Biogeography
FOURTH EDITION

"… that grand subject,
that almost keystone of the laws of creation –
Geographical Distribution."

Charles Darwin,
in a letter to Joseph Dalton Hooker, 1845

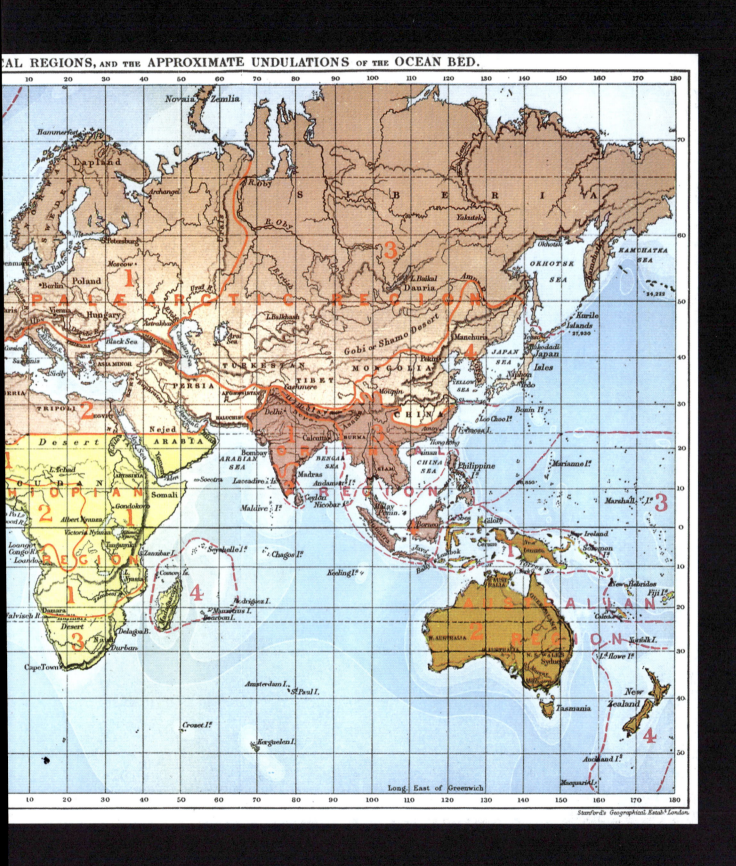

Biogeography

FOURTH EDITION

Mark V. Lomolino
*SUNY College of Environmental
Science and Forestry, Syracuse*

Brett R. Riddle
University of Nevada, Las Vegas

Robert J. Whittaker
University of Oxford

James H. Brown
University of New Mexico

Sinauer Associates, Inc. • Publishers
Sunderland, Massachusetts

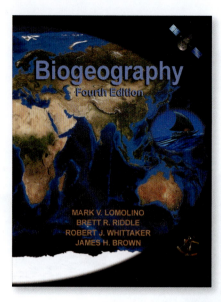

The cover

The cover of the Fourth Edition of *Biogeography* captures the conceptually and visually compelling nature of modern biogeography–its broad historical foundations, which represent the rich legacy of generations of naturalist explorers guided by early maps, compasses and sextants; its application of the most recent technological advances (including those in geo-statistics, remote sensing and satellite imagery) which continually redefine the frontiers of the discipline; and the dynamic geography of humanity, which has and will continue to shape the geography of nature. (Basemap courtesy of Reto Stöckli, NASA Earth Observatory, NASA Goddard Space Flight Center; Sextant image courtesy of The Smithsonian Institution; image of the KODAMA Data Relay Test Satellite courtesy of Japan Aerospace Exploration Agency; a Tongiaki sailing vessel of Tonga © 2010 Herb Kawainui Kane.)

The endpapers

(*Front*) In his classic map of the world's biogeographic regions, Alfred Russel Wallace (1876) sought to divide the landmasses into a hierarchical system of areas reflecting the evolutionary history and affinities among their biotas (numbers identify subregions within each of the principal biogeographic regions). (*Back*) Map of the world illustrating many of the mountain ranges, lakes, island chains, ocean basins, and other topographic and bathymetric features discussed in this text. The basemap is the same as used in the cover and throughout the text (courtesy of Reto Stöckli, NASA Earth Observatory, NASA Goddard Space Flight Center).

Biogeography, Fourth Edition

Copyright © 2010 by Sinauer Associates, Inc. All rights reserved.
This book may not be reproduced in whole or in part without permission from the publisher.
For information, address:

Sinauer Associates, Inc., PO Box 407, Sunderland, MA 01375 U.S.A.
Fax: 413-549-1118
Email: publish@sinauer.com
Internet: www.sinauer.com

Library of Congress Cataloging-in-Publication Data
Lomolino, Mark V., 1953-
Biogeography / Mark V. Lomolino ... [et al.]. -- 4th ed.
   p. cm.
"Rev. ed. of: Biogeography / Mark V. Lomolino, Brett R. Riddle, James H. Brown. 3rd ed."
Includes bibliographical references and index.
ISBN 978-0-87893-494-2 (hardcover)
1. Biogeography. I. Lomolino, Mark V., 1953- II. Lomolino, Mark V., 1953- Biogeography. III. Title.
QH84.B76 2010
578.09--dc22   2010022662

Printed in China

5   4   3

To our families for their love and patience,
to our colleagues for their insights and inspiration, and
to our students for challenging our ideas and redefining the frontiers of science.

Brief Contents

Contents

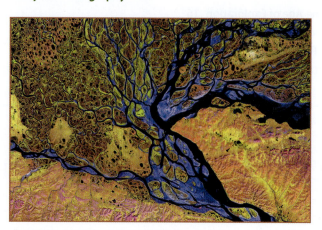

UNIT THREE FUNDAMENTAL BIOGEOGRAPHIC PROCESSES AND EARTH HISTORY 165

UNIT FOUR EVOLUTIONARY HISTORY OF LINEAGES AND BIOTAS 359

UNIT FIVE ECOLOGICAL BIOGEOGRAPHY 507

UNIT SIX CONSERVATION AND THE FRONTIERS OF BIOGEOGRAPHY 695

Preface

Simply defined, biogeography is the geography of nature. Biogeographers study an impressive and sometimes overwhelming diversity of patterns in the geographic variation of Earth's life forms, from molecular variation among geographic populations of the same species, to geographic variation in productivity and diversity of ecosystems, and to the distinctiveness of entire biotas among the continents and ocean basins.

Today, the field is enjoying a renaissance and is recognized as one of science's most insightful and important disciplines, whose conceptual foundations extend back to ancient times. Indeed, knowledge of variation in numbers and types of organisms across areas inhabited by our ancestors was vital to their survival as they searched for particular prey and plants and, in general, adapted to a heterogeneous but geographically predictable environment.

Explorers and naturalists of the seventeenth and eighteenth centuries expanded our understanding of the geographic variation of the natural world from local and regional scales to a global one. They discovered that organisms and biological communities varied in a highly regular fashion along geographic gradients of isolation, elevation, latitude, and increasing area of habitat, and they realized that different regions of the globe—even those with similar climates and environments—were inhabited by different assemblages of species. These very general biogeographic patterns challenged the prevailing views, which held that the Earth's diversity was relatively limited and that its continents, oceans, climates and species were immutable–changing little since the very origins of the planet. Thus, the fascinating geographic signatures of nature ultimately lead to paradigmatic shifts in our understanding of the origins, spread, diversification and extinction of life. From these scientific revolutions entire new disciplines emerged, focusing on the dynamics of the Earth (geology and, in particular, plate tectonics theory), interactions among organisms and their environment (ecology and environmental biology), and heritable changes in the characteristics of populations and species (evolutionary biology).

Ironically, despite its long and distinguished history and its central role in the development of other disciplines, biogeography was not widely recognized as a major discipline in its own right until the latter decades of the twentieth century. The very reasons responsible for its great insights and promise–its holistic and integrative nature–also posed great challenges as few scientists could master the many fields relevant to the diversity of patterns and processes biogeographers study. Indeed, up until the 1980s, the most comprehensive works in biogeography were those written many decades earlier by the "fathers" of the field, including those by Alfred Russel Wallace, Charles Darwin, and Joseph Dalton Hooker. Few universities offered a course in biogeography and many of us, including at least one of the authors of this textbook, never had the opportunity to take a course in the subject.

During the 1970s, modern biogeographical discourse began to take a clearer shape, with first editions of introductory textbooks such as those by C. B. Cox, I. N. Healey, and P. D. Moore (1973, *Biogeography:An Ecological and Evolutionary Approach*) and E. C. Pielou (1979, *Biogeography*), while the flagship disciplinary journal, *Journal of Biogeography*, was launched in 1974. Yet there remained a need for a full-length, modern, and comprehensive textbook in biogeography, a niche that was filled with the publication in 1983 of the First Edition of *Biogeography*, by James H. Brown and Arthur C. Gibson. This was just the book that many of us interested in teaching biogeography had been hoping for as a guide to the breadth and scope of the discipline. Since then, more universities and colleges have offered courses in biogeography, other textbooks and journals have joined the field, and a growing number of scientists have taken to referring to themselves as "biogeographers."

Recognition of the heuristic and applied relevance of the field of biogeography continued to grow at an accelerated rate, and by the 1990s it was clear that the book required a complete revision. The Second Edition of *Biogeography*, authored by James H. Brown and Mark V. Lomolino, emulated the original text in its balanced coverage of the entire breadth of the discipline, its integration of ecological and evolutionary approaches, and its emphasis on general concepts richly illustrated with empirical examples drawn from a wide variety of organisms, ecosystems, and geographic regions. We were pleased to see that, in addition to continuing as a popular textbook, the Second Edition became a general reference and frequently cited source in the primary literature. Although 15 years passed between publication of the First and Second Edition of *Biogeography*, in just five years it was clear to us that another revision was required. With the addition of Brett Riddle to the team for the Third Edition, we again set out to provide a comprehensive and integrative account of the entire field, restructuring the text and updating it with over 1000 new publications from the most exciting frontiers of biogeography.

We published the Third Edition of *Bioeography* at a rather remarkable time within the growth trajectory of the discipline. Notably, the *International Biogeography Society*—formed ca. 2000—was offering biogeographers several venues focused explicitly on themes, concepts, and issues that form the essence of a Modern Biogeography for the Twenty-first Century, including a newsletter (that has recently become the online magazine *Frontiers of Biogeography*), and the opportunity to gather every two years for a conference. During this time, we have seen a steady increase in the wealth of biogeography-related research in a broad range of ecological, evolutionary, conservation, and geography-based journals. The upshot of this modern scientific renaissance was the need for a new encapsulation of the fascinatingly rich and relevant discipline that Biogeography has become. The preparation for the Fourth Edition of the book included the addition of Robert J. Whittaker to the author team, and we set out to capture what we could of the rich diversity of material that has lately been added to the primary literature in all areas of the subject, ranging from developments in phylogenetics and phylogeography, to island biogeography, human biogeography, and conservation biogeography. The use of color illustrations (new to this edition), evaluated and optimized for colorblind readers as well, has transformed the way in which the book sets out to illustrate key concepts and empirical patterns in the geography of nature.

We hope that in some measure the Fourth Edition of *Biogeography* captures the essence of the discipline and conveys our enthusiasm about its past, present, and, mostly importantly, its future to a new generation of students as well as our colleagues in this and all interrelated disciplines. We are confident that their efforts will soon make it necessary to once again update and revise our text with emerging and often transformative advances in our understanding of the geography of nature.

In addition to those who contributed to the previous editions, many additional people–far too many to recognize individually–have helped to revise, update and improve this book. Many readers of the previous editions provided suggestions and corrections that we have incorporated here, and we thank them for their assistance. We also thank Andy Sinauer, Chelsea Holabird, David McIntyre, Azelie Aquadro, Jeff Johnson, Christopher Small, Lou Doucette, and Marie Scavotto for their unfailing professionalism and Herculean efforts to produce a high-quality book on a tight schedule. Finally, we are indebted to all biogeographers and other scientists, from the earliest workers to our contemporaries, for their contributions to the discipline. It is their individual, collaborative, and cumulative contributions that make biogeography so exciting to teach and study.

INTRODUCTION TO THE DISCIPLINE

THE SCIENCE OF BIOGEOGRAPHY

Life varies from place to place in a highly nonrandom and predictable manner. This seemingly simplistic observation is nonetheless one of the most fundamental and most important patterns in nature. Even the earliest human societies were aware that as they expanded their search area, they would encounter a greater number and greater diversity of plants and animals. Those societies living in mountainous regions could see that vegetation changed in an orderly manner as they moved from the lowlands to the summit. The earliest fishing and seafaring societies learned that their catch varied from place to place and from the shallows to deeper waters of the ocean. These same societies (or their descendants) eventually learned that the numbers and diversity of terrestrial organisms increased as their journeys took them from the tiniest to the largest islands, and from environments of temperate regions to those of the tropics (**Figure 1.1**).

Knowledge about the geographic variation of nature across this planet was thus unavoidable and likely essential to survival of these ancient societies. Thousands of years later, scientists would rediscover these patterns and add many additional insights into the geography of nature during the Age of European Exploration, especially from the 1700s into the early 1800s. In order to explain such patterns, these early biogeographers would eventually realize that the Earth, its climate, and its species were dynamic over both space and time. Soon, their explanations for the development and distribution of life would include references to past environments and extinct life forms, as well as competition, predation, and other interactions among species. Thus the geography of nature became the foundation for entirely new fields of science, including geology, meteorology, paleontology, evolution, and ecology.

The early ecologist and evolutionary biologist Ernst Haeckel (1876) once wrote that "the actual value and invincible strength of [Darwin's] Theory of Descent . . . [is] . . . that it explains *all* biological phenomena, that it makes *all* botanical and zoological series of phenomena intelligible in their relations to one another." Later, Theodosius

FIGURE 1.1 Knowledge that the natural world varies across landscapes and seascapes, from small to larger systems, from near to more isolated ones, from the lowlands to montane peaks, and from polar to tropical regions is fundamental to many disciplines of science, and it was likely ancient knowledge essential to the survival of the earliest hominid societies. (Photo of Beardslee Islands courtesy of Kevin White.)

Dobzhansky put it much more succinctly—"Nothing in biology makes sense except in the light of evolution." We do not take issue with either of these visionaries but offer an assertion that is just as bold and important; indeed, it is the fundamental theme of this book and modern biogeography in general. Few patterns in ecology, evolution, conservation biology—and for that matter, most studies of biological diversity—make sense unless viewed in an explicit geographic context.

We know that living things are incredibly diverse. There are probably somewhere between 5 million and 50 million kinds of animals and plants, and a nearly inconceivable multitude of microbes, living on Earth today. Of these, fewer than 2 million have been formally recognized as species and described in the scientific literature. For only a fraction of these do we have adequate information to map their distributions, let alone describe variation in characteristics among individuals and populations across their geographic ranges. The vast majority of species, however, remain as specimens awaiting description in museums, or individuals awaiting discovery in nature. Additional untold millions, perhaps billions, of species that lived at some time in the past are now extinct; only a small fraction of them have been preserved as fossils.

Nearly everywhere on Earth—from the frozen wastelands of Antarctica to the warm, humid rainforests of the tropics; from the cold, dark abyssal depths of the oceans to the near-boiling waters of hot springs; even in rocks several kilometers beneath the Earth's surface—at least some kinds of organisms can be found. But no single species is able to live in all of these places. In fact, almost every species is restricted to a small geographic area and a narrow range of environmental conditions. The spatial patterns of global biodiversity are a consequence of the ways in which the limited geographic ranges of the millions of species overlap and replace each other over the Earth's vast surface.

What Is Biogeography?

Definition

Biogeography is the science that attempts to document and understand spatial patterns of biological diversity. Traditionally, it has been defined as the study of distributions of organisms, both past and present. Modern biogeography, however, now includes studies of all patterns of geographic variation in nature—from genes to entire communities and ecosystems—elements of biological diversity that vary across geographic gradients, including those of area, isolation, latitude, depth, and elevation.

As with any science, biogeography can be characterized by the kinds of questions its practitioners ask. Some of the questions posed by biogeographers include the following:

1. Why are different regions of the globe, even those with similar soils, climates, and other environmental conditions, inhabited by distinct biotas?

2. What abiotic factors (e.g., water chemistry and temperatures) and what biological processes (e.g., predation, competition, and mutualism) limit species distributions, and how does the relative influence of these factors vary across geographic regions, taxa, and time periods?

3. How do the size, shape, and patterns of overlap of geographic ranges vary among taxa, over evolutionary history of particular lineages, or across geographic regions and realms (e.g., among the continents and across ocean basins)?

4. How have historical events such as continental drift, mass extinctions of the dinosaurs and other once-dominant life forms, glacial episodes of the Pleistocene Epoch (all but the most recent 10,000 of the past 2 million years), and more recent periods of climate change (natural or anthropogenic) shaped distributions and patterns of geographic variation of extant biotas?

5. How do the characteristics of entire communities, including their diversity, species composition, and rates of total production and decomposition, vary across the globe?

6. How are isolated oceanic islands colonized, why are there nearly always fewer species on islands than in the same kinds of habitats on continents, and why are these islands often inhabited by evolutionary marvels—flightless birds, "daisies" and other typically herbaceous plants the size of trees, or elephants no larger than domestic pigs?

7. Does the diversity and species composition of relatively recently discovered communities—such as those of hydrothermal vents and cold seep communities of the deepest reaches of the marine realm—exhibit biogeographic patterns similar to those of islands, mountaintops, and other isolated, terrestrial communities?

8. How do the physiological, genetic, morphological, and behavioral characteristics of individuals and populations of a species vary across its geographic range and along geographic gradients such as those of latitude, elevation, and depth?

9. How have the distributions of species—from primordial, unicellular life forms to *Homo sapiens* and its direct ancestors—developed over evolutionary time, and how has the evolution of these lineages been influenced by geographic variation in environments and interactions with regional biotas?

10. How have the geographic dynamics of human civilizations influenced the distributions, evolution, and extinctions of other species, and finally, how will our unrivaled abilities to modify the natural world influence the distributions, diversity, and geographic signature of nature long into the future?

The list of possible questions is nearly endless, but in essence we are asking: *How and why does biological diversity (biodiversity) vary over the surface of the*

Earth? This is the fundamental question of biogeography. The geographic signature of nature provided some of the most fundamental insights for the world's most visionary scientists—from Socrates, Plato, and Aristotle to Darwin and Wallace—and it continues to intrigue scientists and laypersons who are curious about the origins, diversification, and conservation of nature.

Only within the last few decades, however, have scientists begun to call themselves biogeographers and to focus their research primarily on the distributions and geographic variation of living things. Not surprisingly, biogeographers have not answered and will likely never answer all the questions listed above. They have, however, learned a great deal about geographic variation in biodiversity and its underlying explanations. Much of this recent progress has been stimulated by exciting new developments in the related fields of ecology, genetics, systematics, paleontology, and geology; by technological developments in our abilities to visualize and analyze geographically explicit data; and by advances in our capacity and willingness to integrate observations and inferences across this broad range of supporting fields and approaches.

Biogeography is indeed a very broad and integrative science. To be a complete biogeographer, one must acquire and synthesize a tremendous amount of information across a broad range of temporal and spatial scales. But not all aspects of the discipline are equally interesting to everyone, including biogeographers. Given different biases in training, biogeography courses and writings have tended to be uneven in coverage. A common specialization in the past was taxonomic—for example, phytogeographers who studied plants and zoogeographers who studied animals—and within these categories one found specialists in groups at finer taxonomic levels. Although viruses and bacteria play crucial roles in ecological communities and in human welfare, microbial biogeography was until very recently poorly known and rarely discussed. Even in modern times, many biogeographers have provided some intriguing insights by specializing in **historical biogeography**, which is a subdiscipline that attempts to reconstruct the origin, dispersal, and extinction of taxa and biotas. This contrasts with **ecological biogeography**, which attempts to account for present distributions and geographic variation in diversity in terms of interactions between organisms and their physical and biotic environments. Paleoecology is one means of bridging the gap between these two subdisciplines by investigating the relationships between organisms and past environments and using data on both the biotic composition of communities (abundance, diversity, and interactions among component species) and abiotic conditions (ancient climate, soils, water quality, etc.) to reconstruct the evolutionary and geographic development of biotas.

Different biogeographers, whether historical or ecological, continue to utilize different methods for understanding the geography of nature: Some approaches are primarily descriptive, designed to document the ranges of species or spatial variation among their populations, whereas others are mainly conceptual, attempting to develop theoretical models and explanations for these patterns. All of these approaches to the subject are valid and valuable, and discounting or overemphasizing any division or specialization is counterproductive and unnecessary. Instead, we advocate here and throughout this book the most integrative of approaches to this very broad and synthetic discipline. Whereas no researcher or student can become an expert in all areas of biogeography, exposure to a broad spectrum of organisms, methods, and concepts leads to a deeper understanding of salient patterns and their underlying causes. As we hope to show, the various subdisciplines and approaches to exploring and understanding the geography of nature contribute to and complement each other, unifying the science.

Relationships to other sciences and outline of the book

Given the very integrative nature of biogeography, we do not want to draw sharp lines between this field and what we see as integral subjects, as some authors have attempted to do. For example, various authors have recommended that paleontology (the study of fossils and extinct organisms) and ecology be divorced from biogeography; this would make biogeography largely a descriptive, map-making exercise. It would deprive the field of its central role as a synthetic discipline that not only has its own theoretical and empirical approaches, but also readily incorporates conceptual, methodological, and factual advances from many other sciences.

Biogeography is, of course, a synthesis of two fields—biology and geography—so knowledge of at least the fundamental concepts in both disciplines is essential. This is why our treatment devotes considerable space (see Units 2 and 3) to reviewing and developing the geographic, geological, ecological, and evolutionary concepts that form the foundations of both the biological and Earth sciences and are used throughout the book. In addition, one must be acquainted with the major groups of plants, animals, and other life forms and know something about their physiology, anatomy, behavior, genetics, and evolutionary history. These topics are not the subjects of separate chapters but are integrated throughout the text, usually by the device of using different kinds of organisms, with distinctive biological characteristics, to illustrate biogeographic patterns, processes, and concepts. For example, it is only by knowing several critical features of biology and evolutionary history that we can understand why polar bears are limited to the Northern Hemisphere, whereas penguins are limited to the Southern Hemisphere (**Figure 1.2**); or why amphibians and freshwater fishes have only rarely crossed even modest stretches of ocean to colonize islands, whereas birds and bats have done so much more frequently.

Naturally, it is important to know some geography and geology. The locations of continents, mountain ranges, deserts, lakes, major islands and archipelagoes (groups of islands), and seas, during the past as well as the present, are indispensable information for biogeographers, as are past and

FIGURE 1.2 One of the central questions that biogeographers attempt to explain is why geographic ranges differ among species: in this case, why polar bears (*Ursus maritimus*, green area) and penguins (family *Spheniscidae*, 17 species, red area) are limited to the Northern and Southern Hemispheres, respectively.

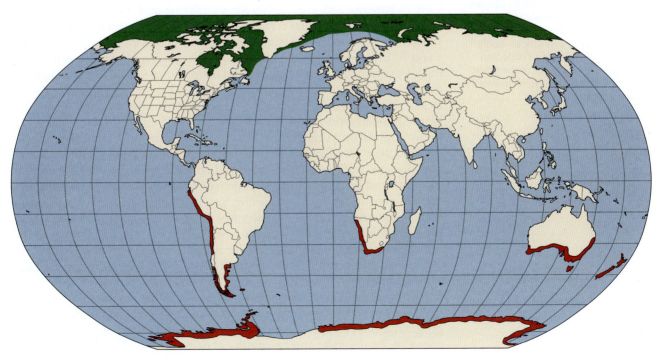

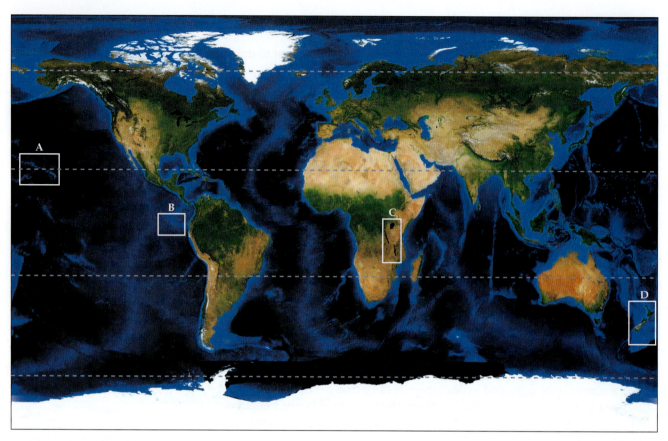

FIGURE 1.3 An illustration of a world locator map to be used throughout this book, in this case indicating locations of (A) the Hawaiian Islands—site of adaptive radiations in many species, including fruit flies, land snails, ferns, flowering plants, and honeycreepers (finches); (B) the Galápagos Islands—famous for Charles Darwin's studies of interisland variation of tortoises and finches; (C) the Great Rift Valley Lakes of East Africa—home to hundreds of endemic cichlids; and (D) New Zealand—islands of the Maori people and flightless birds, including the kiwi and extinct, giant moas.

present climatic regimes, ocean currents, and tides. To remind the reader of the major geographic features of the Earth, we have illustrated many of them on the colored maps of the book's endpapers, and we have included locator maps (**Figure 1.3**) to provide a geographic context for case studies featured in illustrations throughout this book. In addition, Unit 2 provides an overview of patterns and approaches for visualizing and analyzing geographic variation in environmental conditions (see Chapter 3), and basic patterns of distributions of species and entire communities across the globe (see Chapters 4 and 5). Unit 3 describes the fundamental biogeographic processes influencing geographic variation of species and entire biotas—including immigration (see Chapter 6), speciation and extinction (see Chapter 7), geological processes and dynamics of the continents and ocean basins (see Chapter 8), and the biogeographic responses to climate change (see Chapter 9). Unit 4 focuses on the evolutionary and geographic history of biotas, including the geography of diversification (see Chapter 10), the history of lineages (see Chapter 11), and how lineages and biotas have coevolved with the geological development of the Earth and its geographic template (see Chapter 12). Unit 5 explores patterns of geographic variation among terrestrial and aquatic biotas, focusing first on patterns in species diversity (see Chapter 13) and then on the assembly and evolution of isolated biotas (see Chapter 14). The last chapter in this unit (see Chapter 15) explores similar patterns, but in this case focusing on continental and oceanic biotas and introducing and exploring three related subdisciplines—ecogeography, areography, and macroecology. The final section of this book, Unit 6, provides an overview of the status of biological diversity and how the now impressive body of knowledge and the tools of biogeography can be applied to advance our understanding of, and abilities to conserve, the diversity and geography of nature.

Philosophy and basic principles

Most people have a vague and perhaps misleading impression of what science is, how scientists work, and how major scientific advances come about. To put it simply, scientists try to understand the natural world by explaining its enormous diversity and complexity in terms of general patterns and basic laws. Some philosophers and historians of science, viewing its progress with 20/20 hindsight, often suggest that it is possible to provide a recipe for the most effective way to conduct an investigation. Unfortunately, as most practicing scientists know, scientific inquiry is much more like working on a puzzle or being lost in the woods than like baking cookies or following a road map. There are numerous mistakes and frustrations, but also some wonderful surprises. Luck, timing, and trial and error play crucial roles in even the most important scientific advances. Some important discoveries, such as Alfred Wegener's evidence for continental drift, are long ignored or even totally rejected by other scientists. While the progress of science owes much to such admirable human traits as intelligence, creativity, and perseverance, it is also impeded by equally human but less admirable characteristics such as prejudice, jealousy, shortsightedness, arrogance, and stupidity. This is not meant to imply that science has not made great advances in our understanding of the natural world. It has. But like most human activities, the progress of science has followed a much more complex path than is usually portrayed in textbooks. As we shall see, beginning with our next chapter, which discusses the rich history of this field, biogeography is no exception.

Despite the diversity of potentially insightful approaches, we can identify some commonalities in what often prove to be effective strategies for understanding the complexity of nature. In essence, scientists proceed by investigating the relationships between pattern and process. *Pattern* can be defined as nonrandom, repetitive variation of focal elements (e.g., organisms, species, populations, or communities) among units (e.g., regions, entire continents, or ocean basins) or along relevant gradients (e.g., those of area, isolation, elevation, depth, or geographic variation in environmental conditions). The occurrence of pattern in the natural world implies causation by some general process or processes. Science usually advances by the discovery of patterns, then by the development of mechanistic explanations for them, and finally by rigorous testing of those models and more general theories until the ones that are necessary and sufficient to account for the patterns become widely accepted.

Traditional treatments of the philosophy of science usually devote considerable space to distinguishing between inductive reasoning—reasoning from specific observations to general principles—and deductive reasoning—reasoning from general constructs to specific cases. Several influential modern philosophers, especially Karl Popper (1968a,b), have strongly advocated so-called hypothetico-deductive reasoning. Any good scientific theory contains logical assumptions and consequences, and if any of these can be proven wrong, then the theory itself must be flawed or, at the least, in need of major revisions. The hypothetico-deductive method provides a powerful means of testing a theory by setting up alternative, falsifiable hypotheses and testable predictions to distinguish among them. First, an author puts forth a new, tentative idea (hypothesis) stated in clear, simple language, generating predictions that can be tested and potentially falsified by means of experiments or observations. After the statement has withstood the severest empirical tests, it can be considered to be supported or corroborated, but by hypothetico-deductive logic a theory can never be proven true, only falsified (see also Kuhn 1970; Meadows 2001; Carpenter 2002; Graham et al. 2002).

Moreover, just as organisms and species develop over time, theories also undergo their own ontogeny and evolution, sometimes experiencing periods of contraction in the domain of phenomena considered and models employed, alternating with periods of domain expansion (Pickett et al. 2007). Perhaps the most fascinating and, paradoxically, the most important stages in the development of theories are those periods that Thomas Kuhn (1996) describes as scientific crises: when the long-standing accepted explanation (the so-called paradigm) seems so fraught with exceptions and inconsistencies that it no longer serves to guide scientific inquiry and discovery. Classic cases of crises in science include those of Copernicus, who complained that his contemporary astronomers had become so "inconsistent in these [astronomical] investigations . . . that they cannot even explain or observe the constant length of the seasonal year" (see Kuhn 1996: 83–84). Similarly, Einstein described his frustrations during a time of crisis in physics as being "as if the ground had been pulled out from under one, with no firm foundation to be seen anywhere, upon which one could have built." Wolfgang Pauli, in characterizing the crisis in quantum mechanics during the 1920s, admitted to a friend that "at the moment physics is again terribly confused. In any case, it is too difficult for me, and I wish I had been a movie comedian or something of the sort and had never heard of physics." Yet it appears that such crises, as frustrating and painful as they may be, are not just unavoidable but also may be essential to genuine revolutions in science—when a new theory or paradigm finally replaces a long-revered, but now obsolete one. Often, the breakthrough is finally achieved during an epiphany, as Kuhn (1996: 122) observed, "not by deliberation and interpretation, but by a relatively sudden and unstructured event like the gestalt switch. Scientists then often speak of the 'scales falling from the eyes' or of the 'lightning flash' that 'inundates' a previously obscure puzzle."

Throughout the process of scientific discovery, empirical observations and conceptual models are played back and forth against each other, theories are devised and modified, and understanding of the natural world advances, typically through protracted periods of gradual growth interspersed with bursts of revelations—the scientific revolutions that Kuhn describes. This is particularly true of biogeography, in which new empirical observations accumulated until the mass of knowledge and the morass of exceptions and anomalies finally triggered the overthrow of a reigning paradigm for one more capable of explaining our more advanced understanding of the natural world. For example, although Alfred Wegener's ideas of continental drift were rejected for many decades, by the 1970s the data and theory of tectonic plate movements became so overwhelming that the paradigm of the fixity of the continents, which scientists clung to for centuries, was finally overturned. Biogeographers were then freed to explain distributions and geographic variation among biotas using a more integrative theory that was based on the dynamics, not only of species and climates, but of the continents and ocean basins as well (see Chapter 8).

One especially encouraging, albeit rather humbling, aspect of Kuhn's view of scientific revolutions is that they are often led by scientists at relatively young stages in their careers. Einstein wrote his first scientific work—"The Investigation of the State of Aether in Magnetic Fields"—at the age of 15, experienced his famous thought experiment of traveling alongside a beam of light at 16, and completed his *Annus Mirabilis* collection of papers (featuring his theories of quantized light, Brownian motion, special relativity, and the equivalence of matter and energy) while cloistered away in his corner of the patent office at just 26 years old. Alfred Russel Wallace and Charles Darwin were both quite young (25 and 22, respectively) when they set out on their transformative voyages; Edward Wilson was just 30 when he published the first of his two seminal papers on the taxon cycles of island faunas; and Robert MacArthur, already widely respected for his previous research in mathematical ecology, was only 33 when he and Wilson first published their equilibrium theory of island biogeography (see Chapter 13). As Kuhn (1996: 90) put it, "the very young . . . are little committed to the traditional rules of normal science, are particularly likely to see that those rules no longer define a playable game and to conceive another set that can replace them."

Unlike many of the physical and biological sciences, biogeography usually is not an experimental science, at least not in the sense of controlled manipulations of laboratory studies. In very rare but often illuminating cases, a few biogeographers have used experimental techniques to manipulate small systems such as tiny islands, sometimes with spectacular success (e.g., Simberloff and Wilson 1969). However, most important questions in biogeography involve such broad spatial and temporal scales that experimentation is rendered either impractical or unethical. This methodological constraint does not diminish the rigor and value of biogeography, but it does pose major challenges. Other sciences, such as astronomy and geology, face the same problems. As Robert MacArthur once observed, astronomers and physicists, including Copernicus, Galileo, Kepler, and Newton, never moved a planet, but that did not prevent them from advancing our understanding of the motion of celestial bodies (see Brown 1999).

Indeed, despite these practical and ethical limitations, physicists, geologists, and biogeographers continue to provide some fascinating discoveries by utilizing other types of experiments—those where variation among "treatments" and putative driving forces are not manipulated by the scientists themselves but result from natural (or unplanned anthropogenic) manipulations (e.g., differences in the size and degree of isolation of islands). Although such natural and opportunistic experiments lack precise controls, they more than make up for this in realism, spanning the temporal and spatial scales required for the phenomena under study (e.g., drifting of entire continents, colonization of isolated ecosystems, evolution of lineages, and extinction of species and entire biotas). Thus, Wallace and Darwin used the patterns of variation in animal and plant communities among islands of modern-day Indonesia and the Galápagos, respectively, to develop truly transformative insights into the evolution and geography of nature. Islands, in particular, have had a great influence on these and numerous subsequent biogeographers, ecologists, and evolutionists because they represent unrivaled natural experiments—many thousands of replicated systems whose biotas have developed over long periods and across the broad expanses necessary for the fundamental processes to generate their distinctive signatures. Just as in manipulative laboratory experiments, natural experiments must be designed—selecting among all islands available a strategic subset of islands that are similar with respect to all characteristics except those central to the hypothesized causal factors (e.g., the influence of competition, predation, or availability of particular types of prey on evolution of Darwin's finches; see Grant and Weiner 1999).

In dealing with historical aspects of their science (i.e., the development of lineages over space and time), most biogeographers make one critical assumption that is virtually impossible to test: They accept the principle of **uniformitarianism**, or **actualism**. Uniformitarianism is the assumption that the basic physical and biological processes now operating on the Earth today have operated throughout time because they are manifestations of universal scientific laws. This principle is usually attributed to the British geologists James Hutton (1795) and Charles Lyell (1834), who realized that the Earth was much older than had been previously supposed and that its surface was constantly being modified by the same geological processes. In this same spirit, one of Darwin's great insights was the recognition that changes in domesticated plants and animals over historical time resulted from the same process that caused natural changes in organisms over evolutionary time.

As noted by Simpson (1970), acceptance of uniformitarianism has never been universal, in part because some authors have attached additional meanings to the concept. Some have assumed that the term implies that the average intensities of processes have remained approximately constant over time, and that both geological and biological changes are always gradual. Neither of these amendments is necessary or acceptable. We know that some historical events have been of great magnitude but so infrequent that they have never been observed in recorded human history. For example, contrary to some science fiction movies, no humans were living during the Cretaceous Period when dinosaurs still roamed the Earth. Consequently, no one observed the collision of the Earth with an asteroid that presumably eliminated the dinosaurs along with roughly half of all other organisms in existence (see Chapter 8). Yet there is abundant evidence, preserved in ancient rocks, for this extremely rare and unpredictable, but extremely important, event. Scientists now accept that the intensity of forces will vary from time to time and from place to place; only the nature of the laws that control these processes is timeless and constant. The above example—that of the asteroid responsible for the end-Cretaceous extinctions—provides a spectacular demonstration for a paradox common to disciplines of long temporal and vast spatial scales: Many features of the natural world have been strongly influenced by extremely rare events (asteroid impacts that wiped out long-dominant species, volcanic explosions and tsunamis that destroyed entire biotas, or hurricanes that blew entire flocks of birds off course to colonize oceanic islands far outside their normal migration routes). Thus, while highly unpredictable, such extremely rare events have had perhaps unparalleled impacts on the distributions and diversity of life across this planet.

To avoid the unfortunate connotations associated with uniformitarianism, and assumptions that it does not allow for variation in the rates of fundamental processes or the importance of rare events, we will follow George Gaylord Simpson's adoption of the term *actualism* (Simpson 1970; see Gould 1965). Historical biogeographers in particular use this principle to account for present and past distributions, assuming that the processes of speciation, dispersal, and extinction operated in the past by the same mechanisms that they do today (although varying in their rates and relative importance). This premise has become an accepted tool for interpreting the past and predicting the future. The most serious challenge to using this principle is, of course, discovering which timeless processes have operated to produce different patterns of geographic variation among different biotas, regions, and time periods.

Doing Contemporary Biogeography

Biogeography differs from most of the biological disciplines, and from many other sciences, in several important respects. We have mentioned one of them above: Biogeography is, for the most part, a comparative observational science, still quite rigorous and insightful in many respects, but largely restricted from conducting manipulative experiments because it usually deals with phenomena and processes that operate across such broad spatial and long temporal scales.

Another way that biogeography differs from most other sciences is that even the most accomplished researchers depend on data collected by a large number of individuals working across a diversity of regions and time periods. This reliance on a variety of sources of empirical data is inherent and unavoidable, given the integrative nature of the field and the quest of its practitioners to understand patterns of variation in the characteristics of populations, species, lineages, or communities over broad spans of space and time. Thus, although being a biogeographer often carries the sometimes deserved connotation of intrepid scientists embarking on adventurous collecting expeditions to exotic places, most practicing biogeographers obtain more of their data from museums and libraries and, increasingly, from digital resources than from their own fieldwork.

Finally, given the diverse sources of data required to explore regional- to global-scale patterns and their causal explanations, we again emphasize that biogeography is one of the most synthetic sciences—one that puts a premium on integrative and collaborative research (see Wiens and Donoghue 2004). Biogeographers work at the interfaces of several traditional disciplines: ecology, systematics, evolutionary biology, geography, paleontology, and the "Earth sciences" of geology, climatology, limnology, and oceanography. Much of their best work comes from co-opting and modifying existing theory from other fields, or developing entirely new theories that bridge between what were typically viewed as distinct and unrelated disciplines. At first, this might seem to require such exceptionally broad training as to be intimidating to the beginner. However, while some breadth of knowledge is desirable, the interdisciplinary nature of biogeography means that there is much to be gained through collaboration among specialists in different but complementary fields. Indeed, some of the most transformative discoveries and advances of modern science were the fruits of collaborative research. Exemplary cases of transforming science through collaborative syntheses include James Watson and Francis Crick's legendary deciphering of the structure of DNA, Robert MacArthur and Edward Wilson's equilibrium theory of island biogeography, and James H. Brown's collaborations with Brian Mauer, which laid the foundations for macroecology and its approaches for discovering complex, emergent patterns across broad spans of time, space, and biological diversity. Indeed, during the early history of biogeography, the importance of the collaborative and constructive spirit in advancing science was richly evidenced in the long and incisive correspondences among its most distinguished scientists, including those between Charles Darwin, Joseph Dalton Hooker, Charles Lyell, and Alfred Russel Wallace (**Figure 1.4**; see Chapter 2).

Biogeography is accessible to almost anyone with the curiosity and motivation to tackle some of the greatest mysteries of nature. Large research grants, modern laboratory facilities, and even sophisticated statistical techniques—while often desirable—are not required to do state-of-the-art biogeographic research. All that is required is a good idea, access to geographically explicit data, and the creativity to design opportunistic, but nonetheless scientifically rigorous, comparative analyses of those data. This means that even beginning students can do original research. In fact, the authors of this book usually require that each student in our biogeography courses do an original research paper. We are highly gratified, and simultaneously humbled, to report that some of the most interesting and important work discussed in this book and published in distinguished journals began as student research projects in our courses. Students' contributions, along with those of an ever-increasing number of biogeographers, continue to advance our understanding of the

FIGURE 1.4 Darwin benefited from many discussions and lengthy correspondences with other leading naturalists of the day, including Joseph Dalton Hooker and Charles Lyell (sitting and standing across from Darwin, respectively). (Down House, Kent, UK; painting by Evstafieff.)

geography of nature and contribute to the long and distinguished history of this field—one shared with the disciplines of evolution and ecology.

In the next chapter, we recount the historical development of these fields and the fascinating discoveries that were to serve as the foundations for modern research into the variation of life across space and time.

THE HISTORY OF BIOGEOGRAPHY

Biogeography has had a long and distinguished history, one that is inextricably woven into the development of evolutionary biology and ecology. Indeed, the patterns of species distributions and geographic variation among biotas were matters of primary interest to early evolutionary biologists, including the distinguished "fathers" of the field, such as Carolus Linnaeus, Charles Darwin, and Alfred Russel Wallace. The field of ecology, a relatively young offspring of this lineage, grew out of attempts to explain biogeographic patterns in terms of the influence of environmental conditions and interactions among species. But the origin of these three fields—biogeography, evolutionary biology, and ecology—is ancient, dating back well before the Darwinian revolution. Indeed, Aristotle was contemplating many of the questions we continue to ponder today:

> But if rivers come into being and perish, and if the same parts of the Earth are not always moist, the sea also must necessarily change correspondingly. And if in places the sea recedes while in others it encroaches, then evidently the same parts of the Earth as a whole are not always sea, nor always mainland, but in process of time all change. (Meteorologica, ca 355 BC)

Aristotle offered this prophetic view of a dynamic Earth in order to explain variation in the natural world over space and time. He was one of the earliest of a long and prestigious line of scientists to ask the same questions: Where did life come from, and how did it diversify and spread across the globe?

Despite Aristotle's impressive insights, answering these questions would require a much more thorough understanding of the geographic and biological character of the Earth. Thus, the development of biogeography, evolutionary biology, and ecology was tied to the Age of European Exploration. As we shall see below, the early European explorers and naturalists did far more than just label and catalog their specimens. They immediately, perhaps irresistibly, took to the task of comparing biotas among regions and developing expla-

Table 2.1 *Persistent Themes in Biogeography*

1. Classifying geographic regions based on their biotas
2. Reconstructing the historical development of lineages and biotas, including their origin, spread, and diversification
3. Explaining the differences in numbers as well as types of species among geographic areas and along geographic gradients, including those of area, isolation, latitude, elevation, and depth
4. Explaining geographic variation in the characteristics of individuals and populations of closely related species, including trends in morphology, physiology, behavior, genetics, and demography

nations for the similarities and differences they observed. The comparative method served these early naturalists well, and by the eighteenth century the study of biogeography began to crystallize around fundamental patterns of distribution and geographic variation. Here, we trace the development of biogeography from the Age of European Exploration to its current status as a mature and respected science.

Many, if not all, of the themes central to modern biogeography (**Table 2.1**) have their origins in the pre-Darwinian period. This is not to say that biogeography has not advanced tremendously in the past few decades—only that modern biogeographers owe a great debt to those before them who shared the same fascination with, and asked the same types of questions about, the geography of nature. As Newton once replied when asked how it was that he had developed such unparalleled insights in physics, "If it appears that I have been able to see more than those who came before me, perhaps it is because I stood on the shoulders of giants." Like physics, biogeography has a history of "giants," visionary scientists each building on the collective knowledge of those who came before them.

The Age of Exploration

It is hard for us to appreciate that roughly 250 years ago, biologists had described and classified only 1 percent of all the plant and animal species we know today. Biogeography was essentially founded and rapidly accelerated by world exploration and the accompanying discovery of new kinds of organisms, which often baffled even the most distinguished scientists of that era. The arrival of the first specimen of a duck-billed platypus in Britain in 1799 challenged the great anatomists of the day to place it within the accepted, albeit limited, contemporary taxonomy (Moyal 2001). Reactions ranged from accusations that it was a clever hoax created by taxidermists— "exhibiting the perfect resemblance of the beak of a Duck engrafted on the head of a quadruped"—to puzzlement over its affinities to mammals, birds, and reptiles, having an amalgam of traits reminiscent of each. These and other marvels of remote lands led the distinguished naturalist Reverend Sydney Smith to remark in 1819 that "in this remote part of the earth [Australia], Nature . . . seems determined to have a bit of play, and to amuse herself as she pleases."

The great naturalists of the eighteenth century were largely driven by a calling to serve God. The prevailing belief was that the mysteries of creation would be revealed as these scientists developed more complete catalogs of the diversity of life. Up until the mid-eighteenth century, the prevailing world view was one of stasis—the Earth, its climate, and its species were

FIGURE 2.1   The contributions of Carolus Linnaeus (1707–1778) to biology are legendary and include development of the binomial system of nomenclature, the production of many volumes describing identifying characteristics and distributions of hundreds of plant species, and the development of some of the earliest theories on the origins and spread of life across the planet (see Figure 2.2A). (Carl von Linné by Alexander Roslin, 1775, currently owned by and displayed at the Royal Swedish Academy of Sciences.)

immutable (unchanging). However, as the early biogeographers (then called naturalists or simply geologists) returned with their burgeoning wealth of specimens and accounts, two things became clear. First, biologists needed to develop a standardized and systematic scheme to classify the rapidly growing wealth of specimens. Second, it was becoming increasingly obvious that there were too many species to have been accommodated by the biblical Noah's Ark. It was just as difficult for these early biologists to explain how animals and plants—now isolated and perfectly adapted to dramatically different climates and environments—could have coexisted at the landing site of the Ark and then spread to populate all regions of the globe.

One of the most ambitious and visionary of these eighteenth century biologists was Carolus Linnaeus (1707–1778; **Figure 2.1**). He believed that God spoke most clearly to man through the natural world, and he felt that it was his task to methodically describe and catalog the collections of this divine museum. Toward that end, Linnaeus developed a scheme to classify all life: the system of binomial nomenclature that we continue to use today. Linnaeus also set his energies to the task of explaining the origin and spread of life. Like his contemporaries, he believed that the Earth and its species were immutable. He realized that the challenge was not just in explaining the number of species, but in explaining patterns of diversity and distribution as well. The rapidly growing list of species included organisms adapted to environments ranging from the moist tropics to deserts, forests, and tundra. Given that species were immutable, how could they have spread from a single site (Paradise and, later, Noah's landing place) across inhospitable environments to eventually colonize those distant sites with habitat conditions they found, by creation, to be optimal? Linnaeus (1781) hypothesized that all life-forms had originally been placed along the slopes of a "Paradisical Mountain" located near the Equator, where each species was perfectly adapted to the environmental conditions and interacting species of particular "stations" (or habitats) along the mountain slope. Given that the slopes of tropical mountains include a diverse complement of stations—from tropical forests along the lowlands to tundra near their peaks—Linnaeus assumed that all of nature's original species at the time of creation could be accommodated on this one tropical mountain, which he suggested was actually an island—the only land that rose above sea level of this primordial period (roughly 6000 years ago).

Following these origins, rocks and various depositional materials were spewed up onto the land by the great forces of the sea. The land, and ultimately the nascent continents, expanded as the once pan-global sea contracted and then deepened. And as the terrestrial realm expanded, its species—still immutable and restricted to the same stations—dispersed to colonize newly emergent regions—some on the backs or in the fur, feathers, and bellies of others, and many by virtue of a myriad of dispersal mechanisms richly de-

(A)

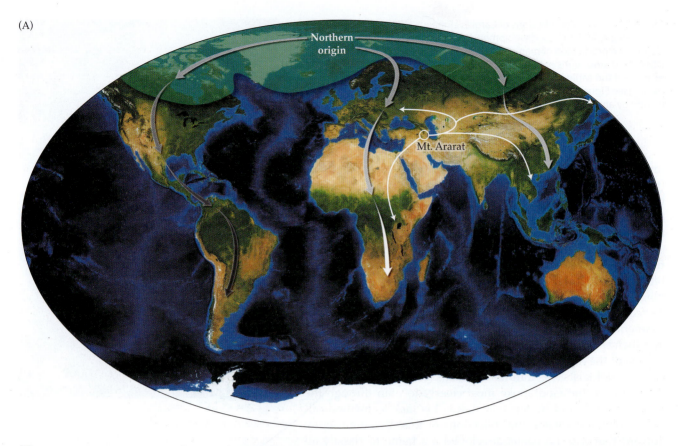

(B)

tailed in the volumes of Linnaeus's treatises. His explanation for the survival and redistribution of these same immutable species following the biblical deluge was analogous to that of his theory of creation. Linnaeus hypothesized that Noah's landing occurred along the slopes of one of the world's tallest mountains—possibly Mount Ararat, which is located near the border of Turkey and Armenia (**Figure 2.2**). Once the Flood receded, these species migrated down from the mountain and spread, to eventually colonize and inhabit their respective stations in different regions of the globe.

◄ **FIGURE 2.2**   (A) Two early hypotheses proposed to account for the diversity and distributions of terrestrial organisms. Carolus Linnaeus hypothesized that terrestrial plants and animals survived the biblical Flood along the slopes of a mountain such as Mount Ararat (B) near the present-day border of Turkey and Armenia, and then spread to suitable environments from there (thin arrows). Buffon, on the other hand, hypothesized that species had a far northern origin during a relatively warm period and then spread southward, changing ("improving" or "degenerating") as they colonized climatically and ecologically diverse landmasses in both the New and Old Worlds (thick arrows). (Photo courtesy of NASA.)

Georges-Louis Leclerc, Comte de Buffon (1707–1788) was a contemporary of Linnaeus, but his studies of living and fossil mammals led him to a very different view of the origin and spread of life. Buffon (1761) noted two problems with Linnaeus's explanation. First, he observed that different portions of the globe—even those with the same climatic and environmental conditions—were often inhabited by distinct kinds of plants and animals. The tropics, in particular, contained a great diversity of unusual organisms, and those of the New World (i.e., in Central and South America) shared few species with those of the Old World tropics (i.e., in Africa). Second, Buffon reasoned that Linnaeus's view of the spread of life required that species migrate across inhospitable habitats following the Flood. Species adapted to mesic montane forests, for example, would have had to migrate across expansive deserts before they could colonize deciduous and coniferous forests to the north. If species were immutable, and therefore incapable of adapting to new environments, then their spread would have been blocked by these environmental barriers. Buffon, therefore, hypothesized that life originated, not on a Paradisical Mountain in the tropics, but in a region of northwestern Europe during an earlier period when climatic conditions were more equable (see Figure 2.2A). He speculated that when climates later cooled, life-forms migrated southward through the northern continents and then on to colonize the Southern Hemisphere of both the New and Old Worlds. During this migration, the populations of the New and Old Worlds were separated and became increasingly modified until tropical biotas of the New and Old Worlds shared few, if any, forms. According to Buffon (1761),

> Man is totally a production of heaven; But the animals, in many respects, are creatures of the Earth only. Those of one continent are not found in another; or, if there are a few exceptions, the animals are so changed that they are hardly to be recognized.

While Buffon's theory of a northern origin of all life-forms may now seem fanciful, it provided two key elements of what would become central parts of modern biogeographic theory. Not only was the Earth (both land and sea) and its climate dynamic, but so were its species; they evolved, or in Buffon's terms, "improved" or "degenerated," as they became further isolated in different regions and environments. The results included survival of the improved forms at the expense of those that either could not change or degenerated (and ultimately died out), a process that Alfred Wallace and Charles Darwin would later call "natural selection." Buffon's views of the process of evolution—degeneration, in particular—were admittedly extreme and betrayed an obvious distaste for the New World and its founding people:

> In this New World, . . . there is some combination of elements and other physical causes, something that opposes amplification of Nature. There are obstacles to development, and perhaps to the formation of large germs . . . [persisting forms that] shrink under the niggardly sky and an unprolific land, thinly peopled with wandering savages. (Buffon 1761)

One very positive outcome of Buffon's otherwise objectionable view of the New World was that it did spur Thomas Jefferson, then President of the United States, to commission one of the most ambitious biogeographic surveys ever conducted—Lewis and Clark's expedition across the central and northwestern portions of the United States.

In response to Buffon's hypothesis of degenerations in the New World, Charles Darwin (1839) observed that "if Buffon had known of the gigantic sloth and armadillo-like animals, and of the lost Pachydermata, he might have said with greater semblance of truth that the creative force in America had lost its power, rather than that it had never possessed great vigour." That is, like other continents prior to colonization by exotic human civilizations, the New World possessed a diversity of great beasts that at least rivaled those more familiar to Buffon, but it lost them during the megafaunal extinctions of the Late Pleistocene and Early Holocene Epochs (see Chapter 9). To his credit, Buffon's many other contributions strongly influenced the early development of biogeography and evolutionary biology, not the least of which was his observation that environmentally similar but isolated regions have distinct assemblages of mammals and birds. This became the first principle of biogeography, known today as **Buffon's law**.

From 1750 to the early 1800s, natural scientists continued to explore the diversity and geography of nature and to write systematic catalogs and general syntheses of their work. One of the most prominent naturalists/collectors of this period was Sir Joseph Banks, who, during a three-year voyage around the world with Captain James Cook on HMS *Endeavor* (1768–1771; **Figure 2.3**), collected some 3600 plant specimens, including over 1000 species not known to science (i.e., in addition to the 6000 species described by Linnaeus in his *Species Plantarum*). The efforts of Banks and many other naturalists/explorers resulted in two important developments. First, they affirmed and generalized Buffon's law. Second, they developed a much more thorough understanding of, and appreciation for, the complexity of the natural world. Banks and his colleagues discovered some interesting exceptions to Buffon's law—specifically, cosmopolitan species. Moreover, they noted other biogeographic patterns, which in their own right would become major themes as biogeography developed.

Johann Reinhold Forster (1729–1798; see Figure 2.3D)—a German pastor and naturalist, provided many fundamental contributions to phytogeography, and to biogeography in general. By his great fortune, and Captain Cook's exhausted willingness to indulge the extravagant demands of Sir Joseph Banks, Forster became ship's naturalist on Cook's second voyage, this time on HMS *Resolution* (1772–1775), on which he was accompanied by his son—Johann Georg Adam Forster (1754–1794), who would also distinguish himself as naturalist and geographer. Forster pieced together hundreds of observations on regional floras to develop the first systematic global view of botanical regions, where each is defined by its distinct plant assemblages (Forster 1778). He found that Buffon's law applied to plants as well as to mammals and birds, and to all regions of the world—not just the tropics. Forster also described the relationship between regional floras and environmental conditions, and how animal associations changed with those of plants—insights that well over a century later would prove foundational to the field of ecology. He also provided some important early insights into what was to become island biogeography and species diversity theory. He noted that insular (or island) communities had fewer plant species than those on the mainland and that the number of species on islands increased with available resources (island area and variety of habitats). Forster also noted the tendency for plant diversity to decrease from the Equator to the poles, a pattern that he attributed to latitudinal trends in surface heat on the Earth.

(A)

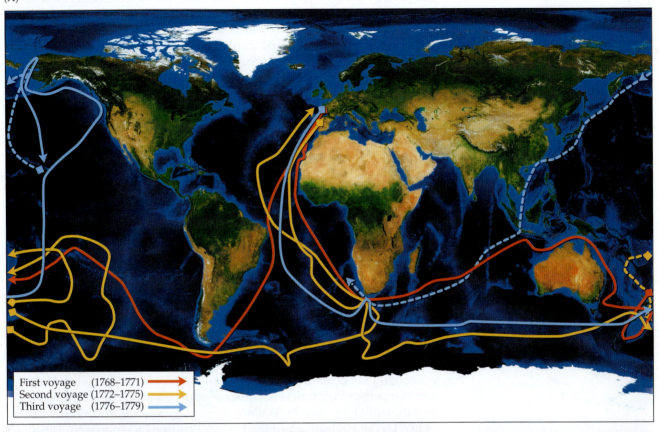

First voyage   (1768–1771)
Second voyage (1772–1775)
Third voyage (1776–1779)

(B)

(C)

(D)

In 1792, Karl Ludwig Willdenow (1765–1812), another German botanist, wrote a major synthesis of plant geography. He not only provided detailed descriptions of the floristic provinces of Europe and how plant geography reflects that of climates, but he also offered a novel explanation for their origin. Rather than one site of creation (or survival during the biblical deluge), Willdenow suggested, there were many sites of origination—isolated mountain chains across the continents that in ancient times were separated by global seas. Each of these mountain refuges was inhabited by a distinct assemblage of locally created plants. As the Flood receded, these plants spread downward to form the floristic regions of the world.

FIGURE 2.3 Explorations of the world by Europeans, including circumnavigational voyages by Captain James Cook (A, B) during the eighteenth century, provided invaluable information on the distributions of plants and animals (dashed line—route of the crew's return to England following Cook's death). Sir Joseph Banks (C) served as ship's naturalist during Cook's first voyage on HMS *Endeavor* (1768–1771), but he was replaced on Cook's second voyage, on HMS *Resolution*, by Johann Reinhold Forster and his son Georg (D). (B portrait by Nathaniel Dance; C portrait by Benjamin West; D depicted here in Tahiti by John Francis Rigaud [1742–1810] in 1780.)

The latter decades of the eighteenth century marked the final tumultuous, but triumphant, revolution in Western science and society known as the Age of Enlightenment. Scientific advancement during this period was inseparably and causally linked to this first true development of a world view of nature, fueled by the fascinating accounts of the variety of lands and life-forms encountered during the Age of European Exploration. Scientists were enjoying a growing appreciation for the diversity and geographic variation of nature—including its landscapes and seascapes, soils and climates, plants and animals, and the characteristics of regional societies and tribal peoples. This ultimately drove intellectuals to free themselves from the shackles of authoritative doctrines and religious dogma to instead use careful, unbiased observation and unfettered rationalism as the primary means of advancing their understanding of all aspects of the natural world. This new, truly holistic view of nature included all phenomena, from the physics of the cosmos to the world's climates, diversities, and distributions of its species to the natural character and innate freedom of its peoples. This was indeed a time of revolutionary intellectual and cultural achievement, when bold practitioners developed integrative views of diverse but essentially interrelated phenomena.

One of the most influential scientists and intellectuals of this period of enlightenment and globalization was Alexander von Humboldt (1769–1859), who is widely recognized as the father of phytogeography and a founder of an eclectic list of other disciplines, including volcanology, anthropology, meteorology, geomagnetism, oceanography, and archaeology—all of which he saw as interrelated and essential means of comprehending the unity and universal harmony of nature. He was a singular visionary and powerful inspiration for many of the greatest scientists, essayists, and artists of the nineteenth century, including Darwin, Wallace, Louis Agassiz, Ralph Waldo Emerson, Henry David Thoreau, and Edgar Allan Poe (see Jackson 2009). He was a student of Willdenow and a colleague and close friend to Georg Forster, both of whom strongly influenced Humboldt's ideas on the interplay between geography, climate, and vegetation. Humboldt (**Figure 2.4A**) was both a meticulous naturalist and an integrative, conceptual scientist—keenly aware that fundamental laws of nature could be discovered through the careful study of spatial variation in climates, soils, and associated life-forms. After studying the works of Pierre André Latreille on arthropods, and Georges Cuvier on reptiles, Humboldt further generalized Buffon's law to include most terrestrial biotas—plants as well as animals. He studied in great detail climatic associations of plant communities, and he invented the isobar and isotherm (lines joining points of equal atmospheric pressure and temperature) to help map and visualize these associations. Humboldt noted that the floristic zonation that Johann and Georg Forster described along latitudinal gradients could also be observed at a more local scale along elevational gradients. After conducting many floristic surveys, including those along the slopes of Mount Teide (3718 m; **Figure 2.4B**) of Tenerife Island (Canary Islands) and Mount Chimborazo of the Andes (2500 m; **Figure 2.4C**), Humboldt concluded that even within regions, plants were distributed in elevational zones, or **floristic belts**, ranging from equatorial tropical equivalents at low elevations to boreal and arctic equivalents at the summits.

While Humboldt's (1807) narrative of his travels through the tropical ("Equinoctial") regions of the New World inspired a generation of scientists—including Darwin, Wallace, and many others—to embark on their own explorations of exotic biotas, it was his *Essay on the Geography of Plants* (Humboldt 1805) that is generally recognized as a genuine watershed in the history of integrative earth sciences, including biogeography and ecology (see Jackson 2009). In addition to his meticulous yet holistic descriptions of

(A)

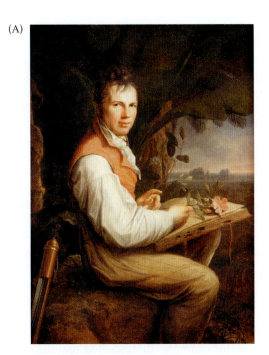

(B)

(C)

FIGURE 2.4 Alexander von Humboldt (A) is widely regarded as the "father of phytogeography" for his detailed and incisive descriptions of the distributions of plants, including those along the slopes of Mount Teide (B) on Tenerife Island in the Canary Achipelago, and those of Mount Chimborazo (C) in the Ecuadorian region of the Andes. (A by Friedrich Georg Weitsch, 1806; B courtesy of Leticia Ochoa; C courtesy of Maury McKinney.)

covariation in climate and vegetation, Humboldt's *Essay* showcases an illustration of these patterns—*The Physical Tableau of the Andes and the Neighboring Countries*—which became a seminal exemplar for visualization in biogeography and ecology (see Chapter 3). As the distinguished paleontologist, glaciologist, and geologist Louis Agassiz (1807–1873) remarked in his eulogy of Humboldt, before him "we had no graphic representation of complex natural phenomena which made them easily comprehensible."

In summary, Humboldt, Forster, Willdenow, and other visionaries of that era were setting a new research agenda—one that called for careful observations of patterns, combined with more integrative explanations for the interplay of processes influencing the geography of nature (in particular, the associations among geography, climates, and plants). One of Humboldt's friends and colleagues, Swiss botanist Augustin P. de Candolle (1778–1841), added another important insight that would prove fundamental to ecology: Not only are organisms influenced by light, heat, and water, but they compete for these resources as well—that is, their distributions are also influenced by interactions with other species. Candolle emphasized the distinction between biotic provinces or regions (which he termed "habitations") and local habitats (or "stations"—a term earlier used by Linnaeus). Later, he returned to Johann and Georg Forster's observations on insular floras, adding that while species number is most strongly influenced by island area, other factors—such as island age, volcanism, climate, and isolation—also influence floristic diversity. Thus, some of the key elements of ecological biogeography and modern ecology were established by the early 1800s. Moreover, in an essay published in 1820, Candolle appears to be the first to write about competition and the struggle for existence, a theme that would prove central to the development of Darwin and Wallace's theory of natural selection.

Biogeography of the Nineteenth Century

By the early 1800s, the first three themes of biogeography (see Table 2.1) were well established. Biogeographers (or "geologists") were studying the distinctiveness of regional biotas, their origin and spread, and the factors responsible for differences in the numbers and kinds of species among local and regional biotas. Buffon's observations on mammals from the New and Old World tropics had been generalized to a law applying to most forms of life. But with this increased appreciation of the complexity and geographic variation of nature, the challenges for biogeography were becoming even greater. Scientists had made great progress in describing fundamental biogeographic patterns, including many that we continue to study today, but explanations for those patterns were wanting. How was it, for example, that isolated regions with nearly identical climates shared so few species? Causal explanations would have to include factors accounting for the similarities as well as the differences among isolated biotas. That is, explanations for Buffon's law would also have to account for the exceptions to it—the cosmopolitan species.

During the next century, the legacy of Buffon's work would be evidenced by a long and continuing succession of explanations for his law, which would lead from a view of a static Earth populated by immutable, cosmopolitan species to one in which the Earth, its climate, and its species were dynamic. It was already well established that in nearly all taxa, the number of local species (at Candolle's "stations") tended to increase with area and to decrease with distance from the Equator, with distance of islands from the mainland, or as we climb from the lowlands to the montane peaks. But why?

To summarize, we see that nearly two centuries ago biogeographers had described many of the more striking patterns we continue to study today, explored the generality of those patterns, and ventured some causal explanations. For the most part, however, they fell short on this last challenge. Their understanding of the mutability of the Earth and its species still lagged far behind their rapidly increasing knowledge of the Earth's geological structure, climatic patterns, and biological diversity. For the field to come of age and to develop more rigorous, testable explanations for its fundamental patterns, it would have to await three important advances of the nineteenth century:

1. a better estimate of the age of the Earth (many early biogeographers were working with an estimate of less than 10,000 years);

2. a better understanding of the dynamic nature of the continents and oceans (i.e., continental drift and plate tectonics);

3. a better understanding of the mechanisms involved in the spread and diversification of species—specifically dispersal, vicariance, extinction, and evolution.

To achieve these advances, biogeography had to draw on new discoveries in geology and paleontology. During the early decades of the nineteenth century, Adolphe Brongniart (1801–1876) and Charles Lyell (1797–1875; **Figure 2.5**), regarded as the fathers of paleobotany and geology, respectively, concluded that the Earth's climate was highly mutable. Both men used the

(A)

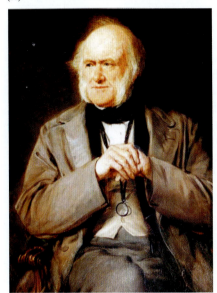

(B)

FIGURE 2.5 (A) Charles Lyell and (B) Adolphe Brongniart, often regarded as the "fathers of geology and paleobotany," respectively. Lyell strongly influenced the development of biogeography in the nineteenth century, largely through his *Principles of Geology*, first published in 1830. (A courtesy of King's College, London; B courtesy of the Council of the Linnaean Society of London.)

fossil record to infer conditions of past climates (Ospovat 1977). They found that many life-forms adapted to tropical climates had once flourished in the now temperate regions of northern Europe. Lyell also documented that sea levels had changed and that the Earth's surface had been transformed by the lifting up and eroding down of mountains. This, he argued, was the only way to account for the existence of marine fossils on mountain slopes. Lyell also provided incontrovertible evidence for the process of extinction. Many fossil forms, once dominant and presumably perfectly adapted to existing climatic conditions, had perished and left no further trace in the fossil record. The causal agent, again, was inferred to be climatic variation and associated changes in sea level. Lyell, however, held firm to the belief that, although extinctions occurred, species were not mutable—at least not to the point that new species arose from existing ones. He also believed that, despite episodes of extinction, the diversity of the Earth on a grand scale remained relatively constant. He resolved this apparent contradiction by suggesting that each episode of extinction was followed by an episode of creation, which established a new set of species perfectly adapted to the altered climatic conditions. Not only were there many sites of creation, but there were many periods of creation as well!

Lyell argued that these great upheavals of the Earth's biotas resulted from physical processes such as mountain building and erosion—processes that had operated continually throughout the history of the Earth. Thus, uniformitarianism (see Chapter 1) replaced earlier catastrophic explanations for the changes in the Earth, its landforms, and its inhabitants. It permitted new thinking about the dynamics of living systems because, after all, their physical and biotic components are historically inseparable. Moreover, Lyell and other geologists such as James Hutton (1726–1797) realized that, given the gradual nature of these geological processes, the Earth must be much older than just a few thousand years. Only with an ancient Earth could they account for the formation and erosion of entire mountains, the submergence of ancient landmasses, and the migration or replacement of entire biotas that were so well documented in the fossil record.

As we shall see, the acceptance of uniformitarianism and the antiquity of the Earth were essential to the theories of both Darwin and Wallace—that organic diversity results from the gradual yet persistent effects of natural selection operating over thousands of generations. Ironically, for most of his long and prestigious career, Lyell rejected the idea that species, like the geological features he studied, were the results or "creations" of physical forces that acted throughout time. Yet, despite his insistent denial of what we now regard as the incontrovertible fact that species arise from other species, Lyell played a key role in the development of biogeography, largely through his treatise entitled *Principles of Geology*. In this massive work he discussed at great length not only the geological dynamics and antiquity of the Earth, but also the geography of terrestrial plants and animals and the distributions of marine algae. This book was essential reading for all serious geologists and naturalists of that era, including the four British scientists discussed below.

Four British scientists

Perhaps most prominent among nineteenth century naturalists were four British scientists: Charles Darwin, Joseph Dalton Hooker, Philip Lutley Sclater, and Alfred Russel Wallace (**Figure 2.6**). Although many other naturalists produced important works during this period, these four are responsible for what would prove to be truly seminal advances in both biogeography and evolutionary biology. They all studied the works of Linnaeus, Buffon, Forster, Candolle, and Lyell. They shared similar experiences as naturalists and explorers—traveling to distant archipelagoes, high mountains, and tropical and temperate regions of the New and Old Worlds. They also shared a common goal: to account for the diversity of life, including the origin, spread, and diversification of biotas. Therefore, in retrospect, it is not surprising that they featured so prominently in the

(A)

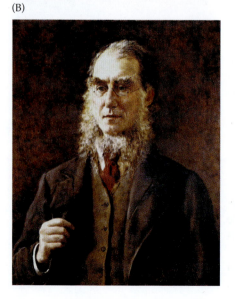

FIGURE 2.6 Four British scientists who, in the mid-nineteenth century, revolutionized our understanding of the history of the Earth and the dynamic distributions and evolution of its organisms: (A) Charles Darwin, (B) Joseph Dalton Hooker, (C) Philip Lutley Sclater, and (D) Alfred Russel Wallace. (A, portrait by George Richmond in the late 1830s; C from Goode 1896; D, portrait by Thomas Sims [ca 1863–1866] © National Portrait Gallery, London.)

(B)

(C)

(D)

development of biogeography and evolutionary biology or that they developed great mutual respect and lasting friendships. The correspondence between these four scientists is, in many cases, as informative and insightful as their formal papers. In their letters to one another, they reveal their shared preoccupation with what we now call biodiversity and their conviction that the key to understanding the natural world was to study the geography of nature. In a letter to Hooker in 1845, Darwin referred to the study of geographic distribution as "that grand subject, that almost keystone of the laws of creation."

With a copy of the first volume of Lyell's *Principles of Geology* in hand, young Charles Darwin set sail in 1831 on a five-year surveying voyage aboard HMS *Beagle*, on which he served as a naturalist and gentleman companion for its captain, Robert Fitzroy (**Figure 2.7**). His travels would take him around the world, where he would visit islands of the Atlantic, Pacific, and Indian Oceans and explore the tropical pampas and the Andes regions of South America. Darwin studied geology, native plants and animals, indigenous peoples, and domesticated animals in an attempt to understand the fundamental laws of nature and their underlying causes. From his diary and collections of specimens, he later published a fascinating account (1839) of his adventures and observations during the voyage of the *Beagle*. Darwin was both intrigued and perplexed by the patterns he observed: the fossils of extinct mammals in Argentina, the presence of seashells at high elevations in the Andes, and the occurrence of unique forms of life on islands. The pat-

FIGURE 2.7 Charles Darwin's voyage on the HMS *Beagle* (1831–1836) was instrumental in the development of his theory of natural selection and the origin of species.

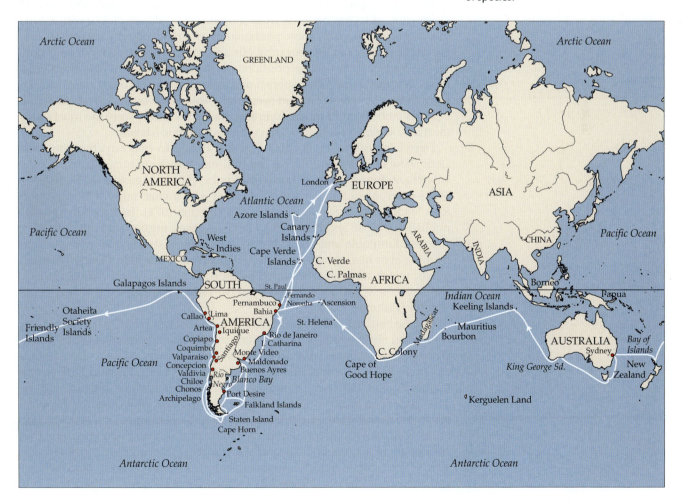

terns of variability in the Galápagos Archipelago, where different forms of tortoises and finches inhabited different islands, suggested to him the idea that geographic isolation facilitates inherited changes within and between populations. On his return to England, Darwin developed his theory of evolution, invoking natural selection as the primary mechanism by which new forms of life arose and are still arising today. This theory ranks as one of the most important scientific advances of all time and is woven into all aspects of biogeography.

The writing and eventual publication of Darwin's theory of evolution by natural selection is itself an interesting story that has been the subject of much review. Darwin drafted a manuscript on the subject in 1845 but withheld the idea from print for 15 years while he continued to amass evidence to support his revolutionary theory. He was finally forced to publish when he received a manuscript from another brilliant scientist, Alfred Russel Wallace, who had independently developed the identical theory based on similar observations of the natural world. The upshot was that a paper by Darwin and one by Wallace were read together before the Linnaean Society of London in 1858; the following year, Darwin's (1859) great book, *On the Origin of Species by Means of Natural Selection, or the Preservation of Favoured Races in the Struggle for Life*, was published and became an immediate best seller.

It is difficult to overemphasize Darwin's contribution to the field of biogeography. He, along with Wallace, provided the basis for understanding changes in the adaptations and distributions of organisms over time as well as space. He proposed that the diversification and adaptation of biotas resulted from natural selection, while the spread and eventual isolation and disjunction of biotas resulted from long-distance dispersal. Darwin argued this latter point perhaps more passionately and more convincingly than anyone in the history of biogeography. His arguments were drawn not only from inferences based on the distributions of isolated biotas, but also from ingenious "experiments" on dispersal and colonization. Through these studies Darwin was able to show that seemingly unlikely events, such as dispersal of seeds embedded in mud clinging to the feet of birds, were the most likely means by which land plants had colonized oceanic islands.

Darwin's arguments on dispersal threatened to overturn the long-held static view of biogeography. In the mid-nineteenth century this older view was championed by Jean Louis Rodolphe (Louis) Agassiz (1807–1873), a Swiss-born paleontologist and systematist who trained most of North America's leading zoologists and geologists (Dexter 1978). In two papers—one published in 1848 and the other in 1855—Agassiz argued that not only were species immutable and static, but so were their distributions, with each remaining at or near its site of creation. However, as a result of Darwin's arguments, which were later bolstered by those of Asa Gray and Alfred Wallace, the static view was abandoned by most biogeographers of the nineteenth century (Fichman 1977).

But the battles to be waged by the dispersalist camp had just begun. They were soon challenged by much more formidable adversaries—the "extensionists," whose ranks were no less prestigious than those of the dispersalists and included such respected scientists as Charles Lyell, Edward Forbes, and Joseph Hooker. Both camps agreed that distributions were dynamic in time and space. The extensionists, however, argued that long-distance dispersal across great and what were then assumed to be permanent barriers was too unlikely to explain distributional dynamics and related phenomena such as cosmopolitan species and disjunct distributions. Rather, they proposed that species had spread across transoceanic, but now submerged, land bridges and ancient continents (**Figure 2.8**). Despite Darwin's great respect for Lyell,

(A)

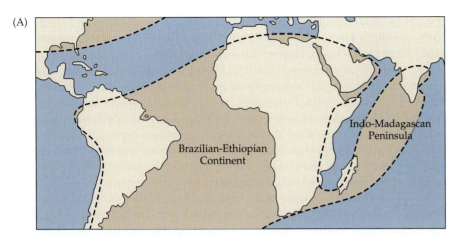

FIGURE 2.8 Hypothetical land bridges were proposed by extensionists during the late nineteenth and early twentieth centuries to account for major disjunctions in the distributions of terrestrial organisms. (After Hallam 1967.)

(B)

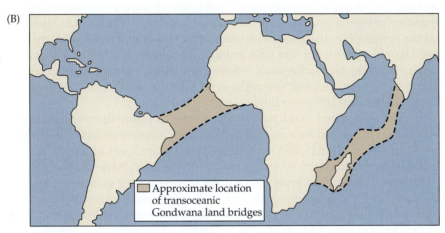

(C)

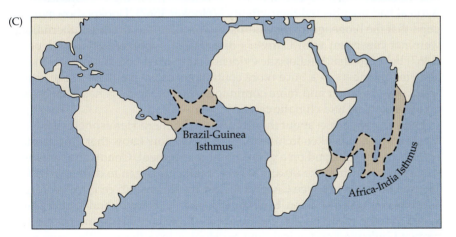

nothing vexed him more than those extensionists who created land bridges "as easy as a cook does pancakes." In a letter to Lyell in 1856, Darwin complained of "the geological strides, which many of your disciples are taking.... If you do not stop this, if there be a lower region of punishment of geologists, I believe, my great master, you will go there."

Despite Darwin's persistent and passionate arguments, the extensionist camp would remain a viable and influential force throughout the latter decades of the nineteenth century. One of its greatest proponents, Joseph Dalton Hooker (see Figure 2.6B), was also a good friend and admirer of both Darwin and Wallace. He was a remarkably ambitious plant collector (Turrill 1953). At the age of 22, he became the assistant surgeon and botanist on an expedition to the Southern Hemisphere led by Sir James Clark Ross on HMS *Erebus* and HMS *Terror* (1839–1843). During this expedition Hooker studied the floras of many southern landmasses, including Australia, Tasmania, Tierra del Fuego, and many archipelagoes in temperate and subantarctic waters. Hooker also traveled to Africa, Syria, India, and North America, where he studied the flora of the Rocky Mountains region with Asa Gray. In addition to his own collections, Hooker studied those of other botanists, especially those from the Galápagos Archipelago and Arctic and Antarctic regions. These collections became the raw material for his analyses of the affinities and probable origins of the flora of each of these regions, group by group.

Upon his return to England in 1843, Hooker formed a friendship with Charles Darwin, whom Hooker had admired from his account of the voyage of the *Beagle*. Within a year of their meeting, Hooker had read Darwin's manuscript on the theory of natural selection—the only person to see the manuscript before Asa Gray in 1857. Hooker shared with Darwin his ideas on the geographic distribution of plants and was one of the few to encourage Darwin to work on, and later publish, *The Origin of Species*. In Darwin's original introduction to the book, Hooker was the only person singled out for acknowledgment. Hooker also influenced Wallace, who dedicated his third book—*Island Life* (1880)—to him.

Hooker recognized that long-distance dispersal across open oceans might account for the occurrence on remote islands of plants with easily dispersed seeds and fruits: "I may add that the large Bean-like seeds of *Entada*, a West Indian climber, are thrown up abundantly on the islands [the Azores] by the Gulf Stream, but never grow into plants, if indeed they ever germinate on their shores. Some years ago a box of these seeds from the Azores was sent to Kew, where many germinated and grew to be fine plants, showing that their immersion during a voyage of nearly 3000 miles [from the West Indies to the Azores] had not affected their vitality" (Hooker 1867). Hooker, however, argued that the biogeographic patterns and peculiarities of the southern floras were not consistent with Darwin's dispersalist hypothesis. Rather, Hooker (1867) believed that the floristic evidence supported "the hypothesis that of all being members of a once more continuous extensive flora, . . . that once spread over a larger and more continuous tract of land, . . . has been broken up by geological and climatic causes."

Hooker was correct about the affinities of Southern Hemisphere plants, although he was mistaken in his explanation. We now know that the ancestors of these plants occurred on a giant southern continent, Gondwana (or Gondwanaland), which broke up and whose fragments began to drift apart about 180 million years ago (see Chapter 8). But the nineteenth century geologists and biogeographers—from Lyell and Hooker to Darwin and Wallace—believed that the relative sizes and positions of the continents and oceans had changed little over geological time. So Hooker hypothesized the emergence and submergence of ancient and undiscovered continents and land bridges to account for the disjunct distributions of closely related plants. Geological evidence for these ancient, transoceanic land bridges never materialized. By the end of the nineteenth century, most geologists and biogeographers (including Lyell) had abandoned the extensionist doctrine.

We must not, however, discount Hooker's contributions. He is still regarded as the founder of causal historical biogeography, and rightly so. He developed and applied many of the principles of what we now call "vicariance biogeography." In addition to emphasizing the importance of a dynamic Earth and changes in climate, Hooker stressed the importance of comparing insular biotas to gain insights into biogeographic processes. He confirmed the earlier observations of Forster and others that insular floras tend to be more depauperate than those on the mainland, and he noted that as island isolation increases, the number of plant species decreases, while the distinctness of the flora increases. He also observed that the most diverse floras tend to be those in lands with the greatest diversity of temperature, light, and other environmental conditions. Hooker drew an analogy between the floras of oceanic islands and those of high mountains, suggesting that both are influenced by the same processes. As we shall see, these observations and principles would be echoed by biogeographers of the twentieth century.

During most of its early history, biogeography was dominated by the contributions of botanists—from Linnaeus and Candolle to Forster and Hooker. It is no great mystery why zoogeography lagged behind phytogeography. There are many more animal species than plant species, and most animals are relatively small and difficult to collect and identify. Over 70 percent of all known species are animals, and most of these are insects. The search for general zoogeographic patterns had to await a better understanding of animal diversity and distributions. By the middle decades of the nineteenth century, however, zoologists had made great strides toward these goals. One of the most prolific and creative intellects of that era was the German scientist, artist, and philosopher Ernst von Haeckel (1834–1919; **Figure 2.9A**). He not only introduced the concept of ecology and the idea that "ontogeny recapitulates phylogeny" but also coined the terms *Protista*, *phylum*, *ecology*, and *phylogeny*, and he described thousands of species of invertebrates. Most relevant were his calls to recognize biogeography (which he termed "chorology") as a new discipline—one that must incorporate the theory of evolution. His maps of phylogenies vividly demonstrated his understanding of evolution as a process that occurs across both space and time (**Figure 2.9B**).

In addition to Darwin and Hooker, two other British zoologists made foundational contributions to biogeography: Philip Lutley Sclater and Alfred Russel Wallace. Sclater (see Figure 2.6C), also a good friend of Darwin, was an eminent ornithologist who described 1067 species, 135 genera, and 2 families of birds. His ambitious career included service as chief executive of-

FIGURE 2.9 (A) Although limited by the absence of information on genetic differences among geographic populations and therefore fraught with inaccuracies, Ernst von Haeckel's 1876 map of the phylogeny and dynamic geography of man (B) demonstrates his understanding that the process of evolution, whether in humans or other lineages, occurs over space as well as time.

(A)

(B)

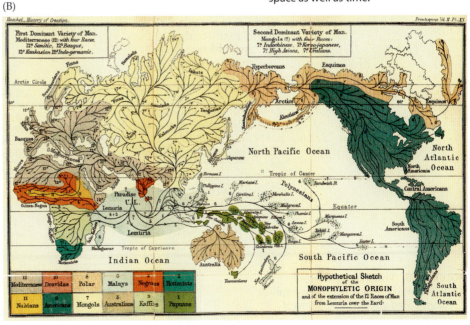

ficer of the Zoological Society of London for over 30 years; he was also a life member of the Royal Geographic Society of London, as well as the first editor of the ornithological journal *Ibis*. Of the hundreds of articles he wrote, only a handful focused on zoogeographic patterns, but they had a major impact on the field.

In 1857 Sclater (1858) read a paper before the Linnaean Society of London entitled "On the General Distribution of the Members of the Class Aves." He proposed a scheme that divided the Earth into biogeographic regions that would reflect "the most natural primary ontological divisions of the Earth's surface." As Sclater acknowledged, many others had developed global schemes of biogeographic divisions, but these earlier systems had given too little regard to the fauna and flora, and too much to arbitrary boundaries such as latitude and longitude. Sclater wished to do much more than just categorize each landmass with a list of species. He wanted to develop a system that would reflect the "ontogeny" (origination and development) of distinctive biotas. He assumed that there were many areas of creation and that most species must have been created in the geographic regions where they now occurred. Therefore, the ontological divisions of the globe could be identified by analyzing the similarity and dissimilarity of biotas. Sclater based his scheme on the group he knew best—birds. He acknowledged, however, that given their impressive dispersal abilities, birds might not be the best group for such an exercise. Accordingly, he limited his work to passerines, which he believed had higher site fidelity than other birds. The system of biogeographic regions Sclater developed is illustrated in **Figure 2.10**. While coarse, it formed the basis for the system of six biogeographic regions we continue to use today.

Whereas Sclater and Darwin spent much of their energies on other subjects, biogeography was Alfred Russel Wallace's (see Figure 2.6D) lifework, and he is considered the father of zoogeography. He developed many of the basic concepts and tenets of the field, combining the insights of others with

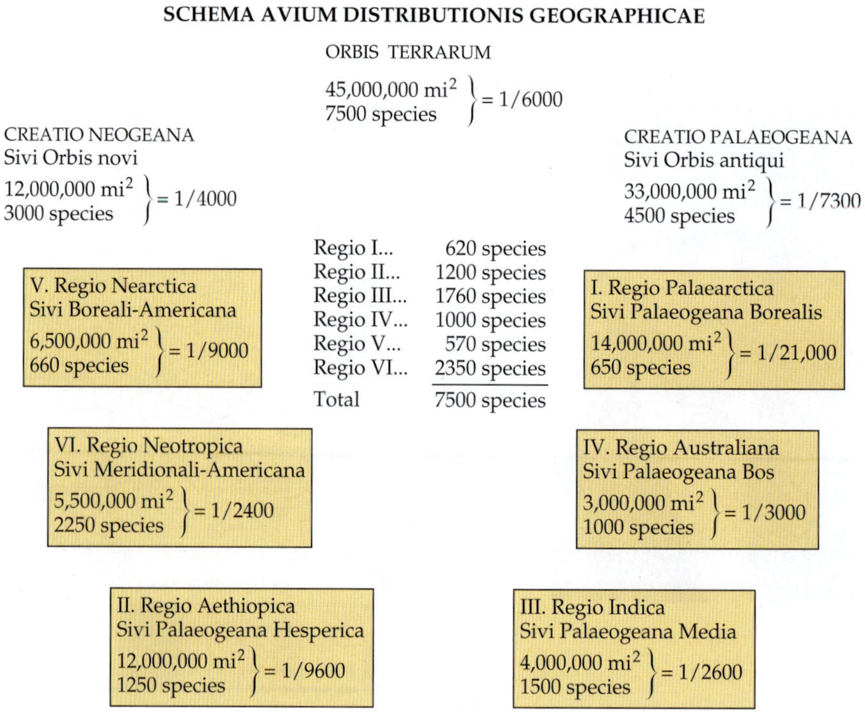

SCHEMA AVIUM DISTRIBUTIONIS GEOGRAPHICAE

ORBIS TERRARUM

45,000,000 mi^2
7500 species $\Big\} = 1/6000$

CREATIO NEOGEANA
Sivi Orbis novi

12,000,000 mi^2
3000 species $\Big\} = 1/4000$

CREATIO PALAEOGEANA
Sivi Orbis antiqui

33,000,000 mi^2
4500 species $\Big\} = 1/7300$

Regio I...	620 species
Regio II...	1200 species
Regio III...	1760 species
Regio IV...	1000 species
Regio V...	570 species
Regio VI...	2350 species
Total	7500 species

V. Regio Nearctica
Sivi Boreali-Americana

6,500,000 mi^2
660 species $\Big\} = 1/9000$

I. Regio Palaearctica
Sivi Palaeogeana Borealis

14,000,000 mi^2
650 species $\Big\} = 1/21,000$

VI. Regio Neotropica
Sivi Meridionali-Americana

5,500,000 mi^2
2250 species $\Big\} = 1/2400$

IV. Regio Australiana
Sivi Palaeogeana Bos

3,000,000 mi^2
1000 species $\Big\} = 1/3000$

II. Regio Aethiopica
Sivi Palaeogeana Hesperica

12,000,000 mi^2
1250 species $\Big\} = 1/9600$

III. Regio Indica
Sivi Palaeogeana Media

4,000,000 mi^2
1500 species $\Big\} = 1/2600$

FIGURE 2.10 Philip Lutely Sclater's (1858) scheme of terrestrial biogeographic regions based on the distributions of passerine birds.

his own experiences and his theory of evolution through natural selection. While Darwin and Sclater wrote only a few papers or chapters on biogeography, Wallace devoted decades to amassing his ideas and accounts in three seminal books: *The Malay Archipelago* (1869—dedicated to Darwin), *The Geographical Distribution of Animals* (1876), and *Island Life* (1880—dedicated to Hooker). We have summarized Wallace's contributions to the field in **Box 2.1**. Many of the concepts enunciated by Wallace were actually introduced by earlier scientists, but Wallace restated, documented, and interpreted them in an evolutionary and biogeographic context. As you read this book, you may wish to refer periodically to this box and note how many of Wallace's ideas are still being investigated by contemporary biogeographers.

Like Darwin and many other young and ambitious scientists of that era, Wallace was captivated by Humboldt's tales of discovery in the New World Tropics. Unlike Humboldt and Darwin, however, Wallace was not from a wealthy family and so had to support his travels by selling specimens of butterflies and birds. Tragically, his first voyage—to Amazonia—ended in disaster when a ship carrying his specimens for sale back to England foundered in the Caribbean. Fortunately, not just for Wallace but for the fields of evolutionary biology and biogeography as well, he recovered and set out on a second voyage, this time an eight-year mission of exploring the diverse

BOX 2.1 *Biogeographic Principles Advocated by Alfred Russel Wallace*

■■ These conclusions are summarized from Wallace's writings and have been verified many times by researchers in the twentieth century:

1. The processes acting today may not be at the same intensity as in the past.
2. Prerequisites for determining biogeographic patterns are detailed knowledge of all distributions of organisms throughout the world, a true and natural classification of organisms, acceptance of the theory of evolution, detailed knowledge of extinct forms, and sufficient knowledge of the ocean floor and stratigraphy to reconstruct past geological connections between landmasses.
3. To analyze the biota of any particular region, one must determine the distributions of its organisms beyond that region as well as the distributions of their closest relatives.
4. Distance by itself does not determine the degree of biogeographic affinity between two regions; widely separated areas may share many similar taxa at the generic or familial level, whereas those very close may show marked differences—even anomalous patterns.

5. Climate has a strong effect on the taxonomic similarity between two regions, but the relationship is not always linear.
6. Long-distance dispersal is not only possible, but also the probable means of colonization of distant islands across ocean barriers; some taxa have a greater capacity to cross such barriers than others.
7. The distributions of organisms not adapted for long-distance dispersal are good evidence of past land connections.
8. The fossil record provides positive evidence for past migrations of organisms.
9. The present biota of an area is strongly influenced by the last series of geological and climatic events; paleoclimatic studies are very important for analyzing extant distribution patterns.
10. Disjunctions of genera show greater antiquity than those of single species, and so forth for higher taxonomic categories.
11. Competition, predation, and other biotic factors play determining roles in the distribution, dispersal, and extinction of animals and plants.

12. In the absence of competition and predation, organisms on isolated landmasses may survive and diversify.
13. Speciation may occur through geographic isolation of populations that subsequently become adapted to local climate and habitat.
14. When two large landmasses are reunited after a long period of separation, extinctions may occur because many organisms will encounter new competitors.
15. Discontinuous ranges may come about through extinction in intermediate areas or through the patchiness of habitats.
16. The islands of the world can be classified into three major biogeographic categories: continental islands recently set off from the mainland, continental islands that were separated from the mainland in relatively ancient times, and distant oceanic islands of volcanic and coralline origin. The biotas of each island type are intimately related to the island's origin.
17. Studies of island biotas are important because the relationships among distribution, speciation, and adaptation are easier to see and comprehend on islands. ■■

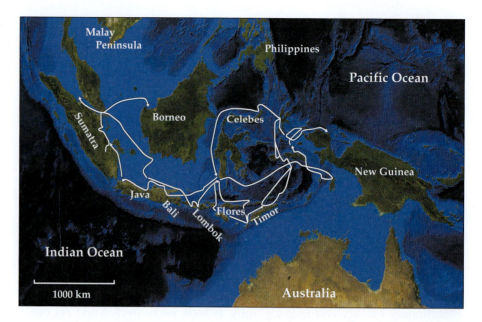

FIGURE 2.11 Alfred Russel Wallace's voyages through the Malay Archipelago (1854–1862) of Indonesia led to major discoveries in evolutionary biology and biogeography.

fauna of the Malay Archipelago, Indonesia (**Figure 2.11**). In addition to independently developing a theory of natural selection equivalent to Darwin's, Wallace was the first person to systematically analyze faunal regions (within Indonesia and across the entire globe) based on the distributions of multiple groups of terrestrial animals. As we see in his map, which forms the front endpaper of this book, Wallace's analysis supported—but greatly expanded upon—Sclater's 1858 scheme (see Figure 2.10). His system was based not just on birds but on vertebrates in general, including nonflying mammals, which because of their limited dispersal abilities, should more precisely reflect the natural divisions of the Earth. Thus, Wallace developed a detailed and very precise map of the Earth's biogeographic regions. His map includes sharp divisions between regions as well as subregions, along with bathymetric divisions reflecting the isolation of different archipelagoes. A distinctive, original contribution was his observation of a sharp faunal gap between the islands of Bali and Lombok in the East Indies, where many species of Southeast Asia reach their distributional limit and are replaced by forms from Australia (Wallace 1860). This biogeographic division among biotas was first named by Thomas Henry Huxley in 1868 and to this day is known as Wallace's line (see Chapter 10; Mayr 1944a; Carlquist 1965).

Other contributions of the nineteenth century

Other scientists of the nineteenth century were also looking for and interpreting important patterns in distributional data. Rather than considering only names and numbers of species, some of these pioneering biogeographers began to analyze geographic variation in the characteristics of individuals and populations (see Table 2.1, theme 4). Chief among these early contributions were the generalized morphogeographic rules of Carl Bergmann (1847), Joel Asaph Allen (1878), and David Starr Jordan (see Chapter 15). **Bergmann's rule** states that in endothermic (warm-blooded) vertebrates, closely related forms from cooler climates tend to have larger body sizes and hence to have smaller surface area-to-volume ratios than races of the same species living in warmer climates. The original explanation for this pattern was that a lower surface area-to-volume ratio helps to conserve body heat in cold environments and, conversely, small size and a relatively large surface area facilitate the dissipation of heat in warmer regions. Along this same line of reasoning, **Allen's rule** states that among endothermic species, limbs and other extremities are shorter and more compact in individuals living in colder climates: Birds and mammals of polar regions tend to be stout with short limbs. A similar phenomenon was reported by D. S. Jordan in 1891 for ectothermic (cold-blooded) teleost fishes inhabiting marine environments. According to **Jordan's law of vertebrae**, as one moves farther from the Equator the vertebrae of teleost fishes become smaller and more numerous. Although the generality and causality of these "rules" have been questioned, these contributions represented pioneering efforts in the study of geographic variation and adaptation. They stimulated the development of the field of physiological ecology and led to important observations on allometry—that is, how traits scale with body size.

In addition to studying geographic variation in the traits of individuals, biogeographers noted that the demographic characteristics of populations also varied across regions. In 1859 Darwin observed that within most genera, species "which range widely over the world are the most diffused in their own country, and are the most numerous in individuals"—in other words, wide-ranging species tend to occur at relatively high densities. This pattern, while not a major emphasis of early biogeography, was rediscovered and documented for a variety of taxa during the twentieth century. The study of how population-level parameters vary along geographic dimensions, and that of allometry and scaling of biological traits, are now central questions of the relatively new disciplinary area termed **macroecology** (see Chapter 15).

Some other early evolutionary "rules" were described by paleontologists who were searching for patterns in the history of life and trying to interpret the fossil record. The theory of **orthogenesis** is an example. This theory states that the evolution of a group continues in only one direction and that this orientation is an intrinsic property of the organism and is not controlled by natural selection. Of course, this theory of evolutionary inertia was used to oppose Darwin's theory of natural selection as the agent of evolutionary change. While few, if any, modern evolutionary biologists believe in orthogenetic trends, evolution is nonetheless conservative and is constrained by the phylogenetic history and preexisting characteristics of lineages (see Chapter 7).

A special type of orthogenesis was described by the American paleontologist Edward Drinker Cope (1840–1897), who noted that lineages of mammals such as horses often exhibit a trend toward increased body size over their evolutionary history. Although there are many exceptions to **Cope's rule**, it does seem that certain advantages of large size have resulted in repeated increases in size in many animal lineages. Large body size, however, also tends to make species susceptible to extinction, so large forms (such as the dinosaurs and many now-extinct groups of giant birds and mammals) die out and are replaced by representatives of new groups, which in turn evolve to a larger size (see Stanley 1975). Borrowing from Jonathan Swift's (1726) wonderful tales of *Gulliver's Travels*, the tendency for smaller life-forms to out-survive large ones during extinction episodes is sometimes termed the Lilliput effect (Twitchett 2007).

Simultaneously with these advances in zoogeography and paleobiogeography, phytogeographers continued to make important contributions to the development of biogeography. In 1860, E. W. Hilgard demonstrated that climatic factors and the activities of plants are directly responsible for converting parent rock into different kinds of soils varying in pH, mineral composition, texture, and so forth (see Chapters 3 and 4). Shortly afterward, a Russian scientist named Vasily Vasilievich Dokuchaev recognized that each soil has a characteristic structure. These two contributions led scientists to understand that the soils of a region are governed in large part by climatic patterns, which influence the breakdown of parent materials, the growth of plants, the decomposition of organic materials, and ultimately the kinds of plants and even animals that occur there.

In the late 1800s, plant researchers in Europe began to develop novel classifications in which plant taxa were grouped according to their external architectural designs or their tolerance of abiotic stresses such as shortages of water or excesses of salts. By the early twentieth century, contributions by the Danish scientists Oscar Drude (1887) and Eugenius Warming (1895, 1905) quickly led to the widely used scheme for classification of plant life by Christen C. Raunkiaer (1934), who defined major types of plants based on the positions of their perennating tissues. An ecological rather than a taxonomic approach was also adopted by the great German phytogeographer Andreas Franz Wilhelm Schimper (1898, 1903), who summarized in elaborate detail the forms and habits of plants from around the world. These foundational contributions led to the development of two vital areas of biogeography and ecology: plant physiological ecology, which seeks to understand how various species are adapted to the habitats in which they are found, and phytosociology, a subdiscipline of plant community ecology that describes how and attempts to explain why certain combinations of plant species, but not others, co-occur in a given habitat (Good 1974).

Early botanists such as Candolle and Humboldt were well aware that different types of vegetation occurred at different elevations, but it was a zoologist, Clinton Hart Merriam, who provided one of the most valuable insights into these broad patterns and is also credited with actually coining the term *biogeography* (Merriam 1892; Ebach and Goujet 2006). Based on his extensive field studies in southwestern North America, Merriam (1894) confirmed that elevational changes in vegetation type and plant species composition are generally equivalent to the latitudinal changes found as one moves from the Equator toward the poles (**Figure 2.12**). He called these belts of similar vegetation "life areas" and, later, "life zones." Although Merriam was not successful in generalizing his concept of life zones to animals and to other regions and continents, he correctly concluded that elevational zonation of vegetation, like latitudinal zonation, is a response of species and communities to environmental gradients of temperature and rainfall.

Nineteenth century biogeographers continued to develop a much more comprehensive understanding of the dynamics of the Earth's climates and how they in turn influence the dynamics of its biotas. Louis Agassiz (1807–1873) studied under Georges Cuvier before emigrating to America. Although, as we noted earlier, Agassiz refused to accept Darwin's theories on the mutability of species, he developed some of the most seminal insights on the dynamics of Earth's climates and its biotas. While visiting the Alps on a return trip to his homeland in 1836, Agassiz studied the dynamics and ecological impacts of alpine glaciers. From these and other observations on distributions of fossils and extant species across the Northern Hemisphere, Agassiz (1840) developed the first comprehensive theory of the "Ice Age" and its influences on Holarctic biogeography. For these contributions, Agassiz is often recognized as the "father of glaciology," and his theory indeed proved fundamental to understanding biogeographic dynamics of plants and animals throughout the Pleistocene (see Chapter 9).

While Agassiz, Cuvier, and other paleontologists were developing a more comprehensive understanding of the geography of nature (past and present) by deciphering the fossil record, other nineteenth century naturalists/ explorers began to turn their attention to a new frontier: the oceans and their biotas. As mentioned earlier, in his *Principles of Geology*, Charles Lyell

(A)

(B)

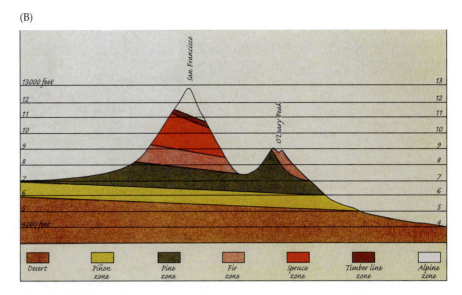

FIGURE 2.12 (A) Clinton Hart Merriam's "life areas," aka "life zones," of North America, which were based on the relationship between climate and vegetation. (B) Elevational distribution of life zones on the San Francisco Peaks of Arizona as viewed from the southeast. (C) Distribution of Merriam's life zones across North America. (Merriam 1890, 1894; photo courtesy of Zenaida Merriam Talbot.)

(1834) discussed the distributions of marine algae. Edward Forbes (1815–1854) produced the first comprehensive work on marine biogeography (1856), in which he divided the marine world into nine horizontal (latitudinal) regions of similar fauna ("homozoic belts"), which he then subdivided into five zones of depth. Echoing the observations of Linnaeus and Forster on elevational and latitudinal variation of terrestrial communities, Forbes (1844) noted that "parallels in depth are equivalent to parallels in latitude." Similarly, Forbes observed that whereas distributions of terrestrial organisms are primarily influenced by climate, soil, and elevation, those of marine organisms are influenced by climate, water chemistry, and depth.

The American geologist James Dwight Dana (1813–1895) expanded on some of Forbes's observations while also providing

(C)

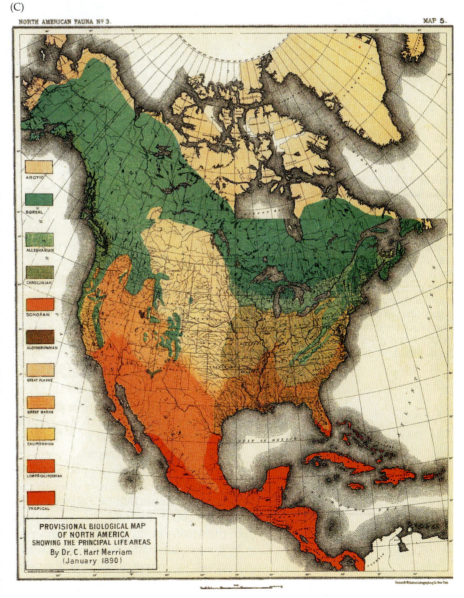

(A)

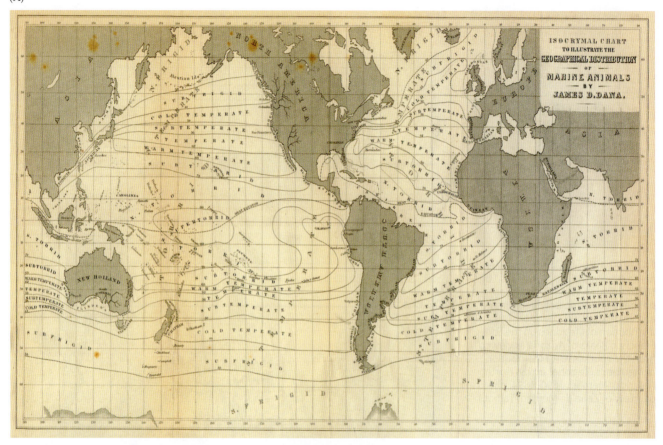

FIGURE 2.13 (A) James Dwight Dana's (1853) scheme of biogeographic regions of marine organisms (*Isocrymal Chart to Illustrate the Geographical Distribution of Marine Animals*) was based on "isocrymes," or lines of equal minimum sea surface temperatures. (B) In contrast, Philip Lutley Sclater's classification of biogeographic regions in the marine realm was based largely on endemic fauna, in particular, marine mammals. Like his scheme for terrestrial regions, this system included six regions and was based on the distributions of geographically localized genera.

some fundamental insights on volcanism, mountain building, the origins of the continents and the seas, and distributions of corals and crustaceans. Dana (1853) showed that distributions of marine organisms could be explained and mapped using "isocrymes" (i.e., lines of equal minimum sea surface temperature; **Figure 2.13A**). Near the end of the nineteenth century, John Murray (1895), G. Pruvot (1896), and Arnold Ortmann (1896) also published important general works on marine geography. In 1897, Philip Sclater (primarily known for his ornithological studies) published a paper on the distributions of marine mammals. Much as in his earlier scheme for terrestrial zoogeographic regions (see Figure 2.10), Sclater divided the marine realm into six regions based on the distributions of geographically localized genera of marine mammals (**Figure 2.13B**; Sclater 1897).

Despite these early contributions, just as zoogeography lagged behind phytogeography, marine biogeography would not come of age until more explorers focused on this final frontier and, in particular, its geological dynamics. A generally accepted system of biogeographic "regions" for the marine realm was not developed until the publication of John Briggs's work in the 1970s, over a century after the terrestrial system of biogeographic regions was established by Sclater and Wallace. While there remained much to learn about the marine frontier, the insights garnered from explorations of the oceans in the twentieth century not only would expand the study of biogeographic patterns to marine life, but would challenge and eventually overturn the long-held view on the permanence of the oceans and continents (see Chapter 8).

(B)

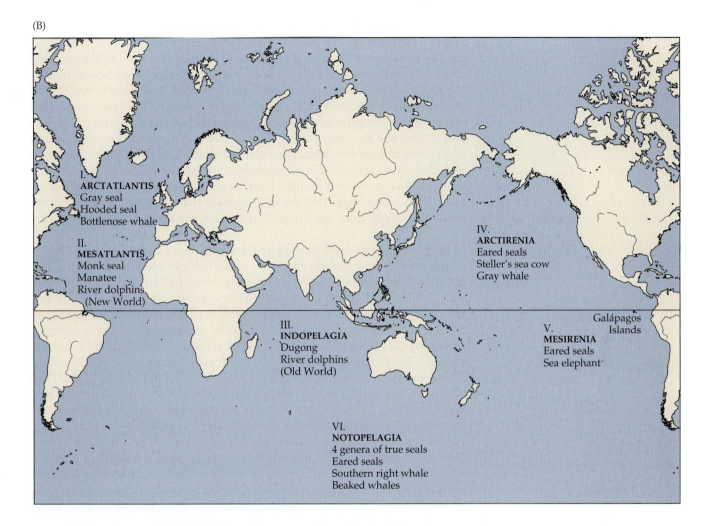

The First Half of the Twentieth Century

From 1900 through the early 1960s, several major trends in research had extraordinary effects on biogeography. Paleontology in particular deserves credit for providing new and fascinating descriptions of faunal changes across the continents. Numerous paleontologists—especially C. Ameghino, W. D. Matthew, G. G. Simpson, E. H. Colbert, A. S. Romer, E. C. Olson, and B. Kurtén—described the origin, dispersal, radiation, and decline of land vertebrates. They showed that new groups increase in number of species, radiate to fill new ecological niches, expand their geographic ranges, and become dominant over and contribute to the extinction of older forms. Thus our present-day faunas have extremely long and complex histories that can be understood only by elucidating the **phylogenies** of the groups and their geographic dynamics across the continents and ocean basins.

As mentioned earlier, explanations for how terrestrial organisms could have spread from one landmass to another were legion. Investigators proposed an incredible number of short-lived, transoceanic land bridges, now vanished; former continents, now sunken; or once-joined continents, now drifted apart. During this period, tempers often flared as investigators debated alternative explanations for how groups arrived in such isolated places as Australia, the Galápagos, and the Hawaiian Islands. Many such explanations are now rejected or considered unlikely, but these efforts infused phylogeny, paleoclimatology, and geology into biogeographic syntheses.

A common question arising out of such inquiries concerned "centers of origin." Where were the cradles of formation and diversification of various groups or biotas? Biogeographers of the early part of the twentieth century returned to the same challenges tackled by their predecessors, but with much-expanded evidence and unparalleled intensity. A common goal of systematic monographs (treatments of the phylogenetic affinities and taxonomic classification of a group) was to propose a probable place of origin for the group and to describe its subsequent spread and diversification. The fossil record was spotty, so such reconstructions were often made from characteristics that included the geographic distributions of contemporary organisms, using arguments that now seem circular. Some authors were bolder and more dogmatic than others regarding the assumptions that could be used to deduce centers of origin from such limited data (see, for example, Matthew 1915; Willis 1922).

Early in the twentieth century, researchers began to investigate patterns of phenotypic and genetic variation within single species. Following the leads of Bergmann and Allen on ecogeographic patterns, Joseph Grinnell, Lee R. Dice, and Bernhard Rensch demonstrated close relationships between geographic and ecological properties of the environment and patterns of morphological variation within and among species. Subsequently, physiological and genetic variations were related to distributions in nature through the pioneering studies of Theodosius Dobzhansky on fruit flies (*Drosophila*) and Jens Clausen, David Keck, and William Hiesey (1948) on plants. By the early 1940s, evolutionary biologists were building on Darwin's synthesis to investigate patterns of geographic variation and to infer the mechanisms responsible for the origin of new species. A long list of scientists contributed to our understanding of the modes of speciation. Ernst Mayr provided especially important contributions in the fields of systematics, evolution, and historical biogeography. Out of this work arose one unifying theme: the biological species concept, which states that a species is definable as a group of populations that is reproductively isolated from all other such groups (Mayr 1942, 1963; see Chapter 7). Moreover, Mayr's studies of patterns in the geographic distributions of species and of the underlying evolutionary mechanism now known as allopatric speciation enabled an important new synthesis in evolutionary biology and biogeography (Mayr 1944a,b, 1965a,b, 1969, 1974).

By the middle of the twentieth century, many authors had produced new and more general syntheses of evolutionary and biogeographic patterns for various taxa and biotas. These included studies focusing on vertebrates by Phillip J. Darlington (1957) and George Gaylord Simpson (1956, 1965); marine zoogeography by Sven Ekman (1953) and Joel W. Hedgpeth (1957); vascular plants by Evgenii V. Wulff (1943), Stanley A. Cain (1944), and Ronald Good (1974 [first published in 1947]); and island biogeography of animals by Philip J. Darlington (1938) and of plants by Sherwin Carlquist (1965). Moreover, an impressive body of literature began to accumulate on ecological biogeography, which was summarized for animals by Richard Hesse, W. C. Allee, and Karl P. Schmidt (1951) and G. Niethammer (1958), and for plants by Pierre Dansereau (1957). These works—and others too numerous to list here—established biogeography as a respected science that could provide insights for other fields, including ecology, evolutionary biology, and conservation biology. In a similar manner, biogeographers would continue to draw upon advances in these complementary fields as well as those in geology. This led to a revitalization that would continue throughout the remainder of the twentieth century.

Biogeography since the 1950s

Three major developments have revitalized biogeography in the last 50 years: the acceptance of plate tectonics, the development of new phylogenetic methods, and advances in our abilities to compile, visualize, and analyze geographically explicit information. Until the 1960s, most biogeographers considered the Earth's crust to be fixed and without lateral movement. As we shall discuss in Chapter 8, a theory of plate tectonics and continental drift was first introduced by Antonio Snider-Pelligrini in 1858. However, his radical and largely unsubstantiated theory was readily dismissed by his contemporaries. It would take another 60 years before the theory resurfaced with the arguments of a German meteorologist, Alfred Lothar Wegener (1912a,b, 1915, 1966), and an American geologist, Frank Bursley Taylor (1910). Wegener's theory of continental drift, published and revised from 1912 until his tragic death in 1930, was based on extensive geological and biological evidence for great movements of the continents. But again, the theory was harshly criticized and rejected by most biogeographers, including distinguished leaders such as George Gaylord Simpson and Philip Jackson Darlington, the latter arguing that it was much easier to move animals than it was to move entire continents.

The theory of continental drift was not generally accepted until the late 1960s, when geological evidence for the process became irrefutable. Once accepted, however, the theory of plate tectonics revolutionized historical biogeography and required scientists to rethink their explanations for many distributional patterns. Joseph Dalton Hooker had not been far off the mark—changes in the connections among landmasses had indeed resulted in important movements and disjunctions of biotas. The connections he proposed, however, rather than resulting from the emergence and drowning of mysterious continents and land bridges, were caused by great movements of existing continents. Regardless of one's interpretation of a distribution, the explanation now had to be consistent with this geological history of the Earth's surface.

Since the 1960s, biologists also have made tremendous strides toward achieving phylogenetic classifications that trace the history and relationships of taxa, thus vastly improving our understanding of how biotas are, and have been, related. Guidelines for reconstructing phylogenies were already available for traditional systematics (Simpson 1961) and for phylogenetic systematics (or cladistics; Hennig 1950), but the issues regarding the history of lineages over space and time became crystallized in 1966 when the work of Willi Hennig was published in English. Phylogenetic research was transformed from a discipline that compared anatomical and other similarities among taxa into one in which the historical geography and diversification of a lineage is reconstructed and the evolutionary relationships among species are quantified.

Early in the historical development of the field, biogeographers attempted to use information on geographic distributions of related species to reconstruct the histories of continents and other landmasses. In the mid-1800s, Asa Gray pioneered research on plant **disjunctions**—cases in which two closely related species are widely separated in space (see Chapter 10). Since then, biogeographers have been fascinated by disjunctions because they can reveal past land or water connections or long-distance dispersal between two regions. Interest in disjunctions as a means of reconstructing the evolution and geographic dynamics of lineages was revitalized in the middle decades of the twentieth century, in part through the writings of L. Croizat (1958, 1960,

1964). With the availability of new phylogenetic approaches, the study of disjunct species—now called **vicariants**—has taken a central position, particularly in historical research. Advances in molecular biology and our ability to assess the affinities among geographically isolated species have continued to provide new and more reliable means of reconstructing the phylogenetic and biogeographic development of lineages and biotas (see Chapters 11 and 12; see also Funk 2004; Briggs 2007).

Up until the 1960s, the emphasis in biogeography had been primarily on these issues of historical biogeography—concentrating mostly on the phylogenies of groups and their means of dispersing into and surviving in different regions and habitats. By the late 1950s, George Evelyn Hutchinson began to focus attention on questions about the processes that determine the diversity of life and the number of species that coexist in local areas or habitats. Ecologists began to emphasize the importance of competition, predation, and mutualism in influencing the distributions of species and their coexistence as ecological communities. Of all the work in ecological biogeography during this period, perhaps the most influential were the collaborations between one of Hutchinson's most accomplished students—Robert H. MacArthur—and an equally distinguished naturalist and evolutionary biologist—Edward O. Wilson. Their seminal paper and subsequent monograph proposed a radically new, dynamic theory to account for the diversity of species on islands (1963, 1967; see Chapter 13). While Agassiz's view of static distributions of continental biotas had long been abandoned by most biogeographers, island biogeographers still assumed that species distributions within archipelagoes changed only very slowly, that is, on an evolutionary but not ecological timescale. MacArthur and Wilson's equilibrium theory of island biogeography challenged this view and quickly became the new paradigm of the field. Their work changed the direction of ecological biogeography by focusing attention on a new set of questions: abstract questions about patterns of distribution and species diversity. And they suggested that these patterns reflected the operation of fundamental biogeographic processes—those leading to species additions (via immigration and speciation) versus species loss (via extinction) (see Chapters 13 and 14).

Questions about species diversity and coexistence, in addition to dominating the fields of ecology and theoretical biogeography, spawned other areas of inquiry such as those on the extent to which the dispersal, establishment, and radiation of a lineage are **stochastic** (i.e., random) or **deterministic** (i.e., predictable if the underlying mechanisms are understood; Raup et al. 1973; Simberloff 1974b; Stanley 1979; Eldredge and Cracraft 1980). Moreover, these abstract questions stimulated experimental testing of biogeographic concepts (e.g., the island defaunation experiments of Simberloff and Wilson 1969) as well as new mathematical ways of quantifying and analyzing observations (e.g., Pielou 1977a, 1979; Upton and Fingleton 1990; Cressie 1991; Manly 1991; Maurer 1994; Gotelli and Graves 1996).

Although some ecologists tended to emphasize the roles of interspecific interactions in influencing the distribution of species and communities (MacArthur and Connell 1966; MacArthur 1972; Whittaker 1975), others emphasized the importance of the abiotic environment in limiting the distributions of individuals and populations and, in turn, determining the diversity of species in different regions. Important advances in techniques and instrumentation since the mid-1960s have permitted a flurry of research in this area, which has resulted in a tremendous amount of information that must be selectively integrated with biogeography. In particular, advances in molecular biology, along with those in computer technology and related techniques, including satellite imagery, geographic information systems (GIS), and spatial statistics (geostatistics), have enabled quantum leaps in our ability to explore, describe, and explain the historical, ecological, and geographic development of the world's biotas (**Figure 2.14**).

In summary, many great scientists have contributed to the development of biogeography. All of them, from Linnaeus, Buffon, and Candolle to Simpson, Croizat, MacArthur, and Wilson, have shared a common goal—understanding the origin, spread, and diversification of biotas. In the mid-1700s, Buffon gave biogeography its first principle, while others tested its generality and broadened the field to explore a diversity of other patterns and their causal forces. As the field progressed, theories of natural selection and evolution, plate tectonics, and immigration/extinction dynamics provided new mechanisms to explain long-standing patterns, while new comparative and experimental methods, phylogenetic analyses, and advances in computer science

(A)

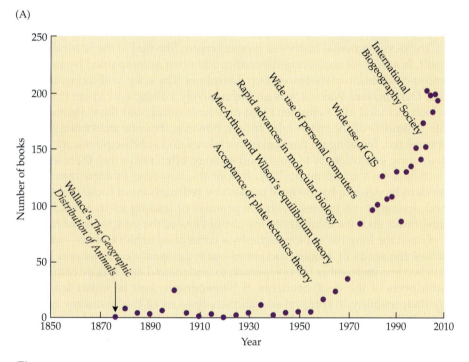

(B)

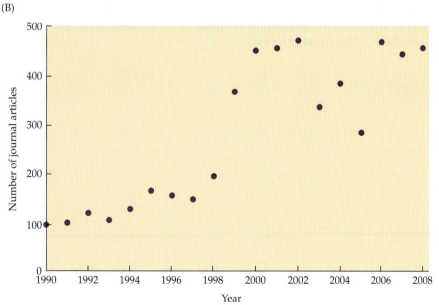

FIGURE 2.14 The maturity and vitality of biogeography is evidenced by the increasing number of publications in the field, including (A) the number of books on the subject published since Alfred Russel Wallace's seminal work, *The Geographical Distributions of Animals* (the only biogeography book published in 1876), and (B) the number of journal articles on biogeography published over the past two decades. (Tallies based on searches using OCLC [Online Computer Library Center, Inc.] databases—WorldCat (books) and FirstSearch (journal articles)—using the keyword *biogeography*, which standardizes the search but likely underestimates publication activities in this field, which often assume but do not explicitly declare *biogeography* as one of the keywords.)

and statistics provided invaluable tools for visualizing and analyzing geographic patterns. By the 1960s, biogeography had finally come of age as a rigorous and respected science. It has continued to build on those gains and flourish in subsequent decades. Since Wallace published his seminal book—*The Geographical Distribution of Animals*—in 1876, the number of publications on biogeography has increased exponentially, with most of the increase taking place in the past few decades (see Figure 2.14). There are now at least four journals whose primary focus is biogeography, and in 2000 the International Biogeography Society (www.biogeography.org) was founded to foster the advancement of all studies of the geography of nature. In recent years, the society's membership has grown to include over 700 scientists from more than 35 countries. The society has contributed to the development of two books—*Foundations of Biogeography*, which describes the historical development of the field (including reproductions of 72 classic works), and *Frontiers of Biogeography*, which attempts to identify some of the most promising directions for advancing the field.

The developmental history of biogeography, like that of the natural world itself, is a story of increasing diversification often accompanied by isolation and specialization of its descendant disciplines. The science of Linnaeus, Buffon, Darwin, and Wallace was an impressively holistic and integrative one, which included all phenomena that may have influenced the distribution of life. Scientists of the twentieth century, however, became increasingly more specialized, focusing on particular disciplines and subdisciplines, including taxonomy, evolutionary biology, and ecology, often restricting themselves to particular taxa and time periods. It appears, however, that the frontiers of modern biogeography are now marked by reticulations of those descendant and divergent disciplines, with the promise of leading us to a much more integrative and, we think, more insightful view of the geography of nature.

Given the long list of biogeography's conceptual achievements—in themselves the seeds of whole disciplines—one can easily comprehend how it has become impossible for one person to understand and follow completely all aspects of the field. Students of biogeography can be either frustrated by their inability to comprehend all the subtleties of this awesome body of knowledge or encouraged by the prospect of using biogeography as a focal point to synthesize many separate disciplines and to acquire a unique perspective on the development, diversity, and distribution of life. In the next chapter we explore the predictable, spatial variation in Earth's physical setting, which forms the **geographic template** for all biogeographic patterns and their underlying, causal processes.

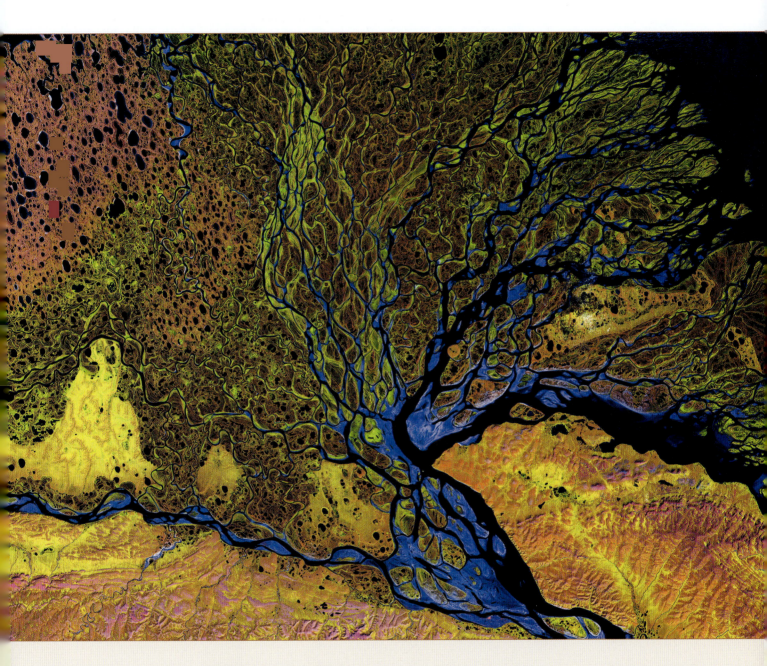

THE GEOGRAPHIC AND ECOLOGICAL FOUNDATIONS OF BIOGEOGRAPHY

TWO

The Chemical and
Ecological Foundations
of Sustainability

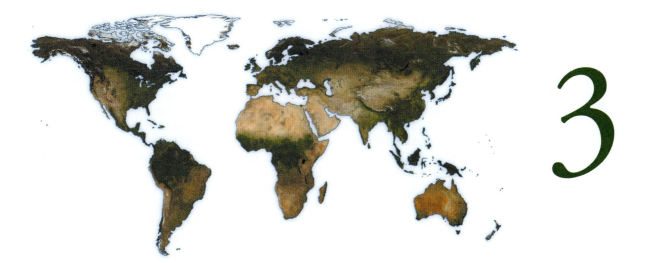

3

THE GEOGRAPHIC TEMPLATE:
Visualization and Analysis of Biogeography Patterns

Definition and Components of the Geographic Template

Organisms can be found almost everywhere on Earth: from the cold, rocky peaks of high mountains to the hot, windswept sand dunes of lowland deserts; from the dark, near-freezing depths of the ocean floor to the steaming waters of hot springs. Some organisms even live around hydrothermal vents in the deep ocean, where temperatures exceed 100° C (but the water does not boil because of the extreme pressure). Yet no single kind of organism lives in all of these places. Each species has a restricted geographic range in which it encounters a limited range of environmental conditions. Polar bears and caribou are confined to the Arctic, whereas palms and corals are rare outside the tropics. There are a few species, such as *Homo sapiens* and the peregrine falcon, that we call cosmopolitan because they are distributed over all continents and over a wide range of latitudes, elevations, climates, and habitats. These species, however, are not only exceptional but also much more limited in distribution than they appear at first glance. Humans and peregrines, for example, are absent from the three-fourths of the Earth that is covered with water; indeed, they are little more than rare visitors to large expanses of the terrestrial realm with extremely harsh environments.

The geographic template

As we will point out in later chapters, we may need to invoke unique historical events or ecological interactions with other organisms to account for the limited geographic ranges of some species, but the most obvious patterns in the distributions of organisms occur in response to variations in the physical environment. In terrestrial habitats, these patterns are largely determined by climate (primarily temperature and precipitation) and soil type. The distributions of aquatic organisms are limited largely by water temperature, salinity, light, and pressure.

As Edward Forbes and other early biogeographers observed (see Chapter 2), climate, soil type, water chemistry, and a long list of other environmental conditions vary in a highly nonrandom manner across geographic gradients of latitude, elevation, depth, and proximity to major landforms such as coastlines and mountain ranges. In addition, regardless of whether we consider the aquatic or terrestrial realms, the environmental variables tend to exhibit strong spatial autocorrelation, or what is sometimes referred to as **distance-decay**, which simply means that similarity of environmental conditions between sites decreases as we compare more distant sites.

Taken together, these nonrandom patterns of spatial variation in environmental conditions constitute a multifactor **geographic template**, which forms the foundations of all biogeographic patterns. That is, most biogeographic patterns ultimately derive from this very regular, spatial variation in environmental conditions. For example, as we move from low to high latitudes, from the Equator to the poles, or from the ocean through estuaries and then upstream, environmental temperatures tend to cool, and this either directly or indirectly influences biotic communities along each of these geographic and environmental gradients (see Chapter 15). Diversity, species composition, and vital processes (e.g., productivity and decomposition) of biotic communities along these gradients change in a highly predictable manner, with similarity among communities being much higher for those located in close proximity along the geographic template.

The conceptual model illustrated in **Figure 3.1** may prove useful in understanding a great diversity of biogeographic patterns. Again, the fundamental layer or foundation for all patterns of variation across space is the geographic template. Organisms can then respond to this highly nonrandom spatial variation through adaptations (both behavioral and physiological), dispersal, or speciation, or if unsuccessful at any of these responses, populations may suffer local extinction. All of these responses, taken together, ultimately determine the geographic distributions and patterns of variation among populations, species, and communities across geographic gradients. To more fully account for the diversity of patterns biogeographers study, this conceptual model must include at least two additional layers of complexity—both of which may be viewed as feedback or interactions among system components. Not only are species distributions influenced by environmental conditions, but those distributions and related patterns are also influenced by interactions among the species themselves (e.g., mutualism, parasitism, and

FIGURE 3.1 All biogeographic patterns are ultimately influenced by the geographic template. Distributions and patterns of geographic variation of life-forms and communities result from (1) the highly nonrandom patterns of spatial variation in environmental characteristics across the Earth (i.e., the geographic template); (2) responses of biotas (including adaptation, dispersal, evolution, or extinction) to this variation; (3) interactions among species (e.g., competitive exclusion and mutualistic interactions); (4) impacts of particular species—"ecosystem engineers" such as beavers, prairie dogs, and humans—on the geographic template; and (5) the temporal dynamics of the geographic template (including the so-called TECO events—plate Tectonics [including orogeny], Eustatic changes in sea level, Climate change, and Oceanographic processes).

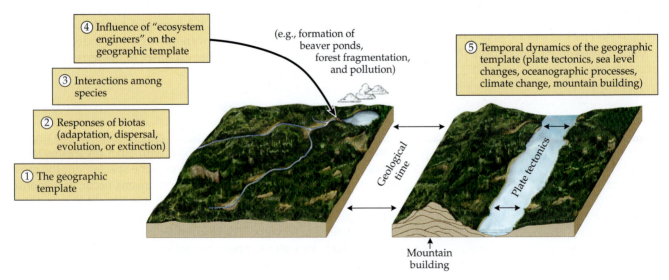

④ Influence of "ecosystem engineers" on the geographic template

③ Interactions among species

② Responses of biotas (adaptation, dispersal, evolution, or extinction)

① The geographic template

(e.g., formation of beaver ponds, forest fragmentation, and pollution)

⑤ Temporal dynamics of the geographic template (plate tectonics, sea level changes, oceanographic processes, climate change, mountain building)

Geological time

Plate tectonics

Mountain building

competitive exclusion; see Chapter 4). Furthermore, some species—often referred to as ecosystem engineers (e.g., beavers, prairie dogs, and humans)—can modify the geographic template itself. Even before our own species rose to become the world's dominant **ecosystem engineer**, microscopic and otherwise "primitive" life-forms fundamentally altered Earth's atmosphere, thereby increasing its oxygen content and its abilities to store heat, modifying its climate, and eventually molding and recasting the geographic template across both the terrestrial and aquatic realms. We still need to introduce one final but nonetheless fascinating layer of complexity to this conceptual model. The geographic template itself is dynamic—not just across space but across time as well. Over the 3.5 billion year history of life on Earth, and indeed even over much shorter timescales, climates have swung dramatically from glacial to interglacial episodes of the Pleistocene (including most of the past 2 million years) and from the much earlier periods of the so-called snowball Earth to the global sauna of the Mid Eocene (roughly 50 million years ago; see Chapter 8). Just as fundamental, and sometimes driving these major climatic shifts, Earth's crust itself is highly dynamic, emerging from the mantle below to drift, split, and collide in a kaleidoscopic jigsaw puzzle known as plate tectonics and continental drift (see Chapter 8).

All of these dynamics in space and time are themselves driven by two great engines, which are powered by two different sources of energy. The energy stored in the Earth's core at the time the solar system was formed is also supplemented by compressional heating due to gravity. A portion of this energy is gradually but continuously being dissipated through the Earth's mantle and crust and ultimately out into space. This transfer of heat energy moves and shapes the Earth's crust, shifting the positions of the crustal plates containing the continents, thrusting up mountains, creating or consuming ocean basins, and causing earthquakes and volcanic eruptions.

The other great engine is driven by the energy of the sun. Radiant energy emitted by the sun strikes the Earth's surface, where it is absorbed and converted into heat, warming the surface of the land and water and the atmosphere above them. The resulting differences in the temperature and density of air and water cause them to move over the Earth's surface, both horizontally and vertically, creating the Earth's major wind patterns and ocean currents. The heating of surface water also causes evaporation, and the resulting water vapor is carried by the air and redeposited as rain or snow. These processes, which are responsible for the Earth's climate and for many physical characteristics of its oceans and fresh waters, are the subject of the following sections of this chapter.

Climate

Solar energy and temperature regimes

SOLAR RADIATION AND LATITUDE. Sunlight sustains life on Earth. Not only does solar energy warm the Earth's surface and makes it habitable, but it also is captured by green plants and converted into chemical forms of energy that power the growth, maintenance, and reproduction of most living things.

According to the principles of thermodynamics, heat is transferred from objects of higher temperature to those of lower temperature by one of three mechanisms: (1) conduction, a direct molecular transfer (especially through solid matter); (2) convection, the mass movement of liquid or gaseous matter; or (3) radiation, the passage of waves through space or matter. Heat flows as radiant energy from the hot sun across the intervening space to the cooler Earth. When incoming solar radiation strikes matter such as water or soil, some of it is absorbed, and the matter is heated. Some solar radiation is ini-

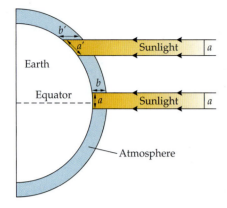

FIGURE 3.2 Average input of solar radiation to the Earth's surface as a function of latitude. Heating is most intense when the sun is directly overhead, when incoming solar radiation strikes perpendicular to the Earth's surface. The higher latitudes are cooler than the tropics because the same quantity of solar radiation is dispersed over a greater surface area (*a'* as opposed to *a*) and passes through a thicker layer of filtering atmosphere (*b'* as opposed to *b*).

tially absorbed by the air, particularly if it contains suspended particles of water or dust (e.g., clouds), but most passes through the sparse matter of the atmosphere and is absorbed by the denser matter of the Earth's surface. This surface is not heated uniformly. Soil, rocks, and plants absorb much of the radiation and may be heated intensely. Although air is heated to some extent by absorption of incoming solar radiation, most of the heating of air occurs at the Earth's surface, where it is warmed by direct contact with warm land and water, by latent heat released by the condensation of water, and by long-wave infrared radiation emitted from the surfaces of warm objects such as leaves and bare soil. In contrast, much of the solar radiation striking the surface of the oceans is reflected back toward the atmosphere, while the rest penetrates and warms layers of the water column below.

The angle of incoming radiant energy relative to the Earth's surface affects the quantity of heat absorbed. The most intense heating occurs when the surface is perpendicular to incident solar radiation, for two reasons: (1) the greatest quantity of energy is delivered to the smallest surface area; and (2) a minimal amount of radiation is absorbed or reflected back into space during passage through the atmosphere, because the distance it travels through air is minimized (**Figure 3.2**). This differential heating of surfaces at different angles to the sun explains why it is usually hotter at midday than at dawn or dusk, why average temperatures in the tropics are higher than at the poles, and why south-facing hillsides are warmer than north-facing ones in the Northern Hemisphere (and the reverse in the Southern Hemisphere).

Because the Earth is tilted 23.5° from vertical on its axis with respect to the sun, solar radiation falls directly, that is perpendicularly, on different parts of the Earth during an annual cycle. This differential heating produces the seasons. The seasons are also characterized by different lengths of day and night. Only at the Equator are there exactly 12 hours of daylight and darkness every 24 hours throughout the year (**Figure 3.3**). At the spring and fall equinoxes

FIGURE 3.3 Seasonal variation in day length with latitude is due to the inclination of the Earth on its axis. At either equinox (A), the sun is directly overhead at the Equator, and all parts of the Earth experience 12 hours of light and 12 hours of darkness each day. At the summer solstice (B) in the Northern Hemisphere, however, the 23.5° angle of inclination causes the sun to be directly over the Tropic of Cancer, while the Arctic Circle and areas farther north experience 24 hours of continuous daylight; at the same time, all regions in the Southern Hemisphere experience less than 12 hours of sunlight per day, and the sun never rises in areas south of the Antarctic Circle.

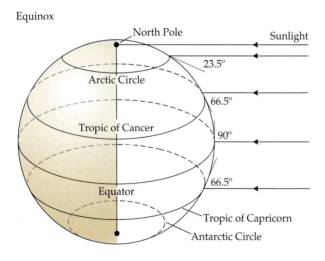

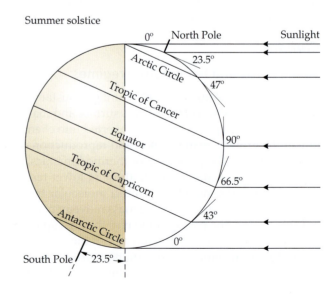

(March 21 and September 22, respectively) the sun's rays fall perpendicularly on the Equator, equatorial latitudes are heated most intensely, and every place on Earth experiences the same day length. At the summer solstice (June 22), sunlight falls directly on the Tropic of Cancer (23.5° N latitude). The Northern Hemisphere is heated most intensely and begins to experience longer days than nights and enjoys summer, while the Southern Hemisphere has winter. At the winter solstice (December 22), the sun shines directly on the Tropic of Capricorn (23.5° S latitude), and the Southern Hemisphere enters its summer, while the Northern Hemisphere experiences winter, cold temperatures, and long nights. The result of all this is that the seasonality of climate increases with increasing latitude, with the Tropics of Cancer and Capricorn marking the northern- and southernmost latitudes, respectively, that receive direct sunlight at least one day each year (i.e., on the north or south summer solstice). At the Arctic and Antarctic Circles (66.5° latitude), there is one day each year of continuous daylight when the sun never sets and one day of continuous darkness—each at a solstice. Although every location on the Earth theoretically experiences the same amount of daylight and darkness over an annual cycle, the sun is never directly overhead at high latitudes; however, considerable solar radiation is absorbed during the long summer days. Temperatures in excess of 30° C are commonly recorded in Alaska. The warmest days typically are in July (after the summer solstice) because of the time lag required to heat the Earth's surface.

THE COOLING EFFECT OF ELEVATION.   The processes just described account for seasonal and latitudinal variation in temperature, but it remains to be explained why it gets colder as we ascend to higher altitudes. The fact that Mount Chimborazo and Mount Kilimanjaro (nearly on the Equator in tropical South America and East Africa, respectively) are capped with permanent ice and snow seems to be in conflict with our intuitive expectation and with our previous discussion on the intensity of solar radiation in the tropics. Mountain peaks are nearer the sun, so why are they cooler than nearby lowlands? The answer lies in the physical and thermal properties of air. As a climber moves up a mountain, the length (and the pressure) of the column of air that lies above the climber decreases. Thus, the density and pressure of air decrease with increasing elevation. When air is blown across the Earth's surface and forced upward over mountains, it expands in response to the reduced pressure. Expanding gases undergo what is called **adiabatic cooling**, a process where they lose heat energy as their molecules move farther apart (temperature essentially is a measure of activity and the frequency of collisions of molecules). The same process occurs in a refrigerator as freon gas expands after leaving the compressor. The rate of adiabatic cooling of dry air is about 10° C per km elevation, as long as no condensation of water vapor and cloud formation occurs.

Higher elevations are also colder because the less dense air allows a higher rate of heat loss by radiation back through the atmosphere. Water vapor and carbon dioxide in the atmosphere would typically absorb radiant heat—the so-called **greenhouse effect**. As the name implies, these gases act like the glass in a greenhouse: They allow the short wavelengths of incoming solar radiation to pass through, but they trap the longer wavelength radiation (infrared, or heat) emitted by surfaces that have been warmed by the sun. The resulting warming effect of greenhouse gases is most pronounced in moist lowland areas, where air is laden with water vapor and carbon dioxide. In contrast, mountains and deserts typically experience extreme daily temperature fluctuations, because the local atmosphere is thinner or drier, respectively.

Winds and rainfall

WIND PATTERNS. Differential heating of the Earth's surface also causes the winds that circulate heat and moisture. As we have already seen, the most intense heating is at the Equator, especially during the equinoxes, when the sun is directly overhead. As this tropical air is heated, it expands, becomes less dense than the surrounding air, and rises. This rising air produces an area of reduced atmospheric pressure over the Equator. Denser air from north and south of the Equator flows into the area of reduced pressure, resulting in surface winds that blow toward the Equator (**Figure 3.4**). Meanwhile, the rising equatorial air cools adiabatically, becomes denser, is pushed away from the Equator by newly warmed rising air, and eventually descends again at about 30° N and S latitude (the Horse Latitudes). This vertical circulation of the atmosphere results in three convective cells (Hadley, Ferrel, and Polar) in each hemisphere, with warm air ascending at the Equator and at about 60°N and S latitude, and cool air descending at about 30°N and S and at the poles. These circulating air masses produce surface winds that typically blow toward the Equator between 0° and 30° and toward the poles between 30° and 60°. In the upper atmosphere between the convective cells are the jet streams—high-speed winds blowing approximately parallel to the Equator.

The surface winds do not blow exactly in a north–south direction; instead, they appear to be deflected toward the east or west by the **Coriolis effect**. Although the Coriolis effect is often called the *Coriolis force*, it is not a force but a straightforward consequence of the law of conservation of angular momentum. Every point on the Earth's surface makes one revolution every 24 hours. Because the circumference of the Earth is about 40,000 km, a point at the Equator moves from west to east at a rate of about 1700 km/h^{-1}. But the parallel lines (really circles) of latitude become increasingly shorter as we move from the Equator to the poles. Therefore, points north or south of the Equator travel a shorter distance each 24-hour rotation of the Earth; that is, they move at a slower rate than points closer to the Equator. Consider what happens at the Equator if you shoot a rocket straight upward. Where does it come down? Right where it was launched; the rocket travels not only up and down but also eastward at a rate of 1700 km/h^{-1}, the same rate as the Earth moving beneath it. Now suppose the rocket is propelled northward away from the Equator. It continues to travel eastward at 1700 km/h^{-1}, but the Earth underneath it moves ever more slowly as the rocket travels farther north, and consequently its path appears to be deflected toward the right. The Coriolis effect describes this tendency of moving objects to veer to the right in the Northern Hemisphere and to the left in the

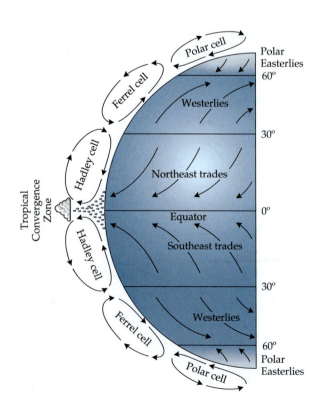

FIGURE 3.4 Relationship between vertical circulation of the atmosphere and wind patterns on the Earth's surface. There are three convective cells (Hadley, Ferrel, and Polar) of ascending and descending air in each hemisphere. As the winds move across the Earth's surface in response to this vertical circulation, they are deflected by the Coriolis effect, which produces easterly trade winds in the tropics, and Westerlies at temperate latitudes. The latitudinal locations of these cells shift with the seasons as the latitude of the most direct sunlight and most intense heating (i.e., the Tropical Convergence Zone) shifts between 23.5° N and 23.5° S (the Tropics of Cancer and Capricorn, respectively).

Southern Hemisphere. The winds approaching the Equator from the Horse Latitudes appear to be deflected to the west and are therefore called northeast or southeast **trade winds**. (Winds are described based on the direction of their *sources*.) Winds blowing toward the poles between about 30° and 60° N and S latitude are called **Westerlies** and are deflected to the east (see Figure 3.4). These winds naturally were very important to commerce in the days of sailing ships, when both the Westerlies and the trade winds (or *trades*) got their names. Ships coming to the New World from Europe traveled south to the Canary Islands and Azores in tropical latitudes to intercept the trades before heading westward, but they returned to Europe at higher latitudes with the Westerlies behind them.

The surface winds, influenced by the Coriolis effect, initiate the major ocean currents. The trade winds push surface water westward at the Equator, whereas the Westerlies produce eastward-moving currents at higher latitudes. Responding to the Coriolis effect, these water masses are deflected toward the east or west, and the net result is that the ocean currents move in great circular gyres—clockwise in the Northern Hemisphere and counterclockwise in the Southern Hemisphere (**Figure 3.5**). Warm currents flow from the tropics along eastern continental margins; as these water masses reach high latitudes, they are cooled, producing cold currents that flow down the western margins.

PRECIPITATION PATTERNS.   By superimposing these patterns of temperature, winds, and ocean currents, we can begin to understand the global distribution of rainfall. We will also need some additional background in physics. As air warms, it can absorb increasing amounts of water vapor evaporated from the land and water. As air cools, it eventually reaches the **dew point**, at which it is saturated with water vapor. Further cooling then results in condensation and the formation of clouds. When the particles of water or ice in clouds become too heavy to remain airborne, rain or snow falls. In the tropics, the cool-

FIGURE 3.5   Main patterns of circulation of the surface currents of the oceans. In each ocean, water moves in great circular gyres, which move clockwise in the Northern Hemisphere and counterclockwise in the Southern Hemisphere. These patterns result in warm currents along the eastern coasts of continents and cold currents along the western coasts. Note the Pacific equatorial countercurrent: the small current along the Equator that flows from west to east, opposite to the gyres, and which strengthens in some years to cause the El Niño phenomenon.

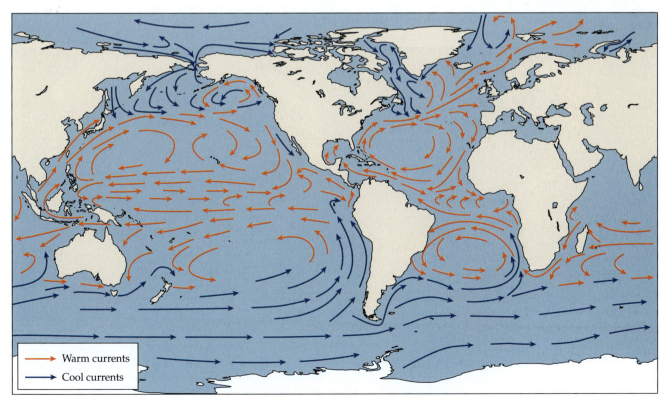

Warm currents
Cool currents

ing of ascending warm air laden with water vapor produces heavy rainfall at low and middle elevations, where rain forests and cloud forests occur. Rainy seasons in the tropics tend to occur when the sun is directly overhead and the most intense heating occurs. The tropical grasslands of Kenya and Tanzania in East Africa, which lie virtually on the Equator but at higher elevations than rain forests, experience two rainy seasons each year, corresponding approximately to the equinoxes (when the **Tropical (Inter-Tropical) Convergence Zone** shifts overhead; see Figure 3.4), and two dry seasons, which correspond to the solstices. In contrast, the area around the Tropic of Cancer in central Mexico has only one principal rainy season—in the summer. Most tropical regions have at least one dry season.

At the Horse Latitudes, where cool air descends from the upper atmosphere, two belts of relatively dry climate encircle the globe. Descending air warms and can therefore absorb more moisture, drying the land. In these belts lie most of the Earth's great deserts (including the Mojave, Sonoran, and Chihuahuan in southwestern North America; the Sahara in North Africa; and the Arid Zone in central Australia; **Figure 3.6**), and adjacent to these deserts are regions of semiarid climates and grassy or shrubby vegetation. Here the seasonality of climate is very marked on the western sides of continents, which experience **Mediterranean climates**. Parts of coastal California, Chile, the Mediterranean region in Europe, southwestern Australia, and southernmost Africa have dry, usually hot summers and mild, rainy winters. In winter, when the land tends to be cooler than the ocean water, the westerly winds bring ashore warm, moisture-laden air; cooling and condensation occur, and fog and rain result. In summer, when the land is warmer than the ocean, the *Westerlies* blowing inland from the cold offshore currents are warmed; this increases their capacity to hold more water vapor and creates the relatively dry climates on land. The effects of cold currents are even more pronounced in localized regions of western South America and southwestern Africa, where they contribute to the formation of coastal deserts that are the driest areas on Earth (Amiran and Wilson 1973).

FIGURE 3.6 Major deserts of the world are not randomly distributed but tend to occur near 30° N or S latitude or along the leeward slopes of mountains, where descending air masses undergo adiabatic warming, increasing their capacity to hold water and drying local environments. (Meigs, 1953.)

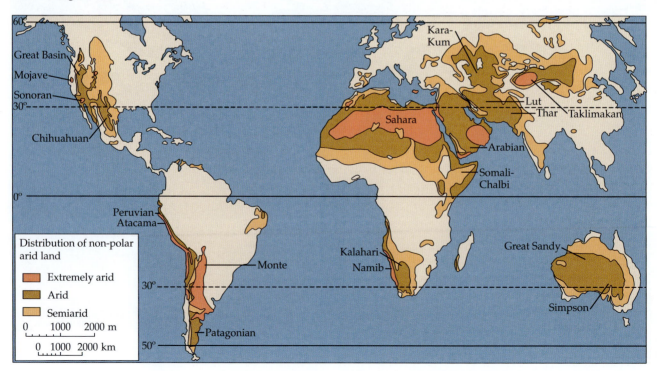

(A)

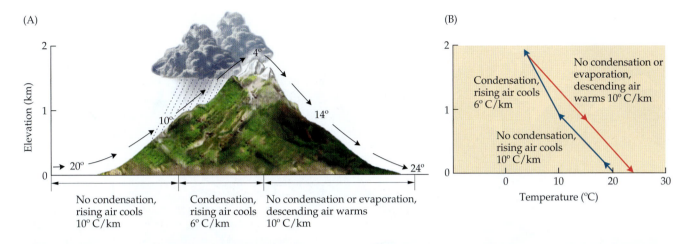

Elevation (km)

20°   10°   4°   14°   24°

No condensation,
rising air cools
10° C/km

Condensation,
rising air cools
6° C/km

No condensation or evaporation,
descending air warms
10° C/km

(B)

Condensation,
rising air cools
6° C/km

No condensation or
evaporation,
descending air
warms 10° C/km

No condensation,
rising air cools
10° C/km

Temperature (°C)

(C)

(D)

Several of the deserts between 30° and 40° N and S latitude are located not only on the western sides of continents, but also on the eastern sides of major mountain ranges. As westerly winds blow over the mountains, they are cooled until eventually the dew point is reached and clouds begin to form. Condensation releases the latent heat of evaporation, so wet air cools adiabatically at a slower rate than dry air—6° C per km of elevation, as opposed to 10° C per km for dry air. As the air continues to rise and cool, most of its moisture falls as precipitation on the western side of the mountain range. When the air passes over the crest and begins to descend, the remaining clouds quickly evaporate, and the dry air warms at the faster rate. This **rain shadow** effect causes the warm, dry climates found on the leeward sides of temperate mountains (**Figure 3.7**). Thus, for example, the Sierra Nevada in California has lush, wet forests of giant sequoias and other conifers on its western slopes but arid woodlands of piñons and junipers on its eastern (leeward) slopes; a bit farther east, with an elevation below sea level, lies Death Valley—the driest place on the North American continent. Similarly, the Monte Desert of South America is in the rain shadow on the eastern side of the Andes in Argentina.

These global patterns of temperature and precipitation frequently are summarized in climatic maps like the one in **Figure 3.8**. Such maps are useful, but they can be misleading because they fail to show the local-scale patterns of spatial and temporal variation that influence the abundance and distribution of organisms.

FIGURE 3.7 Factors causing rain shadow deserts. (A) Air blowing over a mountain cools as it rises, water vapor condenses, and the air loses much of its moisture as rain on the windward side, so the leeward side experiences warm, dry winds. (B) The rate of change in air temperature with elevation is greater for drier air, resulting in warmer, drier conditions on the leeward side than at the same elevation on the windward side. Comparison of vegetation on opposite sides of the central mountain range on the tropical island of Puerto Rico. (C) On the northeastern side, which receives the moisture-laden trade winds, lush rain forests occur. (D) In marked contrast, the southwestern side lies in a rain shadow, has a hot and dry climate, and has cacti and other plants typical of desert regions. (A,B after Flohn 1969; C © Lawrence Sawyer/istock; D © Michele Falzone/AGE Fotostock.)

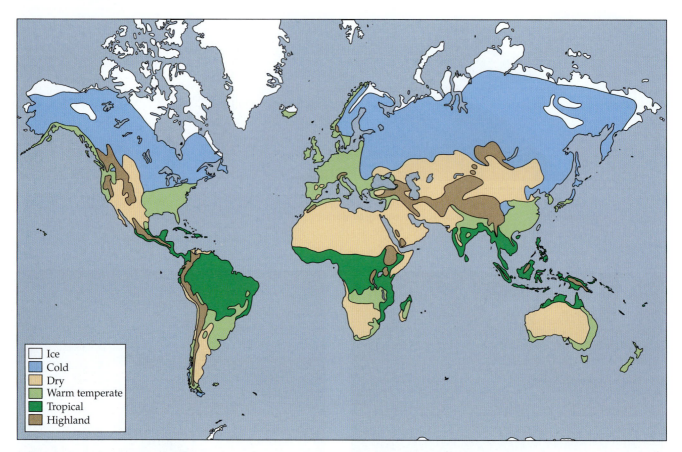

FIGURE 3.8 Major climatic regions of the world. Note that these regions occur in distinct patterns with respect to latitude and the positions of continents, oceans, and mountain ranges. (After Strahler and Strahler 1973.)

Legend:
- Ice
- Cold
- Dry
- Warm temperate
- Tropical
- Highland

SMALL-SCALE SPATIAL AND TEMPORAL VARIATION. The processes that we have just described on a global scale can also produce great climatic variation on a local scale. The effect of mountains is particularly great, as we can illustrate with several examples. From Tucson, Arizona, it is only 25 km by a paved road to the top of Mount Lemmon (2800 m elevation) in the Santa Catalina Mountains. But the climate and the plants at the summit are far more similar to those in northern California and Oregon—1500 km to the north—than to those in the desert just below (**Table 3.1**). Similarly, the spruce–fir forests on the summit of the Great Smoky Mountains in Tennessee are more similar to the boreal forests of northern Canada than to the deciduous forests in the valleys below. Puerto Rico, which lies in the Caribbean Sea at 18° N latitude, is about 150 km long and 50 km wide and has a central mountainous backbone

TABLE 3.1 *The Influence of Elevation on Climate*

Site	Elevation (m)	Temperature (°C)				Mean annual precipitation (cm)
		Mean January	Mean July	Lowest	Highest	
Tuscon, Arizona	745	10.8	30.7	−9.4	46.1	27.3
Mt. Lemmon, Arizona	2791	2.3	17.8	−21.7	32.8	70.0
Salem, Oregon	60	3.2	19.2	−24.4	40.0	104.3

Source: Data from U.S. Weather Bureau.

Note: Two of the sites are near one another in Arizona; the third site is in Oregon. Note that the climate of the high-elevation site in Arizona—Mt. Lemmon—is much more similar to that of Salem, Oregon, 1700 km to the north, than to that of Tucson, only 25 km away but 2000 m lower in elevation.

rising to about 1000 m. The lowlands on the northern and eastern sides are lush and tropical, but much more rain falls at higher elevations on the northeastern slopes, and this is where the best-developed rain forests are found. So much moisture is lost as the northeast trade winds traverse the mountains that the southwestern corner of Puerto Rico is extremely dry; the cacti and shrubby vegetation that occur there remind a visitor of the deserts and tropical thorn forests of western Mexico (see Figure 3.7C, D). Even more dramatic are the combined effects of the cold Humboldt Current and the rain shadow cast by the westward-flowing southeast trades over subtropical regions of the Andes in Peru and Chile. Here, up to 10 m of precipitation per year drenches the tropical rain forests on the eastern slope, while there may be several years in succession with no rain at all in the Atacama Desert on the western slope. Note that because they are located in regions where the prevailing winds come from different directions, the Atacama Desert (10°–15° S, with westward-flowing southeast trades) and the Monte Desert (around 30° S, with eastward-flowing Westerlies) are located on opposite sides of the Andes.

There are also year-to-year and longer-term temporal variations in climate. The entire global system of moving air masses, ocean currents, and patterns of precipitation fluctuates on a five-to-seven-year cycle. The fluctuations appear to be initiated by events in the vast tropical Pacific Ocean (although similar events occur in the tropical Atlantic). This pattern is called the **El Niño Southern Oscillation**, or **ENSO** for short. We are uncertain about its initial cause—perhaps a variation in the output of solar radiation or intrinsic fluctuations in the atmosphere–ocean system. Whatever the ultimate cause, the pattern of tropical ocean circulation changes. While the primary ocean currents are the great hemispheric gyres mentioned above, close inspection of Figure 3.5 will show a small current running west to east right along the Equator. It is called the **equatorial countercurrent** because it runs in the opposite direction of the gyres. It is usually small, as the figure suggests, but in some years it becomes much stronger and pushes warm water away from the Equator up the coasts of North and South America. As the westerly winds pass over this warm water, they pick up moisture and carry it onto the adjacent continents, causing heavy precipitation in the winter when the land is colder than the offshore waters. This phenomenon is called **El Niño** (literally, "little boy" in Spanish, the predominant common language in western South America) because the resulting rains tend to fall around Christmas, the celebration of the birth of the Christ child. El Niño years are the only times that it rains in the extremely arid coastal deserts of South America. The seemingly lifeless Atacama Desert bursts into bloom, as plants that have survived as seeds (dormant in the soil) germinate, grow, and reproduce. In contrast, ENSO events also are characterized by reduced coastal upwelling, which leads to dramatic reductions in nutrients and affects the entire food chain of oceanic communities. Seabirds and other marine organisms along the Pacific coast and in the Galápagos Islands suffer wholesale reproductive failure and mortality due to the unusual rain and reduced upwelling.

Other kinds of temporal variation can also have important biogeographic consequences. For example, a hurricane may pass over a Caribbean island only once in a century on average, yet these rare, unpredictable storms wreak incredible devastation. Hurricanes probably are one of the primary causes of disturbance on Caribbean islands. Such large, infrequent storms may increase or decrease biodiversity; while they can inundate tiny islets, causing extinction of some terrestrial animals and plants, they also clear space in forests and coral reefs, which facilitates the continued existence of competitively inferior species (Spiller et al. 1998). The more general lesson for biogeographers cannot be overstated: Seemingly unpredictable and extremely rare events such as hurricanes, cataclysmic volcanic eruptions, long-distance chance dispersal to isolated oceanic islands, or collision with a wayward asteroid can fundamentally transform the development and distributions of life on Earth.

Soils

Primary succession

Except for the polar ice caps and the perpetually frozen peaks of the tallest mountains, almost all terrestrial environments on Earth can and do support life. Areas of bare rock and other sterile substrates created by volcanic eruptions or other geological events are gradually transformed into habitats capable of supporting living ecological communities by a process called **primary succession**. This process involves the formation of soil, the development of vegetation, and the assembly of a complement of microbial, plant, and animal species.

We cannot understand the distribution of soils without knowledge of the role of climate and organisms in successional processes. The type of vegetation covering a region depends primarily on three ingredients: climate, type of soil, and history of disturbance. For example, three distinct vegetation types (temperate deciduous forest, pine barrens, and salt marsh) occur in northern New Jersey in close proximity to one another but on different soil types (Forman 1979). Moreover, if a mature stand of deciduous forest is destroyed, such as at the hands of humans or by natural fire, it is not reestablished immediately. Instead, certain plant species colonize the area and are in turn replaced by later colonists, beginning with weedy pioneer species and continuing until the mature or **climax** vegetation is reestablished. This process of community development on existing soils (as opposed to that on volcanic ash or bare rock) is called **secondary succession**. Throughout this process, both the microclimate and the soil of the site also change—becoming more favorable for some species and less favorable for others.

Soil formation is both a chemical and a biological process resulting from weathering of rock and the accumulation of organic material from dead and decaying organisms. The process by which new soil is formed from mineral substrates is usually long and complicated. Physical processes such as freezing and thawing, and water and wind erosion, break down the parent rock material. Organisms also play key roles: Lichens hasten the weathering of rock; decaying corpses of plants, animals, and microbes add organic material; the activities of roots and microbes alter the chemical composition of the soil; and burrowing animals mix and aerate it.

Totally organic soils (or **histosols**), such as peat, form in certain unusual environments where cold, acidic, or other conditions inhibit decomposition of accumulating plant and animal debris. In fact, the rate of soil formation varies widely depending largely on the nature of the parent material and the climatic setting. The formation of shallow soils may take thousands of years in Arctic and desert regions, where temperature and moisture regimes are extreme (e.g., McAuliffe 1994). For example, soils only a few centimeters deep cover much of eastern Canada where the retreat of the last Pleistocene ice sheets left bare rock only about 10,000 years ago. In other cases, especially when soils are formed from sand, lava, or **alluvial** materials in regions with warm, moist climates, primary succession can be amazingly rapid. In 1883, the small tropical island of Krakatau in Indonesia experienced an explosive volcanic eruption that exterminated all living things and left only sterile volcanic rock and ash. Organisms rapidly recolonized Krakatau from the large neighboring islands of Java and Sumatra, and by 1934—only 50 years after the eruption—35 cm of soil had been formed and a lush tropical rain for-est containing almost 300 plant species was rapidly developing (Docters van Leeuwen 1936; Thornton 1996; Whittaker 1998).

Formation of major soil types

Anything we write about soils must be a gross oversimplification because both the classification and the distributions of soils are very complex, even controversial. Visit the vast flat plains of the United States or the Ukraine and you will find just one or a few soil types distributed as far as the eye can see, but in other geographic regions—especially mountainous areas—soil and geological maps are mosaics that look like complicated abstract paintings. For example, Great Britain has a series of unusual organic soil types formed in cold, wet environments, as well as soils overlaid onto a complex geological foundation and greatly modified by centuries of human activities.

We can begin to appreciate the diversity and distribution of soils by study-ing the four major processes that produce the primary (or *zonal*) soil types. These so-called **pedogenic regimes** are those that typically occur in habitats characterized by temperate deciduous and coniferous forests (**podzolization**), tropical forests (**laterization**), arid grasslands and shrublands (**calcification**), and waterlogged tundra (**gleization**).

Podzolization occurs at temperate and subarctic latitudes and at high el-evations where temperatures are cool and precipitation is abundant. In such climates plant growth may be substantial, but the low temperatures inhib-it microbial activity, so organic matter, called **humus**, accumulates. As the humus decays, organic acids are released and carried downward (**leached**) through the soil profile by percolating water. The hydrogen ions of these ac-ids tend to replace cations that are important for plant growth, such as cal-cium, potassium, magnesium, and sodium, which are removed by leaching from the soil (**Figure 3.9A**). This process leaves behind a silica-rich upper soil containing oxidized iron and aluminum compounds, but few cations. Co-niferous forests, which thrive in such acidic conditions, are a characteristic vegetation type on podzolic soils.

In the humid tropics, which experience high temperatures and heavy rain-fall, little humus can accumulate, because microbes and other organisms rap-idly break down dead organic material. In the absence of organic acids, oxides of iron and aluminum precipitate to form red clay or a bricklike layer (later-ite). The heavy rainfall causes silica and many cations such as potassium, so-

FIGURE 3.9 Schematic representations of the four major pedogenic regimes showing the resulting soil profiles: (A) podzolization, (B) laterization, (C) calcifi-cation, and (D) gleization. (After Strahler 1975.)

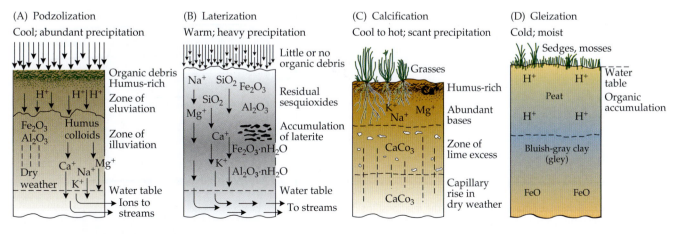

dium, and calcium to be leached out of the soil (**Figure 3.9B**), leaving behind a firm and porous soil with very low fertility. In some areas, if the tropical forest cover is removed, the organic material and its bound nutrients are easily lost and the intense equatorial sun bakes the exposed lateritic soils hard, retarding secondary succession and making the area unsuitable for agriculture.

Calcareous soils typically occur in arid and semiarid environments, particularly in regions where thick layers of calcium carbonate were deposited beneath ancient shallow tropical seas. Where rainfall is relatively low, such that evaporation and transpiration exceed precipitation, cations are generally not leached out. Instead they are carried downward through the soil profile to the depth of greatest water penetration, where they precipitate and form a layer rich in calcium carbonate (**Figure 3.9C**). In desert soils, the scanty rainfall penetrates only a short distance below the surface, where it leaves behind a rocklike layer of calcium carbonate called **caliche** or **petrocalcic horizons**. In regions where precipitation is higher, water and roots penetrate deeper into the soil profile, leading to the formation of deep, fertile soils rich in organic material and essential nutrients such as potassium, nitrogen, and calcium. Such soils are typical of tallgrass and shortgrass prairie habitats, although little of the former remains because these soils are so highly prized for agriculture.

In cold and wet polar regions, gleization is the typical process of soil formation. At the permanently wet (or frozen) surface, where the low temperatures and waterlogged conditions prevent decomposition, acidic organic matter builds up, sometimes forming a layer of peat that can be several meters thick (**Figure 3.9D**). Below this organic upper layer an inorganic layer of grayish clay, containing iron in a partially reduced form, typically accumulates. While few nutrients are lost through leaching, the highly acidic conditions cause nutrients to be bound up in chemical compounds that cannot be used by plants. Thus, gley soils typically support a sparse vegetation of acid-tolerant species.

The above descriptions represent four idealized cases of soil formation processes. Given the complex variation in parent material and climate over the Earth's surface, pedogenic regimes vary in complex but predictable ways. The processes of soil formation and soil types described above occur where the chemical composition of the parent material is typical of the common rock types: sandstone, shale, granite, gneiss, and slate. The soils that are derived from these "typical" rocks are called **zonal soils**. A simplified summary of the relationship between climate and zonal soil type is given in **Figure 3.10**.

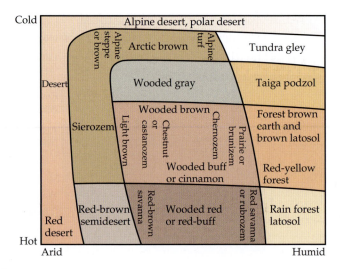

FIGURE 3.10 Schematic diagram depicting the relationships between major soil types and climate, showing that different combinations of temperature and precipitation cause the formation of distinctive soil types. (After Whittaker 1975.)

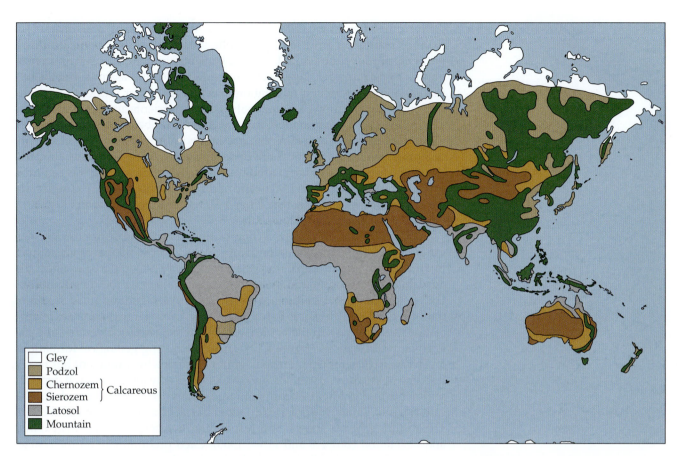

Gley
Podzol
Chernozem } Calcareous
Sierozem
Latosol
Mountain

FIGURE 3.11   World distribution of major soil types. Note the close correlation of these soil types with the climatic zones shown in Figure 3.7, which reflects the influence of temperature and precipitation on soil formation.

The global distribution of zonal soil types (**Figure 3.11**) can also be compared with the global climate map (see Figure 3.8) to demonstrate the close relationship between soils and climate.

Unusual soil types requiring special adaptations

In addition to such zonal soils, there are unusual soil types derived from parent material of unusual chemical composition. Certain rock types such as gypsum, serpentine, and limestone contain unusually high amounts of some compounds and little of others. Serpentine, for example, is particularly deficient in calcium, and gypsum contains an excess of sulfate. Few plant species can tolerate such azonal soils, and the low-diversity plant communities that do grow on such soils have special physiological adaptations for dealing with their unusual chemical composition.

One example of a soil type that requires special adaptations by plants is **halomorphic soil**, which contains very high concentrations of sodium, chlorides, and sulfates. Halomorphic soil typically occurs near the ocean in estuaries and salt marshes, and in arid inland basins where shallow water accumulates and evaporates, leaving behind high concentrations of salts. A small number of specialized **halophytic** (salt-loving) plant species grow in such areas. They include a variety of taxonomic and functional groups, each of which has special adaptations for dealing with the problem of maintaining osmotic and ionic balance in these environments. Some species of mangroves and grasses excrete salts from specialized cells in their leaves, whereas pickleweeds and ice plants store salts in special cells in their succulent leaves.

As mentioned above, highly acidic soil conditions cause essential nutrients—especially nitrogen and phosphorus—to be bound in compounds that plants cannot use. Pitcher plants, sundews, Venus's flytraps, and other insectivorous plants can grow in highly acidic soils or other environments where nutrients are severely limited. These plants obtain their nitrogen and phosphorus by capturing living insects, digesting them, and assimilating the nutrients. A less spectacular adaptation to acidic and other nutrient-poor soils is evergreen vegetation (Beadle 1966). Because nutrients are lost when leaves are dropped, and because more minerals must then be taken up by the roots to produce new leaves, plants can use limited nutrients more efficiently by retaining their leaves for longer periods. In mesic (moist) temperate climates, where the predominant vegetation is usually deciduous forest, it is common to find evergreens growing on acidic and nutrient-poor soils. Examples are the pine barrens of the eastern United States and the eucalyptus forests of Australia (Daubenmire 1978; Beadle 1981).

In addition to their chemical composition, the physical structure of soils can influence the distribution of plant species and the nature of vegetation. In arid regions, for example, the size and porosity of soil particles affect the availability of the limited moisture to plants by affecting the runoff, infiltration, penetration, and binding of water. Thus, even within a small region of uniform climate, differences in soil texture can cause large differences in vegetation. A striking example is provided by the **bajadas** (or extended alluvial fans) of desert regions (**Figure 3.12**). These interesting geological formations are made up of sediments carried out of mountains by infrequent but heavy flooding of the canyons. As the floodwater gradually loses energy, it deposits sediments in a gradient—dropping large, heavy rocks at the mouths of the canyons and small sand- and clay-sized particles at the bottom of the fan. The resulting *bajada* shows a corresponding gradient in water availability and vegetation (Bowers and Lowe 1986). Cacti predominate on the coarse, rocky, well-drained soils high on the *bajada*, where water is available only for short periods during and after rains. These succulents can take up water rapidly through their extensive shallow roots and store it in their expandable tissues. Shrubs and grasses are much more common farther down the *bajada*, where their roots can extract the water held on and among the smaller soil particles.

A somewhat similar situation occurs along the coast of the Gulf of Mexico in the southeastern United States. The uplands have coarse, sandy, well-drained soils that support drought-tolerant, coniferous woodland/savanna vegetation. In contrast, the lowlands have accumulated fine, water-retaining soils, and this—as well as their proximity to the water table—allows them to support much more mesic vegetation. Thus, a person interested in the factors influencing plant distributions and community composition at this local to

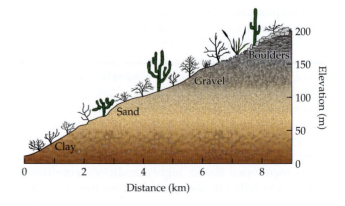

FIGURE 3.12 Schematic representation of the local elevational distribution of soil particle size and vegetation on a desert *bajada* on the Sonoran coast of the Gulf of California (Sea of Cortez). At the upper end of the alluvial fan, where large boulders have been deposited, the vegetation is dominated by cacti and other succulents that can take up water rapidly before it percolates below the root zone. At the lower end, where water infiltration is poor and the existing water is tightly bound by fine clay particles, the vegetation consists of sparse, shallowly rooted shrubs. The greatest water availability, productivity, and species diversity occur at intermediate elevations, where the soils are sandy, infiltration is high, and water is not tightly bound by soil particles.

regional scale must pay particular attention to how subtle characteristics of soil structure affect the runoff, infiltration, and retention of rainwater.

Although we have concentrated here on the relationship between soils and vegetation, soils also affect the distributions of animals—both indirectly by controlling which plant species are present, and directly through the effects of the chemical and physical environment on their life cycles. Many kinds of mammals, reptiles, and invertebrates are restricted to particular types of soils that meet their specialized requirements for burrowing and locomotion. For example, in North American deserts, lizards of the genus *Uma*, the kangaroo rat *Dipodomys deserti*, and the kangaroo mouse *Microdipodops pallidus* are all restricted to dunes and similar patches of deep, sandy soil. Another set of species, including chuckwallas (*Sauromalus obesus*), collared lizards (*Crotophytus collaris*), and rock pocket mice (*Chaetodipus intermedius*) show just the opposite habitat requirement, that of being restricted to rocky hillsides and boulder fields.

Aquatic environments

As anyone who has ever tried to keep tropical fish knows, warm and relatively stable temperatures are essential for their survival and reproduction. Salinity, light, inorganic nutrients, pH, and pressure also play key roles in the distributions of aquatic organisms. Like terrestrial climates, the physical characteristics of water often exhibit predictable patterns along geographic gradients, which can be understood with a basic background in physics.

Stratification

THERMAL STRATIFICATION. When solar radiation strikes water, some is reflected but most penetrates the surface and is ultimately absorbed. Although water may be transparent, it is much denser than air, and its absorption of radiation is rapid. Even in exceptionally clear water, 99 percent of the incident solar radiation is absorbed in the upper 50 to 100 m, and this absorption occurs even more rapidly if many organisms or colloidal substances are suspended in the water column. Longer wavelengths of light are absorbed first; the shorter wavelengths—which have more energy—penetrate farther, giving the depths their characteristic blue color.

This rapid absorption of sunlight by water has two important consequences. First, it means that photosynthesis can occur only in surface waters where the light intensity is sufficiently high (the **photic zone**). Virtually all of the primary production that supports the rich life of oceans and lakes comes from plants living in the upper 10 to 30 m of water. Along shores and in very shallow bodies of water, some species such as kelp are rooted in the substrate. These plants may attain considerable size and structural complexity, and may support diverse communities of organisms. In the open waters that cover much of the globe, however, the primary producers are tiny, often unicellular algae (called **phytoplankton**), which are suspended in the water column. **Zooplankton**, tiny crustaceans and other invertebrates that feed on phytoplankton, migrate vertically on a daily cycle: up into the surface waters at night to feed, and down into the dark, deeper waters during the day to escape predatory fish that rely on light to detect prey.

Second, the rapid absorption of solar radiation by water means that only surface water is heated. Any heat that reaches deeper water must be transferred by conduction or convection by vertical currents. Consequently, deep waters are characteristically cold, even in the tropics. The density of pure water is greatest at 4° C and declines as its temperature rises above or falls below this point. This unusual property of water is significant for the sur-

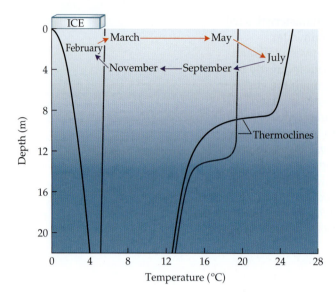

FIGURE 3.13 Vertical temperature profiles of Lake Mendota, Wisconsin, at different dates from summer through winter showing the loss of thermal stratification as the lake cools. In July the thermocline is pronounced and shallow; in September it is less pronounced and deeper; and by November it has disappeared, allowing surface and deep waters to mix during the fall overturn. Because the density of water increases as it cools down to 4° C, but then decreases as water molecules begin to form ice crystals, thermal stratification is typically reversed under the ice cover in winter (February). (After Birge and Juday 1911.)

vival of many temperate and polar organisms because it means that ice floats. Ice provides an insulating layer on the surface that prevents many bodies of water from freezing solid. The presence of salts in water lowers its freezing point, and some organisms are therefore able to exist in unfrozen water below 0° C (de Vries 1971).

A more general consequence of the relationship between water density and temperature is that water tends to acquire stable thermal stratification. When solar radiation heats the water surface above 4° C, the warm surface water becomes lighter than the cool deeper water, so it tends to remain on the surface where it may be heated further and become even less dense. In tropical areas and in temperate climates during the summer, the surfaces of oceans and lakes are usually covered by a thin layer of warm water. Unless these bodies of water are shallow, the deep water below this layer is much colder (sometimes near 4° C). The change in temperature between the surface layer and deeper water is called a **thermocline (Figure 3.13)**. Mixing of the surface water by wave action determines the depth of the thermocline and maintains relatively constant temperatures in the water above it. In small temperate ponds and lakes that do not experience high winds and heavy waves, the thermocline is often so abrupt and shallow that swimmers can feel it by letting their feet dangle a short distance. In large lakes and oceans, where there is more mixing of surface waters, the thermocline is usually deeper and less abrupt.

Tropical lakes and oceans show pronounced permanent stratification of their physical properties, with warm, well-oxygenated, and lighted surface water giving way to frigid, nearly anaerobic, and dark (aphotic) deep water. Oxygen cannot be replenished at great depths where there are no photosynthetic organisms to produce it, and the stable thermal stratification prevents mixing and reoxygenation by surface water. Only a relatively small but fascinating menagerie of organisms can exist in these extreme conditions. The feces and dead bodies of organisms living in the surface waters sink to the depths, taking their mineral nutrients with them. The lack of vertical circulation thus limits the supply of nutrients to the phytoplankton in the photic zone. Consequently, deep tropical lakes are often relatively unproductive and depend on continued input from streams for the nutrients required to support life.

OVERTURN IN TEMPERATE LAKES. The situation is somewhat different in temperate and polar waters. Deep temperate lakes, in particular, undergo dramatic seasonal changes: They develop warm surface temperatures and a pronounced thermocline in summer, but freeze over in winter. Twice each year, in spring and fall, the entire water column attains equal temperature and equal density (see Figure 3.13), the temperature/density stratification is eliminated, and moderate winds may generate waves that then mix deep and shallow water, producing what is called **overturn**. This semiannual mixing carries oxygen downward and returns inorganic nutrients to the surface. Phosphorus and other mineral nutrients may be depleted during the summer, when warm temperatures allow algae to grow and reproduce at high rates; overturn replenishes these nutrients by stimulating the growth of phytoplankton. Temperate lakes, such as the Great Lakes of North America, are often quite productive and support abundant plant and animal life, including valuable commercial fisheries. However, abnormally high nutrient inputs—

often due to runoff from agricultural fields and discharges of inadequately treated sewage—can cause excessive production, rapid algal growth, depletion of oxygen, fish kills, and other environmental problems.

Oceanic circulation

The vertical and horizontal circulation of oceans is more complicated than that of lakes, in part because oceans are so vast, extending through many climatic zones, and in part because salinity affects the density of water. Salts are dissolved solids carried into the oceans by streams and concentrated by evaporation over millions of years. The presence of salts in water increases its density, causing swimmers to experience greater buoyancy in the ocean than in freshwater. Varying salinity and density have important effects on ocean circulation. Rivers and precipitation continually supply freshwater to the surface of the ocean, and this lighter water tends to remain at the surface. If you have ever flown over the mouth of a large, muddy river such as the Mississippi, Thames, or Nile, you may have noticed that its water remains relatively intact, flowing over the denser ocean water for many kilometers out to sea. In polar regions, the input of freshwater to the ocean from rivers and precipitation generally exceeds losses from evaporation, but the reverse is true in the tropics. This pattern creates a somewhat confusing situation because warm tropical surface water tends to become concentrated by evaporation and to increase in density, counteracting to some extent stratification owing to temperature. Conversely, cold polar water—which would be expected to show little stratification—may become somewhat stabilized as low-density freshwater accumulates on the surface.

Vertical circulation occurs in oceans, but the rates of water movement are so slow that a water mass may take hundreds or even thousands of years to travel from the surface to the bottom and back again. Areas of descending water tend to occur at the convergence of warm and cold currents in polar regions, where the colder, denser water sinks under the warmer, lighter water. Areas of rising water, called **upwellings**, are found where ocean currents pass along the steep margins of continents. This happens, for example, along the western coast of North and South America, where there is little continental shelf and the land drops sharply offshore. As the Pacific gyres sweep toward the Equator along these shores, the Coriolis effect and, in tropical latitudes, the easterly trade winds tend to deflect the surface water offshore, and water wells up from the depths to replace it. Because upwelling, like the overturn in lakes, returns nutrients to the surface, productivity tends to be high in areas of upwelling (see the global productivity map of Figure 5.30A). Probably the greatest commercial fishery in the world is located in the zone of upwelling off the coasts of Chile and Peru, making the episodic ENSO events that were discussed earlier in this chapter particularly devastating to local economies.

Surface currents, such as the great hemispheric gyres (see Figure 3.5), are relatively shallow and rapidly moving, so they tend to form discrete water masses, each of which has a characteristic salinity and temperature profile distinct from those of neighboring water masses. Some organisms with limited capacity for locomotion may drift in currents for long distances without leaving a single uniform water mass. Organisms that can move actively to overcome the currents must also be able to tolerate the contrasting physical environments in different water masses.

Although oceanographers have recognized the existence of distinct water masses within the oceans for many years, modern technology has revealed the extent of spatial heterogeneity in shallow ocean waters. For example, investigators from the Woods Hole Oceanographic Institution have studied the

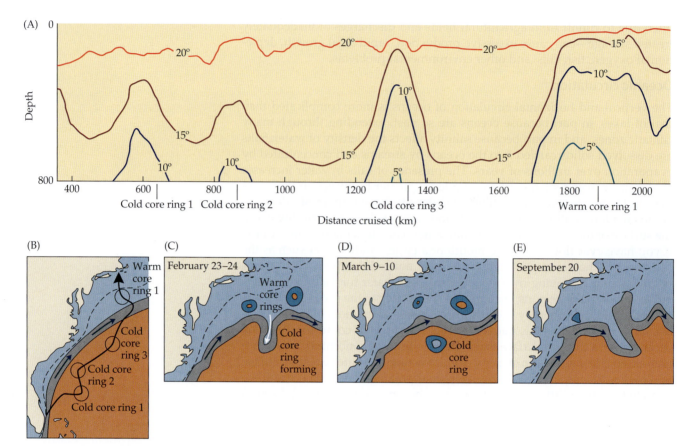

FIGURE 3.14 Local- to regional-scale spatial and temporal heterogeneity of surface waters in the North Atlantic Ocean is caused by meanders of the Gulf Stream (dark gray shading) that create great rings (some as much as 300 km in diameter) of Gulf Stream water which encircle water of different origin. Rings to the north of the Gulf stream encircle a core of water derived from the Sargasso Sea, and thus tend to be relatively warm (around 18°C), while the core of those to the south of the stream tend be cold (< 10°C), drawing their waters from northern regions (see cold core ring forming in part C). (A) Temperature/depth profile recorded by an oceanographic vessel that traveled through several rings, as indicated by the dashed line on map (B). (C–E) Changes in water surface temperatures as mapped by infrared satellite imagery showing the formation, movement, and disappearance of rings. (After Wiebe 1982.)

physical environment and the biota of Gulf Stream "rings" (Wiebe 1976, 1982; Lai and Richardson 1977; Katsman et al. 2003). These rings are small masses of cold or warm water that have broken away from the southern or northern edges of the Gulf Stream to drift through water of contrasting temperature in the North Atlantic. They can be readily seen on infrared satellite images that show sea surface temperatures (**Figure 3.14**). These rings not only have physical environments that are strikingly different from their surroundings, but also contain a unique biota that can persist in these special conditions far from its normal distribution in the Gulf Stream. The possible roles of these floating warm- or cold-water eddies in trans-Atlantic dispersal—both now and in the past—are intriguing subjects for future biogeographic research.

Pressure and salinity

Pressure and salinity vary greatly among aquatic habitats. These variations have major effects on the distributions of organisms, because special physiological adaptations are necessary to tolerate the extremes. As every scuba diver knows, water pressure increases rapidly with depth. It becomes a major problem for organisms in the ocean, where the deepest areas are up to 6 km below the surface. Pressure increases at a rate of about 1 atmosphere (about 1.5 mega Pascals) for every 10 m of depth. In the abyssal depths, pressures are more than 200 times greater than at the surface. Organisms adapted to living in surface waters cannot withstand the pressures of the deep sea, and vice versa.

Variation in salinity is relatively discontinuous. The vast majority of the Earth's water is in the oceans and is therefore highly saline (greater than 34 parts per thousand). In contrast, freshwater lakes, marshes, and rivers, which account for less than 1 percent of the Earth's waters, contain very few dissolved salts. Habitats of intermediate or fluctuating salinity, such as salt marshes and

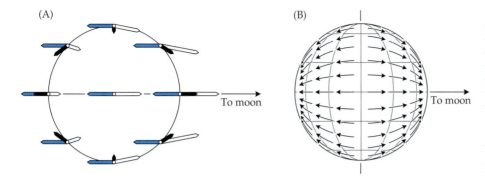

(A)

(B)

To moon

To moon

FIGURE 3.15 (A) Schematic represen-tation of how the centrifugal force of the spinning Earth and the gravitational force of the moon cause the tides. On the side of the Earth closest to the moon, the gravitational force (white) is stronger than the centrifugal force (blue), and the net tidal force (black) tends to draw surface water toward the moon. On the side opposite the moon, the gravitational force is weaker than the centrifugal force, and the net tidal force tends to draw water away from the moon. In between these extremes, the gravitational and centrifugal forces are balanced, and there is essentially no net tidal force. (B) The movement of surface waters in response to these tidal forces.

estuaries, constitute only a tiny fraction of the Earth's aquatic habitats. Conse-quently, most aquatic organisms are physiologically adapted and geographi-cally restricted either to freshwater, where the physiological problem is ob-taining sufficient salts to maintain osmotic balance, or to salt water, where the problem can be eliminating excess salt. Only a few widely tolerant (**euryhaline**) organisms have the special physiological mechanisms required to survive in the widely fluctuating salinities of estuaries and salt marshes.

Tides and the intertidal zone

We can learn a great deal about the factors determining the distributions of organisms by studying environmental gradients: both gradual changes such as variation in light and pressure with depth in lakes and oceans, and rapid changes such as the variation in temperature in the cooling outflow of a hot spring. One of the steepest, best-studied, and most interesting environmental gradients occurs where the ocean meets the land. Along the shore is a nar-row region that is alternately covered and uncovered by seawater. It is called the **intertidal zone** because it experiences a regular pattern of inundation and exposure caused by tides.

Sir Isaac Newton explained how the gravitational influences of the moon and sun interact to cause the global fluctuations in sea level that we call tides. The entire story is complicated, but the main pattern and its mechanism are simple. The tides are flows of surface waters. They occur in response to a net tidal force, which reflects a balance between the centrifugal force of the Earth and moon revolving around their common center of mass, and the gravi-tational forces of the moon and sun (**Figure 3.15**). Because the gravitational force exerted by one object on another is equal to its mass divided by the square of the distance between those objects, the smaller but nearer moon has a greater effect than the sun.

Most shores typically experience both a daily and a monthly tidal cycle: There are two high and two low tides every 24 hours, and there are two pe-riods of extreme tides each month, corresponding to the new and full moons (**Figure 3.16**). During these periods, the moon and sun are in the same plane as the Earth, and their gravitational effects are additive, causing high-ampli-tude tides, or **spring tides**, with the highs occurring at dawn and dusk and the lows near noon and midnight. During the quarter moons, the sun and moon are at right angles to each other (from the perspective of Earth), and their gravitational effects tend to cancel each other, resulting in low-amplitude tides, or **neap tides**.

A distinct community of plant and animal species lives in the intertidal zone. Nearly all aspects of the lives of these organisms are dictated by the cyclical pattern of inundation by seawater at high tide and exposure to des-iccating conditions at low tide. Most species are confined to a very narrow zone of tidal exposure, so their distributions form thin bands running hori-

FIGURE 3.16 A tide calendar for the northern Gulf of California (Sea of Cortez) showing the typical pattern of tides due to the gravitational influences of the moon and sun. Note that there are two high and two low tides each day. There are also two periods of low-amplitude (neap) and high-amplitude (spring) tides each month; the latter correspond to the times of the new and full moons when the gravitational forces of moon and sun are aligned. (Courtesy of D. A. Thomson.)

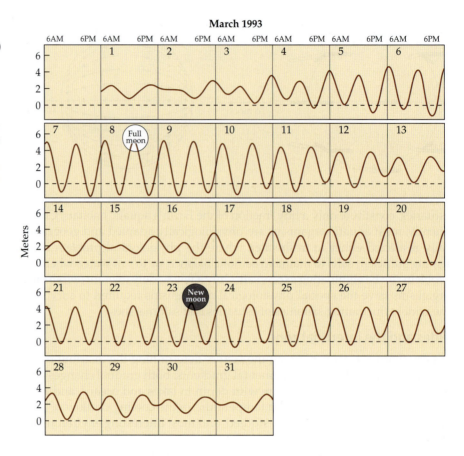

zontally along the shore. As we shall see in Chapter 4, the narrow ranges of species in the intertidal zone (typically only a few centimeters or meters) and the ease with which critical environmental conditions can be manipulated experimentally have produced a wealth of information about the factors limiting the distributions and regulating the diversity of species.

Time

Before discussing important methodological issues associated with describing and then exploring the response of biotas to variation in the geographic template, we first should emphasize one dimension that, although fundamental to all patterns and underlying processes in biogeography, ecology, and evolution, is often overlooked—time. There are at least two reasons why time should be explicitly considered in all attempts to understand the geography of nature. First, and perhaps most obvious, is that all features of the geographic template discussed above—climate, soils, and water chemistry—are temporally dynamic, varying in some complex but important manner over various scales of time from that of days and generations to millennia and geological periods. Second, but perhaps less intuitive, is that time itself influences the abilities of individuals, species, and entire biotas to respond to features of the geographic template even during periods of relative environmental stasis. That is, the fundamental biogeographic processes of dispersal, range expansion, and subsequent extinction or evolution in new regions takes time, such that biogeographic responses often lag far behind the actual dynamics of local to global environments. For example, geographic ranges of many species of plants and animals are still expanding poleward in response to retreat of the glaciers that occurred some 15,000 years ago, while the biogeographic imprint of plate tectonics, in terms of distinctiveness of biotas among the continents, remains indelible hundreds of millions of years since the supercontinent of Pangaea began to split apart (see Chapters 8 and 9).

Two-Dimensional Renderings of the Geographic Template

Early maps and cartography

As we observed in the previous chapter, setting the precise date of any major discovery or innovation in science is bound to be an elusive if not futile exercise. So it is with maps and cartography, whose developmental history is intricately linked with advances in perhaps all physical sciences from geology to astronomy. Maps, at least in their most rudimentary forms, are probably ancient and may well date back to the earliest renderings of game species and elements of their surroundings (e.g., a local river, mountain, or shoreline) that early humans etched into the sands at their feet or onto the walls of their dwellings. The earliest formal biogeographic maps, however, are not known to have existed until the early nineteenth century, perhaps the first of these commissioned by two luminaries in the history of biogeography, evolution, and ecology—Augustin-Pyramus de Candolle (1779–1841) and Jean-Baptiste Lamarck (1744–1829) (see Chapter 2).

In their discussion of the history of biogeographic maps, Ebach and Goujet (2006) make the distinction between maps that serve two fundamental objectives in biogeography: (1) maps of "**chorology**," a term coined by Ernst Haeckel in 1866 to describe the science of the geographic spread of organisms, and (2) **systematic biogeographic maps**, which describe the distinctiveness and/or similarities among biotas from local (provincial) to global scales. We have already discussed examples of **chorological maps** in Chapter 2, including Buffon's description of the spread of life-forms from his hypothesized Northern Origins (see Figure 2.2) and the extensionists' maps of transoceanic land bridges across which biotas were hypothesized to have migrated among the continents (see Figure 2.8). Not surprisingly, one of the earliest and most comprehensive chorological maps was that created by Haeckel himself (see Figure 2.9), which vividly illustrates his hypothesis on the monophyletic origins and migration routes of early humans. Haeckel's map also nicely captures the interdependence between these two types of maps and objectives in biogeography. That is, systematic maps describe the differences and similarities among regions in terms of their biotas (geographic races, species, etc.), while chorological maps provide hypothetical explanations for those patterns (migrations and subsequent evolution of species in isolated and environmentally disparate regions, and the dynamics of land, sea, and climates that affect those migrations). In fact, even the most descriptive biogeographic map, such as Lamarck and Candolle's 1805 map of France's botanical provinces (**Figure 3.17**) or Wallace's world map of biogeographic regions (frontispiece) are both systematic and chorological, illustrating differences among regional biotas and simultaneously, at least implicitly, proposing explanations for those differences based on hypothetical migration paths or barriers to dispersal.

Next, we will briefly introduce the basic methods for creating maps, which are essentially two-dimensional representations (and distortions) of the three-dimensional curved surface of the Earth, and we will describe geographic coordinate systems for identifying particular locations across this surface. We will then explain, again in very basic terms, how scientists have delineated the fundamental unit of biogeography—the geographic range—before showcasing some advances in **visualization** and analysis of spatial pat-

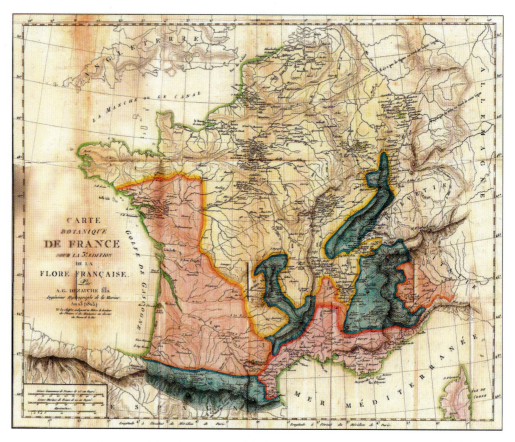

FIGURE 3.17   Perhaps the first systematic, biogeographic map was the *Carte Botanique de France*, which was commissioned by Jean-Baptiste de Lamarck and Augustin-Pyramus de Candolle for the third edition of *Flore Francaise* (Lamarck and Candolle 1805). The five regions are distinguished by color: green = maritime plants; blue = mountain plants; crimson red = Mediterranean plants; yellow = plants widely dispersed throughout much of France and surrounding areas (lacking endemics); vermillion = vegetation that is intermediate or transitional between those of northern plains and southern provinces. (For additional descriptions of this map, see Ebach and Goujet 2006.)

terns—technological advances that provided the tools and new perspectives integral to the revitalization of biogeography that took place during the latter decades of the twentieth century.

Flattening the globe: Projections and geographic coordinate systems

Before we can describe and analyze distributional patterns and search for explanations for those patterns, it seems appropriate that we review the basis for visualizing these patterns (i.e., maps, which are simplifications and, quite often, distortions of the true geographic template). The distortion arises from two challenges: (1) describing and somehow transforming three-dimensional patterns across the curved surface of the Earth onto the two-dimensional plane of a map and (2) locating particular points or areas of interest on that plane. Fortunately, cartographers have been tackling these challenges for centuries, providing us with a wide array of **projections** and **geographic coordinate systems** to address each of these challenges, respectively. As **Figure 3.18** illustrates, developing a map projection is relatively simple in theory, but in practice it turns out to be impossible to accurately represent a three-dimensional entity on a two-dimensional surface without distortion. For relatively small areas, across which the curvature of the Earth is negligible, map distortions are so minor that they can be ignored for most applications. At larger scales, however, map projections can cause substantial distortions of shape, area, distance, and direction. Fortunately, cartographers have developed a diversity of projections with known properties such that we can choose the one most appropriate to our question. For example, conformal, equal-area, equidistant, and true directional projections preserve the shape, area, distances between points, and directions, respectively.

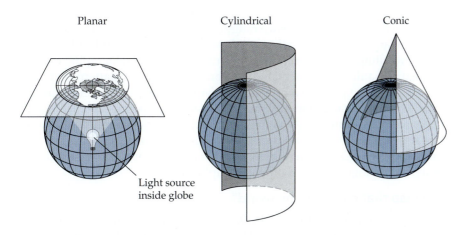

Planar   Cylindrical   Conic

Light source
inside globe

FIGURE 3.18 Although map projections are created using various mathematical formulae to transform spatial data from a curved three-dimensional surface to a flat two-dimensional map, the basic approach is not difficult to understand. If the features of interest (e.g., continental outlines or geographic ranges) are drawn onto a transparent model of the globe, those features can be transformed ("projected") by a light source that passes through the globe and then onto a piece of paper located either inside or outside the globe. By folding the paper (e.g., to form a plane, cylinder, or cone) and by moving it to different positions around the globe, projections can be adjusted to better preserve one or more of the key geographic features (i.e., shape, area, distance, or direction).

Once we have decided on an appropriate projection, we still cannot locate particular points or areas of interest without a geographic coordinate system. The two most commonly used types of coordinate systems are (1) latitude and longitude and (2) the Universal Transverse Mercator (UTM) system. Locations in the UTM system are expressed as number of meters east of any of 60 arbitrary reference meridians ("false eastings") and, in the Northern Hemisphere, number of meters north of the Equator. In the Southern Hemisphere, "eastings" are expressed in the same manner, while "northings" are expressed as number of meters north of an imaginary "false origin" located 10,000 m south of the Equator. With the addition of elevation, each of these systems can better approximate true complexities of the Earth's surface.

Visualization of Biogeographic Patterns

History and exemplars of visualization in biogeography

Visualization is much more than just a means of displaying a particular phenomenon such as a geographic range; it is a creative and insightful process that ideally serves to fundamentally transform our abilities to view and conceptualize the world. As with Galileo's telescope and van Leeuwenhoek's microscope, biogeographic maps and graphical models provide new perspectives on these phenomena and, thus, serve as invaluable tools for exploring the interplay of processes that may be causally related to patterns in the geographic variation of nature. The most familiar visualizations in biogeography are, of course, maps, but the most informative and insightful biogeographic maps are much more than just accurate portrayals of landscapes and seascapes. Indeed, there exists throughout the history of the field a great wealth of transformative visualizations—maps and otherwise—whose diverse forms were limited only by the scientific and artistic ingenuity of their creators. Success in terms of the ability of any visualization to fundamentally alter the views of contemporary scientists and advance the field depends largely on striking a delicate balance between two competing objectives that constitute a fundamental tension in visualizations: achieving accuracy and precision (in describing geographic, environmental, and biotic features) on the one hand, versus simplification and abstraction (to develop a conceptual synthesis of the interplay and interdependence of underlying processes) on the other.

Here, we showcase three of what were arguably the most influential visualizations in the history of biogeography: William Smith's "map that changed the world," Alexander von Humboldt's *Tableau Physique* (the keystone illustration in his *Essay on the Geography of Plants*; see Chapter 2), and Robert H. MacArthur and Edward O. Wilson's graphical model of island biogeogra-

FIGURE 3.19 William Smith (1769–1839), although suffering great tragedy throughout his life, fundamentally transformed the way scientists of the eighteenth and nineteenth centuries viewed the world and the evolutionary dynamics of its rocks and life-forms. (Portrait by Hugues Fourau.)

phy. These three exemplars span a range of approaches in scientific visualization, from largely descriptive (but still fundamentally insightful and aesthetically compelling) to predominately abstract (with little geographic context but great versatility for understanding a broad diversity of patterns in biodiversity). Later, we provide a brief and unavoidably incomplete introduction to the various visualization and analytical tools that became available and, in no small way, contributed to the revitalization of biogeography that took place during the latter decades of the twentieth century. In Unit 4, we will introduce a new generation of mapping that projects three-dimensional renderings of phylogenetic trees onto geographic maps.

"The map that changed the world"

The story of William Smith (1769–1839; **Figure 3.19**) and his "map that changed the world" is a heartrending saga of tragedy and triumph (Winchester 2001). Unlike most of the distinguished scientists of that era—including Linnaeus, Buffon, Lamarck, and Humboldt, who were blessed with the privileges and wealth of aristocracy—Smith was of humble origins. His father, the village blacksmith, died at a young age and left him an orphan to be raised by his uncle. Indefatigable throughout his challenged life, Smith was a voracious, albeit by necessity self-taught, student, taking up the trade of surveying at 18 and eventually becoming an accomplished surveyor in his own right—working for Somerset Coal Canal Company for some eight years until he was dismissed in 1799. By then, at the age of 29, he had amassed a great wealth of information, meticulously recording the changes in rocks and associated fossils both horizontally across the landscapes of England and vertically along the slopes of pit mines. Most importantly, he realized that the series of those rocks and ancient, fossilized life-forms was invariant across England and that they formed what eventually would be recognized as the "geological record" of succession in ages of the Earth and its species.

He drafted the first, rough sketch of his map of England's geological strata in 1801. Unfortunately, he was unable to find consistent employment for most of the next decade, which freed him to travel across the countryside to further develop his map, but also left him unable to fund its substantial production and publication costs. Eventually, he found a benefactor who fully appreciated the map's great promise, and it was finally published in 1815. Rather than a time of heralded triumph, this marked the beginning of some 15 years of professional and personal suffering for Smith. His map was plagiarized and sold for less than the cost of the original. Smith became bankrupt and was sentenced to debtors' prison; he became homeless and was forced to leave London, and his wife fell into insanity. It wasn't until he was in his early 60s and working as an itinerant surveyor that he was finally recognized for his contributions, which truly did transform the way scientists viewed the world (**Figure 3.20**) to that of an ancient but dynamic system where evolution of its biotas is intricately linked to that of the Earth itself.

FIGURE 3.20 William Smith's seminal map, *A Delineation of the Strata of England and Wales with Parts of Scotland*, first published in 1815, is often referred to as "the map that changed the world" (see Winchester 2001). This marked the first time geologists firmly established the connection between geographic variation in rocks and the temporal development of those rocks into distinctive strata—each identified by their unique associations (and evolutionary series) of fossils. In this region, uplift and erosion have exposed the strata along a roughly northwest to southeast gradient of decreasing antiquity (i.e., oldest strata in the upper left-hand corner of this map; see Figure 8.1 for an enlarged image of the key to this map). This colossal visualization (the original comprised 15 separate copperplate engravings and spanned 74 × 105 inches) finally made it possible to fully appreciate the antiquity and dynamics of the Earth and its life-forms, and was thus foundational to the fields of geology, biogeography, ecology, and evolution. (From University of New Hampshire, Durham, NH, www.unh.edu/esci/wmsmith.html.) ▶

A
DELINEATION
OF THE
STRATA
OF
ENGLAND AND WALES,
WITH PART OF
SCOTLAND;
EXHIBITING
THE COLLIERIES AND MINES,
THE MARSHES AND FEN LANDS ORIGINALLY OVERFLOWED BY THE SEA,
AND THE
VARIETIES OF SOIL
ACCORDING TO THE VARIATIONS IN THE SUBSTRATA,
ILLUSTRATED BY THE MOST DESCRIPTIVE NAMES
BY W. SMITH

Summit of Chimborazo

Point on Chimborazo where Messrs Bonpland, Montufar, and Humboldt carried instruments on June 23, 1802...

Summit of Popocatépetl

Highest point (on Corazón) reached by Messrs Bouguer and de la Condamine in 1738

Height of the Teide summit

Height of 7016 meters reached by Mr. Gay-Lussac alone in a balloon (above Paris on September 16, 1804), in order to determine the intensity of the magnetic forces, the quantity of oxygen in the air, and the decrease in heat...

Summit of Cotopaxi

Height of the Orizaba summit or Sidatepetel

Height of the Mont-Blanc reached by M. de Saussure in 1787

Height of the city of Quito

Height of Vesuvius

GÉOGRAPHIE DES PLANTES ÉQUINOXIALES.

Tableau physique des Andes et Pays voisins

Dressé d'après des Observations & des Mesures prises sur les lieux depuis le 10.° degré de latitude boréale jusqu'au 10.° de latitude australe en 1799, 1800, 1801, 1802 et 1803.

PAR

ALEXANDRE DE HUMBOLDT ET AIMÉ BONPLAND.

FIGURE 3.21 Alexander von Humboldt's *Tableau physique des Andes et pays voisins*, published first in 1807, was far more than just a vividly descriptive portrait but a landmark in visualization and conceptualization of the interrelationships among geography, climate and soils, and vegetation. This was the centerpiece illustration of his seminal book, *Essai sur la Geographie des Plantes*. Humboldt's *Essai* demonstrated to generations of scientists, including Charles Darwin and Alfred Russel Wallace, how biogeography, ecology, and evolution could be advanced through meticulous observations complemented by integrative, conceptual explanations based on the interplay of causal processes. Labels within the central plate delineate distributions of particular plant species (along the central and right-hand slopes of the mountain), and labels against the atmospheric background reference key elevations, including those of the world's tallest mountains and the highest elevations reached by climbers and balloonists at that time (translations provide here for five of these). Humboldt's Tableau also included detailed side panels with columns in tables describing elevational gradients in meters; light refraction; distance at which mountains are visible from the sea; elevations measured in various parts of the Earth; electrical phenomena; cultivation of the soil; decrease in gravity; azure color of the sky; decrease in humidity; atmospheric air pressure; scale in toises (a measure approximately equal to 6 feet); air temperature; chemical composition of the atmosphere; elevation of the lowest limit of perpetual snow at various latitudes; scale of animals according to where they live; temperature at which water boils at various altitudes; geological aspects; intensity of light at various elevations. (For a full translation of the *Tableau* and Humboldt's entire *Essai*, see Jackson and Romanowski 2009.) (Courtesy of Cop. Bibliothéque central MNHM Paris 2006.)

Humboldt's *Tableau Physique*

Unlike William Smith, Alexander von Humboldt (1769–1859) was born into wealth and, to his credit, used it well to unselfishly support the scientific and intellectual endeavors of many of his colleagues as well as his own explorations and creations. As we observed in the previous chapter, his masterful works—including his multivolumed *Cosmos*, the narrative of his travels through the equinoctial (tropical) regions of America, and his *Essai sur la Geographie des Plantes*—inspired the generation of scientists that were to establish the fields of biogeography, ecology, and evolution. The centerpiece illustration in Humboldt's 1807 *Essai* (the *Tableau physique des Andes et pays voisins*) is deservedly recognized as one of the hallmarks of visualization and conceptualization in science.

Like William Smith's map, Humboldt's *Tableau* is a colorful illustration of the geographic variation of nature. The *Tableau*, however, is much more conceptualized and abstract—purposely distorting, exaggerating, or altering the juxtaposition of landforms and associated soil types, climates, plants, and animals in order to emphasize the interdependence among these components of the natural world. As Stephen Jackson observes, the *Tableau* is a comprehensive and compelling visualization that illustrates "the first mature, integrated statement of Humboldt's view of a unified nature" (see Jackson and Romanowski 2009. The *Tableau*'s center illustration along with its tabular side panels (**Figure 3.21**) describe the regular and interdependent variation in environmental characteristics and dependent life-forms as one ascends from sea level to the summit of what was then recognized as the world's highest peak—Mount Chimborazo, in the equatorial Andes of South America.

Humboldt's *Essai*, in its entirety, was finally translated into English in 2009, making this masterful work once again accessible to a modern audience who should benefit, not just from its historical perspective, but from its ability to serve as an exemplar for advancing today's frontiers of science. The lesson has come full circle, such that modern biogeographers are now adopting a very holistic approach in many ways similar to that utilized by the founding figures of biogeography, from Humboldt and Smith through Darwin and Wallace: meticulous observation and unbiased interpretation of all relevant phenomena, complemented by integrative, conceptual explanations premised on the interplay of causal processes.

MacArthur and Wilson's graphical model

The final in the three exemplars in the history of visualizations in biogeography is both the most recent and also the most abstract, devoid of color and unquestionably the most limited in its artistic aesthetics. Robert H. MacArthur and Edward O. Wilson's graphical model of their equilibrium theory of island biogeography (MacArthur and Wilson 1963, 1967) is not even, at least explicitly, geographic (**Figure 3.22**). It is not a map, and indeed, their seminal 1967 monograph—which revolutionized the fields of ecology and island biogeography—is notable for its surprising dearth of maps (save for a scale-less map illustrating the process of fragmentation, and a map depicting latitudinal gradients in avian diversity [MacArthur and Wilson 1967: their figures 1 and 37, respectively]).

The abstraction and de-emphasis of geographic context was, of course, purposeful and a central reason for the success of their theory (Losos and Ricklefs 2009). Following the dictum that "beauty is found in simplicity," MacArthur and Wilson's graphical model became one of the most compelling and versatile visualizations and conceptual tools in biogeography and biodiversity research, in general: one that could easily be adapted to explain a variety of phenomena, including the diversity of biotas from microscopic to

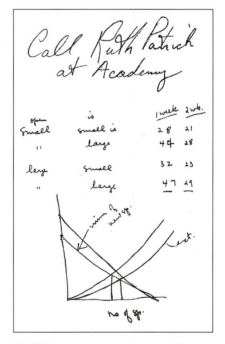

FIGURE 3.22 The graphical equilibrium model, first sketched in a letter from Robert H. MacArthur to Edward O. Wilson in April of 1962, became one of the most compelling features of their theory of island biogeography and was integrally responsible for its rapid acceptance as the new paradigm of the field in the late 1960s. The theory itself and its influence on the development of the field are discussed in Chapter 13 (see also various contributions to the volume on MacArthur and Wilson's theory edited by Losos and Ricklefs 2009). (From the Robert Helmer MacArthur [1930–1972] Papers, Department of Ecology and Evolutionary Biology, Princeton University.)

global scales and from ancient to recent time periods, the extinction dynamics of imperiled species in the face of fragmentation and climate change, and the patterns of variation in linguistic diversity among regional populations of our own species (see Chapter 13).

The spartan nature of MacArthur and Wilson's approach to advancing science was, we believe, a sign of the times: when simplification and reductionism influenced many fields of science and their practitioners' quests for truly transformative approaches to understanding the universal laws of nature. Interestingly, as we remark elsewhere throughout this book, it appears that biogeographers are once again adopting more holistic approaches that may—like those of Humboldt, Darwin, and the founding fathers of the field—strike a balance between empirically realistic and theoretically insightful approaches for understanding the geographic and temporal dynamics of life. Contributing in no small way to the enhanced "vision" and integrative powers of these modern biogeographers are the recent waves of technical advances in our abilities to record geographically explicit information, to visualize it in truly innovative ways, and to develop conceptual and predictive models to advance both basic and applied research.

The GIS revolution

One of the most important innovations that contributed to the revitalization of biogeography during the latter decades of the twentieth century was the advent and eventual wide availability of **geographic information systems** (**GIS**) (**Figure 3.23**). While they can be used to create some aesthetically compelling maps, they are much more than this. Put simply, a GIS is a system of technologically sophisticated but wonderfully accessible tools for visualizing, modifying, and analyzing patterns among spatially referenced observations. In its applications to biogeography, that information is typically in the form of locations of particular biological features—such as the locations of individuals, populations, species, or entire biotas—and the characteristics of those features and the local (geographically referenced) environmental factors that may underlie biogeographic patterns of interest.

There now exists an impressive variety of GIS software, but all of these utilize one or a combination of two georeferencing platforms. In **raster-based GIS**, the georeference platform is a system of cells (typically rectangular, but sometimes hexagonal) that tessellate to form a virtual grid work of geographic units representing the area to be mapped and analyzed. Because each cell has a unique identity, it can be attributed with particular biological, geographic, and environmental features (e.g., a list of species occurring within that location, its elevation, and its climate and soils). The alternative georeferencing platform is a **vector-based GIS**, where each feature or attribute is assigned to a precise and exact coordinate location (most commonly, latitude and longitude, or the Northing and Easting of the Universal Transverse Mercator system described above), or it is assigned to vectors and polygons created by interconnecting lines

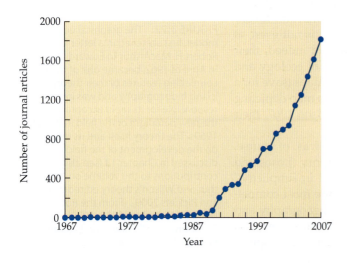

FIGURE 3.23 Since the first general accessibility to geographic information systems (GIS) in the 1980s, their use by scientists has continued to accelerate. Shown here is the number of journal articles with either geographic information system or GIS as keywords. (After Sastre et al. 2009.)

or arcs. Because each system has its advantages (e.g., raster-based GIS's ease of computation and representation of continuous features, and vector-based GIS's relationship to traditional cartography, precision, and ability to represent discontinuous features), modern GIS software programs are sometimes hybrids of these complementary platforms.

This is a highly dynamic and innovative area of technological development, such that any more detailed description—even if it were within the scope of this very general introduction—would likely be outdated before the ink on this page dried. We strongly urge those genuinely serious about studying the geographic variation in nature to add GIS to their tool kits through courses and self-directed studies of more comprehensive accounts of the strategies for utilizing GIS in biogeography (e.g., Burrough 2001; Bernhardsen 2002; Millington et al. 2002; Vogiatzakis 2003; McMaster and Usery 2005; Anselin et al. 2006; Stigall and Lieberman 2006; Harvey 2008; Kidd and Liu 2008).

Cartograms and strategic "distortions"

While it may seem counter-intuitive, one of the most informative means of visualizing geographic variation is through purposeful distortion (e.g., vertical exaggeration of elevations in three-dimensional maps are relatively simple cases of purposeful distortion that have been used in cartography for centuries). This again emphasizes our earlier assertion that maps are far more than just dutiful works of accuracy and precision; they are potentially powerful tools for conceptualizing the interplay among a variety of spatial patterns and their underlying, causal forces. **Cartograms** are examples of such strategic distortion where mapping units (particular grid cells or polygons such as those representing countries or biogeographic regions) are scaled (and distorted), not according to their actual surface area, but in proportion to another theme, such as population density or species diversity. One implicit, but important, consideration is that the cartogram be paired up with a reference map with geographic units drawn to more standard projections (i.e., equal area projections), or in proportion to another theme to be compared with the first.

The utility and versatility in potential applications of cartograms are best appreciated by example. In an illustrative case study, Wake and Vredenburg (2008) use cartograms to demonstrate the highly uneven diversity and endangerment of amphibians across the globe, along with the geographic variation in rates of discovering new species (**Figure 3.24**). In these examples, the highly uneven nature of these variables (i.e., highly disproportionate to the areas of the countries) is also emphasized by monochrome shading and use of alternative colors. The result is a series of visualizations that, although at first somewhat otherworldly in appearance, illustrate geographic variation in biodiversity more clearly than could be done using conventional maps.

Obtaining Geo-Referenced Data

Humboldt's legacy: A global system of observatories

We have one remaining tribute to Alexander von Humboldt (see Chapter 2), for his remarkable legacy also includes a prescient call in 1836 for the establishment of a global system of climatological and geomagnetic observations. This would not only provide the data requisite to generations of scientists who pursued Humboldt's vision of the unity of nature, in its broadest sense, but ultimately would provide invaluable information on the temporal dynam-

FIGURE 3.24 Cartograms utilize special projections to purposefully distort geographic units to emphasize geographic variation in characteristics (themes) of those units. Here, Wake and Vredenburg (2008) utilize cartograms to visualize the geographic variation in (A) diversity (number of species), (B) endangerment (percentage of fauna in the top three categories of threat—critically endangered, endangered, and threatened), and (C) discovery rates (number of species discovered and named during the period of 2004–2007) of amphibians (per area of each country) across the globe. (For a description of the actual projection [density-equalization] methods used to develop these cartograms, see Gastner and Newman 2004.)

(A) Diversity

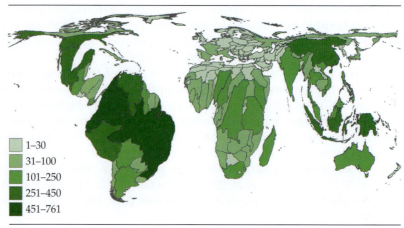

1–30
31–100
101–250
251–450
451–761

(B) Endangerment

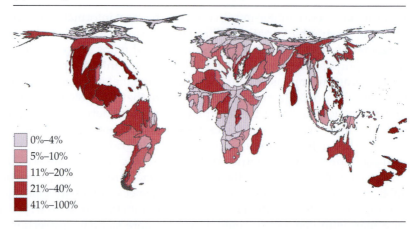

0%–4%
5%–10%
11%–20%
21%–40%
41%–100%

(C) Discovery rates

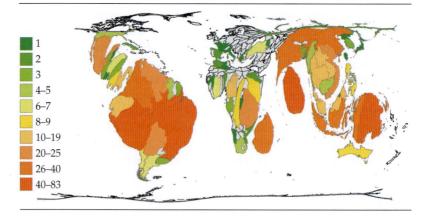

1
2
3
4–5
6–7
8–9
10–19
20–25
26–40
40–83

ics of a diversity of natural and anthropogenic phenomena—not the least of which being ongoing global changes in climates, landscapes, and seascapes.

The current wealth of on-site, globally integrated observatories—for example, Smithsonian Institution Global Earth Observatories (SIGEO; www.sigeo.si.edu/), Ocean Observatories Initiative (OOI; www.oceanleadership.org/programs-and-partnerships/ocean-observing/ooi), and Terrestrial Environmental Observatory (TERENO; www.tereno.net/)—provide critical information to biogeographers using GIS and spatial analyses to investigate the geographic dynamics of biotas and their underlying, environmental drivers. Of course, it is essential that these in situ measurements follow standardized protocols and that resultant databases include **metadata**—detailed descriptions of the methods used to record and store data and the relevant characteristics of those data. Contemporary analyses in biogeography, however, increasingly utilize other systems of complementary databases generated by off-site technology—in particular, via continually advancing innovations in remote sensing.

Remote sensing and satellite imagery

Remote sensing refers to any means of collecting data where the recorder or sensing device is not in direct contact with the area or objects of interest. In biogeographic applications, this typically includes the use of technologically sophisticated sensing devices operated from distant platforms, including aircraft, ships, and satellites. As a result, these systems can generate massive data sets covering areas that are much more extensive and often more continuous in coverage than even the most extensive in situ global observatories, which are by definition point-based and discontinuous.

Modern remote sensing systems are capable of recording an incredible variety of characteristics about focal biota, landscapes, and seascapes, depending on the particular types of electromagnetic radiation (typically ranging through radio, microwave, infrared, visible, and ultraviolet bands) being emitted or reflected from those targets (see overview by Gillespie et al. 2008). Once detected, the sensory information must first be processed, which typically requires using complex systems of mathematical models to interpret the spectral (wavelength) information in terms of relevant features of the biota and regions under study (e.g., the locations, temperatures, or activity states of organisms; the nature of relatively deep geological formations and the bathymetric features of the oceans; the depth and volume of ice sheets; or the rates of productivity of entire ecosystems; see Figure 5.30A).

Interpolation over space and time

As we indicated above, although in situ observations and, for that matter, many forms of remote sensing yield point-based (discontinuous) data, biogeographers often require information that covers a broader span of sites and areas, including those between locations of existing observations. Similarly, information from these sources, especially very remote sensing programs such as the Landsat and Envisat satellite programs, generate images and other information that are limited in temporal extent—often recorded during a time period far removed from that of other data to be used or the period most relevant to intended applications of that imagery.

Fortunately, we can solve this problem of observational discordance by creating more continuous coverages and estimates of values at unmeasured points in space and time using the method of **interpolation**. Spatial and temporal interpolations are equivalent procedures that provide estimates of the expected value of a variable at an unmeasured point in space (or time), based on statistical models that take into account the values of recorded variables at actual observation sites, and their distances to the site whose value is to be interpolated. These estimates can then be weighted such that, rather than being the simple average of all measured values in the vicinity, the estimated value is more strongly influenced by values of sites closer to the site to be interpolated. The result is tremendous variety of spatially and temporally concordant data (measured or derived from interpolations) that can then be subjected to a powerful battery of analytical tools now available to contemporary biogeographers.

Analysis of Biogeographic Patterns

Comprehensive coverage, even at the level of an introductory primer of all of the modern approaches for analyzing geographic patterns, would require many volumes on a variety of topics in geo-statistics (i.e., spatial statistics; see Webster and Oliver 2001; Fortin and Dale 2005; Kent et al. 2006; Rangel et al. 2006; Webster and Oliver 2007; Bivand et al. 2008). It is important, however, that we introduce some of the fundamental considerations and concepts that are especially relevant to, and sometimes confound, analyses of spatial patterns.

As with any statistical analyses, biogeographers must carefully consider the special nature of their data prior to evaluating the significance of any empirical patterns. Because of the extensive spatial and temporal scales inherent in most biogeographic studies, their research approach is often comparative and opportunistic. As a result, requisite data are typically collected by different scientists, often utilizing a variety of sampling protocols across different locations and time periods. As we described above, the latter two challenges can be solved by interpolation of values over space and time, respectively. Interpolations, however, are not observations but derived estimates—thus introducing another level of error to spatial analyses. Of more fundamental concern to nearly all forms of spatial analyses is that geographically explicit data, whether interpolated or genuine observational data, often violate a fundamental assumption of almost all statistical analyses—independence of observations. Ironically, this tendency for points closer in space or time to be more similar (**spatial** and **temporal autocorrelation**) can be viewed as an asset or an actual pattern of interest, as well as a potentially confounding problem (Fortin et al. 1989; Legendre 1993). That is, autocorrelation of adjacent observations is why interpolations work, and indeed the primary objective of many scientific investigations (e.g., those in landscape ecology) is determining the spatial extent of this interdependence (or the distance beyond which observations and relevant phenomena become independent). On the other hand, multiple data points taken within the critical windows of space and time where observations are not independent may repeatedly sample the same observations or phenomena—a problem termed **pseudoreplication** (inflation of the functional sample size). Fortunately, geo-statistical tool kits include methods both for estimating the extents of empirical autocorrelation in space and time and for adjusting statistical analyses to control for these potential problems (e.g., by adjusting inferred probabilities of statistical tests to those based on a smaller number [N] of independent observations, or by adjusting the values of data to be analyzed based on empirical trends in autocorrelation over space and time).

These and other recent advances in spatial analyses and remote sensing, combined with what are sometimes centuries' worth of empirical information on species distributions and their ecological associations, have enabled biogeographers to rigorously explore a wonderful diversity of patterns (see Turner et al. 2003; Loarie et al. 2007; Kark et al. 2008). One of the most fundamental yet challenging of these applications is estimation of what can be viewed as the fundamental unit of biogeography—the geographic range (see Chapter 4). Regardless of how ranges are represented (i.e., as outline, dot, or contour maps; see Figures 4.5–4.8), it is important to realize that all distributional maps are *estimates* of the actual distributions. Fortunately, although biogeographers will never reach perfect consensus on how best to estimate species distributions, alternative approaches share some key features in common: They utilize both empirical data (e.g., on recorded occurrences of the focal species) and remotely acquired and processed information (on soils, climates, or cover of habitats); they employ sophisticated GIS and statistical software to estimate environmental associations of the focal species; they use this information to develop models predicting distributions over a particular area and time period; they test the efficacy of those predictive models using empirical data not utilized in previous steps; and they modify the predictive models until some target efficacy is achieved (see pages 118–119 in Chapter 4).

Finally, it would be misleading for us to imply that the solutions to the challenges of rigorously exploring and assessing patterns in the geography of nature—in features of individuals, ranges of species, or diversity of entire biotas—are to be solved simply by advances in technology. Granted, these advances have provided and will continue to provide us with abilities to visualize and investigate biogeographic phenomena in ways unimagined, even by the most visionary founders of the field, including Humboldt, Darwin, and Wallace. Yet we continue to wrestle with one of the most fundamental questions in biogeography: What actually limits the range of a species? This question is the focal topic of the next chapter, and one relevant to nearly all patterns in the geography of nature.

4

DISTRIBUTIONS OF SPECIES:
Ecological Foundations

Biogeography's Fundamental Unit

As we observed in the previous chapter, the proposition that each species has a unique geographic range is central to all of biogeography. Biogeographers study many phenomena—locations of individual organisms, shifts in the local or regional distribution of a population, present and past distributions of higher taxa, clines of variation in morphology, physiology, behavior, and diversity across geographic gradients—but the ecological processes and historical events that have shaped the ranges of species are directly relevant to nearly all of them.

In exploring the factors and processes influencing this fundamental unit of biogeography—the **geographic range**—we proceed in a hierarchical fashion; first considering the various factors influencing the distributions of individuals, then discussing the additional factors and processes influencing the ranges of populations and entire species. While it seems natural to consider these phenomena within the realm of ecological biogeography, it is important to remember that geographic ranges have been influenced by responses of individuals, populations, and species to environmental conditions throughout their evolutionary history. We conclude this chapter by first previewing later chapters where we discuss derived patterns (e.g., variation of the size, shape, and overlap of ranges over space and time) and then summarizing some of the challenges and recent advances in predicting species distributions under current and future conditions (e.g., alternative scenarios for global climate change).

The Distribution of Individuals

Even the best map can convey only a highly simplified and abstract picture of the geographic distribution of a species. Real units of distribution are the locations of all the individuals of a species. A map depicting these locations would be impossible to prepare for most

FIGURE 4.1   An aerial photograph near the edge of the local distribution of the juniper tree (*Juniperous osteosperma*) in eastern Nevada. Individual trees, which are recognizable as dark spots, generally decrease in both size and abundance as elevation decreases from left to right. Note three things: (1) the overall complexity of the pattern of abundance and the difficulty of defining a precise range boundary; (2) the relatively uniform distribution of plants along an alluvial outwash plain at the top of the photograph; and (3) the patchy distribution of plants on southeast-facing slopes of small hills toward the bottom of the photograph. (Photograph courtesy of U.S. Forest Service.)

kinds of organisms, but we can get some idea of what one would look like from aerial photographs in which we can identify individuals of certain conspicuous species (**Figure 4.1**). Rapoport (1982) prepared a map of the distribution of a distinct, easily recognizable palm tree (*Copernicia alba*) from aerial photographs along a transect through part of its range in Argentina (**Figure 4.2**). As we can see from both the sample aerial photograph (see Figure 4.1) and Rapoport's data (see Figure 4.2), distribution is complex, with individual plants occurring in clumps separated by gaps. As the edge of the range is approached, there is a tendency for individuals to be more sparsely distributed, clumps to be smaller, and gaps between them to be larger.

The clumpy-gappy distribution of individuals across a landscape means that any kind of map of a species' range is not only an abstraction but a scale-dependent abstraction. Imagine that we superimposed a grid on an aerial photograph such as Figure 4.1. Whether an individual will be found in a particular grid cell will depend on the size of that cell: the larger the cell, the higher the probability of its containing at least one individual. This exercise reveals that the edge of a range is also a scale-dependent abstraction. Although in Figure 4.1 there are no individuals in the lower right corner,

FIGURE 4.2   Abundance of the palm tree (*Copernicia alba*) along a 3 km wide transect from the western edge toward the center of its geographic range. Data were taken from aerial photographs on which individual palms were easily recognized by their distinctive shapes. Note that near the western edge of the range, abundance tends to be low and the distribution of trees tends to be patchy (as indicated by values of zero individuals per square kilometer). (After Rapoport 1982.)

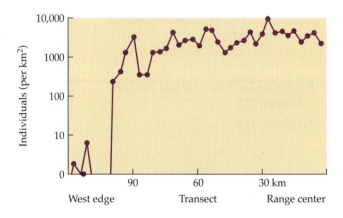

the exact definition of the distributional boundary will depend on the scale at which we connect the locations of individuals to draw an edge. Furthermore, in addition to relatively obvious range boundaries, there are "holes" within the range where no individuals occur. Therefore, Rapoport (1982) has likened the range to a slice of swiss cheese. But even this is an oversimplification because the sizes and locations of the areas where no individuals are considered to occur also depends on the spatial scale of analysis. Perhaps the best representation of the effect of spatial scale on the perceived distribution of a species is still Erickson's (1945) classic distribution maps of the shrub *Clematis fremontii* (**Figure 4.3**).

Even an aerial photograph—thought to be a faithful depiction of the distribution of a species—fails to capture another critical feature of the geographic range because it represents only a snapshot in time. The distribution of any species is dynamic, and any accurate depiction of its range should in theory be constantly updated to reflect the changes that occur as individuals are

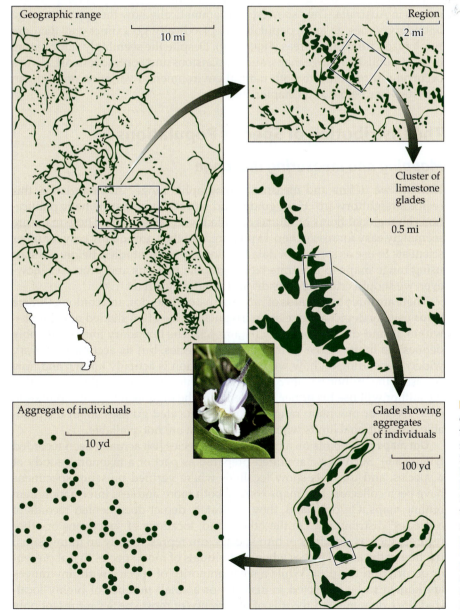

FIGURE 4.3 Erickson's classic depiction of the distribution of Fremont's leather flower (*Clematis fremontii*) within the state of Missouri in the central United States, on different spatial scales. The largest scale shows the geographic range based on known collecting localities. Successively finer scales show distributions of populations. The most local scale shows dispersion of individual plants within a single local population. Note that, at all scales, the distribution is patchy, and areas where plants are found are separated by uninhabited areas. (After Erickson 1945; photo courtesy of Lisa Francis.)

FIGURE 4.4 A schematic diagram showing how abundance and distribution of a hypothetical organism might vary in time and space. Shown are fluctuations in abundance over many years at three different localities (A–C) separated by distances of several kilometers. Note that all three populations fluctuate over time. At locality A, which is presumably at the margin of the local or geographic range of the species, only a few individuals are intermittently present, indicating repeated episodes of local extinction and recolonization. (After Andrewartha and Birch 1954.)

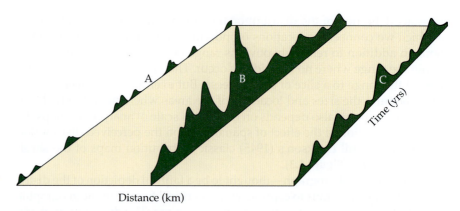

Distance (km)

born, move, and die and as populations colonize new areas and go locally extinct in parts of their former range. Andrewartha and Birch (1954), for example, documented large shifts in the geographic ranges of several species of insects in Australia. They showed diagrammatically how the apparent range boundary varies as local populations episodically go extinct and then are recolonized from other areas (**Figure 4.4**). Despite the seemingly static nature of most published range maps, such expansions and contractions are always occurring in response to both natural environmental variation and human activities (see below and Chapter 16).

The Distribution of Species Populations

Mapping and measuring the range

How do we define and measure a geographic range? At first glance, this seems straightforward. Field guides and more technical systematic publications on regional floras or faunas are filled with **range maps**. These maps are seemingly easy for researchers to prepare, and they are equally easy for other scientists to use as sources of data for biogeographic studies. Before we start using range maps to illustrate biogeographic patterns and processes, however, we should critically consider just what they tell us.

There are three basic kinds of range maps: outline, dot, and contour. **Outline maps** usually depict a range as an irregular area—often shaded or colored—within a hand-drawn boundary (**Figure 4.5**). The boundary line presumably defines the limits of the distribution of a species, but its accuracy can vary widely depending on how well the distribution is actually known and how precisely the author has incorporated this information into the map. Often the author will use his or her knowledge of the organism (e.g., its abilities to cross rivers or mountain ranges) to make educated guesses about the probable distributional limits when adequate data are not available.

Dot maps plot points on a map where a species has actually been recorded (**Figure 4.6**). Such maps are often prepared as part of a taxonomic study of a species, and the dots show localities where verified museum specimens have been collected. Dot maps convey both more and less information than outline maps. On one hand, they accurately depict documented records of a species' distribution. On the other hand, locations of specimens or other records, such as sightings of bird species, can represent only an infinitesimal fraction of the actual places where individuals of most species live at present or occurred in the past. While a small minority of species with tiny ranges are known to be restricted to just one or a small number of highly localized sites, documented records of occurrence of most species represent only

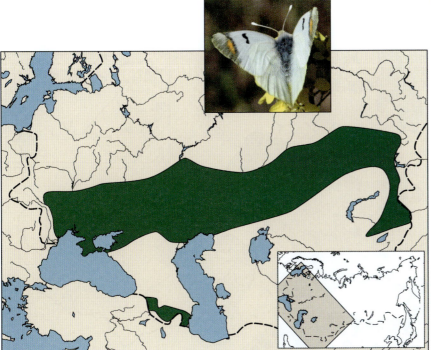

FIGURE 4.5 An example of an outline map of the geographic range of a species—in this case the endangered sooty orange-tip butterfly (*Zegris eupheme*), which occurs in southwestern Asia. The outer boundary has been drawn by hand to include localities where the species is known to occur. (After Borodin et al. 1984; photo © Matt Rowlings, www.eurobutterflies.com.)

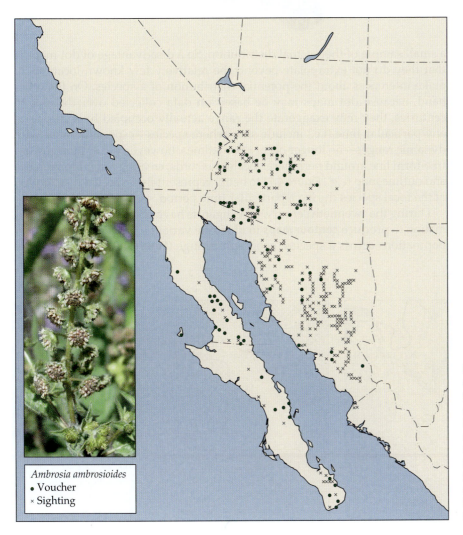

Ambrosia ambrosioides
● Voucher
× Sighting

FIGURE 4.6 An example of a dot map of the geographic range of a species—in this case the canyon ragweed (*Ambrosia ambrosioides*) of the Sonoran Desert. Each circle represents a locality where someone has documented the presence of the species by collecting a voucher specimen and depositing it in an herbarium. Each cross represents an additional record based on a sighting and identification of the plant in the field. (After Turner et al. 1995; photo courtesy of Max Licher.)

FIGURE 4.7 An example of a combination dot and outline map of the geographic range of a species—in this case the endangered southern festoon butterfly (*Zerynthia polyxena*), which is restricted to a small area north of the Black Sea in southern Eurasia. Each dot represents a locality where the species has been recorded. A line has been drawn by hand to include the outermost dots, thereby enclosing the known geographic range. (After Borodin et al. 1984; photo © Matt Rowlings, www.eurobutterflies.com.)

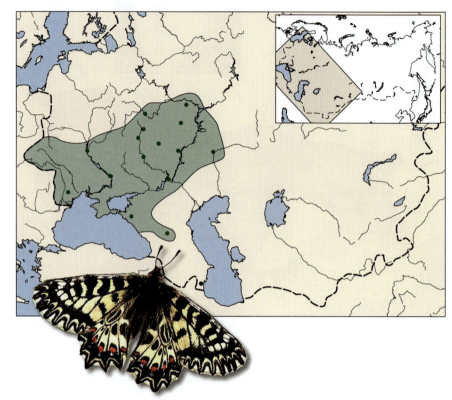

FIGURE 4.8 An example of a contour map of the geographic range of a species—in this case the winter range of the blue jay (*Cyanocitta cristata*), showing geographic variation in abundance. (A) Each contour line (or isocline) indicates a twentieth-percentile class of relative abundance. (B) A three-dimensional landscape depicting relative abundance. Data on abundance come from North American Christmas Bird Counts. Raw data from these census counts (number of birds seen per hour per field party) were entered into a computer program that averaged and smoothed them to estimate abundance between actual census localities in order to draw the maps. (From Root 1988a.)

a small sample of their actual distribution. So a disadvantage of dot maps is that they do not extrapolate beyond the relatively few known locations to make inferences about the potential distribution of a species. On the other hand, because dot maps may be based on data collected over decades to centuries, they may exaggerate the range actually occupied by a species at any particular time (i.e., include sites where species—especially those with dynamic ranges—no longer occur). Sometimes, however, the author draws a free-form line around peripheral location records, creating a combination dot and outline map (e.g., **Figure 4.7**) to better represent the expected distribution of the focal species during a particular time period.

During the twentieth century, investigators have obtained sufficient information to produce **contour maps** to illustrate variation among individuals or populations across a species' geographic range (**Figure 4.8**). Although these

(A)

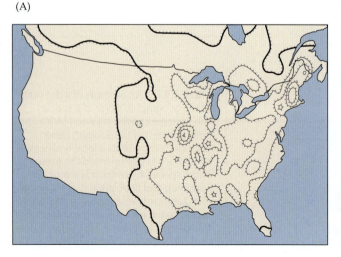

(B)

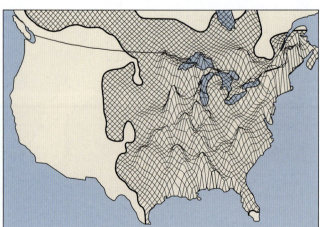

maps convey much more information than either outline or dot maps, they should be interpreted with caution. Accurate information on abundance or other local (individual- to population-level) variables usually is available for only a limited number of fairly widely separated localities, so much of the depicted pattern is based on **interpolation**, thus introducing errors and biases of scale discussed earlier in this chapter. So let the user beware: Even contour maps are highly oversimplified visualizations and estimates of the complex and dynamic structure of ranges.

Population growth and demography

The size of a range, location of its boundaries, and shifting patterns of abundance within those boundaries reflect the influence of environmental conditions on the survival, reproduction, and dispersal of individuals and the dynamics of populations. In 1798, in his seminal *Essay on the Principle of Population*, Thomas Malthus showed that all kinds of organisms have the inherent potential to increase their numbers exponentially. A population increases when the combined rates of birth and immigration exceed the combined rates of death and emigration. We can express this concept mathematically as

$$r = b + i - d - e \tag{4.1}$$

where r is the per capita rate of population growth (if r is positive, the population increases; if r is negative, it decreases); b and d are per capita birth and death rates, respectively; and i and e are the respective per capita rates of immigration from and emigration to other populations. Given unlimited resources and favorable environmental conditions, a population will grow continuously at its maximum possible r. It will increase its numbers as described by the equation

$$dN/dt = rN \tag{4.2}$$

where dN/dt is the rate of change in number of individuals, N, with respect to time, t; and r is the population growth rate, as above. We call this **exponential growth**, and we can describe its rate in terms of the time interval required for a population to double its numbers. If it kept growing exponentially, any species would eventually cover the Earth with its own kind. The time required would depend on r: Bacteria and houseflies would require only a few years, whereas slowly reproducing trees and elephants, with their lower values of r, would take a few thousand years. The global human population has been growing at a nearly exponential rate for the last several thousand years (**Figure 4.9**). Malthus recognized, however, that because resources ultimately limit growth, and because many environments are unsuitable, no organisms actually continue to increase indefinitely at such exponential rates; all populations are eventually limited by one or a combination of variables defining their ecological niche.

Hutchinson's multidimensional niche concept

In 1957 G. Evelyn Hutchinson developed the concept of the multidimensional **ecological niche** in order to conceptualize how environmental conditions limit abundance and distribution. Hutchinson's view of the niche was a modification of the earlier niche concepts of Grinnell (1917) and Elton (1927; see also James et al. 1984; Schoener 1988). Hutchinson realized that over a period of time and over its geographic distribution, every species is limited by a number of environmental factors. He conceptualized a species' environment as a multidimensional space or "hypervolume" in which different axes

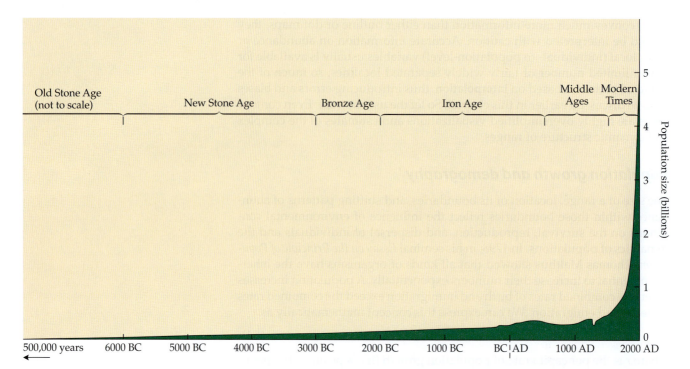

FIGURE 4.9 Estimated growth of the human population over the last 10,000 years. Note the almost continuously increasing, nearly exponential shape of the curve as *Homo sapiens* not only increase in local abundance but also spread over most of the Earth. All populations have the capacity to increase exponentially so long as environmental conditions are not limiting. Colonizing exotic species typically show similar near-exponential growth rates during a period of rapid range expansion. (After Desmond 1965.)

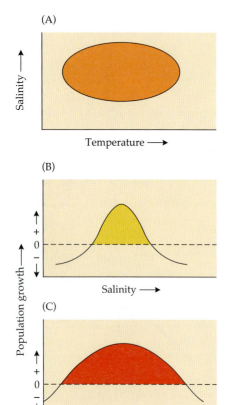

or dimensions represent different environmental variables. The niche of a species can thus be visualized as the combination of these variables that allow individuals to survive and reproduce, and populations to maintain their numbers.

This concept sounds intimidating and is indeed a bit abstract, but the basic idea is very simple (**Figure 4.10**). Imagine the effects of just two environmental variables—say, temperature and salinity—on some aquatic species. If all other conditions are favorable, individual performance and population growth will be limited by the joint effects of these two variables. We can plot out the range of conditions under which population growth will be negative, zero, or positive; the space inside the zero-growth contour, or isocline, represents the niche space of this hypothetical species. In reality, it is almost certain that this aquatic organism would also be limited by other variables, such as dissolved oxygen concentration, a competing species, presence of a predator, or all three. Each of these variables would represent another dimension of the niche, causing the niche space to be a multidimensional volume, which is hard to visualize or draw but not too hard to imagine. It is easy to see that the

FIGURE 4.10 (A) A diagram representing two dimensions—temperature and salinity—of the niche of a hypothetical aquatic species. The shaded area represents the combinations of these two variables—a broad range of temperatures but a narrow range of salinities—under which individuals can survive and reproduce or a population can increase. In (B) and (C), population growth rate is plotted as a function of each variable, with the horizontal dashed line showing the zero value used to plot the elliptical niche space in (A).

niche of each species is unique because it differs at least slightly—sometimes greatly—from all others in the combinations of environmental conditions required for the survival and reproduction of its individuals and the growth of its populations.

The geographic range as a reflection of the niche

The geographic range of a species can be viewed as a spatial reflection of its niche, along with characteristics of the geographic template and the species that influence colonization potential. If species were unlimited by geographic barriers and colonization abilities, they might be expected to occur wherever environmental conditions were suitable. But just as interspecific interactions can limit a species to its **realized niche**, which is only a subset of its entire **fundamental niche**, so geographic barriers and colonization abilities restrict a species to its **realized geographic range**, which is a small subset of its **fundamental geographic range**. Geographic range is limited over broader temporal and spatial scales by limited time and opportunities for dispersal.

One of the earliest studies of the ecological niche of a species is still one of the most complete. Joseph Connell (1961) studied the environmental factors that limit the range of a barnacle (*Chthamalus stellatus*) on the rocky coast of the Isle of Cumbrae in Scotland. As mentioned in Chapter 3, the intertidal zone is that narrow strip of coastline that is alternately inundated by seawater and then exposed between tides. Species are typically restricted to a narrow range of exposures within the intertidal zone. Connell used some elegantly simple field experiments to characterize important variables of the niche of *C. stellatus* and to explain its distribution in the uppermost portion of the intertidal zone (**Figure 4.11**). He showed that the upper edge of the species' range is set by the ability of the barnacles to tolerate the physiological stress of desiccation while exposed during low tides. The lower edge of the species' range is set by interactions with other intertidal organisms, primarily by competition with another barnacle species (*Balanus balanoides*) and secondarily via predation by a snail (*Thais lapillus*). The power of the experimental method is illustrated by the effect of removing *B. balanoides* from small patches of shore—on those plots where its competitor was absent, *C. stellatus* extended its realized range lower into the intertidal zone.

Connell's study pioneered the use of field experiments in ecology. It is also a classic for demonstrating how three niche variables—exposure to desiccation, competition with another barnacle species, and predation by a snail—can largely explain the limited distribution of *C. stellatus* on the rocky shores of the Isle of Cumbrae. Note, however, that these are almost certainly not the only niche variables affecting the distribution of this barnacle: For example, some other factor(s) presumably account for the absence of *C. stellatus* from the sandy and muddy substrates that occur only a short distance from Connell's study site.

FIGURE 4.11 Diagrammatic representation of the effects of interspecific competition and other factors on the distribution of the barnacle (*Chthamalus stellatus*) in the intertidal zone on rocky shores in Scotland. Diagrams on the left show species distribution; the width of each bar indicates population density at that depth. Larvae settle over a wide range, but many die before reaching maturity, leaving adults confined to a much narrower zone. Diagrams on the right indicate the effects of three mortality-causing factors: desiccation between tides, A, which sets the upper limit of distribution; predation by the snail (*Thais*), B; and competition from the barnacle (*Balanus balanoides*), C, which together with A and B set the lower limit. The width of each bar shows the strength of each effect at that depth. (After Connell 1961.)

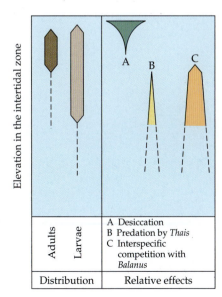

Habitat 1 (Source population)

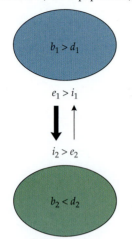

Habitat 2 (Sink population)

FIGURE 4.12 Diagrammatic representation of source and sink habitats, showing the relative magnitude of the four processes that determine growth and persistence of their populations. In the source population (Habitat 1), the birth rate (b_1) is greater than the death rate (d_1), but the population does not increase. Instead, the "excess" individuals disperse, resulting in a higher rate of emigration (e_1) than immigration (i_1). The opposite situation results in the sink population (Habitat 2), which is able to persist despite having a lower birth (b_2) than death rate (d_2), because the rate of immigration (i_2) is sufficiently higher than the rate of emigration (e_2).

While the multidimensional environmental niche provides a conceptual framework for understanding how environmental limiting factors influence both geographic ranges and local population densities of a species, niche variables alone are inadequate to account for all patterns of distribution and abundance. Three complications will be mentioned briefly here, and then considered further in later chapters. First, it is too simplistic to assume that environmental conditions are equally favorable for a species at all localities where it occurs. Some localities may be so favorable that birth rates exceed death rates; these localities can serve as **source habitats**, producing surplus individuals that disperse out to other areas (**Figure 4.12**; Pulliam 1988). Some other localities may be so unfavorable that death rates exceed birth rates, but such **sink habitats** may still be inhabited if they receive a sufficient supply of immigrants to maintain a local population (refer to Equation 4.1 and note the terms *i* and *e*, representing the contributions of immigration and emigration to the population growth rate). One might expect that some of the areas near the border of a species' range would be sink habitats, with environmental conditions so marginal that they would not be able to sustain populations in the absence of immigration. The sea rocket (*Cakile edentula*), a broadly distributed annual plant that lives on coastal sand dunes, provides an example of this pattern. The proportionately few individuals growing in exposed seaward sites have the highest growth rates and produce the majority of seeds, but most plants occur in unfavorable inland sites, where storms have carried large numbers of dispersing fruits (Keddy 1982).

A second complication is that, just as there may be sites where environmental conditions are unfavorable but are nonetheless inhabited, there may also be favorable localities that are uninhabited. Some ecologists would say that there is an unfilled niche for the species in such places, but most prefer to define the niche as a characteristic of organisms (i.e., of species) rather than of places. As mentioned in Chapter 3, a species is likely to be absent from many places where it could live (realized ranges are always smaller than fundamental ranges). Often this is because such favorable sites are isolated from inhabited areas by some combination of distance and unfavorable environmental conditions, such that immigration and establishment are rendered highly unlikely. This situation is common, as demonstrated by the large number of exotic species that have been successful in becoming established in new habitats only after humans have transported them across otherwise formidable barriers to dispersal (see Chapter 16). Biogeographers often invoke "history" to account for situations in which species are absent from apparently suitable areas. Indeed, to understand why a species occurs where it does and not elsewhere, it is necessary to understand both the history of apparently favorable places and the barriers between them, and the history of the species itself. We will return to consider these problems and the relationships between the **history of place** and the **history of species** in Chapters 6, 10, and 11.

Finally, some places may be inhabited only intermittently. Local populations increase and decrease—sometimes to local extinction—as environmental conditions fluctuate or as stochastic events affect their growth. Because most habitats are patchily distributed, it is likely that the ranges of most species include isolated populations separated by uninhabited areas. When a species population is subdivided in this way, it is said to be a **metapopulation**, comprised of multiple **subpopulations** linked together over space and time by infrequent but nonetheless important dispersal. In such cases, some of the subpopulations are likely to go extinct intermittently, especially if they are sink populations. Even source populations in favorable patches of habitat, however, may go extinct by chance. New subpopulations are also likely to be founded by immigration to uninhabited patches, including those vacated by previous local extinction events. We will return to consider the implications of such metapopulation and source–sink dynamics in Chapters 6, 13, and 16. For the moment, it is sufficient to stress that they are especially likely to occur on the periphery of a species' range and to contribute to dynamic shifts in range boundaries.

The relationship between distribution and abundance

The contour maps discussed earlier (see Figure 4.8) imply that there is considerable variation in abundance within a species' range. In fact, most published contour maps underestimate the magnitude of this variation, because of the statistical and graphical methods used to interpolate between data points and construct the maps. The real spatial abundance patterns of nearly all species are extremely heterogeneous—what statisticians would call **clumped** or **aggregated** (Brown et al. 1995). That is, compared with a random distribution of individuals across a landscape, some places have many more individuals, and others have many fewer or none at all.

Although the complications of source–sink dynamics and history discussed above are relevant here, most of this spatial variation in abundance presumably reflects the extent to which the local environment meets the niche requirements of a species. Each species tends to be most abundant where all niche parameters are in the favorable range, and to be rare or absent where one or more environmental factors are strongly limiting (Brown 1984; Hengeveld 1990; Lawton et al. 1994; Brown et al. 1995).

Common species are typically several orders of magnitude more abundant at some sites than at others (**Figure 4.13**). For example, the North American Breeding Bird Surveys conducted from 1966 to 1992 revealed that, on average, only a single red-eyed vireo (*Vireo olivaceus*) was recorded on more than 200 of the annual routes surveyed each year, but more than 100 individuals were counted on 6 routes. One consequence of this highly clumped distribution pattern is that the majority of individuals of a species actually occur in a very small proportion of its geographic range. For the majority of common songbirds in eastern North America, more than half of the individuals occurred at fewer than 20 percent of the sites within their geographic ranges. Of course, rare species may be uncommon throughout their ranges (Rabinowitz et al. 1986; Gaston 1994), but they too typically have patchy distributions and are absent from many, presumably unfavorable, localities.

As we noted earlier, species populations also exhibit substantial, sometimes remarkable, variation over time; most of it presumably reflecting temporal variation in niche parameters. The fluctuations of Australian insect populations in response to climatic variation documented by Andrewartha and Birch (1954) are excellent examples (see Figure 4.4). Migratory locusts of the Old World provide additional, perhaps even more dramatic, examples

FIGURE 4.13 Variation in the abundance of two common songbird species, (A) the red-eyed vireo (*Vireo olivaceus*) and (B) the Carolina wren (*Thryothorus ludovicianus*), among hundreds of census routes distributed throughout their respective geographic ranges in eastern North America. Numbers of birds counted on each standardized census route of the North American Breeding Bird Survey (BBS) between 1966 and 1992 are plotted in rank order, so routes with one bird are on the left and routes with the maximum number of birds recorded are on the right. Note that for both species, fewer than 5 birds (BBS route abundance) were recorded on the vast majority of census routes, but more than 100 birds were counted on at least six routes. This highly clumped pattern of abundance is characteristic of most birds as well as many other organisms. (After Brown et al. 1995.)

(A) Red-eyed vireo

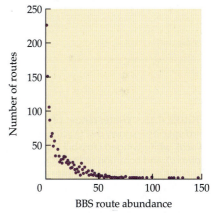

(B) Carolina wren

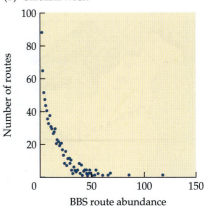

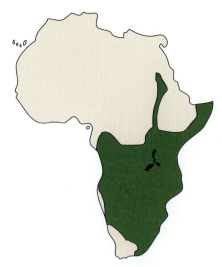

FIGURE 4.14 The temporally shifting range of the red locust (*Nomadacris septemfasciata*) in Africa. The small black areas are source habitats at the core of the range. These outbreak areas are the only places known to sustain permanent populations. Enormous population increases and geographic expansions begin in these areas, and in favorable years the locusts can spread into sink habitats throughout the invasion area (green area), which includes about half the continent. (After Albrecht 1967.)

FIGURE 4.15 Winter ranges of three bird species—snowy owl (*Nyctea scandiaca*), white-winged crossbill (*Loxia leucoptera*), and common redpoll (*Carduelis flammea*)—that normally winter at high latitudes (dark green area) but that disperse far to the south (to the dotted line), greatly expanding their ranges in years of food shortage.

(Waloff 1966; Albrecht 1967; White 1976). Source populations of these grasshoppers persist in limited regions called outbreak sites, where conditions are suitable for their continued survival and reproduction. During periods when weather and food supplies are particularly favorable, these populations increase fantastically, change their morphology and behavior, aggregate into huge swarms, and migrate outward from the outbreak sites to forage over enormous areas. Such plagues of both the African migratory locust (*Locusta migratoria*) and the red locust (*Nomadacris septemfasciata*) have occurred two or three times in the last century, sweeping over most of southern Africa, an area consisting of millions of square kilometers—more than 1000 times the size of the outbreak area from which the locusts originated (**Figure 4.14**).

Similar fluctuations occur in many other regions and for species other than insects. For example, tundra and taiga (coniferous, or boreal, forest; see Chapter 5) spread across northern North America, Europe, and Asia. In those regions, several species of voles and lemmings (mouselike rodents, subfamily Arvicolinae, formerly Microtinae) fluctuate by several orders of magnitude in abundance over a three-to-four-year period (see Finerty 1980; Lidicker 1988; Stenseth and Ims 1993). Unlike locusts, however, these rodents have very limited dispersal abilities, so their geographic ranges do not change very much during these cyclical fluctuations; however, their local patterns of habitat use may shift or expand considerably. Some northern birds such as snowy owls, which feed on these rodents, and crossbills, which feed on conifer seeds that vary markedly in production among years, also show wide fluctuations in abundance. Associated with population dynamics of these birds are large shifts in their winter ranges—sometimes hundreds of kilometers to the north or south—over periods of fluctuating food availability (**Figure 4.15**; Bock and Lepthian 1976).

Superimposed on all of this spatial and temporal variation in abundance are some intriguing spatial patterns. We will illustrate and comment briefly on three of them here, again using data and inferences generated by the North American Breeding Bird Surveys (**Figure 4.16**). First, like many other variables, abundance is autocorrelated in space—which is a technical way of saying that abundances tend to be more similar at sites that are closer together. This is just what we would expect if abundance reflects the suitability of the local environment and if niche variables also exhibit spatial autocorrelation across the geographic template. Second, abundance often varies systematically over a species' geographic range, from relatively high numbers near the center, to zero at the boundaries. In particular, localities near the edges of the range tend to have highly variable but generally low populations, whereas population abundances at sites near the center of the

Snowy owl

White-winged crossbill

Common redpoll

(A) Population abundance   (B) Coefficient of variation

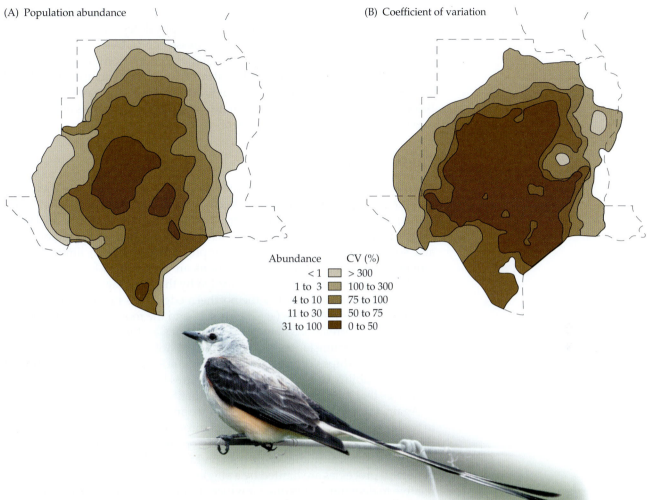

Abundance	CV (%)
< 1	> 300
1 to 3	100 to 300
4 to 10	75 to 100
11 to 30	50 to 75
31 to 100	0 to 50

range tend to be high and exhibit low variability from year to year. This pattern is again consistent with tendencies for environments to be spatially autocorrelated (thus becoming more dissimilar and less optimal as we move away from the more optimal, central sites) and for range boundaries to occur where one or more niche factors become unfavorable. Third, an interesting exception to the previous pattern is that abundance may be high near a range margin that is set by a coastline or other "hard" barrier. Again, this is what would be expected when a particular range boundary is determined by just one niche variable that abruptly becomes unfavorable.

What Limits the Geographic Range?

So far we have talked about niche dimensions and limiting factors only in abstract terms. But what are the environmental variables that set the boundaries of a species' range by limiting the survival, reproduction, and dispersal of its individuals? Our previous discussion of the multidimensional environmental niche, of spatial and temporal variation in abundance, and of Connell's classic study of barnacle distributions in the intertidal zone would suggest that locations of range boundaries are set by multiple environmental factors and that these can include both physical conditions (abiotic factors) and effects of other organisms (biotic factors).

Unfortunately, there have been few studies as comprehensive as Connell's. Most investigators have studied the effects of just one or a few factors in a small area near one edge of a geographic range. This is understand-

FIGURE 4.16 Abundance (A) and population variability (coefficient of variation) (B) for the scissor-tailed flycatcher (*Tyrannus forficata*) in its breeding range of Texas and Oklahoma. These maps illustrate three general tendencies: (1) autocorrelation, such that nearby areas tend to have similar abundances; (2) maximal abundances in one or more areas, usually located near the center of the range; and (3) generally low abundances and relatively high variability at the edge of the range, except where the boundary is set by a coastline. Abundance = average count in North American Breeding Bird Surveys; CV = coefficient of variation of abundance over a 25 year period (compare with Figures 4.8 and 4.2). (From D. Certain, unpublished report; photo courtesy of S. Maslowski, USFWS.)

able because geographic ranges of most species are large, and the scientific expertise of most ecologists is limited (e.g., physiological ecologists study the effects of different kinds of abiotic stresses on individuals, while community ecologists study the different kinds of interactions among species). Nevertheless, we are left with a literature that provides many examples of single limiting factors but little overview and synthesis (but see Gaston 1994; Brown et al. 1996). So first let's examine some of these examples and then try to see whether they reveal any general patterns and processes.

Physical limiting factors

Many widespread species appear to be limited in at least parts of their geographic ranges by physical factors such as temperature regime, water availability, and soil and water chemistry. For example, many Northern Hemisphere plants and animals become increasingly restricted to low elevations and south-facing exposures as they approach the northern limits of their ranges, suggesting that their distributions are determined by ambient temperature. An excellent example is afforded by the giant saguaro cactus (*Carnegiea gigantea*) as it approaches the northern edge of its range in the Sonoran Desert of southern Arizona (see Figure 4.17 and the more detailed discussion below). Such correlations between distributions and local climate, however, provide only circumstantial evidence and do not necessarily indicate direct causal relationships. The species in question might, for example, be limited not by its inability to tolerate low temperatures but by competition from other species that happen to be superior competitors in cold climates.

Again, even if we could tease apart cause and correlation, ranges are likely to be determined by a combination of environmental factors and physiological, behavioral, and ecological capacities of the species. This seems to contradict the classical studies of Justus von Liebig, who suggested in 1840 that biological processes are limited by a single factor—the one in shortest supply relative to demand, or the one for which the organism has the least tolerance. At one time ecologists accepted this idea so completely that they called it **Liebig's law of the minimum** and then spent much time trying to identify *the* single factor that limited the growth of each population. Now, however, as implied by Hutchinson's multidimensional niche concept and demonstrated by numerous empirical studies, we realize that Liebig's concept is too simplistic.

For example, many temperate and arctic birds and mammals appear to be limited by their inability to tolerate cold temperature regimes in the winter. Often this is not because they simply cannot survive at such low temperatures (Dawson and Carey 1976; Root 1988b,c), but rather because changes in solar radiation, precipitation, and temperature reduce primary productivity resource availability while simultaneously increasing resource requirements of these species. Consistent with this more inclusive view of range limitation, many species previously limited to temperate regions are now expanding their ranges poleward in response to the provision of backyard bird feeders and recent, anthropogenic amelioration of winter climates (e.g., Boyd and Nunneley 1964; see Chapter 16).

Another problem in determining the causes of distributional limits is the difficulty of identifying the mechanisms by which environmental factors affect the growth of populations. Cold, for example, is not a single variable, and different aspects of low temperature regimes limit different populations in different ways. Adults of some plant species may be killed by critically low short-term temperatures such as those experienced on a single, exceptionally cold winter night. Other species may be more susceptible to damage from prolonged freezing. Still other species may be limited by cold climates, not because they cannot withstand low winter temperatures, but because the summer growing season is too short to allow for sufficient growth and reproduction.

As mentioned above, one of the best-documented examples of cold temperatures limiting the upper elevational and latitudinal distribution of a species is provided by the saguaro cactus (**Figure 4.17**). This giant multiarmed columnar cactus, which may reach 15 m in height and grow to be 200 years of age, is a conspicuous part of the landscape in much of the Sonoran Desert of southern Arizona and northwestern Mexico. Although it lives where winter nighttime frosts are not infrequent, the saguaro is extremely sensitive to temperatures below –7° C and to prolonged freezing. Individual cacti are often killed by frost damage to their tissues, especially by destruction of the growing shoot tips. Young saguaros are more susceptible to frost damage than are adults, but seedlings typically become established under the canopy of small desert trees, which provide young cacti with a protective microclimate for the first few decades of their lives (Nobel 1980b). These "nurse trees" shield young saguaros from the cold night sky and prevent their freezing in much the same way that frost damage to tomato plants can be prevented by covering them at night with paper or plastic—the loss of heat by infrared radiation to the sky is retarded by these nurse trees. Before they reach reproductive age, saguaros grow above their nurse trees, often killing the trees in the process; but by then they are large enough not to be affected by overnight frosts. Nobel (1978, 1980a) has studied the thermal relations of stems and shoot api-

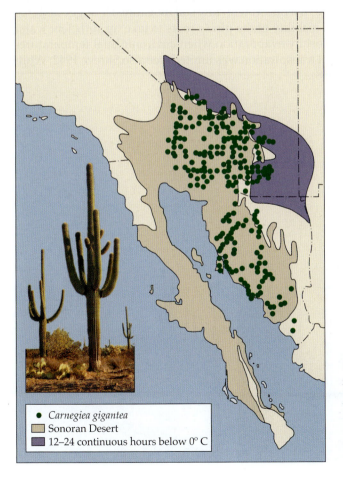

- ● *Carnegiea gigantea*
- ▢ Sonoran Desert
- ▢ 12–24 continuous hours below 0° C

FIGURE 4.17 Distribution of the saguaro cactus (*Carnegiea gigantea*) in relation to winter temperature regime. This cactus, like many other Sonoran Desert plants, is intolerant of prolonged freezing. Note the close correspondence between the northern limit of the saguaro, the northern boundary of the Sonoran Desert, and the region where temperatures remain below 0° C for more than 12 hours. (Data from Hastings and Turner 1965; Hastings et al. 1972; after Shreve, 1964; photo © Dan Eckert/istock.)

ces using computer simulations and direct field measurements. Results show that the large stem diameter of the saguaro enables it to maintain higher minimal temperatures of its apical buds, and thus it has a more northern distribution than related species of columnar cacti.

Steenburgh and Lowe (1976, 1977) studied populations of saguaros at Saguaro National Park outside Tucson, Arizona, near the northeastern and upper elevational limit of the species' range. Extensive mortality of both young and adult plants occurred as a result of exceptionally low temperatures in January of 1937, 1962, 1971, and 1978. The 1971 freeze killed about 10 percent of individual cacti and severely injured an additional 30 percent; many of the injured cacti died during the next few years as a result of microbial infections that started at the site of the frost damage (**Figure 4.18**). These observations of episodic winter kill, together with a close correspondence between the northern and eastern boundaries of the species' range and areas that experience below-freezing temperatures for more than 12 hours at a time (see Figure 4.17), suggest that low winter temperatures directly limit the distribution of saguaros in this region. Evidence for this explanation has been provided by a remarkable set of recent studies that link regeneration and survival of saguaros to global volcanic activity and resultant levels of volcanic dust, which temper extremes of temperature in both winter and summer (Drezner and Balling 2008).

Distributions of many other plant species appear to be limited, not just by low temperatures, but by their interaction with other environmental conditions such as water availability and soil chemistry. Hocker (1956) studied the distribution of loblolly pine (*Pinus taeda*) in the southeastern United States and concluded that the northern and western edges of the pine's range were set by low temperatures in concert with low soil moisture. He suggested that this resulted from the inability of the pine's roots to take up sufficient water to replace quantities lost by evaporation when environmental temperatures were low and stomatal transpiration was limited (see also Shreve 1942; Whittaker and Niering 1968).

FIGURE 4.18 Matched photographs of a stand of saguaro cacti near Redington, Arizona, near the upper elevational and northern edge of the species' range. (A) In 1961. (B) In 1966, showing the loss of one large individual (center foreground) and scars (white patches near tips of arms) on several other cacti as a result of severe frost in 1962. (C) In 1979, showing much additional mortality due to severe frosts in 1971 and 1978; several of the individual cacti still standing are dead or dying. (A and B courtesy of J. R. Hastings; C courtesy of R. M. Turner.)

(A)

(B)

(C)

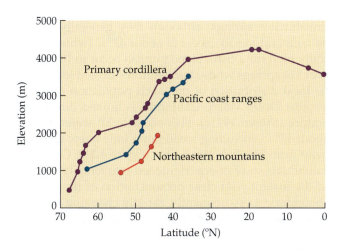

FIGURE 4.19 Relationship of timberline to elevation and latitude in three different major mountain chains in North America. In general, timberline increases in elevation with decreasing latitude reflecting the influence of increasing temperature. Note, however, that the relationship is different in each mountain chain because of other factors such as length of the summer growing season. Note also that along the primary cordillera (mountain chains extending from the northern Rocky Mountains to Panama), there is essentially no change in the elevation of timberline within the tropics and subtropics between about 35° N latitude and the Equator. (After Daubenmire 1978.)

Across the world's mountainous regions, many investigators have attempted to determine the cause of **timberline**, the upper elevational limit of trees on mountains. The large-scale geographic position of timberline seems to be related to the mean or maximum temperature during warm months of the growing season, which varies with latitude and elevation (**Figure 4.19**). Timberline, however, is also influenced locally by other factors such as wind, snow depth, and energy balance (Daubenmire 1978; Stevens and Fox 1991). Trees at timberline in windswept alpine regions are sometimes called krummholz (crooked timber), referring to their dwarfed nature, contorted shape, and lack of branches on their windward side. At timberline, established trees often live for a long time, but their growth is very slow, and successful reproduction and seedling establishment are rare. Bristlecone pines (*Pinus longaeva*), which often grow just below timberline on arid mountains in the southwestern United States, are the oldest known living things, some individuals being more than 3000 years old. Bristlecone pines grow slowly and consequently produce exceptionally hard, dense wood that is highly resistant to decay (**Figure 4.20**). In some places, dead bristlecones form a "fossil timberline" above the timberline of living trees and extending back many millennia, with annual growth rings providing an invaluable record of past climatic conditions and growing seasons (La Marche 1973, 1978; Fritts 1976).

Unlike plants and sessile animals (such as the barnacles discussed earlier), most animals can move to seek out favorable microenvironments and thus avoid the most stressful abiotic conditions. Nevertheless, even highly mobile animals such as fish can be limited directly by physical factors, such as environmental temperatures and water chemistry. Pupfish of the genus *Cyprinodon* are extremely eurythermal and euryhaline; that is, they can tolerate a wide range of temperatures and salinities (Brown 1971c; Brown and Feldmeth 1971; Soltz and Naiman 1978; Varela-Romero et al. 2002). Species of this genus occur in rigorous physical environments, including shallow streams and marshes in deserts, small pools in tidal flats, estuaries, and mangrove swamps, where temperature and salinity may fluctuate widely. Some populations

FIGURE 4.20 A bristlecone pine (*Pinus longaeva*) growing near timberline in the White Mountains of California. As is typical of individuals at the upper elevational edge of the range, this one has much deadwood and a highly contorted growth form. (Photo © Andrew Orlemann/Shutterstock.)

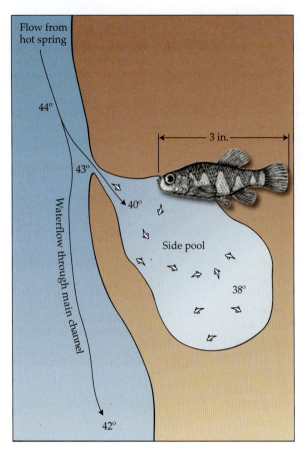

Flow from hot spring

44°

43°

40°

3 in.

Side pool

38°

42°

Waterflow through main channel

FIGURE 4.21 Temperature limits the local distribution of a desert pupfish (*Cyprinodon nevadensis*) in the outflow of a hot spring near Death Valley, California. The fast-flowing main channel is above the lethal temperature of 43° C; fish are trapped but are able to survive in the cooler side pool (enlarged at right). (After Brown 1971c.)

even inhabit hot springs, although they generally cannot tolerate temperatures in excess of about 43° C (which is still amazingly high for a fish). Some pupfish occur in the cooler outlets of hot springs whose temperatures exceed lethal limits. The local distribution of one such population of *C. nevadensis* near Death Valley, California, is limited directly by temperature (Brown 1971c). Fish occur in all waters cooler than about 42° C, including small pools only a few centimeters away from much hotter water (**Figure 4.21**). Occasionally, individuals stray or are frightened into lethally hot water and die instantly—a clear demonstration of the thermal niche boundary! Less frequently, rapidly changing weather conditions cause lethally hot water to flow far downstream from the source, trapping thousands of fish in pools where they are killed when the temperatures rise to over 43° C (see Varela-Romero et al. 2002; Martin and Saiki 2005).

As in many organisms, the adults of pupfish are tolerant of a wider range of physical conditions than are pupfish in early developmental stages. Eggs of *C. nevadensis* develop normally only in water between 20° and 36° C, although adults of this species can withstand temperatures between 0° and 42° C (Shrode 1975). Eggs are also less tolerant than adults of extreme salinity. Consequently, adult pupfish can be found in sink habitats where reproduction is impossible, so long as they can immigrate from nearby source microhabitats that are suitable for egg development. On the other hand, species of *Cyprinodon* are conspicuously absent from some cold springs and other habitats where the adults can grow and survive but where there are no microclimates suitable for earlier stages.

In summary, many factors interact in sometimes complex but important ways to limit the distributions of species across both the terrestrial and aquatic realms. A relatively simple example of such an interaction is the effect of temperature and oxygen concentration on many fish and aquatic invertebrates. As water increases in temperature, its capacity to hold oxygen and other dissolved gases decreases. The resulting combination of high temperature and low oxygen concentration is very stressful to many fish and aquatic invertebrates because high temperatures cause elevated metabolic rates and increased demand for oxygen.

A few additional examples will suffice to illustrate distributional boundaries that can be attributed, at least in part, to such factors as moisture, salinity, and soil chemistry. Many terrestrial plants are limited by low soil moisture at the drier edges of their ranges, just as they are by low temperatures at the colder margins. In nearly all vascular plants, photosynthetic rates decline as soil moisture decreases. Plants can compensate for decreased water uptake through the roots by closing their stomata and reducing transpiration from the leaves, but rates of photosynthesis are reduced concomitantly.

Plants have diverse anatomical, physiological, and phenological (referring to the timing of periodic events such as flowering and seed set) adaptations that enable them to grow in a wide range of temperature, moisture, and light regimes (Bazzaz 1996). For example, species that can grow in full sunlight on dry soils (**xerophytes**) show many specialized mechanisms for keeping their stomates open despite low levels of water in their leaves. In contrast, species from wetter and more shaded environments (**mesophytes**) typically close their stomates when subjected to drought and temperature stress; without evaporative cooling, their leaves suffer high—often fatal—heat loads. On the other hand, xerophytes typically have relatively low rates of photosynthesis when abundant water is available, and they are intolerant of shade. A consequence of this trade-off is that mesophytes are physiologically incapable of growing on dry soils, whereas xerophytes can grow where there is little moisture but are competitively excluded from wetter soils by shading from mesophytes (see Odening et al. 1974) (**Figure 4.22**). These physiological findings provide a mechanistic basis for the conclusion of Forest Shreve (1921) and other early plant ecologists that in gradients of increasing aridity, the limits of plant distributions are determined largely by an inability to tolerate low soil moisture. Widespread diebacks in drought years are commonly observed in local populations at the margins of the range of a species (e.g., Sinclair 1964; Westing 1966). Similar kinds of trade-offs between photosynthetic rate and ability to tolerate low nutrient levels, high salinity, extreme pH, or high concentrations of toxic minerals probably account, at least in part, for the failure of many otherwise widespread plant species to occur locally on soils with these characteristics, while species with special adaptations to these soil types are often restricted to them (Whittaker 1975).

These kinds of physical and chemical factors also limit animal distributions. Because of their osmoregulatory physiology, the vast majority of freshwater fish and invertebrates are intolerant of salinities even approaching the concentration of seawater, whereas marine species usually cannot survive in freshwater. (A notable exception includes the anadromous fishes, e.g., several species of salmon and trout in the genus *Oncorhynchus*, that spend much of their lives in the ocean before migrating long distances up freshwater rivers to spawn.) One consequence of this intolerance is that salt marshes and estuaries are inhabited by neither freshwater nor marine species but, rather, by specialized euryhaline species that can tolerate great—often daily—fluctuations in salinity caused by tides, floods, and storms.

Only a few kinds of specialized organisms occur in those lakes and springs that are even saltier than seawater. Great Salt Lake in Utah, for example, has a salinity approximately seven times that of seawater. It contains no fish and only two macroscopic invertebrates: the pelagic brine shrimp (*Artemia salina*) and the benthic brine fly (*Ephydra cinerea*). Several fish species and numerous invertebrates inhabit the freshwater streams and marshes that empty into the lake. These animals have abundant opportunities to extend their ranges and colonize the lake, but they are prevented from doing so by their inability to tolerate its high salinity. This is dramatically illustrated by the widespread mortality that occurs when occasional windstorms push salt water from the lake into the adjacent marshes.

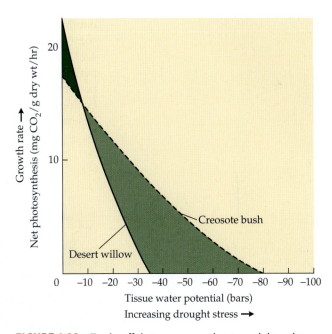

FIGURE 4.22 Trade-offs between growth rate and drought tolerance in two species of desert shrubs: creosote bush (*Larrea tridentate*), which grows in some of the driest North American deserts; and desert willow (*Chilopsis linearis*), which has an overlapping geographic range but is more mesophytic, occurring in microhabitats along watercourses where the soil is permanently moist. Note that under relatively high drought stress (light green region), creosote bush has the higher net photosynthetic rate and is able to grow faster, shade, and competitively exclude desert willow. (After Odening et al. 1974.)

Disturbance and time

Another class of factors that influences both local and geographic distributions of many organisms includes disturbances such as fires, hurricanes, volcanic eruptions, and other agents of sudden widespread disturbance and destruction. These natural disasters are capable of completely destroying habitats and their inhabitants. In regions where such disturbances occur with some frequency, however, they are a natural part of the environment (**Figure 4.23**). In fact, many species are, not only tolerant of, but dependent on these periodic disturbances for their continued existence, and there is a regular pattern of colonization and replacement of species following a disturbance. Ecologists call this process secondary succession. For example, jack pines (*Pinus banksiana*) in eastern North America and lodgepole pines (*P. contorta*) in

(A)

(B)

(C)

FIGURE 4.23   Effect of a natural lightning-caused wildfire on savanna vegetation near Elgin, Arizona. (A) A few days after that fire in June 1975. (B) Several months later, September 1975. (C) Eleven years later, February 1986. Note that while the live oak trees lost most of their leaves as a result of the fire, only a few, such as the one in the center foreground, were killed or suffered major, lasting damage. Fires kill invading shrubs and some trees, maintaining this open, grassy habitat. (Photographs courtesy of R. M. Turner.)

the Rocky Mountains have closed cones that require the heat of periodic forest fires to release their seeds. Somewhat similar cyclical successional processes occur in the chaparral shrublands of coastal California and the coniferous forests of the Rocky Mountain region, which are swept by periodic fires; on the forested islands and offshore coral reefs of the Caribbean region, which are occasionally but inevitably decimated by hurricanes; and in intertidal habitats throughout the world, which are subjected to heavy wave action and sand scouring during major storms.

On the other hand, periodic natural disturbances may be sufficiently severe and frequent to prevent the expansion of some species into areas where they could otherwise survive. In most grasslands, lightning-caused fires are a natural part of the environment. At forest–grassland boundaries, where soil moisture is high enough for woody vegetation to become established, frequent fires may prevent trees and shrubs from extending their ranges (see Figure 4.23). Artificial fire suppression within the last 200 years has contributed to the expansion of forest and shrubland at the expense of prairie and other grassland habitats along the eastern margin of the Great Plains of central North America (e.g., Beilmann and Brenner 1951; Hartnett et al. 1996). A similar phenomenon occurs along the drier margins of the Great Plains in southwestern North America, where fire influences the boundary between arid grassland and desert shrubland. Fire suppression, along with livestock grazing, has played a major role in **desertification** and degradation of grasslands in the southwestern United States, northern Mexico, and other arid regions throughout the world (**Figure 4.24**; Johnston 1963; Bahre 1995). In many of these places where natural fires once burned unchecked over enormous areas, prescribed burns must now be used as a management tool to preserve grassland and prevent the invasion of woody vegetation and exotic species. The tiny reserves of tallgrass prairie in the north central United States, for example, are too small to experience a natural frequency of lightning-caused fires. Now, prescribed burns are essential to prevent local extinction of native prairie plant species.

At more local scales, other causes of disturbance—especially biological ones—can be equally important. Typically, such local-scale disturbance has the effect of removing dominant plants or sessile animals and allowing fast-growing but competitively inferior species to colonize. This creation and filling of gaps results in a patchwork of micro-successional stages that, in total, supports many species. In many tropical and temperate forests, gaps in the canopy caused by the falling of single trees, or even large limbs, create sunny microclimates on the forest floor that are essential for the existence of certain understory species and for the establishment of seedlings of some canopy trees (e.g., tulip tree [*Liriodendron tulipifera*]: Pickett and White 1985). On rocky intertidal shores, such as those of northwestern Washington, the predatory starfish (*Pisaster ocraceous*) removes the competitively dominant mussel (*Mytilus californianus*), creating gaps in the otherwise continuous mussel beds; the gaps can then be colonized by rapidly growing algae and invertebrates (Paine 1966). On some exposed shores, logs banging against the mussel beds can have a similar effect (Dayton 1971). In North American tallgrass and shortgrass prairies, soil disturbance caused by the activities of mammals may be as important as fire for the maintenance of a diverse grassland community. Wallows of bison and mounds of pocket gophers all provide patches of bare, sunny soil that are essential for the establishment and persistence of certain plant species (e.g., Platt 1975; Reichman and Smith 1985; Inouye et al. 1987). Similarly, grazing by prairie dogs (genus *Cynomys*) maintains a dynamic mosaic of shortgrass prairies across broad expanses of the Great Plains Region, providing habitat for hundreds

(A) 1860

(B) 1960

FIGURE 4.24 Contraction of grassland habitat in southern Texas between (A) 1860 and (B) 1960. These changes, which were reconstructed from historical records, are due largely to fire suppression, livestock grazing, and invasion of woody vegetation, especially mesquite (*Prosopis*). (After Johnston 1963.)

of species of plants and terrestrial vertebrates, while the extensive burrow system of the prairie dogs creates microhabitats and refugia for many other animals and plants (Miller et al. 1994, 2000; Stapp 1998; Lomolino and Smith 2003, 2004; Smith and Lomolino 2004).

As we discussed in the previous chapter, time is an integral, although often overlooked, dimension influencing geographic distributions of all species. Even for species long established in areas with suitable and relatively continuous habitat, unless they are powerful dispersers, their realized ranges will fall far short of their fundamental ranges. Striking examples include numerous trees and other plant species that colonized landscapes in temperate regions following retreat of the glaciers during the Early Holocene—some 10,000 years BP (see Chapter 9). Range expansions of these species are still ongoing, lagging centuries to millennia behind expansions of their fundamental ranges (see Figure 9.18; see also discussion of dispersal and range expansion, pages 170–181). Of course, range dynamics also include contractions of the distributions of once-expansive species in response to either natural or anthropogenic disturbance of environmental conditions. While there appear to exist some generalities in qualitative patterns of range contraction, at least during the historic periods (see pages 710–719), its rate may be quite unpredictable and highly variable, with species persisting in some regions from decades to perhaps centuries after environmental conditions have degraded. This variation in the rate of range contraction is likely a function of, among other things, the generation time of the focal species and its ability to sustain populations in areas of marginal conditions by being rescued by dispersal from populations thriving in more suitable habitats ("sink" and "source" sites, respectively; see also discussion of "rescue effects," pages 541–542). Taken together, rather than a static character of any species, geographic ranges should be viewed as complex and ever changing.

Interactions with other organisms

In many cases, geographic distributions are not limited, directly or indirectly, by physical factors; they are limited by other species. Botanical gardens and zoos provide perhaps the most dramatic evidence that individuals can survive, grow, and even reproduce under a much wider range of physical conditions than they encounter anywhere in their natural geographic ranges. The fact that many plants can thrive in these novel ecosystems—but only if they are protected from competing plants, animal herbivores, and microbial pathogens—demonstrates the importance of interspecific interactions in limiting distributions.

There are three major classes of interspecific interactions: competition, predation, and mutualism. All of these can influence the dynamics of populations and limit the geographic ranges of species.

Competition

Competition is a mutually detrimental interaction between individuals. Organisms that share requirements for the same essential resources necessarily compete with each other and suffer reduced growth, survival, and reproduction if the resources are in sufficiently short supply. Plants may compete for light, water, nutrients, pollinators, or physical space, whereas animals most frequently compete for food but also compete for shelter, nesting sites, mates, or living space that contains these resources. These interactions may be purely **exploitative** such that individuals use up resources and make them unavailable to others. Alternatively, competition may be more direct and involve some form of **interference** in which aggressive dominance or active inhibition is used to deny other individuals access to resources. For example, some plants, such as the black walnut (*Juglans nigra*), and some sessile marine animals, such as bryozoans and corals, use a form of chemical warfare called **allelopathy** to defend space from competitors (see Jaksic and Fuentes 1980). In such cases, the interspecific interaction may be referred to as **amensalism** if it is asymmetric enough that it negatively affects the distribution or abundance of only one species.

There is much circumstantial evidence that competition limits geographic ranges; this includes many examples of ecologically similar, closely related species that occupy adjacent but nonoverlapping geographic ranges. Five species of large kangaroo rats (*Dipodomys*) are found in desert and arid grassland habitats in the southwestern United States and northern Mexico, but their geographic ranges do not overlap (**Figure 4.25**). Two species, *D. ingens* and *D. elator*, have isolated (or disjunct) ranges, but *D. spectabilis* shares an extensive border with *D. deserti* in the west and another with *D. nelsoni* in the south. These three species, however, segregate their realized niches at a more local scale: *D. deserti* occurs only in sandy habitats, which lack *D. spectabilis*; *D. nelsoni* inhabits the core of the Chihuahuan desert; while *D. spectablis* is limited to arid grasslands of that region. Although such cases suggest that competition limits distributions by preventing coexistence, the distributions are subject to alternative explanations such as specialization for alternative

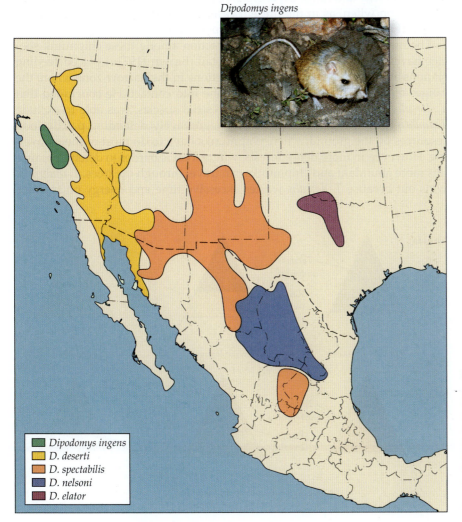

Dipodomys ingens

🟩	*Dipodomys ingens*
🟨	*D. deserti*
🟧	*D. spectabilis*
🟦	*D. nelsoni*
🟥	*D. elator*

FIGURE 4.25 Nonoverlapping geographic ranges of five species of large kangaroo rats (*Dipodomys*) in southwestern North America. These rodents are similar in their ecology, and the fact that their ranges frequently come into contact but rarely overlap suggests that they limit one another's distributions through competition. (After Bowers and Brown 1982; photo courtesy of U. S. Fish and Wildlife Service.)

habitats, and often there is no direct evidence of competitive interactions occurring on the boundaries.

Better evidence for the limiting effects of competition comes from "natural experiments" in which one species, simply by chance, is absent from regions that are apparently suitable for it. If a second species has expanded its range to include habitat types that are normally occupied by the first species, this implies that in areas where the first species is present, the second will be limited by competition. Forests and shrublands of the western United States are inhabited by more than 20 species of chipmunks of the genus *Tamias* (formerly *Eutamias*). In mountains of the Southwest, two or three species typically occur in woodlands and forests, but they are segregated by habitat and elevation. *Tamias dorsalis* inhabits open, xeric woodlands at lower elevations and is replaced in the denser coniferous forests at higher elevations by a species of the *T. quadrivittatus* group. A third species, *T. minimus*, is less restricted to any of these habitats and sometimes is present in low elevation shrublands, in spruce and fir forests, and above timberline on the highest peaks. There are at least 24 isolated desert mountain ranges where appropriate habitats seem to occur and one of these chipmunks species is present but one is absent, apparently because it either never colonized or became extinct sometime in the past. In every case, regardless of which species is absent, the remaining species has expanded its range to include all forested habitats from the edge of the desert to the timberline (Patterson 1980, 1981a; **Figure 4.26**). Such examples of niche and range expansion are particularly convincing evidence of competition when, as in the present case, similar distributional shifts have occurred independently in several different places.

The mechanisms of competitive interaction among these chipmunk species have been investigated in some detail. Brown (1971a) placed feeding stations and observed behavioral interactions in the narrow zone where the ranges of *T. dorsalis* and *T. umbrinus* come into contact. He concluded that *T. dorsalis*, the more aggressive and terrestrial species, was able to exclude *T. umbrinus* from open woodlands by aggressively defending patchy food resources on the ground, where chipmunks have to do most of their traveling. However, in more dense forests, where food is harder to defend because it is more abundant and chipmunks can travel through the trees, *T. umbrinus* wins out because *T. dorsalis* wastes excessive time and energy on fruitless

FIGURE 4.26 Diagrammatic representation of elevational distributions of chipmunks (*Tamias spp.*) on mountain ranges in the southwestern United States. On most ranges, two species are present and their ranges overlap only slightly. On some ranges, however, only a single species occurs, in which case its range has expanded to include nearly all habitats and elevations normally occupied by both species. Such natural experiments are good evidence of competitive exclusion, which in this case has been confirmed by field studies. (Photo © Martha Marks/Shutterstock.)

Tamias spp.

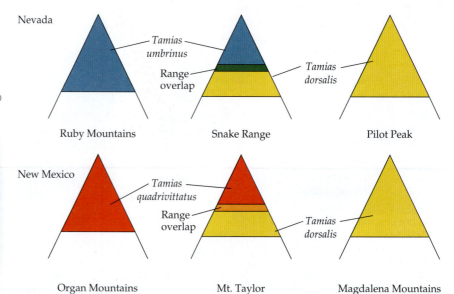

chases. Chappell (1978) studied a more complex situation where the ranges of several species come into contact in the Sierra Nevadas in California. He also found that these mutually exclusive distributions could be attributed to a combination of differences in body size, thermal tolerance, and competition among species, with larger species able to co-opt the more optimal forest and woodland habitats.

Such "natural experiments" implicating competition in limiting geographic ranges are by no means confined to small mammals. Diamond (1975b) gives several examples of niche and range expansion by birds in the absence of competing species on isolated mountain ranges within New Guinea and on nearby islands. Schoener (1970) and Roughgarden (1995; see also Roughgarden et al. 1983; Pacala and Roughgarden 1985) describe several cases among *Anolis* lizards on islands of the West Indies. Perhaps the best example in plants is the segregation of two species of cattails along a gradient of increasing water depth in marshes in central North America. In an elegant series of reciprocal transplant experiments, Grace and Wetzel (1981) showed that *Typha latifolia* was little affected by its congener, *T. angustifolia*. In the absence of its competitor, however, *T. angustifolia* could grow over the entire gradient.

Competition need not involve similar species to limit distributions. In fact, the kinds of pairwise interactions between closely related, ecologically similar species described above probably account for only a small part of the competition experienced by most species. For example, distantly related plants of different growth forms often experience intense, asymmetrical competition (Johansson and Keddy 1991). We have already described how stressful physical conditions and fire limit the distributions of many plant species and vegetation types. The other side of the story is that the plants that are the least tolerant of abiotic stresses (drought, fire, and flooding) are usually superior competitors (Keddy and MacLelan 1990). For example, while some combination of fire, grazing, and drought may prevent trees and shrubs from invading grasslands, these woody plants are superior competitors where these conditions are not too severe. Not only can they prevent establishment by grassland species, but during the last two centuries, as fire frequency has been reduced and grazing regimes altered, these woody plants have aggressively invaded grasslands and replaced herbaceous species (Steinauer and Collins 1996) (see Figure 4.24). Such asymmetrical niche relationships, in which one species is limited by competition while the other is restricted by its inability to withstand disturbance or physical stress, appear to be extremely common among many kinds of organisms—recall Connell's study of barnacles.

Predation

Predation can be defined as any interaction between two species in which one benefits and the other suffers. According to this definition, relationships between herbivores and their food plants (see Harper 1969), parasites and their hosts, and **Batesian mimics** and their models would also be classified as predation. Predator–prey interactions can limit the distribution of either participant because, on one hand, predators may depend on particular prey for food or other benefits necessary to support their own populations, whereas on the other hand, predators may limit prey populations by killing or damaging individuals.

When predators are highly specific, it is obvious that their distributions must depend in part on the availability of appropriate prey. It is hardly surprising that the geographic ranges of many specific parasites and herbivores correspond almost precisely with those of their animal or plant hosts. Thus, the distribution of the checkerspot butterfly (*Euphydryas editha*) in coastal California is limited to the immediate vicinity of patches of serpentine soils (characteristically low in nutrients but high in heavy metal concentrations) to which its host plant (*Plantago hookeriana*) is also restricted (Ehrlich 1961, 1965). Of course, even highly specific predators may range less widely than their hosts because in some areas their populations are limited by other factors in addition to the availability of suitable prey.

It is much more difficult to document cases in which the distributions of prey populations are limited by their predators. Some of the best examples are artificial in that they involve human introductions of predators into regions where they did not originally occur. In some cases these introductions were deliberate attempts to control pest species. Two conspicuously successful examples of such biological control involve drastic reductions in plant populations by introduced herbivores. The prickly pear cactus (*Opuntia stricta*) was introduced into Australia in the mid-1800s to serve as an ornamental garden plant. By the early 1900s the cactus had escaped from cultivation and had become a serious pest on grazing lands. In 1926 Australian scientists introduced a moth (*Cactoblastis cactorum*) whose larva is a specific feeder on *Opuntia* in its native Argentina. By 1940, *O. stricta* had been effectively checked as a pest species in eastern Australia, although small patches of cacti and local populations of the moth remain (Dodd 1959). Unfortunately, many other species of exotic plants are invading Australian habitats, and the search for suitable agents of biological control is ongoing.

Similarly, Klamath weed (*Hypericum perforatum*) was introduced into northwestern North America from Eurasia in about 1900 and became an agricultural pest, but it was subsequently controlled by the introduction of a specific leaf-eating beetle (*Chrysolina quadrigemina*) from its native habitat. In the southern part of its range, Klamath weed persists only in small populations along roadsides and in shady areas, but beetle populations do not do well in colder climates, and the weed is more widely distributed in British Columbia (Huffaker and Kennett 1959; Harris et al. 1969). In the cases of both prickly pear cactus and Klamath weed, specific herbivores have drastically reduced plant populations, resulting in very limited local distributions, but they have not greatly altered the distributions of the weeds on a larger geographic scale.

Perhaps the best examples of complete elimination of prey populations from parts of their geographic ranges by voracious but nonspecific predators are provided by artificial introductions of large predatory fish into certain freshwater habitats. Many of the small native fish of the southwestern United States have suffered great reductions in their geographic ranges, and complete extinction of local populations, as a result of the introduction of large predatory game fish (especially largemouth black bass [*Micropterus salmoides*]) into their habitats (Miller 1961a; see also Braba et al. 1996; Whittier and Kincaid 1999; White and Harvey 2001; Winfield and Hollingworth 2001). Native fish are not adapted to large generalist predators, because they have evolved in isolated lakes, streams, and springs for thousands of years. Zaret and Paine (1973) documented a similar example in Lake Gatun, Panama, where introduction of the predatory fish *Cichla ocellaris* caused the extinction of at least seven species of native fish. Perhaps the most dramatic episodes of predatory exclusion of native fish are the extinctions of some 300 endemic species of cichlids following the introduction of the Nile perch (*Lates niloticus*) into Africa's Rift Valley lakes.

The effects of introduced parasites on native species can be just as devastating. Throughout much of eastern North America, the lake trout (*Salvelinus namaycush*) was a widely distributed and often abundant species, but it is not adapted to the presence of lampreys of the genera *Petromyzon* and *Entosphenus*, which are voracious external parasites. Niagara Falls formerly prevented *Petromyzon* from entering the upper Great Lakes, which supported large populations of lake trout. Construction of the Welland Canal, however, enabled the lamprey to colonize these lakes. The result has been a precipitous decline of lake trout and many other native fish of the upper Great Lakes despite a major effort to control the lamprey and save this valuable commercial fishery.

In the previous examples, the effects of predators on prey populations are particularly clear because we have been able to observe responses to artificial introductions into nonnative ecosystems. Without this historical perspective, however, we would be hard pressed to infer the extent to which prey are limited by their predators. Today, most remaining patches of *Opuntia* in Australia are not infested with *Cactoblastis*. Similarly, it is difficult to observe black bass preying on pupfish of the southwestern United States, because the native fish have already been extirpated from most waters where bass are present. It is likely that many prey populations are limited at least in part by their predators, but it is difficult to obtain convincing evidence for the ghosts of predators past, since the requisite manipulative experiments may have been impossible following extinction of native species.

While the effects of predators (in the traditional sense) have been studied for decades, the effects of internal parasites and disease-causing microbes received much less attention from ecologists and biogeographers until quite recently. It is apparent, however, that they also can limit both local abundances and geographic distributions. There are increasing numbers of examples of domesticated plants and animals that have been able to expand their ranges into previously uninhabitable areas following elimination or control of their parasites or pathogens. Likewise, there are examples of range contraction following the human-aided invasion of new parasites or diseases (Diamond 1997, 2005). One example of the latter is the influence of avian malaria on native Hawaiian birds. Ever since the first humans arrived on the islands, the spectacular endemic Hawaiian avifauna has been suffering extinctions, range contractions, and population reductions (see Chapter 16). An initial wave of extinctions, presumably due primarily to hunting and habitat destruction, followed the arrival of the Polynesians about 1000 years ago. More extinctions followed European settlement, which began less than 200 years ago. Europeans not only caused additional habitat destruction and brought in exotic avian competitors and mammalian predators, but also initiated biological warfare. Along with the exotic birds they brought in as domesticated fowl, game birds, and pets—and sometimes deliberately released—came avian malaria, caused by *Plasmodium*. This parasite has become pandemic at lower elevations, where its persistence and transmission is favored by warm temperatures, availability of mosquito carriers, and relatively resistant exotic avian hosts (e.g., game birds and domesticated fowl). While it is difficult to tease out the multiple interacting influences of all the factors affecting native bird species, avian malaria has clearly played a major role in the extirpation of most endemic Hawaiian birds from the lower elevations, even in areas where relatively undisturbed habitat remains (Warner 1968; van Riper et al. 1986; for an overview of similar effects of parasitism on distributions of birds of the West Indies, see Ricklefs et al. 2008).

Mutualism

The third class of interspecific interactions is **mutualism**, in which each species benefits the other. Examples of mutualistic associations are provided by plants and their animal pollinators and seed dispersers, corals and the photosynthetic zooxanthellae (algae) that live in their tissues, ants and aphids, and cleaner fish and their hosts. Compared with competition and predation, mutualism has been less frequently studied by ecologists and biogeographers (but see Boucher et al. 1984; Boucher 1985; Fleming and Estrada 1993; Rico-Gray and Oliveira 2007; Stadler and Dixon 2008). Much remains to be learned about the effects of these mutually beneficial associations on the abundance and distribution of participating populations.

When mutualistic relationships are obligate for at least one (sometimes referred to as **commensalism**)—and especially for both of the partners, then the interaction must have a major influence on distributions. Some plant populations are dependent on the services of specific pollinators for sexual reproduction. For example, red clover (*Trifolium pratense*) did poorly after being introduced into New Zealand until its pollinator, the bumblebee (*Bombus* spp.), was also introduced (Cumber 1953; Free 1970). Janzen (1966) studied the association between ants of the genus *Pseudomyrmex* and trees and shrubs of the genus *Acacia* in the New World tropics. Although many species of *Acacia* have no ants associated with them, and some species of *Pseudomyrmex* may not be dependent on acacias, the relationship is apparently obligate for numerous species. The trees provide the ants with enlarged thorns, in which the ants build their nests, and with specialized foods rich in sugars, oils, and proteins. In return, the ants attack herbivorous insects and vertebrates that attempt to feed on the trees, and the ants clear away surrounding vegetation, reducing competition from other plants. Such coevolved specializations apparently have made these two mutualists so dependent on each other that they have virtually identical ranges.

These examples of close ecological and biogeographic associations may, however, be exceptional. Most often, relationships among mutualists, at least at the species level, are not so obligatory, and the influence of this interaction on geographic distributions is not so apparent. Usually at least one of the partners does not absolutely require the service of the other, or several different species can supply that service. Thus, most plant–pollinator and plant–seed disperser relationships are not obligately species-specific—at least outside the tropics (Waser et al. 1996).

Figure 4.27 shows the relationship between the geographic ranges of some hummingbird and plant species. The North American geographic ranges of eight hummingbird species are compared with the geographic ranges of some species of plants with red tubular flowers, which the hummingbirds feed upon and pollinate. There is little correspondence between the ranges of particular pairs of hummingbird and plant species, but the distributions are such that, no matter where they occur, the hummingbirds have flowers to feed upon and the plants have hummingbirds to pollinate them (Kodric-Brown and Brown 1979). A similar relationship is seen in the ranges of clownfish (*Amphiprion* spp.) and the sea anemones with which they are associated (**Figure 4.28**). While all clownfish appear to require the protection of anemones, and some kinds of anemones may be unable to exist without clownfish, the particular species are much more independent.

Multiple Interactions

In addition to those cases in which it is possible to isolate the limiting effect of one species on the distribution of another, there are undoubtedly many situations in which ranges are structured by more diffuse and indirect biotic interactions. Such limits may be the result of different, interacting effects of several species (see earlier discussion of competition between woody and herbaceous plants). Robert H. MacArthur (1972) noted that the southern limits of the ranges of many North American bird species apparently could not be attributed to climate (because climate becomes more equable at lower

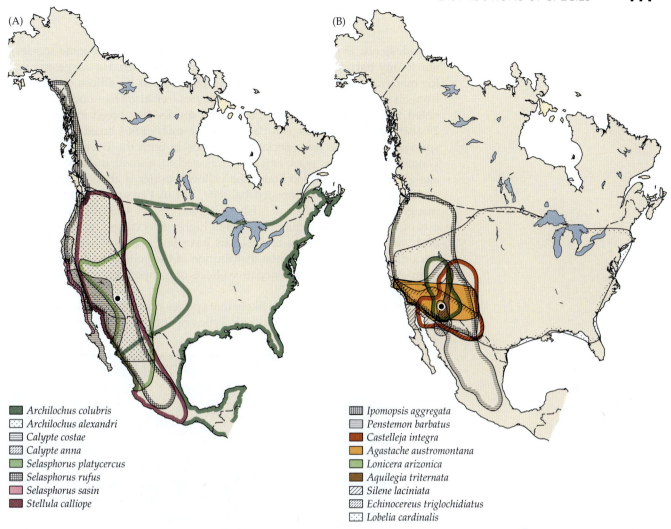

(A)

- ■ *Archilochus colubris*
- ▨ *Archilochus alexandri*
- ▤ *Calypte costae*
- ▨ *Calypte anna*
- ▨ *Selasphorus platycercus*
- ▨ *Selasphorus rufus*
- ▨ *Selasphorus sasin*
- ▨ *Stellula calliope*

(B)

- ▤ *Ipomopsis aggregata*
- ▤ *Penstemon barbatus*
- ■ *Castelleja integra*
- ▨ *Agastache austromontana*
- ▨ *Lonicera arizonica*
- ▨ *Aquilegia triternata*
- ▨ *Silene laciniata*
- ▨ *Echinocereus triglochidiatus*
- ▨ *Lobelia cardinalis*

FIGURE 4.27 Distributions of temperate North American hummingbirds and some of the plants they pollinate while foraging for nectar. (A) Breeding ranges of the eight species of hummingbirds that occur substantially north of the United States–Mexico border. (B) Geographic ranges of the nine species of red tubular flowers commonly visited by hummingbirds at just one site in Arizona (circle). It would be too confusing to plot ranges of the approximately 130 species of flowers used by these hummingbirds throughout their breeding ranges. Note that despite their close mutualistic relationships, there is little relationship between geographic ranges of the specific plants and their pollinators. (From Kodric-Brown and Brown 1979.)

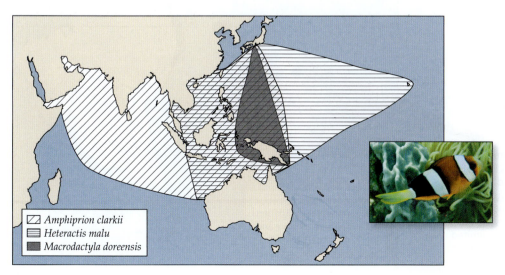

- ▨ *Amphiprion clarkii*
- ▤ *Heteractis malu*
- ▨ *Macrodactyla doreensis*

FIGURE 4.28 Distribution of the Indo-Pacific clownfish (*Amphiprion clarkii*) and two sea anemones (*Heteractis malu* and *Macrodactyla doreensis*) that serve as its hosts. *Amphiprion clarkii* is the only fish that is mutualistic with *H. malu*, but it uses other anemones, including *M. doreensis*. As with the hummingbirds and flowers shown in Figure 4.27, even relatively closely coevolved mutualists may be able to switch partners, and therefore do not necessarily have coincident ranges. (Unpublished data courtesy of D. Dunn; photo © Niels Holm/Shutterstock.)

latitudes), nor to habitat (although it might be important for some species), nor to any particular species of competitor or predator. He suggested that for many of these species, the limiting factor must be **diffuse competition** from an increasing number of tropical species (see also Diamond 1975b). He noted, for example, that of 202 land bird species that breed in Texas, only 29 also breed in Panama, but Panama has a total of 564 breeding land bird species. One of the few species that breeds in both the United States and Panama is the yellow warbler (*Dendroica petechia*), a small, insectivorous foliage gleaner. In the United States the yellow warbler is abundant in a wide variety of shrubby and forested habitats, but in Panama it is restricted to mangrove swamps and small offshore islands (**Figure 4.29**). Because the forests of Panama contain many species of highly specialized foliage-gleaning birds, whereas mangroves and islands have few such species, MacArthur attributed the restricted habitat distribution of the yellow warbler to diffuse competition—the combined negative effects of competition with many other bird species.

In most cases, however, it is likely that species distributions are influenced by the combined effects of multiple types of interspecific interactions, in-

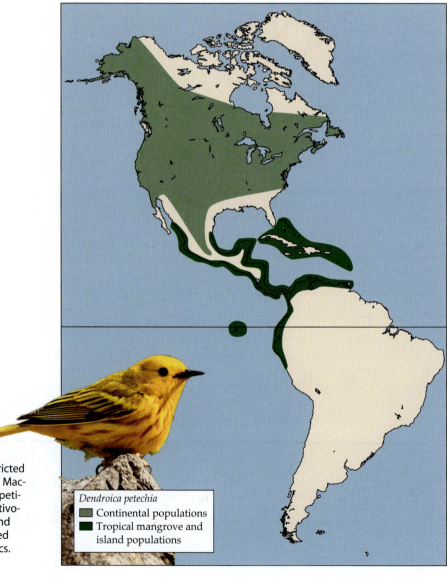

FIGURE 4.29 Geographic range of the yellow warbler (*Dendroica petechia*). This passerine bird is widely distributed over temperate North America, but in the tropics it is restricted to coastal mangrove swamps and islands. MacArthur (1972) suggested that diffuse competition from the many species of small insectivorous birds in most habitats on the mainland of tropical America accounts for the limited distribution of yellow warbler in the tropics. (Photo © Gabor Ruff/Shutterstock.)

Dendroica petechia
Continental populations
Tropical mangrove and island populations

cluding those with a diversity of predators, parasites, diseases, and mutualists throughout the historical development of a species' geographic range. Recall the example of avian malaria that limited the lower elevational distributions of several native bird species in the Hawaiian Islands. In the absence of historical and epidemiological evidence for the role of the malaria parasite, it would be easy to mistakenly attribute restricted distributions of native species solely to competition with introduced birds. In abiotically stressful environments such as salt marshes and intertidal zones, mutualistic interactions may have at least as much influence on the abundance and distribution of species as competition and predation (Bertness 1989, 1991; Bertness and Shumway, 1993).

Careful studies of highly coevolved species at different trophic levels, such as parasites and their hosts or plants and their pollinators, reveal many instances in which such pairs of species have, as we might expect, virtually identical ranges. Most of the species in a community, however, do not interact strongly with one another. The number of possible pairwise interactions between species rapidly increases with the number of species (S) as ($S^2 - S$)/2. If a community contained only 50 species, each species could interact with each of the 49 others, resulting in a total of 1225 possible direct pairwise interactions. Clearly each species cannot be finely adapted to all of the other species with which it coexists. On the other hand, an organism may be involved with just a few other species in strong competitive, predator–prey, or mutualistic interactions that have important influences on its abundance and distribution. In rare cases, a single **keystone species** may strongly influence community structure through its direct, as well as indirect, effects (Paine 1974; Terborgh 1986; Cox et al. 1994; Menge et al. 1994; Paine 1995; Power et al. 1996; Gill 2000).

Interactions among three or more species also may be quite common, and they may sometimes be as influential as two-way interactions. For example, many species may share a common limiting resource. Even where competition is intense, predators can promote the coexistence of otherwise incompatible competitors by preying most heavily on the most abundant prey species, thus preventing it from driving the competitively inferior species to extinction (a phenomenon termed **predator-mediated coexistence**; Holling 1965; Murdoch and Oaten 1975; Roughgarden and Feldman 1975; Caswell 1978; Hsu 1981; Holt et al. 1994). Changes in the population densities of such predators and other keystone species may have pervasive effects that can cascade through the community to affect species across many trophic levels (Carpenter et al. 1985, 1987; Schmitt 1987; Carpenter 1988; Vanni and Findlay 1990; Strauss 1991; Wootton 1992; Holt et al.1994; Jones and Lawton 1994; Menge et al. 1994; Leibold 1996; Knight et al. 2005). For example, in response to the extirpation of sea otters from some regions of their historical range, populations of sea urchins (invertebrate herbivores) increased to the point at which they overgrazed aquatic plants such as kelp, reducing them to a fraction of their original biomass (see Estes and Duggins 1995). On land, the extirpation of wolves and other large predators in Eurasia and North America has also resulted in the ecological release of their herbivorous prey. Populations of deer (*Odocoileus* spp.) and other browsing mammals have increased to the extent that they have become pests. As well as entering suburbs and cities, these animals have denuded the understory of temperate forests, removing important cover and nesting materials, and decreasing the breeding success of native songbirds.

While there is every reason to suspect that these kinds of multispecies interactions also influence distributions and diversity of species on geographic scales, rigorous documentation is usually lacking. Increased emphasis on geographic scale phenomena and nonexperimental methods, capitalizing on the creativity of researchers and the wealth of information amassed during centuries of biological surveys, should lead to a better understanding of the biogeographic consequences of these complex interactions.

A Preview of Derived Patterns

We use the term *derived* here to emphasize that the great diversity of patterns that biogeographers study is a function of variation among or within its most fundamental unit—the geographic range. Once, and only after, we have delineated or at least estimated that unit can we begin to explore how its characteristics (e.g., its size and shape) vary over time, space, and species; how its internal characteristics (the characteristics of individuals and of populations) vary across that range; and how ranges overlap or are disjunctly distributed to produce patterns of special importance to ecologists, evolutionary biologists, and conservation biologists—i.e., global patterns in diversity and endemicity of entire biotas. We return to each of these fascinating and important subjects in later chapters, but provide a brief and, we hope, tantalizing overview of some anticipated patterns here.

Range dynamics

Scientists find range dynamics a compelling subject for both basic and applied reasons. Understanding the distributions of all life-forms on Earth essentially boils down to understanding the history of their geographic dynamics, from dispersal and expansions from their sites of origin, to their decline and range contraction, to the ultimate extinction of their last survivors. The applied value of understanding range dynamics is well appreciated by conservation biologists, who are often challenged with predicting or controlling the spread of a problematic, introduced species while slowing range contraction of the focal, native species.

Studies of empirical patterns of range expansion, particularly those of invasive species, have revealed that their dynamics are far from random but typically exhibit qualitatively similar patterns across a diversity of species, ecosystems, and geographic regions. As we discuss in depth in Chapter 6, initial colonization of an exotic region is often a slow process, often requiring many failed attempts followed by a prolonged period of relative stasis and limited distribution once the invader has established a foothold. After that initial period of establishment, however, the ranges of a successful invader will often enter of phase of rapid expansion until it encounters and is slowed by geographic, climatic, or ecological barriers to dispersal (e.g., oceans or mountain chains, extremes of temperature and humidity, and the absence of a mutualist or presence of a dominant competitor, predator, or pathogen). Similarly, range contraction, while exhibiting substantial variation in rates and other features among species, ecosystems, and regions, is also anticipated to exhibit some general qualitative patterns. For example, based on a general pattern of variation in densities of populations across species' ranges, discussed earlier in this chapter, biogeographers and conservation biologists predicted that ranges of declining species would implode, with the extinction fronts occurring first along their peripheries and then moving toward the centers of the species' historic, geographic ranges. The results of studies conducted over the past few decades provide some very surprising insights into the actual patterns of geographic range collapse—patterns that, while apparently anomalous, may strongly enhance our abilities to conserve the distributions and ecological and geographic context of native species (see Chapter 16).

Beyond these general patterns, our ability to explore and model geographic dynamics of species has now advanced our research from the purely qualitative and exploratory stage to a much more quantitative, spatially explicit stage capable of providing more reliable predictions of geographic dynamics in biological diversity. Moreover, it is well within our reach to utilize the ever-expanding wealth of information and analytical tools to develop prescriptive models, that is, strategies for altering the rates and direction of range dynamics of both invasive and native species (e.g., slowing the advance or redirecting the expansion of a problematic invasive species away from a critical hotspot of native species diversity, or accelerating and directing the reexpansion of a species of high conservation priority to assist its migration in response to future climate change.

Areography: Size, shape, and internal structure of the range

The year 1982 saw the publication of the English language edition of a fascinating little book entitled *Areography*, by the Argentine ecologist and biogeographer Eduardo Rapoport (1975). Working largely in isolation in Latin America, Rapoport showed that simple quantitative analyses of spatial distributions of organisms—such as the maps of geographic ranges in Hall's (1981) *Mammals of North America* or Critchfield and Little's (1966) *Geographic Distribution of Pines of the World*—could reveal fascinating patterns and suggest hypotheses about the mechanisms that limit species' distributions and influence community diversity. Rapoport's book inspired a number of ecological biogeographers to undertake conceptually and methodologically similar studies on variation in the size, shape, and patterns of overlap of geographic ranges—**areography** (**Figure 4.30**). The result was a renewed emphasis in comparative biogeography that was similar in some ways to the resurgence of research in insular biogeography that was stimulated by MacArthur and Wilson's (1963, 1967) seminal theory published some 15 years earlier.

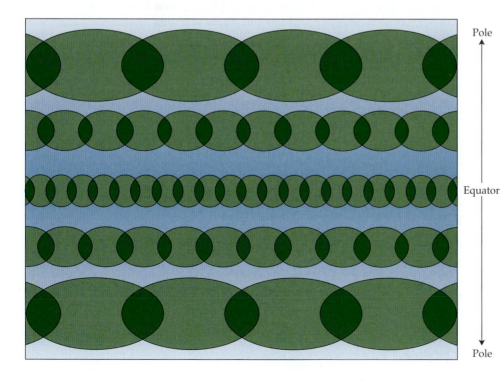

Pole

Equator

Pole

FIGURE 4.30   Idealized, but often reported, areographic patterns include latitudinal gradients in the size, shape, and overlap of geographic ranges (here, ranges of different species are represented by circles and ovals). This idealized map illustrates Rapoport's rule (a tendency for geographic range size to increase with latitude) and the tendencies for geographic ranges to become less numerous and more irregular in shape (often distorted with increased east–west stretch) as we move from the Equator to the poles (see Chapter 15). (Modified from Rapoport 1982: 188.)

Areography remains an active and stimulating research program, exploring all aspects of the structure of geographic ranges (see Gaston 2003).

Rapoport's seminal book also stimulated a generation of biogeographers and ecologists to begin to rigorously explore the "internal structure" of geographic ranges, patterns long anticipated by Darwin and other early scientists but only made tractable with the advent of computers and advances in statistical and spatial data analyses. For example, recall from Figures 4.2 and 4.16 that population densities tend to be higher, but more variable, near the center of a species' range. We will return to reassess the generality and alternative explanations for these and related patterns of variation within and among geographic ranges in Chapter 15.

Ecogeography: Variation in phenotypes across the range

Among the earliest sets of patterns described by the founders of biogeography were the so-called ecogeographic patterns: regular patterns of variation in phenotypic characteristics (traditionally morphological traits) that seemed so general that they were termed "rules." As we noted in Chapter 2, these include the rules of Bergmann, Allen, and Jordan, which describe geographic clines of intraspecific variation in body size, appendage length in mammals and birds, and number and relative size of vertebrae in fish, respectively.

In addition to now being able to conduct more rigorous evaluations of the generality of these patterns, investigate anomalous cases, and search for causal explanations, modern biogeographers have begun to explore a diversity of other geographic clines in phenotypic traits among species' populations, including those of physiological, genetic, and behavioral characteristics. Susan Foster and John Endler's (1999) edited volume *Geographic Variation in Behavior* is an exemplary and provocative example of future research awaiting the next generation of ecogeographers—studies that will continue to advance the shared frontiers of ecology, evolution, and biogeography. Some of the most innately compelling studies of geographic variation in behavior, physiology, and morphology are those conducted by primate biologists, including human biogeographers. While we admit there often tends to be a disjunct between primate biogeographers and those studying analogous patterns in other species, we see no logical justification for this, as each discipline has much to learn from the other. Exemplary case studies include James Bindon and Paul Baker's (1997) reports of Bergmann's rule patterns in humans, John Terrell's (1976, 1986) explanations for geographic variation in linguistic diversity of Pacific Islanders (see Figure 16.35), and the research of Sandy Harcourt and his colleagues (Harcourt 2000; Fuller and Harcourt 2009; Harcourt and Schreier 2009) on geographic variation in species diversity, range size, population density, and body size of primates (see Chapter 15).

Multispecies patterns: Range overlap, richness, and exclusive distributions

The final set of derived patterns we preview here are some of the most commonly studied phenomena of ecology, biodiversity research, and conservation biology, for they lie at the core of all studies in species diversity and composition of local to regional biotas. We refer to such patterns as "multispecies" because these related properties—community richness and composition (the numbers and types of resident species, respectively)—are both determined by patterns of overlap among species' ranges. More simply put, hotspots of species diversity are sites of many overlapping ranges regardless of the total area each occupies (see Figure 4.27), whereas hotspots of endemicity (of special importance to conservation biologists) are sites where the ranges of geographically restricted species (endemics) overlap. Despite the tremendous variation in characteristics influencing species' survival and dispersal abilities, the very regular patterns of variation in the geographic template produce some surprisingly general patterns of geographic variation in species diversity and endemicity (with hotspots of diversity often occurring in the tropics and montane regions, and those of endemicity typically located on large, isolated islands of the tropics and subtropics; see Figure 15.30). These clines have continued to attract the attention of scientists throughout the shared histories of biogeography, evolutionary biology, and ecology and have generated some intense but often illuminating debates over the determinants of the geography and diversity of nature (see Chapter 15).

The converse of patterns in overlap among ranges, that is, gaps and disjunctions in species distributions, are equally compelling to both ecological and historical biogeographers. Even within a species, it is clear that there are gaps in distributions of its populations—a phenomenon aerographers refer to as range porosity. As we pointed out earlier in this chapter, even this measure tends to be highly nonrandom, with porosity appearing to increase from the center to the periphery of a species' range (presumably, along a gradient of decreasing suitability of local environments; **Figure 4.31**). At a higher taxonomic level, gaps or disjunctions in the distributions of related species provide both challenges and clues to historical biogeographers attempting to reconstruct the evolutionary diversification and geographic expansion of that taxon from its ancestral range (Unit 4). Finally, nonoverlapping or exclusive distributions of distantly related but ecological interdependent species (e.g., intense competitors and amensals) can provide key insights into the roles of interspecific interactions in influencing the assembly of local to regional biotas (see Figures 4.25, 4.26, and 14.11). Again, the challenge, of course, is to determine whether these distributions observed in a snapshot of time are actually the result of recent ecological exclusions or a consequence of differences in the fundamental ranges and the evolutionary history and dispersal capacities of the focal group of species.

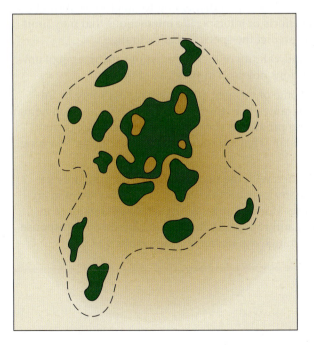

FIGURE 4.31 An idealized map illustrating the anticipated pattern of increased porosity (i.e., gaps between populations of a species) as we move from the center to the periphery of a species' geographic range (range boundary indicated by dashed line; populations represented by green areas). The underlying geographic template is assumed to exhibit spatial autocorrelation, with habitat suitability for this species (tan shading) highest near the center and declining toward the periphery of this region.

Predicting Fundamental and Realized Ranges

As we indicated in the previous chapter, delineating the most fundamental unit of biogeography—the geographic range—is paradoxically one of the field's most challenging exercises. First, our efforts are always limited by the **Wallacean shortfall**—we know precious little about the geographical distributions and geographic variation in the traits of most species. Second, even for those relatively well-studied species (e.g., trees and terrestrial vertebrates), geographic ranges are quite dynamic, while the studies supplying distributional data tend to span many decades and sometimes centuries. Finally, the nature of estimating species' ranges—based on incomplete information gathered for other purposes and with nonstandardized and temporally and spatially heterogeneous protocols—renders the challenge a technologically formidable one. Fortunately, our ability to manage, visualize, and analyze geographic information has advanced markedly over the past few decades; the biogeographers' tool kit now includes a variety of statistically rigorous methods for estimating species' ranges—an approach variously called **ecological niche modeling** or **species distribution modeling**—and, just as important, for predicting range dynamics under alternative scenarios in the future.

While there is an ongoing and sometimes heated but, we think, very healthy debate over alternative approaches for estimating a species' range, all the approaches typically include the following procedures:

1. They utilize both directly acquired (e.g., via field survey) and remotely sensed (e.g., via satellite imagery) information on actual occurrences of the focal species and the environmental conditions (habitat, temperature, precipitation, etc.) at the locations where they occurred.

2. They conduct statistical analyses to determine the species' environment affinities (the suite of conditions under which the species is likely to occur).

3. They utilize spatially explicit statistical models and geographic information systems to predict and map the fundamental range ("environmental space" or "climate space") of the focal species, that is, the area of the landscape or seascape that matches the species' environmental affinities.

4. They then include additional considerations (e.g., historical distributions of the species and the likelihood of it dispersing to other favorable, albeit isolated, locations; the effects of interspecific interactions that may limit or expand its range) to predict the species' realized range.

5. They test the efficacy and validity of these predictions using a set of empirical data on distributions not used in previous steps.

6. Depending on the outcome of this test, predictive models may be improved (by adjusting parameters in the models describing environmental affinities or by including predictors such as additional climate or soil variables in new models) in what are sometimes referred to as "learning routines."

7. Once scientists are satisfied with the performance of their models for estimating current or historic distributions, they can apply them in an actually predictive sense—for example, to predict range dynamics in response to anticipated anthropogenic changes in climate or in response to the introduction and range expansion of problematic species.

We will refrain here from showcasing any particular protocol and software for estimating and predicting species ranges, partially to avoid any apparent bias on our part but also because this is a very active and rapidly advancing front in analytical biogeography, such that any summary would soon be made obsolete. Instead, we refer the reader to some general references and recent syntheses on relevant methodology (special issue of the *Journal of Biogeography* in 2006, Volume 33: 1677–1789; Austin 2007; Kearney and Porter 2009; Kozak et al. 2008). We will integrate insights from these advances into later chapters (in particular, Chapters 11 and 15).

Synthesis

Having presented numerous examples of the kinds of abiotic and biotic factors that limit the geographic ranges of species, it is appropriate for us to draw some general conclusions. First, it may not be productive to search for single limiting factors and simple explanations for the geographic distributions of most species. Not only may a single species be limited by different factors in different parts of its range, but even in one local area, several factors may interact in complex ways to prevent expansion of populations.

Given this complexity, however, there is a hint of one pervasive pattern, which we have alluded to: Many species appear to be limited on one range margin by abiotic stress and on the other by biotic interactions. Although empirical studies and rigorous evaluations of this hypothesis are quite rare, its origins can be traced back to the early biogeographers and ecologists of the nineteenth and twentieth centuries. In 1853, after developing a map that separated the surface waters of the world's oceans into zones based on isocrymes (lines of mean minimum temperature; see Figure 2.13A), James Dana (1853) made the prescient observation that "the cause which limits the distribution of species northward or southward from the Equator is the cold of winter rather than the heat of summer or even the mean temperature of the year." Later, Theodosius Dobzhansky (1950) and Robert H. MacArthur (1972) offered a similar hypothesis for the terrestrial realm: Biotic interactions are more likely to limit abundances and distributions in the tropics, while abiotic stresses are more likely to be limiting at higher latitudes. At regional to local scales, this pattern is particularly apparent in studies of distributions along ecological and environmental gradients, such as those of sessile organisms along the intertidal zone and plants along terrestrial gradients of temperature and moisture (recall Connell's classic studies of barnacle distributions; see Figure 4.11).

We will consider the implications of these relationships and their underlying processes in Chapter 15 when we discuss the correlates and causes of geographic gradients of diversity across oceanic and continental regions. In the next chapter we scale up the ecological hierarchy from a focus on the geographic ranges of particular species to describe the distributions and patterns of geographic variation of entire biotic communities and ecosystems.

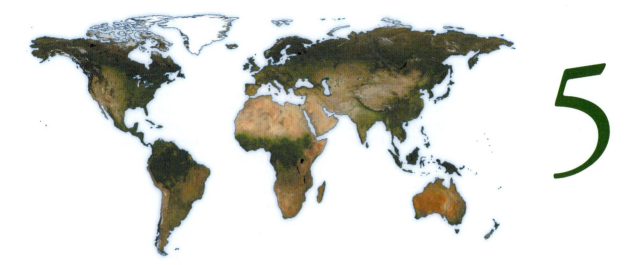

5

DISTRIBUTIONS OF COMMUNITIES

No living thing is so independent that its abundance and distribution are unaffected by other species. In Chapter 3, we observed that not only must species respond to environmental variation across the geographic template, but many also modify it and, in turn, significantly influence distributions of other species. The physical and chemical composition of the Earth's surface has been drastically altered by the activities of organisms, which, among other things, have generated the oxygen in the atmosphere and contributed to the development of soil and modified the chemical composition of the aquatic realm. Organisms vary greatly, however, in the extent to which they are dependent on other organisms. Some autotrophic organisms, such as certain kinds of algae and lichens, not only make their own food from sunlight, carbon dioxide, water, and minerals but also inhabit extremely rigorous physical environments, such as hot springs and bare rocks, where they may encounter and interact directly with few, if any, other species. Heterotrophic organisms, which include all animals, fungi, and some non-photosynthetic vascular plants, cannot make their own food and are dependent on autotrophs for usable energy. Even many photosynthetic plants are directly dependent on specific kinds of organisms, such as nitrogen-fixing bacteria and mycorrhizal fungi (which make available essential mineral nutrients), and insects and vertebrates (which pollinate flowers and disperse fruits and seeds). Thus, as we saw in Chapter 4, because of the myriad of interactions among organisms, species occurring in the same region may tend either to exhibit mutually exclusive distributions at the local scale or to occur together in complex associations within ecological communities. Here, we expand from the focus of Chapter 4—distributions of particular species—to consider the geography of these more inclusive units of biological organization.

Historical and Biogeographic Perspectives

Frederic E. Clements (1916) suggested that a community could be regarded as a type of superorganism with its own life and structure as well as its own spatial and temporal limits. According to this view, individual organisms and species could be analogized to the cells and tissues of an organism, and the process of secondary succession could be likened to the growth and development of an individual. In contrast to Clements's concept of the community as a discrete and highly integrated unit, Henry A. Gleason (1917, 1926) viewed a community as merely the coexistence of relatively independent individuals and species in the same place at the same time. Focusing primarily on plants, Gleason pointed out that the occurrence of species in an area depends primarily on their individual capacities to immigrate to, and to grow in, the local environment.

As is often the case with such debates, both sides made some important points. Communities do have certain emergent properties (analogous to those of individuals) that can be measured and studied. These properties include photosynthesis (total **primary productivity**) and metabolism (community **respiration**), as well as more complex processes associated with the transfer and use of energy and nutrients and with changes in species composition and habitat during succession (e.g., Odum 1969, 1971). We observed in the previous chapter that many species—especially keystone species, ecosystem engineers, and obligate mutualists or hosts of **stenotypic** parasites—strongly influence the abundance and distributions of others. On the other hand, early ecologists amassed much data showing that many species vary in presence and abundance through time and space as if they were independent of other members of their communities (McIntosh 1967; Palmer et al. 1997; Davis et al. 1999; Menge and Lubchenco 2001; Parker 2002).

Even the casual observer will notice that certain kinds of plants tend to occur together in particular climates to create distinctive vegetation types. Ecologists and biogeographers refer to these as **life zones** and **biomes**—or, in more recent years, **ecoregions**—and recognize that specific kinds of animals and microorganisms are associated with these vegetation formations. For example, a broad band of coniferous forest, sometimes referred to facetiously as the "spruce–moose biome," extends around the world at high latitudes in the Northern Hemisphere (**Figure 5.1**). Not only similar vegetation, but also many of the same species and genera of plants, animals, and microbes are distributed over the Old World from Scandinavia to Siberia and across the New World from Alaska to Nova Scotia and Newfoundland. These organ-

FIGURE 5.1 Three groups of organisms inhabiting coniferous forests exhibit similar geographic ranges in North America. Seven species of trees (not shown individually), two species of birds, and three species of mammals are typical inhabitants of the coniferous forests that spread across the northern part of the continent and extend southward at high elevations in the mountains.

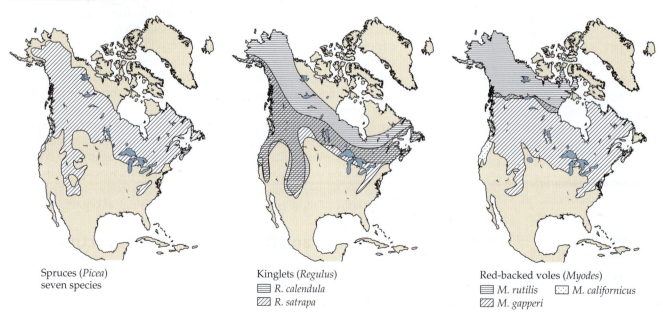

Spruces (*Picea*)
seven species

Kinglets (*Regulus*)
▤ *R. calendula*
▨ *R. satrapa*

Red-backed voles (*Myodes*)
▤ *M. rutilis*   ▦ *M. californicus*
▨ *M. gapperi*

isms generally are adapted to living with one another in similar physical environments. Beyond their general latitudinal limits, their ranges extend further southward together where mountain ranges provide appropriate habitats at high elevations. In North America these coniferous spruce–fir forests extend southward along the Appalachian Mountains in the east and along the Cascades, Sierra Nevadas, and Rocky Mountains in the west. Several kinds of organisms are restricted to these coniferous forest habitats and have similar geographic distributions. Recent advances in historical biogeography and ecology now enable us to assess the causal nature of such broadly overlapping distributions: that is, the degree to which they result from historical legacies of evolution and barriers to dispersal, from coincident but independent environmental affinities, or from ecological interdependence among those species (see Chapters 10, 11, and 12).

Communities and Ecosystems

Definitions

Communities and ecosystems—the highest levels of biotic and ecological organization—are rather arbitrarily defined. As stated above, a community consists of those species that live together in the same place. A member species can be defined either taxonomically or on the basis of more functional ecological criteria, such as life-form or diet. Ecologists study two fundamental properties of communities: community structure (which refers to static properties that include diversity, composition, and biomass of species) and community function (which includes all of the dynamic properties and activities that affect energy flow and nutrient cycling such as photosynthesis, interspecific interactions, and decomposition). The place where member species occur can be designated either by arbitrary boundaries or on the basis of natural topographical features. Thus, we can speak of the fish community in some ecologist's 2 hectare study area on a coral reef on the north shore of Jamaica, of the grazing community (which would include representatives of several invertebrate phyla as well as some fish species) on that entire reef (which might extend uninterrupted over many square kilometers), or of the entire community of all coral reef–inhabiting organisms in the Caribbean Sea (see Ricklefs 2008).

An **ecosystem** includes not only all the species inhabiting a place but also all the features of that place's physical environment and all the interactions between the biotic and abiotic components of the system (Tansley 1935). Because ecosystem ecologists are interested primarily in the exchange of energy, gases, water, and minerals among the biotic and abiotic components of a particular ecosystem, they often try to study naturally confined areas where input and output of energy and materials are restricted and hence easier to control or monitor (e.g., the classic studies of Howard T. Odum [1957], John M. Teal [1957, 1962], and G. Evelyn Hutchinson [1957, 1967, 1975, 1993]). Small, relatively self-contained ecosystems are often called **microcosms** because they represent miniature systems in which most of the ecological processes characteristic of larger ecosystems operate on a reduced scale. A sealed terrarium is a good example of an artificial microcosm, and a small pond is an example of a natural one.

Few ecosystems are as isolated and independent as they might appear. The largest and only completely independent ecosystem is the **biosphere**, which encompasses the entire Earth. The interdependence of all ecosystems is vividly demonstrated by the widespread impact of human activities. Acid rain falls on pristine forests and is carried into lakes far from the source of pollution. Deforestation—especially the cutting of tropical forests—and the burning of fossil fuels are changing the composition of the atmosphere, and perhaps altering the climate, throughout the world. Such linkages also occur naturally, of course, a striking example being that the main mineral source fertilizing the Amazon basin turns out to be dust (about 40 metric tons per annum) from the Sahara; about half of this derives from the Bodélé depression, northeast of Lake Chad—an area about 0.5 percent of the size of the Amazon basin (Koren et al. 2006).

Many ecologists are actively investigating the structure and function of communities and ecosystems. As is usually the case in a rapidly developing field, many important questions remain unanswered, and some tentative conclusions are highly controversial. Because many of these unresolved issues are relevant to biogeography, this situation can be both challenging and frustrating for those trying to account for plant and animal distributions. On one hand, biogeographers have an opportunity to use their own methods and data to make important advances in community ecology. Researchers with a broad biogeographic perspective are in a position to make unique contributions toward understanding how ecological processes influence the numbers and kinds of species that live together in different parts of the Earth. On the other hand, until these ecological processes are better understood, biogeographers will be unable to fully integrate them with the effects of historical events in order to provide a solid conceptual basis for interpreting and predicting distribution patterns (Ricklefs 2008).

Community organization: Energetic considerations

In Chapter 4 we emphasized that each species—the fundamental unit of an ecosystem—has a unique ecological niche that reflects the biotic and abiotic conditions necessary for its survival. How are these individual niches organized to produce complex associations of populations of many species? This is the ultimate question of community ecology, and one for which we can advance only tentative answers. Two characteristics of species that strongly influence their effects on community organization are their body size and trophic status (i.e., whether they are primary producers, herbivores, carnivores, or decomposers.) In fact, if you sought to take a single measurement that would provide the most information on the biology of an organism, you would be best advised to measure its body mass. The larger an organism, the more energy it requires for maintenance, growth, and reproduction. The rate of energy uptake and expenditure of animals at rest (or the **basal metabolic rate**, m) varies with body mass (M) according to the relationship $m = cM^{0.75}$. Although the value of the constant (c) varies somewhat among taxonomic groups, the exponent, or slope (0.75), is very general (**Figure 5.2A**; Hemmingsen 1960; Peters 1983; Calder 1984). The ecological consequences of this simple formula are profound. Because the exponent is positive, total energy requirements increase with body size. It takes about 14,000 times more energy to maintain a 5000 kg elephant than a 15 g mouse. The same applies to all organisms: A mature redwood tree uses much more energy than a strawberry plant. On the other hand, because the exponent is less than 1, the metabolic rate *per unit of mass* of a small organism is greater than that of a larger one (see inset of Figure 5.2A). It requires about 25 times more energy to maintain a gram of mouse than a gram of elephant. Apparently for this reason, almost all rate processes (e.g., cardiac rate and respiration frequency in vertebrates) are accelerated in small organisms, which are more active and have higher reproductive rates and shorter life spans than do large organisms.

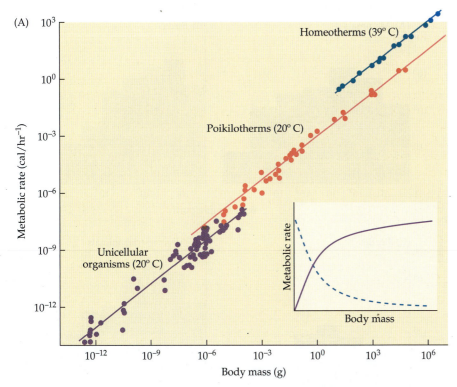

FIGURE 5.2 (A) Relationship between metabolic rate (*m*) and body mass (*M*) for a wide variety of organisms, ranging from unicellular forms to poikilothermic ("cold-blooded") animals to homeothermic birds and mammals. Note that the axes in the principal graph are on a logarithmic scale, so the relationship is described by a power function of the form $m = cM^{0.75}$ where the constant (*c*) varies slightly among the three different groups, but the exponent or slope (0.75) is remarkably constant. The inset illustrates this relationship using arithmetic scales, with per gram metabolic rate (dashed line) decreasing with body size. (B) Because the storage capacities of organisms (e.g., energy and mineral stores, and water volume) increase with body size more rapidly than their requirements, endurance or ability to withstand prolonged periods of stress (vertical arrows) also increase with body size. (A after Hemmingsen 1960.)

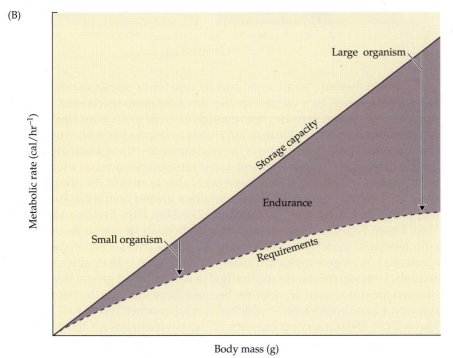

Although metabolic rate and energy requirements increase as less than linear functions of body mass, storage capacities (e.g., energy stored as fat, water volume, mineral stores in bone tissue) increase in direct proportion to mass. Therefore, all else being equal, larger organisms have greater capacities to withstand prolonged stresses such as starvation, dehydration, and subfreezing temperatures (**Figure 5.2B**). Body size also has important ecological consequences because it influences the scale at which organisms use the

FIGURE 5.3 Frequency distribution of body size among species for several different kinds of organisms: (A) terrestrial mammals of the world, (B) land birds of the world, (C) beetles of Great Britain, and (D) all terrestrial animals of the world (approximately). Note that the axes are on a logarithmic scale, so small species are much more numerous than large ones. This pattern is very general and accounts for the obvious fact that insects are much more numerous (in numbers of species as well as individuals) than vertebrates in terrestrial environments. (After May 1988.)

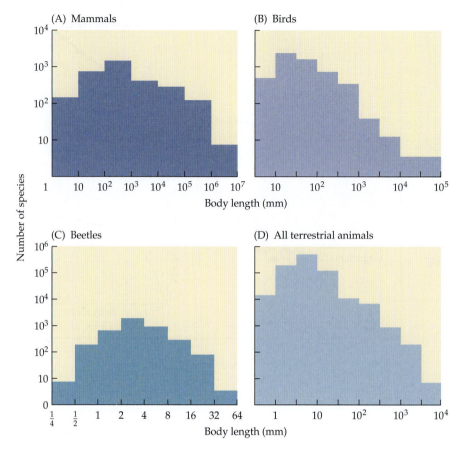

environment. Because small organisms require fewer resources per individual than large ones, they can use smaller areas, be more specialized, and still maintain population densities high enough to avoid extinction. Small organisms are therefore better able than large ones to respond to what Hutchinson (1959) termed the mosaic nature of the environment—the spatial heterogeneity or patchiness of the environment that is most pronounced on a local scale. Consequently, small organisms have been able to divide the environment more finely, so any geographic area contains a greater number of small-bodied species than large ones (Van Valen 1973a; May 1988; Brown and Maurer 1986; **Figure 5.3**). Consider the tremendous diversity of insects as compared with terrestrial vertebrates (about 950,000 versus 28,000 known species, respectively). A biogeographic correlate of this pattern is that, at least among animals, large organisms are constrained to having broad geographic ranges. Reasons for this should be obvious: Because large individuals require more space and have broader niches (McNab 1963; Schoener 1968b; Kelt and Van Vuren 1999, 2001), carrying capacity of the environment for these animals is low, and only large areas can support sufficient numbers of individuals to maintain the species over evolutionary time. By contrast, small areas can support large populations of small organisms, which consequently have a low probability of extinction. Small species may or may not have very restricted ranges (**Figure 5.4**; see also Chapter 15).

The **trophic status** of organisms (or how they acquire energy) also influences the role they play in community structure, where again body size plays an important role: predators tending to be larger than their prey (unless hunting in packs), parasites smaller than their hosts, larger species of plants and animals dominating smaller ones in competition for sunlight or food, respectively. Outside of a handful of fascinating exceptions (e.g., chemosynthetic

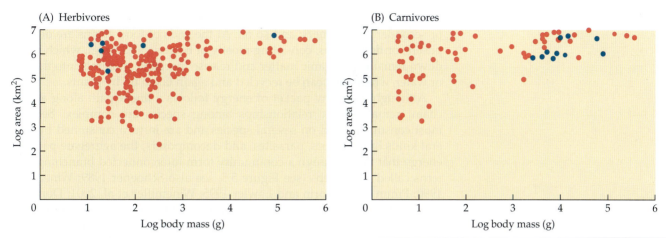

(A) Herbivores

(B) Carnivores

Log area (km²) vs Log body mass (g)

FIGURE 5.4 Relationship between area of geographic range and body mass among North American terrestrial mammals. Note that although there is much variation, the areas of the smallest ranges increase with increasing body mass and are smaller for herbivores (A) than for carnivores (B). Blue circles indicate species that actually have larger ranges than indicated because they range into Central America (only the North American area of the distribution has been measured). (After Brown 1981.)

communities of geothermal vents on the ocean floor), all of the energy used by living things ultimately comes from the sun. Autotrophic green plants use solar radiation, carbon dioxide, water, and minerals to synthesize organic compounds, and they produce oxygen as a by-product. These organic compounds are not only used by plants for making structures and fueling basic metabolism, but are also the sole source of energy for heterotrophic organisms in the community. Solar energy, trapped in organic molecules, is transferred from one species to another and gradually used up as herbivores eat plants, carnivores eat herbivores, and so on. These heterotrophic species oxidize organic compounds to obtain usable energy, and in the process they consume oxygen and release carbon dioxide.

Unidirectional paths of energy that flow between species and through communities are termed **food chains**. Different links in a food chain are called **trophic levels** (**Figure 5.5**). The first level contains green plants, or **primary**

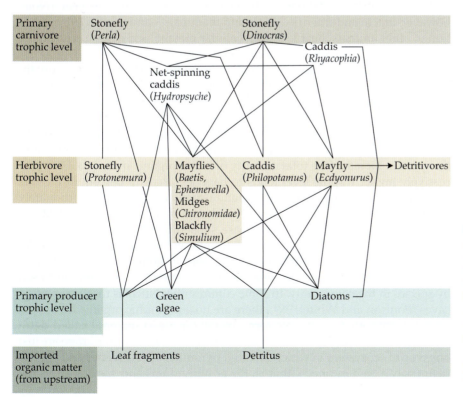

Primary carnivore trophic level

Stonefly (*Perla*)   Stonefly (*Dinocras*)   Caddis (*Rhyacophia*)

Net-spinning caddis (*Hydropsyche*)

Herbivore trophic level

Stonefly (*Protonemura*)   Mayflies (*Baetis, Ephemerella*) Midges (*Chironomidae*) Blackfly (*Simulium*)   Caddis (*Philopotamus*)   Mayfly (*Ecdyonurus*) → Detritivores

Primary producer trophic level

Green algae   Diatoms

Imported organic matter (from upstream)

Leaf fragments   Detritus

FIGURE 5.5 Part of the food web for an aquatic community inhabiting a small stream in Wales. The diagram shows interconnections of food chains to form food webs. There are three trophic levels, but some organisms, such as *Hydropsyche*, which feeds on both plant and animal material, occupy intermediate positions. The many individual species of plants, green algae, and diatoms that make up the primary producer trophic level are not shown individually. (After Jones 1949.)

producers; the second, **herbivores**, or **primary consumers**; the third, **carnivores**, or secondary consumers; and so on. At the ends of food chains are decomposers, or **detritivores**, consisting mostly of bacteria and fungi, which break down remaining organic matter and release inorganic minerals into the soil or water, where they can be recycled and again taken up by plants.

Although each tiny packet of energy follows a linear path along a food chain, actual trophic relationships among species are complex. Because most consumers feed on several species and are in turn consumed by several kinds of predators, parasites, and decomposers, the aggregate paths of energy that flow through a community form interconnected branching patterns called food webs (see Figure 5.5; see also Schoener 1989; Winemiller 1990; Pimm 1991; Morin and Lawler 1995; Winemiller et al. 2001; Dunne et al. 2002; Belgrano et al. 2005; de Ruiter et al. 2005; McClanahan and Branch 2008). As energy is distributed through a community in this fashion, it obeys the laws of thermodynamics, which have important implications for community organization. Simply stated, the **first law of thermodynamics** says that energy is neither created nor destroyed but can be converted from one form to another. During photosynthesis, for example, solar energy is converted into high-energy bonds in the organic molecules of plants, which in turn can be consumed by animals and transformed into ATP to support muscular contractions. The **second law of thermodynamics** states that as energy is converted into different forms, its capacity to perform useful work diminishes, and disorder (entropy) of the system increases. Organisms use energy stored in organic molecules to perform the work of moving, growing, and reproducing. As organic compounds are oxidized to provide energy for these activities, much of the energy is dissipated as heat. In fact, most organisms are surprisingly inefficient. Photosynthetic efficiency of most plants, in terms of conversion of solar energy into chemical energy, typically falls well below 5 percent. Similarly, most animals are able to incorporate only 0.1 percent to 10 percent of the energy they ingest into the organic molecules of their own bodies; the remaining 90 percent to over 99 percent is "lost" as heat. (Note, however, that in endotherms—birds and mammals—and an intriguing selection of other animals, heat lost from cellular metabolism plays a critical role in maintaining relatively warm body temperatures.)

One ecological consequence of these thermodynamic properties of organisms is that substantially smaller quantities of energy are available to successively higher trophic levels. This can be shown diagrammatically by portraying the community as an ecological pyramid (**Figure 5.6**). In any community, green plants have the highest rates of energy uptake, and they usually comprise the largest number of individuals, the greatest number of species, and the greatest **biomass** (the total quantity of living organic material). Each successively higher trophic level tends to have less than 10 percent of the rate of energy uptake of the level below it and usually contains a proportionately lower biomass as well as fewer individuals and species (Lindeman 1942).

Availability of energy must decrease substantially with each successive trophic level in accordance with the laws of thermodynamics, but there are exceptions to the other patterns. Distribution of biomass, individuals, and species among trophic levels depends on how energy is acquired and used by species in the community. In some communities, such as those inhabiting caves (Poulson and White 1969), there are no autotrophic organisms, and all energy is imported in organic form, usually as dead material called **detritus**. Deciduous forests of eastern North America and Western Europe are just one type of ecosystem in which inverted pyramids of abundance and species richness can be described—that is, fewer individuals and species of primary producers than of primary consumers. Here, most of the energy used by

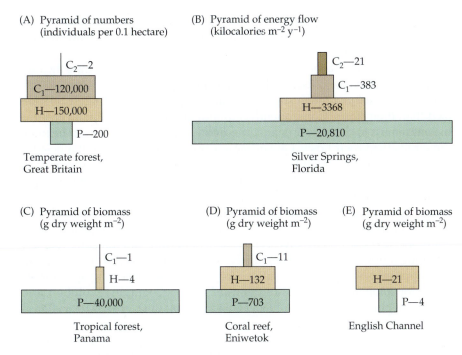

(A) Pyramid of numbers
(individuals per 0.1 hectare)

C_2—2
C_1—120,000
H—150,000
P—200

Temperate forest,
Great Britain

(B) Pyramid of energy flow
(kilocalories m^{-2} y^{-1})

C_2—21
C_1—383
H—3368
P—20,810

Silver Springs,
Florida

(C) Pyramid of biomass
(g dry weight m^{-2})

C_1—1
H—4
P—40,000

Tropical forest,
Panama

(D) Pyramid of biomass
(g dry weight m^{-2})

C_1—11
H—132
P—703

Coral reef,
Eniwetok

(E) Pyramid of biomass
(g dry weight m^{-2})

H—21
P—4

English Channel

FIGURE 5.6 Ecological pyramids of energy flow, biomass, and number of individuals for several different communities. Note that three of the pyramids (B–D) are regular, with each successive trophic level reduced compared with the one below, but two of the pyramids (A, E) are not. The pyramid of individuals for the temperate forest in Great Britain (A) and the pyramid of biomass for the English Channel (E) have fewer individuals and a lower biomass of primary producers than herbivores. P, primary producers; H, herbivores; C1, primary carnivores; C2, secondary carnivores. (After Odum 1971.)

plants is monopolized by a relatively small number of large trees, whereas energy used by herbivores is allocated among many individuals and species of small insects and other invertebrates (Elton 1966; Varley 1970) (see Figure 5.6A). Inverted pyramids of biomass also are possible and, in fact, quite common in some aquatic ecosystems (see Figure 5.6E). In the open ocean, photosynthetic rates of phytoplankton may be so high that these tiny producers can support a biomass of consumers far exceeding their own. This is possible because the consumers depend on rates of energy transfer (i.e., molecules metabolized per unit of time) and not simply biomass per se. Phytoplankton populations are so productive that they can replace themselves many times per day and thus support high rates of herbivory by zooplankton. However, consistent with the first law of thermodynamics, pyramids of energy flow can never be inverted.

Because the **carrying capacity** (measured in units of usable energy) of any area is lower for successively higher trophic levels, organisms occupying these levels exhibit predictable characteristics that affect their ecological roles and geographic distributions. Not only are there fewer species of carnivores than of herbivores and plants, but the carnivores also tend to be larger and more generalized than herbivores. Carnivores usually have to be large enough to overpower their prey. They also tend to feed on several prey species, and to have relatively broad habitat requirements and wide geographic distributions (see Figure 5.4; Rosenzweig 1966; Van Valen 1973a; Wilson 1975; King and Moors 1979; Zaret 1980; Brown 1981; Gittleman 1985). The cougar, or mountain lion (*Puma concolor*), for example, is one of the top carnivores in the New World. It weighs 50 to 100 kg and takes a variety of prey, mostly mammals ranging in size from rabbits to elk. Prior to anthropogenic reductions of its range during the past two centuries, the cougar had the widest geographic distribution of any American mammal, from the Atlantic to the Pacific coasts on both continents. Even today, its range extends from Alaska to the southern tip of South America. The cougar inhabits a diversity of ecosystems, including those in tropical rain forests, coniferous and deciduous hardwood forests, shrublands, and deserts. Other top carnivores, such as the wolf (*Canis lupus*) and jaguar (*Panthera onca*), also have broad geographic and habitat distributions, whereas herbivores of comparable size tend to have more restricted ranges.

Parasites are an exception to these patterns, but they still illustrate the fundamental importance of energetic relationships in community organization. Parasites are not just much smaller but also tend to be more highly specialized than their hosts. Consequently, individual parasites need not consume the entire host to meet their energy requirements, and parasites can be more numerous than their hosts. One or more parasites usually inhabit a host for an extended period, often for the entire life span of either parasite or host. Many parasites are so highly specialized that they infect only one or a few, often taxonomically related, host species. Although these adaptations have allowed parasites to maintain geographic ranges and numbers of individuals roughly comparable to those of their hosts in the trophic level below them, many species have had to solve the problem of finding and infecting sparsely distributed hosts. The elaborate, highly specialized life cycles of many parasites appear to be largely adaptations for locating and infecting appropriate hosts (for some fascinating descriptions of the biogeography, evolution, and ecology of parasites, see Price 1980; Hawkins 1994; Bush et al. 2001; Fallon et al. 2005; Thomas et al. 2009).

These kinds of energetic considerations lead to the prediction that the capacities of geographic areas and their habitats to support many individuals and diverse communities should ultimately depend on their total productivity. Everything else being equal, the higher the fixation rate of sunlight into organic material, the more usable energy should be available to be subdivided among individuals, species, and trophic levels. As we shall see, productivity varies greatly among different habitats, depending on such factors as climate, soil type, water availability, and the influence of human activity (Rosenzweig 1968; Jordan 1971; Lieth 1973; Whittaker and Likens 1973). In general, the predicted relationship between productivity and diversity is observed—at least at a coarse scale of analysis using large sample areas (see Chapter 15). Widespread, highly productive habitats such as tropical rain forests and coral reefs are renowned for their great diversity of specialized species. In contrast, small, isolated areas such as small islands, and widespread, unproductive habitats such as boreal forests and tundra, contain fewer and, for the most part, more generalized species. We will examine such patterns of species diversity in more detail in Chapters 13, 14, and 15.

Distribution of Communities in Space and Time
Spatial patterns

Because species often are adapted to the same physical conditions as other, ecologically similar species, we might expect biotic communities to be distributed as discrete units, with rapid turnover of species as we move along the geographic template from one community to the next. On the other hand, because of competition, we might expect ecologically similar species to adapt to different niches and thus exhibit exclusive, or non-overlapping, distributions.

It is obvious that abrupt changes in the environment, such as those found at the shore of a lake or the edge of a forest, will usually be accompanied by rapid transitions between two different associations of species. If the environmental discontinuity is rapid and severe, most of the species living on each side will be limited almost simultaneously when they encounter inhospitable conditions at the border. If the two habitat types are not too dissimilar, however, the edge (or **ecotone**) may actually contain more species than either pure habitat type (Holland et al. 1991). Species from either side of the boundary may be able to mix and occur together in the narrow area where the two environments meet. If the transition zone is fairly productive and not too narrow, it may even support its own assemblage of organisms uniquely adapted to live there. This is especially likely if the boundary is not simply an ecotone, with conditions intermediate between those on either side, but an environment in which conditions are unique or fluctuate back and forth between those found on either side. The most dramatic example of such a community is provided by the intertidal zone. Special adaptations are required to withstand periodic inundation in seawater followed by exposure to a desiccating terrestrial environment. Along a narrow strip of shore live entire communities of such plants and animals, which show their own local-scale patterns of association and segregation within the vertical gradient of tidal exposure (e.g., Connell 1961, 1975; Menge and Sutherland 1976; Lubchenco and Menge 1978; Lubchenco 1980; Souza et al. 1981; Peterson 1991; Menge et al. 1994).

How are species distributed along environmental gradients, such as the relatively abrupt gradient in exposure within the intertidal zone or the more gradual gradient in climatic conditions caused by latitudinal or elevational changes? Some of the most thorough studies of such patterns, called **coenoclines**, have been those of Robert H. Whittaker and his colleagues on the distribution of tree species on mountainsides in the United States. Their data can be used to evaluate five alternative hypotheses (**Figure 5.7**; see Whittaker 1975):

1. Groups of species exhibit similar ranges along the gradient and are distributed as discrete communities ("superorganisms" in the Clementsian sense), with sharp boundaries between them. This pattern could be caused by competitive exclusion between dominant species and by other

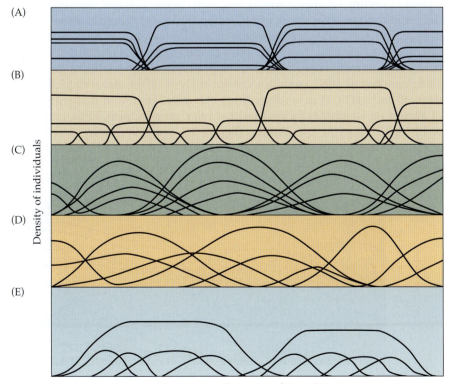

FIGURE 5.7 Five hypothetical coenoclines—patterns of distributions of species along an environmental or geographic gradient. (A) Species are distributed as discrete communities that replace each other abruptly. (B) Species are not segregated into communities, but some sets replace each other abruptly. (C) Species are distributed as discrete communities, which gradually replace each other. (D) Species behave as if they are independent of each other, neither associating in discrete communities nor replacing each other abruptly at transition zones, or "ecotones." (E) Most species are nested within the ranges of a few dominant species, but otherwise occur independently of each other. (After R. H. Whittaker 1975.)

Environmental or geographic gradient

(A) (B) (C) (D) (E)

Density of individuals

species evolving to coexist with the dominants and with one another (Figure 5.7A).

2. Individual species abruptly exclude one another along sharp boundaries, but most species are not closely associated with others to form discrete communities (Figure 5.7B).

3. Much as in hypothesis 1, species form discrete communities, but replacement of communities along the gradient is gradual. This could happen if groups of species evolved to coexist with one another, but competitive exclusion did not cause rapid replacement of species along the gradient (Figure 5.7C).

4. Individual species gradually appear and disappear, seemingly independent of the presence or absence of other species. Species neither competitively exclude each other nor associate to form discrete communities, and species replacement along the gradient is random (Figure 5.7D).

5. Ranges of most species are nested within the ranges of a few dominant species that are overdispersed (non-overlapping) along the gradient. Thus, species distributions may appear highly ordered at geographic scales (with little overlap among assemblages), but random at local scales (Figure 5.7E).

Whittaker and his associates (e.g., Whittaker 1956, 1960; Whittaker and Niering 1965) sampled several sites at varying elevations on mountains. Sample sites were chosen so as to keep soil type, slope, and exposure as constant as possible but to allow natural variation in temperature and rainfall. The results of two such studies (shown in **Figure 5.8**) appear to support the hypothesis that species are distributed as if they were independent of one another, showing no evidence of either abrupt replacements that might be attributed to competitive exclusion or association of species to form discrete communities.

It should be pointed out, however, that Whittaker's methods of collecting and analyzing data may have contributed to his obtaining the patterns shown in Figure 5.8. By censusing relatively large plots and then averaging the results across several mountain slopes, Whittaker would have tended to miss abrupt, local-scale replacements of species owing to competitive exclusion and mediated by the distribution of microsites in a spatially heterogeneous environment (see Davis et al. 1999; Parker 2002). Yeaton (e.g., Yeaton 1981; Yeaton et al. 1981) has made careful analyses of the elevational distributions of pines (*Pinus*) in western North America. He found many instances of abrupt replacements by species of similar growth form, apparently as a result of interspecific competition (**Figure 5.9**; see also Shipley and Keddy 1987). For example, gray pine (*P. sabiniana*) reaches its upper elevational limit

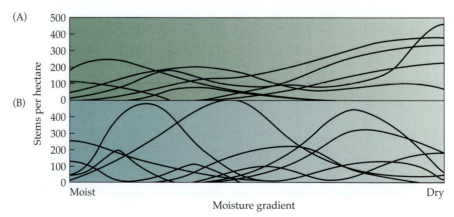

FIGURE 5.8 Coenoclines illustrating the distributions of tree species along two moisture gradients: (A) in the Siskiyou Mountains, Oregon, 760 to 1070 m elevation, and (B) in the Santa Catalina Mountains, Arizona, 1830 to 2140 m elevation. Note that the species replace one another gradually and seemingly independently of one another. The species have narrower elevational ranges along the steeper moisture gradient in Arizona. (After Whittaker 1967.)

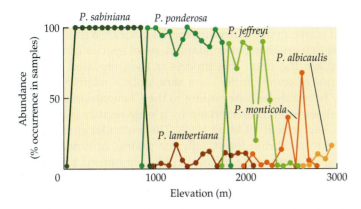

FIGURE 5.9 Elevational distributions of pine (*Pinus* spp.) trees on western slopes of the Sierra Nevadas in California, including (green) species of three-needled pines and (red and orange) species of five-needled pines. The species with the same number of needles are morphologically and ecologically similar, and they overlap little in elevation on sites with similar slopes, exposures, and soil types. Note also that distributions of these pines (three- and five-needled species, combined) suggest some degree of community organization, with *P. ponderosa* and *P. lambertiana* occupying similar elevations, as do *P. jeffreyi* and *P. monticola*, while *P. sabiniana* and *P. albicaulis* exhibit exclusive distributions (at the lowest and highest elevations, respectively). (After Yeaton 1981.)

and is abruptly replaced by ponderosa pine (*P. ponderosa*) at approximately 840 m on southeastern-facing slopes of the Sierra Nevadas near Yosemite National Park. Gray pine is at its greatest density and growth rate in mesic environments at higher elevations just before it is supplanted by ponderosa pine, suggesting that it is competitive exclusion, rather than inability to tolerate the physical environment, that determines the upper limit of its range. Thus, although Whittaker's results accurately reflect the typical distribution of many plant species along environmental gradients, abrupt replacements by competing species can undoubtedly occur in many cases in which ecologically similar or closely related species come into contact (e.g., chipmunks along elevational gradients; see Figure 4.26).

Temporal patterns

Ecological succession

Whether gradual or abrupt, individual species and entire ecological communities replace each other over time, as well as in space. These changes are of great interest both to ecologists, who study relatively short timescales, and to paleontologists, who study very long (geological and evolutionary) time scales. Many early ecologists, especially those studying sessile organisms, have devoted much effort to the study of succession (Cowles 1899, 1901; Cooper 1913; Clements 1916; Beckwith 1954; Monk 1968; Odum 1969; Drury and Nisbet 1973; Horn 1974, 1975, 1981; Connell and Slatyer 1977; Usher 1979; McIntosh 1981; Miles 1987; Facelli and Pickett 1990; Farrell 1991; McCook 1994). Succession, as we saw in Chapter 3, is progressive change in community structure and function over ecological time. Succession is a normal process in any ecosystem in which disturbances repeatedly eliminate entire communities from patches of local habitat. If all soil and organic material is removed—as by a glacier or a volcanic eruption—the process is termed **primary succession**. On the other hand, if a disturbance such as a fire or storm removes most of the living organisms but leaves some signal of the previous system in surviving individuals, or in the seed bank, the process is termed **secondary succession**.

Despite decades of research, succession theory remains a hotly debated topic among ecologists (McIntosh 1999; Walker and Moral 2003; Johnson and Miyanishi 2007; Walker et al. 2007). Much of the controversy echoes the Clementsian–Gleasonian debate over community organization; indeed, much of Clements's work was devoted to the study of succession. Here, the question is whether communities replace each other as interdependent units or as collections of independent species over *temporal* gradients (rather than spatial gradients, as in Figure 5.7). Clements's view, which dominated successional

theory for much of the twentieth century, held that succession was deterministic, predictable, and convergent. Pioneering species colonize a site and then
modify it to the point that another, better adapted, set of species can invade,
outcompete, and replace the first. This process is then repeated through
a progressive sequence of communities, or a **sere**, until a relatively stable
"**climax**" community is established. According to the Clementsian school of
thought, successional seres are characterized by a very regular increase in
biomass, complexity, and stability (Odum 1969). The particular climax community (e.g., lowland tropical rain forest, tallgrass prairie, or broad-leaved
deciduous forest) and its particular sere are largely determined by climate
and soil conditions and therefore are predictable and characteristic of a given
region (or biome).

In contrast to this orderly, deterministic view of succession, the Gleasonian camp observed an impressive diversity of successional seres and alternative climaxes, even for a given region. These ecologists argued that succession may not be the orderly process envisioned by Clements. Rather than
being driven largely by species interactions and autogenic modification of
local conditions, succession may simply reflect the idiosyncratic capacities of
independent species to disperse, establish themselves, and survive at a local
site with a particular combination of environmental conditions.

This debate, while nearly a century old, is not likely to be resolved soon.
Until recently, succession was largely regarded as a botanical phenomenon.
Whether deterministic or stochastic, it was viewed as a sequence of progressive changes in plant communities. Plant ecologists acknowledged that animals exhibited similar successional dynamics, but felt that they did so only
because of their dependence on plants. This view, of course, is problematic.
Grazers, granivores, pollinators, and decomposers—whether animals, fungi,
or microbes—all influence plant community structure and dynamics, as do
those animals serving as seed dispersers, and animal communities also exhibit progressive changes in species composition during primary and secondary succession (e.g., see McCook and Chapman 1997). Indeed, the search
for the driving force, or forces, of ecological succession may require ecologists to look outside their taxonomic biases and study communities defined
by functional—rather than taxonomic—criteria.

As with all debates between two such extreme points of view, it is likely
that the answer to this one lies somewhere in between. Both the purely deterministic Clementsian camp and the stochastic Gleasonian camp agree that
within a region of similar soil and climate, succession is a directional process
that tends to result in suites of species assemblages that have a degree of compositional organization and structure while not being wholly predictable.

Paleoecological perspectives

Paleoecologists and historical biogeographers also study temporal changes
in communities and biotas, albeit over larger geographic areas and longer
time periods. Regional biotas are strongly influenced by the major shifts in
climate and soil conditions (or, in the aquatic realm, by currents and water
chemistry) that have occurred throughout history. Paleobotanists have often
used analyses of fossil pollen to document the reestablishment of eastern
deciduous forest communities in response to climatic and geological changes
that followed the retreat of the last glaciers in North America (**Figure 5.10**;
see Davis 1969, 1976; see also Bernabo and Webb 1977; Grayson 1993; Peng et
al. 1995). Julio Betancourt and his colleagues (e.g., see Betancourt et al. 1990;
Betancourt 2004; Holmgren et al. 2008) analyzed pack rat middens (plant material collected by these rodents and preserved in caves and in rock crevices)
to reconstruct the dynamics of vegetation in arid regions of western North

America over the past 40,000 years. Such studies have revealed that the retreat of glaciers and associated climatic changes resulted in major shifts in terrestrial vegetation far removed from the ice sheets. Communities, however, did not behave as perfectly interdependent units. Apparently, rates of invasion depended in large part on mechanisms of seed dispersal and other aspects of the life history that differed markedly among species. While some groups of species shifted in concert, others became associated with different communities, while still others were unable to adapt and went extinct (see Jackson 2004 and Chapters 9 through 12).

Paleoecologists such as those cited above study the dynamics of biotas over the span of hundreds of years to millennia. Other paleobiologists focus on major upheavals that have occurred on an evolutionary, or geological, timescale (i.e., on the order of millions of years). The taxonomic specialization of most paleontologists and the fragmentary nature of the fossil record pose formidable challenges for those attempting to reconstruct changes in communities that have occurred over geological time (Gray et al. 1981; Behrensmeyer et al. 1992). Much attention has been focused on **mass extinctions**—episodes of relatively abrupt (on the span of a few million years) replacement of virtually entire biotas. Although the exact causes of these catastrophic changes are still hotly debated, many, if not all, must have been triggered by drastic environmental perturbations. In many ways these biotic upheavals are comparable to the succession that occurs after ecological disturbances, or to the abrupt spatial replacement of communities across major environmental discontinuities. These catastrophic events were spectacular, but they were the exceptions—these short pulses of extinction and evolutionary change were interspersed with extremely long spans of more gradual changes in regional biotas.

Throughout geological time, many species have colonized, speciated, and become extinct relatively independently of one another, in a pattern much like that of the spatial replacement of many contemporary species observed along gradual physical gradients. As evidence of this pattern, we can point to the existence in modern communities of recently evolved species alongside "living fossils" (e.g., the ginkgo, horseshoe crab, and coelacanth), forms that have survived virtually unchanged for hundreds of millions of years. However, we must caution against sweeping generalizations. While much of the fossil record suggests independent shifts of species, other evidence may reflect interdependence among species. Mutualism, widespread today (Janzen 1985; Rico-Gray and Oliveira 2007; Stadler and Dixon 2008), must also have been common in the distant past. We know from research on contemporary communities that the loss of a keystone species, or the invasion and establishment of an exotic species, can cause major changes and even wholesale reorganization of communities. It is therefore likely that the evolution, range shifts, and extinctions of such keystone species contributed to the dynamics and upheavals of regional biotas in the past (e.g., see Petuch 1995). The fossil record, riddled as it is with gaps and mysteries, also may hold many insights into the forces structuring ecological communities, both past and present.

(A)

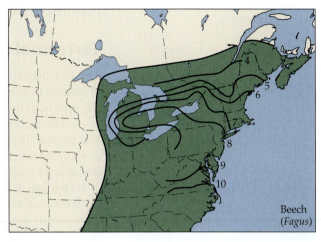

(B)

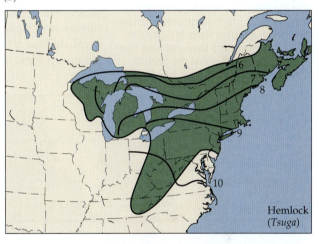

FIGURE 5.10 Reconstruction from fossil pollen records of the recolonization of North America by two tree species, (A) beech (*Fagus*) and (B) hemlock (*Tsuga*), after the last Pleistocene glaciation. Numbered lines indicate the fronts of each species' range at 1000-year intervals BP, showing the progressive northern migration of each species. Note that the migration of these two tree species was quite different, although the northern borders of their present ranges are virtually identical. (After Bernabo and Webb 1977.)

Before delving any further into the past, we now present an overview of contemporary terrestrial biomes and aquatic communities.

Terrestrial Biomes

The fact that communities do not represent perfectly discrete associations of species in either time or space obviously complicates any attempt to classify and map them. Where physical and geographic changes are abrupt, it is relatively easy to recognize distinct community types, but where environmental variation is gradual, we are faced with the problem of dividing the essentially continuous variation in species composition, life-form, and other community traits into a discrete number of arbitrary categories.

In the last few decades, mathematicians and biologists have developed sophisticated multivariate statistical techniques for quantifying the degree of similarity (or difference) between two samples, based on a large number of variables. Many plant and animal ecologists have applied these methods to ecological and biogeographic data (see Pielou 1975, 1979; Omi et al. 1979; Robinove 1979; Rowe 1980; Smith 1983; McCoy et al. 1986; Birks 1987; Cornelius and Reynolds 1991; Bailey 2009). If comparable measurements are available for a large number of communities, these techniques can be used to group them into hierarchical clusters reflecting their similarities in spe-

FIGURE 5.11 World distribution of major terrestrial biomes. Note that the locations of these vegetation types correspond closely to the distribution of climatic regimes and soil types (see Figures 3.6 and 3.11). In some cases, several different vegetation types (e.g., tropical deciduous forest and savanna) have been grouped together so that the general zonal pattern of biomes can be observed.

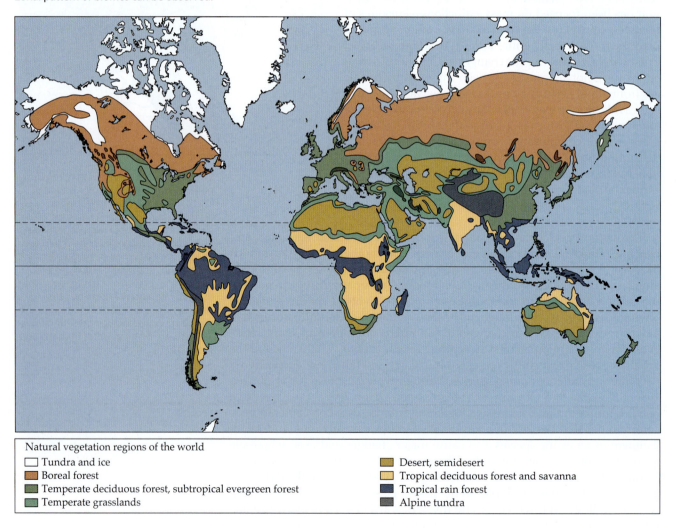

Natural vegetation regions of the world

- Tundra and ice
- Boreal forest
- Temperate deciduous forest, subtropical evergreen forest
- Temperate grasslands
- Desert, semidesert
- Tropical deciduous forest and savanna
- Tropical rain forest
- Alpine tundra

cies composition, life-form, or other attributes of interest. As we shall see in Chapters 11 and 12, these statistical methods are extremely useful for detecting quantitative patterns of floral and faunal resemblance that may suggest the influence of historical geological events, evolution, or contemporary ecological conditions.

Beginning with the pioneering classifications of Schouw (1823) and other early phytogeographers, continuing with Merriam's (1894) life zones (see Figure 2.12) and Herbertson's (1905) natural regions, and extending to the contemporary concept of biomes and ecoregions, ecologists and biogeographers have almost without exception classified terrestrial communities on the basis of the structure (or physiognomy) of the vegetation (**Figure 5.11**). Implicit in all of these classifications is the recognition that life-forms of individual plants, and the resultant three-dimensional architecture of plant communities, reflect the predominant influence of climate and soil on the kinds of plants that occur in a region. Some authors, such as Holdridge (1947) and Dansereau (1957), have attempted to depict these relationships more quantitatively, showing fairly tight relationships between ranges of climatic variables (such as temperature and precipitation) and specific vegetation types (**Figure 5.12**) (Whittaker 1975; see also Leith 1956). Similar climatic regimes do tend to support structurally and functionally similar vegetation in disjunct areas throughout the world. Often these similarities result from convergence—that is, taxonomically unrelated plant species in geographically isolated regions having evolved similar forms and similar ecological roles under the influence of similar selective pressures (see the discussion on convergence of geographically isolated communities in Chapter 10).

There are almost as many different classifications of vegetation types as there are textbooks in ecology and phytogeography. In general, most biogeographers recognize six major forms of terrestrial vegetation:

1. **forest**: a tree-dominated assemblage with a fairly continuous canopy;

2. **woodland**: a tree-dominated assemblage in which individuals are widely spaced, often with grassy areas or low undergrowth between them;

3. **shrubland**: a fairly continuous layer of shrubs, up to several meters high;

4. **grassland**: an assemblage in which grasses and forbs predominate;

5. **scrub**: a mostly shrubby assemblage in which individuals are discrete or widely spaced;

6. **desert**: an assemblage with very sparse plant cover in which most of the ground is bare.

These principal vegetation forms are further classified into a variable number of biomes. In the following sections we describe 12 common terrestrial biomes, whose geo-

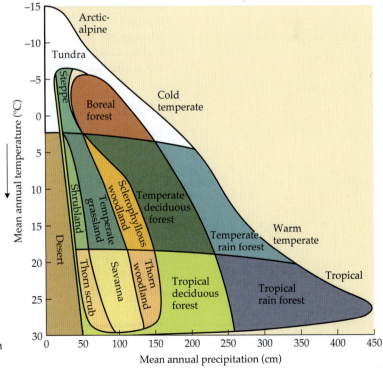

FIGURE 5.12 A climograph, which is a simple diagram quantifying some aspects of the relationships between climate and vegetation types. (After Whittaker 1975.)

TABLE 5.1 *Relationship between Climatic Zone, Soil Type, and Vegetation Communities*

Zonal soil type[a]	Zonal vegetation
Latisols (Oxisols)	Evergreen tropical rain forest (selva)
Latisols (Oxisols)	Tropical deciduous forest or savanna
Chestnut, brown soils, and sierozem (Mollisols, aridisols)	Shortgrass
Desert (Aridisols)	Shrubs or sparse grasses
Mediterranean brown earth	Sclerophyllous woodlands
Red and yellow podzolic (Ultisols)	Coniferous and mixed coniferous–deciduous forest
Brown forest and gray-brown podzolic (Alfisols)	Coniferous forest
Gray-brown podzolic (Alfisols)	Deciduous and mixed coniferous–deciduous forest
Podzolic (Spodosols and associated histosols)	Boreal forest
Tundra humus soils with solifluction (Entisols, inceptisols, and associated histosols)	Tundra (treeless)

Source: Bailey (2009).

[a]Names in parentheses are soil taxonomy orders (USDA Soil Conservation Service 1975).

graphic distributions are mapped in Figure 5.11. Note that the occurrence of these biomes corresponds approximately to the distribution of climatic zones (**Table 5.1**; see Figures 3.8 and 3.11). These latitudinal and elevational patterns reflect the fact that vegetation is highly dependent not only on local climate and underlying soil, but also on the influence of regional climate and topography on soil formation (see Chapter 3). We now consider characteristics of the principal biomes, and then we will conclude this chapter with a global comparison of their salient features and a brief overview of recent advances in ecosystem geography.

Tropical rain forest

Tropical rain forests (**Figure 5.13**) are the richest and most productive of the Earth's terrestrial biomes, covering just 6 percent of the Earth's surface but harboring about 50 percent of its species. Hundreds of tree species may occur in just a few hectares of tropical rain forest, and here one also finds the world's highest diversity of arboreal insects and other invertebrates. The diversity of terrestrial and flying vertebrates is no less impressive, setting the stage for an incredible complexity of biotic interactions.

Tropical rain forests are found at low elevations along tropical latitudes (chiefly 10° N to 10° S) where rainfall is abundant (over 180 cm annually; **Figure 5.14**). Although most tropical rain forests receive some precipitation throughout the year, rainfall tends to be seasonal—marking the periods when the Tropical Convergence Zone (see Figure 3.4) passes over these regions. Temperatures, on the other hand, are nonseasonal—varying little throughout the year and typically ranging between 18° and 32° C. The dominant plants are large, rapidly growing evergreen trees that form a closed canopy at 30 to 50 m. Although the trees are taxonomically diverse, their architecture is often convergent, featuring buttressed bases and smooth, straight trunks, though

FIGURE 5.13 Tropical rain forests often include buttressed trees and lush growth of epiphytes (inset) including orchids, ferns and bromeliads. (Courtesy of Andrew D. Sinauer.)

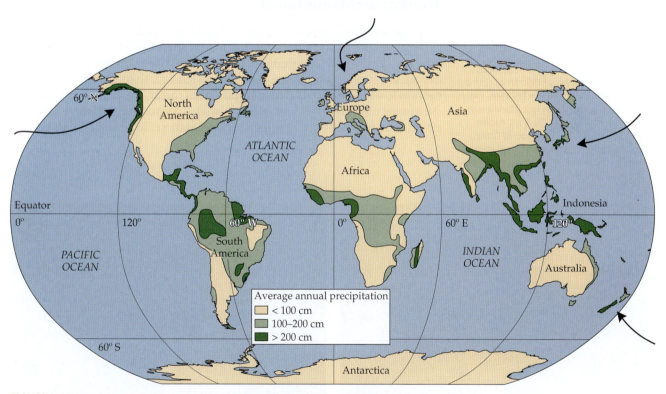

FIGURE 5.14 Global distribution of tropical and temperate rain forests. Temperate rain forests (indicated by arrows) are limited to regions along the coasts of continents and large islands at the midlatitudes, where prevailing winds blow inland, bringing moisture-laden air which, after rising and cooling over coastal mountains, produces heavy precipitation (see Figure 3.7). Tropical rain forests are much more broadly distributed between 10° N and 10° S latitudes, where the principal sources of precipitation are those associated with the Tropical Convergence Zone (see Figure 3.4).

the height and shape of the crowns can be highly variable. The evergreen leaves also tend to be convergent in form: robust and broad with smooth edges. There may be several levels of trees below the uppermost canopy, and palms and other distinctive plants typically occur in the understory. The lush and complex structure of these forests, along with their relatively high humidity, enables dense growths of dependent plants, including a rich diversity of **lianas** (woody vines) and **epiphytes** (orchids, ferns, and in the New World, bromeliads), and **epiphylls** (thin layers of mosses, lichens, and algae that cover many leaves of the dominant trees). Very little light penetrates the dense, multilayered canopy to reach the forest floor, which is usually surprisingly open and devoid of vegetation. Annual plants are conspicuously absent.

Lowland tropical rain forests are the most diverse and productive of the major terrestrial biomes. As a result of high temperatures and high humidity, decomposition of dead organic matter occurs so rapidly that little litter accumulates on the forest floor or in the soils. Many trees have adapted to this environment by developing extensive horizontal root mats—mostly within the upper 20 cm of soil—to capture nutrients released when detritus decomposes. Mycorrhizal fungi, which facilitate the uptake of nutrients, are also closely associated with these root mats. These adaptations, combined with the rapid leaching rates characteristic of tropical soils, are responsible for the paradox of tropical rain forests: One of the world's most productive systems grows on some of the world's poorest soils (Richards 1996).

Tropical deciduous forest

Tropical deciduous forests (**Figure 5.15**) usually occur in hot lowlands outside the equatorial zone (between 10° and 30° latitude) where rainfall is more seasonal (again, tied to migration of the Tropical Convergence Zone), and the dry season more pronounced and more extensive, than in regions of tropical rain forest. Compared with a tropical rain forest, the canopy of a tropical deciduous forest is lower and more open and, because more light reaches the ground, there is often more understory vegetation. To conserve water, many

FIGURE 5.15 (A) Tropical deciduous forests include forest of Pijio trees (*Cavanillesia platanifolia*) of Cerro Blanco, west coast of Ecuador, which lose their leaves during the dry season. (B) A monsoon forest dominated by bamboo in Guilin, China. (A © Pete Oxford/Naturepl.com; B © Haibo Bi/istock.)

(A)

(B)

of the trees and understory plants shed their leaves during the long dry season, although much flowering and fruit maturation may occur at this time.

The dominant vegetation is often called **rain-green forest** because forest trees leaf out during the first heavy rains following the dry season. The most luxuriant form is the **monsoon forest** (**Figure 5.15B**), the layperson's "jungle." Monsoon forest, especially well developed in southern Asia, has many large leaves and dense undergrowth rich in bamboos. These areas are frequently drenched with torrential rainstorms, and some are among the rainiest, but most seasonal, habitats in the world.

Thorn woodland

Tropical and subtropical thorn woodlands (**Figure 5.16**) are low arborescent vegetation types that grow in hot semiarid lowlands. Dominant plants are small spiny or thorny shrubs and trees. Members of the genus *Acacia* and other legumes (Fabaceae) are common in these biomes on all continents. Succulents, such as cacti (Cactaceae) in the New World and convergent forms of the genus *Euphorbia* (Euphorbiaceae) in Africa, are often abundant. Most

(A)

(B)

FIGURE 5.16 (A) An acacia thorn woodland and (B) a thorn scrub of South Africa. (A © Maximilian Weinzierl/Alamy; B courtesy of Andrew D. Sinauer.)

plants lose their leaves during the prolonged dry season, but during the rainy season, trees leaf out and a dense herbaceous understory develops.

Thorn woodlands are often found on drier sites adjacent to tropical deciduous forests. As the climate becomes even drier along a gradient, thorn woodlands give way to **thorn scrub**. A minimum of 30 cm of annual rainfall usually is necessary to establish a thorn scrub, and there is usually a six-month dry season with virtually no rainfall.

Tropical savanna

Tropical savannas (**Figure 5.17**) are biomes dominated by a nearly continuous layer of xerophytic perennial grasses and sedges scattered among fire-resistant trees or shrubs. Savannas usually occur at low to intermediate elevations along tropical latitudes (primarily between 25° N and 25° S). Savannas are characterized by marked seasonality of precipitation, with one or two rainy seasons followed by intense droughts. These weather patterns are largely driven by seasonal shifts in the Tropical Convergence Zone, the zone of most intense solar radiation and convergence of trade winds from the north and south (see Figure 3.4). As the sun shifts between tropical latitudes, the Tropical Convergence Zone shifts as well, passing twice over the Equator, but just once over higher latitudes of the tropics. As a result, equatorial savannas experience two rainy seasons, while those near the limits of the tropics experience a single, longer rainy season. Thus, annual rainfall varies from 30 to 160 cm. Most savannas, however, are strongly influenced by three common factors: seasonally intense precipitation, fire during the dry season, and migratory or seasonal grazing. These dynamic forces combine to make the savanna one of the most spatially and temporally heterogeneous biomes on Earth.

The most extensive savannas are found in intertropical Africa (see Figure 5.11, where they support the most abundant and diverse community of large grazing mammals in the world, as well as a variety of large carnivorous mammals. While easily overlooked, the diversity of smaller herbivores and carnivores living on savannas is no less impressive—typically many times that of their larger counterparts. Savannas have also played an important role in the development of human civilizations. Native peoples inhabiting the savannas of Africa and other continents followed the natural migrations of ungulates,

FIGURE 5.17 Tropical savannas are characterized by sparse trees, and high productivity of grasses and grazing mammals. (© Adrian Assalve/istock.)

which they depended on for food and clothing. Eventually these early "pastoralists" domesticated some species, such as cattle, horses, donkeys, and camels, and moved their herds along with the natural grazers to exploit seasonal shifts in the productivity of the savanna's native communities.

Desert

Hot deserts and semideserts (**Figure 5.18**) occur around the world at low to intermediate elevations, especially within the belts of dry climate from 30° to 40° N and S (the horse latitudes) between the humid tropics and the mesic temperate biomes. Remember from Chapter 3 (see Figure 3.4) that it is at these latitudes that air descends from the upper atmosphere, undergoing adiabatic warming and drying, thus causing these regions to be some of the driest environments on Earth. Rainfall is not only scanty (often less than 25 cm per year) and seasonal, but also highly unpredictable. The evaporative potential of hot desert climates is so strong that most plants have acquired special adaptations that enable them to take up, store, and prevent the loss of water.

The key feature common to deserts is that the amount of rainfall is far less than the evaporative potential, besides being highly unpredictable within, as

(A)

FIGURE 5.18 (A) Hot deserts include a variety of biomes and habitats, including the Sonoran Desert with its diversity of plants such as the large, columnar saguaro cactus (*Carnegiea gigantean*; see Figure 4.17) that are tolerant of drought, but not prolonged periods of freezing temperatures. (B) The sand-dominated ecosystems such as the Sahara desert. (C) The sparsely vegetated and baked soils of the Australian outback. (A © Anton Foltin/Shutterstock; B © Jacques Croizer/istock; C © Sander van Sinttruye/Shutterstock.)

(C)

(B)

well as among, years. Even regions with well over 30 cm of average annual precipitation may be dominated by desert vegetation if located in relatively hot regions with intense evaporation. Some extremely arid regions may not experience any rainfall for several years in a row, and therefore they may have little or no perennial vegetation. In less arid regions, the dominant vegetation consists of widely scattered low shrubs, sometimes interspersed with succulents—cacti, yuccas, agaves, and euphorbs (plants of the genus *Euphorbia*). Where shrubs predominate, the vegetation is usually called desert scrub.

Desert plants possess a variety of adaptations to withstand long, often multiyear periods of drought and to capitalize on the short, unpredictable, but sometimes heavy, rains. Following these brief periods of rain, ephemeral forbs and grasses may grow rapidly to carpet the normally bare ground. Many plants, especially succulents, are able to swell and store water, and they often possess extensive shallow root systems, which act like inverted umbrellas to capture rainfall before it penetrates the soil. Animals also exhibit a variety of adaptations to arid environments, including an ability of many species to derive all the water they need from seeds and other foods, or to remain seasonally or diurnally inactive until environmental conditions become more favorable. Some large desert mammals are able to store substantial amounts of water in their tissues, while birds and bats can avoid the most stressful seasons by practicing migration.

Despite their hardy appearance, desert ecosystems are quite fragile. They show very poor resilience and, once disturbed, may take centuries to recover. On the other hand, overgrazing by livestock and diversion of water for agriculture and other uses have converted otherwise more mesic systems into deserts, albeit atypical ones.

Sclerophyllous woodland

Sclerophyllous woodlands and chaparral (**Figure 5.19**) occur in mild temperate climates with moderate winter precipitation but long, usually hot, dry summers. Sclerophyllous woodlands may also occur in regions with moderate precipitation, but whose sandy soils have little water-holding capacity. This biome includes a broad variety of xeric woodlands ranging from piñon–juniper woodlands and pine barrens to sandhill pine woodlands, sandpine scrub, and pine flatwoods. The dominant plants have sclerophyllous (hard, tough, evergreen) leaves. Sclerophyllous woodlands can be tall, open forests

FIGURE 5.19 Sclerophyllous woodlands include (A) the chaparral of Mendocino County, California, and (B) the fynbos of Cape of Good Hope, South Africa. (A © Alan Tobey/istock; B © Bob Gibbons/ OSF/Photolibrary.com.)

(A)

(B)

that receive over 100 cm of annual rainfall, like the eucalypt woodlands of southwestern Australia, or shorter woodlands that experience less rainfall, like the oak and conifer woodlands of western North America (with evergreen *Quercus*, *Juniperus*, and *Pinus* species).

Those areas that receive less than 60 cm of rainfall per year tend to have low, shrubby vegetation. Sclerophyllous scrublands—called **chaparral**, **matorral**, **maquis**, **fynbos**, or **macchia**—are characteristic of Mediterranean climates (see Chapter 3). Much of their land surface is covered with a dense and almost impenetrable mass of evergreen vegetation only a few meters high. Fires frequently sweep through these hot and dry habitats, burning off the aboveground biomass and apparently playing a major role in preventing the establishment of trees. Shrubs then resprout from their root crowns to reestablish the vegetation.

Subtropical evergreen forest

Subtropical evergreen forests (**Figure 5.20**), some of which have been called oak–laurel forests or montane forests, are common in subtropical mountains at intermediate elevations. These broad-leaved forests cover extensive areas of China and Japan, disjunct areas in the Southern Hemisphere, and much of the southeastern United States. These areas may receive as much as 150 cm of annual rainfall, evenly distributed throughout the year, but subtropical evergreen forests cannot occur where the mean annual temperature is much below 13° C or where severe frosts occur (Wolfe 1979).

FIGURE 5.20   Subtropical evergreen forest of South Island, New Zealand. (© Bob Gibbons/Alamy.)

FIGURE 5.21   A temperate deciduous forest in Maine, USA, during autumn. (© Ken Canning/istock.)

Most of the dominant species consist of dicotyledons with broad, sclerophyllous evergreen leaves, such as laurels (Lauraceae), oaks (*Quercus*, Fagaceae), and magnolias (Magnoliaceae) in the Northern Hemisphere and southern beeches (*Nothofagus*, Nothofagaceae) in the Southern Hemisphere. The canopy is not usually well stratified, and understory plants, especially mosses, may be lush and diverse. A number of temperate broad-leaved deciduous trees occur in these forests, and as the climate becomes colder, broad-leaved evergreens are gradually replaced by deciduous trees or conifers.

Temperate deciduous forest

Temperate deciduous forests (**Figure 5.21**) grow throughout temperate latitudes almost wherever there is sufficient water during the summer growing season to support large trees. They are also called summer-green deciduous forests because they have a definite annual rhythm—the trees are dormant and leafless in the cold and snowy winter, leaf out in the spring, but shed their leaves in the fall as precipitation and temperatures decline. Temperate broad-leaved deciduous forests are extremely variable in their structure and composition across eastern North America, Western Europe, and parts of eastern Asia. In otherwise arid southwestern North America, similar vegetation occurs along permanent watercourses (riparian deciduous woodland). The height and density of the canopy and the importance and composition of the understory vary greatly depending on local climate, soil type, and the frequency of fires. The diversity and coverage of understory plants can be quite high, especially during spring before the trees leaf out. As a result of their extensive accumulation of organic matter and the high capacity of their soils to hold water, temperate deciduous rain forests are much less prone to fire than many other biomes.

In many parts of the Northern Hemisphere, temperate deciduous forests are located next to other arborescent communities, especially temperate evergreen forests, sclerophyllous woodlands, and coniferous forests. Consequently, many phytogeographers recognize a long list of hybrid associations between these communities, such as mixed evergreen–deciduous forest. The trees of temperate forest climax communities grow slowly, and most forests have been significantly affected by logging over the last few centuries.

(A)

(B)

FIGURE 5.22 Temperate rain forest of the Olympic Peninsula, Washington State. (A) Riparian corridor of Olympic National Forest. (B) The world's largest sitka spruce tree, which has a circumference of 17.6 m (58 ft) and is approximately 1000 years old. (Photographs by Mark V. Lomolino.)

Temperate rain forest

Temperate rain forest (**Figure 5.22**) is an uncommon but interesting biome found along the western coasts of continents where precipitation exceeds 150 cm per year and falls during at least 10 months out of each year (see regions identified by arrows in Figure 5.14). Cool temperatures predominate year-round, but these regions are always above freezing, and they experience much fog and high humidity, so there is enough moisture to permit the growth of large evergreen trees. A moderately dry season during the summer inhibits the growth and dominance of deciduous trees. Cool temperatures account for the absence of any true tropical plants, such as palms, and the relatively low number of tree species. Temperate rain forests do not have many kinds of lianas, but the epiphyte diversity is high, consisting of mosses, lichens, epiphyllous fungi, and some ferns. These cool and moist forests are largely uninfluenced by fire. Therefore, while growth rates are relatively slow, temperate rain forests are renowned for possessing some of the world's oldest and largest trees. Their canopies tend to be closed, with many dead standing trees, or "snags," while the humid understory is often covered with lush mats of mosses, ferns, and lichens. Decomposition rates tend to be slow, again due to the relatively low temperatures. Combined with slow growth rates of the dominant tree species, this means that forest development requires many centuries of relatively stable climatic conditions.

The best examples of temperate rain forest include those of New Zealand, Chile, and the Pacific Northwest region of North America. Along these coastal regions, maritime air masses flow inland and then ascend the coastal mountain ranges, cooling and declining in their capacity to hold moisture, and then releasing some 100 to 300 cm of water per annum in the form of rain and fog. Although highly varied among regions, these forests are dominated by immense but slowly growing trees, including *Agathis*, *Eucalyptus*, *Nothofagus*, and *Podocarpus* in the Southern Hemisphere, and *Picea* (spruce) and *Abies* and *Pseudotsuga* (Douglas firs) in the Northern Hemisphere. The understories of these forests are typically covered with lush and diverse mats of mosses, epiphytes, fungi, and ferns.

Temperate grassland

Temperate grasslands (**Figure 5.23**) are situated both geographically and climatically between deserts and temperate forests. While broadly distributed between 30° and 60° latitude, temperate grasslands are most extensive in the interior plains of the Northern Hemisphere. Their location, isolated from the moderating effects of the oceans, causes interior climates to be markedly seasonal, with substantial annual variation in both temperature and rainfall. The vegetation is confined to a single stratum, which is dominated by grasses, sedges, and other herbaceous plants. Vegetation height tends to vary directly with precipitation, both factors decreasing from tall grasslands (veldt of South Africa, *puszta* of Hungary, tallgrass prairie of North America, or pampas of Argentina and Uruguay), to short grasslands (shortgrass prairie or steppes) in colder latitudes, to desert grasslands adjacent to warm arid regions. Even in relatively moist tallgrass prairies, drought, fire, and heavy grazing pressures combine to prevent the establishment of woody plants while favoring the dominance of herbaceous species.

The dominant grasses are perennials with **basal meristems** (growth tissue located in the soil), which make them tolerant to defoliation; indeed, in these species, vegetative growth is stimulated by fire and grazing. Although typically dominated by just a few grass species, temperate grasslands actually harbor a surprising diversity of both plants and animals. Grasses, which may account for over 90 percent of the biomass, typically constitute less than 25 percent of the plant species in grassland ecosystems. Despite the variable and sometimes luxuriant layer of vegetation above the surface, most grassland biomass lies belowground in the extensive root systems of perennial plants. The ratio of belowground to aboveground biomass varies with annual precipitation, ranging from less than 2:1 for arid grasslands to 13:1 for tallgrass prairies (Weigert and Owen 1971). Accordingly, grassland soils

FIGURE 5.23 Bison roaming temperate grasslands of the Great Plains, North America. (© Yin Yang/istock.)

tend to have high accumulations of organic material, which in turn support a rich diversity of soil invertebrates and microbial decomposers. Many vertebrate species, especially some rodents, are completely **fossorial** (burrow dwelling), while others are cursorial grazers (those adapted to run rather than burrow to avoid predation), consuming as much as two-thirds of the aboveground production.

Because of their deep, fertile soils, many temperate grasslands have been converted to agricultural uses. As a result of cultivation or desertification, natural grasslands, which once covered approximately 40 percent of the Earth's surface, have now been reduced to about half of their presettlement range.

Boreal forest

Boreal forests (also called taiga or "swamp forest") occur in a broad band across northern North America, Europe, and Asia in regions with cool temperatures and adequate moisture (**Figure 5.24**). As we saw in Chapter 4 (see Figure 4.19), this biome also extends well southward into temperate and even subtropical regions where mountain slopes provide the requisite cool temperatures for these forests (see Figure 5.1). For example, boreal forests extend down the cordilleras of western North America all the way to southern Mexico; indeed, much of highland Mexico is covered by boreal forest.

Boreal forests, although often thick, are typically dominated by just a few species of coniferous trees, such as spruce (*Picea*), fir (*Abies*), and larch (*Larix*). Because of cool temperatures and waterlogged soils, decomposition rates are relatively slow, resulting in an accumulation of peat and humic acids, which render many soil nutrients unavailable for plant growth. Acidic soils combine with relatively cool temperatures to limit diversity and productivity of the few tree species able to survive these stressful conditions. Often the canopy is not dense, and a well-developed understory of acid-tolerant shrubs, mosses, and lichens may be present in the most mesic sites.

FIGURE 5.24　The boreal coniferous forest (taiga) of Koyukuk National Wildlife Refuge, Alaska. (Photograph courtesy of U.S. Fish and Wildlife Service.)

Tundra

Tundra (**Figure 5.25**) is a treeless biome; *arctic tundra* is found between the boreal forest and the polar ice cap, and *alpine tundra* is found at high elevations on tall mountains. Even more than the boreal forest, tundra is characterized by stressful environmental conditions. Temperatures in the alpine tundra remain below freezing for at least seven months of the year, precipitation is often less than in many hot deserts, and tundra soils tend to be even more nutrient-limited than those of the boreal forest (because of the cool temperatures and, concomitantly, extremely slow decomposition rates). Soils are also saturated with water because of slow evaporation rates and the presence of permafrost (a frozen, impermeable layer of soil that lies at a depth of a meter or less in the summer). Consequently, primary productivity, biomass, and diversity of the tundra are lower than those of almost all other terrestrial biomes.

FIGURE 5.25   (A) Arctic tundra, Denali National Park, Alaska. (B) Alpine tundra, Colorado. (A courtesy of Aaron Collins/USFWS; B © Missing35mm/istock.)

Arctic (including Antarctic) tundra and alpine tundra are all covered with a single dense stratum of vegetation, usually only a few centimeters or decimeters in height. The dominant plants tend to be dwarf perennial shrubs, sedges, grasses, and mosses, and lichens are also an important element of the ground cover. Despite generally low productivity during the rest of the year, tundra plants exhibit bursts of productivity during the short growing season. The lush vegetation is then heavily grazed by migratory or nomadic ungulates, including caribou (*Rangifer tarandus*), muskoxen (*Ovibos moschatus*), and Dall sheep (*Ovis dalli*). Other important herbivores of the tundra include geese (*Branta* spp.), ptarmigan (*Lagopus* spp.), and small mammals (voles and lemmings), whose populations fluctuate dramatically in a complex interaction with the plant community and predators who also migrate during the short growing season to exploit the intense, seasonal pulse of prey populations.

Found above the timberline on mountaintops in the equatorial zone is **tropical alpine scrubland**, with vegetation taller than that of the arctic tundra. The dominant plants are tussock grasses and bizarre, erect rosette perennials with thick stems. These vegetation types are found at elevations above 3300 m in the Andes (**páramo**) in South America, on the upper slopes of the highest mountains in East Africa, and on mountaintops in New Guinea and some other high oceanic islands.

Just like its hot desert counterpart, the tundra is a fragile system. Oil exploration and other human activities are major threats to this delicate environment. Once these activities disturb the permafrost, natural communities may take many decades to recover.

Aquatic Communities

Marine and freshwater ecologists and biogeographers do not classify aquatic communities into categories analogous to those used for terrestrial biomes. For one thing, the relatively simple arrangement of sessile plants growing on a land surface is not comparable to the three-dimensional complexity of the water column. Terrestrial habitats are essentially two-dimensional, in the sense that organisms do not remain permanently suspended above the soil surface. To the extent that a third dimension is present on land, it is formed by the vertical growth of sessile plants and by arboreal and flying animals. The three-dimensional organization of aquatic communities is very different. On one hand, a well-developed structure of attached, vertically growing organisms is absent from most aquatic habitats, although there are obvious exceptions, such as kelp forests, coral reefs, and the submersed vegetation of lakeshores. On the other hand, many aquatic organisms spend much or all of their lives suspended in the third dimension, either drifting passively or swimming actively in the water column.

The physical factors that vary in time and space to affect the abundance and distribution of aquatic organisms are also quite different from those that determine both the terrestrial climate and the organization of terrestrial communities. Because of the high specific heat of water, temperature varies less on a daily, seasonal, and latitudinal basis in aquatic environments than in terrestrial ones. Variations in pressure, salinity, and light are especially important in aquatic systems. Tidal cycles, which fluctuate bimonthly with the phases of the moon, are more important to many marine shore communities than are daily or seasonal cycles (see Figure 3.15).

Oceanographers, limnologists, and aquatic ecologists have developed classification systems for marine and freshwater communities. Like the division of terrestrial communities into biomes, these systems use arbitrary groupings that break up a continuous spectrum of biological associations into a number of convenient categories. Salinity, depth, water movement, and nature of the substrate are physical characteristics that most influence the abundance and distribution of aquatic organisms, and are thus most often used in classifying aquatic communities.

The first major division of aquatic systems is into marine and freshwater communities. On biological as well as geographic grounds, the Earth's bodies of water can be divided into the oceans, which form a huge interconnected water mass covering over 70 percent of the Earth's surface, and the comparatively tiny, highly fragmented lakes, ponds, rivers, and streams, which together cover only a small fraction of the remaining surface of the Earth. The oceans and these various bodies of freshwater differ greatly in salinity. Salt concentration of the oceans varies slightly around 35 parts per thousand, whereas even the hardest freshwaters have salinities of less than 0.5 parts per thousand. As stressed in Chapter 4, this difference in salinity can have dramatic effects on species distributions. Only a tiny fraction of aquatic organisms can live in both salt and fresh water, so salinity effectively divides aquatic communities into two nonoverlapping groups of species. Because of this strong dichotomy, marine and freshwater ecosystems have largely been studied independently by different groups of ecologists (**oceanographers** and **limnologists**, respectively), who have developed different classifications of communities. Accordingly, we describe characteristics of each of the principal divisions of the aquatic realm separately, below.

(A)

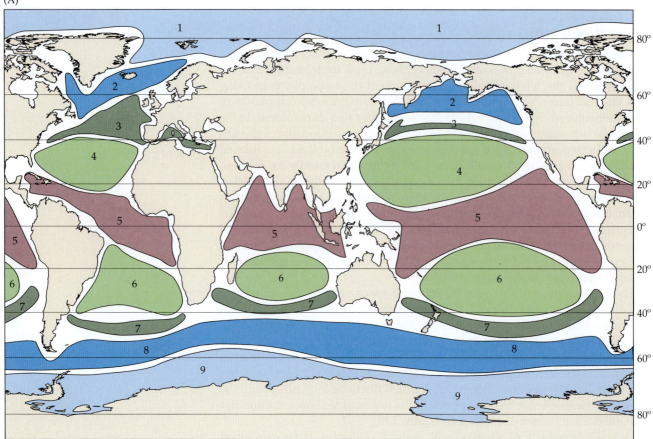

(B)

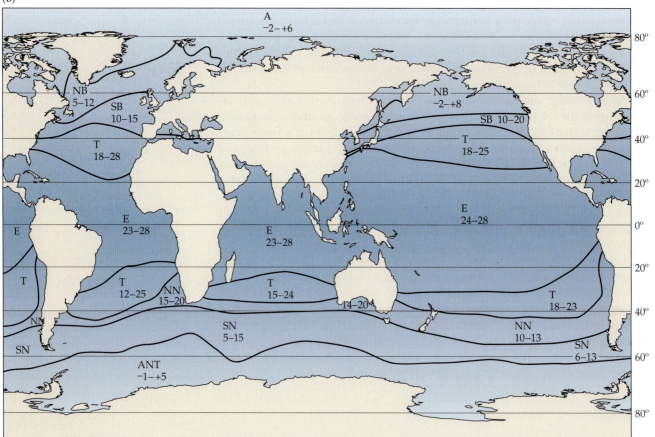

◀ **FIGURE 5.26** Early schemes of biogeographic regions of the marine realm reflect the overriding importance of climate, especially water temperatures and, indirectly, oceanic currents. (A) Biogeographic regions: 1, Arctic; 2, subarctic; 3, northern temperate; 4, northern subtropical; 5, tropical; 6, southern subtropical; 7, southern temperate; 8, subantarctic; and 9, Antarctic. (B) Climatic regions of the oceans based on mean monthly water temperatures (°C): A, Arctic; NB, northern boreal; SB, southern boreal; T, tropical; E, equatorial; NN, northern notal; SN, southern notal; and ANT, Antarctic. (After Rass 1986; see Figure 5.32B for more recent scheme of the world's oceanic ecoregions.)

Marine communities

Compared with terrestrial and freshwater environments, the ocean is immense and essentially continuous, but it is far from homogeneous in terms of its environments and biotic communities. It also comprises the lion's share of the world's ecosystems, with marine waters dominating the Earth's surface and providing a three-dimensional habitat across its entire 1.37 billion cubic kilometer volume (Duxbury 2000).

Organisms live everywhere in the ocean, but the abundances and kinds of life vary greatly depending on the local physical environment. Perhaps the most important features are light, temperature, pressure, and substrate. The current system of biogeographic regions of the marine realm was not generally accepted until the early 1970s (**Figure 5.26A**), roughly a century after Sclater (1858) and Wallace (1876) proposed the system of terrestrial biogeographic regions that we continue to use today. This marine system is primarily based on water temperature; therefore, biogeographic regions of the marine realm encompass broad latitudinal zones and tend to be elliptical because of circular oceanic currents (compare Figure 5.26A and **Figure 5.26B**; see also the map of oceanic surface water currents in Figure 3.5).

Within each of these regions, the ocean can be divided into two vertical zones— the **photic zone** (sometimes euphotic zone, Greek for "well lit") and the **aphotic zone**— based on penetration of sunlight (**Figure 5.27**; see Chapter 3). Because sunlight is gradually absorbed by water with increasing depth, the boundary between these zones is somewhat arbitrary, but it is usually set where light penetration is reduced to between 1 and 10 percent of incident sunlight. The depth of the photic zone increases from coastal waters, where light rarely penetrates more than 30 m because of organisms and inanimate particles suspended in the water column, to the open ocean, where it may extend to a depth of 100 m or more.

The significance of this zonation by light is, of course, that photosynthesis can occur only in the photic zone. Essentially all of the organic energy that sustains marine life is produced in this shallow surface layer of the ocean. Most organisms in the aphotic zone obtain their energy by consuming or-

FIGURE 5.27 Division of marine communities into major zones. Note that these zones are based primarily on water depth, light levels, and relationships between organisms and substrates.

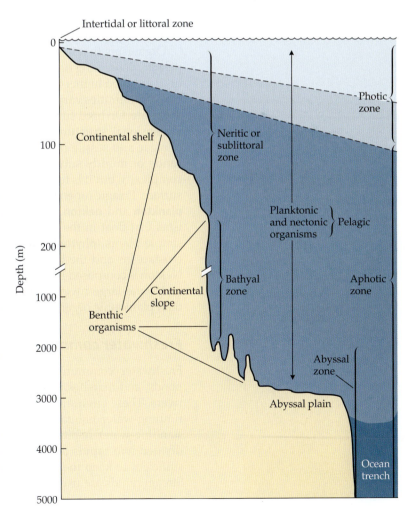

ganic material that is produced in the photic zone and reaches deep water in the forms of feces and dead bodies that sink to the bottom.

In the late 1970s, however, oceanographers discovered entire flourishing communities of organisms, including unique kinds of worms, mussels, and crabs, which do not depend for their energy supply on material from the photic zone. These communities live in highly localized areas such as those on the otherwise barren slopes of the Galápagos rift (eastern tropical Pacific) where submarine hot springs emit hydrogen sulfide. Chemosynthetic bacteria obtain energy by oxidizing hydrogen sulfide, and these unusual autotrophs serve as the base of the food chain for these deepwater, aphotic communities (Jannasch and Wirsen 1980; Karl et al. 1980).

Marine communities are also classified into another set of zones on the basis of **bathymetry** (i.e., depth and configuration of the ocean bottom) (see Figure 5.27). The shallowest zone is the intertidal or **littoral zone**, which occurs on the shore where sea meets land. Although it is inhabited almost exclusively by marine organisms, the intertidal zone is actually an ecotone between land and ocean. Beyond the intertidal zone is the **neritic** or **sublittoral zone**, which encompasses waters of a few meters to about 200 m deep that cover the continental shelves. At the edges of the continental plates are regions of highly varied relief that comprise the **bathyal zone**, with the marine equivalent of mountainsides and canyons, in which the waters rapidly drop away to the great ocean depths. The **abyssal zone**, which constitutes most of the ocean, covers extensive areas in which the water ranges in depth from 2000 to more than 6000 m. Deep ocean waters provide some of the most constant physical environments; they are continually dark, cold (4° C), subject to enormous pressures, and virtually unchanging in chemical composition.

Organisms that inhabit the oceans are often classified as either **benthic** or **pelagic**, depending on whether they are closely associated with the substrate or distributed higher in the water column, respectively (see Figure 5.27). Benthic communities vary greatly in composition depending on the nature of the substrate. On hard substrates, attached benthic organisms often form a three-dimensional structure that varies in complexity from low crusts and turfs of algae and sessile invertebrates to tall "forests" of kelp and coral. On soft sandy or muddy substrates, there is often a comparable three-dimensional complexity, but it is formed by burrowing invertebrates that live beneath the surface. Pelagic (open water) organisms are usually divided into two groups, **plankton** and **nekton**. The former consists of primarily microscopic organisms that float in the water column. The plankton typically includes many simple plants (**phytoplankton**), such as diatoms, and tiny animals or small crustaceans and the larvae of many invertebrates and fishes (**zooplankton**). Nekton is composed of actively swimming animals, including fishes, whales, and some large invertebrates, which usually occupy higher trophic levels than planktonic organisms.

Freshwater communities

Freshwater communities are widely distributed as small, isolated lakes, ponds, and marshes, sometimes connected by long, branching streams and rivers. These environments are usually divided into two categories: **lotic** or running-water habitats, such as springs, streams, and rivers; and **lentic** or standing-water habitats, such as lakes and ponds. Lotic habitats are often divided into rapids (or "riffles") and pools. In the former, water velocity is sufficient to keep the water well oxygenated and the substrate clear of silt. Stream rapids are usually inhabited by organisms that live on the surface of the rocky substrate or swim strongly in the current. Pools are characterized by deep, slowly moving water and silty, often poorly oxygenated, bottoms. Swimming animals are common in stream pools, and many benthic species burrow into the substrate. Although some organic material in streams is manufactured in situ by benthic plants or phytoplankton, most of it is washed in from terrestrial ecosystems of the surrounding watershed.

Lentic habitats are often divided into zones reminiscent of those of the oceans, although somewhat different terms and meanings are applied (**Figure 5.28**). The littoral zone consists of shallow waters where light penetrates to the bottom and rooted aquatic vegetation may be present. Offshore waters are divided into a surface **limnetic zone**, where light penetrates sufficiently for photosynthesis to occur, and a deep **profundal zone** (beyond the depth of effective light penetration). Lakes can be highly productive, supporting extensive food webs based on photosynthesis by attached vegetation in the littoral zone and phytoplankton in the limnetic zone. Productivity is limited largely by the availability of inorganic nutrients, such as phosphorus, which wash in from surrounding watersheds and, in temperate and subarctic lakes, are seasonally replenished from organic material on the lake bottom when thermal stratification disappears and the waters overturn (see Chapter 3). Temperate lakes are often classified as either **eutrophic** or **oligotrophic**. Eutrophic lakes are shallow and highly productive because light penetrates almost to the bottom and vertical circulation of the water column occurs each spring and fall, returning limiting nutrients to the surface and oxygen to the depths. Oligotrophic lakes are characterized by low nutrient input and tend to be deeper than eutrophic lakes. In fact, many oligotrophic lakes are so deep that little or no vertical circulation occurs. As a result, productivity is relatively low, despite the high water clarity of many oligotrophic lakes.

The preceding classification scheme omits a number of freshwater communities. In fact, some classification schemes such as that of the International Ramsar Convention on Wetlands describes some 20 types of natural, freshwater wetlands. **Swamps (marls)** and **marshes (moors)** are two very common types of wetlands that tend to develop on mineral soils and are distinguished

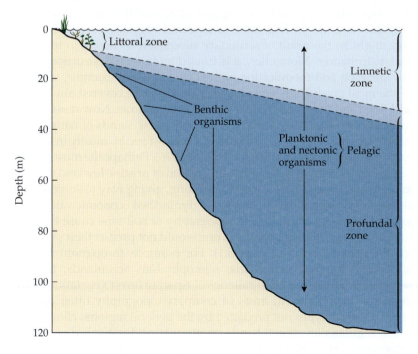

FIGURE 5.28 Division of freshwater lentic habitats into major zones. This classification scheme is similar to that used for marine environments because it is based on variation in light penetration, water depth, and relationships between organisms and substrates; nevertheless, somewhat different terms are used.

primarily by whether their dominant vegetation is woody or herbaceous, respectively. **Peatlands** are freshwater ecosystems that develop in cool temperate areas of the Northern Hemisphere where water drainage is blocked, thus resulting in continuous saturation of soils, which in turns slows decomposition and promotes the accumulation of partially decomposed plant material, called "peat." Two principal types of peatlands include **bogs** (**mires**) and **fens** (**fenlands**). Bogs receive most of their water from precipitation, whereas most of the water flowing into fens is from runoff or groundwater. As a result, fens tend to be less acidic, have higher nutrient levels and support more diverse plant and animal communities. Some other freshwater communities are much rarer, but just as interesting because they represent atypical physical environments that pose special problems for the few organisms that are able to live there. Examples of such harsh environments are hypersaline lakes (undrained lakes, such as Great Salt Lake in North America, the Dead Sea in the Middle East, and the Aral Sea in Central Asia, all of which are much more saline than seawater), underwater caves (which admit no light and contain communities supported entirely by imported organic matter), and hot springs (among the most physically rigorous of all environments).

Another important group of communities occurs in those areas where fresh water, the sea, and the land meet. The **estuaries**, **salt marshes**, and (in tropical regions) **mangrove swamps** that occur at these sites are highly productive ecosystems. They usually contain only a few species that can tolerate the physical rigors of successive exposure to fresh water, seawater, and the terrestrial climate. On the other hand, the sunlight permeating these shallow water ecosystems, the circulation provided by tides, and the input of nutrients from rivers often permits these habitats to support high biomasses and densities of individuals.

A Global Comparison of Biomes and Communities

The biomes described above, and mapped in Figure 5.11, indicate only the general kind of climax vegetation that we would expect to find in a region, based primarily on its climate. However, someone visiting many of the areas on the map might have difficulty finding good stands of the vegetation typical of these biomes and might find some unexpected vegetation types. In some cases, this might be the result of secondary succession occurring in response to natural disturbances. More often, it is caused by human destruction of the original vegetation and modification of the landscape. For example, tallgrass prairie—the most productive temperate grassland—once covered much of Illinois and Iowa and stretched from Saskatchewan to Texas. Now most of this area has been converted to agriculture, and only a few stands of native prairie—totaling less than 5 percent of its original area—remain, mostly in a small number of preserves (see Steinauer and Collins 1996). Perhaps the most significant change now occurring is the rapid destruction of pristine lowland tropical rain forests, whose boundaries are ever shrinking, giving way to successional communities of reduced biotic diversity and diminished economic value.

In some places, local variations in topography or soil type cause types of vegetation to be found in regions where one would not predict their presence based on the general map of Figure 5.11. For example, throughout regions dominated by temperate grasslands, sclerophyllous scrublands, tropical thorn scrub, or deserts—there are galleries of riparian forest vegetation along permanent streams. Similarly, areas of complex topography often contain small areas of biome types not predicted by the general regional climate but arranged in regular, elevational bands on mountainsides. Such diversity of vegetation types contributes greatly to the overall biotic richness of a region

TABLE 5.2 *Net Primary Production and Biomass of Major Kinds of Biomes and Marine Communities*

Biome/Community	Area (10⁶ km²)	Net primary production per unit area (g m⁻² yr⁻¹)	Total net primary production (10⁹ MT yr⁻¹)	Mean biomass per unit area (kg m⁻²)
TERRESTRIAL AND FRESHWATER				
Tropical rain forest	17.0	2000.0	34.00	44.00
Tropical deciduous forest	7.5	1500.0	11.30	36.00
Temperate rain forest	5.0	1300.0	6.40	36.00
Temperate deciduous forest	7.0	1200.0	8.40	30.00
Boreal forest	12.0	800.0	9.50	20.00
Savanna	15.0	700.0	10.40	4.00
Cultivated land	14.0	644.0	9.10	1.10
Woodland and shrubland	8.0	600.0	4.90	6.80
Temperate grassland	9.0	500.0	4.40	1.60
Tundra and alpine meadow	8.0	144.0	1.10	0.67
Desert scrub	18.0	71.0	1.30	0.67
Rock, ice, and sand	24.0	3.3	0.09	0.02
Swamp and marsh	2.0	2500.0	4.90	15.00
Lake and stream	2.5	500.0	1.30	0.02
TOTAL TERRESTRIAL AND FRESHWATER	149.0	720.0	107.09	12.30
MARINE				
Coral reefs and algal beds	0.6	2000.0	1.10	2.00
Estuaries	1.4	1800.0	2.40	1.00
Upwelling zones	0.4	500.0	0.22	0.02
Continental shelf	26.6	360.0	9.60	0.01
Open ocean	332.0	127.0	42.00	0.003
TOTAL MARINE	361.0	153.0	55.32	0.01
WORLD TOTAL	**510.0**	**320.0**	**162.41**	**3.62**

Source: After Whittaker and Likens (1973).

because distributions of many other plants and numerous animal species are strongly influenced by the dominant vegetation.

Given the great diversity of world biomes and communities, it may be instructive to conclude this overview with a global comparison of their salient features (**Table 5.2**). **Net primary productivity** (**NPP**) is a measure of the rate at which solar energy is converted to plant tissue, typically expressed as mass produced per unit of surface area (e.g., g m⁻² yr⁻¹). NPP is one of the most fundamental and important measures of community function, as it represents the energy available to maintain biomass and diversity of almost all forms of life. As we have seen, biomes and communities vary markedly in temperature, precipitation, nutrient availability, and many other factors that influence primary productivity. Consequently, the world's biomes vary markedly in NPP, biomass, and diversity. Tropical rain forests are renowned for their high productivity and have long been considered the most productive of the terrestrial biomes. Recently, however, some ecologists have questioned whether this is generally the case, pointing to the evidence that many rain forests growing on old lateritic soils are strongly nutrient limited and may be unexceptional in their productivity in comparison with temperate forests (see Huston and Wolverton 2009). Moreover, as we see in **Figure 5.29A**, some aquatic systems rival tropical forests in NPP. As we observed in the previous section, productivity tends to be highest in shallow water environments, where high photosynthetic rates are favored by relatively high levels of sunlight and nutrients.

These trends in NPP among biomes are roughly paralleled by trends in biomass: The most productive biomes and communities tend to support the highest density of living tissue (**Figure 5.29B**). Aquatic communities, however, exhibit consistently lower biomass than their terrestrial counterparts. Phytoplankton, the smallest plants, account for approximately 90 percent of

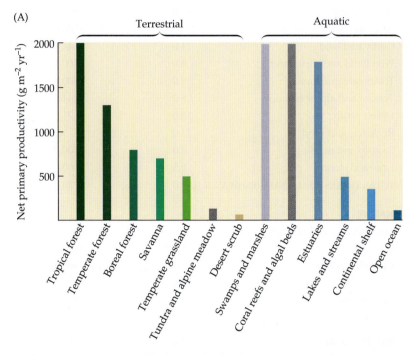

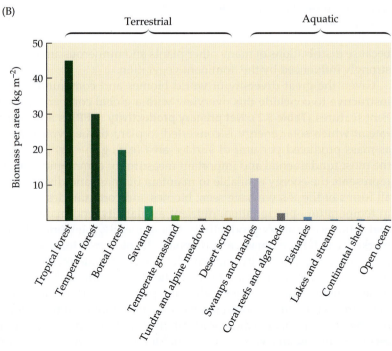

FIGURE 5.29 Comparisons of (A) net primary productivity and (B) biomass among terrestrial and aquatic communities (see Table 5.2). Despite their relatively low biomass, aquatic communities often rival tropical rain forests in productivity on a per area basis.

NPP in aquatic communities. Unlike terrestrial macrophytes, more than a third of whose biomass is photosynthetically inactive tissue, phytoplankton are extremely efficient. Under optimal conditions, solar energy is rapidly converted into phytoplankton tissue and in turn made available to aquatic consumers and decomposers. As mentioned earlier, this high turnover rate, which supports extensive food webs, explains why pyramids of biomass are inverted for some lake and ocean ecosystems (see Figure 5.6E).

We caution that the above estimates of productivity and biomass are averages over space and time. At finer scales, each of these biomes and communities is remarkably heterogeneous. Each is composed of a collection of successional and disturbance stages, all influenced to varying degrees by seasonal changes. Temperate grasslands, meadows, and deserts, for example, can exhibit impressive bursts of productivity during the growing season.

We should also bear in mind that up to now, we have been comparing biomes and communities on a productivity per area basis. Let us now study these systems from a different perspective—a global view. If an astronaut were able to view the Earth through a biologically sensitive lens—one that could distinguish productivity levels—it would look something like the map of **Figure 5.30A**. Both terrestrial and aquatic ecosystems exhibit pronounced latitudinal effects, but other patterns differ markedly between them. On land, NPP tends to be strongly correlated with precipitation and temperature. Thus, except for arid regions along the horse latitudes, terrestrial productivity tends to be highest in tropical and subtropical regions and to decrease as we move toward the poles (but see Huston and Wolverton 2009). In the oceans, however, phytoplankton productivity tends to be most strongly limited by the availability of dissolved nutrients, especially phosphorus and nitrogen. Consequently, hotspots of marine productivity coincide with areas of nutrient input from the discharge of large rivers along continental shelves or upwellings from the nutrient-rich depths of the ocean. As we can see from the map of global productivity (see Figure 5.30A), ocean upwelling tends to occur along the west coasts of continents and at higher southern latitudes.

Reviewing Figure 5.11 makes it quite clear that world distribution of biomes and communities is far from uniform. In fact, the Earth's surface is dominated by open ocean, one of the least productive ecosystems per unit of *surface* area (see Figure 5.29A). Open oceans, however, cover nearly three-fourths of the Earth's surface (see Figure 5.30A). Consequently, when all open oceans are totaled, we find that they account for over one-fourth of the Earth's primary production—a proportion equal to or perhaps slightly exceeding that of all tropical forests (**Figure 5.30B**). On the other hand, some of the most productive communities on a per area basis, such as coral reefs and estuaries, contribute only a minor fraction to Earth's total primary production.

Ecosystem Geography

Primary production, predation, and the great diversity of other processes involving energy flow and nutrient cycling among species, and between species and the abiotic components of their environment, are, of course, not just community properties, but properties of entire ecosystems. **Ecosystem geography** is the branch of science that focuses on how these processes have influenced the distributions of ecosystems. Given the open nature of ecosystems and the need to study them at a broad range of spatial and temporal scales, ecosystem geographers have developed a hierarchical scheme of ecological regions, or ecoregions. This hierarchical, or nested, nature and the focus on processes at a telescoping range of scales, from local sites to landscape mo-

(A)

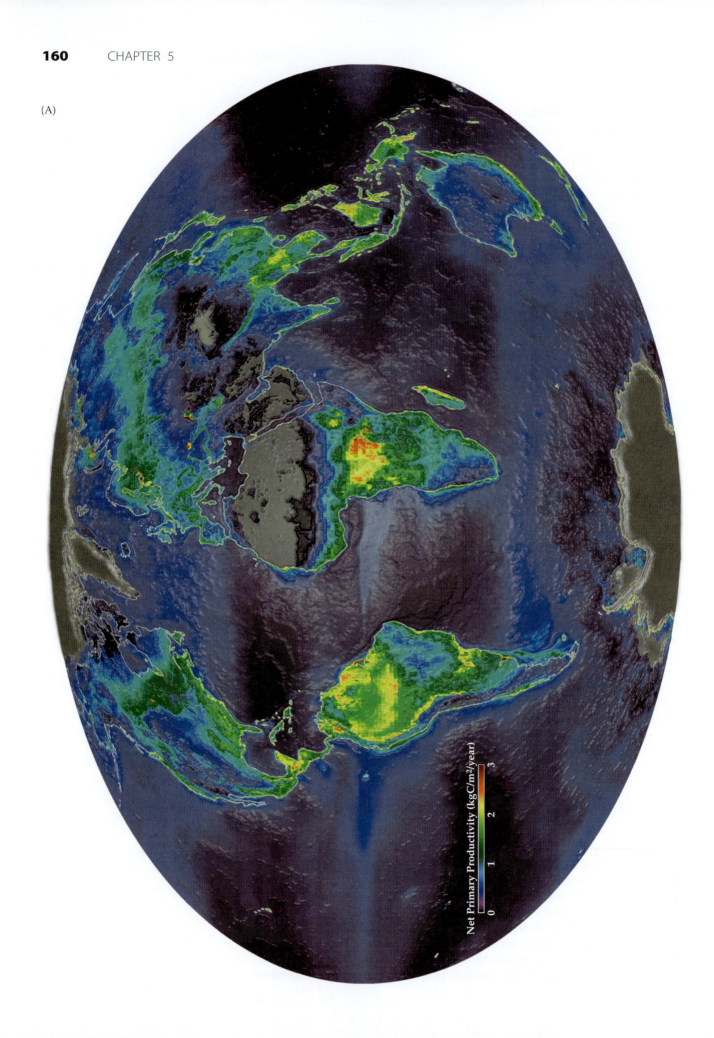

Net Primary Productivity (kgC/m²/year)

0   1   2   3

FIGURE 5.30 (A) Average annual net primary productivity of the Earth during 2002. This map displays carbon metabolism, which is the rate at which plants absorb carbon out of the atmosphere to produce organic matter during photosynthesis. Comparisons among total surface area (B) and total NPP (C) of the world's biomes. Tropical rain forests are so efficient (on a per area basis) at transforming solar energy into plant tissue that, even though they cover just 4 percent of the Earth's surface, they account for roughly one-fourth of the Earth's total primary production. On the other hand, even though open ocean communities are relatively inefficient at fixing solar energy, they rival tropical rain forests in total primary production because they cover nearly three-fourths of the Earth's surface. (A courtesy of NASA's Earth Observatory; B,C data from Whittaker and Likens 1973.)

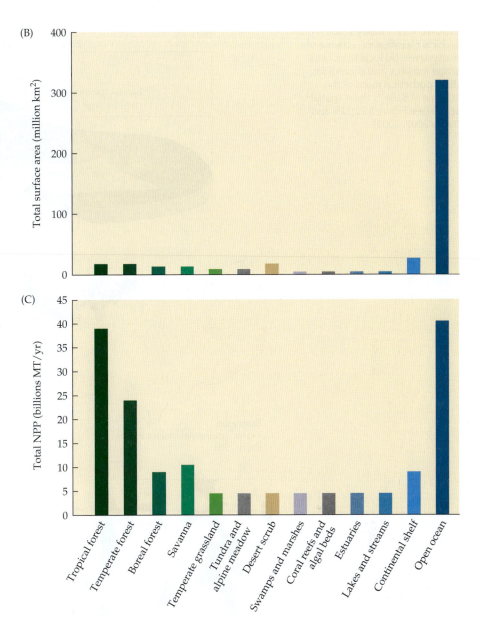

saics and to regional and continental (or ocean basin) scales, are what distinguish the ecoregion approach from traditional schemes based on biomes (**Figure 5.31**). In addition, the ecoregion scheme is truly global, including a scheme of ecological regions for both terrestrial and aquatic realms (**Figure 5.32**). Nested subdivisions of ecoregions (ecoregions forming provinces, provinces within divisions, divisions within domains) each reflect finer-scale differences in climates, which in turn fundamentally influence development of soils and structure and function of ecological communities. In the aquatic realm, knowledge of water currents (surface and vertical currents, which influence water temperature, water chemistry, and salinity) is also used to classify and delineate ecoregions.

Ecoregions, however, are just that—ecological divisions of the Earth based on responses of organisms (on land, especially of plants) to climatic conditions; that is, they are a functionalist solution to the problem of subdividing the world into biological units. Therefore, unlike the system of biogeographic regions first developed by Sclater and Wallace, ecoregions were not devel-

FIGURE 5.31 Bailey's (2009) hierarchical classification scheme of ecosystems: (A) local sites, landscape mosaics, and ecoregions; (B) hypothetical maps of this scheme at Bailey's three spatial scales (see Figure 5.32; see also Bailey 2002, 2009).

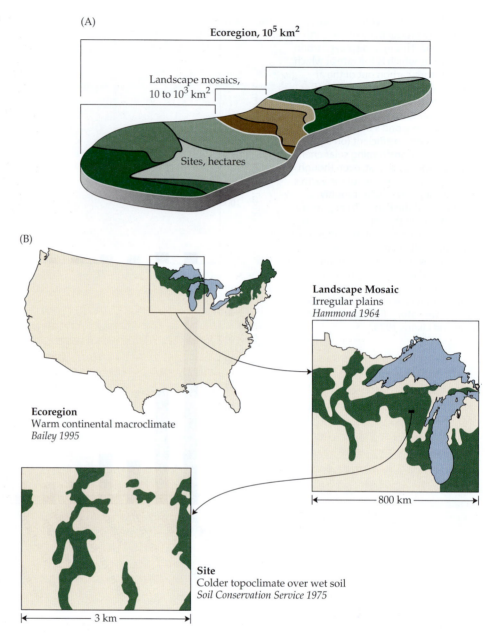

(A)

Ecoregion, 10^5 km^2

Landscape mosaics, 10 to 10^3 km^2

Sites, hectares

(B)

Ecoregion
Warm continental macroclimate
Bailey 1995

Landscape Mosaic
Irregular plains
Hammond 1964

800 km

Site
Colder topoclimate over wet soil
Soil Conservation Service 1975

3 km

oped to reflect the evolutionary history of regions and their biotas. Nonetheless, an increasing number of biogeographers, ecologists, and conservation biologists are coming to rely on the ecoregion approach and its hierarchical maps of the Earth's ecosystems (see Figures 5.31 and 5.32) as valuable tools for understanding and conserving the geography of nature. There is little doubt that the accuracy, resolution, and utility of these maps—and of ecosystem geography, in general—will continue to increase with future advances in satellite imagery, remote sensing, and other means of interpreting spatial variation in climates, oceanic currents, and biotic communities.

FIGURE 5.32 Ecoregions of the terrestrial (A) and marine (B) realms. (From Bailey 2009; ▶ see also Olson et al. 2001, and Spalding et al. 2007 for similar maps of global ecoregions.)

(A)

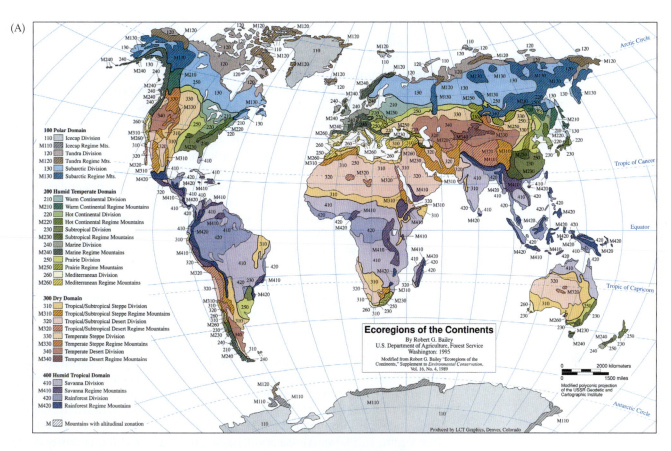

100 Polar Domain
110   Icecap Division
M110 Icecap Regime Mts.
120   Tundra Division
M120 Tundra Regime Mts.
130   Subarctic Division
M130 Subarctic Regime Mts.

200 Humid Temperate Domain
210   Warm Continental Division
M210 Warm Continental Regime Mountains
220   Hot Continental Division
M220 Hot Continental Regime Mountains
230   Subtropical Division
M230 Subtropical Regime Mountains
240   Marine Division
M240 Marine Regime Mountains
250   Prairie Division
M250 Prairie Regime Mountains
260   Mediterranean Division
M260 Mediterranean Regime Mountains

300 Dry Domain
310   Tropical/Subtropical Steppe Division
M310 Tropical/Subtropical Steppe Regime Mountains
320   Tropical/Subtropical Desert Division
M320 Tropical/Subtropical Desert Regime Mountains
330   Temperate Steppe Division
M330 Temperate Steppe Regime Mountains
340   Temperate Desert Division
M340 Temperate Desert Regime Mountains

400 Humid Tropical Domain
410   Savanna Division
M410 Savanna Regime Mountains
420   Rainforest Division
M420 Rainforest Regime Mountains

M Mountains with altitudinal zonation

Ecoregions of the Continents
By Robert G. Bailey
U.S. Department of Agriculture, Forest Service
Washington: 1995
Modified from Robert G. Bailey "Ecoregions of the
Continents," Supplement to *Environmental Conservation*,
Vol. 16, No. 4, 1989

0   2000 kilometers
0   1500 miles
Modified polyconic projection
of the USSR Geodetic and
Cartographic Institute

Produced by LCT Graphics, Denver, Colorado

(B)

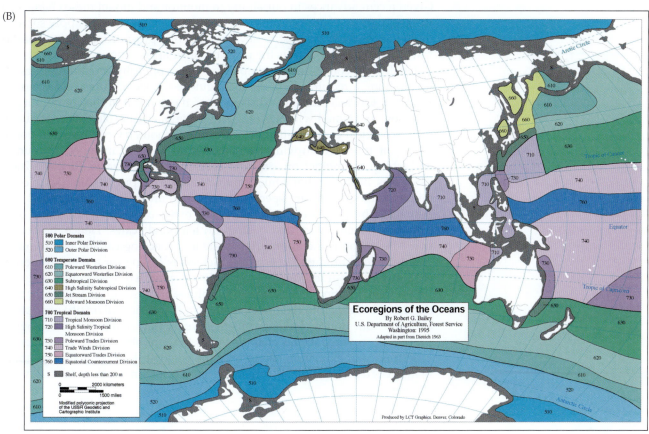

500 Polar Domain
510   Inner Polar Division
520   Outer Polar Division

600 Temperate Domain
610   Poleward Westerlies Division
620   Equatorward Westerlies Division
630   Subtropical Division
640   High Salinity Subtropical Division
650   Jet Stream Division
660   Poleward Monsoon Division

700 Tropical Domain
710   Tropical Monsoon Division
720   High Salinity Tropical
      Monsoon Division
730   Poleward Trades Division
740   Trade Winds Division
750   Equatorward Trades Division
760   Equatorial Countercurrent Division

S   Shelf, depth less than 200 m

0   2000 kilometers
0   1500 miles
Modified polyconic projection
of the USSR Geodetic and
Cartographic Institute

Ecoregions of the Oceans
By Robert G. Bailey
U.S. Department of Agriculture, Forest Service
Washington: 1995
Adapted in part from Dietrich 1963

Produced by LCT Graphics, Denver, Colorado

Finally, we should reemphasize that our current global view of nature, while generally accurate and informative, is really just a snapshot in time. The distribution of biomes and the productivity profile of the Earth have changed dramatically throughout geological time. Since the breakup of Pangaea approximately 180 million years ago, continental drift has caused major changes in global temperatures, precipitation patterns, prevailing winds, and ocean currents. Positions of the continents with respect to latitude and solar radiation have changed substantially as they have drifted, sometimes from one hemisphere to the other (see Chapter 8). These environmental changes have resulted in major shifts and transformations of communities and ecosystems, including extinction of once-dominant ancient biomes and communities, and the creation of novel ones (i.e., those without any analogues in the paleoecological record; Behrensmeyer et al. 1992; Erwin 1993; Betancourt 2004; Jackson 2004). Even as recently as 18,000 years ago—only an instant ago on the geological timescale—most of the northern landmasses were covered by glaciers, often a kilometer or more thick (see Chapter 9). Warm deserts and some other biomes, once relatively rare and limited to lower latitudes and lower elevations, have since expanded at the expense of others. As we shall see in subsequent chapters, these historical changes often leave a lasting imprint on biogeographic patterns (see Chapters 10, 11, and 12).

We can learn much from this history lesson and, just as important, apply it to the future. Even within what has traditionally been viewed as "ecological time"—decades to hundreds of years—human activities have caused significant changes in global climate and, in turn, fundamentally altered the distribution of biomes and ecoregions. As we shall show in the final chapters of this book, **anthropogenic biomes** now dominate many regions of the Earth (see Figure 16.39; see Alessa and Chapin 2008; Ellis and Ramankutty 2008). Given recent advances in our ability to monitor and model these changes in the geography of entire communities and ecosystems, we can apply some of these lessons from the past to development of a prospective view of tomorrow's biogeography. This is the focus of the final chapters.

Finally, we conclude this chapter on the geography of communities by highlighting a recent call by the distinguished biogeographer and ecologist Robert E. Ricklefs (2008) to "disintegrate" the concept of the community as traditionally framed within ecology. In general, most ecologists continue to view ecological communities as local-scale phenomena, composed of populations of species occupying the same locality, whose members encounter and affect one another on a regular basis at that local scale. Ricklefs (2008: 746) notes, however, that while "populations are the primary entities in community ecology," in fact species populations are broadly distributed and influenced by processes operating across extensive spatial scales and over historical (evolutionary and geological) temporal scales. Hence, he argues that "the region is the appropriate scale for an ecological and evolutionary concept of community." In essence, Ricklefs is calling for a reintegration of ecology (with its traditionally local-scale focus) with biogeography and evolution. Reintegration of these three disciplines promises to provide some fundamental insights into the forces influencing biological diversity over space and time. But it also challenges us to develop new syntheses of ecological processes—covered in this chapter—with those of biogeography (dispersal, speciation, and extinction, along with the geological and climatic dynamics of the geographic template), which we will explore in depth in the following chapters.

UNIT THREE

FUNDAMENTAL
BIOGEOGRAPHIC PROCESSES
AND EARTH HISTORY

UNIT THREE

FUNDAMENTAL
BIOGEOGRAPHIC PROCESSES
AND EARTH HISTORY

Previous Page: Mountain peaks of the Himalayas. Courtesy of NASA.

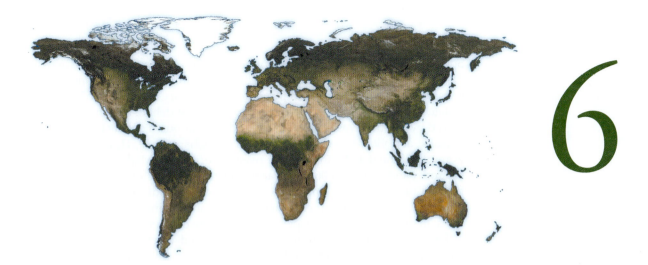

6

DISPERSAL AND IMMIGRATION

Nature has been defined as a principle of motion and change, and it is the subject of our inquiry. We must therefore see that we understand the meaning of motion, for if it were unknown, the meaning of nature too would be unknown. (Aristotle, Physics, III, 1)

There are three fundamental processes in biogeography: evolution, extinction, and dispersal. These are the means by which biotas respond to spatial and temporal dynamics of the geographic template. Thus, all of the biogeographic patterns that we study derive from the effects of these processes. While Aristotle may have been unaware of the importance, or even existence, of evolution and extinction, his writings excerpted above clearly speak to the role of movements.

Throughout the development of biogeography and its maturation as a respected discipline, the relative importance of movement, or **dispersal** (inclusive of **immigration**), has been a subject of great debate. As described in Chapter 2, Charles Darwin was one of the most passionate and persuasive champions of the importance of long-distance dispersal for explaining the occurrence of species in two or more widely separated areas. He and other early dispersalists—including Alfred Russel Wallace and Asa Gray—argued that such distributional disjunctions can best be explained by long-distance dispersal across existing barriers. Their antagonists in this debate were the extensionists, including Charles Lyell, Edward Forbes, and Joseph Hooker, who argued that disjunctions were the result of movements along ancient corridors that have subsequently disappeared. The point of contention was not whether dispersal occurred, but whether range expansion and disjunction required the crossing of extensive barriers.

Evidence for the extensionists' transoceanic land bridges never surfaced. A new explanation for the disjunct distribution of taxa emerged, however, in the twentieth century with Alfred Wegener's proposition of continental drift and subsequent development of plate tectonic theory. No longer was it necessary to propose such rare and unlikely events as long-distance dispersal to account for disjunctions.

Species could simply ride on the continents as they split and dispersed across the surface of the Earth. This splitting of once-continuous populations would serve as a vicariance event and promote evolutionary divergence of the now isolated populations. The dispersalist–extensionist debate had thus been replaced by one that was no less heated and contentious: a debate between dispersalist and vicariance biogeographers (see Chapters 10, 11, and 12; for an example of the debate, in this case over the origins of Madagascar's biotas, see Yoder and Nowak 2006). The earlier debate was transformed into one that focused on the relative importance of long-distance dispersal versus plate tectonics, orogeny, and other vicariant events in shaping biogeographic patterns: in other words, did dispersal occur before or after barriers formed?

Finally, some biogeographers have questioned whether dispersal—over any time frame—has had any lasting influence on biogeographic patterns, either past or present. This view, in perhaps its most extreme form, is summarized by **Beijerinck's Law**, formalized and concisely stated by Baas Beking: "Everything is everywhere, but the environment selects" (Sauer 1988; deWit and Bouvier 2006). Proponents of this theory suggested that, given enough time, dispersal is inevitable for many, if not most, species. Therefore, it was assumed that any differences in dispersal abilities among species are eventually rendered irrelevant and, thus, species exhibit geographically limited ranges simply because they differ in their abilities to adapt to different environments. This view may apply to some life-forms (especially microbes, which Beijerinck and Baas Beking studied), but it seems to fly in the face of some very general biogeographic patterns, including **Buffon's Law**. Beijerinck's view may have been derived from a misinterpretation of G. G. Simpson's suggestion that even exceedingly unlikely events (such as transoceanic dispersal) become more likely if enough trials are performed. Increased likelihood, however, does not mean eventual certainty of colonization by all species to all points across the globe. Indeed, even for microscopic life-forms, the everything-is-everywhere mantra has been increasingly challenged by genetic evidence, such that it cannot be regarded as having the status of a scientific "law" (see Whittaker et al. 2003; Smith and Wilkinson 2007; O'Malley 2008; Perez-del-Olmo et al. 2009). Of course, the other extreme view—that dispersal is the only force influencing isolated communities—is just as problematic.

This debate over the biogeographic relevance of long-distance dispersal is, of course, an important one. However, as with most debates, the truth lies somewhere between the extremes. The relative importance of dispersal, vicariance, and geographic variation in selection regimes varies from one biotic group and region to another, and likely across time periods as well. As Simpson (1980b) put it, "a reasonable biogeographer is neither a vicarist nor a dispersalist, but an eclecticist." Many of today's historical biogeographers acknowledge that biotas may have reticulate phylogenies—that is, biotas may have repeatedly fragmented and merged as dispersal barriers appeared and disappeared through time. Indeed, although it would seem easier to study these three processes—dispersal (immigration), evolution, and extinction—independently, it is becoming increasingly clear that they are interdependent processes. Moreover, one especially important although largely overlooked feature of the dynamics of species is that dispersal abilities also are subject to evolution (Baskett et al. 2007; Ronce 2007). Random mutation, drift, or selection can result in dramatic shifts in abilities of species' populations to cross barriers and further expand their ranges. Some of the most striking cases include the many thousands of species of plants, insects, and birds that evolved greatly reduced dispersal capacity after colonizing isolated, oceanic islands (see Chapter 14). More recently, biogeographers and ecologists are beginning to document the converse phenomenon, where rates of range expansion of some species have accelerated, presumably because they developed altered mechanisms or proclivities for dispersal (e.g., see the review of accelerated expansions of the exotic cane toad across Australia by Urban et al. 2008).

Again, all biogeographic patterns result from the combined effects of these responses to environmental variation across the geographic template—populations disperse to new regions, evolve, and adapt to local environments, or they suffer extinction. As Aristotle observed many centuries ago, however, the geographic template itself is also highly dynamic: "Earth as a whole is not always sea, nor always mainland, but in process of time all change" (Meteorologica, ca 355 BC). Therefore, in order to develop a more comprehensive understanding of the nature and relative importance of dispersal, it is essential that we view both the species and the underlying, geographic template (land or sea) as dynamic in space and time (see Craw et al. 2008).

What Is Dispersal?

All organisms have some capacity to move from their birthplaces to new sites. The movement of offspring away from their parents is a normal part of the life cycle of virtually all life-forms. Often, dispersal is confined to a particular stage of an organism's life history. Higher plants and some aquatic animals are sessile as adults, but in their earlier developmental stages they are capable of traveling long distances from their natal sites. Mobile animals can shift their locations at any time during their lives, but many settle in one place and confine their activities to a limited home range for long periods of time.

Dispersal should not be confused with **dispersion**, an ecological term referring to the spatial distribution of individual organisms within a local population. In biogeography, our focus is primarily on movements of individuals to sites not already occupied by the species (i.e., immigration), because these events allow expansion of the species' distributions toward its fundamental geographic range. This process is distinct from regular seasonal movements between summer and winter ranges, which are termed migrations.

Dispersal as an ecological process

Plants and animals have evolved an incredibly diverse array of dispersal mechanisms that have been the subject of numerous classic studies throughout the histories of biogeography and ecology (see Gadow 1913; Udvardy 1969; Pijl 1972; Carlquist 1974, 1981; Sauer 1988; Stenseth and Lidicker 1992; Thornton 1996; Desbruyeres et al. 2000; Bilton et al. 2001; Bullock and Kenward 2001; Clobert 2001; Figuerola and Green 2002; Jones 2003). Yet in all cases, dispersal is basically an ecological process that is an adaptive part of the life history of every species. Natural selection typically favors individuals that move a modest distance from their natal site. A new location is always likely to be more favorable than an individual's exact birthplace, in part because intraspecific competition between parent and offspring and among siblings is reduced, and in part because the environment—and hence the quality—of the natal site is always changing. On the other hand, most environmental variables exhibit spatial autocorrelation. Thus, as distance from the natal site increases, habitats become more dissimilar, and as a result, would-be colonists are less likely to be well adapted to their new habitats. At biogeographic scales, however, suitable and possibly more optimal conditions can often be found far from the species' realized range, within distant continents or ocean basins—the challenge, of course, is getting there.

Dispersal as a historical biogeographic event

The role of dispersal in biogeography is thus very different from its role as a demographic phenomenon. Although dispersal occurs continually in all species, most of it does not result in any significant change in their geographic distributions. As pointed out in Chapter 4, the geographic ranges of most species are limited by environmental factors, and they thus remain relatively constant over ecological time. Biogeographers are concerned primarily with the exceptional cases: those rare instances in which species shift their ranges by moving over long distances across seemingly insurmountable barriers. This occurs so infrequently that we seldom see it happening, and even less often are we able to study it. Usually biogeographers must look at dispersal as a historical process and must infer the nature and timing of past long-distance movements from indirect evidence, such as distributions of living and fossil forms and genetic similarity among now isolated biotas. Making such inferences is a monumental task. The distribution of every taxon reflects a history of local origin, dispersal, and local extinction extending back to the very origin of life. Patterns of endemism, provincialism, and disjunction of geographic ranges (see Chapter 10) indicate that the dispersal of some groups has been so limited that their histories are indeed reflected in the distributions of their living and fossil representatives. However, reconstructing dispersal history usually requires that we deal with a process operating so infrequently that its effects may often appear to be highly random and idiosyncratic.

The problem of dealing with rare but important events is not unique to biogeography. In many ways the role of successful dispersal in biogeography is analogous to that of beneficial mutations in evolution. Beneficial mutations provide the raw material for evolutionary change, but essentially they occur randomly and so infrequently that they are difficult to observe and study (see Elena et al. 1996). Every inherited feature of an organism has a history comprising a series of unique genetic changes. Trying to reconstruct the history of the mutations that resulted, for example, in the evolution of a reptilian scale into a feather is conceptually similar in many ways to attempting to reconstruct the biogeographic events that led to the present disjunct distribution of large predaceous birds in Africa, Australia, New Zealand, and South America.

We draw this analogy to make two points. First, long-distance dispersal events may be infrequent and somewhat stochastic, but that does not mean that they are unimportant. On the contrary, these movements are among the most important of the events that have shaped present distributions. Second, we cannot afford to ignore the role of dispersal in biogeography just because it is difficult to study. One of the great challenges in understanding the geography and evolution of life is to develop ways of evaluating the influence of rare but important events such as beneficial mutations, asteroid impacts, and long-distance dispersal to archipelagoes as remote as the Galápagos and Hawaii.

Dispersal and Range Expansion

In order to expand its range, a species must be able to (1) travel to a new area, (2) withstand potentially unfavorable conditions during its passage, and (3) establish a viable population upon its arrival. Biogeographers often distinguish among three kinds of dispersal events that can have this result, primarily by the relative rates of movement and range expansion. These three mechanisms of range expansion are called jump dispersal, diffusion, and secular migration (Pielou 1979).

Jump dispersal

There is abundant evidence that many species have undergone long-distance dispersal, also known as **jump dispersal**. Anyone who has built a small pond in the backyard cannot help but be impressed and often vexed by the rate at which populations of aquatic insects, snails, other invertebrates, vascular plants, and algae become established there. The same process of colonization occurs on a much larger geographic scale. A classic example is provided by the Krakatau islands, Indonesia (see locator map in Figure 6.14). In 1883 a volcanic eruption removed two-thirds of the largest island (Rakata) and buried all three islands in a deep blanket of volcanic ash, apparently eliminating life from their land surfaces. Subsequent biological surveys, primarily of birds and plants, documented the rapidity with which new populations of organisms became established on the island (**Figure 6.1**; Docters van Leeuwen 1936; Dammermann 1948; Rawlinson et al. 1992; Bush and Whittaker 1993; Whittaker and Jones 1994; Thornton 1996; Whittaker et al. 1997, 2000). By 1933—only 50 years after the eruption—Krakatau was once again covered with a dense tropical rain forest; indeed, 271 plant species and 31 kinds of birds, as well as numerous invertebrates, were recorded on the island. Recolonization was achieved by both active and passive transport and included both stepping-stone dispersal across island chains and more direct bouts of jump dispersal from source islands, principally from the "mainland" islands of Java and Sumatra, which lie 40 and 30 km, respectively, from Krakatau (see further discussion of the colonization of Krakatau in Chapter 13).

The case of Krakatau is unusual, but only in that the creation of completely virgin habitat was so dramatic that the sources of colonizing organisms could be easily identified and their immigration carefully documented. The same processes have occurred over much longer distances and time periods for other archipelagoes. The Galápagos and Hawaiian archipelagoes lie far out in the Pacific Ocean—972 km west of Ecuador and some 3200 km southwest of the North American mainland, respectively (see the world map in the end cover of this book). These oceanic islands are actually the tops of

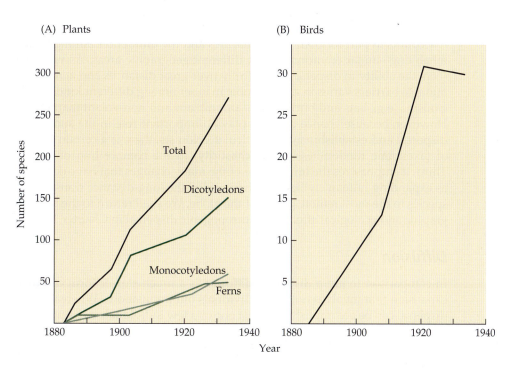

FIGURE 6.1 (A) Plants of the three Krakatau islands of Panjang, Rakata, and Sertung. (B) Birds of Rakata (the largest and best-surveyed island). During the first five decades following the sterilizing 1883 eruption of Krakatau, its remnant islands were rapidly colonized by plant and animal species. Several biological surveys recorded the arrival of colonizing species, which traveled across at least 30 km of ocean. (After MacArthur and Wilson 1967; see also Whittaker and Fernández-Palacios 2007; see also Chapter 13 for an update and synthesis of this classic, natural experiment in long-distance dispersal and dynamic biogeography.)

volcanoes that arose from the ocean floor. Within the life span of the present islands, they have never been much nearer to continents than they are today, providing classic illustrations of how isolated islands have acquired diverse biotas from propagules (see page 203) that dispersed across the ocean.

Some authors continue to downplay the importance of long-distance dispersal in favor of alternative explanations, invoking either land bridges or a long-term history of stepping-stone islands that have come and gone over time as the ocean basins have developed, allowing the possibility of much less dramatic dispersal distances than are apparent from today's atlases. Nevertheless, there is undeniable evidence that many groups of organisms reached distant islands by traveling through the air (sometimes carrying other organisms with them) or floating on the sea (Ridley 1930; Vagvolygi 1975; Baur and Bengtsson 1987; Enckell et al. 1987; J. M. B. Smith et al. 1990; J. M. B. Smith 1994). We also have evidence—both from recent occurrences and from historical patterns—that other species have traveled equally long distances over continental areas (McAtee 1947; Cruden 1966). Moreover, and contrary to assumptions that rare events must also be entirely random, long-distance dispersal may have a strong selective component (Clark 2008). Those species that have special adaptations for long-distance travel have been especially successful at colonizing islands (see Chapters 13 and 14). Both the Galápagos and the Hawaiian Islands have land snails and bats as well as numerous species of trees, insects, and birds. In addition, giant tortoises and native rats inhabit the Galápagos (the latter most likely arriving as waifs on natural rafts of vegetation cast adrift by violent storms). Noteworthy in their absence from these islands are most **non-volant** (nonflying) mammals, amphibians, freshwater fishes, and other forms poorly adapted for dispersal across open oceans.

The acceptance of the reality of long-distance dispersal offers at least three important consequences for biogeographers. First, it can be used judiciously to explain the wide and often discontinuous distributions of many taxa of animals, plants, and microbes. However, just as it would be wrong to dismiss the possibility of long-distance dispersal out of hand, it should also be recognized that not all present-day land–sea configurations provide an accurate guide to the distances and environments that had to be crossed when the species of interest first reached a particular isolated land mass. Second, it accounts in part for both the similarities and the differences among biotas inhabiting similar environments in different geographic areas. As we have seen, the ability to disperse successfully over great distances and across formidable barriers varies in a predictable manner among different kinds of organisms (e.g., bats and birds versus amphibians and non-volant mammals). However, because chance can still play an important role in the successful dispersal and establishment of particular colonists, there is a certain degree of taxonomic randomness (or **stochasticity**) in the composition of biotas. Third, it emphasizes the importance of the many changes that have occurred as expanding human civilizations have aided the long-distance transport of species to the most remote points of the globe. We shall return to the biogeographic and ecological effects of anthropogenic dispersal in Chapter 16.

Diffusion

In comparison to jump dispersal, **diffusion** is a much slower form of range expansion that involves not just individuals but populations. Whereas jump dispersal can be accomplished by just one or a few individuals crossing a barrier within a short period of their life span, diffusion typically is accomplished over generations by populations gradually spreading out from the margins of

FIGURE 6.2 Colonization of the New World by the cattle egret (*Bubulcus ibis*). This heron crossed the South Atlantic from Africa under its own power, becoming established in northeastern South America by the late 1800s. From there it dispersed rapidly, and it is now one of the most widespread and abundant herons in the New World. (After Smith 1974; Osborn 2007; photo courtesy of Suneko/Flickr.)

a species' previous range. However, these two mechanisms of range expansion are closely related, as diffusion often follows the jump dispersal of a species into a distant—but uncolonized—region of hospitable habitat.

An excellent example of both kinds of range expansion is provided by the cattle egret, *Bubulcus ibis* (Crosby 1972). This small heron was once native to Africa, where it still inhabits tropical and subtropical grasslands in association with large herbivorous mammals, foraging for insects and other small animals that are flushed out by grazing herbivores. In the late 1800s, cattle egrets colonized northeastern South America, having dispersed under their own power across the South Atlantic Ocean (**Figure 6.2**). During succeeding decades, descendants of the founding population thrived and spread throughout much of the New World, finding abundant food and habitat as a result of the clearing of tropical forests, primarily for grazing livestock. The cattle egret has now expanded its breeding range northward to the southern United States and has colonized all of the major Caribbean islands.

Other well-known examples of diffusion include the range expansions of starlings (*Sturnus vulgaris*) and house sparrows (*Passer domesticus*) after their intentional introductions into North America (**Figure 6.3A,B**). The American muskrat (*Ondatra zibethica*; **Figure 6.3C**) rapidly expanded its range after introduction into Europe. Opossums (*Didelphis virginiana*) and armadillos (*Dasypus novemcinctus*; **Figure 6.3D**), both natives of South America, continue to expand their ranges northward through North America. The European rabbit (*Oryctalagus cuniculus*; **Figure 6.3E**) and red fox (*Vulpes vulpes*; **Figure 6.3F**), both purposely introduced into Australia in the late 1800s, subsequently expanded their exotic ranges over the next century to occupy most of the

(A) European starling

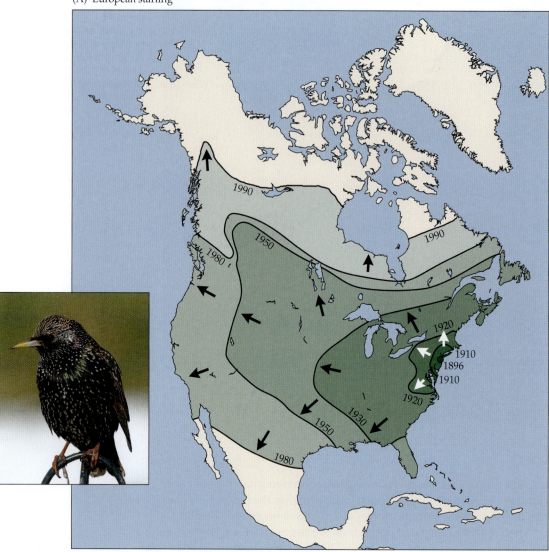

(B) House sparrow

FIGURE 6.3 Range expansions of selected species of animals: (A) European starling (*Sturnus vulgaris*) in North America; (B) house sparrow (*Passer domesticus*) in North America (note that the current populations derive from at least three introductions, including those of 1852 in New York, 1871 in California, and 1873 in Utah); (C) American muskrat (*Ondatra zibethica*) in Europe since its 1905 introduction near Prague; (D) nine-banded armadillo (*Dasypus novemcinctus*) in the southern United States; (E) European rabbit (*Oryctalagus cuniculus*) in Australia; (F) red fox (*Vulpes vulpes*) in Australia; and (G and H) cane toads (*Chaunus* [*Bufo*] *marinus*) in Australia. (A compiled from various sources; see Cabe 1993, photo © Raymond Neil Farrimond/ShutterStock; B after Lowther and Cink 1992, photo © Andrew Howe/istock; C after Van den Bosch et al. 1992, photo courtesy of the U.S. Fish and Wildlife Service; D after Taulman and Robbins 1996, North American Mammals (Smithsonian) Web site, www.mnh.si.edu/mna/ (accessed 2009), photo © Ron Kacmarcik/istock; E after Stodart and Parer 1988, and Edwards et al. 2004, photo © Eduardo Rivero/Shutterstock; F after Dickman 1996, photo courtesy of the National Park Service; G and H after Urban et al. 2008, photo © Art Man/Shutterstock.)

(C) Muskrat

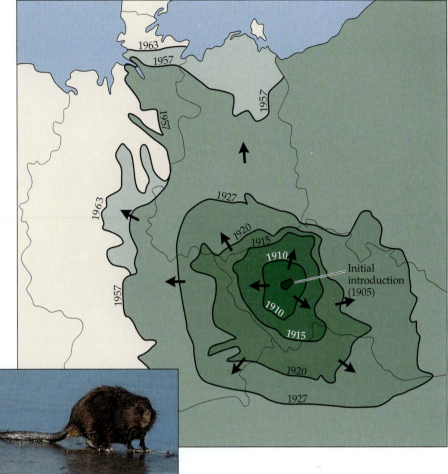

(D) Armadillo

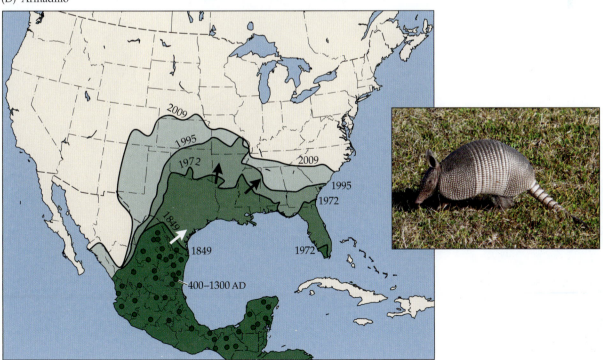

(E) Rabbit

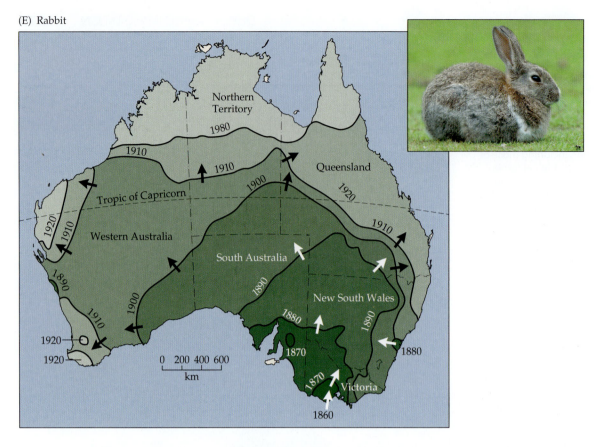

FIGURE 6.3 (*continued*)

(F) Red fox

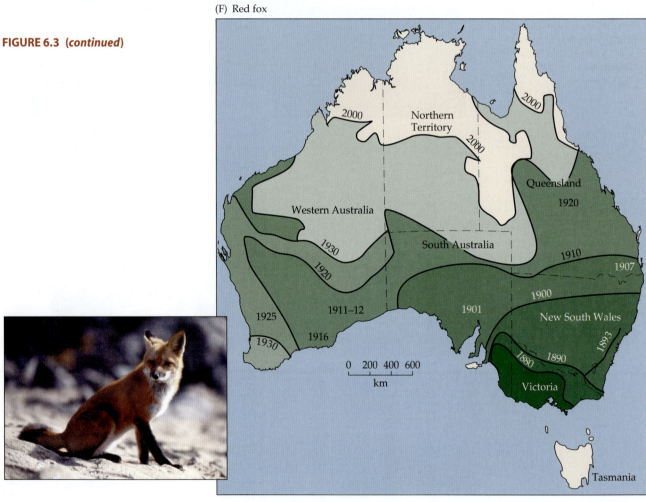

(G) Cane toads (exotic range)

FIGURE 6.3 (*continued*)

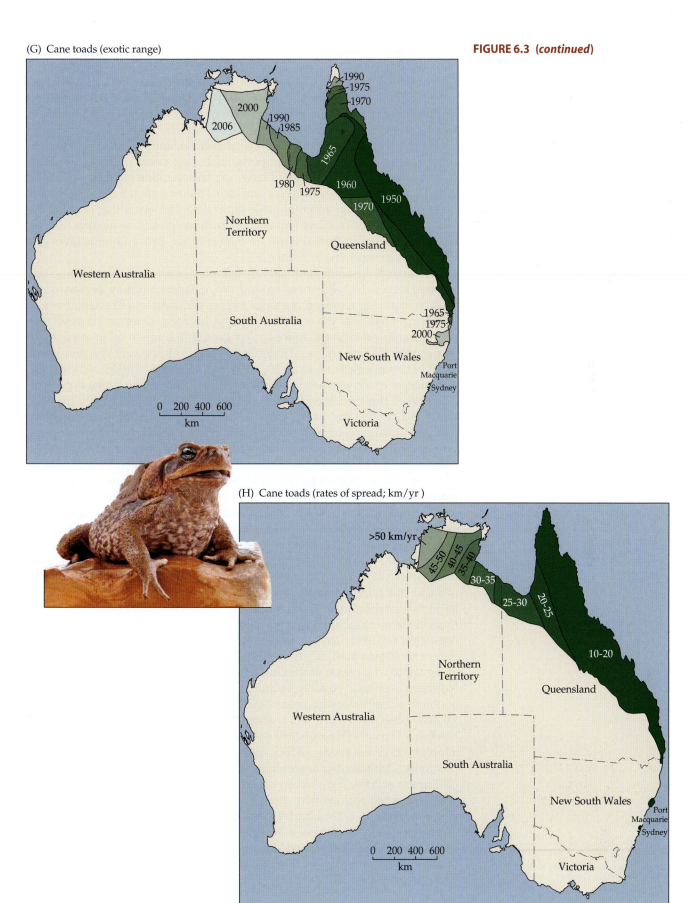

(H) Cane toads (rates of spread; km/yr)

island continent. Cases of diffusion of invertebrates are perhaps all too common, but among them, the most notorious and problematic is the spread of Africanized "killer" bees (*Apis mellifera*) northward, following their introduction into Brazil, and of fire ants (*Solenopsis invicta*), which are continuing their relatively rapid expansion into the southern portion of the United States. In aquatic ecosystems around the globe, many "weeds" such as water hyacinth (*Eichhornia crassipes*), water fern (*Salvinia molesta*), and hydrilla (*Hydrilla verticilata*), as well as many animal pests, including zebra mussels (*Dreissena polymorpha*), have exhibited rapid range expansion following their accidental introduction into lakes and rivers outside of their original ranges. In each of these cases, it was an anthropogenic jump dispersal event that enabled the subsequent waves of range expansion via diffusion across regions that, although far removed from the species' native range, proved highly favorable in their environmental conditions.

Although diffusion is an inherently complex process, it typically proceeds in three stages. Initially, invasion and range expansion may be very slow, usually requiring repeated dispersal events and subsequent adaptations to the characteristics of the ecosystems being invaded. Once an invasive species becomes established, however, its geographic range often expands at an exponential rate and in a roughly symmetrical fashion (Van den Bosch et al. 1992; see also Long et al. 2009 for a comparison of geographic expansion rates of exotic versus native plant species). Eventually, range expansion slows when the species encounters physical, climatic, or ecological barriers, which also tend to distort the shape of the species' geographic range (**Figure 6.4**). At this stage, geographic ranges may remain relatively stable unless environmental conditions become substantially altered or the species somehow crosses another barrier and reinitiates the invasion sequence. One surprising but especially important feature of at least some range expansions in exotic regions is that the species may change, both phenotypically and genetically, as they adapt to their new environs. This may well account for the accelerated rates of range expansion characteristic of the second stage described above, and for a renewed phase of expansion following a period of prolonged stasis. An intriguing case study is that of cane toads (*Chaunus* [*Bufo*] *marinus*) introduced into Australia in the late 1930s in a disastrous attempt to control pests of sugarcane farms of coastal Queensland. Following a relatively protracted period of establishment of initial populations across the numerous sites of introductions, range expansion of these toads then accelerated until they encountered relatively dry environments to the northwest and cooler ones to the south (**Figure 6.3G**). During the 1970s and 1980s, however, populations along the coastal region of the Gulf of Carpenteria expanded into Australia's Northern Territory, possibly by jump dispersal between wetlands that are scattered across otherwise more xeric landscapes. This long-distance dispersal and subsequent acceleration of range expansion across this region of the Northern Territory (**Figure 6.3H**) may have been facilitated, at least in part, by adaptations of the toads—which included altered behavior (e.g., increased proclivity for dispersing or seeking shelter) and development of longer limbs (Phillips et al 2006, 2008; Urban et al. 2008). On the other hand, the biota native to invaded regions may also adapt—adjusting as predators, parasites, or competitors to what Charles Darwin referred to as "the stranger's craft of power." These ecological adjustments of native communities likely contribute to the deceleration and asymmetrical restriction of range boundaries that are characteristic of the late stages of invasion (see also Arim et al. 2006; Skarpaas and Shea 2007).

(A) Fertile Crescent crops

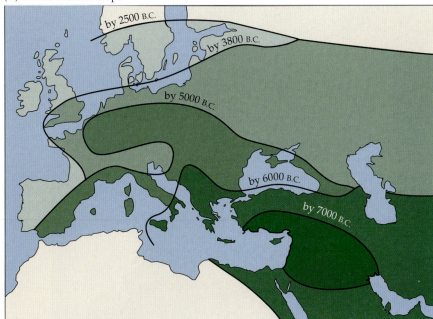

(B) Oaks

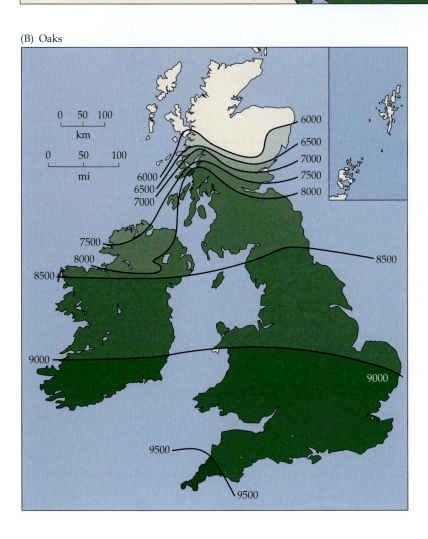

FIGURE 6.4 Range expansion in selected plants. Maps show the spread of (A) "Fertile Crescent" crops across western Eurasia, (B) oaks (*Quercus* spp.) in Great Britain (numbers indicate years BP), (C) elm (*Ulmus* spp.) in Great Britain (numbers indicate years BP), and (D) purple loosestrife (*Lythrum salicaria*) in North America. (A after Diamond 1997; B and C after Birks 1989; D after Thompson et al. 1987; photo courtesy of David McIntyre.)

FIGURE 6.4 (*continued*)

(C) Elm

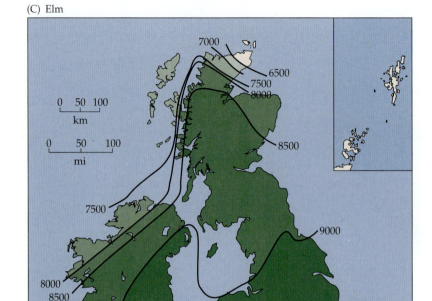

(D) Purple loosestrife

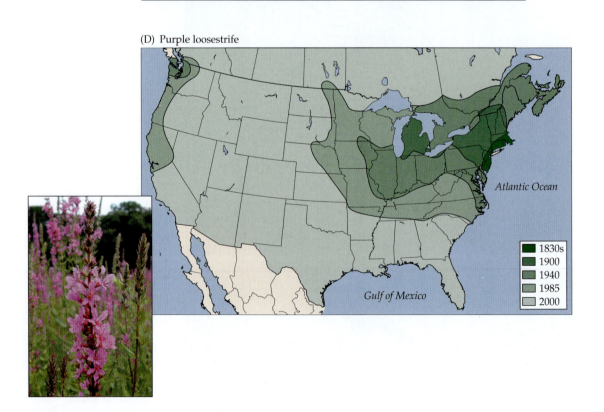

Atlantic Ocean

Gulf of Mexico

■	1830s
■	1900
■	1940
■	1985
■	2000

Secular migration

In contrast to the relatively rapid forms of range expansion discussed above, **secular migration** occurs so slowly—on the order of hundreds of generations—that species have ample opportunity to evolve en route. Although Herbert Louis Mason appears to have coined the term in 1954, the concept of secular migration is an old one, dating back to the early development of biogeography (Pielou 1979). Recall from Chapter 2 that during the eighteenth century, Buffon hypothesized that most life-forms originated in the northern regions of the Old World but then spread to the New World, apparently at a time when the northern continents were much less isolated (see Figure 2.2B). From there, they migrated southward through the New World and, once isolated, they evolved ("degenerated" in Buffon's terms) and became modified such that the New and Old World tropics developed distinct biotas (i.e., Buffon's Law).

While Buffon's explanation for the spread and diversification of terrestrial life-forms (including mammoths and humans!) now seems fanciful, the fossil record provides convincing evidence of evolutionary divergence during range expansion. For example, some now extinct forms of camels spread southward through North America and eventually across the newly formed Central American land bridge during the Pliocene Epoch. Extant products of this secular migration include the guanaco (*Lama guanicoe*) and vicuña (*Vicugna vicugna*) of South America. Similarly, camels and horses spread from the Nearctic region into the Old World. In both cases, the descendants (including camels, zebras, Przewalski's horse, and the Asian wild ass) persisted, although their ancestors were extirpated from their North American homeland.

Mechanisms of Movement

Active dispersal

Organisms can disperse either actively (moving under their own power) or passively (being carried by a physical agent, such as wind or water, or by other organisms). The terms **vagility** and **pagility** are sometimes used to denote the ability of organisms to disperse by active or passive means, respectively. Only a few animals have the capacity to travel long distances under their own power. Strong fliers, such as many birds, bats, and large insects (e.g., dragonflies, some lepidopterans, beetles, and bugs), have the greatest capacity for active, long-distance dispersal. Many of these animals regularly travel hundreds or thousands of kilometers during their seasonal migrations, which are a normal part of their annual life cycles. When stressed or aided by favorable winds, some of these same animals can cover comparable distances during a single flight.

A few examples of normal migratory routes demonstrate the potential of dispersal by flight. The golden plover (*Pluvialis dominica*), a medium-sized shorebird weighing about 150 g, breeds in the Arctic but winters in southern South America, southern Asia, Australia, and the islands of the Pacific. Migrating individuals regularly fly nonstop from Alaska to Hawaii, a distance of 4000 km (**Figure 6.5**; Dorst 1962). The ruby-throated hummingbird (*Archilochus colubris*), one of the smallest birds at just 3.5 g, regularly commutes twice a year nonstop across the Gulf of Mexico en route between its breeding grounds in the eastern United States and its wintering grounds in southern Mexico (Dorst 1962), a distance of 800 km.

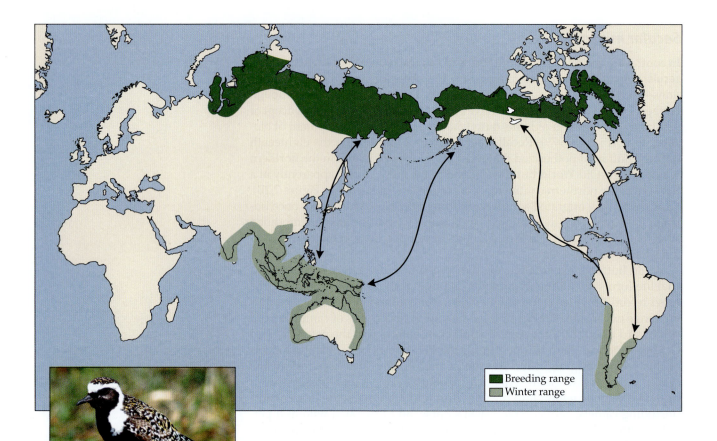

FIGURE 6.5 Migratory routes of the golden plover (*Pluvialis dominica/fulva*), a shorebird that breeds in the Arctic and winters in temperate regions of the Southern Hemisphere. Each year these birds fly prodigious distances, often crossing huge expanses of open ocean. (Photo courtesy of the USFWS.)

In one of his most insightful yet often overlooked papers, Joseph Grinnell (1922) discussed "accidental" (extralimital) occurrences of birds well beyond their documented geographic ranges. In addition to noting that these seemingly rare events are actually commonplace for many species, Grinnell concluded that "the continual wide dissemination of so-called accidentals... provided the mechanism by which each species spreads [i.e., by diffusion] or by which it travels from place to place when this is necessitated by shifting barriers [i.e., by jump dispersal]." Extralimital sightings are well documented for many taxa. Every year a few individuals of bird species native to Europe are seen in eastern North America, and vice versa, usually following severe North Atlantic storms. Hoary bats (*Lasiurus cinereus*, body mass 10–35 g) migrate from northern North America to winter in the Neotropics (Findley and Jones 1964). A species of this bat that is native to the Hawaiian Islands undoubtedly is derived from migrating individuals that went astray.

Monarch butterflies (*Danaus plexippus*) and some dragonflies (*Anax* spp.) migrate distances comparable to those flown by many songbirds—from southern Canada and the northern United States to the southern United States and central Mexico (**Figure 6.6A**). Individual monarch butterflies may fly as far as 375 km in four days and 4000 km during their lifetimes (Urquhart 1960; Brower 1977; Brower and Malcolm 1991). Occurrences of individuals in Cuba and other regions of the Caribbean may indeed represent fatal "accidents," but they might also be the equivalent of genetic "hopeful monsters." These misguided few, or their descendants, may someday found a new insular race or species of monarchs (see gray arrow in Figure 6.6A).

In contrast to these volant species, only a few non-volant animals, such as some of the larger mammals, reptiles, and fishes, are able to disperse substantial distances, by walking or swimming. Active dispersal by these means is generally less effective than by flight, because the animals are forced to

(A)

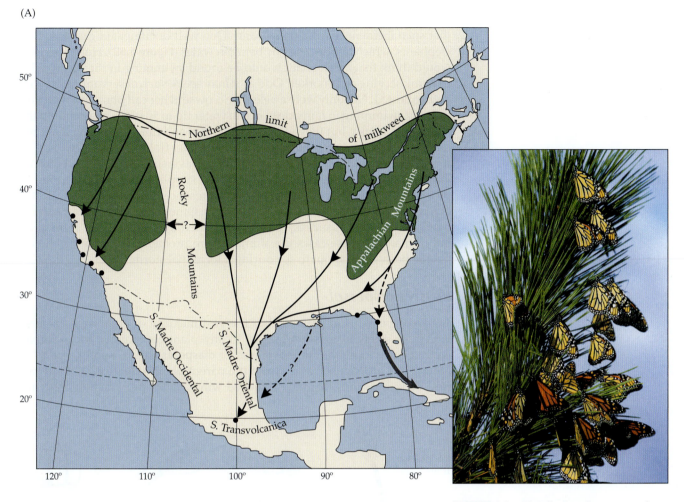

(B)

FIGURE 6.6 (A) Fall migration routes of eastern and western populations of the monarch butterfly (*Danaus plexippus*). Most, if not all, overwintering individuals of the eastern populations gather in spectacular winter congregations in montane forests of the Transvolcanic region of Mexico ("S." in these maps refers to "Sierras," i.e., mountain ranges in Mexico). During the summer, the northern limits of this species are coincident with the northern limits of its host plants, milkweeds (*Asclepias*). (B) Both African and Asian elephants are powerful swimmers, capable of traveling tens to hundreds of kilometers during migrations and dispersal. (A after Brower and Malcolm 1991, photo courtesy of Gene Nieminen, U.S. Fish and Wildlife Service; B courtesy of André Grové.)

walk or swim through unfavorable intervening habitats, whereas flying animals can simply vault barriers. Nevertheless, some large animals—especially aquatic ones such as whales, sharks, predaceous fishes, and sea turtles—have wide, albeit often discontinuous, geographic distributions produced in part by active dispersal.

Stories about swimming terrestrial vertebrates have appeared repeatedly in the literature, and as absurd as some reports may seem, we cannot deny that such incidents occasionally occur. One well-documented case involves elephants, which appear to enjoy being in water (**Figure 6.6B**). However, who would think that an elephant would be found swimming in the ocean, with its trunk serving as a snorkel? Odd as it may seem, such observations are well authenticated. Not only were an elephant cow and her calf photographed voluntarily swimming in fairly deep ocean water up to 50 km to an island off the coast of Sri Lanka, but they were also followed and timed, showing that they made a return trip to Sri Lanka some 3 hours later (Johnson 1978, 1980, 1981). It is thus evident that elephants are able to traverse narrow straits between islands and mainlands in search of new food supplies. Therefore, contrary to previous assertions of extentionists, their presence is not a reliable indicator of ancient but now submerged land bridges. Other large vertebrates, such as tigers and terrestrial snakes, have been spotted over a kilometer from land, but in many of these cases the animals were probably not there by choice, having been washed out to sea during a torrential storm.

Passive dispersal

Active dispersal is impressive, but the vast majority of organisms disperse largely or solely by passive means. In any plant community, for example, we can easily observe the movement of **diaspores** (seeds, spores, fruits, or other plant propagules) away from the parent plants (**Figure 6.7**). Wind carries seeds and fruits that have attached wings, hairs, or parachute-like pro-

FIGURE 6.7 A variety of diaspores from North American plants designed to enable seeds to disperse from the parent plants. Seeds that are carried by wind often bear hairlike appendages (*Rafinesquia*, *Chilopsis*) or papery wings (*Acer*, *Dodonaea*). Seeds that hitchhike on the bodies of animals have barbs (*Krameria*, *Bidens*), sharp spines (*Ambrosia*), or sticky glands (*Plumbago*). Seeds meant to be eaten and later defecated by birds and mammals are enclosed in fleshy fruits (*Solanum*, *Juniperus*, *Rubus*). In an unusual case, seeds of the dwarf mistletoe (*Arceuthobium*) are explosively discharged from a fleshy base.

cesses. Birds and mammals consume fleshy and dry diaspores, scattering some of them during feeding and distributing others later with their feces. Seeds, fruits, and spores of some plants become attached to the feathers or fur of animals and ride as hitchhikers until they are dislodged accidentally or by grooming. Some diaspores are explosively released over short distances, whereas others simply fall to the ground at the base of the parent plant. On the ground, ants, rodents, and birds compete for the fallen seeds and may carry them considerable distances—in some cases many meters and possibly kilometers. Finally, flowing water, tides, and ocean currents may displace diaspores and carry them to distant mainland sites or oceanic islands (Stebbins 1971a; Pijl 1972).

These means of passive dispersal are used not only by plants but also by invertebrates, fungi, and microbes. Some of them are obviously more effective than others. Of course, the distance passively traveled in a generation by a seed, spore, or other disseminule depends on the velocity, distance, and direction of movement of the dispersal agent. Aerial dispersal by winds—or by birds and bats—is especially successful in moving disseminules over long distances. Animal transporters may be nomadic or migratory, leaving their original habitats and covering great distances, or they may be territorial residents that rarely leave their home ranges. Wind tends to have local effects except in violent storms, during which organisms can be widely disseminated. Jet streams are also alleged to be important dispersal agents, especially for tiny spores (Gressitt 1963; Clagg 1966; Kellog and Griffin 2006). Such so-called **aerial plankton** also includes tiny mites, spiders, and insects that do not necessarily have special stages adapted for dispersal, so while their lifeless exoskeletons may drift over great distances, they will not contribute to range expansion.

Even passive mechanisms for dispersal are not necessarily completely haphazard or random. Aerial invertebrates, for example, occupy different strata above the land surface, so it is likely that they differ in their tendency to be wind dispersed (**Figure 6.8**). In addition—all else being equal—small organisms tend to be more effectively transported by wind than large organisms. Interestingly, just the opposite is the case for active immigration and vagility (Jenkins et al. 2007; Nathan et al. 2008). This suggests that size may be one of the many characteristics selected for during immigration, albeit for different extremes—that is, selecting for smaller individuals or propagules during passive dispersal, and for larger individuals during active dispersal (see also the discussion of immigration abilities and selection in Chapter 14). On the other hand, many plant propagules that are passively dispersed by

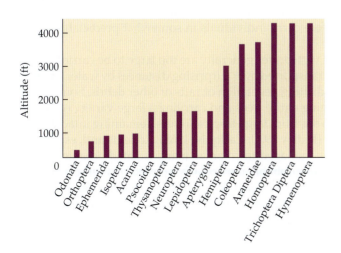

FIGURE 6.8 Aerial distribution of insects, spiders, and mites. Provided they can survive the rigors of high altitudes (especially low oxygen concentrations and temperatures), species found at higher altitudes should have greater dispersal abilities. (Muller 1974, after Glick 1939.)

drifting in seawater (e.g., coconuts) are quite large (Ridley 1930; Pijl 1972), and land snail faunas of many Pacific islands are derived from a disproportionate diversity of "microsnails" (i.e., snail species so small that they drifted as aerial plankton to colonize even the most isolated islands; see Vagvolygi 1975). But all else is not equal. Structural features that promote long-distance dispersal differ markedly among species and taxonomic groups. For bacteria, yeasts, fungi, mosses, and ferns, dispersal is effected by tiny spores that not only are readily transported by wind and water but also are highly resistant to extreme physical environments. They are light enough to float in water or remain aloft in the air for long periods.

Many invertebrates also have small propagules. Often, fertilized eggs or other life history stages encyst to form thick-walled structures that are metabolically inactive and capable of withstanding long periods of desiccation and wide ranges of temperature. Some protozoans, rotifers, worms, and crustaceans disperse over long distances by such means. For example, species of the brine shrimp, genus *Artemia*, live in highly saline pools and lakes throughout the world. When these bodies of water dry up, the shrimps can survive for months or years as encysted fertilized eggs. When the pools refill, the brine shrimps resume their life cycle. Dry eggs of one species of brine shrimp can be purchased at pet stores, and one need only add them to salt water to rear a new generation of *Artemia*. The tiny encysted eggs are also picked up by the wind and dispersed great distances. It is apparently because of this capacity for long-distance dispersal that *Artemia* are found in isolated localities such as Great Salt Lake in the western United States and the Dead Sea in Israel (see Browne and MacDonald 1982; Stephens and Gardner 1999; Abatzopoulos 2002).

Larger propagules are generally heavier and therefore require surface features to keep them aloft in wind currents. We mentioned earlier the adaptations of plants for aerial transport (see Figure 6.7); some invertebrates have equally innovative designs, including gossamer parachutes spun by certain spiders (order Araneae, class Arachnida) and the long dorsal filaments of mealybugs (Hemiptera).

Most marine organisms have free-living juvenile stages that drift near the surface of the water. These tiny planktonic propagules—invertebrate eggs and larvae, fish larvae, algal cells, and the like—move passively with ocean currents. Distances traveled by small versus large forms may differ, but all are controlled ultimately by circulation patterns of the oceans (see Chapter 3). Strong currents may move propagules quickly from one locality to another in a definite direction (see Roughgarden et al. 1987). The wind that drives oceanic circulation also disperses some species, such as the jellyfish (*Vellela* spp., Cnidaria) that are equipped with "sails" that are oriented to the right or left and therefore carry individuals in somewhat predictable directions (**Figure 6.9**).

Although terrestrial vertebrates generally are too large to be carried far by the wind, some can be transported over surprising distances by water, often as passengers on mats of drifting vegetation or rafts of other debris. Sometimes, very large rafts containing entire trees and carrying a large variety of organisms are washed out to sea from large tropical rivers, remaining intact for extremely long distances (Ridley 1930; Carlquist 1965). Rodents and, especially, lizards apparently have managed to colonize many isolated oceanic islands by this means. Lizards are probably particularly favored by their capacity to survive relatively long periods in salt water without sustenance. Although the chance of a particular raft's going ashore on a given island is very small, over sufficiently long periods of time there is a high probability of at least a few such events. It is through the chance occurrence of such individually

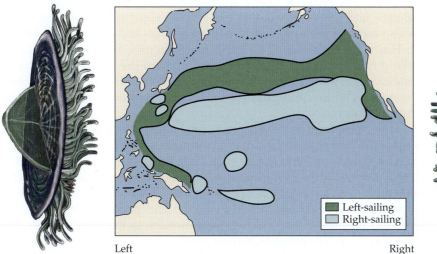

Left   Right

FIGURE 6.9 Like their cnidarian relative the Portuguese man-of-war, individuals of the jellyfish *Velella* spp. ("by-the-wind sailors") possess "sails" that are oriented so that they tend to drift either to the right or to the left of prevailing winds. Each form is restricted to certain regions of the ocean, and it is thought that by sailing in the wind, these animals counteract their tendency to be carried off by ocean currents and thus remain within a limited area. (After Savilov 1961.)

unlikely episodes of long-distance dispersal that oceanic islands and many other isolated habitats acquire much of their distinctive biotas. Implicit in much of this is the influence of habitat preferences on long-distance, passive dispersal. The species most likely to be carried by winds and water currents, or on logs and vegetation mats set adrift, are those inhabiting beachfronts and coastal habitats (Carlquist 1965, 1966, 1974, 1981; MacArthur 1972; J. M. B. Smith et al. 1990; J. M. B. Smith 1994; Guzman and Vargas 2009).

A surprisingly large number of small animals are transported over long distances by other animals. Many examples of this process, called **phoresy**, are obvious. Parasites, for example, may be carried to new areas colonized by their hosts. As Europeans explored and colonized the Earth, they spread a variety of bacterial and viral diseases. In some isolated areas, the native human populations had never been exposed to these microbial pathogens and had evolved no immunity to them. Consequently, diseases such as smallpox, syphilis, and measles had devastating effects when introduced among American Indians and Pacific Islanders (Marks and Beatty 1976; McNeill 1976; Diamond 1997). Conversely, being relatively disease-free may give small founding populations of animals a fitness benefit when they colonize remote islands, by favoring their rapid population expansion and the invasive status of newly colonized species (whether introduced naturally or anthropogenically). This effect has been invoked in the theoretical literature on invasive species and, in particular, to explain avian biogeography of the Caribbean (see Ricklefs and Bermingham 2002; see Chapter 14).

Exozoochory is a process analogous to phoresy, but in this case it refers to dispersal of plant propagules (e.g., sticky or barbed seeds or fruits) that attach to the external tissues of mobile animals (see Figure 6.7). Aquatic birds may transport small invertebrates in their feathers—or, as Charles Darwin demonstrated in one of his classic studies, trapped in mud on their feet—between widely separated lakes and ponds (see also Figuerola and Green 2002). As we noted earlier, other seeds—usually those found in sweet, fleshy fruits—are adapted to pass through the digestive tracts of animals: a process termed **endozoochory**. Because they are resistant to digestive juices, they are still viable when contained in dropped feces; in fact, the germination of certain seeds is enhanced when they are exposed to animal digestive chemicals. Such internal transport is extremely common and is important, not only in the colonization of isolated oceanic islands, but also in dispersal within continental regions from one habitat to another. Birds are the usual agents, but mammals and reptiles (including some now-extinct species of megafauna) can also serve as internal seed dispersers (Janzen and Martin 1982; but see also Howe 1985; Nathan et al. 2008). In the tropics, fruit bats have a particularly important, and to some degree underappreciated, role as agents of endozoochory (e.g., Whittaker and Jones 1994; Shilton et al. 1999). Similarly, insects and mycophagous mammals are important agents of dispersal of fungal spores.

Finally, a few animals are obligatory hitchhikers. Colwell (1973, 1979) and his associates discovered a fascinating example of this in tiny mites that live in certain flowers (see also Poinar et al. 1990; Brown and Wilson 1992; Zeh et al. 1992; Athias-Binche 1993). These mites lack sufficient resistance to desiccation and, thus, are poorly adapted for long-distance dispersal. However, the mites occur in flowers regularly visited and pollinated by hummingbirds, and they are transported between flowers in the birds' beaks. They crawl into the nasal openings, whose protected microclimate enables them to survive and be carried for long distances. At least two species of these specialized flower mites occur on isolated islands of the Greater and Lesser Antilles. Presumably, they originally colonized these islands from the continental mainland of tropical America, riding hundreds of kilometers on the beaks of dispersing hummingbirds.

Despite the remarkable diversity of specific mechanisms and strategies for long-distance dispersal, some generalities are emerging. The likelihood of each particular colonization event is extremely remote; however, given the great abundances of propagules and dispersal vectors (winds, currents, or other species) and given sufficient time periods, long-distance dispersal is quite common. It is often directional and selective in terms of favoring certain species and phenotypes over others. It is thus of paramount importance for understanding the diversity and composition of isolated biotas (Cook and Crisp 2005a; de Queiroz 2005; Cowie and Holland 2006; Leathwick et al. 2008; Nathan et al. 2008).

The Nature of Barriers

Successful long-distance dispersal usually requires that organisms survive for significant periods of time in environments very different from their usual habitats. These unusual environments constitute physical and biological barriers that successful colonists must cross. The effectiveness of such barriers in preventing dispersal depends not only on the nature of the environment, but also on characteristics of the organisms themselves. Of course, these characteristics vary from one taxonomic group to another, so particular barriers may not affect all residents of a region equally. Thus, barriers are species-specific phenomena. Two examples will illustrate this point.

Most freshwater zooplankton have resistant stages that facilitate long-distance dispersal. Consequently, the same species may be found in widely separated localities wherever environmental conditions are similar, such as in the cold temperate lakes of northern North America, Europe, and Asia. On the other hand, fishes inhabiting these same lakes appear to be unable to disperse across terrestrial and oceanic systems. Consequently, the same fish species inhabit only those lakes that have been connected by freshwater at some time in the past. Many isolated alpine lakes that have never had such connections will support a diverse plankton fauna but have no fish at all (unless they have been recently introduced by humans).

Similarly, many of the isolated mountains of arid regions in the southwestern United States are essentially islands of cool, moist forest in a sea of hot,

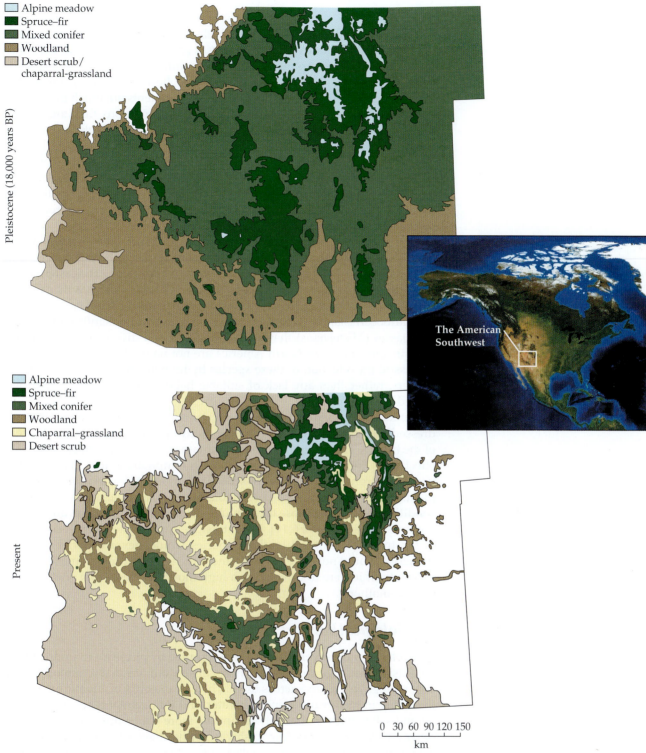

Alpine meadow
Spruce–fir
Mixed conifer
Woodland
Desert scrub/chaparral-grassland

Pleistocene (18,000 years BP)

Alpine meadow
Spruce–fir
Mixed conifer
Woodland
Chaparral–grassland
Desert scrub

Present

The American Southwest

0 30 60 90 120 150
km

FIGURE 6.10 Geographic shifts in vegetation zones of the American Southwest since the most recent glacial maximum. (Elevational shifts in these vegetation zones are depicted in Figure 9.22.) (From Lomolino et al. 1989.)

dry desert. The desert is not a barrier to most birds and bats, which rapidly and repeatedly fly over it to colonize the montane forests. On the other hand, for many small terrestrial mammals and amphibians, which must disperse much more slowly on foot, the desert represents a formidable barrier. It is likely that these animals colonized the most isolated mountains when they were connected by bridges of suitable forest and woodland habitat during the Pleistocene (**Figure 6.10**; see Findley 1969; J. Brown 1978; Patterson 1980; Lomolino et al. 1989; Frey et al. 2007).

Because the effectiveness of a particular kind of barrier depends on both the physical and biotic challenges it poses and the biological characteristics of the organisms attempting to cross it, it is difficult to make generalizations about the nature of barriers. It is usually true, however, that organisms that inhabit temporary or highly fluctuating environments are much more tolerant of extreme or unusual physical and biotic conditions than are species that are confined to more permanent or stable habitats. Plants and animals from fluctuating environments are also more likely to have resistant life history stages, which not only enable them to survive long periods of unfavorable conditions but also can serve as effective propagules. Consequently, the "weedy" species that inhabit unpredictable environments are likely to be better dispersers and less limited by barriers than are species from more permanent or predictable habitats.

A comparable example from the aquatic realm is provided by the pupfish, *Cyprinodon variegatus*, an extremely **euryhaline** and **eurythermal** species that inhabits estuaries, tidal flats, and mangrove swamps along the eastern coast of North America from Cape Cod to Yucatán. This little fish has managed to cross hundreds of kilometers of ocean to colonize similar habitats on many of the Caribbean islands. In contrast, the strictly freshwater sunfishes (*Lepomis*) and basses (*Micropterus*) that are so abundant and diverse in rivers, lakes, and streams of eastern North America are not native to the Caribbean. The successful introduction of these species by humans indicates that saltwater barriers, rather than any lack of suitable habitats on the islands, had prevented their colonization.

Daniel Janzen (1967) made a related point in his insightful paper that addressed the restricted distributions of many species in mountainous areas of the tropics. He pointed out that mountain passes of a given elevation are effectively "higher" in the tropics than in temperate zones. Because temperate environments experience seasonal temperature variations, lowland plants and animals must be able to tolerate a wide range of environmental conditions, including those they would encounter at higher elevations during the summer (**Figure 6.11**). In contrast, because of the thermal constancy of the tropics, a tropical lowland species would never experience—and, therefore, would not have to withstand—the temperature regimes it would encounter if it were to travel over a high mountain pass from one lowland area to another.

Physiological barriers

Probably the most severe barriers are presented by physical environments so far outside the range an organism normally encounters that it cannot survive long enough to disperse across them. The vast majority of aquatic organisms live either in the oceans or in freshwater, and they cannot regulate their water and salt balance sufficiently to survive more than a brief exposure to the other environment. Very few freshwater fishes have successfully colonized across the oceans, but there is a clear advantage to those with diadromous ancestors (i.e., those that regularly migrated between fresh and salt water). The same is true of amphibians, which have permeable skins and are quite intolerant of exposure to salt water. Likewise, many terrestrial plants are unable to withstand prolonged exposure to seawater at any stage of the life cycle, including the seed. As we observed earlier, such barriers can be highly selective—favoring only a limited subset of potential colonists. For example, Hnatiuk (1979) reported that 56 of the 69 terrestrial plant species native to Aldabra Atoll in the western Indian Ocean tolerated total immersion of their seeds in seawater for eight weeks with no inhibition of germination. Not

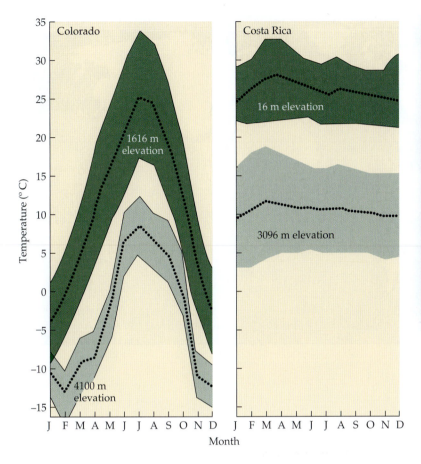

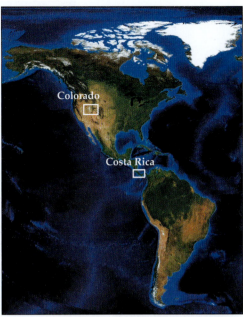

FIGURE 6.11   A given change in elevation tends to be a greater barrier to dispersal in the tropics than at higher latitudes, as shown by these temperature profiles. In tropical regions, sites separated by several thousand meters of elevation usually experience no overlap in temperature, whereas in the temperate zone, winter temperatures at low elevations broadly overlap summer temperatures at much higher sites. (After Janzen 1967.)

only would this tolerance facilitate the survival of these species on tiny islands subject to wave splash during heavy storms, but it would also greatly increase the probability of these species surviving immersion during transport across ocean barriers, thus perhaps explaining how many of them managed to colonize the islands in the first place.

Environmental temperature regimes can also serve as physiological barriers. We have already mentioned the case of mountain passes in the tropics. A similar barrier is created by the tropics themselves, which form a band of high temperatures around the Equator, isolating the cooler temperate and Arctic areas toward either pole. For many cold-adapted organisms—both terrestrial and aquatic—these warm tropical climates are a major barrier to dispersal, resulting in **amphitropical** (sometimes referred to as bipolar or antitropical) distributions of many species (see Figures 10.19 and 10.28). Some groups, such as the Nearctic avian family Alcidae (auks, puffins, guillemots, and murres), are good dispersers within their climatic zone and are broadly distributed throughout one hemisphere (Northern or Southern) but have not managed to cross the tropics to colonize the other hemisphere (**Figure 6.12**). Thus, the two hemispheres are inhabited by different forms with convergently similar morphology and ecology, such as penguins (Spheniscidae) and diving petrels (Pelecanoididae) of the Southern Hemisphere, and auks, auklets and puffins and (Alcidae) of the Northern Hemisphere, respectively.

The nature of barriers may also vary dramatically with the seasons. For example, in temperate regions of North America, large bodies of water serve as effective barriers to the movement of many terrestrial species, including non-volant mammals. However, during winter these waters freeze, providing a seasonal avenue for the dispersal of winter-active species to islands or

FIGURE 6.12   The avian family Alcidae (auks, puffins, and murres) is restricted to cooler regions of the Northern Hemisphere (shaded area), even though all of these birds are winged and most are strong fliers. The tropics appear to constitute a major barrier to the dispersal of this group, although the presence of distantly related but ecologically convergent seabirds in the Antarctic region may also prevent their colonization of the Southern Hemisphere. (After Shuntov 1974; photo © Stockbyte.)

adjacent mainland sites (**Figure 6.13**) (Jackson 1919; Banfield 1954; Lomolino 1984, 1988, 1989, 1993b; Tegelstrom and Hansson 1987).

Ecological and psychological barriers

Dispersing organisms must be able to survive not only the physiological stresses imposed by the environments they traverse, but also the ecological hazards. Just as predation and competition can limit the local abundances of species, they can also prevent successful dispersal and colonization. One would expect biotic interactions to be particularly important components of barriers for animals that neither move very rapidly nor disperse at a resistant stage. Although it is likely that competition and predation limit the dispersal of such organisms, we are not aware of any well-documented examples.

Surprising as it may seem, behavioral or psychological barriers appear to play a major role in preventing the long-distance dispersal of some, and perhaps many, organisms. Most organisms appear to possess mechanisms of **habitat selection**, which is the ability to recognize and respond appropriately to favorable environments. In some animals these traits are so well developed that they strongly inhibit active dispersal across particular habitats and barriers. For example, some forest birds that seem perfectly capable of flying long distances are apparently unwilling to disperse across open habitats, including open waters. Willis (1974) has described species of antbirds (Formicariidae) that have become extinct on Barro Colorado Island and have not recolonized, even though they would have to fly only a few hundred meters across Gatun Lake. MacArthur and colleagues (1972), Diamond (1975b), and others have documented groups of tropical birds, such as New World cotingas and toucans and Old World barbets and pittas, that are strong fliers but that are repeatedly absent or poorly represented on oceanic islands,

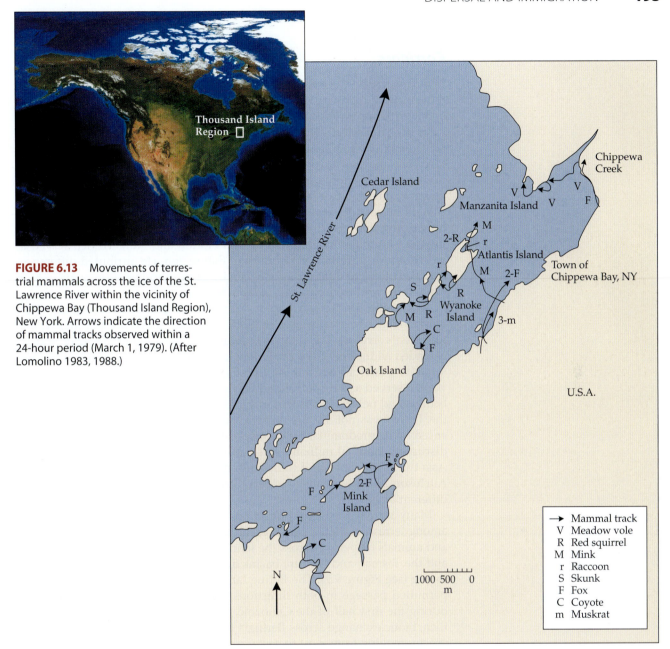

FIGURE 6.13 Movements of terrestrial mammals across the ice of the St. Lawrence River within the vicinity of Chippewa Bay (Thousand Island Region), New York. Arrows indicate the direction of mammal tracks observed within a 24-hour period (March 1, 1979). (After Lomolino 1983, 1988.)

even where good habitat appears to be abundant. Often species restricted to mature rain forest are particularly sedentary, whereas other, even closely related, species that are characteristic of second-growth habitats or edges are good dispersers.

Biotic Exchange and Dispersal Routes

As we emphasized in the preceding section, barriers are species-specific phenomena. Even within a group of related species, a barrier to one is not likely to be a barrier to all. Now we shall broaden our perspective to look at entire biotas and the process of biotic exchange between biogeographic regions. Biogeographers often distinguish three kinds of dispersal routes based on their effects on biotic exchange. Listed in order of increasing resistance to biotic exchange, these include corridors, filters, and sweepstakes routes (Simpson 1940).

Corridors

In biogeography, the term **corridor** refers to a dispersal route that permits the movement of many or most species of a particular taxon from one region to another (Simpson 1936, 1940; see also Udvardy 1969). A biogeographic corridor does not selectively discriminate against any form, and must therefore provide an environment similar to that of the two source areas. A corridor therefore allows a taxonomically balanced assemblage of species to cross from one large source area to another so that each area obtains organisms that are representative of the other. Any differences between biotas on either side of the corridor result either from chance or from ecological and evolutionary responses to different environments. In Simpson's example (1940), the pre-Columbian landscape between Florida and New Mexico served as a corridor for terrestrial vertebrates of that region. He attributed differences between biotas to differences between the subtropical environments of Florida and the arid environments of New Mexico.

As we will discuss in Chapters 8 and 9, the geological record is replete with evidence of great pulses of biotic exchange across ancient marine and terrestrial corridors. The Tethyan Seaway, a circum-equatorial marine system, extended all the way from the Orient and Malaysia to westernmost Europe, separating Africa from Eurasia (see Figure 8.22G,H). For nearly 500 million years it served as a dispersal highway for marine organisms, including both benthic and pelagic forms (Adams and Ager 1967). Exchange of taxa across the Tethys was not always uniform, and it may have been strongly influenced by its predominantly westward-flowing currents. This exchange was disrupted about 60 million years ago when Africa became connected with Asia through Arabia, and India drifted northward to join Asia (see Ali and Aitchison 2008). Now the Mediterranean Sea and Indian Ocean have vastly different biotas.

Many ancient, transient connections have served as corridors for exchange among terrestrial biotas. Most recently, a great diversity of terrestrial plants and animals dispersed across the land bridges of Beringia, the Sunda Shelf, and the Tasman and Arafura basins during the glacial maxima of the Pleistocene (see Figure 9.9). The Bering land bridge between Alaska and Siberia permitted passage in both directions, but it was probably a corridor only during the first half of the Cenozoic, when a mild climate prevailed. Even then, biotic exchange across Beringia tended to be asymmetrical, with more species moving from west to east (i.e., from the relatively diverse regions of eastern Asia and Siberia into Alaska).

Filters

As its name implies, a **filter** is a dispersal route that is more restrictive than a corridor. It selectively blocks the passage of certain forms while allowing those able to tolerate the conditions of the barrier to migrate freely. As a result, colonists tend to represent a biased subset of their respective species pools. Thus, the biotas on the two sides of a filter share many of the same taxonomic or functional groups, but some taxa are conspicuously absent in each.

The Arabian subcontinent is a harsh filter that permits the dispersal of only a limited number of mammals, reptiles, nonpasserine ground birds, invertebrates, and xerophytic plants between northern Africa and central Asia. Organisms that need abundant water, such as freshwater fishes, most amphibians, and forest dwellers of many taxa, are stopped by the deserts in this region. This has not always been true. As mentioned above, for millions of years Arabia and Asia were widely separated by the Tethyan Seaway—a formidable barrier to terrestrial organisms. When the Arabian Peninsula became emergent and contacted Persia, the region intermittently served as a mesic land bridge for movements of Asian taxa into Africa and vice versa, as in the Upper Miocene, when many African forms appeared in Asia for the first time (Cooke 1972). Thus, today's filter has been both more of a corridor and a more formidable barrier in the past, and it might become either in the future.

Filters may be produced by abiotic or biotic factors. They are generally easy to identify because the number of species in certain taxa decreases in a regular manner with distance from the source area. Filters often form transition zones between two biogeographic regions. **Figure 6.14** depicts such a transition zone: a two-way filter formed by the Lesser Sunda Islands that lie between Java and New Guinea. On these islands, reptile species of Oriental origin decrease proportionately as one moves eastward, where Australian groups become dominant; of course, a converse trend is found for Australian groups as one moves westward through Wallacea (i.e., the transition zone bisected by Wallace's line).

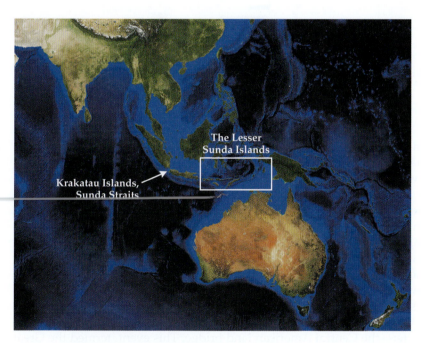

FIGURE 6.14 The Lesser Sunda Islands between Java and New Guinea serve as a two-way filter for the reptilian faunas of southeastern Asia and Australia. Bars quantify the decline in Oriental species and the increase in Australian species going from west to east down the island chain. (After Carlquist 1965.)

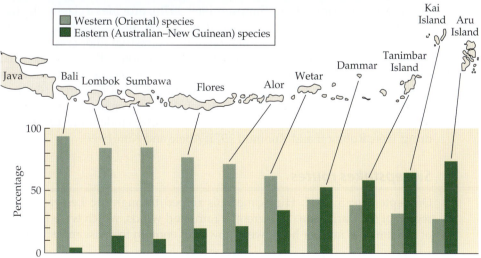

(A)

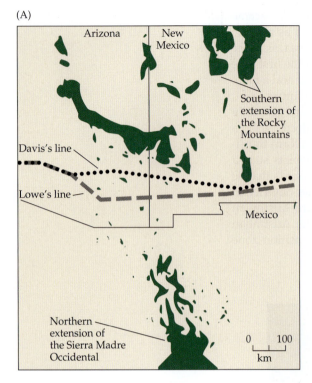

(B)

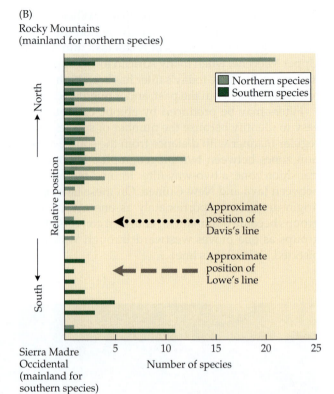

FIGURE 6.15 The desert "sea" of the American Southwest may serve as a transition zone, or two-way filter, for forest and woodland species of reptiles and mammals between the Rocky Mountain "mainland" to the north and the Sierra Madre "mainland" to the south (see Figure 6.10). (A) Two biogeographic lines—Lowe's line and Davis's line—have been proposed to mark the divisions between these regional faunas. (B) Bars quantify how the numbers of species of non-volant mammals decline with distance from their respective source regions. (A after Davis 1996; B after Lomolino and Davis 1997.)

On the opposite side of the globe, biogeographers have discovered a terrestrial version of Wallacea. For most of the Pleistocene, assemblages of forest-dwelling terrestrial vertebrates in western North America have been isolated in two major "mainland" regions—forests of the Rocky Mountains to the north and the Sierra Madre Mountains to the south. As we observed earlier in this chapter, however, intervening habitats have allowed some mixing among these regional biotas, especially during glacial maxima when forests and woodlands expanded at the expense of more xeric habitats (see Figures 6.10 and 9.22). Thus, the xeric landscapes of the American Southwest serve as a two-way filter or transition zone between the two faunal regions (**Figure 6.15**; see Davis 1996; Lomolino and Davis 1997).

In Chapter 10 we will discuss the biotic exchange across another terrestrial filter—the Central American land bridge. This event, termed the Great American Interchange (G. G. Simpson 1969, 1978), provides a classic illustration of the combined effects of dispersal, evolution, and extinction. As many have observed, however, the formation of this terrestrial filter created a barrier to the dispersal of marine organisms. Construction of the Panama Canal reunited the Caribbean Sea and the Pacific Ocean by means of a continuous waterway, but biotic exchange between them has been extremely limited. Low-salinity waters flowing from Gatun Lake serve as an effective physiological barrier to most marine animals (**Figure 6.16**; Hildebrand 1939; Woodring 1966; Rubinoff and Rubinoff 1971; Abele and Kim 1989).

Sweepstakes routes

On September 5, 1995, Hurricane Luis ripped through the Lesser Antilles region of the Caribbean. Its torrential rains and winds, which were in excess of 225 km (140 mi) per hour, were powerful enough to cause massive land-

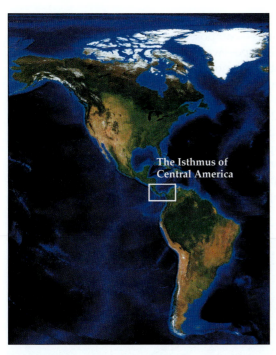

FIGURE 6.16 Map of the Panama Canal (in the Isthmus of Central America) and associated waterways, showing salinity levels (in ppt). Freshwater of Gatun Lake serves as an effective physiological barrier to most marine animals. (After Abele and Kim 1986.)

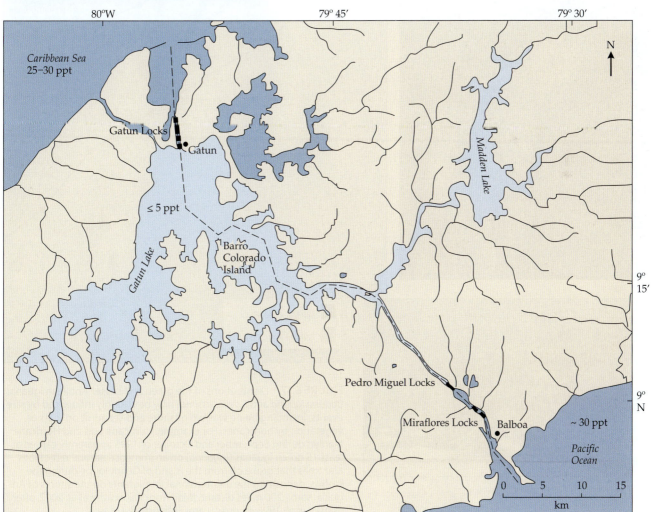

slides and send large tangled mats of vegetation adrift in the sea. Rafting along on one of the largest mats—composed of hundreds of logs and trees torn from the island of Guadeloupe—were at least 15 green iguanas (*Iguana iguana*). Incredibly, these castaways not only survived Hurricane Luis and Hurricane Marilyn—which followed closely on its heels—but also clung to and survived on their raft for a month until it washed ashore along the coast of Anguilla—some 320 km (200 mi) distant (**Figure 6.17**; Censky et al. 1998; see also Raxworthy et al. 2002; Rieppel 2002). The saga of these rafting lizards once again speaks to the importance of rare events (Darlington 1938; Nathan et al. 2008). Again, not everything is going to get everywhere, but things will definitely end up in surprising places and, as a result, such rare events strongly influence the geography of nature.

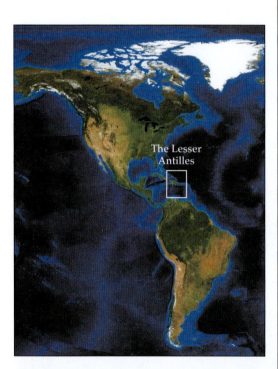

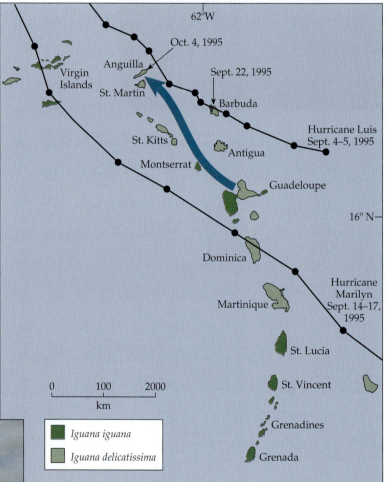

Iguana iguana

FIGURE 6.17   Distributions of the green and Lesser Antillean iguanas (*Iguana iguana* and *I. delicatissima*) may be strongly influenced by rare events, including being washed offshore and set adrift on mats of vegetation by hurricanes. For example, Hurricane Luis and Hurricane Marilyn during September and October of 1995 apparently carried at least 15 green iguanas (*I. iguana*) on an immense natural raft of logs and trees that broke off from the island of Guadeloupe during the storms, depositing the unintentional immigrants on the shores of Anguilla, some 200 miles distant. (Map from Raxworthy et al. 2002, after Censky et al. 1998; photo © Larry Ebbs/istock.)

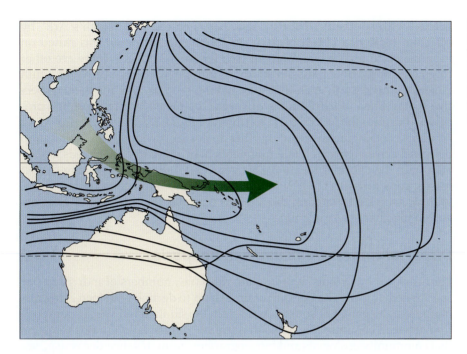

FIGURE 6.18 A classic example of a sweepstakes route. Lines show limits of the distributions of eight different families of land snails in Australasia and the South Pacific. These groups originated in Southeast Asia and then spread southward and eastward to different extents. (After Solem 1981.)

The distinguished paleontologist G. G. Simpson coined the term **sweepstakes route** to refer to barriers that must be crossed by such rare, chance interchanges (i.e., by long-distance, jump dispersals). When people enter a sweepstakes, they do so knowing that although many individuals enter the contest, only one or a handful of lucky ones will win prizes. Yet, because winners are so few and far between, prizes can be very high indeed. In the natural world, propagules continually disperse from established populations into new areas, but only a small fraction are ever successful in founding new populations. The arrival of such castaways on isolated and ecologically impoverished oceanic islands is often followed by adaptive radiations (see Chapter 7). The seemingly chance colonizations of isolated oceanic islands by lizards and many other terrestrial animals are classic cases of interchange across a sweepstakes route. This process is not as random as it may seem, however. Even when organisms cross barriers independently of one another, they may still use the same dispersal routes and be dispersed by the same, prevailing winds, storms, and ocean currents (e. g., sweepstakes dispersal of green iguanas followed the hurricane tracks through the Lesser Antilles). Thus, as we discussed earlier, most of the species inhabiting islands of the South Pacific have Indo-Malayan and, to a lesser extent, Australasian affinities, and the number of taxa shared with these landmasses decreases with increasing distance out into the Pacific (**Figure 6.18**; see also Figure 6.14; Wilson 1959, 1961; Solem 1981). Mainland ancestors of most species inhabiting these remote islands were probably preadapted for long-distance dispersal—so-called waif species. Thus, dispersal across these sweepstakes routes is random only in that it is impossible to predict when it will occur and which of these waif species will actually make it to any given island.

Other means of biotic exchange

When a landmass is shifted from one place to another by seafloor spreading, it carries a biota on board. These species can thus be transferred directly without crossing major barriers. India, which moved from its ancient connection with southern Africa to Asia, has been called a "Noah's Ark" by Malcolm McKenna (1973) because an assemblage of organisms rafted on it en masse to a new environment (see Figure 8.22 G–K). There are other fragments of tectonic plates in southern Europe, western North and South America, and eastern Asia that may have served as arks for their respective biotas. Such redistributions of species by means of plate tectonic processes illustrate the axiom that "Earth and life evolve together," providing accounts of changing distributions that do not rely on long-distance or jump dispersal across barriers. Plate fragments may also carry fossil beds associated with their former locations. Movements of these beds of long-extinct fossils, sometimes termed "Viking funeral ships," may complicate biogeographic analyses by mixing the remains of biotas that may never have occurred together at a single time.

Dispersal curves within and among species

For nearly all organisms, immigration rates (i.e., numbers of individuals or species arriving at a site per unit of time) tend to decline with increasing distance from the parent or source. As a result, the similarity in species composition between communities tends to decrease with isolation (a phenomenon that community ecologists and biogeographers term distance-decay; e.g., see Krasnov et al. 2005; Oliva and Gonzalez 2005; Steinitz et al. 2006). Depending on the mechanism of dispersal, this relationship between immigration rate and distance often takes one of two general forms (**Figure 6.19**). If dispersal is a purely random process with a constant probability or proportion of organisms "dropping out" with each increment of distance, then the dispersal curve should take the form of a negative exponential (see lower, red curve in Figure 6.19B). This pattern is often found among wind-borne propagules and other forms that are passively dispersed.

On the other hand, for actively dispersing organisms or those dispersing on logs or rafts with normally distributed persistence times, the frequency distribution of dispersal abilities may approximate a normal or lognormal distribution (see Figure 6.19A). The precise form of this frequency distribution is not critical; only that most individuals (or species) can disperse beyond some minimal distance, and that few can travel beyond some greater distance. In this case, the dispersal curve describes the number of individuals or species whose dispersal capacity is exceeded as distance or isolation increases. Immigration rates should at first remain relatively high until the distance exceeds the dispersal abilities of the least vagile individuals, then decline more rapidly until the modal dispersal ability is approached, and finally slow to asymptotically approach zero for the most distant sites. In such cases (upper curve in Figure 6.19B), immigration curves should approximate a sigmoidal function (which is similar to what MacArthur and Wilson [1967: 128] termed a "normal" function).

We will return to the biogeographic relevance of dispersal curves in our chapters on island biogeography. It is important to reemphasize here, however, that actual rates of dispersal or arrival at distant sites are influenced by the nature of dispersal barriers as well as by characteristics of the species in question.

Establishing a Colony

If dispersal is to be of biogeographic significance, not only must an organism be able to travel long distances and cross barriers, but it also must be able to establish a viable population upon its arrival at a new site. Several fac-

(A) Dispersal capacity

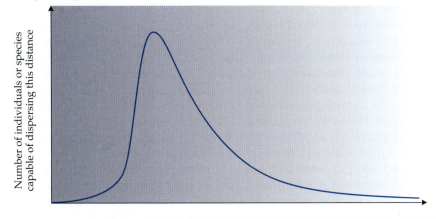

Number of individuals or species capable of dispersing this distance

Dispersal capacity (maximum distance)
units arbitrary

(B) Immigration rate

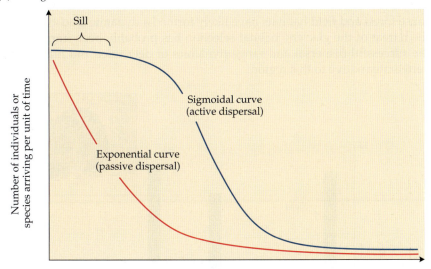

Number of individuals or species arriving per unit of time

Sill

Sigmoidal curve
(active dispersal)

Exponential curve
(passive dispersal)

Isolation (distance from source)

FIGURE 6.19 (A) An idealized lognormal frequency distribution of dispersal capacities among individuals or species. (B) Dispersal curves describe the expected decline in arrival or immigration rate as a function of increasing distance from a source region. If dispersal is a passive and random process with a constant probability that organisms will drop out with each increment of distance, then the dispersal curve should take the form of a negative exponential (red line). On the other hand, if dispersal is active or if organisms are carried on rafts with normally or lognormally distributed persistence times, then the dispersal curve should take the form of a sigmoidal function (blue line), with the extent (range in distance) of the upper threshold increasing with dispersal capacity of the species or faunal group. (After MacArthur and Wilson 1967.)

tors, including interspecific interactions, habitat selection, and reproductive strategies, may play little role in dispersal per se, but they may be of great importance in determining the fate of an immigrant once its journey is over (Silvertown 2004).

Influence of habitat selection

Highly mobile organisms actively seek out favorable environments, which they recognize either instinctively or from having learned the characteristics of their place of origin. Many passively dispersed organisms, such as the planktonic larvae of sessile marine invertebrates, will not settle unless they perceive certain sensory cues indicating a substrate that is suitable for establishment. Even the seeds and spores of plants and the cysts of invertebrates have some capacity for habitat selection. Although these resistant structures are passively dispersed, certain specific conditions of temperature, moisture, light, and other factors are usually required to break their dormancy and initiate growth and development. Unless cues indicating favorable conditions are received, the diaspore remains in the resistant stage and is capable of further dispersal.

Stanley Wecker (1963, 1964; see also Harris 1952) performed a classic study of habitat selection in deer mice (*Peromyscus maniculatus*). He worked with two subspecies that inhabit the north central and northeastern regions of the United States: *Peromyscus maniculatus gracilis*, a form primarily restricted to mature mesic forests of the Alleghenies and northward, and *P. maniculatus bairdi*, which inhabits grasslands of the Midwest (see Figure 7.6, subspecies numbered 45 and 42, respectively). The latter subspecies is of particular interest because it is an animal that originally inhabited the prairies of central North America but extended its range eastward by colonizing agricultural cropland and old fields as the forests were cleared by Europeans (Hooper 1942; Baker 1968). Wecker conducted a series of experiments in which he reared mice in different environments and then tested them in a large outdoor enclosure that was half old field and half forest. Each subspecies tended to wander and explore until it found its appropriate habitat (*P. m. bairdi* in fields, *P. m. gracilis* in forests), whereupon it would establish residence (**Figure 6.20**). These mice used genetically inherited information, reinforced by early experience in their natal habitat, to select the correct environment. Because forest and field habitats have likely formed a dynamic landscape since the retreat of the glaciers, habitat selection has probably been important to the successful dispersal and range dynamics of deer mice and many other vertebrates native to this region.

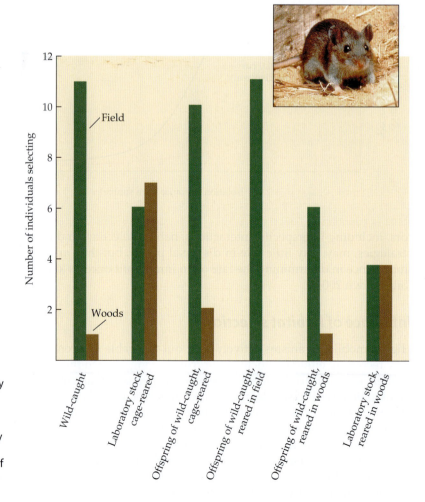

FIGURE 6.20   Habitat selection by the field (grassland) subspecies of the deer mouse (*Peromyscus maniculatus bairdi*) depends on both inherited ability and early experience. Wild-caught individuals and their offspring—especially if they were exposed to field habitat at the time of weaning—spent most of their time in the field end of a large enclosure that covered a field–forest boundary. Many generations in the laboratory or early experience in the woods diminished this preference. (After Wecker 1963; photo courtesy of James Gathany/CDC.)

What constitutes a propagule?

In addition to finding the right kind of habitat upon arrival, successful colonists must be able to reproduce and establish a viable new population. The number and composition of the dispersing units are of critical importance in this regard. In MacArthur and Wilson's (1967) insightful chapter "The Strategy of Colonization," they used the term **propagule** to define the unit necessary to found a new colony, emphasizing that it varied substantially but predictably among species, depending on their reproductive patterns and behavior.

Colonization is a major challenge for the many higher plants and animals that reproduce sexually, because reproduction normally requires the participation of two individuals of different sexes or mating types. Since each case of long-distance dispersal is a rare event, it is especially unlikely that potential mates will arrive sufficiently close to each other in space and time to find each other and reproduce. In contrast, asexually reproducing organisms are often successful colonists because a single individual can found an entire population by fission, budding, or some other asexual means (see Browne and Mac-Donald 1982). Although many microbes and lower plants and animals are capable of sexual reproduction, many of them can also reproduce asexually, as indeed do many higher plants. This includes an impressive diversity of plants that are capable of asexual reproduction by vegetative growth of their rhizomes, stolons, root crowns, or bulbs, and some may even propagate by roots formed from fallen stems or branch tips that come into contact with the soil, eventually becoming separate physiological entities (ramets) from the original rooting point. Similarly, the freshwater zooplankton is often teaming with crustaceans that can go through numerous generations of asexual reproduction after emerging from encysted zygotes (which are produced by a sexual generation that occurs periodically during the life cycle).

Among higher plants and animals, there are some species with reproductive patterns that make them much better colonists than others. A large number of plant species are both hermaphroditic and self-compatible; that is, the same individual has both male and female flower parts and can produce viable seeds by self-fertilization. This is true of many species, such as tumbleweed (*Salsola iberica*), that we normally think of as weeds. Dandelion (*Taraxacum*), another common weed, is **apomictic**: It can produce viable seeds from unfertilized ovules. In all of these kinds of weeds, a single seed can serve as a propagule. Animals also exhibit a marvelous diversity of reproductive strategies. A few species of fish, amphibians, and reptiles are parthenogenetic: The entire population of these species consists solely of females that produce female offspring asexually (White 1973; Cuellar 1977). Some fishes can also change their sex, depending on the sex of other members of their local population.

Although these reproductive patterns do not necessarily mean that the organisms will always be good colonists, they appear to facilitate successful colonization in some species. At least three genera of gecko lizards, for example, contain parthenogenetic species that are distributed primarily on oceanic islands (Cuellar 1977). It takes only a single individual, drifting ashore on a floating log or arriving in an islander's canoe, to establish an entire new population. On the other hand, sexuality, if it is associated with aggressive behavior, may confer a competitive advantage to some invading lizard species over asexual natives (Petren et al. 1993; Case et al. 1994).

Some sexual species have reproductive cycles and social behaviors that facilitate simultaneous colonization by several individuals. In animals with internal fertilization, a propagule can consist of a single female with stored sperm or developing embryos. In plants in which several seeds are normally retained and dispersed in a fruit, the fruit can be the unit of colonization even in self-incompatible, obligately sexual species. Some birds, which have virtually none of the characteristics listed above, are good dispersers and successful colonists, not only because they can fly long distances, but also because they normally travel in flocks, which obviously increases their chances of establishment in new locations. Thus we find that fruit pigeons (*Ducula*)—medium-sized, highly social birds that often travel in flocks far out to sea—are widely distributed in the tropical Pacific (Diamond 1975b) and were even more widespread before aboriginal humans colonized those islands (see Chapter 16; Steadman 1995, 2006).

Among higher plants, just as the dispersal mechanism and breeding system may be important to establishment in a remote isolate, the pollination system may also be crucial to the prospects of establishing a breeding population. An intriguing illustration of the complexities involved in colonization comes from figs (genus *Ficus*, family Moraceae). Each species of fig (a genus of about 850 trees, shrubs, and hemi-epiphytes) depends for pollination on its own particular species of fig wasp (Agaonidae), tiny insects with very short life spans. Fig trees fruit asynchronously throughout the year, so there are always a few fruiting individuals available for the fig wasps. Establishment of a viable population of a fig species, however, depends both on its particular pollinating wasp and on the activities of vertebrates (in particular, birds or fruit bats), which ingest the fruits and disperse fig seeds.

Such tightly constrained mutualisms would seem to suggest high interdependence and low colonization chances. However, as evidenced by their colonization of the Krakatau islands (Indonesia), figs are rather good colonists, and fig wasps have remarkable abilities to locate fruiting trees, such that interpopulation distances of 40 km or more appear to be crossed by both fig seeds and fig wasps with surprising frequency (e.g., Parrish 2002; Zavodna et al. 2005; Shilton and Whittaker 2009; and see also Ahmed et al. 2009 for evidence of fig wasp pollination over distances of 160 km).

Survival in a new ecosystem

Because long-distance dispersal almost inevitably means colonizing a habitat that is not exactly like the one in which the colonists originated, successful colonists must be able to survive physical stresses and ecological challenges of their new environments. Thus, as we observed earlier, organisms from highly fluctuating and unpredictable environments have typically experienced strong selection for both dispersal ability and the capacity to cope with varied environmental conditions. Nonetheless, the biological hazards posed by competitors and predators (including parasites and diseases) should not be underestimated. Organisms from large, continuous areas with diverse ecological communities, such as productive continental habitats, tend to be relatively successful in establishing populations in small, isolated habitats containing few species, such as oceanic islands. Insular species, however, are rarely successful in invading mainland habitats, presumably because they are not adapted to cope with the variety of threats posed by a diverse array of competitors and predators (but see Dávalos 2007; Bellemain and Ricklefs 2008).

When small numbers of individuals are involved in colonization, as is almost always the case in long-distance dispersal, chance becomes very important in determining the course of events. Colonies may fail, for example, because a freak storm or an unlikely predator kills just one or two individuals. Thus, despite the great diversity of traits that enhance prospects of long-distance dispersal, even if a viable propagule arrives in a suitable environment, successful colonization is not assured; in fact, most colonizations fail. We know this from studies of well-documented, human-assisted introductions of hundreds of species of plants and animals—only a small fraction of these (roughly 10%) resulted in establishing a population in the introduced range (see Sax et al. 2005).

Advances in the Study of Dispersal

Throughout the history of the field of biogeography, a majority of its practitioners have acknowledged the fundamental importance of long-distance dispersal, but they have also been profoundly frustrated because, given its rarity, it was one of the most difficult, if not intractable, processes to study. Early students of the discipline would be envious of modern scientists, who can now marshal a diverse tool kit of recently developed technologies for retracing the origins and movements of most life-forms. This includes advances in a variety of disciplines, including molecular biology and genetics, biogeochemistry, and remote sensing and telemetry. As we will describe in detail in Chapters 10 through 12, historical biogeographers can now reconstruct the biogeographic histories of lineages (including not just the routes of long-distance dispersal, but the timing of those events as well) by analyzing genetic similarity of their extant descendants (and sometimes the remains of their ancestors, as well) (de Weerd et al. 2005; Rocha et al. 2005; Perrie and Brownsey 2007; Voigt et al. 2009). Biogeochemists are now capable of conducting analogous reconstructions by measuring the levels of **stable isotopes** in the tissues of individuals that may have dispersed long distances but from sites or during periods previously unknown to the investigators. The ratios of certain stable isotopes to each other (e.g., those of carbon [^{13}C], hydrogen [^{2}H], oxygen [^{18}O], nitrogen [^{15}N], sulfur [^{34}S], and strontium [^{87}Sr]) have characteristic geographical distributions. Therefore, assays for these isotopes in, for example, the feathers of birds, the shells of mollusks and other invertebrates, or the hair and bones of mammals (including living and extinct members of our own species and our ancestors) provide powerful means of retracing the migration paths and long-distance dispersal of individuals, populations, and entire biotas (Hobson 2002, 2005; Moore 2004; Elewa 2005; Cerling et al. 2006; Hobson and Wassenaar 2008; Sexton and Norris 2008).

Most of these techniques, as instructive as they are, constitute indirect means of studying dispersal. That is, they allow us to infer movement paths by connecting the dots between assumed sites of origin and colonization. Precious few methods, such as attaching radio collars to individuals or following the tracks of mammals laid down in the snow of a frozen lake or river, have allowed us to directly study this fundamentally important process. A growing number of scientists in these fields, however, are capitalizing on advances in remote sensing (including applications of various forms of radar, infrared imagery, and ultrasound detection) to visualize the three-dimensional movements of organisms through the aerosphere (the relatively thin portion of the troposphere, closest to the Earth's surface, that supports life). This new discipline and exciting frontier of science—**aeroecology**—holds great promise for unlocking many of the mysteries of dispersal and dynamic biogeography in general (**Figure 6.21**; Kellog and Griffin 2006; see the overview of this emerging discipline by Kunz et al. 2008).

These and other, perhaps unanticipated, advances in our abilities to investigate rare events, such as flights of insects, birds, and bats across sweep-

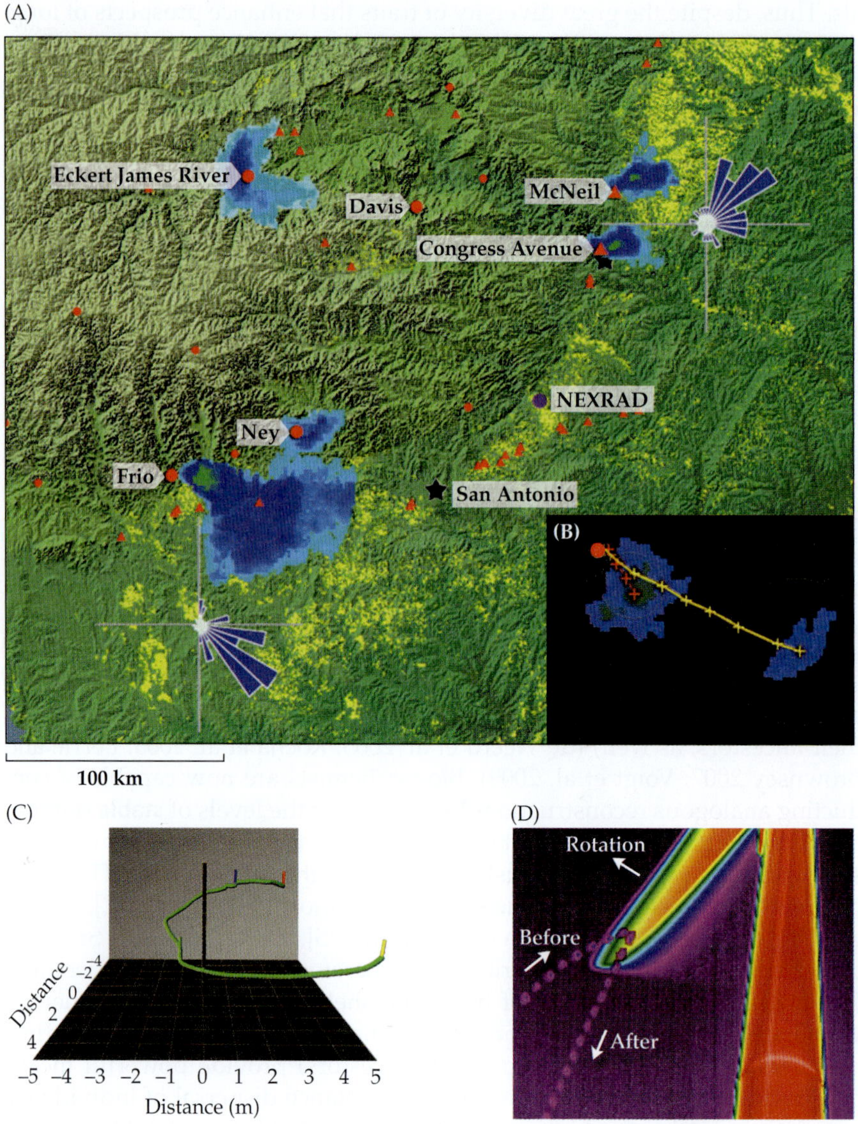

FIGURE 6.21 Applications of technological advances in remote sensing have enabled entire new disciplines, such as aeroecology (see Kunz et al. 2008), and have greatly enhanced our abilities to visualize rare events—such as dispersal of individuals or entire flocks or colonies of birds, bats, or insects—and identify the factors (including anthropogenic structures such as wind turbines) that influence movements. (A) NEXRAD (next generation doppler radar) was used to summarize the movement patterns of bats from caves (red dots) and roosting sites under bridges (red triangles) in the foothills and agricultural plains near San Antonio, Texas. General headings of these movements are shown in blue; agricultural areas are shown in yellow. (B) NEXRAD was used to generate a more local-scale image of Brazilian free-tailed bats (*Tadarida brasiliensis*) dispersing from Frio Cave (located in the lower left quadrant of Panel A) on May 19, 2005. The red circle marks the location of the colony; the first and second series of tracks of their movements are marked by yellow crosses and a yellow line and by red crosses, respectively. (C) Two thermal-imagery cameras were used to create this three-dimensional reconstruction of the flight of a bat pursuing a moth near a streetlight in southeastern New Mexico (the beginning and end of this tracked movement are marked in red and yellow bars, respectively). (D) A series of time-lapse infrared images document the mortality of a bat as it collides with a wind turbine blade at a wind energy center in West Virginia. (A and B from Horn and Kunz 2008; C from Hristov et al. 2008; D from Horn et al. 2008.)

stakes routes, will provide invaluable insights into the nature of dispersal and the ability of organisms to respond to the temporal and spatial dynamics of the geographic template (see Nathan 2005). We will focus on the latter subject (i.e., organisms' response to the dynamics of the Earth and its environments) in Chapters 8 and 9. Before we do that, we will review two other, fundamental biogeographic processes—speciation and extinction—and their influences on the geographic variation of nature, in Chapter 7.

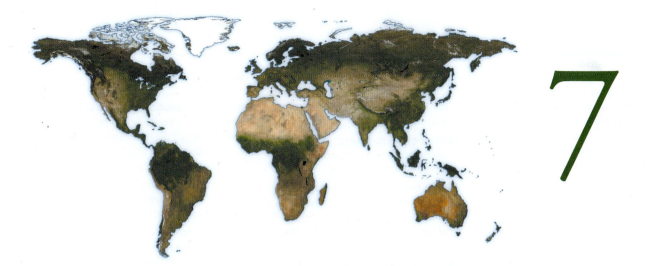

7

SPECIATION AND EXTINCTION

We began the last chapter with the proposition that the three fundamental processes of biogeography are evolution, extinction, and dispersal (inclusive of immigration). In this chapter we take up the first two processes—evolution and extinction—recognizing that both of these processes are strongly influenced by dispersal and that all three have a strong geographic context. Evolution occurs across space as well as across time, and the production of new species (**speciation**) often requires geographic isolation, while **extinction** is the final stage of the long-term, large-scale processes influencing geographic range collapse.

Species have histories that are somewhat analogous to the lives of individuals. New species originate through the multiplication of old species; they survive for varying periods of time, during which they may or may not leave descendants; and they die (extinction). There are, however, important differences between the histories of individuals and the histories of species. Whereas the births and deaths of most individuals are discrete, easily recognizable events, it is often difficult to determine when a new species has come into existence, and when the last individual of a species has died.

Species are composed of groups of populations. Even if a local population goes extinct, the species it belongs to does not (unless that population represented the entire species, e.g., the Devil's Hole pupfish; see Figure 10.4). As such, a species is considered to have temporal and spatial **cohesion** (see **Box 7.1** for definitions or descriptions of several terms new to this chapter). Cohesion can be maintained genetically in species with different sexes, or perhaps ecologically through adaptation to a specific set of niches (Templeton 1981). Whatever the mechanisms involved, it is this property of spatial and temporal cohesion that makes a species an evolving entity with both a discrete beginning and a discrete end, a place of origin and a final refuge, and thus an important entity in evolutionary biology and biogeography. (Note for example that the title of Darwin's most influential work began with "On the Origin of *Species*.")

BOX 7.1   *Glossary of Some Terms Used by Systematists and Evolutionary Biologists*

■■▮ Definitions or descriptions of specialized terminology new to this chapter and used frequently by systematists and evolutionary biologists in their exploration of the nature and history of species and evolutionary lineages—others will be introduced in Chapters 11 and 12.

Apomorphic In a transformation series, the derived character state.

Clade Any monophyletic evolutionary branch in a phylogeny, using synapomorphic characters to support genealogical relationships.

Cladogenesis The process of evolution that produces a series of branching events.

Cladogram A graphical depiction of evolutionary relationships among clades.

Cohesion The array of genetic and ecological components that serve to maintain the integrity of a species; particularly relevant to the cohesion species concept.

Homology A character shared by a group of organsims or taxa due to inheritance from a common ancestor.

Homoplasy A character that appears similar among a group of organisms or taxa, but the similarity is due to parallel or convergent evolution rather than inheritance from a common ancestor.

Mitochondrial DNA (mtDNA) The closed circular DNA of the organelle, transmitted through the cytoplasm from one generation to the next.

Monophyletic Of a group, or clade, of organisms that includes an ancestral taxon and all of its descendant taxa.

Nuclear DNA (nucDNA) The vast majority of DNA in a eukaryotic cell, transmitted through the nucleus from one generation to the next.

Plesiomorphic In a transformation series, the primitive character state.

Reticulate evolution or speciation The formation of new evolutionary lineages or species by the hybridization of dissimilar populations, for example, through interspecific hybridization.

Species concept Any one of some 22 criteria and approaches used to delineate separate species.

Symplesiomorphic In a transformation series, a character state shared by taxa in a focal clade but also shared by other clades at more basal nodes in a phylogeny and therefore not useful in diagnosing the focal group as being monophyletic.

Synapomorphic In a transformation series, a derived character state shared by taxa because of inheritance from a common ancestor and therefore useful in diagnosing a monophyletic group or clade.

Tokogenetic Reticulate (netlike or interwoven) relationships between individuals within a sexual species. ■▮■

As evolving entities, species have a unique set of properties. In sexual organisms, a species includes groups of individuals held together genetically through the process of reproduction, whereby they interbreed and exchange genetic material during each generation. Alternatively, the "clones" in an asexual species remain genetically distinct from one another across generations (**Figure 7.1**). Individuals within a sexual species are said to have a **tokogenetic** or **reticulate** (netlike or interwoven) set of relationships. On the other hand, the process of speciation represents the splitting (**cladogenesis**) of an ancestral species into two or more daughter species—each with its own evo-

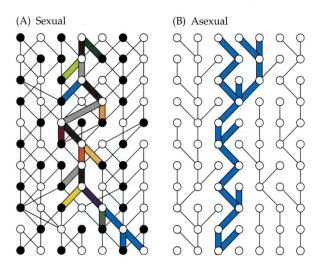

(A) Sexual   (B) Asexual

FIGURE 7.1   Contrasting modes of inheritance between (A) sexual and (B) asexual organisms in species level lineages. In sexual organisms, reproduction produces a complex network of relationships among individuals, whereas "clones" in an asexual species must be held together by some process other than reproduction. Circles represent individuals; black or white circles in A represent different sexes; and solid lines represent parent–offspring reproductive connections. Bold lines trace one set of parent–offspring connections across 11 generations in each case, with different colors noting identical genetic lineages (clones). (After de Queiroz 1998.)

lutionary trajectory. The species, therefore, represents the boundary between a reticulate network of genetic exchange among individuals and **cladogenetic** evolution *among* different species (**Figure 7.2**). Although this demarcation is relatively straightforward, it is also possible for different species to exchange genes through **hybridization**. Sometimes hybridization is restricted to an occasional cross-species mating at a narrow hybrid zone, but at other times it can produce a hybrid species with its own evolutionary trajectory, ecological niche, and geographic structure. Such **reticulate species** are particularly common in plants (Grant 1981), but they also occur in animals (e.g., the edible frog, *Rana esculenta*, is a hybrid species in the Palearctic Region, formed by matings between *R. lessonae* and *R. ridibunda* [see Figure 7.2C]; Spolsky and Uzzell 1986).

Like related individuals, related species possess particular combinations of characters that have been inherited from their common ancestors (**homologies**). Even if a character changes along the way (Darwin's "descent with modification"), it is still recognized as being homologous if it was inherited along an **evolutionary lineage** connecting descendant taxa with their common ancestor. Many characters inherited by descendants from ancestors might have arisen through an evolutionary force other than natural selection. However, some characters that make a lineage distinct represent innovative solutions (**adaptations**) to a problem that may have limited survival and reproduction in other lineages. These sorts of characters often are acquired independently in unrelated species that are faced with similar environmental challenges (e.g., the wings of birds, bats, and butterflies that are required for powered flight) and are called **homoplasies**, or **convergent** characters.

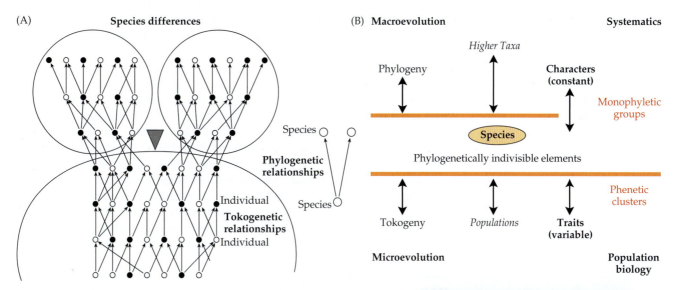

FIGURE 7.2 (A) Simplified version of Hennig's original depiction of the demarcation between ontogenetic (within individuals of a sexual species), tokogenetic (among individuals within a sexual species), and phylogenetic (between different species) relationships. (B) Diagram showing the placement of species in the delineation between tokogenetic relationships across populations (generally the interest of microevolution and population biologists) and phylogenetic relationships (generally the interest of macroevolution and systematists). (C) The edible frog, *Rana esculenta*, is an example of a species derived from hybridization, in this case formed by matings between *R. lessonae* and *R. ridibunda*. (A after Hennig 1966; B after Nixon and Wheeler 1990; photo © Marek R. Swadzba/Shutterstock.)

Further, although a particular lineage may flourish for a period of time and give rise to a diverse group of species, this success is usually ephemeral on a geological time scale. Eventually, the rate of extinction exceeds the rate of speciation, the diversity of the group decreases, and all or nearly all of its representatives go extinct. While the vast majority of species that were at one time alive went extinct without leaving descendants, it's obvious that some lineages have continued to survive in order to produce more species. The vast variety of living and extinct organisms on Earth have been produced through repeated cycles of speciation and extinction, often through rapid bursts of speciation following an abrupt mass extinction, and all set against a background of changes in Earth's geographic template (see Chapters 8 and 9).

Systematics

The science of biology that addresses patterns and processes associated with biological diversity and diversification (**systematics**) consists of two interrelated but distinct subdisciplines, **taxonomy** and **phylogenetics**. Taxonomy is the discipline that assigns names to organisms and classifies biological diversity using Linnaeus's binomials (species and genus) and a hierarchical classification scheme (genera within families, families within orders, etc.; **Table 7.1**). Phylogenetics is the discipline that reconstructs the history of evolutionary relationships between organisms—the study of cladistics is a subdiscipline of phylogenetics that studies cladogenesis (i.e., splitting but not reticulation) that results in separate evolutionary lineages. A number of the **species concepts** summarized below rely on explicit hypotheses of phylogenetic relationships to delineate separate species. However, apart from the few definitions needed to proceed through this discussion, we will defer a formal development of phylogenetic theory and practice until Chapter 11.

What are species?

If the fundamental unit of biogeography is the geographic range of a species, then what is a species, and what criteria shall we use to recognize different species? Biologists have long wrestled with the problem of how to recognize closely related but distinct species (Mayr 1982), and the debate is not over (e.g., Mayden 1997; Wheeler and Meier 2000; de Queiroz 2007). Several reviews have analyzed and summarized the large number of currently debated **species concepts** (e.g., Mayden 1997; Avise 2000; de Queiroz 2007). Each of a large number of modern species concepts (Mayden 1997) has its own strengths and weaknesses. Some may be more applicable to a given kind of organism than to others (e.g., sexual vs. asexual; plant vs. animal; fossil vs.

TABLE 7.1 *The Primary Units in the Linnaean System of Hierarchical Taxonomic Classification of Living Things*[a]

Kingdom	Animalia
Phylum	Chordata
Class	Mammalia
Order	Primates
Family	Hominidae
Genus	*Homo*
Species	*sapiens*

[a]Illustrated by the classification of the human species, *Homo sapiens*.

extant species) and may emphasize a particular aspect of what may turn out to be a more general theoretical framework (Mayden 1997; de Queiroz 2007), and some are considered more difficult to implement in practice than others (Mayden 1997).

Mayden (1997) identified 22 different species concepts but argued that only the **evolutionary species concept** (**ESC**) qualifies as a primary definition of species—all others are secondary concepts that serve as operational criteria for delineating species, but none is sufficiently general to serve as a primary concept (**Figure 7.3**). Although the ESC was originally developed by Simpson (1961), the most modern definition is "an entity composed of organisms that maintains its identity from other such entities through time and over space and that has its own independent evolutionary fate and historical tendencies" (Wiley and Mayden 2000: 73). Although it may seem counterintuitive, Mayden (1997) also suggested that the ESC is the only concept that is not operational, requiring maintenance of an array of secondary concepts to be employed in practice; a few of the more important of these are discussed briefly below and summarized in Figure 7.3.

Under the **morphological species concept** (**MSC**), "species are the smallest groups that are consistently and persistently distinct, and distinguishable by ordinary means" (Cronquist 1978: 15). This is also known as the **classical species concept** because it has, since the beginning of taxonomy, been the main operational criterion used to delineate different species. The main assumption is that morphological differences between species have an underlying genetic basis that is heritable, and it has been applied to asexual as well as sexual species. This criterion fails, however, when speciation occurs without detectable morphological change (e.g., in **sibling** or **cryptic species**) or when morphological traits are highly plastic (i.e., subject to rapid change—with or without genetic changes—as ecological conditions shift) within a single species.

The **biological species concept** (**BSC**) defines a species as "a group of interbreeding natural populations that is reproductively isolated from other such groups" (Mayr and Ashlock 1991: 26). This influential species concept arose from the Modern Synthesis in evolutionary biology that developed in the early twentieth century. Mayr (1963, 1970) considers reproductive isolation as the most fundamental attribute delineating closely related species, because it

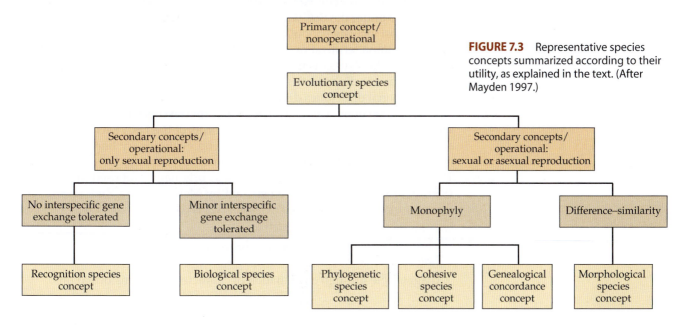

FIGURE 7.3 Representative species concepts summarized according to their utility, as explained in the text. (After Mayden 1997.)

protects a reproductive community of harmonious genotypes that have been molded by evolution to work within a particular ecological context.

As straightforward as it may sound, the criterion of reproductive isolation is difficult to employ. It cannot be applied to fossil organisms. The best one can do in such cases is to determine whether the morphological gaps between specimens are as large as—or larger than—those between living species that are reproductively isolated (i.e., reproductive isolation is inferred from an MSC perspective). The BSC is unwieldy for extant groups as well. The reproductive isolation criterion has little relevance in organisms that are exclusively asexual and do not exchange genetic information with one another (see Figure 7.1). While asexual reproduction is the hallmark of prokaryotes (bacteria, archea), it is also common throughout the fungal, plant, and animal world. Even for sexual species, application of the BSC can be difficult or in some cases impossible to apply. For instance, we know of many examples of **hybridization**: interbreeding that produces viable and fertile offspring, but which occurs between populations that remain distinct genetic and evolutionary units. This is especially problematic in vascular plants, in which hybridization is common between species and genera and has been known to occur even between plants in different families (Levin 1979; Grant 1981; Briggs and Walters 1984; Wyatt 1992). Biologists have argued that hybridization may frequently produce novel evolutionary lineages with equal or higher fitness than the parental lineages (Arnold and Emms 1998), which runs counter to Mayr's original emphasis on the importance of isolating mechanisms for protecting coadapted gene complexes. Finally, spatially isolated populations cannot interbreed under natural conditions, and so their potential to interbreed must be investigated under artificial laboratory conditions. Only in a few cases have such mating tests actually been conducted, because of the time and expense involved and the fact that many organisms cannot easily be reared in the laboratory. Usually, as in the case of fossil species, it has traditionally been left to a systematist to use a surrogate such as the MSC to make an educated guess about whether individuals would have interbred sufficiently freely to form a single population if the isolated populations had come back into contact under natural conditions. Mayden and Wood (1995) discussed ten criteria under which the BSC fails as a general and operational species concept.

One attempt to circumvent the difficulty of identifying reproductive isolating barriers under the BSC was the **recognition species concept** (**RSC**), defined as "that most inclusive population of individual, biparental organisms which share a common fertilization system" (Paterson 1993: 105), which focused on mate recognition systems within a species rather than reproductive barriers between species. Beyond this difference, however, the RSC shares many of the operational difficulties discussed for the BSC.

There are several forms of a **phylogenetic species concept** (**PSC**) that are presented by Mayden (1997) as operational forms of the ESC. Each of these has a common goal of defining species based on evolutionary lineages (clades) derived through the method of phylogenetic systematics (see Chapter 11). The concept of a phylogenetic species was originally proposed in the early 1980s (Cracraft 1983) and, as elucidated by McKitrick and Zink (1988), has three components. Under the PSC, a species is considered as

1. a monophyletic lineage, which was

2. derived through an evolutionary process of descent from an ancestral lineage and is

3. diagnosable through examination of character state transformations.

Those transformations, or derived characters, that uniquely define a monophyletic lineage are called **apomorphic** characters, to distinguish them from primitive, or **plesiomorphic**, characters. This definition considers the reproductive isolation criterion that is fundamental to the BSC as irrelevant, even misleading, in the delineation of species (McKitrick and Zink 1988; Mayden and Wood 1995) and is applicable to both sexual and asexual species.

One criticism of the PSC is that it might lead to an inflation of species diversity if species designations are based on few, and sometimes one, apomorphic character (Agapow et al. 2004). Another criticism is that, at the molecular level, phylogenies derived from independent parts of an organism's genome (e.g., mitochondrial vs. nuclear genes) will not always be exactly concordant in their delineation of species—the so-called gene tree versus species tree problem (Nei 1987), outlined in greater detail in Chapter 11. Both of these issues were considered under another form of the PSC, the **genealogical concordance concept** (**GCC**), which proposes that species should only be recognized as being distinct if independent genetic traits (e.g., a mitochondrial and a nuclear gene) concordantly recognize the same groups of separate populations (Avise and Ball 1990). In practice, most advocates of the PSC do recognize the gene tree versus species tree problem, and they normally do incorporate information from more than a single genetic trait into decisions about species boundaries (e.g., Baum and Shaw 1995; Sites and Marshall 2003).

Perhaps of greatest relevance to asexual organisms is the **cohesion species concept** (**CSC**), which considers species to be groups of populations that maintain cohesion either through genetic or demographic exchangeability (Templeton 1989). Thus, cohesion within a species can be maintained through the exchange of genetic information in sexual species but also through the occupation of a particular ecological niche even when gene exchange is not possible, as in asexual lineages. Cohan (2001) has argued that different bacterial "ecotypes"—populations of cells that might overlap spatially but differ in their ecological niches—qualify as distinct species under an ESC definition, based on the demographic exchangeability criterion of the CSC.

Brown (1995) suggested that the difficulties of defining species reflect fundamental problems that stem from the way in which organisms vary in their traits. Variation among organisms in most traits is **discontinuous**: There are clumps of organisms with similar traits separated by gaps. **Figure 7.4** illustrates this pattern using the classical case of variation in beak size among Darwin's finches that occur on several islands of the Galápagos. This "clumpy-gappy" organization of biological diversity is apparent in nearly all characters of organisms, not only in morphology and genetics, but also in physiology, ecology, and behavior. It reflects something very basic in the genetic and evolutionary processes through which both individuals and populations inherit characteristics from their ancestors. At the individual level, these processes are essentially the same in all organisms, but at the population level, as we saw above (see Figure 7.1), they are fundamentally different in sexual and asexual organisms: The former have two parents and thus reticulate patterns of inheritance, whereas the latter have only one parent and thus only a simple branching pattern of ancestry and descent.

The essence of most species concepts is that they either explicitly or implicitly attempt to identify the points along an evolutionary trajectory at which a particular level of clumpy-gappiness exists because populations have crossed the line between genetically reticulate (tokogenetic) and phylogenetically divergent associations (see Figure 7.2). The concepts differ from one another in the sorts of clumps they seek to recognize, such as morphological

FIGURE 7.4 Distribution of one trait, beak depth, in several populations of Darwin's finches inhabiting different islands of the Galápagos. This discontinuous variation, especially within a single island, illustrates the "clumpy-gappy" distributions of traits that most species concepts try to capture. The pattern of interisland differences also illustrates character displacement, that is, the fact that species tend to be more different from one another where they occur together—presumably reflecting adaptive divergence to avoid competition—than where they do not coexist with closely related forms. (After Lack 1947.)

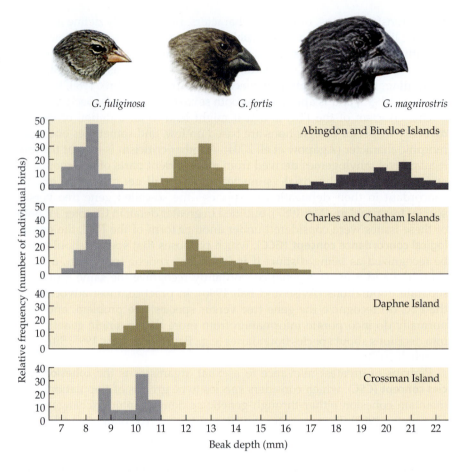

(MSC), genealogical (GCC), phylogenetic (PSC), or reproductive (BSC and RSC). While the same clumps (species) are often recognizable on the basis of distinctive morphological, genetic, physiological, behavioral, and reproductive characteristics, sometimes they are not concordant—the application of different concepts can often yield disparate lists of species. Nowhere is this discrepancy more evident than at the level of gene trees nested within species trees. For very good theoretical reasons that we will address in Chapter 11, we should often expect that following the cessation of reproductive exchanges between populations, some gene trees, such as mitochondrial DNA (mtDNA) in animals, will normally cross the tokogenetic–phylogenetic border much faster than will nuclear gene trees (Moore 1995; Avise 2000; see Figures 11.7 and 11.8).

As argued by Mayden (1997), the most practical approach might be to utilize a particular operational criterion or set of criteria for recognizing the discontinuous organization of biological diversity that consistently associates major clumps with a conceptually sound primary species concept (e.g., the ESC). Similarly, de Queiroz (2007) has argued for a unified species concept recognizing that each separate concept describes a threshold within the ongoing differentiation of one species into two or more separate species (**Figure 7.5**). It is important to recall that the variety of species recognition criteria (secondary species concepts) have been invented by humans as useful surrogates to simplify and classify biological diversity in a fashion that, hopefully, will increasingly be consistent with an evolutionary lineage-based concept of species. Several investigators have argued that different species concepts are most appropriately applied at different stages in the process of speciation

(Harrison 1998; Brooks and McLennan 2002; de Queiroz 2007). Nevertheless, the process of evolutionary change includes such a variety of dynamic interactions among populations that the task of species recognition will always include some degree of subjectivity, although some effort has been made to develop empirical methods for delimiting species in nature (Sites and Marshall 2003). We can, however, take some comfort from the fact that specialists on most taxonomic groups have traditionally recognized separate species at a scale of gappiness that is similar to those discovered through more recent surveys of molecular-level variation (Avise and Walker 1999).

Subspecies, ecotypes, phylogroups, and evolutionarily significant units

Taxonomists, evolutionary biologists, and ecologists recognize that some well-differentiated populations probably do not warrant recognition as distinct species, using any species definition, but might nevertheless warrant formal recognition as distinct taxa. These populations have traditionally been called **subspecies** by animal taxonomists when referring to populations that are morphologically, and presumably genetically, distinct. Typically, a subspecies is given a trinomial Latin name, such as *Peromyscus maniculatus gracilis*, in which the last part is the subspecies designation. *Subspecies* may be considered synonymous with the less formally recognized **geographic races** in discussion of populations with nonoverlapping distributions. Plant taxonomists use the term **variety** rather than *subspecies* in formal taxonomies, and ecologists use **ecotype** to refer to a distinct population that occurs in a particular habitat type (note that Cohan [2001] equated bacterial ecotypes with species). Unlike subspecies, geographic races, and varieties, which normally have nonoverlapping geographic ranges, two or more ecotypes may occur together in the same local area so long as they are restricted to different habitats. Typically, ecotypes have distinctive morphological and physiological traits, which may reflect genetic adaptations to their unique environments.

More recently, with the advent of molecular approaches to examining biological diversity, populations that are found to be genetically coherent have been called **phylogroups** or **evolutionarily significant units** (**ESUs**). The former term has a fairly precise meaning as a reference to distinct, genetically delineated clades often defined using mtDNA in animals or chloroplast DNA (cpDNA) in plants (Avise 2000). The ESU has been defined in several different ways beginning with Ryder (1986) and aligned with the "distinct population segments" amendment to the 1973 U.S. Endangered Species Act (Waples 1991). One way to view ESUs that is most relevant for historical biogeography (see Unit 4) is as a historically isolated set of populations with evidence, often from mtDNA, for divergence into separate evolutionary lineages (Moritz et al. 1995). We will revisit phylogroups and ESUs in Chapter 11.

Higher classifications

Having considered the problem of how species are defined, let's briefly turn to the equally vexing problem of how species are arranged in a taxonomic classification scheme. The great eighteenth century naturalist, Carolus Linnaeus, developed the hierarchical scheme for classifying organisms that is still used today. In the Linnaean classification scheme, species are grouped into genera, genera into families, families into orders, and so on (see Table 7.1). In

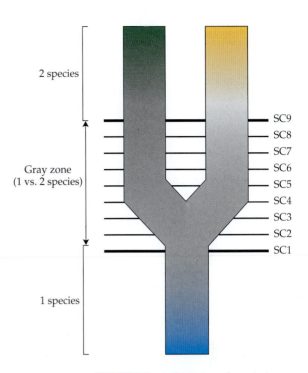

FIGURE 7.5 A depiction of speciation as an extended process of splitting of one ancestral lineage into two distinct evolutionary lineages. Horizontal lines (labeled SC1–SC9) show different species concepts (perhaps those shown in Figure 7.3), which would represent where various lines of evidence (e.g., morphological, mtDNA, multiple genes, reproductive isolation, ecological differentiation) would support existence of two separate species. Below and above this "gray zone," most or all species concepts would concur in recognizing either one or two species. (After de Queiroz 2007.)

Linnaeus's time, when species were thought to have been individually created by God, this classification was simply meant to reflect the fact that species differ in the degree to which they resemble one another.

Now that we know that all living things are descended—through a complex historical pattern of branching lineages—from a common ancestor, most taxonomists want their hierarchical classification schemes to reflect this history of ancestor–descendant relationships. The discipline of **phylogenetic systematics** has had considerable success in developing both a rigorous conceptual framework for recognizing evolutionary lineages with a shared history (Hennig 1966) and methods for reconstructing the evolutionary history of lineages (see Chapter 11). These developments have led a large group of phylogenetic systematists to propose a radical new system of taxonomic classification called the **PhyloCode** (http://www.ohio.edu/phylocode) that gets rid of the Linnaean hierarchical ranking system and instead ties nomenclature explicitly to the phylogenies generated using modern concepts and methods of phylogenetic reconstruction (e.g., as posted on the Tree of Life Web Project, http://tolweb.org/tree). However, while exciting, this proposal remains controversial, and at least for now, it is still necessary to describe some of the attributes of the traditional rank-based system.

Major levels of classification under the Linnaean system are given in Table 7.1. Additional levels are sometimes recognized by taxonomists working on particular groups—examples include superspecies, subgenera, infraorders, and so on (for a good example, see the classification scheme for mammals by McKenna and Bell [1997]). It is usually easy to figure out from their names where these levels are located in the hierarchy.

The hierarchical classification scheme implies that the taxa grouped together at each level are most closely related to one another. Thus, for example, species within the same genus are more closely related to one another than to species in any other genus, as are genera within the same family, and so on. By *more closely related* we mean that they more recently shared a common ancestor (i.e., all included taxa comprise a **monophyletic clade**). One thing that the classification scheme does not imply, however, is that all of the taxa at the same level of classification are equally closely related—there is no explicit reference to phylogenetic relationships among clades nested within a particular genus, family, etc. Thus, if three families are placed in an order and two have split from their common ancestor more recently than the third, this particular hierarchical evolutionary relationship will not be reflected in the classification. Sometimes this system is criticized for not being standardized across taxa—in some groups, such as certain vascular plants and vertebrates, species in the same genus may have diverged only within the last few million years, whereas congeners in other groups may have split tens of millions of years ago. In Chapter 11, we will consider in some detail how systematists reconstruct phylogenetic relationships, including estimates of times of divergence, and what these genealogies imply. In Chapter 12, we will examine why these phylogenies are important in biogeography.

While the Linnaean classification scheme has worked well for over 250 years, its weaknesses (in addition to those already mentioned) are evident now that we have a rapidly expanding, phylogenetically based framework for understanding the "tree of life" and reconstructing its biogeographic history (Cracraft and Donoghue 2004). First, it does not provide for a clade, once given a name, to retain that name should future analyses indicate that it needs to be repositioned on a phylogenetic tree (in a rank-based system, for example, if a genus-level clade is elevated to a family-level rank, then the name of the clade must also be changed). Second, it requires that, should a species be transferred to a different genus, the formal name of the species (which includes the genus name) must also be changed (e.g., the desert pocket mouse, formerly called *Perognathus penicillatus*, is now recognized as *Chaetodipus penicillatus*). While the PhyloCode system seeks to eliminate the conventions of nomenclature that lead to such taxonomic instability, it is quite controversial and, while it was first introduced to the public in 2000, we still do not know if and how rapidly it will gain popularity and be adopted by the systematics community.

Speciation

Although the process of speciation was not widely recognized by that term until the 1930s, a number of researchers prior to the Modern Synthesis were interested in the formation of new species and explored issues that are still being addressed today. They asked whether geographic isolation is required, and they explored the roles of natural selection and genetic drift (Berlocher 1998; Coyne and Orr 2004). Today, speciation is usually seen as a branching process (cladogenesis) in which new kinds of organisms originate from a single ancestral species. However, new species might also arise through hybridization (**reticulate speciation**), and some investigators include the non-branching (anagenetic) process of **phyletic speciation**—the transformation of one ancestral species into a single descendant species—with each major stage in such an evolutionary sequence being called a **chronospecies**. Many examples of phyletic speciation have been cited in the literature, but relatively few have been studied carefully enough that we can be confident that the chronospecies represent an unbroken anagenetic series—and not, in fact, products of cladogenetic events that were followed by rapid extinction of all but one lineage.

The magnitude of this process is staggering, especially when we consider that all species of green land plants that have ever lived ultimately share a common ancestor in a simple green alga that lived over 500 million years ago, that all vertebrates are traceable to some ancient chordate, and that millions of species of insects have evolved in the 400 million years since their ancestor first invaded land. Most of the species in all of these lineages are no longer around—they lived in the past and went extinct. How did this diversity come to be?

Mechanisms of genetic differentiation

Regardless of the geographical mode of speciation (reviewed below), the divergence of an ancestral species into two or more daughter species requires genetic changes among populations, ranging from those that are simply used to diagnose different species (e.g., under the PSC) to those that actually cause reproductive isolation (e.g., under the BSC). Population geneticists have long recognized four primary microevolutionary processes by which fully or partially isolated populations can diverge (or converge)—mutation, genetic drift, natural selection, and gene flow—in genetic composition. Different forms of a gene at the same locus, known as **alleles**, are responsible for differences among individuals in their heritable traits. New alleles arise by mutation, and changes in the frequencies of alleles in populations occur primarily through genetic drift, natural selection, and gene flow.

Mutation

Changes in DNA are called **mutations** and include four kinds of change: **substitution** of single nucleotides; and **deletion**, **insertion**, or **inversion** of one or more nucleotides. Deletions and insertions are often facilitated by **transpos-**

able elements or **horizontal gene transfer**. Any of these events can cause a change in a protein-coding gene, RNA-coding gene, or gene regulatory region of the DNA. New gene copies in multigene families (e.g., globins) can be generated through **unequal crossover** at **recombination**, and damaged genes can be repaired through **gene conversion**. The importance of mutations in the genetic differentiation of populations is that they can be incorporated into the genetic architecture of a population through genetic drift or natural selection. Gene flow will distribute new mutations among populations, whereas geographic or reproductive isolation will prevent that from happening.

Genetic drift

Genetic drift is a relatively weak force because it involves changes in the genetic constitution of a population caused solely by chance. Given sufficient time—which usually means many generations—the frequencies of alleles in a population tend to change randomly as different individuals happen to survive, mate, and produce offspring. Genetic drift has relatively little effect in large populations, but it can have important influences on the evolution of small populations. How small is small is a matter of considerable discussion in evolutionary biology, but we should mention that when population geneticists discuss population size, they are usually more interested in the **effective population size**—the actual number of breeding individuals in a population—which can differ substantially from the **census population size**—the total number of individuals in a population. Factors that can make the effective population much smaller than the census population include social structure, age structure, and transmission of genes from one generation to the next through a single sex (e.g., female transmission of mtDNA in bisexual organisms). If new species start from small founding populations—as many island forms undoubtedly do—then genetic drift may play an important role in their initial differentiation, as we shall see below.

Natural selection

Natural selection, on the other hand, can be a potent force for evolutionary change in both large and small populations. **Natural selection** is the change in a population that occurs because individuals express genetic traits that alter their interactions with their environment so as to enhance their survival and reproduction relative to other individuals in the population. Over many generations, alleles for such adaptive traits tend to increase in frequency at the expense of alleles that confer less fitness. Populations tend to diverge if there is sufficient variation in the environment to select for different characteristics to deal with different environmental conditions. If a new mutation has a strongly positive fitness effect, the rapid fixation of a favorable new mutation, as well as linked alleles that "hitchhike" with it, might trigger a rapid differentiation of populations—a **selective sweep**—such sweeps being of particular importance in the evolution of asexual populations (Cohan 2001).

This fine-tuning of phenotypes to environmental heterogeneity can be readily documented. A classic example involves industrial melanism in moths. From 1848 to 1895, moths in the species *Biston betularia* increased from predominantly pale to 98 percent melanistic phenotypes in accord with increases in soot and SO_2 that darkened the trees the moths rested on during the daytime. Experiments during the 1950s (Kettlewell 1961) provided evidence that this phenotypic change was driven by selection of bird predators for the most conspicuous individuals. Subsequent experiments and data collection have supported the natural selection hypothesis by more rigorously establishing the role of bird predation, by documenting that moth phenotypes reversed in frequencies in association with environmental laws reducing soot and SO_2, by showing that the strength of natural selection is reduced in moth species with high levels of gene flow, and by demonstrating a parallel pattern in *B. betularia* from industrialized regions of northeastern North America (Grant 2009). Rapid evolution by natural selection has also occurred in house sparrows. Since being introduced to North America from Europe less than 200 years ago (Johnston and Selander 1964, 1971; Johnston and Klitz 1977), these enormously successful birds have not only spread to colonize most of the continent (see Figure 6.3B) but have also evolved distinct geographic races. Other examples supporting evolution by natural selection include changes in body size in response to climatic shifts and habitat fragmentation (see Chapters 14, 16, and 17). Changes due to climatic shifts include those seen in pocket gophers (Hadly 1997; Hadly et al. 1998), field mice (Millien and Damuth 2004; Millien 2004), wood rats (Smith et al. 1995; Smith et al. 1998), land snails (Chiba 1998), bivalves (Roy et al. 2001b), and ostracods (Hunt and Roy 2006). Changes due to habitat fragmentation include those seen in tropical forest tree frogs (Neckel-Oliveira and Gascon 2006), rain forest birds (Smith et al. 1997; Fredrickson and Hedrick 2002), and birds and mammals of Denmark (Schmidt and Jensen 2003, 2005; Lomolino and Perault 2007).

Gene flow

Migration, or **gene flow**, often tends to act counter to genetic drift and natural selection and impede genetic divergence. Individuals that migrate to a new area carry their genes with them, and if they subsequently reproduce successfully, their genes are introduced into the local population. Such migration, therefore, tends to have a homogenizing influence, preventing or at least retarding the development of geographically isolated and genetically differentiated populations. An example of this influence has been documented in the case of industrialized melanism in moths (discussed above), wherein the strength of natural selection in a species with relatively high levels of gene flow between populations (*B. betularia*) is less than in the more sedentary species *Gonodontis* (*Odontoptera*) *bidentata* (Grant 2009).

Adaptation and gene flow

The niche of a species is not constant in either space or time. Since natural selection is a universal process that tends to increase the capacity of individuals to survive and reproduce, we would expect populations to adapt to their local environments. Such adaptation might be evidenced by geographic variation within species in physiology, morphology, or behavior. It might also be reflected in adaptive changes in niche relationships over time. We might expect peripheral populations to be able to adapt to the environmental factors that limit their ranges, increase in density, and then colonize adjacent areas (see Baker and Stebbins 1965). Lewontin and Birch (1966) describe an apparent example in the Queensland fruit fly, *Bactrocera* (*Dacus*) *tryoni*. During the last century this species has expanded its range several hundred kilometers to the south along the eastern edge of the continent. These flies are limited by low temperatures, and their expansion has been accompanied by adaptation of the peripheral populations for increasing cold tolerance. But clearly not all species are increasing their ranges in this fashion. A historical perspective suggests that throughout most of the Cenozoic Era, the distributions of some species and higher taxa have indeed increased, but these expansions have been almost equally matched by contractions in the ranges of other organisms. As a result, there appears to have been little change in global biodiversity during this period.

Why don't the peripheral populations of all species adapt to local conditions, resulting in a continual expansion of the range on all margins? In some

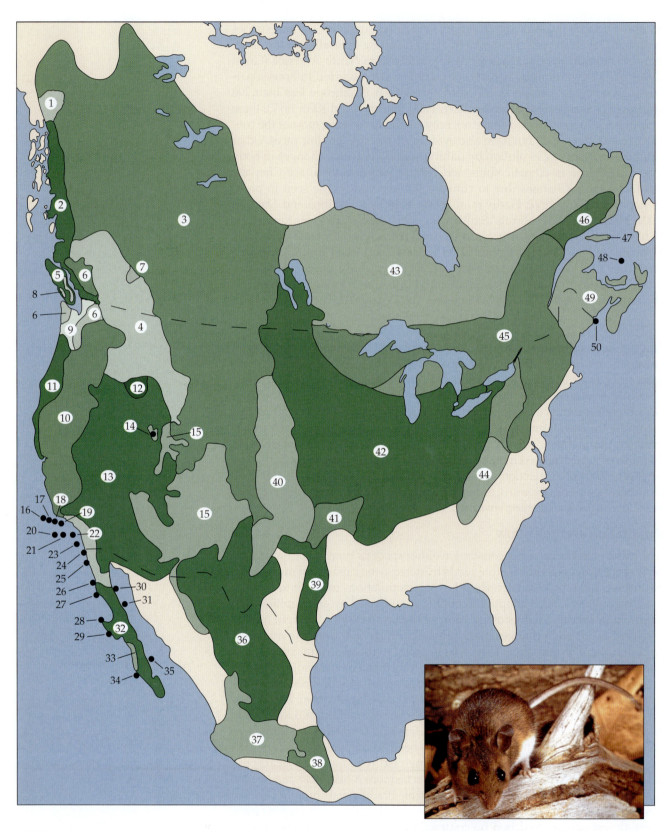

FIGURE 7.6 Geographic variation in the deer mouse (*Peromyscus maniculatus*) is indicated by the subdivision of the species into 50 formally recognized subspecies (which are not necessarily congruent with phylogeographic structure inferred through molecular genetic studies; Dragoo et al. 2006). Each of these geographic races has distinct characteristics—including dorsal coat color, which resembles background color and provides camouflage, and tail and hindfoot length, which are related to climbing ability and habitat structure. (After Hall 1981; photo courtesy of James Gathany/CDC.)

cases, the reason is obvious: There are fundamental constraints that cannot be easily overcome by local adaptation. Terrestrial life that forms along coastlines, for example, cannot simply evolve adaptations for aquatic life and invade the marine realm. This kind of adaptation did happen in the past—as terrestrial ancestors of penguins and whales invaded the sea—but it took millions of years and required that these animals give up their capacity for living on land. But such cases hardly explain why species do not expand their ranges along ecological gradients. Why don't local populations adapt to deal with a limiting physical stress or biological enemy? Why don't they expand the species range by evolving to tolerate just a bit more cold, aridity, predation, or competition?

Genetic, ecological, and biogeographic processes seem to interact to limit the capacity of peripheral populations to adapt to local conditions. Exchange of genes among populations through dispersal and gene flow may prevent local populations from acquiring and maintaining the combinations of genes necessary for continual adaptation. Sufficiently high gene flow can swamp a local population with genes from outside, effectively working against increasing adaptation to local conditions.

The critical question in any particular case of local adaptation is whether gene flow is high enough to overwhelm natural selection, prevent the continual adaptation of peripheral populations, and thereby preclude expansion into new areas of ever more extreme environments. The answer to this question for many organisms is not clear, and it has been the subject of much debate among population geneticists and evolutionary biologists. Certainly, many species show adaptive genetic changes in response to variation in their environments over their geographic ranges. The deer mouse (*Peromyscus maniculatus*), whose enormous geographic range encompasses most of the North American continent, has been the subject of many genetic and ecological studies. There is a great deal of geographic variation in the color and shape of deer mice, which is reflected by the subdivision of the species into many geographic races or subspecies (**Figure 7.6**; see also Figure 6.20). Classic studies by Lee Dice and his students showed that much of this variation reflects adaptation to local environments. Coat color tends to match the local soil color because of strong selection by owls and other predators against contrastingly colored individuals (Dice and Blossom 1937; Dice 1947). Compared with individuals from desert and grassland habitats, animals from forest populations tend to have longer feet and tails because they use these appendages in climbing (Horner 1954). Populations in regions where contrasting habitat types come into close proximity, however, show lower levels of such morphological and behavioral specialization, because of the diluting influence of gene flow (Thompson 1990).

These and similar studies of other organisms support theoretical models that predict that even modest levels of gene flow may be sufficient to preclude local adaptation (Wright 1978). If gene flow could be reduced or eliminated, however, local adaptation could proceed, and further range expansion might be possible. Evidence in support of this conjecture comes from studies of the genetics of Old World burrowing rodents of the genus *Spalax*. Differences in chromosome number among populations of the *S. ehrenbergi* species complex appear to have been important in facilitating their colonization of the most arid deserts of the Middle East by reducing gene flow from populations adapted to more mesic areas (Nevo and Bar-El 1976; **Figure 7.7**). Gene flow between populations with different chromosome numbers and configurations is probably reduced because when individuals with different karyotypes breed, problems with pairing and separation of chromosomes reduce the viability and fertility of their offspring. Patton (e.g., 1969, 1972,

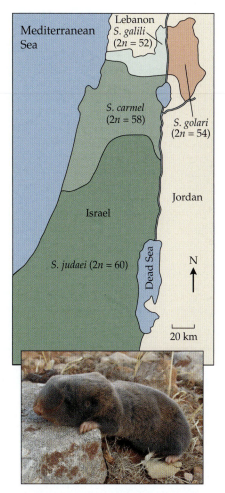

FIGURE 7.7 Distribution of karyotypic races, supported by several other types of nuclear and mitochondrial genetic markers, of burrowing rodents in the Palestine mole rat *Spalax ehrenbergi* species complex. Number of chromosomes increases along a gradient of increasing aridity from north to south (and also from west to east) in the Middle East. Note that these "chromosomal races" replace each other with virtually no overlap, suggesting that reduced gene flow between them has facilitated adaptation and expansion into increasingly arid environments, ultimately resulting in the evolution of several new species. (After Karanth et al. 2004; photo courtesy of Alenka Krystufek.)

1985) describes other examples of how local and regional ecological conditions, natural selection, and genetic barriers to gene flow interact to influence the distributions of small mammals.

Chromosomal rearrangements also have been implicated in limiting the distributions of plants. Populations at the geographic margins of a species range, as well as those inhabiting extreme soil types or climates, are often characterized by major chromosomal changes, especially polyploidy (Stebbins 1971b). Although these rearrangements of genetic material may themselves confer specific adaptations, their general effect is to reduce the frequency of crossing with individuals from other populations and thereby reduce—or completely block—gene flow, permitting adaptation to the local environment and facilitating colonization of new areas.

Ultimately, of course, the capacity of populations to adapt to new environments and to colonize new habitats is limited. Although some species are more widely distributed and more tolerant of varying conditions than others, there are no superorganisms that occur everywhere. Adaptation inevitably involves trade-offs and compromises. In order to tolerate the physical conditions and deal with the biotic interactions in some environments, a species must sacrifice its ability to do well in others. Using a combination of "common garden" and breeding experiments on yarrow (*Achillea millefolium*), Clausen and colleagues (1940, 1947, 1948) elegantly demonstrated the interacting roles of genetic isolation, local selection, and trade-offs in determining the degree of adaptive differentiation of local populations and allowing this species to have a wide distribution along an elevational gradient in the Sierra Nevada Mountains of California (**Figure 7.8**). Again, gene flow can prevent populations from adapting to different local environments, but when gene flow between populations is interrupted—as it is during the process of speciation—then over evolutionary time, populations can and often do diverge in response to natural selection, adapt to widely different conditions, and expand their ranges to occupy new habitats and geographic areas.

Geographic variation

There is often a geographic component to genetic divergence, because both genetic drift and natural selection are facilitated, and gene flow is impeded, by geographic isolation. Genetic drift can be an important force in small,

FIGURE 7.8 Morphological differentiation of the yarrow, *Achillea millefolium*, along an elevational gradient in the Sierra Nevada Mountains of California. Note the distinctive characteristics of different populations of the plant from different places along the gradient. Since these plants were all grown in the identical environments of a "common garden," we can infer that the differences among them are genetic. Presumably, these differences reflect local adaptation due to a combination of natural selection and reduced gene flow. (After Clausen et al. 1948.)

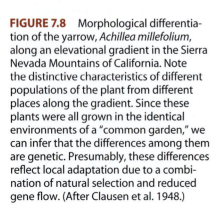

isolated populations, such as those that inhabit small outlying patches at the periphery of a species' range or those that have recently been founded by long-distance colonization. A population started by a few colonizing individuals usually contains only a small random sample of the alleles present in a much larger ancestral population. Genetic drift that occurs during such a **population bottleneck** has been termed the **founder effect**. Mayr (1942) suggested that this process accounts for the apparently random differences often found among bird populations on different islands, each of which was probably derived from a few successful colonists (**Figure 7.9**). Arguments that the founder effect can play a major role in speciation by converting this initially random genetic sampling into different species through both genetic drift and natural selection of small colonizing populations (e.g., Carson 1971; Templeton 1980a; Carson and Templeton 1984) have been challenged by theoretical arguments (Barton and Charlesworth 1984).

Whether or not coupled to founder events, geographic separation of populations facilitates genetic differentiation by natural selection. Different environmental regimes tend to select for different traits, and spatially isolated populations are likely to occur in different environments. Under these

FIGURE 7.9 Divergence of populations of the monarch flycatcher (*Monarcha castaneiventris* complex) on the Solomon Islands east of New Guinea. These seemingly chance differences among populations were interpreted by Mayr (1942) as reflecting the founder effect: the random sampling of genes present in the ancestral population due to the small number of individuals in the initial colonizing populations. Updated phylogenetic analyses and reconstructions of Pleistocene sea levels suggest that the story is more complex than envisioned originally by Mayr, with speciation resulting from a combination of founder events and vicariance owing to island isolation during Pleistocene sea level fluctuations. Mate recognition experiments suggest that both plumage color and song variation are important in forming premating reproductive isolating mechanisms. (From Uy et al. 2009.)

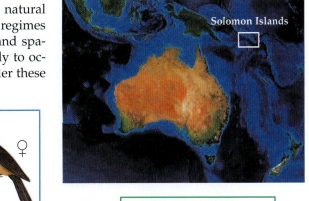

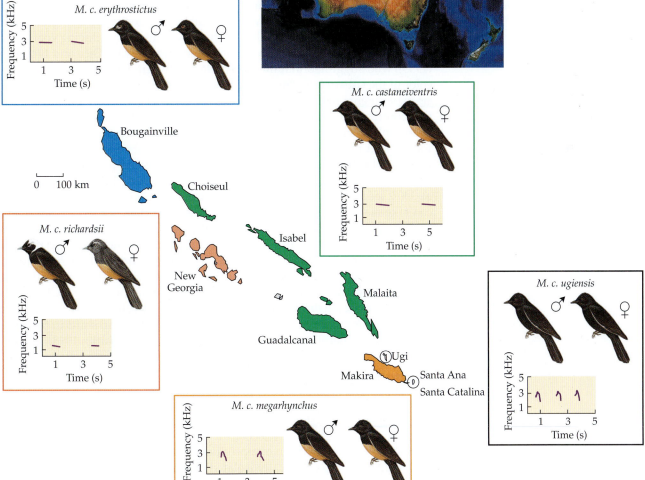

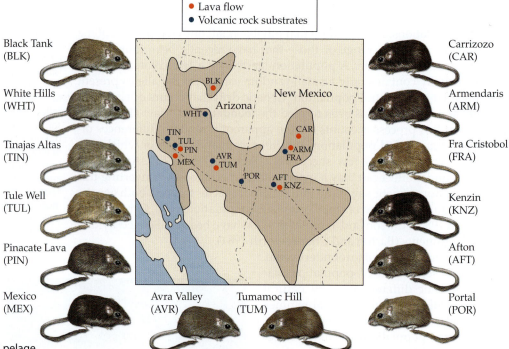

FIGURE 7.10 Typical dorsal pelage color variation representing either lava flows or volcanic rock substrates across the geographic distributions of the rock pocket mouse, *Chaetodipus intermedius*, in the southwestern North American deserts. (After Hoekstra et al. 2005.)

conditions, natural selection can facilitate a process termed **ecological specia-tion** (Schluter 1996; Schluter and Conte 2009). The effects of natural selec-tion associated with spatial environmental heterogeneity are exemplified by data on desert rodents. The role of natural selection in producing coat color variation in desert mammals was suggested originally by Dice and Blossom (1933), who suggested that the close match between dorsal fur color and the color of the substrate is the consequence of selection by predators (e.g., clas-sic experiments by Dice [1947] showed that owls selectively captured mice that contrasted with their background). More recently, mutations at specific genes linked with dramatic coat color differences among populations of the rock pocket mouse (*Chaetodipus intermedius*) that inhabit either lava bed or lighter soil substrates (**Figure 7.10**) have been shown to have more likely re-sulted from natural selection rather than phylogenetic history (Hoekstra et al. 2005; see also Mullen and Hoekstra 2008). Patterns of geographic variation can take many forms (see Chapter 15). The term **cline** is used to describe a gradual change in one or more features along a single environmental or other geographic gradient. Many birds and mammals exhibit clinal variation in clutch or litter sizes with latitude and elevation (**Figure 7.11**). Such variation presumably reflects the adaptation of life history traits to environments that differ in temperature, seasonality, productivity, and other factors. Clines can also develop in a **hybrid zone**, which marks a relatively narrow geographic zone where one or more phenotypes or alleles exhibit rapid shifts in frequen-cies between otherwise isolated populations or species (Butlin 1998).

Allopatric speciation

Geographic isolation has historically been considered the most frequent, if not the necessary, first step in the speciation process (Mayr 1942, 1963), be-cause it is the simplest way to cut off gene flow between populations and ultimately lead to formation of reproductive barriers. Several possible modes of **allopatric** (or geographically isolated) **speciation**, along with two alterna-tives to complete geographic isolation, are described below.

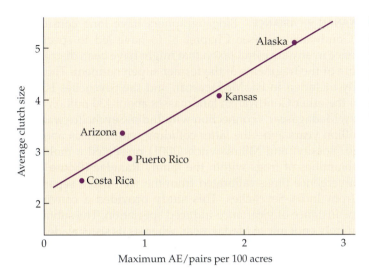

FIGURE 7.11 An example of a latitudinal cline in clutch size in breeding birds. In this study, average clutch size increases along a latitudinal gradient in relation to the ratio of maximum actual evapotranspiration (AE) to density of breeding birds. AE is proportional to primary production, and so the positive correlation of clutch size with the ratio of AE to breeding bird density can be interpreted as a response to resource availability. (After Ricklefs 1980.)

If the environment is heterogeneous, a geographically widespread ancestral population will tend to develop regional genetic differences in response to either natural selection or genetic drift. Because of barriers that limit dispersal, free gene flow from one end of the range to the other rarely occurs. Thus, populations living in environmentally distinct and geographically separated regions tend to become somewhat differentiated from one another, but some gene flow maintains the genetic cohesiveness of the species. Regional variation in house sparrows and deer mice, and the clinal variation in litter and clutch sizes mentioned above, are examples of such differentiation.

If, however, isolation of the regional populations becomes sufficient to cut off or drastically reduce the cohesive gene flow between them, then they can become independent evolutionary units over time. Without dispersal and gene flow, isolated populations tend to diverge. Divergence proceeds more rapidly when effective population sizes are small, or when substantially different environments subject isolates to different selective pressures. For example, Darwin called attention to the morphological differences among the giant tortoises of the Galápagos, which are obviously descended from a common ancestor but presently occur on different islands. On some of the most arid islands, where treelike cacti are abundant, the endemic tortoises have evolved long necks and forelimbs and distinctively shaped shells that allow them to reach up high to feed on these plants. On wetter islands, where the tortoises feed mostly on lower vegetation, they have more generalized body forms. We can distinguish two general means by which such isolation can occur.

Allopatric speciation mode I: Vicariance

At one extreme, some environmental change can create a barrier to dispersal somewhere within the range of an ancestral species, isolating previously connected and interbreeding populations. Rising sea levels, for example, can isolate an island on a continental shelf; tectonic events can cause part of a continent to split off and drift away; conversely, landmasses can drift together to isolate formerly continuous oceans. Such changes are called **vicariant events**, and they usually isolate relatively large sets of populations (**Figure 7.12**). We will see later that these sorts of events are important in biogeography because they also tend to isolate many different, codistributed species at the same time.

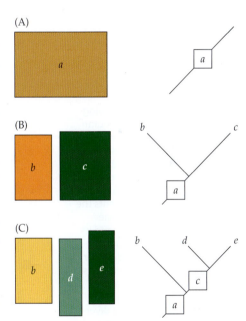

FIGURE 7.12 An illustration of the process of allopatric speciation mode I, and resulting phylogenetic patterns. (A) An ancestral species, *a*, widely distributed throughout a geographical area, is subdivided by the formation of a geographic barrier (vicariance). (B) With gene flow prevented, populations diverge from one another over time and eventually can be recognized as separate species, *b* and *c*. (C) Another geographic barrier subsequently subdivides species *c*, and the isolated populations diverge in geographic isolation to form species *d* and *e*. (After Brooks and McLennan 2002.)

EXAMPLE: GONDWANAN VICARIANCE. Biogeographers have long considered the breakup of Gondwanaland (see Chapter 8) to have led to speciation caused by vicariance in any ancestral taxon that might have been distributed across two or more of the fragments of the former supercontinent. The fragment that initially included India, Madagascar, and the Seychelles Islands broke away from the other Gondwanan landmasses about 130 million years ago. Subsequently, Madagascar's connection with India–Seychelles was severed about 88 million years ago—the latter colliding with Eurasia about 56 million years ago. Bossuyt and Milinkovitch (2001) presented a molecular phylogeny of subfamilies of frogs within the family Ranidae (**Figure 7.13**), with estimated divergence times calibrated from the well-dated separation of Madagascar from India–Seychelles. This study provided a convincing case of at least one instance of vicariance-induced speciation. The subfamily, Mantellinae, is entirely endemic to Madagascar, whereas its sister clade, Rhacophorinae, has its greatest diversity in the Oriental region and most likely was a member of the ancestral India–Seychelles biota prior to the suturing of India and Eurasia early in the Cenozoic.

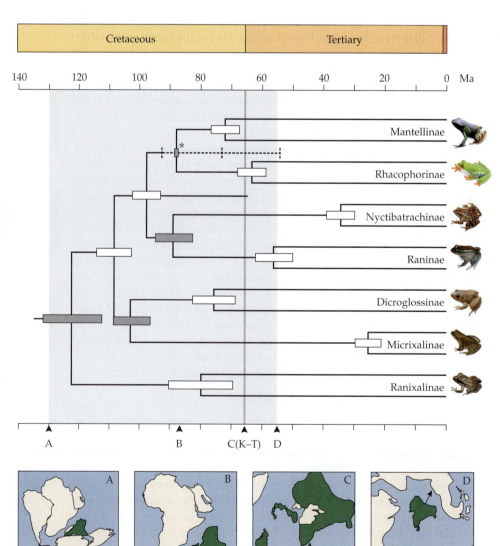

FIGURE 7.13 Molecular phylogeny for subfamilies of the anuran family Ranidae. The node with the asterisk represents the well-dated geological separation of Madagascar from the India–Seychelles landmass about 88 million years ago. This node also represents the separation of the Mantellinae—endemic to Madagascar—and its sister clade the Rhacophorinae, which was most likely a member of the ancestral India–Seychelles biota as it drifted northward. The close association between calibrated times of lineage divergence (see discussion of molecular clocks in Chapter 11) and times of separation of landmasses provides an example of vicariance-induced speciation. Diagrams A–D below the phylogeny, and the shaded area of the phylogeny, represent configurations of several Gondwanan landmasses between the time of separation of Madagascar–India–Seychelles about 130 million years ago (A) and the time of India's collision with Eurasia about 56 million years ago (D). (After Bossuyt and Milinkovitch 2001.)

Allopatric speciation mode II: Peripheral isolates, or founder events

At the other extreme, individuals may disperse across an existing barrier to colonize a previously uninhabited region. Such occasional and random events, called **jump dispersal** (see Chapter 6) or **founder events**, are the way in which oceanic islands and many other isolated patches of habitat come to be colonized. Typically involved is a small initial population, sometimes only one or a few individuals (**Figure 7.14**). Unlike vicariance, these events tend not to happen simultaneously across multiple species. As indicated above, mechanisms and rates of initial genetic divergence may differ depending on the mode of isolation. Genetic drift may play a greater role relative to selection, and divergence may be more rapid, at least initially, in allopatric mode II relative to mode I speciation (Carson 1971; Bush 1975; Templeton 1980a, 1981).

EXAMPLE: ALLOPATRIC SPECIATION IN THE GALÁPAGOS ARCHIPELAGO. Despite uncertainties about the details, there can be little doubt that geographic isolation has been a common mode of speciation in many groups. The finches and giant tortoises that Darwin observed in the Galápagos Archipelago provide examples of populations in different stages of the process. As we have seen, there are morphological differences among populations of tortoises on each of the large islands. These populations have diverged from a single ancestor that originally colonized the older islands in the archipelago (represented today by Española and San Cristóbal), probably from South America about 2 to 3 million years ago (Caccone et al. 2002). Their descendants subsequently dispersed, most likely by drifting along with the prevailing northwesterly ocean currents, to establish populations on the different islands as the younger islands formed (**Figure 7.15**). Two of the larger islands, Isabela and Santa Cruz, were each colonized more than once at different times and from different source islands, and there appear to be no cases of younger populations on younger islands recolonizing older islands. Cases in which islands have been colonized more than once appear to have resulted in some islands having more than one named taxon. Taxonomists, however, have not yet determined just how many should be considered distinct subspecies based on morphological and molecular differences among populations, although some populations of Galápagos tortoises might be recognizable as distinct subspecies of *Geochelone nigra* (Russello et al. 2005).

On the other hand, in Darwin's finches, the final stages of speciation are represented. Here again, all of the species are believed to be derived from a single ancestral population that colonized the archipelago between 2 and 3 million years ago, probably—somewhat surprisingly—

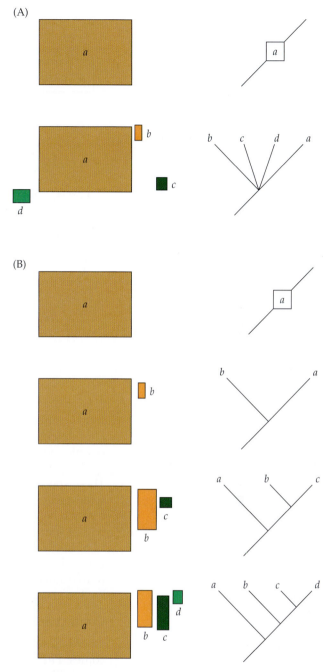

FIGURE 7.14 Illustration of the process of allopatric speciation mode II, and resulting phylogenetic patterns. (A) Random dispersal of members of the geographically widespread species *a* to peripheral areas initiates new, geographically isolated populations. Over time, the new peripherally isolated populations will diverge to form new species *b*, *c*, and *d*. (B) Peripheral isolates are again formed by founding events, but now the process of dispersal is temporally and spatially sequential such that the first population isolated to become new species *b* is adjacent to ancestral species *a*; then members of species *b* disperse to the next peripheral area and found the next new species *c*; and finally, dispersing members of species *c* found new species *d*. (After Brooks and McLennan 2002.)

FIGURE 7.15 Proposed phylogeographic history of Galápagos tortoises on the larger islands of the Galápagos Archipelago. The colonization and divergence history of tortoise populations is congruent with the geological ages of the islands, which are drifting across a hotspot in the Earth's crust (currently located under the western island of Fernandina). Mainland progenitors from South America first colonized the older islands of San Cristóbal and Española. Intraisland dispersal among different volcanoes on Isabela likely occurred after originally isolated volcanic islands coalesced into a single island. Arrows represent likely colonization events: solid arrows represent natural colonization events, and dashed arrows are possible translocations caused by humans. Numbers indicate approximate temporal order of colonization. Note that there were at least two independent natural colonizations of the islands of Santa Cruz and Isabela. (After Caccone et al. 2002 and Parent et al. 2008.)

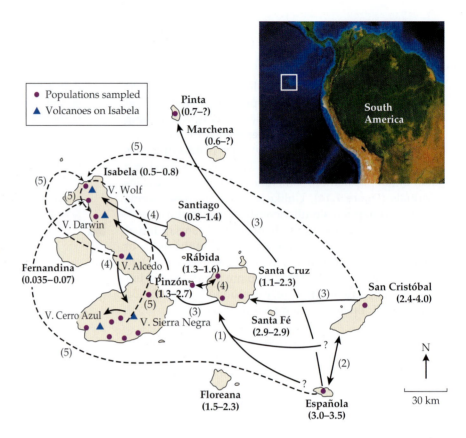

from Caribbean islands (Burns et al. 2002). In the case of the finches, however, after diverging in isolation on different islands or in different habitats, some populations have successfully reinvaded already inhabited areas, so several species (as many as ten on certain islands) now coexist on a single island. These species have not only evolved specific mating behaviors that prevent widespread interspecific hybridization, but have also diverged in morphology and behavior to exploit different ecological niches (**Figure 7.16**). The process of speciation is an ongoing one, and so there are also documented cases of hybridization between forms that are not yet completely reproductively isolated; for example, Grant and colleagues (2004) characterized the pattern of hybridization between *Geospiza fortis* and *G. scandens* (see Figure 7.16), as well as across a range of Darwin's finches (Grant et al. 2005).

Contact and reinforcement

Once isolates have formed and differentiated, there are several possible outcomes. First, they may come back into contact with one another. This may occur either through the disappearance of a geographic barrier or through dispersal back across it. If the populations reestablish contact, there are three possible outcomes:

1. the two populations may not interbreed, because they have developed prezygotic isolating mechanisms, or they may fail to produce fertile offspring if they do interbreed (postzygotic isolating mechanisms), in which case, reproductive isolation is complete, speciation has occurred, and they might begin to share the same geographic range but as two separate species;

2. the two populations may interbreed extensively, producing fertile hybrids that then backcross to parental populations to the extent that the populations gradually merge genetically and their differentiation breaks down; or

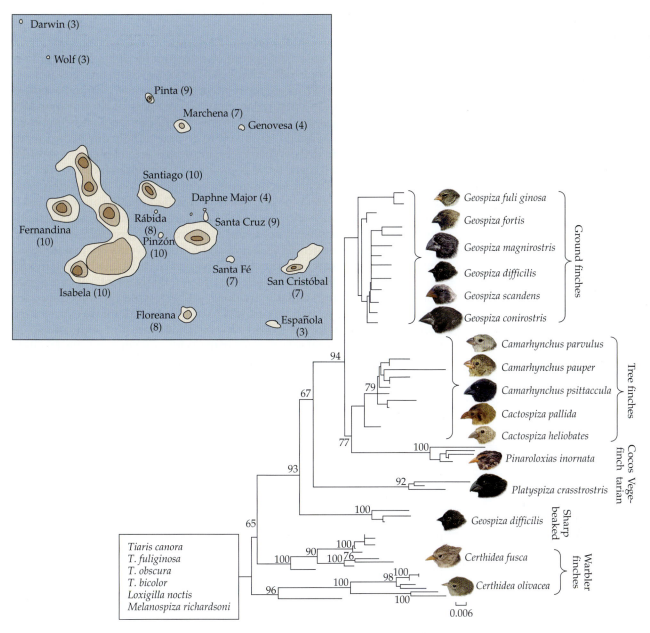

FIGURE 7.16 Adaptive radiation in Galápagos finches based on mitochondrial DNA analysis, showing the diversity of beak shapes and diets. A single ancestor from within a clade that includes six species in three genera colonized the archipelago. Subsequent allopatric speciation events, due to repeated episodes of colonization and divergence, mostly among different islands, produced six genera and 15 species as well as the Cocos Island finch, *Pinaroloxias inornata*. Note that not all morphological species are monophyletic lineages at the molecular level (e.g., *Geospiza difficilis* occurs at two different places on the tree). Following largely allopatric speciation, subsequent dispersal has led to as many as 10 breeding species in sympatry on a given island (see inset map; number of sympatric, morphological species in parentheses), which provides ample opportunities for interspecific hybridization, for example, between *Geospiza fortis* and *G. scandens*. (After Petren et al. 2005 and Burns et al. 2002.)

3. the two populations may not have developed prezygotic isolating mechanisms and may in fact hybridize, but the hybrids may be less fit beyond a narrow hybrid zone than the offspring of within-population matings. In this last case, selection favoring those individuals that choose mates from within their own population may be sufficiently strong to lead to reproductive isolation and completion of the speciation process.

This process of selection for prezygotic mechanisms that promotes within-population matings is termed **reinforcement** (Dobzhansky 1937). A final scenario is one in which the isolated populations never do come into contact. In this case, reproductive isolation may take a long time to develop, and as mentioned above, it may be difficult for a systematist to apply a biological species concept when deciding whether they should be considered different species.

While there is broad agreement that speciation often occurs through geographic isolation, there is much less consensus about the details of the process (Howard and Berlocher 1998; Coyne and Orr 2004). How much do mechanisms of speciation vary among different kinds of organisms and even among different speciation events within lineages of closely related organisms? What are the relative frequencies of vicariant events—with isolation between initially large populations less likely to experience rapid genetic change—and founder events—with initial isolates composed of only a few individuals and thus likely to diverge rapidly due to the founder effect and genetic drift? When vicariant events are involved, how often are the isolates small, peripheral populations (**microvicariance**) as opposed to large fragments of a once-continuous population? How important is gene flow in maintaining cohesion and preventing differentiation among populations? How important is the genetic inertia of large populations and the influence of similar selective environments in retarding divergence? Conversely, what are the roles of genetic isolation and founder events relative to divergent selective pressures in promoting divergence of isolated populations? To what extent is there active selection for "reinforcement" of prezygotic isolating mechanisms so that individuals actively avoid interbreeding with members of closely related species, and to what extent might we expect these traits to move away from a hybrid zone back into ancestral populations? To what extent does sexual selection for mates with exaggerated traits facilitate the evolution of prezygotic reproductive isolation? There is much disagreement among evolutionary and systematic biologists on the answers to these questions, but progress is being made (Howard and Berlocher 1998; Coyne and Orr 2004). Rather than searching for one universal process of speciation, we must recognize the enormous variation in the process due to the special characteristics of different kinds of organisms and the different historical and environmental contexts in which speciation occurs.

Sympatric and parapatric speciation

Although many evolutionary biologists once maintained that speciation in most organisms has occurred primarily or solely as a result of geographic isolation, most now accept that speciation can occur within spatially overlapping populations (Coyne and Orr 2004; Bolnick and Fitzpatrick 2007). This process is called **sympatric speciation** if the geographic overlap of populations of the ancestral species is extensive (**Figure 7.17A**); it is called **parapatric speciation** if the populations are largely allopatric but overlap in a narrow zone where depressed fitness of interpopulation matings leads to selection for isolating mechanisms (**Figure 7.17B**). In the decades immediately following the modern synthesis, there was a tendency to regard allopatric speciation as the general process and sympatric speciation as the exception. This

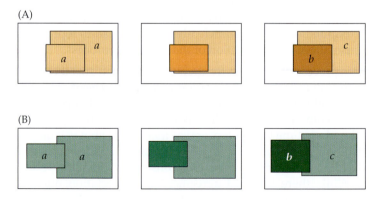

FIGURE 7.17 Illustration of sympatric and parapatric speciation. In sympatric speciation (A), two populations of ancestral species *a* overlap extensively when population differentiation begins and maintain extensive contact throughout the speciation process until new species (*b* and *c*) are recognized. Alternatively, in parapatric speciation (B), overlap occurs only along a narrow zone of contact between populations of ancestral species *a* and throughout the course of the speciation process. (After Brooks and McLennan 2002.)

assumption was largely due to the enormous influence and powerful arguments of Ernst Mayr (e.g., 1942, 1963), who argued for the near universality of allopatric speciation. The pendulum began to swing back in the 1980s as increasing evidence for sympatric speciation accumulated, particularly evidence for speciation associated with chromosomal changes in plants (Briggs and Walters 1984).

Sympatric speciation, however, has been difficult to demonstrate for at least three reasons. First, current overlap in geographic distributions between two closely related species does not necessarily mean that their ranges overlapped at the time of speciation—jump dispersal or range expansion might have brought them together following speciation in allopatry; therefore, in putative cases of sympatric speciation, biologists should be able to rule that an allopatric phase is highly unlikely. Second, defining the ranges of species as sympatric has often depended on the spatial scale being considered—two species might be considered sympatric at a regional scale but perhaps be geographically separated into discrete habitats at a local scale; therefore, biologists should be able to determine that home ranges of breeding members of the ancestral species did indeed overlap. Third, the best cases of sympatric speciation might be between closely related species that have not had the time to experience appreciable range shifts, and prior to use of molecular systematics, these were often the most difficult taxa to delineate as being separate species. In short, Bolnick and Fitzpatrick (2007) listed 63 putative cases of sympatric speciation (mostly animal examples), with only a small handful of those being generally accepted as good examples, and most of those are sister species or species flocks of fishes endemic to crater or postglacial lakes. Two primary mechanisms of sympatric speciation have been proposed.

Disruptive selection

If strong selective pressures cause a population to adapt to two or more different environmental regimes or niches, they can progressively pull the population apart and eventually result in speciation. Bush (1975), Price (1980), and others (see papers in Howard and Berlocher 1998) have argued that sympatric speciation by **disruptive selection** may be common in certain groups of phytophagous (herbivorous) insects and in animal parasites that are highly specialized for specific host species. Even in plants, a strong case has been presented that disruptive selection generated two species of palms on the remote and small Lord Howe Island (Savolainen et al. 2006). Among these organisms, successful colonization of a new kind of host must be a rare event, but when it occurs, the colonists are immediately subjected to selection for the ability to survive and reproduce in a drastically different environment. Usually this selection pressure is intensified by counter-evolution of the host to escape from or reject the parasite. Selection to meet the challenge of a new host could potentially lead to the rapid differentiation of totally sympatric populations. Initially, matings between insect or parasite host-specific races would generate offspring that had reduced fitness in either host niche, which would favor selection for evolution of a **prezygotic isolating mechanism** and completion of the speciation process. A parapatric model of speciation proposed by Endler (1977; see also Slatkin 1973; Rosenzweig 1978) suggested that disruptive selection, acting along an environmental gradient, could gradually sharpen clinal variation until a single ancestral population fragmented into two or more species. Disruptive selection might also underlie sympatric speciation, but in the form of sexual selection rather than natural selection; this seems to be the case for several African cichlid radiations (Kocher 2004).

Chromosomal changes

Sympatric or parapatric speciation (also called **stasipatric** speciation by White 1978) can also occur through chromosomal changes. Chance rearrangements of the genetic material of a parent during meiosis, or of an embryo during fertilization or early development, can sometimes change the number of chromosomes or the sequence of genes on chromosomes. Changes in chromosome number are of two kinds: **aneuploidy**, in which a single chromosome breaks or fuses with another to change the total number by plus or minus one; or **polyploidy**, in which an entire additional set of chromosomes is passed on, changing the number by some multiple (e.g., a doubling or tripling). In other cases, the chromosome number remains the same, but some of the genetic material is either rearranged within a chromosome (**inversion**) or transferred to another chromosome (**translocation**).

In diploid organisms, precise pairing during meiosis of the genes and chromosomes inherited from each parent usually is necessary to ensure the transmission of a complete set of genes to each gamete, and hence to produce viable offspring. Consequently, mutant individuals with new chromosomal arrangements often have impaired fertility when they mate with an individual having the original chromosomal arrangement, and they may be able to reproduce only by mating with another individual having the new arrangement. For this reason, it is obviously difficult for a population with a new arrangement to become established. However, once established—especially in a small, isolated, inbred population—the new type is genetically isolated from its parental population and can diverge rapidly as a new species. Navarro and Barton (2003) have recently incorporated this ephemeral isolation step into an updated and more realistic "allo-parapatric" model.

Sympatric speciation by way of polyploidy appears to have occurred frequently in some groups of organisms, especially plants (e.g., Stebbins 1971b; de Wet 1979; Lewis 1979; Briggs and Walters 1984). There are many documented ways in which polyploidy has been achieved in plants, but one such process is considered especially common (de Wet 1979). Diploid ($2n$) organisms have two sets of chromosomes and produce haploid (n) gametes by meiosis. Occasionally, a female gamete is formed without undergoing meiosis and remains diploid ($2n$). This unreduced gamete can then fuse with a haploid pollen grain (n) to produce a triploid ($3n$) plant; this will then produce triploid gametes because of complications during meiosis. In the next generation, a triploid female gamete can fuse with a haploid pollen grain to yield a tetraploid ($4n$) zygote, which can survive and produce fertile offspring from diploid ($2n$) gametes, either by self-fertilization or by crossing with other rare tetraploids in the population. The resulting tetraploid population is immediately genetically isolated from the diploid population. Interestingly, tetraploidy has also been discovered in a mammal—the red viscacha rat, *Tympanoctomys barrerae* (Octodontidae)—in Argentina (Gallardo et al. 1999).

Polyploidy can occur either within a population, called **autopolyploidy**, or as a result of hybridization between different but usually closely related populations or species, called **allopolyploidy**. Allopolyploidy is thought by many researchers to be more common. Because chromosomes from different species may not pair and segregate properly, interspecific hybridization often results in abnormalities in the meiotic process that can facilitate the process described above. Allopolyploids may not only arise more frequently than autopolyploids, but may also be more likely to become established. In addition to possessing a larger genome than either of the parental species, allopolyploids tend to be intermediate in their characteristics, which enables them to be superior competitors in certain habitats.

TABLE 7.2 *Distribution of More than 1400 Cichlid Species in 12 African Lakes, Representing about 60 Percent of the Global Cichlid Species Richness*

Location	Number of known species	Estimated age of basin (Myrs)	Major radiating lineages
Malawi	600	8.6	Haplochromine
Victoria	>500	0.4	Haplochromine
Tanganyika	180	~20	Several
Edward	60	2.0	Haplochromine
Kivu	16	5.0	Haplochromine
Barombi Mbo	11	~1	Tilapiine
Kyoga	>10	0.4	Haplochromine
Albert	10	2.0	Haplochromine
Bermin	9	0.8	Tilapiine
Ejagham	7	0.01	Tilapiine
Mweru	6	0.35	Haplochromine
Natron	5	1.0	Tilapiine

Source: After Turner 2007.

Examples: Sympatric speciation in isolated lakes

Some of the most compelling cases of possible sympatric speciation in animals involve the divergence and adaptive radiation of fishes in isolated lakes (Bolnick and Fitzpatrick 2007). Examples include species flocks of cichlids in the Great Lakes of Africa's Rift Valley and in the giant Cuatro Ciénegas spring system in northern Mexico; whitefishes in the Great Lakes of North America; sculpins in Lake Baikal in Siberia; herrings in Scandanavian lakes; pupfishes in Lago Chichancanab in Yucatán; and sticklebacks in lakes in British Columbia. These cases differ in when the isolation occurred, how much differentiation has taken place, and the strength of the case for sympatric speciation.

The cichlids of Africa's Rift Valley lakes (**Table 7.2**) have been isolated for tens of thousands to millions of years, although shallow Lake Victoria may have been completely dry in the Late Pleistocene (until only about 14,500 years ago: Seehausen 2002; see also Fryer and Iles 1972; Greenwood 1974, 1984; Kaufman and Ochumba 1993; Johnson et al. 1996; Kornfield and Smith 2000). Even if complete desiccation did not occur, this lake is probably no more than 0.5 to 1 million years old (Martens et al. 1994), and so the fact that it now supports a fauna of more than 500 species of endemic cichlids suggests that under appropriate conditions, speciation and adaptive radiation can be extremely rapid.

Species flocks of these cichlids have diversified morphologically and have specialized behaviorally and ecologically into hundreds of species from a few founding lineages. Each of the species flocks in separate lakes fill a variety of different feeding and habitat niches (**Figure 7.18**). Lake Malawi alone contains over 300 species in a group of rockfishes called the *mbuna*, and many of these have ecological counterparts in completely separate adaptive radiations in Lake Victoria (the *mbipi*) and Lake Tanganyika. There are herbivores

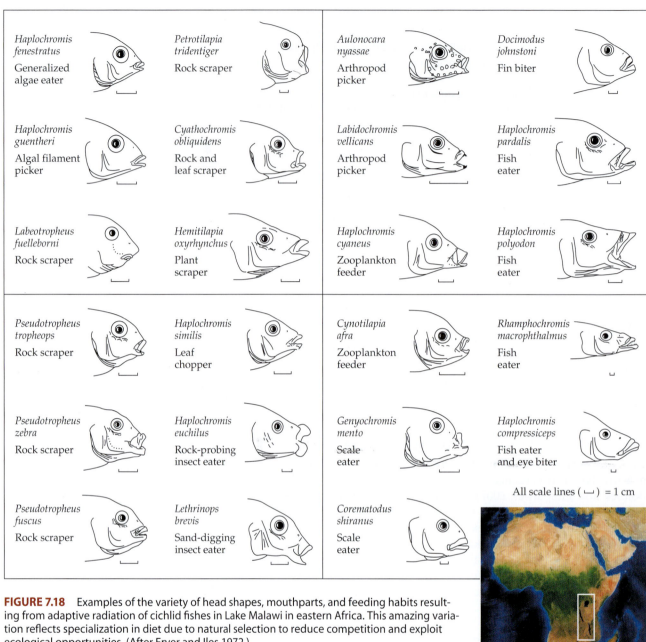

FIGURE 7.18 Examples of the variety of head shapes, mouthparts, and feeding habits resulting from adaptive radiation of cichlid fishes in Lake Malawi in eastern Africa. This amazing variation reflects specialization in diet due to natural selection to reduce competition and exploit ecological opportunities. (After Fryer and Iles 1972.)

and carnivores; species with mouths and teeth adapted for catching tiny zooplankton, crushing snails, and eating other fishes whole; and even forms specialized to feed just on the fins, scales, or eyeballs of other fishes. Some species are largely pelagic and others are specialists in either sandy or rocky shoreline habitats. Variation in color patterns and behavior are related to courtship and mating displays, which form strong premating isolating mechanisms that mask a high frequency of hybridization in laboratory matings (Turner 2007). Speciation could also be enhanced by population structure in which adults are strongly philopatric to breeding sites, and genetic data suggest low levels of gene flow among local populations (Kocher 2004).

While hypotheses of sympatric speciation through ecological and sexual selection mechanisms are viable, it is still difficult to rule against an alternative involving allopatric phases in the diversification of many of these species flocks. In the larger lakes, for example, changing lake levels probably formed isolated refugia in the deeper basins, and a wealth of geographically localized habitats are contained within each of the larger lakes. In some cases, multiple colonization events might have led to hybrid swarms with adaptive superiority over the cichlids. We do know at this point that in many of the lakes (Victoria, Malawi, Barombi Mbo) there is a monophyletic assemblage of species (Verheyen et al. 2003; Kocher 2004; Schliewen and Klee 2004), which provides support for sympatric speciation hypotheses. This case is particularly strong for the 11 species in the small Cameroonian crater lake Barombi Mbo, whose diversification satisfies each of the aforementioned criteria—previous allopatry unlikely, intrabasin microgeographic isolation unlikely, and all species representing a monophyletic lineage relative to a probable ancestor that inhabits nearby streams (Turner 2007).

At the opposite extreme are the pupfishes of Yucatán's Lago Chichancanab and the sticklebacks of British Columbia, which have been isolated in small lakes for only a few thousand years as a result of changing sea and lake levels since the Pleistocene (see Chapter 9). Lago Chichancanab contains seven putative species of *Cyprinodon* (**Figure 7.19**), which have distinctive morphologies related to diet and also exhibit a degree of genetic and behavioral divergence that points to either complete or partial reproductive isolation (Horstkotte and Strecker 2005; Strecker 2006; Plath and Strecker 2008). Small lakes in coastal British Columbia typically contain two forms of sticklebacks—benthic and pelagic—which exhibit morphological adaptations for swimming and feeding along the bottom or in open water, respectively (McPhail 1994). While these pupfishes and sticklebacks have diverged much less than the African cichlids, they nevertheless provide an excellent opportunity to study the roles of ecological and evolutionary processes in speciation and adaptation (e.g., Schluter 2000; see discussion of taxon cycles in Chapter 14).

These cases suggest that sympatric speciation may be more common than most evolutionary biologists have suspected, although they also indicate the

FIGURE 7.19 Morphology and divergence of feeding behaviors in six forms of pupfishes (*Cyprinodon*) in Lago Chichancanab on the Yucatán Peninsula of Mexico. These forms have diverged, most likely sympatrically, within no more than the 8000 years since the isolation of the saline lake from the sea. Genetic and behavioral studies suggest that the speciation process is generally not complete (with the exception of *C. maya*) and that some of these forms still interbreed to some extent. Differences in size and shape reflect dietary differences and suggest that strong selection for trophic and behavioral differentiation is driving sympatric speciation. (After Horstkotte and Strecker 2005.)

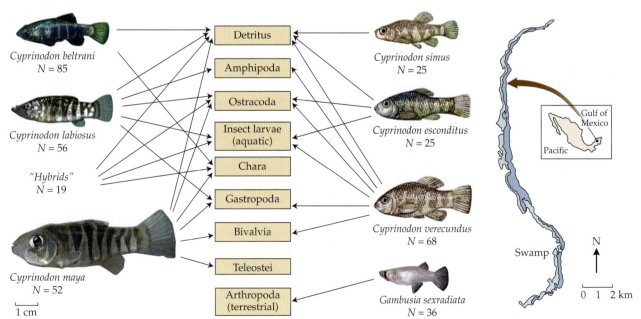

difficulties involved in ruling out alternative scenarios. For example, Lago Chichancanab is composed of a series of deeper basins separated by shallower ridges (see Figure 7.19), and so we still cannot rule out the possibility of allopatric phases during Holocene droughts (Hodell et al. 2005). They also suggest that either disruptive selection, to exploit different ecological niches, or strong sexual selection may often be sufficiently powerful forces to produce speciation, even in the absence of geographic isolation. Nevertheless, well-documented cases of sympatric speciation are still scarce (Coyne and Orr 2004; Bolnick and Fitzpatrick 2007), and an analysis that examined the geography of speciation in a phylogenetic context within several clades of birds, fishes, and insects (Barraclough and Vogler 2000) concluded that the most predominant form of speciation was an allopatric mode.

Diversification

Ecological differentiation

Once new species have formed, what happens to them? Immediately after a speciation event, the resulting species are often quite similar to each other. Above, we pointed out that they are most likely to diverge if they are subjected to different environments with different selective regimes. Ecological differentiation increases the likelihood that the species will be able to occupy overlapping geographic ranges. Thus, it facilitates the buildup within a region of a biota of closely related, sympatric species. Gause (1934) showed in laboratory experiments with protozoans of the genus *Paramecium* that two species with identical resource requirements could not persist in the same environment: One species was eventually outcompeted and driven to extinction. Ecologists have generalized this phenomenon and termed it the principle of **competitive exclusion** (see Hardin 1960; Miller 1967; Hutchinson 1978).

According to this principle, two or more *resource-limited* species cannot coexist in a *stable environment* without segregating their *realized niches*. The important point is not simply that identical species cannot coexist. Rather, the competitive exclusion principle describes how species *can* coexist—they may be limited by something other than resources (e.g., by predators or parasites); they may inhabit unstable environments where floods, fires, or other disturbances limit populations of competitors before they can exclude each other; or they may inhabit different regions of their shared, fundamental geographic range (see Chapter 4).

A biogeographic corollary of the principle of competitive exclusion is that species that are extremely similar in their niches tend to have nonoverlapping geographic distributions, whereas species that coexist in the same area and habitat tend to differ substantially in their resource use. Perhaps the most striking example of this phenomenon is provided by so-called **sibling**, or **cryptic**, **species**: species that are genetically distinct but extremely similar in their morphology and ecology. Examples include species that have recently formed through chromosomal changes, as well as many examples that have only recently been detected using modern molecular methods (examples given in Avise 2000)—they may be particularly abundant in the sea (Knowlton 1993). Sibling species often exhibit almost perfectly abutting, but not strongly overlapping (i.e., parapatric), geographic ranges, as demonstrated by numerous examples in both plants and animals (**Figure 7.20**). Even related species that are no longer extremely similar may competitively exclude each other from local habitats or extensive geographic areas. Examples include the kangaroo rats and chipmunks mentioned in Chapter 4 (see Figures 4.25 and 4.26). Several species of *Plethodon* salamanders have ranges

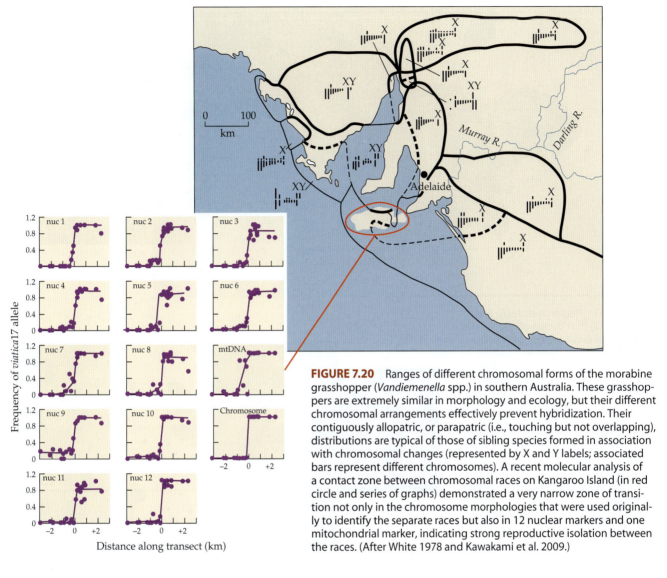

FIGURE 7.20 Ranges of different chromosomal forms of the morabine grasshopper (*Vandiemenella* spp.) in southern Australia. These grasshoppers are extremely similar in morphology and ecology, but their different chromosomal arrangements effectively prevent hybridization. Their contiguously allopatric, or parapatric (i.e., touching but not overlapping), distributions are typical of those of sibling species formed in association with chromosomal changes (represented by X and Y labels; associated bars represent different chromosomes). A recent molecular analysis of a contact zone between chromosomal races on Kangaroo Island (in red circle and series of graphs) demonstrated a very narrow zone of transition not only in the chromosome morphologies that were used originally to identify the separate races but also in 12 nuclear markers and one mitochondrial marker, indicating strong reproductive isolation between the races. (After White 1978 and Kawakami et al. 2009.)

in eastern North America that come into contact but do not overlap. There is good evidence that the parapatric ranges of some of the pairs of species may be maintained through competitive exclusion based on differences in body size and behavioral aggression (**Figure 7.21**; Griffis and Jaeger 1998; Myers and Adams 2008).

Conversely, numerous field studies show that when closely related species do coexist in nature, they often differ substantially in their use of limiting resources (e.g., MacArthur 1958, 1972; Cody and Diamond 1975). Often these niche differences are reflected in pronounced morphological, physiological, or behavioral differences. Galápagos finches provide excellent examples of this phenomenon. As the finches have reinvaded inhabited islands after speciating, they have diverged morphologically, behaviorally, and ecologically from other sympatric species. This process, called **character displacement** (Brown and Wilson 1956), has resulted in species' being more different where they coexist than where they live allopatrically. As we have seen (see Figure 7.4), character displacement in these finches is most apparent in the size of the beak, which enables coexisting forms to specialize on different kinds of foods (Lack 1947; Abbott et al. 1977; Schluter and Grant 1984; Schluter et al. 1985; Grant 1986; Grant and Grant 1989). Some intermediate phenotypes

(A)

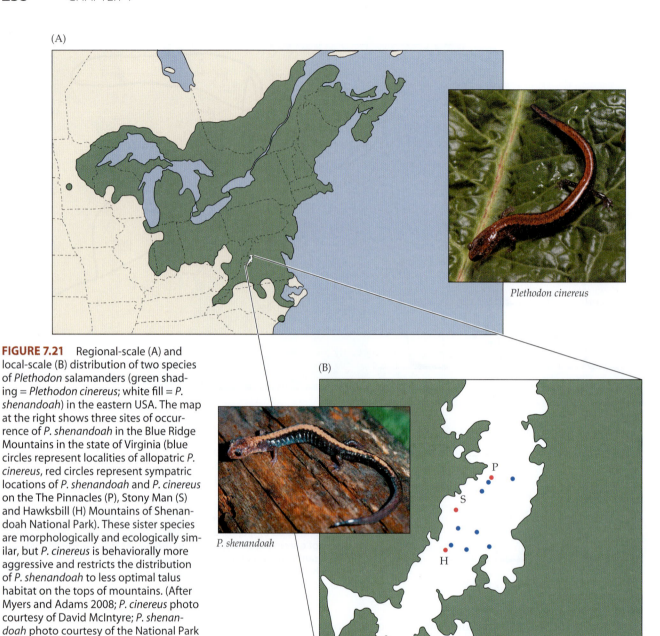

FIGURE 7.21 Regional-scale (A) and local-scale (B) distribution of two species of *Plethodon* salamanders (green shading = *Plethodon cinereus*; white fill = *P. shenandoah*) in the eastern USA. The map at the right shows three sites of occurrence of *P. shenandoah* in the Blue Ridge Mountains in the state of Virginia (blue circles represent localities of allopatric *P. cinereus*, red circles represent sympatric locations of *P. shenandoah* and *P. cinereus* on the The Pinnacles (P), Stony Man (S) and Hawksbill (H) Mountains of Shenandoah National Park). These sister species are morphologically and ecologically similar, but *P. cinereus* is behaviorally more aggressive and restricts the distribution of *P. shenandoah* to less optimal talus habitat on the tops of mountains. (After Myers and Adams 2008; *P. cinereus* photo courtesy of David McIntyre; *P. shenandoah* photo courtesy of the National Park Service.)

Plethodon cinereus

(B)

P. shenandoah

do occur, but strong selection against interspecific hybrids and other deviant phenotypes during periods of food shortage maintains the clumpy-gappy distribution of distinctive sympatric species (see Figure 7.4).

Adaptive radiation

Adaptive radiation is the diversification of species to fill a wide variety of ecological niches. It occurs when a single ancestral species gives rise, through repeated episodes of speciation, to numerous kinds of descendants that become or remain sympatric. Often, these bursts of speciation are thought to be initiated by the development of a **key innovation** in the ancestral species. A well-characterized example of key innovations leading to adaptive radiations is the evolution of the tentacular mouthparts of yucca moths (Prodoxidae), which led to diversification of over 25 extant species that pollinate yuccas

(Agavaceae) in western North America (Pellmyr and Krenn 2002). Another well-characterized example is the decoupling of the food-gathering and food-processing functions of the oral and elaborate pharyngeal jaws associated with radiation of well over 1500 species in the diverse group of cichlid fishes in tropical lakes (Hulsey et al. 2006; Turner 2007).

Today, looking at the variety of living things, we can find numerous examples of successful lineages that have radiated to produce diversity at many levels. We can, for example, consider the adaptive radiation of Hawaiian honeycreepers (Drepanidinae), a group of small perching birds that probably colonized the archipelago within the last few million years (Amadon 1950; Raikow 1976; Tarr and Fleischer 1995). Or we can examine the major radiation of placental mammals that produced many of the existing mammalian orders (e.g., Lilligraven 1972), which occurred during the Cenozoic Era after the mass extinction event that eliminated most of the giant reptiles. In all such cases, the basic ecological opportunities are created either by an adaptive (key) innovation or an environmental change, such as colonization of a new area or the extinction of competing species. These opportunities are exploited as an ancestral form initiates repeated speciation events, producing specialists that fill numerous ecological niches (Schluter 2000).

One of the most dramatic examples of adaptive radiation, as we discussed above, is provided by the cichlid fishes of the lakes of East Africa (See Table 7.2 and Figure 7.18). Key innovations underlying the explosive and parallel evolution of species flocks in different lakes probably include a combination of behavioral, morphological, and sexually selected traits (Salzburger et al. 2005). Once a radiation begins within a lake, it can progress along a course in which the forces underlying diversification are progressive, beginning with adaptation to alternative rocky or sandy habitas, followed by diversification via natural selection of mouthparts for feeding specialization, and continuing to the proliferation of male color patterns via sexual selection (**Figure 7.22**).

Islands, archipelagoes, and lakes provide many examples of adaptive radiations, some that could have occurred in a sequential pattern similar to the

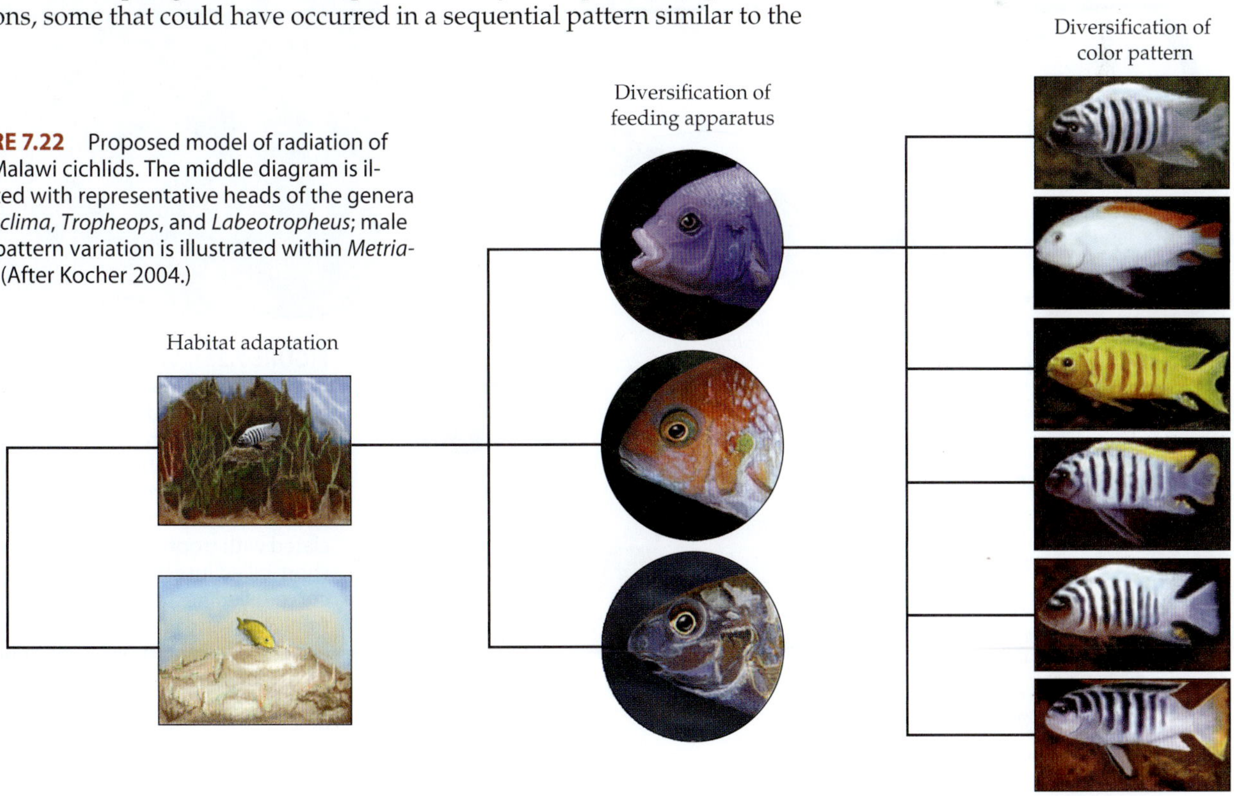

FIGURE 7.22 Proposed model of radiation of Lake Malawi cichlids. The middle diagram is illustrated with representative heads of the genera *Metriaclima*, *Tropheops*, and *Labeotropheus*; male color pattern variation is illustrated within *Metriaclima*. (After Kocher 2004.)

one described above for the Malawi cichlids (Streelman and Danley 2003). Madagascar, with its long history of isolation from Africa and other southern continents (see Chapter 8), has been the site of several spectacular radiations. These include not only the well-publicized lemurs (Primates) but also the less well-known tenrecs (a morphologically, behaviorally, and ecologically diverse endemic family, Tenrecidae), vanga shrikes (a similarly diverse group of perching birds belonging to the endemic family Vangidae), leaf-tailed geckos (genus *Uroplatus*), and mantellid frogs (family Mantellidae, including forms convergent with Neotropical poison-dart and tree frogs). Among the many examples in the Hawaiian Islands (see also Chapter 12; Wagner and Funk 1995) are Hawaiian honeycreepers of the endemic subfamily Drepanidinae (Pratt 2003); the Hawaiian silversword alliance (Baldwin and Robichaux 1995); and the picture-winged *Drosophila* (Carson and Kaneshiro 1976). There

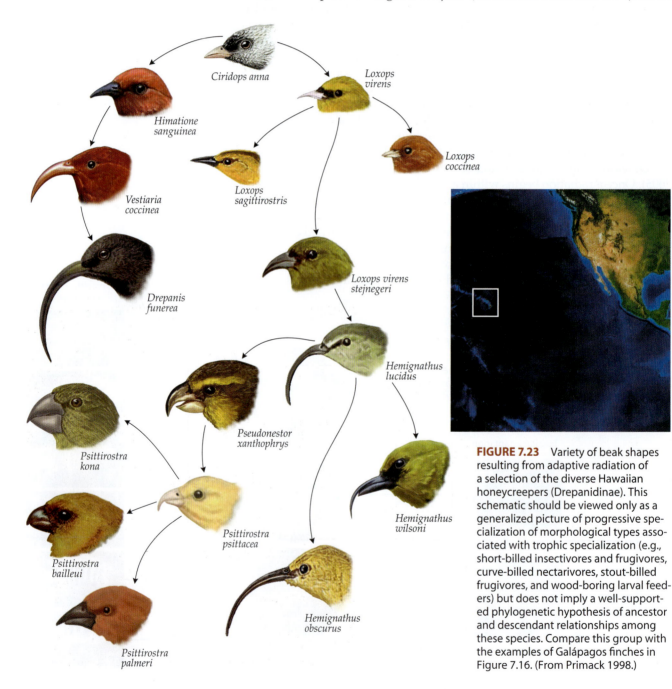

FIGURE 7.23 Variety of beak shapes resulting from adaptive radiation of a selection of the diverse Hawaiian honeycreepers (Drepanidinae). This schematic should be viewed only as a generalized picture of progressive specialization of morphological types associated with trophic specialization (e.g., short-billed insectivores and frugivores, curve-billed nectarivores, stout-billed frugivores, and wood-boring larval feeders) but does not imply a well-supported phylogenetic hypothesis of ancestor and descendant relationships among these species. Compare this group with the examples of Galápagos finches in Figure 7.16. (From Primack 1998.)

were about 33 species of honeycreepers when Europeans first visited Hawaii, although at least 10 have since gone extinct, and most of the others are endangered. They are descended from a single common ancestor, now thought to be a cardueline finch that colonized from North America, probably less than 5 million years ago (Tarr and Fleischer 1995). The lineage radiated to produce an amazing variety of species, which differ conspicuously in the sizes and shapes of their beaks (**Figure 7.23**) and to a lesser extent in color pattern. As in Darwin's finches, ecological differentiation to exploit different niches—especially food sources—seems to have been the primary process that led to radiation of the honeycreepers (Amadon 1950; Raikow 1976; Tarr and Fleischer 1995).

While many examples come from isolated habitats, such as islands and lakes, comparable adaptive radiations have occurred in other groups in oceans and on continents. They have also taken place on many different temporal and spatial scales. An ancient radiation occurred in the marine realm in the cryptically colored marine anglerfishes, in the order Lophiiformes (**Figure 7.24**), producing eight extant families and over 300 species, including the shallow-water frogfishes (*Antennarius* and *Antennatus*), the open-ocean sargassum fish (*Histrio histrio*), the stingraylike batfishes (Ogcocephalidae), and a variety of deep-water bioluminescent seadevils (suborder Ceratioidea). In plants, a group that has radiated over the past 20 million years on the North American continent is the phlox family (Polemoniaceae). Within this single family, flower form and color have become amazingly variable as different species have adapted to be pollinated by hawk moths, bees, butterflies, flies, bats, or hummingbirds, and some have become specialized for self-pollination (**Figure 7.25**). Australia, the most isolated continent, is the site of many spectacular radiations: marsupial mammals, lizards of the genera *Ctenotus* and *Varanus*, and plants of the genera *Eucalyptus*, *Melaleuca*, and *Acacia*.

Whereas many genera and families exhibit a fascinating variety of ways of life, many others are extremely monotonous, differing mainly in minute

(A)

(B)

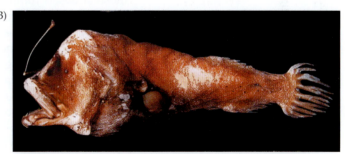

(C)

FIGURE 7.24 Examples of the diverse and ancient adaptive radiation of anglerfishes (Lophiiformes). A diagnostic feature and possible key innovation is the luring apparatus on the tip of the snout, produced through modification of the first dorsal-fin spine. (A) Frogfish (*Antennarius pictus*); (B) sargassum fish (*Histrio histrio*); (C) *Himantolophus* sp., "football fish." (A courtesy of John E. Randall; B courtesy of T. W. Pietsch/University of Washington; C © David Shale/Naturepl.com).

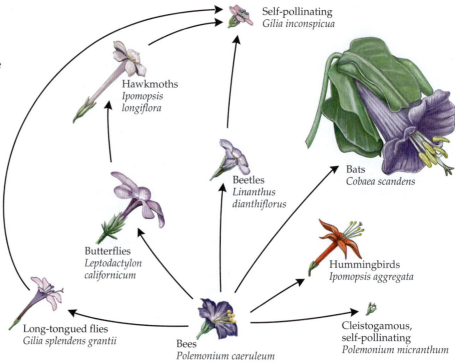

FIGURE 7.25 Adaptive radiation in the phlox family, Polemoniaceae, showing diversity of flower form reflecting different modes of pollination. A generalized bee-pollinated ancestor is believed to have given rise to the other forms; most coevolved for mutualistic relationships with particular kinds of animals, but some specialized for self-pollination. Plants in other families that have specialized to use the same kinds of pollinators often have convergently evolved similar flower sizes, shapes, colors, odors, nectar, and/or pollen rewards. (After Ehrlich and Holm 1963.)

Self-pollinating
Gilia inconspicua

Hawkmoths
Ipomopsis longiflora

Bats
Cobaea scandens

Beetles
Linanthus dianthiflorus

Butterflies
Leptodactylon californicum

Hummingbirds
Ipomopsis aggregata

Long-tongued flies
Gilia splendens grantii

Bees
Polemonium caeruleum

Cleistogamous, self-pollinating
Polemonium micranthum

structural details. Taxa that have changed only slightly over evolutionary time apparently are good at what they do and have not had the genetic flexibility or ecological opportunity to shift adaptive strategies. The wide variation in rates and degrees of divergence makes it difficult to generalize about the diversification of higher taxa and the influence of geography on adaptive radiation. The many examples of spectacular adaptive radiations on islands seem to show the importance of long historical isolation in what were intially species-poor and disharmonic ecosystems (i.e., those colonized by just a few groups of organims with relatively high dispersal abilities; see Chapter 14). Equally impressive radiations in less isolated settings, however—such as those of desert rodents of the family Heteromyidae in southwestern North America; sigmodontine rodents in South America; darters of the fish genus *Etheostoma* in the Mississippi River drainage; lizards in the *Liolaemus darwinii* complex in South America; and the spiny, succulent plants of the family Cactaceae in the arid regions of North and South America—suggest that long episodes of geological and ecological transformation may induce adaptive radiation on less geographically isolated landscapes. Comparative studies that are attempting to assess the influence of phylogenetic relationships and evolutionary constraints (Felsenstein 1985; Harvey and Pagel 1991; Brooks and McLennan 2002) are making important contributions to our understanding of the enormous differences in rates of speciation and adaptive differentiation among lineages.

Extinction

Ecological processes

Although all living organisms represent a continuous evolutionary lineage extending billions of years back to the origin of life, the ultimate fate of every species is extinction. This can be appreciated by taking a brief glance at the fossil record. Earth was teeming with life 100 million years ago. Both terrestrial and aquatic habitats were occupied by diverse biotas that formed complex ecological communities. However, the species and genera—and many of the families and orders—that were dominant then have been eliminated or drastically reduced by extinction, and they have been replaced by new lineages. On land, dinosaurs and other reptilian groups have been replaced by birds and mammals, while ferns and gymnosperms have been largely supplanted by angiosperms. In the oceans, cephalopod mollusks have been supplanted by teleost fishes, while icthyosaurs and mesosaurs (reptiles) have been replaced by dolphins, whales, seals, and sea lions (mammals). Extinctions have apparently occurred continuously throughout the history of life, although the fossil record also catalogs occasional episodes of widespread disaster when much of the Earth's biota was wiped out, apparently by rapid and drastic environmental change (Raup and Sepkoski 1982). Not surprisingly, extinctions have had a major influence, not only on the kinds of organisms in existence at any given time, but also on the geographic distributions of those now-extinct forms and the contemporary lineages that are descended from them.

Several authors have likened the evolutionary history of life to a continual race with no winners, only losers—those species that become extinct. This view is probably best expressed in Van Valen's (1973b) **Red Queen hypothesis**, named for the Red Queen in Lewis Carroll's *Through the Looking Glass* who said, "It takes all the running you can do to keep in the same place." The idea is that a species must continually evolve in order to keep pace with an environment that is perpetually changing, not just because abiotic conditions are shifting, but also because all the other species are evolving, altering the availability of resources and the patterns and processes of biotic interactions. Those species that cannot keep up with the changes become extinct, but others do well temporarily and speciate to produce new forms.

Van Valen points out that the probability that a species will become extinct appears to be independent of its evolutionary age (although more recent work suggests that the interplay between speciation and extinction dynamics does lead to an overall association of species richness with clade age [McPeek and Brown 2007]) but not of its taxonomic and ecological status. Certain taxonomic and ecological groups have consistently higher rates of extinction than others. For example, apparently because of their lower extinction rates, small and herbivorous mammals are found on more and smaller islands than large or carnivorous species (Brown 1971b; Heaney 1986; Lawlor 1986). This appears to be a general pattern (see Van Valen 1973a). Somewhat similarly, differences in diversity and duration in the fossil record among lineages of marine invertebrates are correlated with characteristics of their life history, as we shall see below.

While we speak of extinction as the end of a species, the processes that begin a species on the path to extinction typically occur at the scale of individual populations. Some researchers have developed mathematical models to predict a population's vulnerability to extinction, based on its demographic characteristics (MacArthur and Wilson 1967; Richter-Dyn and Goel 1971; Leigh 1981; Gilpin and Hanski 1991; Hanski 2005; Hanski and Gaggiotti 2008). All populations experience fluctuations in size as a result of variations in environmental conditions and the activities of their enemies. When populations become very small, however, purely chance factors, such as random variations in the sex ratio, can also affect their abundance and result in what are termed stochastic extinctions. Mathematical models show that, in general, the smaller a population becomes and the lower its ratio of births to deaths (see Equation 4.1) and the longer it remains at low numbers, the more vulnerable it is to extinction. A population with a low birth rate, especially when coupled with a high death rate, cannot recover rapidly from a temporary reduction in numbers. Long-term overall population size is probably the most important factor. Models suggest that probability of extinction increases nonlinearly as population size decreases, and it becomes very high when population size becomes, and remains, very low—say, fewer than 100 indi-

FIGURE 7.26 Output of a mathematical model showing how estimated time to extinction depends on two demographic characteristics of a population: equilibrial population size, or carrying capacity, (*K*), and the ratio of birth rate (*b*) to death rate (*d*). The graphs show that the probability of extinction is high (expected time to extinction is low) when populations are small and birth rates are low relative to death rates, but the probability of extinction decreases rapidly as these demographic parameters increase. (After MacArthur and Wilson 1967.)

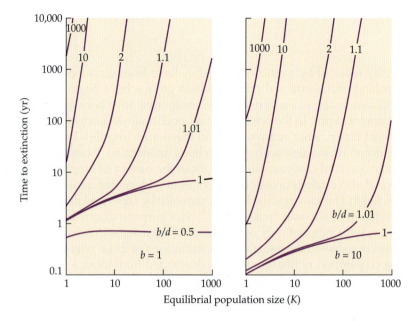

viduals (**Figure 7.26**). These mathematical models are based on the intrinsic demographic characteristics of populations, but it is important to recognize that changes in extrinsic environmental conditions are likely to be the cause of changes in those characteristics that ultimately lead to extinction.

Information on how intrinsic demographic and extrinsic ecological factors interact to cause extinction is difficult to obtain and interpret, because extinctions of species, with the exception of those caused by humans, are rarely observed. Researchers have gained valuable insights, however, by studying the effects of these factors on the turnover of small, isolated subpopulations within a metapopulation (Gilpin and Hanski 1991; Moilanen et al. 1998; Kreuzer and Hunty 2003; Franken and Hik 2004; see also Chapters 4 and 16). One particularly well-documented example is provided by the work of Andrew Smith (1974, 1980) on the American pika (*Ochotona princeps*). The American pika is a small relative of rabbits; it lives in rock slides and boulder fields in the mountainous regions of western North America. Smith carefully monitored pikas that had colonized the rock piles left by a mining operation in the Sierra Nevada Mountains of California. These mine tailings functioned as habitat islands for the pikas in a sea of sagebrush habitat. Smith was able

TABLE 7.3 *Estimated Time to Extinction for Pika Subpopulations*

Population size (number of individuals) *K*	Birth rate (per capita per year) *b*	Death rate (per capita per year) *d*	Time to extinction (years) *E*
1	0.35	0.00	2.9
2	0.35	1.63	6.9
3	0.35	2.44	46.2
4	0.35	2.84	405.1
5	0.35	3.15	3751.5

Source: Data from Smith 1974, 1980.

Note: The populations represented here occupy habitat islands of varying size. Per capita birth and death rates were determined from the age structure of this population. A mathematical model, based on the model of MacArthur and Wilson (1967), predicts that only very small populations (*K* < 3 individuals) should have measurable rates of extinction owing to ordinary random demographic fluctuations. This prediction was confirmed by Smith's subsequent measurement of population turnover on these same islands during a 5-year period.

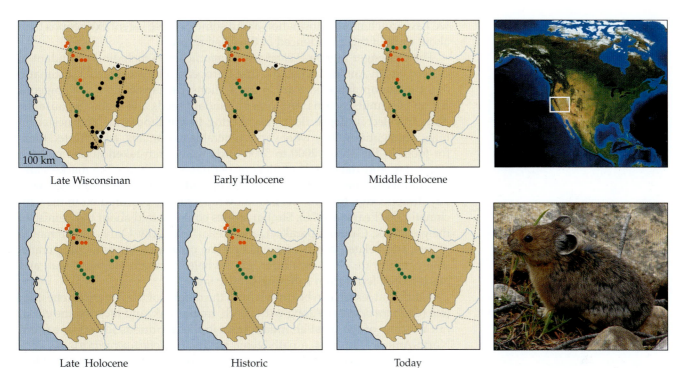

Late Wisconsinan   Early Holocene   Middle Holocene

Late Holocene   Historic   Today

FIGURE 7.27 A summary of Holocene extinctions of pikas (*Ochotona princeps*) in the Great Basin, USA. Red circles = records of historic populations that are now extinct; green circles = extant populations; black circles = extinctions in recent times. (After Grayson 2005; photo courtesy of Sevenstar/Wikipedia.)

to document extinctions of subpopulations by censusing the rock piles, initially in 1972 and then again in 1977. The repeat censuses documented both extinctions of previously existing subpopulations and colonizations of previously uninhabited rock piles. From data on the birth and death rates of pikas and the sizes and degrees of isolation of the rock piles, Smith and colleagues developed a stochastic model that quite accurately predicted the frequency of extinction and colonization events as a function of these factors (Moilanen et al. 1998; see also Beever et al. 2003) (**Table 7.3**). Somewhat similarly, Lomolino (1984, 1993b) and Crowell (1986) documented the extinction and immigration of small mammals on islands in the St. Lawrence River, in Lake Huron, and in the Gulf of Maine. Even though these studies are concerned with the turnover of small, isolated subpopulations rather than extinctions of entire species, they illustrate how both intrinsic demographic characteristics and extrinsic environmental conditions influence the extinction process. In the pika example, we now have a good record (Beever et al. 2003; Grayson 2005) documenting historical extinctions on increasingly isolated mountaintops in the Great Basin as climates warmed following the Pleistocene and, consequently, extrinsic environmental conditions shifted the dynamics in favor of extinctions of subpopulations without colonizations, particularly at lower and hotter elevations (**Figure 7.27**).

Recent extinctions

Over the last 200 years, humans have caused the extinctions of thousands of species; moreover, we are undoubtedly unaware of many more species of microbes and small animals and plants that have disappeared. On the other hand, the demise of some larger, more spectacular organisms is well documented (see Chapter 9). Recent reductions in populations, contractions of geographic ranges, and extinctions of species caused by humans will be considered in more detail in Chapter 16. We will present a few examples here, however, to illustrate some of the processes involved in extinctions, because we have much more information about some of these recent extinctions than about most of those evidenced in the fossil record.

The passenger pigeon (*Ectopistes migratorius*) was incredibly abundant in eastern North America when the first European colonists arrived. Estimates of the total population size are in the millions, perhaps billions. The pigeons fed on beechnuts, acorns, and other abundant seeds and fruits. Because the birds traveled in dense flocks numbering in the thousands and nested in huge aggregations, they were very vulnerable to humans, who hunted them for food. In the 1870s, 2000 to 3000 birds were often taken in one net in a single day, and 100 barrels of pigeons per day were shipped to markets and restaurants in New York City for weeks on end. By 1890, the birds had virtually disappeared. In 1914, the last known passenger pigeon died in the Cincinnati Zoo (Pearson 1936). Similar stories could be told about the demise of the great auk, Carolina parakeet, and Stellar's sea cow. Only a combination of luck and belated conservation action has prevented the whooping crane, trumpeter swan, bison, sea otter, gray whale, northern elephant seal, and black-footed ferret from suffering similar fates.

A different lesson can be drawn from the demise of the American chestnut tree (*Castanea dentata*). Along with beech, maple, oak, and hickory, the chestnut was one of the most abundant trees in the deciduous forests of eastern North America. In 1904, a pathogenic fungus (*Endothia parasitica*) was accidentally introduced, apparently from Asia, where it attacks a related but less susceptible species of chestnut. The disease spread rapidly (Metcalf and Collins 1911), and within 40 years, mature chestnuts were virtually eliminated from their entire range (see map in Figure 16.12). In some areas, scattered small trees that have sprouted from surviving rootstock can still be found. Unfortunately, these are usually attacked by the disease and killed before they can reach reproductive size, so it is questionable whether the chestnut can much longer avoid absolute extinction. Similar cases of biological warfare due to introduced pathogens are the near extinction of the American elm due to Dutch elm disease, of native Hawaiian birds due to avian malaria (see Chapter 4), and of native peoples on oceanic islands, and perhaps even on the American continents, due to smallpox, measles, diphtheria, and other diseases brought by European invaders.

One of the best-documented cases of local extinction of multiple species is the loss of bird species from Barro Colorado Island in Panama (see Figure 6.16). Prior to the early 1900s, the island did not exist; it was simply a hill in a tract of tropical lowland forest. During the construction of the Panama Canal, the Chagras River was dammed, and the rising waters of Gatun Lake covered the adjacent lowland areas, creating Barro Colorado, an island of about 16 km^2. Despite its relatively large size, Barro Colorado no longer contains many of its original bird species. Because the Smithsonian Institution has operated a long-term biological research station on the island, its biota is well known, and some of the extinctions are well documented. Using these records, Willis (1974) calculated conservatively that at least 17 species had been lost from a total land bird fauna in excess of 150 species. More recently, by comparing the avifauna of the island with that of a mainland site of similar size and habitat, Karr (1982) concluded that at least 50 species have become extinct on the island since its isolation. Although habitat changes may have contributed to the demise of some forms, most of the missing species were those normally occurring at low densities, and it appears that when the small populations present on the island died out, they were not replaced by colonists from across the water gap. By 1996, up to 65 species extinctions were recorded (Robinson 1999)—many being interior forest inhabitants—indicating that the island avifauna continued to "relax" to a lower level of species richness.

The above examples illustrate the vulnerability of many species to environmental changes caused by modern humans (see also Chapter 16). However, humans do not need to have guns, agriculture, or European culture to cause extinctions. Aboriginal humans almost certainly played a major role in many extinctions documented in the fossil record, as we will see in Chapter 9. On the other hand, the capacity of some species to recover from low numbers is amazing. Two species of marine mammals that inhabit the Pacific coast of North America—the northern elephant seal (*Mirouga angustirostris*) and the sea otter (*Enhydra lutris*)—were hunted almost to extinction for their fat and fur, respectively. Tiny populations managed to escape detection, however, and once protected, increased rapidly to produce large, healthy populations. Another Pacific marine mammal, the eastern Pacific population of the gray whale (*Eschrichtius robustus*), has been steadily increasing following the protection of its breeding grounds off Baja California and the cessation of hunting (Hobbs et al. 2004), and the gray whale was removed from the U.S. Endangered Species List in 1994. Several bird species that were hunted to near extinction in the late 1800s for feathers to supply the millinery industry, including trumpeter swans, sandhill cranes, and great and snowy egrets, are again abundant.

Extinctions in the fossil record

The fossil record provides abundant evidence of extinctions, but it often provides only tantalizing clues to their causes. In many cases, the fossil record is complete enough to show that many diverse species became extinct virtually simultaneously over relatively short periods of geological time. We can infer that each **mass extinction** was caused by some drastic, widespread environmental change, but the exact nature of the perturbation and its effect on the organisms concerned is often difficult to deduce clearly. Long-standing, vigorous debates about the causes of mass extinctions mark paleontological literature. The six largest mass extinctions recognized in the fossil record have consequences beyond the elimination of large portions of the standing biological diversity. First, they have often been considered important precursors to new cycles of adaptive radiation because they created new opportunities for speciation of surviving lineages to fill the array of ecological niches left unoccupied in their wakes. Second, it often appears that even if a few species in a group that was extremely successful prior to a mass extinction event happen to survive, they seldom give rise to another equally successful radiation within that group—a phenomenon called "dead clade walking" by Jablonski (2002).

One of the most recent episodes of mass extinction was the disappearance of the Pleistocene megafauna of North and South America between 15,000 and 8000 years ago. The cause of this dramatic event is still the subject of controversy, as we will see in Chapter 9. Originally, paleontologists favored hypotheses invoking climatic change. The idea that aboriginal humans played a major role, advanced by Paul Martin in the 1960s and known as the overkill hypothesis, was initially rejected by most paleontologists. Multiple lines of accumulating evidence, however, have shifted the balance, causing many scientists to conclude that humans almost certainly played a significant—and perhaps pivotal—role in the extinctions of megafauna (e.g., Martin and Klein 1984; Owen-Smith 1987, 1989; Barnosky et al. 2004). Examples of additional extinctions caused or threatened by modern humans will be discussed in detail in Chapter 16.

Causes of other mass extinctions that occurred even earlier have also been highly controversial. Two of the most dramatic episodes saw the disappearance of dinosaurs and many other groups of terrestrial and marine organ-

(A)   (B)

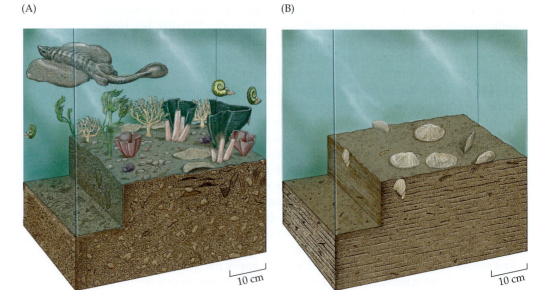

FIGURE 7.28 Reconstructions of the ancient seabed in southern China (A) before and (B) after the Permo-Triassic mass extinction. Before the crisis, the seabed included more than 100 species of reef dwellers and burrowing infauna. Species richness was reduced to 4 or 5 species following the event. (From Benton and Twitchett 2003; artwork © J. Sibbick.)

isms at the end of the Cretaceous period, about 65 million years ago, and of even more species and higher taxa—particularly marine organisms—at the boundary between the Permian and the Triassic periods, about 250 million years ago (see Raup 1979; Raup and Sepkoski 1982; Jablonski 1991; Benton and Twitchett 2003). Raup (1979) and Jablonski (1991) estimated that 96 percent of all marine species then in existence went extinct at the end of the Permian (**Figure 7.28**). Causes most often proposed for these and other mass extinctions are large, rapid changes in the climate of the Earth: either global cooling, with resulting continental glaciation and exposure of continental shelves, or global warming, with associated rising sea levels and inundation of continental shelves (e.g., Stanley 1984). Some evidence points to mass volcanic activity with associated global warming as a cause of the Permo-Triassic event (Benton and Twitchett 2003). But other explanations, including continental drift (e.g., Schopf 1974) and reorganization of interaction in ecological communities (Bak 1996), have had their proponents.

While the causes of the Permo-Triassic event remain poorly understood, there is now little doubt about the cause of the Cretaceous–Tertiary event. Its discovery is one of the most exciting detective stories in recent Earth sciences, rivaling the development of plate tectonic theory. It has long been apparent that the end of the Cretaceous was a time of major change. The Cretaceous–Tertiary (K-T) boundary was defined based on dramatic changes in the fossil record. Not only did major groups of previously dominant organisms (such as dinosaurs on land and ammonites in the oceans) go extinct, but there were also major changes in the abundance and species diversity of the lineages that survived (e.g., dramatic increases in birds and mammals on land, and teleost fishes in the oceans; and decreases in marine cephalopod mollusks). Until the early 1980s, these changes were generally thought to have occurred over several million years and to have been caused by climatic change.

In 1980, three scientists at the University of California at Berkeley suggested a radically new hypothesis: an asteroid impact (Alvarez et al. 1980, 1984). Initially, this sounded like an idea from a science fiction movie and was ridiculed by many established Earth scientists. Nevertheless, the group, led by Walter Alvarez, presented intriguing data and a testable hypothesis. The asteroid impact hypothesis was based on the occurrence in rock strata that spanned the K-T boundary of a layer highly enriched in the element iridium (Ir). Iridium is rare in the Earth's crust but is often present in high concentrations in materials of extraterrestrial origin (meteorites and asteroids). The Alvarez group hypothesized that the iridium-enriched layer was produced by the dust injected into the atmosphere and circulated around the Earth following the impact of a large asteroid. The dust presumably blocked solar radiation and resulted in rapid global cooling, which in turn caused the extinction of dinosaurs and other lineages.

A strength of the Alvarez hypothesis was that it made testable predictions (Alvarez et al. 1984). First, it predicted that when an iridium-enriched layer occurred near the K-T boundary, it would always separate Cretaceous rocks below from Tertiary strata above. This prediction led to efforts to find stratigraphic sequences spanning the K-T boundary, to search for an iridium-enriched stratum, and to date the underlying and overlying rocks. Many independent studies of localities throughout the world confirmed the prediction: When present, the iridium-enriched layer marked the K-T boundary. Second, the hypothesis predicted that the vast majority of extinctions should have occurred right at the K-T boundary. Previous interpretations of fossil remains had suggested that some taxa disappeared millions of years before or after the K-T boundary. More-accurate dating of fossil remains, however—especially with reference to the telltale iridium-enriched layer, when present—increasingly supported the hypothesis: Extinctions were concentrated during a very brief period. Third, an asteroid impact and resulting global cooling would be expected to lead to differential extinction or survival of different kinds of organisms, depending on their life histories and ecologies. And, indeed, the K-T extinctions were highly selective. Those groups with capacities for body temperature regulation, such as endothermic birds and mammals, and those with life history stages that would be resistant to brief but intense cold stress, such as seed plants, insects, and freshwater invertebrates, suffered fewer extinctions than many other kinds of organisms.

Finally, the hypothesis predicted that the crater formed by the asteroid impact might actually be discovered, and it suggested that the thickness of the iridium-enriched stratum should provide clues to its location. Obtaining evidence bearing directly on this prediction initially seemed to be a long shot. No known meteorite craters had the appropriate combination of size and age to be good candidates. Further, since most of the Earth's surface is ocean and most of the seafloor is young, there was a high probability that the impact occurred in the ocean and that the crater had long since been subducted into an ocean trench (see Chapter 8). However, mapping of increasing data on the iridium-enriched stratum showed a clear geographic pattern, with the thickest layers in the Caribbean region. Then came the definitive find: the discovery of a crater—of the right size and age—on the continental shelf off the coast of Yucatán, termed the Chicxulub crater (Swisher et al. 1992).

So we now know that an asteroid did strike Earth about 65 million years ago. It appears to have caused the extinction of dinosaurs, ammonites, and many other organisms and in so doing, to have opened the way for the adaptive radiation of mammals, teleost fishes, and other lineages and their rise to their present dominance. While the major features of this event are now well documented, many pieces of the story are not. How much was the Earth cooled, and how long did the cold period last? Beyond the differences in life history and ecology mentioned above, what determined which species and clades died out and which survived? Clearly, even such an unprecedented and unpredictable event as the impact of an extraterrestrial body did not cause "random" extinctions. Instead, it acted as a severe filter, eliminating many lineages, allowing others to pass through, and causing revolutionary changes in the composition of the Earth's biota that have endured for 65 million years.

Macroevolution

Species delineate two sometimes very different approaches to investigating evolutionary patterns and processes—**macroevolution** above and **microevolution** below the species level. Observations of changes in organisms in the fossil record, as well as those inferred from syntheses of morphological studies and molecular systematics, have given us new insights into the patterns and processes of macroevolution and their biogeographic context.

Until the 1970s, most biologists assumed that evolution usually proceeds gradually through successive incorporation of relatively small genetic changes as populations respond to environmental change in space and time. However, many eminent evolutionary biologists (e.g., Simpson 1944; Mayr 1963) emphasized that evolutionary rates can vary widely, even within the same evolutionary lineage. Simpson categorized evolutionary lineages into **bradytelic** (slower-evolving than typical), **tachytelic** (faster-evolving than typical), or **horotelic** (typical) rate classes. Mayr and others also suggested that major, rapid changes can occur during the process of speciation.

The prevailing microevolutionary view was the product of the Modern Synthesis that dominated evolutionary thought for most of the twentieth century. The new synthesis was concerned primarily with how the characteristics of populations change as a result of the genetic mechanisms of natural selection, mutation, genetic drift, and gene flow that we described earlier in this chapter. Microevolution has traditionally been studied in contemporary organisms, either by observing experimentally induced changes in laboratory populations of organisms, such as microbes and fruit flies, or by drawing inferences from comparative studies of populations in the field. A long-running debate among evolutionary biologists has been whether—and to what degree—these microevolutionary processes can adequately explain a variety of evolutionary changes observed above the species level, including the origins of evolutionary novelties (e.g., C_4 metabolism in plants, tetrapod limbs and endothermy in vertebrates, compound eyes in arthropods), the differential rates of evolution and success of lineages through time, and the influences of a dynamic Earth history (see Chapters 8 and 9) on patterns of diversification.

Evolution in the fossil record

In the 1970s, a group of young paleontologists that included Niles Eldredge, the late Stephen J. Gould, Steven M. Stanley, and Elisabeth Vrba began to emphasize a variety of kinds of evolutionary changes recorded in fossil remains that are not easily explained through simple extrapolation of microevolutionary processes. Because of the long time frame typically involved, these changes tend to be large and include speciation and extinction events as well as fundamental changes in morphology. The paleontologists thus resurrected the concept of macroevolution. Eldredge and his colleagues suggested that macroevolution occurs primarily as a result of two processes, which they called punctuated equilibrium and species selection (Eldredge and Gould 1972; Stanley 1979).

The fossil record indicates that the evolution of a lineage usually consists of long periods of virtually no change (**stasis**) interspersed with relatively brief periods of rapid change, which often appear to be associated with speciation events. These bursts of rapid change are often accompanied by morphological innovations, geographic range expansion, and exploitation of new areas of ecological niche space. This pattern of highly variable rates of evolution is referred to as **punctuated equilibrium**. For example, one of the most profound events in the history of life was the Cambrian Explosion that defined the beginning of the Phanerozoic Eon and Paleozoic Era (see Table 8.1) and marked the origination of most major body plans in modern oceans and the extinction of many earlier groups. The idea of punctuated bursts of evolutionary change is supported by data from the fossil record—especially that of marine and freshwater invertebrates—in which new fossil species often appear abruptly and then persist for millions of years with virtually no morphological change. These bursts of rapid change often followed episodes of mass extinction, suggesting that the removal of prior occupants from ecological niche space provides a stimulus for the adaptive radiation of new forms to fill that space (Raup 1994; Jablonski 2002). Evidence of rapid change associated with speciation events comes, for example, from P. G. Williamson's (1981) work on the molluscan fauna of Lake Turkana Basin in eastern Africa. In a stratigraphic sequence representing several million years of fossil deposition, several lineages of mollusks showed long periods of virtually no change in shell morphology, punctuated by rapid shifts (**Figure 7.29**).

FIGURE 7.29 "Punctuated" equilibrium in the evolution of fossil mollusks in Lake Turkana Basin in eastern Africa. The diagram depicts the reconstructed history of shell morphology in several genera. Dotted lines indicate inferred changes during periods for which no fossils have been recovered. Note that most lineages exhibited long periods of virtual stasis followed by rapid, substantial change. The latter, punctuational events (marked by either a T for lake transgression or R for lake regression along the right margin) often occurred in one of two ways: virtually simultaneously in several different lineages, suggesting major environmental changes, such as shifts in lake level owing to climatic change (confirmed by other evidence); or in association with speciation events, which often left one species virtually unchanged while the other species diverged substantially. (After Williamson 1981b.)

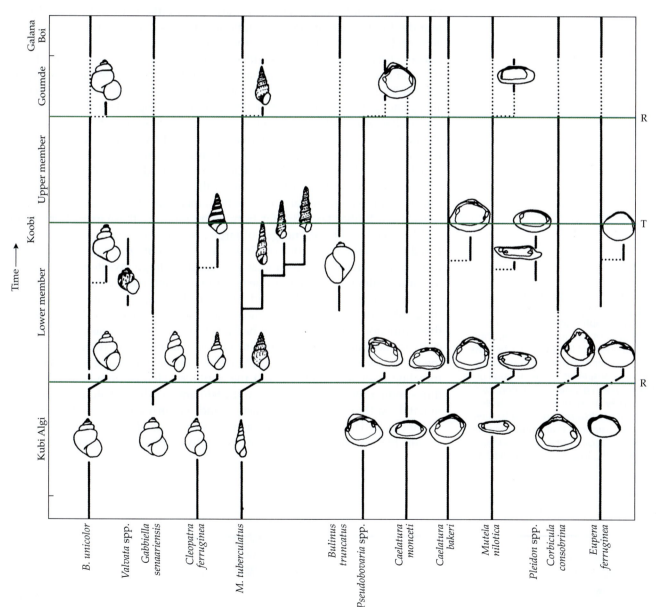

These changes took place within a time interval of less than 50,000 years—so rapidly with respect to the fossil record that often no shells with intermediate characteristics were found. Many of these shifts were accompanied by the splitting of lineages, and they often occurred in conjunction with rapid, climatically mediated changes in lake level. A high frequency of punctuated equilibrium is also detectable in benthic (e.g., the bryozoans *Metrarabdotos* and *Stylopoma*) and planktonic (e.g., the foraminiferan *Globorotalia*) lineages that existed in marine and brackish water environments during the Neogene of the last 25 million years (Jackson and Cheetham 1999).

Species selection

The other macroevolutionary process evidenced in the fossil record has been called **species selection**. The idea is simple—just as natural selection is a process of evolutionary change caused by differential survival and reproduction of individual organisms with certain heritable characteristics, so species selection is a process of evolutionary change caused by the differential survival and speciation of *species* with particular heritable traits (Stanley 1979). In order for this to be a viable evolutionary process, species must have emergent and heritable traits beyond those molded by microevolutionary processes acting upon individuals within a species. Two traits in particular, body size and geographic range size, have received much attention in this regard (Gould 2002; **Figure 7.30**). The demonstration, however, of species selection acting independently of selective processes operating below the level of the species has proven difficult. Even so, paleontologists have shown that major episodes of extinction and speciation in the fossil record have been selective with respect to the characteristics of species. For example, the asteroid that struck the Earth at the end of the Cretaceous Period about 65 million years BP (resulting in the Chicxulub crater off the coast of the Yucatán Peninsula) caused wholesale death and destruction, including the extinction of many species and higher taxa. But far from being random, these extinctions were highly selective. Entire lineages of large animals were completely exterminated (e.g., mass extinctions of dinosaurs, several other groups of giant terrestrial and marine reptiles, and several kinds of large invertebrates, including ammonites), while many lineages of small animals (e.g., insects, small mammals, and teleost fishes) and vascular plants survived and became the progenitors of many of today's most successful taxa. The selective nature of this extinction event apparently had a strong geographic bias as well, especially immediately following the impact—it was particularly pronounced to the north, owing to the fact that the meteor hit the Earth at a low trajectory, traveling over the Antarctic and the Southern Hemisphere before impact, thus sending the incendiary heat and a tsunami northward along the epicontinental sea that separated the eastern and western parts of North America (Flannery 2001). Similarly, in major episodes of speciation, some species with certain morphological (e.g., body size), physiological, life history, or geo-

FIGURE 7.30 Hypothetical example of species selection. The species-level trait "geographic range size" is more expansive for species 2, with populations on four adjacent islands (A–D), because it has greater dispersal potential than species 1 or species 3, even though both 1 and 3 reach greater population sizes when they co-occur with 2. However, this trait predisposes species 2 to survive volcanic eruptions that would drive the whole of species 1 or 3 to local extinction on any of the islands. (After Gould 2002.)

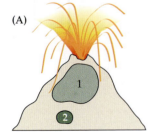 (A)

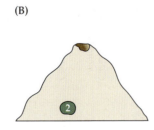 (B)

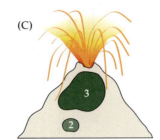 (C)

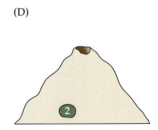 (D)

graphic range size characteristics have produced many more descendant species than have other ancestral species with contrasting characteristics.

Episodic events such as mass extinctions followed by explosive adaptive radiations have been particularly important in the historical pattern of speciation and extinction, and thus in the taxonomic diversity and geographic distribution of living things. The aftermath of the K-T extinction event makes a point that is also illustrated by other parts of the fossil record: Different lineages of organisms have not been equally successful, and even the most successful lineages have been dominant during different periods of Earth history and have prospered for different lengths of time. Some lineages have radiated rapidly to leave many descendants; others have survived virtually unchanged for millions of years; still others have disappeared quickly, leaving no descendants. Paleontologists and biologists have long recognized that certain lineages appear to possess particular traits that result in high speciation rates or low extinction rates, often leading to episodes of adaptive radiation and relative evolutionary success, compared with those groups that die out or barely maintain themselves.

Species selection and individual selection should be viewed, not as totally different biological processes, but as the consequences of generally similar ecological and genetic processes operating at different levels of biological organization. The same conditions (e.g., rapid environmental changes) that cause large differences in the birth and death rates of individuals of different genotypes, and thus result in rapid evolution by individual selection, are also likely to lead to differential multiplication and survival of species, and thus result in rapid evolution by species selection. The most important requirement for species selection to operate is that species-level traits are heritable. So, for example, if two unrelated marine benthic invertebrate lineages shared a trait such as having pelagic larvae (whereas other lineages within each of their clades did not) and we discovered that those lineages speciated more rapidly than other members of each clade, then we might suspect that selection favoring the trait "pelagic larvae" led to the differential production of species having that trait in two evolutionarily independent lineages.

The role of historical contingency

To some extent the fate of evolutionary lineages is a matter of chance and opportunity—it depends on the presence of a species with particular traits in a favorable place at an opportune time. As emphasized above, many adaptive radiations begin when a population either colonizes a new area or evolves a key innovation that substantially increases its fitness in a particular environment (Simpson 1952a). In either case, the species is suddenly presented with new ecological opportunities, which stimulate further rapid evolution and speciation. The Cenozoic radiations of mammals illustrate the influence of both of the above factors. During the Early Mesozoic, ancestors of modern mammals developed several innovations. These include precise articulation between jaws and crania, and specialized teeth—both traits conferring sophisticated efficiency in processing food—as well as an erect posture that increased efficiency in locomotion, a large brain, and probably endothermic temperature regulation. During the Late Mesozoic, ancestors of modern placental mammals acquired several new traits, including higher metabolic rates; improved temperature regulation; highly developed tactile, auditory, and olfactory systems; and new mechanisms of nourishing their developing young. These traits represented major advances over their reptilian ancestors and over other kinds of primitive mammals, including monotremes and some now extinct groups. However, it is possible that even with this array of innovations, mammals would not have become the dominant terrestrial vertebrates of the Cenozoic without the abrupt change in "rules" brought about by the K-T mass extinction event:

> We may reasonably conjecture that, absent this ultimate random bolt from the blue, dinosaurs would still dominate the habitats of large terrestrial vertebrates, and mammals would still be rat-sized creatures living in the ecological interstices of their world . . . at least in one cogent interpretation, certain marks of our ancestral incompetence—persistently small size in a dinosaurian world, for example—suddenly turned into a crucial and fortuitous advantage under the different rules of K-T impact, while the former source of triumph for dinosaurs may have spelled their doom under these same newly imposed rules. (Gould 2002: 1320)

Extinction of the dinosaurs and other vertebrate groups at the end of the Cretaceous eliminated the rivals of placental mammals and provided them with new ecological opportunities. Immediately afterward, in the Early Cenozoic, a relatively small number of lineages of placental mammals radiated explosively (**Figure 7.31**). They not only "replaced" the extinct reptilian groups, but also diversified at the expense of the other mammalian lineages, such as the "primitive" docodonts, and the rodentlike multituberculates that had survived the K-T mass extinction (Luo 2007).

The role of contingency in historical events is further illustrated by later Cenozoic history of mammals on several southern continents. Radiation of

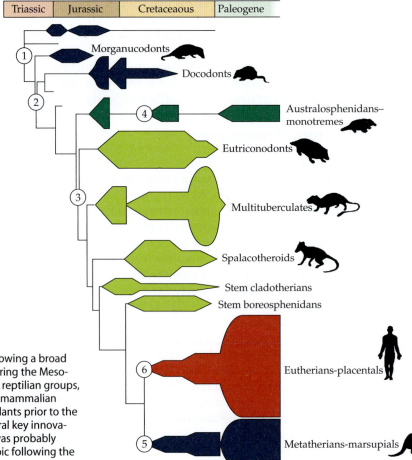

FIGURE 7.31   Early radiation of mammals, showing a broad range of diversity of clades and ecomorphs during the Mesozoic, thus their coexistence with the dominant reptilian groups, including the dinosaurs. While most Mesozoic mammalian lineages went extinct without leaving descendants prior to the Cenozoic, the coalesced accumulation of several key innovations in stem placental and marsupial clades was probably critical to their proliferation during the Cenozoic following the K-T extinction event. (After Gould 2002.)

placental mammals occurred throughout the world except in Australia. On this isolated continent, surrounded by ocean barriers, egg-laying monotremes and pouched marsupials were the norm. In the absence of dominant reptiles, marsupials underwent their own adaptive radiations. South America had a mixture of marsupial and basal placental groups for much of the Cenozoic, but was subsequently colonized by many other groups of placental mammals from North America (and perhaps Africa), and many of the original South American taxa became extinct (see Chapter 10). Australia has remained isolated right up to the present. Two groups of egg-laying monotremes (duck-billed platypus and echidna) survived, and marsupials underwent an extensive radiation, giving rise to many families and diverse morphological and ecological types that are amazingly convergent with placental forms (but no flying or marine forms, likely because of functional morphological constraints) (see Figure 10.38). These include a sand-burrowing "mole" (Notoryctidae); small to medium-sized insectivores and carnivores—some called marsupial "mice," "cats," and "wolves" (Dasyuridae); numbats (Myrmecobiidae) that converge on anteaters and aardvarks, possums, and gliders; koalas (Phalangeridae); bandicoots (Paramelidae); wombats (Phascolomyidae); and a wide variety of wallabies and kangaroos (Macropodidae). Prior to the Pleistocene and the arrival of humans and their dogs (dingoes) and other mammals that the humans imported, only two orders of placental land mammals—rodents and bats—had managed to colonize Australia from Asia, presumably across water barriers. The rodents diversified dramatically, after at least two different colonization events, to produce about 80 species, most belonging to endemic genera. These include water rats (*Hydromys*), desert hopping mice (*Notomys*), and stick nest builders (*Leporillus*), which are ecologically and morphologically similar to North American muskrats (*Ondatra*), kangaroo rats (*Dipodomys*), and pack rats (*Neotoma*), respectively (Keast 1972a,b,c).

In the absence of a historical event that eliminated competitors or provided access to a new area, radiations often were slower, and the buildup of diversity took longer. Many such radiations apparently occurred when some evolutionary innovation gave an ancestral species an advantage over coexisting organisms. Despite its apparent superiority, however, it usually took considerable time for the founding lineage to speciate, diversify ecologically, and supplant other groups of organisms that were already present. An example is provided by the neogastropods, a group of specialized, predatory marine snails that originated in the Cretaceous 130 million years ago, which have gradually diversified to become a dominant invertebrate group. Another is provided by angiosperm plants, which also originated in the Cretaceous. They possessed a number of structural and functional innovations in reproductive biology that, among other things, allowed them to evolve mutualistic associations with the animals that pollinate their flowers and disperse their seeds. Despite these advantages, however, it took until the mid-Cenozoic—about 100 million years—for angiosperms to largely supplant the previously dominant ferns and gymnosperms (**Figure 7.32**; Niklas et al. 1983; Knoll 1986; Taylor 1996).

The importance of evolutionary innovations, or at least particular combinations of adaptive traits, is illustrated by macroevolutionary and biogeographic patterns in mollusks. There is an interesting relationship between speciation rate, geographic distribution, and mode of larval dispersal among these groups (Jackson 1974; Hansen 1980; Jablonski and Lutz 1980; Jablonski 1982; Jablonski 1987).

FIGURE 7.32 The relatively gradual radiation of angiosperms as illustrated by changing composition of fossil floras representing past communities. Although innovations in reproductive biology apparently gave angiosperms advantages over the previously dominant gymnosperms, the angiosperms' rise to dominance took over 100 million years. Angiosperms have maintained much higher diversity than their gymnosperm progenitors, although some gymnosperms, such as conifers and cycads, still survive today. (After Knoll 1986.)

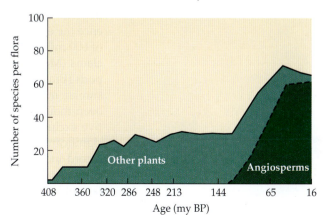

FIGURE 7.33 Relationship between mode of larval dispersal, extent of geographic range, and survival time in the fossil record of Late Cretaceous mollusks on the east coast of North America. Note that species whose larvae could disperse long distances in the plankton (A) tended to have larger geographic ranges and lower extinction rates (longer survival times) than those species whose larvae settled close to their parents (B). (After Jablonski 1982.)

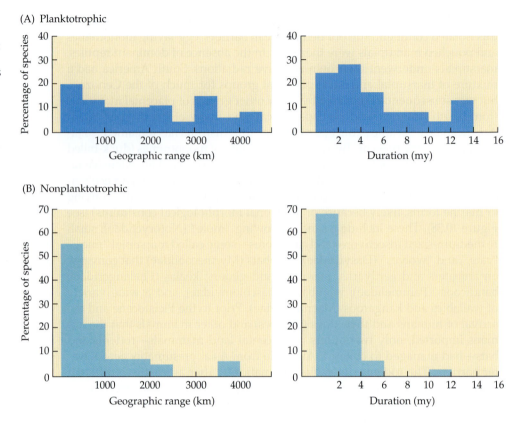

(A) Planktotrophic

(B) Nonplanktotrophic

Some groups have planktotrophic larvae—juvenile stages that drift passively in the ocean—whereas others brood their offspring or have other specializations that do not involve their young being dispersed in the plankton. Mode of larval dispersal can be determined for fossil as well as for recent forms because an individual's original larval shell is preserved as part of its adult shell, which is readily fossilized. Molluscan species with planktotrophic larvae tend to have larger geographic ranges (**Figure 7.33A**) and lower speciation rates than related species whose young rarely disperse far from their parents. Limited dispersal of offspring presumably reduces gene flow, enables populations to adapt to local conditions, and thus facilitates genetic differentiation and allopatric speciation. On the other hand, because small, specialized populations are also particularly vulnerable to extinction, nonplanktotrophic species tend to persist in the fossil record for shorter periods of time than do those with planktotrophic larvae (**Figure 7.33B**).

As implied by this simple example, patterns of species extinction, which depend in part on the characteristics of the organisms themselves, can influence the evolutionary histories of lineages by species selection. Over a period of 600 million years, one group of marine bottom-dwelling invertebrates, the clams (Mollusca, class Bivalvia), has been replacing a phylum, the brachiopods (Brachiopoda; **Figure 7.34**). Because these two groups have superficially similar morphology, feeding habits, and habitat requirements, it had long been thought that the clams were supplanting the brachiopods by competitive exclusion (Elliot 1951). In fact, species of the two groups may compete significantly where they occur together, but careful examination of the fossil record indicates that the increase of clams and the dramatic decline of brachiopods were also facilitated by mass extinctions that eliminated many more taxa of brachiopods than of clams (Gould and Calloway 1980). Valentine and Jablonski (1982) suggest that brachiopods are differentially susceptible to

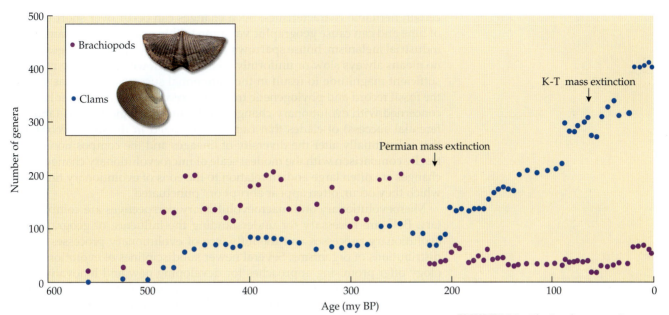

FIGURE 7.34 The "replacement," over a 600 million year period, of brachiopods (phylum Brachiopoda) by clams (phylum Mollusca, class Bivalvia). This shift in dominance has often been attributed to competition because the two groups are generally similar in morphology and ecology. While competition may have played an important role, so did several mass extinction events: After both the Permian and the K-T mass extinctions, brachiopods declined in diversity, while clams increased. (After Gould and Calloway 1980; photos courtesy of David McIntyre and Andrew D. Sinauer.)

catastrophic extinction because they do not have planktotrophic larvae, but instead brood their young or produce larvae that spend only a brief time in the plankton. As noted above, marine invertebrates with nonplanktotrophic larvae tend to have restricted geographic distributions and high extinction rates. Brachiopods have largely been eliminated from shallow-water habitats, where they are exposed to environmental fluctuations, but some forms persist in deep waters, where their mode of reproduction is less disadvantageous.

The fossil record documents the origination and extinction of lineages at all taxonomic levels, from species to phyla. Many of these lineages radiated to produce considerable morphological and ecological diversity, and some spread throughout the world. But success was almost invariably fleeting: Most groups declined in diversity and ultimately went extinct. Sometimes their fates were due primarily to chance events, such as random extinctions of small populations or mass extinctions caused by the collision of an asteroid with Earth. Often, however, their biological attributes, shaped by the microevolutionary force of natural selection and the macroevolutionary force of species selection, played key roles. The more we learn about the history of life, the more we appreciate the interplay between the biological and abiotic environmental processes that have shaped the patterns of speciation and extinction. In the next two chapters we describe how this dynamic arena for speciation and extinction has been influenced by fundamental changes in Earth's geographic and geologic template.

Micro- and macroevolution: Toward a synthesis

When they first aired their ideas of punctuated equilibrium in the 1970s, proponents of macroevolution appeared to be using data from the fossil record to challenge the traditional view of microevolution that had come from the modern synthesis. But as is often the case in such apparent controversies, neither side has been proved right or wrong; rather, we have discovered that, for the most part, these apparent antagonists were actually talking about different things. In this case, the primary difference is one of scale. *Microevolution* is concerned with evolutionary changes within populations that occur as a result of the differential births, deaths, and movements of individuals with certain heritable characteristics. Such changes can occur over short periods of time and can cause geographic variation within a species (see examples of industrial melanism, house sparrows, and clines above). Such changes are by no means always slow or uniformly continuous; however, they are rarely of sufficient magnitude to result in the differential success of clades as seen in the fossil record and phylogenetic trees. *Macroevolution*, on the other hand, is concerned with evolutionary changes in the morphological forms (and differential success) of clades that can be recognized in the fossil record and that substantially alter the diversity of lineages and the composition of biotas. By comparison with the modest scale of microevolutionary change, these changes are often large and, in relation to the eons of evolutionary time over which they occur, often appear abrupt or "punctuated."

Microevolutionary and macroevolutionary perspectives are complementary. Both are necessary for understanding the influence of geography on evolutionary processes and the influence of evolutionary processes on the distributions of organisms. Several newer subdisciplines of evolutionary biology offer promising approaches for developing new and innovative syntheses across micro- and macroevolutionary arenas (phylogeography, Avise 2000; evo-devo biology, Goodman and Coughlin 2000; for an excellent review of synthetic approaches that incorporate both micro- and macroevolution, see Jablonski 2000). We will illustrate the need for complementary perspectives in Unit 4, with emphasis on the "macroevolutionary" processes of speciation and extinction. Yet, we will show that a deep understanding of these processes also requires a microevolutionary perspective, focusing on how populations respond to environmental change in both space and time.

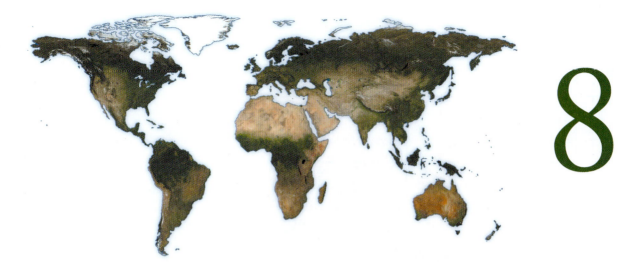

8

THE CHANGING EARTH

Throughout the history of life on Earth, countless generations of organisms either responded to the temporal and spatial dynamics of the geographic template (via evolution or dispersal) or they suffered extinction. Paleontologists sometimes refer to the sum total of Earth's dynamic history as TECO events, for plate *T*ectonic, *E*ustasic (global fluctuations in sea levels), *C*limatic change, and *O*ceanographic events (Rosen 1984). As shown in this and the next chapter, these processes are interdependent, with drifting plates often causing major changes in global sea level and climates, and creating volcanic islands and mountain chains. The relatively recent acceptance of plate tectonics theory (i.e., that Earth's landmasses and ocean basins are in a constant state of flux) is one of science's most important advances—providing a much more comprehensive understanding of the "history of place" and knowledge that is essential for understanding the biogeography and evolutionary history of all life-forms (see Chapter 2 and Chapters 10 to 12).

The Geological Timescale

Anyone studying historical biogeography needs to be familiar with the timescale used to date the history of the Earth (**Table 8.1**). Beginning with the revelations and painstaking reconstructions of William Smith, including his *A Delineation of the Strata of England and Wales with Parts of Scotland* in 1815—"the map that changed the world" (**Figure 8.1**; see also Figure 3.20 and Winchester 2001), early geologists recognized that each layer in a stratigraphic column contains a unique assemblage of fossils characteristic of a particular time span. These assemblages could, therefore, be used to correlate ages of rock strata in one region with those in distant localities. The most reliable fossils for use in such correlations were wide-ranging species whose lifestyles were largely independent of small-scale patchiness in the environment—especially species that were freely dispersed among marine habitats by currents—and that evolved rapidly so as to have

TABLE 8.1 *The Geological Time Scale (All Numbers are In Millions of Years)*

Precambrian

Eon	Era	Period	Age (Ma)
PRECAMBRIAN — PROTEROZOIC	Neoproterozoic	Ediacaran	542
		Cryogenian	~635
		Tonian	850
	Mesoproterozoic	Stenian	1000
		Ectasian	1200
		Calymmian	1400
	Paleoproterozoic	Statherian	1600
		Orosirian	1800
		Rhyacian	2050
		Siderian	2300
PRECAMBRIAN — ARCHEAN	Neoarchean		2500
	Mesoarchean		2800
	Paleoarchean		3200
	Eoarchean		3600
	HADEAN (informal)		4000
			~4600

Paleozoic

Eon	Era	Period	Epoch	Age (Ma)
PHANEROZOIC	PALEOZOIC	DEVONIAN	Upper	359.2 ±2.5
			Middle	385.3 ±2.6
			Lower	397.5 ±2.7
		SILURIAN	Pridoli	416.0 ±2.8
			Ludlow	418.7 ±2.7
			Wenlock	422.9 ±2.5
			Llandovery	428.2 ±2.3
		ORDOVICIAN	Upper	443.7 ±1.5
			Middle	460.9 ±1.6
			Lower	471.8 ±1.6
		CAMBRIAN	Furongian	488.3 ±1.7
			Series 3	~499
			Series 2	~510
			Terreneuvian	~521
				542.0 ±1.0

Mesozoic – Paleozoic

Eon	Era	Period	Epoch	Age (Ma)
PHANEROZOIC	MESOZOIC	JURASSIC	Upper	145.5 ±4.0
			Middle	161.2 ±4.0
			Lower	175.6 ±2.0
		TRIASSIC	Upper	199.6 ±0.6
			Middle	~228.7
			Lower	~245.9
	PALEOZOIC	PERMIAN	Lopingian	251.0 ±0.4
			Guadalupian	260.4 ±0.7
			Cisuralian	270.6 ±0.7
		CARBONIFEROUS (Pennsylvanian)	Upper	299.0 ±0.8
			Middle	307.2 ±1.0
			Lower	311.7 ±1.1
		CARBONIFEROUS (Mississippian)	Upper	318.1 ±1.3
			Middle	328.3 ±1.6
			Lower	345.3 ±2.1
				359.2 ±2.5

Cenozoic – Mesozoic

Eon	Era	Period	Epoch	Age (Ma)
PHANEROZOIC	CENOZOIC	QUATERNARY	Holocene	0.0117
			Pleistocene	2.588
		NEOGENE	Pliocene	5.332
			Miocene	23.03
		PALEOGENE	Oligocene	33.9 ±0.1
			Eocene	55.8 ±0.2
			Paleocene	65.5 ±0.3
	MESOZOIC	CRETACEOUS	Upper	99.6 ±0.9
			Lower	145.5 ±4.0

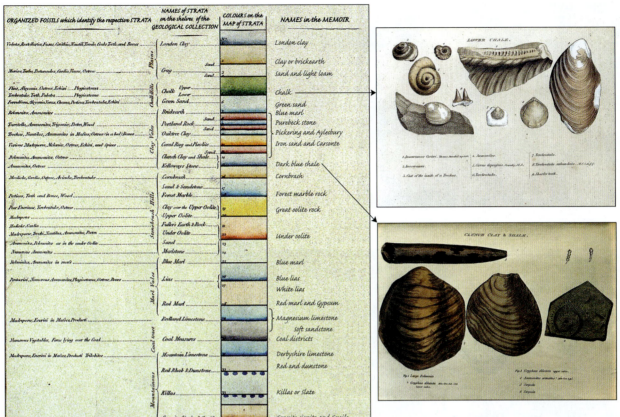

FIGURE 8.1 William Smith's geological "Strata Identified by Organized Fossils" and representative diagrams of fossils he used to delineate strata (shown here are Smith's plates from the Lower Chalk, and Clunch Clay and Shale strata). Coloring of the strata is keyed to colors on Smith's *A Delineation of the Strata of England and Wales with Parts of Scotland* shown in Figure 3.21.

a relatively short temporal duration. These forms are called **index**, or **guide**, **fossils**. Examples (**Figure 8.2**) include planktonic microfossils such as calcareous foraminiferans and siliceous radiolarians (phylum Protozoa) of the Cenozoic Era; chitinous colonial graptolites (phylum Hemichordata)—floating animals of the Ordovician and Silurian Periods; trilobites of the Cambrian and Ordovician Periods; the enigmatic conodonts (hard mouthparts from extinct eel-like vertebrates) of the Paleozoic Era; and swimming ammonoid cephalopods (phylum Mollusca) of the Paleozoic and Mesozoic Eras, whose buoyant calcareous shells probably drifted with currents after their death. On continents, widespread fossils found in coal beds of the Upper Carboniferous Period included pollen, spores, and macrofossils of distinctive vascular plants that thrived in the expansive swamps and forests of that period.

Estimating time

At first, correlations of fossils around the world gave only relative estimates of the ages of various rocks and fossils; early scientists had no specific knowledge of the actual dates of rocks. Carolus Linnaeus and his eighteenth century colleagues accepted the biblical doctrine that the Earth was just a few thousand years old. One century later, Alfred Russel Wallace (1880) used an estimate of 400 million years for the absolute age of the Earth. In the twentieth century, the discovery of radioactive materials finally led to more reliable and accurate dating procedures. Radioactive elements are unstable and they decay, forming stable atoms through a series of intermediate unstable products; during this disintegration process, atomic particles are released. The rate of decay can be quantified and expressed as a half-life—the amount of

(A)

(B)

200 μm

(C)

(D)

(E)

1 mm

FIGURE 8.2 Representative "index fossils" used by paleontologists to chronologically correlate rock strata from different geographic localities. (A) An array of fossil forminifera. (B) Ordovician graptolite (*Amplexograptus* sp.). (C) A Cambrian trilobite (*Elrathii kingii*). (D) An array of conodonts from the Devonian. (E) Jurassic ammonite, *Kosmoceras* sp. (A,D,E courtesy of David McIntyre; B courtesy of Mark A. Wilson; C © Russell Shively/Shutterstock.)

time needed for half of the radioactive material to decay to the stable element. Thus, by calculating the ratio of radioactive element to stable end product, one can determine, within limits, the age of a sample. By using radioactive isotopes of uranium and thorium, whose stable end product is lead, scientists have pushed back the estimate of Earth's age to 4.6 billion years. The age of the oldest known fossils is controversial, but several lines of evidence point to photoautotrophic prokaryotes in chert deposits from Australia and South Africa during the Archean Eon, as early as 3.5 billion years before the present (BP) (Schopf et al. 2007).

The potassium–argon method is a valuable technique for dating Phanerozoic rocks. Radioactive potassium ($^{40}K^{19}$) decays to stable calcium ($^{40}Ca^{20}$) and the inert gas argon ($^{40}Ar^{18}$); the half-life of potassium 40 is 1.25 billion years. A major problem with this technique, however, is that argon gas will escape from rock heated above 300° C—the temperature that would be reached during metamorphism—therefore making this method not wholly reliable for dating rocks older than 100 million years. In such cases, measurement of the decay of rubidium 87 (^{87}Rb) to strontium 86 (^{86}Sr) is used.

For very recent material, radiocarbon dating is extensively used. Carbon 14 decays to nitrogen 14 at a fairly rapid rate (half-life is 5730 ± 40 years). After 50,000 years, so little radiocarbon remains that detection is very difficult. For dating fossils formed within the last 10,000 years, radiocarbon dates can be calibrated with tree ring analysis because many trees in temperate latitudes form an annual ring during each growing season. Paleoclimatic reconstructions are also possible because the width of a fossil growth ring is correlated with the length of the growing season and the availability of water (Fritts

1976; Bradley 1985). Other advances include accelerator mass spectrometry dating, which requires only minute amounts of organic carbon, that is, < 2 mg. Cosmogenic dating is used to determine the length of exposure of an object at the Earth's surface, whereas optically stimulated luminescence and thermoluminescence can be used for materials with very low concentrations of uranium, thorium, and potassium. Each of these techniques is now often used to provide high-quality and high-precision estimates of age within a range of 100 to 200,000 years (see Jackson 2004 and references therein; see also Stanley 2009).

In combination, these methods have been used to estimate the times of major geological and evolutionary events during the past 542 million years, a period known as the Phanerozoic Eon (age of obvious, or multicellular, life; see Table 8.1). The geological timescale is hierarchical, with each division among eons, eras, periods, or epochs marking transitions among geological strata and embedded fossil assemblages. Given the great difficulty of dating such ancient events, it should not be surprising that precise dates are not universally accepted. For example, the scale accepted in 1971 divided the Cretaceous Period into 12 equal epochs of 6 million years apiece, but radioisotope dating has revealed that some epochs were longer and others quite short (Baldwin et al. 1974). Reexamination of Triassic deposits may eventually lead to a drastic shortening of that period and to increased time spans for the Jurassic and Permian. Accurate dates within the Mesozoic are crucial because the early evolution and dispersal of major lineages of land vertebrates and seed plants, as well as the extinctions of certain marine groups and radiations of others, occurred during that era.

Traditionally, the Cenozoic Era was divided into the Tertiary and Quaternary Periods. However, the Tertiary covered some 63 million years, while the Quaternary lasted just 1.8 million years. This traditional scheme has now been replaced by a newer one that divides the Cenozoic into three periods of more similar duration: the Paleogene (65 to 23 million years BP), the Neogene (23 to 2.6 million years BP), and the Quaternary (2.6 million years BP to the present; see Table 8.1).

The Theory of Continental Drift

No contribution to biogeography has had more of an impact than the theory of continental drift. This theory developed from a highly speculative idea in the early 1900s to a well-established fact by the 1960s (see Briggs 1987; Strahler 1998; Scotese 2004). Simply defined, the theory of continental drift states that continents and portions of continents have rafted across the surface of the globe on the viscous upper mantle beneath the Earth's crust. Thus the Earth's crust is not composed of fixed ocean basins and continents, as was assumed by an overwhelming consensus of scientists up until the 1950s, but instead it is a dynamic landscape in which once-distant lands are now in juxtaposition, and other once-attached lands are now widely separated.

Evidence in favor of crustal movements is conclusive, and within the last few decades the theory of continental drift has given rise to a respected science. Today's more comprehensive theory—referred to as plate tectonics—explains the origin and destruction of Earth's plates as well as their lateral movement, or drift. However, throughout the long history of the field, biogeographers, from Linnaeus through Lyell, Darwin, and Wallace and to Darlington as recently as the 1950s, insisted on the fixity of the continents and great oceans. Lyell, Cuvier, and others had long ago established that the land and sea exhibited great vertical fluctuations throughout the geological record. But the notion that great masses of the Earth's crust could drift, collide,

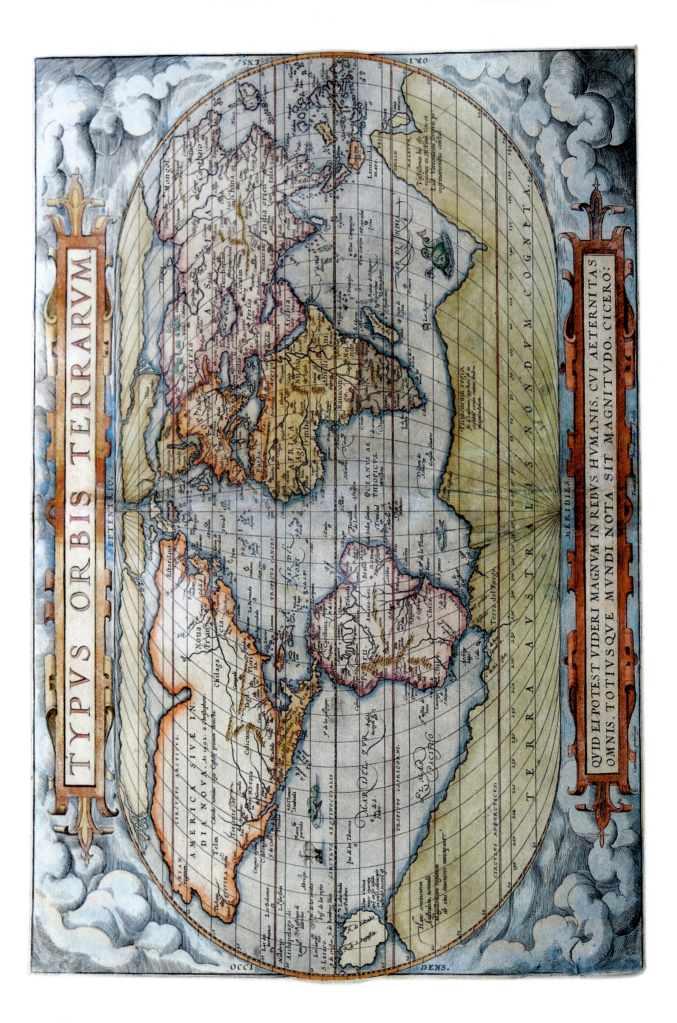

◀ **FIGURE 8.3** Map of the world printed in 1570 by Abraham Ortelius in his *Theatrum Orbis Terrarum*. While the outline of South America was very inaccurate, Ortelius was able to postulate an early theory of continental drift by observing the fit between eastern South American and western African coastlines.

and separate like ice on a partially frozen river was a tough sell indeed, as we will see below when we review the history of this bold idea.

Scholars have searched diligently to determine who first proposed a theory of continental drift. It has been suggested that the curious matching of coastlines between South America and Africa was first noticed by Sir Francis Bacon in 1620 and by the Compte de Buffon in 1749. The German theologian Theodor Christoph Lilienthal probably deserves credit for actually commenting on the "fit" of the continents in 1756 (Romm 1992). Lilienthal's observation, however, was preceded by nearly 200 years by the Flemish geographer and cartographer Abraham Ortelius (1527–1598), who speculated in his *Thesaurus Geographicus* in 1596 that the Americas were "torn away from Europe and Africa . . . by earthquakes and floods. . . . The vestiges of the rupture reveal themselves, if someone brings forward a map of the world and considers carefully the coasts of the three [continents]" (**Figure 8.3**).

In 1782, Benjamin Franklin postulated that the Earth's crust was like a shell floating on a fluid interior and that broken pieces of the shell could move about as the fluid flowed. In the early decades of the nineteenth century, Sir Charles Lyell realized that fossils found in Europe indicated that a tropical climate had once prevailed in that region, and he suggested that the Earth experienced cycles of global climatic change (Lyell 1834). According to Lyell, these changes were triggered by vertical shifts in the Earth's crust, which rose above sea level in one region of the world while sinking in another. Like Franklin, Lyell also postulated that the Earth's crust was composed of one rigid shell, which at times rotated independently of the fluid mass of molten matter below. When the crust rotated such that most of Earth's landmasses were concentrated near the Equator, global warming occurred. During other "great seasons of the Earth," landmasses were concentrated near the poles, resulting in what Lyell called Earth's "great winter," which was accentuated by glacial episodes. Lyell's theory of geoclimatic cycles, albeit novel, had little if any impact on the field, perhaps largely because he insisted that while the Earth's crust shifted vertically and at times rotated as one rigid shell during these great cycles, the continents (plates) changed little in size, shape, or position relative to one another (**Figure 8.4**). Lyell, one of the most persuasive and revered scientists of his time, never made his theory of Earth's great seasons one of his crusades. As we shall see, his model bore only a trivial resemblance to the modern theory of continental drift and plate tectonics.

On the other hand, one of Lyell's contemporaries, Antonio Snider-Pelligrini (1858), may have been the first to persuasively demonstrate the geometric fit of the coastlines of continents on opposite sides of the Atlantic Ocean, and to use rock formations and fossil distributions to argue cogently that they once formed a supercontinent that subsequently split apart (**Figure 8.5**). Snider-Pelligrini's theory, however, was quite rudimentary and included the suggestion that separation of the continents was somehow caused by Noah's Flood.

Throughout the remainder of the nineteenth century and indeed into the early decades of the twentieth century, geologists had little understanding of the past relationships of continents. In 1908, Frank Bursley Taylor, an American geologist and astronomer, presented a detailed model (published in 1910) in which the continents were hypothesized to move, distorting crustal materials into mountain ranges and island chains (**Figure 8.6**). Most of Tay-

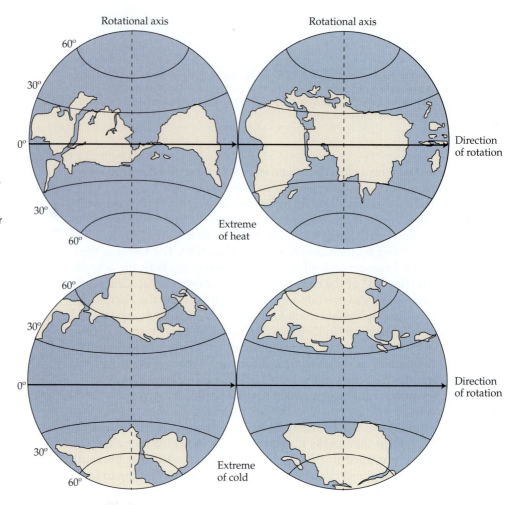

FIGURE 8.4 Lyell's theory of cycles of global climatic change. In the third edition of his *Principles of Geology*, Lyell suggested that although the Earth's crust remained rigid, it rolled on its rotational axis (dashed lines) at various times such that the continents (which absorb more solar radiation than the oceans) were concentrated near the Equator during some periods (i.e., those at "Extreme of heat") and closer to the poles during others (i.e., those at "Extreme of cold").

lor's research focused on glacial history of the Great Lakes region of North America during the Pleistocene, but he was also strongly interested in the formation of mountain chains, island archipelagoes, and the solar system. Taylor postulated that recent glacial periods were caused by movements and massing of the continents near the poles and that, during earlier periods, mountain chains and island arcs formed along the forward margins of moving continents, which also opened ocean basins behind them (Taylor 1928). Taylor also perceptively suggested movement of South America and Africa away from the mid-Atlantic ridge. Although Taylor's ideas were innovative, he did not correctly perceive the directions of continental movements, and it is not surprising, given his interests in astronomy, that he wrongly attributed these movements to tidal and rotation forces of the Earth, moon, and sun.

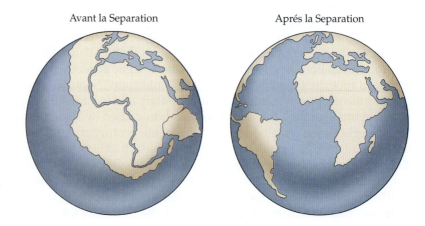

FIGURE 8.5 Antonio Snider-Pellegrini produced in 1858 a "pre-Wegenerian" depiction of continental drift that accounted for the zipperlike fit of the coasts of the Americas and Africa across the Atlantic Ocean.

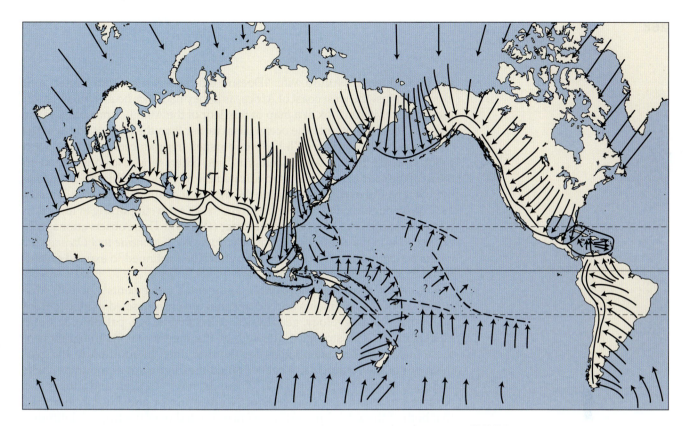

Later research would identify more plausible causes of these movements and would confirm that continental movement, crustal upheaval, and the formation of ocean basins are indeed intimately related.

Wegener's theory

Alfred Lothar Wegener (**Figure 8.7**), a German meteorologist, conceived and championed the theory of continental drift, which he presented with admittedly scanty evidence at first, but which anticipated much of our current knowledge. Wegener developed his ideas on continental displacement in 1910, independently of Taylor, after visiting tectonically active areas of Greenland as well as observing on a world map the congruence of opposite coastlines across the Atlantic. His enthusiasm for this idea was encapsulated in a letter

FIGURE 8.6 An early model of continental drift proposed by Frank B. Taylor (1910). This scenario suggests a general drift of major landmasses toward the Equator that Taylor referred to as crustal creep. Taylor used position and orientation of island arcs and mountain chains to infer direction and magnitude of movement (summarized by direction and size of arrows, with uncertainties shown as question marks).

(A)

(B)

FIGURE 8.7 Two views of Alfred Lothar Wegener (1880–1930), who developed the ideas leading to the modern science of plate tectonics and its confirmation of continental drift. The first figure shows Wegener in 1910, shortly before the first publications of his theory of continental drift in 1912. The second figure shows Wegener shortly before his death during a mission to rescue colleagues in Greenland in 1930. (From Schwarzbach 1980; A courtesy of Deutsches Museum, Munich; B photo by K. Herdemerten, courtesy of the Alfred Wegener Institute.)

to Else Köppen, his future wife: "Doesn't the east coast of South America fit exactly against the west coast of Africa, as if they had once been joined? The fit is even better if you look at a map of the floor of the Atlantic and compare the edges of the drop-off into the ocean basin rather than the current edges of the continents. This is an idea I'll have to pursue" (reproduced in Schwarzbach 1980). Unlike Ortelius, Snider-Pelligrini, and others who also commented on the fit of the continents, Wegener dedicated the remainder of his life to advancing what would eventually become the theory of continental drift. In January 1912, Wegener unveiled his working hypothesis, with supporting evidence, in two oral reports, published later that year (1912a,b). These observations were expanded into his classic book, *Die Entstehung der Kontinente und Ozeane* (*The Origin of Continents and Oceans*) (1915), which he continued to expand on and revise over the following two decades.

Wegener's theory on horizontal continental movements not only discussed all of the continents and oceans, but also synthesized evidence from many disciplines: geology, geophysics, paleoclimatology, paleontology, and biogeography. The strongest attribute of the theory was its integration of these many types of phenomena for the first time. In addition to the geometric fit of the continents, Wegener noted the alignment of mountain belts and rock strata on opposite sides of the Atlantic. As Lyell and others had observed much earlier, coal beds of North America and Europe indicated that both of these landmasses were once situated over the tropical latitudes. Similarly, Wegener observed that glacial deposits, or **tillites**, in what are now subtropical Africa and South America suggest that they were once displaced poleward. In addition, the tillites of North America seemed continuous with those of Europe. Many anomalous biogeographic patterns, such as the highly disjunct occurrence of extant marsupials in South America and Australia, were easily explained by the theory that these continents were once physically connected.

Like many revolutionary ideas, Wegener's were well ahead of his time and, consequently, were ignored by most scientists and ridiculed by others. At first, his conclusions were accepted by just a few geologists and biogeographers, most notably those in the Southern Hemisphere and Europe. After all, the idea of ancient connections and biotic affinities among the southern biotas was not new to students of the southern biotas. Joseph Dalton Hooker had suggested this in 1853 after reviewing distribution patterns of plants among the southern continents and archipelagoes:

> Enough is here given to show that many of the peculiarities of each of these the three great areas of land in the southern latitudes . . . is agreeable with the hypothesis of all being members of a once more extensive flora, which has been broken up by geological and climatic causes. (Hooker 1853)

As we now know, Hooker postulated intermittent connections and dispersal across transoceanic land bridges, but geological evidence for his extensionist theory never materialized. Then again, while Wegener had assembled an impressive interdisciplinary array of evidence to support his theory of continental drift, development of a plausible mechanism of continental movements would not be forthcoming for several decades more. The second edition (1920) of Wegener's treatise received some attention, albeit negative, when it was criticized by several prominent geologists, but wide knowledge of the theory came only after the third edition (1922) was translated into five languages, including English (1924). A fourth edition (1929), the one generally used now, contained much more information, but throughout the various editions, the substance of Wegener's ideas remained the same. The following are some of his pertinent conclusions:

1. Continental rocks, called sial (granitic rocks composed largely of silicon and aluminum), are fundamentally different, less dense, thicker, and less magnetized than those of the ocean floor (basaltic rocks, called sima—consisting primarily of silicon and magnesium). The lighter sialic blocks—the continents—float on a layer of viscous, fluid mantle.

2. Major landmasses of the Earth were once united as a single supercontinent, Pangaea. Pangaea broke into smaller continental plates, which moved apart as they floated on the mantle. Breakup of Pangaea began in the Mesozoic, but North America was still connected with Europe in the north until the Late Tertiary or even the Quaternary (**Figure 8.8**).

3. Breakup of Pangaea began as a rift valley, which gradually widened into an ocean, apparently by adding materials to the continental margins. Midoceanic ridges mark where opposite continents were once joined, and ocean trenches formed as the continental blocks moved. The distributions of major earthquake centers and regions of active volcanism and orogeny (mountain building) are related to movements of these blocks.

4. The continental blocks have essentially retained their initial outlines, except in regions of mountain building, so the manner in which the continents were once joined can be seen by matching up their present margins. When this is done, similarities in the stratigraphy, fossils, and reconstructed paleoclimates of now-distant landmasses demonstrate that those blocks were once united. These patterns are inconsistent with any explanation that assumes fixed positions of continents and ocean basins.

5. Rates of movement for certain landmasses range between 0.3 and 36 meters per year—the fastest being Greenland, which may have separated from Europe only 50,000 to 100,000 years ago.

6. Radioactive heating in the mantle may be a primary cause of block movement, but other forces are probably involved. Whatever the causal processes, they are gradual and not catastrophic.

Early opposition to continental drift

Wegener's ideas were clearly prescient, and they formed the basis of plate tectonics, the modern theory of continental drift (see Briggs 1987). Yet, despite Wegener's cogent and persistent arguments, continental drift was not generally accepted until the early 1960s, some 50 years after he and Taylor first published their ideas. Why was the theory resisted for so long? The history of this debate serves as an important lesson in the nature of scientific revolutions. Strong criticism of Wegener's theory arose as soon as translated volumes made it available to most geologists and biogeographers in the mid-1920s. Some scientists resisted the new idea because it conflicted with their preconceived ideas of fixed continents and a solid Earth, and because it was proposed in the shadows of World War I by a German meteorologist—a man who was not part of the geological establishment of Western Europe. Others, particularly in America, resisted it because of its sheer magnitude and audacity.

Other scientists opposed Wegener's theory on much more objective and defensible grounds. Although it would

FIGURE 8.8 Wegener's (1929) model of continental drift. Wegener envisioned that the continents, including shallow seas (not differentiated here as in his original map)—initially united into the one giant landmass of Pangaea—had moved apart during the Eocene and Late Quaternary. In Wegener's time, the geological epochs and periods were thought to have been more recent than has been indicated by modern dating methods. Nevertheless, comparison with Figure 8.22E and Box 8.1, Figure A shows that Wegener's view was very similar to current reconstructions of continental movement. (After Wegener 1966.)

Upper Carboniferous

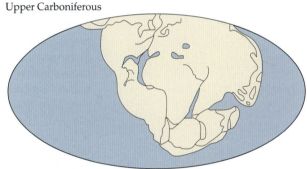

Eocene

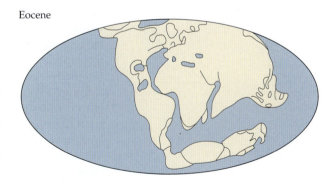

Lower Quaternary

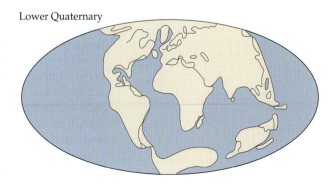

eventually prove to be one of science's most important paradigms, Wegener's theory suffered from at least three shortcomings. First, the nature of scientific revolutions is such that new theories that challenge long-held paradigms are resisted until evidence clearly shows them to be more parsimonious. The prevailing attitude in biogeography and paleontology was first expressed by Alexander du Toit: "Geological evidence *almost entirely* must decide the probability of this hypothesis" (1927: 118). Wegener's theory included too many assumptions that remained unsubstantiated by available geological evidence. Paleontologists in particular were unconvinced by the biogeographic and fossil evidence marshaled to support the continental drift model. For example, in 1943, G. G. Simpson published an analysis of past and present mammalian distributions to show how these data fit alternative scenarios of past intercontinental connections; that is, most known patterns of distributions during the Cenozoic Era could be explained without invoking continental drift (Simpson 1943a). It was difficult, using paleontological data, to overturn long-accepted theories based on ancient land bridges and periodic floods. Although Wegener's thesis was plausible, the theory of continental drift would not be accepted until enough unambiguous new evidence was collected to make it the most parsimonious explanation for geological and biogeographic patterns.

A second shortcoming of Wegener's theory is that it contained many factual errors. The theory aroused skeptical interest in many fields, and scientists in each discipline recognized major inconsistencies in Wegener's presentations. These inconsistencies had to be resolved. Even du Toit—Wegener's strongest proponent, who published two books (1927, 1937) in favor of the theory—had to concede unquestionable errors by Wegener. For example, Wegener proposed that plates move at an incredibly high speed—as rapidly as 36 m per year. His theory might have been much more palatable if his estimates were closer to what we now believe to be true—that is, rates of 2 to 12 cm per year. Wegener's error, however, is understandable, as he and his colleagues were working with what they believed to be a much younger Earth than we now assume and were limited to very crude methods of measuring displacement rates. In addition to correcting these errors, Wegener and his followers would have to gather much more evidence from a variety of disciplines in biogeography and geology (especially marine geology) and to actually test their model.

Finally—and this was perhaps more critical than any of the other shortcomings—Wegener's theory was criticized for lacking a plausible mechanism. How could the plates—rigid and enormous masses of rock—move about, and what force or forces could drive such movements? Wegener's insights in this area, however, are largely overlooked by science historians. In fact, Wegener did indeed discuss three potential driving forces, or "displacing forces" (see Wegener 1929: 167–179). Not surprisingly, given that he was a meteorologist and an astronomer, two of these forces involved celestial phenomena (effects of centrifugal forces on the Earth's surface, and combined effects of gravitational fields of the Earth, moon, and sun). The third force Wegener discussed was one developed in detail by British geologist Arthur Holmes—convective currents of molten rock beneath the Earth's crust—which, indeed, is the same ultimate force assumed by today's model of plate tectonics. Wegener admitted that his ideas on causal mechanisms were speculative and that "the problem of forces which have produced and are producing continental drift is still in its infancy" (Wegener 1929: 179). Speculation, however, may be a normal and perhaps an essential part of scientific revolutions. As Darwin once wrote in a letter to Wallace (1857), "I am a firm believer that, without speculation, there is no good and original observation."

Soon after completing the fourth and final revision of his book, Wegener set out on an expedition in 1930 and returned to Greenland to document its purportedly rapid movement and, by so doing, to verify his theory of continental drift. In fact, his view of a dynamic Earth with rapidly drifting landmasses was largely stimulated by his earlier expeditions to Greenland in 1906 and 1912. It is one of science's great and tragic ironies that, as Pascual Jordan observed, Wegener "died in a snowstorm on the very Greenland expedition he undertook to verify his theory" (Jordan 1971).

Evidence for continental drift

Providing convincing evidence for Wegener's theory was to take nearly five decades of research conducted by many teams of geologists, paleontologists, and biogeographers. From the first time Wegener proposed the supercontinent of Pangaea, his opponents criticized the liberties he took to achieve a "good fit." A good reconstruction was finally achieved when S. W. Carey (1955, 1958), an Australian geologist, used plasticine shapes of landmasses sliding over a globe. Nonetheless, the fit was not widely accepted until three researchers (Bullard et al. 1965) combined computer mapping techniques and statistical analyses to test continental fits. Their analyses showed that the continents do fit together if one uses submarine contours of the continental shelves to delineate the margins of continental plates (see also Smith and Hallam 1970). In retrospect, as shown by the letter quoted earlier (p. 270) from Wegener to his wife Else, Wegener perhaps should have been given more credit for coming to the same conclusion some five decades earlier.

Other stratigraphic, paleoclimatic, and paleontological evidence gradually accumulated to support the theory of continental drift. Along with these important discoveries (summarized in **Box 8.1**), some of the most compelling evidence in favor of the theory of continental drift was provided by marine geologists, working initially under a secrecy imposed by the military during World War II.

Marine geology

After World War II, a second generation of scientists, who had not been directly involved in the initial debates, made some important discoveries about ocean basins and rock magnetism that encouraged reexamination of the ideas and evidence advanced by Wegener and du Toit (see Briggs 1987). When Wegener proposed his ideas, very little was known about the structure of the ocean floor. On the basis of loose samples obtained by dredging, geologists suspected that the ocean floor was composed of basalt (sima), but no one had actually taken core samples of the deep basins. Sialic continental rocks, however, were well known. Echo soundings from several transoceanic expeditions had portrayed ocean bottoms as smooth structures (abyssal plains) lying 4 to 6 km beneath the ocean surface. A midoceanic ridge was known only in the Atlantic Ocean. Finally, deep cuts in the ocean floor (known as trenches) had been found on the ocean sides of island arcs, and they were known to display unusual gravitational properties.

Oceanographic research was just beginning to accelerate before the outbreak of World War II, when charting ocean topography became a practical goal. During the war, Harry H. Hess—a marine geologist who was using an echo sounder as he sailed aboard a U.S. troop transport—discovered some flat-topped submarine volcanoes 3000 to 4000 m high. Peaked submarine volcanoes, called **seamounts**, had been previously identified. The new structures, which Hess later named **guyots** in honor of a Princeton geologist, were thought to be volcanic islands that had formed above the ocean surface, later

BOX 8.1 *Stratigraphic, Paleoclimatic, and Paleontological Discoveries that Contributed to the Acceptance of the Theory of Continental Drift*

■■■ By the 1970s, newer stratigraphic, paleoclimatic, and paleontological data combined with evidence provided originally by Wegener and du Toit to provide convincing support for the theory of continental drift. In particular, many pieces of evidence came together to support the matchup of continents—particularly the southern continents—in their former positions as part of Pangaea.

Stratigraphic evidence

Topographic features, including mountains, oceanic ridges, and island chains, along with specific rock strata (e.g., Precambrian shields and flood basalts) and fossil deposits, were found to be aligned along Wegener's hypothesized connections of the now fragmented portions of Gondwanaland (**Figure A**)

(Hurley 1968; Hurley and Rand 1969). In addition, on each of the now isolated southern continents, rocks from the Late Paleozoic and Early Mesozoic contained the same stratigraphic sequence (see Allard and Hurst 1969): glacial sediments, coal beds, and sand dunes and other desert deposits, all overlain by a layer of volcanic rock.

Paleoclimatic evidence

All of the continents in the Southern Hemisphere have Late Paleozoic glacial deposits (tillites) in their southernmost regions. Moreover, as glaciers move, they scour the underlying rocks, leaving deep scratches that mark the direction of their movements. If we plot these glacial lines on a map with the southern continents in their current positions, the patterns appear

quite confounding (**Figure B1**). Not only do the glaciers appear to have been situated in what are now some relatively warm latitudes, but many appear to have risen out of the sea. This perplexing anomaly disappears when the same glacial lines are plotted on a reconstruction of Gondwanaland as it was during the Permian Period (**Figure B2**).

Paleontological evidence

The Late Paleozoic glacial deposits of the southern continents are covered with Permian rocks bearing the so-called *Glossopteris* flora (Schopf 1970a,b). These arborescent gymnosperms are presumed to have been adapted to (and therefore indicative of) temperate climates because they had deciduous leaves and conspicuous growth rings in their wood (Schopf 1976). When the occurrence of the *Glossopteris* flora is plotted on a map of Pangaea (Figure B), the points circumscribe a discrete region of Wegener's Gondwanaland, thought to be correlated with the margins of the glaciers.

FIGURE A (1) Distributions of Precambrian shields (hatched areas), illustrating how they match up into separate northern and southern units if the continents are reassembled as they were before the breakup of Pangaea. (2) Although substantially isolated today, flood basalts similar to those shown in the inset also serve as evidence of the previous connections of the southern continents in the former Gondwanaland. (1 after Hurley and Rand 1969; 2 after Storey 1995; photo courtesy of Williamborg/Wikipedia.)

(1)

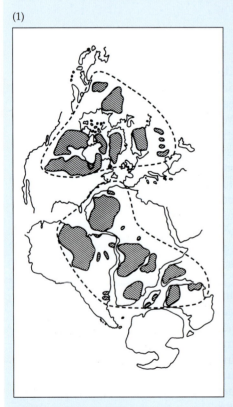

(2)

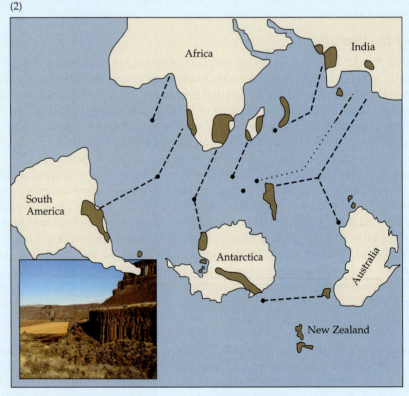

BOX 8.1 *(continued)*

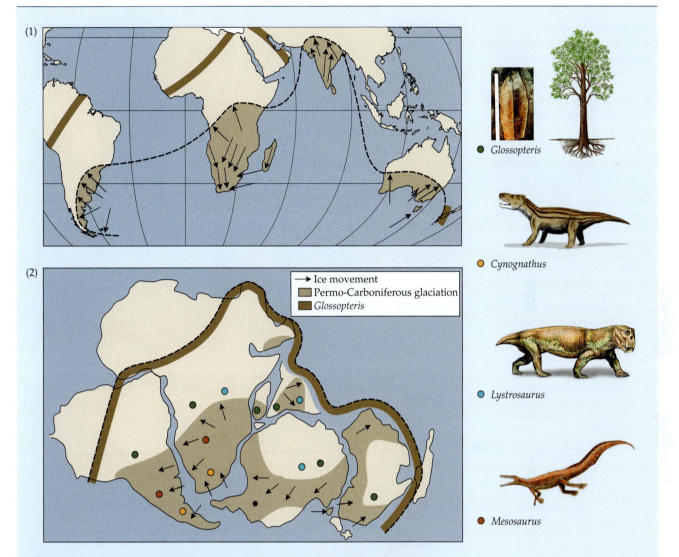

FIGURE B Several lines of paleontological evidence for continental drift are found on the southern continents. Glaciers carved lines in the underlying rock material, marking their location and direction of movement (arrows). The *Glossopteris* flora (or "tongue ferns") included more than 20 species of plants that grew along the margins of the glaciers. (1) The origin and directions of glacial movement (shaded area with arrows) and the distributions of fossils of *Glossopteris* (darker shading), as well as the mammal-like reptiles *Lystrosaurus* and *Cynognathus* and the marine reptile *Mesosaurus*, are difficult to explain based on the current positions of the southern continents. (2) These patterns, however, are consistent with reconstructions of Gondwanaland as it was during the Permian Period. (1 after Stanley 1987; 2 after Windley 1977; *Glossopteris* photo courtesy of the U. S. Geological Survey; *Cynognathus* and *Mesosaurus* courtesy of Nobu Tamura/Palaeocritti; *Lystrosaurus* © John Sibbick.)

In 1969 D. H. Elliot and E. H. Colbert unearthed the first tetrapod fossils found in Antarctica (Elliot et al. 1970) from mudstone and volcanic sandstone of the Early Triassic. Additional finds provided convincing evidence that many of these bones belonged to *Lystrosaurus*, a mammal-like reptile also found in rocks of similar age in the Karoo of southern Africa and the Panchet Formation of southern India. The reconstruction of Gondwanaland explained many such biogeographic anomalies in the distributions of extant as well as fossil assemblages (Figure B2). Many vertebrates underwent major radiations during the Permian, when the proximity of the continents allowed their rapid spread across what are now isolated landmasses. Like basaltic rocks, glacial tillites, and fossil assemblages, many of these extant forms exhibit disjunct distributions. The extensive breaks in their current ranges are consistent with the former connections

Continued on next page

BOX 8.1 *(continued)*

of Pangaea and its subcontinents, although more recent phylogenetic analyses have in some cases indicated a more complex history of Gondwanan vicariance and subsequent inter-continental dispersal (**Figure C**). ▮▮▮

FIGURE C   The disjunct distributions of some living taxa have often been taken as evidence that their ancestral forms radiated across Gondwanaland prior to its breakup. Examples include (1) southern temperate beetles of the tribe Migadopini of the family Carabidae, (2) nontropical freshwater fishes of the superfamily Galaxioidea, (3) plants of the family Proteaceae, and (4) clawed aquatic frogs of the family Pipidae, with distributions in tropical South America and tropical Africa, suggesting a common ancestor that was once distributed in western Gondwanaland. For several of these groups, more recent morphological or molecular phylogenetic studies indicate a more complex history, including a mixture of both Gondwanan vicariance and subsequent intercontinental dispersal events (e.g., Waters et al. 2000 for Galaxioidea; Barker et al. 2007 for Proteaceae; Cannatella and Trueb 1988 for Pipidae). (1 after Darlington 1965; 2 after Berra 1981; 3 after Johnson and Briggs 1975; 4 after Savage 1973.)

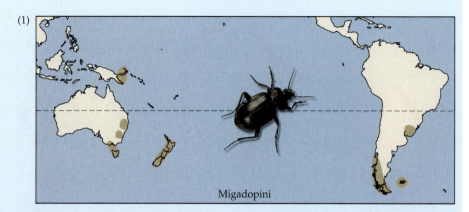

(1) Migadopini

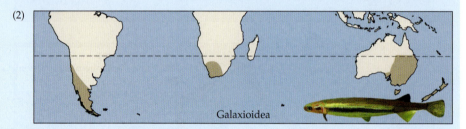

(2) Galaxioidea

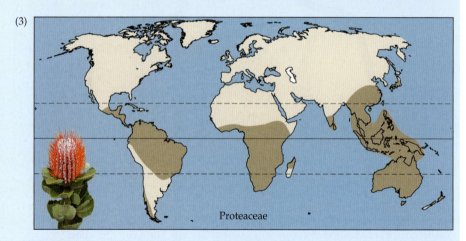

(3) Proteaceae

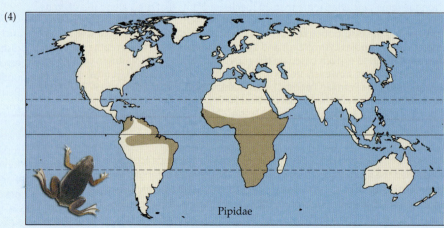

(4) Pipidae

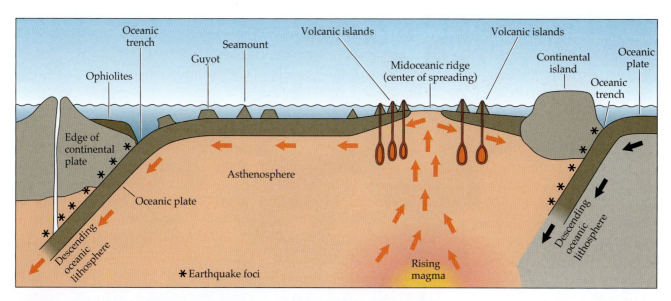

FIGURE 8.9   A highly simplified model of seafloor spreading that depicts how oceanic plates are pushed apart at the center of spreading by the upwelling of magma from the mantle, which causes the plates to slide away from a midoceanic ridge over the viscous asthenosphere. Magma may also produce volcanic islands near the spreading center, but as a point on the plate is displaced from the ridge, it also descends to 4–6 km below sea level, and the islands become submerged. These submerged volcanic structures (seamounts and guyots) eventually disappear into an oceanic trench where the oceanic plate meets another plate. In the case illustrated, the heavier oceanic plate descends beneath the lighter continental plate, which causes the metamorphosis of the surface material on the oceanic plate into ophiolites and their deposition on the continent, the consumption of volcanic islands, and eventually the remelting of the plate itself. Asterisks indicate epicenters of earthquakes resulting from contact of two plates (the Benioff zone).

became truncated by wave action, and finally sunk to 1 or 2 km beneath the waves (**Figure 8.9**). Guyots are common in the northern and western Pacific Ocean, and as we shall see shortly, they figured prominently in the development of models of continental drift.

Following the war, marine exploration blossomed because of generous funding by Allied navies. Important discoveries were made using new deep-sediment piston corers and explosive charges. From samples obtained using these new techniques, geologists learned that under recent sediments, all ocean floors are composed of basalt and that this basement is young, for the most part dating back only to the Jurassic (150 million years BP; **Figure 8.10**). Thus the modern oceans are considerably younger than the continents, whose ancient foundations, called **cratons** or **Precambrian shields**, date to the Archean Eon more than 3 billion years ago (**Figure 8.11**). Prior to the formation of large continents, the earliest continental crust developed from basaltic material that melted when oceanic plates were subducted into the very hot mantle, forming sodium-rich feldspar granitoid rocks that then rose to the surface to form the **protocontinents** early in the Archean Eon. As the Earth cooled later in the Archean, protocontinents were sutured together into larger continents between about 2.7 and 2.3 billion years ago. Subsequently, except for the the remaining Archean cratons, destruction and creation of continental crust has happened at similar rates such that the approximate volume has remained about the same since Late Archean time (Stanley 2009).

By the mid-1950s a team of scientists had recognized that the submarine mountain ranges that bisect the oceans are really segments of a continuous global system of **midoceanic ridges** 65,000 km long (Figure 8.10). This system

Age of Oceanic Lithosphere

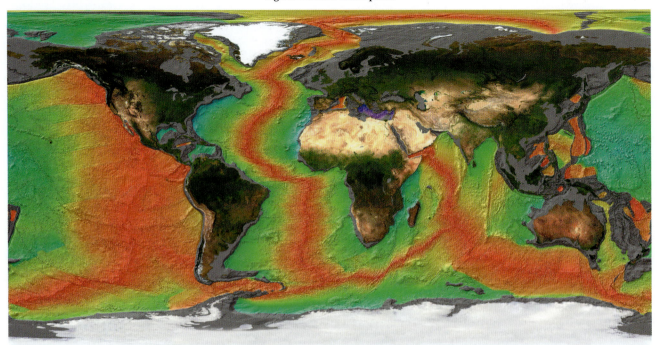

Million years

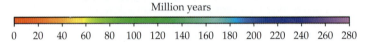

0 20 40 60 80 100 120 140 160 180 200 220 240 260 280

FIGURE 8.10 Global system of mid-oceanic ridges, which mark regions of seafloor spreading, showing age of the oceanic lithosphere. Note the regularity in the pattern in which the ages increase with distance from the midoceanic ridges. (After Muller et al. 2008, courtesy of Mr. Elliot Lim, CIRES & NOAA/NGDC.)

is marked by a central rift valley, which is closely associated with a zone of frequent shallow earthquakes. At that time, new instruments measured remarkably high temperatures in these rifts, suggesting that molten mantle material was being released there. As anticipated by Wegener, geologists of the 1950s began to interpret midoceanic ridges as zones where the oceans expand, establishing the concept of **seafloor spreading**.

Located far from the ridges, oceanic **trenches** are so deep that until the 1950s, most knowledge of them was obtained by taking soundings. Oceanic trenches are V-shaped troughs about 10 km deep (see Figure 8.9). Through the use of seismic refraction techniques, marine geologists learned that the

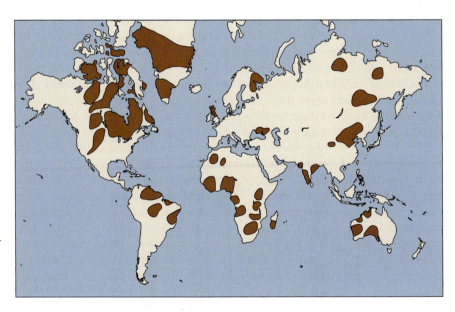

FIGURE 8.11 Distribution of Archean cratons, some representing the protocontinents that sutured together during the Proterozoic Eon to form larger cratons that became the sources of modern continents. (After Stanley 2009.)

Earth's crust is extremely thin in trenches and that the heat flow beneath the trenches is half that found in the abyssal plain, implying that heat is consumed in trenches. Gravity measurements in the trenches are lower than in any other place on Earth. Geologists therefore postulated that it is here, in the oceanic trenches, where the crust is pulled downward and its material reincorporated into the mantle.

Surrounding the Pacific Ocean is a belt of volcanism and earthquake activity known as the Ring of Fire. It was proposed that the subduction (downward movement) of crustal materials in the trenches was the direct cause of these violent geological events. Hugo Benioff (1954) provided the first convincing evidence for this hypothesis. By plotting positions and depths of earthquake epicenters in the vicinity of the trenches, he demonstrated that the epicenters closest to trenches are shallow and those farther away are progressively deeper. Epicenters are aligned along a zone dipping downward at about 45° behind a trench, indicating that earthquakes are caused as a cold, rigid crustal slab descends into the mantle (see also Calvert et al. 1995). These zones are now termed **Benioff zones** (see asterisks in Figure 8.9).

Paleomagnetism and the emergence of a mechanism

Studies of **paleomagnetism** provided additional evidence for seafloor spreading. Convective flows of molten material from inside the Earth's core and through its mantle generate magnetic fields that permeate the planet (Rai et al. 2002; Dehant et al. 2003). Paleomagnetism refers to the orientation of magnetized crystals at the time of mineral formation (i.e., when molten rock solidifies). Rocks containing iron and titanium oxides become magnetized as they solidify and cool, and this magnetization is reflected in their crystalline structure, which remains "frozen" in the rock, oriented as a fossil compass in the direction and **declination** (the downward orientation of magnetic lines, which varies with latitude) of the then-prevailing magnetic field. This high-temperature magnetization, referred to as **remnant magnetism**, is very stable unless the rocks are reheated to extremely high temperatures (their **Curie points**). Hence, by measuring the direction and declination of remnant magnetism in cooled lavas, it is possible to determine the relationship of any landmass to the magnetic poles at the time the rock was formed; and by using triangulation and computer techniques, it is possible to reconstruct the positions of landmasses relative to the Equator and to one another (**Figure 8.12**).

FIGURE 8.12 (A) The Earth acts as a great bar magnet. Because its magnetic fields are oriented toward its core (as well as poleward), latitudinal position can be read as declination in a compass needle. The same phenomenon also influences the orientation of crystals during the formation of magnetically active rock, thus recording the latitudinal position of the rock when it was formed. (B) Such paleomagnetic information can be used to reconstruct the positions and movements of the continents, such as the movements of Gondwanaland relative to the South Pole during the Paleozoic Era. Gondwanaland (including some present-day equatorial regions) drifted over the South Pole twice—in the Late Ordovician and in the Late Devonian. (After Stanley 1987.)

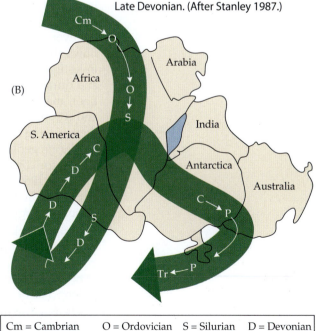

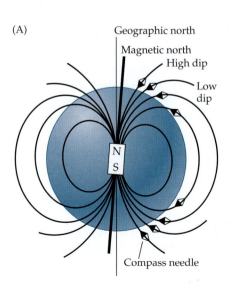

Cm = Cambrian	O = Ordovician	S = Silurian	D = Devonian
C = Carboniferous	P = Permian	Tr = Triassic	

FIGURE 8.13   Shifts in the orientation and latitudinal positions of Labrador, Africa and Australia between the Triassic (200 million years BP, dashed outline) and the present (solid outline) are revealed by paleomagnetism. (After Pielou 1979.)

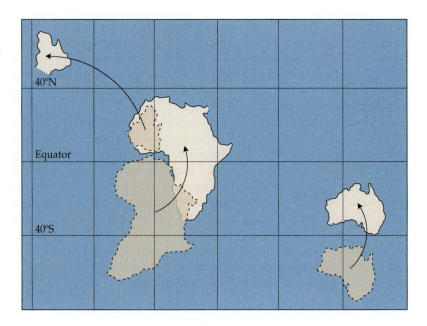

In the early 1950s the British physicist P. M. S. Blackett invented a new supersensitive magnetometer (magnetic detector) that could be used to determine continental orientation throughout geological history. First, the magnetometer was used to show that the British Isles had rotated 34° clockwise since the Triassic (Clegg et al. 1954). A major breakthrough came when other British scientists (Creer et al. 1954, 1957; Runcorn 1956) analyzed geological strata in Europe and North America and provided strong evidence that the two continents had once been joined but had later drifted apart. Subsequent studies around the world reaffirmed the necessity of continental movement to explain existing paleomagnetic patterns (Irving 1956, 1959; Runcorn 1962; **Figure 8.13**).

At the beginning of the twentieth century in central France, Bernard Bruhnes (1906) first discovered **magnetic reversals** (anomalies) when he found lavas that were magnetized in a direction opposite that in recently formed ones. Such patterns reflect reversals of the Earth's magnetic field, which occur every 10^4 to 10^6 years and appear to be generated by changes in magma flows through the mantle (Buffett 2000). Since Bruhnes's discovery, many investigators have found additional geological evidence for both the number and durations of these reversals (**Figure 8.14**; see Lowrie 1997). On the ocean floor, magnetic reversals are recorded by alternating patterns of normally and reversely magnetized basalt that appear as **magnetic stripes** that retain their spacing and shapes for long distances (**Figure 8.15**).

Marine geologists were the first to perceive the significance of magnetic stripes for continental drift theory. Two seminal insights were provided in the early 1960s, first by Harry H. Hess (1962) and then confirmed by Frederick Vine and Drummond Matthews (1963) through discovery of several important properties of ocean floors:

1. Basaltic rocks at the midoceanic ridges have normal field (present-day) magnetic properties.

2. Widths of alternating magnetic stripes on opposite sides of a ridge are often roughly symmetrical, and the stripes are generally parallel to the long axis of the ridge (see Figure 8.15).

3. The banding pattern, if not the actual widths of the stripes, of any one ocean closely matches that of others, and ocean patterns correspond approximately to reversal timetables from terrestrial lava flows.

FIGURE 8.14 The history of magnetic reversals through time (in million years ago, Ma), demonstrating their temporal irregularities (e.g., note the very long durations of normal magnetic polarity during the Mid Cretaceous). (After Lowrie 1997.)

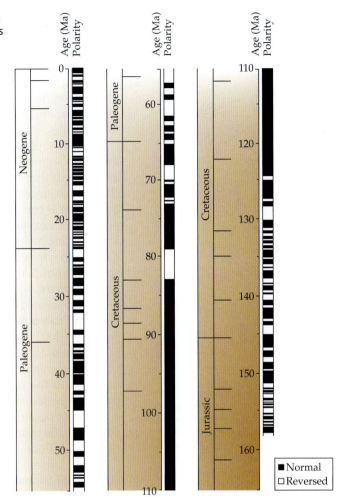

Hess proposed the first model of seafloor spreading to account for major tectonic events (see Figures 8.9 and 8.15). He presented his synthesis at Princeton University in 1960, but it was not published for general readership until 1962. By that time, R. S. Dietz (1961) had published similar but less detailed accounts of global continental movements. Hess and Dietz hypothesized that the oceans are formed by addition of material and spreading at the midoceanic ridges. Moving away from the ridges, basalt along the ocean floor increases in age and, as confirmed by Vine and Matthews, is marked by magnetic stripes, which record the polarity of the prevailing magnetic field. The stripes tend to be highly symmetrical on opposite sides of a ridge. In contrast, differences in stripe widths indicate either that the periods of normal or reversed polarity were of different lengths or that the rate of seafloor spreading varies over time—and that it is not uniform across different oceans or even different parts of the same ocean.

With this model of seafloor spreading, Wegener's theory finally had a plausible mechanism. The initial theory of continental drift was now included in a more comprehensive theory, plate tectonics, which included not just lateral movements of the continents and ocean basins but their origin and destruction as well.

In Great Britain, most geologists who studied global tectonics were converted to the theory of continental drift by 1964, convinced especially by the soundness of Hess's model and the new synthesis that included Vine and Matthews's contributions. Acceptance in North America, however, lagged behind by several years. Meanwhile, widely circulated articles appearing in *Scientific American*, *Science*, and *Nature* did much to make the entire scientific community aware of the rebirth of Wegenerism. Not only were young scientists introduced to the latest evidence, but they then helped to create a wave of acceptance following the mid-1960s. The theory of plate tectonics is now firmly established as a unifying paradigm for much of geology, paleontology, and biogeography.

FIGURE 8.15 During seafloor spreading, reversals in the Earth's magnetic field are recorded as the magnetically sensitive, iron-rich crust cools. Differences in the widths of the magnetic stripes (normal polarity in green, reversed polarity in brown) match known differences in the duration of these polarity episodes (see Figure 8.14), confirming the seafloor spreading hypothesis originally proposed by Harry H. Hess (1962). (After Stanley 2009.)

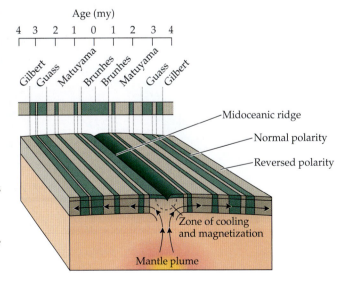

The Current Model

The theory of plate tectonics remains an active and exciting area of research (see Briggs 1987; Oreskes 2001; Bercovici 2003; Scotese 2004; Stanley 2009). Following its general acceptance, geologists, paleontologists, and biogeographers were free to focus their attention on more intriguing questions relating to the temporal and spatial patterns of plate dynamics and potential causal mechanisms.

Lateral movements of the plates result from a complex interaction among the Earth's crust, its underlying mantle, and its core—which is the site of intense heat that drives plate movement (**Figure 8.16**). This immense heat includes heat stored from when the Earth first formed, plus heat generated by gravitational compression and nuclear reactions (primarily from radioactive potassium, uranium, and thorium) deep in the core. Plates are roughly 100 km thick and are composed of a relatively thin, rigid layer of crust, which adheres to the upper layer of the mantle, and together they comprise the **lithosphere** (**Figure 8.17**). The mantle also includes a deeper, more fluid layer, the **asthenosphere**, which is composed primarily of molten material. Plate movements are caused by a combination of forces, including **ridge push**, **mantle drag**, and **slab pull** (Kerr 1995; Deplus 2001; Price 2001; Conrad and Lithgow-Bertelloni 2002; Bercovici 2003). Ultimately, all of these forces are generated by heat and convective forces deep inside the Earth. Ridge push occurs at the midoceanic ridges, where **magma** (molten rock) upwells from the asthenosphere to the surface. It is believed that the parent rock of the mantle is partially melted and the basaltic portion is then brought to the surface. The addition of basaltic magma at the center of the ridge causes the older rocks on either side to spread—to be literally pushed apart. Thus, ridge push is the cause of seafloor spreading in Hess's model (see Figure 8.9).

Mantle upwelling is part of a convective cell system that also includes lateral flow of the mantle beneath the plates and downward flow of cooler rock

FIGURE 8.16   The current model of plate tectonics includes the possibility that at least three forces may be responsible for crustal movements: (1) ridge push, or the force generated by molten rock rising from the Earth's core through the mantle at the midoceanic ridges; (2) mantle drag, the tendency of the crust to ride the mantle much like boxes on a conveyor belt; and (3) slab pull, the force generated as subducting crust tends to pull trailing crust after it along the surface. (After Stanley 1987.)

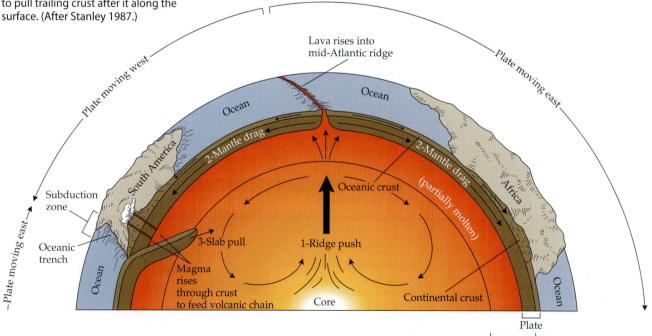

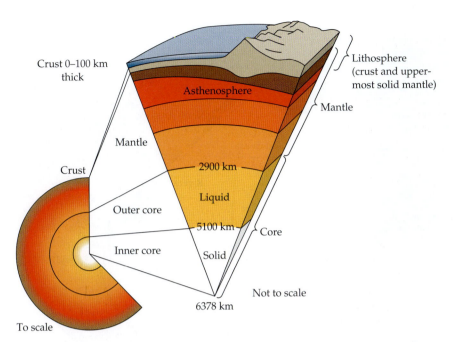

Crust 0–100 km thick

Mantle

Crust

Asthenosphere

Mantle

2900 km

Liquid

Outer core

5100 km

Inner core

Solid

6378 km

Lithosphere (crust and upper-most solid mantle)

Mantle

Core

Not to scale

To scale

FIGURE 8.17   Cross-sectional view of the Earth and its major layers.

toward the Earth's core (see Figure 8.16). Lateral flow and friction between the mantle and the overlying plate create a dragging force much like that of a conveyor belt (i.e., mantle drag). However, much, if not most, of the drifting force may be generated at **subduction zones**, where dense oceanic plates are pulled deep into the magma, eventually contributing to the convective gyre of molten rock. As the leading edge of the subducted plate descends, it pulls the rest of the plate laterally toward the subduction zone (see Kerr 1995 and references therein).

The relative importance of these three forces—ridge push, mantle drag, and slab pull—appears to vary markedly across tectonic regions and geological periods (Deplus 2001; Bercovici 2003; Conrad and Lithgow-Berteiloni 2002). Computer modeling conducted by Lithgow-Bertelloni and Richards (1995) suggests that slab pull may at times account for over 90 percent of net tectonic forces. On the other hand, slab pull cannot account for movement of those continental plates, such as the South American Plate, that lack subduction zones (actually, the Pacific Plate is being subducted under the western edge of the South American Plate). Obviously, this central issue—the underlying mechanisms of plate tectonics—will remain an active and hotly debated question for some time.

Despite remaining uncertainties about ultimate causal mechanisms, geologists have developed a sound understanding of plate configurations, plate movements and interactions, and related phenomena, including earthquakes and volcanism. While the number of plates has varied continually throughout geological time, 19 major plates are currently recognized (**Figure 8.18**). These plates range in size from the Gorda Plate (roughly 750 km^2) to the Pacific Plate (with an estimated area of over 100 million km^2). Rate of lateral movement also varies markedly among the plates, with some, such as western portions of the Pacific Plate, drifting as rapidly as 5 cm per year, while others appear fixed.

The biogeographic relevance of plate tectonics becomes immediately evident when we compare plate configurations with Wallace's map of biogeographic regions (compare Figures 8.18 and the endpapers): Biogeographic regions are defined by biotas and portions of the Earth's crust that share both evolutionary and tectonic histories, with representative assemblages on each plate evolving

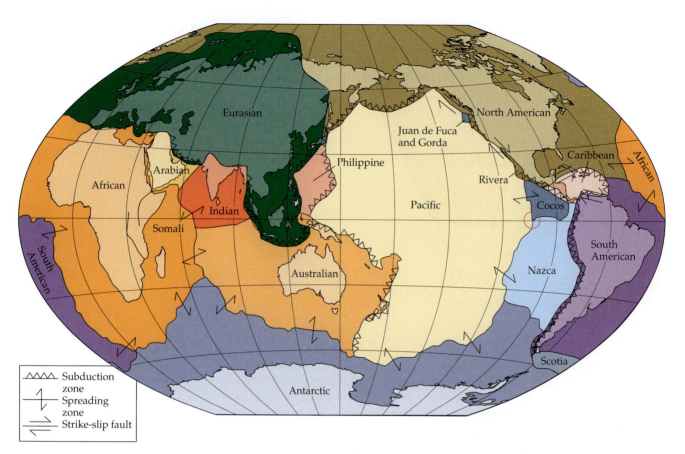

Subduction zone

Spreading zone

Strike-slip fault

FIGURE 8.18 Earth's major tectonic plates, showing the different kinds of plate boundary interactions described in the text. The triple junction of the Nazca, Pacific, and Cocos plates has been circled to mark the zone of volcanic activity that resulted in the formation of the Galápagos Archipelago.

in isolation from those on other plates. In Chapter 12 we explore in depth the relationships between history of the Earth and that of its biotas.

While geological history of the Earth's plates must be profoundly complex, **plate boundaries** take three basic forms: **divergent boundaries** associated with **spreading zones**, **convergent boundaries** associated with **collision** and **subduction zones**, and **transform boundaries** associated with major **fault zones** and **strike-slip faults** (see Figure 8.18). Other tectonic phenomena, including earthquakes, volcanism, and the formation of mountains and island arcs, are closely associated with these interactions among plates. For example, the Himalaya Mountains formed 60 million years ago when the Indian Plate drifted across the Equator to collide with the Eurasian Plate (see Briggs 2003a). These remain the world's tallest mountains. On the other hand, older mountain ranges, such as the Appalachians of eastern North America, which formed over 300 million years ago when North America collided with northwestern Africa, have been reduced by long-term erosion (in this case, such that the tallest mountains in the Appalachians are less than 2000 m; see Scotese 2004).

As noted above, midoceanic ridges mark the sites where two plates are drifting apart. Spreading zones, however, are not confined to oceanic plates. On continents, plates also diverge to form **rift zones**, which in the past created the Red Sea and the great, deep lakes of the Baikal Rift Zone and the East African Rift Valley. The latter region remains a tectonically active area marked by frequent earthquakes. Rather than continually and gradually sliding away from spreading centers, plates tend to resist moving until tectonic forces finally exceed some threshold, creating powerful bursts of movement interspersed with relatively long periods of stasis. Spreading zones also are marked by volcanic activity where magma rises to the surface (see Figure 8.9). Along ridge systems, volcanoes may become emergent as oceanic islands. Once formed, such an island is eventually carried away from the ridge and down a slope to the abyssal plain. This movement, along with accumulation of what often amounts to many tons of coral and carbonates, decreases elevation of the island relative to sea level and eventually draws it beneath the surface, making it into a submarine seamount. In Hess's classic model, wave action wears an island flat to form a guyot; however, some evidence suggests that flat-topped guyots may actually be formed that way without ever having been emergent. If the seafloor spreading model is correct, the ocean floor and its associated chains of volcanic islands, seamounts, and guyots should be youngest at ridges and oldest—and sink deepest below sea level—as it approaches subduction trenches. Beginning with Tarling (1962) and Wilson (1963b), various researchers have shown that these predictions are correct (**Figure 8.19**).

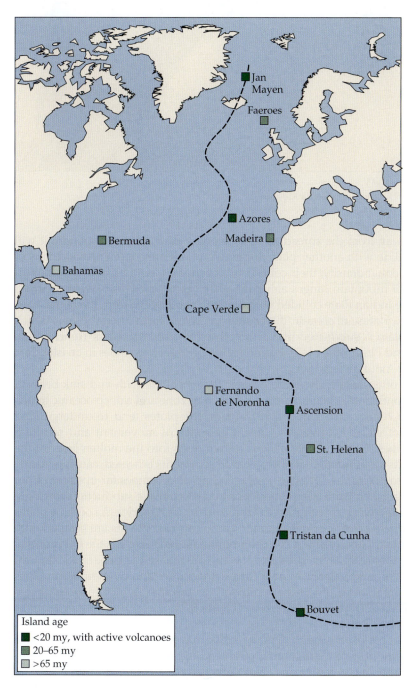

FIGURE 8.19 Because new crust is added along the spreading zones of midoceanic ridges, the age (in millions of years, my) of islands and other crustal features tends to increase with increasing distance from these ridges, as demonstrated by islands at different distances from the mid-Atlantic ridge (dashed line). (After Pielou 1979.)

Island age
■ <20 my, with active volcanoes
■ 20–65 my
□ >65 my

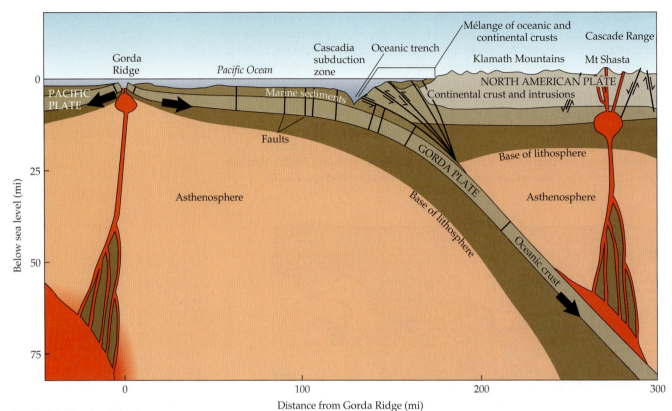

FIGURE 8.20 A subduction zone, where a relatively heavy oceanic plate slides beneath a lighter continental plate. Here, seafloor spreading causes the Gorda Plate to be subducted beneath the continental crust of the Pacific Northwest region of North America. This, in turn, results in faults and terranes (accumulations of oceanic material; see Figure 8.21) at the subduction zone and volcanic activity deep beneath the continental plate, forming the Cascade Range.

Far away from the spreading zone, along a plate's leading margin, it will often collide with another plate. If the two plates are continental and thus of roughly equal density, their collision will cause violent uplifting and the formation of mountain ranges along the plate boundary. This is what occurred when the Indian Plate collided with the Eurasian Plate to form the Himalayas. If the two plates are oceanic, their convergence can form a deep oceanic trench as one plate is subducted under the other (e.g., the Marianas Trench at the Pacific and Philippine plates boundary) or produce volcanoes in an **island arc** (e.g., the Aleutian Islands).

More often, relatively dense oceanic plates collide with and sink beneath lighter continental plates to form a subduction zone and a deep oceanic trench (**Figure 8.20**; see also Figure 8.16). Again, the plates tend to undergo long periods of stasis followed by violent episodes of movement and resulting earthquakes. As an oceanic plate is drawn deep into the molten layer of the mantle, its accumulation of relatively wet sediments is heated, causing mantle plumes to rise to the surface and form a ring of volcanoes far upstream from the superficial layers of the subduction zone. Extensive subduction zones are found along western margins of North and South America. As oceanic plates are subducted over geological time, seamounts and guyots are scraped onto continental plates, forming coastal mountain ranges such as the Olympics of Washington State. Thus, subduction zones are marked with parallel bands of earthquakes and volcanism near their active edges and by mountain ranges and accumulations of marine sediments, as **terranes**, which increase in age with distance from the active zone (**Figure 8.21**; see also Coney et al. 1980; Calvert et al. 1995).

Finally, plates of roughly equal density may slide and grind against each other, without either subducting or spreading, to form a transform boundary along a major fault zone. Not surprisingly, transform boundaries are

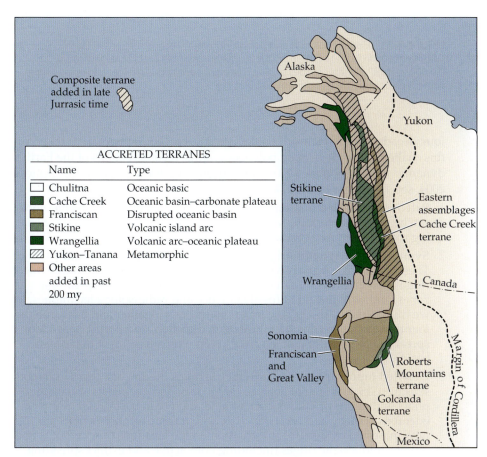

FIGURE 8.21 Terranes, or fragments of lithosphere that acrete onto other plates, mark the locations of earlier subduction zones along the west coast of North America, beginning with the suturing of the Golconda and Sonomia terranes to the Roberts Mountains terrane in the Triassic Period; those shown in Canada were sutured during the Late Jurassic. This process occurred as oceanic plates slipped beneath the relatively buoyant continental plates but islands, seamounts, and other superficial features were scraped off and added to the continental plate. During the 1970s, these as well as accretions across the Pacific basin were used as evidence for existence of an ancient continent, "Pacifica" (Nur and Ben-Avraham 1977), which has subsequently been discredited upon recognition that they are not of continental origin. (After Stanley 2009 from original in Saleeby 1983).

characterized by high seismic activity as plates grind against each other. In southern California, the North American and Pacific plates grind past each other along the San Andreas fault. While this may be the most notorious example, transform boundaries are quite common. Some are closely associated with major plate boundaries (e.g., those along the margins of the Scotia and Antarctic plates, along the southern tip of South America), while others are found at great distances from plate boundaries (e.g., the Altyn Tagh fault of western China).

Earth's Tectonic History

The tectonic processes described above have recurred throughout Earth's 4.6 billion year history and will certainly continue to modify the geographic template as well as subsequent distributions of future biotas (Briggs 1987; Richards et al. 2000; Scotese 2004). It is hard to overemphasize the effects of plate tectonics on biogeographic patterns of virtually all organisms. The origin, spread, and radiation of many taxa took place when the Earth's surface was dramatically different from its current profile. These processes—along with the many extinctions evidenced in the fossil record—were often associated with the collision, separation, or destruction of continents or oceanic basins, which altered opportunities for biotic exchange or affected climate on regional to global scales (Harris 2002).

In the following sections, we summarize the current understanding of Earth's tectonic history, focusing on the dynamics of its continents and marine basins during the Phanerozoic. As you can imagine, the detective work involved in this 542 million year reconstruction, while fascinating, was a tremendous challenge. It is without doubt one of the most fundamental and insightful advances in our understanding of the natural world. Over the past four decades, many teams of geologists and biogeographers have taken up this challenge and developed paleogeographic reconstructions of the Earth's dynamic history (**Figure 8.22**) drawing on four lines of evidence (Scotese 2004; see also Burke et al. 1977; Dewey 1977):

1. **Paleomagnetic declinations**: As discussed earlier, magnetic declinations can be used to determine latitude of sites where iron- and titanium-bearing rocks cooled and crystallized (see Figure 8.12A).

2. **Symmetrical magnetic stripes ("magnetic anomalies") of ocean basins**: These magnetic reversals recorded in rocks on either side of midoceanic ridges (see Figure 8.15) can be used to "match up" ancient spreading zones, and with the aid of fossil evidence and radioisotope techniques, we can virtually "re-zip" the ocean floors back together and back into the mantle and thus reconstruct the extent, direction, and timing of seafloor spreading (and drifting of landmasses on opposite sides of that ocean basin).

3. **Detailed topographic and bathymetric maps**: Recent advances in remote sensing, GIS, and spatial analyses have provided geologists and geographers with maps of Earth's current land and ocean basins with unprecedented accuracy and precision. Paleogeographic reconstructions are now informed by detailed information on the locations of seamounts and guyots, midoceanic ridges (including both active and extinct spreading zones), deep sea trenches, linear fracture zones (indicating directions of drift of previously joined landmasses), mountain chains, linear archipelagoes (e.g., Emperor and Hawaiian archipelagoes; see Figure 8.29), and tracks of flood basalts (magma from hotspot plumes that burned through the lithosphere and poured onto the continents; see Box 8.1, Figure A).

4. **Lithologic indicators of climate**: As discussed in Chapter 3, Earth's soil and rock formations are strongly influenced by climatic conditions; coals form where bauxite occurs in wet and warm (i.e., tropical) climates, tillites where cool and wet conditions prevail (especially along the edges of glaciers), and evaporates and calcretes under warm and dry conditions, respectively. Thus, even though continental plates may drift from pole to pole, their rock formations have recorded their previous climates and, indirectly, their general latitudinal positions when they formed. Again, these rocks and associated tectonic events can be aged based on fossils and radiometric techniques.

Tectonic History of the Continents

Unlike the oceanic lithosphere, which is continually, albeit slowly, recycled into great gyres of the molten material of the mantle, continental lithosphere is less dense and, therefore, not readily subducted. Thus, whereas the age of ocean basins rarely exceeds 150 million years, continental lithosphere is truly ancient—with the Archean cratons (see Figure 8.11) perhaps as much as 3.8 billion years old (Scotese 2004). As a result, reconstructions of the tectonic history of the continents can be much more extensive than that of ocean basins. In the following sections, we will summarize paleogeographic history of continental plates and terranes, starting with the formation of the most recent supercontinent—Pangaea—realizing, however, that this is just a small portion of the long and complex tectonic history of these landmasses.

(A) Late Proterozoic (600 million years BP)

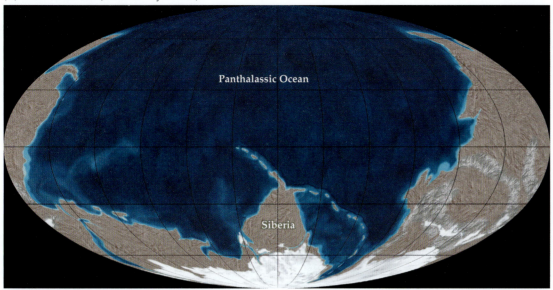

(B) Early Cambrian (540 million years BP)

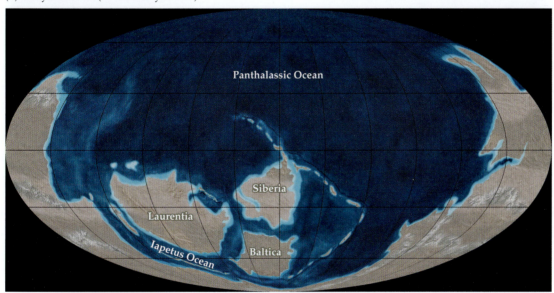

(C) Silurian (430 million years BP)

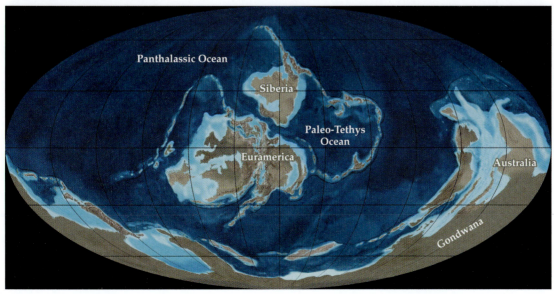

(D) Late Carboniferous (300 million years BP)

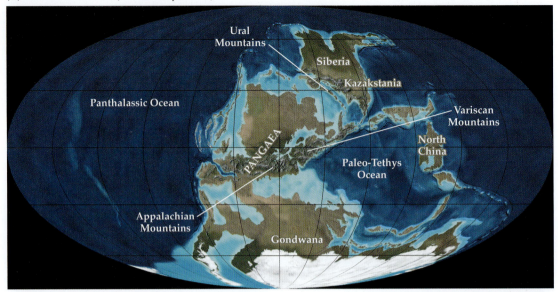

(E) Early to Middle Triassic (240 million years BP)

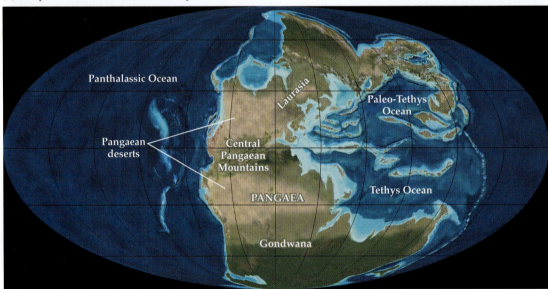

(F) Middle Jurassic (170 million years BP)

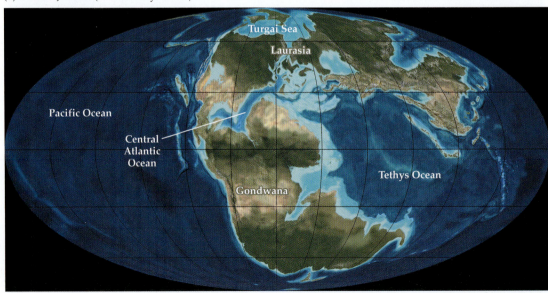

(G) Early Cretaceous (120 million years BP)

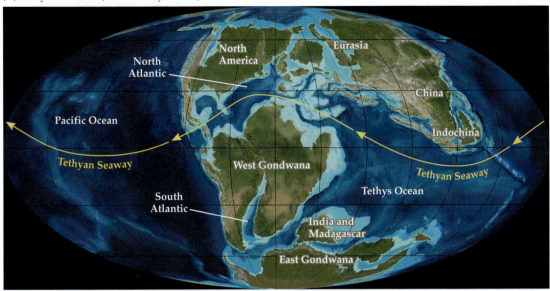

(H) Late Cretaceous (90 million years BP)

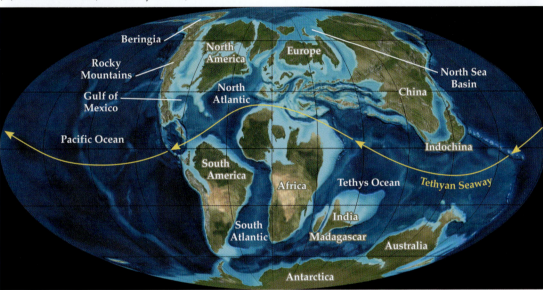

(I) Cretaceous-Tertiary boundary (65 million years BP)

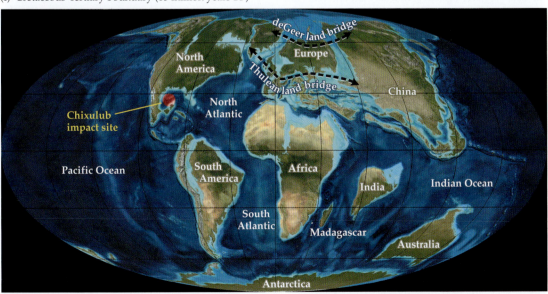

(J) Eocene (50 million years BP)

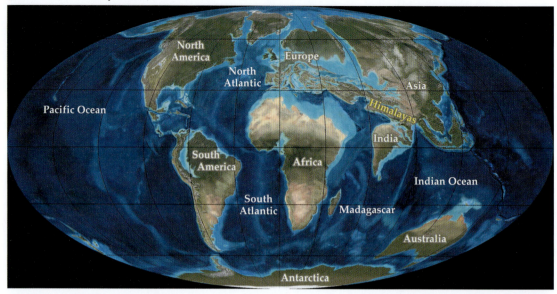

(K) Miocene (20 million years BP)

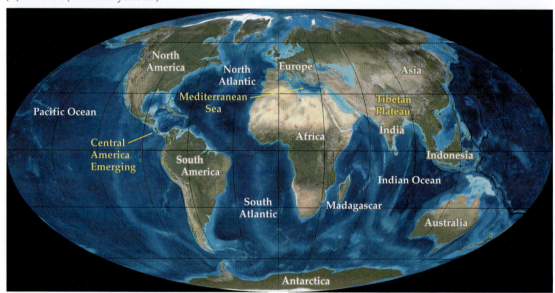

(L) Late Pleistocene glacial period (50 thousand years BP)

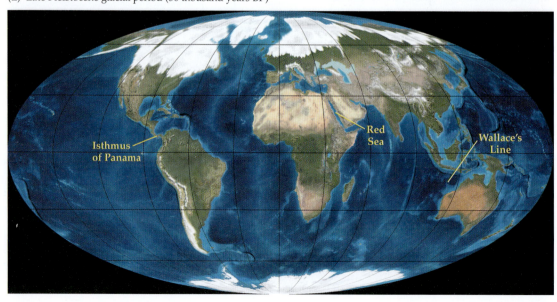

Gondwanaland, Laurasia, and the formation of Pangaea

Early supporters of the theory of continental drift originally deduced that prior to the Mesozoic, all landmasses had been united in a supercontinent—Pangaea—which occupied one-third of the Earth's surface before separating into the northern supercontinent of **Laurasia** and southern supercontinent of **Gondwanaland** (or **Gondwana**). Now paleogeographers have not only confirmed this but tell us that other global-scale supercontinents existed prior to this. For example, the supercontinent of **Rodinia** formed about 1.1 billion years BP and later broke apart during the Late Proterozoic, separating three fragments (Laurentia, Siberia, and Baltica) from the main continent as the Iapetus Ocean expanded (**Figures 8.22A,B**), and Pangaea was a relatively temporary structure that probably existed as a single unit only during Late Paleozoic and Early Mesozoic times (**Figure 8.22D**). Its northern half, Laurasia, had a very complex early history that was substantially different from that of the southern half, Gondwanaland.

Gondwanaland included the foundations of present-day South America, Africa, Madagascar, Arabia, India, Australia, Tasmania, New Guinea, New Zealand, New Caledonia, and Antarctica. It was, by far, the most ancient of the Pangaean landmasses, forming some 650 million years BP during the Precambrian Era (see Figure 8.22A). From this time through the middle Ordovician Period (about 475 million years BP), Gondwanaland remained a relatively continuous supercontinent in the Southern Hemisphere, although it drifted substantially between the Equator and the South Pole (see Figure 8.12B). If one examines the history of Queensland, Australia—now at 12° S latitude—one discovers that it was located near the North Pole in the Proterozoic (1 billion years BP), at the Equator in the Silurian (**Figure 8.22C**), and at 40° S in the Mesozoic (**Figures 8.22H–K**; Embleton 1973). The great antiquity of Gondwanaland accounts in part for the similarity of the now-isolated biotas of the Southern Hemisphere, patterns that strongly influenced historical biogeography since the early writings of Joseph Dalton Hooker (see Chapter 2).

In contrast to its ancient southern counterpart, the history of Laurasia is much more recent. Pre-Laurasian landmasses remained isolated until the Early Devonian (about 400 million years BP), when the precursors of northern Europe (Baltica), which once occupied subantarctic latitudes, drifted northward across the Equator to collide with the precursors of North America (Laurentia) and the northern part of Siberia, forming the continent of Euramerica (see Figure 8.22C). The precursors of China and southern Siberia remained isolated in subtropical latitudes of the Northern Hemisphere, far from Gondwanaland, which was then situated near the South Pole. During the Carboniferous Period, Gondwanaland began drifting northward, and it eventually united with Euramerica about 300 million years BP to begin the formation of Pangaea (see Figure 8.22D). Later, during the Early Permian (about 270 million years BP), western Asia collided with Europe to form the Ural Mountains and unite the bulk of Pangaea. At this time, eastern Asia (largely derived from accreted remnants of Gondwanaland: Metcalfe 1999) still remained isolated as several large islands, which partially divided the world's oceans and encircled the Paleo-Tethys Ocean along the eastern margin of Pangaea (**Figure 8.22E**). By the Late Permian (250 million years BP), these landmasses consolidated to form Pangaea—a great, continuous landmass stretching from pole to pole.

Although this fact is often overlooked, coincident with the formation of Pangaea, marine basins (e.g., Panthalassian, Paleo–Tethys, and Tethys oceans) were united to form one global ocean, **Panthalassa**. Thus, the Permian was a time of great connectivity and exchange among both terrestrial and marine biotas, many of which had evolved in isolation for hundreds of millions of years. On the other hand, while this was a time of interchange among long-isolated biotas, Triassic landscapes (and, likely, seascapes as well) were not without barriers and ancient, biogeographic divisions. Formidable barriers to biotic interchange during the Triassic included extensive mountain ranges (e.g., the Central Pangaean Mountain Range of the Early Triassic) and the subtropical Pangaean deserts, which at times stretched from central North America to the Amazon basin and west-central Africa (see Figure 8.22E).

The breakup of Pangaea

Great evolutionary radiations of most terrestrial families and many orders, especially among vertebrates, seed plants, and insects, occurred after the Mid Mesozoic breakup of Pangaea and Panthalassa (**Figures 8.22F–G**). In fact, the diversification of many taxa may have been a direct outcome of this break-up and the resultant isolation of terrestrial and marine biotas. As continents fragmented and drifted toward different latitudes and climates, reduced gene flow and altered selective pressures allowed rapid speciation and radiation of their rafting biotas. Again, the history of tectonic events provides invaluable insights for interpreting biogeographic patterns, both past and present.

The breakup of Laurasia and its rifting from Gondwanaland

The breakup of Pangaea was actually initiated in the Early to Mid Jurassic (about 180 million years BP) when the Turgai Sea (see Figure 8.22F) expanded southward from the Arctic to split Eurasia from North America (i.e., eastern from western Laurasia). By the Late Jurassic, a shallow "Atlantic Sea" developed and began to separate North America from Eurasia (see Figure 8.22G–H). By Early Cretaceous time, active seafloor spreading had begun to open the Atlantic, and the remaining narrow land bridges between Laurasia and Gondwanaland (near Central America and north Africa) were broken, transforming the Tethys Sea into a circum-equatorial seaway. This drifting also marked the origin of the Gulf of Mexico as North and South America separated, while extensive volcanic activity along the eastern margins of the African Plate formed the initial basins of the Indian Ocean. Pangaea had fragmented into three great landmasses (eastern and western Gondwanaland in the Southern Hemisphere, and a more diffuse Laurasia in the north), and these continued to split into smaller and increasingly more isolated continents as the numerous, incipient ocean basins continued to expand throughout the Jurassic to Mid Cretaceous Periods (roughly 215 to 100 million BP).

By the Late Cretaceous (75 million years BP), an intermittent land connection between Laurasian landmasses had been reestablished across the Bering Strait (Fiorillo 2008). Thus, at about the same time that one land connection between eastern North America and western Europe was severed, a new one formed between western North America and Asia, permitting continued exchange of organisms. This connection, Beringia, persisted through most of the Cenozoic, although an uninterrupted land bridge was not continually present (see Chapter 9). Transient connections among the remnants of Laurasia (between North America and Europe) also existed across the Greenland–Scotland ridge during the Early Tertiary (i.e., the Thulian and DeGeer routes). Paleontological evidence indicates that these connections across the North Atlantic were important corridors for the migration of animals and plants (McKenna 1972a,b; Sclater and Tapscott 1979; Barron et al. 1981). Earlier, biotic exchange also occurred across transient land connections that often linked the British Isles with Greenland and Canada during the Early Cretaceous (Hallam and Sellwood 1976). As a result of this long history of exchange between the Nearctic and Palearctic regions, they share many species of plants and animals and have often been coupled as a superregion, the Holarctic.

Other areas of the original Laurasian supercontinent had a dynamic Mesozoic history. In Europe, great chalk deposits formed in the shallow North Sea basin during the Late Cretaceous. As North America drifted away from Europe, its southeastern coast (east of the Appalachian Mountains, including land from New Jersey to Yucatán, Mexico), became submerged to form a great sedimentary basin. The Atlantic Gulf coast was under water from the Late Jurassic until the Eocene, so all of the forms living there today are more recent colonists. Similarly, the North American Interior Seaway stretched from Alaska to the Gulf of Mexico (**Figure 8.23**), as evidenced by an abundance of marine fossils across the interior of North America. To the south, it now appears that the Caribbean Plate formed and then expanded as the

FIGURE 8.23 Reconstruction of the North American Interior Seaway, a major epeiric transgression that reached its maximum extent during the Late Cretaceous Period. (From Ron Blakey.)

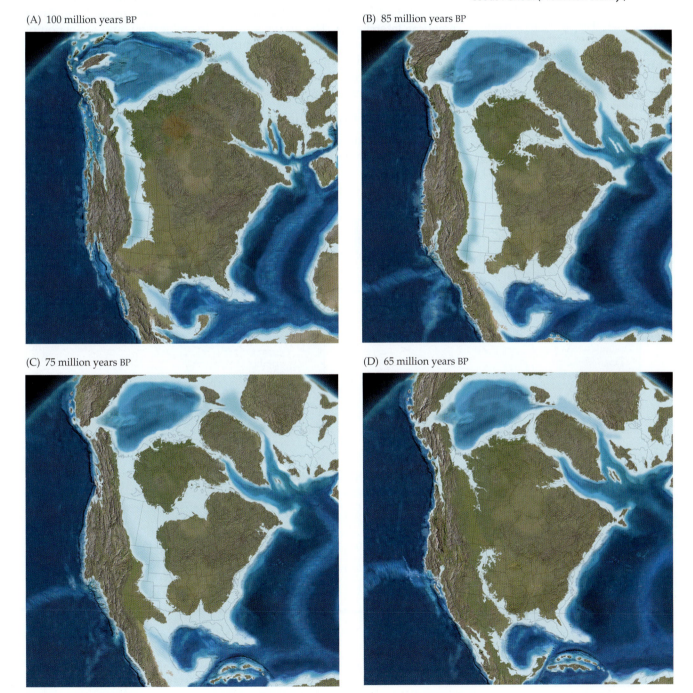

(A) 100 million years BP

(B) 85 million years BP

(C) 75 million years BP

(D) 65 million years BP

North American and South American plates separated during the Late Jurassic (about 150 million years BP; **Figure 8.24**). The Greater and Lesser Antilles formed de novo as a result of tectonic and volcanic activity, and not as fragments from the mainlands of North or South America (see below). Therefore, it appears that most of the biota of these islands, including a large number of vertebrates that went extinct during the Holocene, colonized these islands during multiple episodes of over-water immigrations (see Briggs 1995; Meschede and Frisch 1998; Hedges 1996, 2001).

The final punctuation mark to the tectonic development of this region of the New World was the emergence of the Central American land bridge (see Figures 8.22K–L; Iturralde-Vinent 2006), which allowed biotic exchange between the long-isolated Nearctic and Neotropical regions, something the distinguished paleontologist George Gaylord Simpson termed the **Great American Interchange**. The origin, spread, diversification, and extinction of species across these two regions and throughout the Caribbean is indeed a complex but fascinating account of the relationships between the dynamics of the Earth and its climates, oceans, and biotas, and so we further explore the current understanding of these events in Chapter 10.

The breakup of Gondwanaland

Roughly simultaneously with the initial breakup of Laurasia, the landmasses of Gondwanaland began to separate. According to the review by Storey

FIGURE 8.24 Paleogeographic reconstructions summarizing the development of the Caribbean region from the Late Jurassic to Early Eocene. (A) Opening of the Caribbean (= western Tethyan) Seaway. (B–D) Progressive development and eastward drift of the Caribbean Plate, highlighting both the Antilles volcanic arc system and the Central American volcanic arc system. Later Cenozoic reconstructions continue in Figures 10.22 and 10.34. (After Iturralde-Vinent 2006.)

(A) Late Jurrasic 150–145 Ma

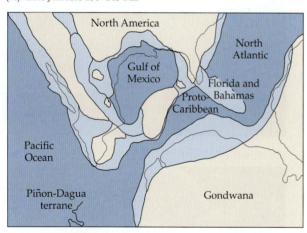

(B) Early Cretaceous 125–120 Ma

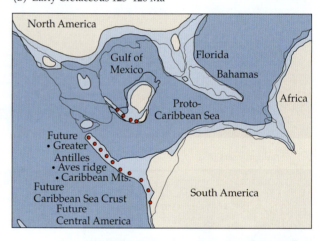

(C) Upper Cretaceous ~70 Ma

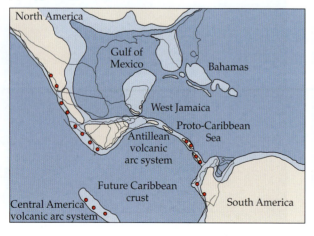

(D) Early Eocene ~55 Ma

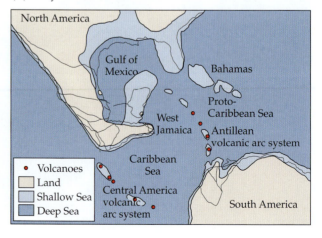

(1995), separation of Gondwanaland (**Figure 8.25A**) proceeded in three stages. First, during the Early Jurassic (about 180 million years BP), rifting opened the Somali, Mozambique, and Weddell seas between eastern and western Gondwanaland. This initial rifting was followed by seafloor spreading, which united these relatively narrow seas and split Africa and South America from the remainder of Gondwanaland (Madagascar, India, Australia, Antarctica, and New Zealand) about 160 million years BP. At this time, Madagascar and India were adjacent to the northeastern margin of Africa, while Australia was located to the south along the eastern margin of Antarctica in a position rotated approximately 90° clockwise from its present position (**Figure 8.25B**).

During the second stage of the breakup, Madagascar remained joined to India as this landmass separated from the Antarctica–Australia landmass by a narrow seaway (about 155 million years BP; Briggs 2003a). Active seafloor spreading had begun by about 130 million years BP as India–Madagascar separated from its connection to Antarctica (**Figure 8.25C**). Also at about this time, South America and Africa rifted apart, thus forming four Gondwanan landmasses (South America, Africa, Madagascar–India, and Antarctica–Australia–New Zealand). Madagascar then separated from India and the Seychelles landmass approximately 90 million years BP. This process of global-scale fragmentation of Gondwanaland continued during the Late Cretaceous and throughout the Paleocene (100 to 58 million years BP), when oceanic basins such as the Mascarene continued to expand, which in this case separated Madagascar and the Seychelles archipelagoes from India (see Figures 8.22G–J). On the other hand, extensive drifting of continents also resulted in collisions and mixing of biotas that had been isolated since the initial rifting

FIGURE 8.25 Stages in the breakup of Gondwanaland. Colors outline different plates as they pull apart along rifts through a time interval of about 100 million years (my) during the Jurassic and Cretaceous. One line of evidence for this reconstruction comes from linking of flood basalts across landmasses, as shown in Box 8.1, Figure A2. (After Storey 1995).

(A)
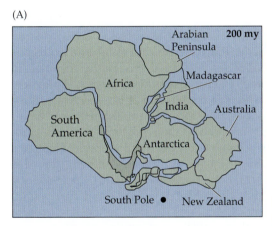

(C)
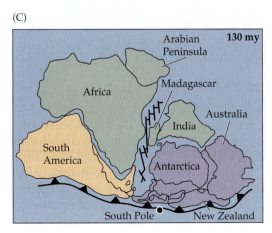

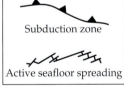

(B)
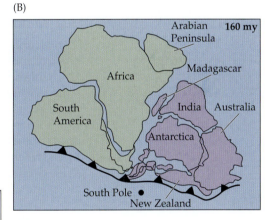

(D)
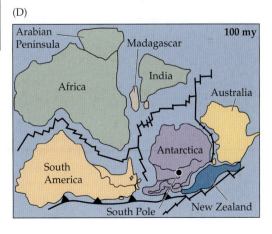

of Pangaea. India's migration northward was very rapid indeed, averaging 15 to 20 cm per year, and by about 55 million years BP, it had collided with Eurasia (Scotese 2004). The period of biotic interchange may, however, have been quite punctuated since the reconnection of these remnants of Pangaea was soon followed by formation of one of the world's most formidable barriers to interchange among terrestrial biotas—the Himalayas—when India collided with the Eurasian Plate between 60 and 50 million years BP (see Figure 8.22J; Butler 1995; Briggs 2003a).

Much farther to the south, Australia and New Zealand rifted from Antarctica about 100 million years ago (**Figure 8.25D**), and extensive seafloor spreading continued to disperse these and other remnants of Gondwanaland during the Late Cretaceous Period. New Zealand retained a connection to Antarctica and what is now the eastern margin of Australia until about 90 million years BP. New Zealand broke away, first from Australia and then from Antarctica, about 80 million years BP, drifting farther east of Australia during development of the Tasman Sea. While other remnants of Gondwanaland were drifting toward the lower latitudes, Antarctica continued its slow poleward journey. By the Miocene (24 million years BP), Antarctica was situated over the South Pole, triggering the development of the great polar icecap.

Thus, these reconstructions of the tectonic history of Gondwanaland go a long way toward explaining both the similarities among biotas of now isolated remnants (e.g., those of India, the Seychelles, and Madagascar as well as those of New Zealand and Australia) and the distinctiveness and endemicity of these biotas. Yet, there are still details that are controversial, for which biogeographic evidence can prove illuminating (e.g., Noonan and Chippendale 2006). We will further explore the reciprocal nature of geological and biogeographic evidence in more detail in Chapters 11 and 12.

Cenozoic tectonics

Throughout much of the Cenozoic Era, extensive drifting and collisions of long-isolated continental landmasses continued, frequently resulting in lithospheric uplift and formation of extensive mountain ranges. While new ocean basins expanded, others were subducted, and ancient seaways, such as the circum-equatorial Tethys, were transformed into isolated seas. In addition to uplift of the Himalayas (discussed above), collision of drifting landmasses during the Mid and Late Cenozoic formed the Pyrenees Mountains (when landmasses now occupied by France and Spain collided), the Alps (formed by the collision of Italy with France and Switzerland), the Hellenide and Dinaride mountains (formed by the collision of Greece and Turkey with the Balkan States), and the Zagros Mountains (formed by the collision of Arabia with Iran). Again, each of these collisions of continental plates were likely followed by waves of relatively rapid biotic interchange, which likely decreased as uplifting proceeded to replace an oceanic barrier with a mountainous one. Two of the most biogeographically important collisions of long-isolated regions are illustrated in more detail below.

History of Central America and the Antilles

The history of Mesoamerica and the Caribbean region is a complex and intriguing case study. The reasons for this are obvious. First, biogeographers need to know how and when biotic exchange occurred between the Nearctic and Neotropical regions following the breakup of Pangaea. Second, we need to know whether the rich biota of the Caribbean islands, especially the Greater Antilles, arrived there exclusively by long-distance dispersal, as suggested by Darlington (1938) and many others, or whether the islands are remnants of an ancient landmass that at one time was continuous with the mainland (i.e., either North or South America).

The development of the Caribbean region continues to be controversial, but the "Single arc Pacific origin model" reviewed here is among the most clearly developed and associated with biogeographic evidence (Iturralde-Vinent 2006). Following the formation of Pangaea, the precursors of North America, South America, and Africa remained connected until Pangaean rifting in the Middle and Late Jurassic created the Tethyan (or Caribbean) Seaway—a continuous equatorial marine connection between the Pacific and emerging Atlantic oceans (see Figure 8.24A). During the Early Cretaceous (120 to 140 million years BP), a chain of volcanic islands began to emerge along the eastern edge of the Caribbean Plate. By the Late Cretaceous, this plate had drifted eastward so that an Antilles volcanic arc system formed a stepping-stone route for limited exchange of some terrestrial organisms between the Nearctic and Neotropical regions (see Figure 8.24C). The Antilles volcanic arc system continued to drift eastward along the leading edge of the Caribbean Plate to eventually form the Greater Antilles, the Caribbean Mountains of Venezuela, the Lesser Antilles, including the ABC (Aruba, Bonaire, and Curaçao) islands off the coast of Venezuela, and the Aves–Greater Antilles Ridge (see Figures 10.22 and 10.34). The core islands of the Greater Antilles had achieved their present positions by Eocene times (58 million years BP), and the Caribbean Plate collided with the Bahamas Platform by 55 million years BP (see Figure 8.24D).

Formation of the current land bridge between North and South America was an important event for New World biogeography. Waves of migration across this land bridge, called the **Great American Interchange**, profoundly affected the diversity and composition of the Nearctic and Neotropical faunas (see Chapter 10). Like the Antilles volcanic arc system, the Central American land bridge may have first emerged during the Late Cretaceous as the Central America volanic arc system (far west of its current position), which subsequently drifted eastward with the boundary between the Cocos and Caribbean plates (see Figure 8.24B–D).

The final emergence of the Central American land bridge in the Neogene was produced by coalescence of islands in the arc system as the Cocos, Nazca, and Caribbean plates converged (see Figures 10.22 and 10.34). By the Late Miocene (10 to 5 million years BP), the Central American archipelago provided a stepping-stone route for the dispersal of a variety of terrestrial organisms. Approximately 3.5 million years BP, the archipelago finally fused to close the Panamanian Isthmus and form the current Central American land bridge between the Nearctic and Neotropical regions (see Stehli and Webb 1985; Briggs 1994, 1995).

History of the Indo-Australian Region and Wallacea

The region of the world that lies at the junction of the Australian, Eurasian, Indian, Pacific, and Philippine Plates (see Figure 8.18) holds a particular fascination, as well as considerable historical signficance for biogeographers. It is here that Alfred Russel Wallace developed some of his greatest insights into the geography of life (see summary in Box 2.1) during his eight years of island-hopping and collecting across the Malay Archipelago. Wallace's keen powers of observation lead him to surmise correctly that two great biotas came together in a narrow transition zone, which he placed between the islands of Borneo and Bali to the west, and Sulawesi and Lombock to the east. He was also aware of the complexities of this faunal transition, which created difficulties for him, for example, in deciding whether Sulawesi belonged to

FIGURE 8.26 Reconstructions of plate and landmass positions and development of accretionary landmass that would develop into modern islands as the Indonesian portion of the Eurasian Plate, the eastward-drifting Pacific Plate, and the northward-drifting Australian Plate collided 45 to 5 million years BP (see Figures 8.18 and 8.27 for present configuration of island and continental landmasses and plate boundaries). Lighter shades represent currently submarine parts of continental plates. (After Hall 2002.)

(A) Middle Eocene 45 Ma

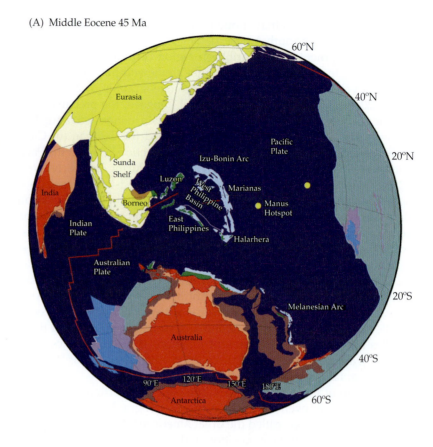

(B) End Oligocene 25 Ma

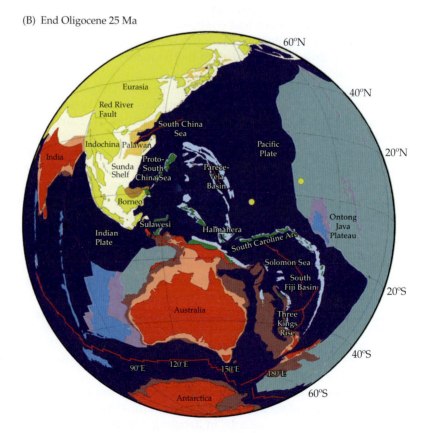

(C) End Miocene 5 Ma

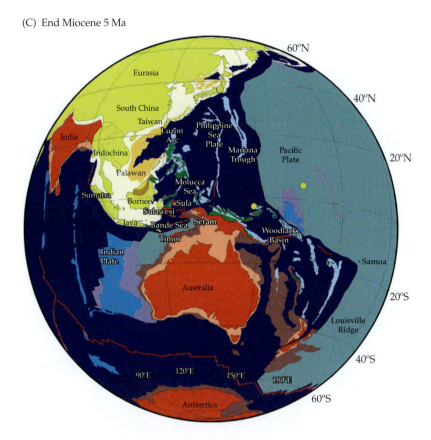

the west or east of Wallace's line (named as such in 1868 by T. H. Huxley; see Figures 9.9 and 10.14). We wonder how he might have addressed this complexity if he had known that Sulawesi represents a relatively recent suturing of landmasses at one of the most dynamic of the Earth's plate boundaries.

Fragments of Gondwanaland, including the Australian Plate, have drifted northward throughout the Late Cretaceous and Cenozoic. While the resulting collision of the Indian and Eurasian Plates produced massive mountain ranges and plateaus, the results from the Australian Plate collision with the Sundaland region of the Eurasian Plate were very different, primarily because it represented a meeting of oceanic with continental margins and, thus, led to the development of several expansive subduction zones (**Figure 8.26**), complete with long trenches and island arc volcanoes (**Figure 8.27**). During this time, the Pacific Plate was subducting from the east, and the ongoing collision of India was deforming Asia. The modern abundance of South Pacific islands (including the Philippines) thus represent a mosaic of island arc volcanoes or accreted terranes sloughed from leading edges of subducting plates. Of interest to biogeographers is that the current geography of Southeast Asia, including Wallacea, was largely in place by about 10 million years BP (Hall 2001), although volcanic activity along plate boundaries and Pleistocene sea level fluctuations continue to provide new opportunities for dispersal, vicariance, and evolution.

Tectonic development of marine basins and island chains

We commented earlier on the unfortunate tendency of biogeographers to ignore tectonics of the marine realm, except as it relates to biogeography and paleontology of terrestrial biotas. Yet the marine realm, in addition to creat-

(A) Middle Oligocene 30 Ma

(B) Early Miocene 20 Ma

(C) Late Miocene 10 Ma

(D) Present 0 Ma

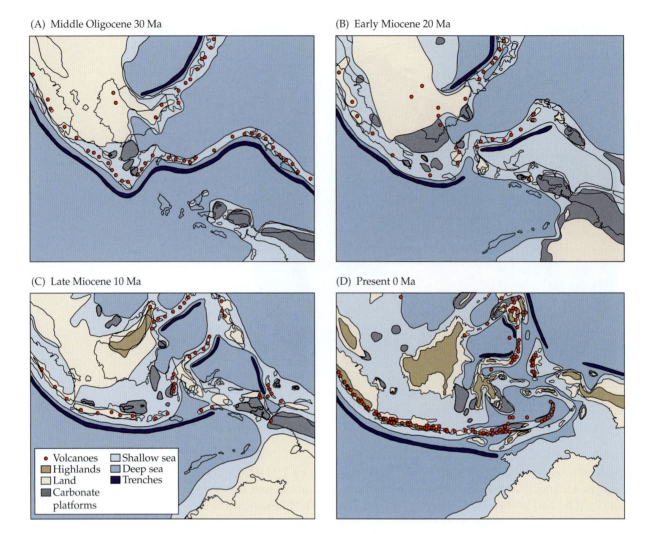

- Volcanoes
- Highlands
- Land
- Carbonate platforms
- Shallow sea
- Deep sea
- Trenches

FIGURE 8.27 A reconstruction of tectonic events centered on the Wallacean region, from the Middle Oligocene to the present, as it formed between the Indonesian portion of the Eurasian Plate, the Pacific Plate drifting from the east, and the Australian Plate as it drifted northward from Gondwanaland (see Figure 8.26 for larger-scale reconstruction). The many islands in the region have formed either as volcanic arcs along subduction zones or as accretionary landmasses formed along plate boundaries and sometimes coalesced into larger islands (e.g., Sulawesi). (After Hall 2001.)

ing major barriers to dispersal for terrestrial organisms and influencing climates on continents and islands, constitutes most of the biosphere. As noted earlier, when the Earth's landmasses were consolidated to form Pangaea, its marine basins also were joined to form one world ocean—Panthalassa. As Pangaean landmasses split apart, seafloor spreading created new marine basins at the expense of Panthalassa, which was fragmented and isolated by the drifting continents. Just as terrestrial climates are strongly influenced by adjacent oceans, changes in the positions of continents changed paleocurrents dramatically, strongly influencing the temperatures and basic chemistry of marine waters. Often throughout the Earth's history, these changes in climates and currents produced glacial cycles, which lowered sea levels during their peak phases and raised sea levels to flood the continents during interglacial periods.

Again, a sound understanding of the unique tectonic history of the marine realm is essential for anyone studying biogeography of its biotas. While continental plates have drifted, broken apart, and united, their total expanse has changed very little over the past 650 million years. In contrast, the ocean floor is continually created at the midoceanic ridges and consumed at the trenches. Consequently, most of the present ocean floor is less than 50 million years old. In the following sections, we summarize some of the major events associated with the formation and disintegration of marine basins and shallow, epeiric seas, and then we will discuss associated phenomena, including the formation of island chains and the dynamics of marine currents.

Epeiric seas

Epeiric seas, also called epicontinental seas, are formed when sea levels rise and oceans flood continental plates. We happen to be living today in one of the driest periods in the history of the Earth. Anyone who has hunted for fossils in the interior of the United States or Eurasia, however, knows that vast seas covered these areas many times, leaving the fossilized remains of marine organisms and thick deposits of limestone, formed from the calcium carbonate of ancient coral reefs.

Epeiric seas usually act as barriers to terrestrial organisms, subdividing a landmass into smaller emergent regions. In the Early Cretaceous (about 120 million years BP), for example, epeiric seas divided Australia into two subcontinents and drowned a vast portion of its northeastern subcontinent (Gurnis et al. 1998). Such events had important biogeographic consequences for biotic differentiation and extinction (see Chapters 10, 11, and 12). Some of the more interesting examples of marine transgressions occurring elsewhere during the Cretaceous include the North American Interior Seaway, which extended from the Gulf of Mexico as far west as Arizona (see Figure 8.23), and the subdivision of mainland Africa (see Figure 8.22H). The North American Interior Seaway dried up by the Paleocene (about 65 million years BP), while the Turgai Sea (which isolated Europe from Asia) dried up during the Oligocene (about 30 million years BP), again allowing biotic exchange among terrestrial biotas of Europe and Asia.

On a smaller scale, we can find important changes in river systems. In South America, the Amazon had a complex history, changing its drainage several times, from westward in the Early Cenozoic to its present eastward drainage into the Atlantic following uplift of the Andes. River systems such as the Mississippi and the St. Lawrence in the United States also have changed course many times. In the western United States, the Great Basin, which includes much of Utah, Nevada, and eastern California, comprises a series of **endorheic basins**—numerous alkali sinks, dry lakes, and river beds with no external drainage to the ocean. However, during earlier Cenozoic times, when geological configurations were quite different from today, there were connections to the Colorado River in the south, the Columbia River (via the Snake River) in the north, and the Pacific Ocean to the west (Smith et al. 2002).

Formation of the Mediterranean and Red Seas

After being freed from North America and South America, Africa began to swing counterclockwise toward Eurasia, and it eventually closed the formerly extensive Tethyan Seaway, leaving as its modern remnants the Mediterranean, Black, Caspian, and Aral seas. A bridge was formed between Asia and Africa through Arabia following their collision in the Mid Tertiary (35 million years BP; Jolivet and Faccenna 2000), which not only created the Zagros Mountains of Iran but also confined the Mediterranean Sea along its eastern margin. As Africa approached southern Europe, a number of deformations were initiated, including the counterclockwise rotation of Italy, which produced its characteristic diagonal orientation (McElhinny 1973a; see also Hsü 1972; Biju-Duval and Montadert 1977; Bureau de Recherches Géologiques et Minières 1980a,b; Ghebreab 1998). By the Early Miocene (24 million years BP), the Straits of Gibraltar had nearly closed, causing desiccation and a 6 percent increase in salinity in the Mediterranean (see Figure 8.22K). Seafloor spreading then accelerated during the Early Miocene (21 to 25 million years BP). The Straits of Gibraltar rifted apart during the Late Miocene to open the Mediterranean Sea to the Atlantic. Rifting proceeded more rapidly in the south, resulting in an angular separation of the African and Arabian plates hinged along their northern juncture. This created a large basin continuous with the Indian Ocean to the south but isolated from the Mediterranean. Near the end of the Miocene, between 5.96 and 5.33 million years BP, the Mediterranean experienced a complete or nearly complete closure from the Atlantic, creating a massive dessication event known as the **Messinian salinity crisis** (Krijgsman et al. 1999; Duggen et al. 2003). Not only would this event set a maximum time frame for subsequent marine biogeographic patterns in the Mediterranean, but it also would expand terrestrial dispersal routes between northern Africa and southern Europe.

While the Mediterranean formed through the closure of a once-expansive, circum-equatorial seaway, the Red Sea formed de novo through updoming and rifting of continental landmasses (Wegener 1929; Coleman 1993), which began approximately 30 million years BP (**Figure 8.28**; Ghebreab 1998). The rift system that created the Red Sea seems continuous with Africa's Great Rift Valley. If current trends continue, the tectonic events that created the Red Sea

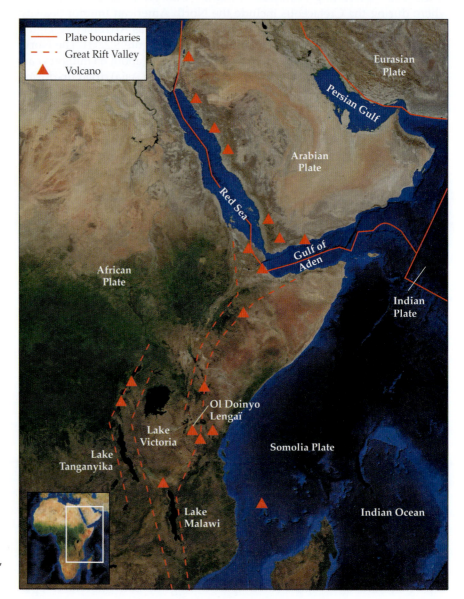

FIGURE 8.28 Tectonic activity in the Great Rift Valley system of eastern Africa. The Great Rift Valley and its Great Lakes formed as the African Plate separated and rotated away from the Somali Plate. This rift represents one arm of a triple junction meeting at the Afar Triangle. In this case, each of three plates is spreading away from the others, with volcanoes, the Red Sea, and the Gulf of Aden forming along plate boundaries.

may be repeated in eastern Africa, with its great lakes eventually connecting to form an extensive inland seaway continuous with the Indian Ocean. The biota of eastern Africa would then drift in isolation, perhaps replicating the original isolation of a similar-sized area that became the island of Madagascar, providing a fascinating, natural biogeographic experiment for future biogeographers.

Dynamics of the Pacific Ocean

Early articles on continental drift focused mainly on continental plates, but soon considerable attention was focused on oceanic plates in the Pacific basin, which contain relatively little emergent land. It may surprise you to learn that our largest ocean, the Pacific, has been getting smaller due to subduction under plates to the west and north (see Figures 8.18 and 8.27); indeed, the Pacific is a remnant of the continuous global ocean, Panthalassa, which once covered two-thirds of the globe when Pangaea existed.

HOTSPOTS AND TRIPLE JUNCTIONS. In the basic model of seafloor spreading, volcanic islands are formed either at midoceanic ridges, in chains perpendicular to the ridges, or behind oceanic trenches, as parallel arcs of islands (see Figure 8.27). In the Pacific Ocean, however, there is no single central ridge, and many of the islands are not associated with ridges or trenches. The Hawaiian Islands are a perfect example: This narrow, linear island chain is located far from any ridge or trench. The oldest emergent island is Midway, and the youngest is Hawaii, which remains volcanically active. J. Tuzo Wilson (1963a) proposed a unique mechanism to explain this pattern, which he called a **hotspot**: a fixed weak spot in the mantle at which magma is released. As the oceanic plate passes over this hotspot, volcanoes are produced at the surface, causing the formation of islands (**Figure 8.29**). The spot is very narrow, so the islands form in a chain, which is linear as long as the plate moves strictly in one direction. The youngest islands in the chain are emergent, and the progressively older ones gradually become submerged (see also Jarrard and Clague 1977; Wagner and Funk 1995; Neall and Trewick 2008).

FIGURE 8.29 A hotspot has produced the volcanic Hawaiian Islands and the Emperor seamount chain in the central Pacific. The hotspot, presently located near the Hawaiian Deep, is presumably responsible for the intense volcanic activity on the island of Hawaii and current development of the new island of Loihi forming southeast of Hawaii. Paleogeologists generally agree that these islands were created as their locations on the Pacific Plate successively drifted over the hotspot (but see alternative arguments cited by Neall and Trewick 2008). The older islands to the west and north were formed first; at the northernmost terminus, the Meiji Seamount is next in line to be subducted into the Aleutian Trench. The sharp bend between the Emperor and Hawaiian chains is dated to 47 million years BP and, while originally interpreted as a change in direction of the drifting Pacific Plate, most recently has been considered to more likely reflect the drifting location of the hotspot itself. (After Neall and Trewick 2008.)

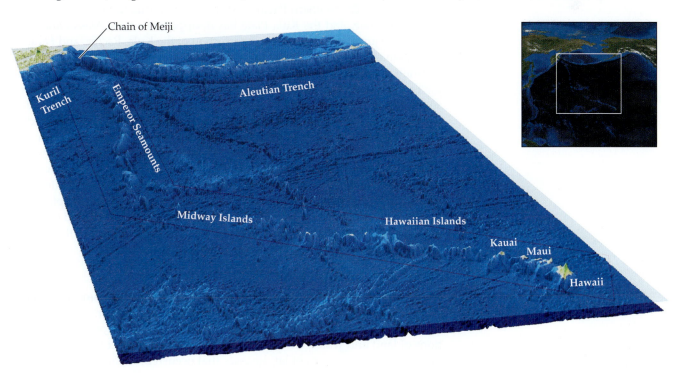

One nice feature of Wilson's model was that it explained the occurrence of most islands in the eastern Pacific. The Hawaiian Islands—many of which are still emergent—are really only the most recent part of a long series extending to the Aleutian Trench, including the submerged Emperor seamount chain (Morgan 1972a; Dalrymple et al. 1973; Neall and Trewick 2008). The seamount closest to the trench is naturally the oldest (about 80 million years BP). Morgan (1972a,b) noted that three very similar chains of emergent and then submerged volcanoes occur in the Pacific basin: the Tuamoto Archipelago (including Easter Island) and the Line Islands chain; the Austral chain and the Marshall–Gilbert–Tuvalu island chain; and a chain from the Cobb seamount opposite Washington through the Gulf of Alaska. Another interesting quality of three of these four chains is that the recent islands are all roughly parallel; then there is a sharp bend (elbow), so that all the older sections are parallel again (see Figure 8.29). This is the pattern you would expect if a rigid plate started moving in a different direction as it passed over a stationary hotspot, although a viable alternative explanation is that the hotspot itself has drifted southeastward over time (Neall and Trewick 2008). The Hawaiian–Emperor elbow is dated at 47 million years BP. Consequently, we can use the distance and orientation of the chains to reconstruct their movements through geological time.

Volcanic islands may also be formed at a **triple junction**, a place where three plates rest against each other to form a complex of trenches (McKenzie and Morgan 1969; see also Henstock and Levander 2003). Volcanoes are often associated with triple junctions. Because each plate has its own rate of subduction into a trench, the triple junction and its associated volcanic activity can shift position through time, forming an island arc. An example of an archipelago formed in this manner is the Galápagos, which originated near the intersection of the Nazca, Pacific, and Cocos plates (see the circled triple junction in Figure 8.18).

Inferred history of the Pacific Basin

Using the data on hotspots and triple junctions, many researchers have attempted to reconstruct the early history of the Pacific basin. The Phoenix Plate was located in the southern Pacific and disappeared due to the enlargement of the Pacific Plate, and the Kula Plate has disappeared under Alaska and Siberia. A triple junction existed at the intersection of the Kula, Pacific, and Farallon plates, which migrated to come into contact with the North American Plate; at 150 million years BP, this junction was just east of the Hawaiian hotspot. Most of the Farallon Plate has been subducted under western North America and is responsible for much of the mountain building there, but small pieces of this plate remain in the vicinity of the Pacific Northwest (the Juan de Fuca Plate and its southern extension, the Gorda Plate) and opposite central mainland Mexico (the Rivera Plate; see Figure 8.18). The Pacific Plate also became fused with a section of mainland Mexico to become Baja California, while its northward movement began the formation of the Gulf of California. This gulf has been spreading over the last 4 million years. The convergence of the Cocos and Caribbean plates has produced the present mountainous backbone of Central America, and the meeting of the Nazca and South America plates induced the formation of the Andean Cordillera, mostly since the Oligocene (24 million years BP).

Paleoclimates and paleocirculations

As discussed in Chapter 3, global circulation patterns of wind and ocean currents are controlled not only by the equatorial-to-polar gradients of solar radiation and temperature, but also by the relative proportions and distributions of land and water. Given the great shifts, splittings, collisions, submergences, and emergences of landmasses that occurred, oceanic currents—and thus regional to global climates—must have changed dramatically through the Phanerozoic Eon, strongly affecting both marine and terrestrial biotas (Harris 2002; Zachos et al. 2001; **Table 8.2**).

Following the final Jurassic breakup of Pangaea, the surface currents and gyres in the Tethys Sea and throughout Panthalassa were replaced by circulation through a warm, tropical, circum-equatorial Tethyan Seaway, which formed during the Early Cretaceous (see Figure 8.22E–G). Later, the Tethyan Seaway was interrupted (about 49 million years BP) when the Indian Plate began to collide with Eurasia, and warm tropical waters were again deflected to high latitudes, initiating a global cooling trend. During the Early Oligocene (about 33 million years BP), Australia and South America had separated enough from Antarctica to establish cold currents among those three continents, deflecting relatively warm currents in the South Pacific, South Atlantic, and Indian oceans, which had until then moderated the water temperatures and coastal environments of southern landmasses (**Figure 8.30**). As it continued to separate from South America and Australia during the Oligocene, Antarctica drifted farther poleward, and cold currents strengthened. Thus, the warm circum-equatorial currents of the Late Cretaceous were replaced by cold circum-Antarctic currents, setting the stage for regional extinctions of thermally intolerant marine invertebrates, buildup of an ice cap on Australia, intensification of the latitudinal gradient in temperature, and global cycles of glaciation and climatic change.

FIGURE 8.30 Continental drift strongly influenced paleocirculations and paleoclimates (see also Tables 8.2 and 8.4). (A) Northward drift of India interrupts the Cretaceous circum-equatorial Tethyan Seaway (see Figure 8.22H) during the Paleocene, deflecting ocean circulation to south of Africa. (B–D) The separation of Australia and South America from Antarctica established the cold Antarctic Circumpolar Current as deep water flowed through the Drake Passage (DP) between Antarctica and South America and over South Tasman Rise (STR) between Australia and Antarctica beginning sometime during the Eocene and Early Oligocene, resulting in the onset of global cooling. (After Lawver and Gahagan 2003.)

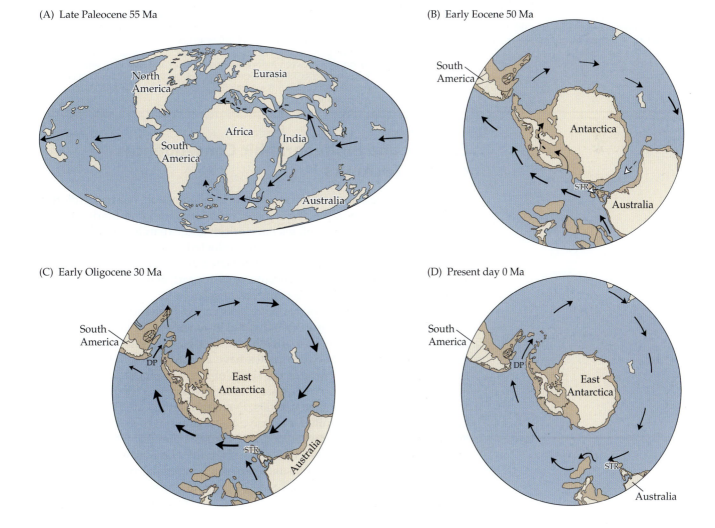

(A) Late Paleocene 55 Ma

(B) Early Eocene 50 Ma

(C) Early Oligocene 30 Ma

(D) Present day 0 Ma

TABLE 8.2 *Key Tectonic Events of the Cenozoic and Their Effects on Marine Waters and Paleocurrents*

Event	Effects
PHYSICAL ISOLATION OF ANTARCTICA	
Early Eocene (50 million years BP)—full deep water separation of South Tasman Rise	Almost complete deep-water isolation of Antarctica. First indications of global cooling at 50–40 million years BP and significant 2° C temperature drops in both the late Middle Eocene and Middle-late Eocene boundary. Further isolation of Antarctic marine biota.
Eocene–Oligocene boundary (37 million years BP)	Major cooling of both surface and bottom waters by 5° C in 75–100 thousand years BP. Onset of widespread Antarctic glaciation.
Opening of Drake Passage (~36–23 million years BP)	Precise time of opening uncertain, but deep-water connections probably not established until 23–28 million years BP. Almost complete isolation of Antarctic marine biota.
Mid Miocene (15 million years BP)—full establishment of Antarctic Circumpolar Current	Development of Polar Frontal Zone. Latitudinal temperature gradient similar to that of today.
CLOSURE OF TETHYAN SEAWAY IN THE MIDDLE EAST	
End of the Cretaceous period (~75–65 million years BP)—vast circumpolar-equatorial tropical ocean	Major westerly flowing equatorial current system. Some faunal differentiation in this vast ocean, but no indications of clear high-diversity foci.
Paleogene (65–23 million years BP)—continuity of the tropical Tethyan Ocean.	Largely homogeneous tropical fauna. Major pulse in coral reef development at end of Oligocene (~23 million years BP), but even then there were marked similarities between western Tethys (Mediterranean) and Caribbean/Gulf of Mexico.
Early Miocene (20 million years BP)—closure of Tethyan Seaway by the northward movement of Africa/Arabia landmass	Westerly flowing tropical current drastically curtailed. Mediterranean Sea excluded from reef belt. Caribbean and eastern Pacific regions become progressively isolated. This marks the beginning of the distinction between Indo-West Pacific (IWP) and Atlantic Caribbean–East Pacific (ACEP) foci. Further development through the Neogene (<20 million years BP) sees relative impoverishment of ACEP and enrichment of IWP.

Climatic and Biogeographic Consequences of Plate Tectonics

Plate tectonics, perhaps more than any other phenomenon, has had profound effects on the biogeographic patterns of both terrestrial and marine biotas. Over geological time, new plates have arisen and expanded at the expense of more ancient ones, which have been subducted and consumed within the deeper layers of the Earth's mantle. Plates have varied dramatically in shape and size and, during their existence, have split apart, slid against each other, or collided to feed the mantle's convective cycle.

Surface features of the plates have also varied tremendously over time and space. Many plates were often divided by extensive shallow seas, which harbored a great diversity of marine life while isolating terrestrial biotas. At various times, bridges between plates have created routes for great biotic invasions. Also, as illustrated in the series of maps in Figure 8.22, extensive mountain ranges often served as barriers to lowland species or as corridors for those adapted to montane habitats. Even during the Late Permian, when most landmasses were united, extensive mountain ranges along the convergence zone between Gondwanaland and Laurasia may have effectively isolated their respective biotas. As noted earlier, the Himalayas and the Urals also formed as a result of collisions between continental plates, effectively isolating many species on either side of these topographic barriers. Other mountain ranges formed when oceanic plates were subducted beneath con-

TABLE 8.2 *(continued)*

Event	Effects
COLLISION OF AUSTRALIA/NEW GUINEA WITH SOUTHEAST ASIA	
Beginning of Cenozoic era (~65 million years BP)—Australia/New Guinea separated from S E Asia by deep-water gateway (~3000 km wide)	Single tropical (Tethyan) ocean; no differentiation in Indian and Pacific oceans.
Paleogene (65–23 million years BP)—progressive closure of Indo-Pacific gateway; northward subduction of Indian-Australian lithosphere beneath the Sunda-Java-Sulawesi arcs	
Mid-Oligocene (30 million years BP)—gap narrowed substantially but still a clear deep water passage formed by oceanic crust	
Latest Oligocene (25 million years BP)—major changes in plate boundaries when New Guinea passive margin collides with leading edge of eastern Philippines—Halmahera-New Guinea arc system	
Early Miocene (20 million years BP)—continent-arc collision closes deep water passage between Indian and Pacific oceans	Major reorganization of tropical current systems; new shallow water habitats appear in Indonesian region.
Early–late Miocene (20–10 million years BP)—continued northward movement of Australia/New Guinea; rotation of several plate boundaries and formation of tectonic provinces that are recognizable today.	Widespread growth of coral reefs in IWP region; huge rise in the numbers of reef and reef-associated taxa; many modern genera and species evolve.
Later Neogene (10 million years BP–present)—continued northern movement of Australia/New Guinea block; in Early Pliocene (4 million years BP)—a critical point when close contact is made with the island of Halmahara.	Warm South Pacific waters deflected eastward at the Halmahera eddy to form Northern Equatorial Countercurrent. Thus warm waters in Indonesian through-flow are replaced by relatively cold ones from the North Pacific. These changes affect heat balance between east and west Pacific and help promote onset of Northern Hemisphere glaciation.
UPLIFT OF CENTRAL AMERICAN ISTHMUS (CAI)	
Mid Miocene (~15–13 million years BP)—sedimentary evidence of earliest phases of shallowing.	Still a deep water connection through CAI.
Latest Miocene–Early Pliocene (6–4 million years BP)—continued shallowing to <100 m depth.	Major effect on oceanic circulation; Gulf Stream begins to deflect warm, shallow waters northward along eastern U.S. seaboard. Initially leads to some warming in northern mid to high latitudes; however, this water is slightly hypersaline and sinks when it reaches the highest latitudes to form the North Atlantic Deep Water. This then spreads into both the South Atlantic and Pacific Oceans to initiate a major "conveyor belt" of deep ocean circulation.
Mid Pliocene (3.6 million years BP)—further closure of CAI	North Atlantic thermohaline circulation system intensifies; Arctic Ocean effectively isolated from warm Atlantic. This leads eventually to the onset of Northern Hemisphere glaciation at 2.5 million years BP
Late Pliocene (~3 million years BP)—complete closure of CAI	Final separation of shallow water Atlantic–Caribbean and eastern Pacific provinces. A further important effect of the closure of CAI is the reverse flow of water through the Bering Strait; this in turn influences Pliocene–Pleistocene patterns of thermohaline circulation in Arctic and North Atlantic oceans.

Source: After Crame 2004.

tinental plates, which caused great, upward buckling of the land surface (as in the Andes) or the welding of seamounts and guyots onto the continental plate (as in the Olympic Mountains). Finally, other mountain ranges, including island chains, formed as a result of volcanic activity associated with subduction zones (as in the island arcs between the converging Australian, Eurasian, and Pacific plates; see Figure 8.27) or as plates drifted over hotspots in the mantle (as in the Hawaiian Island chain; see Figure 8.29).

Biogeographic consequences of these tectonic events are manyfold, but most stem from the effects of plate movements on the configurations of landmasses and marine basins, which determine degrees of biotic isolation and

TABLE 8.3 *Summary of Proposed Causes of the "Big Five[a]" Events as well as Lesser Phanerozoic Mass Extinction Events*

Event	Bolide impact	Volcanism	Cooling	Warming	Marine regression	Anoxia/Marine transgression
Late Precambrian						Strong
Late Early Cambrian					Strong	
Late Cambrian biomeres			Possible			Strong
End-Ordovician			Possible		Possible	Possible
Frasnian–Famennian			Possible		Possible	Possible
Devonian–Carboniferous			Possible			
Late Guadaloupian					Possible	
End-Permian		Strong		Strong	Strong	Possible
End-Triassic		Strong		Possible	Strong	Possible
Early Toarcian		Strong		Possible		Strong
Cenomanian–Turonian				Possible		Strong
End-Cretaceous	Strong	Strong	Possible	Possible	Strong	Possible
End-Paleocene		Possible		Strong		Strong
Late Eocene	Possible		Strong			

Strong link	Possible link

Source: After Hallam 2004.
[a]"Big Five" events are in bold font.

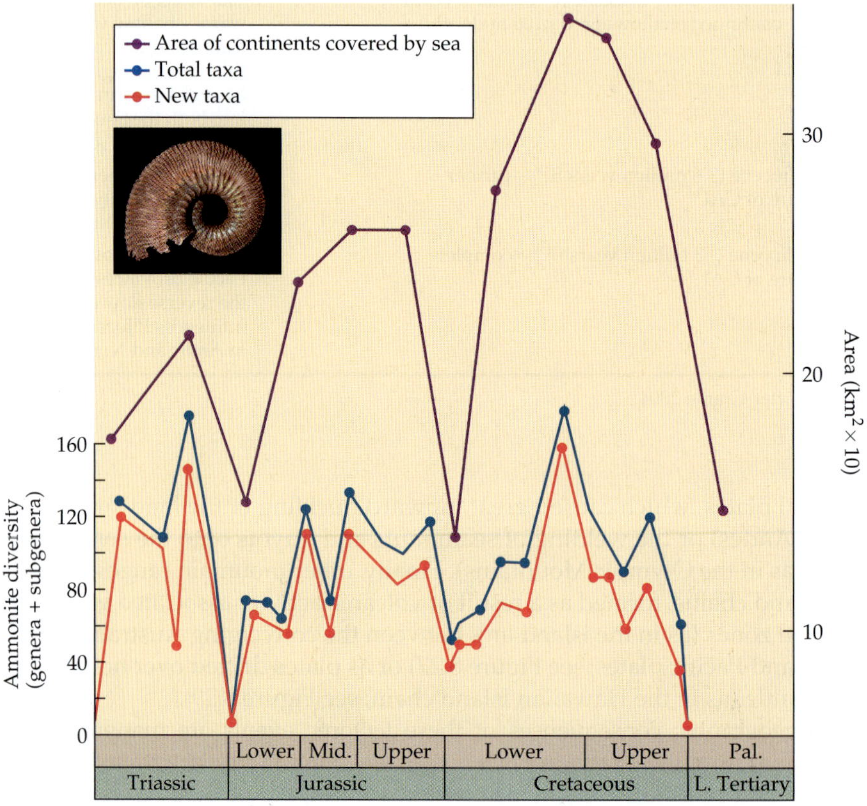

FIGURE 8.31 Many extinction events in the marine realm may be linked to regression and transgression cycles in sea level. The transgressions increased—sometimes dramatically (see Figure 8.23)—the area of the continents covered by shallow epeiric seas. The diversity of the once-dominant ammonites, for example, appears to be directly correlated with the area of Earth's epeiric seas. (After Hallam 1983; photo courtesy of D. McIntyre.)

opportunities for biotic exchange or, more indirectly, influence regional and global climates (see Chapters 9 and 12). One remarkable outcome of tectonics is the phenomenon of mass extinctions, with many such events having been attributed to tectonically driven cycles of regression and transgression of marine waters across continental epeiric seas (Hallam 2004). Regression cyles lead to vast decreases in area of shallow water environments, whereas transgressions can bring masses of anoxic water across vast expanses of shallow seas, and each of these events can result in mass extinctions, perhaps in conjunction with additional potential drivers of high-impact environmental change (**Table 8.3**).

Since the early development of the field, biogeographers have been well aware of the effects of area, isolation, and latitudinal position on diversity. Across a great variety of taxa, time periods, and ecosystems, diversity increases with area, while diversity and similarity among biotas decrease with increasing isolation. As Hallam (1983) has shown, the diversity of ammonites—a once-dominant but now extinct group of marine invertebrates—was strongly correlated with the area of epeiric seas (**Figure 8.31**). Other paleontological work with marine invertebrates shows that when shallow-water marine communities on either side of the North Atlantic drifted apart, their similarity (i.e., the proportion of shared species) decreased (**Figure 8.32**; Fallaw 1979).

Precipitation patterns, paleocurrents (see Figure 8.30), wind, and temperature also must have varied tremendously as continents split, converged, or drifted between the poles. On a global scale, the proportion of landmasses at different latitudes has been nearly reversed since the Silurian (425 BP), when most landmasses were situated south of the Equator (**Figure 8.33**). When landmasses drifted over the poles, they triggered cycles of glaciation and changes in sea level (**Table 8.4**). Glacial cycles of the Pleistocene are now familiar to most biologists, but these events were dwarfed by the magnitude of some earlier glaciations and fluctuations in sea level. Similarly, the formation

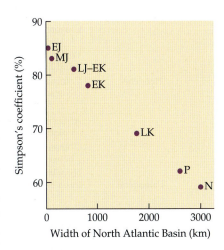

FIGURE 8.32 Seafloor spreading in the North Atlantic caused separation and divergence of shallow-water marine invertebrates during the Mesozoic and Cenozoic (Fallaw 1979). As the North Atlantic widened and the coastal environments on either side became more isolated, the taxonomic similarity of their biotas decreased. Simpson's coefficient of similarity = 100 × (the number of taxa in common/the number in the sample with the smallest number of taxa). EJ, Early Jurassic; MJ, Mid Jurassic; LJ–EK, Late Jurassic to Early Cretaceous; EK, Early Cretaceous; LK, Late Cretaceous; P, Paleogene; N, Neogene. (After Hallam 1994.)

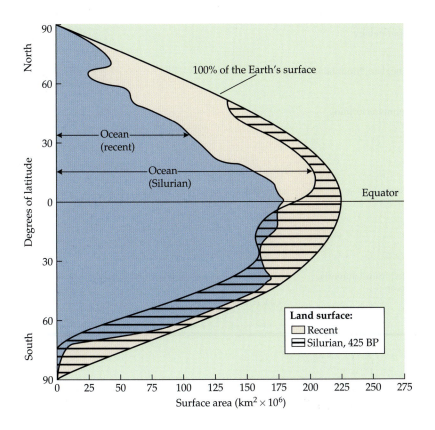

FIGURE 8.33 Continental drift has substantially altered the latitudinal distribution of landmasses and, in turn, has caused dramatic changes in regional and global climates. (After Rosenzweig 1995.)

TABLE 8.4 *Summary of the Tectonic, Climatic, and Biotic Events of the Phanerozoic and Latest Precambrian*

MY BP	Period	Tectonic events	Marine events
		Closure of Panamanian Isthmus	Complete separation of Atlantic/ Caribbean from eastern Pacific; flow through Bering Strait reverses
		Wallacea forms between Australia and Eurasia	Messinian salinity crisis; Red Sea forms; Arctic Ocean effectively isolated from warm Atlantic
	Neogene	Colorado Plateau uplifting; Andean and Sierra Nevada Mountains uplifting rapidly	Northern Equatorial Countercurrent forms; deep ocean "conveyor belt" forms
			Deep water connection through Central American Isthmus
		Africa collides with Eurasia; Central American volcanic arc system coalescing	Tethyan Seaway closes; Mediterranean Sea forms
23		Tibetan Plateau uplifting; deep water passage between Indian and Pacific oceans closes	New shallow water habitats in Indonesian region Drake Passage opens; Antarctic Circumpolar Current forming
		Himalayas uplifting; Australia drifts north rapidly	
	Paleogene	India begins to collide with Eurasia	Deep water separation of South Tasman Rise
			Closure of Indo-Pacific gateway
65		Australia drifts away from Antarctica; India and Seychelles drift north rapidly	Single tropical (Tethyan) Ocean
		India and Seychelles separate from Madagascar; Beringia forms	Interior seaways fragment most continents
	Cretaceous	South America and Africa separation complete; Australia begins separation from Antarctica	
		India–Madagascar separate from Antarctica; South America and Africa rift from south to north	
146			
		West and East Gondwanaland separation complete	Circum-equatorial Tethyan Seaway opens
	Jurassic	Pangaea breakup complete	
			Central Atlantic Ocean and Turgai Sea form
200		Pangaea rifting begins	
	Triassic		
251			Opening of Tethys Ocean
	Permian	Deserts in western Pangaea separated by Central Pangaean Mountains	
299	Pennsylvanian (Upper Carboniferous)	Western half of Pangaea forming; Ural Mountains forming	
318	Mississippian (Lower Carboniferous)	Appalachian and Variscan Mountains forming	
359			
	Devonian		
416			
	Silurian	Euramerica continent forms; Australia near the Equator	Panthalassic and Paleo-Tethys oceans First jawed fishes
444			
	Ordovician		
488			
	Cambrian		
542		Gondwanaland forms, separated from Laurentia, Baltica, Siberia by Iapetus Ocean	Continents flooded with shallow seas
	Proterozoic	Rodinia breakup	Glaciation, icehouse climate
		Rodinia formation	
2500			
	Archean		
4000			

Source: Sea levels after Hallam 1997.

TABLE 8.4 *(continued)*

Global sea level	**Biotic events**
	Late Pleistocene Mass Extinction: plankton, marine invertebrates, land mammals
	Global expansion of C_4 grasslands; first hominids
	Great increase in Indo–West Pacific coral reefs and associated taxa Indo–West Pacific vs. Atlantic/Caribbean–East Pacific marine differentiation
	Major coral reef development
	Adaptive radiation of mammals *End-Cretaceous Mass Extinction*: foraminifera, coccolithophorids, benthic invertebrates, cephalopods, dinosaurs Adaptive radiation of angiosperms
	First angiosperms (flowering plants); conifers become dominant gymnosperms First birds (e.g., *Archaeopteryx*)
	Diversification of marine reptiles, including plesiosaurs, ichthyosaurs, crocodiles Adaptive radiation of ammonoid and belemnoid marine mollusks; bony fishes First dinoflagellates and calcareous nannoplankton *End-Triassic Mass Extinction*: reefs, ammonites, conodonts, benthic bivalves and brachiopods; terrestrial plants First flying vertebrates (pterosaurs) Forests dominated by ferns and gymnosperms First mammals First dinosaurs, turtles, marine reptiles, and frogs *End-Permian Mass Extinction*: reefs, echinoderms, brachiopods, foraminifera; terrestrial vertebrates and insects
	First mammal-like reptiles
	First reptiles
	Equatorial coal-swamp forests
	Late Devonian Mass Extinction: reefs, marine invertebrates and vertebrates First terrestrial forests; first terrestrial tetrapod vertebrates First insects First vascular land plants
	End-Ordovician Mass Extinction: graptolites, conodonts, trilobites, brachiopods First jawless fishes First vertebrates
	Great radiation of marine invertebrates with skeletons
	Adaptive radiation of soft-bodied invertebrates
	Prokaryotes only Origins of life

Note: Shading marks icehouse climate, periods of global cooling with (dark blue) or without (light blue) evidence for glaciation.

of Pangaea isolated most of its interior land area from the buffering effects of the ocean, increasing temperature fluctuations and aridity and, in turn, causing the drying of shallow seas.

We have tried to summarize the major tectonic, climatic (including currents and sea levels), and biotic events of the Phanerozoic in Tables 8.2 and 8.4. Admittedly, this account is far from complete or precise, but it should be instructive. While the mass extinction at the end of the Cretaceous was caused by an asteroid impact, most other major extinction events appear to have been associated with tectonic events (Table 8.3; see also Ward 2001; Hallam 2004). The challenge of establishing cause and effect across geological time remains a daunting one indeed, but we should note that Cuvier himself attributed mass extinctions to great changes in relative sea level. We should, however, add a note of caution: Mass extinctions, as spectacular and intriguing as they appear, tend to attract a disproportionate—and perhaps unwarranted—amount of attention. As Stanley (1987) noted, only a small minority of all extinct species were victims of the major mass extinctions: "Most of the global turnover of species was very gradual, not catastrophic." On the other hand, tectonic events also set the stage for major radiations of new species, which are evident throughout the fossil record. As noted earlier, many terrestrial vertebrates evolved and spread across the relatively continuous landmasses of Pangaea during the Late Permian. However, their major radiations did not take place until Pangaea split, isolating its biotas, which were then free (if they survived) to diversify under different selective pressures and in the absence of gene flow. Few contemporary scientists doubt that tectonic, climatic, sea level, and biogeographic phenomena are causally related (see Fischer 1984).

Our extensive discussions of plate tectonics, ecology, and evolution in the early chapters of this book are recognition of their fundamental importance to biogeography. Again, all of these disciplines are intricately related, and any distinction we attempt to make is surely an arbitrary one. In the next chapter, we will take a closer look at the relationship between climate and biogeographic events during a relatively thin slice of time—the Pleistocene Epoch (roughly, the past 2 million years). In Chapters 10, 11, and 12, we will return to the theme that is central to both historical biogeography and evolutionary biology—that the Earth and its life-forms have evolved in concert (see Lieberman 2000), and that the best approach to understanding the geography of nature, past and present, is to develop a better understanding of the histories of place and of life.

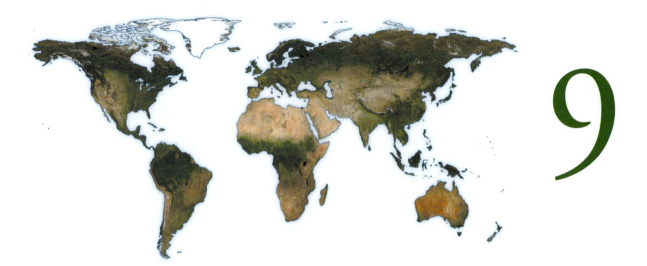

GLACIATION AND BIOGEOGRAPHIC DYNAMICS OF THE PLEISTOCENE

Just as layers of rocks and fossils provided William Smith and subsequent generations of geologists with a wonderfully rich record of Earth's geological and evolutionary history (Chapter 3), the layers of sediment along the ocean floors and of ice that now comprise the Arctic and Antarctic ice caps provide an invaluable chronicle of Earth's climatic history—one that has been marked by great fluxes in global temperatures, precipitation, atmospheric and oceanic currents, and indeed, all conditions of terrestrial and marine environments. These changing conditions are recorded in various forms of proxy data that can be extracted from deep cores drilled through the sediments and from accumulations of ice. Among these proxies, some of the most important indicators of past climates are dust particles (which provide a measure of the relative aridity of regional to global climates), along with tiny organisms (e.g., planktonic foraminifera, which provide samples of the chemical composition of ocean waters) and air bubbles (which provide samples of atmospheric gases, e.g., CO_2 and other greenhouse gases, and the isotopic composition of water).

The isotopic composition of these samples of sea life, water, and the atmosphere, in particular, can serve as invaluable proxies of past climates. Recall that oxygen and hydrogen occur in alternative forms as stable isotopes—molecules having the characteristic number of protons for that element (oxygen with 8 and hydrogen with 1) but varying in the number of neutrons and, therefore, their atomic weights. The most common stable isotopes of oxygen are ^{16}O (99.75 percent) and ^{18}O (0.20 percent), with 8 and 10 neutrons, respectively. The most common isotopes of hydrogen are ^1H (99.8 percent) and ^2H (0.01 percent), with 0 and 1 neutron, respectively. Because the lighter isotopes evaporate more readily, increased evaporation during warming periods results in increased ratios of heavy to light isotopes (^{18}O / ^{16}O; ^2H / ^1H) in seawater, especially at the Equator, and in reduced ratios of these isotopes (greater deficiencies of heavy isotopes) in atmospheric vapor and in rain and snow, this effect intensifying toward the poles as water repeatedly cycles through the hydrological cycle. The isoto-

FIGURE 9.1 Changes in average global temperatures of the Earth at three different time scales—(A) Phanerozoic Eon, (B) Cenozoic Era, and (C) Pleistocene Epoch—based on δ^{18}O, that is, the ratio of two isotopes of oxygen ([18]O / [16]O) in ice cores. (D) While the drivers of these changes varied depending on the time period, the climatic dynamics of the Pleistocene appear to have been largely a function of changes in Earth's orbit, which influenced the planet's heat budget. (A–C after Veizer et al. 1999; Zachos et al. 2001; Lisiecki and Raymo 2005; D from Bloom 2010, after Laskar et al. 2004 and Huybers 2006.)

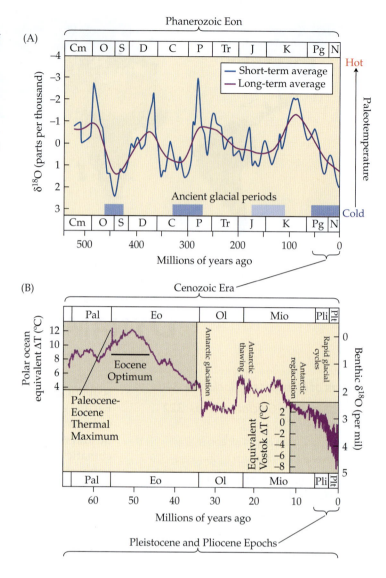

pic signatures are then fixed in the shells of marine organisms as they die and drift down to join the benthic sediments, and they are also captured in ice as it accumulates in layers within the polar ice caps.

These isotopic proxies, along with a diverse meteorological tool kit of other climatic indicators, including accumulations of various types of dust and sediments on land and beneath the sea, geomorphic (ancient landforms) and other direct effects of glaciers, the relative widths of tree rings, and the depths and latitudinal positions of ancient coral beds and other biological indicators of changes in ocean currents, temperatures, and past sea levels (see reviews in Anderson et al. 2007; Bloom 2010), enable reconstructions of Earth's climatic history through the Phanerozoic Eon—the age of obvious life. This record (**Figure 9.1**) reveals major swings in temperature and all other features of ancient climates that must have strongly influenced the diversity and distributions of life across the planet. The dynamics of paleotemperatures, in particular, include a mix of irregular and regular or cyclical signals, the latter with periodicities ranging from the familiar seasonal to millennial and multi-millennial time scales. While explanations for the Earth's climatic history over this broad time period are understandably quite complex, the principal drivers of climate change distill down to those factors influencing the temporal and spatial variation of insolation (the amount of solar radia-

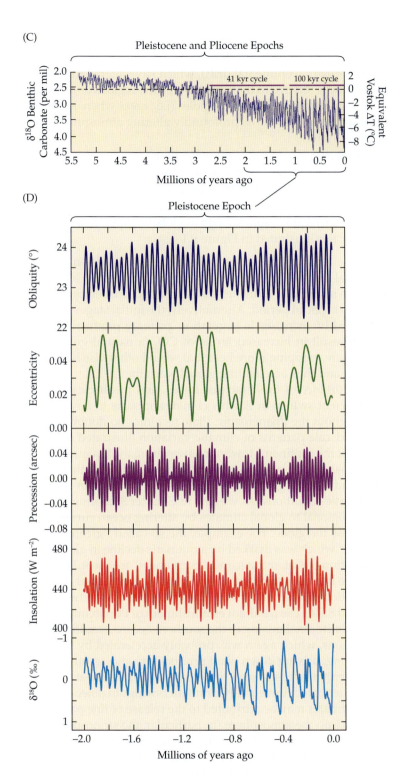

tion absorbed by the Earth's surfaces) and the subsequent redistribution of that heat across the atmosphere and oceans. Thus, these factors include the following:

1. Variation in radiation emitted by the sun

2. Variation in the absorptive and reflective nature of the atmosphere, which at times in the past has been strongly influenced by volcanic activity, and at other times by the impact of asteroids

3. Differences in the absorptive and reflective nature of Earth's surfaces (soil, vegetation, water, snow, and ice)

4. Plate tectonic processes (continental drift), which have altered the positions of land surfaces relative to the Equator (i.e., the regions of greatest absorption of solar radiation) and to the poles (i.e., areas where ice may accumulate and persist), as well as profoundly altering atmospheric and ocean circulation systems

5. Variation in the geometry of Earth's orbit, which have affected its heat budget by altering both the proximity of our planet to the sun and the intensity of solar radiation received by different regions of the Earth

6. Feedback interactions among the above factors and processes, which likely account for incredibly rapid swings in global temperatures obvious at all times scales illustrated in Figure 9.1

Such global climatic feedback processes include the accumulation of snow and ice during initial stages of cooling, which in turn increase the albedo, or reflectivity, of Earth's surfaces, reducing insolation and accelerating global cooling transition to a glacial period. Alternatively, the initial stages of global warming are marked by increases in vegetation, primary productivity, and respiration, which result in increased levels of CO_2 (a natural greenhouse gas) in the atmosphere, trapping more heat that would otherwise have been lost from the planet and thus accelerating global warming.

The relative importance of these factors, forces, and feedback processes in influencing the great swings in Earth's climates appears to have varied substantially throughout the Phanerozoic. Plate tectonics no doubt strongly influenced the broadscale patterns of climatic variation illustrated in Figure 9.1A, with drifting of the continents toward or away from the Equator, as well as the varying availability of land at or near the poles, affecting the Earth's total heat budget and altering geographic and seasonal variation in temperature and other climatic conditions. As we saw in Chapter 8, continental drift can also influence regional to global climates by altering ocean currents that distribute heat across the globe. For example, the so-called "greenhouse to icehouse transition," which was marked by a dramatic decline in global temperatures and initiation of Antarctica's now "permanent" ice sheet 24 million years ago, may have been triggered, at least in part, by continental drift that completed the isolation of Antarctica from other remnants of Gondwanaland (see Figures 9.1B and 8.30). The resultant establishment of the circum-Antarctic current reduced latitudinal mixing of the southern oceans, resulting in the onset of the glaciations, which again was probably accelerated by feedback processes (e.g., cooling, driving down plant productivity and, in turn, reducing the levels of greenhouse gases; see Prothero et al. 2002; Ivany et al. 2006). Some of the more attenuated, but equally dramatic, swings in paleotemperatures and ancient climates may be attributed to the release of fine particulate matter and of gases into the atmosphere (in particular, the stratosphere), either from asteroids colliding with the Earth or from surges in volcanic activity at regional to global scales. Recall that the end-Cretaceous extinction of the dinosaurs may have been caused by an asteroid impact that darkened the skies and brought on a "global winter" that lasted for decades. On the other hand, one of the likely contributors to the spike in global temperatures about 55.8 million years ago—the Paleocene–Eocene Thermal Maximum (see Figure 9.1 B)—was increased volcanic activity near Greenland and other regions of the Northern Hemisphere. In addition to spewing more CO_2 into the atmosphere, volcanic activity and its initial warming of waters in the northern oceans may have released methane previously trapped in shallow-water sediments, again increasing the total volume of greenhouse gases and accelerating global warming. Volcanic activity and this CO_2- and methane-driven greenhouse feedback may have also contributed to the subsequent, and more prolonged, thermal maximum of the Eocene (the Eocene Optimum; see Figure 9.1B).

Yet these geological events, and in particular continental drift, were not the principal drivers for the more regular cycles of climate change that characterize the latter stages of this climatic record, that is, the Quaternary Period. The Quaternary consists of two periods, the Pleistocene, which is currently adjudged by the International Union of Geological Sciences to have begun about 2.588 million years ago (+/− 5000 years), and the present interglacial, or Holocene, which commenced about 12,000 years BP. The period of time from about 2.6 million years ago to the present has simply been too short for appreciable drifting of the continents. This is not to say that plate tectonics was unimportant and, indeed, it appears that the positions of the continents during this period were precursory to the impact of other drivers of climatic change during the Pleistocene. Recall from Chapter 8 that although the total surface of Earth's terrestrial plates has changed little throughout the Phanerozoic Era, their relative position has (see Figure 8.33). Currently, approximately 65 percent of the planet's land surface lies north of the Equator, creating a strong asymmetry of heat budgets between the Northern and Southern Hemispheres (land absorbs more solar radiation than does sea surface). The configuration of ocean basins, which in turn influenced circulation of heat energy through the hydrosphere, also contributed to the heightened instability of global climate during the Pleistocene.

The most likely explanation for the overall pattern of heating and cooling during the Pleistocene is the cyclical changes in Earth's orbit about the sun that are driven by changes in the positions of the planets and their magnetic pull on the Earth (**Figure 9.2**). This hypothesis dates back to early works by mid-nineteenth century scientists, including the French mathematician Joseph Alphonse Adhemar (1842) and the self-educated Scottish scientist James Croll (1875). These early ideas were then further developed by the Serbian mathematician and geophysicist Milutin Milankovitch during the early decades of the twentieth century (Milankovitch 1920, 1930). The principal features of the Earth's orbit that affect its total heat budget and the temporal variation and geographic distribution of that heat are its eccentricity (how its orbit deviates from that of a perfect circle), its obliquity (the tilt of the Earth on its axis), and what is termed the precession of the equinoxes (the orientation of the Earth's axis in the solar system, which affects whether periods of most direct solar radiation—summer solstice for a particular hemisphere—occur when it is at perihelion, aphelion, or some intermediate position in its orbit around the sun; see Figure 9.2). As Milankovitch hypothesized, the repeated but comparatively brief events of global warming during the Pleistocene correspond with periods when the summer solstice in the Northern Hemisphere (which has most of the land and therefore absorbs more solar radiation and reradiates it as heat) occured when the Earth was closest to the sun (i.e., in perihelion; see Figure 9.2D). This orbital configuration would also create the conditions for relatively cold winters (winter solstice in the Northern Hemisphere when the Earth was farthest from the sun), with the cold reducing evaporative infusion of moisture into the atmosphere and, thus, reducing snowfall. The overall result was snow and ice melt during summers in the Northern Hemisphere that far exceeded accumulation during winter, thus triggering the onset of the current interglacial period (see Kohler et al. 2010; Lourens et al. 2010). Calculations indicate that, taken separately, the so-called astronomical forcing we have discussed is insufficient to account for the magnitude and rapidity of the changes captured in the

FIGURE 9.2   (A–C) Changes in the characteristics of Earth's orbit around the sun were among the key drivers of global climate change through the Phanerozoic Eon, especially during the Pleistocene Epoch when approximately 65 percent of Earth's land surfaces (which absorb substantially more solar radiation than sea surfaces) had drifted to the Northern Hemisphere. (D) As Milutin Milankovitch hypothesized in the early 1900s, when the Northern Hemisphere experienced its winter solstice at aphelion of the Earth's orbit and its summer solstice at perihelion, winter snowfall was exceeded by summer snowmelt, marking the onset of an interglacial such as the one that began some 10,500 years ago. Eventually, when the characteristics of Earth's orbit combine to reverse these conditions, another glacial period will begin. (Portrait of Milankovitch by Paja Jovanovic.)

(A) Eccentricity (ellipticity of orbit)
    Cyclic period ~ 96,000

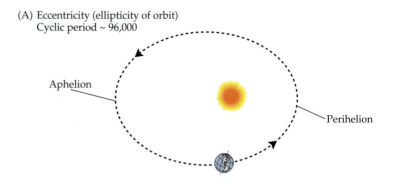

Eccentricity = (A− P)/(P + A)		Distance to sun (millions km)	
		Perihelion	**Aphelion**
Minimum	0.005	148.9	150.4
Current	0.0167	147.2	152.2
Maximum	0.0617	140.4	158.9

(B) Obliquity (orbit tilt)
    Cyclic period ~ 41,000

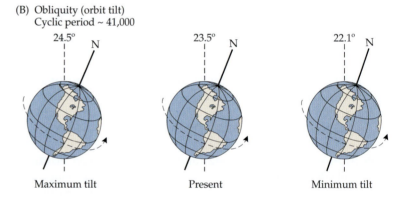

(C) Precession
    Cyclic period ~ 22,000

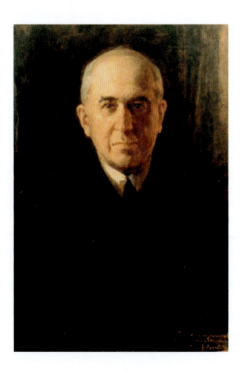

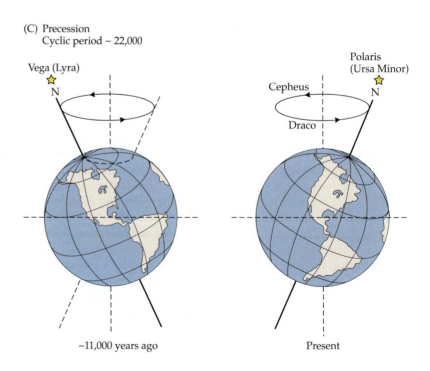

(D) Precession of the solstices and equinoxes

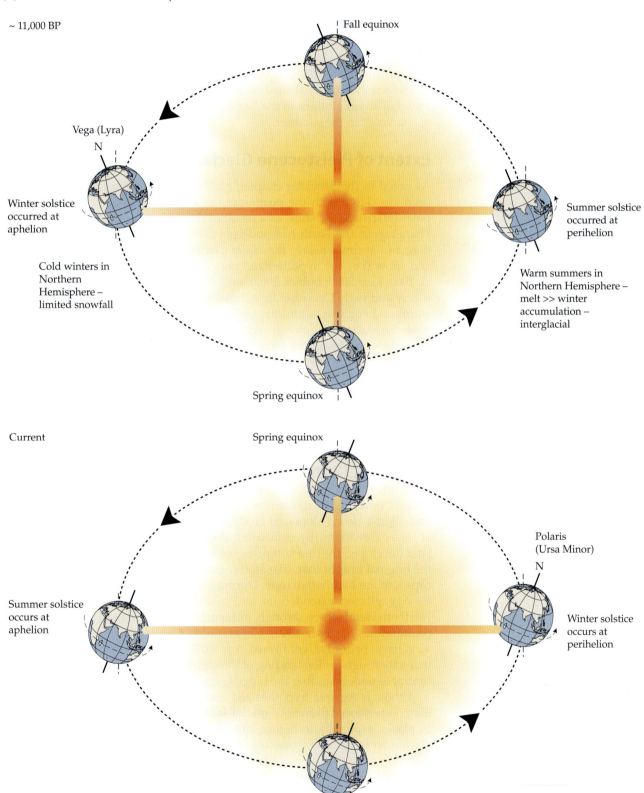

~ 11,000 BP

Fall equinox

Vega (Lyra)
N

Winter solstice
occurred at
aphelion

Summer solstice
occurred at
perihelion

Cold winters in
Northern
Hemisphere –
limited snowfall

Warm summers in
Northern Hemisphere –
melt >> winter
accumulation –
interglacial

Spring equinox

Current

Spring equinox

Polaris
(Ursa Minor)
N

Summer solstice
occurs at
aphelion

Winter solstice
occurs at
perihelion

Fall equinox

proxy records. Thus, explanations for the dramatic swings between glacial and interglacial conditions also include feedback effects (e.g., reduced albedo as the ice sheets waned, increased production of CO_2 as plant productivity increased). Although the full explanation is a bit more complex than this, with eccentricity, obliquity, and precession varying over unequal periods of approximately 96,000, 41,000, and 22,000 years, respectively, interglacial conditions eventually waned as patterns of insolation reversed to produce snowfall accumulations in the Northern Hemisphere that outpaced the rates of snowmelt in the summer.

Extent of Pleistocene Glaciation

As noted in the previous chapter, glacial events within Earth history have been associated with the positioning of large landmasses over or near the poles (Harris 2002). Thus, as a result of the drifting of continents during the Phanerozoic, the Earth has undergone several periods of extensive glaciation, and at times nearly the entire planet may have been covered with ice and snow (i.e., the hypothesized periods of "snowball Earth"; see Hyde et al. 2000; Kerr 2000; Lubick 2002; see Table 8.2). Throughout much of the Mesozoic and early Cenozoic Eras, however, the global climate was relatively warm and equable, with little variation across seasons or latitudes. In contrast, the glacial–interglacial cycles that characterized the late Neogene were dramatic episodes in the biogeographic and evolutionary history of Cenozoic biotas, many of which developed and radiated during an age known as the Mid-Eocene "sauna" (Beard 2002; Bowen et al. 2002).

The origins of Pleistocene biogeography may be traced back to Swiss-born (Jean) Louis Rodolphe Agassiz. After studying comparative anatomy under the renowned paleontologist and systematist Georges Cuvier, Agassiz emigrated to America, but he returned to his homeland in 1836 to study glaciers along the slopes of the Alps. Agassiz was struck, not just by the immense mass of the glaciers, but also by their dynamics—sometimes nearly imperceptible, sometimes cataclysmic. From this, he inferred that glaciers are, over long periods of time, highly dynamic—constantly growing at the higher elevations and sliding down slopes to erode soils and strongly influence vegetation at lower elevations. By combining these and related observations, Agassiz later developed the first comprehensive theory of glaciation, which included the original insight that, like alpine glaciers, those forming Earth's polar ice caps also are highly dynamic. During an earlier period, which he called the Ice Age, glaciers extended far into subarctic and temperate latitudes of the Northern Hemisphere, strongly influencing distributions of plant and animal communities (Agassiz 1840). Generations of paleontologists and climatologists have, of course, confirmed Agassiz's prescient inferences and the fact that a thorough understanding of the climatic dynamics of the Pleistocene is central to understanding the biogeographic dynamics of its biotas.

Although Agassiz hypothesized one "Ice Age," it is now clear that the Earth experienced numerous **glacial–interglacial cycles** during the Pleistocene. In places, especially in higher latitudes of the Northern Hemisphere, these glaciers were incredibly massive sheets of ice. They often were 2 to 3 km thick, and their mass was so great that it deformed the underlying lithosphere by 200 to 300 m. At their maximum extent, these ice sheets covered up to a third of the Earth's land surface. At the height of the most recent glacial period, ice sheets in the Northern Hemisphere extended from the Arctic southward to cover most of North America and central Asia to approximately 45° N latitude (**Figure 9.3**). During these periods of glacial maxima, prevailing winds shifted, sometimes blowing wet oceanic air masses deep into the interior of some con-

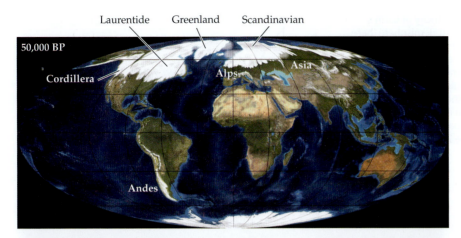

FIGURE 9.3 Variations in Earth's orbit around the sun frequently combined to produce rapid declines in insolation to the Northern Hemisphere which, in turn, triggered periods of rapid and extensive glaciation, with ice sheets as much as 3 km deep and extending far into temperate latitudes, and to even lower latitudes across the mountainous regions of northern America, Europe, and Asia (see Figure 9.4). (Courtesy of Ron Blakey, Northern Arizona University.)

tinents; therefore, glacial maxima in now arid regions are also referred to as **glacio-pluvial** (ice-rain) periods. In contrast, tropical regions tended to be drier during glacial maxima. Glacial periods alternated with **interglacial** periods, when climate warmed and ice sheets retreated. As the recent paleontological record tells us, these events had profound effects on both terrestrial and marine biotas, including those far removed from the glaciers.

Most authors focus on Pleistocene glaciation episodes in the Northern Hemisphere because over 80 percent of glacial ice occurred there (**Figure 9.4**). In the Southern Hemisphere, where much less land occurs in temperate and

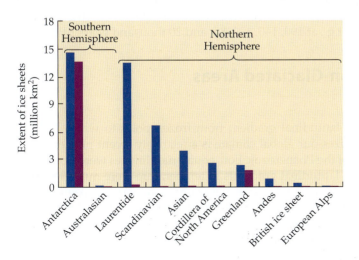

FIGURE 9.4 Although previous periods of glaciation during the Pleistocene may have been even more extensive (covering as much as 45 million km² of Earth's land surface), the last glacial maximum of around 18,000 BP was a major event, with ice sheets covering roughly 40 million km² (see Figure 9.3). The asymmetry in glaciation between Southern and Northern Hemispheres derives largely from the current distribution of approximately 65 percent of the planet's land surfaces (which heat and cool more rapidly than the oceans) north of the Equator. Red = current coverage of glaciers; blue = that during the last glacial maximum. (See Figure 9.3.) (Modified from Anderson et al. 2007, after Smithson et al. 2002.)

FIGURE 9.5 The relationship between eccentricity of Earth's orbit, solar radiation absorbed by the Northern Hemisphere (here, insolation at 65° N), and glaciation is evident in the reconstruction of these cyclical changes during the Late Pleistocene Epoch (200,000 years ago to present). Simulation models of future changes in Earth's orbit can be used to predict changes in these parameters over the next 120,000 years, although the duration of the interglacial predicted by this simulation is uncharacteristically long (throughout the Pliestocene, interglacials seldom lasted more than 10,000 years). (After Berger and Loutre 2002.)

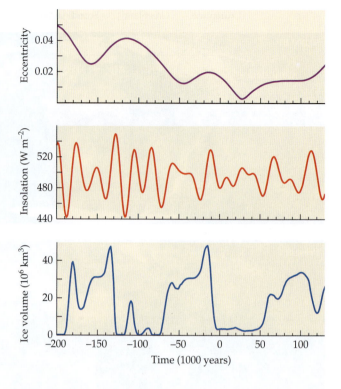

subarctic regions, glaciation was mostly confined to high elevations at high southern latitudes, such as the Central Plateau of Tasmania and the New Zealand Alps (see Flint 1971). The Andean cordillera was glaciated, but the greatest ice coverage was in Chile and Argentina. Mainland Australia was unglaciated except for the Victorian Alps, and Africa lacked glaciation except in the Atlas Mountains of the extreme northwestern corner and in the highest mountains of eastern Africa. At the peak of the last glacial maximum, ice sheets covered approximately 40 million km^2 of Earth's land surface and, assumiung an average of over 2 km thickness, their total volume may be estimated at about 84 million km^3. During the current interglacial, ice sheets have all but disappeared except for those of Greenland and Antarctica (see Figure 9.4), dwindling to just over 15 million km^2 surface area and a volume of just 32 million km^3 (Anderson et al. 2007; after Yokoyama et al. 2000). The relationships between characteristics of the Earth's orbit about the sun, insolation in the Northern Hemisphere, and glaciation are illustrated in the graphs of **Figure 9.5**. Note in particular that each period of rapid expansion in volume of the glaciers was preceded by a rapid decline in Northern Hemisphere insolation (e.g., at 190, 140, 120, 80, and 20 thousand years BP).

Effects on Non-Glaciated Areas

Temperature

The steep thermal latitudinal gradient from frozen poles to warm Equator that now characterizes our global climate is a relatively recent phenomenon. Throughout most of the Phanerozoic Eon, equatorial climates were tropical, as they are today, but latitudinal gradients in temperature were much less pronounced, and terrestrial climates were generally more equable. Beginning in the Miocene, global climates gradually began to cool and become drier. Extension of the Antarctic ice sheet and intensification of oceanic and atmospheric circulation during the Mid-Miocene (15 million years ago) established a strong

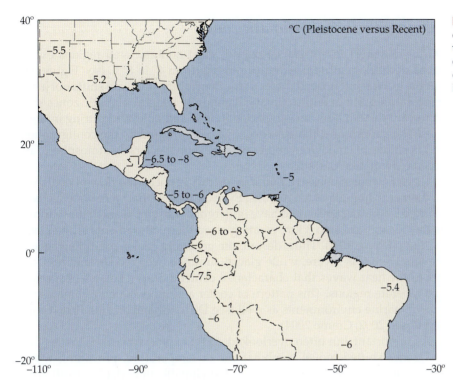

FIGURE 9.6 Glacial cycles of the Pleistocene influenced regional climates far from the edges of the glaciers. Temperatures over much of North and South America, for example, ranged from 4° to 8° C cooler during the Wisconsin. (After Stute et al. 1995.)

latitudinal thermal gradient that intensified during interglacial periods. During glacial maxima of the Pleistocene, average air temperatures were 4° to 8° C cooler than during interglacial periods (**Figure 9.6**, and see Figure 9.1C).

Due to the high heat capacity of water, global temperature of the oceans varied by just 2° or 3° C between glacial and interglacial periods. On the other hand, atmospheric and oceanic circulation changed dramatically, and climatic zones shifted substantially in latitude and elevation with each climatic flush. In mountainous regions, snow lines shifted as much as 1000 m in elevation between glacial and interglacial periods (Behrensmeyer et al. 1992). Before exploring some of their biogeographic and evolutionary consequences, we summarize some of the major climatic changes associated with glacial–interglacial cycles.

As Figures 9.1 and 9.5 indicate, glacial conditions prevailed throughout most of the Pleistocene, with intermittent interglacials accounting for less than 10 percent of this epoch. In addition to subdividing the period into glacials and interglacials, Quaternary scientists also recognize rather shorter excursions within them, whereby brief but particularly cool or warm periods are termed stadials and interstadials, respectively (interstadials during glacial periods may have been comparable in temperature to interglacials). A closer look at the thermal record for the Pleistocene and Holocene confirms that climatic change in and out of glacial conditions can be remarkably rapid (see Jackson 2004). Since the peak of the most recent glaciation—the "Wisconsin" Glaciation in North America or the "Würm" in Europe—global temperatures increased by approximately 4.5° C, with most of that increase occurring within a span of just 3000 years (17,000 to 14,000; see Figure 9.1). Again, remarkably rapid cooling and warming events took place, the most dramatic recent event being the Younger Dryas, in which a large part of the Northern Hemisphere plunged back into and then out of glacial conditions within just decades because of positive feedback mechanisms in the global climate system (Adams et al. 1999).

Geographic shifts in climatic zones

During the most recent glacial period, most non-glaciated regions experienced declines in air temperatures ranging from 4° to 8° C (see Figure 9.6). But the glacial–interglacial cycles of the Pleistocene involved shifts not just in temperatures, but in entire climatic regimes. In general, climatic zones shifted toward the Equator during glacial periods and poleward during interglacials. The pattern of shifts, however, was complicated by the configurations and positions of land, large bodies of water, and glaciers on atmospheric and oceanic circulation patterns.

Glaciers were so massive that they greatly reduced the flow of cool polar air masses to non-glaciated regions. Even air masses that were not effectively blocked by the glaciers would have undergone substantial adiabatic warming as they descended 2 to 3 km, creating steep thermal gradients at the glacial margins (**Figure 9.7A**). Thus, despite the generally cooler conditions, glacial winters were less severe, while glacial summers were cooler and less subject to the heat waves that characterize contemporary climates—especially in temperate regions. This pattern of cooler and more equable climates was true of marine environments as well (see D'Hondt and Arthur 1996; Crame and Rosen 2002; Crame 2004).

An important, but often overlooked, effect of glacial cycles is that as thermal zones shifted, novel combinations of temperature, prevailing winds, ocean currents, and precipitation created climatic and edaphic (soil) zones that lack any contemporary analogues (Jackson 2004). The glaciers caused dramatic changes in prevailing winds and ocean currents, which in turn strongly influenced regional climates. During the Wisconsin, the jet stream of North America was split and diverged around the glacier, and an anticyclonic (clockwise) circulation pattern was established over the Laurentide Ice Sheet (see Figure 9.7). As a result, while average global temperature of the oceans cooled by just 2° to 3° C, surface temperatures of waters in the North Atlantic cooled by as much as 10° C (see also Figure 9.12). Along the southwestern edge of the glacier, relatively dry easterly winds from the interior caused desiccation of lakes in the American Northwest. While cold and dry conditions characterized the glacial climates of other temperate regions, including Europe, other regions were wetter during the glacio-pluvial periods (see Van der Hammen et al. 1971; Wells 1979; Spaulding and Graumlich 1986). For example, far south of North America's glaciers, prevailing wester-

FIGURE 9.7 (A) As relatively cool winds descended the 2 to 3 km high faces of the glaciers, they were adiabatically warmed, thus moderating climatic conditions adjacent to the glaciers. (B) The Laurentide Ice Sheet (18,000–9000 years ago) caused major changes in prevailing winds, including shifts in the jet stream and Westerlies over North America, which strongly influenced regional climates, bringing cold air masses to the Northeast, dry air to the Pacific Northwest, and heavy precipitation (blue arrows) to what today is the aridlands region of the American Southwest. (B after Gates 1993.)

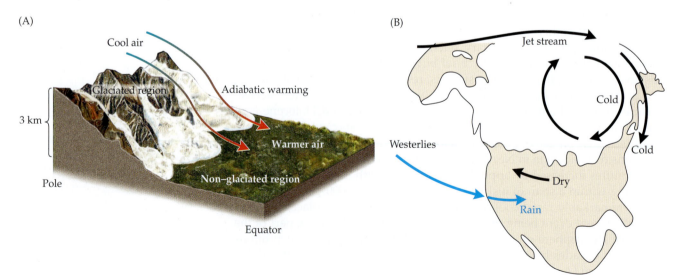

ly winds brought saturated oceanic air masses, causing elevated water levels in lakes of the American Southwest (**Figure 9.7B**).

Paleoclimatic data also reveal a strong association between glacial events and monsoonal circulation and precipitation. **Monsoons** are driven by differential solar heating of adjacent land and sea during hot summers in tropical and subtropical latitudes. Because of its lower heat capacity, land heats much more rapidly than water. Ascending air masses over land create convective cycles that draw oceanic air masses inland, causing heavy summer precipitation—monsoon rains. Pollen records and global circulation models indicate that the strongest monsoonal events of the Pleistocene coincided with periods of peak summer solar radiation (roughly 126,000, 104,000, 82,000, and 10,000 years BP; Gates 1993). Conversely, monsoons were weakest during glacial maxima, often resulting in aridification of otherwise moist tropical regions. In fact, by reducing evaporative input to the hydrosphere, glacial periods are also characterized by general expansion of deserts and other arid ecosystems across the globe (**Figure 9.8**).

In summary, the world we see outside our windows is not typical of the conditions that animals and plants lived and evolved under for the past 2 million years. The current very warm period was perhaps equaled during only 10 percent of the Pleistocene. While climatic zones tended to shift in latitude and elevation with glacial cycles, the specific nature of these zones varied markedly, often resulting in novel combinations of temperature, precipitation, and atmospheric or oceanic circulation. Not only were climatic zones altered in both character and location, they frequently were displaced to occupy new soil regimes. Finally, while most climatic zones shifted toward the Equator during glacial periods, tropical climates had little opportunity for geographic shifts. The inevitable reduction of what we now view as tropical climatic zones (i.e., the "tropical squeeze") raises an intriguing question: How is it that not only did the biotas adapted to these zones survive the cooler climates of glacial cycles, but many maintained or developed a diversity unparalleled by that of any other region on Earth? Later in this chapter

FIGURE 9.8 Because glacial periods reduce evaporative input of water into the atmosphere, they are often marked by regional to global aridification and the expansion of deserts and other xeric ecosystems. Thus, the extent of sand dune areas was much greater during the last glacial maximum than today. (Anderson et al. 2007.)

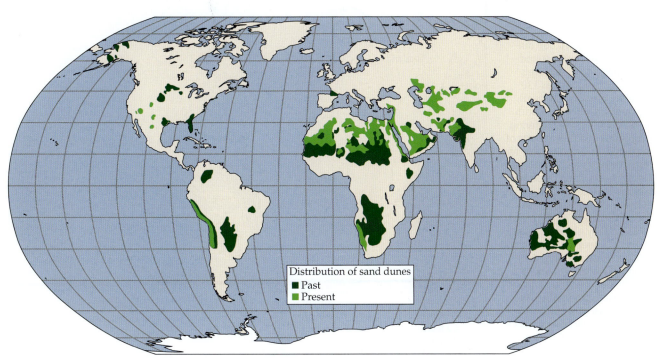

Distribution of sand dunes
■ Past
■ Present

and in Chapter 15, we return to this fascinating paradox, along with other biogeographic consequences of glacial events. First, however, we need to focus on one of the most pervasive and most important phenomena associated with glacial cycles: regional and global changes in sea level.

Sea level changes during the Pleistocene

Throughout the Pleistocene, sea levels fluctuated dramatically on both global and regional scales. **Eustatic** changes are global fluctuations in sea level result-

(A)

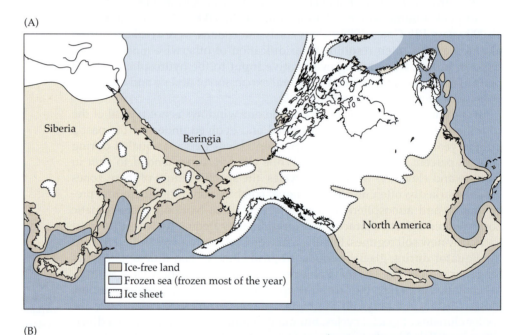

(B)

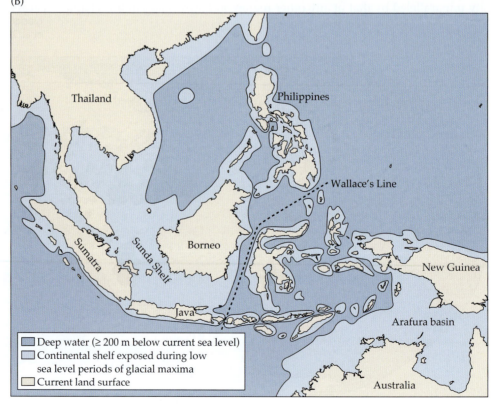

FIGURE 9.9 Glaciation during the Pleistocene resulted in the lowering of sea levels by 100 m to as much as 130 m below their current levels, although changes were both asynchronous and of differing magnitude in different parts of the world. As a result, many terrestrial regions and associated biotas now isolated by oceanic barriers were connected during glacial maxima. (A) Beringia connected North America and Asia. (B) Many islands of Indonesia were connected to mainland Asia and Australia. The island region of Wallacea, marking a division between the biotas of Southeast Asia and Australia, coincides with the division between these glacial landmasses. (A after Pielou 1991; B after Heaney 1991, 2004.)

ing from changes in the volume of water in the oceans, such as those resulting from the locking up or release of massive amounts of water in glacial period ice sheets. In contrast, **isostatic** changes in sea level occur when portions of the Earth's crust rise or sink into the less buoyant asthenosphere, causing local or relative changes in sea level even when global (eustatic) levels remain unchanged. Crustal depression or rebounding due to ice sheet formation or melting are termed glacio-isostacy. In some areas of the high northern latitudes, the sheer mass of 2 to 3 km high glaciers caused crustal downwarping by over 300 m. As we shall see, both eustatic and isostatic changes during the Pleistocene strongly influenced the distributions and diversity of biotas.

During the most recent glacial maximum, nearly one-third of all land in the Northern Hemisphere was covered with thick glaciers, and great fields of sea ice occurred in both northern and southern polar regions (see Figure 9.3). In consequence of the former, a large volume of water was removed from the ocean, probably exceeding the equivalent of 50 million km^3 of ice—more than the present-day ice sheets in Antarctica and Greenland combined. The resulting drop of about 120 m in sea level during the Wisconsin Glaciation was of a similar order to the maximum of about 130 for the Pleistocene as a whole. The current interglacial is not the warmest one on record for the Quaternary, and there is evidence that some earlier interglacials featured periods of slightly higher sea levels, about 25 m higher, than at present. Thus, the total amplitude of global sea level changes during the Pleistocene may have been about 155 m, although it should be understood that there can be pronounced local and regional discrepancies from the average trends in sea level, which not only makes calculating global figures hazardous, but also means that they may be an inadequate guide to past land connections in particular regions.

Despite the relative stability of the tectonic plates during this thin slice of the geological record, Earth's biogeographic profile may have been transformed as much during the Quaternary as during any period in its history. For example, at the end of the last glacial, sea levels were about 100 m or so lower than today, and there was a substantial lag period between the initiation of the postglacial climate regime and the melting of the ice sheets, meaning that large areas of continental shelf were terra firma, allowing easy exchange of plant and animal species by land bridge connections to what are now isolated islands—these connections being maintained for hundreds to thousands of years, depending on the local bathymetry (**Figure 9.9**). The biogeographic relevance of these events is difficult to overstate. Wallace's line, for example, which separates the Oriental and Australian biogeographic regions, coincides with the division between the glacial land bridges of the Sunda Shelf and Australia–New Guinea–Tasmania.

In **Figure 9.10** we summarize global, or eustatic, changes in sea level. For many regions, however, isostatic changes may have been just as large. Also, just as eustatic increases in sea level entering the Holocene lagged behind climate warming, isostatic changes tend to lag behind eustatic changes. It takes many centuries for downwarped crust to rebound after glacial recession. Consequently, as temperatures increase and glacial meltwater accumulates, rising seas often spill over onto the downwarped regions of continents, creating extensive shallow seas. During the Early Holocene (about 12,000 to 10,000 years BP), for example, the Saint Lawrence River Valley and Great Lakes of North America were inundated with marine waters from the Atlantic (**Figure 9.11A**). This saltwater corridor provided a dispersal avenue for shoreline flora of the Atlantic coast (e.g., beach plants including *Ammophila breviligulata*, *Cakile edentula*, and *Euphorbia polygonifolia*; coastal bog species such as *Xyris caroliniana*; aquatic macrophytes such as *Utricularia purpurea*: Pielou 1991). Subse-

FIGURE 9.10 Variation in global sea levels across three different scales of time. (A) The Phanerozoic record reveals that sea levels were typically much higher than in recent times. (B) The record throughout the Pleistocene illustrates the repeated and very rapid transitions between full glacial and interglacial conditions and the resultant declines and rises in sea levels. (C) These dynamics are also evident over more abbreviated time scales, as illustrated here for the period since the last major interglacial at about 140,000 years BP. In addition to these global (or eustatic) changes in sea levels, regional sea levels also varied substantially as Earth's crust rose and sank in the asthenosphere. Such fluctuations in sea level can occur even when global levels remain unchanged. (A from Lambeck and Chappell 2001, after Hallam 1984; B after Dyer 1986; C after Hopkins et al. 1982.)

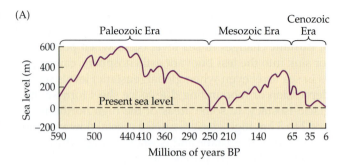

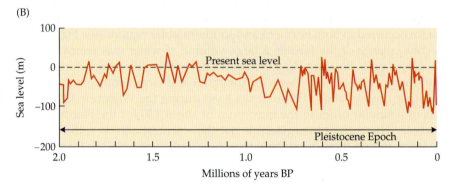

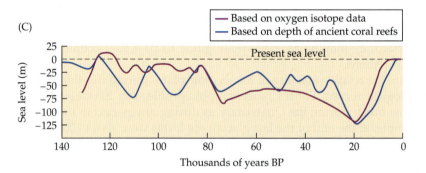

quent rebound of Earth's crust by as much as 275 m drained this corridor and transformed it into the freshwater Saint Lawrence River, which drains the Great Lakes. Events such as this may account for the disjunct ranges of many species adapted to coastal environments (**Figure 9.11B**).

In summary, glacial events and associated changes in sea level altered dispersal routes among non-glaciated regions far from the ice sheets and long after the glaciers receded. During glacial maxima, the once-continuous ecosystems of northern regions were often isolated across widely scattered pockets of unglaciated land, while other, long-isolated biotas dispersed across formerly submerged land bridges. Although these land bridges provided dispersal routes for terrestrial biotas, they created barriers for marine life. Glacial cycles thus created great, alternating waves of biotic exchange and isolation (vicariance) among both terrestrial and marine biotas. Before exploring biogeographic responses to these changes in the geographic template, we reemphasize that, as Stephen Jackson (2004: 52) puts it, "the environment of each glacial period and interglacial period was unique . . . there is no modal condition for the past century, the past millennium, the Holocene, or the Quaternary."

(A)

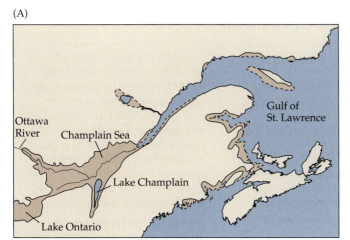

(B)

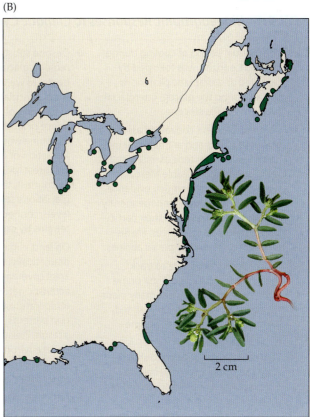

FIGURE 9.11 During the most recent glacial retreat, sea levels began to rise relatively rapidly, while previously glaciated land surfaces were slow to rebound from the downwarping force of the massive glaciers. As a result, marine waters spilled over onto these low-lying regions of the continents, creating shallow seas and providing saltwater corridors for the migration of plants and animals. (A) About 10,000 years BP, shallow seas covered the Saint Lawrence River valley, Lake Champlain, and the Ottawa River, forming the Champlain Sea (dashed line depicts modern coastline; shading represents area covered by shallow seas). (B) The existence of such transient corridors explains the disjunct distributions of many shallow-water, marine, and estuarine species, such as the seaside spurge (*Chamaesyce polygonifolia* [formerly *Euphorbia*]; distribution indicated by dark areas) along the coastlines of eastern North America and the Great Lakes. (After Pielou 1979; photo courtesy of Michael Terry.)

Biogeographic Responses to Climatic Cycles of the Pleistocene

The biogeographic dynamics of Pleistocene biotas were triggered by three fundamental changes in their environments:

1. Changes in the location, extent, and configuration of their prime habitats
2. Changes in the nature of climatic and environmental zones
3. Formation and dissolution of dispersal routes

The responses of biotas, long adapted to relatively stable and equable climates, also were of three types:

1. Some species were able to "float" with their optimal habitat as it shifted across latitude or altitude.
2. Other species remained where they were and adapted to altered local environments.
3. Still other species underwent range reduction and eventual extinction.

As we shall see, the environmental stresses of the glacial–interglacial cycles triggered all three of these responses. The specific nature of biogeographic responses varied considerably among species and geographic regions, but we have summarized some of their salient features in **Box 9.1**. We emphasize that, while such a synopsis may be useful, a richer understanding of these truly complex and fascinating events can be developed only by thoroughly exploring individual case studies (see Unit 5; see also Lieberman 2000).

BOX 9.1   *Biogeographic Responses to Climatic Cycles of the Pleistocene*

1. The gradual period of cooling during the Mid Cenozoic was followed by repeated and dramatic climatic reversals during the glacial–interglacial cycles of the Pleistocene.

2. Communities and coevolved assemblages of plants and animals that may have persisted for tens of millions of years during the equable Mesozoic were disrupted, with much evidence of species' responding independently of one another based on their particular physiological tolerances, life history strategies, and dispersal abilities.

3. Many species were able to track the geographic shifts of their prime climates and habitats, but many plant species, especially, typically lagged behind, often by centuries, sometimes by millennia.

4. Vegetation zones tended to shift toward the Equator (or lower elevations) during glacial periods and toward the poles (or higher elevations) during interglacials, but the shifts were complicated and strongly influenced by geographic features (e.g., mountains, ocean basins, prevailing winds, and proximity to the ice sheets).

5. In general, open-canopied biomes (tundras, savannas, grasslands, and prairies) expanded during glacial maxima at the expense of closed biomes (i.e., forests). These trends were reversed during periods of global warming, but again, rates of shifts varied substantially among biomes, as did the particular species composition of each biome and community.

6. Despite substantial variation among regions, glacial climates tended to be dry as well as cool. On the other hand, postglacial warming resulted in flooding of coastlines, submergence of land bridges, transgression of marine waters onto land, and formation of extensive shallow seas and great, postglacial lakes and rivers. By contrast, the peaks of the major glaciations saw great drops in sea levels, connecting previously isolated terrains by dry land.

7. On land, climatic zones changed dramatically, not only in location and areal coverage, but also in their characteristic nature (i.e., combinations of temperature, seasonality, precipitation patterns, and soil conditions). As a result, Pleistocene events created novel environments, fostering development of novel communities, while other communities disappeared.

8. Although there was much variation within taxonomic groups, plants tended to shift more slowly than animals. The geographic dynamics of species during the Pleistocene created many isolated populations, in some cases promoting evolutionary divergence and diversification of certain biotas.

9. Many plants and animals that were unable to track their shifting environments were able to remain in situ by adapting to altered conditions, while still others adapted and evolved en route—that is, during shifts to other regions.

10. The remaining species, unable to shift or adapt, went extinct. During the initial cycles of climatic reversals, extinctions were much more common among plants than animals. This may have been a consequence of the comparatively limited ability of plants to disperse and the decoupling of associations among plants and between plants and animals that served as pollinators, parasites, and herbivores.

11. In contrast, until the most recent glacial cycles, animal extinctions were relatively few, and many groups, especially the large herbivores and carnivores, underwent major radiations.

12. The tables were turned, however, during the more recent glacial cycles, which witnessed waves of extinctions of many animals, especially larger ones, while comparatively few plants suffered extinctions in the Late Pleistocene. It appears that the initial climatic reversals may have "weeded out" most of the intolerant plants, leaving behind those more capable of dispersing with, or adapting to, climatic reversals.

13. During the most recent glacial cycle, large mammals may have become too specialized on the then-waning glacial habitats (especially steppes and savannas). Alternatively, these "megafaunal" extinctions may have resulted from biotic exchanges associated with glacial events (including invasions by humans during periods of low sea levels), which caused mass anthropogenic extinctions of the naive native fauna (see Chapter 16). ∎∎∎

Biogeographic responses of terrestrial biotas

Many of the trends summarized in Box 9.1 are illustrated in the vegetative dynamics of the most recent deglaciation (see Figures 9.13–9.18). Climatic changes associated with the glacial maximum caused the general expansion of steppes, savannas, and other open-canopied terrestrial ecosystems at the expense of closed ecosystems, especially tropical rain forests. In general, biomes have shifted from 10° to 20° in latitude between glacial and interglacial periods yet have tended to occupy the same relative positions across latitudes or elevations (i.e., a sequence from tundra and boreal forest to savanna and tropical rain forest). This is because the climatic belts of the Earth (see Figure 3.8), then as now, created zonal patterns of vegetation, although the zones were compressed during glacial episodes. In the marine realm,

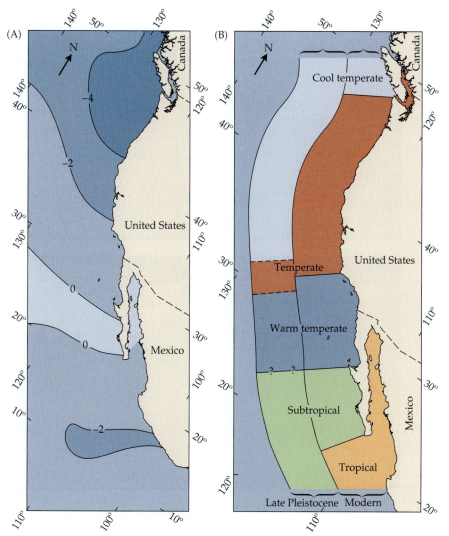

FIGURE 9.12   Shifts in (A) February sea surface temperatures (temperatures are in °C; minus sign indicates cooler temperature during the Pleistocene) and (B) marine biotic provinces along the west coast of North America between 18,000 years BP and the present. Note that sea surface temperatures and nearshore marine provinces have remained relatively stable in the waters off southern California and northern Baja, while those in other regions have varied substantially during the same period. (After Fields et al. 1993.)

latitudinal shifts in isotherms and biogeographic patterns have tended to be substantial in the mid-latitudes (35° to 55°), but relatively minor at lower latitudes (**Figure 9.12**).

Shifts in climatic zones and biomes, however, were complicated by currents and topographic features, including mountain ranges, large rivers, and other bodies of water. In Europe, the southward shift of some biomes and biotas during the most recent glacial maximum was blocked by the Alps, Pyrenees, and Mediterranean Sea. In contrast, the rivers and mountain ranges of North America that run north–south facilitated extensions of high-latitude biomes deep into subtemperate and subtropical latitudes. During the Wisconsin maximum (about 18,000 years BP), boreal forests and tundra penetrated deep into the interior of the continent along both the Mississippi River Valley and the Appalachian Mountains (**Figure 9.13**). Similar extensions of boreal forests and otherwise high-altitude biomes occurred along the Rocky Mountains of western North America; the Carpathian, Ural, and Atlay Mountains of Eurasia; the Great Dividing Range of Australia; and the Andes of South America.

In **Figure 9.14**, pollen profiles of the Andes in Colombia show how each of the vegetation zones has shifted upward since the most recent glacial maximum (see Flenley 1979b). The lower tropical, sub-Andean, and Andean

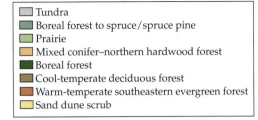

FIGURE 9.13 Shifts in vegetation zones of eastern North America during the most recent deglaciation. Note the great southern expansion of boreal forest and tundra along the Mississippi Valley and Appalachian Mountains during the Wisconsin (18,000 BP). (After Gates 1993.)

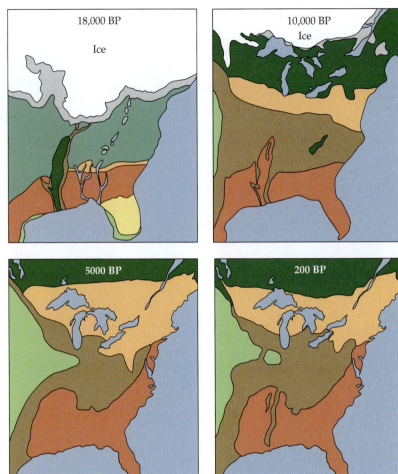

forest types (including the tropical rain forest) have much wider elevational amplitudes now than they did 14,000 years ago. Subpáramo has remained about the same, but **páramo** and superpáramo are elevationally compressed. Because upper vegetation types occur closer to narrow mountain peaks, the total area covered by these biomes has fluctuated greatly, causing extinctions when populations were isolated in restricted montane areas. In these mountainous regions, elevational shifts ranged from 150 to 1500 m between glacial and interglacial periods, and typically were much more rapid than latitudinal shifts.

Figure 9.15 tracks the probable upper elevational limit of subtropical forests in three equatorial regions (East Africa, New Guinea, and South America). Each graph suggests several drastic elevational shifts for these forests during the last 33,000 years. The curves are not identical, but in each case the upper elevational limit of tropical forest began to decrease gradually beginning about 29,000 to 27,000 years BP, but then it reversed and increased sharply about 16,000 to 15,000 years BP. Flenley (1979a,b) also noted that in Africa, South America, Indo-Malaya, and New Guinea, montane vegetational zones were depressed by about the same amounts (1000 to 1500 m) at roughly the same times.

The area just south of New Guinea, the state of Queensland in northeastern Australia, exhibited an exceptional and rapid change from predominantly sclerophyllous woodland to rain forest between the Late Pleistocene and

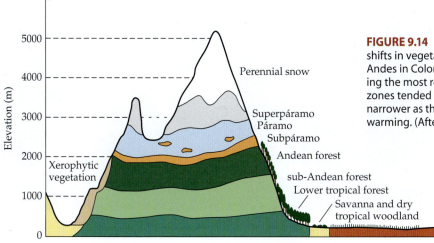

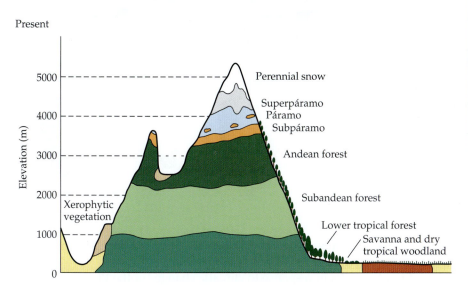

FIGURE 9.14 A simplified description of elevational shifts in vegetation zones in the eastern cordillera of the Andes in Colombia in response to climatic change following the most recent glacial maximum. Note that while all zones tended to shift in concert, the upper zones became narrower as they shifted upward in response to global warming. (After Flenley 1979a.)

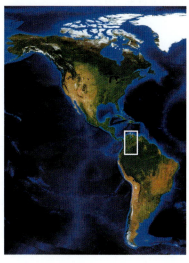

6000 years BP (**Figure 9.16**). The latter point helps to illustrate how difficult it is to extrapolate trends in terrestrial paleoclimatology and to understand vegetational history merely by studying present-day vegetation (**Figure 9.17**). Again, the responses to glacial conditions were not uniform across the globe. While African and Amazonian rain forests contracted in response to glacial

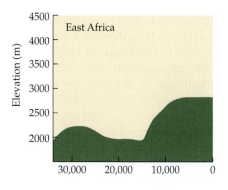

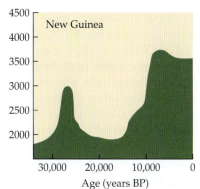

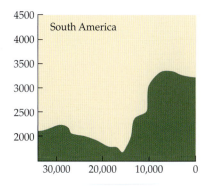

FIGURE 9.15 Shifts in the upper elevational limits of tropical forests in three different regions (New Guinea, East Africa, and South America) during the past 33,000 years, as inferred from a number of published pollen diagrams. Note that these elevational shifts were roughly synchronous, although the exact upper elevational boundary was somewhat different in each region. (After Flenley 1979a.)

FIGURE 9.16 Comparisons of distributions of vegetation zones during the most recent glacial maximum (18,000 years BP) and 3000 years BP (i.e., before significant disturbance by humans) for three regions of the Southern Hemisphere: (A) Australia, (B) New Zealand, and (C) southern South America. (After Markgraf et al. 1995.)

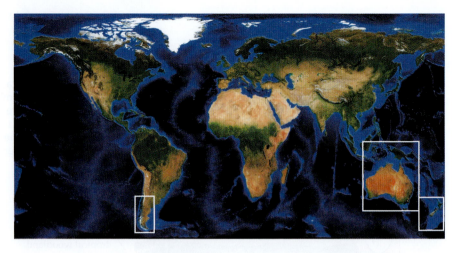

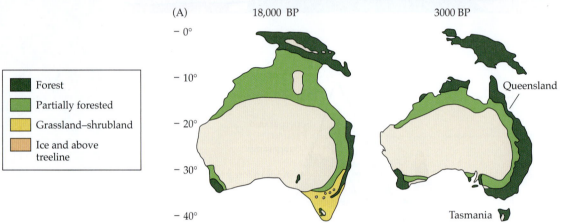

■	Forest
■	Partially forested
■	Grassland–shrubland
■	Ice and above treeline

aridity, those of Sumatra remained intact (see Flenley 1979a; Maloney 1980; see also Figure 9.13; but see Colinvaux et al. 2000).

Geographic shifts in response to climatic changes of the Pleistocene were even more complex for individual species. Rather than responding as discrete units to shifts in climates, biomes and communities often disintegrated, with many, if not most, species responding in an individualistic manner (i.e., consistent with Gleasonian views of community organization; see Chapter 5). Shifts in species ranges were influenced by both extrinsic and intrinsic factors. Extrinsic factors included climate, soils, prevailing winds, ocean currents, and topographic features such as those discussed above. Colonization of deglaciated areas of North America by beech provides an exemplary case study of the influence of these factors. According to Margaret Davis (1986), after beech became established along the southeastern shore of Lake Michigan (about 7000 years BP), it took another 1000 to 2000 years to reach the far side of the lake. Birds, including the now extinct passenger pigeon (*Ectopistes migratorius*), may have played an important role in carrying the seeds of beech and many other species across such geographic barriers. The potential for range shifts also may have been strongly influenced by other interspecific interactions, especially competition, predation, and parasitism.

Individualistic responses to climatic change among species derive largely from their considerable differences in ability to respond to these extrinsic factors. Physiological tolerances, behavior, life history strategies, intrinsic rates of increase, capacities for evolutionary adaptations, and dispersal abilities all vary markedly among species, even those occupying the same community (see also Davis and Shaw 2001). Pollen analyses and other paleontological records document considerable interspecific variation among tree species in rates of migration (**Table 9.1**). Average rates of range extension during the Ho-

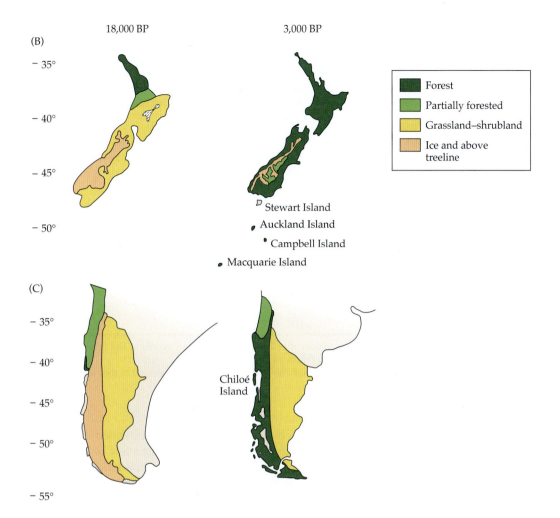

(B)

18,000 BP   3,000 BP

- 35°
- 40°
- 45°
- 50°

Forest
Partially forested
Grassland–shrubland
Ice and above treeline

Stewart Island
Auckland Island
Campbell Island
Macquarie Island

(C)

- 35°
- 40°
- 45°
- 50°
- 55°

Chiloé Island

locene in North American trees varied from 100 to 400 m per year. As a result, some species now recognized as dominants may be relatively recent arrivals in contemporary communities. The American chestnut (*Castanea dentata*), for example, was a dominant species in oak–chestnut forests of the Appalachian Mountains for over 5000 years, but it was a relatively recent invader of similar forests in Connecticut (arriving about 2000 years BP). In western North Amer-

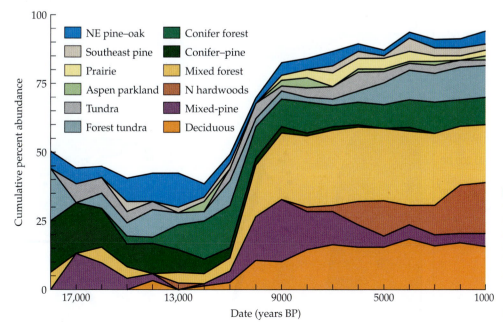

NE pine–oak   Conifer forest
Southeast pine   Conifer–pine
Prairie   Mixed forest
Aspen parkland   N hardwoods
Tundra   Mixed-pine
Forest tundra   Deciduous

Cumulative percent abundance

Date (years BP)

FIGURE 9.17 Variation in the relative abundance of different types of vegetative communities in North America since the most recent glacial maximum (based on samples of fossil pollen). Note that the most rapid changes occurred between 14,000 and 10,000 years BP. (After Shane and Cushing 1991.)

TABLE 9.1 *Early Estimates of Average Rates of Holocene Range Extensions of Trees (m/yr) following Glacial Recession*

Species	North America	European mainland	British Isles
Pines	300–400	1500	100–700
Oak	350	150–500	350–500 (50 near northern limit)
Elm	250	500–1000	550 (100 near northern limit)
Beech	200	200–300	100–200
Hazelnut	—	1500	500
Alder	—	500–2000	500–600 (50–150 near northern limit)
Basswood	—	300–500	450–500 (50–100 near northern limit)
Ash	—	200–500	50–200
Spruce, larch, balsam, fir, maples, hemlock, and hickory	200–250	—	—
Chestnut	100	—	—

Source: North American data after Davis 1981; European mainland and British Isles data after Huntley and Birks 1983; Birks 1989.

Note: For most of these species, the rate of range extension slowed as they approached their northern limits.

FIGURE 9.18 Rates of migration of trees following glacial recession varied substantially among species and were strongly influenced by extrinsic factors such as topographic features and prevailing winds. (A) Inland varieties of the lodgepole pine (*Pinus contorta* var. *latifolia*) have expanded their range northward over the past 12,000 years and may be continuing this range expansion in modern times. Dots indicate northern range boundaries at various times, based on pollen samples. (B) Northward range expansion of the white spruce (*Picea glauca*) during retreat of the Laurentide glacier was strongly influenced by prevailing winds (blue arrows). The species moved rapidly once it reached the western edge of the glacier, where north-flowing winds aided its range expansion. Dots indicate locations of sample points. Long arrows indicate slow migration; short arrows, rapid migration. (After Pielou 1991.)

ica, lodgepole pine (*Pinus contorta* var. *latifolia*) has extended its range at about 200 m per year, reaching southern Alaska just a few centuries ago, and may still be expanding its range northward (**Figure 9.18A**).

Rates of range expansion can vary considerably within, as well as among, species. At the end of the Wisconsin in North America, for example, white spruce (*Picea glauca*) migrated northward along both the eastern and western margins of the retreating Laurentide glacier. From 14,000 to 7000 years BP, it migrated along the eastern edge of the ice sheet at about 300 m per year. In contrast, its migration along the western edge of the ice sheet, where north-flowing winds aided its dispersal, was nearly an order of magnitude faster (**Figure 9.18B**).

Species varied, not only in their rates of range expansion, but in direction as well, with many species shifting as much in longitude as

(A)

Wisconsin ice sheet

400 BP
1100 BP
2500 BP
5000 BP
5000 BP
8000 BP
8000 BP
10,700 BP
11,200 BP
12,000 BP

(B)

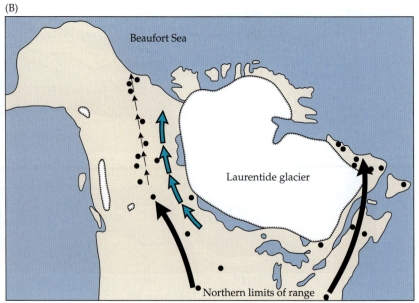

Beaufort Sea

Laurentide glacier

Northern limits of range

in latitude. For example, the geographic range of the eastern chipmunk (*Tamias striatus*) shifted to the northeast, that of the northern pocket gopher shifted to the west (**Figure 9.19**), and the range of the papershell piñon (*Pinus remota*) contracted 300 km to the south (Graham 1986; Graham et al. 1996; Lanner and Van Devender 1998). In a similar manner, many plants and animals inhabiting mountainous regions exhibited individualistic shifts in elevation (e.g., see Jackson and Whitehead 1991; Jackson 2004). The overall result of these independent responses was a great reshuffling of community

FIGURE 9.19   Geographic range shifts in four species of rodents during the Holocene. Shaded areas represent present range; dots indicate locations of late Pleistocene fossils. Note that these range shifts differed in extent, with (A) collared lemmings (*Dicrostonyx*) shifting farther northward than (B) brown lemmings (*Lemmus*); species also shifted in direction, with the above species shifting northward while (C) eastern chipmunks (*Tamias striatus*) shifted toward the northeast and (D) northern pocket gophers (*Thomomys talpoides*) shifted toward the west. (After Graham 1986.)

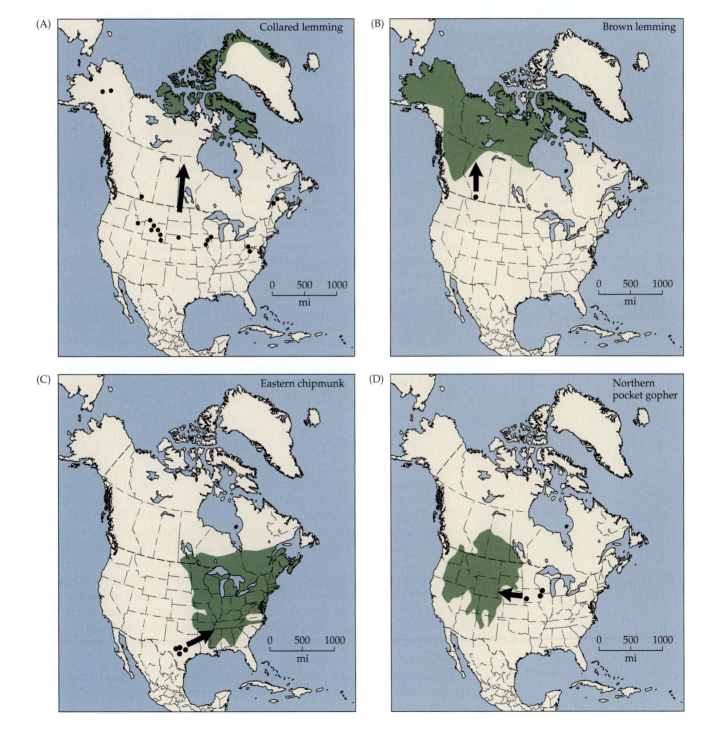

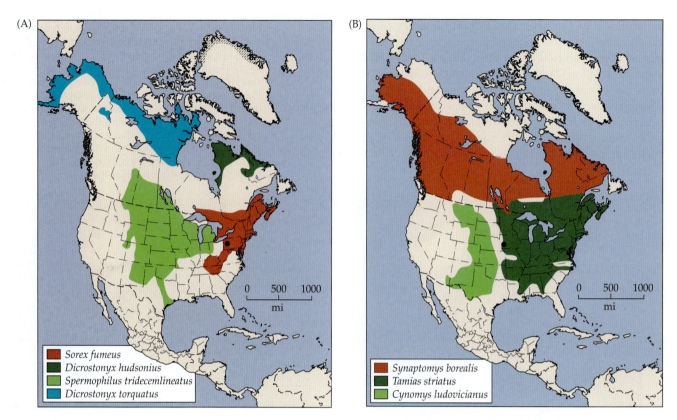

(A)

🟥	*Sorex fumeus*
🟩	*Dicrostonyx hudsonius*
🟢	*Spermophilus tridecemlineatus*
🔵	*Dicrostonyx torquatus*

(B)

🟥	*Synaptomys borealis*
🟩	*Tamias striatus*
🟢	*Cynomys ludovicianus*

0   500   1000
mi

FIGURE 9.20 As a result of the differences in extent and direction of range shifts during the Holocene, species that co-occurred during the most recent glacial maximum often exhibit disjunct ranges today. (A–D) In each map, black dots indicate coincident occurrences of the named species during the late Pleistocene, while shaded areas indicate their current ranges. (E–F) Quantitative analysis of range shifts in North American mammals (exclusive of regions in Canada and Beringia) indicate that, counter to simple predictions of southward shifts during glacial advances and northward shifts during glacial recessions, geographic range shifts often were in other directions and were quite varied among species (arrows indicate the relative frequency of species whose range centroid shifted in the direction indicated, N = the total number of species included in each analysis, D = the median distance the centroid of the ranges shifted, and asterisks indicate the expected direction of range shifts under climate change). (A–D after Graham 1986; E–F modified from Lyons 2003.)

composition from one climatic reversal to the next. Again, animals tended to shift much more rapidly than plants, apparently influenced more strongly by habitat structure than by abundances of particular plant species. Range shifts of some and perhaps many plant species may have been slowed by relatively limited dispersal abilities of symbiotic fungi—in particular, large-spored, belowground fruiting mycorrhizal fungi (Wilkinson 1998). Thus, glacial reversals often caused decoupling of animal and plant communities. In the subtemperate latitudes of the American West (30° to 40° N latitude), for example, glacial and contemporary communities typically share less than one-third of their plant species. While biomes shifted in a somewhat predictable manner, the species composition of each system varied significantly, often with communities of one glacial stage having no comparable analogue in the next (**Figure 9.20A–D**).

While such novel, or non-analog communities (i.e., those with species compositions novel to that particular stage) were common, community disintegration was not total or universal, despite significant shuffling of spe-

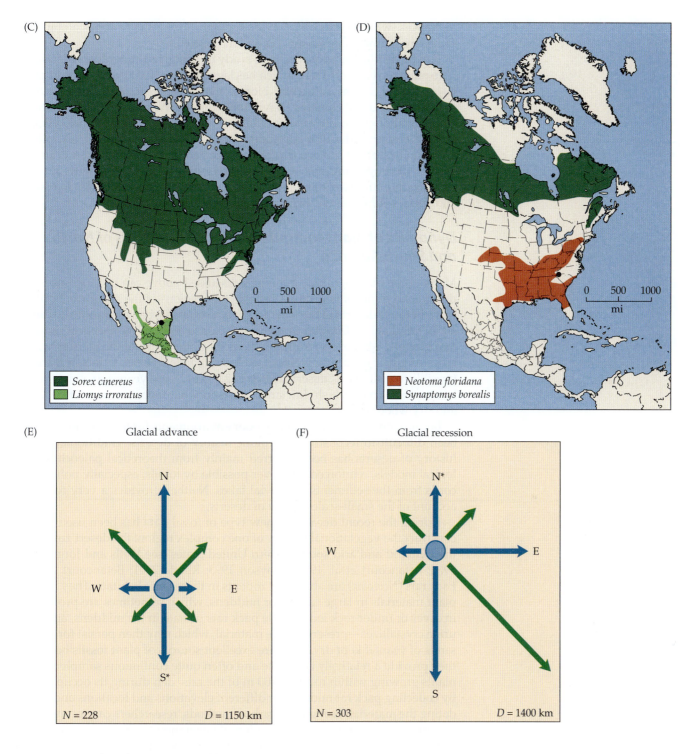

cies among glacial stages. S. Kathleen Lyons' (2003) quantitative analyses of Late Pleistocene to Holocene range shifts in North American mammals are consistent with the case studies illustrated in Figures 9.20 and 9.21A–D. That is, geographic range shifts in these species often exceeded 1000 km, varying substantially in direction and often counter to simplistic expectations of responses to glacial advance and recession (**Figure 9.20E,F**). These range dynamics, however, were far from random, with species often shifting in similar directions—reflecting either an ecological interdependence among species (e.g., predators following prey) or shared environmental and climatic tolerances, with ecologically similar species shifting their ranges along the same environmental gradients (Lyons 2003, 2005).

Finally, as illustrated for the case study of postglacial migration of lodgepole pine (see Figure 9.18A), which has lagged some 12,000 years behind glacial recession, communities and biomes may never reach equilibrium with their contemporary climates (Davis 1986; but see Wright 1976; Webb 1987). Animals and, especially, plants often lag far behind climatic zones, which shift at least an order of magnitude more rapidly (Gates 1993). Thus, plant communities of the Holocene often are viewed as ephemeral collections of species, sharing many species with communities of previous stages, but often subject to turnover, reintegration, and persistent disequilibria in diversity and species composition (Behrensmeyer et al. 1992).

Dynamics of plant communities in the aridlands of North and South America

Many of the biogeographic dynamics discussed above can be illustrated by detailed accounts of vegetative shifts in arid regions of the American Southwest. Because deserts of the Northern Hemisphere lie between about 30° and 40° N latitude (see Chapter 3), they were not covered by ice sheets or mountain glaciers. Yet these and other regions far removed from the glaciers were still strongly influenced by glacial cycles of the Pleistocene. In fact, if it were possible to visit these "desert" regions 12,000 to 18,000 years ago, you would be surprised to find lush, lowland forests in what today are seas of deserts and dry woodlands.

Because deserts are notoriously poor environments for fossilization, it has been difficult to reconstruct the biotic history of deserts. Traditionally, the history of deserts has been inferred mainly from theoretical paleoclimatological models, reinforced wherever possible by fossils, especially pollen records from the sediments of pluvial lakes. Neither provides a very reliable account of the small-scale history of deserts.

During the recent decades, a new type of fossil data has been used to reconstruct the vegetational history of one complex region: the desert zones of the semiarid and arid southwestern United States (see Wells and Jorgensen 1964; Betancourt et al. 1990; Grayson 1993; Rhode 2001; Betancourt 2004). Pack rats (*Neotoma*) are abundant rodents in these xeric habitats. They hoard plant materials in large caches, or middens, which sometimes are protected in caves or under rock ledges. The pack rats urinate in the middens, and the urine crystallizes—preserving the material, which may then persist for thousands of years if kept dry. They are excellent sources of plant fossils because they provide a relatively complete and often quite continuous sample of the plants growing within roughly 100 m of the rats' den during its occupation. By collecting pack rat middens at different elevations and locations and then dating the materials using radiocarbon methods, researchers can reconstruct the shifts in elevation and composition of vegetation types and deduce past climatic regimes (i.e., those back to ca 40,000 years BP).

Data from pack rat middens of the American Southwest have provided important insights into the vegetive dynamics of this region (Van Devender 1977; Van Devender and Spaulding 1979; Wells 1979; Betancourt et al. 1990; Betancourt 2004; see also Fritts 1976). In addition, paleoecologists now analyze middens from a growing diversity of other mammals, including other rodents such as ground squirrels (*Spermophilus*), stick-nest rats (genus *Leporillus*), chinchilla rats (genus *Abrocoma*), and hyraxes (order Hyracoidea), thus allowing reconstructions of Pleistocene and Holocene climates across other regions of North America as well as midden sites in Australia,

FIGURE 9.21 (A) Elevational shifts in vegetation zones in the mountainous region of the American Southwest near southern Arizona. (B) Vegetation of the U.S. Southwest during the last glacial period compared with (C) modern vegetation. (A after Lomolino et al. 1989; Merriam 1890; B and C from Betancourt 2004, after Swetnam et al. 1999.)

(A)

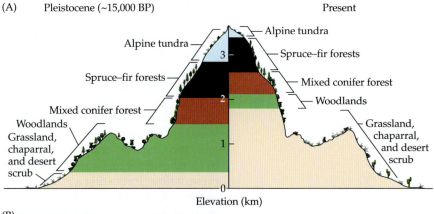

(B)

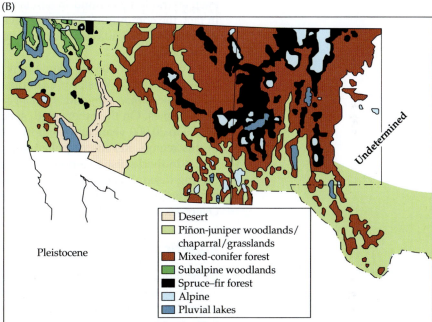

South America, and Africa (Latorre et al. 2002; Pearson and Betancourt 2002; Betancourt 2004; Emslie et al. 2005; Webeck and Pearson 2005; Gil-Romera et al. 2007; Zazula et al. 2007; Holmgren et al. 2008; Chase et al. 2009; Fisher et al. 2009).

These studies have revealed major elevational shifts in vegetation over the last 20,000 years. As explained earlier, climates in the present day aridlands were substantially cooler and wetter during the most recent (Wisconsin) glacial maximum. During this period, vegetation zones across the American Southwest were displaced as much as 500 to 1000 m below their present limits (**Figure 9.21**), with most of the shift occurring within the last 8000 to 12,000 years. Analyses of mid-dens from the Atacama Desert along the Pacific slope of the Andes (18° to 27° S latitude) in-

(C)

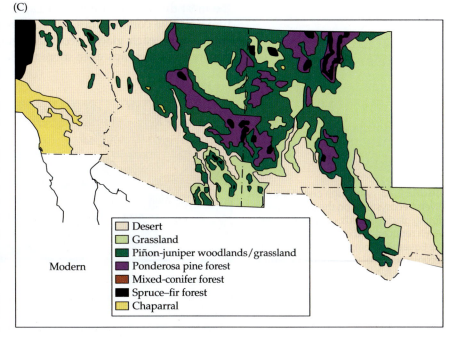

dicate similar elevational shifts in plant communities between 14,000 and 10,500 BP (Latorre et al. 2002). It appears, however, that vegetative responses to climatic dynamics of the Pleistocene were not universal; for example, rodent middens east of the Andes in Argentina's Monte Desert revealed surprising stability, with little latitudinal displacement of floras during the Early Holocene (Betancourt 2004).

Even in those regions with significant shifts of biomes, plant communities did not simply move as entire, integrated entities up and down the mountains. Rather, they changed dramatically in species composition. During the most recent glacial maximum (21,000 to 15,000 years BP), most plant species in North America's Grand Canyon occurred 600 to 1000 m lower than at the present time, indicating a much cooler and wetter climate (Cole 1982). Many of the species that inhabited areas along the rim of the Grand Canyon, however, are no longer found in similar communities in the same region today. In fact, the community contained several species that are presently found in northeastern Nevada and northwestern Utah, at least 500 km to the north (Cole 1982).

Similarly, many of the plants now dominant in the coniferous forests of the intermountain West, for example, ponderosa pine (*Pinus ponderosa*) and piñon pine (*Pinus* subgenus *Ducampopinus* spp.), were rare and restricted in glacial times. On the other hand, species that were much more widespread 10,000 to 30,000 years ago are now narrowly distributed or no longer occur in this region. Results of pollen analyses over the same time periods also reveal individualistic responses to climatic fluxes of the Pleistocene, again driving disequilibrium and disruption of vegetative communities that once characterized glacial maxima (Davis 1986, 1994; Jackson et al. 1997; see also Overpeck et al 1992; Williams et al. 2003).

Aquatic systems

Distributions and diversities of natural communities across the marine realm as well as freshwater communities were strongly influenced by changes in water temperatures, ocean currents, and changes in water chemistry that were associated with each glacial–interglacial cycle. For example, some of the most diverse ecosystems on Earth, but also some of the most temperature-sensitive communities, are coral reefs, whose distributions waxed and waned with each glacial recession and subsequent advance, respectively (**Figure 9.22**). Perhaps even more fundamentally influenced by the climatic

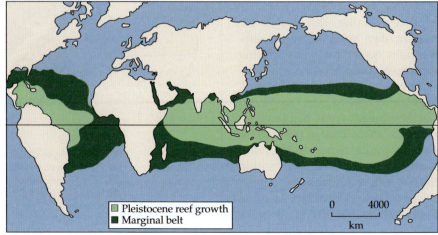

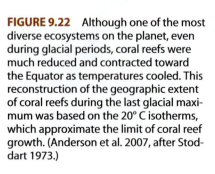

FIGURE 9.22   Although one of the most diverse ecosystems on the planet, even during glacial periods, coral reefs were much reduced and contracted toward the Equator as temperatures cooled. This reconstruction of the geographic extent of coral reefs during the last glacial maximum was based on the 20° C isotherms, which approximate the limit of coral reef growth. (Anderson et al. 2007, after Stoddart 1973.)

Pleistocene reef growth
Marginal belt

0   4000

km

cycles of the Pleistocene, however, were the diversity of freshwater systems, many of which were actually created and at times cataclysmically destroyed by the advances and recessions of glaciers across the mid to high latitudes.

Proglacial and postglacial lakes

No other force of lake formation can compare with the glacial activity of the Pleistocene (Hutchinson 1957; Wetzel 1975). As glaciers melted, many shallow-water marine systems were decimated by rising waters. Throughout the world, great volumes of meltwater created rivers and lakes of such magnitude that they dwarf their contemporary analogues. Modern Lake Superior, the world's largest freshwater lake, is less than a fourth the size (surface area) of postglacial Lake Agassiz (**Figure 9.23A,B**), appropriately named after the scientist who developed the first comprehensive theory of the Ice Ages—(Jean) Louis Rodolphe Agassiz. Lake Agassiz covered approximately 350,000 km^2 of North America during the Early Holocene, influencing regional climates and possibly contributing to the very rapid cooling of the Younger Dryas Period (12,800 to 11,600 BP; see Hostetler et al. 2000). Like Lake Agassiz, most other postglacial lakes following the last glacial maximum developed between 12,000 and 11,000 years BP and peaked in extent about 10,000 to 9000 years BP.

While many of these lakes persisted for several millennia, their demise was often catastrophic and spectacular. Even during their northerly retreat, the great mass of glaciers continued to downwarp the lithosphere, causing glacial meltwater to flow toward, and be blocked by, remnants of the retreating glacier (i.e., the great ice dams; see Figure 9.23A). As climates continued to warm during the Holocene, ice dams eventually gave way in explosive outbursts. For example, postglacial Lake Missoula, equivalent in size to modern-day Lake Ontario, emptied its 2000 km^3 volume of water in less than two weeks, scouring deep channels into the basalt flows on the Columbia Plateau to create the dramatic scablands topography of eastern Washington State (McPhail and Lindsey 1986; cited in Pielou 1991). As cataclysmic as this must have been, it appears relatively minor in comparison with the breaching of Lake Agassiz's ice dam, which occurred around 8000 BP and poured some 163,000 km^3 of freshwater (about seven times the total volume of the modern Great Lakes) into the Tyrrell Sea (Hudson Bay) and the Atlantic Ocean (see Figure 9.24A; Leverington and Teller 2002, 2003; Teller et al. 2002).

Glacial and postglacial lakes, however, were not limited to the Pleistocene and Holocene but likely were common features dominating regional landscapes during earlier glacial–interglacial cycles as well. The sediments now lying beneath the English Channel still bear the impacts of repeated floods generated by the collapse of glacial lakes, the first of these having developed about 450,000 years ago (**Figure 9.23C**; Gupta et al. 2007). In this case, the waters of the Rhine and Thames drainages were blocked at their northern extent by the North Sea and Scandinavian ice sheets as they advanced southward. As this proglacial lake formed, its southern discharge was blocked by a rock dam that once lay above Dover Strait. Lake waters eventually rose above the dam, eroding it to the point that it eventually burst with a cataclysmic release of lake waters that rivaled that of any other known megaflood (with discharge rates estimated at between 0.2 and 1.0×10^6 m^3 s^{-1}) and carved out the channel at around 400,000 BP (Gupta et al. 2007).

Cryogenic lakes

Some of today's lake basins were created when retreating glaciers left great blocks of ice in their wakes. The ice blocks often persisted for many centuries, leaving their imprints as prairie potholes in temperate regions of the Hol-

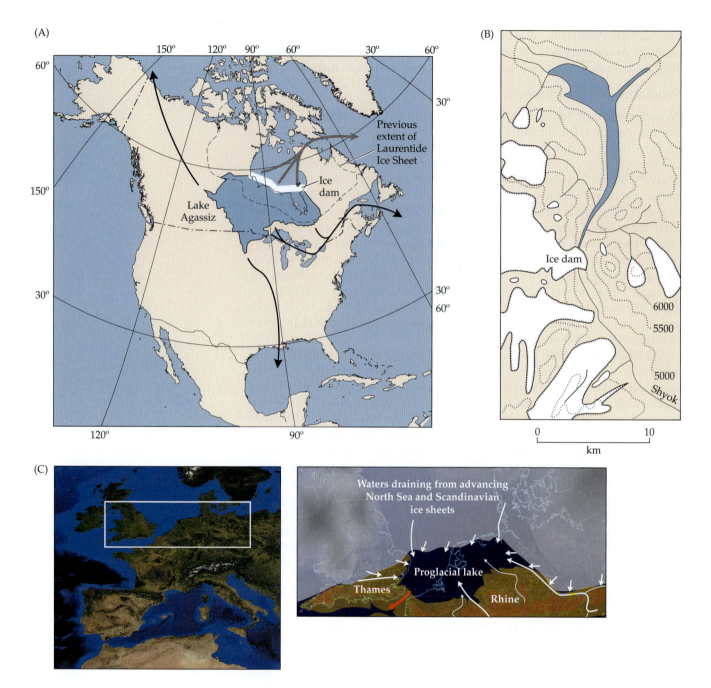

FIGURE 9.23 Many of today's lakes were formed by glacial activity. (A) Glacial Lake Agassiz, approximately 9000 BP, with its outflows (arrows), including its catastrophic release (gray arrows) of some 163,000 km³ of freshwater into the Tyrrell Sea (Hudson Bay) and the Atlantic Ocean once its ice dam burst around 8000 BP. (B) Shyok Ice Lake along the western margin of the Himalayas, an example of an extant, postglacial lake formed when retreating glaciers acted as dams, with glacial meltwater accumulating in valleys carved by previous glacial activity. (C) This proglacial lake formed as glaciers expanded southward from the North Sea and Scandinavia and blocked the Rhine and Thames river drainages (white arrows) between 450,000 and 425,000 years ago. Lake levels continued to rise and eventually rose above and then eroded through (red arrow) the rock dam that once lay over the Dover Strait, releasing a megaflood that carved out the English Channel at about 400,000 BP. (D) Formation of kettle lakes: 1, retreating ice sheet leaves stagnant ice blocks on the outwash plain; 2, lakes are formed in the outwash and in the till by melting blocks of ice; 3, a large, partly buried ice block melts, causing irregular sliding of the outwash originally covering the sides of the block. (E) Formation of a glacial plunge pool lake. Glacial meltwater sometimes formed large rivers that flowed over the surface of a retreating glacier, then plunged as much as 3 km down the face of the glacier to carve out a roughly circular lake basin below. (A after Teller et al. 2002; B and D after Hutchinson 1959; C from Gibbard 2007, after Gupta et al. 2007.)

(D)

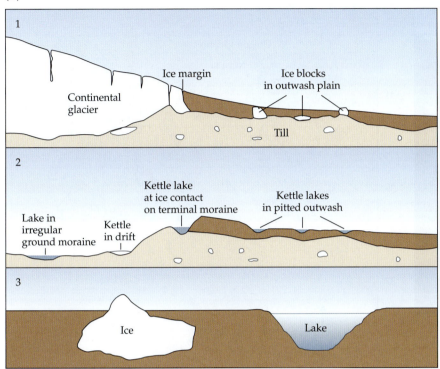

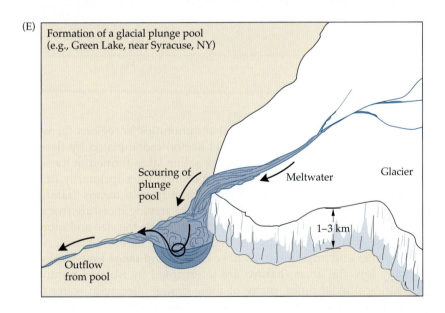

(E)

arctic. The largest blocks of ice created deep and persistent depressions that later filled to form **kettle lakes** (**Figure 9.23D**). The legacy of glacial recession can also be seen in the relatively deep, roughly circular type of lake called a **plunge pool**, which was formed by glacial meltwaters that flowed over the surface of a glacier before plunging off the edge to carve a basin in the Earth some 2 to 3 km below (e.g., Green Lake near Syracuse, New York; **Figure 9.23E**). Countless other lakes were formed by the scouring action of glaciers and meltwater, which formed basins, many of them dammed by the deposition of rock debris along glacial moraines (e.g., Lake Mendota and Devil's Lake of Wisconsin, and the Finger Lakes of New York).

(A)

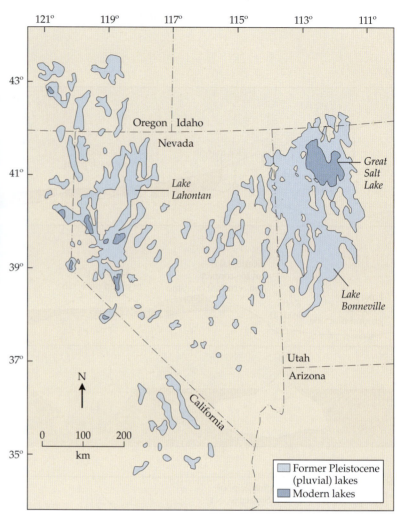

FIGURE 9.24 (A) Distribution of pluvial lakes in western North America during the most recent (Wisconsin) glacial maximum. During glacio-pluvial periods, most of the arid region of the continent experienced a wetter and cooler climate, and lakes and marshes filled what are now desert valleys. Potential range expansions and contractions of hypothetical warm stenothermal (B) and cool stenothermal (C and D) marine taxa. During a glacial period, cool stenothermal taxa could penetrate from one hemisphere to the other (middle diagram, D) and could possibly achieve bipolar distributions when waters rewarm and their ranges contract toward the poles. (A after Benson and Thompson 1987; Elias 1997; B–D after Crame 1993.)

Pluvial lakes in arid regions

Glacial cycles affected plant and animal communities in regions far withdrawn from glaciers, including those that are now dominated by deserts. During pluvial periods, large freshwater or saline lakes formed in these regions because of a combination of low evaporation rates (primarily due to lower temperatures) and high precipitation rates (*pluvial* means "rainfall") sometimes fueled by shifts in prevailing winds carrying moisture from the oceans (see Figure 9.7). Few biogeographers appreciate the size and number of pluvial lakes that existed in places that now have desert climates. Aridlands of Nevada and adjacent areas have a **basin-and-range topography**, in which many low, flat areas (**basins**) are interrupted by isolated mountain ranges. During pluvial times, many of these basins filled with water (**Figure 9.24A**). The largest such water body was Lake Bonneville in Utah and parts of Nevada and Idaho. At times it contained freshwater and drained northward into the Snake and Columbia Rivers. In the Middle Wisconsin, this lake was 330 m deep, had an area exceeding 50,000 km² (slightly smaller than present-day Lake Michigan), and supported a diverse freshwater community, including cutthroat trout (*Salmo clarki*). The present Great Salt Lake is a small remnant of Lake Bonneville. Nevada, southern Oregon, eastern California, southeastern Arizona, and southwestern New Mexico had numerous large and small pluvial lakes. Yet most of these lakes had evaporated by 10,000 years BP. One of these lake basins, Death Valley, with the lowest elevation in North America (93 m *below* sea level), now contains perhaps the most extreme desert on the continent.

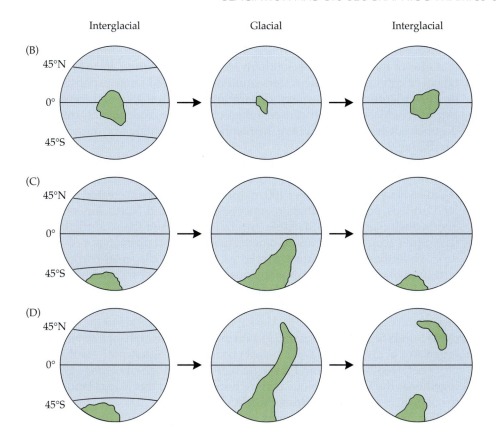

Pluvial lakes were also present in what are now deserts on other continents. Their remnants—saline lakes and dry lake beds—are abundant in the Atacama Desert of northern Chile, the Monte Desert of Argentina, many areas in interior Australia, the region of the Dead Sea in the Middle East (ancient Lake Lisan), the Kalahari Desert of southern Africa, and many places in arid and semiarid Asia (Flint 1971). The western portion of the Sahara was still relatively mesic until 5000 years ago. Perhaps the most remarkable of these remnants of glacio-pluvial periods is Lake Chad. This lake, now just 16,000 km^2, was once 950 km long and covered over 300,000 km^2 (the present size of the Caspian Sea), including a significant portion of the southern Sahara. Lake Chad remained at this maximum size from 22,000 to 8500 years BP.

The disappearance of pluvial lakes during the Holocene had several profound biogeographic effects. It caused the wholesale extinction of many plants and animals living in or around these bodies of water. In addition, the dissection of large lakes into smaller, isolated units led either to local extinctions or to the vicariant speciation of surviving forms, as with the pupfishes (*Cyprinodon*) of the southwestern United States (see Figures 7.19 and 10.4; Miller 1961b; Smith 1981).

Biotic Exchange and Glacial Cycles

Figures 9.14 and 9.21 demonstrate how downward shifts of high-elevation vegetation (primarily forests) during glacial maxima could have created avenues of dispersal: Such a lowering would allow a plant or animal species, previously isolated in montane forests of individual peaks, to cross ridges and migrate along mountain ranges and mesic lowlands. Glacial cycles may have had similar effects on the distributions of many marine organisms (**Figure 9.24B–D**; see Figures 9.12 and 9.22). In the case of cold-water stenothermal species, relatively warm tropical waters serve as an effective physiological barrier to dispersal, limiting their distributions to the middle or high latitudes of the Northern or Southern Hemisphere. During glacial maxima, however, cooling of marine waters could have allowed range expansion into the lower latitudes. Subsequent rewarming of tropical waters during interglacials could have again caused range contraction and possibly result in disjunct northern and southern hemispheric distributions of these species (see Figure 9.24D).

As noted earlier, eustatic and isostatic changes in sea level greatly altered opportunities for biotic exchange for both terrestrial and marine biotas. Drops in sea level by sometimes over 100 m during the Wisconsin created extensive land bridges, such as Beringia (connecting Siberia and North America), the Sunda Shelf (connecting Malaysia and Indonesia), and the Arafura Sea and Bass Strait (connecting New Guinea, Australia, and Tasmania; see Figure 9.9). While these land bridges served as important dispersal corridors for terrestrial organisms, they simultaneously eliminated or fragmented marine biotas. In most cases, however, biotic exchange tended to be asymmetrical, with more species migrating from larger (species-rich) to smaller areas than vice versa (e.g., from Siberia to Alaska; from Southeast Asia to the "islands" of the Sunda Shelf; from Australia to Tasmania). Biotic exchange of terrestrial organisms across Beringia (see below) contributed significantly to the similarity between Nearctic and Palearctic biotas, which are sometimes grouped together as the Holarctic biota. Our own species used this glacial land bridge to colonize North America from Siberia.

Biotic exchange among regions of the marine realm often tended to be asymmetrical, depending on the size and diversity of each species pool and the ocean currents and other factors influencing dispersal (see Briggs 2004b,c; Vermeij 2004). While Beringia served as a dispersal corridor for terrestrial organisms during glacial maxima, the Bering Strait was an important corridor for the dispersal of marine life during interglacial periods. Again, more species migrated from the larger, more species-rich region—that is, from the Pacific basin northward. For example, Durham and MacNeil (1967) reported that during the Late Cenozoic, 125 species of marine invertebrates invaded the Arctic–Atlantic region from the Pacific, while no more than 16 species colonized in the reverse direction (see also Vermeij 1991a,b).

In Chapters 6 and 8 we discussed the tectonic events that ultimately formed a Central American land bridge between North and South America (about 2.5–3.5 million years ago). The resultant waves of biotic exchange between Nearctic and Neotropical biotas, referred to as the **Great American Biotic Interchange**, were made possible not just by tectonic events but by eustatic changes and vegetative shifts associated with glacial cycles. During glacial maxima, the lowering of sea levels increased both the area and elevation of the Central American landbridge. Perhaps just as important, the relatively dry conditions that prevailed during glacial maxima caused savannas to expand toward the Equator and form a continuous habitat corridor for the migration of many species adapted to these open habitats (savannas and shortgrass prairie; see Webb 1991). We shall return to the profound effects of these waves of biotic interchange in Chapter 10.

Glacial Refugia

Nunataks are refugia that persisted within, or adjacent to, ice sheets. A nunatak might occur along the continental periphery of a glacier, internally as an isolated pocket along a mountain range (e.g., in a snow shadow or along a steep, ice-free slope), along a coastline where bluffs were too steep for glaciers, or between adjacent ice sheets. Geologists and biogeographers have identified several possible "internal" nunataks that may have served as important refugia for a diversity of species during the last glacial maximum. The Laurentide Ice Sheet covered most of northeastern and north central North America (**Figure 9.25**). Non-glaciated regions within the North American ice sheets include the so-called **driftless area** in southern Wisconsin and adjacent Illinois and Iowa, an elliptical area that was bypassed by the glacial front. A number of narrow nunataks may also have occurred in intermountain valleys between the Laurentide and Cordilleran ice sheets (see Figure 9.7B).

In contrast to the questionable importance of these relatively small, internal nunataks, those along the periphery of glaciers may well have provided refugia for a great many species. In North America, at least three large refugia

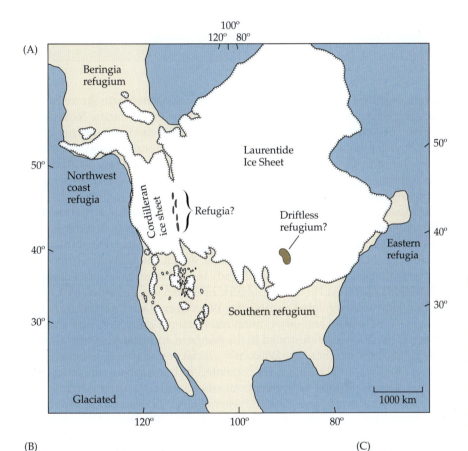

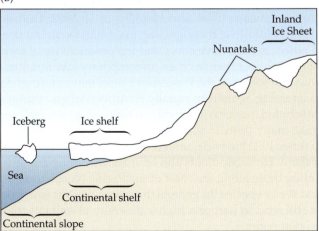

FIGURE 9.25 (A) Even during the Wisconsin glacial maximum, ice-free refugia may have occurred between the Laurentide and Cordilleran ice sheets and in a region called the Driftless Area. (B and C) Ice-free areas in mountainous regions along the Pacific coast may also have served as refugia—and possibly as migration corridors—for plants and animals during full glacial conditions. (A after Rogers et al. 1991; B and C after Pielou 1991.)

FIGURE 9.26 The high frequency of endemic plant species in central Alaska was used by Hultén (1937) to advance the idea that a large part of Beringia remained unglaciated and served as a refugium for Arctic forms during the Pleistocene. Each contour line indicates the number of narrow endemics present in that region.

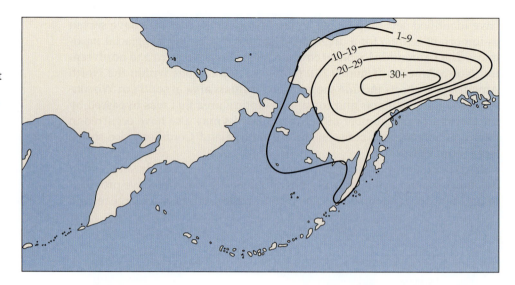

persisted north of the glaciers during the Wisconsin in North America: the expansive iceless areas of Beringia, coastal regions of the Pacific Northwest, and much of Nova Scotia (see Figure 9.25). Many species that were unable to disperse with their shifting habitats persisted, and in some cases diversified, in these refugia. Thus, many Holarctic biogeographic patterns bear witness to the importance of these glacial refugia in patterns of diversity, endemism, and phylogenetic relatedness among extant species.

The thesis that Beringia was a Pleistocene refugium was clearly presented by the phytogeographer E. Hultén (1937) (**Figure 9.26**). Hultén's work on this subject is classic, and one of the best efforts to document plant distributions on a broad scale. His detailed distribution maps, which stimulated an important creative synthesis, should be required study for all biogeographers. Although Hultén's idea was ahead of its time, several lines of evidence helped to confirm that Beringia was vegetated during the Wisconsin glacial maximum. First, Colinvaux (1981, 1996) and others analyzed fossil pollen records and showed that mesic tundra was widespread in the northern and central portions of the land bridge at the height of the Wisconsin glacial period (see Elias et al. 1996). Second, Hopkins and Smith (1981) presented rather strong evidence for the occurrence of deciduous dicotyledonous trees and larch (*Larix*, Pinaceae) in the Yukon at this same time. Finally, Weber et al. (1981) and others excavated Late Pleistocene mammalian fossils (aged from 40,000 years BP) in the interior of Alaska, finding a great diversity of large ungulates (woolly mammoth, woolly rhino, barren-ground and forest musk oxen, mountain sheep, steppe antelope, reindeer, horse, and camel; see also Hopkins et al. 1982; Guthrie 1990). Although it is counterintuitive, these grazing mammals occupied Beringia during a period when its vegetative cover and the productivity of their preferred forage were much lower than in other contemporary communities. The waves of extinctions that followed the Wisconsin glacial maximum, when ungulate forage was increasing, remain an equally controversial and intriguing paradox and are attributed, perhaps justifiably, to humans, as we shall see below. While it may take many years to solve this mystery, it is clear that the Bering land bridge, at times 1000 km wide, served as an important refuge and dispersal corridor for many Holarctic plants and animals, including humans. Smaller refugia, including those along the coast of the Pacific Northwest (see Figure 9.25), supported fewer species. In general, however, regions formerly within glacial refugia still tend to harbor a higher diversity of animals than those in formerly glaciated regions.

In the absence of gene flow among refugia, populations diverged genetically, often to the specific level in small mammals or to the subspecific level in larger species. Examples of such potential divergence between populations of small mammals in Beringia and those in regions south of the glaciers have included northern and southern red-backed voles, tundra and Arctic shrews, and Arctic and Columbian ground squirrels (see Hoffman 1971; Nadler et al. 1973, 1978). Between large mammals they include the divergence of moose (*Alces alces gigas*) and Beringian Dall's sheep from their southern counterparts (Peterson 1955; Korobytsina et al. 1974). See Weksler et al. (2010) for a synopsis of phylogeographic evidence that in some cases supports these patterns, while in other cases refutes them. In contrast, smaller refugia of the Pacific Northwest not only supported fewer species, but their populations underwent less divergence than those inhabiting Beringia. Shrews, deer mice, voles, marmots, chipmunks, brown bears, and ermines exhibited modest divergence, especially those isolated on islands of the Queen Charlotte Archipelago (see Banfield 1961; Hoffmann et al. 1979; Hoffman 1981; Riddle 1996; Paetkau et al. 1998).

Anthropological data also bear strong witness to the legacy of glacial refugia. Biogeographic patterns in the dental traits, genetic markers, blood proteins, and linguistic diversity of indigenous peoples appear to parallel those found in other mammals. Diversity in these traits (e.g., number of languages per unit of area) tends to remain higher where refugia occurred, and highest in large refugia (i.e., higher south of the ice sheets than in Beringia, and higher in Beringia than in the northwestern refugia; see Rogers et al. 1991).

While we have focused on North America, this brief discussion also highlights that refugia were not peculiar to North America. Biogeographic patterns and, in particular, phylogenetic analyses have identified locations of principal glacial refugia in the peninsula regions of southern Europe, along with smaller ("cryptic") refugia that may have persisted along the snowless intermountain regions and along the coastal regions and edges of the Scandinavian and North Sea ice sheets (**Figure 9.27**; see Hewitt 1999; Willis and Whittaker 2000; Stewart and Lister 2001; Provan and Bennett 2008). In summary, the repeated formation and dissolution of refugia, dispersal corridors, and barriers throughout the Pleistocene left a lasting imprint on distributions and ecological and genetic diversity of nearly all biotas, including those far from the glaciers and our own species as well (see Chapter 16; see also Stewart and Lister 2001; Forster 2004; Gamble et al. 2004; Hewitt 2004; Willis et al. 2004; Provan and Bennett 2008; Fagan 2009). Moreover, there may have

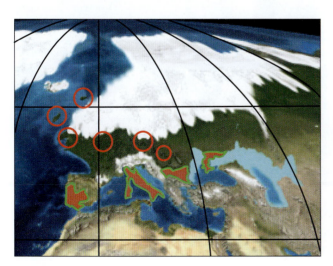

FIGURE 9.27   Lowland regions of the Iberian, Italian, and Balkan Peninsulas were principal, glacial refugia (red shading) for a diversity of plants and animals as they dispersed southward of the advancing ice sheets during glacial periods of the Pleistocene. Phylogenetic studies are now revealing isolated, so called cryptic, refugia (red circles) in regions that, by virtue of the prevailing winds and their proximity to warm ocean currents or mountain ranges, remained snowless during glacial maxima. (After Stewart and Lister 2001; Provan and Bennett 2008; base map courtesy of Ron Blakey, Northern Arizona University.)

(A) Saber-toothed cats and imperial mammoths of North America

(B) Saber-toothed cats and litoptern ungulates of South America

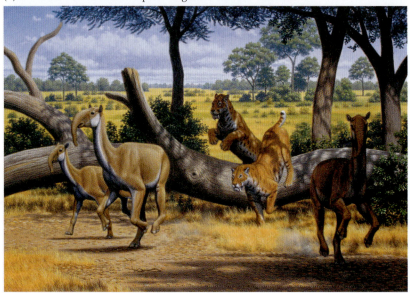

(C) Marsupial lions and short-faced kangaroos of Australia

FIGURE 9.28 Megafaunal mammals, reptiles and birds were widespread across all major landmasses, except Antarctica, until the latter stages of the Pleistocene and early Holocene Epochs. (Courtesy of Mauricio Antón, http://www.mauricioanton.com/)

been only relatively small areas globally in which reiterative expansion and contraction of species ranges and resorting of assemblages as a result of oscillations in climate throughout the Pleistocene have not been evident. We return to this intriguing subject—the evolutionary legacies of dynamics in the geographic template during the Pleistocene and earlier periods—in the next two chapters.

Extinctions of the Pleistocene Megafauna

For many of us, the term *megafauna* conjures up vivid images of Africa's elephants, hippos, rhinos, and giraffes, herds of swift-footed antelope, and the powerful predators that hunt them. This menagerie of great beasts in the Old World tropics was the basis for the eighteenth century naturalist Comte de Buffon's criticism of the New World, whose fauna was, by comparison, diminutive and "degenerated" (Buffon 1766). Thomas Jefferson, then president of the United States, was understandably keen on disproving Buffon's criticisms, and so he appointed Lewis and Clark to lead the Corps of Discovery to, among other things, prove that North America had at least some beasts that rivaled those of the Old World tropics.

As it turned out, Buffon was correct, but only partially; while naturalists failed to discover the great beasts that Jefferson was hoping for, paleontologists would soon unearth incontrovertible evidence that North and South America, and indeed other continents and certain large islands as well, were once inhabited by a rich diversity of huge mammals, birds, and reptiles (**Figure 9.28**). By the middle decades of the nineteenth century, it became clear to natural scientists, including Charles Darwin and Alfred Russel Wallace, that the native megafauna experienced mass extinctions on nearly every major landmass, save for Africa.

> It is impossible to reflect on the changed state of the American continent without the deepest astonishment. Formerly it must have swarmed with great monsters: now we find mere pigmies, compared with the antecedent, allied races."... "If Buffon had known of the gigantic sloth and armadillo-like animals, and of the lost Pachydermata, he might have said with greater semblance of truth that the creative force in America had lost its power, rather than that it had never possessed great vigour. (Darwin 1839: 448)

> We live in a zoologically impoverished world, from which all the hugest, and fiercest, and strangest forms have recently disappeared. (Wallace 1876: 150)

Given our descriptions throughout this chapter, it seems logical to assume that the principal drivers of these megafaunal extinctions were the 20 or so climatic upheavals that characterize the Pleistocene. Over the latter decades of the last century, however, paleobiologists and biogeographers discovered features of the megafaunal extinctions that seemed inconsistent with the climate-based hypotheses—namely, the species selectivity and temporal–spatial signature of this geologically very recent wave of extinctions.

First, megafaunal extinctions were just that—extinctions of the largest mammals, reptiles, and birds, while many if not most species in the smaller size classes were spared (**Figure 9.29**). It is not readily intuitive why the largest vertebrates would be at a disadvantage under climate change, although hypotheses based on fragmentation of the once-expansive ecosystems required by these megaherbivores and their predators may have some merit. Two other, interrelated features of the extinctions of the Pleistocene megafauna, however, seem especially problematic for a climate-based hypothesis. The latter hypothesis predicts that extinctions should be roughly synchronous across the globe and that they are most likely to have occurred during the initial

(A) All continents combined

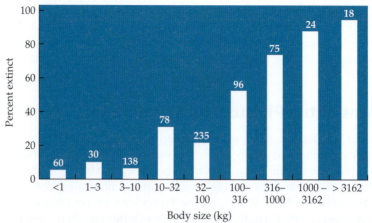

(B)

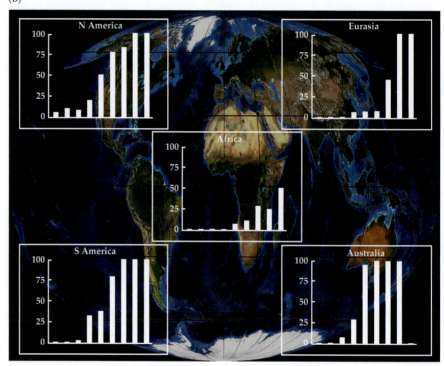

cycles of climate change, or at least during a stage that is especially severe. Yet the megafaunal collapse occurred at different times on different continents, and during the last stages of the Pleistocene or during the Holocene. Why would this diversity of huge vertebrates, which survived and often diversified during the previous 20 or so glacial–interglacial cycles, suffer something close to annihilation in the most recent one? The intensity of the last glacial maximum and the subsequent warming period were not remarkable in comparison with those during earlier periods, and extinctions occurred at different glacial–interglacial stages across different continents (**Figure 9.30**). A final anomaly that challenges climate-based explanations for megafaunal extinctions is Africa; why were most of the large vertebrates spared there but destroyed across all other landmasses across the globe?

The singular nature of Africa's surviving megafauna may have provided the essential clue, and one that Jared Diamond—with the aid of the world's

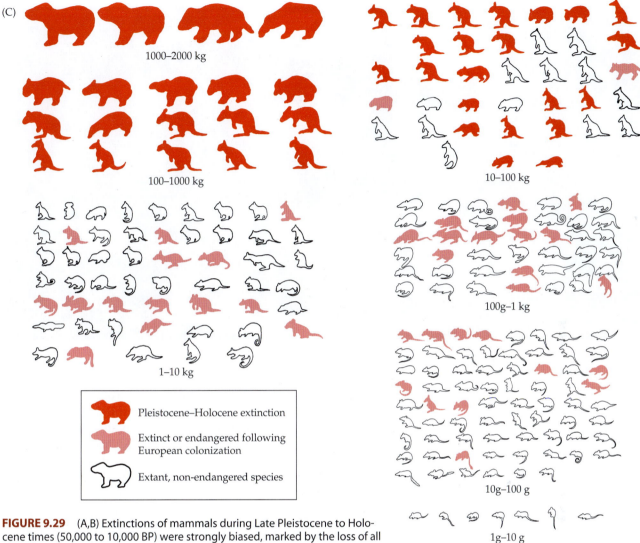

(C)

1000–2000 kg

100–1000 kg

10–100 kg

1–10 kg

100g–1 kg

10g–100 g

1g–10 g

Pleistocene–Holocene extinction

Extinct or endangered following European colonization

Extant, non-endangered species

FIGURE 9.29 (A,B) Extinctions of mammals during Late Pleistocene to Holocene times (50,000 to 10,000 BP) were strongly biased, marked by the loss of all megafauna and most of the very large mammals, except in Africa, while most of the small mammals survived to recent periods. (Body size categories in the graphs of part B are the same as those in A; the number above a column in A indicates the number of species in that size category known to exist at the beginning of this period.) (C) Selective extinctions of large, megafaunal mammals in Australia during the Pleistocene and Holocene. Shown are outline drawings of species known to occur in these regions when they were colonized by aboriginal humans. Species suffering extinction during the Pleistocene and early Holocene are shown in dark red, while those that became extinct or endangered following European colonization are shaded pink (open outlines indicate extant, non-endangered species). (A, B source data from Koch and Barnosky 2006, base map courtesy of Ron Blakey, Northern Arizona University; C information courtesy of T. F. Flannery, art based on original drawings by Tish Ennis, Australian Museum, Sydney.)

most famous detective, Sherlock Holmes—seized on to explain this mystery of the lost megafauna.

In Sir Arthur Conan Doyle's *Silver Blaze,* Holmes called attention to "the curious incident of the dog in the night-time." When the stable owner observed that "the dog did nothing in the night-time," Holmes pounced on him—"that was the curious incident"—for it indicated that the stables pet was familiar with the "intruder." As Diamond (1984) hypothesized, the "intruder" in the case of megafaunal extinctions during the Pleistocene was

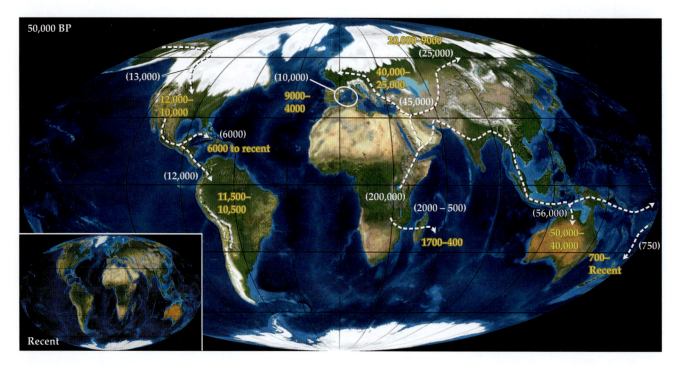

FIGURE 9.30 The collapse of mega-faunal assemblages of mammals, birds, and reptiles were asynchronous across the globe (approximate dates of extinction periods in years BP are in yellow), but they began soon after the arrival of ecologically significant humans (dates in parentheses and white). (Extinction chronologies after MacPhee and Marx 1997; Burney et al. 2003; Fiedel 2009; MacPhee 2009. Chronologies of colonization by populations of ecologically significant humans simplified from Figure 16.32; base maps courtesy of Ron Blakey, Northern Arizona University.)

Homo sapiens. As populations of ecosystem-transforming humans colonized different landmasses, they carried with them the powers of an advanced novel hunter, a developing ecosystem engineer capable of transforming native landscapes with fire, and a global-scale vector of vertebrate pests and disease (Martin 1967, 2005; Flannery 1994; Diamond 1997; MacPhee and Marx 1997; Alroy 1999, 2001; Haynes 2009a). Although basing the following observations on the fragility of insular biotas, Darwin's comments apply equally as well to megafaunal extinctions on the continents:

> We may infer from these facts, what havoc the introduction of any new beast of prey must cause in a country, before the instincts of the indigenous inhabitants have become adapted to the stranger's craft of power. (Darwin 1860: Chapter 17)

As powerful and once-dominant as they were, the megafauna were ecologically naive—lacking any adaptations to the "strangers' craft of power" on all continents save for Africa, where ecologically significant humans evolved and likely coevolved with the native biota. The notion of an ecologically naive biota has developed more broadly into a common explanation for why the influences of a biotic invasion on native faunas—for example, the Great American Biotic Interchange mentioned previously and discussed in Chapter 10—tend often to be highly asymmetric in nature. As early humans in Africa developed each new tool or hunting strategy, the native megafauna had the opportunity to adapt in a continuing arms race that is characteristic of most predator–prey relationships. But once early tribes migrated out of Africa, they encountered ecologically naive prey and quickly (i.e., within centuries to millennia) overwhelmed them. Accordingly, the sequence of megafaunal collapses across the globe is consistent with human colonization of those distant lands—some collapses occurring during glacial periods, transitional stages, or interglacials (**Figure 9.31**), but on the heels of colonization by ecologically significant humans.

Because these events involved some of the most charismatic species ever to exist, and because the above hypothesis challenges the notion of the "no-

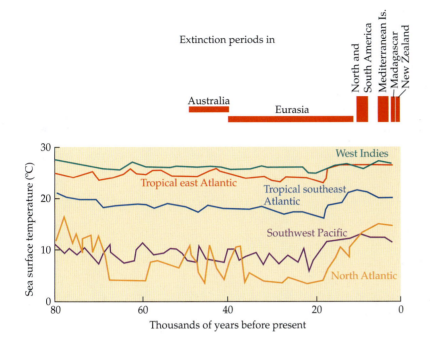

FIGURE 9.31 Extinctions of megafaunal mammals, birds, and reptiles on different landmasses were far from synchronous (extinction periods are indicated by labeled bars at the top of the chart); instead they occurred during different stages of the glacial–interglacial cycles (here indicated by sea surface temperatures in different oceanic regions). (Modified from Koch and Barnosky 2006; extinction chronologies after MacPhee and Marx 1997; Burney et al. 2003; Fiedel 2009; MacPhee 2009.)

ble savage" assumed to have lived in harmony with the environment, it is understandable that this is a subject of continuing, and often much heated, debate (e.g., see Graham 1986, 2001; Graham et al. 1996; Grayson 2001; Brook and Bowman 2002; Grayson and Meltzer 2002, 2003; Lyons et al. 2004; Burney and Flannery 2005, 2006; Surovell et al. 2005; Koch and Barnosky 2006; MacFadden 2006; Wroe et al. 2006; Gillespie 2008; Carrasco et al. 2009; various chapters in Haynes 2009b; Roberts and Brook 2010). Although scientists in this, or any arena, for that matter, may never reach complete consensus, continuing research in this area should provide valuable insights, not only on the nature and causes of extinctions of the Pleistocene megafauna, but on causes of imperilment and the most promising strategies for conserving the world's surviving megafauna. We take up this subject, insights from biogeography for conserving biological diversity, in Chapter 16.

In the following chapters (Unit 4), we dig deeper into the paleontological record to reconstruct the geologic, climatic, and evolutionary processes that influenced the origins, diversification, and distributions of life on Earth. To paraphrase George Gaylord Simpson, our ability to understand and effectively curtail the ongoing wave of extinctions may well depend on our ability to learn the lessons of these prehistoric extinctions. Finally and just as important, ecologists, conservation biologists, and others studying extant biotas should bear in mind that the many waves of megafaunal extinctions during the Pleistocene may have fundamentally changed the geography of nature. Given the diversity, numbers, and presumed ecological dominance of the thousands of now vanished species, many if not most of the patterns and processes we study are those of highly altered biotas.

UNIT **FOUR**

EVOLUTIONARY HISTORY OF LINEAGES AND BIOTAS

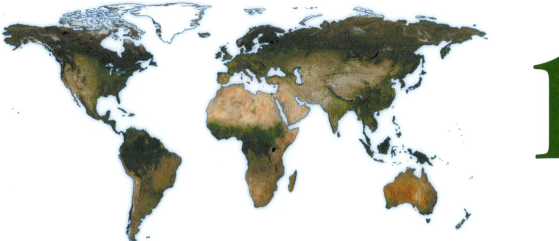

THE GEOGRAPHY OF DIVERSIFICATION

In Unit 3, we learned that the evolutionary fates and distributions of species have been dynamic throughout the history of life on Earth. New species have continuously evolved from ancestral species, and occasionally, large numbers of species have been eliminated in an episode of mass extinction followed by the evolution of entirely new sorts of species. Species distributions often have changed, either through jump dispersal of founding individuals or populations across a barrier, or through more gradual range expansion over continuous expanses of suitable habitats. Some of the most profound changes in distributions have come when entire groups of species have crossed from one biogeographic region to another following the erosion of a barrier. Additionally, we learned that the geology of the Earth itself changes continuously through time and that at various times and particularly over the past several million years, climatic cycles have produced an exceedingly dynamic ecological arena. In order to avoid extinction, species and biotas have had to respond to frequent and unpredictable changes in habitat structure and distribution through dispersal or adaptation. In short, while evolutionary diversification occurs over time, it also has a distinct spatial dimension, and as a way of thinking about the contributions of both, we recognize here two kinds of "histories" that have shaped the characteristics of contemporary biotas, which we will call the *history of place* and the *history of lineage* (Brown 1995).

The **history of place** is the history of the Earth itself: the changes in geography, geology, climate, and other environmental characteristics that are extrinsic to the particular organism, lineage, or biota being studied. We like to think of the history of place as the environmental and geographic template that each organism, lineage, or biota has experienced during its own unique evolutionary history. Thus, the history of place includes the geological and climatic history of the Earth; changes in soils, climate, and oceanographic and limnological conditions; and the varying composition of biotic communities. Past environments may have influenced a contemporary lineage in many ways by influencing the abundance, distribution, and adaptive evolutionary changes of its ancestors. Even the most distantly related kinds of organisms that lived together in the past at a particular place—and hence, experienced some features of a common environment—share some history of place.

The **history of lineage** is the series of changes that have occurred in the intrinsic characteristics of organisms, species, or higher taxa over generations of evolutionary descent. These are changes in heritable characteristics, derived from and constrained by the characteristics of the organisms' ancestors. Only to the extent that organisms share common ancestors do they share a history of lineage. The history of place strongly influences the history of lineage, because characteristics of past environments (e.g., geology, climate, and biotic composition) undoubtedly influenced the survival, distribution, and diversification of all lineages that occurred in those places. But the converse is less true. The history of lineage influences the history of place only to the extent that the activities of particular organisms altered the past environment for themselves and other lineages. While many kinds of organisms substantially modify their environments—the building of reefs and atolls by corals, and the modification of the climate of the Amazon basin by rain forest trees are just two examples—usually these influences are diffuse and cannot be attributed to just one lineage or taxonomic group.

In this chapter, we describe the fundamental biogeographic patterns that emerge from a history of diversification, dispersal, and extinction of lineages and biotas on a geologically and climatically dynamic Earth. We first consider general patterns of endemism, provincialism, and disjunction and the abiotic and biotic features that function to maintain geographically separate biotas. We then consider what happens when barriers between previously separate biotas erode, opening corridors for biotic interchange. We finish this chapter with a discussion of several general evolutionary trends that tend to arise when lineages and biotas have long been separated by geographic barriers. In Chapter 11, we present an overview of approaches to reconstructing the history of lineages, and then in Chapter 12 we build on that discussion by introducing the variety of approaches available for reconstructing historical relationships between lineages, biotas, and places.

The Fundamental Geographic Patterns

The most pervasive feature of geographic distributions is the fact that they have limits. No species is completely **cosmopolitan** (organisms that are widely distributed throughout the world), and most species and genera, and even many families and orders, are confined to restricted regions, such as a single continent or ocean—the term **endemic** means occurring in one geographic place and nowhere else, and it can be applied at any taxonomic scale. Many distributions are extremely limited. The minute, redfinned blue-eye (*Scaturiginichthys vermeilipinnis*), the sole species in the fish genus *Scaturiginichthys*, lives in a handful of tiny springs in arid western Queensland, Australia (Ivantsoff et al. 1991), which it shares with another equally restricted species, the Edgbaston goby (*Chlamydogobius squamigenus*; Larson 1995; **Figure 10.1**). On the other hand, even very depauparate taxa can also have broad distributions. The avian family Opisthocomidae contains only a single species, the bizarre hoatzin (*Opisthocomus hoazin*)—a nearly flightless, leaf-eating bird— but in contrast to the redfinned blue-eye, the hoatzin occurs across a fairly large expanse of the Amazon and Orinoco river drainages of northern South America (see Figure 10.1). More diverse taxa tend to be endemic within larger areas. For example, the avian family Furnariidae (ovenbirds), with more than 65 genera and 300 species, is endemic to South and Central America plus sev-

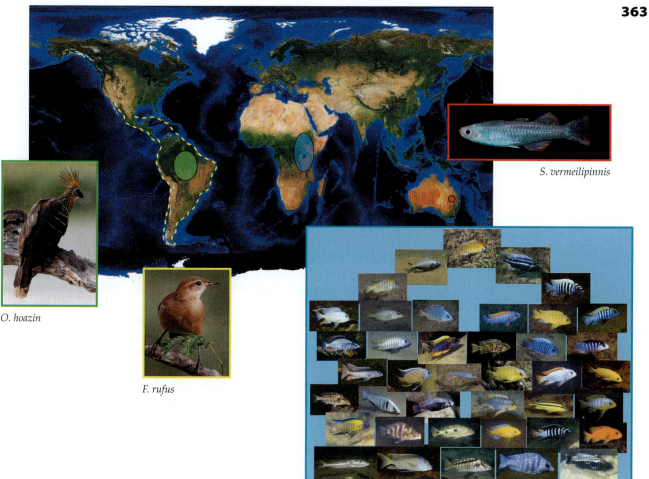

S. vermeilipinnis

O. hoazin

F. rufus

eral neighboring islands (one species extends north to northern Mexico; see Figure 10.1). Several large fish families, including catfish eels (Plotosidae), are confined to the Indo-Pacific region. In South and Central America, there are over 50 families and subfamilies of flowering plants that occur nowhere else; many of these are small taxa, but they also include several extremely diverse families. Cacti (Cactaceae) and bromeliads (Bromeliaceae) would be endemic to the New World if one species of each had not recently crossed the ocean by natural means (long-distance dispersal) and become established in Africa. However, there are notable exceptions, with even some diverse taxa having very restricted distributions. We have already discussed, for example, the extreme diversity of cichlid fishes in the Great Lakes of the African Rift Valley (see Chapter 7; see Figure 10.1).

Endemics are not distributed randomly, but tend to overlap with other geographically restricted species in certain regions, a phenomenon that is called **provincialism**. Australia, southern Africa, Madagascar, New Zealand, New Caledonia, Wallacea, and southern California, for example, are terrestrial, and the Indo–West Pacific (or Coral Triangle) and Caribbean are marine hotspots of endemism, each containing a large percentage of endemic species and numerous endemic higher taxa. Alternatively, other regions such as Europe, northern North America, and the southern Atlantic Ocean share much of their biotas with other areas. Often, different groups of plants and animals occur, not only in the same ocean or on the same continent or island, but also in the same localities and habitats within those regions (recall the redfinned blue-eye and the Edgbaston goby examples given above; see Figure 10.1). These coincident distributions of endemics often do not coincide precisely with the present boundaries of continents and oceans, and they certainly do not always coincide with obvious characteristics of abiotic and biotic envi-

FIGURE 10.1 Endemism occurs across a broad range of geographic scales and is often decoupled from taxonomic diversity or phylogenetic distinctness. The hoatzin (*Opisthocomus hoazin*) represents a highly unique species and genus of bird in the Amazon and Orinoco river basins of northern South America (green area). The ovenbirds (Furnariidae: the Rufous hornero, *Furnarius rufus*, shown here) represent a diverse Neotropical family of suboscines with over 65 genera and 300 species (yellow dashed area). The African cichlids represent a complex of over 1400 very closely related species in 12 African lakes (blue area) (see Chapter 7) The redfinned blue-eye (*Scaturiginichthys vermeilipinnis*) represents a single species in the genus and occurs with the Edgbaston goby (*Chlamydogobius squamigenus*) in a handful of desert springs in a localized spring complex in Queensland, Australia (red area). (*O. hoazin* © Marshall Bruce/istock; *F. rufus* courtesy of Wagner Machado Carlos Lemes; cichlids from Turner, 2007; *S. vermeilipinnis* courtesy of Cady Bryce.)

ronments. As pointed out in Chapters 4, 6 and 16, the many successful introductions of species by humans demonstrate that species can thrive in regions far from their native habitats. These introductions emphasize the unique influences of historical events in determining where organisms occur today. In many cases, the spread of taxa from regions in which they evolved has been blocked by barriers to dispersal. In other cases, one species of a once widespread group has persisted in a limited area after its representatives in other regions have become extinct.

Disjunctions are those cases in which populations, closely related species, or higher taxa occur in multiple but widely separated regions, but are absent from intervening areas. Disjunct distributions reflect past events: The disjunct forms dispersed long distances over geographic barriers were carried to distant sites aboard crustal plates as they drifted apart, or they are the surviving remnants of a once-widespread taxon. Usually the disjuncts are morphologically similar and inhabit similar environments. The evergreen southern beeches (*Nothofagus*), for example, grow in wet, cool temperate forests in South America, New Zealand, New Caledonia, New Guinea, New Britain, and Australia (**Figure 10.2**). However, disjuncts that are phylogenetically related are not always biologically and ecologically similar. The flightless ratite birds (superorder Paleognathae) have a disjunct Southern Hemisphere distribution similar to that of the southern beeches and other groups (both living and known only from fossils) that once occurred on Gondwanaland (see Box 14.1, Figure C). Most ratites, such as the ostriches of Africa, emus of Australia, and rheas of South America, inhabit open, relatively arid habitats. The cassowaries of New Guinea and northeastern Australia, however, occur in tropical rain forests. The ecologically and morphologically most divergent

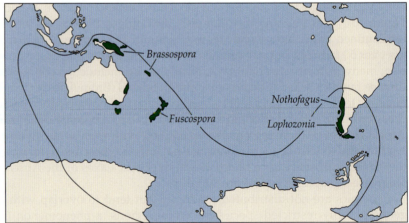

FIGURE 10.2 Recent (dark shading) and fossil (outline) distribution of the Southern Hemisphere beeches, *Nothofagus* (Fagaceae), and four extant subgenera within this genus shown on the map and summarized in graph format. This distribution is disjunct and also probably relictual to some degree, with representatives of the genus presently occurring on many of the widely dispersed fragments of the ancient southern continent of Gondwanaland, although recent molecular-based studies have indicated instances of historical dispersal between landmasses following the breakup of Gondwanaland (Cook and Crisp 2005b; Knapp et al. 2005). Also, note that the fossil record of the genus in some of these places extends back to the Mesozoic and that fossil localities also include Antarctica. As with many other Gondwanan relicts, no representatives of the genus presently occur in Africa, Madagascar, or India. (After Schlinger 1974; Dettmann et al. 1990; Heads 2006.)

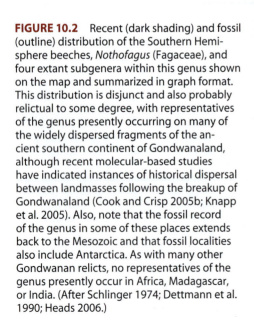

Geological periods and epochs	NGU New Guinea	NCA New Caledonia	AUS Australia	NZE New Zealand	ANT Antarctica	SAM Chile and Argentina
Recent		■		■		■
Pliocene						
Upper Miocene	■					
Lower Miocene			■	■		■
Oligocene			■	■	■	
Eocene			■			
Paleocene			■			
Upper Cretaceous			■	■		■

(A)

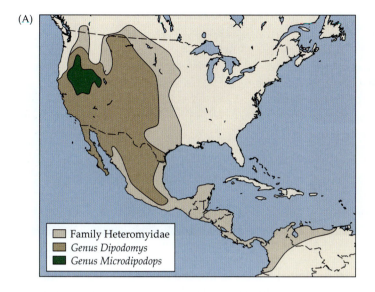

Family Heteromyidae
Genus Dipodomys
Genus Microdipodops

(B)

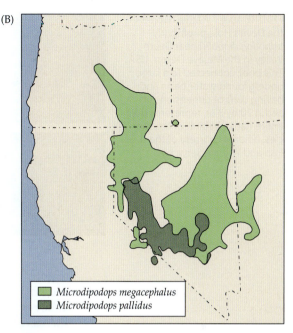

Microdipodops megacephalus
Microdipodops pallidus

(C)

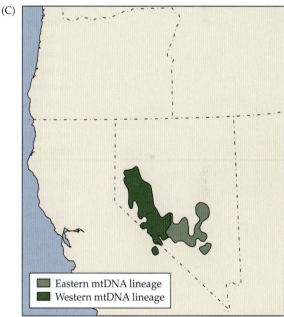

Eastern mtDNA lineage
Western mtDNA lineage

of the ratites are the strange, "mammal-like" kiwis (family Apterygidae), which live in wet temperate forests of New Zealand, are relatively small and nocturnal, feed on earthworms, and nest in burrows.

Endemism and Cosmopolitanism

Because the term *endemic* simply means occurring nowhere else, organisms can be endemic to a geographic location on a variety of spatial scales and at different taxonomic levels. Organisms can be endemic to a location for three different reasons: (1) because they originated in that place and never dispersed, (2) because their entire range has shifted in locality subsequent to origination, or (3) because they now survive in only a small part of their once-expansive former range.

Taxonomic categories are hierarchical (see Table 7.1), so the distributions of lower taxa within a higher taxon are also organized in a hierarchical fashion. An order, for example, contains a nested set of families, genera, and species that represent the historical phylogenetic branching pattern of a single evolutionary lineage. Similarly, the geographic range of an order contains within its boundaries the ranges of all of its families, genera, and species in a cumulative series. For this reason, the lowest taxonomic categories—species and genera—tend to be more narrowly endemic than the higher taxa, such as families and orders, of which they are members. **Figure 10.3** provides an example. The rodent family Heteromyidae, containing the pocket mice, kangaroo rats, and kangaroo mice, is endemic to western North America, Central America, and northernmost South America. Each of the six extant genera of heteromyids has a more restricted range; the genus *Microdipodops* (kangaroo mice), for example, occurs only in the Great Basin of the western United States. The genus *Dipodomys* (kangaroo rats) has a much wider distribution, and its species vary greatly in their ranges, from *D. ordii*, which occurs in most of the cold desert and arid grassland

FIGURE 10.3 Hierarchical patterns of endemism in the rodent family Heteromyidae, which includes kangaroo rats, four genera of pocket mice, and kangaroo mice. (A) The entire family is endemic to the New World, ranging from southern Canada to northern South America. Distributions of its six genera vary in extent, from the kangaroo rats (*Dipodomys*), which range over most of western North America, to the kangaroo mice (*Microdipodops*), which are endemic to the western portion of the Basin and Range physiographic province. (B) Within the genus *Microdipodops*, two extant species (*M. megacephalus* and *M. pallidus*) have largely nonoverlapping distributions within the Great Basin. (C) Within *M. pallidus*, two distinct evolutionary lineages (based on mitochondrial DNA [mtDNA] analysis) have largely nonoverlapping distributions within southeastern and western parts of the Great Basin. (B and C after Hafner et al. 2008.)

FIGURE 10.4   Native habitat of the Devil's Hole pupfish (*Cyprinodon diabolis*) near Death Valley in the Mojave Desert, USA. This pool, connected to a large underground water basin and having surface dimensions of about 20 by 3 m, represents the entire native habitat of the species. (Photos courtesy of the U. S. Fish and Wildlife Service.)

regions of the western United States, to the endangered *D. ingens*, which is endemic to an area of a few thousand square kilometers in the San Joaquin Valley of California (see Figure 4.25). We note here that nontaxonomic units such as phylogroups and evolutionarily significant units—defined in Chapter 7 and discussed in more detail in Chapter 11—can also be described in the context of their pattern of endemism.

Many species and genera, and even some families and orders, are entirely restricted to tiny islands or equally small patches of terrestrial or aquatic habitats. The entire native population of the Devil's Hole pupfish (*Cyprinodon diabolis*) numbers fewer than 600 individuals, which are confined to a spring pool measuring 20 by 3 m in the Mojave Desert of Nevada, just east of Death Valley (**Figure 10.4**; see also Figure 4.21). Some remarkable plant endemics occur in California, especially on the Channel Islands off the southern coast (**Figure 10.5**). The only population of the distinctive shrub *Munzothamnus blairii* (Asteraceae) lives on the tiny island of San Clemente. The entire known population of a species of the Catalina Island mountain mahogany, *Cercocarpus traskiae* (Rosaceae), consists of four individuals growing in one small canyon on Santa Catalina Island. Also on Santa Catalina, San Clemente, Santa Rosa, and Santa Cruz islands lives the Catalina ironwood (*Lyonothamnus floribundus asplenifolius*; Rosaceae), an elegant evergreen tree whose fossils are known from the mainland, including some from Death Valley, during the Miocene. Even some flying organisms can have narrow ranges. The todies (Todidae), for example, are a family of tiny birds entirely restricted to a few West Indian islands; and the highly specialized Mexican fishing bat, *Myotis vivesi* (Vespertilionidae), is almost entirely restricted to quiet lagoons on islands and along mainland coasts in the Sea of Cortés of Mexico.

Of course, isolated islands, such as Madagascar and New Zealand, are famous for their endemics. Some of the endemic groups that have radiated on Madagascar are mentioned in Chapters 7 and 12. Perhaps the most famous endemic reptile is the tuatara (*Sphenodon punctatus*) of New Zealand, a lizardlike species that is the sole surviving representative of the order Rhynchocephalia, the sister clade of all other extant lizards and snakes. Widespread on continents in the Mesozoic, the tuatara persists only on a few small islands that have not been reached by introduced rats and cats. Not so lucky was the Stephen's Island wren (*Xenicus lyalli*), which occurred only on one small islet in the strait between North and South Islands of New Zealand. The original size of its population is unknown because the only individuals known to science were "collected" by the lighthouse keeper's cat, which apparently single-handedly exterminated the last population of this species.

Castilleja mollis

Cercocarpus traskiae

Urocyon littoralis

Other peculiar New Zealand endemics indicating a long history of isolation of these islands from the rest of the world include the previously mentioned kiwis, the now-extinct moas, and the New Zealand short-tailed bats (Mystacinidae), which forage on the ground and inhabit burrows (see Chapter 14).

In contrast to such narrow endemics are cosmopolitan taxa, organisms that are widely distributed throughout the world. No genus, species, or family is truly cosmopolitan, although our own species, *Homo sapiens*, and certain microorganisms come close. Some plant, animal, and microbe species are now widely distributed because humans have intentionally or inadvertently introduced them throughout the world. Terrestrial organisms that have achieved nearly worldwide distributions by natural means include the peregrine falcon (*Falco peregrinus*), the diverse plant genus *Senecio* (ragworts and groundsels), and the bat family Vespertilionidae (**Figure 10.6**). Although they do not occur on Antarctica and on some remote islands, these and other exceptionally widespread taxa are often said to be cosmopolitan. Numerous minute animals and plants, such as protozoans, algae, and fungi, also

FIGURE 10.5 The entire geographic range of a taxon can be very small, illustrated here with known natural distributions of several species that reside on one or more of the Channel Islands along the southern coast of California, USA. (*U. littoralis* courtesy of the National Park Service; *C. traskiae* courtesy of Stickpen/Wikipedia; *C. mollis* courtesy of Sarah Chaney, National Park Service.)

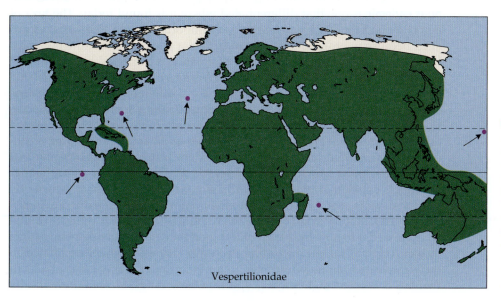

FIGURE 10.6 The nearly cosmopolitan distribution of the bat family Vespertilionidae, which includes some 35 genera and 318 species. Representatives of this group occur on all of the continents except Antarctica and have colonized isolated archipelagoes (indicated by arrows and purple dots) such as Hawaii, the Galápagos, and the Azores. (After Koopman and Jones 1970.)

have extremely broad ranges because their resistant life stages are dispersed widely by water or wind (see Wilkinson 2001; Finlay 2002).

Even in the sea—where any organism could, theoretically, swim around the world unimpeded by land barriers—there are relatively few species or genera that are actually found in all oceans. The most notable exceptions are certain whales and some invertebrates that have been widely dispersed by the whales or by ships. On the other hand, some kinds of freshwater organisms are surprisingly widely distributed, especially considering that other groups, such as fishes and mollusks, contain some of the most narrowly distributed endemics. Most freshwater plant families, and even genera and species, tend to have broad ranges. Some species, including many duckweeds (Lemnaceae), some aquatic ferns (*Azolia*, *Salvinia*, *Marsilea*), water milfoil (*Myriophyllum*), and hornwort (*Ceratophyllum*), are nearly cosmopolitan. Similarly, several genera and even species of freshwater zooplankton (e.g., the water flea, *Daphnia*, and other tiny invertebrates, such as rotifers and tardigrades) are distributed worldwide, or nearly so. The key to these wide distributions appears to be dispersal (see Chapter 6). Freshwater plants are dispersed largely by wading and swimming birds, which carry the seeds or plantlets from one pool or lake to another. The widely distributed freshwater zooplankters have resistant life history stages, often in the form of fertilized eggs that are capable of surviving long periods of desiccation as they are transported by waterbirds or blown like dust in the wind.

Again, few species are truly cosmopolitan. However, many families and genera with exceptionally wide distributions contain at least some species that also have very large geographic ranges, indicative of their broad ecological tolerances and their capacities for dispersing long distances with or without human assistance. On the other hand, many high-ranking taxa, such as orders and classes, are essentially cosmopolitan, because the ecological diversity within these groups is broad enough to include forms that can exist in a variety of terrestrial or aquatic habitats, and also because these groups are old enough either to have had historical opportunities to colonize most parts of the world or to have occupied ancestral landmasses (e.g., Gondwanaland or Laurasia) prior to their fragmentation into the modern landmasses.

The origins of endemics

The origins of endemics are indicated by a variety of terms. An **autochthonous endemic** is one that differentiated where it is found today. The taxa of autochthonous endemics are particularly important in the analysis of vicariance, because in order to infer that a particular geographic feature became a barrier to dispersal within the range of an ancestral species, one must have confidence that current geographic distributions of sister taxa on either side of that barrier represent their original areas of differentiation (see allopatric speciation mode I and Figure 7.12 in Chapter 7, and a more detailed discussion in Chapter 12). A geographic area that contains two or more nonrelated autochthonous endemics is formally defined as an **area of endemism**, a concept of central importance to much of modern historical biogeography that will be developed formally in Chapter 12. Alternatively, an **allochthonous endemic** is one that originated in a different location from where it currently survives. Prime examples of allochthonous endemics include the previously mentioned Catalina ironwood on the Channel Islands and other taxa that have crossed into new biogeographic regions and subsequently gone extinct in their original ranges, including peccaries (*Tayassu*), which originated in the Nearctic but are now restricted to the Neotropics. Many cases of allochthonous endemics might include species whose ranges shifted dramatically in response to Pleistocene climate changes (several examples are provided in Chapter 9).

There are two kinds of relicts, taxonomic and biogeographic. **Taxonomic relicts** are the sole survivors of once-diverse taxonomic groups, whereas **biogeographic relicts** are the narrowly endemic descendants of once-widespread taxa. Often the two categories coincide, which is often true for organisms called **living fossils**. One example discussed previously is the tuatara (*Sphenodon punctatus*) of New Zealand. Another is the monito del monte (*Dromiciops gliroides*), which is the only surviving member of an entire order of marsupials (Microbiotheria) and is restricted to southern beech forests of Chile and Argentina. Still another is the coelacanth (*Latimeria*), known only from the deep waters of the tropical Indian Ocean. This "primitive" fish is the only living member of the lobe-finned fishes, the crossopterygians, a group that was widely distributed in freshwater habitats as well as in oceans and shallow epicontinental seas in the Paleozoic. An example of a plant relict is the ginkgo (*Ginkgo biloba*, Ginkgoales), a gymnosperm native to a small region in eastern China and the sole survivor of a group of primitive conifers that was quite diverse in the Mesozoic. Today the ginkgo is a widely distributed ornamental tree, valued for its unusual, aesthetically pleasing form; its ability to tolerate drought, poor soil, and air pollution; and its supposed therapeutic qualities.

The terms **paleoendemic** and **neoendemic** are used to identify old and recently formed endemic species, respectively. One can see immediately that the use of such terms requires judgments, usually subjective ones, about the origins of endemics. In the previous paragraph we mentioned several examples of ancient relicts, all of which could also be considered paleoendemics.

Also easy to identify are some very recent endemics, especially those of the Quaternary Period (1.8 to 0.1 million years BP). As pointed out in Chapter 9, the Pleistocene was a time of great climatic and biogeographic change. Within just the last 10,000 years, many species ranges have shifted dramatically in response to the warming climate and retreating ice sheets. The ranges of many once widespread species, especially those of cool mountain climates, have contracted as climates warmed, such that now only small, isolated populations are found. An example is the bristlecone pine (*Pinus longaeva*), which is now restricted to arid, rocky microenvironments just below timberline on a few isolated mountains in the Great Basin of California and Nevada, but was widely distributed at lower elevations during the most recent glacial period. The saltbush genus *Atriplex* is widely distributed in western North America, but several endemic polyploid forms occur in distinctive habitats surrounding Great Salt Lake in Utah. Since they occur in areas that were covered by the water of Pleistocene Lake Bonneville—of which the Great Salt Lake is a small remnant—they have almost certainly formed within the last few thousand years (Stutz 1978). On the other hand, many desert-dwelling species of plants and animals have expanded their distributions greatly since the beginning of the Holocene—often into valleys that were previously covered with large lakes. At the same time, boreal forest and tundra species have reinvaded large areas of northern North America and Eurasia that were previously glaciated. Species restricted to these newly colonized areas, whether their ranges are large or small, most certainly speciated very recently—including the sticklebacks and lake whitefish in lakes across Canada (Bernatchez and Wilson 1998; Schluter 2000)—and thus are good examples of neoendemics.

We summarize this overview of endemism with a recognition of the complex interweaving of biotic and abiotic processes that occured throughout the history of both lineages and places, to determine why a taxon exists where it does today. On the one hand, historical events must be invoked to explain how the taxon became confined to its present range and to reconstruct the geographic origin, spread, and contraction of the taxon. As one looks for the influence of historical events, such as the formation of barriers by drifting continents, changing sea levels, glaciation, and shifting ocean currents (i.e., TECO events), it must also be kept in mind that responses to these events are often taxon-specific. As a result, range dynamics—even in response to the same events—often vary markedly among species, creating novel combinations of species while sometimes causing extinctions of others at the hands of ecological interactions with novel competitors, predators, and parasites. Thus, while investigators may search for satisfying, simplistic explanations for the origins of endemics, what they often find is a complex picture, explained in part by Earth history and in part by past and present ecological processes—sometimes confounded in ways that are hard to disentangle, although our ability to do so is getting better all the time (see Chapter 12).

Provincialism

Terrestrial regions and provinces

As we observed earlier, when the ranges of organisms are examined closely, we see that endemic forms are neither randomly nor uniformly distributed across the Earth, but instead are clumped in particular regions. Three patterns are apparent. First, the most closely related species tend to have overlapping or adjacent ranges within restricted parts of continents or oceans. Second, completely unrelated higher taxa—for example, certain plant and animal orders and classes—often show similar patterns of endemism. Third, a small but significant number of taxa have markedly disjunct ranges, with species living in widely separated areas on different continents or islands. The first two patterns make it possible for biogeographers to identify circumscribed regions of the Earth's surface that share common, taxonomically distinctive biotas. The third pattern encourages us to search for historical explanations for how these organisms came to be distributed among such widely separated regions.

Provincialism was one of the first general features of land plant and animal distributions noted by such famous nineteenth century biologists as the phytogeographers Joakim Frederik Schouw (1823), Joseph Dalton Hooker (1853), and Alphonse de Candolle (1855), and the zoogeographers Philip Lutley Sclater (1858) and Alfred Russel Wallace (1876). When these naturalist/explorers traveled among different continents, they were struck by the differences in their biotas (e.g., Buffon's law described in Chapter 2). The limited distributions of distinctive endemic forms suggested a history of local origin and limited dispersal. As discussed in Chapter 2, much biogeographic research has been devoted to identifying centers of origin—areas where one or more groups of organisms originated and began their initial diversification (e.g., Udvardy 1969; Nelson and Platnick 1981; Funk 2004; see also Chapter 12). There have been complementary efforts to find evidence of both historical barriers that blocked the exchange of organisms between adjacent regions and of historical corridors that allowed dispersal between currently isolated regions. The result has been a division of the Earth into a hierarchy of regions reflecting patterns of faunal and floral similarities. In order of decreasing size, the common subdivisions are usually referred to as **realms** or **regions**, **subregions**, **provinces**, and **districts**.

The largest units are the most general. As mentioned in Chapter 2, the division of the Earth into biogeographic regions (see Figure 2.10) recognizes that these areas contain endemic and closely related taxa in many different groups. Sclater divided the Earth into six large units to describe the world distributions of bird families and genera. These were essentially the same six zoogeographic regions later adopted by Wallace in his major treatise on zoogeography (see front endpapers). In the Northern Hemisphere, landmasses north of the tropical zone are called the **Holarctic**, which is composed of the **Nearctic** (North America) and **Palearctic** (Eurasia and northernmost Africa) regions. The remaining regions are primarily tropical: the **Neotropical** (Central and South America, and the West Indies), the **Ethiopian** (Africa south of the Sahara; and Madagascar), and the **Oriental** (Southeast Asia and the adjacent continental islands). The many oceanic islands positioned between the Sunda and Sahul shelves harbor a large number of endemic species and are often referred to collectively as Wallacea. The remaining faunas of the oceanic islands in the Pacific basin are anomalous in this classification scheme because either they contain a small number of taxa from adjacent continents and have relatively few unique groups, or they have their own unique sets of adaptive radiations (e.g., the Philippine and Hawaiian archipelagoes).

These large units have often been divided into subunits that appear to form geographically nested subsets of distributional congruence. For example, Wallace (1876) subdivided the Australian Region—typically recognized as distinct based on a terrestrial biota characterized by endemic marsupials (kangaroos, koalas, wombats, possums, etc.) and placental mammals, and unique plants including eucalypts (Myrtaceae) and proteads (Proteaceae)—into Austro-Malayan, Australian, Polynesian, and New Zealand subregions (see front endpaper). Within Australia, biogeographers have classically developed modifications of W. Baldwin Spencer's 1896 scheme in recognizing at least four major subregions (**Figure 10.7**; see Keast 1981). The first subregion is the Eyrean Province, named for the giant Lake Eyre Basin, encompassing the entire central two-thirds of the continent. It is now a vast arid and semiarid area without major mountain ranges or other internal barriers. The other subregions include three provinces that comprise the wetter fringe of Australia. The northern portion, called the Torresian Province, is a warm tropical belt that contains animals and plants with close affinities to those in New Guinea and sometimes also Southeast Asia. Their isolation is relatively recent, often only about 10,000 years old. During glacial episodes of the Pleistocene, when low sea levels created a land bridge across the present Torres Strait (see Figure 9.9), many of these organisms ranged continuously from Australia to New Guinea. The southeastern portion of Australia is called the Bassian Province and links Victoria with Tasmania across the Bass Strait, which became a continuous land bridge during Pleistocene glacial maxima. This province is inhabited by animals and plants adapted to cool, mesic temperature climates, including the southern beech forests, which contain *Nothofagus* and other relicts dating back to the ancient southern continent of Gondwanaland. The other province, sometimes called Westralia, includes the southwestern corner of the continent, where great numbers of endemic forms reside. Many of the taxa shared between the Mediterranean climatic regions of Australia and South Africa, such as certain groups in the plant family Proteaceae, occur in Westralia.

The majority of articles and books on the distributions of organisms now use the six regions of Sclater and Wallace as fundamental descriptors of animal distributions, and this classification scheme

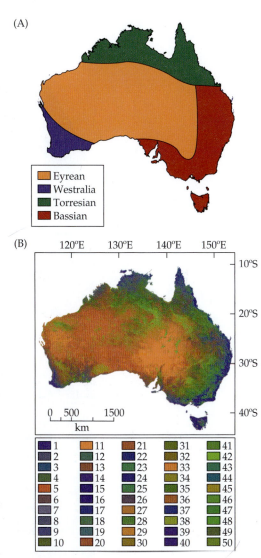

FIGURE 10.7 Classical biogeographic provinces of Australia. (A) Australia is typically divided into four subregions: the Eyrean province, which includes the entire arid center of the continent; and three subregions that include the strip of more moist, mostly forested environments around the northern, eastern, and southwestern periphery, designated the Torresian province, the Bassian province, and Westralia, respectively. The last three provinces are rich in endemics and in groups with representatives on other southern continents. The Bassian province shares southern beeches (*Nothofagus*) with New Zealand, New Caledonia, and South America, while Westralia shares several disjunct plant groups with the Mediterranean climatic region of South Africa. (B) Distribution of 50 primary productivity regimes of Australia, illustrating broad similarities with the subregions based on species distributions. (After Keast 1981; Johnson and Briggs 1975; Mackey et al. 2008.)

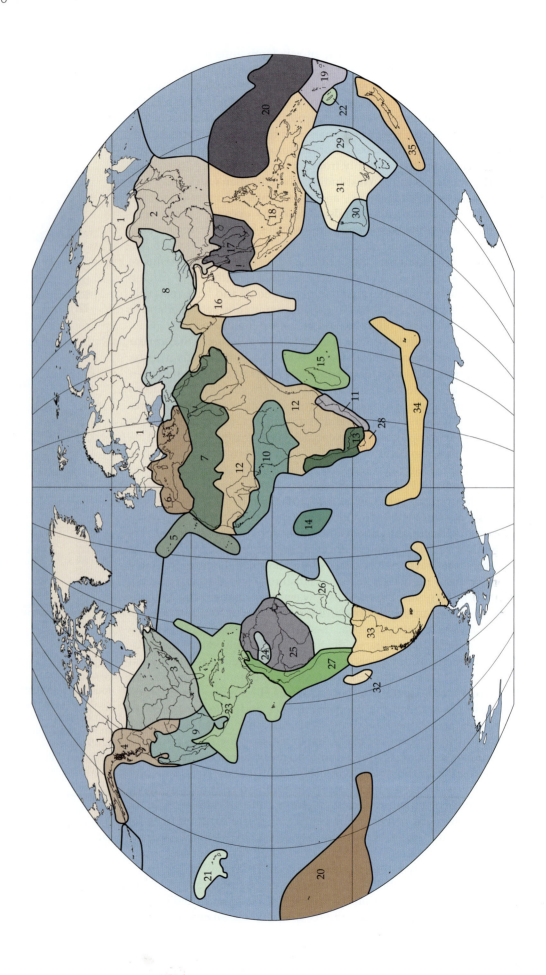

has become one of the primary empirical foundations of biogeography, even though, as we will see later in this chapter, biogeographers continue to refine the delineations between these fundamental terrestrial regions. This is a tribute to the early zoogeographers who created it, who were working with woefully incomplete information on distributions and who were using classifications that did not accurately reflect phylogenetic relationships. Moreover, these pioneering scientists were working without an understanding of the underlying events in Earth history that have created barriers to dispersal or corridors for biotic movements. These barriers and corridors explain many of the similarities and differences that exist between the great biogeographic regions, as exemplified by Wallace's speculations on Earth history as he pondered the abrupt biogeographic transition across Wallace's Line: "I believe the western part to be a separated portion of continental Asia, the eastern the fragmentary prolongation of a former Pacific continent" (Wallace's letter to Henry Bates in 1858, as reported by Berry 2002).

The Sclater–Wallace classification scheme is basically similar to those used for plants, but there are a number of significant differences (**Figure 10.8**). Disregarding for the moment the fact that phytogeographers call the provinces of vertebrates "regions" and call the districts "provinces," there are other important differences that reflect the distinctive characteristics of plants. Vegetation types are tightly restricted by abiotic factors, such as soils, temperature, and rainfall. Consequently, regions of endemism are more clearly defined by climatic and other physical barriers for plants than for animals, which are often able to surmount climatic barriers by dispersal or by physiological and behavioral adaptations to tolerate stressful conditions. Hence, the tip of South Africa, which has a Mediterranean climate, is a very distinctive phytogeographic region. This small area has an exceedingly rich flora, and about 90 percent of its species are endemic (Goldblatt and Manning 2000; Verboom et al. 2009). In contrast, most animal groups of southernmost South Africa are not restricted to that zone and are not spectacularly diverse (Werger and van Bruggen 1978). South America has been divided into numerous provinces based primarily on the distributions of plant taxa. As elsewhere, these provinces are characterized by particular combinations of temperature, precipitation, and soil. However, Cabrera and Willink (1973), who defined the provinces shown in **Figure 10.9**, pointed out that distinctive animal species also are largely restricted to or exceptionally abundant in those regions.

Biogeographers continue to work toward dividing continents into more or less natural biotic provinces. While these units are based primarily on shared evolutionary histories rather than the ecological similarities used to construct biomes (discussed in Chapter 5), there are often underlying relationships between the two (see Figure 10.7) that provide a basis for developing more synthetic mapping schemes such as ecoregions (see Figure 5.32). The need to develop a sound biogeographic basis for conservation planning (Ladle and Whittaker 2011) has motivated construction of ever more detailed schemes of biogeographic regionalization that combine components of floral and faunal distribution with abiotic features of the landscape. For example, the Interim Biogeographic Regionalisation of Australia has subdivided the continent into a mosaic of 50 primary productivity regimes (see Figure 10.7B), 85 bioregions and 403 subregions used by the Collaborative Austra-

◀ **FIGURE 10.8** Division of the world into biogeographic regions based on the distributions of land plants. While the major regions have been subdivided to a greater extent, comparison with Figure 2.10 and with Wallace's map in the endpapers shows the close correspondence between the divisions used for plants and animals. (After Takhtajan 1986.)

FIGURE 10.9 Division of South America into biogeographic provinces based primarily on the distributions of land plants. Note that the configuration of the provinces corresponds closely to geological and climatic features, such as the Andes, showing the influence of climate and soil on the distributions of plant groups and vegetation types. (After Cabrera and Willink 1973.)

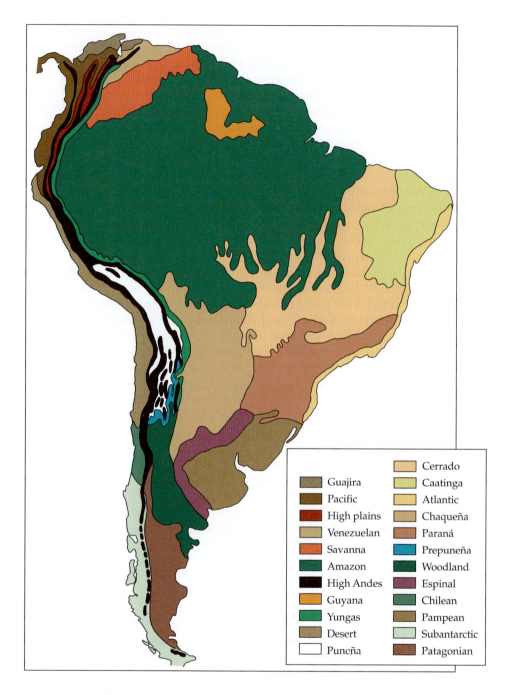

Legend:

Guajira		Cerrado	
Pacific		Caatinga	
High plains		Atlantic	
Venezuelan		Chaqueña	
Savanna		Paraná	
Amazon		Prepuneña	
High Andes		Woodland	
Guyana		Espinal	
Yungas		Chilean	
Desert		Pampean	
Puncña		Subantarctic	
		Patagonian	

lian Protected Area Database to assign to each an explicit level of biodiversity protection (for comparison to regions of North America based on plant and animal distributions, see **Figure 10.10**).

Continental taxa have been stranded on certain landmasses because their ranges have been limited by both abiotic and biotic factors (see Chapter 4), and because major barriers to dispersal have restricted them to those landmasses for millions of years (see Chapters 8 and 12). We shall illustrate patterns of endemism in two groups (land birds and angiosperm plants) in two regions with long histories of isolation from other continents—Australia and South America (**Box 10.1**). Despite the enormous differences in lifestyle and dispersal ability between birds and plants, it is readily apparent that these historically isolated southern continents are centers of endemism for both groups.

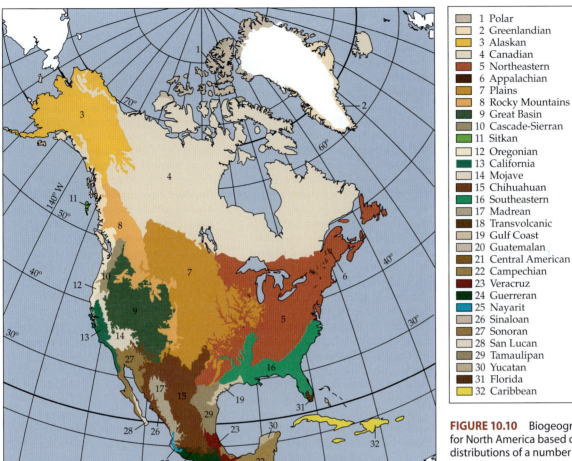

1 Polar
2 Greenlandian
3 Alaskan
4 Canadian
5 Northeastern
6 Appalachian
7 Plains
8 Rocky Mountains
9 Great Basin
10 Cascade-Sierran
11 Sitkan
12 Oregonian
13 California
14 Mojave
15 Chihuahuan
16 Southeastern
17 Madrean
18 Transvolcanic
19 Gulf Coast
20 Guatemalan
21 Central American
22 Campechian
23 Veracruz
24 Guerreran
25 Nayarit
26 Sinaloan
27 Sonoran
28 San Lucan
29 Tamaulipan
30 Yucatan
31 Florida
32 Caribbean

FIGURE 10.10 Biogeographic provinces for North America based on combined distributions of a number of plant and animal taxa. Note that, as for South America (see Figure 10.9), the province boundaries largely conform to geological and climatic features. As such, there are more provinces in the western half of North America, with its extreme topographic and climatic complexity, than in the eastern, more homogeneous, half of the continent. Compare with Figure 10.11 and notice that the boundaries of the desert provinces are in close accord with one another but not identical. (After D. E. Brown et al. 1998.)

By itself, however, the observation that these two regions contain endemic taxa is of limited interest. Even randomly selected areas would be expected to contain some endemics, given the fact that most taxa have limited distributions. The recognition of formal biogeographic provinces implies that each province contains distinctive endemic taxa that share a history of having long been isolated by some combination of unique ecological conditions and barriers to dispersal, but most biogeographic units have not been quantitatively analyzed to show that they do indeed contain assemblages with just this sort of a shared history. Much quantitative characterization of patterns of endemism still remains to be done, although several studies demonstrate how this is being accomplished for selected groups and regions (e.g., Crisp et al. 2001; Morrone 2006; Slatyer et al. 2007). Another factor that has historically impeded a better quantification of endemism within biogeographic provinces is the quality of phylogenetic data used to identify distinct taxa. Consider, for example, passerine birds in the Australian Region. Prior to the novel hypotheses presented by Sibley and Ahlquist (1985) using DNA-DNA hybridization techniques, many species were considered to be closely related to species elsewhere in the world. We now know that many of these species instead are distantly related, and they probably represent a number of separate and old lineages that originated on the Australian plate after separation

BOX 10.1 *Endemic Birds and Plants of South America and Australia*

■■▮ South America and Australia share some important features of Earth history and biogeography. Both were parts of the giant Southern Hemisphere continent of Gondwanaland that drifted apart in the Mesozoic, and both were completely isolated island continents for most of the Tertiary (see Figure 8.22). By the Late Tertiary, however, continental drift had carried the northern, tropical parts of both South America and Australia into close proximity to large northern continents that were once part of the giant continent of Laurasia (see Figures 8.24, 8.26, and 8.27). South America joined with North America at the very narrow Central American isthmus about 2.5 to 3.5 million years BP. Australia and New Guinea, while part of the same plate and joined to each other by land connections for most of their history, drifted close to southeastern Asia (especially when the Sunda Shelf was exposed in the Pleistocene), but never established land connections with it.

Both South America and Australia contain many endemic taxa. Here we use patterns of endemism in angiosperm plants and birds to illustrate some features of their biogeography that reflect the long histories of isolation of these continents. Much attention has been given to the few groups, such as southern beeches (*Nothofagus*) and giant flightless birds (ratites, superorder Paleognathae), that are shared between the two continents and have been long thought to be Gondwanan in origin. Many of the endemic groups,

however, are greatly differentiated, and their nearest presumed relatives do not necessarily occur on other southern continents. For example, the presence of volant paleognaths in the Paleogene of North America and Europe (Dyke 2003) might bring into question the assumption that ratites originated on Gondwanaland. At the level of families, approximately 27 plant taxa are endemic, or nearly so, to South and Central America, and more than 18 are endemic, or nearly so, to Australia and New Guinea. Each of the "nearly endemic" taxa has one or a few close relatives that occur in a different biogeographic region, apparently as a result of fairly recent long-distance dispersal: to Africa in the case of the South American Mayacaceae (bog moss), Rapateaceae (rapateads), Bromeliaceae (bromeliads), and Vochysiaceae (*Vochysia*, etc.); and to Melanesian or Polynesian islands in the case of the Australian Corynocarpaceae (karaka), Davidsoniaceae (davidson), and Trimeniaceae (*Trimenia*).

Despite large differences in their capacities for dispersal and other traits, similar patterns of endemism occur in birds. If we consider only land birds that are considered endemic as breeders, about 20 families containing a total of about 3121 species are endemic, or nearly so, to South and Central America, including the very diverse New World suboscines with over 1000 species (flycatchers, manakins, ovenbirds, woodcreepers, antbirds). This compares with approximately 18 families

containing more than 1415 species that are endemic breeders, or nearly so, in Australasia. As with plants, while no South American lineages of birds apparently have recently dispersed across the Atlantic to Africa, several groups of Australian birds, including the Psittacidae (parrots, lorikeets, cockatoos), Pachycephalinae (whistlers, shrike-thrushes, pitohuis), Meliphagidae (honeyeaters), and Artamidae (wood swallows), have colonized Pacific islands. Another pattern of endemism seen in both plants and birds is exhibited by a large number of additional taxa that probably originated in South America, but have spread northward to colonize at least some of the Central American isthmus. In addition, many more taxa are most diverse in South America, and almost certainly originated there, but they include several species or genera that have spread varying distances into subtropical and even temperate North America. These include the plants (such as those whose ranges are shown in Figure 10.28) that are arid habitat disjuncts and the main groups of birds (e.g., flycatchers, vireos, wood warblers, and tanagers) that breed in temperate North America but migrate to winter in the tropics of Central and South America. Thus, South American and Australian plants and birds illustrate the point that often (but not always: see text and Chapter 12) very different groups of organisms show strikingly similar patterns of endemism. ▮■■

from Gondwanaland (Barker et al. 2004; Jonsson and Fjeldså 2006). The striking levels of morphological similarity to non-Australian passerines represent either convergent evolution or retention of ancestral characteristics.

Regardless of difficulties in the quantitative and phylogenetic characterization of biogeographic regions, at least four lines of evidence suggest that many of them do indeed indicate the pervasive influence of geography, geology, and climate on the historical origin, diversification, and spread of lineages—and that the legacy of this history of place and of lineage is preserved in the distributions of contemporary taxa.

Congruence between taxonomic and biogeographic hierarchies

A convincing line of evidence for the interdependence of histories of place and lineage is the correspondence between the hierarchy of taxonomic categories and the hierarchy of biogeographic regions, with the clearest patterns occurring at the largest scales. Thus, higher taxa (orders and families) tend to have fairly broad distributions within a continental landmass, presumably reflecting the ancient origin and long confinement of major lineages, while more recently differentiated forms (genera and species) tend to be confined to small areas nested within those regions. As pointed out earlier, nested patterns are required by the hierarchical classifications of both organisms and biogeographic regions, so this apparent pattern could be an artifact if based merely on taxonomic classifications. Nevertheless, the relationships between the two hierarchies suggest that the progressively finer biogeographic subdivisions reflect a history of less ancient and less formidable barriers between more recently differentiated lineages (more on this in Chapter 12).

Lineage congruence

The second line of evidence that biogeographic subdivisions reflect the long-standing influence of geological events and climatic patterns is that the boundaries of provinces, determined independently for different groups of organisms, tend to coincide. We have already described, in the preceeding pages and in Chapter 2, how the division of the Earth into the six major terrestrial biogeographic regions was originally developed for birds by Sclater, then generalized to terrestrial mammals by Wallace, and finally extended to land plants and, by implication, all terrestrial organisms. The same kind of generality across different taxa often applies to the finer divisions as well. The North American deserts, for example, have long been divided into four provinces—the Great Basin, Mojave, Sonoran, and Chihuahuan (**Figure 10.11**)—based originally on the pioneering floristic studies of Forest Shreve (1942). This division reflects the distributions of many endemic plants and animals, which are typically restricted to only one or a subset of the four

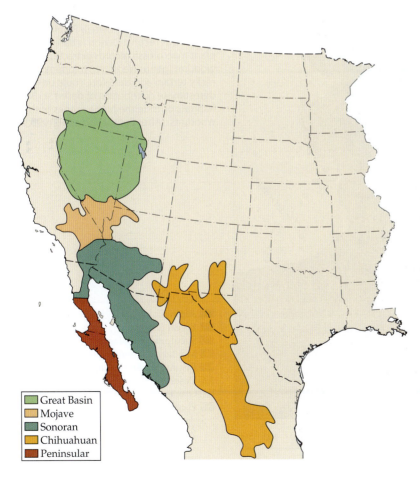

Great Basin
Mojave
Sonoran
Chihuahuan
Peninsular

FIGURE 10.11 Traditional (Shreve 1942) division of arid North America includes four desert provinces: Chihuahuan (yellow), Sonoran (blue), Mojave (orange), and Great Basin (green). The first three are relatively hot, low-elevation deserts, distinguished primarily by an east–west gradient of seasonality of precipitation: Rain falls in summer in the Chihuahuan, in both summer and winter in the Sonoran, and in winter in the Mojave. The Great Basin, due to its higher latitude and elevation, is a cold desert, or shrub-steppe. Each province has distinctive endemic plants and animals. An evolutionary biogeographic analysis using molecular phylogenies in cladistic and phylogeographic analyses for a cadre of terrestrial vertebrates and plants demonstrated a high frequency of endemism on the Baja California Peninsula, and thus supported the recognition of a peninsular desert (red) distinct from the Sonoran Desert. (After Riddle and Hafner 2006a.)

deserts, but are fairly wide-ranging within the desert(s) where they occur. Indeed, molecular-based studies (e.g., Riddle et al. 2000; Zink et al. 2001; Nason et al. 2002; Riddle and Hafner 2006a) have revealed even finer subdivisions (see Figure 10.11); for example, the Baja California Peninsula—the southern half, in particular—harbors a high number of endemic lineages not occurring elsewhere within the Sonoran Desert. These patterns of endemism can in turn be attributed to the unique histories and environments of the deserts. On one hand, during the glacial periods of the Pleistocene, the climate of southwestern North America was cooler and wetter than at present, and the deserts contracted to lowland basins isolated by barriers of grassland and woodland at higher elevations (**Figure 10.12**). On the other hand, even during the interglacial periods, when these barriers were removed, the distinctive precipitation and temperature regime of each desert limited distributions and prevented wholesale mixing of the desert biotas. Prior to the Pleistocene, geological events during the Miocene and Pliocene transformed the landscape in southwestern North America into the complex basins and plateaus bracketed by mountain ranges and seas that delineate each of the deserts today. The result is that each desert has distinctive endemic plants and animals. For example, the saguaro (*Carnegia gigantea*), the Sonoran Desert toad (*Bufo alvarius*), the regal horned lizard (*Phrynosoma solare*), Bendire's thrasher (*Toxostoma bendirei*), and Bailey's pocket mouse (*Chaetodipus baileyi*)

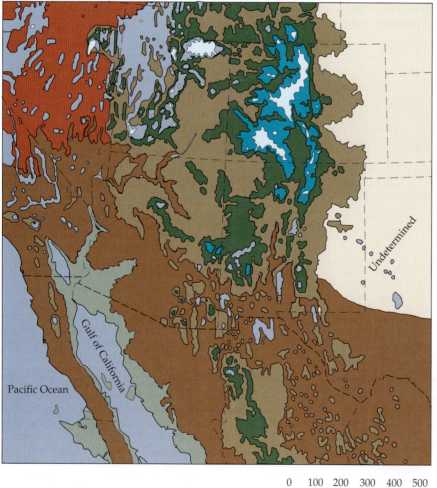

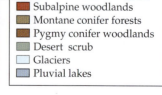

FIGURE 10.12 Reconstruction of Late Pleistocene vegetation in southwestern North America. Note, in particular, the restricted distribution of desert scrub vegetation (contrast with distribution of modern deserts shown in Figure 10.11). This reconstruction is based on plant macrofossil records extracted from woodrat (genus *Neotoma*) middens in rocky terrain and therefore could underestimate distribution of desert scrub habitats outside the home ranges of the woodrats. (After Betancourt et al. 1990.)

■ Alpine
■ Boreal forests
■ Subalpine woodlands
■ Montane conifer forests
■ Pygmy conifer woodlands
■ Desert scrub
□ Glaciers
■ Pluvial lakes

0   100 200 300 400 500
km

are but a few of the many plant and animal species that are endemic to the Sonoran Desert of southern Arizona and northwestern Mexico. Such congruence in patterns of endemism across different lineages of organisms suggests that they have responded similarly to geographic variation in important geological, topographic, and climatic features of the Earth (we will return to this topic in Chapter 12).

Freshwater organisms include forms with both some of the most cosmopolitan and some of the most narrowly endemic distributions. The former, as mentioned above, tend to have life history stages that can withstand desiccation and are readily dispersed. The latter tend to lack such resistant stages and to be equally intolerant of exposure to air and to seawater. As a consequence of their limited dispersal abilities, some freshwater groups, such as freshwater fish and mollusks, show clear patterns of endemism that correspond to the Sclater–Wallace biogeographic scheme. Furthermore, these same taxa typically show pronounced patterns of provincialism and endemism within continents as well. Each major river drainage or lake basin has distinctive, well-differentiated endemic forms (e.g., for fish, see **Figure 10.13**; Hocutt and Wiley 1986; Mayden 1992b). Thus, each of the world's great river systems, such as the Amazon, Congo, Volga, and Mississippi, has a diverse fish fauna that contains many endemic species and genera. And such endemism is found even at much smaller spatial scales. Each of the Great Lakes of Africa's Rift Valley, such as Victoria, Tanganyika, and Malawi, has been a major center of speciation (see Chapter 7) and supports unique lineages containing many closely related species of both fish and mollusks. These same groups are spectacularly diverse, not only in the Mississippi River itself, but also in many of its tributaries. Especially in the Ozark and Appalachian highlands, even the smaller drainages contain endemic forms of darters (*Etheostoma*), shiners (*Notropis*), and clams. Giant Lake Baikal in Siberia contains not only endemic fish and invertebrates, but also the unique Baikal seal (*Phoca sibirica*) that colonized sometime during the Pleistocene when the now landlocked lake drained into the Arctic Ocean. In the Mojave Desert in the vicinity of Death Valley, thermal springs within the drainage of the Pleistocene Amargosa River contain a diverse array of endemic snail lineages (family Hydrobiidae) in the genera *Tryonia* (Hershler et al. 1999) and *Pyrgulopsis* (Liu et al. 2003), and fish in the genus *Cyprinodon* (Echelle and Dowling 1992). Thus, because of the inability of most freshwater fish and certain invertebrates to disperse across either land or sea, their distributions probably reflect historical events more faithfully than do those of most other organisms (see Berra 2001). These groups offer fruitful systems for research on speciation and diversification, as well as on many aspects of historical biogeography.

Biogeographic lines

Another observation that reflects the pervasiveness of provincialism is the rapid turnover of many taxa at the boundaries between regions. Traditionally, biogeographers have drawn lines to fairly precisely define the limits of regional biotas. Although these biogeographic lines are usually derived for one taxon at a time, they often coincide with geological or climatic barriers that have prevented the dispersal of a diversity of other organisms. A prime example of such a sharp boundary is Wallace's line, drawn between the Indonesian islands of Borneo and Sulawesi (Celebes), on the west, and Bali and Lombok, on the east, to mark the boundary between the Oriental and Australian Regions (based primarily on Wallace's observations on the turnover of vertebrates, especially mammals and birds). Not all taxa show distributional boundaries corresponding precisely to Wallace's line, and other lines have been described to accommodate them (**Figure 10.14**). Nevertheless, two

FIGURE 10.13 Biogeographic provinces for North American freshwater fishes. Note that most of these provinces correspond to major drainage basins, such as those for the Great Lakes and the Mississippi and Colorado rivers. The two numbers given are the number of families and species, respectively, and the values in parentheses are the percentages of the species that are endemic to each province. (After Burr and Mayden 1992.)

things about Wallace's line are noteworthy. First, most of the other lines deviate only slightly from the boundary described by Wallace. Second, Wallace's line corresponds almost exactly to the outer limit of the Sunda Shelf, the part of the continental shelf of Southeast Asia that was intermittently exposed by lowered sea levels during the Pleistocene (see Figure 9.9; Hall 1998). Thus, although Wallace did not realize this, his line corresponds to the deep water that marks the limit of historical land connections among the major East Indian islands, and between them and the Southeast Asian mainland (see also

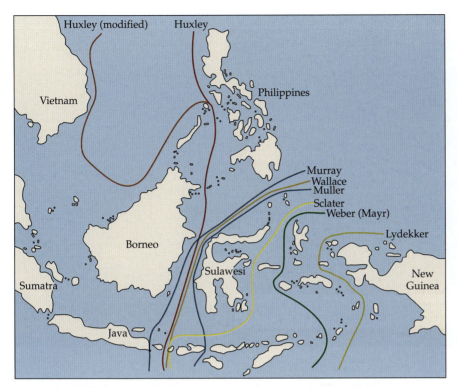

FIGURE 10.14 The biogeographic lines drawn by Wallace (and later biogeographers) to mark the boundary between the Oriental and Australian regions. The locations of the various lines indicate that different taxa have dispersed to different degrees into the islands of the East Indies from their continents of origin. The approximate limits of the continental shelves are given by Lydekker's line in the east and Huxley's modified line in the west. Much of the variation has to do with emphasis on distributional limits of different taxa; for example, Wallace and Huxley emphasized birds, while Murray and Sclater emphasized mammals. Weber's line was favored by Mayr (1944a) because in his analysis it delineated a faunal balance in which over 50 percent of the biota was Oriental to the west, and over 50 percent was Australian to the east. (After Simpson 1977.)

Figure 9.9, and Box 16.1, Figure B). Lydekker's line (see Figure 10.14), just west of New Guinea, corresponds to the edge of the Sahul Shelf, which was exposed in the Pleistocene as part of the supercontinent that included both Australia and New Guinea. It marks the northwestern limit of the ranges of many species of the Australian Region. In between these two lines are islands surrounded by deep water that were not connected by Pleistocene land bridges to either Southeast Asia or Australia–New Guinea and which contain an extraordinarily rich mixture of endemic species derived from ancestors from both regions that managed to disperse across the water barriers.

Boundaries between other biogeographic regions are also open to interpretation. The challenges of delineating such boundaries usually reflect a combination of two interrelated phenomena. First, either the historical, geological, and climatic isolation of the regions was originally not so discrete, or it became less discrete following a major geological transformation allowing more dispersal between regions and causing the apparent boundary to be blurred. Second, different groups of organisms responded somewhat differently to the same historical events and climatic patterns, resulting in a lack of consistency across taxa. These two factors have conspired to make it difficult to draw a single line defining the boundary between most regions, including the Nearctic and Neotropical regions, and indeed, biogeographers are now more inclined to speak of biogeographic transition zones than of discrete lines—for example, the Mexican Transition Zone for insects as delineated by Halffter (1987) and Morrone (2006).

The Isthmus of Panama is the location of the historical separation between the Nearctic and Neotropical regions. It marks where the last marine barrier, separating the formerly isolated continents of North and South America, closed about 3.5 million years ago. When the Central American land bridge emerged as a continuous corridor, many organisms invaded and dispersed along it, an event that the distinguished paleobiologist G. G. Simpson called the Great American Interchange (see Biotic Interchange later in this chap-

FIGURE 10.15 Northern limits of the ranges of Neotropical mammal families (red lines) and southern limits of the ranges of Nearctic families (yellow lines). Similarly complex patterns occur in nearly all taxa (see Figure 10.16). Note that the transition between these historically isolated mammal faunas occurs over a wide area, making it difficult to draw a definitive line separating the two biogeographic regions. A prominent feature of this transition zone is the remarkably complex topography, including a mosaic of upland and lowland habitats that serve to filter dispersing taxa according to habitat constraints. Within this mosaic, several prominent physical features have been postulated to form historical or ongoing barriers to dispersal in different taxa, including (IP) the Central American–South American transition across the Isthmus of Panama, (NL) the Nicaraguan lowlands, (IT) the Isthmus of Tehuantepec, (TMVB) the Trans-Mexican Volcanic Belt, (SMOC) the Sierra Madre Occidental and Mexican plateau, (SMOR) the Sierra Madre Oriental, and (SC) the Sea of Cortez.

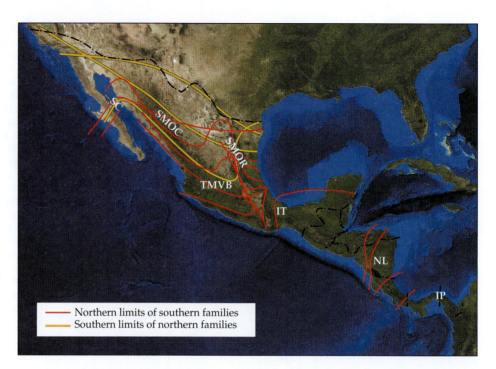

Northern limits of southern families
Southern limits of northern families

ter and in Chapter 12). Because the climate and habitats of Central America and southern Mexico are tropical, there was much biotic exchange between this area and tropical regions of northern South America. Thus, even though Central America and southern Mexico have many endemic forms, they are usually recognized as a subregion of the Neotropics (see front endpaper). But the problem of where to draw the line remains. Often it is placed at the Isthmus of Tehuantepec, which marks both the approximate northernmost extension of tropical rain forest, and a major lowland gap in the mountain cordilleras running the length of western North America. By no means do all Neotropical taxa have their northern limits there, however. **Figures 10.15** and **10.16**, which plot the northernmost and southernmost limits of the ranges of terrestrial mammals and freshwater fishes, respectively, illustrate the difficulty of defining the line separating the Nearctic and Neotropical regions. Each family has colonized southward or northward from its historical origin in North or South America in a somewhat different pattern, reflecting some combination of its capacity for dispersal and its tolerance for environmental conditions.

Congruence between the histories of lineage and of place

While the tendency of range boundaries of species and higher taxa to be coincident facilitates the objective recognition of biogeographic provinces, two more interesting questions can be asked: To what extent are the distributions of taxa a reflection of the history of the landmasses or bodies of water in which these organisms now live? And to what extent can they be used to reconstruct the history of geographic changes on the Earth's surface? We will consider the relationships between phylogenetic, biogeographic, and Earth history in much more detail in Chapter 12, but finish our overview of terrestrial biogeographic regions in this chapter by presenting an update by Morrone (2002, 2009) that incorporates information from cladistic phylogenetic analyses and plate tectonics to create a new, evolutionarily based view of global provinciality (**Figure 10.17**). While this depiction looks very similar to the original Sclater–Wallace schemes, it recognizes that several continents—

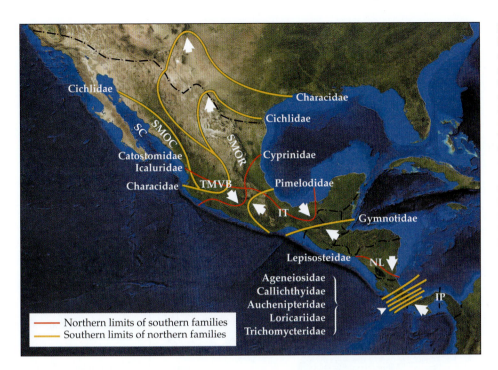

FIGURE 10.16 Distributional limits of freshwater fish families of South American (yellow lines) and North American (red lines) origin. Only two species of obligately freshwater fishes of South American origin have reached the United States, and North American forms extend no farther south than Costa Rica. Note that, as in the case of the mammals shown in Figure 10.15, no single line can be drawn to separate unambiguously the Neotropical and Nearctic faunas. See legend of Figure 10.15 for key to putative barrier names. (After Miller 1966.)

in particular, South America, Australia, and Africa—include a combination of distinct assemblages with very different biogeographic and tectonic histories. Specifically, the biotas on each of these southern continents are split into separate components—the Holotropical and Austral Realms—that share biogeographic affinities to either an eastern or western part of the ancestral Gondwanaland supercontinent.

That an association should exist between plate tectonics and the development of global patterns of provinciality might be fairly obvious. However, the history of provinciality on a single continent is more difficult to ascertain because the timing and magnitude of geological events are often more difficult to reconstruct, and until recently, their influence on biotic distributions

FIGURE 10.17 Global biogeographic realms and regions based on evolutionary biogeographic analyses. The distinction between eastern versus western Gondwanaland biotas results in a subdivision of the traditional Sclater and Wallace Neotropical, Ethiopian, and Australian regions into separate regions. (After Morrone 2002, 2009.)

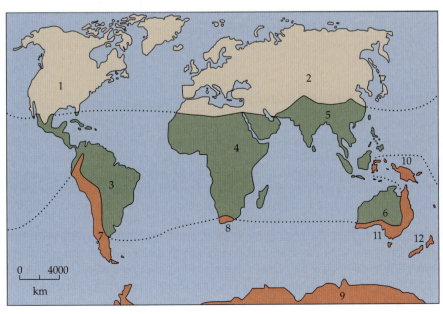

Holarctic Realm (= Laurasia)
 1 Nearctic Region
 2 Palearctic Region

Holotropical Realm (= eastern Gondwana)
 3 Neotropical Region
 4 Afrotropical Region
 5 Oriental Region
 6 Australotropical Region

Austral Realm (= western Gondwana)
 7 Andean Region
 8 Cape or Afrotemperate Region
 9 Antarctic Region
 10 Neoguinean Region
 11 Australotemperate Region
 12 Neozelandic Region

FIGURE 10.18 An example of changes in provinciality through time. Mean turnover index (A) in large mammalian carnivore and in herbivore species between eight sampling regions (B) in western North America from 44 million years ago (beginning of sampling interval 0–1) to the Recent (end of sampling interval 24–25). Changes indicate extent of provinciality, which remains more or less constant from Late Oligocene through Late Miocene time, followed by a general trend toward decreasing provinciality (e.g., a greater proportion of species that range across two or more sampling regions). (After Van Valkenburgh and Janis 1998.)

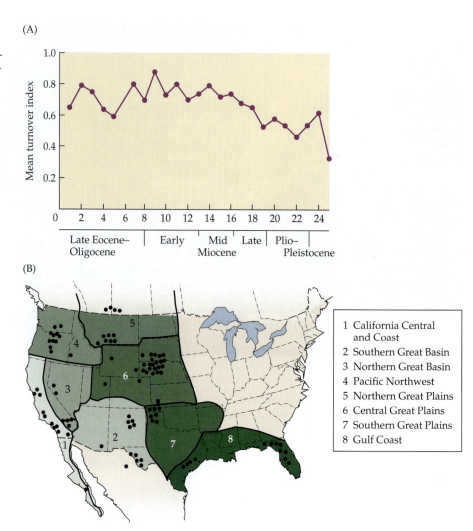

1   California Central and Coast
2   Southern Great Basin
3   Northern Great Basin
4   Pacific Northwest
5   Northern Great Plains
6   Central Great Plains
7   Southern Great Plains
8   Gulf Coast

could only be inferred indirectly from a geographically and taxonomically detailed fossil record (see, for example, **Figure 10.18**). Much progress has been made in the past two decades, however, by exploiting a different kind of data—the molecular phylogenies of populations within species and across related species—to infer the roles of intracontinental, geological, and climatic histories on the biogeographic and evolutionary histories of extant lineages (see Chapters 11 and 12).

Marine regions and provinces

Numerous biogeographers have attempted to define regions and provinces in the oceans. Some of these classifications, however, are really ecological characterizations based on such criteria as water temperature, depth, and substrate. Such classifications often emphasize vertical rather than horizontal divisions of the three-dimensional marine realm—and understandably so, because there is much less difference among the composition of marine biotas in different oceans than among terrestrial biotas on different continents. Not only are the oceans more interconnected than the continents, but many marine forms have tiny planktonic eggs or larvae that are passively and widely dispersed by ocean currents. Most marine plants and animals are cosmopolitan at the familial level, and many even at the generic level. Consequently, it has traditionally been difficult to detect patterns in the distributions of marine taxa that clearly reflect the histories of the present water masses, although such patterns are beginning to emerge as more taxa are surveyed at a molecular genetic level (e.g., Barber et al. 2000; Lourie and Vincent 2004). This is especially true of open-water, or pelagic, organisms, both because there are few physical barriers to dispersal and because a variety of organisms can disperse passively along with ocean currents. A good example of the latter process is provided by the West Wind Drift in the southern oceans (Waters 2008). Benthic plants and animals that live close to, on the surface of, or buried in marine substrates tend to have more restricted distributions and, consequently, to exhibit more provincialism.

Part of the problem in reconstructing the histories of oceanic biotas is that many ocean taxa have had a very long and complicated history, even though (remember from Chapter 8) the Earth's oceans are much younger than the blocks of continental crust. By studying fossil marine invertebrates of the Paleozoic and Mesozoic, we can see that during certain periods there was very little provincialism in some, if not the majority, of the taxa. For example, in the Silurian when Earth was dominated by a nearly continuous ocean (see Figure 8.22C), many genera of even benthic forms, such as brachiopods, gastropods (Boucot and Johnson 1973), and graptolites (Berry 1973), had nearly cosmopolitan distributions. On the other hand, the majority of phyla contain distinctive taxa that, at least during some periods, were restricted in distribution to some early ocean, such as the Tethys of the early Triassic (see Figure 8.22E; Hallam 1973a; Gray and Boucot 1979). So, marine invertebrates show an accordion effect as landmasses fragment and coalesce, with ranges expanding and contracting, to produce alternating stages of cosmopolitanism and provincialism, respectively. This fact should cause us to be cautious in searching for legacies of ancient (i.e., pre-Cenozoic) Earth history in the distributions of present-day marine organisms.

Some history is preserved, however, in the distributions of certain taxa, especially those benthic forms with limited dispersal ability. An obvious horizontal pattern is latitudinal variation in species diversity and composition (see Chapter 15). Several biogeographers have described marine provinces distributed latitudinally along continental coastlines that are separated by zones of rapidly changing species composition (e.g., Valentine 1966; Pielou 1977b; Horn and Allen 1978; **Figure 10.19**). However, these latitudinal patterns appear to be primarily the result of environmental gradients in temperature and correlated abiotic limiting factors, rather than a legacy of unique historical events (see Chapter 7). Thus, the locations of the boundaries between provinces have shifted latitudinally with changing ocean temperatures during the Pleistocene (Valentine 1961, 1989; Valentine and Jablonski 1982), and the taxonomic composition of local assemblages has changed as species have expanded and contracted their ranges (Enquist et al. 1995).

There is, however, marine provincialism that more clearly reflects the influence of tectonic and oceanographic history on the origin and distribution of lineages. Warm tropical oceans have served as significant barriers to dispersal for many cold-adapted temperate and Arctic organisms. Thus, the high-latitude seas represent centers of endemism and speciation for groups such as fishes, seabirds (penguins and auks), and marine mammals (whales and seals). Other polar groups do not exhibit such provincialism, apparently because they have frequently dispersed across the Equator in deep water, where temperatures are also very cold (an obvious impossibility for air-breathing marine birds and mammals). Some marine organisms exhibit amphitropical disjunctions: ranges that include the cold waters of both the Northern and Southern Hemispheres, but not the warm tropical oceans in between. **Figure 10.20** shows two such examples for whales, but similar pat-

List of Provinces within Realms:

Arctic Realm
1. Arctic (no provinces identified)

Temperate Northern Atlantic Realm
2. Northern European Seas
3. Lusitanian
4. Mediterranean Sea
5. Cold Temperate Northwest Atlantic
6. Warm Temperate Northwest Atlantic
7. Black Sea

Temperate Northern Pacific
8. Cold Temperate Northwest Pacific
9. Warm Temperate Northwest Pacific
10. Cold Temperate Northeast Pacific
11. Warm Temperate Northeast Pacific

Tropical Atlantic Realm
12. Tropical Northwestern Atlantic
13. North Brazil Shelf
14. Tropical Southwestern Atlantic
15. St. Helena and Ascension Islands
16. West African Transition
17. Gulf of Guinea

Western Indo-Pacific
18. Red Sea and Gulf of Aden
19. Somali/Arabian
20. Western Indian Ocean
21. West and South Indian Shelf
22. Central Indian Ocean Islands
23. Bay of Bengal
24. Andaman

Central Indo-Pacific
25. South China Sea
26. Sunda Shelf
27. Java Transitional
28. South Kuroshio
29. Tropical Northwestern Pacific
30. Western Coral Triangle
31. Eastern Coral Triangle
32. Sahul Shelf
33. Northeast Australian Shelf
34. Northwest Australian Shelf
35. Tropical Southwestern Pacific
36. Lord Howe and Norfolk Islands

Eastern Indo-Pacific
37. Hawaii
38. Marshall, Gilbert, and Ellis Islands
39. Central Polynesia Cook Islands
40. Southeast Polynesia
41. Marquesas
42. Easter Island

Tropical Eastern Pacific
43. Tropical East Pacific
44. Galápagos

Temperate South America
45. Warm Temperate Southeastern Pacific
46. Juan Fernández and Desventuradas
47. Warm Temperate Southwestern Atlantic
48. Magellanic
49. Tristan Gough

Temperate Southern Africa
50. Benguela
51. Agulhas
52. Amsterdam–St Paul

Temperate Australasia
53. Northern New Zealand
54. Southern New Zealand
55. East Central Australian Shelf
56. Southeast Australian Shelf
57. Southwest Australian Shelf
58. West Central Australian Shelf

Southern Ocean
59. Subantarctic Islands
60. Scotia Sea
61. Continental High Antarctic
62. Subantarctic New Zealand

(A)

Temperate Northern Pacific

Central Indo-Pacific

Arctic

Temperate Northern Atlantic

Western Indo-Pacific

Temperate Australasia

Temperate Southern Africa

Southern Ocean

Temperate South America

Tropical Eastern Pacific

Temperate Northern Pacific

Eastern Indo-Pacific

(B)

◀ **FIGURE 10.19** Spalding et al.'s (2007) classification scheme for biogeographic regions of the marine realm is a nested system of (A) 12 realms, (B) 62 provinces and 232 ecoregions (see their Box 1 for a full list of ecoregions). Outlines in each map denote divisions among ecoregions. (From Spalding et al. 2007.)

terns are also found in fishes and invertebrates. Within the vicinity of Wallacea, several molecular studies have shown a surprising historical subdivision between Pacific and Indian Ocean species, including butterfly fish in the genus *Chaetodon* (McMillan and Palumbi 1995), the mantis shrimp (*Haptosquilla pulchella*; Barber et al. 2000), and sea horses (*Hippocampus*; Lourie et al. 2005).

Some shallow-water and coral reef organisms show significant differentiation among different oceans or coastal regions. Coral reef fish faunas are quite variable in composition, and certain areas in the Indo-Pacific and the Caribbean not only are centers of high species diversity, but also contain a number of endemic species and genera (Briggs 1974; Mora et al. 2003; Briggs 2004b; Carpenter and Springer 2005; Hoeksema 2007; Bellwood and Meyer 2009). There is also significant endemism at the generic level among the corals and mollusks. Fossils indicate that many genera of these invertebrate groups were widely distributed in the Mesozoic, which means that many of today's narrowly restricted forms are relicts of once circumtropical forms (Vermeij 1978). Recent information suggests that many shallow-water marine organisms, even some of those with planktonic larval stages, do not disperse as far as once believed (Gaines and Bertness 1992; Palumbi 1997).

On the other hand, we are just beginning to learn about the biology and distribution of deep-water benthic organisms, including those inhabiting hydrothermal vents, cold water seeps, and the vast abyssal plains that constitute the majority of the ocean floor. At least some of these organisms seem to show remarkably little differentiation or endemism, suggesting that they may have enormous powers of dispersal, although the vent and seep faunas appear to exhibit greater levels of differentiation, providing for the delineation of distinct hydrothermal vent provinces (**Figure 10.21**). Clearly, the relationship between marine biogeography and historical tectonic events will be a fruitful area for future research as we explore one of the Earth's last frontiers.

FIGURE 10.20 Amphitropical distributions of (A) the long-finned pilot whale (*Globicephala melas*) and (B) the right whale (*Eubalaena glacialis*). Warm tropical waters are powerful barriers to dispersal, isolating populations of these cetaceans in the Northern and Southern Hemispheres and apparently preventing the pilot whales from colonizing the North Pacific. (After Martin 1990.)

(A)

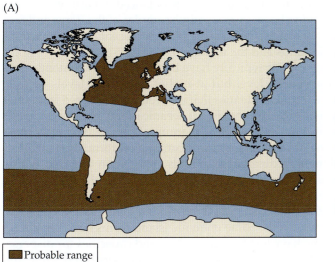

■ Probable range

(B)

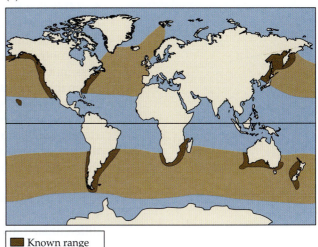

■ Known range
■ Probable range

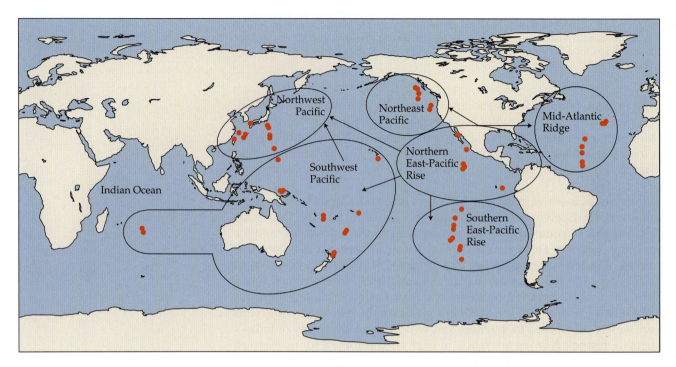

FIGURE 10.21 Depiction of biogeographic provinces at deep sea hydrothermal vents based on distributions and relationships among specialized vent biotas. (After Bachraty et al. 2009.)

Classifying islands

Since Wallace (1876, 1880), biologists have attempted to classify islands as either continental or oceanic to reflect whether they were once connected to continental biotas by a land bridge (**Table 10.1**). Initially, such classifications were made on the basis of the composition of the insular biotas instead of information on their actual geological origin, which was lacking for most regions. As we shall see in Chapter 13, both kinds of islands typically have substantially fewer species than are found in comparable habitats on nearby continents. Continental islands, as their name implies, support plants and animals that are closely related to forms on the nearby mainland; these forms are often so little differentiated as to be considered populations of the same species. Further, continental islands usually have kinds of organisms that are poor over-water dispersers, such as large-seeded plants, amphibians, and non-volant terrestrial mammals. Thus, they are balanced or harmonic (Chapter 14) in that they tend to have representative samples of the floras and faunas of nearby continents. These two features suggest recent land connections between continental islands and the mainland.

In contrast, oceanic islands typically have biotas of lower taxonomic richness, not only at the species level but at higher taxonomic levels as well. Often these forms are insular endemics that are well differentiated from their apparent relatives on nearby continents. An oceanic island, or an archipelago of many such islands, typically supports a group of species obviously more closely related to one another than to any continental form. This suggests that not only evolutionary divergence, but also speciation occurred within the island or archipelago. Further, the limited taxonomic richness of oceanic islands is strongly biased in favor of the kinds of organisms that are relatively good over-water dispersers, such as flying animals and plants with small wind- or bird-dispersed seeds. Some endemic island forms have become so differentiated that they have lost their structures and capacities for dispersal (e.g., parachutes, hooks, and awns have been lost from seeds; reduction and loss of wings have occurred in insects and birds; see Chapter 14 and Carl-

TABLE 10.1 *Some Biogeographically Interesting Islands Classified According to Their Modes of Origin*

FULLY OCEANIC ISLANDS

Totally volcanic islands of fairly recent origin that have emerged from the ocean floor and have never been connected to any continent by a landbridge.

Midoceanic island chains or clusters formed from hot spots (HS) or along fracture zones (FZ) within an oceanic plate
Austral-Cook Island chain (HS or FZ); Carolines (HS); Clipperton Island (FZ); Galápagos Islands (FZ, Carnegie Ridge); Hawaiian Islands (HS); Kodiac Bowie Island chain (HS); Marquesas (HS); Society-Phoenix Island chain (HS, but some contribution by the Tonga Trench)

Island arcs formed in association with trenches
Aleutians (may have been part of the Bering Landbridge in the Cenozoic); Lesser Antilles; Lesser Sunda Islands; Marianas; New Hebrides; Ryukyus (may have been associated with neighboring islands); Solomons; Tonga and Kermadec

Islands formed at presently spreading midoceanic ridges
Ascension Island; Azores (some islands have continental rocks); Faeroes; Gough Island; Tristan da Cunha

CONTINENTAL ISLANDS

Formed as part of a continent and subsequently separated from the landmass. Some of these have added oceanic material since they were formed.

Islands permanently separated from the mainland since the split was initiated (time of final separation in parenthesis)
Greater Antilles (80 million years BP); Kerguelen Island (Upper Cretaceous); Madagascar (ca. 100 million years BP); New Caledonia (ca. 50 million years BP); New Zealand (80–90 million years BP); Seychelles (65 million years BP); South Georgia (45 million years BP)

Island groups with connections of some islands, but not others, to the mainland
Canary Islands

Islands most recently connected to some mainland in the Pleistocene by land or an ice sheet
British Isles; Ceylon; Falklands; Greater Sunda Islands; Japan and Sakhalin; Newfoundland and Greenland; New Guinea (with Australia); Taiwan; Tasmania

quist 1965, 1974). Their phylogenetic affinities, however, suggest that these organisms lost their dispersal abilities secondarily, after their ancestors had colonized the islands.

From Wallace's time until the 1960s, this division between continental and oceanic islands, based on their biotic composition, appeared to work well. It was presumed to reflect geological history in a straightforward way: Continental islands, located on the continental shelves, were presumed to have had recent land connections to the mainland during periods of lowered sea levels; whereas oceanic islands, located in deeper waters, were presumed never to have been connected to the mainland, but to have been pushed up from the seafloor, usually as a result of volcanism. As knowledge of plate tectonics and insular geology accumulated, however, this clear dichotomy between continental and oceanic islands was called into question. Some islands, isolated from all continents by oceans hundreds of kilometers wide and hundreds of fathoms deep, and long thought to be oceanic, were found to contain continental (sialic or andesitic) rocks. And, as tectonic plate movements were reconstructed, it was realized that these islands had once been part of continental plates. Madagascar, New Zealand, and New Caledonia, for example, were recognized as fragments of the ancient supercontinent of Gondwanaland, and the islands of the Greater Antilles were possibly connected via a continuous land bridge (along with the Aves Ridge) with northern South America (**Figure 10.22**; Iturralde-Vinent 2006). In the wake of this accumulating geological evidence for ancient continental connections, the biological and biogeographic evidence of isolation was reexamined for these islands (as was the case for continents; see Chapter 8). Many elements of their floras and faunas may indeed be descendants of over-water colonists (e.g.,

Upper Eocene (37–35 Ma)

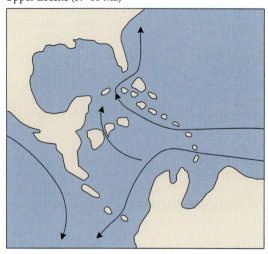

Upper Eocene–Lower Oligocene (35–32 Ma)

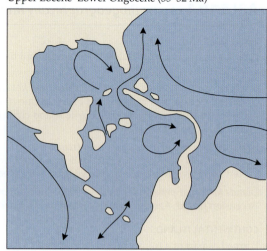

Uplift of GAARlandia

Upper Oligocene–Mid Miocene (32–9 Ma)

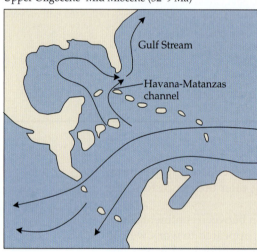

Gulf Stream

Havana-Matanzas channel

Drowning of GAARlandia

Upper Miocene (9–7 Ma)

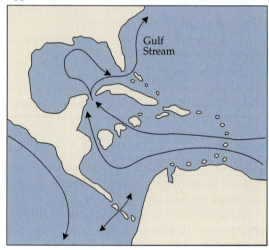

Gulf Stream

General uplift and closure of the Havana–Matanzas Channel

Miocene–Pliocene (7–3.7 Ma)

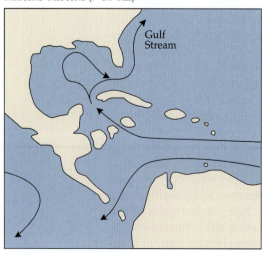

Gulf Stream

Drowning of the Panamanian area

Pliocene–Holocene (3.7–0 Ma)

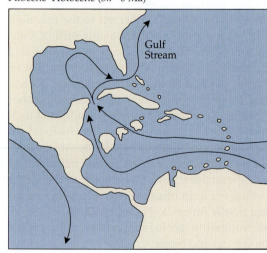

Gulf Stream

Closure of the Panamanian Isthmus

◀ **FIGURE 10.22** Paleoceanographic reconstructions of latest Eocene to recent land configurations and connections of the Caribbean region (continues the Mesozoic and earlier Cenozoic reconstructions in Figure 8.24). Note the Upper Eocene–Lower Oligocene continuous land connection (known as the Greater Antilles–Aves Ridge, GAARlandia) that extended from the northern coast of South America to at least a portion of Cuba, and may have included a portion of Jamaica (see Figure 10.34). The eventual drowning of GAARlandia, leaving a series of disjunct islands, would represent an opportunity for vicariance rather than long-distance dispersal and would make it difficult to classify islands of the Greater Antilles as either oceanic or continental in origin. (After Iturralde-Vinent 2006.)

Lack 1976). Often, however, their biotas contain telltale elements, ignored or misinterpreted by earlier biogeographers, that have until quite recently been assumed to more likely be relictual survivors of ancient continental floras and faunas. Examples include the southern beeches (*Nothofagus*) and fresh-water galaxioid fishes of New Zealand and New Caledonia, the elephant birds of Madagascar, and the moas and kiwis of New Zea-land, all often recognized as remnants of the Gondwanan biota that owe their disjunct distributions to vicariance (but see discussion in Chapter 12 of recent findings supporting a significant role as well for dispersal across continents and islands of a former Gondwanaland).

There are also examples of the converse situation for presumably continental islands. On one hand, increasing information on bathymetry has permitted much more ac-curate estimation of the timing and extent of past land con-nections. Of particular importance is the depth of the wa-ter between island and continent in relation to Pleistocene fluctuations in sea levels. The 100 m depth contour is now known to mark the approximate sea level during the most recent glacial period (see Chapter 9). Islands isolated by shallower depths can usually be assumed to have had their continental land connections severed within the last 10,000 years. This assumption is generally consistent with the bio-geographic evidence, such as the occurrence of a diverse and only moderately differentiated terrestrial mammal fau-na on the islands of Great Britain and the Sunda Shelf (e.g., Java and Sumatra), which had extensive land connections with Europe and Southeast Asia, respectively, during the latest Pleistocene.

On the other hand, we cannot simply assume that about 10,000 years ago, all continental islands had their mainland connections severed and that this stopped all biotic inter-change. Just as with oceanic islands, both the geological and biogeographic histories of continental islands can be complex. One example concerns the more isolated islands of the East Indies, especially Borneo and the Philippines (Heaney 1985, 1986, 1990). Both geological and biogeo-graphic evidence suggest that Borneo and a subset of the Philippine islands had past land connections to the islands of the Sunda Shelf and to the mainland of Southeast Asia, but not during the latest stages of the Pleistocene (**Figure 10.23**). Thus, these islands contain much more differentiat-ed endemic land plants, terrestrial mammals, and other or-ganisms than other islands on the Sunda Shelf. Within the Phillipine archipelago, many current islands were joined

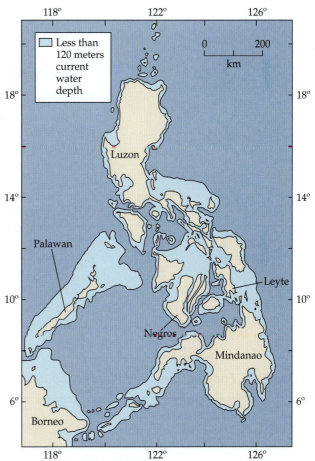

FIGURE 10.23 Map of the eastern Sunda Shelf and the Philippines, showing the apparent extent of land above sea level in the Late Pleistocene. Note that while many of the currently isolated islands (lighter shading) were connected and the straits between others were much narrower than at present, there were still no land connections between the Philippines and Borneo (although they were separated by only a 12 km strait and may have been connected about 165,000 years ago: Esselstyn et al. 2009) or the rest of the exposed Sunda Shelf, and thus to the Southeast Asian main-land. (After Heaney 1986.)

during the Late Pleistocene into several much larger landmasses (see Figure 10.23), whereas a number of smaller islands of volcanic origin have never had such connections (see also Box 16.1).

Also, just because most continental islands have not had land connections with the mainland within the last 10,000 years does not necessarily mean that there has not been more recent biotic interchange. Forms with well-developed capabilities for over-water dispersal, such as bats and some birds, tend to be especially well represented, even on tiny islands, and to exhibit virtually no evidence of evolutionary divergence. For example, the continental island of Trinidad, separated from Venezuela by two shallow straits, both less than 25 km wide, is famous in bird-watching circles for its scarlet ibis. These hardly constitute an isolated population, however, because many of the birds commute on a daily basis to feed on the South American mainland. While this is an extreme example, many kinds of organisms have continued to disperse from the mainland after the isolation of continental islands by rising sea levels at the end of the Pleistocene. Such over-water immigration serves to recolonize the islands after extinction events and to prevent divergence and extinction of the insular populations (see discussion of "rescue effects" in Chapter 13).

Quantifying similarity among biotas

Early biogeographers, and many of their successors, defined biogeographic provinces and regions subjectively, based on their intuition about how to interpret the geographic patterns. This does not necessarily mean that their classifications are unreliable; in fact, the human brain has an exceptional capacity for recognizing patterns. The biogeographic regions first defined subjectively by Sclater almost certainly include real patterns that could be defined objectively by modern quantitative analyses. During the last several decades, however, quantitative techniques have been applied increasingly to systematics and biogeography in order to make the process of classifying organisms and biogeographic regions more rigorous, objective, and repeatable.

In principle, quantitative techniques are simple. First, the items to be classified are described using objective criteria. In the case of biogeographic studies, the data usually consist of complete lists of the relevant taxa that occur at specific sites or within given areas. Sometimes other data, such as average

---

BOX 10.2 *Endemic Birds and Plants of South America and Australia*

Jaccard	$\dfrac{C}{N_1 + N_2 - C}$	Second Kulczynski	$\dfrac{C(N_1 + N_2)}{2(N_1 N_2)}$	Braun-Blanquet	$\dfrac{C}{N_2}$
Simple matching	$\dfrac{C + A}{N_1 + N_2 - C + A}$	Otsuka	$\dfrac{C}{\sqrt{N_1 N_2}}$	Fager	$\dfrac{C}{\sqrt{N_1 N_2}} - \dfrac{1}{2\sqrt{N_2}}$
Dice (Bray-Curtis; Sorenson)	$\dfrac{2C}{N_1 + N_2}$	Correlation ratio	$\dfrac{C^2}{N_1 N_2}$	
First Kulczynski	$\dfrac{C}{N_1 + N_2 - 2C}$	Simpson	$\dfrac{C}{N_1}$	

Note: A = the number of taxa absent in both sites being compared (but present in others); C = the number present in both sites; N_1 = the number present only in the first site; N_2 = the number present only in the second site (when the first unit contains fewer taxa). (After Cheatham and Hazel 1969.) ▮▮

TABLE 10.2 *Flessa's (1979) Classic Matrix of Similarity Coefficients (Simpson Index as Percent) between the Mammalian Faunas of Various Regions*

	North America	West Indies	South America	Africa	Madagascar	Eurasia	SE Asian islands	Philippines	New Guinea	Australia
North America		40	55	8	9	19	8	9	6	6
West Indies	67		33	11	9	11	7	7	9	11
South America	81	73		3	7	4	7	6	4	3
Africa	31	27	25		30	25	21	27	17	12
Madagascar	38	27	35	65		32	26	22	22	17
Eurasia	48	27	36	80	69		75	64	25	14
SE Asian islands	37	20	32	82	63	92		73	30	18
Philippines	40	20	32	88	50	96	100		26	18
New Guinea	36	21	36	64	50	64	79	64		46
Australia	22	20	22	67	38	50	61	50	93	

Source: After Flessa et al. 1979. Courtesy of The Geological Society of America.

Note: Values above the diagonal are similarities at the generic level; values below the diagonal are similarities at the familial level.

abundance or area of geographic range, are available for each taxon for each region. Several mathematical techniques can then be used to quantify the similarity between each pair of biotas. The most commonly used measures are the similarity indexes (**Box 10.2**), because they all can be computed from simple presence–absence data. These indexes differ primarily in the extent to which they incorporate taxa that are present in both regions, in the range of values they can assume, and how they behave mathematically (i.e., how variations in presence–absence patterns are combined to produce a single number). The Jaccard and Simpson similarity indexes are the two that have probably been most frequently used in biogeographic analyses. Similarity values for each pair of biotas can be conveniently expressed in matrix form (**Table 10.2**). Once the similarity between each pair of biotas has been computed, a quantitative clustering method is typically used to organize the biotas of multiple regions in groups that reflect their distinctiveness (see example in **Figure 10.24**). There are several such clustering techniques, each of which has unique procedures for making mathematical computations and each of which can give somewhat different results.

Most applications of quantitative methods to define biogeographic provinces have been limited to a single group of organisms within a limited geographic region. Examples include the division of Australia into ten provinces based on avian distributions (Kikkawa and Pearse 1969), analyses of similarities among mammalian faunas on the different landmasses (Flessa et al. 1979; Flessa 1980, 1981), and the division of the North American deserts into a hierarchy of biogeographic regions based on plant distributions (McLaughlin 1989, 1992). Such studies illustrate the practicalities of applying quantitative methods to biogeographic problems, but are so limited in taxonomic scope that their results may not reflect more general patterns shared by other taxonomic groups.

As demonstrated in earlier studies, independent analyses of the distributions of several groups in the same region are potentially more informative. Connor and Simberloff (1978) compared the similarities among the Galápagos Islands for both land birds and land plants, and they found both similarities and differences between the two groups, which presumably reflect the influences of such factors as the geological histories and ecological environ-

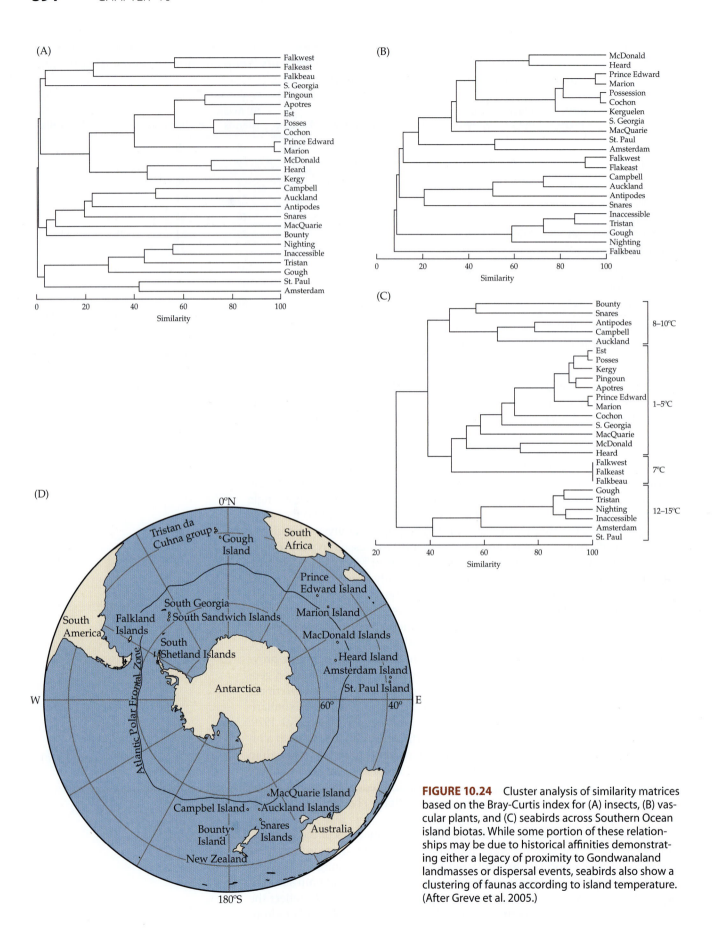

FIGURE 10.24 Cluster analysis of similarity matrices based on the Bray-Curtis index for (A) insects, (B) vascular plants, and (C) seabirds across Southern Ocean island biotas. While some portion of these relationships may be due to historical affinities demonstrating either a legacy of proximity to Gondwanaland landmasses or dispersal events, seabirds also show a clustering of faunas according to island temperature. (After Greve et al. 2005.)

(A)

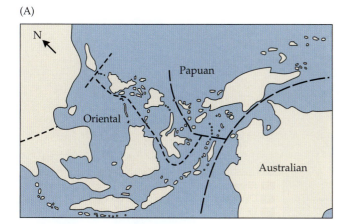

(B)

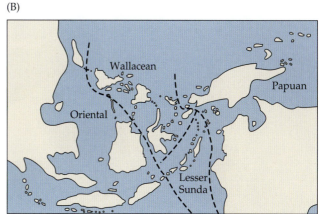

(C)

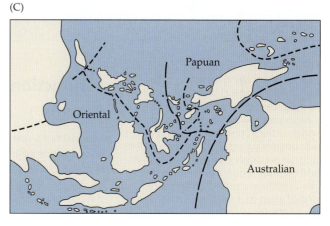

ments of the islands and the dispersal capabilities and population biologies of the organisms. An ambitious and early use of quantitative methods was Holloway and Jardine's (1968) analysis of the distributions of butterflies, birds, and bats in the Indo-Australian area (**Figure 10.25**). They measured similarities among different islands at the species level and made inferences about past trends of dispersal and speciation. They concluded that all three of these groups of flying organisms dispersed predominantly eastward out of Southeast Asia, but that bats showed a different pattern of faunal differentiation than birds and butterflies. Subsequently, Nelson and Platnick (1981) reanalyzed the original data of Holloway and Jardine, applied the alternative approach of cladistic biogeographic analysis (see Chapter 12), and made somewhat different interpretations.

Interestingly, the use of similarity indexes and clustering methods to define biogeographic provinces has declined since the 1970s. There appear to be two reasons for this. First, as indicated above, at levels finer than the Wallace–Sclater regions, the provinces defined in this way for one group usually do not generalize well to other taxa (see Figure 10.24). This should not be surprising. Different kinds of organisms originated and spread during different historical periods, when the Earth's geography and climate were different, and their patterns of speciation and dispersal have been affected by their different intrinsic characteristics, such as dispersal modes and life histories, and by different environmental factors, such as biogeographic corridors and barriers. Second, efforts to more rigorously characterize and interpret the legacy of biogeographic history have seen methods based on biotic similarity in species composition of extant biotas supplanted by methods based explicitly on phylogenetic relationships of those species. These phylogenetic-based methods have made it clear that many, if not all, biogeographic provinces are composite or reticulate areas, composed of a mixture of taxa with evolutionary connections to more than one other province (see South American, African, and Australian examples in Figure 10.17). Indexes provide only a single value summarizing overall similarity among areas, and are thus of limited utility in revealing the details of a complex reticulated biogeographic history. We will examine the phylogenetically based approaches in the next two chapters.

FIGURE 10.25 Biogeographic regions and subregions in the Indo-Australian area defined for (A) birds, (B) bats, and (C) butterflies, based on the quantitative analyses of Holloway and Jardine (1968). Note that the divisions for birds and butterflies are virtually identical, but those for bats are quite different. The differences among the lines reinforce the difficulties earlier biogeographers had in delineating the boundary between the Oriental and Australian regions, as shown in Figure 10.14. (After Holloway and Jardine 1968.)

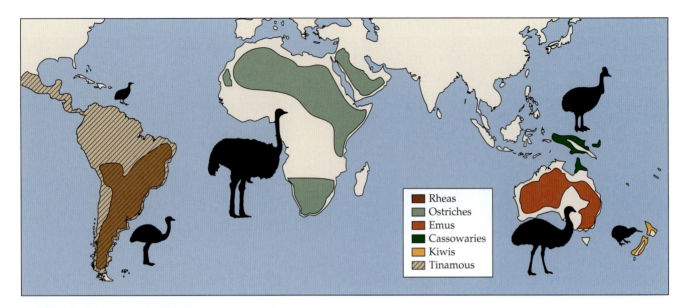

FIGURE 10.26 The disjunct distribution of the surviving members of the bird lineage that includes the tinamous and flightless ratites: ostriches, emus, cassowaries, kiwis, and rheas. The current widely disjunct distribution of this group was originally considered to reflect their original occurrence on Gondwanaland and their subsequent isolation as the ancient southern continent fragmented and drifted apart (but see text and Harshman et al. 2008 for alternative historical reconstruction). Extinct Gondwanan taxa in this group with a good fossil record include giant flightless moas on New Zealand and elephant birds on Madagascar.

Disjunction

Patterns

Disjunctions are those distributions in which closely related organisms (e.g., populations of the same species, or species of the same genus) live in widely separated areas. There are many classic examples. Many related forms occur on some combination of the Southern Hemisphere landmasses of Australia, Africa, South America, and sometimes New Zealand and Madagascar: These disjuncts include southern beeches (*Nothofagus*), trees and shrubs of the genus *Acacia*, several groups of mayflies, lungfishes, galaxioid fishes, clawed frogs (family Pipidae), ratite birds (emus, cassowaries, ostriches, rheas, tinamous, kiwis, and extinct moas and elephant birds; **Figure 10.26**), and marsupial (pouched) mammals. Lungless salamanders (family Plethodontidae) are broadly distributed in the wetter parts of the New World; the only representative in the Old World, the genus *Hydromantes*, occurs in southern

FIGURE 10.27 The disjunct distribution of lungless salamanders (Plethodontidae). Note that this ancient group has many isolated relicts in North America, and even one in the Mediterranean region of France, Italy, and Sardinia. (After Wake 1966; photo courtesy of David McIntyre.)

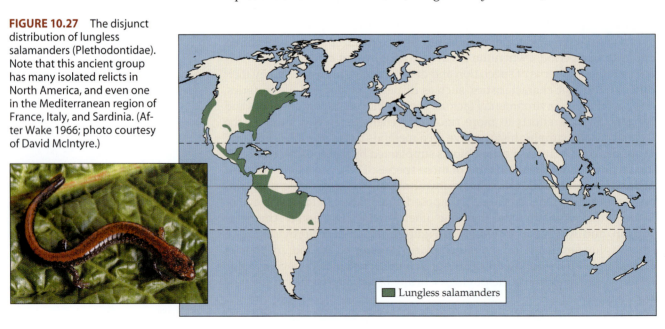

Europe (France, Italy, and Sardinia), but also occurs in California (**Figure 10.27**). Many of the common genera and even species of plants in the deserts of southwestern North America, including creosote bush (*Larrea*), mesquite (*Prosopis*), and paloverde (*Cercidium*), also occur in the desert regions of southern South America, but not in the extensive tropical and montane areas in between. Other examples of amphitropical distributions in terrestrial organisms are shown in **Figure 10.28**; see also Figures 10.20 and 9.24B–D.

There are disjuncts on smaller scales as well. Plants provide some of the best examples, perhaps because they often have very specific environmental requirements and are able to persist for long periods in small isolated patches as long as conditions are suitable. The sweet gum tree, *Liquidambar*, occurs in the deciduous forests of the eastern United States and also in a series of isolated localities in central and southern Mexico. The Black Belt and Jackson Prairies in the southeastern United States are isolated grassland habitats that occur on unusual soils. They contain a number of disjunct plants whose nearest populations occur far to the west on the Great Plains (Schuster and

FIGURE 10.28 Examples of amphitropical, disjunct distributions of plant species in North and South America. These examples represent only a small fraction of the disjunct distributions of closely related plant species, mostly in arid regions, on the two continents. These disjunct distributions raise interesting questions about the historical, geological, and climatic events that have allowed these plants to disperse across the tropics.

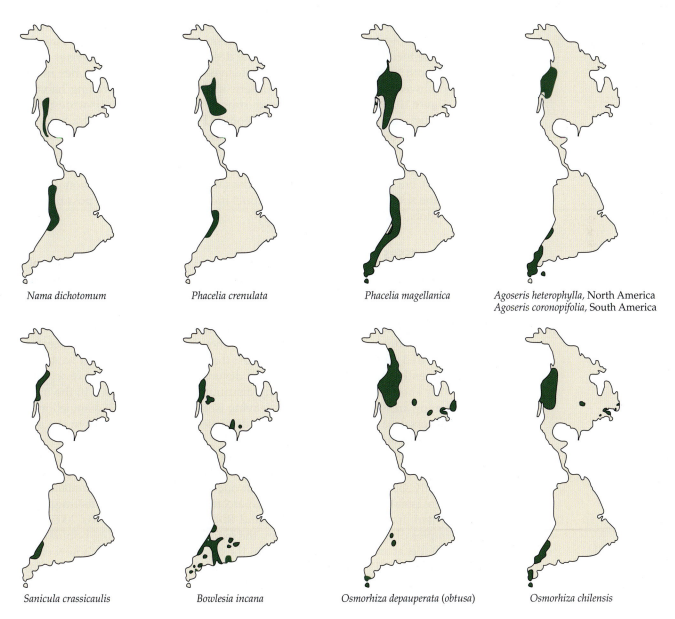

Nama dichotomum

Phacelia crenulata

Phacelia magellanica

Agoseris heterophylla, North America
Agoseris coronopifolia, South America

Sanicula crassicaulis

Bowlesia incana

Osmorhiza depauperata (obtusa)

Osmorhiza chilensis

FIGURE 10.29 The disjunct distribution of the lizard genus *Uma*, which is restricted to dunes of wind-blown sand in the desert region of southwestern North America (photo of *U. scoparia*, the Mojave fringe-toed lizard). Many other plants and animals that are restricted to specific kinds of spatially isolated habitats, such as mountaintops, lakes and springs, and patches of unusual soil types, show similar patterns of small-scale disjunct distributions. (Photo © Shutterstock.)

Uma scoparia

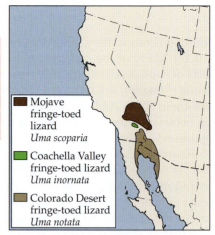
- ■ Mojave fringe-toed lizard
  Uma scoparia
- ■ Coachella Valley fringe-toed lizard
  Uma inornata
- ■ Colorado Desert fringe-toed lizard
  Uma notata

McDaniel 1973). Some wetland plants found along the Atlantic coastal plain of eastern Canada also occur as isolated populations well inland in central Ontario (Keddy and Wisheu 1989).

Animals also exhibit fine-scale disjunctions. Like plants, many kinds of insects have been able to disperse to, and persist in, widely separated areas that meet their narrow niche requirements. Vertebrate examples are the fringe-toed lizard genus *Uma* and the desert kangaroo rat (*Dipodomys deserti*), both of which occur almost exclusively on isolated sand dunes in the southwestern North American deserts and not in the intervening habitats (**Figure 10.29**). The eastern collared lizard (*Crotaphytus collaris*) is widespread throughout the Chihuahuan and Colorado Plateau deserts, but also has a set of highly disjunct populations in arid, rocky habitat patches in the Missouri Ozarks. The checkerspot butterfly, *Euphydryas editha*, occurs in many isolated populations widely dispersed across western North America (Ehrlich et al. 1975), but many of the isolates near the southern and lower elevational limits of the species' range have gone extinct in the last few decades, possibly due to global warming (Parmesan 1996).

Processes

Since the early nineteenth century, explaining disjunct distributions has been a major goal of biogeography. These distributions raise an obvious challenge: If the organisms are indeed closely related, then how did they get from where their common ancestor occurred to their widely separated ranges? There can be only three possible answers: (1) disjunction by tectonics—their ancestors occurred on pieces of the Earth's crust that were once united, but have subsequently split and drifted apart; (2) disjunction by intervening extinction—their ancestors were once broadly distributed, but populations in the intervening areas have gone extinct, leaving isolated relicts; or (3) disjunction by dispersal—at least one lineage has dispersed a long distance from the area where its ancestor(s) originally occurred.

Examples of all three mechanisms have been found. The Southern Hemisphere disjuncts mentioned above have until recently been considered as mostly vicariant relicts of forms that once occurred on Gondwanaland and were isolated when that ancient supercontinent broke up and its fragments drifted apart (but see Sanmartín and Ronquist 2004; de Queiroz 2005). Examples of disjunctions caused by the extinction of intervening populations include, at an intercontinental scale, the camels and tapirs, which now occur only in South America and Eurasia, but are found in North America as fossils. Many of the organisms that exhibit more fine-scale disjunctions, including the collared lizards mentioned above, are relicts left by the extinction of intervening populations due to changes in climate or other environmental conditions since the Pleistocene. Many fishes found in now isolated desert springs, and mammals restricted to the cool, mesic habitats on isolated mountain ranges, had wider, more continuous ranges in the Pleistocene when their habitats were more continuous (see Chapter 9). Some disjuncts are the result of long-distance dispersal—either by natural means, such as the cattle egrets that a few decades ago crossed the South Atlantic Ocean to colonize South America from Africa (see Figure 6.2), or by human transport, such as the exotic European house sparrows and rabbits that are now established in such outposts as North America, Australia, New Zealand, and southern South America (see Figure 6.3).

There are also examples of forms once thought to be disjuncts that are now known not to be close relatives at all. Thus, "porcupines," rodents characterized by certain features of the skull and by hairs modified to form distinctive defensive quills, were once thought to be disjuncts between the New and Old Worlds. These animals occur in Africa, southern Europe and Asia, and North and South America, but are absent from all of temperate Eurasia. Subsequent systematic research showed, however, that New and Old World porcupines are not closely related; they arose independently from primitive rat-like rodents, retaining unspecialized features of the skull and evolving quills by convergence. Similarly, the anteaters of the tropical New World and the pangolins (or scaly anteaters) of tropical Africa and Asia were once thought to be related disjuncts and were placed in the same order, Edentata. These animals share several characteristics, especially anatomical features of the skull and tongue that are related to their exclusively insectivorous diet, but these have arisen by convergence. These "anteaters" actually are unrelated members of two different orders, Pilosa (which includes the New World anteaters, armadillos, and sloths) and Pholidota (which includes only the Old World pangolins).

The revelation that some organisms with seemingly disjunct distributions are not in fact close relatives underscores the need for good systematics. Patterns of cosmopolitanism, provincialism, and disjunction raise important questions in historical biogeography. The discipline's most fundamental questions include how the origin, diversification, and spread of a particular kind of organism was influenced by the dynamic geographic, geological, climatic, and biotic features of the Earth. But unless we have an accurate assessment of the phylogenetic relationships among species and higher taxa, as well as phylogenetic and population genetic relationships among populations within species, we do not even know which biogeographic questions to ask. Those are the subjects of the next chapter. We will finish this chapter with three final questions: What abiotic and biotic attributes serve to keep distinct biotas separated from one another over time? What happens when barriers erode between previously separated biotas? And what are some of the evolutionary patterns we see following long-term isolation of species and biotas?

Maintenance of Distinct Biotas

Here we consider briefly how distinctiveness between biotas is maintained. Given the present land bridges connecting Africa and Eurasia, as well as North and South America, and the frequent Pleistocene connections between North America and Eurasia, why hasn't biotic interchange been more complete? What processes are responsible for the preservation of biogeographic provincialism, especially in organisms that are good dispersers?

At one level, the answer is fairly straightforward. Since biotic interchange is due to dispersal and subsequent ecological success, the maintenance of provincialism must be due largely to some combination of continued isolation, resistance to invasion, and the role of evolutionary history in ensuring that migratory species predictably move between the same wintering and breeding regions each year. Each of these factors is important.

Barriers between biogeographic regions

Most biogeographic regions are isolated in the sense that they are separated by barriers to dispersal. For example, inspection of Wallace's map of biogeographic regions (see front endpapers) immediately reveals that even where the terrestrial biogeographic realms are connected by land bridges, these are either narrow isthmuses, harsh deserts, or high mountains. Furthermore, the geographic ranges of the majority of species within a region do not extend to the boundary and, therefore, fall short of the land bridge. In order to move between regions, the majority of species, including distinctive endemic forms, would have to disperse long distances through unfavorable habitats. Consequently, the opportunity for interchange between regions is limited to a small fraction of the biota. For example, even though the Bering land bridge was exposed by the lowered sea levels that prevailed during most of the last 2 million years, it allowed only limited exchange between North America and Eurasia. It did provide a dispersal corridor, but only for those organisms whose ranges extended into steppe, coniferous forest, and tundra habitats at high latitudes. Some species, such as bears, wolves, ermines, caribou, moose, and lemmings among mammals, and ravens, hawks, owls, ptarmigan, siskins, and crossbills among birds, dispersed freely and now occur on both continents. In contrast, many groups of North American and Eurasian amphibians, reptiles, birds, and mammals that did not inhabit these high-latitude environments did not disperse across the Bering land bridge.

Resistance to invasion

Although it is tricky to document, it also appears that the biotas of large landmasses are relatively resistant to invasion. When faced with a diverse native biota that has evolved to tolerate an abiotic environment and coevolved to withstand existing biotic interactions, it is difficult for invading species to become established. Evidence in support of this hypothesis comes from the fate of exotic species; the majority of introductions fail (for individual case histories, see Chapter 16 and Elton 1958; Udvardy 1969; Drake et al. 1989; Hengeveld 1989; Sax et al. 2005). Another factor involved in resistance may be one or more forms of biotic inertia—the retention of native species such as long-lived forest trees, or of soil chemistries produced by allelopathic plants, long after changes of environmental conditions that favor the invader have occurred (Von Holle et al. 2003).

Numerous Old World plant and animal species have become locally abundant and geographically widespread in North America within the last four centuries. Although the success of these invaders is impressive, they represent only a small fraction of the species that have been intentionally or accidentally introduced. The vast majority of introduced populations have gone extinct, and many others have not spread far beyond the sites of their introduction. An example is the Eurasian tree sparrow (*Passer montanus*): More than a century after its establishment, it is still confined to a small area near St. Louis, Missouri.

Furthermore, of the hundreds of Eurasian plants established in North America, most can be classified as weeds, species that occur in successional habitats created primarily by human disturbance. Of the many introduced insects, the majority of successful species are crop pests, associated with introduced plant species, or confined largely to disturbed habitats. This pattern holds even for vertebrates. Many Eurasian birds have been introduced into North America, but only a few have become established. The two amazingly successful introduced bird species that have spread to cover most of the North American continent—the house sparrow (*Passer domesticus*) and the starling (*Sturnus vulgaris*; see Figure 6.3A,B)—are largely commensal with humans, using artificial structures for nesting sites and using urban and agricultural habitats for food resources. Mammalian examples include the house mouse (*Mus musculus*), the Norway rat (*Rattus norvegicus*), and the roof rat (*Rattus rattus*). The success of these Eurasian exotics in North America and the much lower success of New World species in the Old World suggests that most of the Eurasian forms are adapted to occupy niches that are dependent on human activity, and they exploit similar niches in North America. As evidence of this, we note not only that the exotics have been generally unsuccessful at invading undisturbed native habitats, but also that it is difficult to point with confidence to the extinction of a native continental species owing to replacement in its niche by a naturally invading competitor.

While Elton (1958) suggested that diverse biotas that have radiated to fill many niches may be relatively resistant to invasion, several lines of evidence indicate that native species richness alone is not a robust predictor of community invasability (Sax et al. 2007). First, comparisons of local and regional species richness suggest that many communities appear not to be saturated. Second, in experiments at small scales, even species-rich communities can often still be invaded by more species, increasing total species richness. Third, inspection of invasions versus extinctions at regional scales suggest that many regions can accommodate more species through invasion than are lost through extinction. So, an emerging perspective is that if communities are often not saturated, competition may not be a globally important process in confering biotic resistance to invasion (Sax et al. 2007).

Avian migration and provincialism

Birds, with their capacity to disperse by flight over great distances and formidable geographic barriers, would seem to be one of the groups of organisms least likely to exhibit biogeographic provincialism. Indeed, a few bird species, such as the peregrine falcon, osprey, and barn owl, are virtually cosmopolitan, and birds have managed to colonize even the most remote of oceanic islands (see Chapter 14). Remember, however, that Sclater (1858) originally divided the world into biogeographic provinces based on the distributions of passerine birds. This seems counterintuitive.

An examination of avian distributions on a global scale reveals that observed provincialism is due largely to two very different "functional groups" of birds. On one hand, birds with very limited powers of dispersal and specialized adaptations for particular environments often exhibit high degrees of endemism. Such birds include the flightless ratites (ostriches, emus, cassowaries, kiwis, and rheas) of the southern continents; the weakly flying tinamous, hoatzins, and curassows of tropical America; secretary birds and turacos of Africa; megapodes, lyrebirds, bowerbirds, and birds of paradise of Australia; and turkeys of North America. They also include highly specialized birds such as nectar-feeding hummingbirds of the New World, sunbirds of Africa and southern Asia, and honeyeaters of Australasia, as well as the woodcreepers, ovenbirds, antbirds, and manakins of South America; the honeyguides, wood hoopoes, and weaver finches of Africa; and the lorikeets, cockatoos, wood swallows, whistlers, and logrunners of Australia. Two additional obser-

FIGURE 10.30 Major avian migratory flyways. As the plotted routes suggest, in both the New and Old Worlds, many bird species travel hundreds and even thousands of kilometers twice each year, commuting between breeding grounds in subarctic and temperate regions to wintering grounds that are located mostly in the tropics. Very few of these birds, however, cross between the New and Old Worlds. (After McClure 1974 and Baker 1978.)

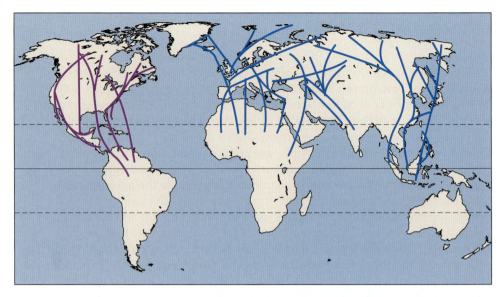

vations are worth making. First, many of these groups are endemic to one of the southern continents, presumably reflecting their long history of geographic isolation. Second, other groups of birds that might seem to be equally poor fliers or equally specialized are surprisingly widespread. Examples include the large, rarely flying bustards (Africa, Europe, Asia, and Australia) and the marsh-living, rarely flying rails (cosmopolitan).

The other "functional group" that contributes importantly to biogeographic provincialism consists of small land birds that are long-distance migrants (Bohning-Gaese et al. 1998). These include the hummingbirds, tyrant fly-

(A) Arctic warbler (*Phylloscopus borealis*)

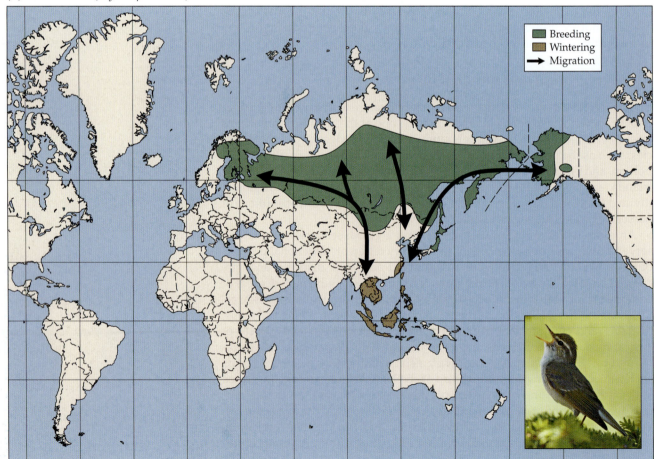

catchers, vireos, wood warblers, tanagers, orioles, blackbirds, and emberizine buntings and sparrows, which are all families or subfamilies endemic to the New World, and also the true flycatchers and warblers, which are restricted to the Old World. These subfamilies and families of birds include many species that migrate twice each year between their breeding grounds, in the subarctic and temperate regions of either North America or Eurasia, and their very distant wintering grounds in tropical areas of South and Central America or Africa, southern Asia, and Australia (**Figure 10.30**). With very few exceptions, representatives of these groups do not occur in both the New and Old Worlds. The exceptions prove the rule: A few species of Old World warblers breed in the Aleutian Islands and even on the Alaskan mainland but migrate to winter in tropical Asia or Australasia (**Figure 10.31**).

Also surprising is the fact that similarly small land birds that are either nonmigratory or only short-distance migrants show exactly the opposite pattern: Many families and subfamilies contain genera that occur broadly across both North America and Eurasia. These groups include woodpeckers, nuthatches, creepers, wrens, crows, tits, finches, thrushes, and waxwings (**Figure 10.32**). These relatively sedentary avian taxa have distributions very similar to those of land mammals, insects, and both coniferous and angiosperm plants: Closely related species occur across Arctic, subarctic, and cool temperate regions of both Eurasia and North America. Apparently such Holarctic distribu-

FIGURE 10.31   Maps showing the breeding ranges, wintering ranges, and migration routes (arrows) of two passerine bird species: (A) the Arctic warbler (*Phylloscopus borealis*), which breeds in western Alaska, and (B) the northern wheatear (*Oenanthe oenanthe*), which breeds in Alaska, northern Canada, and Greenland. Both of these species have colonized the New World, but show their Eurasian origins by migrating to winter in the Old World tropics. Other species with migratory routes similar to the Arctic warbler include the eastern yellow wagtail (*Motacilla tschutschensis*) and the bluethroat (*Luscinia svecica*). (A © Markus Varesvuo/Naturepl.com; B © Sue Robinson/Shutterstock.)

(B) Northern wheatear (*Oenanthe oenanthe*)

(A) Species

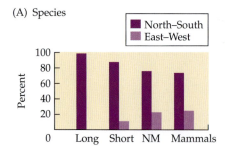

(B) Genera

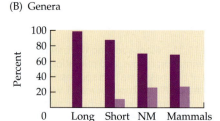

(C) Families

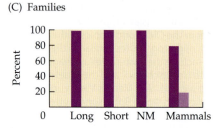

FIGURE 10.32 Comparison of the distributions between hemispheres of long-distance migrant (Long), short-distance migrant (Short), and nonmigratory (NM) birds and terrestrial non-volant mammals at three levels of taxonomic classification: (A) species, (B) genera, and (C) families. North–South (dark purple bars) indicates that the geographic range of the taxon includes both North and South America in the New World, or both Europe and Africa in the Old World; East–West (light purple bars) indicates that the range of the taxon includes both North America and Eurasia in the Northern Hemisphere. Note that the long-distance migrant birds have exclusively north–south ranges, indicating that they have not colonized between the Old World and the New World; in contrast, resident birds resemble non-volant mammals in having a substantial number of taxa with East–West distributions, indicating colonization between North America and Eurasia. (After Bohning-Gaese et al. 1998.)

tions reflect a long history of close proximity, including repeated land connections and resulting biotic exchanges, most recently across the Bering land bridge during the Pleistocene (see above, and Chapters 9 and 12).

Why are the long-distance migrants so different? What has prevented interchange between the New and Old Worlds? At least three interrelated factors seem to be involved. First, traits associated with long-distance migration may actually make it difficult for migrants to colonize a new continent. An initially small founding population not only must become established in a suitable breeding area, it must also find a new, distant wintering ground and a route there and back. The difficulty of developing new migratory routes appears to be so severe as to make exchange between New and Old Worlds highly improbable. Support for this hypothesis comes from the observation that the land bird faunas of isolated islands are constituted primarily of species derived from nonmigrants and short-distance migrants, rather than long-distance migratory ancestors (L. Gonzalez-Guzman, pers. comm.). Long-distance migrants can and do reach distant places, but they are unlikely to stay there or find their way back there to breed.

Second, this pattern seems to have a historical component. Lineages that contain most of the long-distance migrants appear to be of tropical origin. Migration seems to be an adaptation that allowed highly mobile birds to rear their young using the seasonal pulse of productivity that is available for only a few months at higher latitudes. Thus, most of the long-distance migrants feed their nestlings on insects that are active, abundant, and accessible only during the warm months. This is especially true in the New World, where, as noted above, the families and subfamilies containing most of the long-distance migrants are of South American origin.

Third, despite the broad geographic extent of their range throughout the year, the only relevant range in terms of provincialism is that of the breeding season, that is, when genetic material is exchanged. Thus, similar to the habitat philopatry of endemic cichlids of Africa's Rift Valley lakes (see Chapter 7), the requirement for genetic exchange constrains these migratory birds to visiting the same breeding grounds each year and thus restricting the geographic arena for the evolution of migratory routes.

The special case of birds illustrates how evolutionary constraints and ecological factors can interact to maintain the historical legacy of biogeographic provincialism. Even though the continents may be connected by land bridges that would seem to provide dispersal routes, and even though organisms may possess traits that would seem to permit long-distance dispersal, the actual interchange and mixing of long-isolated biotas has been relatively limited. The result is that the influence of ancient Earth history, especially of tectonic events, is preserved in the distributions of contemporary forms, even such vagile ones as migratory birds.

Biotic Interchange

The opposite of biotic segregation or biogeographic provincialism is biotic interchange. The fate of artificially introduced species provides some limited insight into the kinds of interactions that can occur when representatives of one formerly isolated biota come into contact with those of another. This has happened naturally in the past, not only when individual species have managed to disperse across barriers, but also when the barriers themselves have been reduced or abolished, bringing two distinct, previously isolated biotas into direct contact. Such contacts have occurred many times as continents have drifted over the Earth and new land and water connections have been formed (Vermeij 1978, 1991a,b; **Figure 10.33**). Unfortunately, the record of their

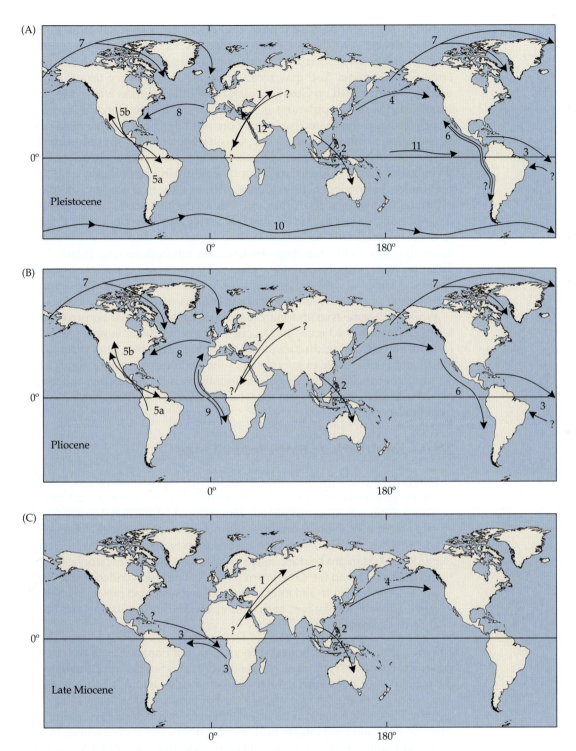

FIGURE 10.33 Major episodes of terrestrial and marine biotic interchange during three time intervals: (A) Pleistocene–Recent, (B) Pliocene, and (C) Late Miocene. Note that, in several cases, interchange occurred along similar routes during more than one time frame. Arrows depict predominant direction of dispersal during interchange, and question marks indicate uncertainty about directionality. 1, terrestrial, between Africa and Asia; 2, terrestrial, from Southeast Asia to Australia and New Guinea; 3, marine, across tropical Atlantic; 4, marine, across North Pacific; 5a, Great American Interchange for lowland rain forest species; 5b, Great American Interchange for savanna and upland species; 6, marine, transequatorial in east Pacific; 7, marine, transarctic; 8, marine, across North Atlantic; 9, marine, transequatorial eastern Atlantic; 10, marine, circum-Antarctic; 11, marine, tropical Pacific; 12, Lessepsian, trans-Suez. (After Vermeij 1991b.)

results is often poor because it must be pieced together, largely from limited fossil evidence. For example, when the Indian plate collided with southern Asia, it presumably brought a distinct Gondwanan biota into contact with a large Eurasian biota. You may recall from our discussion of taxa that were originally distributed on Gondwanaland that little mention was made of relictual Gondwanan populations in India. Was the Indian biota derived from Gondwanaland eliminated by interactions with the more diverse biota of the larger Eurasian landmass? This is a plausible explanation, but unfortunately, the fossil inhabitants of the Indian subcontinent are so poorly known that there is little direct evidence to support or refute this hypothesis. In fact, recent molecular phylogenetic evidence, with its increased accuracy in tracing evolutionary histories and, often, good estimates of divergence times, offers the intriguing possibility that some of what we have assumed to be lineages of northern origin instead originated on Gondwanaland and subsequently have diversified after reaching the northern landmasses (see Figure 7.13). This also appears to be true for oscine passerine birds that probably originated in eastern Gondwanaland–Australia prior to achieving a widespread distribution across the northern continents (Jonsson and Fjeldså 2006).

While both terrestrial and marine biotic interchange have occurred many times in the past (see Figure 10.33), one of the better documented events, the focus of the following discussion, is the interchange that took place when North and South America joined across the Isthmus of Panama about 2.5 to 3.5 million years BP, providing a continuous corridor for dispersal between Nearctic and Neotropical biotas that had evolved for millions of years apart from one another.

The Great American Biotic Interchange

During the age of mammals, the Cenozoic in geological terms, South America was an island continent. Its land mammals then evolved in almost complete isolation, as in an experiment with a closed population. To make the experiment even more instructive, the isolation was not quite complete, and while it continued two alien groups of mammals were nevertheless introduced, as might be done to study perturbation in a laboratory experiment. Finally, to top the experiment off the isolation was ended, and there was extensive mixture and interaction between what had previously been quite different populations, each with its own ecological variety and balance. (G. G. Simpson 1980)

The ecological and evolutionary experiment that Simpson was referring to was the exchange of mammals between North and South America following the formation of the Central American land bridge about 2.5 to 3.5 million years BP. This event, known as the Great American Interchange—or, more recently, the Great American Biotic Interchange—provides a fascinating case study of the combined effects of dispersal, interspecific interaction, extinction, and evolution on biological diversity. Most of the available information concerns mammals, in part because they left a rich fossil record. In fact, mammals figured prominently very early in the recognition of this event, when Wallace wrote in 1876 that "we have here unmistakable evidence of an extensive immigration…and the fact that no such migration had occurred for countless preceding ages, proves that some great barrier to the entrance of terrestrial mammalia which had previously existed, must for a time have been removed." Wallace was even more prophetic when he used his knowledge of both terrestrial and marine taxa to postulate that it was "almost certain that the union of North and South America is comparatively a recent occurrence, and that during the Miocene and Pliocene periods, they were separated by a wide arm of the sea…[and that] when the evidence of both land and sea animals support each other as they do here, the conclusions arrived at are almost as certain as if we had (as we no doubt some day shall have) geological proof of these successive subsidences." As more information becomes available for other taxa, a story is emerging in which the effects of the interchange were either quite different or rather similar in other terrestrial organisms, such as amphibians, reptiles, birds, and plants.

Historical background

Mammals evolved from reptilian ancestors during the Mesozoic, beginning about 220 million years ago, when the continents were united in the single great landmass of Pangaea. By the time Pangaea began to break up, mammals had spread over the continents and diversified into the modern lineages (monotremes, marsupials, placentals). Compared with the ruling reptiles, however, they remained a small component of the biota until the mass extinctions at the end of the Cretaceous, about 65 million years BP. This biotic upheaval, which saw the extinction of many terrestrial and marine lineages (see Chapter 8), including the dinosaurs, was followed by rapid diversification of birds, mammals, and some other surviving groups.

The most important geological, climatic, and biotic events associated with the history of South American mammals are summarized in **Table 10.3** (see also Stehli and Webb 1985). South America was a part of Gondwanaland until about 160 million years BP but then began drifting apart and remained a giant island continent until about 2.5 to 3.5 million years BP. During this period of what Simpson called "splendid isolation," a distinct endemic land mammal fauna evolved. At least one monotreme (a platypus), several groups of marsupials, and at least one lineage of eutherian mammals speciated and differentiated to produce a morphologically and ecologically diverse fauna that included large carnivores (even a marsupial saber-toothed "tiger") and giant herbivores. This radiation was in some ways comparable to, but even more diverse than, that which occurred on the other island continents formed by the breakup of Gondwanaland—Australia and Madagascar.

TABLE 10.3 *Summary of Tectonic, Climatic, and Biotic Events Associated with the Great American Biotic Interchange*

220–160 million years BP	South America remains connected to the rest of Gondwanaland and to North America. The origin, spread, and diversification of mammals and birds across the landmasses.
140–75 million years BP	South America becomes isolated. Its biota evolves and diverges on a separate landmass.
140–70 million years BP	Increasingly connected transient stepping-stone, sweepstakes route between South and North America, possibly ending in a land bridge along Antillean volcanic arc system.
~65 million years BP	Bolide impact at Chicxulub coincides with widespread extinctions.
~55–3.5 million years BP	Several arcs of islands, including the Antilles volcanic arc system as it moves east and forms the Greater Antilles–Aves Ridge system, and the Central American archipelago serve as transient stepping-stone, sweepstakes routes. Limited biotic exchange among Nearctic and Neotropical regions.
~2.5–3.5 million years BP	Emergence of the Central American land bridge (closure of the Bolivar Trench) provides a filter dispersal route for terrestrial forms, but a barrier for marine organisms.
2–0 million years BP	Lowering of sea level and extension of savanna and other open-habitat biomes during glacial maxima open a corridor or filter route for biotic exchange. Subsequent invasions and diversification of invaders, extinctions of invaders and natives.

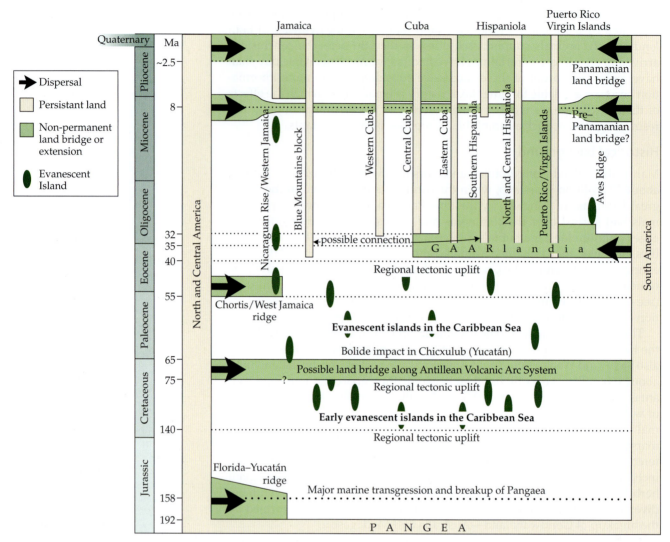

FIGURE 10.34 A paleogeographical scenario for the Caribbean region spanning a Middle Mesozoic to Late Cenozoic time frame, summarizing detailed reconstructions shown in Figures 8.24 and 10.22. Locations and numbers of the "evanescent islands" are intended to be only conjectural, and illustrate the increasing difficulty in reconstructing accurate depictions of paleogeography at older times. Light green shading represents possible land connection between two or more terrestrial blocks, including the Greater Antilles–Aves Ridge (GAARlandia), that now form the modern islands of Jamaica, Cuba, Hispaniola, Puerto Rico, and Aves Island. Notice that in addition to the known Late Pliocene Panamanian land bridge that resulted in the Great American Biotic Interchange, several conjectured earlier interconnections between North and South America are also shown. (After Iturralde-Vinent 2006.)

The biogeographic isolation of South America was not complete, however. As Simpson noted in the above quote, it was probably interrupted by one or two transient periods of limited exchange, although presence and configurations of former archipelagoes and land bridges in the Caribbean region are difficult to reconstruct accurately for times prior to the last several million years (**Figure 10.34**; see also Figures 8.24 and 10.22; Iturralde-Vinent 2006). During these periods, the ancestors of the present South American primates, edentates (armadillos, sloths, and anteaters), caviomorph rodents (porcupines, capybaras, pacas, agoutis, guinea pigs, chinchillas, and other forms), and possibly sigmodontine rodents (New World rats and mice) colonized the continent. In each of these groups, speciation and adaptive radiation subsequently produced multiple families and genera from a few founding lineages.

The isolation of South America ended dramatically about 2.5 to 3.5 million years BP when the Central American uplift formed the land connection to North America that still exists today. Since its formation, however, the Central American land bridge has served more as a filter than as a highway (**Figure 10.35**). David Webb (1991) has suggested that dispersal across the land bridge was strongly influenced by the climatic cycles of the Pleistocene (see Chapter 9). Interchange was greater during glacial periods, when savanna

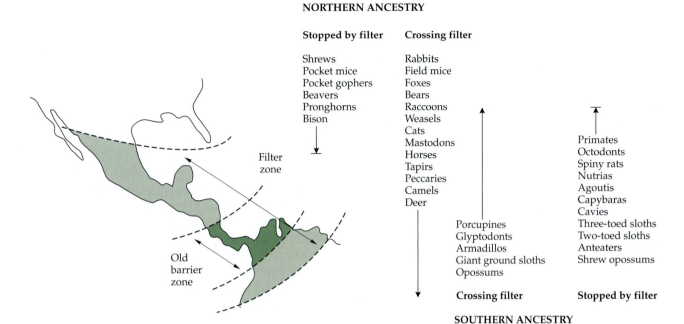

FIGURE 10.35 Map showing the location of the Central American land bridge, with lists of the mammalian families of both North and South American origin that either crossed through the filter of tropical lowland habitats in Central America during the Great American Biotic Interchange, to colonize temperate regions of the other continent, or were stopped in or near the filter. Note the asymmetry, with more groups of North American origin passing through the filter and more families of South American origin stopped by the filter. (After Simpson 1950.)

habitats expanded to cover much of Central and northern South America, than during interglacial periods, such as the present, when most of the land bridge was covered with tropical forest.

Patterns and consequences of mammalian faunal exchange

The fossil records of North and South America over the last 10 million years reveal the magnitude of the Great American Biotic Interchange. The diversity of the North American fauna, at the generic level, remained virtually unchanged despite the invasion of some South American forms, such as the extant porcupines, opossums, and armadillos and the now extinct hippo-sized glyptodonts and bear-sized giant ground sloths. The overall diversity of the South American fauna increased because of the invasion and establishment of lineages from North America. The many North American groups that crossed the Isthmus of Panama and became established in South America included not only the extant shrews, rabbits, squirrels, dogs, bears, raccoons, weasels, cats, deer, peccaries, tapirs, and camels, but also the now extinct mastodons and horses. The magnitude of change was even greater than the final figures for generic or familial diversity suggest, however, because several endemic South American forms went extinct. These included several kinds of large marsupial carnivores and even larger eutherian (true placental) herbivores.

The imbalance of the interchange can be seen in the fact that about half of contemporary South American species are derived from North American ancestors, whereas only about 10 percent of North American species are of South American origin (see Figure 10.35). The northern forms appear to have had three advantages that contributed to this imbalance:

1. *They were better migrators.* While many northern forms (listed above and in Figure 10.35) crossed the land bridge and invaded deep into South America, fewer taxa (porcupines, opossums, armadillos, and ground sloths) of ancient South American ancestry managed to colonize deep into temperate North America, and of these, the ground sloths are now extinct. Interestingly, however, several other groups of South American ancestry (primates, sloths, anteaters, and other kinds of marsupials and caviomorph

rodents) have colonized northward across the Isthmus of Panama but extend no farther than the tropical forests of southern Mexico.

2. *They were better survivors and speciators.* Most of the northern forms that invaded South America not only survived (mastodons and horses were exceptions), but also diversified there, giving rise to such distinctive forms as kinkajous, coatimundis, giant otters, swamp deer, guanacos, and vicuñas. The mouselike sigmodontine rodents, which probably colonized South America by island-hopping before the completion of the land bridge, underwent an explosive adaptive radiation and are now the most diverse group of mammals on the continent, comprising about 61 genera and 299 species (Marshall 1979; Webb and Marshall 1982; Reig 1989; Engel et al. 1998). This diversity is all the more impressive when viewed from two perspectives. First, while the ancestral sigmodontine may have originated in Mexico or Central America, the vast majority of extant diversity in the South American sigmodontines arose within South America following dispersal of one or a handful of ancestral species to that continent. Second, the neotomine-peromyscine rodents of North and Central America, sister group to the South American sigmodontines, represent a highly successful radiation in their own right (including the deer mouse shown in Figure 7.6) yet, with 18 genera and 124 species, are far less diverse than the sigmodontines. Furthermore, over the last 2.5 million years following the final connection of the land bridge, while lineages of North American ancestry have been diversifying, the ancient South American lineages have been dwindling away because speciation has not kept pace with extinction.

3. *They were better competitors.* In the face of differential colonization, survival, and speciation of North American forms, it is hard to avoid the conclusion that they have been, on average, superior competitors. Clearly, the North American forms not only have increased in generic and species diversity at the expense of their more ancient South American counterparts, but they also have radiated to usurp their ecological roles. For example, all of the large carnivores and herbivores, and the vast majority of the mouselike rodents in South America today, are descendants of northern invaders.

Some authors have questioned whether such differential ecological and evolutionary success can be attributed to "competition." Indeed, not all of the interactions are strictly competitive. Predation, parasitism, and disease probably play a significant role. Furthermore, much of the competition that does occur may be diffuse and involve many species rather than simple pairwise interactions. It is also clear that not all mammals of South American ancestry are competitively inferior: Witness the success of porcupines, opossums, and armadillos in invading North America. The latter two species have actually been expanding their ranges northward in recent decades, as seen in Figure 6.3D for armadillos.

Elizabeth Vrba (1999) presented a novel alternative to the competition model that she termed the "habitat theory." Briefly, she combined evidence of long-term climate change (a general cooling trend over the past 2.5 million years) and associated shrinkage of tropical forests, in favor of more open and arid savanna habitats extending between northern South America and Central America, with ecological data on the sorts of mammals that crossed the land bridge. From this evidence she made some predictions for the observed immigration imbalance without invoking competitive displacement. The first prediction is that, because the northern biota contained a greater number of savanna habitat specialists, one would expect greater migration success southward solely because savanna habitats were more continuously distributed across the land bridge. The second prediction is that the extinction imbalance toward South American natives could be attributed to a higher extinction rate among the closed-forest specialists after climatically induced fragmentation of forest habitats, in favor of more open savanna in northern South America. The final prediction is that the greater speciation rates of northern invaders into South America were due primarily to their containing a greater proportion of habitat specialists, prone to frequent cycles of habitat vicariance over the past 2.5 million years. Some evidence for this prediction comes from the great diversity of sigmodontine rodent species, which appear to have speciated in areas with high topographic diversity, within the higher valleys and plains of the Andes, and in the deserts and savannas of the southwestern part of the continent (see Figure 10.9).

Returning to Simpson's analogy, the Great American Biotic Interchange constituted a vast natural experiment. By the movement of continents, the mammalian fauna of South America were first allowed to evolve in "splendid isolation," and then brought into contact with a more diverse fauna—one that had evolved on the "World Continent" that included the larger North American landmass and its frequent interchanges with the fauna of the Old World. The outcome of the experiment is clear. Taken as a whole, the North American mammalian fauna proved superior, and they differentially "replaced" much, though by no means all, of the original South American fauna.

Inter-american exchange in other vertebrates

Information on historical exchanges between North and South America in vertebrate groups other than non-volant mammals is scanty because the fossil record of these groups is relatively poor, although several studies have made considerable progress by using a molecular genetics approach. It is also more difficult to interpret, because some of these groups, such as reptiles and especially birds and bats, are better over-water colonists than non-volant mammals. This means that some level of faunal interchange between the two continents probably occurred continuously throughout the Cenozoic, rather than being concentrated in the last 2.5 million years following the completion of the Central American land bridge (Vuilleumier 1984, 1985; Estes and Baez 1985; Vanzolini and Heyer 1985). For plants, considerable dispersal between the continents appears to have occurred prior to the completion of the land bridge (Cody et al. in press).

Two patterns are noteworthy. First, there is not a clear dichotomy between an ancient South American fauna, which dates back to the isolation of Gondwanaland, and relatively recent invaders. Instead, the South American herpetofauna and avifauna appear to have been assembled more gradually. They are made up of a mixture of ancient Gondwanan forms, lineages that colonized across water during the Early and Mid Cenozoic, and forms that crossed the Central American land bridge (or the island chain that preceded it; see Figures 8.24, 10.22) during the last several million years. Thus, at least for terrestrial reptiles, amphibians, and birds, it is difficult to distinguish easily between South American "natives" and North American "invaders."

Second, the Late Tertiary invasion of South America by North American forms, so clearly seen in mammals, is less obvious in many other groups. The exchange of birds appears to have been much more balanced than that of mammals. While some groups, such as pigeons, owls, woodpeckers, and jays, colonized South America from the north, other groups, such as hummingbirds, tyrant flycatchers, vireos, wood warblers, blackbirds, orioles, tanagers, and emberizine buntings (grosbeaks and sparrows), moved in the opposite direction. Even so, several patterns of Pliocene exchange and diversification appear to be shared by birds and mammals—birds also responded with increased dispersal across the isthmus after it was completely formed, and the pattern of subsequent diversification was strongly asymmetrical in favor of North American forms invading and then radiating in South America (Smith and Klicka, in press). An interesting feature of the North American avifauna is that the Neotropical migrants, which make up the majority of breeding passerines in most temperate habitats, are virtually all of South American ancestry.

The dominance of the North American fauna by lineages of South American origin is even more pronounced in reptiles and amphibians. Thus, South America has no salamanders and perhaps only one species of frog that colonized from North America. In contrast, of the North American frog fauna, only three families (Ascaphidae, Pelobatidae, and Ranidae) are of northern origin, whereas four (Bufonidae, Hylidae, Leptodactylidae, and Microhylidae) are of South American origin. Similarly, in reptiles, the predominant movement has been from South America to North America, although this pattern has sometimes been complicated by secondary centers of speciation and diversification in Central America and Mexico. Thus, for example, the two large families of New World lizards, Iguanidae and Teidae, and most of the snakes have dispersed northward from South America to North America.

The completion of the Central American land bridge resulted in some limited interchange of freshwater fish (Miller 1966; Rosen 1975; Bussing 1985; Bermingham and Martin 1998). Again, the predominant direction of dispersal was from south to north. Because of their requirement for freshwater connections in order to disperse, however, the invasion of primary freshwater forms has been limited. Several South American lineages have reached as far north as central Mexico (Miller 1966), whereas no North American group has made it farther south than Costa Rica (Bussing 1985).

To summarize, the Great American Biotic Interchange, that wonderful natural experiment so thoroughly documented by Simpson and later researchers, shows a clear pattern: Terrestrial mammals of North American ancestry were better dispersers, survivors, and speciators. Their differential success was due to some currently unknown combination of the contingency of habitat distributions across the land bridge and competitive superiority. Over the last 2.5 to 3.5 million years, since the completion of the Central American land bridge, they have invaded South America and largely supplanted the ancient South American groups. Either similar (e.g., in birds) or completely different patterns, but usually also with unbalanced exchange, occurred in other vertebrates.

The Lessepsian exchange: The Suez Canal

Continuing with Simpson's analogy of biotic interchange as a natural experiment, humans have unintentionally performed analogous manipulation by bringing into contact two previously long-isolated marine biotas. One such experiment began in 1869 with the completion of the Suez Canal between the Mediterranean Sea and the Red Sea. The resulting biotic exchange has been called the Lessepsian migration or exchange, named in honor of the chief engineer of this project, Ferdinand de Lesseps (Por 1971, 1975, 1978, 1990). Although the hypersaline lakes through which the canal passes constitute an impassable barrier for many marine forms, an increasing number of taxa have been able to disperse between the two seas. Again, the biotic exchange has been very unbalanced. More than 50 species of fish, 20 species of decapod crustaceans, and 40 species of mollusks have colonized the eastern Mediterranean from the Red Sea, but it is hard to document cases of migration in the opposite direction.

Three explanations for this unidirectional dispersal have been proposed, and all three factors probably contribute to successful colonization by Red

Sea forms. First, the Gulf of Suez, at the southern end of the canal, is itself more saline than ocean water, so the species that occur there may have been preadapted to cross the barriers formed by the hypersaline lakes. Second, most of the successful migrants have populations in shallow sandy or muddy bottom habitats in the Indian Ocean, into which the Red Sea opens. Such habitat affinities also appear to be preadaptations, because they facilitate movement through similar habitats in the Suez Canal. Finally, species of the Indo-Pacific biota may be more resistant to predation than their Mediterranean counterparts, competitively superior to them, or both. The Red Sea is an arm of the Indian Ocean, which contains a far more diverse biota than the Mediterranean. Biotic interactions are implicated by observations of declining populations of some endemic Mediterranean species in the eastern part of the sea (e.g., along the coast of Israel) where Red Sea colonists have become well established. (For a much more complete discussion of the history and results of the Lessepsian exchange, see Por 1971, 1975; Vermeij 1978, 1991b; Golani 1993; Galil 2000; Gofas and Zenetos 2003; Bradai et al. 2004; Shefer et al. 2004; Bianchi 2007.)

One final point is warranted. The differential extinctions of South American mammals in the face of North American invaders and the unbalanced exchange of marine forms through the Suez Canal are both consistent with a general pattern noted by many biogeographers. As early as 1915, W. D. Matthew (see also Willis 1922) noted that organisms from diverse biotas on large landmasses are best able to successfully invade smaller areas and replace the native organisms. Simpson (1950) argued that the northern invaders were better competitors than their South American counterparts because they had a long history of being tested and surviving within the more diverse World Continent: "When ecological vicars met, one or the other generally became extinct. . . . Those [northern invaders] extant in the Plio-Pleistocene were the ones that had been successful in a long series of competitive episodes. They were specialists in invasion and in meeting competitive invaders." Darlington (1957, 1959b) reemphasized this point, although he argued that the successful forms usually originated in tropical regions, whereas Matthew had thought they came from temperate climates. This point may be difficult to resolve, because the climates of regions that are now temperate or tropical have changed greatly over their geological history (see Chapters 8 and 9). Nevertheless, the success of organisms from large, diverse biotas in colonizing small, isolated regions containing fewer native species seems to be a consistent phenomenon in biogeography. The interactions that have been seen on a continental scale fit the pattern of invasions of the island continent of Australia from larger continents, as well as the natural and human-assisted colonization of many islands by continental forms (see Chapters 13, 14, and 16).

The Divergence and Convergence of Isolated Biotas

We have considered what happens when long-isolated biotas are brought into contact. Now let's examine the opposite situation. Are there detectable general evolutionary trends that tend to arise when lineages and biotas become separated by persistent geographic barriers?

Divergence

Given sufficient time since being geographically isolated, and particularly if ecological conditions change as well, populations and species might be expected to diverge in a variety of morphological, physiological, behavioral, and life history traits. Let's return to the event in Earth history—the formation of the Central American land bridge—that brought Neotropical and Nearctic terrestrial biotas together, but let's now examine how this event influenced the evolution of marine biotas that were then completely isolated between tropical Atlantic and Pacific Oceans.

The fragility of the isthmus as a barrier became particularly apparent in the 1960s, when the possibility of constructing a sea level canal across the Isthmus of Panama was being seriously considered. The present Panama Canal, constructed in the early part of the twentieth century, incorporates a large body of fresh water, Gatun Lake, and uses a series of locks to raise and lower ships as they traverse the isthmus. The freshwater effectively prevents interchange between most elements of the Pacific and Caribbean tropical marine biotas. The construction of a sea level canal would constitute a biogeographic experiment of gigantic proportions. Its effect on marine biotas would be analogous to the influence of the original establishment of the isthmus on terrestrial forms, with the exception that the Caribbean and Pacific biotas have been isolated in entirety for only about 2.5 to 3.5 million years, whereas North and South America had been separated for at least 135 million years.

Controversy surrounding the possible ecological effects of an interchange of species as a result of a sea level canal stimulated much research on the similarities and differences between the Caribbean and eastern tropical Pacific biotas. These studies revealed major differences, especially in species richness, between the marine faunas of the two regions (Briggs 1968, 1974; Rubinoff 1968; Porter 1972, 1974; Vermeij 1978). Most groups are more diverse in the Pacific; examples include most major taxa of mollusks, crabs, and echinoderms. There are exceptions, however. The Caribbean has about 900 species of shallow-water and coral reef fish, compared with only about 650 species in the eastern Pacific. Sea grasses and their specialized animal fauna are abundant, widespread, and diverse in the Caribbean, but virtually absent from the eastern Pacific, where suitable, highly productive shallow-water habitats are not extensive.

In many groups, including fishes, sea urchins, gastropods, crabs, and isopods, there are closely related, so-called geminate sister species on either side of the isthmus (Jordan 1908; Vermeij 1978). The rates of divergence of some of these forms have been of considerable interest to systematists and evolutionists (e.g., Rubinoff and Rubinoff 1971; Knowlton and Weight 1998; Lessios 1998; Marko 2002; Hickerson et al. 2006). The time of isolation is often presumed to be the same for all groups, although there might be a broader range of times of isolation in reality than would be expected under the assumption that all species pairs were isolated during the final stages of closure of the isthmus (Knowlton and Weight 1998; Lessios 1998). Although there has been some differentiation, most of these species pairs remain similar in morphology—often masking a surprising level of evolutionary divergence and speciation (Craig et al. 2004; Marko and Moran 2009)—and, presumably, in their ecological niches. This suggests that competition between such species pairs might prevent much interchange across a sea level canal. On the other hand, if some forms were superior competitors and able to invade the other ocean, competitive exclusion might result in the extinction of some sister species without greatly affecting the diversity of the biotas on each side of the isthmus.

The absence of complementary species in a few exceptional groups has caused more concern. Two Pacific taxa in particular that do not have close relatives or obvious ecological counterparts in the Caribbean are the yellow-bellied sea snake, *Pelamis platurus* (see Figure 12.2), and the crown-of-thorns starfish, *Acanthaster planci*. The former is highly venomous, whereas the latter feeds voraciously on certain corals and occasionally devastates reefs in its native region (Chester 1969). It is possible that *Acanthaster* might wreak even greater havoc if it were able to colonize the rich Caribbean reefs, where the coral species have had no opportunity to evolve resistance to it (Porter 1972; Vermeij 1978).

Convergence

Most of what we have said thus far would cause one to expect that geographically isolated organisms should diverge. Much of morphologically based phylogenetic systematics and vicariance biogeography is based on the assumption that genetically and geographically isolated lineages tend to become more different as they evolve independently of each other. It need not be true, however, particularly of ecological characteristics. During the past decade a number of investigators have emphasized what is perhaps a surprising degree of **niche conservatism**, described as "the tendency of species to retain ancestral characteristics" (Wiens and Graham 2005) influencing processes ranging from allopatric speciation (Kozak and Wiens 2006) to community assembly (Moen et al. 2009; Pennington et al. 2009) and conservation of range sizes across episodes of profound climate change (Hadly et al. 2009).

Nevertheless, if groups are isolated in regions of different area, geology, and climate, these differences in the physical environment will tend to promote ecological divergence. If the physical environments are similar, however, distantly related organisms may independently evolve similar adaptations. This phenomenon is called convergent evolution, and it can occur on many levels. Within species, it may be restricted to a few traits, or it can involve essentially the entire organism, resulting in convergence in morphology, physiology, and behavior as unrelated forms specialize for similar niches. Convergence can also occur at the level of entire biotas, resulting in geographically isolated ecological communities with similar structures and functions.

Convergence at the species level

It is possible to find examples of geographically isolated, distantly related pairs of species that are spectacularly similar. Some of the best examples occur in plants, especially those living in regions where similarly stressful abiotic environments have selected for similar form and function. In Chapter 3 we discussed the Mediterranean climates that occur at about 30° latitude where cold ocean currents flow down the west coasts of continents; these climates are found in the Mediterranean region, southwestern Africa, southwestern Australia, Chile, and California. The latter two regions were the sites of extensive comparative ecological studies by the International Biological Program in the early 1970s (Mooney 1977). Analyses of the anatomy and physiology of the woody plants revealed many similarities. The dominant plants in both *matorral* habitat in Chile and chaparral in California are shrubs with small to medium-sized, thick (sclerophyllous), evergreen leaves. It is possible to identify many pairs of species that are extremely similar in growth form and leaf morphology (**Figure 10.36**). In addition, these species usually have similar physiological and life history adaptations to photosynthesize, grow during the cool winter rainy season to minimize water loss and survive through the hot summer dry season, and regenerate rapidly—usually from vegetative sprouts from the stumps, but sometimes by seed—after frequent wildfires (Mooney et al. 1977).

Plants from desert regions throughout the world provide equally spectacular examples of convergent form and function. Examples are exhibited in several botanical gardens, including in the Arizona-Sonora Desert Museum near Tucson, Arizona, and in Kew Gardens outside London, England. Dif-

FIGURE 10.36 Convergence in leaf morphology of distantly related plant species from evergreen shrub habitats in Mediterranean climates in four widely separated regions: California, Chile, Sardinia (Mediterranean), and South Africa. Presumably, the similarities—not only in the sizes and shapes of the leaves shown here, but also in their physiological characteristics—reflect convergence: the independent evolution of similar traits in response to natural selection for similar adaptations to similar environments. (After Cody and Mooney 1978.)

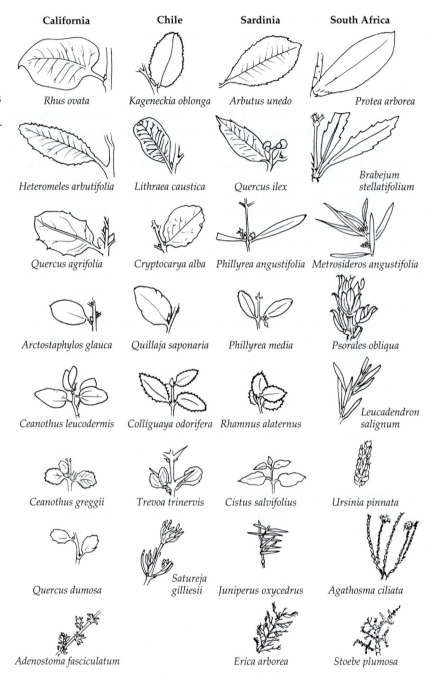

ferent genera of succulent thorny cacti (family Cactaceae) from North and South America are remarkably similar, not only to each other, but also to the much more distantly related euphorbias (family Ephorbiaceae) of Africa. Another convergent theme is succulents with whorls of tough, pointed leaves: agaves in North America, terrestrial bromeliads in South America, and aloes in Africa (each representing a different family). And, the similarity does not stop at superficial resemblances among life forms, but extends to details of cellular anatomy, physiology, and biochemistry. For example, these desert succulents all share a special form of photosynthesis, called crassulacean acid metabolism (CAM). This adaptation enables them to conserve water by opening their stomata at night when temperatures and rates of evaporative water loss are low, taking up CO_2, storing carbon in the form of organic acids, and then completing photosynthesis with their stomata closed during the day when the sun shines.

These examples of convergence are convincing. Some geographically isolated plant species in different families that live in similar environments are much more similar to one another than to more closely related species that occur on the same continents, but in different environments. Furthermore, their similarities clearly represent evolutionary adaptations to deal in similar ways with similar kinds of abiotic environmental conditions. It should be emphasized, however, that these convergent plants are not strikingly similar in all of their characteristics. Naturally, they each retain distinctive traits indicating their divergent ancestry. In addition, they often differ considerably in reproductive biology, exhibiting divergent forms and functions of flowers, fruits, and seeds that reflect adaptations for different agents of pollination and seed dispersal.

Examples of convergence among geographically isolated animal species are equally spectacular. As in plants, many come from desert regions. The mammalian order Rodentia has a virtually cosmopolitan distribution, but many rodent families have much more restricted distributions. Each of the biogeographic realms has large areas of desert and semiarid habitat, and each of these has a distinct, highly specialized desert rodent fauna made up of different families that have independently evolved to fill a variety of niches. The most striking case of convergence is perhaps the independent evolution of forms with elongated hind legs; long, tufted tails; and bipedal hopping (saltatorial or ricochetal) locomotion in different families on several continents (**Figure 10.37**; Mares 1976, 1993a). Some of these rodents also share other adaptations, including light-colored pelages to match their backgrounds; enlarged ear cavities (auditory bullae), perhaps to detect predators; short,

FIGURE 10.37 Apparent convergence, at least in morphology, of five genera of rodents and one extinct marsupial from deserts throughout the world. All of the rodents are derived from unspecialized mouselike ancestors, and have independently evolved long hind limbs, short forelimbs, long tufted tails, light brown dorsal and white ventral pelage, and bipedal hopping locomotion. Some of them share other morphological, physiological, and behavioral characteristics, but some of them also differ conspicuously in body mass, ear length, diet, and other characteristics. (After Mares 1993a.)

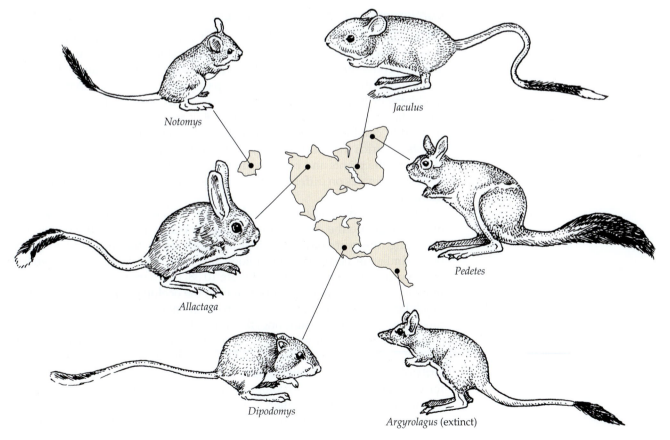

Notomys

Jaculus

Allactaga

Pedetes

Dipodomys

Argyrolagus (extinct)

long-clawed front legs for digging burrows and collecting food; and urine-concentrating kidneys for maintaining water balance on a dry diet. On the other hand, as in the plants, their convergence does not extend to all traits. Thus, the North American kangaroo rats and kangaroo mice are highly specialized seed eaters, whereas the similar-looking hopping mice in Australia and jerboas in North Africa, the Middle East, and central Asia eat some seeds, but also include substantial amounts of other items, such as insects, tubers, and leaves, in their diets. Furthermore, in these different deserts, some of the rodent species that are not very similar in external morphology are much more similar in diet and, as a consequence, in digestive and excretory physiology (Mares 1976; Mares and Lacher 1987; Mares 1993a,b; Kelt et al. 1999).

Similar generalizations could be drawn for other taxa, such as the superficially similar snakes and lizards from arid regions of North America, Africa, Asia, and Australia. The pronghorn or "antelope" of the plains of North America is the sole representative of an endemic family (Antilocapridae) that is convergent in morphology and behavior with the true antelopes (family Bovidae) of the grasslands of Africa. Several species of toucans (family Ramphastidae) of the New World tropics are superficially similar to some species of hornbills (family Bucerotidae) of the Old World. Representatives of both families have striking black and white plumage and very large, light, keel-shaped bills adapted for feeding on fruit. Toucans and hornbills differ conspicuously, however, in their reproductive biology, with only hornbills showing the distinctive behavior of males building mud walls to imprison their mates in nesting cavities. The European fire salamander (*Salamandra salamandra*) and North American tiger salamander (*Ambystoma tigrinum*) are similar in their large size, robust body shape, poison glands in the skin, and striking yellow-on-black warning coloration, but they differ in other aspects of their ecology and behavior.

Convergence of entire assemblages?

While such convergence between isolated species and groups may be extremely precise, some investigators have suggested that there can be at least as much convergence at the level of entire biotas. In the literature, there are many illustrations purporting to show pairs of similar species that make up the biotas of isolated regions with similar environments. The mammals of Australia and North America, shown in **Figure 10.38**, are perhaps the most frequently cited examples, but others are the mammals inhabiting tropical forests of Africa (or Asia) and South America (e.g., Bourliere 1973; Eisenberg 1981) and the plants and birds living in semiarid (including Mediterranean) or desert climates on different continents (e.g., Cody 1968, 1973; Solbrig 1972; Mooney 1977).

Such illustrations are not intended to be misleading. Nevertheless, they often exaggerate the degree of overall resemblance between the biotas. For one thing, the pairs of species that are depicted in the drawings may not be as similar in body size, physiology, behavior, and ecology as in general body form. Thus, for example, in Figure 10.38, the marsupial "cat" is smaller, more insectivorous, and less arboreal than the ocelot; the flying phalanger is much larger than, and differs in diet from, the flying squirrel; the wombat is much larger than the groundhog; the numbat, or marsupial anteater, is much smaller than the giant anteater; and the marsupial "mole" inhabits sandy deserts, unlike any true mole. In addition, such figures rarely show the majority of the species in the two biotas, most of which are not similar. Australia, for example, has no close ecological equivalents of many North American mammals, including mountain lions, bison, weasels, skunks, prairie dogs, and beavers; similarly, North America has no species that is really similar

Placentals   Marsupials

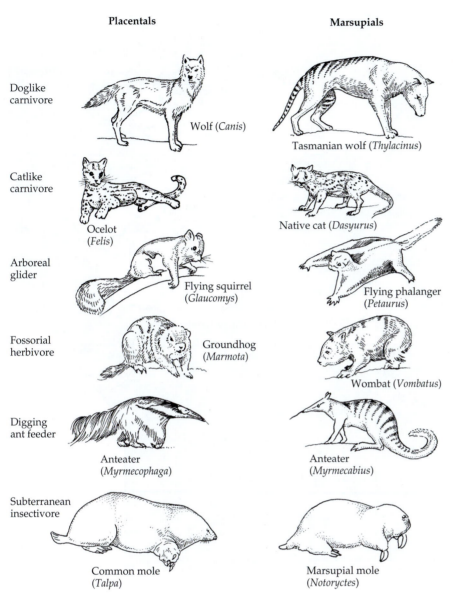

Doglike carnivore

Wolf (*Canis*)

Tasmanian wolf (*Thylacinus*)

Catlike carnivore

Ocelot (*Felis*)

Native cat (*Dasyurus*)

Arboreal glider

Flying squirrel (*Glaucomys*)

Flying phalanger (*Petaurus*)

Fossorial herbivore

Groundhog (*Marmota*)

Wombat (*Vombatus*)

Digging ant feeder

Anteater (*Myrmecophaga*)

Anteater (*Myrmecabius*)

Subterranean insectivore

Common mole (*Talpa*)

Marsupial mole (*Notoryctes*)

FIGURE 10.38 Drawings of pairs of species of North American and Australian mammals purporting to show convergence. Figures such as this one are somewhat misleading. As described in more detail in the text, the species paired here are often not drawn to the same scale, and some of those that look alike do not have similar ecological niches. (After Begon et al. 1986.)

to the duck-billed platypus, spiny anteater, bandicoot, koala, or the numerous small and medium-sized wallabies. And, finally, other elements of the North American and Australian biotas exhibit even less convergence than the mammals. North America has no real equivalents of Australia's parrots and emus among birds, or the eucalypts and acacias among plants, and its lizard fauna is much less diverse (Pianka 1986). Australia has no arid-zone plants that are similar to the cacti and agaves of the North American deserts. In fact, after visiting both North America and Australia, many naturalists begin to question the dogma of convergence.

It is not really surprising that the differences in these two biotas far outweigh their similarities. Comparison of the geography, ecology, and geology of the two continents reveals that their environments are in fact very different in several respects. First, as illustrated in **Figure 10.39**, Australia is much more tropical than North America. The Tropic of Capricorn passes through the very center of Australia just north of Alice Springs, whereas the Tropic of Cancer passes just north of Mazatlan, Mexico. Compared with North America, Australia has extensive older geological formations (Paleozoic and

FIGURE 10.39 A map in which the locations of Australia and North America have been juxtaposed while their relative latitudinal positions have been maintained. Note that while Australia is about the same size as the United States, its latitudinal position overlaps more with that of Mexico. The locations of the arid habitats that are often suggested to contain convergent species or ecological communities are, on average, much more tropical in Australia than in North America. (Australia has been inverted in order to maintain its orientation relative to the Equator.)

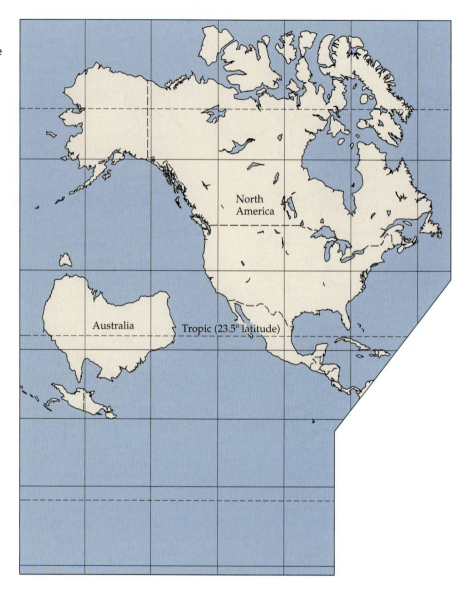

Mesozoic rather than Cenozoic), little recent tectonic activity, more eroded landscapes, and hence much less elevational relief (the highest mountain is only 2229 m, compared with 6194 m in North America). As a result of their long exposure to erosion and leaching, Australian soils are extremely low in nutrients. The low availability of essential resources has strongly influenced many aspects of the ecology and evolution of Australian plants, and the plants in turn have affected the animals. Given these major differences in their geological and geographic histories and current environments, it seems naive of biogeographers, evolutionary biologists, and ecologists to expect to find any substantial degree of convergence between the biotas of these two continents. Similar caution should be exercised before uncritically accepting other examples of expected or claimed convergence at the level of whole floras or faunas (e.g., Cody and Mooney 1978; Lomolino 1993a; Kelt et al. 1999).

Overview

So, we end by emphasizing the complexity evident in the diversification and assembly of biotas. On one hand, as noted by such early biogeographers as Buffon, de Candolle, Sclater, and Wallace, and confirmed by many more recent investigators, the biota of each of the major landmasses is made up of distinct kinds of plants and animals. To a large extent these differences reflect geological histories of isolation and the presence of different lineages, each with its unique constraints and potentials to diversify and adapt to abiotic and biotic environments. We are beginning to discover, for example, the surprisingly fundamental role of Gondwanaland in the early diversification of many now-common Northern Hemisphere lineages (e.g., amphibians: Bossuyt and Milinkovitch 2001; modern birds: Cracraft 2001, Barker et al. 2004; placental mammals: Eizirik et al. 2001, Wildman et al. 2007; and grasses: Bremer 2002). The pattern and process of diversification has been additionally influenced by differences, both small and large, in the environmental settings, which have favored the differential diversification of some lineages and the evolution of certain traits.

On the other hand, the biotas of the major continents are also similar in some respects. They share some taxonomic groups—and even some species are cosmopolitan, or nearly so. Some of these shared lineages have persisted from the time before the continents were isolated. Others have spread more recently, either colonizing across long-standing barriers or dispersing at times in the past when bridges of habitat permitted biotic interchange. Similar environments on different continents have facilitated the colonization and persistence of closely related organisms with similar requirements, as well as the convergent evolution of distantly related forms to use similar environments in similar ways. The same processes can be extended, albeit at some notably different spatial and temporal scales, to marine environments (Wainright et al. 2002; Johannesson 2003; Fratini et al. 2005).

The diversity of life on Earth reflects the outcome of opposing forces promoting both convergence and divergence among biotas. Evolutionary conservatism, phylogenetic constraints, gene flow, and similar environments limit the rates and directions of diversification, and thus tend to maintain similarities among biotas. Different environmental conditions, geographic isolation, evolutionary innovations, adaptation, and speciation enhance the rates and directions of divergence and diversification, and thus they tend to promote differences among biotas. Nowhere is the complex interplay of these opposing forces more evident than in the diversity of the plants and animals inhabiting biogeographic regions. Reconstruction of the developmental stages in diversification (splitting) and interchange (reticulation) among regional biotas is the subject of our next two chapters.

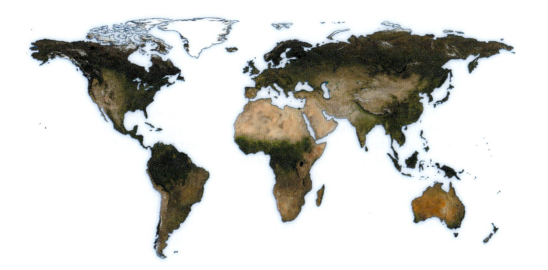

11

RECONSTRUCTING THE HISTORY OF LINEAGES

At the beginning of the previous chapter, we introduced the basic differences between the "history of lineage" and the "history of place." Each of these can be, and has often been, investigated without reference to the other, but an important goal of historical biogeography is to discover causal relationships between lineage and place histories. Our ability to reconstruct such an association can only be accurate to the extent that the underlying hypotheses of history are accurate. We have described in detail many of the important events in the history of place in Chapters 8 and 9. We have also introduced the basic concepts of systematics in Chapter 7, but left until now a detailed discussion about how we reconstruct and interpret the histories of lineages.

The importance of phylogenetic systematics to biogeography has grown rapidly in the past several decades, in part because of the increasing availability of high-quality phylogenetic trees, in part because of the increasing sophistication of phylogenetically based analyses employed to assess the relative influence of vicariance and dispersal events in biogeographic history, and in part because of an expanded technical and analytical capacity to address biogeographic history at relatively recent time frames (i.e., during the past 2 million years). Within this last development, we have also witnessed an exciting synthesis of phylogenetics with population genetics into phylogeography, the approach in biogeography that studies the geographic distributions of genealogical lineages within species and among closely related species. Additionally, while fossils have always been used in the reconstruction of lineage histories, they assume a renewed importance, particularly in the estimation of times of lineage divergence, when used in combination with molecular approaches to constructing phylogenetic trees. This chapter reviews the basic principles of phylogenetic systematics, phylogeography, and the role of fossils in biogeography after a brief review of the development of modern systematics. The histories of lineage and place will then be considered together in our formal discussion of modern historical biogeography in Chapter 12.

Classifying Biodiversity and Inferring Evolutionary Relationships

The hierarchical system of biological classification used traditionally to classify biodiversity dates back to the eighteenth century Swedish naturalist Linnaeus, but he believed that each kind of plant and animal had been specially created by God. In the latter half of the nineteenth century, Darwin placed Linnaeus's scheme of classification within an evolutionary framework, arguing that species are nested within genera, genera within families, and so forth because of a history of descent with modification within evolutionary lineages: "the view that an arrangement is only so far natural as it is genealogical" (Darwin 1859). In fact, the only illustration in *The Origin of Species* was a hypothetical phylogeny (**Figure 11.1**). But while Darwin's observation formed the conceptual foundation for the modern science of systematics, this discipline has wrestled over the past century with basic questions about how to define and diagnose natural groups, and what sorts of characters and procedures can best be used to reconstruct them. Fortunately, much of the heated controversy that infused this discipline throughout the 1960s and into the 1980s has subsided, and tremendous progress continues in both the quality and quantity of phylogenetic hypotheses that are becoming available for use in biogeography. Developments in four principal areas have had a profound impact on our ability to reconstruct the history of lineages: (1) general acceptance and application of the concepts of phylogenetic systematics established by Willi Hennig (1966), (2) utilization of abundant characters within a DNA sequence, (3) continual improvements in our understanding of the tempo and mode of evolution of those characters, and (4) bridging the threshold between the interspecific and intraspecific boundaries by using phylogenetic

FIGURE 11.1 A hypothetical phylogeny, the only illustration Darwin included in *The Origin of Species*. Darwin used this diagram to illustrate the principal of descent with modification from a common ancestor and his view that a natural system of classification was one in which "propinquity of descent . . . is the bond, hidden as it is by various degrees of modification, which is partially revealed to us by our classifications." (After Darwin 1859.)

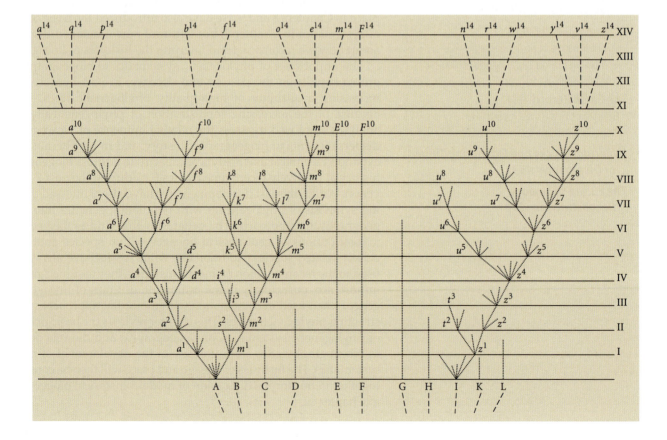

and population genetic approaches to investigate the phylogeography of populations within species.

Systematics

As pointed out in Chapter 2, humans have long recognized that some kinds of organisms are more similar to each other than they are to others, and we have tried to capture these patterns of similarity by naming and classifying living things. Aboriginal peoples have names for the plants and animals in their environments, and these names typically recognize degrees of difference, with classifications incorporating functional information based on the roles each play in the lives of the humans, but nevertheless arriving at a taxonomy surprisingly similar to one developed by trained taxonomists (e.g., Patton et al. 1982). While Darwin provided the conceptual foundation for linking the history of lineages into a nested hierarchy of ancestor–descendant evolutionary relationships, early systematists classified biodiversity for another century without developing a clearly defined methodology for doing so. Beginning in the 1950s, three very different schools of classification emerged—**evolutionary systematics**, **numerical phenetics**, and **phylogenetic systematics**, or **cladistics** (Mayr 1988). Each school claimed to have developed a robust set of methods for constructing a natural classification, but their underlying conceptual principles differed greatly and resulted in heated controversies and debates throughout the 1960s and into the 1980s. Although the vast majority of systematists now employ some form of phylogenetic systematics, it is worthwhile to review briefly the first two methods, both because each enjoyed great influence in systematics at various times during the 1960s and 1970s, and because some of the methods that arose originally within a phenetics framework are continued into modern systematics within a phylogenetic framework.

Evolutionary systematics

Recall from Chapter 7 that modern systematics has two related goals—to classify biological diversity (taxonomy) and to recover patterns of ancestor–descendant relationships (phylogenetics). Prior to the widespread acceptance of the methods and logic formalized by Willi Hennig into cladistics, two founders of the Modern Synthesis—Ernst Mayr (1942) and George Gaylord Simpson (1945, 1961a), among others (Dobzhansky 1951; Rensch 1960)— were leading advocates for the use of two kinds of evolutionary information in the classification of organisms. On one hand, evolutionary classifications were consistent with Hennigian logic in using the best estimates of branching sequences representing the hierarchy of ancestor–descendant relationships. However, they went a step further and sought to include information on degree of divergence between taxa, as well. In the language of phylogenetics, these classifications therefore had a tendency to produce **paraphyletic** groups—those that contain a common ancestor and some, but not all, of its descendants. Many long-recognized taxa are actually paraphyletic and therefore considered artificial rather than natural groups under phylogenetic systematics. Consider, for example, the amniotic vertebrates (reptiles, birds, and mammals). While evolutionary classifications have long placed reptiles, birds, and mammals into separate classes (Reptilia, Aves, Mammalia; **Figure 11.2A**), detailed character analyses—primarily from morphological studies of fossil and extant specimens, but supported by molecular analyses—have generally provided strong support for a basal amniote divergence into two monophyletic clades. The first branch is a sauropsid clade that includes liv-

FIGURE 11.2   Contrast between (A) an evolutionary classification and reconstructed phylogeny of living vertebrates from the classic work of A. S. Romer and (B) a phylogenetic classification (cladogram) showing nested monophyletic groups within the Vertebrata. In (A) the thickness of the branches indicates the approximate diversity of the groups through time. The degree (rate) of evolutionary divergence from a common ancestor as well as the time of the split are taken into account in grouping organisms into taxa. Note that the cartilaginous fishes (Chondrichthyes), bony fishes (Osteichthyes), amphibians, reptiles, birds (Aves), and mammals are all recognized as equivalent units and given class rank. In (B) note that reptiles (Lepidosauria, Crocodilia, and Chelonia) are identified as the monophyletic archosaur clade, but would become a paraphyletic group if Aves were removed from the Sauropsida and placed in its own category. Shared-derived (synapomorphic) characters that support the inclusion of Aves in a particular group include (1) loss of the medial centrale bone of the pes, (2) dorsal and lateral temporal openings and suborbital fenestra, (3) hooked fifth metatarsal and foot directed forward for much of the stride. (A after Romer 1966; B modified from Carroll 1988.)

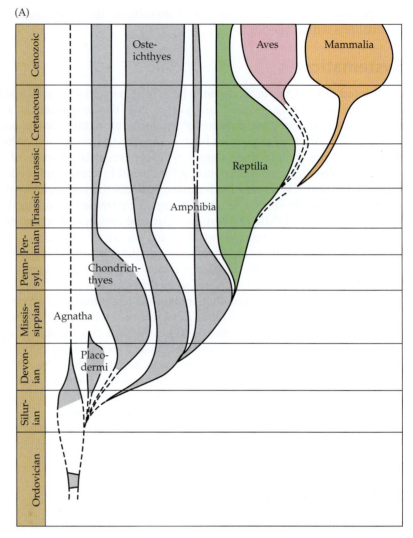

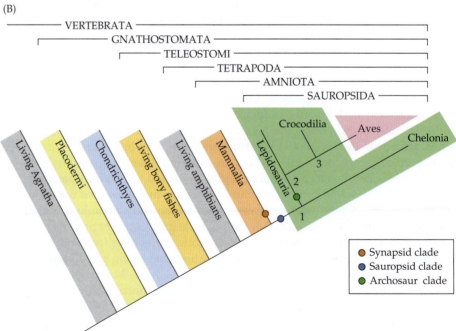

ing reptiles, birds, and the extinct dinosaurs (and nested within this clade is an archosaur clade that contains just the dinosaurs, crocodilians, and birds); the second branch is a synapsid clade that includes mammals and the extinct mammal-like reptiles (compare **Figure 11.2B** with Figure 11.2A).

Numerical phenetics

Another problem with evolutionary classifications was that they tended to be highly subjective. Different systematists, using their own schemes for choosing and weighting characters, produced different classifications for the same groups—and then argued about which one reflected the "true" evolutionary relationships. In response, some biologists questioned the scientific legitimacy and relevance of systematics (Ehrlich 1964), and others attempted to develop a more objective and quantitative process of classification called numerical phenetics or numerical taxonomy (Sneath and Sokal 1973). As an alternative to attempting to construct explicitly evolutionary classifications, numerical taxonomists advocated purely *phenetic* classifications, which attempted to use degree of overall similarity derived through quantitative analyses of phenotypic traits to cluster taxa into hierarchical units. Although the practitioners of this approach had hoped to be able to arrive at natural classifications through use of many characters, these classifications frequently grouped unrelated taxa into the same taxon because of an undue influence of traits that were similar because of convergent or parallel evolution, or because rates of evolution differed among taxa. As with evolutionary systematics, the original form of numerical phenetics has been largely replaced by phylogenetic systematics. However, traces of this school are still found in modern systematic procedures that use quantitative measures of similarity to produce distance matrices that are analyzed within a phylogenetic framework. An important legacy of numerical phenetics survives in disciplines outside of systematics, such as ecology and sociology, that employ a variety of procedures based on ordination and clustering methods to examine attributes and similarities among communities and populations.

Phylogenetic systematics

While Darwin provided great insight by defining a "natural" classification as one that grouped taxa according to their evolutionary affinities, the systematics community did not have a straightforward procedure for objectively coding this fundamental property into phylogenetic trees until the mid-twentieth century. Willi Hennig, a German entomologist, solved this problem with a landmark work first published in German (1950) but having much greater impact after its translation into English (1966). Hennig's genius was to recognize the logical consequences of a feature of the evolutionary process that had been known for a long time. Organisms are very complex systems; they comprise many parts and processes that must interact in many ways to produce a functional individual that can survive and reproduce. Thus, the kinds of changes that can occur over generations of evolution—and also during the ontogenetic development of individual organisms—are highly constrained. New structures and functions are almost never created de novo. Instead, they are obtained by modifying already existing structures and functions. The history of these changes is recorded in the similarities and differences in the complex characteristics of related organisms—in the extent to which the characteristics of their common ancestors have been modified by subsequent additions, losses, and transformations. Hennig showed how one could deduce a history of ancestral connections between descendant taxa by ordering the transformation of ancestral to derived states of a **character** (any heritable and observable part or attribute of an organism) into a **transformation series**. From this simple insight, Hennig developed an elegantly simple yet logical method for reconstructing the history of a lineage and assessing the nested hierarchy of relationships among its taxa. It is this pattern that is captured within a cladogram, and the construction of cladograms that reflect the true history of cladogenesis is the basic goal of phylogenetic systematics.

Anatomy of phylogenetic systematics

The essence of Hennig's logic is easy to understand. First, a natural taxon is defined as a **monophyletic group** (a clade), which includes all descendant taxa and their common ancestor (**Figure 11.3A,B**). Any other grouping of taxa comprises an artificial group. For example, as we introduced above, a group that includes an ancestor and some—but not all—descendant taxa is paraphyletic (**Figure 11.3C,D**); a **polyphyletic** group includes taxa that trace back through two or more separate ancestors before reaching a common ancestor (**Figure 11.3E,F**). Second, a monophyletic group is supportable through evidence derived in an analysis of the pattern of distribution of different states of a character across taxa. A character that exists in a particular state in two or more taxa is said to be **homologous** if it is shared because of its inheritance from a common ancestor; if it is shared because of convergent or parallel evolution, it is considered a **homoplasy**. Only a homologous character can be used to support a monophyletic group, but not all homologous characters are equally

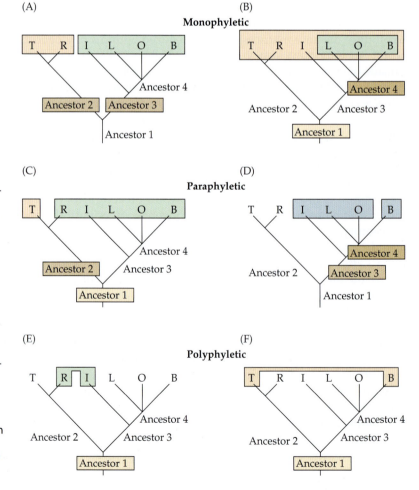

FIGURE 11.3 Cladograms illustrating monophyletic, paraphyletic, and polyphyletic groups. Capital letters represent species, and each group of species included within a box represents a taxonomically recognized group. The groups in the top two diagrams are monophyletic because they include an ancestor and all of its descendants. For example, on tree A one group includes ancestor 2 + species T + R; another group includes ancestor 3 + species I + L + O + B. Groups in the center row are paraphyletic because each contains an ancestor and some, but not all, of its descendants. For example, on tree C the large group is paraphyletic because it includes the common ancestor 1 + species R + I + L + O + B but excludes species T; if the group were revised to include species T, it would become a monophyletic group. Groups in the bottom diagrams are polyphyletic because we need to pass through one or more ancestors before arriving at a common ancestor for the group. For example, on tree E the group includes ancestor 1 + species R + I, but we need to count back through ancestor 2 and ancestor 3 before arriving at the ancestor shared by species R + I. (After Brooks and McLennan 2002.)

valid for doing so, and in order to understand which are valid, we first need to introduce two more terms. As a character evolves between ancestral and descendant taxa, its form in the ancestor is said to be **plesiomorphic** (primitive), while the transformed state in the descendants is called an **apomorphic** (derived) character. Now, we can define those "special homologies" that can be used to diagnose monophyletic groups—these are called **synapomorphic** (shared-derived) characters, while homologous characters that diagnose a group of taxa more inclusive than the focal monophyletic group are called **symplesiomorphic** (shared-primitive) relative to the focal group. For a simple example, look again at Figure 11.2B. If our focal monophyletic group is the archosaur clade, two characters (dorsal and lateral temporal openings, suborbital fenestra) are synapomorphies, while the character that defines and is a synapomorphy for the more inclusive sauropsid clade (loss of the medial centrale bone of the pes) would be a symplesiomorphy for the archosaur clade.

How do we infer which state of a character is plesiomorphic and which is apomorphic (i.e., elucidate the correct order of changes in a transformation series)? The most straightforward approach is to inspect its transformation within a group of taxa considered to comprise a monophyletic group (an **ingroup**) with reference to one or more distantly related **outgroups** (**Figure 11.4**). The most useful outgroup for this purpose is the next most closely related group to the ingroup and is called the **sister group**. The logic here, again, is simple—that state of the character that occurs in both the sister group (or another closely related outgroup, if the sister group is unknown) and one or more ingroup taxa is inferred to be the plesiomorphic state. Obviously, this will not always be the correct inference—there will always be a chance that character states shared by ingroup and outgroup taxa are homoplasies. However, systematists typically include many characters in an analysis, recognize that some of those characters will likely be homoplasies, and employ methods that account for homoplasy while producing the "best" phylogenetic trees possible. In the simplest case, the best tree is the one that requires the fewest character-state changes to group all ingroup taxa into a phylogeny—this is called the most **parsimonious** tree.

Overall, there are only a few basic principles and rules that one needs to know in order to understand the basic logic for building trees under the principles and rules of Hennigian argumentation (**Box 11.1**). Once the basic terminology, principles, and rules are understood, one can begin to understand the relationship between cladogenesis and character evolution that forms the basis of Hennigian logic (**Box 11.2**).

Methods in phylogenetic systematics

While the logic of Hennig's paradigm is quite simple, the current array of competing methods and computer programs available to produce and analyze phylogenetic trees can be intimidating to beginning and experienced systematists alike. A short list of the more popular approaches, each implemented in multiple readily available software packages, includes distance matrix methods and the character-based methods consisting of maximum parsimony, maximum likelihood, and Bayesian inference. Distance matrix methods first produce a quantitative estimate of difference between each pair of operational taxonomic units (OTUs) and then use one or more clustering methods (e.g., UPGMA or neighbor-joining) to generate a tree called a phenogram. In contrast, the maximum parsimony method uses discrete character states directly to find the tree, or set of trees, that require the smallest number of character-state changes to represent relationships among OTUs. Both maximum likelihood and Bayesian methods allow users to choose an explicit, and often rather complex, model of the evolutionary process—the

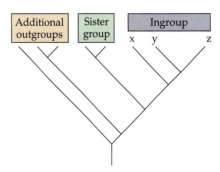

FIGURE 11.4 A simple phylogenetic tree illustrating the concepts of ingroup, outgroups, and the special outgroup that is called the sister group to the ingroup. The ingroup is the focal set of taxa within a monophyletic group, whereas one or more outgroups are typically included in an analysis in order to root the tree and provide an ancestor–descendant orientation to the character transformation series. (After Brooks and McLennan 2002.)

BOX 11.1 *The Principles and Rules of Hennigian Logic*

Hennig's auxiliary principle

Never presume convergent or parallel evolution; always presume homology in the absence of contrary evidence.

Significance: This principle establishes a testable hypothesis about character evolution. It does not state that convergent evolution, which produces homoplasious characters, is nonexistent or even particularly uncommon. It simply establishes the logical basis of employing putatively homologous characters in phylogenetic analysis.

Relative apomorphy rule (outgroup comparison)

Homologous characters found within the members of a monophyletic group that are shared with members of the sister group are plesiomorphic, while homologous characters found only in the ingroup are apomorphic.

Significance: This rule establishes the basis, through outgroup comparison (with the sister group being the most important

outgroup), of polarizing the transformation series of character evolution in the ingroup. This rule forms the foundation for stating the following rule.

Grouping rule

Only synapomorphies (shared special homologies) provide evidence of common ancestry relationships. Symplesiomorphies (shared general homologies), autapomorphies (unique homologies), and homoplasies (convergences and parallelisms) are useless in this procedure.

Significance: A homologous character that is a synapomorphy in reference to a particular clade will become a symplesiomorphy in reference to clades nested within that clade.

Inclusion/exclusion rule

The information from two transformation series can be combined into a single hypothesis of relationship if that information allows for the complete inclusion (or the complete exclusion) of groups that were formed by the separate

transformation series. Partial overlap of groupings necessarily generates two or more hypotheses of relationship.

Significance: This rule establishes the logic behind being able to say that character states from one transformation series either are or are not completely consistent with one another regarding the groupings of taxa that are supported by synapomorphies. More often than not, phylogenetic analyses will produce more than one tree because conflicts between relationships supported by different characters do not allow them to be combined on a single phylogenetic tree. One must add more homologous characters to the analysis to further assess degree of support for the alternative hypotheses, or one must collapse conflicting branches into unresolved polychotomous nodes on a tree that represents a consensus between the conflicting hypotheses. ∎∎∎

Source: Adapted from Brooks and McLennan 2002.

former then searches for the tree that maximizes the probability of observing the data given the tree, whereas the latter searches for the most probable tree given the data under a posterior probabilities framework. These methods are popular in molecular studies, where it is often possible to understand with a fair degree of precision the different rates of evolution in different "partitions" of a genome (see discussion below). A good resource for entry into the available methods of phylogenetic analysis, including a large array of methods for statistical analysis and for comparing trees for degrees of congruence, can be found at http://evolution.genetics.washington.edu/phylip/software.html. Good tutorials on phylogenetic systematics include Wiley et al. (1991, available online at http://www.archive.org/details/compleatcladistp00wile), Skelton and Smith (2002), Hall (2007), and the online resource provided by the Museum of Paleontology at the University of California, Berkeley, at www.ucmp.berkeley.edu/clad/clad4.html.

We turn now to a more detailed discussion of two recent revolutions that have had a profound impact on biogeography—molecular systematics and phylogeography.

Molecular Systematics

In 1953, Watson and Crick elucidated the molecular structure of DNA, the molecule that encodes and transmits genetic information. Systematists soon realized that just as the structural and functional characteristics of organisms could be used to reconstruct phylogenetic relationships, the molecules that code for those characteristics also could be used to infer two patterns

BOX 11.2 *The Basis of Hennig's Paradigm: A Hypothetical Example of Cladogenesis and Cladogram Construction*

■■■ The following example and diagrams will illustrate Hennig's insight into how the branching pattern of phylogeny leaves its legacy in the derived characteristics of contemporary forms. We begin with a group of four species, numbered 1–4, and six characters, lettered A–F, each of which can exhibit one of two states. The first thing we need to do is to determine which of the character states are ancestral (or plesiomorphic) and which are derived (or apomorphic). To do this, we need to identify an outgroup, a species closely related to species 1–4 but whose lineage diverged at an earlier time (preferably, the sister group to the ingroup). In this case, we choose species 0 to be the outgroup. Each character has two discrete states, such as A or A′, and for simplicity in coding we assign each state that occurs in the outgroup the nonprime value, which, through application of the relative apomorphy rule (see Box 11.1) makes those states plesiomorphic and those with prime signs apomorphic (i.e., the transformation series is A → A′). The matrix showing the distribution of character states among the five species is shown below. Note that we will adhere to Hennig's auxiliary principle until the inclusion/exclusion rule gives us reason to consider alternative hypotheses.

Now let's look at the first character, for which all ingroup species share the

derived state. The tree for the ingroup based on this character is the following:

Likewise, the tree depicting change in character B is this:

Proceeding in a similar way to characters C and D, we get this tree:

It should be apparent that we can apply the inclusion/exclusion rule at this point to derive the following tree, considering the first four characters all together:

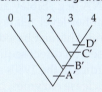

Character state E′ for the next character occurs in only one ingroup taxon (species 2) and is therefore an autapomorphy and therefore uninformative with regard to sister lineage relationships according to the grouping rule.

Finally, when we get to the last character, we find that the derived state specifies the following tree:

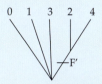

According to the inclusion/exclusion rule, this tree is not combinable with the preceding one, and given the otherwise strong character support for that tree, the simplest way to explain this discrepancy is that F′ represents a convergence or homoplasy. Under that hypothesis, we arrive at the final tree using all characters:

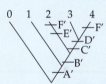

Note that a particular character state can be either a synapomorphy or a symplesiomorphy, depending on the focal group in question. For example, B′ is a synapomorphy for the group (species 2, species 3, species 4), but it is a symplesiomorphy for the group (species 3, species 4).

This example is highly simplified. It considers only a few characters, and although they change from plesiomorphic to apomorphic states, there is only one case of reversal or independent evolution (homoplasy) of character states. ■■■

Taxon	Character transformation series					
Species 0 (outgroup)	A	B	C	D	E	F
Species 1	A′	B	C	D	E	F
Species 2	A′	B′	C	D	E′	F′
Species 3	A′	B′	C′	D′	E	F
Species 4	A′	B′	C′	D′	E	F′

of great interest to biogeographers—the phylogenetic relationships between taxa, and the times of their divergence (Sarich and Wilson 1967). The utilization of molecular data, primarily in the form of DNA sequences, within the theoretical framework of molecular evolution is often treated as a separate subdiscipline in systematics called **molecular phylogenetics**, or **molecular systematics**. The role of molecular systematics in biogeography has increased vastly since its inception during the 1970s.

Evolution of methods in molecular systematics

Several early approaches to estimating changes in the structure of DNA are indirect. One of these methods estimates evolutionary distances between taxa from immunological assays (e.g., Sarich and Wilson 1967). A protein such as albumin is used to assay the strength of antibody–antigen binding with a group of "complement" proteins, using a reference and test species in a procedure called micro-complement fixation. The amount of complement "fixed" is related to the number of amino acid replacement substitutions between proteins from different taxa, and so can be used as a measure of phylogenetic relatedness based on a single measure of overall similarity. Biogeographers began implementing this approach to assessing geographic variation among populations and species during the 1970s (Gill 1976; Higgins 1977). One of the most popular of the early methods that is still used is protein electrophoresis, which takes advantage of differences in the mobility of soluble proteins migrating across a gel medium in an electrical gradient to infer amino acid substitutions. Those amino acid changes that cause differences in protein charge lead to changes in the rates of migration of the protein across the gradient and can thus be scored as character-state changes (i.e., those that can be analyzed using either distance or parsimony approaches; e.g., Selander et al. 1971). If a protein is inherited in a Mendelian fashion, this method can also be used to examine attributes of genetic structure within and among populations and species (Canestrelli et al. 2006; Izawa et al. 2007; Pauly et al. 2007), and test for hybridization between species at contact zones (Sequeira et al. 2008; Schreiber 2009). Some of the more novel phylogenetic surprises, particularly with regard to relationships across higher taxa (e.g., families or orders of birds), came from DNA-DNA hybridization, which uses the thermal melting profiles of "heteroduplex" double-strand DNA built from DNAs from different taxa that are allowed to form a double-stranded pair as an index of relatedness. The more different the strands are, the lower the temperatures at which bonds are broken between heteroduplex double-stranded DNA (e.g., Sibley and Ahlquist 1990). Many of the novel and controversial higher-level relationships among avian taxa that were proposed by Sibley and Ahlquist based on DNA-DNA hybridization have since been rejected, based on direct sequencing of DNA (e.g., Barker et al. 2002).

Many studies in the 1980s and 1990s (reviewed in Avise 2000) used restriction endonuclease enzymes, produced from recombinant bacterial DNA, that cut the DNA at specific (usually four-, five-, or six-base) sequences (e.g., the restriction enzyme *Eco*RI recognizes the sequence 5′-GAATTC-3′ and cuts a double-strand DNA everywhere this sequence occurs). If a point mutation occurs at any of the six nucleotide bases in a recognition sequence, the restriction enzyme no longer recognizes it as a site to cut. The DNA fragments are passed through a gel medium in an electrical gradient, and length of fragment determines the rate of migration through the gel. The DNA fragment profiles are then used to infer character-state changes, and as with the isozymes produced in protein electrophoresis, these profiles can be analyzed with either distance or parsimony approaches in phylogenetic trees and population genetic studies.

The vast majority of biogeographic work using restriction sites was done using the small, circular genome found in animal mitochondria (plant chloroplast DNA was also employed, but not as successfully until direct sequencing methods became easier to use: Avise 2004). This was also the approach that motivated the development of the discipline of phylogeography in the 1980s, as described below. An enormous advance in technology in the late 1980s—the polymerase chain reaction (PCR)—motivated a new generation of phylogenetic and population genetic studies. Using PCR, an investigation can begin with very small amounts of DNA and with DNA that has been degraded into small fragments (as happens, for example, in the DNA of museum specimens and fossils), because the method uses a process of cycling through multiple rounds of amplifying the original DNA into many thousands of copies (see Avise 2004). A target region of DNA, and even whole mitochondrial genomes, that have been amplified through PCR can be sequenced directly, for tens or hundreds of individuals, within a few weeks of laboratory work. The vast majority of molecular biogeographic studies today—whether they examine changes in a DNA sequence directly or use a newer generation of indirect methods (e.g., microsatellites and amplified fragment length polymorphisms [AFLPs])—begin with PCR amplification of the DNA.

Molecular characters and properties of molecular evolution

Mechanisms of genetic differentiation were reviewed briefly in Chapter 7. The nucleotide bases and their abbreviatons are adenine (A), guanine (G), cytosine (C), and thymine (T). The most important form of mutational change for use in phylogenetic analysis is nucleotide base substitution at a single position in a DNA sequence, although both insertion and **deletion** of one or more bases can also be used as characters. Should a substitution occur in a protein-coding sequence, the change in the three-base codon that translates DNA information into an amino acid will often not lead to a change in the amino acid (a synonymous or silent mutation) because of the "redundancy" built within the genetic code, although sometimes it will result in an amino acid change (a nonsynonymous or amino acid replacement mutation). If it alters the activity of a protein, a nonsynonymous mutation is open to being filtered by natural selection and therefore is less likely than a synonymous mutation to become fixed within a population. Within a codon, the rates of DNA substitution differ according to position, with third-base substitutions being generally more frequent than first- or second-base position substitutions, because the genetic code dictates that third-base substitutions are less likely to alter the amino acid. Often, because a substitution, insertion, or deletion in a noncoding segment of DNA is less likely to lead to a major change in a phenotype (and therefore is under less selective constraint), the fixation of mutations in this type of DNA is more frequent than in coding DNA. There is also a bias in substitution rates related to the chemical composition of the bases—substitutions between purines (A and G) and pyrimidines (C and T) are called transition substitutions and are generally more frequent than those that substitute a purine with a pyrimidine or vice versa, which are called transversion substitutions.

The simplest model of nucleotide base substitution assumes that a mutation at any particular base pair along a DNA sequence is equally as likely as one at any other base pair, which forms the basis for the common maximum parsimony procedure where all characters are given the same weight in a phylogenetic analysis. However, variation in rates of evolution across the various partitions in a genome offer the prospect of being able to develop models—ranging from fairly simple to quite complex—of the evolutionary process. One popular approach is to use a maximum likelihood approach to search for the model that best describes the number of partitions that are evolving at different rates (e.g., Posada and Crandall 1998; Sullivan and Joyce 2005) and then use a maximum likelihood or Bayesian approach to construct phylogenetic trees under the chosen model. Accessible discussions of the various models, their assumptions, advantages, and limitations are provided in the tutorial by Hall (2007).

FIGURE 11.5 Morphological phylogeny placing the 66–68 million year old Cretaceous fossil bird, *Vegavis iiai*, from western Antarctica, within the waterfowl (Anseriformes); it is most closely related to the living lineage that includes the true ducks (Anatidae). Other living waterfowl lineages represented are the magpie goose (*Anseranas*), two screamers (*Anhima* and *Chauna*), chickenlike birds (Galliformes), and tinamous (Tinamiformes). Given that at least four nodes leading to modern lineages subtend the node uniting *Vegavis* with Anatidae and *Presbyornis* (another extinct wading bird), at least five divergence events leading to modern bird lineages are inferred here to have occurred prior to the end of the Cretaceous. Note that while previous fossil evidence had supported the idea of an explosive evolution of modern bird lineages following the Cretaceous–Tertiary mass extinction episode (e.g., Feduccia 2003), accumulating molecular evidence, with calibrated molecular clock estimates of divergence times, has consistently indicated a Cretaceous time frame for the origination of many modern lineages (e.g., Cooper and Penny 1997; van Tuinen and Hedges 2001; Harrison et al. 2004). (After Clarke et al. 2005.)

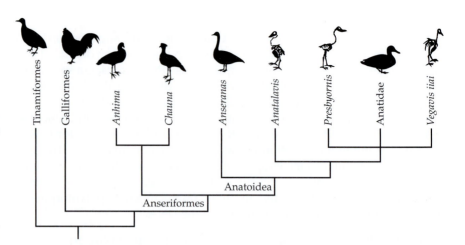

Molecular versus morphological characters

From the beginning of the utilization of molecular characters in systematics several decades ago, there have been controversies over the relative merits of molecular versus morphological data in reconstructing evolutionary relationships. Certainly, paleontologists are almost always limited to using morphological characters (Lieberman 2000), although studies using ancient DNA often can lead to novel interpretations of lineage histories (e.g., Cooper et al. 2001; Haddrath and Baker 2001; Kuch et al. 2002; Shapiro et al. 2004; Dalen et al. 2007; Valdiosera et al. 2007; Debruyne et al. 2008), and new fossil finds continue to be utilized in concert with molecular systematics to transform our knowledge of the ages and geography of early diversification in modern groups (**Figure 11.5**). Nevertheless, as evidenced by the growth in popularity of several journals devoted to publishing molecular-based studies (e.g., *Molecular Phylogenetics and Evolution* and *Molecular Ecology*), and recent papers arguing for more sophisticated uses of molecular phylogenies in historical biogeography (Donoghue and Moore 2003), there is really no question that molecular phylogenetics has assumed an enormous importance in modern biology. The vast number of characters available (even the very small animal mitochondrial genome contains over 16,000 nucleotide bases) provides the opportunity to include from hundreds to thousands of characters in a phylogenetic study. Many of those characters are likely to be evolving in a neutral or near-neutral fashion, without the strong natural selection component that could result in convergent evolution. Another attractive feature is that characters can be grouped into different categories or "partitions" that differ in their rates of evolution (**Table 11.1**). Thus, slowly evolving characters can be used to reconstruct older events in a phylogeny, whereas faster-evolving characters can be used to address more recent phylogenies and population histories (**Figure 11.6**). Some of the character partitions could also be evolving in a clocklike fashion, an important attribute that has been exploited to estimate both absolute and relative times of divergence in a phylogeny.

And now, we are entering a new era in molecular systematics in which many thousands of base pairs—in some cases, whole genomes—can be sequenced rapidly and inexpensively. The approach to systematics that samples many genes from across a genome to reconstruct evolutionary histories has been called **phylogenomics** (Delsuc et al. 2005) and has the potential to resolve long-standing controversies in biogeography, for example, the influence of the Earth's plate tectonic history on the origins of ratite birds (Harshman et al. 2008) and major clades of placental mammals (Hallström et al. 2007; Wildman et al. 2007).

TABLE 11.1 *Representative Rates of Molecular Evolution of an Array of Different DNA Markers*

DNA markers	Organism	Average of divergence (%/my)	Reference
Nuclear DNA			
Nonsynonymous sites	Mammals	0.15	Li 1997
	Drosophila	0.38	Li 1997
	Plant (monocot)	0.014	Li 1997
Synonymous (silent) sites	Mammals	0.7	Li 1997
	Drosophila	3.12	Li 1997
	Plant (monocot)	0.114	Li 1997
Intron	Mammals	0.7	Li 1997; Li and Graur 1991
Chloroplast DNA			
Nonsynonymous sites	Plant (angiosperm)	0.004–0.01	Li 1997
Synonymous (silent) sites	Plant (angiosperm)	0.024–0.116	Li 1997
Mitochondrial DNA			
Protein-coding region	Mammals	2.0	Brown et al 1979; Pesole et al. 1999
	Drosophila	2.0	DeSalle et al. 1987
COI	*Alpheus* (shrimps)	1.4	Knowlton and Weigt 1998
D-loop	Human	14	Horai et al. 1995
	Human	17.5	Tamura and Nei 1993
	Human	23.6	Stoneking et al. 1992
	Human	260	Howell et al. 1998
	Human	270	Parsons and Holland 1998
Nonsynonymous sites	Plant (angiosperm)	0.004–0.008	Li 1997
Synonymous (silent) sites	Plant (angiosperm)	0.01–0.042	Li 1997

Source: Hewitt 2001.

Note: The wide discrepancy of rates provides an opportunity to choose a marker that is likely to be most informative within the temporal window of a particular biogeographic study.

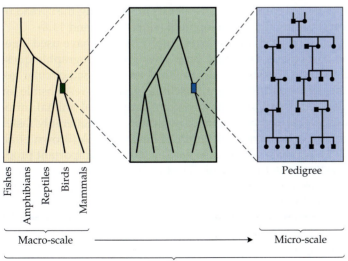

FIGURE 11.6 An illustration of the hierarchical nature of phylogeny. Embedded within the phylogeny of the traditional classes of vertebrates are any number of possible phylogenies for groups within each class; shown here is one possible phylogeny across some set of mammalian groups, but we could imagine those groups as anything from taxonomic orders to species and even groups of populations within a single species. Shown here also is the extended genetic pedigree of a familial lineage, completing the continuum of genetic connectivity, from macro to micro scale. (After Avise 2004.)

The variation in substitution rates across different partitions of a DNA sequence and different genomes results in some distinct advantages to using molecular systematics in biogeography. In particular, it has allowed investigators to target a partition or genome that is most likely to be informative within a given time frame of biogeographic history. Thus, molecular systematic studies can address questions ranging from relationships among taxa originally isolated during the fragmentation of Gondwanaland millions of years ago (e.g., Ortí and Meyer 1997; Wildman et al. 2007) to the isolation of populations resulting from Late Pleistocene habitat fragmentation only a few thousand years ago (Knowles 2009).

Yet, even given this long list of potential advantages of molecular characters, systematists have not abandoned morphological characters entirely, recognizing that good phylogenetic characters (those with little homoplasy) do not rest entirely within the molecular realm—notice in Figure 11.2B that synapomorphic characters that diagnose major vertebrate clades are morphological. Sometimes systematists employ the **total evidence** approach, incorporating both molecular and morphological characters into a phylogenetic analysis. In other cases, phylogenies are constructed from molecular characters, and then morphological characters are "mapped" onto the molecular phylogeny as a means of testing hypotheses of morphological evolution. Either way, there is little question that morphological characters will continue to play an important role in phylogenetic systematics well into the foreseeable future.

Phylogeography

The other revolution of the late 1980s took full advantage of the potential of molecular phylogenetics and population genetics, but differed from the traditional realm of phylogenetic systematics in a very important way. While the vast majority of phylogenetically based studies in biogeography had been conducted among taxa at or above the species level, the new discipline of phylogeography was defined explicitly as "a field of study concerned with the principles and processes governing the geographic distributions of genealogical lineages, especially those within and among closely related species" (Avise 2000). In other words, whereas biogeography has typically concentrated on species as the most fundamental units of analysis, and phylogenetic trees were built among species or higher taxa, the newly developing molecular approaches were providing a basis for mapping the spread of lineages (**gene trees**) during their development (**Figures 11.7** and **11.8**).

In animals, the vast majority of phylogeographic studies have used a genome with some particularly favorable properties for biogeographic questions—the small (about 16,500 base pairs in vertebrates), closed circular genome found in the mitochondrion, the intracellular organelle responsible for cellular respiration. Perhaps the most important property of animal mitochondrial DNA (mtDNA) is that its rate of evolution is considerably faster than the average rate of sequence evolution of nuclear DNA (Brown et al. 1979; see Table 11.1). Animal mtDNA also has several other important properties that increase its utility for intraspecific studies: It is usually passed on clonally, without recombination, from one generation to the next, thus making it possible to treat the entire molecule—protein-coding, tRNA, and rRNA genes, as well as noncoding sequences—as a single "locus"; it exists as a haploid rather than diploid genome, thus different genotypes are typically called **haplotypes**; and it is usually transmitted between generations exclusively through a female (matrilineal) pedigree. These properties combine to give mtDNA a much smaller effective population size than nuclear DNA.

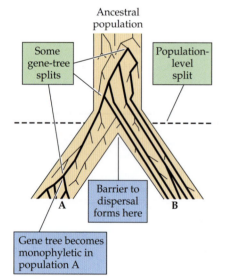

FIGURE 11.7 Gene trees embedded within population trees. Note that the gene tree divergence between populations will not always be concordant with the time of population divergence, and that the process of ancestral lineage sorting can occur at very different rates in different populations. In this case, the gene tree in population A is monophyletic with respect to population B, but it did not become so until well after the time of the population split. Population B is still paraphyletic with respect to population A. At any point in time prior to the coalescence of the gene tree in population 1 to monophyly, populations A and B would have been polyphyletic. (After Avise 2000.)

FIGURE 11.8 An illustration of a gene tree that differs topologically from the "true" phylogenetic relationships among species. In the top figure, species D and E are allied with C, but the gene tree in D and E is allied with F. This is because population B retains ancestral polymorphism in the time interval t_1 to t_2 and across the speciation event between C and D + E. As this interval becomes shorter (i.e., time between consecutive branching events becomes reduced), the probability of a discordance between species and gene tree topologies increases. The bottom figure shows that once such a discordance arises, it becomes permanently frozen in place, so this must always be considered a possible source of error when gene trees are used to infer species histories from sampling of extant lineages (G, H, and I). (After Avise 2000.)

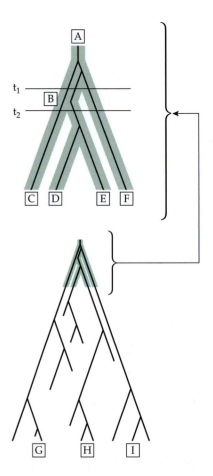

One consequence of this reduced effective population size is that within a population, as DNA is transmitted from parents to offspring across multiple generations, ancestral mtDNA diversity is lost more rapidly than nuclear DNA through a process called **lineage sorting**, the stochastic extinction of ancestral haplotype lineages within a population. Thus, following an event that isolates populations previously connected through gene flow (see Figure 11.7), the mtDNA diversity in each now isolated population will **coalesce** to a single mtDNA haplotype much faster than will nuclear DNA. In fact, following the subdivision of an ancestral population into geographically separated populations, these populations are theoretically expected to go from a polyphyletic stage, through a paraphyletic stage, and eventually to a **reciprocally monophyletic** stage with respect to their matriarchal ancestry (compare Figure 11.7 with Figure 11.3, and refer to the definitions of these terms in the glossary). So, with a high rate of mutation rapidly generating new haplotypes and with a rapid rate of lineage sorting (extinction) of ancestral haplotypes in a population, animal mtDNA will develop a signal of divergence—ultimately reciprocal monophyly—among isolated populations sooner after their initial isolation than nuclear DNA will (Moore 1995). In plants, most phylogeographic studies have used chloroplast DNA (cpDNA) rather than mtDNA, and while the overall rate of evolution is much slower than in animal mtDNA, several gene regions are evolving at a relatively rapid rate; thus, the number of studies demonstrating phylogeographic structure in cpDNA (e.g., Soltis et al. 1997; Soltis et al. 2006) have increased steadily since the advent of PCR-based sequencing methods.

The dual nature of phylogeography

An interesting feature of phylogeography is that it lies at the intersection between two traditionally separate sets of questions in evolutionary biology. As discussed in Chapter 7, macroevolution is typically concerned with evolutionary patterns at and above the species level, and it is generally approached using the tools of phylogenetics and paleontology. Alternatively, in a microevolution framework, evolutionary questions focus on individuals and populations within a species and are often addressed using the tools of population genetics. Phylogeography incorporates a variety of phylogenetic, as well as population genetic, approaches and methods to address questions that span the traditional boundaries between the macroevolutionary and microevolutionary arenas (**Figure 11.9**). Drawing from this variety of methods allows phylogeographers to address a very broad range of possible patterns of phylogeographic architecture (**Figure 11.10**), which is expected to arise over a broad range of time frames and as a consequence of a variety of possible vicariance and dispersal histories.

Phylogeographic patterns exhibiting larger genetic gaps between **phylogroups** (groups of closely related haplotypes separated from other phylo-

FIGURE 11.9 Depiction of phylogeography as a bridge between traditionally separate concerns of (A) microevolution, addressed using population genetics methods, and (B) macroevolution, addressed using historical biogeographic approaches. (After Riddle and Hafner 2004.)

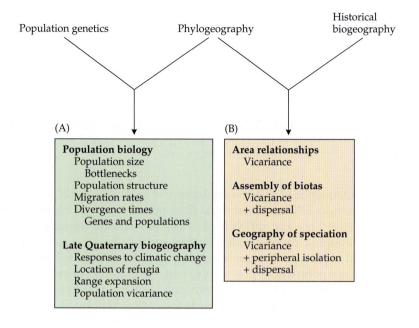

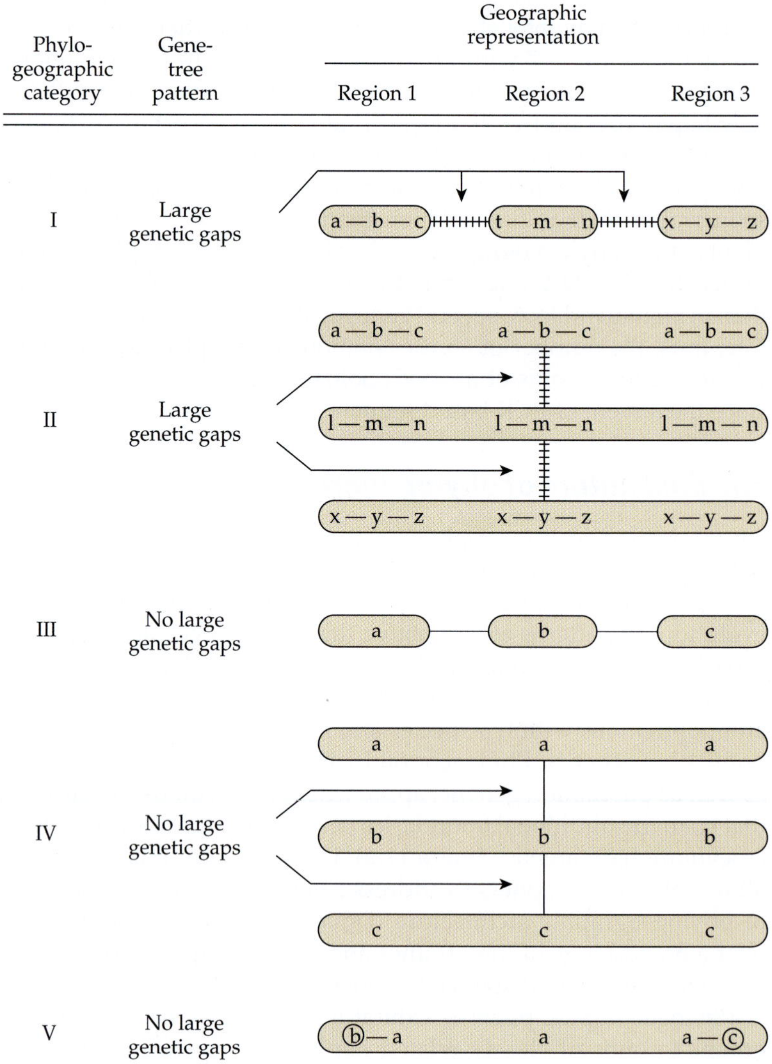

FIGURE 11.10 Illustration of five possible categories of phylogeographic patterns. Ovals and circles encompass mtDNA haplotypes (denoted by letters) or groups of closely related haplotypes, separated from other such groups within a phylogenetic network. Each slash on a line connecting groups of haplotypes into a network represents a single mutation step; lines without slashes represent a single mutational step between connected groups. Category I occurs when haplotype groups are strongly divergent and geographically separated from one another. Category II demonstrates similar levels of divergence but without geographic separation, as might occur with dispersal following erosion of a long-term barrier to dispersal between populations. Categories III and IV are similar to I and II, respectively, but occur within a much shallower time frame and possibly result from life history rather than biogeographic attributes, or they occur in a species with a much slower rate of mtDNA evolution. Category V represents a species with historically continuous gene flow among populations but with some rare haplotypes that are contained within a single population, which might be new mutations that have yet to spread beyond their population of origin. (From Avise 2000.)

(A)

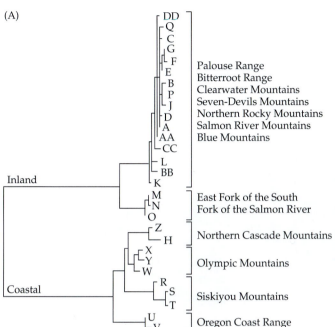

Palouse Range
Bitterroot Range
Clearwater Mountains
Seven-Devils Mountains
Northern Rocky Mountains
Salmon River Mountains
Blue Mountains

East Fork of the South
Fork of the Salmon River

Northern Cascade Mountains

Olympic Mountains

Siskiyou Mountains

Oregon Coast Range

Inland

Coastal

(B)

FIGURE 11.11 (A) Mitochondrial DNA phylogenetic tree for mtDNA haplotypes (capital letters) drawn from populations of the tailed frog, *Ascaphus truei*, from the Pacific Northwest, USA. Haplotypes form two deeply divergent, reciprocally monophyletic, and geographically disjunct clades—(B) an inland group and a coastal group. With exception of the split between haplotype group M + N + O and all other inland haplotypes, the coastal clade has generally deeper points of coalescence among subclades, suggesting older splits between populations from different mountain ranges than is shown for the inland clade. This phylogeny was used to propose that the inland and coastal populations represent two separate, morphologically cryptic species. (From Nielson et al. 2001.)

groups by relatively large genetic gaps), such as those representing categories I and II in Figure 11.10, should be amenable to a phylogenetic analysis using methods described previously in this chapter. This is a simple matter of treating each unique haplotype in a data set as an OTU and employing one or more of the many phylogenetic algorithms—either distance matrix or discrete character based—capable of producing a phylogenetic tree of haplotype relationships (**Figure 11.11**). When variable haplotypes are closely related and groups of populations are not clustered within the clearly reciprocally monophyletic phylogroups shown in Figure 11.11, one can still often depict an informative picture of evolutionary history among populations or closely related species (e.g., categories III, IV, and V in Figure 11.10) by using an unrooted phylogenetic network to summarize haplotype relationships. Given the generally slower rates of evolution and lineage sorting relative to mtDNA, nuclear DNA genes are often portrayed as networks rather than bifurcating trees (**Figure 11.12**).

There are, however, several important caveats that arise when using gene trees to infer the evolutionary and biogeographic history of populations or closely related species. First, a gene tree embedded within a population tree can, under a variety of historical demograph-

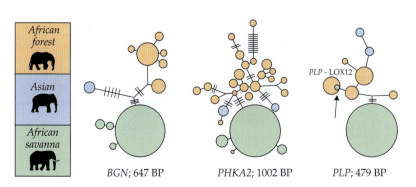

BGN; 647 BP   *PHKA2*; 1002 BP   *PLP*; 479 BP

African
forest

Asian

African
savanna

PLP - LOX12

FIGURE 11.12 Three unrooted haplotype networks within and among two species of African (genus *Loxodonta*) and one species of Asian (*Elephas*) elephant. Each network represents a separate gene from the nuclear DNA. Each slash across a line connecting haplotypes represents a single mutational step (lines without a slash represent one mutational step between haplotypes). Note that haplotypes from all species cluster together in all cases except one (for *PLP*, the two savanna elephant individuals at the arrow carried a haplotype that was otherwise common in the forest species). The pattern of haplotype clustering provides evidence of genetic divergence among the three species. (After Roca et al. 2005.)

FIGURE 11.13 Using multiple gene trees under a statistical phylogeographic approach to reconstruct the pattern of Pleistocene speciation in grasshoppers within the genus *Melanoplus*. (A) Combined distributions of species in the Montanus and Indigens groups are distributed across the northern Rocky Mountains (note the current isolation of some populations from the more connected mountain chain). (B) Phylogenies from one mitochondrial gene (COI) and three nuclear genes are discordant with one another because of different patterns of incomplete lineage sorting in each gene, but coalescent-based approaches can be used to estimate the underlying phylogeny across the four species. (After Knowles and Otte 2000; Knowles 2001, 2009.)

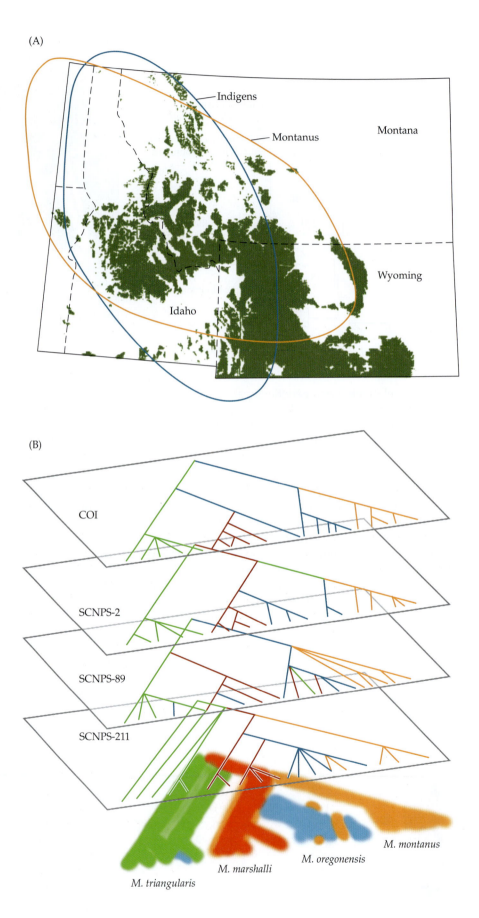

ic scenarios (e.g., number of breeding individuals in a population), be expected to trace to a most recent common ancestral haplotype at a time well before the divergence time for the populations themselves (see Figure 11.7) such that the gene tree might overestimate population divergence time considerably. Second, because of the stochastic nature of the lineage-sorting process and because of the strong influence of demographic factors on it, the confidence intervals around an estimate of population divergence time from a gene tree can be so large that the process can exclude few realistic scenarios (Edwards and Beerli 2000). Third, again because of the stochastic nature of the lineage-sorting process, a gene tree will not always trace the true historical sequence of population differentiation (see Figure 11.8), and an incorrect trace can happen if lineage sorting does not produce reciprocal monophyly in the period of time separating sequential branching events, the likelihood of which increases with decreasing time between branching events. Phylogeographic studies increasingly attempt to solve each of these issues by using estimates of divergence across multiple unlinked gene trees to infer the history of population or species divergence (**Figure 11.13**; Knowles 2009; Hickerson et al. 2010).

We will return to a more focused discussion of the applications of phylogeography in reconstructing biogeographic histories in Chapter 12.

Combining phylogeography and ecological niche modeling

An interesting new approach that has emerged in the past few years combines phylogeography with ecological niche models (ENMs, aka species distribution models); these are two very different approaches to examining the distributional changes in a species in reponse to the climatic oscillations across the latest glacial to interglacial transition. The ENMs are generally constructed in a GIS framework (see Chapter 3) using an array of climatic variables assumed to be important drivers of habitat distributions. Once ENMs predicting extant distributions are generated, they can be transferred through time to model past distributions—most commony during the Late Glacial Maximum about 21,000 years ago—using models of climatic conditions during the previous time frame (Hijmans and Graham 2006). The rationale generally given for using ENMs is that they provide a model of changes in distribution that are independent of the genetically based reconstructions generated through phylogeographic analyses, thereby providing a means of either validating the reliability of phylogeographic reconstructions (e.g., Waltari et al. 2007) or going further and generating alternative hypotheses to be addressed using genetic data (Richards et al. 2007). The questions being addressed using this approach are varied, but some interesting examples include reconstructions of Late Pleistocene refugia within a biodiversity hotspot (**Figure 11.14**; Carnaval et al. 2009) and delimiting geographically and ecologically distinct, but phenotypically cryptic species (Rissler and Apodaca 2007).

The Fossil Record

The task of reconstructing the evolutionary histories of organisms has by long tradition been carried out by two different groups of scientists. Most cladograms are produced by systematists, and many of those are based solely on molecular data and include only contemporary organisms, but with expanding possibilities for incorporating extinct taxa through ancient DNA sequencing. At the same time, paleobiologists have been studying the evolutionary histories of organisms directly from the fossil record, often producing cladograms based solely on fossil evidence. Fossils, the preserved remains of or-

(A)

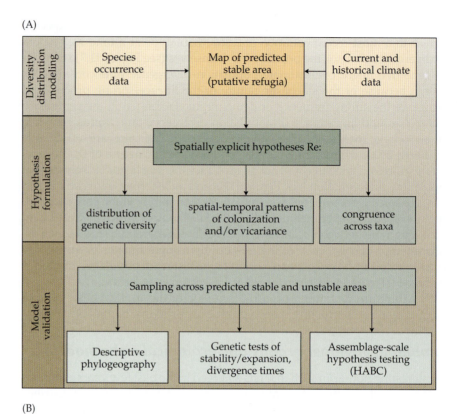

(B)

(1) (2) (3)

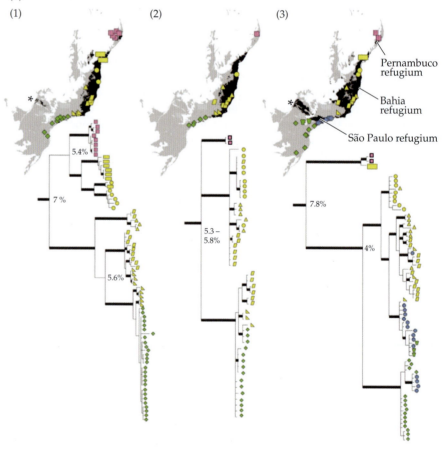

Pernambuco
refugium

Bahia
refugium

São Paulo refugium

FIGURE 11.14 Combining phylogeography and ecological niche (or species distribution) modeling to infer Late Pleistocene forest refugia in the Brazilian Atlantic Forest biodiversity hotspot. (A) In this study, species distribution modeling was first used to develop predictions of putative refugia, which were then used to develop testable hypotheses of phylogeographic patterns within and across species. (B) Summarized species distribution models of Late Pleistocene putative refugia (black polygons) and molecular phylogenies within three species of tree frogs in the genus *Hypsiboas*: A = *H. albomarginatus*; B = *H. semilineatus*; C = *H. faber*. The distribution models predict loss of habitat in the south but retention of large expanses of habitats in the north; the phylogeographic results are largely congruent—deep divergences between yellow and pink clades are generally located within areas predicted to have served as separate refugia, whereas shallow genetic divergences (green haplotypes) are consistent with recent range expansion into previously unsuitable habitats in the south. (After Carnaval et al. 2009.)

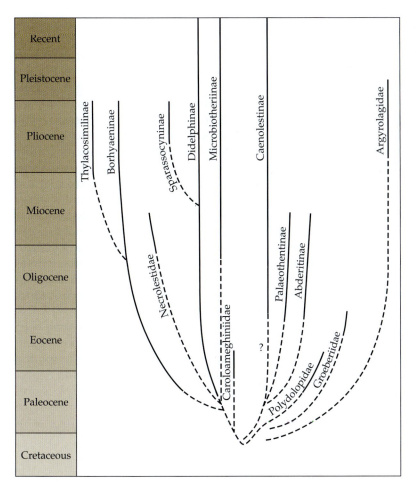

FIGURE 11.15 A phylogenetic hypothesis for the family-level relationships and evolutionary history of South American marsupials. The fossil record shows that this group had a diverse radiation on the isolated South American continent during the Early Cenozoic, but only three families have survived to the present. The solid lines show the known time spans of the families in the fossil record; the dashed lines show their inferred durations and relationships to other families. This hypothetical reconstruction uses a modification of cladistic methods, based on detailed information on the living forms supplemented by limited data on morphology and dates obtained from the fossils. A classification that employed molecular techniques would show only the relationships among the three extant families (and their relationships to extant Australian marsupials) but miss much of the rich historical diversification of this group in South America. (After Patterson and Pascual 1972.)

ganisms that lived at various times in the past, provide the most direct factual evidence of evolution, but the fossil record has even greater importance for modern systematics and biogeography. Fossils provide an important means of calibrating absolute rates of molecular evolution so that molecular phylogenies can be used to estimate divergence times (Gandolfo et al. 2008), and they provide often important evidence of character evolution (Wiens 2009; Edgecombe 2010). Fossils can also either corroborate or contradict estimates of divergence times based on molecular trees (see Figure 11.5). Long before the advent of molecular systematics, fossils provided a window into evolutionary history that could never come from examination of relationships among extant organisms alone. For example, regardless of how accurate a phylogeny based only on extant lineages might be, it cannot address directly the often considerable history of lineage extinction in a group without examination of the fossil record (**Figure 11.15**). Furthermore, simulation studies have shown that biogeographic inferences that rely entirely on cladograms produced using only extant taxa can often be inaccurate (Lieberman 2002).

Limitations of the fossil record

Although it might seem more straightforward to rely on the fossil record than on cladograms constructed only from extant organisms, the fossil record has its own set of practical and logical problems. The known fossil record is incomplete, and its interpretation can be difficult. There are many uncertainties in dating fossils, assigning them to taxonomic groups, understanding how the living organisms were structured and how they functioned, and reconstructing their paleoenvironments.

Only a minute fraction of all the organisms and species that have ever lived have been found as fossils. **Taphonomy**, a subdiscipline of paleontology, is concerned with the processes by which remains of living things become fossilized and the ways in which these processes can bias the fossil record or cause problems of interpretation (e.g., Behrensmeyer et al. 1992). Animals lacking hard tissues, and plants without durable chemicals in their cell walls, are poorly represented in the fossil record, because they are readily decomposed. Some organisms are known only from their distinctive chemical remains, such as the limestone formed from coral reefs and the coal and oil formed from undecomposed biomass in ancient swamps and shallow seas. Some animals, especially soft-bodied forms, are known only from their tracks or burrows, called **trace fossils**. Of those fossils that are formed, many are destroyed by erosion or tectonic processes, and many others are inaccessible, hidden away in deep strata. The fossil record of relatively recent times is the most complete (**Figure 11.16**), one reason why so much more is known about organisms and their environments during the Pleistocene (see Chapter 9) than during earlier periods of Earth history (see Chapter 8).

Even when organisms have been fossilized and those fossils have been discovered, problems of interpretation remain. Most fossils are deposited underwater and preserved in sediments. Remains may have been transported by wind or by the currents of streams, rivers, lakes, and oceans, and deposited far from where the organisms originally lived. A striking example is pro-

(A) Area

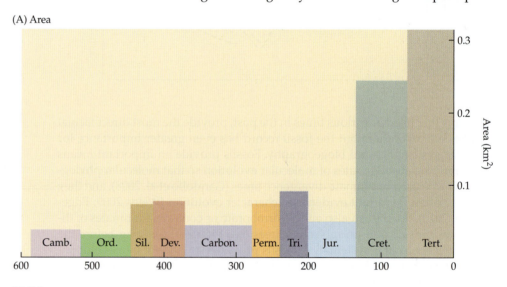

(B) Volume

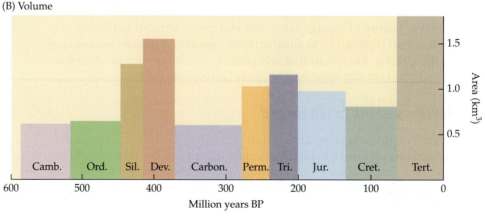

Million years BP

FIGURE 11.16   Estimated (A) area and (B) volume of potentially fossil-bearing sedimentary geological strata of varying ages (here measured as area or volume of rock per year of the past preserved). Note that the area of exposed rocks decreases with age, introducing an important sampling bias into the fossil record. Although some ancient strata are extensive and rich in fossils, in general there is simply less opportunity to find fossils from older geological periods. (After Raup 1976.)

vided by the remains of recently deceased fish that are frequently washed up on the shores of the Great Salt Lake. A naive observer might easily conclude that these fish live in the lake. In fact, the reverse is true: The fish die in the lake. These fish are killed when storms push the hypersaline water of the lake into the freshwater marshes and streams where they live, and their bodies float into the lake and are eventually washed ashore.

One consequence of such long-distance transport of remains before fossilization, therefore, is that species that occur together in fossil deposits may not have actually coexisted locally in the same communities (e.g., the "Viking funeral ships" mentioned in Chapter 6). Pleistocene fossils of small vertebrates from river bottom and cave deposits often represent more species than occur together in single habitats today, probably in part because they have been collected from larger areas. It is easy to imagine how a river could collect sediments from a large basin, but it might be thought that a dry cave would contain fossils only of the animals from its immediate vicinity. Many of the small vertebrate fossils preserved in caves, however, were brought in by owls, which then regurgitated the bones of prey captured in their territories, which may have included several square kilometers in area and a variety of habitat types. So, the take-home message seems to be a disappointing one: The fossil record is biased in favor of easily preserved organisms, and it is temporally and geographically uneven, still greatly unexplored, and sometimes difficult to interpret.

Biogeographic implications of fossils

Despite these limitations, a great deal of information about the history of life on Earth can only or best be obtained from fossils. It is only fossils that record with certainty the kinds of organisms that occurred at particular times and places in the past. Fossils provide the ages when different taxonomic lineages lived (**Table 11.2**) and when they occurred in particular areas. Because of the incomplete nature of the record, however, the organisms may

TABLE 11.2 *Oldest Known Fossils of Selected Taxa*

Taxon	Earliest undisputed fossil	Million years BP	Period
Vascular land plants	*Cooksonia*	L. Silurian	428
Land animals	*Pneumodesmus*	L. Silurian	428
Insects	*Rhyniognatha*	U. Silurian	412
Lungfish	*Youngolepis* and *Diabolepis*	U. Siluarian	412
Amphibians	*Acanthostega*	U. Devonian	375
Conifers		Mississippian	330
Reptiles	*Hylonomus*	Pennsylvanian	320
Flying insects (and natural flight)		Pennsylvanian	320
Dinosaurs	*Eoraptor*	M. Triassic	228
Turtles	*Proganochelys*	L. Jurassic	210
Mammals	*Morganucodon*	L. Jurassic	200
Birds	*Archaeopteryx*	U. Jurassic	150
Angiosperms (flowering plants)	*Archaefructus*	L. Cretaceous	142
Monotremes (egg-laying mammals)	*Steropodon* and *Kollikodon*	L. Cretaceous	100
Bats	*Icaronycteris*	L. Ecocene	50
Hominids	*Australopithecus*	L. Pliocene	4

Source: Stratigraphic Commission 2003 Time Chart.

have been present for unknown lengths of time before and after their known fossils were preserved. Fossils document some of the enormous diversity of prehistoric life and some of the important events in the histories of lineages: speciations and radiations, expansions and contractions of ranges, and extinctions. Finally, inference about the sorts of environments that organisms could have—and perhaps more importantly, could not have—occupied can inform paleogeologists about the geographic locations of land masses at the times of fossilization.

Fossil histories of lineages

Fossils can provide insights into the history of a lineage that cannot be obtained from cladistic reconstructions based solely on recent material (Wiens 2009; Edgecombe 2010). They can supply data on the characteristics and character states of earlier representatives of the lineage, some of which may have been the ancestors of contemporary forms. But even when the fossils are not on the direct branches leading to living representatives, they can show the unique combinations of traits that these organisms possessed. Sometimes the traits present in extinct lineages can be spectacular and unpredicted until a new fossil is discovered. For example, an Early Cretaceous birdlike dinosaur in the genus *Microraptor*, discovered in China about a decade ago, had both forelimb and hindlimb "wings" (Xu et al. 2003), and several other "feathered dinosaur" taxa from the same time period are known (Norell and Xu 2005). While recent analyses suggest that *Archaeopteryx* may not have been, as was once thought, a direct ancestor of living birds (Feduccia 1996; Norell and Xu 2005), its exquisitely preserved Jurassic fossils show the special combination of avian and reptilian traits found in a bird that lived 150 million years BP. One of the most exquisite mammalian fossils is of the earliest known bat, *Onychonycteris finneyi*, which shows unequivocally that the lineage that includes all extant species of bats had already evolved the capacity for powered flight by at least 52.5 million years BP (**Figure 11.17**). Knowing the characteristics of earlier representatives permits tests of cladistic assumptions about the kinds and directions of change in character states that occurred during the evolution of a lineage. For example, in a recent study it has been possible to include another ancient bat, *Icaronycteris*, as well as several other fossil and extant taxa, into a phylogenetic analysis of bat relationships (Sim-

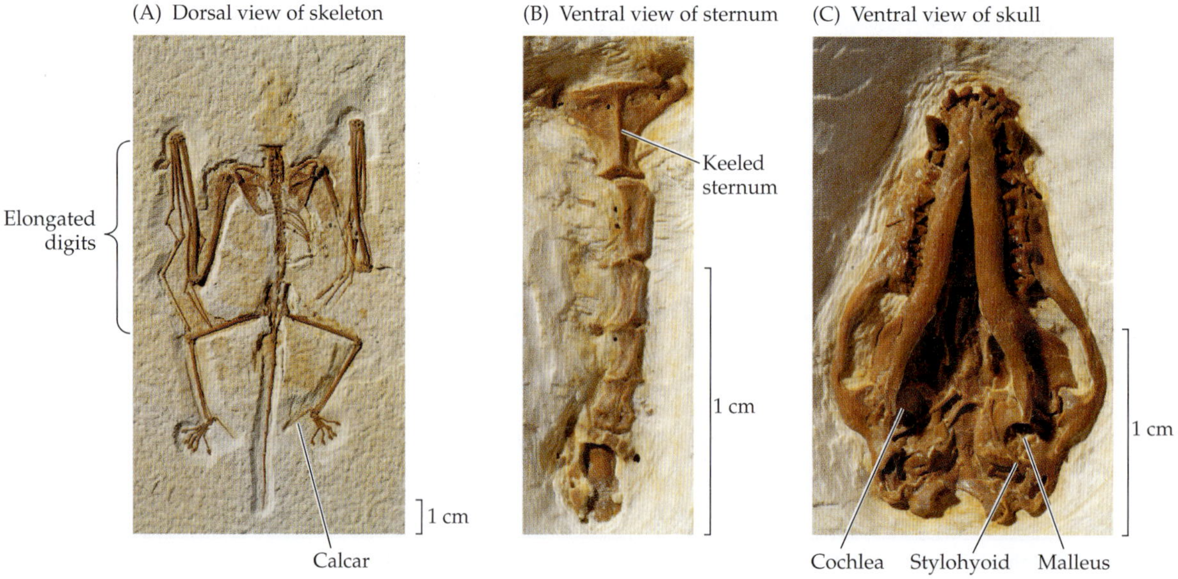

FIGURE 11.17 Morphological characteristics of the earliest known bat, *Onychonycteris finneyi*, provide strong evidence that flight had evolved at least 52.5 million years BP, and before bats developed echolocation. Morphological evidence for flight include (A) the elongated digits of the forelimbs to form the chiropteran namesake ("hand wings") and (B) the keeled sternum, which provides an expanded and reinforced point for attachment of flight muscles. This species, however, also retained some ancestral features shared with non-volant mammals, including relatively long hind limbs and claws on all digits (assisting quadrapedal walking and climbing along limbs), a relatively small stylohyoid (C) (which supports the tongue and larynx, i.e., structures generating ultrasound), and relatively small malleus and cochlea (the latter structures involved in mechanical transmission and neurologic reception of ultrasound, respectively). (After Simmons et al. 2008; photos courtesy of Nancy B. Simmons.)

(A) Dorsal view of skeleton

Elongated digits

Calcar

1 cm

(B) Ventral view of sternum

Keeled sternum

1 cm

(C) Ventral view of skull

1 cm

Cochlea   Stylohyoid   Malleus

mons and Geisler 1998) that addressed the origins of echolocation and foraging patterns.

In addition, fossils can be dated to provide fairly precise information on when prehistoric representatives of a lineage showing particular degrees of differentiation lived on Earth. As we discuss further below, such dating allows cladograms to be calibrated with respect to real time. When fossils are used to calibrate rates of molecular evolution, this calibration is invaluable because the timing and rates of diversification have been very different among different plant and animal groups (see Table 11.1) and even among different subclades within the same group. Such information can have important implications for biogeography. For example, in a fossil deposit dated about 15 million years BP from the Mid Cenozoic of the Magdalena River basin in Colombia, South America, Lundberg and colleagues (1986) found the exquisitely preserved remains of a fish that was identified as a living species, *Colossoma macropomum*, which currently occurs in the Orinoco and Amazon basins. The discovery of fossils of this and other fishes on the western side of the Andes, where they no longer occur, allowed minimum dates to be assigned to events of speciation and extinction, as well as to dispersal between now isolated watersheds (see also Lundberg and Chernoff 1992).

Past distributions

Another unique biogeographic contribution of fossils is their ability to document localities of past occurrence. There are countless examples of fossil representatives of lineages being found far outside the current geographic range of the group, and often such occurrences provide key answers to interesting biogeographic puzzles. Of course, it must be borne in mind that when the fossils were deposited, the environment may have been very different than it is today. Some fossils have been transported on drifting crustal plates ("Viking funeral ships") to locations far from where the organisms originally lived. Thus, for example, the discovery in Antarctica of an ever-growing number of fossils, including of the plants *Nothofagus* and *Glossopteris* (Schopf 1970a,b, 1976); the reptiles *Lystrosaurus*, *Thrinaxodon*, and *Procolophon* (Elliot et al. 1970); and marsupials (Woodburne and Zinmeister 1982; Goin and Carlini 1995; Woodburne and Case 1996), not only provides increasing evidence of the rich biota that continent once shared with other continents of Gondwanaland, but it also indicates that Antarctica did not always occur at polar latitudes and was once much warmer than it is today (see Figure 8.12B).

Fossils show that many now narrowly endemic forms are relicts of much more diverse and widely distributed lineages. Examples of such fossils include a member of the avian family Todidae, which is now restricted to the Greater Antilles but inhabited Wyoming during the Oligocene (about 30 million years BP: Olson 1976), and a platypus (a representative of the order of egg-laying mammals, Monotremata) whose sole surviving species is now restricted to Australia, although its ancestors once occurred in Argentina during the Paleocene (about 60 million years BP: Pascual et al. 1992). Cenozoic fossils of tapirs and camels from temperate North America show how intervening populations went extinct to create the present disjunct distributions between South America and Asia (**Figure 11.18**). The extent of the historic range contractions that can be documented using the fossil record can be impressive. We know that some "living fossils," such as the superficially lizardlike tuatara (*Sphenodon punctatus*) now found only on a few small islands of New Zealand (see Chapter 10), are members of once-widespread groups. Another spectacular example is the fish family Ceratodontidae, now represented by a single species, *Neoceratodus forsteri*, in Queensland, Australia, but

FIGURE 11.18 Fossil localities of camelids (family Camelidae) from the Pleistocene in western North America. The records show that these mammals were widely distributed south of the continental ice sheet (shaded area) until they went extinct abruptly—probably due to human influences—less than 20,000 years ago. The extinctions left a disjunct distribution of camelids, with guanacos and vicuñas in South America, and dromedary and Bactrian camels in the Old World. (After Hay 1927.)

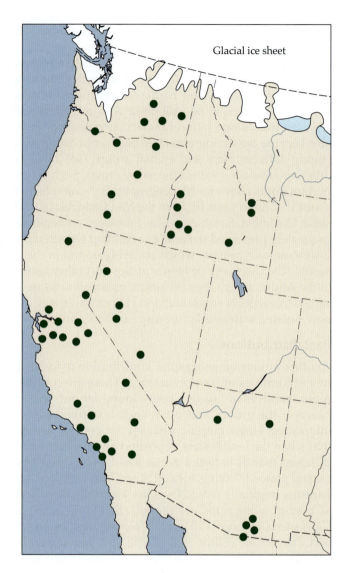

Glacial ice sheet

known from fossils found on all of the continents except Antarctica (**Figure 11.19**). Many cladistic or phylogenetic biogeographers are still attempting to reconstruct past geographic distributions of lineages without using fossil evidence (see Chapter 12), but as new fossil finds are made, they promise to provide strong evidence to support or reject these hypotheses (Lieberman 2002, 2003).

Paleoecology and paleoclimatology

Finally, fossils and the other materials that are preserved with them provide invaluable information on the nature of past environments. Some of this information comes from the physical and chemical composition of fossil-bearing rocks. Their particle size and structure indicate the nature of the substrate and the surrounding geological formations where the remains were preserved. Their chemical composition can indicate whether the environment was marine, freshwater, or terrestrial, and anoxic or oxygen-rich, and it can often give some idea of the temperature. For example, the fossilized remains of ancient coral reefs form distinctive limestone rocks, and since reef-building corals require water temperatures above about 20° C and sunlight for their symbiotic algae, these strata indicate the past occurrence of shallow

(A)

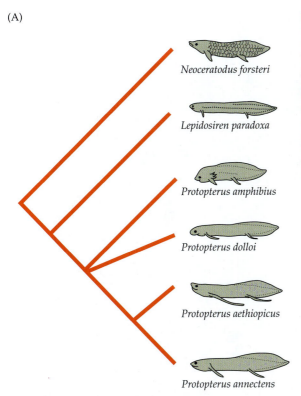

Neoceratodus forsteri

Lepidosiren paradoxa

Protopterus amphibius

Protopterus dolloi

Protopterus aethiopicus

Protopterus annectens

FIGURE 11.19 Information from the fossil record is essential for interpreting biogeographic history. The shaded areas show the current relict distribution of six extant species of lungfish (Lepidoseriformes) on three southern continents: *Neoceratodus forsteri* on Australia, *Lepidosiren paradoxa* in South America, and four species of *Protopterus* in Africa. (A) A molecular phylogeny among all extant species. (B) The circles show fossil localities for one of the three lungfish families, Ceratodontidae. This family contains just one surviving species (*Neoceratodus fosteri*), which is restricted to Queensland, in northeastern Australia, but was distributed nearly worldwide in the Mesozoic. Accurate dating of the fossils is obviously crucial to determining how tectonic events, such as the breakup of Gondwanaland, affected the distribution of the lungfish. (A after Tokita et al. 2005; B after Sterba 1966; Keast 1977.)

(B)

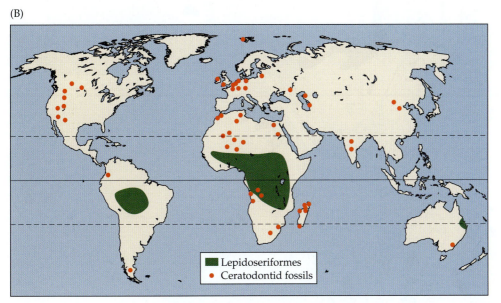

■ Lepidoseriformes
● Ceratodontid fossils

tropical and subtropical seas (**Figure 11.20**). The orientation of crystals due to paleomagnetism indicates the latitude of liquid volcanic rocks when they were cooling (see Chapter 8). Mapping of the past positions of the drifting continents and oceans relative to the Equator has enabled paleobiologists to determine that the latitudinal gradient of species diversity is an ancient feature of the biogeography of the planet, dating back at least 100 million years (see Chapter 15; see also Stehli 1968; Stehli et al. 1969; Stehli and Wells 1971; Crane and Lidgard 1989; Crame 2004).

Remains of organisms preserved together in the same strata often can provide abundant information on the climate, vegetation, and biotic communi-

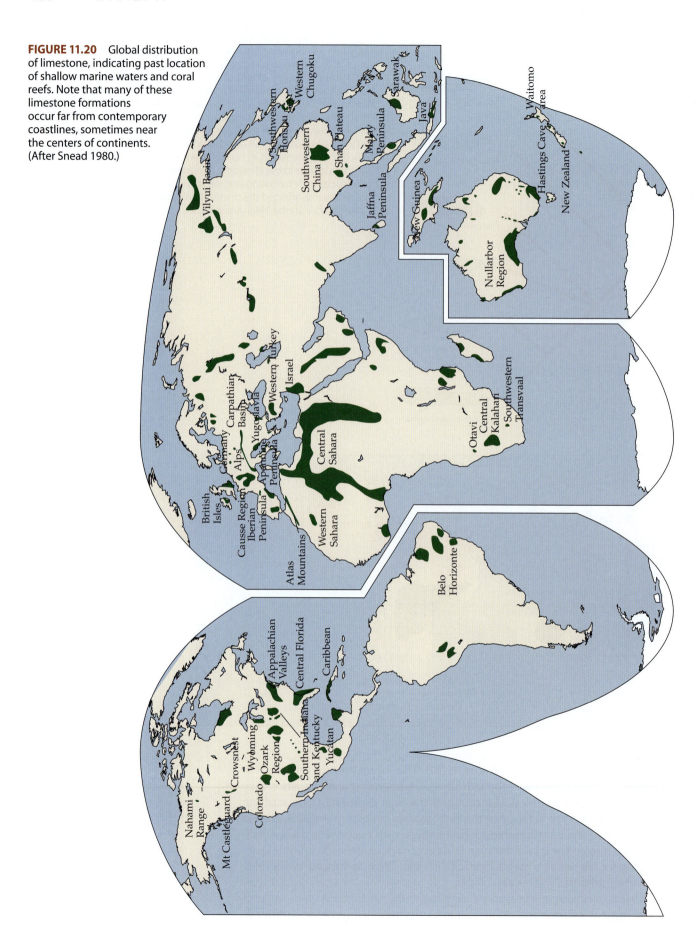

FIGURE 11.20 Global distribution of limestone, indicating past location of shallow marine waters and coral reefs. Note that many of these limestone formations occur far from contemporary coastlines, sometimes near the centers of continents. (After Snead 1980.)

ties during the times these organisms were alive. Perhaps the fossils most informative about paleoecology are **catastrophic death assemblages**. The lives and environment of the people inhabiting the Roman cities of Pompeii and Herculaneum are frozen in time, exquisitely preserved by the ash spewed out by the eruption of Mount Vesuvius in 79 AD. Similarly, many soft-bottom marine assemblages were preserved in situ when they were covered by large quantities of sediment. Individuals were trapped in their exact locations, providing an invaluable record of the depth and orientation of burrows, the size and age structure of populations, and the species composition and spatial relations of communities. Extensive, exquisitely preserved fossil deposits in the northeastern corner of Nebraska preserve a catastrophic death assemblage of the terrestrial mammals that inhabited the area during the Miocene, and that died over a three- to five-week period about 12 million years ago as a consequence of volcanic ash that drifted eastward from a giant volcano far to the west in the state of Idaho. Along the Florida Gulf Coast, a catastrophic event—possibly a red tide—killed and deposited the skeletons of over a hundred individuals of a single species of now extinct cormorant during the Late Pliocene (Emslie and Morgan 1994).

Pleistocene assemblages provide a wealth of paleoecological information. Not only is there a large quantity of well-preserved fossil material, but the fact that most of the plants and animals were very similar to their living descendants enables us to make strong inferences about their behavior and ecology. Not only do data on oxygen isotope ratios in ancient ice and fossils provide information on global, regional, and local temperatures (see Chapter 9), but the kinds of organisms and their characteristics also allow for reconstructions of past climates. Fossil pollen from lakes and bogs and plant fragments from wood rat middens provide so much information on the occurrences and abundances of still-extant plant species that it is possible to reconstruct the structure and composition of Pleistocene plant communities and the climatic regimes in which they lived. The diets of extinct Shasta ground sloths, *Nothrotheriops shastensis*—large ground-dwelling herbivores that went extinct in North America about 11,000 years ago—have been reconstructed by sequencing ancient plant DNA from fossilized fecal remains (Poinar et al. 1998). From plant and animal fossil deposits, it is also possible to document the shifts in geographic ranges and changes in composition of communities as a consequence of the advances and retreats of glaciers and associated changes in climate during the Pleistocene. For example, since the maximum of the Wisconsin glacial period about 20,000 years ago, some small mammals in North America have shifted their ranges hundreds of kilometers, and not always consistently in a northward direction, as would be expected if the climate had simply warmed up uniformly. As a consequence, Pleistocene communities contained combinations of species strikingly different from contemporary communities (see Chapter 9). In South America, the post-Pleistocene range shift of a species of leaf-eared mouse, *Phyllotis limatus*, has been inferred by sequencing ancient DNA from hairs left in the midden currently occupied by another species of rodent, demonstrating its former occurrence in a region where it no longer occurs (**Figure 11.21**).

Fossils can provide convincing evidence of the role of biotic interactions in the past. Many researchers have used shifts in dominance over time between certain functionally similar but distantly related groups of organisms to infer that competition has played a major role in the history of these lineages; examples include comprehensive dominance shifts in brachiopods and mollusks (see Elliot 1951; but see also Gould and Calloway 1980), gymnosperm and angiosperm plants (Knoll 1986), and multituberculate and rodent mammalian herbivores (Simpson 1953; Lilligraven 1979). Pollen profiles and wood

FIGURE 11.21   (A) Geographic distribution and (B) mtDNA phylogeny for a group of closely related species of leaf-eared mouse, genus *Phyllotis*, that are distributed throughout the Andean region of western South America. The operational taxonomic units (OTUs) labeled "midden consensus" and "midden clone" were derived from rodent feces preserved in an 11,700 year old rodent midden within the boxed area on the map, located in the Atacama Desert. Note that from the phylogeny we can infer that (1) *P. limatus* was the species that occupied this region 11,700 years ago (and probably created the midden), and (2) the current southern limit of its distribution is about 100 km to the north of this location. (After Kuch et al. 2002; lithograph by George Robert Waterhouse.)

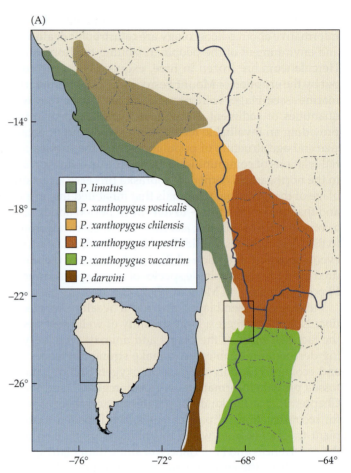

(A)

P. limatus
P. xanthopygus posticalis
P. xanthopygus chilensis
P. xanthopygus rupestris
P. xanthopygus vaccarum
P. darwini

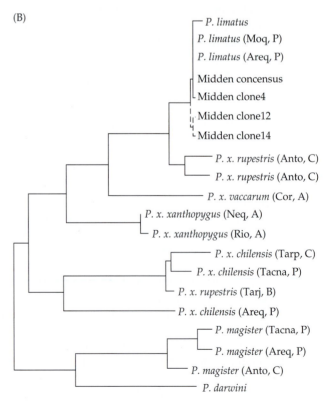

(B)

rat middens provide evidence of successional dynamics and reorganization of past plant communities in response to shifts in climate (e.g., Delcourt and Delcourt 1977, 1994, 1996; Davis 1986, 1994; Woods and Davis 1989).

In the marine realm, Vermeij (1974, 1978) has used patterns of co-occurrence of different groups of carnivorous and herbivorous mollusks and evolutionary trends in shell form to document the history of the coevolutionary arms race between predators and prey throughout most of the Phanerozoic. But such coevolution proceeds slowly, at least in mollusks. The drill holes made by predatory snails in the shells of their prey are clearly visible on well-preserved fossils. The sizes and locations of 100,000 year old drill marks are identical to those found on contemporary shells, indicating that there has been no detectable change in predatory behavior over that period (Tull and Bohning-Gaese 1993). Fossils are also providing evidence of the role of predation by aboriginal humans in the extinctions of other vertebrates (Barnosky et al. 2004). The correspondence between the time of arrival of humans and wholesale extinctions of native birds and mammals on islands and continents shows the effect that an invading predator can have on its prey, and it demonstrates that humans have been having a major influence on global biodiversity for tens of thousands of years (see Chapter 9).

Molecular Clocks and Estimating Times of Divergence

The idea that proteins and DNA could be evolving at a sufficiently constant rate to provide a metric for dating branching events in the history of lineages was first proposed by Zuckerkandl and Pauling (1965). The notion that evolutionary events could be dated even when fossils are unavailable captured the imagination of biologists and, in particular, was attractive to those interested in addressing questions about biogeographic histories. Some of the estimates of divergence times were often surprising. For example, Maxson and Roberts (1984) used a molecular clock for albumin evolution to postulate that speciation in several Australian frog lineages occurred through Miocene and Pliocene vicariance (roughly between 12 million and 4 million years BP, respectively), millions of years earlier than required by the then prevailing model of multiple episodes of eastern-to-western dispersal during Pleistocene glaciations.

Molecular clocks have always been controversial (e.g., Graur and Martin 2004; Heads 2005), but they continue to play an important role in modern biogeography for several reasons. First, we have increasingly robust methods for using molecular dating approaches to estimating times of divergence (reviewed in Kumar 2005; Rutschmann 2006). If a fossil record exists for a group of interest, and if those fossils provide good dates for one or more divergence points in a molecular phylogeny, one can calibrate an "absolute rate" of molecular evolution (**Figure 11.22**). However, even without the possibility of calibration with fossils, the **relative rate test** (Sarich and Wilson 1973) provides a means of testing for rate heterogeneity among lineages in a phylogeny. If a relative rate test indicated a homogeneous rate of evolution in the group, we could employ an estimate of absolute rate, perhaps derived from a related group that did have a good fossil record, to estimate actual time of divergence—with the added assumption, of course, that rates were similar between the focal and related groups. Second, studies (reviewed in Arbogast et al. 2002) have shown how important it is to address the often considerable rate variation that occurs among different classes of nucleotide sites in a DNA sequence when estimating rates, and this can be accomplished by employing maximum likelihood or Bayesian methods of choosing a model of molecular evolution that takes into account the appropriate pattern of

(A)

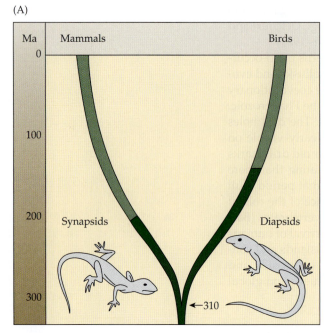

(B)

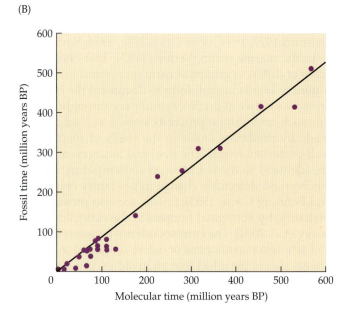

FIGURE 11.22   (A) Calibration of molecular divergence times from the fossil record requires one or more well-dated and diagnostic fossils that can be used to fix the time of a cladogenetic divergence event. In this case, synapsids (the stem amniote lineage [darker shading] leading to mammals [lighter shading]) and diapsids (the stem amniote lineage [darker shading] leading to birds [lighter shading]) have diagnostic cranial morphologies that allow paleontologists to infer a "first appearance" of each in the fossil record at 310 million years ago. (B) This point was then used to estimate rates of molecular divergence across 658 different nuclear genes (details of estimation methods provided in original article), and the composite estimates derived from all genes agreed remarkably well with estimates derived directly from the fossil record. (After Kumar and Hedges 1998.)

rate heterogeneity across character partitions. Third, several maximum likelihood and Bayesian procedures are now available for estimating absolute rates and times of divergence, even when among-lineage rate heterogeneity does in fact occur in a group or when large confidence intervals surround depositional or cladistic information for some of the fossils used to calibrate times of divergence (e.g., Sanderson 2002; Drummond and Rambaut 2007; **Figure 11.23**). Fourth, often for purposes of testing alternative biogeographic hypotheses, it might not be necessary for estimated rates of divergence to be particularly precise. For example, the corridor known as the Bering land bridge opened and closed dispersal routes between the Palearctic and Nearctic as early as 20 million years ago (see Chapter 9) and at least two more times before the Pleistocene, and then it continued to do so during glacial climatic periods throughout the Pleistocene up until about 10,000 years ago (see Figure 12.16). If one is interested in asking whether Beringian dispersal and subsequent divergence of a group occurred several million years ago versus during the Late Pleistocene only a few thousand years ago, a molecular clock would not need to be particularly precise to rule out one or the other of these alternative divergence times.

The Future of Lineage Reconstruction

We conclude this chapter by noting that knowledge of the history of lineages, and our technical and analytical sophistication to gain this knowledge, is advancing along several exciting fronts. Systematists have widely incorporated methods of phylogenetic reconstruction that simultaneously estimate parameters such as variance in rates of molecular evolution across genes, lineages, and fossil calibration points. New technologies provide cost-effective approaches for surveying thousands of genes—often whole genomes—across many individuals and taxa. Phylogeography has added the approaches and methods of population genetics to reconstruct the complex history of gene trees in closely related lineages (Knowles 2009; Hickerson et al. 2010), within a time frame of the past several thousand to several million years. Paleobiologists and paleoecologists have been using new finds and more informed

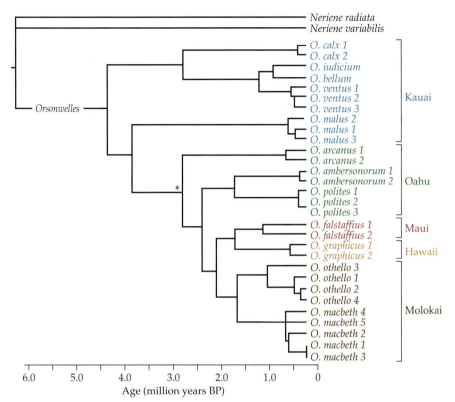

FIGURE 11.23 Phylogeny and estimated divergence times derived from molecular data for 12 of the 13 known species of the endemic Hawaiian linyphiid spider genus *Orsonwelles* (see Figure 12.13 for distributions). Branch lengths differed significantly in the original phylogeny such that a simple molecular clock could not be used to estimate divergence times. Therefore, a nonparametric rate-smoothing procedure (Sanderson 2002) was used to estimate divergence times after calibration of the rates of divergence with the divergence event at the node marked by an asterisk, a geologically well-dated age of origination of the Koolau range in Oahu, and the point of separation of species on Kauai from those on all other islands. (After Hormiga et al. 2003.)

interpretations of fossil remains to document what kinds of organisms lived in particular places at different times in the history of the Earth, along with their ecological associations and the habitats they occupied, and quantitative methods have been developed to assess the incompleteness and accuracy of a fossil record (Kidwell and Holland 2002; Foote 2003). Until quite recently, however, there had been far too little communication between these groups of scientists, and too little synthesis of their findings, even though the need has been recognized for some time (e.g., Grande 1985; Lieberman 2003). In part, this was because the scientists themselves historically came from different backgrounds—systematists and phylogeographers from biology, and paleontologists from geology—and because they presented their results in different journals and meetings. But in part, it was because they had often treated each other as rivals and with suspicion, rather than as mutualists who could bring different backgrounds, techniques, and perspectives to a common endeavor: the formidable task of reconstructing the history of life on Earth. At no time was this suspicion more apparent than when molecular systematists suggested times of divergence that were at odds with the known fossil record.

Fortunately, a detectable mutualism between disciplines has emerged. First, paleobiogeographers are employing modern phylogenetic methods and theory in reconstructions of evolutionary and biogeographic histories (Lieberman 2003). Second, a new bridge between molecular systematics and paleontology combines ancient and modern DNA sequences in phylogenetic analyses (Ramakrishnan and Hadly 2010). Third, as discussed in this chapter, fossils are increasingly sought after by molecular systematists for calibration of molecular clocks. Fourth, while molecular estimates of divergence times still continue to often be at odds with the known fossil record, many of the more sophisticated estimates appear to be quite consistent with fossil dates (see Figure 11.22). Finally, phylogeography is conceptually positioned to build bridges between modern and paleoecological perspectives on the assembly and disassembly of biotas during Late Neogene time frames (Riddle and Hafner 2004). The task of reconstructing biogeographic histories is formidable, but we are enthusiastic about the growing potential to do so, given the progress in making available the needed robust lineage histories. In the next chapter, we will turn more directly to an exploration of historical biogeographic methods and what they are revealing about the history of life on Earth.

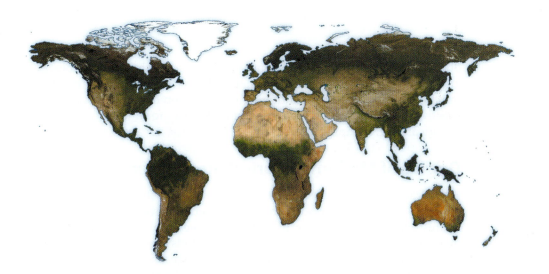

RECONSTRUCTING THE HISTORY OF BIOTAS

As we discussed in Chapter 2, de Candolle first differentiated the historical from the ecological determinants of geographic distributions, but not until the 1960s were these attributes turned into two generally separate research agendas. As a result, historical and ecological biogeography subsequently developed along very separate routes and concentrated on largely, but not exclusively, separate kinds of questions. For much of the twentieth century, the prevailing paradigm in historical biogeography was one in which certain regions of the Earth were considered to be particularly important centers of origination of new species, and from these centers new forms would disperse to far corners of the globe (see, for example, Figure 2.2A).

Beginning in the latter half of the twentieth century, the center of origin paradigm was largely replaced with one in which historical biogeography became increasingly more narrowly focused on how the distribution and diversification of lineages changed through tectonic events and vicariance. A complementary endeavor was the development of hypotheses about Earth's tectonic history based on the dynamic geography of organisms.

Historical biogeography is again undergoing a rapid transformation with the infusion of new approaches and methods, some of which we reviewed in Chapter 11 and others that we present new in this chapter, and all offering previously unattainable opportunities to analyze biogeographic structure across multiple, codistributed lineages. New data on phylogeny, distribution, and diversity—largely based on DNA sequencing—are increasingly being combined with innovative approaches to visualizing lineage histories on a map.

This chapter is divided into four parts. First, we trace the development of concepts and approaches in historical biogeography over the past century. Second, we review some of the many currently available approaches and methods in historical biogeography—some that have been around for a while and some that have emerged more recently. Third, we feature applications of several of these methods that provide some fascinating insights into the biogeographic histories of terrestrial and marine lineages and biotas. Finally, we provide a synopsis on the continuing reintegration of once-divergent perspectives and approaches that are paving the way for a transformation of historical biogeography.

Origins of Modern Historical Biogeography

Historical biogeography serves as an interesting case study in scientific revolutions—in the questions that biogeographers believe can and should be addressed, in the mechanisms underlying changes in biotic distributions and diversification, and in the approaches and methods available (Funk 2004). In the previous chapter, we reviewed approaches to reconstructing the patterns and timing of diversification in lineages using phylogenetic (and particularly molecular) systematics and paleobiology. Phylogeography, in particular, is an advance that allows us to address questions of historical biogeography at the more recent temporal, and often finer, spatial scales encompassed by the climatic fluctuations of the Pleistocene.

We have discussed in several previous chapters the profound influence that advances in plate tectonics theory during the 1950s and 1960s had on biogeographers—no longer was it necessary to construct scenarios of biogeographic history (e.g., ancient land bridges between continents) under an assumption that the continents had largely remained static over the course of Earth history. If the climate of the Earth and the positions of the continents had remained fixed over time, this would drastically limit the kinds of hypotheses that could be advanced about the causes of endemism, provincialism, and disjunction. For example, there would be only one likely explanation for the disjunct distribution of a terrestrial organism between South America and Australia: that the ancestors originally occurred on one of those continents, and that individuals dispersed over long distances across extensive oceanic barriers to colonize the other continent. On a dynamic Earth, however, it was now possible to invoke an alternative hypothesis: that the ancestors originally inhabited Gondwanaland, and vicariance accounted for disjunct distributions and the diversification of distinct biotas on the previously connected but now isolated landmasses. Of course, there are more complicated versions of both hypotheses, perhaps involving past occurrences of the ancestral lineage on the northern continents and extinction of those populations, with some combination of over-water dispersal and continental drift being required to explain the current distribution. Using information from phylogenetic reconstruction, plate tectonics, and the fossil record, it should be possible to erect and evaluate an array of alternative hypotheses.

Not all historical biogeographers "think alike" about the questions to address and appropriate methods to employ in a biogeographic study (van Veller et al. 2003; Riddle and Hafner 2004; Ebach and Morrone 2005; Morrone 2009). The late nineteenth and twentieth centuries were times of great change in philosophical and methodological approaches in historical biogeography. Early in the twentieth century, W. D. Matthew developed the center of origin–dispersal model from Darwin and Wallace's **evolutionary biogeography** and their original concept of "dispersal over a permanent geography." Throughout the next four decades, this model continued as the ruling paradigm of historical biogeography, largely owing to the many insights of George Gaylord Simpson (1940), Philip Darlington (1957), and their colleagues.

Early efforts: Determining centers of origin and directions of dispersal

Since the time of Buffon and de Candolle (see Chapter 2), biogeographers have realized that distributions of organisms have shifted over time. This realization led naturally to efforts to determine the birthplace—or center of origin—of each taxon. Two general questions motivated the search for centers of origin. First, investigators wanted to know whether certain geographic regions served as cradles for the evolution of new kinds of organisms, and which special features allowed these lineages to be successful. Second, biogeographers wanted to understand how biotas have been assembled: where taxa started from, what routes they followed to disperse around the world, and what factors produced present patterns of endemism, provincialism, disjunction, and diversity. These are still among the goals of historical biogeography, but fortunately, the methods used to infer the origins and movements of lineages have improved greatly, and they have been enhanced through incorporation of detailed models of Earth history (e.g., plate tectonics) and their influence on dynamic biogeography and diversification of lineages.

Initially, biogeographers tried to develop simple rules for determining the center of origin for a taxon. For example, Adams (1902, 1909) listed ten criteria that could help to identify such centers. Some authors insisted categorically on using a single criterion to the exclusion of others. Matthew (1915), for instance, followed Buffon in believing that centers of mammalian origin were in the Holarctic (**Figure 12.1**; see also Figure 2.2). Matthew argued that new, successful forms arose in response to the challenges of temperate climates; these forms eventually supplanted their progenitors and other lineages, forcing them into peripheral habitats, then to lower latitudes, and eventually into the Southern Hemisphere. This "boreal superiority" notion was considered by Simpson (1940) as finding some support in the asymmetric effects of the Great American Biotic Interchange between northern and southern continents (see Chapter 10).

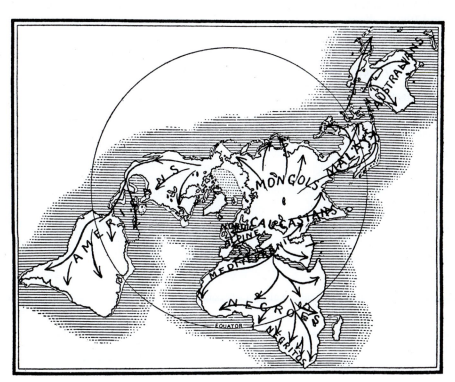

FIGURE 12.1 An example of Matthew's arguments for a center of origin for mammalian speciation in the Holarctic region of the Northern Hemisphere, with subsequent dispersal around the world. This figure shows his scenario for the principal races of humans as they were understood in the early 1900s. In his 1915 book *Climate and Evolution*, Matthew presented a scenario in which consecutive waves of more advanced forms would evolve in the Northern Hemisphere and displace more primitive forms as they dispersed away from the center of origin. (From Matthew 1915.)

Matthew (1915) thought that the center of origin is where the most derived forms reside. This is in direct opposition to Hennig's (1966) **progression rule**, which holds that an ancestral population remains at or near the point of origin, and that progressively more derived forms are found at distances farther away from the center, in a "stepping-stone" pattern of sequential dispersal and speciation events. In reality, either explanation may be correct for different groups, because displacement of one form by another depends in part on how they disperse, speciate, and interact with their biotic and abiotic environments. An important difference between the two ideas, however, is that Matthew's was based on a model of evolutionary asymmetry in which environmental challenges at one geographic area drove the improvement of superior traits in new species that then dispersed outward and drove more primitive forms to extinction. In contrast, Hennig's model was based solely on a model of dispersal, isolation, and speciation in peripheral populations, without presumption of evolutionary asymmetry, and thus it was testable through phylogenetic analysis.

A thorough evaluation of the challenge of determining centers of origin was made by the phytogeographer Stanley Cain (1944). He listed the 13 criteria in use at the time (**Table 12.1**) and argued convincingly that none by itself could reliably identify the center of origin. Some authors had claimed, for example, that the center of origin should be where the greatest number of species in a group reside. This assumption is, of course, invalid if the majority of forms inhabit a region of secondary radiation, as do the heaths of South Africa (*Erica* spp.), with 658 species in that country's Cape Region (Linder 2003), or the hundreds of species of *Drosophila* on the Hawaiian Islands (see Chapter 14). Cain also argued against using the location of a primitive form, or the earliest fossil, as an absolute criterion. Primitive forms often survive in isolated regions located far from their original ranges and containing few competing species (e.g., paleoendemics such as the tuatara in New Zealand; other examples of such relicts are described in Chapter 10).

An example is sea snakes. To give the flavor of the sort of center of origin scenario that was popular prior to conceptual advances beginning in the 1970s and 1980s, let's examine one case in which a number of Cain's criteria point to

TABLE 12.1   *Criteria Used and Abused for Indicating Center of Origin of a Taxon*

1. Location of greatest differentiation of a type (greatest number of species)
2. Location of dominance or greatest abundance of individuals (most successful area)
3. Location of synthetic or closely related forms (primitive and closely related forms)
4. Location of maximum size of individuals
5. Location of greatest productiveness and relative stability (of crops)
6. Continuity and convergence of lines of dispersal (lines of migration that converge on a single point)
7. Location of least dependence on a restricted habitat (generalist)
8. Continuity and directness of individual variation or modifications radiating from the center of origin along highways of dispersal (clines)
9. Direction indicated by geographic affinities (e.g., all Southern Hemisphere)
10. Direction indicated by the annual migration routes of birds
11. Direction indicated by seasonal appearance (i.e., seasonal preferences are historically conserved)
12. Increase in the number of dominant genes toward the centers of origin
13. Center indicated by the concentricity of progressive equiformal areas (i.e., numerous groups are concentrated in centers, and numbers decrease gradually outward)

Source: After Cain 1944.

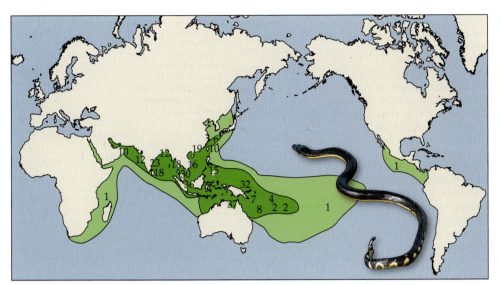

FIGURE 12.2   The current distribution and diversity (in numbers of species) of sea snakes (tribe Hydrophiini). All but one species are confined to the densely shaded area. The exceptional, very wide-ranging species *Pelamis platurus* occurs in this region but also in the more lightly shaded areas. The concentration of species and lineages in the Indo-Pacific oceans and the occurrence of the seemingly most closely related outgroup in swamps in Australia led Cogger (1975) and others to hypothesize that sea snakes spread from a center of origin in the Australia–New Guinea region. (After Heatwole 1999; photo courtesy of Aloaiza/Wikipedia.)

a similar conclusion. Consider the true sea snakes (tribe Hydrophiini) of the Indo-Pacific region, a group whose fossil record is unknown. They are venomous marine predators and members of the family Elapidae, which includes many Australasian snakes as well as cobras, kraits, coral snakes, and mambas. At least 60 species of true sea snakes in 16 genera are known from tropical and subtropical coastal habitats along reefs in the western Pacific and Indian oceans, with related species inhabiting brackish inlets, rivers, and terrestrial habitats in Australia. Several related species have entered and adapted to freshwater habitats in the Philippines and the Solomon Islands (**Figure 12.2**; Dunson 1975; Heatwole 1999). The species with the widest distribution is the pelagic form *Pelamis platurus* (see Figure 12.2); its range includes the geographic range of the entire family, and extends across the tropical Pacific to the west coast of the Americas, from Mexico to Ecuador. Postulating a center of origin for sea snakes was made easier for earlier investigators because these marine reptiles possess several useful attributes: (1) their inability to tolerate cold water (below 20° C) has confined them to the Indian and Pacific oceans, (2) they appear to be a young tribe (of Cenozoic age), (3) their closest relatives—the terrestrial members of the subfamily Hydrophiinae—overlap with them in distribution in the Australasian region, and (4) the lines of morphological specialization appear to be relatively unbroken, thus seemingly reducing problems of tracing geographic patterns obscured by major extinctions.

Cogger (1975) concluded that the Hydrophiini originated in the Australia–New Guinea region—the region of highest hydrophiin diversity, with more than 30 species (see Figure 12.2). Most genera of Hydrophiini are classified in the *Hydrophis* group, and most of these taxa are Australian (Cain's first criterion, see Table 12.1). Here too live what Cogger believed to be their putative elapid ancestors (*Rhinoplocephalus* and *Drepanodontis*), swamp-inhabiting Australian snakes, as well as the most primitive and unspecialized hydrophiins such as *Hydrelaps* and *Parahydrophis*. These putative basal genera in the Hydrophiini radiation share several traits with the elapids (Cain's third criterion) and are notably weak swimmers (McDowell 1969, 1972, 1974; Dunson 1975). Genera that range to the northwest of Australia often have at least one representative in the Australian coral reefs as well. Finally, the sea krait genus *Laticauda* (**Figure 12.3**) was considered by Cogger (1975) and Voris (1977) to be an early offshoot (perhaps a separate tribe) but still closely related to true sea snakes—as such, its occurrence around New Guinea pro-

FIGURE 12.3 Molecular phylogenies for the sea snakes and their terrestrial relatives. Numbers along branches refer to levels of statistical support from different methods of reconstructing the phylogenies. (A) The true sea snakes, Hydrophiini, form a monophyletic group (represented by four species in this phylogeny) in the family Elapidae. Note that two earlier assumptions about relationships (Cogger 1975; Voris 1977) are shown here to have been wrong. First, the sea kraits (genus *Laticauda*) are not as closely related to the Hydrophiini as was postulated; rather they are a sister clade to the rest of the Elapidae. Second, the closest relatives to the true sea snakes are in the genus *Hemiaspsis*, not *Rhinoplocephalus*. (B) A more detailed phylogeny of several species in the Hydrophiini disproves another earlier assumption: the so-called "primitive" genera *Hydrelaps* and *Parahydrophis* are not outside the more specialized radiation of Hydrophiini, but instead are nested well within it. (A after Sanders et al. 2008; B after Lukoschek and Keogh 2006.)

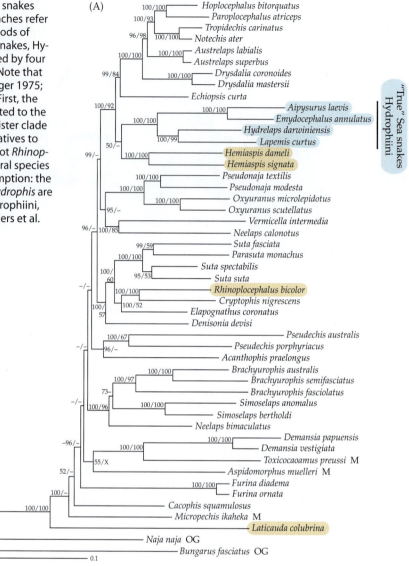

vides yet another degree of support for this as the center of origin of true sea snakes under Cain's third criterion.

Scenarios like the above are just that: stories concocted to explain most or all of the known facts about a group that might be relevant to its history. In a sense, such scenarios are hypotheses because many specific parts of the story can potentially be falsified. For example, the above scenario would be rejected, either by the discovery of a fossil of appropriate age and with plesiomorphic characters indicating hydrophiin origination along the west coast of Africa, or by a robust phylogeny that rejects one or more of the postulated relationships between true sea snake and their closest relatives. Indeed, this scenario allows us to make a critical observation about the importance of good phylogenetic data in historical biogeography. For example, recent studies of Hydrophiinae evolution (Scanlon and Lee 2004; Sanders et al. 2008) have demonstrated that sea kraits are the sister taxon of the entire clade that includes the very diverse terrestrial radiation of Hydrophiinae in Australasia, as well as true sea snakes—the sister group to true sea snakes appears to be a viviparous (live-bearing), terrestrial Australian snake (see Figure 12.3A). Second, in a recent molecular phylogeny for true sea snakes that includes two

(B)

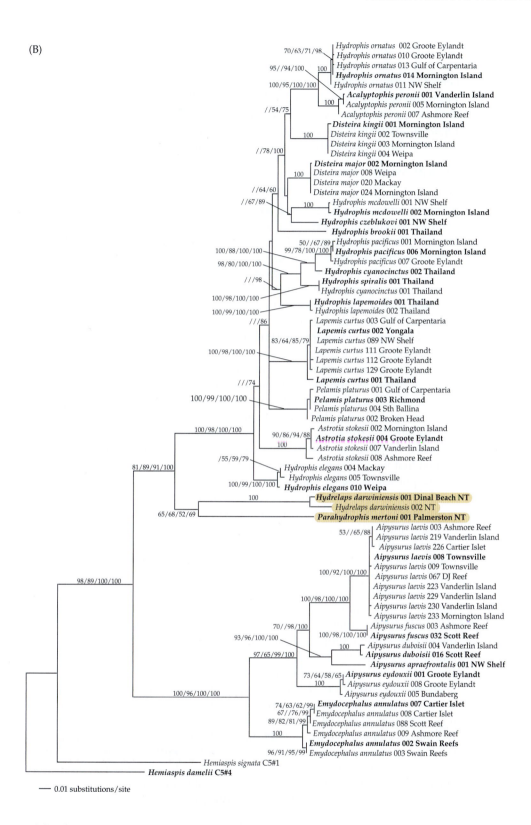

of the three mentioned above (Lukoschek and Keogh 2006), morphologically primitive genera of these species are not basal to all remaining sea snakes, but rather sister to the most diverse and most recent adaptive radiation, the *Hydrophis* group (see Figure 12.3B). Both of these examples demonstrate the value of a robust phylogeny and specifically demonstrate the weaknesses associated with earlier uses of Cain's third criterion. And so, we conclude that earlier authors, with no reference to a rigorous phylogenetic analysis or to fossil evidence, had precious little direct evidence to support the above center of origin scenario for sea snakes. This is not to say, however, that as more and better evidence becomes available, all or parts of this scenario will not be supported; for example, Scanlon and Lee (2004) used an updated phylogenetic analysis to conclude that hydrophiin sea snakes had "their origin… almost certainly somewhere along the Australian coast."

Critical issues

Why did we go into so much detail about the center of origin concept? First, inferring centers of origin using some or all of the criteria in Table 12.1 was common practice in historical biogeography up until the 1970s and 1980s—and it is still practiced, at least implicitly. When other evidence is not available and a group is diverse and widely distributed in one region but has a few disjunct species in another, the former region is often assumed to have been inhabited for a long period and the latter region to have been more recently colonized. For example, several clades of small birds, including hummingbirds (Trochilidae), flycatchers (Tyrannidae), wrens (Troglodytidae), tanagers (Thraupinae), blackbirds (Icterinae), and sparrows (Emberizinae) that occur in temperate North America but are much more diverse in the Neotropics, are often assumed to have originated in South America and to have subsequently colonized North America as the two continents drifted into proximity during the Cenozoic.

Second, scenarios about centers of origin and directions of subsequent dispersal have historically been used to make sweeping generalizations about the innovations that triggered the success of lineages, and about global patterns of diversity. We have mentioned that first Buffon (1761) and then Matthew (1915) argued that new lineages and new innovations arose predominantly as adaptive responses to the climatic rigors of the temperate zones and then spread southward to thrive and diversify in the milder tropics. The exact opposite position was advanced by Darlington (1959b), who argued that new lineages and new innovations were more likely to have originated in the tropics and spread toward the poles. He suggested that the greater diversity of species and higher taxa at low latitudes meant an increased probability that some of these "evolutionary experiments" would be successful (see Chapter 15).

These questions about the origins of lineages and their unique attributes remain of great inherent interest. Biogeographers, systematists, and ecologists still debate whether the tropics are the source or the last refuge of the distinctive combinations of traits that characterize lineages (Gaston and Blackburn 1996; Wiens and Donoghue 2004). A related example is the discussion among marine paleobiologists of whether the rates of origination of new lineages and extinction of existing ones have been higher in the more abiotically stressful onshore environments or in the more benign ones offshore (e.g., Jablonski et al. 1983; Miller 1989; Jablonski and Bottjer 1990, 1991).

Third, we now have much better methods and a greater wealth of data to use in reconstructing biogeographic histories. Currently, as we will explore later in this chapter, more integrative approaches in historical biogeography are emerging and replacing the last few decades of discord and controversy (e.g., Nelson and Rosen 1981; Ebach and Humphries 2002; van Veller et al. 2003).

From center of origin–dispersal to vicariance

To appreciate the evolution of the ideas that led to modern approaches, it is helpful to see what the opponents to center of origin–dispersal scenarios were reacting against. It is clear that historical biogeography, based largely on studies of single taxa and plagued by flawed concepts of centers of origin and dogmatic ideas about dispersal (see Chapters 2 and 6), was in need of more rigorous approaches.

Croizat's panbiogeography

As is often the case with revolutions, the rapid changes in historical biogeography that took place largely in the 1970s were triggered by a few individuals with radical ideas. And like the bizarre evolutionary changes in organisms that have occurred on islands, novel and divergent ideas often seem to arise in isolated circumstances. Thus it was that Leon Croizat, an unorthodox plant biologist who labored for most of his career in relative obscurity in Venezuela, provided the spark that ignited the revolution in historical biogeography. Croizat (1952, 1958, 1960, 1964), like many phytogeographers before him, recognized that the present limited distributions of narrowly endemic, disjunct taxa are the relicts of ancestral taxa—and often of entire floras—that were more broadly distributed in the past. Unlike most previous researchers, who typically studied just one taxon, Croizat compared the distributions of different groups. He realized that very distantly related kinds of organisms often exhibited similar disjunctions, and that these distributional patterns were legacies of historical events that influenced many different lineages in the same way. Amassing data on disjunctions from all over the world, Croizat developed an approach that he termed **panbiogeography**.

Croizat argued that each pattern of multiple disjunctions reflected the fragmentation of a biota that originally inhabited interconnected regions. He plotted the ranges of narrowly endemic species on a map and then drew lines, called **tracks**, connecting the distributions of the most closely related taxa. Croizat then superimposed the maps for multiple taxa and noted where the positions of their tracks coincided. He called the resulting coincident tracks **generalized tracks (Figure 12.4)**. He inferred that such tracks indicate historical connections—that they were pathways connecting the isolated fragments of a formerly continuous biota. By plotting all the generalized tracks on a map, he could theoretically recreate the way in which regional biotas had developed in time and over space.

When first published, Croizat's ideas received little favorable response. For one thing, the basic concept of historical subdivision and isolation of

FIGURE 12.4 A composite drawing of the major and moderately important "tracks" drawn by Croizat (1952, 1958, 1960, 1964) to show hypothesized past connections between the distributions of terrestrial organisms that currently occur in widely separated regions. These tracks were derived by plotting the distributions of endemic species within several taxa and then drawing lines to connect regions that share disjunct endemics in several taxonomic groups. Circles identify five major "biogeographic nodes"—intersections between two or more tracks. The tracks radiating out from Madagascar, for example, show that its endemics have close affinities with disjuncts in Africa, India, Australia, and the East Indies. While many of Croizat's tracks correspond to past land connections, others are highly problematic. Note, for example, the lines connecting northeastern North America and northwestern Europe with islands and continents in the South Atlantic Ocean.

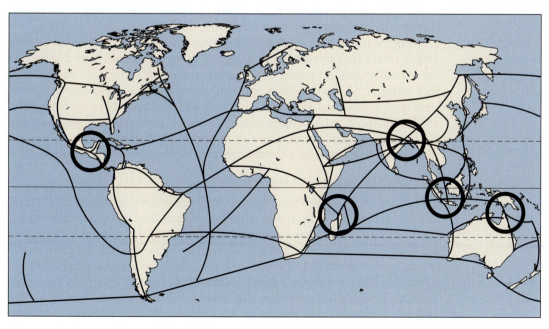

formerly widespread biotas was hardly new. For another, Croizat arrogantly dismissed alternative views while zealously promoting his own. In addition, some of his ideas and examples were so extreme that many of his contemporaries did not take his work seriously. His books contained many technical errors, and some of his purported disjunctions were based on questionable systematics. He categorically rejected dispersal over long distances and across barriers to account for the generalized tracks because he reasoned that "dispersal must be orderly and continuous in time and space" (1958). He argued that long-distance dispersal could not have been a major mechanism of geographic range expansion, even for the colonization of distant oceanic islands. And while continental drift would seemingly have provided a mechanism to explain many of the observed disjunctions, Croizat—at least in his earlier writings—denounced Wegenerism (see Chapter 8). Rather than moving continents and oceans, he kept their locations fixed and invoked ephemeral land and water bridges to explain most of the tracks connecting disjunctions, such as those among the southern continents.

During the 1980s and 1990s, panbiogeographic theory and methods continued to be developed, primarily by a group of New Zealand biogeographers (e.g., Craw 1982, 1988; Heads 1984; Craw et al. 1999), although most biogeographers considered Croizat's approach flawed by its unrealistic and idiosyncratic assumptions (Nelson and Platnick 1981; Parenti 1981; C. Patterson 1981; Humphries and Parenti 1986; Page 1990) and, most importantly, by the fact that he never used a phylogeny. Morrone and Crisci (1995) suggested a more synthetic perspective, considering panbiogeography a potentially valuable first step within a more integrative historical biogeography. Prior to this, however, an influential paper senior-authored by Croizat (Croizat et al. 1974) attempted to conjoin Croizat's track analysis with Hennig's rigorous phylogenetic methods into a new approach that the authors called **vicariance biogeography** (see below). Later, Croizat would write a scathing denial of his contribution to this paper, and criticize Nelson at length for blending phylogenetics and track analysis (Croizat 1982). Nevertheless, Croizat made an important contribution to the field. He emphasized the need for formalized methods of inferring past distributions from present ones. His search for distributional patterns that hold across many different kinds of organisms was a major departure from much previous work, which had been content to confine itself to a single taxon and to propose ad hoc events and taxon-specific mechanisms to account for its distribution.

While he went to the extreme of insisting that his methods be applied uniformly across all organisms without regard for their dispersal mechanisms and times of divergence, Croizat was reacting to the lack of rigor in the methods that many of his predecessors and contemporaries had been using to infer centers of origin, mechanisms of dispersal, and directions of range expansion, and thus to develop historical scenarios to explain contemporary distributions. Croizat rightly pointed out that even though these narratives may be correct, many of them are ad hoc, untested constructs of human imagination.

Brundin's phylogenetic biogeography

Croizat criticized evolutionary biogeographers for lacking an explicit method and for the "storytelling" flavor of their explanations, but by denying the importance of robust phylogenies in historical biogeographic reconstruction, he fell short of developing a general method for biogeography. Instead, it was left to Lars Brundin (1966, 1967, 1972, and 1988) to introduce the power of Hennig's phylogenetics to a new generation of biogeographers. Long before Brundin, Joseph Dalton Hooker had developed an interest in explaining the disjunct distributions of temperate floras on southern continents, including

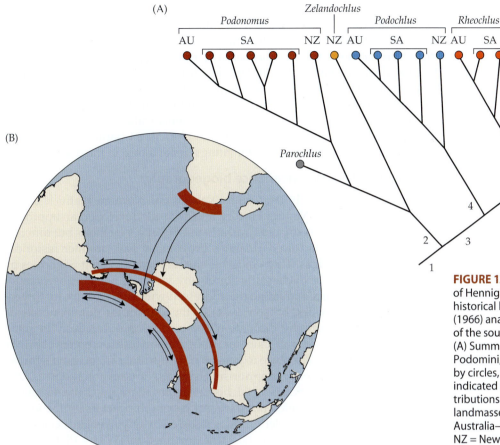

FIGURE 12.5 One of the first uses of Hennig's phylogenetic methods in historical biogeography was in Brundin's (1966) analysis of the chironomid midges of the southern temperate forests. (A) Summary phylogeny of the tribe Podomini, with the species indicated by circles, their grouping into genera indicated above the circles, and their distributions among Southern Hemisphere landmasses indicated as follows: AU = Australia–Tasmania; SA = South America; NZ = New Zealand. Note that there are multiple cases in which species in New Zealand are a sister group to species in South America and Australia–Tasmania, and that the Australia–Tasmania species are always the sister group to an array of South American species (which are always a paraphyletic rather than monophyletic group, if Australian species are not included). (B) Diagram Brundin used to summarize his full set of analyses of several midge tribes. Current landmasses are shown as having two general tracks across Antarctica—a bolder, more dominant track across west Antarctica between New Zealand and South America (with "dispersal" events going in both directions) and a lesser track across east Antarctica between South America and Australia–Tasmania (with dispersal in only one direction into the latter landmass). Phylogenetic trees also supported a track connecting these landmasses with South Africa. (After Brundin 1966.)

South America, New Zealand, and Australia–Tasmania (see Chapters 2 and 6). Although Hooker (1866) had considered these distributions to represent a separate center of evolution in the Southern Hemisphere, the scenarios of dispersal from northern to southern regions of the world favored by evolutionary biogeographers, including Darwin, Matthew, and Darlington, were still a prevailing theme in historical biogeography into the 1960s.

Brundin clearly was showing his frustration with one of the last of the evolutionary biogeographers, Philip Darlington, when he accused him of being "largely unsuccessful because he has misunderstood the method of approach in biogeography. It is plainly meaningless to speculate . . . without knowledge of the reliability and meaning of the basic data and by perpetual neglect of the principles of phylogenetic reasoning" (Brundin 1966). In his classic 1966 monograph, Brundin argued that a phylogenetic analysis based on Hennigian logic (see Boxes 11.1 and 11.2) was a fundamental component of historical biogeographic analysis. His study involved a detailed phylogenetic analysis of the chironomid midges that occupied high-elevation cold-water streams throughout the world, including many species in southern temperate forests that occur on many of the landmasses in the Southern Hemisphere.

Brundin produced detailed phylogenetic trees for several tribes of midges, with an example shown in **Figure 12.5A**). Multiple clades showed a basal split between New Zealand, which was then followed by a South American versus Australia–Tasmania split, leading Brundin to conclude decisively that the southern end of the world was indeed, as predicted by Hooker, an important center of evolution rather than just a receiver of dispersers from the north. Brundin interpreted this history by imagining Antarctica to have played a dual role (**Figure 12.5B**)—both as an earlier "dispersal" route between New Zealand and South America via west Antarctica, and as a more recent route between South America and Australia–Tasmania via east Antarctica. Note that even though Australia–Tasmania is now geographically much closer to New Zealand than to South America, Brundin's powerful phylogenetic methods allowed him to make a bold and robust conclusion that its temperate forest biota had a closer historical affinity with the latter continent. In fact, Brundin was one of the first biogeographers to take full advantage of the plate tectonics revolution in geology of the 1960s (see Chapter 8) to interpret biogeographic histories, thereby initiating a renaissance in the discipline.

Nelson and Platnick's vicariance biogeography

Beginning in the early 1970s, Gareth Nelson and Norman Platnick of the American Museum of Natural History in New York developed an approach that combined elements of both Croizat's and Hennig's methods but differed from Brundin, who had used Hennig's Progression Rule to infer historical scenarios for the midges. These biogeographers became ardent advocates of the point of view that dispersal histories—by virtue of being idiosyncratic, lineage-specific events—could not be addressed within a rigorous, hypothetico-deductive framework. Nelson (1974) set the tone for the dominant role of vicariance in biogeography throughout the next several decades: "In summary, I reject as aprioristic all 'clues' or 'rules' used to resolve centers of origin and dispersal without reference to general patterns of vicariance and sympatry. . . . I include as a rejectable apriorism Hennig's 'Progression Rule.' . . . Unencumbered by aprioristic dispersal, historical biogeography is the discovery and interpretation, with reference to causal geographic factors, of the vicariance shown by the monophyletic groups resolved by phylogenetic ('cladistic') systematics." The vicariance method was described in 1978 by Platnick and Nelson (1978), the same year that Donn Rosen published his classic vicariance biogeography study of the poeciliid fish genera *Heterandria* and *Xiphophorus* (Rosen 1978). Other descriptions of the method included those by Nelson and Platnick (1981) and, with some modifications, Wiley (1981, 1988).

The basic reasoning behind vicariance biogeography was simple. Platnick and Nelson formalized the observations made earlier by Brundin that historical explanations for disjunct distributions of related organisms fall into two classes: dispersal hypotheses (which they called **dispersal biogeography**), in which organisms are assumed to have migrated across preexisting barriers, and vicariance hypotheses (vicariance biogeography), in which the formation of new barriers is assumed to have fragmented the ranges of once continuously distributed taxa. Extant distributions are usually inadequate, not only to determine whether a given barrier was in existence before or after the migration of a particular taxon to its present disjunct areas, but also to reconstruct the direction of the migration or the sequence of barrier formation. If, however, the organisms inhabit three or more disjunct areas, techniques of cladistic systematics (see Chapter 11) can be used to determine the sequence of branching in the lineage (**Figure 12.6**).

If we can assume that the speciation events in a lineage were caused by geographic isolation, then the phylogenetic relationships within the lineage also indicate the relative times of initial spatial separation of the now disjunct groups. If we can further assume that these ancient geographic separations have been preserved and are reflected in the current distributions of the species of the lineage, then the cladistic phylogeny provides not only a phylogenetic hypothesis about the historical ancestor–descendant relationships among the taxa, but also a biogeographic hypothesis about the historical relationships among geographic localities. Both phylogenetic and biogeographic

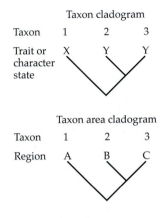

Taxon cladogram

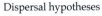

Taxon area cladogram

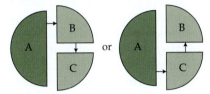

FIGURE 12.6 A hypothetical cladogram showing reconstructed phylogenetic relationships among three taxa (numbers), supported by synapomorphic character state Y, and corresponding dispersal and vicariance hypotheses to account for their current distributions among three areas (A, B, and C). Note that even without any extinction or multiple colonization events, there are two dispersal hypotheses that account equally well for the known phylogenetic and area relationships, but only one vicariance hypothesis.

Dispersal hypotheses

Vicariance hypothesis

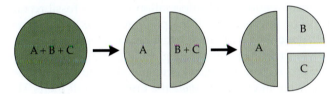

reconstructions are hypotheses about the historical branching of a taxonomic lineage in time and space. The biogeographic reconstruction is called a **taxon area cladogram** because of its precise logical analogy to a cladistic reconstruction of phylogenetic relationships.

Figure 12.6 shows a hypothetical example of a taxon area cladogram. Three taxa are distributed in three disjunct areas. The arrangement of the diagram indicates the relationships among the taxa as revealed by cladistic analysis: Taxa 2 and 3 are more closely related to each other than either is to taxon 1. From this we can infer that taxa 2 and 3 share not only a more recent common ancestor, but also a more recent common geographic range than either taxon does with taxon 1. We can now advance two kinds of biogeographic hypotheses that are consistent with these relationships. One dispersal hypothesis would have the ancestral population inhabiting area A, with a propagule dispersing to area B and then another propagule dispersing from area B to area C. But because taxa 2 and 3 are symmetrically related to each other, the colonization order could just as easily have been from area A to C to B, but dispersal from either area B or C to A is inconsistent with the phylogenetic diagram. Alternatively, a vicariance hypothesis would have the ancestral taxon inhabiting all three areas—A, B, and C—followed by the formation of a barrier to isolate areas B and C (still interconnected) from area A, and finally by the formation of a barrier to isolate area B from area C.

How would we test these hypotheses? Initially, the American Museum group wanted, like Croizat, to reject out of hand any hypothesis that called for long-distance dispersal. Platnick and Nelson (1978) and others tried to justify this seemingly arbitrary procedure by suggesting that dispersal hypotheses are extremely difficult to falsify, even with information from the fossil record. They argued that an episode of long-distance or barrier-crossing dispersal is an event of low probability and predictability. Each episode is likely to represent an independent event in which a propagule of one taxon crosses the barrier and colonizes the isolated region; several different kinds of organisms are unlikely to disperse together or simultaneously. Further, if a species could disperse across a barrier once, presumably it could have done so repeatedly, and this would invalidate the assumption that the historical separation created by the original allopatric speciation event has been preserved. It is easy to show that if we allow for the possibility of repeated episodes of colonization and extinction, a wide variety of histories can produce the same taxon area cladogram, and most of them will not preserve the geographic history of past speciation events (**Figure 12.7**).

The American Museum group argued that a vicariance hypothesis is easier to falsify than a dispersal hypothesis. Provided that barriers—once formed by a vicariance event—cannot be crossed, a vicariance hypothesis makes explicit predictions about the historical connections among areas. In particular, the formation of a new barrier is likely to isolate populations of many different kinds of organisms at approximately the same time. Therefore, a vicariance hypothesis would be supported if multiple taxa exhibit similar patterns of endemism, as Croizat emphasized. A vicariance hypothesis would be falsified if it was incompatible with data on the geological and climatic history of the Earth, if it was incompatible with information from the fossil record on past distributions of the lineage(s), or if the phylogenies of two or more codistributed taxa each generated a different taxon area cladogram.

In the case of our hypothetical example in Figure 12.6, the vicariance hypothesis would be falsified if clear geological evidence showed that the barriers between the areas were in existence *before* the taxa occurred there. The vicariance hypothesis also would be falsified by the occurrence of a fossil of taxon 2 or 3 in area A, or one of taxon 3 in area B or of 2 in C, or if a molecular phylogeny indicated that divergence dates between taxa were more recent than the ages of barrier formation. Perhaps the strongest test, however, could be conducted by constructing taxon area cladograms for other extant groups with endemic taxa in areas A, B, and C. If all the lineages were in existence

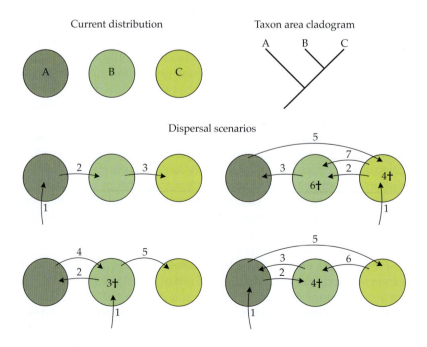

FIGURE 12.7 Hypothesized sequences of colonization and extinction events that are all consistent with a taxon area cladogram in which taxa in areas B and C are more closely related to each other than either is to a taxon in area A. The numbers show the sequence of events, arrows indicate colonizations, and crosses indicate extinctions. In the sequence at the lower left, for example, (1) the ancestor colonizes area B, (2) a population disperses to area A, (3) the original population in area B goes extinct, (4) dispersal from area A recolonizes area B, and (5) dispersal from area B colonizes area C. The scenarios on the right are even more complex but still consistent with the taxon area cladogram. The four scenarios shown here are only a subset of the possible sequences that are consistent with the cladogram, demonstrating how repeated colonization and extinction events can erase the evidence of earlier biogeographic history.

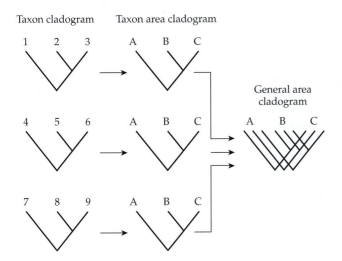

Taxon cladogram   Taxon area cladogram

General area
cladogram

FIGURE 12.8 The main steps in several historical biogeography approaches listed in Table 12.2 and discussed in text, including cladistic biogeography, phylogenetic biogeography II, and comparative phylogeography. Here, the general area cladogram reflects a congruent sequence of speciation events among three codistributed taxa across three areas of endemism, which is more likely to have resulted from a history of vicariance than from a series of unique jump dispersal events in each taxon. Perfect congruence across taxon area cladograms such as that shown here rarely is found in real data sets, even if vicariance did play a role in the biogeographic history of a biota (see Box 10.2). (After Crisci et al. 2003.)

at the same time, one would expect them to have responded similarly to the sequential fragmentation of their ranges. Therefore, they should exhibit similar or congruent taxon area cladograms, reflecting similar responses to the same sequence of barrier-forming geological and/or climatic events (**Figure 12.8**; **Box 12.1**).

Of course, there is some probability that taxon area cladograms will be congruent by chance alone. These probabilities can be determined, however, because they depend on the number of areas and taxa being considered. As the number of areas—and in particular, the number of independent taxonomic groups included in the analysis—increases, the probability of observing congruent cladograms simply by chance diminishes rapidly. Calculating these probabilities may not be as simple as it may seem, but the long-proposed possibility of testing the degree of congruence of taxon area cladograms in a rigorous statistical framework (Simberloff et al. 1981a,b) has been implemented in several of the more recent methods (e.g., Page 2003).

Beyond Vicariance Biogeography and Simple Vicariance

Fundamental questions and issues in modern historical biogeography

The vicariance biogeography advocated by the American Museum group represented an important advance over Croizat's panbiogeography because it demanded that historical biogeography be conducted within a rigorous hypothesis-testing framework, but it had some of the same, and some other unique, weaknesses. In their quest for analytical rigor, its supporters created some highly restrictive and simplifying rules or assumptions that reduced its relevance in a world full of complexity:

1. Long-distance dispersal could not be incorporated into a strict cladistic/ vicariant framework, and it was assumed to be sufficiently infrequent to overwhelm the signal of vicariance in a general area cladogram. However, we presented many examples of long-distance dispersal in Chapter 6, and we will review later in this chapter the emerging evidence for its profound influence throughout the historical assembly of modern biotas.

2. The advocates also assumed that geographic isolation and speciation arising through vicariance is preserved in the spatial configurations of

BOX 12.1 *Defining and Delineating Areas of Endemism*

◼◼▮ In Chapter 10, we discussed the concept of endemism in detail and its importance in biogeography. In historical biogeography, an area of endemism is generally considered the fundamental unit of analysis in cladistic-based approaches. Several decades of historical biogeography have been developed upon the premise that "the most elementary questions of historical biogeography concern areas of endemism and their relationships" (Nelson and Platnick 1981). Clearly, the importance of understanding relationships among areas, based on the taxa that occupied them, was associated with the idea that vicariance, followed by "allopatric speciation mode I" (see Figure 7.12), would produce congruent relationships across co-distributed taxa (see Figure 12.8).

Delimiting areas of endemism would seem to be an easy thing to do. After all, at the simplest level, they merely represent geographic areas where two or more endemic taxa share overlapping, or congruent, distributions. But we know that it is rare that the distributions of two or more taxa overlap exactly, except in cases where distributional limits are set by very discrete abiotic boundaries (e.g., lakes or islands), and biogeographers still are debating how to define and delineate them. How much or how little overlap in ranges, or sympatry, should we accept in order to delimit an area of endemism? Or should some criterion other than sympatry be applied? Recently proposed definitions emphasize one of three criteria to define an area of endemism: (1) degree of distributional overlap, or sympatry; (2) barriers between separate areas resulting from vicariance; and (3) as an operational extension of criterion 2, phylogenetic congruence between co-distributed taxa and their sister taxa in the area on the other side of the barrier. For example:

Platnick's definition "At the minimum, it would seem that an area of endemism can be defined by the congruent distributional limits of two or more species. Obviously 'congruent' in this context does not demand complete agreement on those limits, at all possible scales of mapping, but *relatively extensive sympatry* [italics added] at some scale must surely be the fundamental requirement" (Platnick 1991; see also Morrone and Crisci 1995; Linder 2001).

Hausdorf's definition "Areas of endemism can be defined as areas *delimited by barriers* [italics added], the appearance of which entails the formation of species restricted by these barriers" (Hausdorf 2002).

Harold and Mooi's definition An area of endemism is "a geographic region comprising the distribution of *two or more taxa that exhibit a phylogenetic and distributional congruence* [italics added] and having their respective relatives occurring in other such-defined regions" (Harold and Mooi 1994).

So, the definition of an area of endemism can range from requiring extensive sympatry (Platnick's definition) to little or none, with the main criterion being derived from a vicariance model of barrier formation and subsequent speciation (Hausdorf's and Harold and Mooi's definitions). The approaches to actually delineating areas of endemism are as diverse as are the definitions, ranging from the strongly geopolitical (e.g., "historically persistent Gondwanan landmasses according to paleogeographic reconstructions" [Sanmartín and Ronquist 2004]; see figure below) to character-based quadrat approaches to delineate areas of endemism based on the distributions of taxa within a region (Morrone 1994; Szumik and Goloboff 2004).

In the Gondwanan example, the approach to delimiting areas would be an example of using Hausdorf's definition, because the appearance of barriers following the fragmentation of Gondwana into separate landmasses is a more important criterion than the "extensive sympatry" of any taxa at smaller scales within each landmass. The areas delimited based on this criterion are the following:

- Africa, south of the Sahara
- Madagascar, including several Indian Ocean islands
- India, including Nepal, Tibet, and Sri Lanka
- Australia and Tasmania
- New Zealand, including subantarctic islands on the same continental block
- New Caledonia
- New Guinea, including the Solomon and New Hebrides islands
- Southern South America
- Northern South America

These areas of endemism are informative to historical biogeographers because they have arisen from a well-understood sequence of historical fragmentation of Gondwana landmasses (see area cladogram, at left), which forms a basis for addressing the relative importance of vicariance and dispersal in the biogeographic history of these regions.

In many cases, however, the physical discreteness between areas is not so clear-cut, and so other methods need to be employed to delineate areas of endemism. A number of approaches and methods have been proposed

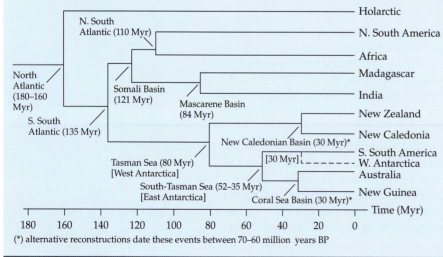

(*) alternative reconstructions date these events between 70–60 million years BP

BOX 12.1 *(continued)*

(e.g., see Morrone 1994; Linder 2001), including some that employ sophisticated statistical or simulation methods (Hausdorf and Hennig 2003; Mast and Nyffeler 2003; Szumik and Goloboff 2004).

Example: Morrone's (1994) Quadrat-PAE approach to delimiting areas of endemism This method relies on the use of Parsimony Analysis of Endemicity (PAE) (Rosen 1988) to delimit areas of endemism. There are five steps in the analysis, explained in greater detail by Morrone (1994) and Crisci et al. (2003):

1. Lay a grid of quadrats across the study area. The actual sizes of the grids are determined by the investigator and depend on the scale of desired resolution.

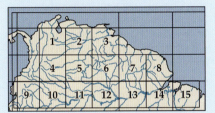

2. Determine the geographic distributions of taxa across the study area. Any number of unrelated taxa, at any taxonomic hierarchy, can be used, with the constraint that the members of each taxon are a monophyletic group.

3. Construct a data matrix summarizing distributions of taxa-by-quadrats,

assigning to a quadrat a 1 if a taxon is present in it and a 0 if a taxon is absent. Construct a hypothetical quadrat with all taxa absent (coded as 0 in all quadrats) as an outgroup.

Quadrats	Species 1	2	3	n
1	1	0	0	-	1
2	1	0	0	-	1
3	1	1	0	-	1
-	-	-	-	-	-
15	0	0	0	-	0

4. Perform a maximum-parsimony cladistic analysis on the matrix and construct a tree.

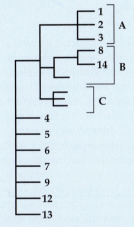

5. Select areas of endemism based on results of the cladistic analysis. In

this case, three quadrats (1 + 2 + 3) are supported by several taxa as a monophyletic clade and are selected to represent area of endemism A. The same reasoning leads to recognition of areas B and C. This analysis results in the delineation of three areas of endemism.

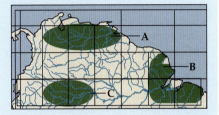

Note that seven of the original quadrats were not grouped into any of the areas; this may happen for several reasons. First, quadrats may not contain any of the species in the data set—these regions may represent current dispersal barriers between areas of endemism. Second, a quadrat may contain only a single endemic species found nowhere else and therefore not share any synapomorphies with other quadrats. Finally, one or more quadrats may have a mixture of some species that otherwise occur in one area of endemism and other species that occur in another area—as may happen with dispersal out of an area following the erosion of a previous dispersal barrier. ∎

current geographic ranges. But note the cautionary message we must take from the magnitude of relatively recent changes in distributions described in Chapters 9 and 16, and later in this chapter.

3. Finally, the advocates argued that dispersal is nearly always a single-lineage, idiosyncratic process and, therefore, not amenable to an analytical and hypothesis-testing approach. But recent advances in paleobiogeography and phylogeography have shown how the process of **geo-dispersal** can also produce biogeographic congruence, and thus it is testable in a cladistic framework (Lieberman 2004).

The vicariance biogeography of the 1970s and early 1980s was renamed **cladistic biogeography** in the mid-1980s, and is thoroughly reviewed by Chris Humphries and Lynn Parenti in their 1999 book. With the growth of power in computers in the 1980s, a quantitative implementation of the original method of **component analysis** introduced by Nelson and Platnick (1981) was incorporated by Rod Page (1989) into his software, COMPONENT 1.5. Efforts to develop better approaches to resolving the general vicariance structure (see Figure 12.8) shared by two or more co-distributed lineages with widespread taxa, redundant distributions, or missing areas (**Box 12.2**) continued into the 1990s with Nelson and Ladiges's (1992, 1995) development of a method called **three area (item) statement analysis** (TAS or TASS), and then one called **paralogy-free subtrees** (Nelson and Ladiges 1996), also implemented using TASS software. These and other methods are addressed thoroughly, with examples, by Crisci et al. (2003).

BOX 12.2 *Processes That Reduce the Generality of the General Area Cladogram*

The main goal of cladistic biogeography, and one of the goals of several other approaches in Table 12.2, including phylogenetic biogeography II, is to derive a general area cladogram that summarizes congruent area relationships among a set of co-distributed lineages. In a world where only vicariance produced disjunct distributions, where all speciation resulted from vicariance, and where there was no extinction, general area cladograms would be simple to interpret—the taxon area cladograms of all co-distributed taxa would be perfectly congruent, as shown in Figure 12.8. We know, however, that this idealistic world does not actually exist, and instead, species sometimes do not speciate when a vicariant event leads to allopatric speciation in other co-distributed lineages; that organisms do sometimes disperse across barriers and establish populations on the other side, and may then speciate; and that species do go extinct. Adding to this complexity, areas of endemism are often difficult to delineate with precision (see Box 12.1) and can include lineages that have undergone episodes of sympatric or embedded allopatric speciation prior to a vicariant event. Biogeographers recognize three categories of patterns in taxon area cladograms that reduce their congruence with an underlying general area cladogram.

Suppose the general area cladogram summarized the vicariant history across a set of five areas, A–E, as shown below:

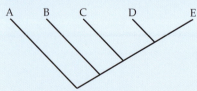

There are three ways in which a taxon cladogram can be less than perfectly congruent with this general area cladogram.

A taxon area cladogram with a *widespread species* that occurred in both areas A and B, but without a species endemic to either area alone, would look like this:

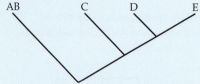

This pattern introduces a source of ambiguity into the interpretation of biogeographic history. One explanation might be that the widespread species did not speciate at the vicariant event that separated area 1 from ancestral area (B + C + D + E). Another might be that the widespread species did, in fact, speciate through vicariance but then dispersed to the other area without undergoing yet another round of speciation. Many biogeographers suggest that we cannot know from only this cladogram which of these explanations might be valid, and so we need to make one of three *assumptions*, called assumptions 0, 1, and 2, that each provide a set of rules about how to remove the ambiguities (see van Veller et al. 1999; Crisci et al. 2003) so that the general pattern of vicariance emerges. Other biogeographers prefer to analyze the widespread lineages using methods that allow them to infer ancestral distributions and dispersal histories in addition to the general vicariance pattern (see discussion in text).

Another source of ambiguity is introduced through the presence of *redundant distributions*, where an area occurs more than once on an area cladogram. The area cladogram below shows two cases of redundant distributions: area B appears once at a more basal, and again at a more derived, position; area D is repeated serially. For area B, we might explain the repetition by a dispersal of the taxon that inhabited

area E at the time of vicariance into area B followed by a subsequent speciation event. Alternatively, for area D, we might explain the repetition by invoking a sympatric speciation event in area D prior to vicariance between areas D and E, as shown above in the general area cladogram. Again, biogeographers differ about whether to remove or analyze redundant distributions. Note here that area B has two different relationships with the other areas. For example, with regard to area E, in one case two vicariant events separate areas B and E. In the other case, areas B and E share a sister area relationship. This is an example of a *reticulate history*, where one area has historical connections with more than one other area (also sometimes called *composite areas* by biogeographers).

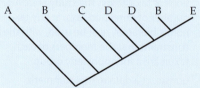

Finally, *missing areas* describe the situation where not every area on the general area cladogram is occupied by a member of a particular taxon, illustrated in the taxon area cladogram below where no member of the group occurs in area B.

This source of ambiguity may be explained as *primitively absent*—the ancestral taxon never got to the area and so was not a member of the biota in that area at the time of vicariance—or it may be explained as *extinction* sometime subsequent to vicariance.

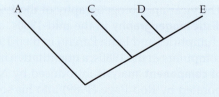

TABLE 12.2 *A Selection of the More Historically Important or Currently Popular Approaches and Methods in Historical Biogeography Updated from Crisci et al. 2003[a]*

Approaches	Goal	Selected methods	Original authors and general references
Descriptive biogeography	Biogeographic regions	Comparing species lists	Sclater 1858
	Biogeographic history		Hooker 1844–1860
Evolutionary biogeography I	Dispersal from center of origin	See Table 12.1	Matthew 1915; Cain 1944
Phylogenetic biogeography I	Dispersal or vicariant history	Phylogenetic systematics	Hennig 1966; Brundin 1966
Ancestral areas analysis	Area(s) of origin prior to dispersal		Bremer 1992, 1995
		Weighted ancestral areas analysis	Hausdorf 1998
Panbiogeography	Generalized tracks on a dynamic Earth	Track analysis	Croizat 1958
		Graph-based track analysis	Page 1987
Cladistic (vicariance)	Vicariance on a dynamic Earth		Nelson 1974
		Reduced area cladogram	Rosen 1978
		Component analysis	Nelson and Platnick 1981; Page 1989; Humphries and Parenti 1999
		Three-item (area) statement (TASS)	Nelson and Ladiges 1992
		Paralogy-free subtrees	Nelson and Ladiges 1996
Phylogenetic biogeography II	Vicariance, dispersal, geography of speciation	Brooks parsimony analysis (BPA)	Wiley 1981
		Primary and secondary BPA	Brooks et al. 2001; van Veller and Brooks 2001
		Phylogenetic analysis for comparing trees (PACT)	Wojcicki and Brooks 2005
Endemicity analysis	Natural distribution patterns of taxa among areas of endemism	Parsimony analysis of endemicity (PAE)	Craw 1988; B. Rosen 1988; Morrone 1994
Parsimony event-based methods	Benefit/cost modeling of events (e.g., dispersal, vicariance, extinction, sympatric speciation)		Ronquist and Nylin 1990
		Dispersal–vicariance analysis (DIVA)	Ronquist 1997
		Parsimony-based tree fitting	Page 1994; Ronquist 2002
Parametric event-based methods	Geographic range evolution	Disperal, extinction, and cladogenesis (DEC)	Ree et al. 2005; Ree and Smith 2008
		Bayesian DIVA	Nylander et al. 2008
Parametric island dispersal biogeography	Dispersal history across islands	Bayesian island biogeography	Sanmartín et al. 2008
Phylogeography	Geography of genealogical lineages		Avise et al. 1987; Avise 2000
		Phylogeny of gene trees	Various
		Nested-clade phylogeographic analysis (NCPA)	Templeton et al. 1995; Templeton 2004
		Coalescent-based approaches	Various; Knowles 2003
		Comparative phylogeography	Zink 1996; Bermingham and Mortiz 1998; Arbogast and Kenagy 2001
Evolutionary biogeography II	Stepwise approach to develop geobiotic scenarios	Various	Morrone 2009

[a]This list differs from Crisci et al. 2003 by adding newer approaches; distinguishing older and newer uses of "phylogenetic biogeography" as I and II, respectively, and separating the latter from cladistic biogeography; and distinguishing older and newer uses of "evolutionary biogeography" as I and II, respectively.

While much of the original vicariance biogeography has largely been replaced by an array of different approaches and methods (**Table 12.2**; Crisci et al. 2003), most of them retain three of its key features: (1) its emphasis on rigorous logic and hypothesis testing, (2) its reliance on robust phylogenetic hypotheses, and (3) the delineation of areas of endemism (see Box 12.1). Many of these approaches depart from vicariance biogeography largely in their willingness to address a wider range of hypotheses and mechanisms, including long-distance dispersal, geodispersal, and pseudocongruence. Dan Brooks and his colleagues (van Veller and Brooks 2001; van Veller et al. 2003; Brooks 2004) have argued that what traditionally was considered a single research program in cladistic biogeography actually included two different goals. One of these was what we described above as vicariance, or cladistic, biogeography, the goal being to identify the causal association between changes in the configuration and connectivity of areas of the Earth and the emergence of differentiated biotas as a result of that Earth history. In this vein, cladistic biogeographers treated all area relationships on a cladogram that resulted from some process other than vicariance as "noise" that needed to be removed from the analysis (see Box 12.2). They proposed to do so by creating a set of a priori assumptions—called Assumptions 1, 2 (Nelson and Platnick 1981), and 0 (Wiley 1988; Zandee and Roos 1987)—each based on the processes that were assumed to generate the noise, including dispersal, sympatric speciation, and extinction. Implementing one or more of these assumptions would allow investigators to "find" the vicariance structure embedded in a general area cladogram (see Figure 12.8) that also contained nonvicariance relationships. A detailed treatment of the rather complex literature attempting to explain, understand, and implement Assumptions 0, 1, and 2 would take more time than we want to spend on the topic here, but good places to explore this topic include van Veller et al. (1999) and Crisci et al. (2003).

The other goal embedded within cladistic biogeography, according to Brooks et al. (2004), was to try to understand the geography of speciation and historical assembly of biotas. Here, rather than trying to remove the "noise" of taxa incongruent with the vicariance-driven general area cladogram, events including post-speciation dispersal, peripheral isolates speciation (see Figure 7.14), sympatric speciation (see Figure 7.17), and extinction were considered parts of the whole of biotic histories and, therefore, worth trying to understand if historical biogeographers wanted to contribute significantly to evolutionary biology. These investigators proposed that this latter research program be decoupled from cladistic biogeography and called "phylogenetic biogeography." Notice that this is also the name of Brundin's earlier approach, described above, but the methods being currently developed under the rubric are quite different from those employed by Brundin. Therefore, to reduce confusion, we will call Brundin's approach "**phylogenetic biogeography I**" (see Table 12.2) and that of Brooks et al. "**phylogenetic biogeography II**."

Three additional examples demonstrate how far modern historical biogeography has drifted away from the strict vicariance biogeography of the American Museum group. First, Bremer's (1992: 440) ancestral area analysis is based on assumptions similar to those of dispersal biogeography, "namely that each group originated in a more or less small ancestral area, *the center of origin* [italics added], and that the distribution areas have subsequently expanded by dispersal" (Hausdorf 1998: 445). In other words, here is a quantitative, phylogenetically based set of methods that have explicitly resurrected the concept of center of origin–dispersal so loathed by the vicariance biogeographers! Of course, Bremer's method was based on a much more rigorous footing than Matthew, Simpson, and Darlington's approach—including cladistic procedures and delineation of areas of endemism. In a second example of how historical biogeography has diversified in the past decade, one of the so-called event-based methods (see Table 12.2) is actually called **dispersal–vicariance analysis** (**DIVA**), which has been the inspiration for development of a new generation of analyses that seek to estimate a history of *geographic range evolution* (Ree and Sanmartín 2009). Finally, many of the popular methods in phylogeography are designed to sort allopatric fragmentation from range expansion (i.e., *dispersal*) histories among intraspecific populations (Crandall and Templeton 1996).

Different approaches to the same question, or different questions?

Crisci et al. (2003), in their introduction to historical biogeography, point out that various distinct questions often have unknowingly been embedded within the seemingly single discipline of historical biogeography, leading to confusion in choosing the appropriate methods of analysis. They suggest that approaches and methods have developed along two separate lines. Several are designed to reconstruct the history of lineages across areas for a single lineage at a time, an approach they call **taxon biogeography**. The cladogram of Brundin's midges in Figure 12.5A provides a good example of the early use of this approach. Ancestral areas analysis and event-based methods that emphasize geographic range evolution are often used in this fashion. Recent methods here include (see Table 12.2) DIVA; parsimony-based tree fitting; and *d*ispersal, *e*xtinction, and *c*ladogenesis (DEC) analysis. Many studies using phylogeographic methods are concerned with a single taxon. The ultimate goal, of course, could certainly be to compare separate lineage histories with one another or with a geological area cladogram and assess the level of congruence across them for evidence of vicariance or dispersal histories.

Alternatively, an investigator doing **area biogeography** might wish to produce a general area cladogram (see Box 12.2) because it is the historical relationships among the areas and the biotas inhabiting those areas that are of interest to the biogeographers. Here the investigator might use a component analysis (see Table 12.2) to produce a **consensus area cladogram** depicting vicariant history; a **reconciled-tree** approach (Page 1994) estimating the number of vicariance, dispersal, and extinction events; or a **Brooks parsimony analysis** (**BPA**) (see below) or **phylogenetic analysis for comparing trees** (**PACT**) analysis to address the geography of speciation.

In summary, the questions in historical biogeography are extremely diverse—ranging from the reconstruction of an ancient Gondwanan biota, through colonization and diversification on oceanic island archipelagoes, to other questions that concern biotic responses to the influences of Late Pleistocene glaciation. As such, it perhaps is not surprising that the available approaches and methods are equally diverse (see Table 12.2; Crisci et al. 2003; Morrone 2009). Even so, Juan J. Morrone has attempted to reconcile this diversity into a research program with the ultimate goal of producing a **geobiotic scenario** (Evolutionary Biogeography II in Table 12.2). Morrone's evolutionary biogeography "integrates distributional, phylogenetic, molecular, and paleontological data in order to discover biogeographic patterns and assess the historical changes that have shaped them, following a stepwise approach." The common ground lies in treating each approach as just one part of a series of steps (**Figure 12.9**) that, for example, would employ panbiogeography and a method to postulate areas of endemism to identify **biotic components**, which "are sets of spatiotemporally integrated taxa that coexist in given areas" (Morrone 2009: 3). Cladistic biogeography is then used to classify the nested pattern of biotic components into biogeographic regions (e.g., realm, region, dominion, province, and district). Meanwhile, molecular

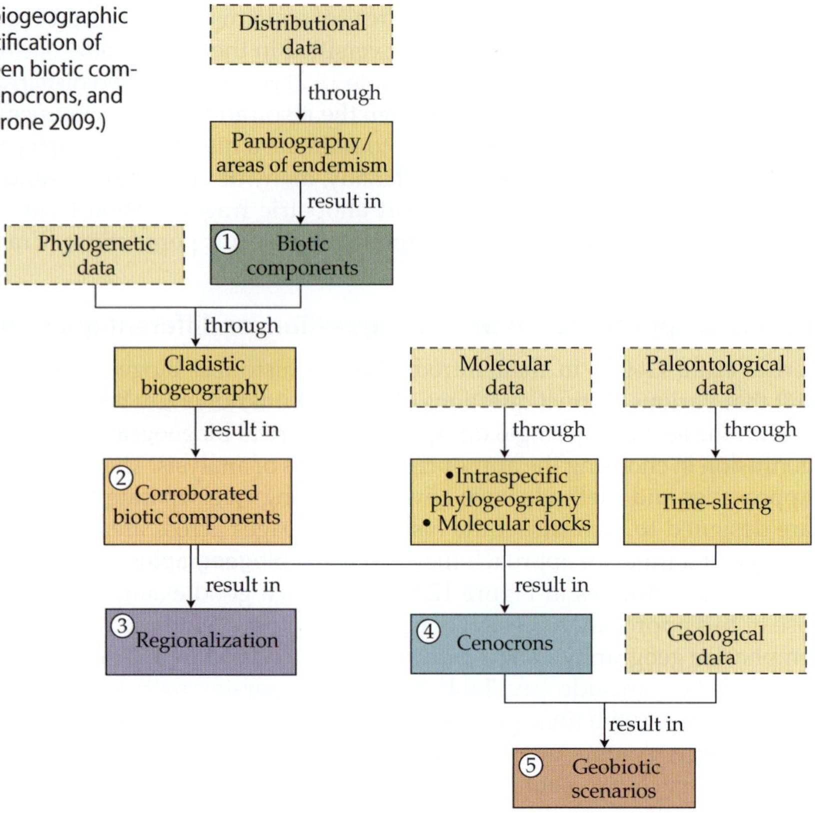

FIGURE 12.9 Five steps within an evolutionary biogeographic analysis as envisioned by Morrone (2009): (1) identification of biotic components, (2) testing relationships between biotic components, (3) regionalization, (4) identification of cenocrons, and (5) construction of a geobiotic scenario. (After Morrone 2009.)

phylogenetics, phylogeography, and paleontology are employed to identify **cenocrons**, which are "sets of taxa that share the same biogeographic history, constituting identifiable subsets within a biotic component by their common biotic origin and evolutionary history from a diachronic perspective" (Morrone 2009: 18). The combination of information derived from each of these steps leads to the ultimate goal of generation of a geobiotic scenario. Clearly, modern historical biogeography has become a much broader endeavor than was advocated by the earlier vicariance biogeographers, although not without a certain level of remaining discord over its appropriate role in evolutionary biology (e.g., Humphries 2000; Ebach and Humphries 2002).

How complex, really, are lineage and biotic histories?

If the relationships among biotas in different areas of the world were mostly the product of either a simple vicariance or predictable dispersal history (e.g., via the progression rule), most of the many available methods would find the same general area cladogram and, given accurate phylogenies, it would not be particularly difficult to reconstruct the history of biotas. Let's look in some detail at an example of a system with a very simple geological history—the Hawaiian Archipelago.

Warren Wagner and Vicki Funk (1995) edited an excellent book that brought together much of the phylogenetic and biogeographic research on the biota of the Hawaiian Islands at that time. We now have available a number of more recent molecular phylogenetic and phylogeographic studies of this biota as well (e.g., see Figure 11.23), allowing for an updated synthesis (Cowie and Holland 2008). As we saw in Chapter 8, the Hawaiian Archipelago has a dramatically simple geological history resulting from the Pacific Plate's drifting over a hotspot now located at the southeastern end of the island chain, currently beneath the island of Hawaii and another volcano

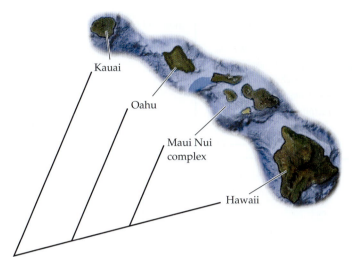

FIGURE 12.10 The area cladogram predicted for the modern large islands of Hawaii based on a simple progression rule that takes into account the age of formation of each island. The Maui Nui complex has been a single island for most of its existence, separating into different islands only during the higher sea levels of interglacial periods. This is the pattern of phylogenetic relationships that would be expected if a lineage colonized each island in turn as it was formed by the emergence of a new set of volcanoes from the sea. (After Cowie and Holland 2008.)

(Loihi Seamount) to the southeast, which is growing but still submerged below the ocean (see Figure 8.29). The formation of the islands began 75 to 80 million years ago, but there may have been times since then when there was little or no emergent land. The oldest of the present major islands is Kauai, the northwestern-most island, which was formed about 5.1 million years ago. The ages of the islands decrease down the chain to the southeast. The islands of Molokai, Lanai, Kahoolawe, and Maui have been united into a single "Maui Nui complex" more often than not, splitting apart first about 600,000 years ago and undergoing cycles of coalesence and isolation with subsequent glacial-interglacial cycles (Price and Elliott-Fisk 2004).

Given this simple geological history, we can develop a straightforward geological area cladogram (**Figure 12.10**) to use as a hypothesis of taxon relationships under the following assumptions: that a taxon originally colonized either Kauai, the oldest present island, or an older island to the northwest of Kauai that is now submerged, and that, as newer islands were formed, each was colonized from an ancestral population on the adjacent and older island up the chain. This prediction should look very much like the progression rule we discussed earlier in this chapter (and shown in Figure 7.14), and is one process that would lead to a high level of congruence across the co-distributed lineages analyzed by Funk and Wagner (1995), and Cowie and Holland (2008). Deviations from this predicted pattern could, however, result from (1) multiple colonizations of the islands from other regions, (2) back-colonization from younger to already inhabited and older islands, (3) colonization of nonadjacent islands (which could also be interpreted as extinctions of populations on islands in between), (4) in situ speciation on a single island (which might occur via sympatric or micro-allopatric modes), or (5) more recent colonization of the archipelago itself. So what do the results of the phylogenetic and biogeographic analyses show? Cowie and Holland (2008) summarized our current state of knowledge as follows:

1. It has been demonstrated that many members of the Hawaiian fauna form a monophyletic assemblage that arose from a single ancestral colonization of the islands, including bees in the subgenus *Nesoprosopis*, the spider genus *Havaika*, the Hawaiian drosophilid flies (**Figure 12.11**), and the Hawaiian honeycreepers (Drepanidinae). However, other groups appear to have colonized the islands more than once, including Hawaiian tetragnathid spiders, spiders in the genus *Theridion*, and the land snail family Succineidae (**Figure 12.12**).

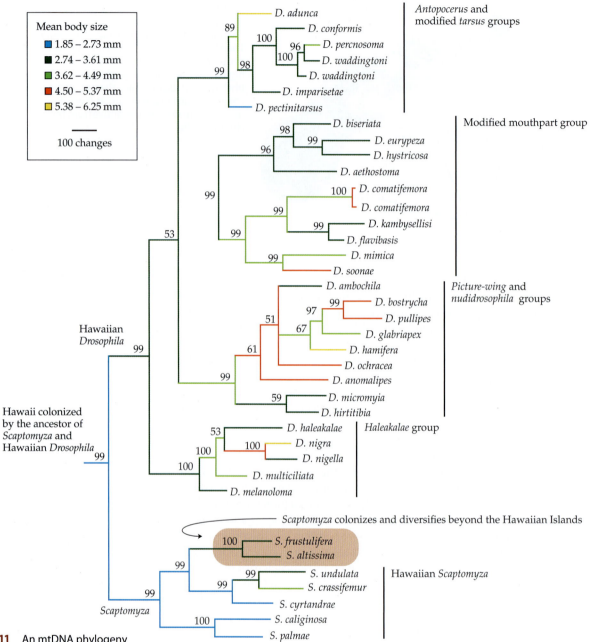

FIGURE 12.11 An mtDNA phylogeny for the Hawaiian Drosophilidae. A larger phylogeny incorporating many more taxa in Drosophilidae from around the world strongly supports a single colonization of Hawaii by an ancestor to two extant clades: Hawaiian *Drosophila* and the genus *Scaptomyza*. A surprising result of this analysis is that *Scaptomyza* colonized a number of distant islands and continents after originating in Hawaii—relatively small body size may have been a contributing factor in their capacity to disperse long distances. (After O'Grady and DeSalle 2008.)

2. The simple progression rule prediction generated from the geological area cladogram is partially supported as a general rule, although even when this pattern can be detected, such as in two studies of endemic snail lineages (**Figure 12.13**; see also Figure 12.12), there are frequent episodes of back-colonization to older islands (e.g., bees in the genus *Hylaeus*) and in situ speciation on a single island (e.g., spiders in the "spiny leg" clade of *Tetragnatha* and the spider genus *Orsonwelles* [**Figure 12.14** and see Figure 11.23]). It is important to note, however, that although speciation occurred *within* an island and not just among islands, geographic isolation still may have played an important role in the differentiation of the populations. All of Hawaii's large islands have a great deal of topographic relief and habitat heterogeneity, including mountain ranges, large rivers, and other land features that may serve

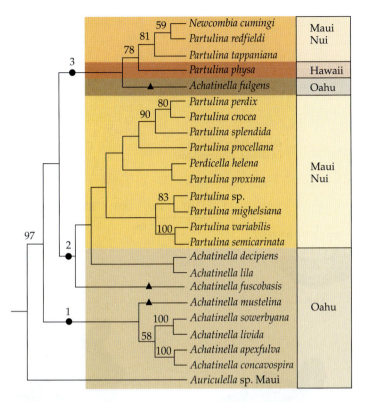

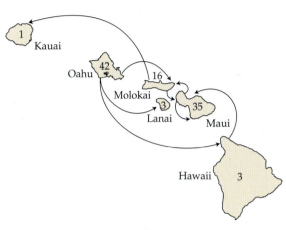

FIGURE 12.12 An mtDNA phylogeny for the Hawaiian achatinelline tree snails (number on each island indicates the diversity of endemic species). A single colonization of the Hawaiian Islands, most likely the island of Oahu, is supported by the phylogeny. Subsequent colonizations of the younger islands of the Maui Nui complex and Hawaii from Oahu are consistent with the progression rule (see Figure 12.10), with a back colonization of Kauai postulated, based on a fossil specimen from that island. (After Cowie and Holland 2008.)

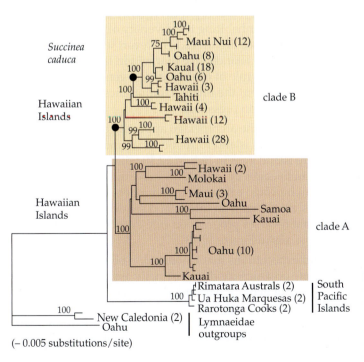

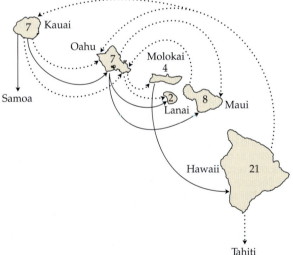

FIGURE 12.13 A molecular phylogeny for the Hawaiian succineid amber snails (number on each island indicates the diversity of endemic species). The phylogeny of clade A (solid lines on map) postulates an original colonization of Kauai, the oldest of the "high islands," and subsequent colonizations of progressively younger islands is consistent with the progression rule (see Figure 12.10). However, clade B (dotted lines) shows a more complex pattern, beginning with an original colonization of Hawaii, the youngest of these islands, and with two separate colonizations of South Pacific islands from Hawaii. (After Cowie and Holland 2008.)

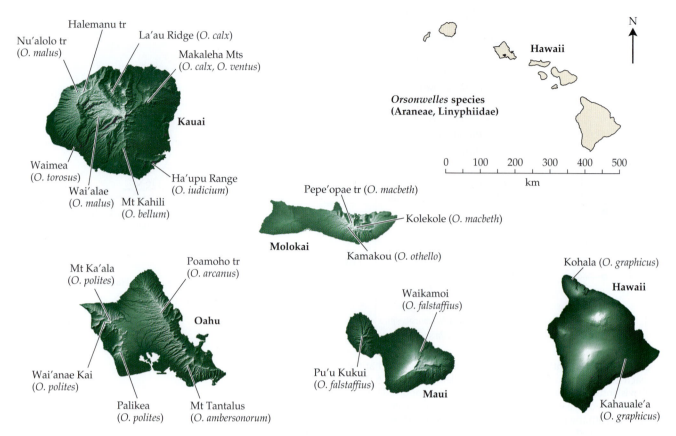

FIGURE 12.14 The geographic distribution of species of *Orsonwelles* spiders across the Hawaiian Islands. Although phylogenetic reconstructions (see Figure 11.23) suggest that a number of speciation events occurred within a single island, there is only one case of a completely sympatric distribution (*O. calx* and *O. ventus* in the Makaleha Mountains of Kauai). Even here, the phylogeny is indecisive about whether this is a result of a sympatric mode of speciation (followed by dispersal and divergence between *O. ventus* and the ancestor of *O. bellum* + *O. iudicium*) or of dispersal in the opposite direction resulting in secondary sympatry. Because of the allopatric distributions of all other species on different mountains, a micro-allopatric speciation mode is generally more likely than sympatric speciation. (After Hormiga et al. 2003.)

as barriers to dispersal. For organisms that disperse as poorly as some insular plants and invertebrates, this topographic heterogeneity promotes **micro-allopatric** speciation and rapid divergence and adaptive radiation among populations inhabiting the diversity of environments found on these large islands (see Figure 12.14). Nevertheless, the high frequency of speciation within islands, like that within lakes, raises other important questions about the role of ecological and genetic processes in speciation—and especially about the relative importance of geographic isolation and divergent selection pressures.

3. A number of colonizations of the archipelago appear to be far too young to adhere to the time frames of island formation that underly the progression rule (compare island ages in Figure 12.10 with molecular-based ages of some Hawaiian lineages in **Table 12.3**)—obviously, more recent colonizations of the archipelago can occur on younger islands just as easily as on older islands, resulting in an interisland sequence of colonization and speciation that reverses progression rule expectations (see clade B of succineid amber snails in Figure 12.13).

4. Contrary to long-held expectations that colonization between continental landmasses and remote oceanic islands is distinctly asymmetrical (see Chapter 14 for examples of evolutionary trends on islands that reduce the colonization abilities of species), and given the remoteness of the Hawaiian Islands, little colonization *from* them back to other landmasses, including continents, might have been anticipated. It is therefore surprising that several molecular phylogenies show clear evidence of a long history of evolutionary radiation on the islands, followed by one or more subsequent long-distance colonization events. The most interesting of

TABLE 12.3 *Estimated Ages of the Most Recent Common Ancestors (MRCA) of Some Hawaiian Lineages[a]*

Lineage	Type of organism	Number of species	Age (Ma)	Source
Hawaiian fruit flies (Drosophilidae)	Insect	ca 1000	26	Russo et al. 1995
Hawaiian lobelioids (Campanulaceae)	Plant	125	15	Givnish et al. 1996
Megalagrion damselflies (Coenagrionidae)	Insect	23	9.6	Jordan et al. 2003
Silversword alliance (Asteraceae)	Plant	28	5.1	Baldwin and Sanderson 1998
Laysan duck, *Anas laysanensis* (Anatidae)	Bird	1	<5	Fleischer and McIntosh 2001
Hawaiian crows, *Corvus hawaiiensis* + other spp.? (Corvidae)	Bird	1 + ?	<4.2	Fleischer and McIntosh 2001
Hawaiian honeycreepers, Drepanidinae (Fringillidae)	Bird	ca 50	4–5	Fleischer et al. 1998
Viola spp. (Violaceae)	Plant	6	3.7	Ballard and Sytsma 2000
Flightless Anseriformes, moa-nalos (Anatidae)	Bird	4	<3.6	Sorenson et al. 1999
Hawaiian thrushes, *Myadestes* spp. (Muscicapidae)	Bird	5	<3.35	Fleischer and McIntosh 2001
Kokia spp. (Malvaceae)	Plant	4	<3	Seelanan et al. 1997
Flightless rails, *Porzana sandwicensis* + other spp.? (Rallidae)	Bird	1 + ?	<2.95	Fleischer and McIntosh 2001
Geranium spp. (Geraniaceae)	Plant	6	2	Funk and Wagner 1995
Hesperomannia spp. (Asteraceae)	Plant	4	1.81–4.91	Kim et al. 1998
Flightless ibises, *Apteribis* spp. (Plataleidae)	Bird	2	<1.6	Fleischer and McIntosh 2001
Hawaiian duck, *Anas wyvilliana* (Anatidae)	Bird	1	<1.5	Fleischer and McIntosh 2001
Flightless rails, *Porzana palmeri* + other spp.? (Rallidae)	Bird	1 + ?	<1.05	Fleischer and McIntosh 2001
Hawaiian geese, *Branta* spp. (Anatidae)	Bird	3	<1	Fleischer and McIntosh 2001
Hawaiian black-necked stilt, *Himantopus mexicanus knudsenii* (Recurvirostridae)	Bird	1	<0.75	Fleischer and McIntosh 2001
Hawaiian hawk, *Buteo solitarius* (Accipitridae)	Bird	1	<0.7	Fleischer and McIntosh 2001
Tetramolopium spp. (Asteraceae)	Plant	11	0.6–0.7	Lowrey 1995
Metrosideros spp. (Myrtaceae)	Plant	5	0.5–1.0	Wright et al. 2001

[a]Refer to Price and Clague 2002 for methods used to calculate divergence times. Note that several lineages are older than the oldest present large island (Kauai, 5.1 million years old), suggesting an initial colonization of an older, now submerged island. The MRCA of other lineages is considerably younger, implying relatively recent colonization events.

these events documented thus far occurred within the Hawaiian drosophilid flies. *Scaptomyza* represents one of two highly successful clades that diversified on the Hawaiian Islands (see Figure 12.11), but many species also inhabit several continents, as well as oceanic islands as distant from Hawaii as Tristan da Cunha in the South Atlantic. O'Grady and DeSalle (2008) provided clear evidence that *Scaptomyza* originated on the Hawaiian Islands and that the widespread distribution of this genus has resulted from one or more colonizations from Hawaii (see Figure 12.11). Indeed, two additional instances of colonization from Hawaii back to other parts of the world appear to have occurred in the succineid amber snails (see Figure 12.13), one to Tahiti and another to Samoa.

The perils of ignoring time

We have seen in the Hawaiian example that, even in a region with a comparatively simple geological history, phylogenies for co-distributed lineages can conform in part to predictions that arise from the geological area cladogram, but can also generate a diversity of incongruent taxon area cladograms, owing to a complex history of dispersal, vicariance, extinction, and speciation within and among areas. Yet, even if we found perfect congruence across a set of taxon area cladograms, leading us to the provisional conclusion that all lineages shared a single history of simple vicariance, we might still have arrived at the wrong conclusion. Comparisons of hypothetical taxon area cladograms such as those in **Figure 12.15** may appear to be relatively straightforward, but that is true only for those clades that diversified at roughly the same times (i.e., the left side of Figure 12.15). Either the divergence events are both topologically and temporally congruent (clades 1 and 2 in the upper left box), indicating that the two lineages share a single biogeographic history; or they are geographically incongruent (clades 5 and 6 in the lower left box), indicating that they do not share the same history. However, the patterns shown in the boxes on the right in this figure (i.e., those for clades that diversified at different times) are likely to lead us to the wrong conclusions, unless of course our approach explicitly incorporates this asynchrony in clade diversification. Note that clades 3 and 4 in the upper right comparison appear to exhibit perfect topological congruence, yet this is an artifact of not incorporating differences in divergence times for these lineages. Without accurate information on temporal coincidence of different clades, such patterns of cryptic biogeographic incongruence (called **pseudo-congruence**; Cunningham and Collins 1994) may generate erroneous conclusions about colonization history and evolution of these lineages (Hunn and Upchurch 2001). On the other hand, as more and more molecular phylogenies incorporate robust estimates of divergence times (see the discussion on molecular clocks in Chapter 11), we may discover that pseudo-congruent patterns are relatively common in parts of the world that have experienced temporally layered cycles of formation and erosion of dispersal barriers—first isolating, then allowing movement of, a succession of lineages between the same set of

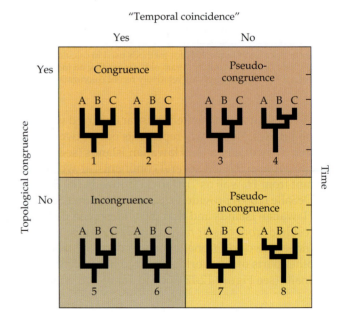

FIGURE 12.15 Four hypothetical sets of taxon area cladograms for two lineages distributed across three areas (A, B, and C) in each comparison. See text for discussion. (After Donoghue and Moore 2003.)

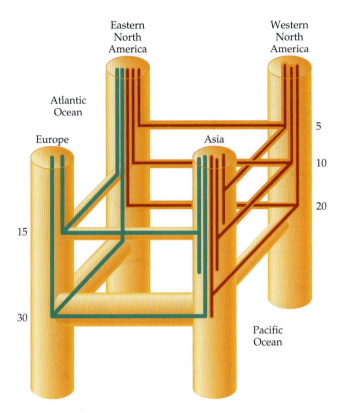

FIGURE 12.16 A depiction of area cladograms that summarizes historical tracks inferred from phylogenetic analyses of seven plant lineages distributed across four Northern Hemisphere areas of endemism. Each track traces one of two postulated intercontinental dispersal routes: (1) a Beringian route across the Pacific Ocean, with evidence presented here of dispersal routes forming at three different time frames (numbers are in millions of years) or (2) a North Atlantic route, with two different time frames for dispersal shown here (note also that these times are more recent than the postulated Eocene deGeer and Thulean land bridge routes across the North Atlantic as shown in Figure 8.22I, implying episodes of transoceanic dispersal). The temporal component of this complex biogeographic history was inferred by estimating divergence times from molecular phylogenies, and it demonstrates pseudocongruence embedded within topologically congruent sets of area cladograms. (After Donoghue and Moore 2003.)

areas (**Figure 12.16**; Donoghue and Moore 2003). Thus, even when the phylogenies of two codistributed lineages suggest a shared history of area relationships (see Figure 12.8), we should use caution in concluding that this is the true history unless we have estimates of the time frame of isolation and divergence.

Reticulate area relationships: The end result of complex histories

As we continue to develop new tools to reconstruct the history of biotas—including molecular-based phylogenies that can provide estimates of the timing of historical vicariance and dispersal events, improved reconstructions of geological histories, and newer and more sophisticated analytical approaches (e.g., the parametric methods listed in Table 12.2)—we are approaching an empirically broad-based, new synthesis that holds that biotic histories are often complex rather than simple. It is becoming increasingly more clear that two or more co-distributed lineages often have sister taxa in an array of different areas (indicating a variety of dispersal as well as vicariant histories in the assemblage of that biota), or even in the same area (suggesting sympatric or embedded allopatric speciation within an area).

Consider an example that addresses the prospect for reticulation, given changing land configurations across separate geographic areas through a series of time slices. In **Figure 12.17**, the Mediterranean region is reconstructed across four time slices beginning in the Middle-Late Eocene. Although simplified, this example illustrates shifting dispersal routes and possible vicariance events as plate tectonics drives a change in geographic connectivity between four regions. The result of such a history is that the biota of any one of these four regions could be expected to share a complex set of historical affinities with surrounding regions, and so a single taxon area cladogram, with each area represented only once, could not capture the complete history of area relationships.

FIGURE 12.17 Changing probabilities for dispersal and vicariance during four time intervals based on a simplified reconstruction of Cenozoic tectonic development of the Mediterranean region. (A) Four separate areas are identified: northwestern (NWM), northeastern (NEM), southeastern (SEM), and southwestern (SWM). Arrows on maps represent likely dispersal routes (solid lines more likely; dashed lines are less likely). (B) Dispersal parameters between regions during each time interval are summarized as numbered estimates of probability (d = most likely dispersal route; any fraction of d is less likely). For example, if we focus on area SEM, note that it is postulated to have a stronger biotic affinity with area SWM prior to the Middle Miocene, when it switches to having stronger affinities with area NEM as the Arabian and East European platforms become sutured together and the opening of the Red Sea increasingly isolates the Arabian and North African Platforms. A taxon area cladogram would therefore reconstruct a reticulate set of affinities between SEM and areas SWM and NEM. (After Ree and Sanmartín 2009.)

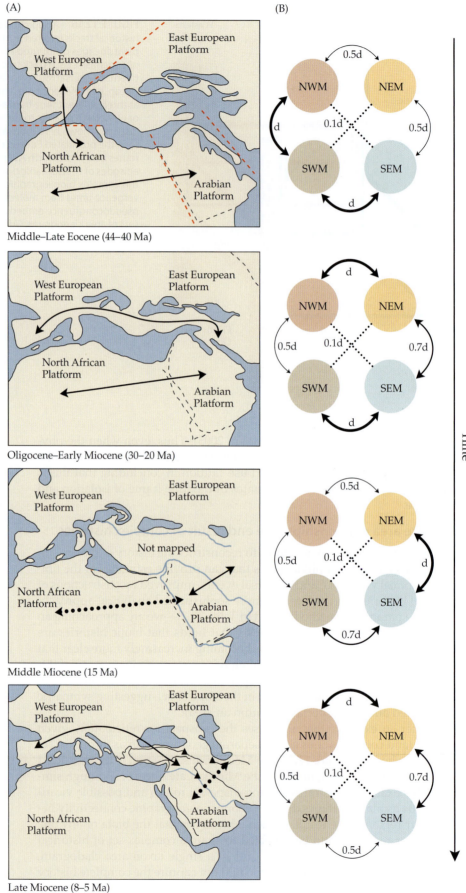

(A)

Middle–Late Eocene (44–40 Ma)

Oligocene–Early Miocene (30–20 Ma)

Middle Miocene (15 Ma)

Late Miocene (8–5 Ma)

(B)

Time

Next, we explore in greater detail two distinctly different methods developed to reconstruct complex biogeographic histories—a phylogenetic biogeography II method and an event-based method (see Table 12.2)—and then return to where we left off in Chapter 11 in our discussion of phylogeography.

Three approaches to reconstructing area and lineage histories

Within the past decade, a number of creative approaches have been developed—several mentioned above and listed in Table 12.2—that begin to resolve conceptual and methodological issues that became apparent in response to the strict form of vicariance biogeography advocated by the American Museum group. Our intent is not to provide an exhaustive review of these approaches and methods (for that, see Crisci et al. 2003; Morrone 2009; Ree and Sanmartín 2009) but, rather, to compare two innovative but very different approaches. Indeed, new methods continue to appear so frequently that one of those considered in detail here (Brooks parsimony analysis, BPA) is already being replaced by a conceptually related method called phylogenetic analysis for comparing trees (PACT; Wojcicki and Brooks 2004, 2005). Even so, the logic behind the approach is unchanged and can be explored by looking at the BPA implementation.

Brooks parsimony analysis shares a goal with the methods that are considered "area biogeography" approaches—of producing a general area cladogram to describe the vicariant or geo-dispersal "backbone" in the biotic relationships among areas of endemism (see Lieberman 2003, 2004). BPA parts ways with several of these methods, however, in providing insights into the unique, lineage-specific patterns that arise from processes such as post-vicariance dispersal among areas, peripheral isolates speciation, sympatric speciation, and extinction (van Veller and Brooks 2001; Brooks and McLennan 2002; van Veller et al. 2003; Brooks 2004). It does so by implementing a two-step procedure: primary BPA and secondary BPA (**Box 12.3**). The end result is a general area cladogram that contains, in addition to the general area cladogram produced through primary BPA analysis, some number of duplicated areas generated through secondary BPA that describe those area relationships that conflicted with the general area cladogram.

In contrast, the most widely used event-based method, dispersal–vicariance analysis (DIVA) departs from BPA and the cladistic biogeography methods by not starting with a general area cladogram (Ronquist 1997). Indeed, even in the worst case scenario of reticulated areas with no discernable hierarchical relationships (**Figure 12.18**), DIVA can be used to estimate episodes of vicariance and dispersal. Although taxon distributions among areas of endemism are still used, as are phylogenies, each taxon area

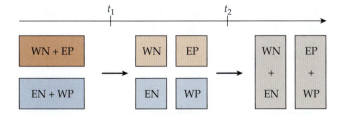

FIGURE 12.18 A "worst case scenario" of reticulate area relationships. Four Holarctic deciduous forest areas are recognized: WN = western Nearctic, EN = eastern Nearctic, WP = western Palearctic, EP = eastern Palearctic. Prior to time 1 (t_1), two ancestral areas connecting regions between current continents are recognized based on paleogeographic reconstructions. At t_1, each of these areas is fragmented by a vicariance event, producing four separate areas. Subsequently, at t_2, geological events remove dispersal barriers between the two areas in each of two pairs of areas, but now, areas are connected within, rather than between, continents. Ronquist (1997) argued that DIVA is more appropriate than one of the hierarchical approaches to reconstructing biogeographic histories in such cases. (After Ronquist 1997; originally described in Enghoff 1996.)

BOX 12.3 *Primary and Secondary Brooks Parsimony Analysis*

■■■ While numerous approaches are available for reconstructing biogeograpic histories (see Table 12.2 and discussion in text), Brooks parsimony analysis (BPA) is a method that allows us to illustrate clearly the principal kinds of histories that could be embedded within a general area cladogram, including vicariance, peripheral isolate speciation, postspeciation dispersal, and sympatric (i.e., within-area) speciation. BPA starts, as with any of the cladistic biogeographic methods, with a phylogeny for each lineage. We do not show the details of the phylogenetic analysis here and assume that it is an accurate depiction of relationships within each lineage. The next step is to summarize the distributions of each taxon within a set of areas of endemism.

Below, we show hypothetical phylogenies for three genera (*A*, *B*, *C*), each containing four species, and their distributions in areas *W*, *X*, *Y*, and *Z*. This is a highly simplified example, where each species occupies only one area, each area is represented only once on each phylogeny, and the topologies of the three taxon area cladograms are all perfectly congruent. As such, the general area cladogram summarizes all of the information contained in each separate taxon area cladogram, with no incongruence, a hypothesis of a simple vicariant history. The numbers on the taxon phylogenies represent "characters" coding the phylogenies for the *primary BPA*. Once each phylogeny is coded as such, they are all compiled into a single binary matrix (1 = character present; 0 = character absent; ? = taxon absent from area [does not apply in this example]):

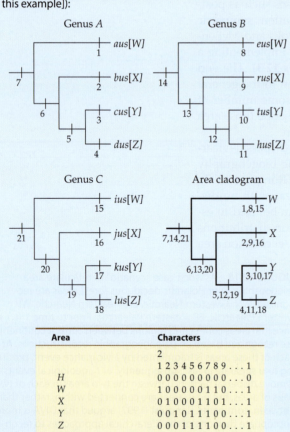

Area	Characters
	$\quad$ 2
	1 2 3 4 5 6 7 8 9 . . . 1
H	0 0 0 0 0 0 0 0 0 . . . 0
W	1 0 0 0 0 0 0 1 1 . . . 1
X	0 1 0 0 0 1 1 0 1 . . . 1
Y	0 0 1 0 1 1 1 0 0 . . . 1
Z	0 0 0 1 1 1 1 0 0 . . . 1

Source: After Bouchard and Brooks 2004.

This matrix includes an additional "hypothetical area" *H*, which includes all zeros and which is a methodological necessity in order to root the tree with an outgroup for phylogenetic analysis—the next step in the procedure. If all character-state changes for all taxon area cladograms conform to a single general area cladogram with no incongruence—that is, no *homoplasy* (see Chapter 11)—then the analysis is completed.

However, this will rarely be the case in real biotas with complex biogeographic histories. *Secondary BPA* was therefore developed as an approach to resolving homoplasy on the general area cladogram produced in the primary BPA. Suppose we add a fourth lineage, genus *D*, to the biogeographic analysis and that it matches closely the pattern of area relationships shown by all of the first three genera, but is not quite a perfect match. Perhaps *D. nus* crossed a barrier between areas *X* and *W* and speciated (*peripheral isolate speciation*; see taxon area cladogram below):

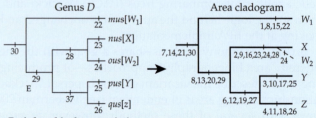

Peripheral isolate speciation

In this case, a new binary matrix could be constructed that treats area *W*, which now has a component that conflicts with and will produce homoplasy in the general area cladogram shown above, as *two separate areas*, W_1 and W_2. This is called the *area duplication convention*, which simply states that areas causing homoplasy on the general area cladogram are to be duplicated until the homoplasy is removed (explained more formally in Brooks and McLennan 2002; Brooks 2004). Parsimony analysis of this new matrix will produce the new general area cladogram shown above on the right. Note that it preserves the information suggesting a "vicariant backbone" to the historical relationships among areas, but now it also postulates a subsequent peripheral isolate speciation event between area *X* and duplicated area *W* (W_2).

As we discussed in the text, several other processes can lead to incongruence between any one of the taxon area cladograms and the general area cladogram. Below, we show the case of a *sympatric speciation* event in area *X*, resulting in the sister species of *D. nus*, which is *D. ous*. This event does not result in itself in incongruence with the general area cladogram, although note the addition of characters on the branch leading to *X*:

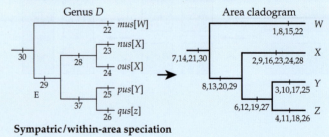

Sympatric/within-area speciation

BOX 12.3 *(continued)*

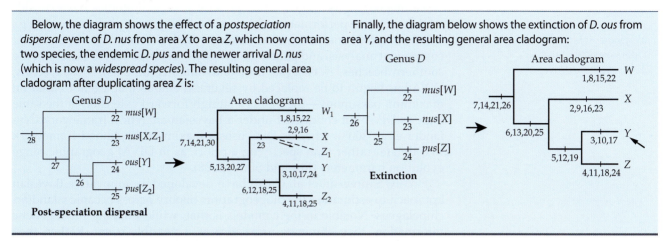

Below, the diagram shows the effect of a *postspeciation dispersal* event of *D. nus* from area *X* to area *Z*, which now contains two species, the endemic *D. pus* and the newer arrival *D. nus* (which is now a *widespread species*). The resulting general area cladogram after duplicating area *Z* is:

Genus *D* — Post-speciation dispersal

Area cladogram

Finally, the diagram below shows the extinction of *D. ous* from area *Y*, and the resulting general area cladogram:

Genus *D* — Extinction

Area cladogram

cladogram is analyzed separately to develop a "taxon biogeography." The idea here is to establish a "cost" for each of several kinds of biogeographic "events"—the more likely the event is considered to be, or considered to be shared across co-distributed taxa, the lower the cost assigned. The basic assumption used to build the cost matrix is that most speciation is allopatric and produced through vicariance. Thus, costs are usually assigned as follows: Two kinds of events—vicariance and sympatric (or embedded allopatric) speciation events—within a single area are assigned a cost of 0, whereas two other kinds of events—extinction and random dispersal—are assigned a cost of 1. The algorithm searches for the distribution of ancestral areas at each node on the tree that minimizes the overall cost of events. An example of a DIVA search for the ancestral area of a currently widespread and diverse taxon is shown in **Figure 12.19**. While DIVA analyzes each taxon area cladogram separately, results from each analysis can be evalu-

FIGURE 12.19 An example of a dispersal–vicariance analysis used to infer the ancestral area for lime swallowtails, *Papilio demoleus* group, including the species *demodocus*, *demoleus*, *erithonioides* (*erl*), *grosesmithi* (*gro*), and *morondavana* (*mor*). Currently, populations of this group are distributed across landmasses bordering the Indian Ocean, but with the highest species diversity on Madagascar. This DIVA analysis of a molecular phylogeny for the mtDNA cytochrome oxidase I and II genes reconstructed Madagascar as the most likely area of origin for the group and suggested several independent dispersal events: one underlying speciation between a Madagascar species and one in Africa, the other between a different Madagascar species and one that is widespread in the Oriental region and Australia. This result was also produced through analysis of phylogenies based on morphology and two different nuclear DNA genes. (After Zakharov et al. 2004; photo courtesy of Hapa/Wikimedia.)

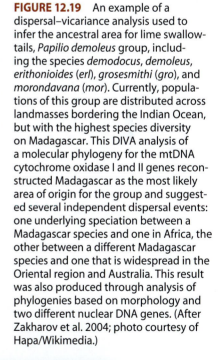

A Madagascar
B Africa
C Oriental Region
D Australia

ated collectively to identify any general, shared patterns. As such, there are cases in which either DIVA or BPA could be a method of choice to use in an "area biogeography" study. More generally, several studies have considered the relative performance of different methods on a given data set (e.g., Crisci et al. 2003 compared ancestral areas analysis, reconciled trees, DIVA, component analysis, and panbiogeography to reconstruct the history of the southern beeches, *Nothofagus*, on Gondwanan continents). Most recently, DIVA has begun to be replaced by several methods that are not based on maximum parsimony—the first is a straightforward extension of the same logic as DIVA but conducted under a Bayesian analytical framework (Nylander et al. 2008), whereas DEC uses a maximum likelihood framework to estimate (rather than specifying a priori, as in DIVA) geographic range evolution parameters (Ree and Smith 2008).

Finally, Sanmartín et al. (2008) have developed an innovative Bayesian approach to estimating the biogeographic history across oceanic island archipelagoes. Notable in their models is that, while utilizing phylogenetic information, they eliminate vicariance as a possible event. Rather, they model inter-island dispersal probability as a function of both island proximity (e.g., in stepping-stone models) and island carrying capacity to infer the predominant historical pattern of dispersal and divergence across islands. Using 13 plant and animal clades from the Canary Islands—an ocean archipelago with a relatively well-known biogeography (Emerson 2002)—as a test case, they suggest that (1) dispersal has been primarily between adjacent island groups, possibly with the central group of islands (Gran Canaria, Tenerife, La Gomera) serving as a center of diversification and source of colonists, and (2) the central group has the highest carrying capacity, and the eastern group (Fuerteventura, Lanzarote) has the lowest carrying capacity. Possible extensions of this dispersal-only approach might include other regions in which dispersal is assumed to have been a pervasive feature of the biogeograhic history, such as within vast expanses of the marine realm (Briggs 2006) or certain mountain archipelago systems (Lomolino et al. 1989).

Phylogeography, again

In Chapter 11, we introduced the general approach of phylogeography, which has revolutionized "temporally shallow" historical biogeography by providing us with tools to reconstruct phylogenetic and population genetic structure across gene lineages within species and among closely related species. However, while we discussed some of the unique features that allow us to reconstruct lineage histories that trace back to relatively recent times—within the most recent thousands to several millions of years of Earth's history—we did not elaborate on the role of phylogeography in reconstructing biogeographic histories, and so we pick up the discussion at that point.

Reconstructing shallow biogeographic histories

The most revolutionary aspect of phylogeography is the way it has brought the scope of historical biogeography into a time frame spanning the past several thousand to several million years, a period that experienced tremendous climatic oscillations with dramatic changes in the locations and composition of biotic assemblages (see Chapter 9). Not long ago, our best clues to how particular species and biotas responded to this dramatic climatic forcing came almost exclusively from fossils (e.g., pollen deposition in lakes, wood rat

(A) *Crotalus atrox*

(B) *Polioptila californica*

FIGURE 12.20 Two examples of overlap between ecological niche modeling (ENM) (green) and phylogeographic (outlines) reconstructions of Late Pleistocene refugia for (A) the western diamondback rattlesnake (*Crotalus atrox*) and (B) the California gnatcatcher (*Polioptila californica*). While both methods tend to predict similar refugia locations, the fit is much better for the gnatcatcher, while the ENM method overpredicts the number of refugia locations (based on phylogeographic inferences) for the rattlesnake. (After Waltari et al. 2007; *C. atrox* © Dave Rodriguez/istock; *P. californica* courtesy of Peter Knapp.)

midden contents, and hard parts of animals and plants in sediments). While the importance of fossils has not faded, we now can employ phylogeographic analyses as a complementary—and in some situations, superior—form of discovery about the effect of Pleistocene paleoclimatic cycles on distributions of species and biotas. Increasingly, phylogeographers are turning to synthetic approaches that incorporate ecological niche (species distribution) modeling to infer geographic range shifts from Late Pleistocene time (**Figure 12.20**, and see Figure 11.14). Moreover, we can address the challenging subject of the time frames and geographic context of speciation during the past several million years, including an assessment of the importance of the Pleistocene on the formation of extant species (Klicka and Zink 1997, 1999; Johnson and Cicero 2004; Weir and Schluter 2004; Zink et al. 2004; Lovette 2005). In addition, we now can examine phylogeographic structure across two or more co-distributed species, using the approach called **comparative phylogeography** (Bermingham and Moritz 1998; Arbogast and Kenagy 2001), to assess the relative contributions of vicariance and dispersal to the assembly of continental, island, and marine biotas.

Earlier in this chapter, when we compared different approaches to addressing complex biogeographic histories, we emphasized the rapid accumulation of new and innovative methods. This point is no less true for phylogeography. Between the inception of phylogeography in 1987 (Avise et al. 1987) and its maturation into a widely recognized subdiscipline of biogeography (Avise 2000), phylogeography was biased toward the description of biogeographic patterns based on mtDNA gene trees in animals (and fewer but important cpDNA studies for plants). When gene trees for populations form discrete, reciprocally monophyletic clades that are genetically distinct and geographically disjunct, it is often quite clear that populations within each of the separate clades have had long histories of allopatric isolation and divergence (e.g., see the coastal vs. inland clades of the tailed frog in Figure 11.11, but see also Irwin 2002 for an interesting alternative view).

Now look at the phylogeny for the red-tailed chipmunk (*Neotamias ruficaudus*), a small mammal that occurs in the Pacific Northwest region of North America (**Figure 12.21**). Without undue controversy, we might infer through visual inspection of the tree that populations in the western clade and eastern clade have very likely experienced a history of allopatric isolation and divergence of sufficient duration to result in the clearly reciprocally monophyletic clades. Now, as we work our way up closer to the tips of the phylogeny, we still find several well-supported clades embedded within either the western or eastern clades, but the times of divergence and degree of

FIGURE 12.21 An mtDNA gene tree for the red-tailed chipmunk (*Neotamias ruficaudus*). This phylogeny sorts intraspecific genetic variation (mtDNA haplotypes labeled at tips of branches) into two well-supported clades (eastern and western) that delineate geographically distinct portions of the species' distribution (see Figure 12.22B, panel 1). Numbers along branches refer to several common measures of support for the robustness of a clade (see original reference for description); unlabelled branches represent outgroups used to root the tree. The key in the upper left box refers to nested clades derived through a nested-clade phylogeographic analysis (see Figure 12.22A and description in Box 12.4), with shading corresponding to clade distributions shown in Figure 12.22B, panels 2 and 3. (After Good and Sullivan 2001; photo courtesy of Noah Reid.)

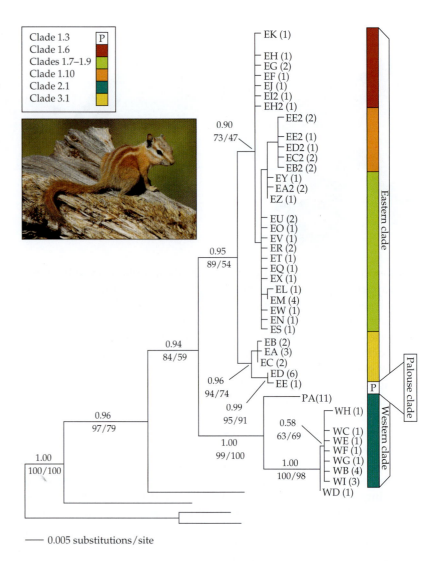

separation between these clades becomes very shallow and their geographic separation tends to be less distinct (**Figure 12.22B**). We have two choices at this point. On one hand, we can conclude that little more can be inferred about geographic structure in this species once we get beyond the inference of historical isolation of the western and eastern clades. On the other hand, we can ask whether another approach might be available that would allow us to assess, first, whether there is a significant, nonrandom association of genetic variation and geography and, if there is, to determine what historical processes—allopatric isolation, recent range expansion, or simply a relationship between spatial proximity of isolated populations and rates of gene flow among them—might have generated the association. Increasingly, phylogeographers are finding innovative ways to pursue this latter alternative within the emerging framework of **statistical phylogeography** (**Box 12.4**).

Comparative phylogeography

Another aspect to phylogeography that has experienced an accelerating growth in popularity, although the term was beginning to be used nearly a decade ago (Zink 1996; Bermingham and Moritz 1998), is "comparative phylogeography," which describes the explicit comparison of phylogeographic structure across two or more co-distributed lineages (Arbogast and Kenagy 2001). Reasons to do so are essentially those that we described previously,

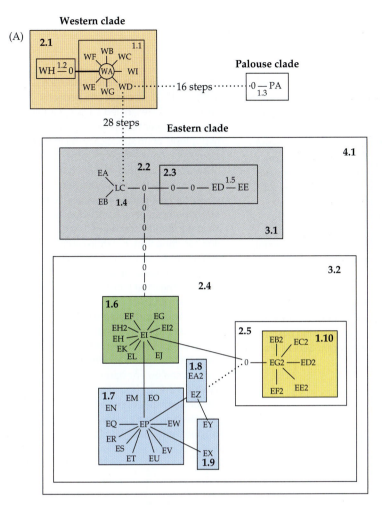

FIGURE 12.22 (A) A nested-clade phylogeographic analysis created from an mtDNA haplotype network for the red-tailed chipmunk (*Neotamias ruficaudus*). The procedure for converting a haplotype network into a series of nested clades is described briefly in Box 12.4, and in more detail elsewhere (e.g., Avise 2000). Each branch connecting haplotypes represents a single mutation step; zeros represent unsampled haplotypes, inserted in order to make connections between haplotypes that differ by more than a single mutation step. Compare with rooted phylogeny in Figure 12.21, and note that the western and Palouse clades are too divergent from the eastern clade and from each other to be nested into a single box by the nesting algorithm. (B) Geographic distributions for (1) the species; (2) sampling localities (with frequencies of nested clades at each location shown); and (3) nested clades (showing the largely complete separation of the western and Palouse clades from those that are nested within the eastern clade). (After Good and Sullivan 2001.)

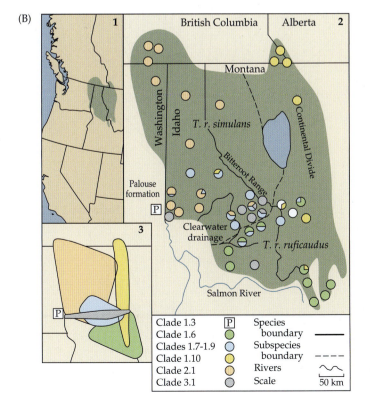

BOX 12.4 *Statistical Phylogeography*

■■■ The example given in this chapter for the red-tailed chipmunk (see Figures 12.21 and 12.22) illustrates a situation that has motivated phylogeographers to call for—and develop—a rigorous statistical framework for phylogeographic research and, in many cases, to frame questions within explicitly stated alternative hypotheses about the geographic history of populations. The development of statistical phylogeography includes two complementary, yet conceptually different, approaches (Knowles and Maddison 2002; Knowles 2004; Templeton 2004; Carstens et al. 2005). The first approach, developed in a number of papers over the past decade by Alan Templeton, Keith Crandall, and their coworkers, is called **nested-clade analysis**

(**NCA**) (Templeton et al. 1995) or, more recently, **nested-clade phylogeographic analysis** (**NCPA**) (Templeton 2002). This method begins by building an *unrooted statistical phylogenetic network among haplotypes* (i.e., a haplotype network that is similar to the gene tree phylogenies shown in Figures 11.11 and 12.21, but without a "root" and, therefore, no outgroup; see Figure 11.12). A statistical approach, such as that implemented in the program TCS (Clement et al. 2000), is used to determine the most likely relationships among different haplotypes in the network. The next step is to convert this phylogenetic network into a hierarchical series of nested clades (see Figure 12.22), generally using nesting rules within the program GeoDis (Posada

et al. 2000). For each nested clade (e.g., in Figure 12.22, clades in one perfectly nested series are labeled 4.1, 3.1, 2.2, and 1.4), the program addresses the "null hypothesis" of no statistically meaningful association between geography and the history of populations within a particular clade. Possible reasons underlying a statistically supported alternative to the null hypothesis of no association include allopatric fragmentation, restricted gene flow resulting from isolation by distance, and recent range expansions—all evaluated a posteriori using an inference key, with the most recent version in Templeton (2004).

A second approach to statistical phylogeography differs from NCPA in the development of plausible alternative hypotheses a priori and then use of various coalescent-based methods (**Table 12.4**) to statistically address the probabilities of the alternative models (see figure below). Several phylogeographers who favor this

Consider this example of an a priori approach to statistical phylogeography: (A) Sampling localities in the northern Rocky Mountain montane forest "sky island" archipelago for the grasshopper species *Melanoplus oregonensis*. Some of this area was glaciated during the Late Pleistocene, and the species could have recolonized areas that were formerly under ice from either a single refugium or multiple refugia. Alternative historical scenarios can be developed into two alternative hypotheses (B) to explain the distribution of genetic diversity within and among populations (see original reference for details). If populations were recolonized from a single ancestral source population, any of the populations might be expected to contain a set of genotypes drawn at random from the overall diversity within the species, as shown in the top tree. Alternatively, if post-Pleistocene recolonization came from more than one Late Pleistocene refugial population, ancestral lineage sorting between the refugial populations prior to establishment of post-Pleistocene geographic distributions would result in the pattern shown in the bottom figure (after Knowles 2001).

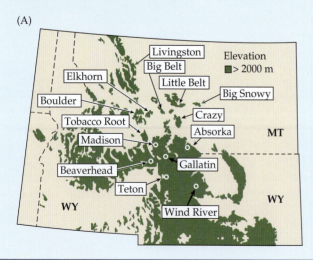

(A)

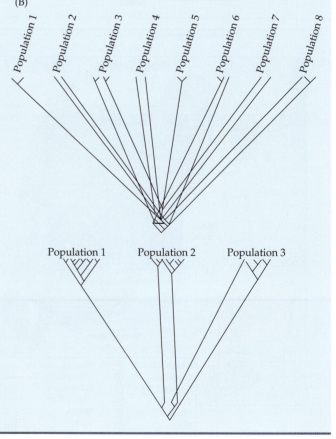

(B)

BOX 12.4 *(continued)*

approach also argue that a modeling approach should be used to determine the robustness of support for one or another of the alternative models (e.g., using the Mesquite program; Maddison and Maddison 2000). Another argument made by investigators favoring use of coalescent-based methods is that phylogeographers working with groups that have particularly shallow divergence times need to worry about the stochastic nature of the sorting of ancestral genetic polymorphism within and among recently diverged populations (see Figures 11.7, 11.8, and 11.13), and many methods, including those listed in the table below, do this.

The term "statistical phylogeography" has emerged only within the past few years, and so its recognition as an approach distinct from an "older" form of phylogeography is relatively recent. However, innovative methods using coalescent approaches and efficient Bayesian computer programs are appearing regularly (reviewed in Hickerson et al. 2010). ▮▮▮

TABLE 12.4 *Examples of Programs and Biogeographic Utility of Coalescent-Based Statistical Phylogeography*

Program	Data type	Biogeographic utility	Source	Reference
LAMARC 2.0	Sequence, SNP, microsatellites, electrophoretic	Estimate exponential population expansion, migration rates	http://evolution.genetics.washington.edu/lamarc/index.html	Kuhner 2006
GENETREE	Sequence	Migration and growth rates in structured populations	http://www.stats.ox.ac.uk/~griff/software.html	Griffiths and Tavaré 1994
BATWING	SNPs, microsatellites	Model population divergence, size, growth, and mutation rates	http://www.mas.ncl.ac.uk/~nijw/#batwing	Wilson et al. 2003
MDIV	Sequence	Estimate divergence times and migration rates between two populations; test divergence with gene flow models	http://people.binf.ku.dk/rasmus/webpage/mdiv.html	Nielsen and Wakeley 2001
IM	Sequence	Estimate migration under isolation with migration model	http://genfaculty.rutgers.edu/hey/software	Hey and Nielsen 2007
BEAST	Sequence	Estimate divergence dates, and population size and growth	http://beast.bio.ed.ac.uk/Main_Page	Drummond and Rambaut 2007
STRUCTURE	Microsatellites, RFLPs, AFLPs, SNPs	Infer population structure, test models of admixture	http://pritch.bsd.uchicago.edu/structure.html	Pritchard et al. 2000
MESQUITE	Sequence	Test models of population divergence, including vicariance, fragmentation, and isolation by distance	http://mesquiteproject.org/mesquite/mesquite.html	Maddison and Maddison 2000
msBAYES	Sequence	Test for simultaneous divergence or colonization across multiple co-distributed taxa	http://msbayes.sourceforge.net/	Hickerson et al. 2007

and they can be traced at least as far back as the 1970s and the American Museum group of vicariance biogeographers. As they asserted, if we are interested in revealing the general historical patterns in a biota—those that arise mainly through vicariance or geo-dispersal—we must examine more than a single co-distributed lineage for a signal of congruence. This is true whether the lineages are higher taxa, species, or gene trees. As such, one implementation of a comparative approach in phylogeography is really noth-

FIGURE 12.23 A depiction of the relationship between comparative phylogeography and historical biogeography, in which phylogroups with taxon phylogenies can be used in the same fashion as species or higher taxonomic units traditionally have, in order to postulate area histories and vicariant events (arrows). Each symbol in the taxon phylogenies depicts a reciprocally monophyletic set of alleles (haplotypes) occurring within, and among, a set of populations. (After Riddle and Hafner 2004.)

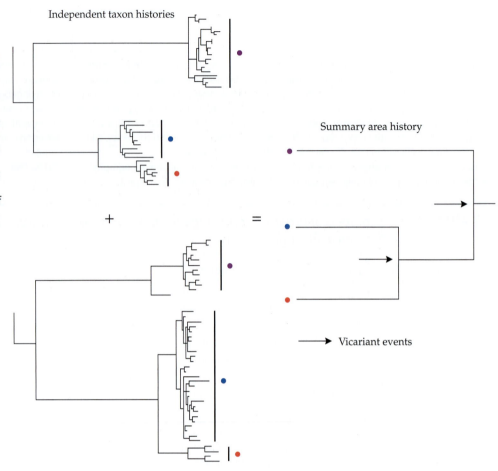

Independent taxon histories

Summary area history

+

=

Vicariant events

ing more than co-opting the logical framework of cladistic biogeography (**Figure 12.23**) or phylogenetic biogeography II (Riddle and Hafner 2004, 2006a). Comparative phylogeographic studies are elucidating shallow biogeographic histories in terrestrial and marine environments across the Earth (**Figure 12.24**), although such studies are still relatively rare in the Southern Hemisphere (Beheregaray 2008).

While the goals of comparative phylogeography and phylogenetic biogeography II are broadly overlapping, each nevertheless brings a different, but complementary, set of strengths (and weaknesses) to a historical biogeographic study. The former approach offers strengths that derive from the sampling and analytical methods that make phylogeography a powerful means of examining relatively shallow and intraspecific histories, including the alternative histories of allopatric fragmentation versus range expansion. On the other hand, the latter approach offers the means to sort general from individualistic events in the vicariance, dispersal, and extinction histories of co-distributed species (through primary and secondary BPA and, more recently, using PACT). Riddle and Hafner (2004, 2006a) argued that the tree generated through BPA analysis could be treated as a set of hypotheses about events such as vicariance, jump dispersal, and range expansion, allowing an investigator to return from a phylogenetic biogeographic to a phylogeographic framework in order to further test those hypotheses. On the other hand, comparative approaches in phylogeography are also being employed even when the phylogenetic tree is down-weighted in favor of one or more of the statistical population genetic approaches summarized in Box 12.4 (Wares 2002; Lourie et al. 2005).

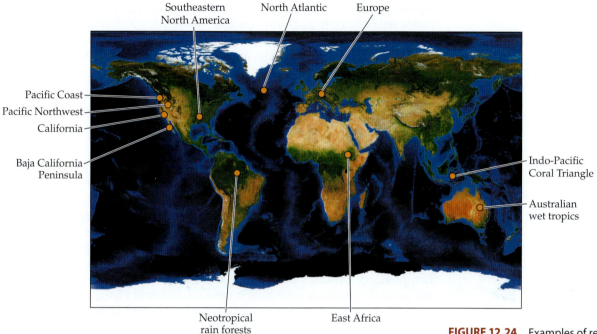

Southeastern North America

North Atlantic

Europe

Pacific Coast

Pacific Northwest

California

Baja California Peninsula

Indo-Pacific Coral Triangle

Australian wet tropics

Neotropical rain forests

East Africa

FIGURE 12.24 Examples of regions in which comparative phylogeographic studies have demonstrated novel historical biogeographic structure. Original references are summarized in Hickerson et al. (2010).

Final thoughts on phylogeography within historical biogeography

Phylogeography has enjoyed an explosive rate of growth and continues to increase in popularity (19 hits in an ISI Web of Science search using topic term "phylogeograph*" for 1994; 470 a decade later in 2004; 1079 just five years later in 2009; and 6757 total to date). Clearly, this approach is here to stay and has become a dominant force in modern historical biogeography. A major reason for the growing popularity of phylogeography is that it offers the potential to attract researchers from disparate disciplines that traditionally focused on a particular question using different approaches, perhaps without even acknowledging the overlap with other disciplines and approaches. For example, Wares (2002) demonstrated how a phylogeographic approach could be employed to address biotic and abiotic processes driving the assembly of a marine intertidal community in the Atlantic Ocean, and the continuing integration of phylogeography with community ecology is quite promising (Riddle et al. 2008; Hickerson et al. 2010).

Because phylogeography is almost always focused on variation within species and between closely related species, it lies temporally at the intersection between recent ecological and historical processes. It therefore can be used to address the relative roles of geological events, in situ diversification, colonization, and extinction in the biogeography of biotas assembled through the recent or ongoing process of colonization, as might be the case for a number of insular systems (Emerson 2002; Heaney et al. 2005; Cowie and Holland 2008). Phylogeography also provides a valuable tool for conservation biologists, enabling them to objectively identify evolutionarily significant units (ESUs) and the geographic processes that underlie their histories and those of other populations, species, and biotas of high conservation priority.

The lesson for the current—and, perhaps more importantly, the next—generation of biogeographers seems clear. Regardless of whether they are most concerned with developing new theory in biogeography, or applying those lessons to conserving the distributions and diversities of native biotas, students should learn as much as they can about the rich tapestry that is modern historical biogeography. Historical biogeography is undergoing a major renaissance (e.g., Donoghue and Moore 2003; Riddle 2005; Wiens and Donoghue 2004), and phylogeography has become an integral component of this new biogeography (Riddle and Hafner 2004, 2006b; Riddle et al. 2008; Hickerson et al. 2010).

What Are We Learning about Lineage and Biotic Histories?

This chapter began with an extended discussion of historical biogeography during the early- to mid-twentieth century. This period was dominated by a tradition of using one or more criteria (see Table 12.1) to "locate" a center of origin, generally in the Northern Hemisphere, and to propose scenarios for the dispersal of species away from that center, sometimes including untested notions of waves of derived species supplanting more primitive forms as they advanced out of the center. The death knell for this approach was two-pronged: Croizat's panbiogeography, with its emphasis on discovering the general patterns of distribution on one hand; and Hennig's phylogenetic methods, with their insistence on discovering monophyletic groups and the ancestor–descendant cladogenetic sequence, on the other hand. Brundin applied both methods in his classic study in the late 1960s and interpreted his results with the benefit of the recent revelations of plate tectonic theory. The 1970s saw the remolding of phylogenetics and track analysis into a form of biogeography that narrowed the field to a search for the general vicariant backbone shared by a set of taxon area cladograms.

While most of today's historical biogeographers find this adherence to "vicariance only" unnecessarily and unrealistically narrow, vicariance biogeography provided a conceptual and methodological foundation for many of the approaches we use today, including those that incorporate methods to estimate dispersal, sympatric speciation, and extinction (see Table 12.2). Finally, from the arenas of molecular evolution and population biology, phylogeography emerged two decades ago and continues to mature into a remarkably popular aspect of modern historical biogeography.

Clearly, historical biogeography has experienced a series of important transformations over the past half century. Yet, until relatively recently, precious few data sets were available for addressing the history of terrestrial and marine biotas, and a good deal of the effort in historical biogeography focused on the "performance" of different approaches using exemplar data sets; most notable among these was Donn E. Rosen's (1978, 1979) poeciliid fish genera, *Heterandria* and *Xiphophorus*, from the uplands of Guatemala. Fortunately, all this changed with key technological advances of the past decade, including the increasing ease of obtaining DNA sequence and other forms of molecular data; the analytical power of sophisticated phylogenetic, population genetic, and biogeographic algorithms; and the availability of data from multiple co-distributed taxa, providing opportunities for exactly the kinds of comparative investigations required to sort general from individualistic biogeographic histories. These breakthroughs have greatly enhanced the analyses of biotic histories in both terrestrial and marine systems—covering a wide range of "deep" as well as "shallow" time frames, and comparisons of biotas within, as well as among, the continents, oceans, and island archipelagoes as well. The burgeoning number of publications from these studies is both encouraging and sometimes daunting, with the number and sophistication of publications increasing each year (e.g., see recent issues of these and other journals: *Evolution, Journal of Biogeography, Molecular Ecology, Molecular Phylogenetics and Evolution, Biological Journal of the Linnean Society, Proceedings of the Royal Society of London*, and *Systematic Biology*).

Earlier in this chapter and elsewhere in this book (see Chapters 7 and 11), we featured a variety of examples of modern, molecular-based, biogeographic analyses of either single lineages or multiple co-distributed taxa (e.g., for the Hawaiian Archipelago). Here, we highlight a handful of intriguing studies that integrate information from a number of co-distributed taxa and demonstrate how modern historical biogeography is poised to produce synthetic and, in many cases, perhaps surprising insights about the histories of biotas.

Histories in Gondwanaland

From the beginning of modern historical biogeography, the plate tectonics model gave biogeographers one very clear exemplar system that should demonstrate a history of vicariance—the breakup of the continent of Gondwanaland. The timing and sequence of fragmentation of landmasses from the ancient Gondwanan continent are well known and provide for the construction of a geological area cladogram that offers explicit predictions about the topology of taxon area cladograms for lineages that diversified in accordance with a vicariance model (see the first figure in Box 12.1). These lineages would have included the ratite and allied birds in the subclass Paleognathae, the chironomid midges studied by Brundin, the southern beeches (*Nothofagus*), and a number of lineages of fishes and reptiles.

What evidence could we use to reject vicariance in favor of dispersal for any of these Gondwanan lineages? First and most obvious, taxon area cladograms that are incongruent with the geological area cladogram would provide a reason to reject vicariance in favor of dispersal. Second, as we discussed in Chapter 11, molecular data could be used to estimate the absolute and relative times of divergence, and they would provide a strong argument against vicariance if an estimated divergence time was younger than the time of area fragmentation. Third, vicariance could be rejected by phylogenetically informative fossils discovered beyond the Gondawanan landmasses where they are predicted to occur according to the geological area cladogram.

The results of a number of recent analyses are pointing to a surprising result—that transoceanic dispersal has played a far greater role in the biogeographic history of the Southern Hemisphere than had been predicted from the Gondwanan vicariance model. De Queiroz (2005) summarized many examples of disjunctions of a broad range of organisms, including primates, chameleons, frogs, and many genera of plants, distributed among landmasses across the Earth, many of Gondwanan origin (**Figure 12.25**). In each case, the disjunct distributions between sister taxa were interpreted as products of transoceanic dispersal, based on incongruence between molecular-based estimates of divergence times and geological estimates of the ages of fragmentation of landmasses. Bergh and Linder (2009), Hopper et al. (2009), and Saquet et al. (2009) have provided evidence derived primarily from time-calibrated molecular trees for multiple cases of transoceanic plant dispersal between the Cape Floristic Region of southern Africa and Australia. In another study, Sanmartín and Ronquist (2004) used a large data set, including 54 animal (insect, fish, reptile, and mammal) cladograms and 19 plant taxon area cladograms in a parsimony-based tree-fitting analysis. Their analyses indicated that, overall, animal distributions are more congruent with the fragmentation sequence of Gondwanaland than are those of plants (**Figure 12.26**). A dramatic case of incongruence in plants, for example, in-

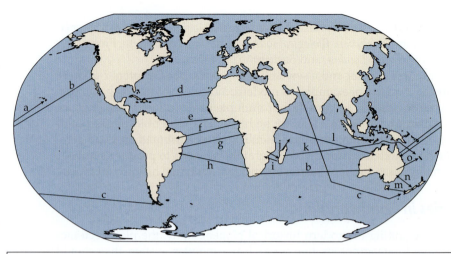

FIGURE 12.25 Examples of transoceanic dispersal, derived mainly from recent molecular phylogenies with estimates of divergence time. A strong case for dispersal rather than vicariance is made when the phylogeny suggests a divergence date between two lineages that is much younger than predicted from a geological area cladogram, such as that shown in Box 12.1 for Gondwana. Arrows on lines indicate direction of dispersal, a line with two filled arrows indicates bidirectional dispersal, and unfilled arrows indicate uncertainty about direction. (After de Queiroz 2005.)

a *Scaevola* (Angiospermae: Goodeniaceae, three episodes of dispersal)
b *Lepidium* (Angiospermae: Brassicaceae)
c *Myosotis* (Angiospermae: Boraginaceae)
d *Tarentola* geckos from Africa to Cuba
e *Maschalocephalus* (Angiospermae: Rapateaceae)
f Monkeys (Platyrrhinii)
g Melastomes (Angiospermae: Melastomataceae)

h *Gossypium* (Angiospermae: Malvaceae)
i Chameleons, three episodes of dispersal
j Several frog genera
k *Acridocarpus* (Angospermae: Malpighiaceae)
l Baobab trees (Angiospermae: Bombacaceae)
m 200 plant species
n Many plant taxa
o *Nemuaron* (Angiospermae: Atherospermataceae)

(A)

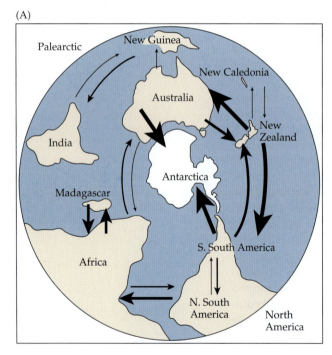

(B)

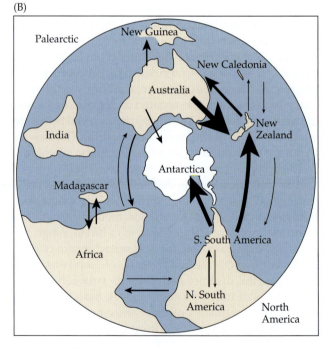

FIGURE 12.26 Analysis of historical dispersal events between major landmasses of Gondwanaland, inferred from a parsimony-based analysis of (A) 54 animal and (B) 19 plant cladograms. The width of the arrows is proportional to the frequency of a particular route (details of analysis provided in original reference). For the animals in (A), the thick arrows connecting Australia and southern South America via Antarctica, as well as the one connecting New Zealand and southern South America, are consistent with the vicariance model of area fragmentation (see first figure in Box 12.1), suggesting that "dispersal" between these areas occurred prior to the breakup of the ancient continent. How-

ever, one could argue the same thing for the high frequency of Madagascar and Africa dispersals, but many of these are now considered to have resulted from postvicariance dispersal events (see discussion in text and Figure 12.27). For the plants in (B), the signal of transoceanic dispersal is stronger than for animals, particularly in the very high frequency of dispersal from Australia to New Zealand, clearly incongruent with the geological cladogram. Note also the weak connections between northern and southern South America, the latter having much stronger historical affinities with other southern landmasses. (After Sanmartín and Ronquist 2004.)

volves the modern flora of New Zealand, which may have originated in large part, if not in total, via long-distance dispersal following the near disappearance of exposed land in New Zealand during the Oligocene (37–23 million years BP; Pole 1994; Winkworth et al. 2002). Generally, their results suggest that plants have dispersed more frequently and more recently than animals among landmasses in the Southern Hemisphere, and patterns of plant dispersal appear to be partially, but not entirely, consistent with predictions if the West Wind Drift and Antarctic Circumpolar Current are important dispersal agents (Sanmartín et al. 2007). But even in animals, with their better overall fit to a Gondwanan vicariance model (see Figure 12.25), several long-held presumptions about purely vicariant histories appear to be inconsistent with molecular estimates of divergence dates. For example, divergence dates between African and Malagasy chameleons (Raxworthy et al. 2002), frogs (Vences et al. 2003), plants (Renner 2004), primates (Yoder and Yang 2004), and carnivores (**Figure 12.27**; Yoder et al. 2003) appear to be much younger than the geological estimate of about 120 million years BP for the separation of Africa from a Madascar–India landmass. This appears to be strong evidence for a history of multiple colonization events between Africa and Madagascar via sweepstakes dispersal, rather than vicariance.

We emphasize here the accumulating evidence of an important role for dispersal in the historical assembly of Southern Hemisphere biotas (McGlone 2005), but stress also that Sanmartín and Ronquist (2004) found congruence between some of their taxon area cladograms and the geological cladogram, supporting vicariance as a component of biotic history as well. Interpreting the histories for any group of taxa can still be controversial (e.g., Briggs 2003a,b,c; Sparks and Smith 2005).

We can mention two additional insights about the dynamic biogeography of Gondwanaland that have emerged from recent studies. The first provides substantial support for the reticulate nature of South America, with Andean and southern parts of South America aligned historically with Australia and New Zealand, and northern (tropical) South America showing greater affinities to the Holarctic and, to some degree, Africa (see Sanmartín and Ronquist 2004; see Figure 12.26). This reticulated history formed the basis for Morrone's (2002) subdividing South America into separate biogeographic

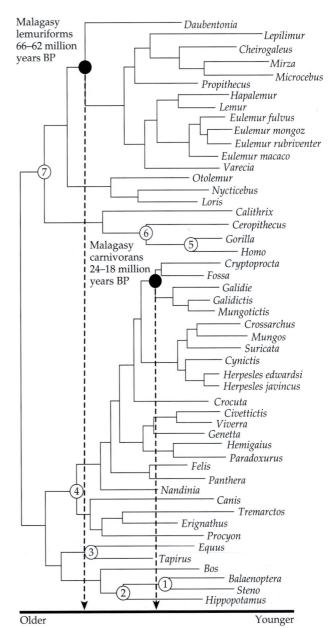

FIGURE 12.27 Molecular phylogeny comparing the ages of divergence of Malagasy primates (*Daubentonia*, the aye-aye; and a number of genera of lemurs) and carnivores (three genera of Malagasy "mongooses"; *Fossa*, the Malagasy civet; and *Cryptoprocta*, the fossa). In each case, the Malagasy clade is monophyletic with a common ancestor at the black filled circle. Open circles with numbers are fossil-based calibration points used to estimate divergence times, which for the primates (66–62 million years BP) and carnivores (24–18 million years BP) postdate the geologically estimated time of separation of Africa and an ancestral Madagascar–India landmass (about 121 million years BP) and the separation of Madagascar from India (about 88 million years BP). This suggests colonization by ancestors of both clades by over-water "sweepstakes" dispersal. (After Yoder et al. 2003.)

regions (see Figure 10.17). Second, Gondwanaland appears to have played a surprisingly important role in the early diversification of a number of major groups of vertebrates previously thought to have originated in Laurasia, including neognathine birds (Cracraft 2001; Barker et al. 2004), ranid frogs (Bossuyt and Milinkovitch 2001; see Figure 7.13), and placental mammals (Eizirik et al. 2001; for similar inferences for grasses, see Bremer 2002).

Histories in the Holarctic

While the tectonic history of Gondwanaland can be summarized concisely into a geological area cladogram with a minimal number of area reticulations (see the first figure in Box 12.1), the geological history of connections and biotic interchange between Laurasian landmasses was much more complex throughout the Cenozoic. For example, although biogeographers have recognized only four broad areas of endemism for temperate deciduous forests—two in the Nearctic (eastern North America and western North America), and two in the Palearctic (Europe and eastern Asia; **Figure 12.28A**)—there has been a long history of debate about the historical sequence of connections between these areas. That these areas are likely to be highly reticulated is suggested by the inferred history of connections within and across the Holarctic continents, summarized as follows:

1. Western and eastern Nearctic landmasses were separated by an epicontinental sea until the earliest Cenozoic Epoch, the Paleocene (roughly 65 million years BP; see Figure 8.23).

2. Multiple Beringian connections formed between the eastern Palearctic

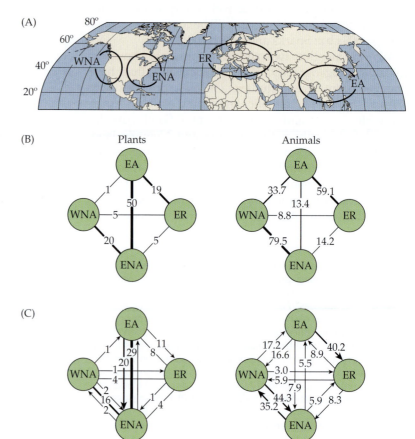

FIGURE 12.28 (A) Holarctic areas of endemism in plants and animals across eastern North America (ENA), western North America (WNA), Europe (ER), and Eastern Asia (EA). (B) A comparison of disjunct patterns of distribution (numbers indicate the relative number out of 100 disjunctions between two areas, with line thickness proportional to that number), and (C) inferred ancestral areas and direction of movement (numbers and line thickness as in B, analyzed using animal data from Sanmartín et al. (2001). See text for discussion. (After Donoghue and Smith 2004.)

and western Nearctic landmasses (see Figure 12.16) during the Cenozoic (culminating in the important Late Pleistocene connections; see Chapter 9).

3. At least two Tertiary connections—the deGeer and Thulean land bridges—formed between the western Palearctic and eastern Nearctic across the North Atlantic (around 50 to 60 million years BP; see Figure 8.22I).

The fossil evidence has previously been interpreted as demonstrating that ancient forests and taxa were widespread across Laurasia prior to its complex Cenozoic geological history and that the current differences in species composition between areas are due primarily to extinction of ancestrally widespread taxa (Wolfe 1975; Tiffney 1985; Tiffney and Manchester 2001).

An increasing number of molecular phylogenetic data sets, many with estimates of divergence times, are becoming available for temperate deciduous forest plant and animal taxa distributed across the four recognized areas of endemism. Donoghue and Smith (2004) used dispersal–vicariance analysis to compare taxon area cladograms from 66 plant clades with the 57 animal clades analyzed by Sanmartín et al. (2001). As was the case in the Southern Hemisphere (Sanmartín and Ronquist 2004), the relative roles of dispersal and vicariance and patterns of dispersal among the Holarctic areas of endemism appear to differ between plants and animals, particularly in the historical relationship of eastern Asia and the Nearctic areas (as summarized in **Figures 12.28B,C**). That is, plants share a higher frequency of disjunct distributions of sister taxa between eastern Asia and eastern North America, and animals share more disjunct distributions between eastern Asia and western North America. Furthermore, again mirroring the Southern Hemisphere, there appears to be a higher frequency of more recent intercontinental dispersal events in plants than in animals in the Northern Hemisphere (but see Donoghue and Smith 2004). Contrary to the "widespread ancient forest" model preferred by paleontologists, these and other studies support a history of multiple episodes of dispersal and vicariance between Palearctic and Nearctic areas during the Tertiary, primarily via a Beringian route (also supported in another study by estimated divergence dates on a molecular phylogeny for squirrels; Mercer and Roth 2003), but also to some degree via a Northern Atlantic route (see Figure 12.16).

Histories in, and just before, the ice ages

Phylogeographic—and particularly comparative phylogeographic—studies have begun to reveal much about the responses of lineages and biotas to the dramatic climatic oscillations of the Pleistocene (Hewitt 2004). Yet, in a slightly expanded time frame, many of the extant species and genera were also members of pre-Pleistocene, Pliocene, and Miocene biotas. We know that the Earth experienced dramatic geological and climatic changes during these epochs, including uplifting of mountains and plateaus, and closure of the Panamanian land bridge. Debate continues regarding the relative importance of ice ages versus earlier events regarding the origin of extant species and assembly of modern biotas (e.g., Johnson and Cicero 2004; Weir and Schluter 2004; Zink et al. 2004). Nevertheless, phylogeographic studies are clearly suggesting that the originations of many extant species and regional biotas date to pre-Pleistocene times, in a wide range of biogeographic regions and biomes, including the tropical forests of northeastern Australia, central Africa, and northern South America (Moritz et al. 2000); the Mexican Neovolcanic Plateau (Hulsey et al. 2004); the conifer forests of the Pacific Northwest in North America (Carstens et al. 2005); and the southwestern deserts of North America (Riddle et al. 2000; Zink et al. 2000; Riddle and Hafner 2006a).

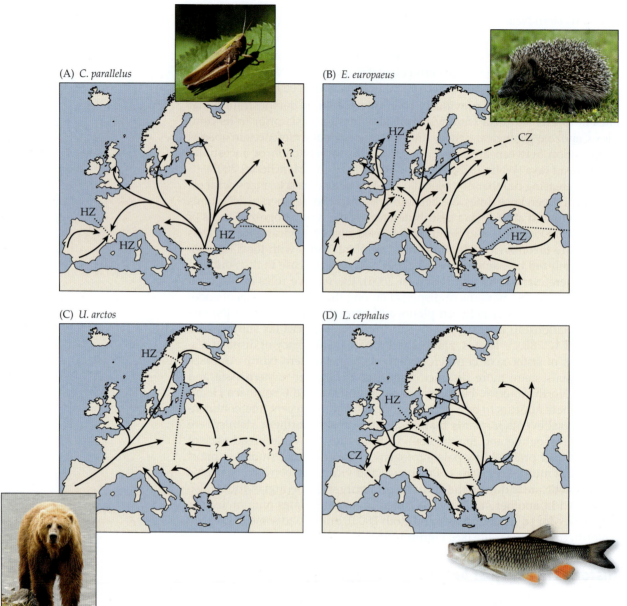

FIGURE 12.29 Four paradigms of postglacial colonization from Late Pleistocene southern refugia in the Palearctic as inferred from mtDNA phylogeographic studies. CZ and HZ are contact zones and hybrid zones, respectively, between lineages expanding from different refugia. The exemplars representing each of the four patterns here are for (A) grasshopper (*Chorthippus parallelus*), (B) hedgehog (*Erinaceus europaeus*), (C) brown bear (*Ursus arctos*), and (D) chub (*Leuciscus cephalus*). (After Hewitt 2004; *C. parallelus* © Birute Vijeikiene/Shutterstock; *E. europaeus* © JurgaR/istock; *U. arctos* courtesy of the U.S. Fish and Wildlife Service; *L. cephalus* © Fedor Kondratenko/istock.)

Within the Pleistocene, comparative phylogeography has provided interesting insights on the temporal cohesiveness of biotas across one or more glacial-interglacial climatic oscillations. For example, in the western Palearctic, Hewitt (2004) synthesized responses across many lineages in Europe into cohesive subsets of taxa whose ranges retreated during glacial periods to one or more southern, unglaciated refugia (e.g., Iberian, Italian, Balkan), followed by northward range expansions after retreat of the glaciers (**Figure 12.29**; see also Figure 9.28). Subsequent phylogeographic studies, combined with fossil evidence, suggest a number of other vertebrates and plants persisted in more northern areas of Europe during the Late Pleistocene (Bhagwat and Willis 2008)—taxon geographic responses to glacial-interglacial cycles appear to have been more complex than originally conceived by Hewitt (2004). Often, separate "phylogroups" within a species that are assigned to a particular refugium are sufficiently distinct that their isolation and divergence likely extended deeper into the Pleistocene than just the latest of the 20 or so glacial-interglacial cycles. Farther to the east, Beringia served as a refugium for mammals, birds, plants, and invertebrates (Waltari et al. 2007; DeChaine 2008).

The Continuing Transformation of Historical Biogeography

As we close this chapter, and this unit of the book, we hope that it is clear that historical biogeography has blossomed into a productive and energetic discipline with the power to offer some amazing insights about the geography of lineage and biotic diversification. We highlighted just a few studies demonstrating the tremendous progress being made in understanding the history of Earth's biotas, and we emphasized terrestrial systems, but recognize that much progress is being made in testing alternative hypotheses and elucidating the histories of marine lineages and biotas as well (e.g., Barber et al. 2000; Santini and Winterbottom 2002; Briggs 2003a, 2004a,b; Lourie and Vincent 2004; Meyer et al. 2005; Bellwood and Meyer 2009). Although we discussed the fascinating biogeographic history of the Hawaiian Islands in some detail, we could just as well have featured a growing number of other interesting studies of biogeographic and evolutionary experiments on oceanic archipelagoes (e.g., Cook et al. 2001; Emerson 2002; Heaney et al. 2005). Finally, we barely mentioned recent studies that are advancing the paleobiogeographies of long-extinct lineages, ranging from Paleozoic trilobites (Lieberman 2003, 2004) to Mesozoic dinosaurs (Upchurch et al. 2002).

Yet despite its great progress, especially in the past several decades, historical biogeography still has large hurdles to overcome if even more important advances are to be made. Even though the large array of modern approaches illustrated throughout this chapter suggest that this is a discipline rich in theory and methods, historical biogeographers still are concerned about whether methods are sophisticated enough to unravel histories that are full of complexity, with reticulated biotas more often than not integrating multiple episodes of vicariance-driven speciation, dispersal, extinction, and sympatric speciation across time frames spanning a few thousand to many millions of years of Earth's history. New methods continue to appear (e.g., Wojcicki and Brooks 2005; Hickerson et al. 2007; Ree and Smith 2008; Sanmartín et al. 2008), and what diverged to form distinct disciplines and methods are now merging into more synthetic approaches in which different methods are employed to address different questions at sequential stages in an analysis (Althoff and Pellmyr 2002; Riddle and Hafner 2004, 2006a; Morrone 2009). Visualization has always been an important component of biogeography (see Chapter 3), no less so in historical biogeography which is poised to incorporate GIS-based geophylogenies (**Figure 12.30**) into the methods toolbox. Finally, along with continuing advances in methods, and growth in numbers and variety of lineages and biotas available for analyses, we are encouraged by the ongoing re-integration of the historical biogeographic perspective into broader and more insightful ecological and evolutionary arenas (e.g., see Wiens and Donoghue 2004, summarized in Figure 15.50; Hickerson et al. 2010).

FIGURE 12.30 An example of a geophylogeny, using information on the mtDNA phylogeography of brown bears (*Ursus arctos*) across Europe (Taberlet and Bouvet 1994; Saarma et al. 2007). This example illustrates the process of converting a phylogenetic tree into a graph incorporating spatial and temporal information. In doing so, choices need to be made on where to position terminal nodes and geographic ranges (current distributions), internal nodes (ancestral distributions prior to a cladogenic event), and branches connecting nodes. A complete description of this process is given by Kidd (in press). (After Kidd in press; photo courtesy of Malene Thyssen.)

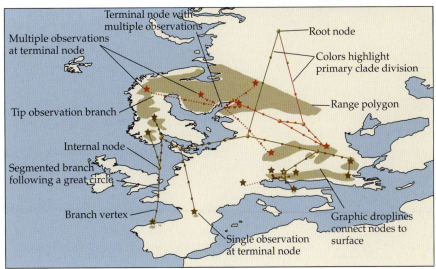

UNIT **FIVE**

ECOLOGICAL BIOGEOGRAPHY

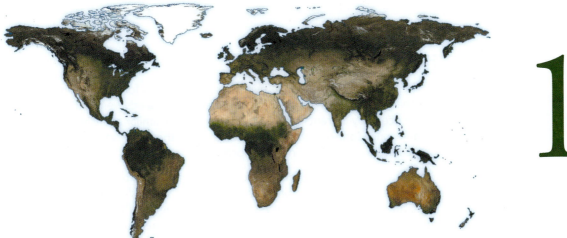

13

ISLAND BIOGEOGRAPHY:
Patterns in Species Richness

Islands have always had a great influence on biogeography, far out of proportion to the tiny fraction of the Earth's surface that they cover (Powledge 2003). The reason for this is straightforward: Islands and other insular habitats, such as mountaintops, springs, lakes, and caves, are ideal subjects for natural experiments. They are well defined, relatively simple, isolated, and numerous—often occurring in archipelagoes of tens or hundreds of islands.

Just as conditions can be varied in artificial manipulative experiments, islands can vary in features such as their area, isolation, or presence of predators and competitors. In this way, the effects of each factor on community properties can be assessed. Despite the limitations of such natural experiments, islands have an important advantage over artificial manipulations in that they provide geographical-scale experimental systems, and many formed so long ago that there has been time for evolutionary responses of their biotas. All one needs to do is to gather the data and interpret them correctly. This, however, has not been an easy task. Basic data on insular biotas are still being compiled, and their interpretation has often been the subject of controversy. Like any other scientific study, a natural experiment must be well "designed." That is, out of all the islands we could study, we must prudently select a subsample of islands that minimizes or controls variation in all factors except those central to our working hypothesis.

Of course, the challenges of designing rigorous and insightful experiments, gathering and analyzing data, and debating and testing alternative methods and ideas are common to all scientific experiments—whether natural or manipulative. As we shall see, biogeographers have utilized both approaches to gain insights into the processes influencing insular communities and, in turn, have obtained important insights into the forces structuring mainland communities as well.

Historical Background

> There are only two possible hypotheses to account for the stocking of an oceanic island with plants from a continent: either seeds were carried across the ocean by currents, or the winds, or birds, or similar agencies; or the islands once formed part of the continent, and the plants spread over intermediate land that has since disappeared. (Joseph Dalton Hooker 1866)

Since the early work of Johann Reinhold Forster and his colleagues during the eighteenth and early nineteenth centuries, studies of islands, mountaintops, and other isolated ecosystems have strongly influenced the development of biogeography. Forster contributed fundamental insights to what was to become island biogeography theory, including the very general tendencies for insular floras to have fewer species than those of the mainland, and for diversity of plants to increase with available resources (i.e., island area and variety of habitats). As we saw in Chapter 2, the influence of islands on biogeography, as well as on evolutionary biology and ecology, began in earnest in the early nineteenth century when various European nations undertook to explore, map, and study the world. Just as Johann Forster and Joseph Banks sailed with Captain James Cook in the late 1700s, other naturalists frequently accompanied voyages of exploration during the nineteenth century. The best of these naturalists—notably, Wallace, Darwin, and Hooker—not only described and collected specimens of what they found, but also noted patterns in nature and sought explanations for them. Some of the clearest patterns were apparent among the various islands of oceanic archipelagoes. Wallace's travels in the East Indies, Darwin's experiences in the Galápagos, and Hooker's explorations in the Southern Ocean had profound effects on the thinking of these scientists and, consequently, on the ideas about evolution and related areas of environmental biology that revolutionized scientific thought in the mid-1800s.

Another revolution occurred in the mid-1900s with the integration of concepts from ecology, evolution, and biogeography. It is difficult to pinpoint the exact beginning of this endeavor, but islands again played a central role. One of the pioneers was David Lack (1947, 1976), who, early in his career, conducted a classic study of the evolution and ecology of Darwin's finches on the Galápagos Archipelago, and later investigated the distribution and ecology of birds in the West Indies. As Lack had followed Darwin to the Galápagos, another ornithologist, Ernst Mayr (1942, 1963), followed Wallace to the East Indies and returned to make major contributions to the understanding of speciation (see also Mayr and Diamond 2001). Another pioneer was G. Evelyn Hutchinson (1958, 1959, 1967), who also traveled widely but studied lakes rather than terrestrial islands. Hutchinson and his students demonstrated the utility of mathematical models as complements to field research and manipulative experiments. In his 1959 paper "Homage to Santa Rosalia, or why are there so many kinds of animals?" Hutchinson called attention to the problem of how to explain geographic variation in the diversity of species. This problem has since remained a key focus of research in ecological biogeography and community ecology.

If any single contribution can be said to have triggered the recent revolution in ecological biogeography, however, it was Robert H. MacArthur and Edward O. Wilson's equilibrium theory of island biogeography (1963, 1967; **Figure 13.1**). MacArthur had been a student of Hutchinson's at Yale. His doctoral dissertation (1958) was a classic study of competition and coexistence in several closely related species of warblers. After completing his degree, he did postdoctoral work in Britain with Lack and then held professorships

(A)

(B)

at the University of Pennsylvania and at Princeton University. Wilson, who has spent his entire career at Harvard, began it as a systematist and biogeographer. Strongly influenced by Mayr, he had worked extensively on the origins and relationships of the ants on islands of the East Indies and South Pacific and had already developed a dynamic explanation, the taxon cycle model, for their island biogeographic patterns (Wilson 1959, 1961). He was also a coauthor of the classic paper on character displacement among competing species (Brown and Wilson 1956). Both men had extensive experience with islands: MacArthur in the montane islands of the southwestern United States, in the West Indies, and in small islands off the coasts of Maine and Panama; Wilson in the East Indies, Polynesia, and the Florida Keys. MacArthur died of cancer in 1972 at the age of 42, but had already produced many theoretical papers on population and community ecology that still motivate research in those fields today. Wilson continued to work on social insects, especially ants, but his interests have shifted from systematics and biogeography to animal behavior (Wilson 1975) and, most recently, to conservation of biodiversity (Wilson 1988, 1992, 1994).

MacArthur and Wilson's equilibrium theory represented a radical change in biogeographic thought. Prior to their work, investigators had focused on historical problems concerning relationships between floras and faunas of islands and continents. As indicated by the above quote from Joseph Dalton Hooker, much of the emphasis within island biogeography was placed within debates about historical connections between regions and whether island endemics were stranded relicts (paleoendemics) or in situ evolutionary products (neoendemics). Hence, much of the work was concerned with idiographic questions such as where a particular taxonomic group of organisms originated and how, as a result of subsequent dispersal, speciation, and extinction in particular regions, its diversity and distribution changed.

Given this big-picture biogeographic focus, island biotas were typically discussed as if fairly static in composition unless modified by long-term evolutionary processes. Insular community structure was largely represented as resulting from unique immigration and extinction events, and species number was determined by the limited number of niches available on each island and the difficulty mainland forms had colonizing small, remote islands. Once a species arrived, it either found adequate resources for survival or failed to establish a population. Large islands had more species because they had more resources and a greater variety of niches that could support a greater diversity of species. Islands closer to the mainland had more species because they were within the colonization abilities of more species. Such a focus is evident, for example, in Hooker's 1866 essay on island floras, although it is interesting to note that he highlighted the subsidence of oceanic islands as a principal cause of rarity and extinction of old species, because subsidence leads to diminished habitat availability, intensifies competition, and generates a form of ecological collapse through decrease in populations of winged pollinating insects. Hooker's remarks highlighted an awareness of the potential importance of island environmental dynamics for biological dynamics some hundred years before the equilibrium theory was written. Another such illustration comes in the early literature on the Krakatau Islands, which self-sterilized in volcanic eruptions in 1883, as botanists and zoologists tracked the colonization and succession of species and communities, occasionally noting how some species established a temporary presence here and there, appearing and disappearing, sometimes repeatedly, as habitats came and went (see, e.g., Docters van Leeuwen 1936).

These two examples serve to illustrate that pre-1960s island biogeographers did not have a restrictively static view of island life, ignorant of all short-term natural dynamics, but it is true to say that their focus was on the big time–space questions of origins and relationships. All this was to change with the publication of MacArthur and Wilson's dynamic equilibrium model of island biogeography, not so much because it overturned a dominant paradigmatic structure among island biologists, but rather because in postulating a simple, general, process-based theory, MacArthur and Wilson opened up an entirely new research program—equivalent to opening a marvelous new suite of laboratories complete with all the equipment and materials any number of research teams could wish for.

From page 249 of E. O. Wilson's (1994) *Naturalist*, wherein Wilson reconstructs the exchange of ideas that led to their famous graphical model at the heart of their theory of island biogeography as if it were a single conversation, the following brief excerpts depict the thinking at the core of it:

> Wilson: "Here's another piece in the puzzle. I've found that as new ant species spread out from Asia and Australia onto the islands between them, such as New Guinea and Fiji, they eliminate other ones that settled there earlier.... So there seems to be a balance of Nature down to the level of the species."

> MacArthur: "Yes, a species equilibrium. It looks as though each island can hold just so many species, so if one species colonizes the island, an older resident has to go extinct. Let's treat the whole thing as if it were a physical process. Think of the island as filling up with species from an empty state up to the limit. That's just a metaphor, but it might get us somewhere."

By starting with the essential processes at work, MacArthur and Wilson deliberately departed from the classic historical approach and were able to pose radically new kinds of questions. They searched for general patterns across different taxa, in the hope that such patterns would have general ecological explanations rather than idiographic and historical ones. Although they knew their theory was simplifying reality, they viewed this as the route to generating testable general hypotheses at the intersection of population and community ecology.

Island Patterns

MacArthur and Wilson's theory embraces two very general and long-known patterns in island biogeography: the tendency for the number of species to increase with island area and the tendency for the number of species to decrease with island isolation (see Chapter 2). Perhaps the greatest inspiration for the equilibrium theory, however, came from Wilson's work on ants, which led him to the inference that immigrations and extinctions were relatively frequent phenomena, even in ecological time. MacArthur and Wilson's innovation was to recognize the common themes that underlie these observations (the species–area relationship, the species–isolation relationship, and species turnover) and to propose a single, unifying theory to account for them.

The species–area relationship

> Theories, like islands, are often reached by stepping stones. The species–area curves are such stepping stones. (MacArthur and Wilson 1967: 8)

Thomas Schoener (1976) described the species–area relationship as "one of community ecology's few laws." Indeed, it is one of the most general, best-documented patterns in nature (Forster 1778; de Candolle 1855; Watson 1859; Jaccard 1902, 1908; Arrhenius 1921; Brenner 1921; Gleason 1922, 1926; Kreft et al. 2008). Regardless of the taxonomic group or type of ecosystem being considered, species number tends to increase with increasing area. This relationship, however, is not linear; richness increases less rapidly for larger islands (**Figure 13.2**).

The Swedish ecologist Olof Arrhenius was the first to propose a mathematical generalization of this fundamental pattern or relationship between species number and area, in articles published in 1920 and 1921. Rather than being concerned with islands per se, his aim was to devise a standardized means of comparing the diversity and assessing the validity of floristic associations, districts, and communities:

> Using this formula one is enabled to find a standard for the relative richness or poorness of a floral district. The species in an association are distributed according to the laws of probability. The number of species increases continuously as the area increases, and the plant associations pass into each other quite continuously. (Arrhenius 1921: 99)

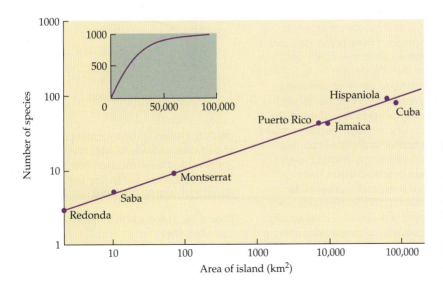

FIGURE 13.2   The empirical relationship between number of species (S) and island area (A) for reptiles and amphibians of the West Indies, plotted from the original data of Darlington (1957). Note that both axes are logarithmic, and the points are well fitted by a straight line and the equation $S = cA^z$, where c and z are fitted values. Inset depicts this relationship in arithmetic space. (After MacArthur and Wilson 1967.)

Although his analysis was directed at debates within **phytosociology**, his insight is rightly recognized as foundational within island biogeography. The Arrhenius equation, or power model, can simply be expressed as

$$S = cA^z$$

where S = species number (or richness), c is a fitted constant, A = island area, and z is another fitted parameter that represents the slope when both S and A are plotted on logarithmic scales.

It was this model of the species–area relationship that Wilson (1961) used in his analysis of Melanesian ants, and with which MacArthur and Wilson

BOX 13.1 *Interpretations and Comparisons of Constants in the Species–Area Relationship: An Additional Caution*

■■❚ The species–area relationship is one of the most important and most frequently studied patterns in biogeography and, according to Schoener, "one of community ecology's few universal regularities" (1986: 560). Yet, the utility of the most common model describing this relationship—the power model ($S = cA^z$, where S is species richness, A is area, and c and z are fitted constants)—is debatable. The strongest controversy stems not from the "fit" provided by this model but from the biological relevance of its exponent, z (Connor and McCoy 1979; Sugihara 1981; Abbott 1983; see also Dengler 2009).

Many misinterpretations can result from the frequent but unfortunate reference to the z and c values as the slope and intercept of the species–area relationship, respectively. The power function, however, "intercepts" at the origin (i.e., when $A = 0$, $S = 0$). Just as important, z values represent the slopes of the relationship between $\log(S)$ and $\log(A)$, not species richness and area. By themselves, z values do not indicate how rapidly S increases with A. For this, we require the values of both parameters in the power model, c and z (Gould 1979).

Although this is not a mathematically startling revelation, it is not rare to see, for example, a high z value equated with a rapid increase in S with increasing A. Yet, such an inference is valid only if c values for the archipelagoes and faunas under study are equal. On the contrary, c values vary considerably (often by an order of magnitude or more among archipelagoes or taxa), whereas z values tend to be conservative (typically ranging

FIGURES A–C The effects of varying the values of c and z on the species–area relationship. The model in all cases is $S = cA^z$. (A) Effects of varying the value of c (z is held constant at 0.25). (B) Effects of varying the value of z (c is held constant at 1). (C) Effects of varying both c and z. Note that species richness increases more rapidly for the upper curve despite its substantially lower in value.

from 0.15 to 0.35; Gould 1979: his Table 1; Wright 1981: his Table 1). This fact led Gould (1979) to suggest that we draw inferences from comparisons of c values for archipelagoes with approximately equal z values (analogous to analysis of covariance for log-transformed data).

The effects of varying one of these parameters, c or z, while holding the other constant are illustrated in **Figures A** and **B**. Note that the species–area relationship (arithmetic scale) is strongly influenced by relatively modest variation in c, but is comparatively insensitive to the variations in z, typical of natural communities. Furthermore, if these parameters vary simultaneously, as they surely do in nature, then the slope of the species–area relationship (again on an arithmetic scale) may actually be lower for studies reporting higher z values (**Figure C**). In summary, although the power model may continue to provide important insights into factors affecting species richness in isolated biotas, there remains considerable potential for misinterpretation. Studies comparing constants of the power model among archipelagoes, or comparisons with

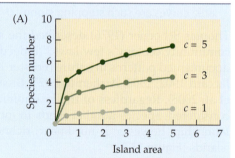

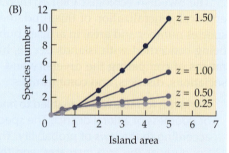

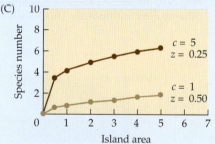

predicted values of these constants, should take measures to avoid the statistical problems discussed here and elsewhere (Connor and McCoy 1979; Gould 1979; Martin 1981). ■■❚

Source: Excerpted and modified from Lomolino 1989.

(1963) began their seminal contribution to island theory. Typically, this relationship is linearized by taking the log of both sides of the equation:

$$\log(S) = \log(c) + z\log(A)$$

The c and z values can now be easily estimated using simple linear regression of log-transformed data. In **Box 13.1** we discuss the biological relevance and potential misinterpretations of c and z.

While the power model appears to be the most commonly used formula for the species–area relationship, a semi-logarithmic model also is frequently used, especially among plant ecologists, and for some data sets it provides better fits than the power model. In 1922, Henry A. Gleason used the following formula to study species–area relationships of plant communities:

$$S = d + k\log(A)$$

As this formula implies, d represents the intercept, and k represents the slope of the line when S (richness) is plotted against the log of A (area).

Frank Preston, an engineer by profession and a naturalist by avocation, made a number of seminal contributions to the mathematical development of ecology between the late 1940s and early 1960s. In his foundational 1962 paper, he noted that the species–area relationship of islands was a special case of the general multiplicative increase in the number of species with an increase in the area sampled. He suggested that this was a consequence of what he termed the canonical lognormal distribution of the number of individuals among species (see also Williams 1953, 1964). In any region, only a few species are extremely common, and most are moderately, or very, rare. Therefore, the distribution of the number of individuals among species, when plotted on a logarithmic abscissa (x-axis), is fairly well fitted by a normal, bell-shaped curve (**Figure 13.3**). Sometimes this curve is cut off on the left-hand side, and Preston suggested that this happens when the sample is so small that some of the rarest species are not observed. It is this effect that shapes the species–area curve. As progressively larger areas are sampled, one obtains not only more individuals but also more species, because some of the new individuals will be representatives of rare species that had not yet been seen. Additionally, larger areas will tend to incorporate new kinds of habitats and therefore add specialized species that are restricted to those environments.

Preston pointed out that small, isolated islands have fewer species per unit of area and higher z values for the species–area curve than do sample areas of comparable size within large

FIGURE 13.3 The relative abundances of species within a local biota often fit a lognormal distribution; in other words, the frequency distribution approximates a normal curve when abundance is plotted on a logarithmic scale. Note that often the left-hand tail of the distribution is cut off by what Preston called a veil line. Because the axis is logarithmic, this curve shows that every community contains more rare species than common ones. The data are from a bird census in Maryland (Preston 1957) and a count of moths at a light trap in England (Williams 1953).

(A)

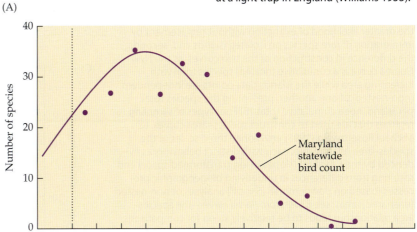

Maryland statewide bird count

(B)

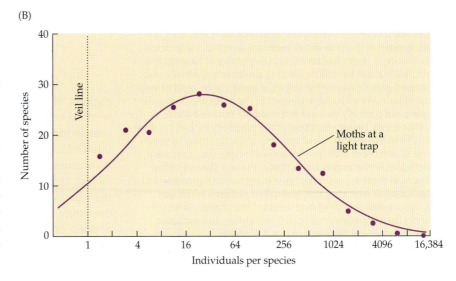

Moths at a light trap

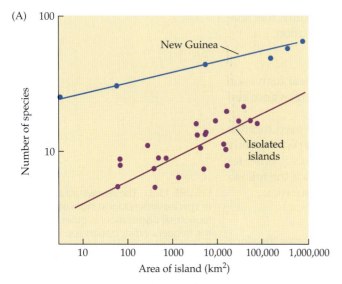

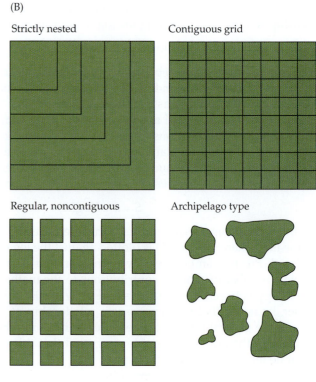

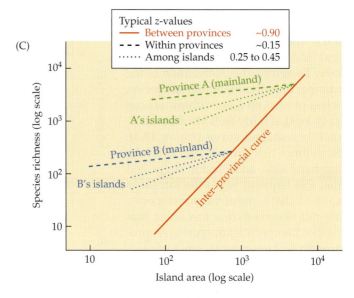

FIGURE 13.4 (A) The slope of the species–area relationship in log-log space is much steeper for isolated islands than for sample areas of different sizes within a single large landmass. These data are for ponerine and cerapachyine ants on the Moluccan and Melanesian islands (lower curve) and in regions of increasing size on New Guinea (upper curve). The difference between the two curves can be attributed to two factors: first, the greater likelihood of extinction without replacement by immigration of rare species on isolated islands and, second, the difference in construction of the two curves, as the New Guinea curve is a cumulative (nested) species total with increasing area while the islands data points are independent (non-nested) of one another, presenting the species total per island, not a cumulative total. (B) Four types of sampling regimes that are used to construct species–area curves. (C) Michael Rosenzweig (1995) envisioned three scales of the species–area relationship, each with distinct behaviors as evidenced by their *z* values (slopes in log-transformed space) and resulting from differences in underlying biogeographic processes (based on power model [$S = cA^z$] analyses of non-nested [unshared] areas; see B): *Interprovincial curve*—the species–area relationship tends to be relatively steep for comparisons among provinces (e.g., separate regions of continents or entire continents) because they are evolutionarily independent and share relatively few species. *Within-province* (*mainland*) *curves*—at this scale, communities are derived from a common pool of species (evolved within that province) and, thus, the increase in richness as larger areas are sampled is a function primarily of ecological processes (availability of habitats and niches) but not evolution, yielding a relatively shallow curve. *Among-island curves*—the slopes of these curves tend to vary systematically with characteristics of the biota and archipelagoes, but typically they are intermediate between those at other scales, being a function of the tendency for habitat diversity to increase with island area and of the interplay between the processes of immigration and extinction (i.e., consistent with MacArthur and Wilson's [1967] theory). Overall, differences in these curves reflect processes operating at different temporal as well as spatial scales: The patterns are driven by processes operating over varying time scales—relatively short for within-province (mainland) curves and evolutionary (hundreds of generations) for the interprovincial curve. (A after Wilson 1961; B modified from Whittaker and Fernández-Palacios 2007, after Scheiner 2003; C after Rosenzweig 1995.)

regions of continuous habitat on continents (**Figure 13.4**; see Schoener 1974; Sugihara 1981; Rosenzweig 1995). The reason for this should be intuitively apparent: Small, isolated islands have fewer species than comparable areas on a continent because if a species becomes too rare on an island, it is likely to become extinct, whereas on a continent its population can be sustained at low levels by the exchange of individuals between local areas. The effect of such extinctions is much more severe on small islands than on larger ones, resulting in the steeper slope of the logarithmic species–area curve.

Preston's (1962) observations were particularly insightful and prescient in the way that they linked two general phenomena together: the species abundance distribution (SAD) (sometimes rank abundance distribution) and the species–area relationship, suggesting that variation in isolation of the system could influence the form of the SAD and in turn the parameters of the species–area relationship (see Box 13.1). Hundreds of papers have since been published on species–area relationships, and continuing efforts have been made to determine the best mathematical model, without final resolution. This reflects the fact that data sets of widely differing properties are being analyzed and to differing ends (see Tjørve 2009). Whereas the above-mentioned models are still the most popular, there are other models of varying complexity that can be applied, depending for example on whether the goal is to linearize the relationship across many different data sets for subsequent comparative analysis, to test a specific hypothesis about the form of the relationship, or to account for the greatest amount of variance in a particular data set (e.g., Kalmar and Currie 2006). In addition, there may be systematic differences in model fits depending on (1) whether the data are from samples or true isolates and (2) whether curves are constructed from nested sets of samples (to form species accumulation curves) or non-nested sets with the species count for each sample being tallied independently of other sample sets (Whittaker and Fernández-Palacios 2007; Tjørve 2009; Tjørve and Turner 2009; and see Figure 13.4B).

The species–isolation relationship

Since the early 1800s, it has been well known that isolated islands support fewer species than islands that are located nearer to continents. Assuming that the decline in species richness results from a decline in dispersal opportunities with isolation, the form of the species–isolation relationship should be a consequence of dispersal curves for the pool of species (potential colonists from the mainland; see Chapter 6). Therefore, for a variety of taxa and ecosystems, species richness should decline as a negative exponential or sigmoidal function of isolation (**Figure 13.5**; see also Figure 6.19).

With appropriate transformations of one or both axes (i.e., richness and isolation), this relationship can be linearized to allow statistical analyses and comparisons among studies (e.g., $S = k_1 e^{-k_2(I)}$, or $S = k_1 e^{-k_2(I \cdot I)}$, where S = richness, k_1 and k_2 are fitted constants, and I = isolation). Many studies report significant species–isolation relationships, especially when island isolation varies substantially and when the effects of island area are statistically "controlled" (e.g., by using correlation or regression analysis of residuals after ac-

FIGURE 13.5 Two of the most common patterns in nature—the species–area (A) and the species–isolation (B) relationships. The log-transformed equivalents of these relationships are presented in the insets (S = species richness; A = island area; and c and z are fitted constants for the power model of the species–area relationship). Note that the species–area relationship will be difficult to detect (slope near 0) if biogeographic surveys are limited to the larger islands (i.e., Region a; see also Figure 13.23). Similarly, the species–isolation relationship will be difficult to detect if surveys are restricted to the very near or very distant islands (Regions b and c; see also Figures 6.19 and 13.28A).

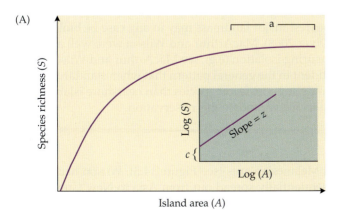

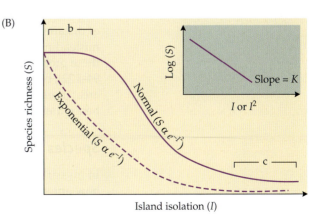

(A)

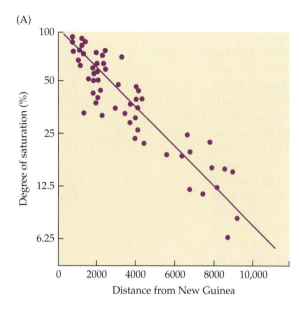

(B)

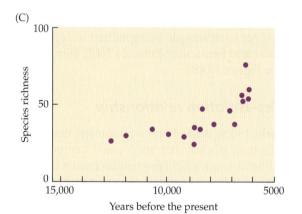

(C)

FIGURE 13.6 A sample of graphs illustrating the effects of isolation on the species richness of various insular biotas (see also Figure 13.28 and 13.30). (A) Resident land birds in the Moluccan and Melanesian archipelagoes. Here species richness is expressed as saturation, which is the species richness of isolated islands expressed as a percentage of that found on an island of equivalent size, but closer to New Guinea. (B) Non-volant mammals of the Thousand Island Region of the St. Lawrence River, New York. Here the ordinate equals the residuals about the species–area relationship, that is, the difference between the observed insular species richness and that predicted for an island of its size based on the regression model: $\hat{S} = 6.51(A^{0.305})$. (C) Species richness of lizards on land bridge islands in the Gulf of California. Here, species richness is graphed as a function of time since isolation of these land bridge islands following rising sea levels of the most recent glacial recession. (A after Diamond 1972; B after Lomolino 1982; C after Wilcox 1978.)

counting for the effects of area on species richness; **Figure 13.6A,B**). However, the species–isolation relationship remains less clearly established than the species–area relationship. This difference may derive, in part, from the tendency for many studies of islands, such as those of a single archipelago, to include a broad range in island area but only a limited range in isolation. Moreover, in assessing isolation, there is a problem that arises from the clumped distribution of islands: Should isolation be measured as the distance to the mainland, the nearest large island that could potentially act as a source pool for the target island, or simply the nearest island of any size? In addition, given the importance of predominant ocean or wind currents and of bird migration routes in influencing the frequency of dispersal, simple measures of distance from the nearest major landmass may in any case be biologically misleading (e.g., see Diver 2008). MacArthur and Wilson were well aware of these problems in measuring isolation (e.g., see MacArthur and Wilson 1967: 94–122), but were confident in the general pattern that when standardized by area, more-distant islands support fewer species than nearshore islands, and hence their theory had to accommodate this observation.

Species turnover

In his earlier work on Melanesian ants (see Figure 14.3), Wilson had shown how ant species had sequentially spread out from Asia to New Guinea and

BOX 13.2 *Independent "Discovery" of the Equilibrium Theory of Island Biogeography*

■■I A correlation of this kind [between number of species and logarithm of area of an island] is as interesting as it is unexpected, for it suggests the existence of an equilibrium value for the number of species in a given island, a value which acts as a limit to the size of the fauna. The processes which determine the equilibrium value for an island of given size must be, on the one hand, the extinction of species, and, on the other hand, the formation of new species within the island, and the immigration of new species from outside it.

The above quotation appears on page 117 of Eugene G. Munroe's (1948) doctoral thesis on the distribution of butterflies in the West Indies. Munroe's work is noteworthy because he did not have just some vague, poorly articulated notion of species equilibrium. He clearly presented the empirical species–area relationship that stimulated his inductive discovery, investigated the generality of this pattern, and developed verbal and mathematical models to explain it. Brown and Lomolino (1989) recognized his articulation of the concepts at the core of MacArthur and Wilson's model as constituting the independent "discovery" of the dynamic equilibrium theory of island biogeography. However, apart from his unpublished doctoral thesis, of which this theory constituted but a few pages, Munroe published only a three-paragraph abstract of his theory, resulting from a conference presentation (Munroe 1953). The first paragraph, quoted above, points out the generality of the semilogarithmic species–area relationship, but then Munroe goes on to present a mathematical model:

The actual form of the curves is that of a shallow sigmoid, with the equation

$$F = k'A^* [iL/(i + kp)]$$

where F = number of species in the fauna at equilibrium; L = number of species in surrounding lands capable of immigrating into the island; i = the probability of any one species actually immigrating; p = the probability of extinction of a single pair of one species; and A = the area of the island, to which the population number of each species is assumed to be directly proportional. (Munroe 1953: 53)

MacArthur and Wilson (1967: 26, Figure 11 and Equation 3-1) use a similar, but somewhat simpler, expression within their monograph:

$$S = (IP)/(E + I)$$

where S = the equilibrium number of species; I = the initial immigration rate, if the island was empty of species; P = the number of species in the species pool available to colonize; and E = the extinction rate if P species were present on the island.

Another ecologist, Frank Preston, went a great deal further than Munroe in developing a dynamic equilibrium model in a series of papers culminating in his seminal and monographic 1962 two-part paper in *Ecology*. Preston's work provided an analysis linking the abundance distribution of individuals within populations to the emergent species richness patterns of samples and isolates (see main text). Preston saw that there was a continuum of connectivity of natural systems such that samples within a fully contiguous system and truly remote islands form the end points. Preston's theory was articulated in a fully fledged mathematical form, with accompanying analysis of numerous data sets, and the parallel with MacArthur and Wilson's independently derived theory is remarkable. In illustration, consider the following excerpt from Preston's papers (1962a,b: 411, Part II):

On isolated islands we must have an approximation to internal equilibrium and presumably to a self-contained canonical distribution [species abundance distribution] and, since an island can hold only a limited number of individuals, the number of species will be very small. But on the mainland a small area is not in internal equilibrium; it is in equilibrium with areas across its boundaries and is a sample of a vastly larger area. This is a matter that does not seem to have been considered previously in studies of Species–Area relationships.

Preston goes on to remark in consolidation of this point that "according to my view, the number of individuals sets a limit to the number of species, and in an 'isolate,' virtually specifies the number of species within rather narrow limits."

Preston explains that in a sample of 100 acres, there will be many species of breeding birds represented by a single pair and that while they may be replenished from elsewhere in a sample, the species can be expected to fail to persist in an isolate. Should the contiguous area of habitat suddenly be greatly reduced, then inevitably the species count must fall as a consequence. In short, Preston invokes changes in population movements and numbers resulting from changes in isolation and area of the system, with the emergent outcome being a readjusted internal equilibrium. MacArthur and Wilson (1963) comment on the remarkable parallels in thinking in Preston's paper, which appeared after they had worked through the mathematical arguments surrounding their core model but in advance of its publication. Regarding Preston's "100 acres" passage they wrote (1963: 382–383):

This point of view agrees with our own. However, the author apparently missed the precise distance effect and his model is consequently not predictive in the direction we are attempting. His model is, however, more accurate in its account of relative abundance.

What distinguishes Preston's seminal macroecological contribution and MacArthur and Wilson's 1967 monograph from Munroe's work and other such contributions that could be considered to have stumbled on the same core ideas is the way in which they develop a comprehensive theory in verbal, mathematical, and graphical form, articulating not only the core assumptions and concepts but also deriving the novel predictions that are the hallmark of the big advances in scientific thinking. ■■■

the Melanesian islands, shifting habitats as they did so, with some species seemingly competitively excluding others from former territories as they spread. Wilson (1959, 1961) noted how his inferences of ant evolutionary biogeography were supported by data showing the "startling ease" with which the Krakatau islands were recolonized by ponerine ant species, filling up to levels that by comparison with his Melanesian and Moluccan islands appeared "near-saturation" within just 50 years of the apparent total sterilization of the islands by the volcanic eruptions of 1883. The rapidity of their colonization is perhaps not entirely remarkable. These islands are located in the Sunda Strait, roughly 30–40 km respectively from Sumatra and Java (**Figure 13.7**) and closer still to some other small islands, and they have also been subject to frequent boat visits by scientists, fishermen, and others, which has aided the natural colonization processes of plant and animal species. In fact, as we shall discuss later, the degree of "saturation" of the Krakatau fauna and flora turns out to be dependent on which taxa you look at.

Notwithstanding, as the earlier exchange recounted in Wilson's book *Naturalist* demonstrates, he and MacArthur had seized on the insight that colonization events may be rapid and recurrent, and from this they inferred that the tendency to acquire new forms must be balanced by other processes that limit persistence of established ones. They may not have been entirely alone in deriving these insights (**Box 13.2**), but only they developed them into a fully fledged theory of island biogeography.

The Equilibrium Theory of Island Biogeography

MacArthur and Wilson produced a single theory to explain what they considered to be the three basic characteristics of insular biotas: the species–area relationship, the species–isolation relationship, and species turnover. They proposed that the number of species inhabiting an island represents a dynamic equilibrium between opposing rates of immigration and extinction (**Table 13.1**). The equilibrium is termed "dynamic" because immigration and extinction are held to be recurrent, opposing processes, maintaining a relatively stable species number despite ongoing changes in species composition.

The equilibrium model can be presented graphically by plotting immigration and extinction rates as a function of the number of species present on an island (**Figure 13.8**). The number of species on the island (S) can range from zero to a hypothetical maximum, P, the number in the pool of species that is

TABLE 13.1   *Key Terms Used in MacArthur and Wilson's Equilibrium Theory of Island Biogeography*

Term	Definition
Colonization	The relatively lengthy persistence of an immigrant species on an island, especially where breeding and population increase are accomplished
Extinction	The total disappearance of a species from an island (which does not preclude recolonization)
Immigration	The process of arrival of a propagule (see below) on an island not occupied by the species (implying nothing concerning the subsequent duration of the propagule or of its descendants)
Propagule	The minimum number of individuals of a species capable of successfully colonizing a habitable island (the minimal unit required—a single mated female, an adult female and a male, or a whole social group, depending on the species)

Source: After MacArthur and Wilson 1967: 185–191, glossary.

Note: In practice, very few studies in island biogeography have direct measurements of immigration rates, because immigration is a difficult process to measure and observations rarely qualify for the definition given here. Hence, it is often the case that data presented as immigration rates are really closer to being colonization rates as originally defined.

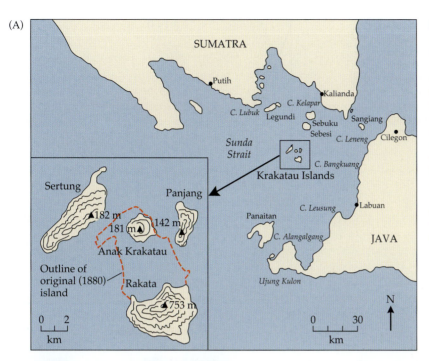

(A)

FIGURE 13.7 The Krakatau islands are located in the Sunda Strait between the Indonesian islands of Sumatra and Java. Rakata, Sertung, and Panjang are remnants of at least two caldera collapse events, with the largest island, Rakata, losing two-thirds of its land area in the most recent collapse of 1883. The eruptions culminating on August 27, 1883, are generally thought to have eliminated all life from the three islands. Within just 50 years, however, extensive forests developed on the larger islands (B–D). Anak Krakatau (E–G) is a relatively young island that emerged from a submarine volcanic vent in 1930: It is already substantially higher than indicated in this 1983 map. (After Whittaker and Jones 1994; photos courtesy of Robert J. Whittaker.)

FIGURE 13.8 A simple model in which the number of species inhabiting an island represents an equilibrium between opposing rates of colonization (immigration) and extinction. Note that the immigration rate declines and the extinction rate increases as the number of species increases from zero to P, the number in the mainland species pool. The point of intersection of the two curves represents a stable equilibrium, because if the number of species is displaced from ($\hat{S}$) to either higher (S'') or lower (S') numbers, it will return to equilibrium (arrows).

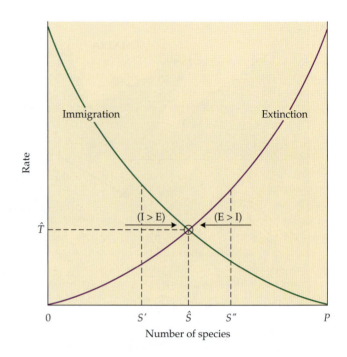

available to colonize the island from a nearby continent or other source area. Next, we need to consider the shapes of the curves representing colonization (immigration) and extinction rates. Here we can start, as MacArthur and Wilson originally did (above), with the "thought experiment" of an empty island. The immigration rate (defined as the rate of arrival of propagules of species not already present on the island) must decline from some maximum value, when the island is empty, toward zero, should the island ever contain all the species in the pool. As the number of resident species grows, there remain fewer new species on the mainland to colonize the island. Conversely, the extinction rate (defined as the rate of loss of existing insular species) should increase from zero, when there are no species present on the island to become extinct, to ever higher values, as the island fills with both individuals and species, with each new addition meaning that more species will have tiny population sizes, increasingly vulnerable to extinction. This argument assumes a fixed carrying capacity for the number of species on an island, a point we return to in the final section of this chapter, on the frontiers of island biogeography.

At some number of species between zero and P, the lines representing the immigration and extinction rates cross. At this point the two rates are exactly equal, resulting in an equilibrial number of species, $\hat{S}$, and an equilibrial rate of species turnover, $\hat{T}$. This point represents a stable equilibrium because if the number of species is perturbed from this value, it should (at least theoretically) always return. For example, suppose that a natural disaster, such as a hurricane, causes the extinction of several insular species, temporarily reducing the number of species from $\hat{S}$ to S' (see Figure 13.8). Then, the immigration rate will exceed the extinction rate, and the island will accumulate species until it has again reached $\hat{S}$. On the other hand, a departure above the long-term equilibrium value might result from a temporary elevation of the immigration rate (e.g., due to lowered sea level or a shift in weather conditions and wind/ocean currents) or from a temporary amelioration of climate (e.g., increased rainfall linked to an El Niño event) resulting in an elevated carrying capacity. In such circumstances, S may shift briefly from $\hat{S}$ to a larger number, S'', then, as normal conditions return, the extinction rate will be greater than the immigration rate, and species will be lost until $\hat{S}$ is restored.

Now let us incorporate the effects of island size and isolation into this model. MacArthur and Wilson assumed that the size of an island would affect only the extinction rate. Although they recognized that a large island would provide a larger target for dispersing propagules than a small one, they reasoned that such an effect on immigration rate would be insignificant compared with the importance of island size in extinction. Greater area allows more individuals and hence more species to persist, hence in the graphical model shown in **Figure 13.9**, the rate at which extinction rate rises to match immigration rate is suppressed in a large island compared with a small one, and its equilibrium point is shifted to the right: Larger islands hold more species at equilibrium and display lower rates of turnover than smaller ones of identical isolation.

MacArthur and Wilson used similar logic to show how immigration curves would be influenced by isolation. They assumed that the distance of an island from the source pool would affect only the immigration rate. No matter what the mechanism of dispersal, if a barrier exerts a filtering effect, then the probability that an organism will cross the barrier decreases as the width of the barrier increases. Hence, not only must the immigration rate for an empty island start at a lower point the more remote the island is, but the rate of decline in immigration rate over time will be slower. This is shown diagrammatically in Figure 13.9, from which it can be seen that the model predicts differing equilibrium richness and turnover combinations as functions of varying isolation and area of islands. The resultant graph depicts four intersections of immigration and extinction rate curves, one for each combination of island size (large = L and small = S) and distance (near = N and far = F). The number of species at equilibrium is predicted to be in the order $S_{SF} < S_{SN} \sim S_{LF} < S_{LN}$. Imagine that these four curves specify the extremes

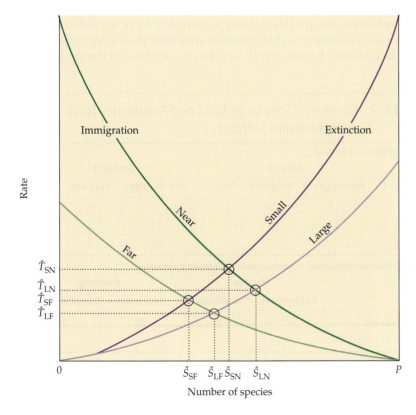

FIGURE 13.9 MacArthur and Wilson's (1963) equilibrium model of island biogeography, showing the effects of island size (different extinction rate curves) and isolation (different immigration rate curves) on the equilibrium number of species ($\hat{S}$) and rate of species turnover ($\hat{T}$). Intersections of the curves for islands of different combinations of size and distance can be used to predict the relative numbers of species and turnover rates at equilibrium.

of isolation and area for a given archipelago; then it will follow that all other islands must have curves that fall within the bounds of the lines shown on the figure. While the thought experiment allows us to follow the curves toward these equilibrium points, once they are reached, only minor deviations from equilibrium may then occur. This means that the extensions of the immigration and extinction curves to the right of the equilibrium points are irrelevant: Immigration and extinction rates should remain at approximately the same levels indefinitely.

The model is thus based on three core premises: (1) the number of species increases with area, (2) the number of species decreases with isolation, and (3) there is a continual turnover of species; the first was universally accepted, the second was generally accepted, and the third was less widely acknowledged at the time the theory was proposed. Given this, much attention has since focused on whether islands do indeed demonstrate measurable rates of equilibrial turnover. Like any good theory, however, the model goes beyond its initial premises to make additional predictions that follow as a consequence of the logic of the model. Specifically, it predicts the following order of turnover rates at equilibrium: $\hat{T}_{SN} > \hat{T}_{SF} \sim \hat{T}_{LN} > \hat{T}_{LF}$. The model also predicts the relative rates at which islands of different sizes and degrees of isolation should return to equilibrium if the biota is perturbed. For example, a near island should return to equilibrium more rapidly than a distant island of the same size because it should have a higher immigration rate, but the same extinction rate function.

Having derived their theoretical model, MacArthur and Wilson (1963) turned to a readily available test data set: that derived from the recolonization history of bird species to the Krakatau islands. Following the defaunation event of 1883, scientists had documented the return of plant and animal life through periodic survey visits, allowing species lists to be compiled for three time points (**Table 13.2**). These data appeared to show a rapid return, by around 1921, to a level of bird species richness close to that expected for the region for islands of their area. By 1932–1934 some turnover in composition had been recorded, but species number appeared to have remained close to the 1921 level—observations that accorded well with the theory. Subsequent work on the Krakatau Islands has shown that this first test of the equilibrium

TABLE 13.2 *Number of Species of Land and Freshwater Birds on Rakata and Sertung*

Number of species found	Rakata			Sertung		
	Nonmigrant	**Migrant**	**Total**	**Nonmigrant**	**Migrant**	**Total**
1908	13	0	13	1	0	1
1919–1921	27	4	31	27	2	29
1932–1934	27	3	30	29	5	34

Number of extinctions and colonizations between censuses	Rakata		Sertung	
	Extinctions	**Colonizations**	**Extinctions**	**Colonizations**
1908 to 1919–1921	2	20	0	28
1919–1921 to 1932–1934	5	4	2	7

Source: After MacArthur and Wilson 1967.

Note: The number of species increased from the census of 1883 to that of 1919–1921 and then remained relatively constant despite extinction of some species and colonization of others.

model was not as definitive as it initially seemed. First, it was based on few surveys of limited duration and efficiency, so the data were rather crude approximations, and second, much of the turnover was due to the widespread replacement of open habitats with closed forest over the course of the 1920s, subsequent to which species numbers of birds and other taxa (e.g., butterflies, plants) have increased further (Bush and Whittaker 1991). Nonetheless, the results were sufficient to encourage further tests of the model.

The description we have given of the MacArthur and Wilson (1963, 1967) theory has focused on the most commonly invoked core properties of their model, but in fact there is a great deal more to their theory. They set out to develop a body of theory building from population processes on ecological time scales through to evolutionary implications and outcomes, a framework that could account for biogeographic patterns across all insular systems. Indeed, it has largely been forgotten that the first formulaic statement of the equilibrium condition (MacArthur and Wilson 1963: 378, their equation 2), was as follows:

$$\Delta s = M + G - D$$

where s is the number of species on an island, M is the number of species successfully immigrating to the island per year, G is the number of new species being added per year by local speciation (not including immigrant species that merely diverge to species level without multiplying), and D is the number of species dying out per year. At equilibrium,

$$M + G = D$$

This statement recognizes that remote oceanic islands receive new colonists so rarely that the evolutionary radiation of species can become a prominent pattern, supplementing immigration as a source of new species. Indeed, the logic of the equilibrium theory generates the prediction that there should be a **radiation zone** on archipelagoes toward the outer limits of the distance a particular taxon can successfully cross (see also Wilson's taxon cycle: Wilson 1959, 1961; Chapter 14). In these remote archipelagoes, so few species of the taxon make landfall that many potential niches remain unfilled. Not only does the provision of empty niche space allow the adaptive radiation of colonist forms on particular islands, but these autochthonous endemic species are then exchanged back and forth within the archipelago as the rates of species exchange among the islands of an archipelago dwarf the rate of immigration from beyond the confines of the archipelago.

Strengths and weaknesses of the theory

Like most important new ideas, MacArthur and Wilson's equilibrium theory elicited a mixed response from other scientists. It generated both swift criticism (e.g., Sauer 1969) and enthusiasm, but more important was that it stimulated a new wave of research in ecological biogeography, focused on islands and other isolates and designed to evaluate or elaborate on MacArthur and Wilson's model. In a review of the equilibrium theory written only a decade after the theory was first published, Simberloff (1974a) cited 121 references, and the pace of research has continued to accelerate, with their 1967 monograph remaining a key point of reference in the primary research literature. Indeed, Schoener (2010) reports that the 1967 monograph was cited over 2000 times in papers recorded in the Science Citation Index in the period 2000–2007.

Several features of the equilibrium theory contributed to its generally favorable reception. Its elegantly simple graphical presentation made the essential elements of the model accessible to a wide audience, including those with minimal mathematical training. Moreover, the equilibrium theory not only introduced stimulating new ideas that helped to bridge the gap between traditional biogeography and ecology, it also presented them in the form of a model that made clear, testable predictions. Many mathematical models, especially those used in ecology, are not empirically operational—that is, they do not explicitly indicate what observations would be necessary to test the theories and reject the models. Such models can still be valuable, however, if they serve a heuristic function—if they cause one to think about a problem in new and more precise ways. MacArthur and Wilson's model certainly plays such a heuristic role, but it also indicates the kind of data that are necessary for a rigorous test. It predicts qualitative trends (increases or decreases) in numbers of species and turnover rates with island size and isolation. These predictions can be tested using simple lists of species inhabiting various archipelagoes at different times. Such lists can be compiled by original fieldwork or by mining previous survey results from the published literature. The only other data required—the areas of the islands and distances to probable source areas—can be obtained from standard maps.

The simplicity of the theory, however, has also been the basis for much criticism. Some have argued that it is so simple as to be useless because it obscures, rather than clarifies, the patterns and processes that make island biogeography interesting (e.g., Sauer 1969; Lack 1970; Carlquist 1974; Gilbert 1980; Williamson 1989). In view of the research the theory has stimulated and the increased understanding of island distributions that has resulted, much of this criticism seems unwarranted. All models are intended to help investigators understand nature by presenting a simplified, abstracted concept of a more complex reality. They inevitably sacrifice a certain amount of precision for clarity and generality. Usually, new theories are presented initially in a very general, incomplete form and are refined as a result of subsequent empirical and theoretical research.

To their credit, MacArthur and Wilson acknowledged some important weaknesses of their theory:

> First, we still know very little about the precise shape of the extinction and immigration curves, so that few numerical predictions can yet be made. Second, and more importantly, the model puts rather too simple an interpretation on the process by making an artificially clear-cut distinction between immigration and extinction. . . . The third difficulty stems from the assumption that extinction and immigration curves have fairly regular shapes for different faunas and different islands and for different times on the same islands. When a new set of curves must be derived for a new situation, the model loses much of its virtue. Deviant cases do occur—for example the Krakatau flora. The extinction curves, furthermore, have a pronounced genetic component: rarity affects gene frequencies, and genetic as well as ecological causes of extinction act in concert. (MacArthur and Wilson 1967: 64–65)

Others were quick to echo these concerns and to identify additional shortcomings of the theory. As you will see from the points listed below, most of these criticisms suggest that the theory is insufficient or incomplete, yet they do not take issue with its basic tenets. That is, insular species richness is a dynamic product of *recurrent* immigration, local extinction, and in situ speciation (at least on remote islands).

Criticisms of MacArthur and Wilson's theory include the following:

1. **Interspecific differences and interactions among species**   The basic model assumes that the identities and characteristics of particular species can be ignored. Immigration, extinction, and turnover are viewed as highly stochastic processes: New species immigrate and existing ones die out more or less at random, and only the approximate number of species

remains the same. It is thus very much a macroecological model, focused on the emergent statistical properties of the system. The implicit assumption that ecological processes—including interspecific interactions—do not determine *which* species can coexist on a particular island is at least technically incorrect. Then again, the equilibrium model was developed to explain patterns in species *richness*, not species *composition*.

2. **Interdependence of immigration and extinction**   The equilibrium theory treats immigration and extinction as independent processes. This is probably justified, given that immigration is defined as the arrival of propagules of new species. However, as the Krakatau system demonstrates, in the buildup toward equilibrium, much turnover is successional, with new recruits driving pulses of extinction through resulting habitat turnover (e.g., Bush and Whittaker 1991). In addition, islands close to source areas should receive high numbers of propagules both of immigrating ("new") species and of those already present on the island, with the latter frequently "rescuing" declining populations from extinction. In a similar manner, area may affect immigration as well as extinction rates because larger islands may intercept more potential colonists (see the later sections on rescue and target area effects).

3. **Biogeographically meaningful measures of isolation**   It may be difficult to identify the source of an island biota without careful investigation of the systematics and historical distribution of the species that are present. Indeed, the species inhabiting a single island may be derived from several sources, including over-water dispersal from continents and other islands, past connections with other landmasses, and in situ speciation. If insular species are acquired by immigration from multiple sources, the theory can potentially be modified to deal with this complication. For example, MacArthur and Wilson considered stepping-stone colonization, in which a species disperses from one island to the next, down a chain of islands. In most cases, when measures of isolation are modified to better reflect the heterogeneity of dispersal barriers, currents, and ecological affinities of the focal species, the effects of isolation and immigrations become clearer (Taylor 1987; Lomolino et al. 1989; Lomolino 1994b; Lomolino and Perault 2000; Lomolino and Smith 2004; Walter 2004; Diver 2008; see Figure 16.33). Put otherwise, the failure to detect a significant effect of isolation on richness of insular communities may be more of a reflection of the difficulties in devising biogeographically meaningful measures of isolation than proof that insular community structure is not influenced by immigration. Walter (2004) suggests that there is a need to capture what geographers distinguish by the word *place* (rather than *space*), that is, measures of area and isolation that better represent the ecological and biogeographic history of places. He terms this concept the "eigenplace." However, the strength and beauty of MacArthur and Wilson's theory is that it is an attempt to find a general model that accounts for the first order pattern in the data, and the challenge for those interested in bringing history of place into the frame of reference is to do so without returning the endeavor to a narrative idiographic (place-by-place) form of biogeography.

4. **Biogeographically meaningful measures of island area**   Total island area provides only a very general and indirect measure of the capacity of islands to support individuals and species. Even for the same island or archipelago, carrying capacity will vary substantially for different groups of species depending on their habitat requirements. In addition, although the extent of most habitat types increases with increasing island size, so usually does the diversity of habitats. Larger islands tend to have higher

mountains, more aquatic habitats, and so on, as well as larger areas of most of the vegetation types found on small islands. Consequently, some of the increase in species diversity with island size may be owing to the addition of specialists whose habitat requirements are met only on large islands. In this case, a more elaborate model, which also incorporates specific habitat variables, should predict patterns of insular species diversity and distribution better than area alone (e.g., Power 1972; Johnson 1975; Triantis et al. 2003, 2008).

5. **The importance of speciation**   It is sometimes stated that the theory ignores speciation. Although in its most basic form and in the familiar graphical representation, MacArthur and Wilson do set the role of speciation to one side, this is not in fact a fair criticism of the theory. They explicitly recognize that speciation becomes of increasing importance on very large and isolated islands on which immigration rate is too low to fill the potentially available niche space. At issue, however, is the applicability of the assumption of a dynamic equilibrium in remote archipelagoes, given the relatively slow rate at which species number can increase as a function of in situ speciation (see Whittaker and Fernández-Palacios 2007).

6. **Disturbance in ecological to geological time scales**   Many insular biotas may not be in equilibrium between opposing processes of immigration and extinction. Rather, the number of species may increase or decrease over evolutionary time, or with major environmental disturbances such as hurricanes and volcanic eruptions. This is particularly likely when immigration occurs on approximately the same time scale as speciation, that is, at a slow pace relative to the geological and climatic events that create, change, and destroy islands. Here, insular communities may never reach or only fleetingly reach equilibrium (Heaney 2000; Whittaker 2000; Whittaker and Fernández-Palacios 2007). An equilibrium theory may still be useful for interpreting these cases, however, because it may be instructive to consider these biotas as approaching a new equilibrium or as relaxing or rebounding toward the original equilibrium following a historical perturbation.

These criticisms and challenges should be kept in mind as we proceed to discuss patterns in insular community structure in this and the following chapter. We will see many cases in which the model, in its original form, is inadequate to account for observed patterns (see Haila et al. 1982; Haila 1986; Case 1987; Williamson 1989). It will also be clear, however, that in following MacArthur and Wilson's general approach, in testing and sometimes rejecting their dynamic theory of island biogeography, we have learned a great deal. As we noted earlier, some biogeographers have drawn a parallel between the heuristic value of MacArthur and Wilson's equilibrium theory and that of the Hardy–Weinberg equilibrium model for gene frequencies (see Heaney 2000, 2007; Heaney and Vermeij 2004). The value of the latter model is not simply that it accurately predicts gene frequencies at equilibrium (a condition seldom achieved in natural populations), but that it identifies populations that deviate from equilibrium and the reasons for this (e.g., genetic drift and natural selection). As MacArthur and Wilson (1967: 19–21) observed, "a perfect balance between immigration and extinction might never be reached . . . but to the extent that the assumption of a balance has enabled us to make certain valid new predictions, the equilibrium concept is useful." Again, given its paradigmatic influence on nearly all disciplines of biogeography and ecology, their theory served the field well. But over the four decades since MacArthur and Wilson first proposed the model, our understanding of the complexity of nature may have advanced beyond the conceptual envelope of the theory. Therefore, after reviewing some of the important tests of the model and related patterns in insular diversity, we conclude this chapter with a preview of some recent attempts to advance or replace this long-reigning paradigm of island biogeography (see also Lomolino and Brown 2009 for a review of the historical development of island biogeography theory).

Tests of the model

Early studies that "tested" and purported to support the equilibrium model fell into three categories: further "snapshot" analyses of species richness variation, tests of changing numbers and turnover rates through time, and experimental manipulations. Most common were analyses of data on insular distributions of various taxa at one point in time. Many of these simply confirmed for more groups of organisms and more archipelagoes the relationship between number of species, island size, and island isolation pointed out by MacArthur and Wilson (e.g., Hamilton et al. 1964; Hamilton and Armstrong 1965; Johnson et al. 1968; Johnson and Raven 1973). Others described similar patterns for other insular habitats, such as mountaintops (Vuilleumier 1970) and caves (Culver 1970; Vuilleumier 1973). Although these studies typically provided general support for the model, they cannot be viewed as rigorous tests of the theory, because they described patterns that were common to alternative hypotheses (e.g., the species–habitat diversity hypothesis). On the other hand, research on biotic turnover, arguably the key to testing MacArthur and Wilson's theory (Gilbert 1980), provided especially important insights into the assembly and dynamics of insular communities.

Biogeographers will continue to gain valuable insights, not just from comparisons among archipelagoes, but also by comparing the community dynamics among functionally different groups of species. For example, Thomas Schoener (1983) discovered some intriguing patterns in his review of 21 studies of turnover across a variety of organisms, ranging from protozoans and plants to terrestrial arthropods and vertebrates. Not only did turnover tend to be lower on larger islands, but it also decreased with the generation time of the organisms (**Figure 13.10**; see also Schoener 2010).

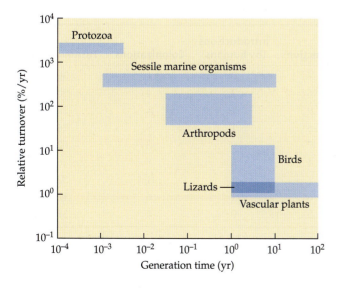

FIGURE 13.10   Relative turnover rates tend to be lower for organisms with longer generation times. (After Schoener 1983.)

Estimates of turnover on land bridge islands

In 1969, Jared M. Diamond reported a more direct test of one crucial prediction: continual turnover in species composition resulting from the balance of colonizations and extinctions. Diamond censused the avifauna of the Channel Islands off the coast of southern California again (**Figure 13.11**) almost exactly 50 years after Howell (1917) had published a detailed account of the birds known to breed on each island at the turn of the century. Comparison of the two censuses revealed striking differences in species composition (**Table 13.3**), although there had been relatively little change in the number of species breeding on each island. In 1968, Diamond observed a number of species not known to breed there prior to 1917, and there was about an equal number of species present 50 years earlier that had apparently disappeared. From these observations, he concluded that at least 20 to 60 percent of the bird species on each island had turned over since 1917. He pointed out that actual turnover rates could well have been even higher; some species might have colonized and become extinct during the intervening 50 years (Diamond 1969). This form or error is termed **cryptoturnover**, and it, along with **pseudoturnover** (cases where species occurrences were missed in an earlier survey and therefore counted as immigrations [turnover] if detected in a later survey), may often confound our interpretations of community dynamics on islands (see Whittaker and Fernández-Palacios 2007). Although there were not enough Channel Island communities to test rigorously for the relationship between turnover rate and either island size or isolation, Diamond noted that apparent turnover appeared to be greatest on the islands with the fewest species (i.e., those that were relatively small and/or isolated; see Table 13.3).

Diamond's study, widely cited as supporting the equilibrium model (e.g., MacArthur 1972), was also vigorously challenged. Lynch and Johnson (1974) pointed out that most of the thoroughly documented changes in the avifauna could be attributed directly to human influence, and some of the other apparent changes may have resulted from errors in conducting and interpreting the census. In particular, most of the extinctions involved the disappearance of large birds of prey, including the osprey (*Pandion haliaetus*), bald eagle (*Haliaeetus leucocephalus*), and peregrine falcon (*Falco peregrinus*), all of which

TABLE 13.3   *Turnover of Breeding Land Bird Species on the California Channel Islands between 1917 and 1968*

Island	Area (km²)	Distance to mainland (km)	Number of species 1917	Number of species 1968	Extinctions	Introductions (by humans)	Colonizations	Turnover (%)[a]
Los Coronados	2.6	13	11	11	4	0	4	36
San Nicholas	57	98	11	11	6	2	4	50
San Clemente	145	79	28	24	9	1	4	25
Santa Catalina	194	32	30	34	6	1	9	24
Santa Barbara	2.6	61	10	6	7	0	3	62
San Miguel	36	42	11	15	4	0	8	46
Santa Rosa	218	44	14	25	1	1	11	32
Santa Cruz	249	31	36	37	6	1	6	17
Anacapa	2.9	21	15	14	5	0	4	31

Source: Diamond 1969.

[a]Turnover rate, expressed as percentage of the resident species per 51 years, is calculated as 100 (extinctions + colonizations)/(1917 species + 1968 species – introductions).

FIGURE 13.11 Locations of selected study areas referred to in this chapter: A, California Channel Islands; B, the montane and intermontane region of the American Southwest, including the Sierra Nevada Mountain Range, Great Basin, and Rocky Mountains; C, Revillagigedo Archipelago; D, Thousand Island Region; E, the Florida Keys, defaunation experiments; F, Mona Island, Puerto Rico; G, Barro Colorado Island, Panama; H, Lago Guri, Venezuela; I, Farne Islands, Great Britain; J, Krakatau islands, Indonesia; K, Kapinga-marangi Atoll, Micronesia; L, islands of the Great Barrier Reef Province, Australia.

were almost certainly eliminated by pesticide poisoning. Many colonizations were a result of the immigration of house sparrows (*Passer domesticus*) and starlings (*Sturnus vulgaris*), which colonized as part of a general regional range expansion after introduction into North America from Europe. Turnover involving sparrows and starlings, and probably other species as well, had also been influenced by anthropogenic habitat changes. Lynch and Johnson concluded that, contrary to Diamond's claims, there was little evidence for "natural" turnover of breeding birds on the Channel Islands within the 50 year period (see also Walter 2000).

This debate, however, was far from resolved. Jones and Diamond (1976) surveyed the birds of the California Channel Islands—especially Santa Catalina—for several years in succession and found year-to-year changes in the breeding status of several species. In a separate but related study, Diamond and May (1976) analyzed many years of careful records of birds breeding on the small Farne Islands off the coast of Great Britain. Several species were reported to have been present only sporadically during the years for which data were available. In both of these studies, however, the "turnovers" (or pseudoturnovers) involved species that were migratory or highly nomadic. Their presence or absence on an island may have been more a matter of the arrival and departure of highly mobile individuals, for which a few miles of water represent no significant barrier, than of the establishment and extinction of real resident populations.

Terborgh and Faaborg (1973) reported apparent turnover in the avifauna of Mona, a small island west of Puerto Rico in the West Indies, as a result of comparisons between an early survey and their own subsequent census. Again, however, the island had been changed substantially by human activities, especially by the effects of introduced goats. Also in contrast to the theory, Lack (1976) reported the relative stasis of birds of Jamaica over more than 200 years of recorded history. On this large Caribbean island, with a resident land avifauna numbering 65 species, there have been just two extinctions and one colonization, and all of these could be attributed directly to human influence (see also Abbott 1983). Studies of turnover on some other isolated oceanic islands, such as those of the avifauna of the Revillagigedo Archipelago, located approximately 400 km south of Baja California, also reveal relative stability and thus seem consistent with Lack's theory of ecological impoverishment on islands (Walter 1998). However, MacArthur and Wilson's theory also predicts relatively low turnover rates on such isolated islands, once they have accumulated their biotas.

Krakatau revisited: Turnover on volcanically active islands

While the species–area and species–isolation relationships fueled the development of the equilibrium theory, biotic surveys of the Krakatau Islands (see Figure 13.7) provided vital clues to the dynamics of insular biotas and the first temporally structured test of the theory. Surveys of the three original Krakatau islands (Rakata, Sertung, and Panjang) in the first 50 years following the 1883 "sterilization event" revealed that they experienced frequent immigration events and some extinctions as species numbers on the islands built up. Bird species numbers appeared to be close to a dynamic equilibrium by the 1920s, but plant species numbers continued to climb. Unfortunately, these surveys were far from standardized, involved varying teams and intersurvey intervals and, because of regional political and security circumstances, they essentially ceased in 1934 (with the exception of limited work in 1951–1952). It was only with the approach of the centenary of the 1883 eruptions that serious survey efforts for plants and several groups of animals began again in 1979 (Whittaker et al. 1989) in a new intensive phase of work that spanned some two decades and involved several teams of scientists (summarized by Thornton 1996).

To fully appreciate the strengths and limitations of the Krakatau research, it is important to understand that survey work on the islands is hampered by significant practical difficulties. For example, there is only one freshwater source on the archipelago, safe landing points exist at only a couple of points for each of the three older islands (see Figure 13.7), and fieldwork is largely constrained to short bursts in the dry season, when the seas and climatic conditions are reasonable. The terrain resembles a fossilized badlands, resulting from swift downcutting and dissection of the soft volcanic ash that initially covered the islands and destroyed any species that survived the explosive eruption. The interiors thus consist of closely packed, steep-sided gullies, many of 20–30 m depth, snaking back and forth in intricate, sometimes un-scalable and disorientating, networks (see Figure 13.7). While surveys of birds benefit from the dispersed nature of their populations, their mobility, and their vocalization, the notion of a complete survey of less mobile taxa, especially plants, on such large and challenging islands is fanciful. Only by compiling lists of species from a series of surveys over a number of field seasons can we be confident that we have a good (but still not complete) list. When this is done, it becomes evident that the strong species-abundance pattern inherent in natural communities confounds attempts to measure turn-over. It is nearly always the rare species whose presence in the lists flickers on and off—species of plants found once or twice beyond the strandline (the area at the top of a beach where debris is deposited) or around a temporary lagoon, such that they are simultaneously on the very edge of presence and of detectability. When a species of fig known from a particular large specimen deep in the interior of upland Rakata disappears for a number of surveys, only to reappear as a large specimen deep in the interior of upland Rakata, the parsimonious explanation of course is that it was there all the time and it was just that no one took the particular turn in the gully network to the right spot. So it is that the analysis of turnover on Krakatau is bedeviled with such problems of pseudoturnover, cryptoturnover, and the genuine comings and goings of species for which the islands present only temporary footholds: the ephemeral members of the species lists. The following points should be read in the light of these problems.

Following the effective sterilization of the islands in August 1883, the process of establishment and succession began via aerial and oceanic dispersal: the first terrestrial colonist detected being a spider. Seeds of sea-dispersed plants colonized the strandlines, and wind-dispersed pioneers began the process of reclothing the interior, initially with ferns, grasses, and herbs. In time, the first restrictively animal-dispersed species of plants arrived, their seeds swallowed by birds and fruit bats prior to embarking on the journeys that took them to Krakatau. These plants included the shrubs and trees that would predominantly clothe the interiors of the islands, as the herbaceous community types gave way to forest over the course of the first three decades of the twentieth century (Whittaker et al. 1989; Whittaker and Jones 1994). As substantial areas of forest became available during the 1920s, the area available to open habitat species crashed, while niche space suitable to the large regional pool of obligate forest dwellers shot up (see Figure 13.7). The emergent outcome of this key phase in the succession is most reliably documented in the data for the largest and best-surveyed island, Rakata. And it took the form of a substantial peak in successful immigration, followed in short order by a peak in extinction as a number of open habitat species either truly disappeared or dropped to such low numbers that they were no longer readily detectable (Bush and Whittaker 1991; Whittaker et al. 2000; **Figure 13.12**). For

FIGURE 13.12 Immigration (*I*) and extinction (*E*) rates (as number of species/year) of animals and plants on Rakata, the largest of the Krakatau islands, as a function of insular species richness (*S*). The different taxa show varying degrees of fit with the expectations of declining immigration, rising extinction rates, and smoothly curved colonization curve required by MacArthur and Wilson's theory (see text). (After Thornton et al. 1993.)

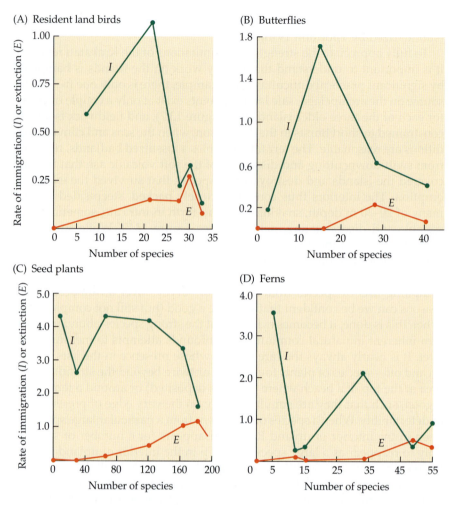

those few taxa where we have reasonably good data, that is, birds, butterflies, and higher plants, the apparent extinction rate has subsequently fallen.

The most recent analyses are based on surveys undertaken between 1979 and about 1984 (Thornton et al. 1990, 1993; Bush and Whittaker 1991; Thornton 1996), with additional plant data collected up to 1994 (see Whittaker et al. 2000). These data, at face value, suggest that resident land bird species numbers have drifted up since the 1930s, but are perhaps today approaching equilibrium (**Figure 13.13**). Butterfly numbers present on the islands have increased substantially since the 1930s, although open habitat specialists (e.g., many Hesperiidae—the "skippers") have not fared so well. While it is tempting to see dynamic equilibria in some of the invertebrate data (e.g., see Odonata and Hesperiidae in Figure 13.13), we have to be cautious about the completeness and comparability of the different surveys and of the role of substantial changes in habitat that may also have driven species turnover (Bush and Whittaker 1991).

The data for reptiles might also appear to indicate some form of dynamic equilibrium, but in fact there have been only two recorded extinctions of stabilized reptile populations, one due to habitat changes linked to coastal erosion and to succession, and the other attributed to catastrophic damage to habitat by eruptions of the new volcano, Anak Krakatau, which formed within the center of the caldera in 1930 (see Figure 13.7). For plants, Bush and Whittaker's (1991) analyses of the data for Rakata up to 1983 fail to show stabilization in species numbers, which continued to climb, while the additional

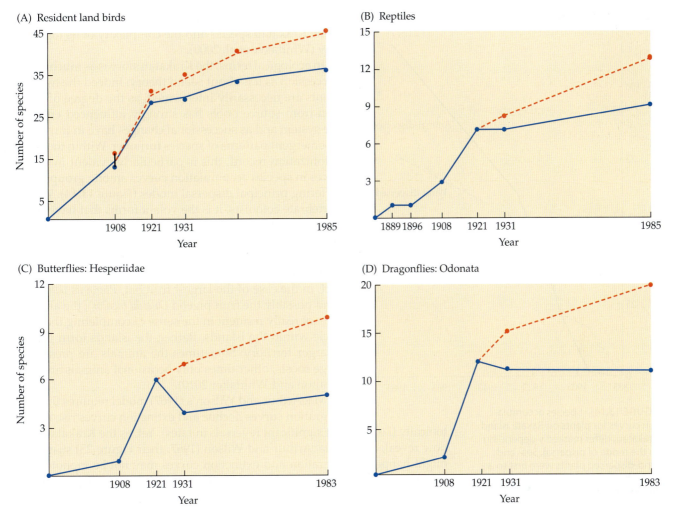

FIGURE 13.13 Colonization curves for various animal taxa in the Krakatau islands: (A) resident land birds, (B) reptiles, (C) butterflies (Hesperiidae), and (D) dragonflies (Odonata). Solid lines indicate actual species numbers at the time the islands were surveyed; dashed lines indicate cumulative number of species recorded (including those present in previous surveys but now absent). (After Thornton et al. 1990.)

survey work undertaken in the late 1980s and early 1990s serves to confirm this trend and also to eat away at the list of species losses (see Whittaker et al. 2000). Indeed, the fairly intensive survey efforts of this period served in part to demonstrate that as more effort goes into a particular survey, more rare species are recorded, driving down estimates of extinction in relation to earlier surveys (see Nilsson and Nilsson 1985 for a quantitative study of similar survey problems in a system of islands in Sweden). Similarly, interpreting the immigration rate data also has its challenges. Recall that MacArthur and Wilson (1967) criticized their own model for "making an artificially clear-cut distinction between immigration and extinction." Immigration is defined as "the process of arrival of a propagule on an island not occupied by that species" (MacArthur and Wilson 1967: 188, 190). Determining what constitutes a propagule is problematic, even assuming that we manage to record the arrival of the potential propagule in the first place! Biotic surveys such as those conducted on the Krakatau Islands, while insightful, typically measure accumulation (or establishment) rates, not immigration rates in this original, strict sense.

What is needed is a better-designed experiment, which means that when recording future Krakataus, we should use far more carefully standardized surveys in which reliable estimates of species abundance are part of the protocol. The effort involved is such that it might be possible for some particularly conspicuous animal taxa, such as birds, but it would be an enormous

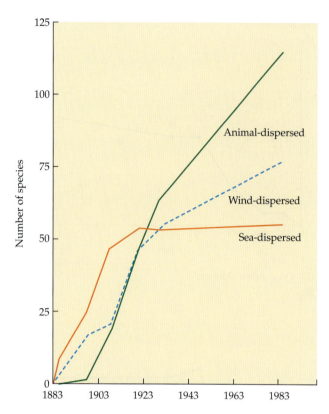

FIGURE 13.14 Species accumulation curves for plants on Rakata Island (Krakatau) differ markedly depending on the mode of dispersal. Sea- and wind-dispersed plants were first to arrive, whereas animal-dispersed plants did not accumulate until successional changes provided suitable habitat for their animal transporters. By the 1920s, however, animal-dispersed plants were accumulating much more rapidly than wind- and sea-dispersed forms. (After Bush and Whittaker 1991.)

FIGURE 13.15 Species–area relationship for birds (A) and mammals (B) on the Krakatau islands (x^1–x^8) and undisturbed oceanic islands of the Sunda Shelf (dark blue circles). Superscripts for birds and volant mammals (bats): 1 = Rakata, 2 = Sertung, 3 = Panjang, 4 = Anak Krakatau (based on total land area), 5 = Anak Krakatau (based on vegetated area only). Superscripts for non-volant mammals (rats): 6 = Rakata, 7 = Sertung, 8 = Panjang. Note that species richness of birds and bats on the Krakatau islands may have rebounded to pre-eruption levels, whereas richness of non-volant mammals remains relatively low in comparison to undisturbed islands of the same area. Anak Krakatau, which emerged from the sea in 1930, still lacked terrestrial mammals in the early 1980s. (After Thornton et al. 1990.)

undertaking for plants, in light of the genuine rarity of some Krakatau species and the complexity of its topography (Whittaker et al. 2000).

In ecological terms, the Krakatau story is as much about succession as it is about island theory: Indeed the two elements are necessarily intertwined, as the degree of isolation from source pools has strongly influenced entry to the system and as successional changes have in turn driven emergent patterns of species turnover. Within the plant colonization record, this is particularly evident in differences in species accumulation curves among groups with differing principal dispersal modes (**Figure 13.14**). Within vertebrate colonists, it is the volant forms that have come closest to establishing species richness consistent with the regional species–area relationships of undisturbed islands (**Figure 13.15**) and for which some degree of continuing species turnover appears detectable (Thornton 1996). In fact, in the case of the largest fruit bat, *Pteropus vampyrus* (and possibly the fruit pigeon *Ducula bicolor*), the animals are not really resident in the sense of completing their life cycle within the islands. Rather, the islands form part of a larger territory: Sometimes the animals are present in large roosts; other times only occasional animals are seen (Shilton and Whittaker 2010).

MacArthur and Wilson's (1967) model requires that immigration curves decline and extinction curves rise monotonically (i.e., without significant reversals in rates). Yet, as the Krakatau data now reveal, and as MacArthur and Wilson (1967: their Figure 23) suspected might happen, immigration and extinction curves may not be monotonic if major successional changes occur. Some species are strongly dependent on other pioneer species to create niches for them, or as key pollinators and agents of dispersal. This seems to be the case for butterflies, land birds, and ferns, whose accumulation rates increased substantially during forest development (1908 to 1931; see Figure 13.13). While we may modify the diagrammatic representation of the equilibrium model to accommodate the major period of successional turnover, as shown in **Figure 13.16**, the suspicion remains that the subsequent behavior of the extinction curve remains stubbornly out of line with the expected trend; rather, as suggested by Bush and Whittaker (1991), approach to equilibrium seems almost entirely driven by declining immigration rate, without detectable increase in rate of extinction. Bear in mind, however, that we cannot be entirely confident that the data really are reveal-

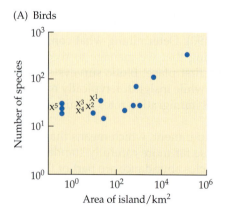

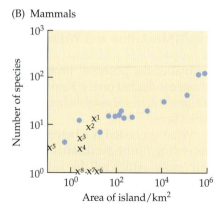

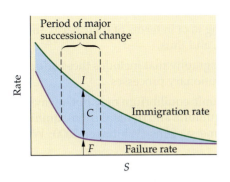

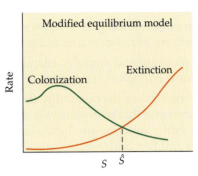

FIGURE 13.16 The effects of succession can be included in an equilibrium model by noting how it alters the likelihood that immigrants will survive and establish a breeding population. S = species richness; $\hat{S}$ = equilibrial species richness; I = immigration rate (number of propagules of new species arriving per unit of time); F = failure rate (number of propagules failing to establish a breeding population per unit of time); and C = colonization rate (= $I - F$). (After MacArthur and Wilson 1967.)

ing the true picture about the arrival of potential viable colonists and their subsequent fates (Whittaker et al. 2000).

As a natural experimental system, the process of succession on Krakatau is a complicating feature, but it is also a general one that theory must accommodate: Newly created islands or those subject to catastrophic damage occur all over the world—all will be sites of ecological succession. In addition, Krakatau has experienced renewed volcanic activity, which has significantly and repeatedly damaged the developing forest cover on Panjang and Sertung islands (but not yet those of Rakata). The islands have also been subject to significant coastal attrition through erosion. These processes have undoubtedly had a role in some of the species turnover recorded over the last 100 years. In describing these dynamics, Bush and Whittaker (1991) noted that the scale and frequency of disturbances affecting Krakatau were significant in the context of the relatively slow dynamics of later successional systems (**Figure 13.17**). If individual forest trees can live and thus hold their ground for many decades, how long must the environment remain stable for ecological dynamics to truly equilibrate? If the approach to equilibrium is as slow, perhaps in the order of hundreds of years, then disturbances such as hurricanes and volcanic eruptions, which occur on similar time scales, may prevent the attainment of a fixed dynamic equilibrium for particular taxa on many island systems.

In summary, Krakatau has provided a number of useful insights pointing to both strengths and weaknesses of the theory as a heuristic device and as a predictive model. The findings from this system point to the need to develop

1. a more sophisticated representation of successional dynamics in the early phases of recolonization,

FIGURE 13.17 For some insular biotas, major disturbances such as volcanic eruptions and hurricanes may occur so frequently that the insular communities seldom achieve a dynamic equilibrium (S_{eq}) as envisioned by MacArthur and Wilson. (A) The shading indicates when immigration rate (I) exceeds extinction rate (E) and therefore when species richness (B) should increase. (After Bush and Whittaker 1991; Whittaker 1995.)

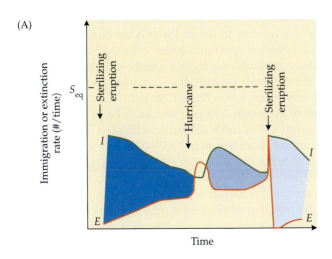

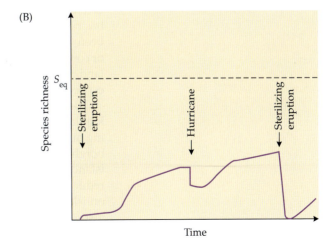

2. an understanding of differences among taxa in terms of their paces and patterns of colonization and turnover, and thus,

3. a more general model of island biogeography that includes these additional ecological, geological, and evolutionary processes.

The work raises the possibility that dynamic equilibrium may not always be attained and that both relatively static equilibria and dynamic non-equilibria may better describe the behavior of some systems (Whittaker 2000; Whittaker and Fernández-Palacios 2007).

We will return to some emerging, more integrative approaches to island biogeography theory in the final section of this chapter.

Turnover on recently created anthropogenic islands

Human activities have fragmented and insularized once-massive and continuous ecosystems across the globe. The effects of these activities on biodiversity are the focus of Chapter 16, but we should acknowledge here that the relevance of island theory for understanding the biotic dynamics of fragmented ecosystems was not lost on MacArthur and Wilson, who featured a classic illustration of the reduction and fragmentation of woodland between 1831 and 1950 in Cadiz Township, Wisconsin, as the first figure in their seminal monograph (MacArthur and Wilson 1967: their Figure 1). Here, we consider the biogeographic dynamics of two anthropogenic archipelagoes created by the flooding of mountainous areas in the American tropics.

Between 1911 and 1914, the Chagras River was dammed to create the Panama Canal and Gatun Lake, in turn flooding lowland areas and transforming forested hilltops into islands. Repeated biotic surveys have been conducted to trace the relaxation (decline toward a new equilibrium) of species numbers on these islands, especially those on the largest island, Barro Colorado (area = 1600 ha). Surveys conducted during the 1970s and early 1980s revealed that about 45 bird species considered probable residents at the point of isolation had disappeared (see Willis 1974; Karr 1982, 1990; Wright 1985). A recent survey, conducted between 1994 and 1996, reported 218 species as having been observed from the island or the waters immediately around it, including 5 new records, none of which were thought to be of breeding species. Consistent with the equilibrium theory and Brown's (1971b) relaxation model (see pages 546–550), the rate of species loss has declined (Robinson 1999), especially for forest interior birds, although edge habitat species have fared poorly as a result of successional loss of open habitats. However, species extinctions—especially for forest interior birds and others with limited abilities, or propensities, for dispersal—continued to exceed colonizations. Overall, the isolation of the hilltops to form Barro Colorado and other islands has been followed by about a century in which the process of relaxation has been the dominant signal. Whether species richness eventually equilibrates or continues to decline likely depends on whether forests of the nearby mainland are maintained, as these forests provide the source of the resident and especially the transient birds observed on Barro Colorado Island.

The second study system considered in this section is Lago Guri, a 4300 km^2 lake created when Caroni Valley in Venezuela was flooded in 1986, creating hundreds of islands ranging from less than 1 to 760 ha. Work by John Terborgh and his colleagues on birds suggests that relaxation on newly created islands within this lake is fastest in its early stages, immediately after area is reduced and isolation increases (Terborgh et al. 1997a). Avian communities on islands of roughly 1 ha may have achieved a new dynamic equilibrium just seven years after the islands were formed (Terborgh et al. 1997b). In contrast, bird species richness still appears to be declining on the larger islands,

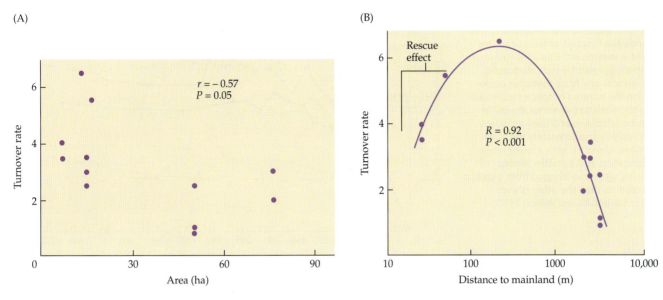

FIGURE 13.18   Consistent with the predictions of the equilibrium theory, turnover rates of birds on islands of Gatun Lake, Panama, tend to decrease with island area (A) and with isolation (B). Yet, because frequent immigrations can rescue otherwise dwindling populations from extinction, turnover rates may also be lower on near islands than on distant islands (e.g., the three near islands in B; see also Figure 13.21). (After Wright 1985.)

a result that is also consistent with the predictions of the equilibrium theory. One especially important discovery was that the relaxation and extinction process was highly selective, resulting in marked faunal imbalances, including the paucity of pollinators and seed dispersers and hyperabundances of populations of persisting species, including selected amphibians, reptiles, ants, spiders, small rodents, and howler monkeys (Terborgh and Estes 2010). The overall result has been a trophic cascade—a sequential collapse of many populations (e.g., plants due to hyperabundance of herbivores and paucity of pollinators and seed dispersers) and release of other species populations in the absence of their natural predators and competitors.

The salient observations with respect to species richness dynamics of both of these anthropogenically shaped systems are consistent with MacArthur and Wilson's theory: Extinctions and immigrations are recurrent, and turnover is lowest on the largest and most isolated islands (**Figure 13.18**). Numerous other studies report a similar inverse relationship between turnover rates and island area or isolation (Brown 1971b; Diamond 1972; Wilcox 1978, 1980; Terborgh and Winter 1980; Nilsson and Nilsson 1982; Heaney 1984, 1986). Although quantitative data on relaxation are limited, it appears that species richness declines most rapidly during the early stages following the loss of area and the gain in isolation (MacArthur and Wilson 1967; Terborgh 1974; Wilcox 1980; see also Lovejoy et al. 1986; Robinson et al. 1992; Kattan et al. 1994; Stouffer and Bierregaard 1995; Lambert et al. 2003). As predicted by the equilibrium theory, relaxation rates then tend to slow as species richness approaches the new equilibrium. Tracking the results of these and similar opportunistic "experiments" is certain to provide additional insights into the dynamic forces structuring insular communities.

Experimental defaunation

The most rigorous, early tests of MacArthur and Wilson's theory were the defaunation experiments conducted by Wilson and his student Daniel Simberloff (see Figure 13.11; Simberloff and Wilson 1969, 1970; Wilson and Simberloff 1969). This study has become justly famous as an example of the use of controlled, manipulative experimentation to test theoretical models in biogeography and ecology. The basic design was simple: All species of arthropods were eliminated from tiny islets of red mangrove (*Rhizophora mangle*) in the Florida Keys, and the subsequent changes were monitored closely. Sim-

FIGURE 13.19 Recolonization by terrestrial arthropods of four small mangrove islands as a function of time since the fauna was removed. The initial number of species present is indicated along the vertical axis. Note that after defaunation the number of species increases rapidly, tends to overshoot the initial number, declines, and then increases gradually to approximately the initial number. Island E1, with a lower rate of colonization and a smaller number of species, was more isolated from a source of colonists than the other islands. (After Simberloff and Wilson 1970.)

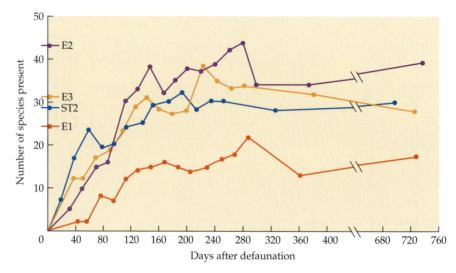

berloff and Wilson hired an exterminator, who used methyl bromide gas to kill all insects, spiders, mites, and other terrestrial animals while leaving the mangrove vegetation virtually undamaged. This was a drastic but effective perturbation. Recolonization, monitored by careful surveys, was surprisingly rapid (**Figure 13.19**). Within less than a year, all but the most distant islands had recovered their initial numbers of species. In fact, the numbers of species increased rapidly and appeared to overshoot the initial numbers before declining and stabilizing close to the initial values. Furthermore, there was a great deal of turnover, even after the number of species had stopped changing significantly. Individual species colonized and disappeared, sometimes repeatedly, during the short-term study. The high turnover rates were not surprising, given the proximity of these islands to the mainland (0.002 to 1.2 km), their small size (75 to 250 m^2), and the lack of intervening dry land.

Thus, Simberloff and Wilson's results strongly supported several predictions of the equilibrium model (see also Molles 1978; Rey 1981; Hockin 1982; Strong and Rey 1982). Although there were too few islands to test rigorously for the predicted relationships of initial colonization rate, equilibrium turnover rate, and equilibrium number of species to island size and isolation, the results were generally consistent with the predictions. For example, the most isolated island (E1 in Figure 13.19) had the fewest species and the lowest rate of recolonization. Simberloff and Wilson also suggested that developing communities may pass through three—and possibly four—types of dynamic equilibria (**Figure 13.20**; Wilson 1969). First, species may tend to accumulate rapidly to reach a "noninteractive" equilibrium, with most species occurring at relatively low population levels—perhaps too low to inhibit populations of other species. Populations may later increase to the point at which interspecific interactions can cause local extinctions of some species, resulting in a reduced or "interactive" equilibrium. Simberloff and Wilson's studies also suggest that succession or ecological sorting may generate a subsequent increase in species richness toward an "assortative" equilibrium (i.e., one with a combination of species more likely to coexist). Finally, on the largest and most isolated islands, species richness may continue to increase until extinction rates balance the combined effects of immigration plus speciation ("evolutionary" equilibrium). Thus, these experimental tests led to a graphical modification (see Figure 13.20) of the basic MacArthur–Wilson model to incorporate the importance of ecological interactions among species, aspects of the dynamics of species composition as well as species richness, and the

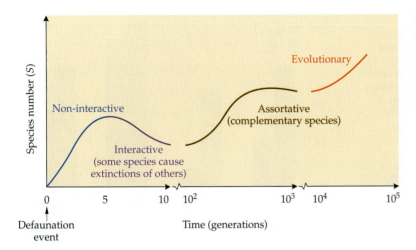

FIGURE 13.20 As an empty island accumulates species, its insular communities may pass through a series of equilibria reflecting demographic, ecological, and evolutionary processes, with the time scale for the process of assortative adjustment longer than for the initial equilibrium, and that for the evolutionary adjustment even longer; hence the breaks in scale indicated on the *x*-axis. (Modified from Simberloff and Wilson 1969, 1970; Wilson 1969.)

influence over time of all three, fundamental biogeographic processes—immigration, extinction, and speciation.

Additional patterns in insular species richness

As we discussed earlier, the equilibrium theory assumes that extinction is affected only by area and that immigration is affected only by isolation. Here, we consider violations of these assumptions that may warrant modification of the original model.

The rescue effect

Another study of arthropods on isolated patches of vegetation points to a potentially important problem with MacArthur and Wilson's equilibrium model. James H. Brown and Astrid Kodric-Brown (1977) censused arthropods (mostly insects and spiders) on individual thistle (*Cirsium neomexicanum*) plants growing in desert shrubland in southeastern Arizona. Although the intervening habitat may have been suitable for some of the arthropod species, the thistle plants constituted isolated patches of favorable habitat. Brown and Kodric-Brown counted the individuals and species of arthropods on the plants at five-day intervals. Their results confirmed several major predictions of the MacArthur–Wilson model. The number of individuals and species increased with plant size and decreased with increasing distance from the nearest plants. Although the arthropods did not maintain real populations on the plants, there was a dynamic equilibrium between the rates of arrival and disappearance. Defaunated thistles were reinhabited rapidly, those near other thistles more quickly than isolated plants; plants closely surrounded by other plants regained 94 percent of their original arthropod biota in 24 hours, whereas isolated plants acquired only 67 percent of the initial number of species in the same period. The turnover of individuals and species was higher on small plants than on large ones.

All of these results were consistent with the predictions of the equilibrium theory. However, contrary to the model's prediction, turnover rates were lower on plants in close proximity to others than on isolated ones (**Figure 13.21**). This single exceptional result is important, because it suggests a problem with the model that may be as relevant for organisms on real islands. The most likely explanation for all of the results taken together is that there is an insular equilibrium maintained by opposing mechanisms of immigration and extinction, as envisioned by MacArthur and Wilson, but the factors affecting the arrival of new species are not independent of those influencing

FIGURE 13.21   Turnover of arthropod species on individual thistle plants. Method: In order to study the relationships between turnover of "insular" populations, Brown and Kodric-Brown (1977) divided the plants into objective size and isolation categories on the basis of the number of flowers and the number of other plants in the immediate vicinity. Turnover rate was calculated as the number of species present only in the first census, plus the number of species present only in the second census, divided by the total number of species present in both censuses. Key finding: Although turnover rates are lower on larger "islands" of thistle plants, turnover also was lower on less isolated plants (compare Large-far to Large-near islands, and Small-far to Small-near islands). The latter result was inconsistent with MacArthur and Wilson's equilibrium theory (see Figure 13.8), but is an important result that Brown and Kodric-Brown termed the rescue effect.

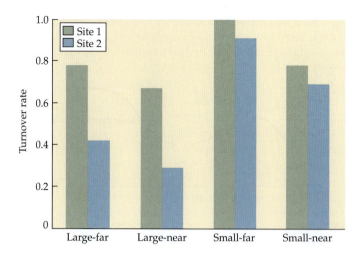

the extinction of species already present. Proximity to a source of immigrants increases the immigration rate of all species, and a continual influx of individuals belonging to species already present tends to prevent the disappearance of those species.

In the case of arthropods on thistles, this **rescue effect** is probably simply statistical: High rates of immigration reduce the probability that a species will temporarily be absent and, hence, recorded as a turnover. On real islands, however, immigrants may rescue populations from extinction by contributing to the breeding stock and by injecting new genetic variability to counteract the deleterious effects of inbreeding, which can be severe in small, isolated populations. In a separate study, Andrew T. Smith (1980) corroborated the rescue effect of immigrants on extinction rates, showing that populations of pikas (*Ochotona princeps*, rat-sized mammals similar in appearance to small rabbits) inhabiting isolated rockslides in North America had higher turnover rates than those near a source of colonists. Other researchers have subsequently reported, or more commonly merely inferred, rescue effects for a variety of organisms (e.g., Wright 1985; Laurance 1990; see Figure 13.18B), and it has become a fundamental feature of metapopulation theory (Gilpin and Hanski 1991; Hanski and Gilpin 1997; Hanski 1999).

The rescue effect was not anticipated by MacArthur and Wilson, but it is easy to modify their model slightly to incorporate it. Drawing different extinction rate curves, as well as different colonization rate curves, for near and far islands (**Figure 13.22A,B**) allows us to take into account the decrease in the extinction rate on near (vs. distant) islands that is due to the rescue effect. Note that when this is done, it may reverse the order of the equilibrium turnover rates ($T_F > T_N$) from that predicted by MacArthur and Wilson's original model, but the predicted relationship for the equilibrium number of species remains unchanged ($S_N > S_F$).

Brown and Kodric-Brown's study emphasizes the importance of testing all predictions of a model as well as critically evaluating its basic assumptions. Unless this is done, investigators risk misinterpreting data that are merely consistent with the model as corroborating evidence, and they miss an opportunity to advance the theory.

The target area effect

Just as extinction rates may be influenced by island isolation, immigration rates may be influenced by island area because larger islands are more likely to be detected by active immigrators and have larger perimeters with which to intercept passive dispersers (Gilpin and Diamond 1976; Buckley and

(A) MacArthur and Wilson's model

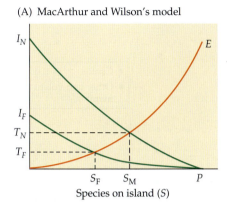

Species on island (S)

(B) Rescue effect

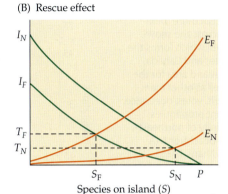

Species on island (S)

I =	Immigration rate
E =	Extinction rate
T =	Turnover rate
S =	Species richness
P =	Richness of the source or "pool" biota
s =	Small island
L =	Large island
N =	Near island
F =	Far island

(C) MacArthur and Wilson's model

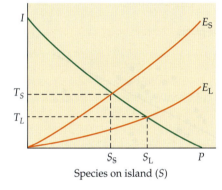

Species on island (S)

(D) Target area effect

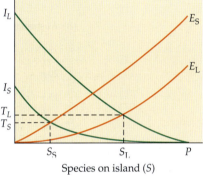

Species on island (S)

FIGURE 13.22 Two modifications of MacArthur and Wilson's equilibrium theory of island biogeography (A and C) can be made, to take observed patterns into account. The rescue effect (B) refers to the reduction in extinction (and therefore turnover) rates on near islands because recruitment of immigrants can supplement otherwise dwindling populations. The target area effect (D) refers to the tendency for immigration rates to be higher on larger islands. (After Gotelli 1991.)

Knedlhans 1986). Again, this is a violation of one of the assumptions of the equilibrium model.

In illustration, Hanski and Peltonen (1988) found that colonization rates of shrews on islets in a Finnish lake increased with island area. While this suggests a target area effect for active immigrators, it is important to note that colonization includes the survival and establishment of breeding populations, as well as immigration (see Table 13.1). Lomolino (1990) was able to conduct a more direct test of the target area hypothesis by tracking the movements of terrestrial mammals across the ice-covered St. Lawrence River of North America during winter. The characteristics of tracks in the snow allowed the identification of species and the mapping of actual movements among islands and sites along the mainland. Immigration rates of both small and large mammals were significantly and positively correlated with island area (**Figure 13.23A**).

In a very different set of studies of plant propagules drifting onto the beaches of oceanic islands in the Great Barrier Reef Province of Australia, Buckley and Knedlhans (1986) also found that immigration rates increased with island size (**Figure 13.23B**). Thus, there is direct evidence, albeit limited, that at least some of the species–area relationship can be attributed to the target area effect. Again, this calls for a slight modification of the equilibrium model, one that alters relative turnover rates but not predicted rank order patterns in species richness (see **Figure 13.22C,D**; see also Hamilton et al. 1964; Connor and McCoy 1979; Coleman et al. 1982; McGuinness 1984).

Unfortunately, if both the rescue effect and target area effect are operative and significant for an insular system, it effectively means that immigration and extinction are each influenced by area and isolation, and hence turnover patterns may depart unpredictably from those anticipated by the MacAr-

FIGURE 13.23 Two examples of the target area effect—the tendency for larger islands to attract or intercept more immigrants. (A) Non-volant mammals dispersing across the snow-covered ice to islands of the St. Lawrence River, which forms the border between the northeastern United States and Canada. Immigration rates increase significantly with island area for the canids (primarily coyotes and red foxes) as well as other mammals (including raccoons, weasels, red squirrels, voles, deer mice, and shrews). (B) The diversity of seaborne plant propagules found along the beaches of islands in the vicinity of northeastern Australia increases with the length of the beachfront. (A after Lomolino 1990; B after Buckley and Knedlhans 1986.)

(A) Non-volant mammals

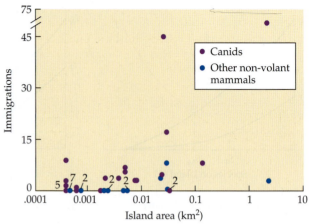

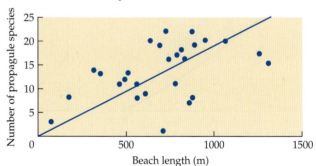

(B) Australian seaborne plants

thur–Wilson model. Accepting these modifications increases the generality of the model at the expense of its simplicity and predictive power.

The small island effect

In their 1967 monograph, MacArthur and Wilson remarked that the species–area relationships of very small islands may be "truly anomalous." They referred to Niering's (1963) study of higher plants on islands of the Kapingamarangi Atoll, Micronesia, which indicated that species richness appeared to be independent of area for the smaller islands (those of less than 3 acres; **Figure 13.24**). Other studies have also reported this so-called **small island effect**, al-

FIGURE 13.24 The small island effect refers to the tendency of the species richness of some insular faunas to remain relatively low and independent of area for the smallest islands. This pattern was found among the higher plants of the Kapingamarangi Atoll, Micronesia. (After MacArthur and Wilson 1967; data from Niering 1963.)

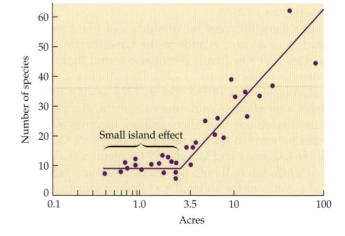

though it is not always detected even when the islands concerned include very small ones (see Whitehead and Jones 1969; Dunn and Loehle 1988; Barrett et al. 2003; Panitsa et al. 2006). Lomolino and Weiser (2001) assessed the relative importance of the small island effect for just over 100 insular biotas. They found small island effects to be common and reported that the range in island areas over which it is detectable varies in a manner consistent with the resource requirements, immigration abilities, and degree of isolation of these biotas. The upper limit of the effect tended to be highest for species groups with relatively high resource requirements and low dispersal abilities, and for biotas of more isolated archipelagoes. More recent work has raised questions over whether the use of log-log data transformation followed by the application of break-point regression might lead to the erroneous detection of small island effects. However, deployment of alternative transformations, regression techniques, and null model analysis indicates that small island effects do indeed occur in many island systems (see Williamson et al. 2001, 2002; Gentile and Argano 2005; Panitsa et al. 2006; Triantis et al. 2006; Burns et al. 2009; Sfenthourakis and Triantis 2009).

Lomolino (2000c, 2002) has suggested that perhaps the most important and most general inference from these studies is that the species–area relationship may be fundamentally different from the pattern we have been describing for well over a century. The relationship may actually be sigmoidal, and if so, future studies in island biogeography may well focus on three fundamentally different realms of the species–area relationship (**Figure 13.25**; Lomolino 2000c, 2002; Lomolino and Weiser 2001; Williamson et al. 2001, 2002; but see also Gentile and Argano 2005):

1. small islands where species richness varies independent of area, but with idiographic differences among islands and with catastrophic events such as hurricanes (see also Barrett et al. 2003);

2. islands beyond the upper limit of small island effects where richness varies in a more deterministic and predictable manner with island area and associated, ecological factors;

3. islands large enough to provide the internal geographic isolation (large rivers, mountains, and other barriers within islands) necessary for in situ speciation.

Similarly, for reasons discussed in Chapter 6 (see Figure 6.19) and earlier in this chapter (see Figures 13.3 and 13.5B), the species–isolation relationship may also be sigmoidal—this time, a negative sigmoidal. Taken together, these patterns suggest that species richness tends to decline most rapidly

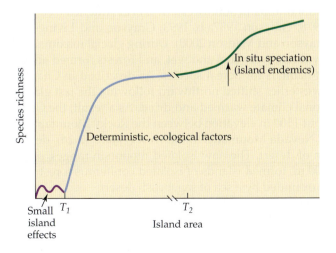

FIGURE 13.25 A general model for the species–area relationship proposed by Lomolino (2000a,b) that includes scale-dependent changes in the principal factors structuring insular communities, including (1) small island effects and idiographic differences among islands and effects of hurricanes and other stochastic extinction forces—predominantly on the small islands (i.e., those < the ecological threshold T_1); (2) more deterministic, ecological factors associated with habitat diversity, carrying capacity, and extinction/immigration dynamics as envisioned by MacArthur and Wilson (1967)—on islands of intermediate size; and (3) in situ speciation—on the relatively large islands (i.e., those > the evolutionary threshold T_2).

across intermediate levels of area or isolation, presumably those levels that approximate the modal resource requirements or dispersal distances of the species pool.

Nonequilibrium biotas

Comparing observed patterns of insular species diversity with the predictions of the MacArthur–Wilson model reveals many cases in which the diversity clearly does not achieve equilibrium between contemporary rates of immigration and extinction. The number of species on these islands is not remaining approximately constant; instead, it is either increasing or decreasing steadily in response to major historical events. Hence, MacArthur and Wilson's approach to island biogeography has, perhaps paradoxically, proven valuable for detecting those systems that are characterized by nonequilibrium biotas. Here, we feature three case studies of nonequilbria, which were created by dynamics in climates, landscapes, and sea levels associated with glacial cycles of the Pleistocene.

Pleistocene refugia

As pointed out in Chapter 9, changes in climate and sea level during the Pleistocene caused major shifts in the distributions of organisms on the continents, especially terrestrial and freshwater species. Their effects on certain insular habitats also were quite profound. In some cases, isolated habitat islands were connected by habitat bridges, permitting a free interchange of biotas; in other instances, new islands or habitat islands were created by the intrusion of formidable barriers to dispersal.

The legacy of these Pleistocene perturbations is still apparent in many insular distributions. Many islands and patches of isolated habitat were formed by rising sea levels and climatic changes at the end of the Pleistocene. Sometimes the fragmentation of extensive areas of once-continuous habitat left small islands oversaturated with species and, following the course of biotic relaxation, diversity began to decrease as certain species were eliminated by extinction. In other cases, completely new insular habitats were created and began to acquire species by colonization and speciation. The extent to which contemporary biotas still show the effects of these historical changes depends largely on the kinds of barriers isolating these post-Pleistocene islands and on the ability of different kinds of organisms to disperse across them.

In one of the first attempts to test the applicability of the MacArthur–Wilson model to the distributions of organisms among habitat islands, Brown (1971b, 1978) studied the small nonflying mammals—and later, the birds—inhabiting isolated mountaintops in southwestern North America (see also Johnson 1975; Behle 1978; Grayson 1987a,b, 1993; Grayson and Livingston 1993; Grayson 2000; Grayson and Madson 2000). During glacial maxima, the climate of this region was relatively cool and wet, and forests formed an expansive "mainland" of habitats for boreal species. As described in Chapter 9, the moisture-laden westerlies shifted northward following the last glacial recession, and the regional climate warmed and dried. As a result, the Great Basin, which lies at about 1500 m elevation between the Rocky Mountains to the east and the Sierra Nevada Mountains to the west (**Figure 13.26**), developed into a vast region of desert. The now isolated mountain ranges, some rising to more than 3000 m, formed islands of isolated coniferous forest and other mesic habitats surrounded by a sea of sagebrush desert.

These isolated mountaintops are now inhabited by a number of boreal (northern forest) mammal and bird species that are restricted to the cool, moist habitats of higher elevations. The patterns of diversity of these mam-

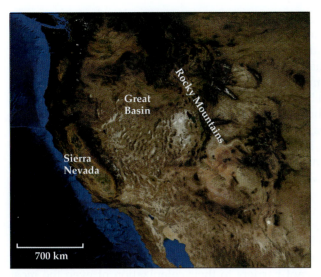

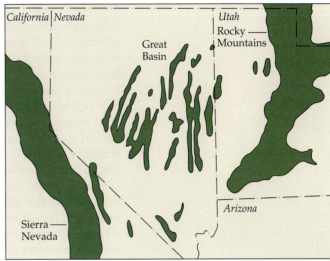

mals and birds exhibit both similarities and differences when compared with each other and with the biotas of oceanic islands. As predicted by MacArthur and Wilson's theory, the number of boreal forest species increases with the size (area) of the mountain range (**Figure 13.27**). However, there is no detectable effect of isolation by distance on the diversity of either taxon.

Brown (1971b) proposed that the present boreal mammal populations of these montane forest islands are relicts, vicariant remnants of once-widespread distributions during the Pleistocene. This model is consistent with plant fossils, which show that as recently as 10,000 to 12,000 years ago, the climate of the Great Basin was cooler and wetter, and the vegetation zones were shifted several hundred meters below their present elevations (Wells and Berger 1967; Wells 1976, 1979; Thompson and Mead 1982). This shift would have connected several presently isolated habitats across the entire Great Basin, permitting all forest islands to be colonized by all species for which appropriate habitat bridges existed. At the end of the Pleistocene, however, the cool, mesic habitats shrank back to higher elevations, completely isolating on the mountaintops those boreal mammal species unable to disperse across the desert valleys (see Grayson 1993). After being isolated, some insular populations became extinct, reducing the diversity of the boreal biota—especially on small mountaintops—but relictual populations of other species have survived until the present.

FIGURE 13.26 Isolated mountain ranges of the Great Basin in western North America are islands of cool, mesic forest habitat in a sea of sagebrush desert. The ranges shown, with peaks mostly higher than 3000 m, lie between two montane "mainlands": part of the Rocky Mountains to the east, and the Sierra Nevada Mountains to the west. The desert valleys between the mountains are readily crossed by birds, but they are virtually absolute barriers to dispersal of many small mammal species. (After Barbour and Brown 1974; photo courtesy of Reto Stockli, NASA Earth Observatory.)

FIGURE 13.27 Species–area relationships for the boreal resident birds (A) and small terrestrial mammals (B) inhabiting the isolated mountain ranges of the Great Basin. Birds continually recolonize these mountains, whereas the mammals are relicts of more widespread populations that existed during the Pleistocene. (After Brown 1978.)

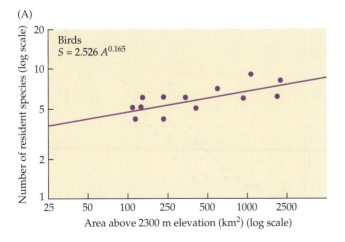

(A) Birds
$S = 2.526\, A^{0.165}$

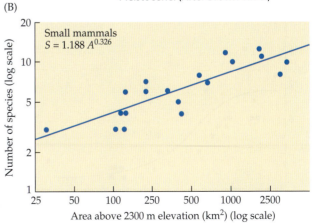

(B) Small mammals
$S = 1.188\, A^{0.326}$

Brown's relaxation model proposes that, in the absence of post-Pleistocene immigration, the mammalian faunas of the isolated mountaintops have been relaxing toward an equilibrium of zero species. Small islands have had high extinction rates and have lost most of their fauna in 10,000 years, whereas large islands still retain most of their original species. One additional observation supports this relaxation model. Late Pleistocene or more recent fossils of boreal species, including some forms still found on some of the larger isolated mountain ranges, have been found in the intermontane valleys and on several mountaintops, indicating that these species were indeed once present and have become extinct within the last 12,000 years (Grayson 1981, 1987a,b, 1993; Thompson and Mead 1982).

Paradoxically, birds also fail to exhibit significant species–isolation relationships, in this case because they are excellent dispersers. In contrast to mammals, boreal bird species appear to be present on all mountain ranges where there are sufficient areas of suitable habitat (Johnson 1975; Behle 1978). Boreal birds have been observed flying across the desert valleys, which appear to pose no significant barriers to their colonization of even the most isolated mountaintops. A high immigration rate apparently continually replenishes bird populations on many small islands. Thus, the distribution of boreal birds may well represent a dynamic equilibrium. However, even this is not quite the sort of equilibrium predicted by MacArthur and Wilson, because the immigration and extinction processes appear to be highly deterministic instead of stochastic. The rate of colonization is so high that habitats are almost completely saturated with those species that can live there. When, on rare occasions, an insular population does become extinct, it probably is replaced rapidly by individuals of the same (or ecologically similar) species.

Other studies also indicate that relaxation rates of supersaturated communities are inversely proportional to island area and tend to be most rapid during the period immediately following isolation (e.g., see Diamond 1972; Wilcox 1978, 1980; Terborgh et al. 1997). It may, however, be invalid to assume that all habitat islands are inhabited by fauna in perpetual states of relaxation. Even Brown's interpretation of the Great Basin mammals may need slight modification, because some species may occasionally disperse across xeric habitats (see Grayson and Livingston 1993; Lawlor 1998; Waltari and Guralnick 2009).

Still, it is likely that most species of forest mammals in the Great Basin, especially in the more xeric, southern reaches of this region, are completely isolated by the intervening deserts (see Lomolino and Davis 1997). In contrast, intermountain habitats in the American Southwest, which lie south of

FIGURE 13.28 (A) Species–isolation and (B) species–area relationships for non-volant forest mammals on 27 montane "islands" in the American Southwest (primarily Arizona, New Mexico, southern Colorado, and southern Utah). Because these mesic forests are isolated by "seas" of xeric woodlands and deserts, their biotic communities often exhibit biogeographic patterns similar to those of species on true islands. (After Lomolino et al. 1989.)

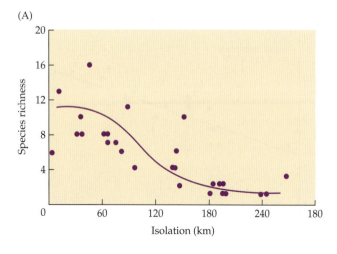

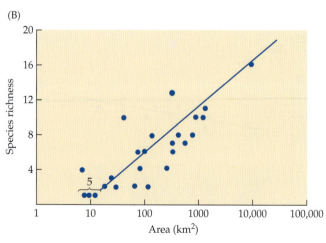

the Great Basin, may well represent filters, not perfect barriers, for the movements of many forest mammals (see Figure 9.22). The key difference between these two regions is that most mountain forests of the American Southwest are isolated by woodlands, not deserts. As a result, post-Pleistocene immigrations are likely for many, if not most, species, and mammalian communities may have approached a dynamic equilibrium as envisioned by MacArthur and Wilson. As illustrated in **Figure 13.28**, the species richness of forest mammals in this region is significantly correlated with both isolation and area. In addition, intermountain dispersal and post-Pleistocene colonization have been documented for a number of these species (**Figure 13.29**; Davis and Dunford 1987; Davis and Brown 1989; Lomolino et al. 1989; Davis and Callahan 1992; Brown and Davis 1995; Lomolino and Davis 1997).

Again, however, we caution against overgeneralizing from these studies. Patterson (1980, 1984, 1995) has provided convincing evidence that at least some of these communities, especially those on the more isolated mountains in southern New Mexico, may truly be Pleistocene relicts (see also Frey et al. 2007). Just as Brown observed in the Great Basin, some of the very distant islands of the American Southwest are surrounded by deserts as well as woodlands, making post-Pleistocene immigration very unlikely, if not impossible, for many mammal species. Thus, mammalian communities on the more isolated mountains of this region still may indeed be relaxing from supersaturated levels of richness in the absence of contemporary immigrations.

We highlight two concluding points about the North American montane mammal system. First, heterogeneous landscapes and species pools render simple measures of geographic isolation inadequate for assessing immigra-

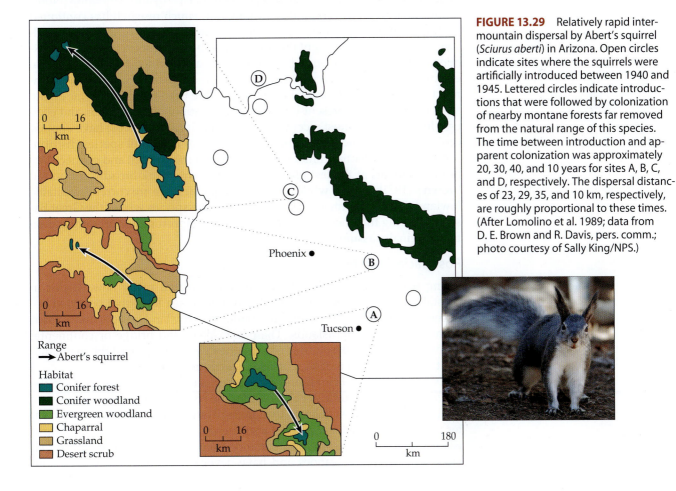

FIGURE 13.29 Relatively rapid intermountain dispersal by Abert's squirrel (*Sciurus aberti*) in Arizona. Open circles indicate sites where the squirrels were artificially introduced between 1940 and 1945. Lettered circles indicate introductions that were followed by colonization of nearby montane forests far removed from the natural range of this species. The time between introduction and apparent colonization was approximately 20, 30, 40, and 10 years for sites A, B, C, and D, respectively. The dispersal distances of 23, 29, 35, and 10 km, respectively, are roughly proportional to these times. (After Lomolino et al. 1989; data from D. E. Brown and R. Davis, pers. comm.; photo courtesy of Sally King/NPS.)

Range
→ Abert's squirrel

Habitat
■ Conifer forest
■ Conifer woodland
■ Evergreen woodland
■ Chaparral
■ Grassland
■ Desert scrub

FIGURE 13.30 Much like that of their counterparts on oceanic islands, the species richness of birds in montane forest islands near the Andes in Venezuela, Colombia, and Ecuador decreases with isolation. (After Nores 1995.)

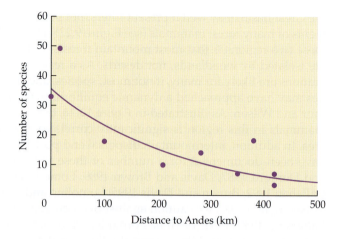

tion potential. Immigration filters vary both within and across archipelagoes. Thus, species richness of mammals of the American Southwest is more strongly correlated with extent of intervening deserts versus the overall distance between montane islands and their potential sources (see Lomolino et al. 1989). Also, as we noted above, comparisons across different archipelagoes may reveal scale or context dependency in the outcomes. Great Basin birds appear to be uninfluenced by isolation; that is, most islands may fall well within their immigration abilities. In other, more isolated regions, avian species richness declines significantly with isolation (e.g., montane birds of southern California and Baja California [Kratter 1992] and of the Andes of Venezuela, Colombia, and Ecuador [Vuilleumier 1970] and Argentina [Nores 1995]; **Figure 13.30**). Second, it is striking how much research has continued to flow from an original analysis intended as a test of MacArthur and Wilson's theory. Brown's insight that there might be a long-term process of relaxation and overall a nonequilibrium pattern has provided both a theoretical extension and a provocative hypothesis for the dynamics of regional mammal fauna. Subsequent work has integrated the island biogeographic analyses with analyses of present-day and fossil distributions, behavior, and ecology. In illustration, most recently, Waltari and Guralnick (2009) have applied ecological niche modeling techniques to explore how well species distributions fill their potential climate niche space and the amount and connectivity of their habitats at the last glacial maximum. This work broadly supported Brown's (1971b) thesis, while also indicating that there are currently suitable lowland dispersal routes for at least some of the species within the Great Basin, indicating that there may indeed be more opportunities for immigration than originally assumed.

Postglacial land bridge islands

Rising sea levels since the Pleistocene have created many new islands in coastal regions. The approximate 120 m rise in sea level that occurred between 18,000 and 10,000 years ago inundated many land bridges and created numerous continental islands. These are called land bridge or continental islands because, unlike oceanic islands, they were once part of the mainland, or at least connected to it by a bridge of terrestrial habitats.

Diamond (1972, 1975b) found that the influence of past connections to the mainland of New Guinea is apparent in the composition of the avifauna of the islands off its coast. Islands that are separated from the mainland by water less than 200 m deep support a greater number of species than oceanic islands of comparable size and distance from New Guinea (i.e., those that arose as undersea volcanoes and have never been connected to the mainland). Although the land bridge islands lack many bird species found in comparable habitats on New Guinea, they have several kinds of birds that are found on the mainland but never occur on the oceanic islands. Of course, there are still other species that occur on both land bridge and oceanic islands. Because these birds obviously have crossed water barriers to colonize the oceanic islands, investigators cannot be sure whether their populations on the continental islands are Pleistocene relicts or whether they have been replenished by subsequent immigration (see also Lawlor 1983, 1986).

Post-Pleistocene dynamics of freshwater faunas

The present distribution of freshwater fishes in southwestern North America reflects a history of aquatic habitat connections during the cooler and wetter climate of the Pleistocene, followed by increasing aridity, isolation, and subsequent extinctions (Hubbs and Miller 1948; Miller 1948; Smith 1978). Hard as it may be to believe, only a few thousand years ago, Death Valley, now in the most arid part of the North American desert, was almost completely filled by a large lake, which was supplied by a major system of permanent rivers and springs (see Figure 9.25). At least three genera of fishes inhabited this basin, and they have persisted as relictual populations in isolated springs. In the case of these fishes, there can be no doubt about the susceptibility of small, isolated populations to extinction. Many have disappeared within the last few years as humans have diverted water or introduced exotic species of competing or predatory fishes.

In most of the examples discussed so far, whether aquatic or terrestrial, the biotas of insular habitats isolated since the Pleistocene are relictual. Once-widespread habitats containing diverse biotas were diminished in size and became fragmented. As soon as these islands formed, they were oversaturated with species and, as a result, they have been losing species by extinction ever since. In contrast, some isolated habitats, as well as areas scoured by glaciers, begin with few, if any, species and remain undersaturated long after glaciers have receded (see Chapter 9).

Many lakes, including those of northern Eurasia and the Great Lakes and the Finger Lakes of northeastern North America, were gouged out by the advancing continental ice sheets and then filled with water as the glaciers retreated (see Figure 9.23A). Some groups, such as certain algae, protozoans, and invertebrates that use cysts or other effective means of long-distance dispersal, may have achieved an equilibrium between rates of colonization and extinction in these habitats over a relatively short period of time. In contrast, other less vagile animals, such as fish (Smith 1981) and some mollusks, may never achieve an equilibrium level of species richness. The fish faunas of many of these lakes appear to be undersaturated and still gradually increasing in diversity (Barbour and Brown 1974; see also Smith 1979; Browne 1981; Holland and Jain 1981; Tonn and Magnunson 1982; Brown and Dinsmore 1988; Hugueny 1989; Watters 1992; Elmberg et al. 1994; Oberdorff et al. 1997).

Only a handful of fish species can survive for more than a few minutes out of water at any stage of their life cycle, so they can colonize new areas only when connections of aquatic habitat are present. They usually colonize lakes from rivers and streams, but most lotic waters contain only a few species that can also be successful in lentic environments. Many glacier-formed alpine lakes lack native fishes entirely because they have never been connected by suitable habitat bridges to waters inhabited by fishes. The fact that several introduced species thrive in some of these lakes is further evidence that the absence of fish is a consequence of barriers to dispersal. Many of those glacier-formed lakes that do contain fishes still appear to be undersaturated with species. This is particularly true of the very large lakes, such as the Great

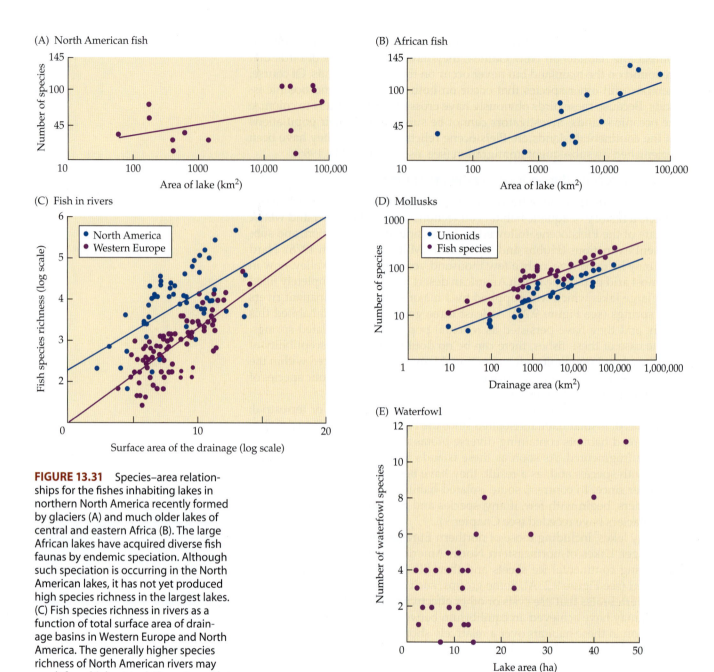

FIGURE 13.31 Species–area relationships for the fishes inhabiting lakes in northern North America recently formed by glaciers (A) and much older lakes of central and eastern Africa (B). The large African lakes have acquired diverse fish faunas by endemic speciation. Although such speciation is occurring in the North American lakes, it has not yet produced high species richness in the largest lakes. (C) Fish species richness in rivers as a function of total surface area of drainage basins in Western Europe and North America. The generally higher species richness of North American rivers may result from higher speciation rates or the tendency for many North American rivers to flow north-to-south, which thus has provided an opportunity for dispersal south of advancing glaciers and therefore higher persistence during periods of glaciation. (D and E) Just as species diversity of fish increases with area of lakes or catchment basins of rivers, diversity of dependent species (e.g., parasitic unionid mollusks of the Ohio River system, and waterfowl in boreal lakes of Finland and Sweden) also increases with surface area of their aquatic ecosystems. (A and B after Barbour and Brown 1974; C after Oberdorff et al. 1997; D after Watters 1992; E after Elmberg et al. 1994.)

Lakes in North America and Lake Baikal in the Russian Republic. There were not enough species in the rivers and streams draining these lakes to fill the niches available to fishes. Some taxa, such as the ciscoes, or whitefishes (*Coregonus*), in the Great Lakes, have been diversifying by endemic speciation and adaptive radiation, but there has not been sufficient time since the Pleistocene to achieve an equilibrium between speciation and extinction.

This conclusion is supported by a comparison of species–area relationships for fishes inhabiting the glacier-formed lakes in temperate North America and the much older lakes of tropical Africa (**Figure 13.31**). As mentioned in Chapter 7, the relatively high diversity of Africa's Rift Valley lakes derives primarily from the spectacular endemic speciation and adaptive radiation of cichlids in the larger lakes (Fryer and Iles 1972; Greenwood 1974; Meyer 1993). Lake Victoria, the largest lake in Africa, now harbors over 300 endemic species of cichlids, yet it appears to have dried up completely during the late Pleistocene (Johnson et al. 1996; Seehausen 2002). Following recolonization, speciation of these fishes must have proceeded at an extremely rapid rate indeed, possibly reaching a dynamic equilibrium before the relatively recent surge in extinctions due to introduced species and other anthropogenic activities. Finally, the relative time scale of underlying glacio-pluvial events holds some key insights for interpreting and predicting the biogeographic dynamics of insular systems. Recall from Chapter 9 that interglacial periods are relatively short compared with glacial periods (often just 10,000–30,000 years vs. 75,000–300,000 years, respectively). Thus, on a geological time scale, both relaxation of relictual communities and accumulation of species in environments cleared by glaciers are likely to be incomplete phenomena. Paleoecologists or biogeographers of future millennia may well reverse our original question and ask, To what extent do distribution patterns during glacial periods represent the legacy of the relatively short pulses of interglacial events?

Frontiers of Island Biogeography

While the equilibrium theory served island biogeography well as the discipline's paradigm since the 1960s, framing much empirical work and many theoretical advances, by the end of the twentieth century there was growing dissatisfaction with the theory, stemming from its omission of some fundamental processes (in particular ecological interactions, evolution, and geological and climatic disturbances) and the scale dependence of those processes. The importance of scale has long been appreciated and is, for instance, evident in Wilson's (1959, 1961) and Preston's (1962) seminal papers, and in the discussions above on the possible nonlinearities of the species richness relationships. In a diagrammatic model similar to that shown in **Figure 13.32**, Haila (1990) argued that the domain of the basic MacArthur–Wilson model was essentially restricted to the intermediate scales of space and time, with decreasing relevance outside this conceptual space. Such conceptual, scale-structured models allow us to recognize that there is order to where the dynamic equilibrium model works and where it fails. Its failures may be not so

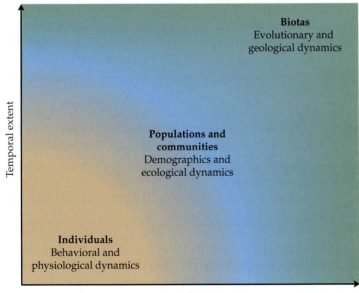

FIGURE 13.32   A scheme of the scale dependence of processes (behavioral and physiological, demographic and ecological, evolutionary and geological) affecting different levels of biotic organization (individuals, populations and communities, and biotas). MacArthur and Wilson's model appears to be most relevant for archipelagoes corresponding to intermediate ranges in temporal and spatial extent. (Modified from Haila 1990.)

FIGURE 13.33 A conceptualization of the major theories of island biogeography based on their assumptions of the dynamics (turnover in species composition) and equilibrium characteristics (relative stability in richness) of insular communities. The *dynamic equilibrium* condition corresponds to MacArthur and Wilson's (1963, 1967) theory; the *static, ecological equilibrium* condition equates to David Lack's (e.g., 1976) ideas on ecological (habitat) saturation of insular birds; the *relaxation, nonequilibrium* condition was inferred from James H. Brown's (1971b) research on mammals and birds of montane forests of the American Southwest; the *evolutionary, nonequilibrium* condition characterizes the biota of very large and remote islands such as Hawaii and the Philippines; and the *disturbance, non-equilibrium* condition is expected for biotas that are subject to episodic environmental (geological) disturbances such as Krakatau and other volcanically active, oceanic archipelagoes. Considering a single taxon, different positions in this conceptual diagram may correspond to different islands or archipelagoes, while different taxa in the same island group may also correspond to different conditions of biotic dynamics and equilibrium. (Modified from Whittaker 1998, 2000; see also Vermeij 2004; Whittaker 2004; Whitaker and Fernández-Palacios 2007; Whittaker et al. 2008, 2010.)

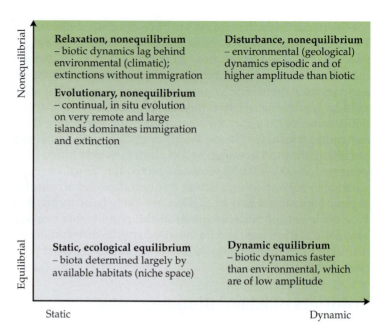

much because the processes it invokes are inoperative, but because the biotic dynamics are dominated by stronger signals from other processes (in particular, evolution, ecological interactions, and disturbance, depending on the spatial and temporal scales being considered). We can represent this idea in an alternative diagrammatic form, proposing that the dynamic equilibrium condition is merely one alternative condition applying to some systems, but not all (**Figure 13.33**). Where a system falls in this conceptual space, whether it is an equilibrium or nonequilibrium system, and whether it is dynamic (with measurable turnover of species) or relatively static in ecological time (i.e., on a time frame of years or decades) is dependent on the magnitudes and rates of each of the fundamental processes influencing insular communities.

We may now be witnessing the emergence of a new synthesis in island biogeography theory—one that includes some elements of MacArthur and Wilson's classic model, but greatly expands the conceptual framework of island theory by explicitly integrating into new theory the following (this list is simplified from that compiled by Lomolino et al. 2009; see references therein):

• The fundamental processes (immigration, extinction, and speciation) are scale dependent.

• These fundamental processes not only interact to affect the structure of insular communities, but also covary among islands and archipelagoes (e.g., immigration decreases while speciation increases with isolation).

• The fundamental capacities of species (their abilities to colonize and to survive, evolve, and dominate other species on islands) covary among species (e.g., when ordered by body size, growth form, or metabolic/photosynthetic pathway and trophic strategies).

• Each of these fundamental processes and capacities, and ultimately the structure of insular communities, will be influenced by feedback in the form of ecological interactions among species and by microevolution and speciation.

- There is history of place; that is, while individuals and species have developmental histories, similarly islands and archipelagoes have a geological ontogeny that, outside of the cataclysmic events that sterilize islands, follows a broadly predictable progression of changes in their physical characteristics that, in turn, influence each of the fundamental biogeographic processes.

This final component of an emerging, more integrative theory of island biogeography is the thesis of **the general dynamic model** of oceanic island biogeography, which was developed by Robert J. Whittaker and his colleagues (Whittaker et al. 2007, 2008, 2010). Their most central observation is that oceanic islands are geologically rather ephemeral, building from the ocean floor and then eroding and subsiding to disappear again in only a few million years (Whittaker and Fernández-Palacios 2007; and see Hooker 1866). Thus, island area, habitat diversity, and all associated factors influencing the carrying capacity of oceanic islands are far from constant, but they vary in a generally predictable manner over geological time.

Figure 13.34 differs from the original MacArthur and Wilson graphic (see Figures 13.8 and 13.9) in placing time on the ordinate, and incorporating species richness as a dependent variable in the modified, graphical model. It postulates that an island initially emerges, grows through continued volcanism to a large size and elevational range, and then begins to decline in area and elevation as erosion and subsidence become dominant (e.g., see Price and Clague 2002 for an account of the geological ontogeny of the Hawaiian archipelago). Hence the theoretical carrying capacity of the island and, consequently, species richness display hump-shaped trends through time.

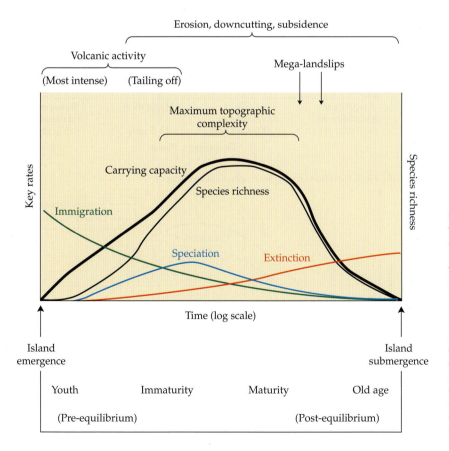

FIGURE 13.34 A graphical representation of the geological ontogeny and biotic dynamics of a single volcanic island, which illustrates the key processes and properties described by the general dynamic model of oceanic island biogeography. Species richness among islands and taxa should vary with the characteristics (e.g., isolation, area and carrying capacity of the islands; and the resource requirements, dispersal abilities, and reproductive rates of the species) affecting the fundamental biogeographic processes (immigration, extinction, and speciation) that, in turn, are affected by geological processes (volcanic activity, erosion, downcutting, and subsidence). (Redrawn from Whittaker et al. 2008.)

However, because immigration to remote islands is limited by the small size of the local archipelagic pool and the great distance to mainland sources, immigration cannot readily fill the potential carrying capacity, and therefore a gap develops between carrying capacity and species richness. This latter condition is indicative of vacant niche space and increased opportunities for speciation and adaptive radiation on larger and more remote islands. Whether we envisage the carrying capacity as a fixed function of area or an adjustable target that can climb over time through increasing specialization in the biota, at some stage in this idealized island's ontogeny, the decline in its size must begin to drive the carrying capacity down once again, pulling the realized richness down with it in a prolonged process of species relaxation, simultaneously reducing opportunities for speciation.

This model of the geological ontogeny of oceanic islands illustrates how it may be possible to advance theory by expanding the conceptual domain of existing models. Of course, it is not unique in this respect, and other theories of island biogeography also promise to contribute to this new synthesis by integrating the salient properties listed above (including scale dependence,

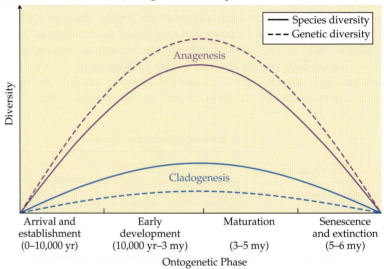

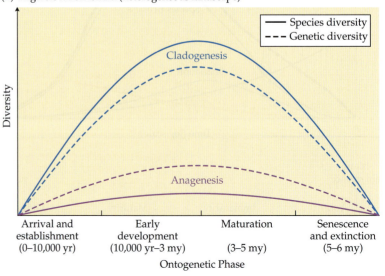

FIGURE 13.35 Stuessy's model of the ontogeny of insular floras explains how the processes of cladogenesis and anagenesis vary with characteristics of the species, and the ecological heterogeneity and geological age of the islands, to influence specific diversity (species richness) and genetic diversity of their assemblages. The biotic dynamics of the first phase are largely influenced by isolation of the islands and dispersal abilities of the species as they immigrate and establish populations. In the second phase, the number of endemic species increases substantially due to cladogenesis, especially on high elevation (ecologically heterogeneous) islands (B), while anagenesis contributes to rapid increases in genetic and, to a lesser degree, specific diversity on low elevation islands (A) (my = million years). Erosion and subsidence reduce surface area and ecological heterogeneity of the islands during the third phase, causing extinctions and substantial declines in both specific and genetic diversity. These processes of decline continue during the fourth and final stage until subsidence of the island and extinction of its species are complete. (After Stuessy 2007.)

interaction, covariation, and feedback among processes, and the histories of place and of species) into the emerging conceptual landscape of island theory. Stuessy's (2007) model of the specific and genetic ontogeny of insular floras (**Figure 13.35**), Heaney's (2000) model of dynamic disequilibria and phylogenesis (phylogenetic diversification; **Figure 13.36**), and Lomolino's species- and process-based hierarchical model (Lomolino 1999, 2000b; Lomolino et al. 2009) are just three examples of alternative but complementary approaches to developing a more integrative theory of island biogeography (see also Losos and Schluter 2000; Sismondo 2000; Ricklefs and Bermingham 2002, 2004; Hanski 2004; Kalmar and Currie 2006; Kadmon and Allouche 2007; Stuessy 2007; Kreft et al. 2008; Scheiner and Willig 2008; Lomolino et al. 2009; and Losos and Ricklefs 2010).

In summary, while we find it difficult to predict the final form of a new paradigm of island biogeography, we are confident that a new synthesis is not far off—one that will continue to expand the domain of current theory to explain other features of the assembly and disassembly of island biotas, including patterns in species composition and the many ecological and evolutionary marvels of island life—subjects we focus on in the following chapter.

(A)

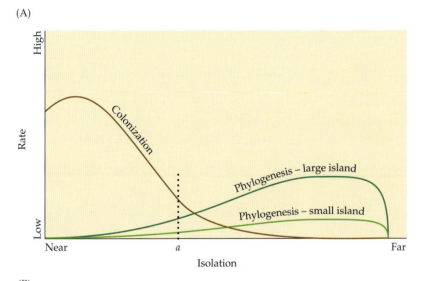

(B)

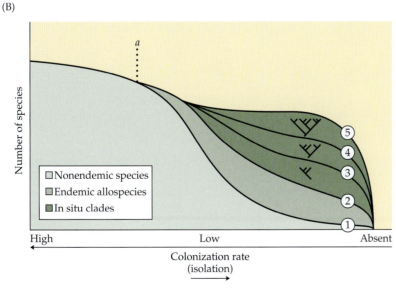

FIGURE 13.36 Heaney's model of dynamic disequilibria and phylogenesis (phylogenetic diversification) is based, in part, on the assertion that evolutionary diversification of insular biotas cannot be treated simply as "another form of colonization." On the contrary, because this process interacts in a complex manner with both immigration and extinction, it should be modeled as a separate, albeit integral, process driving the biotic dynamics of insular biotas. (A) This graph depicts how the rates of colonization and phylogenesis should vary with isolation and area of the islands (*a* marks the approximate degree of isolation at which gene flow for this hypothetical, focal taxon is equal to one individual per generation). (B) This graph illustrates how species diversity and endemicity should vary with colonization rates among islands of different degrees of isolation (*x*-axis; *a* as above), and as phylogenesis progresses over time (curves labeled 1–5 representing increasingly later stages in evolutionary development of these biotas). As the rate of gene flow declines with increasing isolation, phylogenesis becomes increasingly important, and the relative number of endemic species increases. Over time, phylogenesis will produce endemic clades (represented by branching symbols in the shaded areas of the curves) that become progressively more rich in species, especially on the larger islands. (After Heaney 2000.)

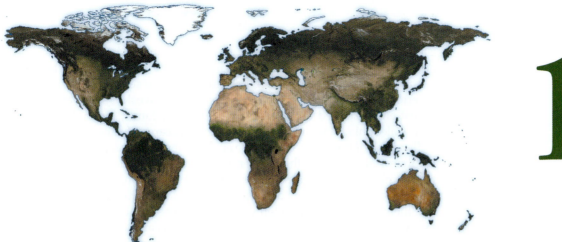

14

ISLAND BIOGEOGRAPHY:
Assembly and Evolution of Insular Biotas

Scientists, and for that matter poets, artists, and likely most humans, find islands innately compelling (Losos and Ricklefs 2009, 2010). The flames of intrigue, however, burn even more intense when their explorations reveal the myriad marvels of island life, including such paradoxical novelties as flightless birds, ground-foraging bats, sunflower "plants" that become woody shrubs and trees, the classical oxymora of giant shrews and pygmy mammoths, and hawks so tame that Darwin could knock them off their perch with the butt of his gun. Most, if not all, of these marvels of island life ultimately derive from the most fundamental property of islands—their isolation—and how, in turn, this has altered the ecological and evolutionary arena for lifeforms that have colonized these remote and remarkable ecosystems.

While Chapter 13 explained why insular biotas tend to be depauperate in comparison to their mainland counterparts, we now explore other equally general and perhaps even more remarkable characteristics of remote islands and archipelagoes, including patterns in similarity and distinctiveness of entire communities and the processes influencing the ecological dynamics and evolutionary development of their component species.

Beyond Richness: The Nature of Insular Biotas

Endemism

Islands, taken collectively, harbor a highly disproportionate number of endemic species. For example, if New Guinea is considered the world's largest island, then islands constitute roughly 3 percent of the Earth's land surface, but some 17 percent of today's endemic avifauna are insular (and this number was certainly much higher before humans colonized oceanic islands, causing wholesale extinction of their native avifaunas; see Chapter 16). Similarly, about 13 percent of the world's plant species are endemic to just ten island archipelagoes, with hotspots of plant endemism typically concentrated on remote islands and archipelagoes, including the Hawaiian, Galápagos, and Macaronesian Islands, St. Helena, Juan Fernández, and archipelagoes of the Caribbean (Whittaker and Fernández-Palacios 2007). We see a similar pattern for invertebrates, where, for example, 9 percent of the world's snails are endemic to just eight isolated archipelagoes (Hawaiian Islands, Japan, Madagascar, New Caledonia, Madeira, Canary Islands, Mascarene Islands, and Rapa).

By virtue of their size, distinctive environmental character and, in particular, their isolation in both space and time, remote islands represent a fascinating collection of novel selective regimes. Populations of plant and animal species on remote islands are typically founded by just a few individuals, and their populations are often subject to crashes and bottlenecks caused by unpredictable events such as hurricanes and volcanic eruptions. Hence, the ecological and evolutionary development of insular biotas often reflects the influence of both stochastic and selective processes. As a result, islands are rightly recognized as hotspots of endemism and evolutionary marvels.

As inherent in MacArthur and Wilson's (1967) equilibrium model, with increasing isolation, the rate of immigration from the mainland species pool gradually diminishes to a point where in situ speciation within an island or archipelago becomes the primary means of increasing richness. Both immigration from extra-archipelagic sources and local extinction oppose the development of distinct biotas; speciation is unlikely to occur in the face of rapid immigration and significant gene flow from the mainland, nor can it occur if insular populations do not survive long enough for significant evolutionary divergence. Hence, the two geographic factors most strongly associated with insular species richness (island area and island isolation) are also correlates of endemism.

Endemism increases with isolation because there are few founding events on very remote island systems; colonists encounter novel environments possessing unexploited ecological niches, and they experience little or no subsequent gene flow from mainland source pools. Endemism should also be correlated with island area, because large islands have a larger resource base, a greater variety of environments (and hence niches), and relatively low extinction rates (i.e., persistence times are long enough to allow evolutionary divergence). In addition, larger islands are more likely to include rivers, mountain chains, and other topographic features that provide within-island barriers to genetic exchange among local populations. In fact, most insular biotas exhibit just this pattern (i.e., endemism increases with area and isolation; **Figure 14.1A,B**), and typically, the greatest degree of evolutionary radiation occurs in the context of archipelagoes rather than isolated individual islands. This in turn suggests that the exchange of autochthonous products between a group of large and remote islands (such as Hawaii and the Philippines) may be a significant feature of the evolutionary divergence of their biotas from the mainland source pools (MacArthur and Wilson 1967; Whittaker and Fernández-Palacios 2007). The propensity of spectacular radiations of lineages on particularly remote islands, toward the edge of the dispersal limits of different taxa, led MacArthur and Wilson (1967) to coin the term **radiation zone** to describe the pattern.

Thus, the distinctive nature of very isolated biotas results, not just from strongly filtered immigration, but also from in situ speciation, which, through radiations of the limited successful colonists, can accentuate ecological distinctiveness among insular biotas (see Gillespie and Roderick 2002). For example, as we consider increasingly more isolated archipelagoes of the Pacific Ocean ranging eastward from New Guinea and Melanesia, the relative dominance of ants attenuates to the point where they are totally absent from most archipelagoes east of Samoa unless introduced by humans (Wil-

(A)

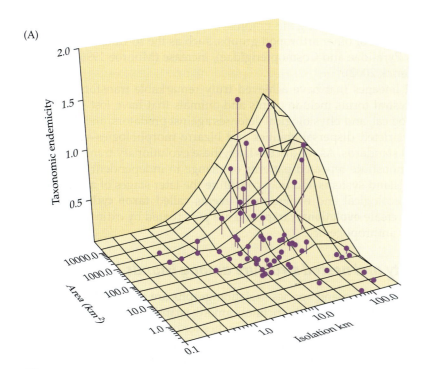

(B)

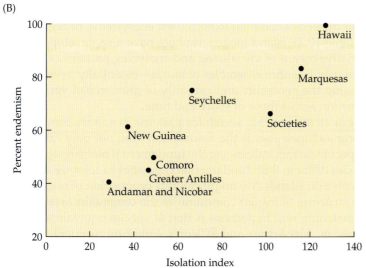

(C)

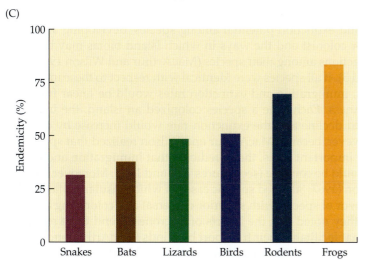

FIGURE 14.1 (A) The degree of endemism of insular avifauna of northern Melanesia increases with both island area and island isolation (this index expresses endemism at the levels of subspecies to genera, with increasing weight for endemism at higher taxonomic levels). (B) Endemism of spiders in the genus *Tetraghnatha* increases with isolation of islands in the Pacific Ocean (here, endemicity is calculated as percentage of species in this genus that are endemic to a particular island; isolation is calculated as the sum of the square root distances to the nearest larger island, nearest archipelago, and nearest continent). (C) Endemism of northern Melanesian vertebrates (as percent of species within each taxon) is highest for those groups with limited vagilities—in particular, rodents and frogs. (A after Mayr and Diamond 2001; B after Gillespie and Roderick 2002; C after Mayr and Diamond 2001.)

son and Taylor 1967). Along this gradient of increasing isolation, the richness and endemism of other arthropod groups, such as the so-called micromoths (families Pyralidae and Cosmopterigidae), increase (Munroe 1996; Gillespie and Roderick 2002).

Island lineages that have achieved truly remarkable transformations of their ancestral forms include plants and animals that have lost behavioral, morphological, and physiological defenses against predators, and those with highly restricted dispersal abilities and bizarre morphologies—at least by mainland standards. As we discuss later, these evolutionary transformations often demonstrate repeated patterns of change in independent lineages on different island systems and may constitute the later stages of a predictable series of ecological and evolutionary events called **taxon cycles**—changes that often create **evolutionary traps** and are punctuated by extinctions, either natural or anthropogenic.

Similarity and nested structure among islands

Just as species–area and species–isolation graphs are powerful tools for investigating patterns in species richness among islands, ecological biogeographers have developed insightful approaches for comparing structure of ecological communities among islands and across other ecosystems. In Chapter 10 we summarized a number of relatively simple formulae developed to characterize the similarity of two communities (primarily by assessing the relative number of species common to both sites, ecosystems, or regions; see Box 10.2 and Table 10.2). Island biogeographers have also developed some elegant and creative means of visualizing and analyzing patterns of similarity, and dissimilarity, of entire ensembles of biotas—essentially developing a means of assessing the generality and causality of gradients of variation in species composition across those of space and time.

At the heart of all studies that search for patterns in species composition with and among archipelagoes is the assumption that, not only do islands differ with respect to factors influencing the fundamental biogeographic processes, but species differ in their fundamental capacities to colonize and then survive and evolve on islands. We might call this proposition of nonrandom patterns in the structure of insular communities the **community assembly hypothesis**. The matching null hypothesis is that of species equivalence. In its simplest expression, MacArthur and Wilson's equilibrium model assumes species equivalence, with each species in a mainland pool having a chance to immigrate and, once on an island, to persist. However, within their 1967 monograph, they go on to address the effects of interspecific differences in immigration and extinction potential among species, developing a portrait of the superior colonist and the ways in which island biotas may be structured by interactions among their species (MacArthur and Wilson 1967: their Chapters 4 and 5). If all species were identical with respect to these important characteristics, immigration and extinction rates would be linear functions of species richness. Each time a species colonized an island, the immigration rate would decline and the extinction rate would increase by constant amounts. However, because MacArthur and Wilson realized that species do differ in some important traits, they predicted that immigration and extinction curves should be concave (see Figures 13.8 and 13.9). Moreover, MacArthur and Wilson (1967: 80) cited specific cases in which insular biotas tend to be biased in favor of species with superior dispersal and colonizing abilities. For example, they noted that the insects of Micronesia tend to be relatively small, a trait that facilitates their transport by winds (MacArthur and Wilson 1967, after Gressitt 1954). Similarly, Wollaston's (1877) earlier work had dem-

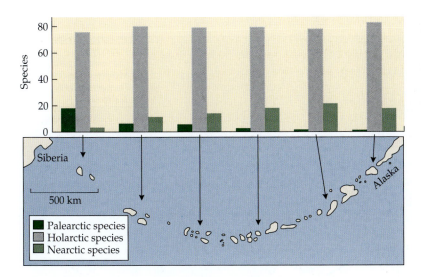

FIGURE 14.2 Much like the Lesser Sunda Islands (see Figure 6.14), the waters between Siberia and Alaska form an oceanic double filter, or transition zone, characterized here by the reciprocal attenuation of land plants on the Aleutian Islands. (After Williamson 1981.)

onstrated that insect colonists of Saint Helena Island in the Atlantic Ocean were predominately wood borers or species preadapted to rafting on logs or vegetation mats. As these and many other studies demonstrated, oceans act as filters, allowing the selective passage of certain types of species. Often, these oceanic filters form transition zones, characterized by differences in the attenuation rates of species groups that differ in dispersal ability (**Figure 14.2**; see also Figures 14.8, 6.14 and 6.15). The selective nature of immigration is also evident for some islands that are quite near the coast (e.g., Whittaker et al. 1997) and indeed for habitat islands embedded in a matrix of altered land cover, suggesting that filtering through isolation may be both general and subtle in its effects (e.g., Watson et al. 2005).

Simberloff and Wilson's (1969, 1970; Simberloff 1978) classic defaunation experiments on invertebrates of mangrove islands not only tested some critical predictions of the equilibrium theory, but also revealed that as these islands were colonized, they tended to converge on their original species composition. Leston's earlier (1957) comparisons of insect communities on isolated islands of the Atlantic and Pacific Oceans also revealed that these communities are convergent with one another and tend to form regular subsets of the mainland biota. The tendency for communities on species-poor islands to form regular subsets of those on species-rich islands is known as **nestedness**. Numerous studies have confirmed that nestedness of insular biotas is an extremely common phenomenon among archipelagoes of islands, mountaintops, and other isolated ecosystems (**Figure 14.3**), reflecting differences among both islands and species in factors influencing immigration to, and extinction and evolution on, islands (Wright et al. 1996; but see Ulrich et al. 2009 for a critique of the analyses used, suggesting that nestedness may be less frequent than hitherto understood).

The earliest documented description of patterns of nested assembly of insular biotas we are aware of was by Darlington (1957), who presented a graphical model termed the **immigrant pattern**, which asserted that the driving force for nestedness was selective immigration (**Figure 14.4A**). Interestingly, Darlington's model of nestedness apparently ignored the possibility that the pattern can also result from selective extinctions. Instead, his **relict pattern** predicted that, while smaller islands should have fewer species, their biotas should be random draws of the mainland pool. However, the capacity to persist, given limited insular resources, may differ dramatically and

FIGURE 14.3   The selective nature of immigration is revealed in the geographic nestedness of many insular biotas. Both ants (A) and birds (B) of the western Pacific exhibit analogous patterns of distributions extending eastward from their sources in New Guinea and Indonesia. (B) Lines indicate the eastern range limits of different groups of land and freshwater breeding birds. (A after Wilson 1959; B from Williamson 1981, after Firth and Davidson 1945.)

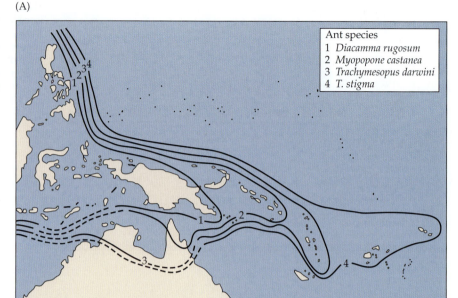

(A)

Ant species
1 *Diacamma rugosum*
2 *Myopopone castanea*
3 *Trachymesopus darwini*
4 *T. stigma*

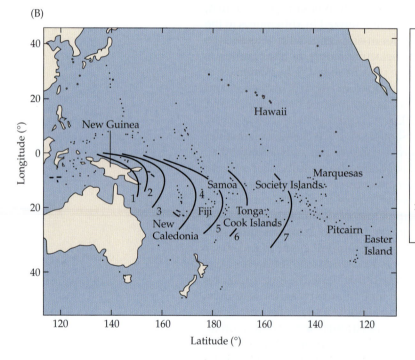

(B)

1 Pelicans, storks, larks, pipits, birds of paradise, and nine others

2 Cassowaries, quails, and pheasants

3 Owls, rollers, hornbills, drongos, and six others

4 Grebes, cormorants, ospreys, crows, and three others

5 Hawks, falcons, turkeys, and wood swallows

6 Ducks, thrushes, waxbills, and four others

7 Barn owls, swallows, and starlings

> 7 Herons, rails, pigeons, parrots, cuckoos, swifts, kingfishers, warblers, and flycatchers

predictably among species, so following the increasing insularization of a habitat (e.g., by sea level rise or ongoing habitat fragmentation), particular groups of species may be excluded from smaller islands. Therefore, regular patterns in species composition on islands may result from extinctions as well as immigrations. For example, Brown's **relaxation model** of island biogeography holds that, as the areas of islands decrease, their communities should converge on a similar set of species (i.e., those with minimal resource requirements; **Figure 14.4B**). Thus, across a great variety of archipelagoes, insular communities tend to form nested subsets when ordered by either increasing isolation or decreasing area, consistent with the hypothesis of se-

(A)

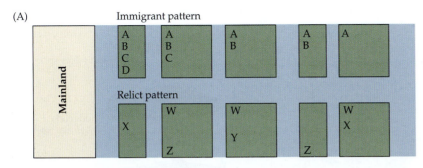

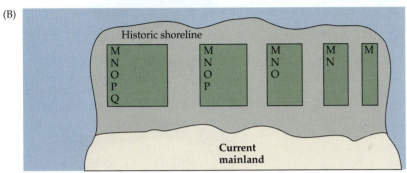

(B)

FIGURE 14.4   Nestedness of insular communities may result from species-selective immigration (the immigrant pattern, A) or extinction (B). Note that Darlington's "relict pattern" implies that the partial extinction of faunas during ecological relaxation might be random (A). Most recent studies, however, suggest that both immigration and extinction are (or can be) selective processes. (A after Darlington 1957.)

lective immigration or selective extinction, respectively (**Figure 14.5**). Colonization studies of volcanically active islands, such as Krakatau (see discussion on Krakatau in Chapter 13) and also Long Island and its inland freshwater island Motmot, confirm the selective nature of immigration and colonization, and they demonstrate how natural disturbances can set the stage for nonrandom assembly of insular communities. On the other hand, activities of modern human societies, including habitat destruction and fragmentation, make it clear that anthropogenic extinctions also can be highly selective and result in nonrandom collapse of native communities—a process known as **community disassembly** (Fox 1987; Mikkelson 1993; Lomolino and Perault 2000; see Chapter 16).

Although nestedness attracted some attention throughout the 1960s and 1970s, it wasn't until 1984 that Bruce Patterson and Wirt Atmar developed the first statistically rigorous approach for analyzing nested subsets (Patterson and Atmar 1986). Their important paper triggered a flurry of studies that confirmed the generality of nestedness, provided alternative statistical approaches, and began to explore the causality of nestedness among taxonomic groups and across archipelagoes (see Kitchener et al. 1980; Lazell 1983; Schoener and Schoener 1983; Patterson 1987, 1990; Cutler 1991; Patterson and Brown 1991; Simberloff and Martin 1991; Wright and Reeves 1992; Atmar and Patterson 1993; Doak and Mills 1994; Cook 1995; Cook and Quinn 1995; Kadmon 1995; Lomolino 1996; McLain and Pratt 1999; Fischer and Lindenmayer 2002; Loo et al. 2002; Hausdorf 2003; Bascompte et al. 2003; Guimaraes and Guimaraes 2006; Carstensen and Olesen 2009).

As this area of research continues to mature, it is certain to contribute substantially to our understanding of the forces contributing to the nonrandom assembly of insular communities. As you may have noted, however, the foregoing discussion of community assembly has focused on selective immigrations and extinctions while largely ignoring the other fundamental biogeographic process—evolution. Indeed, by producing species that are unique to different islands, in situ speciation reduces both similarity and nestedness among islands (see Carstensen and Olesen 2009). Speciation is, of

FIGURE 14.5 The distributions of five fish taxa among 28 springs in the Dalhousie basin of South Australia illustrate the nested subset pattern. Here, nestedness is nearly perfect, in that, with the exception of the taxon that is absent from spring G, the faunas of the more depauperate springs are regular subsets of those found in the richer springs. (After Brown and Kodric-Brown 1993.)

(A)

(B)

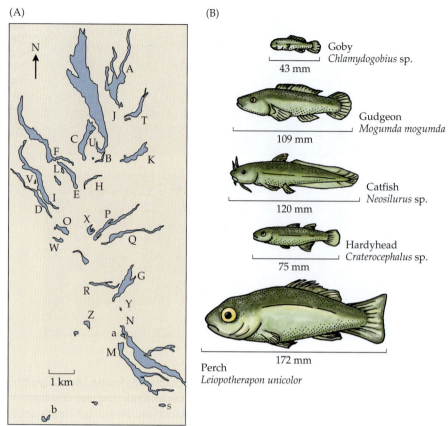

(C)

	Spring (see map in A)																												Number of springs
Species	A	B	C	D	E	F	G	H	I	J	K	L	M	N	O	P	Q	R	S	T	U	V	W	X	Y	Z	a	b	
Goby	X	X	X	X	X	X	X	X	X	X	X	X	X	X	X	X	X	X	X	X	X	X	X	X	X	X	X	X	28
Gudgeon	X	X	X	X	X	X	X	X	X	X	X	X	X	X	X	X	X	X	X										19
Catfish	X	X	X	X	X	X	X	X	X	X	X	X	X	X															14
Hardyhead	X	X	X	X	X	X			X	X	X																		9
Perch	X	X	X	X	X	X	X																						7
Number of species	5	5	5	5	5	5	5	4	4	4	4	3	3	3	3	2	2	2	2	2	1	1	1	1	1	1	1	1	

course, the process responsible for one of the most distinctive characteristics of remote insular biotas—their high endemism. Speciation is also responsible for intensifying an equally distinctive feature of these communities—their disharmonic nature, that is, the fact that remote islands lack whole groups of species characteristic of otherwise comparable mainland ecosystems, as a result of the filtering effect of large stretches of ocean.

Distributions of particular species

Nestedness metrics in essence provide a summary statistic of one form of emergent compositional pattern in island biotas. Further insight can be derived by analyzing the distributions of each constituent species separately. Perhaps the simplest way of doing so is to quantify the **incidence functions** of species in the form of the proportion of islands occupied as a function of the same key properties, island area and island isolation. Analyses of species incidence functions have revealed a variety of interesting patterns (exemplified more fully below) and often serve to indicate that species distributions across islands and habitat islands are highly structured by area, isolation, and (for habitat islands) the properties of the intervening landscapes.

Ecologists have developed an alternative theoretical framework for understanding the dynamics of this species- and landscape-scale biogeography; it is called **metapopulation theory**. The concept of metapopulations has been attributed by some to the influential book by Andrewartha and Birch (1954) *The Distribution and Abundance of Animals*, but both the term and the first metapopulation models were introduced by Richard Levins (1970: 105) to describe the notion of "a population of populations which go extinct locally and recolonize" (see historical account by Hanski 2010). From roots independent of the MacArthur–Wilson theory, Levins took the idea of immigration and extinction to the species level, exploring the idea that the continued persistence of a species across a series of isolates could be the dynamic outcome of a process of small populations failing in a particular patch (local extinction) but being resupplied from a population from another nearby patch. In this way, while each patch can follow its own individual population trajectory, the probabilities of long-term persistence across the patches are interdependent: hence the term *metapopulation*. Basically, metapopulation studies attempt to estimate the proportion of islands, or "patches," that must be occupied to ensure the survival of the interacting populations of a species—its metapopulation. Alternatively, given information on the actual proportion of patches occupied, metapopulation models can be used to estimate the time to extinction of the metapopulation (see Gilpin and Hanski 1991; Hanski 1999; Hanski and Gagiotti 2004).

As posited at a community level in the equilibrium theory, metapopulation models explore the assumptions (1) that the probability or frequency of immigration decreases with increasing isolation and (2) that the frequency of extinction decreases as island area increases. Both metapopulation studies and incidence function studies suggest that for some study systems, one or the other and sometimes both apply (see Lomolino 1986, 1997; Watson et al. 2005). Where both apply, the result can be in the form of thresholds of minimum area for persistence and maximum tolerable isolation, while in other cases there appear to be interactive effects, or **compensatory effects**, of immigration and extinction. The basic rule is that a focal species will occur on those islands where its immigration rate exceeds its extinction rate. Thus, the focal species is expected to occur on some isolated islands, but only those islands that are large enough that extinction rates are low enough to compensate for infrequent immigrations. Conversely, the species may be common on small islands if they are close enough to the mainland that high immigration rates compensate for frequent extinctions. The overall result is that minimal area requirements to maintain populations of the focal species should increase as isolation increases. When this **insular distribution function** is plotted on a graph depicting island area and isolation, the intercept can be interpreted as a measure of resource requirements, while the slope is an inverse measure of the immigration ability of the focal species (**Figure 14.6**; see also Alatalo 1982; Hanski 1986; Peltonen et al. 1989; Peltonen and Hanski 1991; Thiollay 1998; Lomolino 2000a,b; Ovaskainen and Hanski 2003; Watson et al. 2005; Frick et al. 2008; Presley and Willig 2008; Lomolino et al. 2009).

This species-based approach to island biogeography emerged in parallel to the whole-system–level island theory of MacArthur and Wilson and shares key assumptions and concepts. Analyses at the species level can easily be drawn into idiosyncrasies that lack generality; however, many of the

(A)

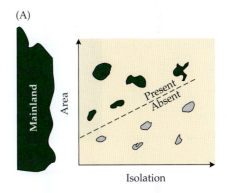

(B) Masked shrew (*Sorex cinereus*)

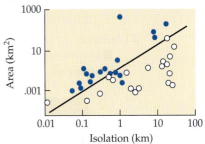

FIGURE 14.6 Insular distributions of species should be influenced by the combined effects of immigration and extinction. Species populations may persist on small islands if they are close enough to the mainland to experience frequent immigrations. Because immigration rates should decrease with isolation, the minimum area required to maintain insular populations of a focal species should increase with isolation. The predicted pattern (A) has been observed for a variety of organisms, including small mammals, fishes, and butterflies. (After Lomolino 1986, 1999.)

emergent patterns community ecologists study result from, not in spite of, differences among species. For example, a high degree of nestedness implies that insular distributions vary in a very regular manner among species (see Figure 14.5; see also Patterson 1984). If species were equivalent, nestedness would be a rare phenomenon indeed.

Insular distribution functions, in addition to providing tools for exploring patterns in assembly of insular communities, highlight the need to study some critical questions. As MacArthur and Wilson (1967) lamented, we still know relatively little about interspecific differences in immigration and extinction probabilities. The general tendency for insular communities to be nested along gradients of the geographic template, including those of isolation and area, suggests some very regular patterns of variation in immigra-

FIGURE 14.7   Species composition of terrestrial vertebrates on islands of northern Melanesia illustrate how some biotas may be relatively harmonic (A and B), with insular communities similar to those of the mainland with respect to relative contributions of component taxa (here, expressed as percentages of species within different families or orders), while others may be highly disharmonic (C and D). (After Mayr and Diamond 2001.)

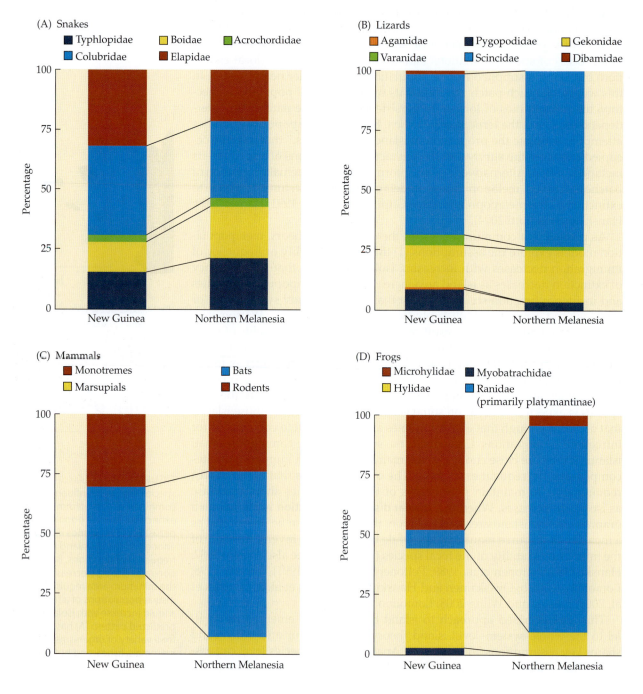

tion abilities and resource requirements—but what are those patterns, and which particular traits of species are correlated with their fundamental capacities to immigrate to and survive and evolve on islands? One of the most likely candidates for a very general if not universal correlate of these capacities for island life is size—either stature and growth form of plants or body mass of animals. Accordingly, we spend substantial time discussing evolutionary shifts in size of insular biotas later in this chapter.

The depauperate and disharmonic character of isolated biotas

As we discuss in subsequent sections of this chapter, development of the special character of insular biotas is often associated with release from ecological pressures that shaped the character evolution of ancestors in the much more diverse, mainland communities. But the ecological and evolutionary marvels of island life are not just products of **ecological release** in species-poor or depauperate environments; they are also the result of ecological **character displacement** from the few colonists and their descendants that have over time come to dominate insular communities. Limiting our inquiries to just one of these distinguishing features of insular biotas—their depauperate nature or their highly endemic, distinctive character—would render many of the resultant novelties (e.g., the gargantuan size of moas and other endemic insular birds—see Box 14.1) inexplicable.

Biogeographers often use the terms **harmonic** and **disharmonic** to characterize insular biotas. Harmonic, or **balanced**, biotas are assemblages that are similar in relative composition to the source biota. They may have fewer species, but the proportions of species in each taxon or ecological category are roughly the same as on the mainland (**Figure 14.7A**). Again, this is the null prediction, with disharmony being the opposing hypothesis. Where species are not equivalent with respect to their abilities to colonize and survive and evolve on islands, we would predict nonrandom patterns in species composition, not just differences between insular and mainland communities, but regular trends and differences among insular communities as well. Given their relatively superior capacities to colonize or survive and speciate on islands, certain types of organisms should be overrepresented in insular communities, and this bias should increase as islands become either smaller or more isolated (**Figure 14.8**). While there is much to learn, island biogeographers have made great strides in identifying the principal forces driving the assembly of isolated biotas, including selective immigrations, extinctions, interspecific interactions, and speciation.

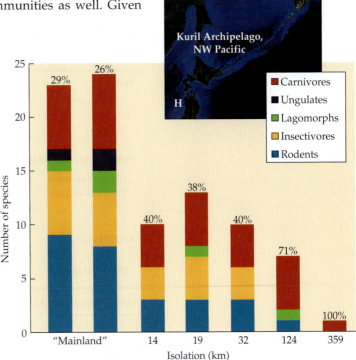

FIGURE 14.8 As a result of differences in the immigration abilities among species, the disharmonic nature of insular biotas increases with isolation. In this case, the percentage of the mammalian fauna of the Kuril Archipelago that are carnivores increases with isolation from the main islands of Hokkaido (H) to the southwest and Kamchatka (K) to the northeast (data excludes islands < 200 km², which are too small to support populations of most of these carnivores). (After Hoekstra and Fagan 1998.)

Forces Assembling Insular Biotas

Immigrant selection

With all due respects to Beijerinck and Baas Becking (see page 168), everything is not everywhere and, especially within the realm of island biogeography theory, intervening waters, by definition, act as barriers or at least filters (sensu Simpson 1940; see Chapter 6). Thus, it becomes axiomatic that along a gradient of increasing isolation (i.e., over greater distances or across more inhospitable barriers), not only should species richness decline, but the disharmonic nature of insular communities should increase. Not only is there predictable variation in immigration abilities among species groups, but even within each major taxon or functional group, particular species often differ markedly in their abilities and propensities to colonize remote islands (see Figures 14.7 and 14.8; **Table 14.1**).

To the extent that these interspecific differences are predictable, they will translate into predictable, nonrandom patterns of species distributions and community structure among islands. As identified by A. R. Wallace (1880) in his classic book *Island Life*, birds, bats, and flying insects tend to be well represented on distant oceanic islands, whereas few, if any, native species of non-volant mammals, amphibians, and freshwater fishes are present there (see Carlquist 1974; J. Smith 1994). Similarly, dispersal of plants to islands is carried out principally by wind, ocean transport, and zoochory, with the relative importance of each means varying as a function of distance, size of island, and other aspects of geography, such as position in relationship to active routes of movement by birds and (in the tropics) fruit bats (see, e.g., Whittaker and Jones 1994; Whittaker and Fernández-Palacios 2007). The predominance of avian-dispersed plants on isolated islands is probably a reflection of the relative efficiency of **zoochory** (transport by animals) rather than sheer number of propagules. That is, while more propagules may be dispersed passively by winds and water (**anemochory** and **hydrochory**, respectively), only a tiny fraction of those potential colonists will reach land.

TABLE 14.1 *Some Examples of Reported Cases of Long-Distance Dispersal by Terrestrial Animals*

REPTILES	
Freshwater turtles	Possible 200 miles (322 km) to Madagascar
Snakes	500 miles (805 km) or more to the Galápagos
Lizards	1000 miles (1609 km) to New Zealand; perhaps more than 1000 miles (1609 km) for geckos
TERRESTRIAL MAMMALS	
Large mammals	Perhaps 30 miles (48 km) or more[a]
Small mammals	Possibly 200 miles (322 km) to Madagascar for civets and insectivores
Rodents	500 miles (805 km) to the Galápagos
AMPHIBIANS	Perhaps 500 miles (805 km) to the Seychelles; perhaps 1000 miles (1609 km) to New Zealand
LAND MOLLUSKS	More than 2000 miles (3218 km) in Polynesia (to Juan Fernandez Island)
INSECTS AND SPIDERS	More than 2000 miles (3218 km)

Source: Wenner and Johnson 1980.

[a]Elephants have been estimated to swim up to 48 km (Johnson 1978; see Figure 6.6B). Also, a semiaquatic hippopotamus occurred on Madagascar.

Indeed, the chance that a randomly dispersed propagule will reach an island decreases as an exponential function of its isolation (see Figure 6.19B). In contrast, avian dispersal to islands typically is a nonrandom, directed, and therefore highly efficient phenomenon. Thus, plants and other avian-dispersed "hitchhikers" have a much higher probability of colonizing oceanic islands.

In an analogous fashion, mammals can be divided into two groups—bats and non-volant forms—on the basis of their over-water dispersal ability (e.g., see Lawlor 1986). Bats have naturally colonized such distant outposts as New Zealand, New Caledonia, and the Canary and Hawaiian Islands, and they are represented by 32 genera on the Greater and Lesser Antilles. In contrast, native non-volant mammals are completely absent from New Zealand, New Caledonia, the Hawaiian archipelago, and the Canaries (although there was one native species, a giant rat, which became extinct following human colonization; see Bocherens et al. 2006). Only 26 genera are known from the Antilles (all but 5 of which have become extinct since the Late Pleistocene, owing, at least in part, to human activity; see Morgan and Woods 1986; Woods and Sergile 2001). For those groups that lack flight or are otherwise incapable of being carried by winds, salt tolerance is an especially important factor influencing oceanic distributions (see Dunson and Mazotti 1989). Freshwater fishes and amphibians are so limited by their intolerance of salt water that their presence was often taken as an indication that an island was once connected to the mainland. However, it is now clear that nearly all species of freshwater fish found on isolated, oceanic islands either were introduced by humans or are descendants of marine species that, over evolutionary time, developed abilities to invade brackish—and then, freshwater—systems. New Zealand, for example, has no native **primary division freshwater fishes**, but introduced trout thrive in its rivers. Among amphibians, the ranids and bufonids tend to be the dispersal champions (Meyers 1953). Crab-eating frogs (*Rana cancrivora*) and giant marine toads (*Bufo marina*) are especially well adapted for oceanic transport, with high salinity tolerance in both adults and tadpoles (Inger 1954; Udvardy 1969). Consequently, these species are much more widely distributed on oceanic islands than their freshwater counterparts (see also Neill 1958). Likewise, the disharmonic nature of frog communities across Melanesia (see Figure 14.7D) results primarily from the relatively high diversity of platymantines, which lay terrestrial eggs, forego the tadpole stage, and therefore have relatively high tolerance for salt water during dispersal.

Among the mollusks, slugs seem equivalent to freshwater fishes and hygrophylic amphibians in their intolerance of saline waters and their generally limited distributions on oceanic islands. In contrast, land snails adapted to arid conditions, and those small enough to be transported by the winds (so-called microsnails), are often significant components of oceanic biotas (Solem 1959; Vagvolgyi 1975; see also Jaenike 1978).

Among non-volant mammals, relatively few insectivores and small rodents are found on islands along coastal and inland bodies of water, presumably because of their limited immigration abilities (Crowell 1986; Peltonen et al. 1989). In regions where waters freeze during at least part of the year, insular distributions are strongly influenced by the ability of species to remain active and withstand subfreezing temperatures during winter. In contrast to hibernators and other seasonally inactive species, those that remain active during winter are disproportionately common on islands. Absent, or surprisingly rare, are chipmunks, ground squirrels, woodchucks, jumping mice, and other species unable or unlikely to utilize the winter ice cover as a seasonal avenue for migration (see Tegelstrom and Hansson 1987; Lomolino 1988, 1993b; Peltonen et al. 1989). Even among winter-active mammals, larger species generally are more capable of withstanding winter conditions and, thus, are often better at dispersing across ice to colonize islands in winter. Non-volant mammals of the Kuril Islands, which stretch from Hokkaido—the northernmost island of Japan—to the southern tip of the Kamchatka Peninsula, provide a nice illustration of selection for immigration abilities. As we move away from the mainland, carnivores (primarily red fox, brown bear, and mustelids) become increasingly more important components of the mammalian fauna, presumably because their large size confers greater immigration abilities (Hoekstra and Fagan 1998; see Figure 14.8).

Selective extinctions

Despite the tremendous variation among insular environments, they all tend to be similar in that they provide limited resources, and as island size decreases, the amount of resources and their replenishment rate decreases. Thus, insular communities should be biased, not just in favor of good dispersers, but also in favor of those species that require less energy to maintain their populations. Although this prediction is far from a rule, the results of many studies are consistent with it. The record of extinctions since the Pleistocene, in particular (see Chapter 9), suggests that extinction has by no means been an entirely random process. As we discussed in Chapter 7, the fossil record is replete with evidence for species selection, and the **Lilliputian effect**—disproportionate survival of the smaller taxa during an extinction event—is frequently evidenced in the geological record (Twitchett 2006, 2007). During the last 10,000 years, some types of species have become locally extinct much more frequently than others, and their differential susceptibility to extinction appears to be related to their ecological characteristics and, ultimately, body size as well. This is consistent with the idea that the ecological characteristics of species determine insular carrying capacities, and in turn, these equilibrium population densities largely determine the probability of extinction (e.g., Brown 1971b, 1978, 1981; Van Valen 1973a,b). In fact, it should be possible to predict the relative abilities of species to persist on isolated land bridge islands based on some easily measured morphological and ecological traits. Animals of large body size, carnivorous diet, and specialized habitat requirements should

FIGURE 14.9   All else being equal, species with relatively low resource requirements tend to be disproportionately more common on islands. The small mammal communities of mountaintop forest "islands" of the Great Basin (A) and of New Mexico and Arizona (B) both tend to be dominated by small, generalist herbivores, presumably because—in comparison with mammals with higher resource requirements—they can maintain higher populations on the same area and, therefore, are less prone to extinction. (A after Brown 1971b, 1978; B after Patterson 1984.)

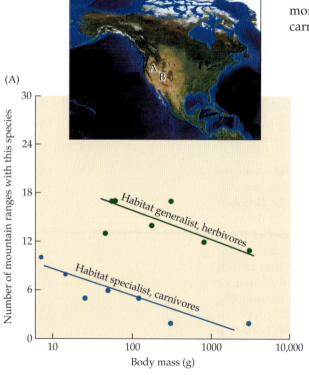

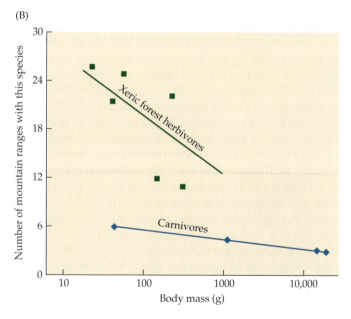

experience higher extinction rates than species that are smaller, herbivorous, or more generalized in habitat use.

This predicted pattern is precisely what is observed for terrestrial mammals on isolated, montane "islands" in western North America (**Figure 14.9**). Small herbivores that are habitat generalists make up the limited subset of species that are found on virtually all mountain ranges, whereas carnivores and herbivores of larger body size or specialized habitat requirements have become extinct on many montane islands and persist on only the largest ones. This pattern is particularly impressive because it is repeated on two different, geographically isolated sets of mountaintop islands in North America: those of the Great Basin (Brown 1971b) and those farther south in the American Southwest (Brown 1978; Patterson 1984). In Chapter 9, we noted that vertebrate extinctions during the Late Quaternary also tended to selectively remove large, carnivorous, or specialist species—that is, species with relatively high resource demands and low reproductive rates. The record of historical extinctions and lists of currently endangered species also bear witness to the vulnerability of large animals (see Chapter 16).

While many studies suggest that extinction is selective, there remains much debate over the traits most closely associated with extinction risk. One reason why it is so difficult to provide an "identikit" portrait of the vulnerable species is that nearly all species are influenced by combinations of threats whose effects vary with the particular environmental and ecological context. Recent developments in analytical methods, such as classification and regression tree analyses (see Olden et al. 2008), promise to provide further insight into the combined and contextual nature of factors (geographic, morphological, ecological, or otherwise) associated with extinction risk across a diversity of ecosystems and life-forms (**Figure 14.10**).

FIGURE 14.10 As this classification tree illustrates, the threatened status of today's living mammals is a function of the interactive and contextual effects of both geographical factors (size of geographic ranges) and biological factors (including body size, relative reproductive rates, and mode of locomotion and foraging). Each split in this tree identifies the factors that are associated with threat across all, or within particular, groups of mammals (traits contributing to relatively high risk are positioned on the right branches and enclosed in red rectangles). (Simplified from Davidson et al. 2009.)

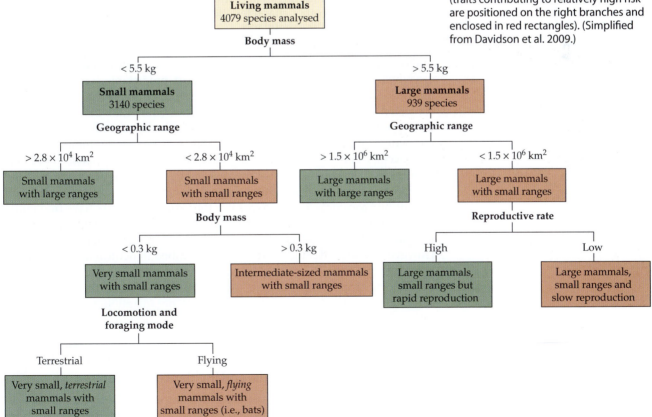

Ecological selection, assembly rules, and ecological release

Ecological communities are among the most complex systems in existence, not just because of the diversity of species and environmental settings, but also because of system feedback—that is, each species can strongly influence the abilities of others to disperse, colonize, and form new species. This has, of course, been a central assumption of ecology—that the distributions and abundances of species populations are strongly influenced by ecological interactions.

As reviewed earlier, nestedness provides one emergent form of nonrandom pattern in island communities that indicates a significant role for species traits in determining community membership via differential immigration, colonization, or extinction. There are, however, other nonrandom forms of insular community structure that have been identified, some of which have been strongly suggestive of an important role for interspecific interactions, either between trophic levels (predator–prey, parasitism, pollinator mutualisms, etc.) or within a single trophic level via competitive mechanisms. Three types of biogeographic patterns that have been hypothesized to be caused by interspecific competition are listed below and discussed in detail in the subsequent sections:

1. Ecologically similar species tend to exhibit mutually exclusive distributions, seldom if ever occurring together on the same island. A corollary of this pattern is the prediction that species that do coexist tend to be more dissimilar in ecologically relevant traits than would be expected by chance.

2. In comparison to conspecific populations on the mainland, populations occurring without close competitors on species-poor islands often exhibit relatively high densities.

3. Insular populations tend to exhibit ecological release, characterized by significantly broader niches and shifts to habitats, feeding strategies, activity periods, or other characteristics that would be considered atypical for the species on the mainland.

The role of competition in generating such patterns has been a subject of intense debate within the island biogeographic literature, and as remarked elsewhere, "we may conclude that the most predictable feature of island assembly rules is that any claim of evidence of competitive effects will be contested" (Whittaker and Fernández-Palacios 2007: 126). In large measure, there are two reasons for this. First is the difficulty of entirely eliminating alternative hypotheses that could account for the nonrandomness identified. Second is the concern that, in effect, the patterns are not so much real features as false constructs that the researcher has (without realizing it) imposed upon the data in searching through complex data sets for evidence of community assembly. Efforts to resolve the resulting debates have frequently focused around the use of so-called null models, which attempt to simulate the patterns that might arise if communities were constructed at random. However, these modeling efforts, better described in neutral terms as "simulation models," are themselves plagued by the problem of unintentionally introducing structure to the "null" world. We here describe the concepts and some classic examples of insular community organization, without complicating the account by detailed evaluation of the criticisms that have often followed (for further discussion see, e.g., Gotelli and McCabe 2002; Whittaker and Fernández-Palacios 2007: their Chapter 5).

Distributions, checkerboards, and incidence functions

In the absence of direct experiments on the competitive exclusion of one species by another, a pattern of mutually exclusive distributions of two or more

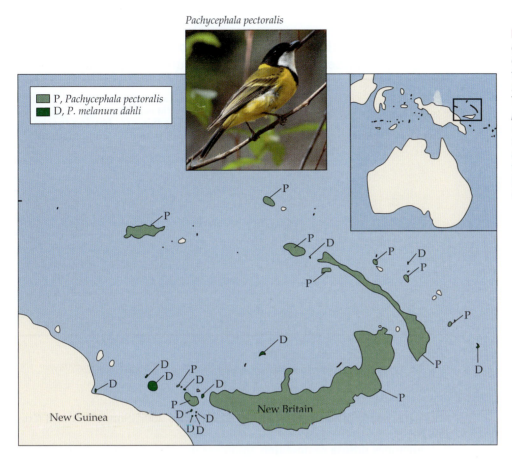

Pachycephala pectoralis

FIGURE 14.11 Mutually exclusive, or checkerboard, distributions of *Pachycephala* flycatchers on the Bismarck Archipelago off New Guinea. There are two species, *P. pectoralis* and *P. melanura dahli*. Most islands have one of these species, no islands have both, and a few islands (especially the smallest) have neither. (After Diamond 1975b; photo courtesy of Lostandcold/Flickr.)

species with similar resource requirements can serve as empirical evidence for competition. Archipelagoes consisting of many similar islands provide numerous examples of such patterns. Diamond (1975b) described several mutually exclusive distributions—which he called checkerboards—for pairs of congeneric bird species inhabiting the Bismarck Archipelago east of New Guinea. The flycatcher *Pachycephala melanura dahli*, for example, is found on 18 islands, and its congener *P. pectoralis* on 11 islands, but the two species do not occur together on the same island (**Figure 14.11**).

Other negative associations resulting from interspecific competition may be more complex than simple checkerboards. Species that do not occur together on small islands may coexist on larger ones, suggesting that they competitively exclude each other where carrying capacities are low, but are able to subdivide the more plentiful resources of larger islands (see Schoener and Adler 1991). In still other cases, only certain combinations of three or more species may be able to coexist, whereas other combinations appear to be "forbidden," or at least much less common, because of competitive exclusion. Diamond (1975b) cites another example from the Bismarck Archipelago—that of cuckoo doves in the genera *Macropygia* and *Reinwardtoena*. There are four species in this group, so there are 15 possible combinations of species that would give biotas of from one to four species (**Table 14.2**). Only 6 of these combinations are actually observed, however, suggesting that certain sets of species are incompatible (see also McFarlane 1989; Gotelli and Graves 1996; Gotelli 2000, 2004). Although debated in the literature, Gotelli and McCabe's (2002) meta-analysis of species distributions matrices confirms the assertions of Diamond's (1975b) general assembly rules: Plant and animal communities exhibit fewer species combinations, more checkerboard patterns, and less species co-occurrence than expected by chance (for further evaluation and

TABLE 14.2 *Combinations of Four Cuckoo Dove Species That Are Present or Absent in the Bismarck Archipelago*

Number of species	Observed combination		Missing species combinations
	Species combinations	**Number of islands inhabited by this species combination**	
1	A	3	N
	M	8	R
2	A, M	5	A, N
	A, R	4	M, N
	M, R	2	N, R
3	A, N, R	5	A, M, N
			A, M, R
			M, N, R
4	None		A, M, N, R

Source: After Diamond 1975b.

Note: A = *Macropygia amboinensis*; M = *M. mackinlayi*; N = *M. nigrirostris*; R = *Reinwardtoena* superspecies.

criticisms, see Whittaker and Fernández-Palacios 2007; Simberloff and Collins 2010).

Sometimes **diffuse competition** among many species, rather than interactions between just two closely related ones, has been implicated in determining insular distributions. Again, perhaps the best examples come from Diamond's (1974, 1975b) studies of the birds of the Bismarck Archipelago. He noted that certain species, which he called **supertramps**, are usually found only on small and/or isolated islands containing few other bird species (**Figure 14.12**). Supertramps are also among the few species that have recolonized islands in early stages of ecosystem succession following recent extensive disturbance by volcanic eruptions. By plotting incidence functions—the proportion of islands inhabited by a given species as a function of the number of other species present—Diamond was able to view what he described as the "'fingerprint' of the distributional strategy of a species," thus quantifying some interesting distributional patterns (note incidence can be plotted against other variables, such as island area and isolation, to explore other biogeographic patterns and their causal processes). Using such graphs, supertramps such as the flycatcher *Monarcha cinerascens* and the honeyeater *Myzomela pammelaena* are readily distinguished as species that appear to be excellent colonists but poor competitors in diverse bird communities (see Figure 14.12). Other species, including a starling (*Aplonis metallica*) and the incubator bird (*Megapodius freycinet*), appear to be both capable colonists and good competitors, as they are found with relatively high frequency on all but the smallest islands. Still other species, especially birds with large body sizes, specialized diets, or restricted habitat requirements, such as hawks (e.g., *Accipiter brachyurus*) and herons (e.g., *Egretta intermedia*), are restricted to the largest islands, which presumably offer more resources but also contain more bird species (see also Peltonen and Hanski 1991; Taylor 1991; Hanski 1992).

A similar approach involves testing the prediction that ecologically similar species coexist on islands less frequently than expected by chance. Several investigators have noted that islands contain relatively few congeners, and few

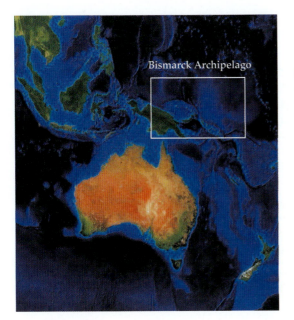

FIGURE 14.12 Incidence functions for various bird species in the Bismarck Archipelago. Note that some species called supertramps, such as *Monarcha cinerascens* and *Myzomela pammelaena*, occur on most islands with few other bird species but are absent from islands with diverse avifaunas; other species, such as *Aplonis metallica* and *Megapodius freycinet*, are found on all but the smallest islands; and still other species, including *Accipiter brachyurus* and *Egretta intermedia*, occur on only the largest islands with many other bird species. (After Diamond 1975b.)

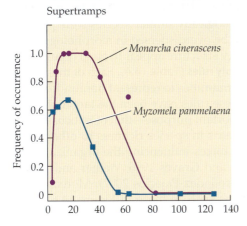
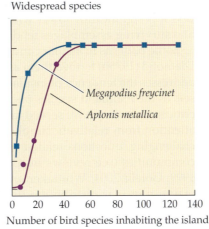
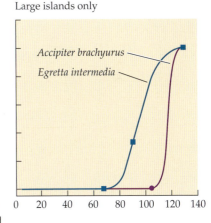

species that are similar in ecologically important morphological traits, such as bill size in birds (e.g., Grant 1965, 1966b, 1968, 1986; MacArthur 1972; Diamond 1973, 1975b; Lack 1973, 1976; see review in Gotelli 2004). In analysis of these patterns, the critical question is whether the differences among co-occurring insular species are greater than would be expected if the island biotas were assembled at random from the mainland species pool. Some analyses that have addressed this problem have produced results that are consistent with the competition hypothesis. For example, in the case of both Darwin's finches in the Galápagos Islands (Grant and Abbott 1980; Hendrickson 1981; Case and Sidell 1983) and hummingbirds in the Antilles (Lack 1973, 1976; Case et al. 1983; Brown and Bowers 1985), it appears that species that coexist on the same island tend to be more different in bill and wing measurements than would be expected if the species associated at random (**Table 14.3**). Actually, the pattern is even more dramatic than Table 14.3 shows, because when species of similar size occur on the same island, they tend to be segregated by elevation and habitat.

Pimm and Moulton's studies on birds introduced to Tahiti and the Hawaiian Islands indicated that establishment of populations on these islands was strongly influenced by interspecific competition from morphologically and ecologically similar species (**Figure 14.13**; Moulton and Pimm 1983, 1986, 1987; Pimm et al. 1988; Moulton 1993). Later eco-morphological studies on birds introduced to Bermuda provided similar results (Lockwood and Moult-

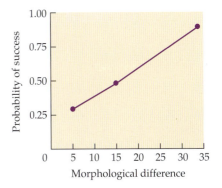

FIGURE 14.13 The probability that introduced species of birds will successfully establish an insular population is directly correlated with the degree to which they differ morphologically from the resident Hawaiian avifauna. Morphological difference is represented here by the difference in bill length between an introduced bird species and its most similar resident congener. (After Moulton and Pimm 1986.)

TABLE 14.3 *Coexistence of Pairs of Hummingbird Species of Varying Size Combinations on the Islands of the Greater and Lesser Antilles*

	Both small	One small, one large	Both large
Observed species pairs	6	27	16
Species pairs expected from a random distribution	14.3	18.7	24.3

Source: Data from Brown and Bowers 1984.

Note: The bill length of every other species was expressed as a ratio relative to the bill length of the smallest species, and the species with ratios less than or greater than 1.8 were categorized as small or large, respectively. Note that coexisting pairs of species tend to be of different sizes; the probability that this will occur by chance is less than 0.02.

on 1994; see also Simberloff and Boecklen 1991; Moulton 1993). Invasion success was significantly lower when ecologically similar species were present, and surviving assemblages tended to be overdispersed in morphology.

Some patterns in insular community structure appear to be so regular that Diamond (1975b) termed them **assembly rules** (**Table 14.4**). Each of these rules addresses the tendency of insular communities to represent nonrandom subsets of the mainland species pool. These rules have been the target of much criticism, yet they have served an important purpose—that of refocusing our attention from relatively simple patterns in species richness to other, more challenging, questions related to the forces influencing the organization of communities (see reviews by Gotelli and Graves 1996; Weiher and Keddy 1999; Gotelli 2004; Whittaker and Fernández-Palacios 2007). Moreover, we may use insular distribution functions such as those in Figure 14.6 to put Diamond's assembly rules—and, in particular, checkerboard patterns—in a more explicit, geographic context. The resultant patterns (**Figure 14.14**) clearly illustrate how both interspecific interactions and limited immigration abilities may combine to restrict a species' realized geographic range to a subset of its fundamental range (see Chapter 4). To complete the picture, Simberloff and Collins (2010) provide a persuasive discussion of how these related patterns—checkerboards and exclusive distribution functions—may

TABLE 14.4 *Diamond's Assembly Rules for Insular Communities*

1. *If one considers all the combinations that can be formed from a group of related species, only certain ones of these combinations exist in nature.*
2. *These permissible combinations resist invaders that would transform them into a forbidden combination.*
3. *A combination that is stable on a large or species-rich island may be unstable on a small or species-poor island.*
4. *On a small or species-poor island, a combination may resist invaders that would be incorporated on a larger or more species-rich island.*
5. *Some pairs of species never coexist, either by themselves or as part of a larger combination.*
6. *Some pairs of species that form an unstable combination by themselves may form part of a stable larger combination.*
7. *Conversely, some combinations that are composed entirely of stable subcombinations are themselves unstable.*

Source: Diamond 1975b.

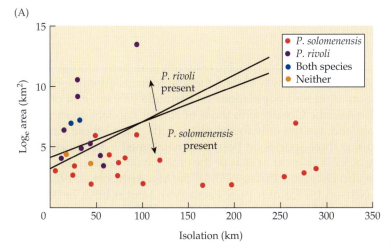

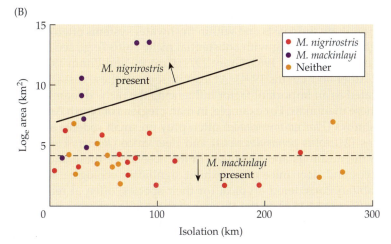

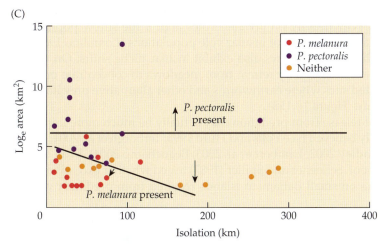

FIGURE 14.14 Insular distribution functions for three of Diamond's (1975b) avian guilds of the Bismarck Archipelago reveal that species distributions are likely influenced by the combined effects of interspecific interactions and interspecific differences in immigration abilities and resource requirements. While the potential or fundamental ranges of these birds may overlap substantially, interspecific interactions appear to limit realized distributions of at least some of these species to a subset of the archipelago. Diamond's checkerboards, or exclusive distributions, appear to result from species segregating their realized distributions over different ranges of isolation and area; (A) fruit pigeon guild, (B) cuckoo dove guild, and (C) gleaning flycatcher guild. (After Lomolino 2000b; see also Simberloff and Collins 2010.)

also be generated by speciation and subsequent radiations from isolated centers of origin, thus providing an interpretation for the birds of the Solomon Islands in strong contrast to the original assembly-rules model proposed by Diamond (e.g., 1975b).

Density compensation and niche shifts on islands

If competition influences the organization of communities by limiting species to only part of their fundamental niches, one would expect populations to expand their niches and increase in density on islands where they interact with few other species. Both of these phenomena, niche expansion (or shifts) and increased densities, appear to be relatively common among insular populations. Crowell (1962) was among the first to focus attention on these patterns by comparing bird populations on Bermuda (900 km east of North Carolina in the Atlantic Ocean) with those of similar habitats on the North American mainland. He not only found niche shifts among these birds, but also observed that three species on Bermuda maintained a combined population density at least as great as that of the entire avian community on the mainland (see also Diamond 1970a, 1973, 1975b; MacArthur et al. 1972, 1973; Yeaton 1974; Nilsson 1977; Kohn 1978; Faeth 1984; Martin 1992; Terborgh et al. 1997; Terborgh et al. 2001; Gaston and Matter 2002; Lambert et al. 2003; Jiang 2007).

This latter phenomenon, termed **density compensation**, is often attributed to relatively low levels of interspecific competition, but it may just as easily result from the paucity of insular predators. Small rodents and shrews inhabiting islands along the coasts and within freshwater systems often exhibit relatively high densities along with substantial niche shifts (e.g., see Hatt et al. 1928; Bishop and Delaney 1963; Grant 1971; Crowell and Pimm 1976; Lomolino 1984; Adler and Levins 1994). At least in some of these cases, predation is implicated as an important factor influencing the niche dynamics and densities of these mammals. For example, in many regions of North America, short-tailed shrews (*Blarina brevicauda*) often prey on small rodents such as meadow voles (*Microtus pennsylvanicus*) and deer mice (*Peromyscus* spp.), especially on the nestlings and juveniles. In the absence of these and other predators, voles and deer mice often exhibit significant ecological release, occupying habitats considered atypical for their species, often at relatively high densities (**Figure 14.15A**). In the absence of short-tailed shrews, meadow voles—typically grassland specialists—expand their niches to occupy all available habitats at relatively high densities. Experimental introductions of short-tailed shrews onto islands of the St. Lawrence River confirmed that these shrews prey heavily on juvenile voles and, if the shrew populations persist, can cause extinction of insular vole populations (Lomolino 1984). Interestingly, short-tailed shrews also exhibit insular niche shifts. On islands where voles are absent, they retreat from meadows to concentrate on more heavily forested sites where earthworms and other alternative prey items are more abundant (**Figure 14.15B**).

The role of release from predation may actually be much more important than originally believed (see Polis and Hurd 1995; Terborgh et al. 2001; Rodda and Dean-Bradley 2002; Lambert et al. 2003; Beauchamp 2004). In an elegant field experiment, Schoener and Spiller (1987) tested whether excessive densities of spiders on small Bahamian islands resulted from predatory release.

FIGURE 14.15 Insular populations often show evidence of niche shifts, exhibiting relatively high densities and occupying habitats considered atypical for the species on the mainland. Such shifts may result from the paucity of competitors or predators on islands. (A) In the Thousand Islands region of the St. Lawrence River, meadow voles (*Microtus pennsylvanicus*) exhibit significant niche shifts on islands lacking one of their predators, the short-tailed shrew (*Blarina brevicauda*). (B) Conversely, on islands lacking meadow voles, the distribution of short-tailed shrews contracts to habitats where the availability of alternative prey items (primarily macroinvertebrates in forest soils) remains relatively high.

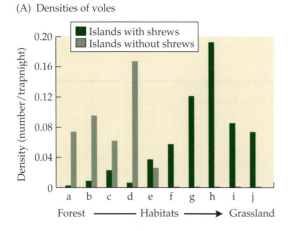

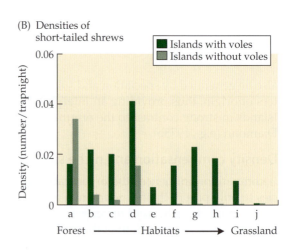

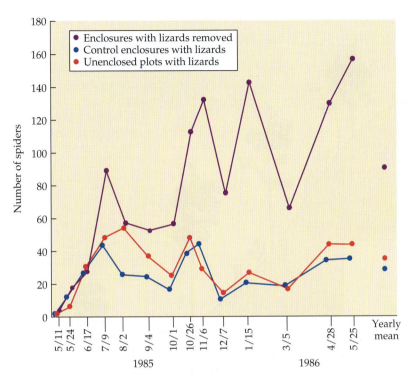

FIGURE 14.16 On Bahamian islands lacking predatory lizards, spiders often exhibit predatory release, reaching densities about ten times as high as on islands inhabited by these lizards. Experimental removal of lizards from enclosed plots demonstrated that population densities of spiders are strongly influenced by predation. (After Schoener and Spiller 1987.)

The primary predators of these spiders are lizards (*Anolis* spp. and *Ameiva* spp.). On islands lacking lizards, spider densities were about ten times as high as on islands where those predators were present. Schoener and Spiller tested the predatory release hypothesis by removing lizards from 84 one-meter-square enclosed plots and monitoring spider densities on experimental and control plots for just over a year. Spider density rapidly increased in the absence of lizards, reaching a level approximately 2.5 times that of the control plots by the end of the experiment (**Figure 14.16**).

Schoener and Spiller's study is just one in a long and distinguished series of studies on the ecology and evolution of West Indian lizards. No field biologist can visit a small Caribbean island without being impressed by the incredible abundance of *Anolis* lizards. These small reptiles seem to be everywhere, from the ground to the tops of the tallest trees, from disturbed habitats along roadsides and in cities, to pristine native forests. Indeed, these lizards are much more abundant on most islands than they are anywhere on the tropical American mainland. E. E. Williams of Harvard University and his students T. W. Schoener, G. C. Gorman, J. Roughgarden, B. Lister, and R. Holt, and their students in turn, have studied the evolution, ecology, and biogeography of the Caribbean *Anolis* in great detail (Losos 2009; Losos and Schneider 2009). Although many distributional patterns have been well documented, the underlying causal processes have proved more difficult to demonstrate convincingly.

Anolis is but one of several important lizard genera on the mainland of tropical America, but it is by far the most abundant genus of vertebrates on the Caribbean iIslands. On the large islands of the Greater Antilles, a few colonizing species have given rise by adaptive radiation to a diverse *Anolis* fauna (Losos 2009; Losos and Ricklefs 2009). Hispaniola, the second largest and ecologically most diverse island, has at least 35 species, which were probably derived from four separate invasions (Williams 1976, 1983). These species occupy a variety of ecological niches. They range from tiny insecti-

FIGURE 14.17 Diagrammatic representation of the habitats occupied by different *Anolis* species on the northern part of Hispaniola. Their niches differ in elevation, vegetation type, perch height, and position along a gradient from sunlight to shade. The species indicated here—only a fraction of at least 35 species that inhabit the island—have been produced largely by speciation and adaptive radiation within the island. (After Williams 1983.)

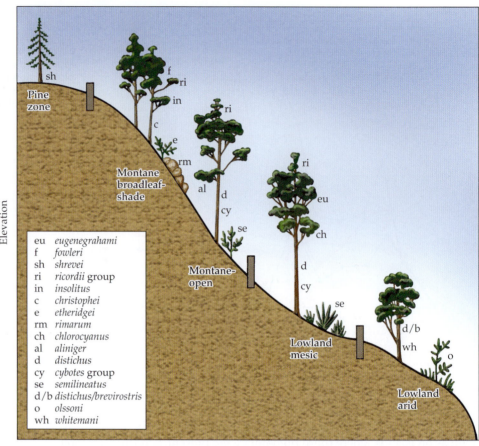

eu *eugenegrahami*
f *fowleri*
sh *shrevei*
ri *ricordii* group
in *insolitus*
c *christophei*
e *etheridgei*
rm *rimarum*
ch *chlorocyanus*
al *aliniger*
d *distichus*
cy *cybotes* group
se *semilineatus*
d/b *distichus/brevirostris*
o *olssoni*
wh *whitemani*

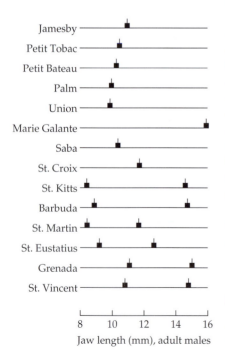

FIGURE 14.18 Body sizes of *Anolis* lizards of the Lesser Antilles. Note that all of these islands have either one or two species. When two species co-occur, they tend to be displaced in size, whereas when only one species is present, it tends to be of intermediate size. (After Roughgarden 1974; Roughgarden et al. 1983; Roughgarden and Fuentes 1997.)

vores to large carnivores, with head and body lengths ranging from 40 mm to more than 200 mm. They are morphologically, physiologically, and behaviorally specialized for distinctive habitats and microenvironments—from sunny sites to deep shade, from open ground and rocks to grasslands and scrub forests, to different layers in the complex vegetation of mature tropical and montane forests (**Figure 14.17**). In contrast, the small islands of the Lesser Antilles each have only one or two generalized species (Roughgarden and Fuentes 1977; Roughgarden et al. 1983). When two species coexist on a small island, they differ in body size, prey size, and habitat; but when only one species is present, it is intermediate in size, takes a wide range of prey, and occupies virtually all habitats (**Figure 14.18**). Clearly, the fundamental niche of an *Anolis* species that has evolved in the absence of congeners is very broad. Some of the observed niche expansion of *Anolis* on small islands may represent behaviorally mediated ecological responses in the absence of competing species. Nevertheless, most of the niche dynamics in Caribbean species represent ecological and evolutionary adaptations to communities containing different numbers and kinds of other species, and these are reflected in morphological, physiological, and behavioral changes (see Roughgarden 1995; Losos and Queiroz 1997; Knox et al. 2001).

Variation in densities and niche characteristics of *Anolis* between islands and between habitats within islands has also been investigated, especially by Schoener (1968a, 1975), Lister (1976a,b), and Holt (1977). They have shown that when a species occurs on an island or in habitats where there are few other lizards, increases in niche breadth often parallel increases in density (**Figure 14.19**). Again, it is tempting to attribute these patterns of density compensation and niche expansion to a release from competition with other *Anolis* species, but this assumption may be unwarranted. The small islands where the lizards show the most spectacular increases in density contain not only fewer *Anolis* species, but lower diversity of most taxa. Thus, *Anolis* populations are potentially released from competition, not only with congeners, but also with insectivorous arthropods (e.g., spiders), birds, frogs, and lizards of other genera, and released from predation by birds and other lizards. It is also possible that small islands support higher standing stocks of insect prey. Holt studied *Anolis* on islands located off the coast of Trinidad and concluded that direct competition with other *Anolis* or other lizard species could not account for the observed density changes. He suggested that the absence of predatory birds might be the most important of the various factors contributing to increased *Anolis* density and niche breadth on small islands. Thus, while ecological release and high densities among insular populations are often attributed to reduced competition, they are more generally a consequence of the depauperate nature and reduced intensity of interspecific interactions.

As Clegg (2010) has observed, niche shifts and expansions are also common for insular birds and may be linked to evolutionary changes in body size, because diet and competitive interactions tend to be strongly influenced by body size of the competitors (Diamond 1970a; Cox and Ricklefs 1977; Blondel et al. 1988; Scott et al. 2003; Schlotfeldt and Kleindorfer 2006).

Density overcompensation

One puzzling aspect of insular density compensation is that the total population densities of a few species inhabiting a small island may actually exceed the combined densities of a much greater number of species occupying similar habitats on the mainland. This phenomenon, called **density overcompensation** or **excess density compensation**, is fairly well documented, especially for birds on small oceanic islands. In fact, Crowell's (1962) studies, mentioned earlier in this chapter, revealed that just 10 species of small passerine birds on Bermuda together maintained a population not just equal to, but about 1.5 times greater than the combined densities of 20 to 30 species on the mainland. Subsequent studies by MacArthur and colleagues (1972) on the Pearl Islands south of Panama, by Diamond (1970b, 1975b) on the Bismarcks and other archipelagoes north and east of New Guinea, and by Emlen (1978) on the Bahamas have described similar patterns. Case (1975) documented density overcompensation among lizards on the islands in the Gulf of California compared with the mainland of Baja California and Sonora, and Adler, Terborgh, and their colleagues have reported the same phenomenon for mammals and birds on the anthropogenic islands of Gatun Lake, Panama, and Lago Guri, Venezuela (Adler 1996; Terborgh et al. 1997; Terborgh et al. 2001; Lambert et al. 2003; Terborgh 2010).

Several explanations have been proposed for such density compensation and overcompensation:

1. Because vertebrates of large body size and higher trophic levels tend to be absent from small islands, the same resources can support substantially larger populations of smaller species. Note, however, that Crowell's

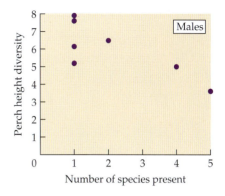

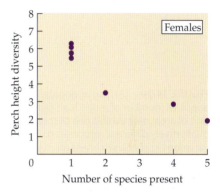

FIGURE 14.19 Relationship between the diversity of heights of perches used by males and females of *Anolis sangrei* and the number of other *Anolis* species occurring in the same habitat on the same island. Note that when *A. sangrei* is the only species present, it uses a wide variety of perch heights, but the breadth of this niche dimension contracts and *A. sangrei* is excluded from arboreal habitats as it encounters an increasing number of coexisting congeneric species. (After Lister 1976a.)

(1962) study found that biomass, as well as density, was substantially higher among insular birds than among mainland birds.

2. Population densities of, for instance, birds reflect release from competition not only with missing bird species but also with other taxa, such as mammals and amphibians that use similar foods and other resources but are even less common on oceanic islands than birds because they are poor over-water dispersers.

3. Inflated population densities on islands reflect the absence or paucity of predators and parasites, which also tend to be poorly represented on oceanic islands (Grant 1966a; MacArthur et al. 1972; Case 1975; George 1987; McLaughlin and Roughgarden 1989; Fallon et al. 2003).

4. Oceanic islands may be more productive, at least in terms of foods and other resources required by smaller vertebrates (i.e., small mammals, lizards, and birds; see Case 1975).

5. On oceanic islands, renewable food resources are harvested at rates closer to their maximum sustained yields than on the mainland, where intense competition can lead to overexploitation and lower productivity of resources.

6. Populations can become more finely adapted to their local environment and, hence, attain higher densities on isolated islands than on continents, where extensive gene flow among populations occupying different habitats tends to prevent specialization for more efficient use of local resources (Emlen 1978, 1979).

7. The surrounding waters may prevent population losses that would otherwise result from emigration of individuals into marginal habitats (the fence effect; see Krebs et al. 1969; MacArthur et al. 1972; Emlen 1979; Ostfeld 1994).

This is a long but not all-inclusive list. As is often the case with the most general patterns in ecology and biogeography, these patterns may result from a combination of convergent mechanisms, rather than just one unique, overriding mechanism.

Patterns of combined densities of insular species, including compensation and overcompensation, may be summarized in a graphical model, as put forward by Wright (1980) (**Figure 14.20A**). This model recognizes that individual species may respond in contrasting ways. Thus, while the overall pattern exhibited by a particular insular community may be density compensation, some of its species are likely to exhibit atypically low populations, while densities of others are very much higher than found in comparative mainland settings. For example, Crowell (1962) found that common crows, house

FIGURE 14.20 (A) Insular population densities (number of individuals per unit of area) may exhibit one of three qualitatively different responses to variations in species richness (or island area). If population densities are not influenced by competition, predation, and other interactions among species, then the densities of all species combined should increase in proportion to the total number of species (density stasis). On the other hand, if interspecific interactions do limit insular populations, they may exhibit density compensation or overcompensation on species-poor islands (i.e., populations of one or a few insular inhabitants may equal or exceed those of a much richer assemblage of species). (B) Population densities of individual species also may exhibit three qualitatively different responses to differences in species richness (or island area) among islands: *density enhancement* (sensu Jaenike 1978)—on very small, species-poor islands, population levels of a species may fall below those of its conspecifics on the mainland but increase as island area increases and environmental conditions become more suitable; *density inflation*—population levels may continue to increase, with increasing area, above their mainland levels, until interspecific interactions begin to regulate population densities of the focal species; *density stasis*—at intermediate to high levels of species richness, many species may regulate one another's densities such that few, if any, exhibit any consistent trend of population density with species richness or island area. (A after Wright 1980.)

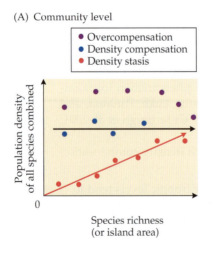
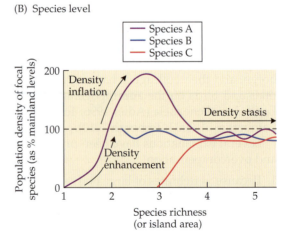

sparrows, and starlings were extremely rare in all but farmland habitats on Bermuda, while catbird, cardinal, and white-eyed vireo populations were 3 to 30 times higher in comparison with the same habitats on the mainland. Similarly, in a study of Caribbean insects, Janzen (1973) found that generalists, especially the Homoptera, tended to exhibit density compensation on islands, while other insects did not.

A species-based alternative to Wright's model introduces an additional density pattern termed **density enhancement** (**Figure 14.20B**; see Jaenike 1978). Population densities of some, and perhaps most, species on very small islands may indeed be much lower than their typical densities on the mainland, but increase with island area. Larger islands not only support more species—some of which may be mutualists or commensals—but also tend to provide less harsh, more stable, and otherwise more favorable habitats for particular species. Alternatively, population densities of at least some species may remain relatively stable over a broad range of species richness—a pattern that Williamson (1981) termed **density stasis**. In fact, the same species may exhibit all three patterns—density enhancement on tiny islands, density inflation on islands of intermediate size, and density stasis on larger, more species-rich islands (see Figure 14.20B, species A). Perhaps density inflation also varies systematically among species: highest for supertramps and lowest for those species restricted to the richest, and largest, islands. We admit that these are largely untested predictions, but the pertinent data seem to be available, at least for some insular communities.

In summary, studies on the ecological responses of insular populations have been, and will certainly continue to be, fertile ground for research on the forces influencing community structure on islands and mainland ecosystems as well.

Adaptive radiations

We now consider the final factor that may contribute to the disproportionate incidence of particular types of species on islands—speciation. As we have seen in Chapter 7, only a select group of species possess the fortuitous combination of ecological capacities and evolutionary machinery to both persist and rapidly speciate on islands. Adaptive radiations of the cichlids of Africa's Rift Valley lakes, and *Drosophila*, honeycreepers, ferns, and asters (Compositae) of the Hawaiian Islands are legendary and help us identify the intrinsic species traits that confer high **evolvability** on remote islands. Among these traits are those associated with the capacity to generate genetic variation, including short generation times, high reproductive rates, and high mutation rates (see Brookfield 2009).

Although the most striking illustrations of evolution on islands involve the radiation of several ecologically distinct species from a single ancestor, not all insular speciation events take this form and, even on remote islands rich in endemics, not all native species evolve into endemics. First, an island endemic may be the sole representative of its lineage on the island and may have diverged (through anagenesis) from its mainland ancestor without forming multiple species on the island or archipelago. Second, some lineages speciate repeatedly within an archipelago, or even within a single island, without any obvious degree of adaptive change, apparently as the result of a series of micro-allopatric speciation events. This process—termed **nonadaptive radiation**—results from founder events and genetic drift within small isolates. As it can be very difficult to determine the degree to which a radiation is adaptive, we should regard adaptive and nonadaptive radiation as conceptual end points of a continuum. If there is doubt as to the mechanisms involved, it may be safer to describe such patterns as archipelago radiations, thus describing the geographical context of the pattern without invoking the mechanism involved (for further exposition see Stuessy et al. 2006; Whittaker and Fernández-Palacios 2007). Finally, even for those archipelagoes that provide examples of spectacular radiations, it is typically the case that only a small minority of lineages have done so, with most colonists producing only one or two native species. We still know much less than we would like to about the combinations of traits that predispose some lineages to radiate while others fail to.

Evolution of Island Life

The above processes, especially those occurring during adaptive radiations of insular biotas, influence more than just the number of component species (i.e., richness), but can fundamentally transform the endemic biota of remote islands—often to the point that they develop some truly bizarre characters associated with dispersing within (but not among) islands and adapting to their singular, and relatively simple, ecological environments. Thus, as productions of evolution under highly altered conditions, the marvels of island life are products of natural experiments in evolution—geographic-scale and long-term "manipulations" constituting dramatic shifts and reversals in selective pressures in comparison to those characteristic of mainland environments. When taken together, these evolutionary artifacts comprise an invaluable Rosetta stone for solving one of the more central mysteries of nature—how biological diversity develops over time and space.

We now discuss how such reversals in natural selection can account for some remarkable changes in the fundamental capacities of insular species, that is, their capacities and propensities for long-distance dispersal; their strategies for developing, surviving, and reproducing on islands; and their abilities to cope with and adapt to a limited and highly altered assemblage of other species. We then conclude our discussions on the evolution of island life with an exploration of how these reversals in natural selection are evidenced in the transformation of one of the most fundamental characters of all life-forms—their size.

Dispersal denied: Sticking to the wreck

> As with mariners shipwrecked near a coast, it would have been better for the good swimmers if they had been able to swim still further, whereas it would have been better for the bad swimmers if they had not been able to swim at all and had stuck to the wreck. (Darwin 1859: 177)

With this one metaphor, Charles Darwin describes how reversals in natural selection between mainland and insular environments can explain a truly fascinating paradox of insular biotas: Although they are typically biased in favor of taxa with superior immigration abilities (e.g., ferns, birds, and microsnails), many insular forms, especially those on the most isolated oceanic islands, have little or no ability to disperse to other islands. As the legendary paleontologist Sir Richard Owen once remarked, there is no greater anomaly in nature than a bird that cannot fly (Darwin 1859). Yet, flightless birds and insects are relatively common on many oceanic islands, as are other animals and plants with little ability or propensity for dispersal. And herein lies the paradox: How could these relatively sedentary forms have colonized such remote ecosystems?

During Darwin's day, the extensionists may have used these observations to bolster their argument that ancient land bridges once connected even the most isolated archipelagoes, including New Zealand with its great moas, with the continents (see also Wallace 1893). As we saw in Chapter 2, nothing "vexed" Darwin more than these post hoc scenarios of the Earth's dynamism. Instead, he offered an explanation embodied in the second part of the above quote. There was no doubt, at least in Darwin's mind, that these peculiar insular forms arrived via long-distance dispersal, but were then transformed by evolution. Unfortunately, Darwin attributed some of these patterns to what was termed the "law of disuse." For example, he believed "that the nearly wingless condition of several birds, species which now inhabit several oceanic islands tenanted by no beast of prey, has been caused by disuse." He acknowledged, however, that the flightlessness of many insular forms, such as 200 of the 550 described beetles of Madeira Island, was "wholly, or mainly due to natural selection." The key point is that the selective forces operating on a potential immigrant may be entirely different from those operating on its insular descendants (see Carlquist 1974). In Darwin's metaphor, selection during immigration favors "good swimmers," but once they (or their descendants) have survived the "wreck" (colonized the island), selection then favors "bad swimmers"—those less likely to swim off or be carried away by winds and lost at sea.

Birds

Interestingly, flightlessness has never been documented in two of the major, or once dominant, volant taxa—bats and pterosaurs. New Zealand's only endemic mammals, short-tailed bats (Mystacinidae) are clearly the most terrestrial of all bats, with incisors specialized for burrowing and with well-developed claws and wing membranes that roll up like furled sails (and, in the case of the extinct greater short-tailed bat, *Mystacina robusta*, folded into pouchlike flaps of skin) to enable adept movement along the forest floor. Yet, these species still retained the ability, albeit somewhat limited, to fly.

On the other hand, derived insular flightlessness has likely evolved thousands of times in birds and is reported in at least ten orders, including kiwis (Apterygiformes); moas (Dinornithiformes); cormorants (Pelecaniformes); ibises (Ciconiiformes); ducks and geese (Anseriformes); kagus, rails, and gallinules (Gruiformes); the dodo, solitaires, and Fiji pigeon (Columbiformes); parrots (Psittaciformes); owls (Strigiformes); and wrens (Passeriformes) (McNab 2002). On New Zealand, some 25 to 35 percent of the terrestrial and freshwater birds are—or were, in the case of extirpated forms—flightless. Similarly, 24 percent (20 species) of Hawaii's endemic birds were flightless. The long list of now-extinct flightless birds includes many flightless giants such as the dodo (*Raphus cucullatus*) of Mauritius, the solitaires (*Raphus solitarius* and *Pezophaps solitaria*) of Réunion and Rodriguez, the elephant birds (Aepyornithidae) of Madagascar, and about 9 species of moas (Dinornithidae) of New Zealand (see review by McNab 1994a). Flightlessness is especially common among the rails, having evolved independently within at least 11 groups of rails across hundreds of islands (Steadman 1989, 1995; Diamond 1991b). In fact, evolution of derived flightlessness may occur very rapidly, on the order of perhaps just a few centuries and, while this may seem astounding, extrapolations based on paleontological evidence suggest that, before human colonization, most Pacific islands were inhabited by at least one species of flightless rail (see Olson 1973; Steadman and Olson 1985; Steadman 1986, 1989; Silkas et al. 2001; Steadman and Martin 2003; Steadman 2006).

Currently, the most widely accepted explanation for the evolution of flightlessness in birds involves selective pressures associated with the absence of predators and with limited resources on islands (Diamond 1991b; McNab 1994b, 2002; see also McCall et al. 1998). Under reduced pressure from predators and, for that matter, from competitors as well, insular populations would be likely to undergo ecological release. Evolutionary responses could include changes that would conserve energy, such as reduced size overall, or a reduction in metabolically expensive tissues, such as the otherwise large flight muscles. On the mainland, such selection to conserve energy is of course countered by the selective advantage of being able to escape ground-dwelling predators. On many islands, however, this is not the case, as "reduction in flight muscle brings great energy savings with little penalty" (Diamond 1991b). Thus, at least for the rails, flightlessness was probably achieved primarily by reduction in mass of flight muscle, with perhaps a modest reduction in overall mass as well.

Evolution of flightlessness in other birds, however, is often associated with an increase in body size (Morton 1978; Livezey 1993). This suggests an alternative path to the loss of flight: Many insular birds may have simply outgrown their wings. In the absence of mammals and other large terrestrial vertebrates, selective pressures may have promoted increased body size overall, without a compensatory increase in flight muscle mass (see the section Transforming Life's Most Fundamental Character: Size, below, on the evolution of insular body size). The result was the moas, dodo, solitaires, and elephant birds—avian giants no more capable of flight than we are. Indeed, these and other flightless insular birds may have taken over the large herbivore niches left vacant by mammals—shifting their diets; developing more drab, cryptic coloration; and increasing in size such that they eventually lost their power of flight (see Livezey 1993; Baker et al. 1995; Trewick 1996). A related, biogeographic pattern supports McNab's (1994b, 2002) energetics-based explanation for the evolution of flightlessness. As he observed, derived flightlessness is relatively common in birds on tropical islands, much less common in those of temperate regions, and absent in those of islands in higher latitudes. In the last case, energetic demands and the need to migrate to warmer regions during winter render flightlessness untenable.

These evolutionary transformations of thousands of insular endemics provide poignant illustrations of both the marvels and the perils of island life. Flightless birds that became adapted and highly specialized for insular environments were also the first to suffer extinctions following colonization by humans and their ground-dwelling commensals (see Chapter 16). Such was the case for (possibly) thousands of flightless rails and other insular birds endemic to remote islands across the Pacific (Steadman and Martin 2003). Similarly, of the three species of ground-dwelling, short-tailed bats discussed above, one (*Mystacina tuberculata*) is critically endangered, and another—the poorest flyer (*M. robusta*)—is extinct.

Insects and other invertebrates

The flightless beetles of Madeira Island in the Atlantic Ocean are just one of the countless groups of insular insects whose powers of flight have been lost, or greatly reduced. In addition to Coleoptera, derived flightlessness has been reported in nearly all orders of insects across a variety of habitats, including those at high altitudes, in deserts, and on islands (Wagner and Liebherr 1992; Roff 1990; Kotze et al. 2000; Peck 2006). Insular forms that have lost the power of flight include butterflies (Lepidoptera); flies (Diptera); ants, bees, and wasps (Hymenoptera); mayflies (Ephemeroptera); bugs (Homoptera); and crickets, grasshoppers, and wetas (Orthoptera). Flightlessness and reduced dispersal abilities of insects appear to be most prevalent on the most isolated archipelagoes. For example, 40 percent of the native insects of Campbell Island (the southernmost of New Zealand's subantarctic islands) exhibit flightlessness or at least some degree of wing reduction; 90 percent of Tristan de Cunha's (south Atlantic) endemic beetles have reduced wings; and 94 percent of New Zealand's moths and butterflies have limited powers of flight (Williamson 1981; Gressitt 1982; Howarth 1990).

While the evolution of flightlessness in insects of xeric mainland habitats may have involved energy and water conservation (Roff 1986, 1990), it is likely that other factors were much more important for development of flightlessness on islands. Like many of the avian examples discussed above, the wetas (Orthoptera) of New Zealand represent another fascinating case of ecological release and evolutionary convergence on a mammalian niche (**Figure 14.21**). In the absence of ground-dwelling mammals, wetas have lost the power of flight, increasing in size and occupying the geophilous niche otherwise occupied by many rodents. Similarly, in the absence of predators along the surfaces of their wetland habitats in Madagascar, some mayflies (genus *Cheirogenesia*) have wings that are so reduced that they are incapable of flight, and instead the adults use these oarlike appendages to skim the water surface while feeding (Ruffieux et al. 1998). Roff (1990) undertook a statistical analysis and found no evidence of an association between oceanic islands and insect flightlessness. His analysis, however, did not take account of a reduced flight prevalence and merely tested for a dichotomy between volant and non-volant forms; therefore it may not be the final word on the propensity to reduced flight in island insects.

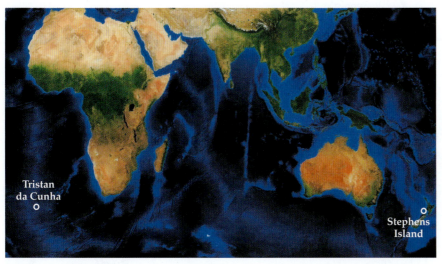

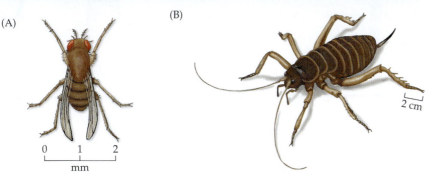

FIGURE 14.21 Two examples of flightless insular insects: (A) *Scaptomyza frustulifera*, a drosophilid from Tristan da Cunha, and (B) the Stephens Island weta, *Deinacrida rugosa*, a flightless orthopteran from Stephens and Mana Islands, Cook Strait, New Zealand. The weta may be an ecological equivalent of rodents. It has been extirpated from the New Zealand mainland and many offshore islands (which lacked native non-volant mammals) by introduced rats and other ground-dwelling mammals. (A after Williamson 1981; B after Collins and Thomas 1991.)

Insular plants

Reduced dispersal ability also is remarkably common in insular plants. In his comprehensive summary on this subject, Carlquist (1974: 429) wrote the following:

> While working with evolutionary phenomena in the Hawaiian Islands and other Polynesian islands, I observed types of fruits and seeds that seemed incongruously poorly adapted at dispersal. The floras of these islands must

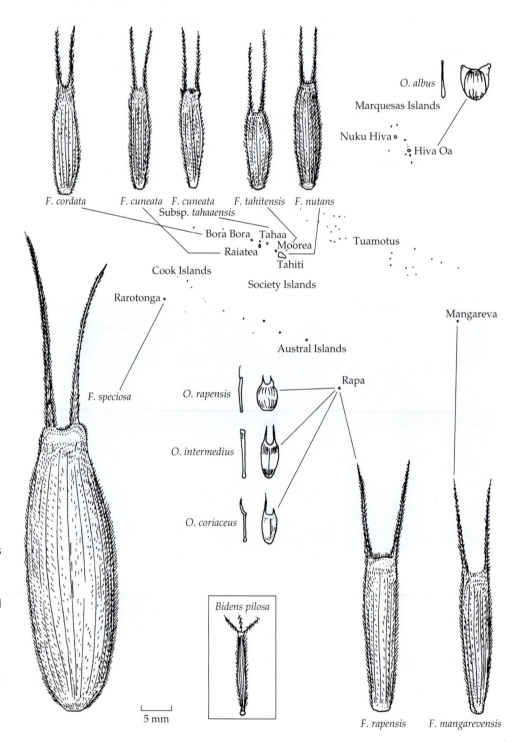

FIGURE 14.22 On islands, many plant species exhibit traits reflecting reduced dispersal ability, including fruits and seeds that are heavier, less buoyant, and less resistant to seawater, as well as a reduction in spines, wings, and structures that would otherwise facilitate lift or attachment to animal dispersers. Fruits of insular forms of *Fitchia* and *Oparanthus* (Asteraceae) are shown; *Bidens pilosa*, which exhibits a more typical mainland morphology, is shown in the inset for comparison.

have arrived by long-distance dispersal, yet during evolution on the island areas various groups of plants have lost their dispersibility.

Many of the changes that result in reduced dispersal ability in plants are associated with shifts from adaptations for dispersing to and colonizing habitats along the beachhead, to traits better suited for living in interior, mesic forests. Precinctiveness appears to be just as important for insular plants as it is for insular insects. Accordingly, both seeds and fruits tend to become heavier, less buoyant, and less resistant to seawater, and structures that facilitate air lift or attachment to animals (spines, wings, and hooks) are greatly reduced (**Figure 14.22**; Carlquist 1974). Morphological changes associated with reduced dispersal ability in insular plants, including insular woodiness—discussed later in this chapter (**Figure 14.23**)—may evolve very rapidly, perhaps in less than a decade (Cody and Overton 1996; see also Millien 2006 for analogous reports of rapid evolution in insular mammals).

Finally, many cases of reduced dispersal ability among insular plants and insects may result from a type of ecological lottery that characterizes most islands. Given that islands are depauperate by nature, plants that depend on particular animals for dispersal (endozoochory) or animals that depend on larger animals for dispersal (**phoresy**) may find themselves stranded (see Whittaker et al. 1997 re: plant dispersal to Krakatau). Luck may still be with them, however, if they can adapt and find a niche somewhere on the "wreck," and stick to it.

Release, displacement, and evolving ecologies

While it is convenient and all too common to make a distinction between ecological and evolutionary realms—that their most defining phenomena are driven by distinct processes operating over very different and nonoverlapping scales of space and time, such assertions likely impede the development of a genuinely integrative understanding of the spatial and temporal dynamics of life. From the dimensions of the proboscis of the giant sphingid

FIGURE 14.23 Evolution of treelike stature and woodiness in insular plants is seen in (A) the silversword (*Dubautia reticulata*) of the Hawaiian Islands and (B) the cactus tree (*Opuntia echios*) of the Galápagos. (A courtesy of Gerry Carr, see Carr et al. 2003; B courtesy of A. Sinauer.)

(A)

(B)

moth—which Darwin (1862) intuited is uniquely adapted to tap the nectar and simultaneously pollinate the 25 cm long, tubular flowers of Madagascar's Christmas orchid (*Angraecum sesquipedale*)—to the teeth, claws, and limb muscles of the big cats, to the brains that allowed hominids to dominate other life-forms, the morphological, physiological, and behavioral traits influencing all ecological interactions evolve, and evolution in turn is driven to a large degree by ecological interactions.

Thus, the species-poor and disharmonic nature of remote islands presents special challenges and opportunities for colonists if they are to survive and maintain populations and ultimately evolve on these islands. In the next section, we spend considerable time discussing some fascinating evolutionary trends in one of life's most fundamental traits—the size of organisms, which strongly influences all of their capacities, including their abilities to adapt to, and sometimes dominate, other species. As Rosemarie Gillespie and her colleagues have demonstrated, however, the interplay between ecological and evolutionary processes is not limited to those associated with body size (Gillespie 2004; Gillespie and Baldwin 2009; see also Roughgarden 1972; Roderick 1997; Schluter 2000; Webb et al. 2002; Silvertown 2004; Silvertown et al. 2005; Trusty et al. 2005; Holt 2010; Vellend and Orrock 2010). On islands, species must adapt not only to the absence of typical **symbionts** (in the general sense, including mutualists, competitors, predators, parasites, commensals, and amensals), but also to the presence and sometimes superabundance of other species, including some bizarre products of evolution in isolation. Despite their simplicity in comparison to mainland biotas, **Red Queens** often run wild on islands as both ecological character displacement and ecological release shape the development of their biotas. Here we feature the development of three insular phenomena—**super-generalists**, unusual pollinators and dispersers, and **ecological naiveté**—to illustrate the interplay between ecological and evolutionary processes on islands.

As we remarked elsewhere in this chapter, niche expansion is a common response of insular populations to the paucity of competitors and predators on islands (see Figure 14.15). This phenomenon is especially well developed in networks of insular plants and their pollinators and dispersers. In contrast to the diverse and intensely competitive mainland assemblages of birds, bees, and butterflies, which may put a premium on specialization, the lucky few nectar-feeding pollinators that colonize islands often serve a disproportionately high diversity of insular plants. Endemic bees of the Canary Islands include *Bombus canariensis* and *Anthophora allouardii*, which visit nearly half of the plant species of their respective communities (laurel forests and subdesert shrubland). Likewise, an endemic *Halictus* bee of Flores, in the Azores, and *Xylocopa darwini* of the Galápagos islands are known to visit 60 percent and 77 percent of the native insular plants, respectively (Olesen et al. 2002; Olesen and Jordano 2002).

As Olesen and Valido (2003) have also shown, super-generalists on islands often include lizards, which rarely serve as pollinators and dispersal agents on the mainland. Of the world's lizard species, some 95 percent of those known to visit flowers (and presumably serve as pollinators) and 63 percent of those known to feed on fruit (dispersing their seeds) are insular forms (**Figure 14.24**). These interrelated island phenomena—of super-generalists and unusual pollinators and dispersal agents—likely result from density overcompensation of lizard populations and lower predation risks, both of which are associated with the absence or paucity of non-volant mammals, arthropods, and other competitors, pollinators, and predators on remote islands.

Finally, the simplified and distinctive ecological regimes of remote islands set the stage for one of the most perilous characteristics of insular plants and

FIGURE 14.24 The only pollinator, frugivore, and seed disperser of the critically endangered plant *Roussea simplex* of Mauritius (Indian Ocean), which is the sole member of the family Rousseaceae and whose numbers have fallen to less than 100 individuals, is the blue-tailed day gecko (*Phelsuma cepediana*). The plant's endangered status may have been exacerbated by an invasive ant that drives the geckos from the flowers and fruits. (Hansen and Müller 2009a,b; photos courtesy of Dennis Hansen.)

animals—their ecological naiveté. In the absence of pressures from the diversity of mainland herbivores, predators, and parasites, the thorns and other physical and chemical defenses of plants are highly reduced (Bowen and van Vuren 1997; Givnish 1998; van Vuren and Bowen 1999), and many endemic island birds, reptiles, and bats are marked by a totally or substantially reduced ability or propensity to fly or run from predators, or even recognize them as such (remember the case of Darwin's ability to approach a Galápagos hawk and knock it off its perch with the butt of his gun). This ecological naiveté or insular tameness explains why a highly disproportionate number of plant extinctions during the past 500 years were insular forms (a subject we take up again in Chapter 16).

Transforming life's most fundamental character: Size

Perhaps the most persistent evolutionary theme is diversification. Barring catastrophes, and indeed despite repeated and sometimes global-scale mass extinctions throughout the geological record, life has exhibited an incredible resilience and capacity for diversification. This indefatigable march to higher levels of diversity was not just simply in numbers of life-forms but in their life processes and distinguishing characteristics as well.

Arguably most important among the defining traits of all species is their size, because it affects resource requirements and all aspects of how those resources are utilized for survival and, ultimately, converted into offspring. All physiological, behavioral, and ecological characteristics of organisms and, in turn, all of the fundamental biogeographic capacities of species (i.e., for colonizing, maintaining populations, and evolving on islands) are strongly influenced by size—height, biomass, and growth form in plants, and body mass in animals (Haldane 1926; Calder 1974, 1984; Brown and Maurer 1986; Brown 1995; Bonner 2006). This, however, raises an intriguing paradox. Natural selection should reduce variation of those traits most strongly associated with fitness (in this case, size). For a given **bauplän** (i.e., growth form in plants, body plan in animals) and energetic and trophic strategy (e.g., floating, in an aquatic autotrophic macrophyte; flying, in a nectarivorous ectotherm; quadrupedal locomotion, in a carnivorous homeotherm), we might predict that there is a particular size that is optimal for the current environmental conditions. There is both theoretical and empirical support for the hypothesis of optimal body size (see Damuth 1991, 1993; Maurer et al. 1992; Brown et al. 1993; Marquet and Taper 1998; Boback and Guyer 2003; Palkovacs 2003; Lomolino 2005; Lomolino et al. 2006; Palombo 2007; Simard et al. 2008).

Yet if we accept the logic of these arguments, how is it that size exhibits such tremendous variation among life-forms? For example, adult body mass varies by 8 orders of magnitude across all extant mammals, and 4 to 5 orders of magnitude for species within the mammalian orders Rodentia, Chiroptera, and Carnivora. Similarly, body masses of extant birds, reptiles, and amphibians range over 4.5, 6.0, and 6.5 orders of magnitude, respectively. Invertebrates and plants are no exceptions: Particular groups of insects range from less than 0.01 cm to over 11 cm in length, and plants range from tiny angiosperms (e.g., duckweeds at less than 0.05 mm and 0.02 g) and gymnosperms (the pygmy cycad, *Zamia pygmaea*, with fronds of just 5 cm) to the largest living trees—the giant sequoias (*Sequoiadendron giganteum*), which tower to nearly 100 m.

The solution to this paradox lies in understanding the driving forces of biological diversification, and islands have once again provided some of the most central insights. Evolution can be an autocatalytic process; that is, diversification may be driven, at least in part, by the species themselves, as ecological interactions put a premium on developing alternative means of acquiring and processing resources and converting them into offspring. Therefore, while there may be an optimal size for each functional group of species, the optimum for each species should differ depending on the characteristics and capacities of others in the community. Accordingly, ecological diversification in species-rich, mainland communities has likely been one of the key forces responsible for a general trend in body size known as Cope's rule. Edward Drinker Cope was an American paleontologist who reported that certain lineages of vertebrates, such as horses, tend to increase in body size over their evolutionary history (**Figure 14.25**; Alroy 1998; Kingsolver and Pfennig 2004; Hone et al. 2008). The key to understanding this appears, however, to be body size diversification: Because there tend to be many more small than larger species, ancestral forms tend to be small, and thus diversification in size will often present itself as a trend of increasing size (but the reverse is often reported as well, i.e., when the ancestral forms are relatively large). David Hone and his colleagues provide an interesting illustration of this for body size evolution in different lineages of dinosaurs. Those with ancestors smaller than 7.8 m in body length exhibited trends consistent with Cope's rule (increasing in size during diversification of the lineage), while those with ancestors larger than this intermediate and perhaps "optimal" size exhibited the opposite trend (Hone et al. 2005).

The relevance of islands for understanding these and related phenomena in variation of size across space and time is that they offer natural and scale-appropriate manipulations of one of the putative driving forces for evolutionary diversification—ecological interactions. The hypothesis of ecological diversification predicts that on remote and ecologically simple (species-poor) islands, the body sizes of insular populations should converge on the optimal size for that functional group (again, the latter based on *bauplän* and energetic and trophic strategy). This is indeed the pattern that has been reported

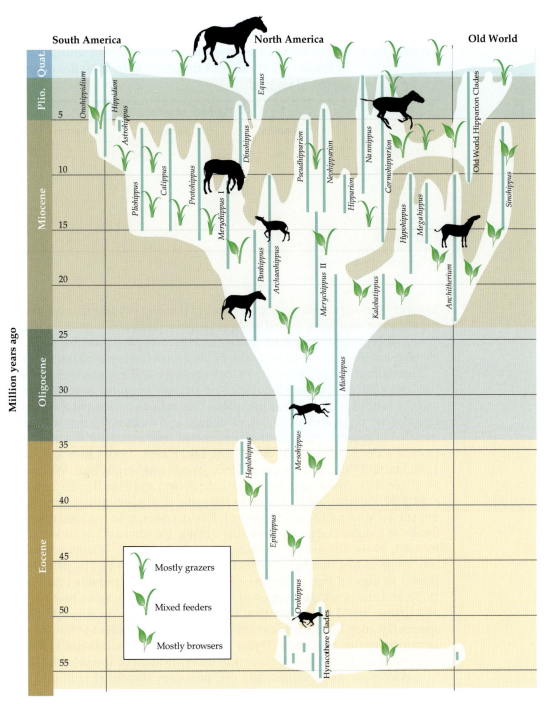

South America   **North America**   **Old World**

Million years ago

Quat.
Plio.
5
Miocene
10
15
20
25
Oligocene
30
35
Eocene
40
45
50
55

Onohippidium
Hippidion
Astrohippus
Equus
Pliohippus
Calippus
Protohippus
Dinohippus
Pseudhipparion
Neohipparion
Nannippus
Cormohipparion
Old World Hipparion Clades
Merychippus I
Hipparion
Hypohippus
Megahippus
Sinohippus
Pandhippus
Archaeohippus
Merychippus II
Kalobatippus
Anchitherium
Miohippus
Haplohippus
Mesohippus
Epihippus
Orohippus
Hyracothere Clades

Mostly grazers

Mixed feeders

Mostly browsers

FIGURE 14.25 Diversification of horses (family Equidae) is a classic case of Cope's rule, an evolutionary increase in body size of the members of a lineage over time. This apparently general pattern is actually part of a syndrome of adaptive radiation and diversification in a variety of morphological and ecological characteristics and geographic distributions of lineages. Silhouettes show the changes in general shape and relative body size of representative species of horses over the past 55 million years. (From MacFadden 2005.)

for a wide variety of life-forms, including living and ancient vertebrates and possibly invertebrates and plants as well.

The nature of the island rule

As we have seen with Cope's rule, general patterns in variation of life-forms over space and time are sometimes elevated, at least in name, to the status of "rules" (see Chapter 15). The label is, of course, an overstatement, but one that may serve a valuable purpose. It is human nature, especially sharpened among natural scientists, to challenge any pattern purported to be a rule of nature. Research programs are then designed to evaluate the generality of these patterns,

TABLE 14.5   *Taxonomic Patterns in the Relative Sizes of Subspecies of Insular Mammals as Compared with Their Relatives on the Mainland*

	Number of subspecies		
	Smaller	**Same**	**Larger**
Marsupials	0	1	3
Insectivores	4	4	1
Lagomorphs	6	1	1
Rodents	6	3	60
Carnivores	13	1	1
Artiodactyls	9	2	0

Source: Foster 1964.

to reject or modify their description, and to develop causal explanations for the apparent anomalies as well as those patterns consistent with the putative "rule." This process of scientific debate and discovery has provided some intriguing insights into evolution of one of life's most fundamental traits—size.

Throughout the early history of paleontology and biogeography, scientists reported fascinating discoveries of strange creatures from the fossil beds of various islands across the globe. This included mice and shrews the size of small dogs, and ungulates, elephants, and mammoths that somehow dwarfed to less than 10 percent of the mass of their mainland ancestors (see Hooijer 1976; Sondaar 1977; Roth 1990; Lister 1989, 1993; Lister and Bahn 1994; Martin 1995; Vartanyan et al. 1995; Woods and Sergile 2001; White and MacPhee 2001; Agusti and Anton 2002; Anderson and Handley 2002; de Vos et al. 2007; Palombo 2008). The first systematic and quantitative synthesis on this subject was published in 1964 by J. Bristol Foster, who reported different trends in insular body size evolution among the mammalian orders (Foster 1963, 1964). Carnivores—including canids and felids—exhibited dwarfism, while rodents tended to exhibit gigantism (**Table 14.5**). The pattern appeared so general that it later became known as the **island rule** (Van Valen 1973a,b). Later work has suggested that the island rule may be best viewed as a graded trend from gigantism in smaller species to dwarfism in larger species (Heaney 1978; Lomolino 1985). This pattern is evident within, as well as among, taxonomic orders (**Figure 14.26A**) and clearly consistent with predictions of the hypothesis of ecological diversification of species-rich biotas: In species-poor communities, selective pressures associated with interspecific interactions are relaxed, and body size converges on an intermediate, and presumed optimal, size.

The hypothesized optimal size can be estimated empirically by simple linear regression (estimating where the trend line intersects the horizontal lines of Figure 14.26, which correspond to when body size of insular populations equals that of the mainland ancestors). Based on this analysis, the putative optimal size for non-volant, terrestrial mammals on islands lies somewhere between 200 and 500 g: roughly the size of a squirrel (see Maurer et al. 1992). As the reader will be aware, however, the class Mammalia includes some quite disparate groups, that is, those with markedly different *baupläne* and energetic (thermoregulatory) and trophic strategies. Thus, it is more appropriate to estimate optimal size for each functional group of species, which yields some interesting and more insightful results. Estimated optimal sizes of extant terrestrial mammals (those of Figure 14.26A) range from approxi-

1111111

(A)

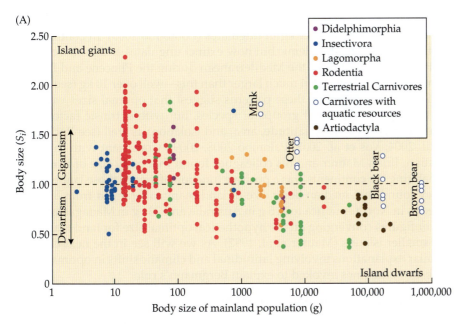

(B)

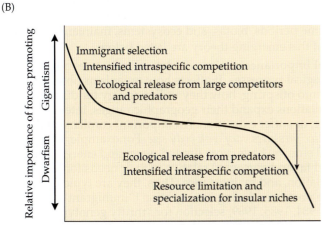

Body size of populations on the mainland

(C)

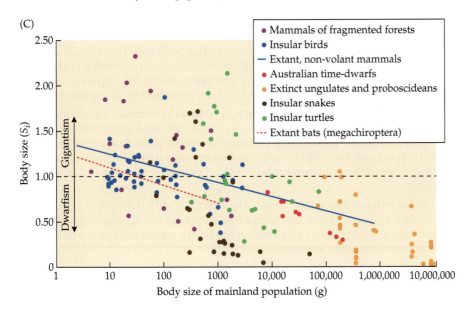

Body size of mainland population (g)

FIGURE 14.26 (A) The island rule refers to the graded trend from gigantism in small species of mammals (e.g., mice and voles) to dwarfism in large species (e.g., canids, ungulates, and elephants). S_i is body size of insular populations of non-volant mammals expressed as a proportion of the body mass of their mainland relative or ancestor. (B) This pattern may result from the combined effects of ecological release and selection for better immigrators, which tend to select for gigantism in small mammals, and from resource limitation, which tends to select for dwarfism, especially in larger species. (C) Body size trends for a variety of other vertebrates also exhibit graded trends from gigantism in the smaller species to dwarfism in the larger species, and are thus consistent with the general trend for insular mammals (blue line from trend in graph A, for non-volant mammals). (A and B after Lomolino 1985; C after Flannery 1994, Australian time-dwarfs; Clegg and Owens 2002, birds; Boback and Guyer 2003, snakes; Schmidt and Jensen 2003, mammals of fragmented forests; N. Karraker, pers. comm. 2004, turtles; Raia and Meiri 2006, M. R. Palombo, pers. comm. 2009, extinct ungulates and proboscideans; see also Rodriguez et al. 2004; Palombo 2007; Krzanowski 1967, bats.)

mately 0.3 kg in rodents, and 0.4 kg in terrestrial carnivores, to 2.1 kg in rabbits and hares, and 6.9 kg in ungulates (Lomolino 2005). In contrast, the optimal body size of endemic but now extinct ungulates and proboscideans that inhabited islands of the Mediterranean during the Pliocene and Pleistocene appears to have been much larger than this—well over 10 kg (see Figure 14.26C). Taken together, these patterns suggest that optimal sizes vary, not just among fundamental groups of species, but with extrinsic factors (climate, carrying capacity, latitude, area, and isolation) as well (see Rodriguez et al. 2004; Vos et al. 2007; Palombo 2008, 2009). Consistent with this suggestion, Brian Maurer and his colleagues found that body size of mammals inhabiting the continents was a positive function of area of those landmasses. They found that median body size of land mammals ranged from 85 g for species in North America to 216 g in Australia and 231 g in Madagascar—the latter figures converging on the hypothetical optimal size for land mammals on smaller islands (Maurer et al. 1992). As we will see in the next chapter, similar explanations involving latitudinal and climatic shifts in optimal body size are sometimes invoked to explain Bergmann's rule.

A causal explanation and general model

The patterns and ideas discussed above may form the basis of a very general theory of body size variation across space and time, that is, one applicable to a broad variety of time periods, biotas, and ecosystems, mainland as well as insular. The tenets of this emerging theory are listed below:

1. **The fundamental importance of size**   Body size influences all physiological and ecological processes. It affects how organisms obtain energy and transform it into offspring, how they utilize energy for dispersal, and how in turn populations expand their geographic ranges, interact with each other, evolve, or suffer extinction. An important corollary of this tenet is that *selective pressures vary with body size*. For example, smaller species (mice and shrews) have more limited water and energy stores (in the form of stored fat and glycogen) relative to their needs for locomotion, thermoregulation, and maintenance. Thus, they are typically under stronger selective pressures during immigration ("immigrant selection," sensu Lomolino 1984, 1985, 1989; selection for "thrifty genotypes" in Pacific Islanders, sensu Neel 1962; Bindon and Baker 1997) and during periods of harsh winter conditions (ice, snow, and subfreezing temperature), famine, and drought than are larger species.

2. **Optimal size**   For each *bauplän* (i.e., functionally distinct body plan) and energetic and trophic strategy, there exists a body size that results in the combination of physiological characteristics, life history traits, and fundamental capacities of individuals and populations (to disperse, survive, and evolve) that is optimal for the current environmental conditions.

3. **Effects of ecological interactions and natural selection**   Evolutionary diversification is driven by, among other things, ecological interactions. Thus, on species-poor islands, natural selection should cause body size to converge on the physiological optimum for a given functional group of species (again, defined by *bauplän* and energetic and trophic strategies).

4. **Variation over space and time**   The optimal body size will also vary, not just among species, but with environmental conditions (e.g., climate, primary productivity, and carrying capacity), which vary in a predictable manner over space and time.

The model illustrated in **Figure 14.27** provides a general explanation for the island rule—one that emphasizes how selective forces (including immigrant selection, ecological release, intraspecific interactions, and resource limitation) vary with ancestral body size. This model may be expanded to include

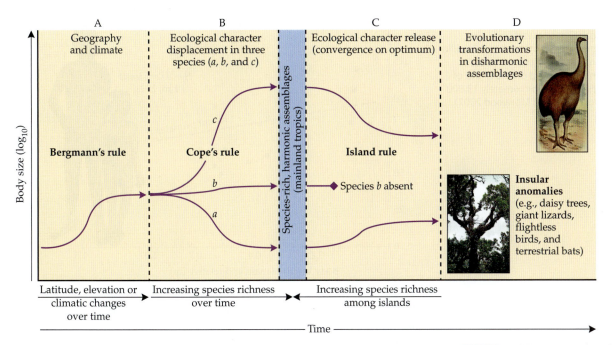

FIGURE 14.27 A general model of body size evolution over space and time. This model assumes that there exists an optimal body size for a particular *bauplän* and trophic and energetic strategy (e.g., a quadrupedal, cursorial herbivorous endotherm), but the optimum for each species within that functional group varies with geographic and climatic conditions (Bergmann's rule, column A) and with ecological interactions among species (i.e., the optimal body size depends on that of other species). Resultant, ecological character displacement in species-rich assemblages is assumed to be associated with diversification of a lineage over time (column B) and, therefore, with Cope's rule, while ecological character release likely has the opposite effect, resulting in convergence in body size in species-poor assemblages (the island rule, column C). Finally, members of depauperate and highly disharmonic assemblages may undergo evolutionary transformations (column D) to alternative *baupläne*, energetic and trophic strategies as a result of the combined effects of character displacement (from relatively rich, but typically small, insular birds and reptiles), and ecological release (in the absence of non-volant mammals that often dominate mainland communities). (From D. F. Sax and M. V. Lomolino, unpublished report, 2009.)

the effects of the disharmonic nature of islands and thus account for a broader range of phenomena including those associated with Bergmann's rule, Cope's rule, the island rule, and insular anomalies such as giant flightless birds that far exceed the intermediate, and presumed optimal, size of flying birds on the mainland. The value of such general models is that they inform us on not just the predicted empirical patterns, but also on what corollary and anomalous patterns we should expect and why; and they serve to identify the future research that is most likely to provide valuable information for understanding the geographic and evolutionary dynamics of body size.

Empirical patterns: Generality of the island rule

Mammals

As illustrated in the charts of Figure 14.26, the island rule appears to be a general phenomenon for the class Mammalia and, while exhibiting substantial variation among orders, most mammals appear to exhibit the expected graded trend from gigantism in the smaller species to dwarfism in the larger ones (we consider apparently anomalous patterns later in this chapter). Additional studies (Lomolino 2005) indicate that the graded trend is evident within numerous mammalian families, in particular those of the species-rich order rodentia. Still other research provides evidence for island rule patterns in other groups of extant mammals, including primates (**Figure 14.28**; Bromham and Cardillo 2007; see also Magnanou et al. 2005; White and Searle 2006; Price and Phillimore 2007; Meiri et al. 2008). Perhaps one of the most sensational of these research findings was the discovery of the fossilized skeletons of *Homo floresiensis*, a relative and possible descendant of *Homo erectus*, which inhabited the remote Indonesian island of Flores between 100,000 and 13,000 BP (Brown et al. 2004; Morwood et al. 2004; Van Oosterzee and Morwood 2007). Despite some initial debate over the condition of the first described specimen and a continuing disagreement as to its status as a distinct species (e.g., see Köhler et al. 2007; Niven 2007; Lyras et al. 2009), analysis of additional specimens indicates that the Flores hominid may well have been an insular dwarf. According to Bromham and Cardillo (2007), *Homo floresiensis* was just over

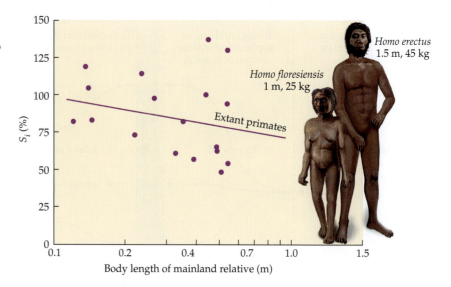

FIGURE 14.28 The island rule's graded trend from gigantism to dwarfism is also exhibited by primates (Bromham and Cardillo 2007), including *Homo floresiensis* (artist's reconstruction) which lived on the island of Flores, Indonesia, until approximately 13,000 BP (see Van Oosterzee and Morwood 2007).

half the size (on a body mass basis) of Indonesian *Homo erectus*, a putative ancestor of the Flores hominid.

It appears that island rule patterns are not restricted to terrestrial mammals, as bats (at least the megachiroptera) also exhibit the predicted, graded trend in insular body size (red dashed line in Figure 14.26C). Nor is the pattern a recent phenomenon, but one repeatedly reported for paleo-insular mammals. As we indicated earlier, the fossil record appears replete with specimens of bizarre insular mammals—giant mice and shrews, and pygmy sloths, ungulates, and mammoths—consistent with the island rule (**Figure 14.29**; see Anderson and Handley 2002; de Vos et al. 2007; Palombo 2008).

As you shall see below, these remarkable shifts in the size of insular lifeforms appear to be a general feature of evolution on remote, depauperate, and disharmonic archipelagoes and for a diversity of taxa in addition to mammals, including other vertebrates, invertebrates, and even plants.

The island rule in other animals

In comparison to other insular faunas, birds and reptiles represent perhaps unrivaled subjects for ecological and evolutionary studies. In comparison to mammals, birds and reptiles are more widely distributed on islands and are generally easier to observe and collect. Some early studies suggested that insular populations of birds and reptiles exhibit trends in body size that are not consistent with the island rule. The equivocal nature of these "patterns" may derive, paradoxically, from the wealth of information available for insular reptiles and birds. Measurements of body size have been taken from populations over a great variety of islands, which differ substantially in area, isolation, climate, and other factors that may influence the evolution of insular body size. Moreover, these islands are often much more isolated than those inhabited by mammals, thus making it more difficult to identify the source populations suitable for comparative studies. Yet, some general trends have emerged in recent years.

Although earlier studies reported that birds exhibited a general trend toward increasing bill size on islands (Lack 1947; Grant 1965; Blondel 2000; but see Clegg and Owens 2002), it was unclear whether they exhibited trends in body size consistent with the island rule (at least when wing length was used as a surrogate of body size; Grant 1965). On the other hand, some relatively small birds such as wrens—and, possibly, fruit pigeons—tend to exhibit gigantism on islands (Williamson 1981; McNab 1994a,b), while larger birds

(A)

(a) **Majorca and Menorca**
Giant dormouse, shrew; small bovids

(b) **Corsica**
Large shrews, rodents
Dwarf deer (2), canid; endemic otter

(c) **Sardinia**
Large shrews, rodents, lagomorphs
Small macaque, hyaena, suid, bovids
Dwarf mammoths, deer, canid
Endemic mustelid; giant, small otters

(d) **Sicily**
Large shrew; giant, small dormouse
Dwarf elephant, hippo; small deer, bovids

(e) **Malta**
Giant dormouse
Dwarf elephant, deer, hippo

(f) **Southern Calabria**
Small elephant, small deer

(g) **Gargano**
Giant hedgehog

(h) **Kephalinia**
Pygmy hippo

(i) **Kythira**
Pygmy elephant; dwarf deer

(j) **Crete**
Giant rodents; small hippo
Dwarf mammoth, deer (6)

(k) **Rhodes, Tilos, Dilos, Naxos**
Pygmy elephant

(l) **Cyprus**
Small elephant

(m) **Kasos and Karpathos**
Dwarf deer

750 km

FIGURE 14.29   (A) A summary of body size trends in mammals inhabiting islands of the Mediterranean Sea during the Pliocene and Pleistocene. Two extreme examples of these insular marvels include (B) the dwarf proboscidean, *Palaeoloxodon* (*Elephas*) *falconeri*, of Sicily during the Pleistocene, and (C) the giant insectivore, *Deinogaleryx*, of Gargano (currently a peninsula, but a paleo-island of eastern Italy during the Miocene). (A after Attenborough 1987.)

(B) *Palaeoloxodon* (*Elephas*) *falconeri*

90 cm

(C) *Deinogaleryx*

75 cm

such as rails, ducks, and ratites tend toward insular dwarfism (Wallace 1857; Greenway 1967; Lack 1974; Weller 1980). Cassowaries are large by most avian standards, but as Wallace noted in 1857, the New Guinea form (*Casuarius bennetti*) is small compared with its relative on the Australian mainland (*C. casuarius*; body lengths of 52 vs. 65 inches, respectively). Emus (*Dromaeius novaehollandiae*) of the small islands of Bass Strait are much smaller than those of Tasmania and the Australian mainland (Greenway 1967). Thus, the trends began to suggest a general pattern.

Clegg and Owens (2002) have provided a comprehensive assessment of morphological variation in insular birds, revealing that both bill length and body mass of a sample of 110 insular populations vary in a manner consistent with the island rule (i.e., a graded trend from gigantism in smaller birds to dwarfism in the larger birds, with a possible "optimal" mass of approximately 125 g; see Figure 14.26C; see also Cassey and Blackburn 2004; Clegg et al. 2008; Mathys and Lockwood 2009; Olson et al. 2009; Clegg 2010). Similar to explanations discussed above, Clegg and Owens attribute these patterns to energetic constraints, interspecific interactions, and physiological optimization for insular environments.

Insular reptiles show a diversity of patterns in body size evolution. Much as Foster (1964) first reported for mammals, different reptilian orders and families appear to exhibit different evolutionary tendencies. Gigantism, for example, is common in insular iguanids, herbivorous lizards, whiptails, tiger snakes, and possibly tortoises (Mertens 1934; Case 1978; Case and Bolger 1991; Case and Schwaner 1993; Petren and Case 1997; Caccone et al. 1999; Barahona et al. 2000), while rattlesnakes tend to be dwarfed on islands (Schwaner and Sarre 1990). Body size of insular populations of the lizard *Uta stansburiana* increases as the number of resident competitor species decreases, suggesting that, just as has been postulated for insular rodents, these lizards tend to exhibit gigantism in the absence of competitors (Soule 1966; see also Case and Bolger 1991). Other studies report different trends even in some of the same groups of species (e.g., in snakes, see Boback 2003; Boback and Guyer 2003; Keogh et al. 2005). The patterns for lizards seem equivocal and vary with diet—omnivores and herbivores exhibit a trend consistent with the island rule, but carnivorous lizards appear to exhibit the opposite trend, and none of these are statistically significant (Meiri 2007). As may have been the case for the remarkable case of body size evolution in moas (**Box 14.1**), ecological character release may be at play here, with increased size in already large lizards allowing them to avoid competition from other lizards (and birds) while becoming more capable of exploiting resources made available in the absence of mammalian carnivores (Meiri 2007; Pafilis et al. 2009; Thomas et al. 2009).

In fact, it is not clear whether the "giant" reptiles of some oceanic islands are actually giants or dwarfs. This situation is reminiscent of the case of many insular mammoths. Compared with almost all other mammals—except, of course, their Pleistocene ancestors—they seemed gigantic. Yet, we know that they underwent dramatic and relatively rapid dwarfism on islands large enough to support their populations but lacking predators. Now, consider the extant reptilian "mammoths": The land tortoises of the Galápagos in the Pacific and Aldabra islands in the western Indian Ocean are indeed the largest extant tortoises—but are they larger than their ancestors? The Galápagos tortoises may have increased in size following colonization by a relatively small ancestor (Caccone et al. 1999). On the other hand, the "giant" tortoises of the Aldabras may actually be insular dwarfs. As recently as 500 years ago, tortoises occurred on at least 20 islands in the Indian Ocean—including Madagascar, which harbored two species. According to Arnold (1979), the supposed ancestor of the Aldabra Island tortoise was a Madagascan species that was at least 115 cm in length—larger than the biggest Aldabra species, which was 105 cm. One Madagascan species reached 122 cm in body length, rivaling the extant giant tortoises, while the two species of Rodriguez Island tortoise were just 85 cm and 42 cm. As Williamson (1981) observed, we cannot be cer-

BOX 14.1   *New Zealand's Moas: Four Times Anomalous*

Sticking to the wreck

■■I Charles Darwin's metaphor of shipwrecked mariners (page 586) was developed, in part, to explain what the renowned paleontologist and comparative anatomist of his day—Sir Richard Owen—described as one of nature's greatest anomalies—a bird that cannot fly. Owen's achievements included development of the concept of the archetype—the primitive and simple ancestral design of vertebrates which, as Darwin theorized, was modified by natural selection during evolutionary development of these lineages. In fact, it was Owen who first described the ancient bird—*Archaeopteryx*, coined the term

dinosaur, and examined a moa bone and formally announced the former existence of these great, flightless birds in 1940.

Throughout the rest of his illustrious and incredibly productive career, Owen would publish and present over 50 additional reports on the anatomy of these bizarre, graviportal birds (see Worthy and Holdaway 2002). Perhaps to his great consternation, Owen's anomaly of flightless birds on oceanic islands was solved by a scientist he frequently and harshly criticized, often anonymously— Charles Darwin. The moas and kiwis of New Zealand, the dodo of Mauritius, and the solitaires of nearby Réunion, along with many hundreds of other flightless insular birds, were in Darwin's metaphor,

the shipwrecked mariners—grounded by natural selection to stick to the wreck. Provided the benefit of another 150 years of discovery, we now believe that other factors, especially diversification of New Zealand's habitats and ecological release from the pressures to avoid mammalian predators on the mainland, also played an important role in the evolution of flightlessness on islands (Bunce et al. 2009) (**Figure A**).

Drifters, swimmers, or flyers

As our understanding of the processes influencing island life has advanced in the generations since Owen and Darwin, we have uncovered other anomalies and paradoxes—initially just as perplexing, but at the same time providing invaluable opportunities to develop a better understanding of the geographic and temporal dynamics of life. Paleontologists and biogeographers came to recognize the moas as members of a broad-ranging group of flightless birds—the ratites, which among other distinguishing anatomical features, lack a keel on their sternum (a structure that serves to anchor major flight muscles).

This then presented a second, and equally intriguing, anomaly of New Zealand's great flightless birds, and of ratites inhabiting remote landmasses in general. How is it that ratites colonized oceanic islands? The first plausible answer would be a long time coming, but eventually biogeographers came to understand that the ancestors of New Zealand's giant moas and the five species of the hen-sized kiwis could just raft with pieces of the continents as they drifted across the globe. Whereas the extensionists would only conjure up another land bridge to explain disjunct distributions of landlocked species such as ratites, biogeographers in the post-Wegenerian era would attribute these great disjunctions to the drifting and separation of Earth's tectonic plates (see

Continued on next page

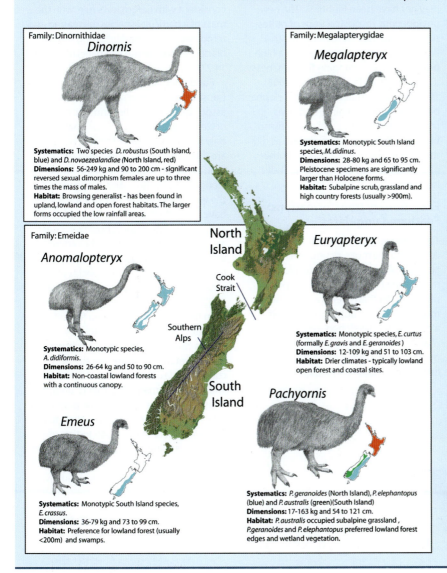

Family: Dinornithidae
Dinornis

Systematics: Two species *D. robustus* (South Island, blue) and *D. novaezealandiae* (North Island, red)
Dimensions: 56-249 kg and 90 to 200 cm - significant reversed sexual dimorphism females are up to three times the mass of males.
Habitat: Browsing generalist - has been found in upland, lowland and open forest habitats. The larger forms occupied the low rainfall areas.

Family: Megalapterygidae
Megalapteryx

Systematics: Monotypic South Island species, *M. didinus.*
Dimensions: 28-80 kg and 65 to 95 cm. Pleistocene specimens are significantly larger than Holocene forms.
Habitat: Subalpine scrub, grassland and high country forests (usually >900m).

Family: Emeidae
Anomalopteryx

Systematics: Monotypic species, *A. didiformis.*
Dimensions: 26-64 kg and 50 to 90 cm.
Habitat: Non-coastal lowland forests with a continuous canopy.

North Island

Cook Strait

Southern Alps

South Island

Euryapteryx

Systematics: Monotypic species, *E. curtus* (formally *E. gravis* and *E. geranoides*)
Dimensions: 12-109 kg and 51 to 103 cm.
Habitat: Drier climates - typically lowland open forest and coastal sites.

Emeus

Systematics: Monotypic South Island species, *E. crassus.*
Dimensions: 36-79 kg and 73 to 99 cm.
Habitat: Preference for lowland forest (usually <200m) and swamps.

Pachyornis

Systematics: *P. geranoides* (North Island), *P. elephantopus* (blue) and *P. australis* (green)(South Island)
Dimensions: 17-163 kg and 54 to 121 cm.
Habitat: *P. australis* occupied subalpine grassland , *P.geranoides* and *P. elephantopus* preferred lowland forest edges and wetland vegetation.

FIGURE A   A summary of the dimensions, geographic distributions, and revised systematics of moas based on recent mitochondrial phylogenetic analyses. (From Bunce et al. 2009.)

BOX 14.1 *(continued)*

Chapters 2 and 8). All seemed well until molecular biology provided us with the abilities to estimate divergence times of ratite lineages (including moas and kiwis), which produced results that seemed to fly in the face of the geological history of New Zealand.

It appears that the moas split from the ratite tree just when they should, that is, at about 80 million years BP, which matches current estimates of when New Zealand separated from the other southern subcontinents of Gondwanaland. Other ratites, however, diverged *after* their respective landmasses split from Laurasia and Gondwanaland (see Waters and Craw 2006; Dickison 2007; Harshman et al. 2008). For example, the evolutionary split between kiwis and other ratites apparently occurred at about 60 million years BP; that is, some 20 million years *after* New Zealand separated from Gondwanaland. The

contention that their ratite ancestors were flightless seems especially problematic given that some paleogeographic reconstructions indicate that most and possibly all of New Zealand may have been submerged for part of the period following its separation from the rest of Gondwanaland (see Cooper and Cooper 1995; Waters and Craw 2006; Bunce et al. 2009) (**Figure B**). Even if there remained some above-water refuge during this period, how could the presumed flightless ancestors of kiwis have dispersed across the approximately 2000 km of oceanic barrier from Australia to colonize New Zealand?

The answer is that they didn't, or more to the point, the kiwis' ancestor may not have been flightless. The new phylogenetic reconstructions of ratites suggest that they were not as much a flightless group as a lineage with a high propensity for developing flightlessness (Dickison 2007; Harshman et al. 2008). Thus, it appears that ancestors of elephant birds, the dodo, solitaires, and kiwis (and possibly moas as well) were flying birds that then became flightless, as did thousands of species of rails and other oceanic birds that evolved in splendid isolation.

Evolutionary transformations: Feathered ungulates

The flightless rails do provide a striking contrast to the moas, raising a third anomaly of avian evolution on islands. The rails were comparatively small, typically much smaller than even the smallest moa, which stood approximately half a meter tall and weighed over 25 kg (Worthy and Holdaway 2002). The largest moa towered to well over 3 m tall and is estimated to have weighed up to 250 kg.

How can the general explanation for the island rule—one of convergence toward an intermediate and presumed optimal size—explain a bird larger than nearly all birds that ever existed? As the last panel of Figure 14.27 implies, it is not that moas evolved toward the optimum size for a bird, but that they evolved toward that of an ungulate. They experienced an evolutionary transformation in their *bauplän* and trophic strategies—taking on the size and other features that enabled them to better exploit resources then made available in the absence of mammalian browsers and grazers. Worthy and Holdaway (2002) note, however, body sizes of moas (within and among species) varied over space and time with climatic conditions in a manner consistent with Bergmann's rule, with the mass of a particular species decreasing as temperatures warmed following the last glacial period (see the first panel of Figure 14.27).

Fundamental to this evolutionary transformation were selective pressures associated with the disharmonic nature of New Zealand's biota. Recall that the only native mammals of this archipelago are three species of short-tailed bats that, although capable of flight, forage along the forest floor and, along with the wetas (see Figure 14.21), appear to have converged on the niche of rodents and shrews. The islands, however, boast an impressive diversity of other animals, including over a thousand species of land snails and about 250 species of native birds (see Worthy and Holdaway 2002; Davies 2003). Ecological pressures from the rich, native avifauna (including predators such as Haast's Eagle, as well as competitors) likely contributed to the evolution of immense size and convergence on the niche of large grazing and browsing mammals, that is, ungulates. Thus, rather than "defying" the island rule, moas provide invaluable clues to its fundamental causation, and to that of body size evolution in general. In addition to the apparent influence of climatic conditions and diversity and productivity

FIGURE B   Kiwis and moas represent the extremes of body size in flightless birds of New Zealand. (Moa print by L. W. Rothschild; kiwi print by G. D. Rowley.)

BOX 14.1 *(continued)*

FIGURE C Haast's Eagle (*Harpagornis moorei*) is the largest known, albeit now extinct, eagle. Weighing up to 19 kg and with talons as long as 60 mm, it preyed on New Zealand's largest prey, including the moas. (Courtesy of John Megahan.)

of habitats, body size evolution is strongly influenced by ecological interactions or, in the case of New Zealand's moas and other biotas evolving in isolation, by ecological release (in the absence of terrestrial mammals) and character displacement (from the pressures of competing with, or avoiding, the very rich assemblage of avian competitors and predators; **Figure C**).

The liabilities of great size

The story of New Zealand's moas raises a final anomaly, one evidenced by the all too common collapse of other marvels of evolution in splendid isolation. Megafaunal species—from moas and elephant birds to mammoths and mastodons—were ecologically dominant ecosystem engineers, disproportionately influencing the structure and function of their native communities. Yet, whether examining the dynamics of the fossil record or the record of historic extinctions, we find that it is not the

powerful giants but the meek and diminutive that inherit the Earth following extinction events. During their evolutionary development, the moas and megafauna rose to ecological dominance under the long-prevailing environmental conditions. Ultimately, however, conditions changed, and those once dominant giants became entangled in evolutionary traps triggered by dynamics of climates, landscapes, and seascapes or by the arrival of novel species with teeth and claws, or spears and fire (Flannery 1994).

New Zealand was one of the last archipelagoes of relatively large islands to be colonized by humans, largely because of its geographic isolation and what, at the time humans set out, were unfavorable prevailing currents, which carried the potential colonists from Australia far north of the islands. This archipelago was finally colonized about 800 BP by Polynesians sailing from eastern regions of the Pacific (see Chapter 16).

Within just a few decades of establishing populations across the main islands, the Maori's formidable hunting skills, along with their use of fire to clear and manage habitats and their introduction of Polynesian rats to New Zealand, began to take their toll on the native wildlife. Moa extinctions occurred in a highly regular and predictable order, generally proceeding from the largest, first, to the next largest, and so on down until the last moa (the upland moa, *Megalapteryx didinus*; **Figure D**) disappeared from the islands about 600 years BP, ending the remarkable 80 million year history of these giant flightless birds in only three centuries of human history. ▮▮▮

FIGURE D An artist's reconstruction of the upland moa (*Megalapteryx didinus*), which may have been the last moa species to survive, inhabiting South Island until about 500 BP. (Illustrated by George Edward Lodge.)

tain whether large size evolved after the islands were colonized or whether it simply made dispersal easier for the initial colonist from Madagascar.

Even monitor lizards—including the famous Komodo dragon—have proved to be a perplexing case study in insular evolution. The Komodo is, after all, the world's largest living lizard, measuring up to 3 m in length and tipping the scales at up to 150 kg (Auffenberg 1981). Even when just extant forms are considered, however, the trend for monitors is equivocal, with some insular forms tending toward gigantism while others appear to be dwarfs (see also Gould and MacFadden 2004; Meiri 2007). Yet, all of these forms are tiny when compared with a monitor that inhabited Australia during the Pleistocene—a 7 m, 600 kg behemoth (Rich 1985). Until sound phylogenies are available and the recently extinct megafauna are taken into account, we must admit the possibility that the extant Indonesian "dragons" are actually insular dwarfs!

The above difficulties, while challenging, are probably not insurmountable. The influence of island characteristics on trends in insular body size can be statistically controlled so that we can better focus on the influence of differences among species (e.g., trophic category, or size of presumed ancestor). In addition, phylogenetic and paleobiogeographic analyses can be used to estimate the most likely ancestral state—in this case, the relative size of the immediate mainland ancestor (see Pianka 1995; Miles and Dunham 1996). Recent studies again reveal some interesting, emerging patterns within particular groups of reptiles. Nancy Karraker's studies of insular turtles and tortoises indicate the expected trend from gigantism in the smaller species to dwarfism in the larger species, with an optimal size of approximately 20 to 25 cm (maximum carapace length; see Figure 14.26C; see also Aponte et al. 2003). Boback and Guyer's (2003) recent analysis of body size variation of insular snakes reveals a similar pattern, in this case with an optimal length of approximately 80 cm (see Figure 14.26C). Boback and Guyer also present an interesting analysis of the influence of island area on species composition. While relatively large islands are inhabited by many species spanning a broad range of body sizes, as island area decreases, species richness decreases and species composition converges on those few snakes of the putative optimal body size (i.e., those with lengths between 60 and 120 cm; **Figure 14.30**).

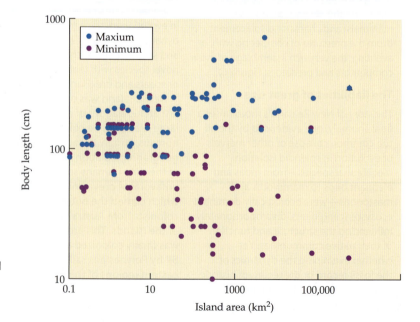

FIGURE 14.30 As we consider assemblages of insular snakes across a gradient of islands from the most to least isolated, species richness declines and species composition converges on snakes of intermediate size (from 80 to 120 cm long; see Figure 14.26). (After Boback and Guyer 2003.)

Very limited but intriguing evidence also suggests that the process responsible for island rule patterns in some groups of extant reptiles may also have been functioning in the past to influence body size of the largest animals to walk the Earth—dinosaurs (**Figure 14.31**). Finally, while it appears that only one study has systematically searched for these patterns in the marine realm,

(A)

Europasaurus

1 m

(B)

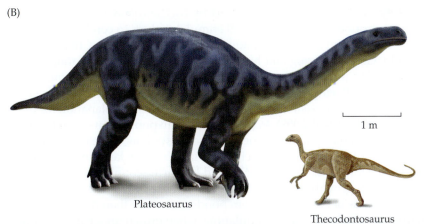

Plateosaurus

Thecodontosaurus

1 m

(C)

FIGURE 14.31 Insular dwarfism may also have occurred in the largest animals ever to walk the Earth—sauropod and hadrosaur dinosaurs. (A) A color reconstruction of a Europasaurus, which was a sauropod that inhabited paleo-islands of the Lower Saxon Basin (Northern Germany) during the Late Jurassic Period. (B) Thecodontosaurus was an early dinosaur of the Late Triassic that inhabited islands of the region of present-day southwestern Great Britain. Adults were just 2 m in length, compared with the species' mainland equivalent Plateosaurus, with a body approximately five times as long (color reconstructions). (C) A reconstruction of the paleo-archipelagoes of western Eurasia during the Late Triassic (about 220 million years BP; see also Jianu and Weishampel 1999). (A after Sander et al. 2006; B after Whiteside and Marshall 2008; C after Ron C. Blakey.)

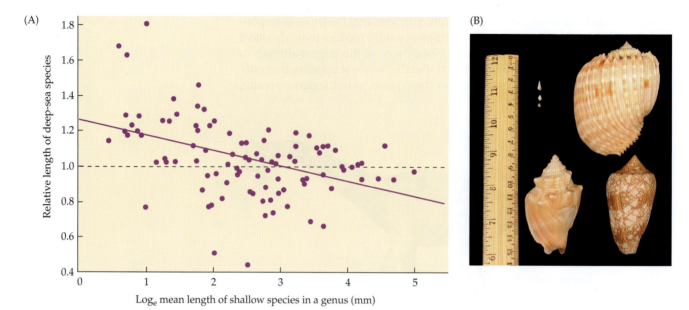

(A)

Relative length of deep-sea species

Log$_e$ mean length of shallow species in a genus (mm)

(B)

FIGURE 14.32 Marine gastropods appear to exhibit a geographic trend in body size analogous to that of the island rule. (A) Like true islands, deep water habitats of the marine realm are isolated and smaller than those along the continental slope, and gastropods exhibit a graded trend in relative body size (as mean length of the deep-sea species divided by that of its shallow-water relative) from gigantism in the smaller species to dwarfism in the larger species. (B) A comparison between three medium-sized shallow-water snails (right and bottom) and small deep-sea snails (upper left). (A after McClain et al. 2006; B courtesy of Craig McClain.)

it is likely that the patterns discovered by McClain and his colleagues for marine gastropods (**Figure 14.32**) will stimulate others to conduct similar research on the island rule in other aquatic taxa and ecosystems. In fact, recent studies by Gabor Herczeg and his colleagues on body size variation among Fennoscandian populations of the nine-spined stickleback (*Pungitius pungitius*) provide both empirical evidence for island rule patterns and experimental results that are consistent with the ecological, causal explanation for those patterns (Herczeg et al. 2009). In comparison with populations from the more diverse (i.e., mainland-like) marine and lake ecosystems, those from relatively small (species-poor) ponds were characterized by very large individuals (35 to 40 percent longer than their marine and lacustrine counterparts). Common garden (pond) growth studies indicated that these differences were genetic, with body mass of fish derived from pond stock more than doubling that of fish from the marine and lake populations when grown under similar conditions during a 36 week period. Opportunistic experiments where predatory fish were introduced to some ponds inhabited by sticklebacks, which then remained relatively small in comparison with other ponds, indicated that it was ecological release that likely drove these patterns (i.e., consistent with the model illustrated in Figure 14.27).

Woodiness and gigantism in plants

> Islands often possess trees or bushes belonging to orders which elsewhere include only herbaceous species. (Darwin 1859: 392)

On the most isolated archipelagoes such as the Galápagos and Hawaiian Islands, otherwise diminutive and herbaceous plants are often transformed into woody "giants" (see Figure 14.23). Across most of their extensive geographic range, *Opuntia* cacti typically are shrubs no greater than a meter or so tall. Yet on remote islands such as Santa Cruz in the Galápagos, *Opuntia* have become trees as much as 10 m high (*Opuntia echios*; see Bohle et al. 1996; Grant 1998, 2001). Among Hawaii's most spectacular plants are the silverswords, woody relatives of sunflowers, whose flowering stalks can tower up to 2 m (see Baldwin and Robichaux 1995).

Woodiness, or arborescence and lignification, is essential for tree-form stature, and evolution of these features on islands has occurred repeatedly in several families of angiosperms (including Apiaceae, Amaranthaceae, Asteraceae, Boraginaceae, Brassicaceae, Campanulaceae, Chenopodiaceae, Crassulaceae, Euphorbiaceae, and Lobeliaceae; Carlquist 1965, 1974; Bohle et al. 1996; Givnish 1998; Panero et al. 1999; Fairfield et al. 2004). This phenomenon is particularly common in the daisy family (Asteraceae), evolving independently in different tribes from numerous islands, including St. Helena and those of the Canary, Juan Fernandez, and Hawaiian Islands (Grant 1998). Woodiness and gigantism of otherwise herbaceous plants have also evolved in the high-elevation habitat "islands" of the Andes and East African highlands (Carlquist 1974; Knox et al. 1993).

Evolution of woodiness on islands is typically associated with ecological shifts, from herbaceous ancestors that occupied open and early successional habitats to arborescent forms that have invaded and created a secondary "forest" niche, left vacant by relatively poorly dispersing trees that seldom colonize isolated archipelagoes. In fact, Darwin (1859) speculated that the limited dispersal abilities and paucity of trees on islands was a key factor contributing to evolution of secondary woodiness on islands. Wallace (1878) also proposed an ecological explanation for this phenomenon, noting that woodiness also connotes greater longevity and a concomitant higher probability of reproductive success on isolated islands, which are notorious for their paucity of pollinators (see also Bohle et al. 1996). It appears that this may also account for the general tendency of insular plants to have relatively drab and otherwise inconspicuous flowers and for their dependence on generalized pollinators, self-fertilization, or wind pollination—especially in dioecious plants (Baker 1955; Carlquist 1974; Ehrendorfer 1979; Sakai et al. 1995; Barrett 1998).

Sherwin Carlquist (1974), on the other hand, downplayed these ecological explanations for insular woodiness in favor of a climatic one. Essentially, he hypothesized that insular climates tend to be relatively moderate, thus favoring species with a different suite of life history strategies. In MacArthur's terms, these plants shifted or evolved from r-selected strategies to K-selected ones (including increased size, slower growth rate, and increased longevity) once their populations approached and adapted to carrying capacities of their insular environments. Alternatively, Bohle and his colleagues (Bohle et al. 1996) modified an earlier hypothesis of Wallace (1878), suggesting that the key selective advantage of woodiness was longevity and the concomitant increase in time and opportunities for cross-pollination—which becomes particularly challenging given the paucity of natural pollinators on remote islands.

Despite these continuing debates over causal mechanisms, evolution of woodiness seems to be a relatively common phenomenon for herbaceous plants on remote islands, and one closely associated with subsequent adaptive radiations of the derived, woody forms (Bohle et al. 1996; Carine et al. 2010). It is, thus, interesting to note that although some of these derived insular forms may undergo evolutionary reversals in growth forms (i.e., back to the ancestral, herbaceous form), there appears to be no well-documented cases of woody invaders from the continents evolving into herbaceous forms on islands. That is, in contrast to the generally symmetrical response of convergence in body size exhibited by many insular animals, evolution of growth form in insular plants seems asymmetrical—herbaceous forms toward woodiness and "gigantism," but not the reverse. The cause of such evolutionary novelties in insular plants is probably complex, but again, it may stem from the disharmonic nature of insular biotas. In comparison with those of large trees and other woody plants, the propagules of small herbaceous plants are typically more easily transported by winds and water currents. Thus, isolated islands are colonized by a highly disproportionate number of small herbaceous species. Just as important, isolated islands often lack large herbivorous mammals, including browsers and folivores, which tend to feed most heavily on woody plants. Under such reduced herbivory but intense competition with other herbaceous species, plants often undergo both ecological release and character displacement—some evolving to occupy the niches of shrubs and trees that were incapable of colonizing islands. The few woody invaders that colonize such islands, however, experience no such release in the face of well-established and rich herbaceous assemblages.

In summary, the generality of these patterns most probably results from a variety of factors, which are ultimately associated with two fundamental properties of remote communities—their depauperate and disharmonic nature. These evolutionary shifts in the distinctiveness of flowers, development of woodiness and tree stature, loss of plant defenses against herbivores, and modification for new pollinators, frugivores, and seed dispersers are analogous to the shifts in appearance, *baupläne*, and life history strategies of insular animals, and perhaps they have similar explanations. They are predictable responses to highly altered ecological selective regimes.

Corollary patterns

One of the best tests of a general model is not just how well it explains the focal pattern, but how well it predicts and explains corollary patterns and anomalous cases, as well. In the case of the island rule, predicted corollary patterns include those of body size variation with island area, isolation, and ecological diversity. An incomplete survey of some relevant literature by Lomolino (2005) turned up numerous cases of correlations between insular body size of terrestrial vertebrates and area or isolation of the islands—presumably because these variables are typically strongly correlated with carrying capacity, immigrant selection, genetic isolation, or species richness (see also Meiri 2008). For an overwhelming majority of those cases yielding statistically significant results, body size increased with both island area (in 12 of 14 studies) and isolation (in 12 of 13 studies surveyed by Lomolino 2005: his Table 3; **Figure 14.33**; see also Meiri 2008). It would seem premature, however, to conclude that these are very general patterns until more integrative assessments (simultaneously analyzing the influence of the principal geographic, climatic, and ecological variables hypothesized to influence insular body size) have been conducted.

Exceptions, anomalies, and evolutionary transformations

Carnivorous mammals may represent a highly instructive group of apparent exceptions to the island rule. There is still some debate as to whether these mammals exhibit the predicted graded trend, depending on whether species of extreme size are included, what measure of body size (tooth or skull dimensions) is used, and the assumptions (one- or two-tailed) of the statistical tests used (see Meiri et al. 2004a; Goltsman et al. 2005; Lomolino 2005: his Figure 2; Meiri et al. 2009). Most researchers would agree that even if the empirical pattern for insular carnivores is statistically significant, there remains much residual variation about the trend. We argue that this is in fact consistent with a general model of body size variation across space and time. If the fundamental driving force for body size evolution on islands is release from ecological interactions, then carnivores are expected to be exceptional. By definition, mammalian carnivores require islands that are large enough and close enough to the mainland to support diverse populations of prey species; they seldom occur on species-poor islands where the ecological mechanisms of the island rule operate most strongly.

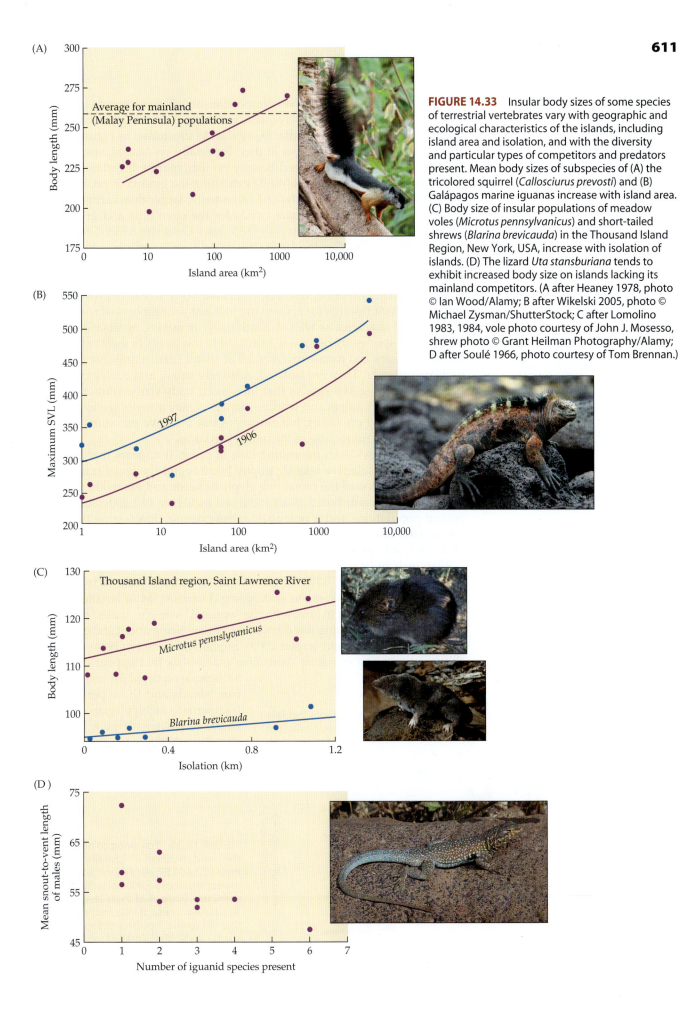

FIGURE 14.33 Insular body sizes of some species of terrestrial vertebrates vary with geographic and ecological characteristics of the islands, including island area and isolation, and with the diversity and particular types of competitors and predators present. Mean body sizes of subspecies of (A) the tricolored squirrel (*Callosciurus prevosti*) and (B) Galápagos marine iguanas increase with island area. (C) Body size of insular populations of meadow voles (*Microtus pennsylvanicus*) and short-tailed shrews (*Blarina brevicauda*) in the Thousand Island Region, New York, USA, increase with isolation of islands. (D) The lizard *Uta stansburiana* tends to exhibit increased body size on islands lacking its mainland competitors. (A after Heaney 1978, photo © Ian Wood/Alamy; B after Wikelski 2005, photo © Michael Zysman/ShutterStock; C after Lomolino 1983, 1984, vole photo courtesy of John J. Mosesso, shrew photo © Grant Heilman Photography/Alamy; D after Soulé 1966, photo courtesy of Tom Brennan.)

Carnivorous mammals are very opportunistic, often shifting their diets on islands to feed more heavily on aquatic resources. Accordingly, a substantial share of their variation in insular body size, as illustrated in Figure 14.26A, can be attributed to differences in diet. Especially on islands, species such as minks, otters, and bears feed heavily on fish, marine invertebrates, and carcasses of marine mammals. While exhibiting the graded trend, they tend to be relatively large on islands, and consistently larger than carnivores that rely on terrestrial prey (compare trends for the open blue and green filled circles of Figure 14.26A; Lomolino 1983; after Goldman 1935; Cowan and Guiget 1956; see also Gordon 1986).

Heuristic and applied value of the island rule

Beyond its heuristic value, research on insular body size may provide especially important insights for conserving, not just the numbers of species, but their natural character as well. A continually increasing number of today's endangered species are becoming restricted to fragments of their native habitats, and to zoos and nature reserves, all of which share many characteristics with true islands: They are isolated, relatively small, and ecologically simple—often lacking key species that influenced body size evolution of their ancestors. Local hunters and wildlife managers may attempt to simulate the effects of missing predators, but anthropogenic selective pressures may differ substantially from those of the native predators.

Elephants provide an interesting case in point. It appears that the combined effects of selective take by trophy hunters and the ivory trade, along with insularization of the remaining elephant herds and exclusion of predators, may have contributed to an increased incidence of tusklessness in elephant populations and a potential, anthropogenic downsizing of Earth's largest land mammals (Jachmann et al. 1995; Lee and Moss 1995; Lomolino et al. 2001; Sukumar 2003).

Mammals in heavily fragmented forests of Europe also exhibit variation in body size consistent with the island rule: In just 175 years of isolation, it appears that larger species are undergoing dwarfism, while smaller species are tending to increase in size, with the "optimal" size (i.e., that of mammals that are neither increasing nor decreasing in size) being about 280 g (see Figure 14.26C; Schmidt and Jensen 2003). Similarly, Flannery (1994) describes shifts in body size of native Australian marsupials that appear consistent with those reported for mammals on true islands and those persisting in fragmented archipelagoes of their native ecosystems (**Figure 14.34**; see also Price 2008). Since the colonization of Australia by aborigines about

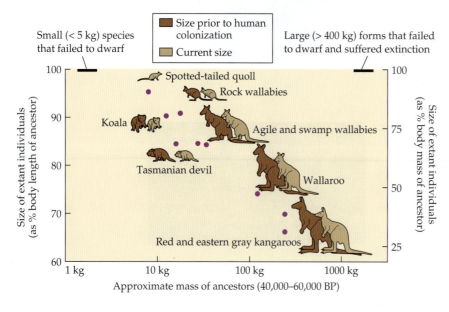

FIGURE 14.34 Following colonization of the island continent of Australia approximately 55,000 BP, all of its megafaunal marsupials (those > 400 kg) suffered extinction, while all of those less than 5 kg survived. Those of intermediate size that did survive, however, did so in altered states—exhibiting a graded trend of increasing dwarfism for the larger species. (After Flannery 1994.)

55,000 BP, all of Australia's largest native marsupials (those > 400 kg) have become extinct, but those of intermediate size that survived have decreased in size (e.g., red and gray kangaroos declining in mass by roughly 30 percent; for reports of possible anthropogenic downsizing in a variety of other animals, see Smith et al. 1997; Sumner et al. 1999; Lomolino et al. 2001; Fredrickson and Hedrick 2002; Aponte et al. 2003; Pertoldi et al. 2005; Neckel-Oliveira and Gascon 2006; Lomolino and Perault 2007). While comprehensive studies across a broader diversity of systems and species are lacking, the possibility of anthropogenic changes in body mass of native species certainly warrants increased attention from biogeographers and conservation biologists.

Taxon Cycles and "A Biogeography of the Species"

E. O. Wilson's epiphany

As we remarked in the early chapters of this book, evolution and the other processes influencing biological diversity make little sense unless viewed in a geographic context. One compelling illustration was that of E. O. Wilson's discovery of a new theory of island biogeography. At the heart of Wilson's scientific epiphany was the geography of nature—maps describing the geographic distributions of Melanesian ants (e.g., Figure 14.3A). Wilson recounted in his autobiography (1994: 214–215):

> It dawned on me that the whole cycle of evolution, from expansion and invasion to evolution into endemic status and finally into either retreat or renewed expansion, was a microcosm of the worldwide cycle envisioned by Matthew and Darlington. To find the same biogeographic pattern in miniature was a surprise then. . . . It came within a few minutes one January morning in 1959 as I sat in my first-floor office . . . sorting my newly sketched maps into different possible sequences—early evolution to late evolution. . . . Discovery of the cycle of advance and retreat was followed immediately by recognition of another ecological cycle. . . . I knew I had a candidate for a new principle of biogeography.

In this passage, Wilson mentions the work of William Diller Matthew and Philip J. Darlington, who, together with George Gaylord Simpson, advanced a theory of the great ecological, evolutionary, and biogeographic cycles of life across the continents. The Earth, its land and sea, its climate, and its species were dynamic, expanding from centers of origin and diversity, dispersing across new regions, and then adapting, evolving, and in most cases, suffering eventual extinction. One key feature of this twentieth century articulation of the center of origin–dispersal–adaptation (CODA) tradition (developed in particular by Darwin and Wallace) was the phenomenon of dominance among species and biotas, with new invading biotas driving the long-term residents of a region to extinction. From inspection of his maps, Wilson was able to adapt this theory of the dynamics of continental biotas to develop a new and truly transformative theory on the dynamics of insular biotas—the theory of **taxon cycles**.

The theory of taxon cycles

As outlined earlier, oceanic barriers severely limit the immigration of propagules, but those species that manage to gain a foothold have a high probability of success. In G. G. Simpson's words, they are the winners of an ecological and evolutionary sweepstakes (see Chapter 6). Wilson saw this infrequent, but continual, establishment of lucky colonists as a driving force for

---

TABLE 14.6 *Stages of the Taxon Cycle*

STAGE I—INITIAL EXPANSION: The initial stage of colonization and establishment of populations of a species across an archipelago before its insular populations have differentiated from one another or from the source population on the mainland. In this stage, the species has a relatively continuous range across the archipelago but exhibits little geographic variation. Such "invaders" are often broad-niched species from marginal habitats on the mainland and are therefore preadapted for marginal habitats along beachfronts and other habitats occupying the island's periphery.

STAGE II—ECOLOGICAL AND EVOLUTIONARY SPECIALIZATION: The insular populations have differentiated to the point at which they may represent endemic subspecies, or even species. The populations have invaded and become adapted to habitats within the island's interior, and they often exhibit associated changes such as reduced dispersal ability, shifts in body size, and increased specialization. Some insular populations, however, have become extinct, so the range of the taxon has contracted and its distribution is now spotty.

STAGE III—INITIAL CONTRACTION: Differentiation and range contraction have continued to the point that the taxon now comprises just a few relictual, endemic species whose populations are highly specialized and restricted to interior habitats.

STAGE IV—SINGLE ISLAND ENDEMICS: The ranges of the relictual populations have contracted further, both within and among islands of the archipelago. As a result of their extreme specialization, perhaps hastened by competition with new invaders (Stage I species), relictual populations disappear from the archipelago, presumably to be replaced by other, Stage III species.

Source: After Wilson 1959 and 1961.

---

sequential phases of expansions and contractions of species' ranges—which he termed taxon cycles (Wilson 1959, 1961).

According to this theory, insular species evolve through a series of stages from newly arrived and broadly distributed colonists—indistinguishable from their mainland relatives—to highly differentiated and ecologically specialized endemics, which ultimately become extinct (**Table 14.6**). This process is termed a "cycle" because once insular populations begin to differentiate and adapt to island life, they appear to be doomed to extinction and replacement by new colonists from the mainland. As Wilson proposed, descendants of the colonizing populations become entangled in an "evolutionary trap" because beachfront habitats are not just marginal but also ephemeral, subject to succession, chronic deterioration, or wholesale destruction from storms and other catastrophic events (Wilson 1959: 141). Wilson credited Ernst Mayr (1942, 1954) and Philip Darlington (1957) for developing the term and original concept of "evolutionary traps," which has more recently been co-opted by conservation biologists (e.g., to explain the precarious status of sea turtles whose hatchlings have been genetically programmed to move toward the moonlight when they emerge from the beach sands, but suffer fatalities when drawn to artificial lighting of nearby highways; Schlaepfer et al. 2002; Keeler and Chew 2008).

As Ricklefs and Bermingham (2002) have observed, taxon cycles are not expected to characterize all biotas, only those occurring on islands where isolation approximates dispersal abilities of the focal taxon; for less-isolated archipelagoes, immigration and gene flow will overwhelm ecological and genetic differentiation, whereas for the very distant archipelagoes, too few colonization events occur to drive a repeated cycle. Wilson's seminal studies of ants focused on an insular fauna that was in just the right geographic zone for these patterns to develop—the islands of Melanesia, north and east of New Guinea in the western Pacific (see Figure 14.3A). By analyzing dif-

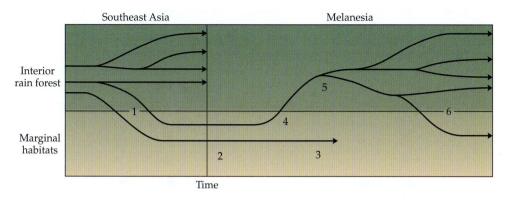

FIGURE 14.35 Ecological changes accompanying the taxon cycle in the ants of Melanesia. Species of forest ancestry that have secondarily invaded marginal (usually disturbed or coastal) habitats of mainland Southeast Asia (1) tend to be good dispersers and to colonize similar habitats on the islands of Melanesia (2). These populations become extinct fairly rapidly (3), or else they invade interior rain forest habitats (4), where they may undergo differentiation and adaptive radiation (5). Sometimes they give rise to forms that secondarily invade marginal habitats and disperse in stepping-stone fashion to more distant islands (6). (After Wilson 1959.)

ferences in distributions and ecological associations of ants in a snapshot of time, Wilson inferred what appeared to be stages in the dynamic expansion and taxonomic differentiation of species. He also analyzed the distributions of these species across habitats within and among islands, from which he inferred successive changes in the niches of these ants as they evolved through the taxon cycle. Those species that are good over-water colonists typically occur in coastal or disturbed habitats of New Guinea; recently arrived, undifferentiated populations are found in similar habitats on the Melanesian islands. As they differentiate in isolation, however, the ants also change their ecological requirements and expand into interior habitats, such as native forests (**Figures 14.35** and **14.36**). They are replaced by a new wave of colonists occupying the beaches and disturbed habitats. Highly differentiated endemic forms, representing the final extant stage of the taxon cycle, typically are restricted to just one island and to a narrow range of environments, usually rain forest or montane forest deep in the island's interior.

Although Wilson's theory was deterministic, he did acknowledge that hurricanes and other stochastic events influenced the stages and dynamics of the taxon cycle—a phenomenon he termed "*faunal drift*" (his italics). As he hypothesized, "the composition of small local faunas varies in an unpredictable manner, that is, there is a subjective element of randomness" (Wilson 1961: 172). Wilson's theory incorporates a number of mechanisms and patterns of insular biogeography, including those associated with colonization, ecological release, niche shifts, character release and displacement, reduced dispersal ability, shifts in insular body size, range contraction, endemicity, and extinction. In its dynamic character, and its emphasis on repeated immigration and extinction (the latter driven by interactions within the fauna), the taxon cycle model was antecedent to the equilibrium theory soon to be developed by MacArthur and Wilson (1963, 1967). Later, Ricklefs and Cox (1972, 1978) reported

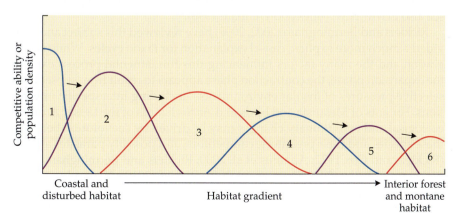

FIGURE 14.36 A graphic model of the processes that are thought to be involved in the insular taxon cycle. Numbers are as in Figure 14.35. The cycle is driven by the colonization of generalized species, which then evolve to become more specialized, sacrificing competitive ability and experiencing reduced population density within particular habitats. Pushed further along the habitat gradient by superior competitors, species in the terminal stages are forced to specialize even more, but evolutionary constraints prevent them from becoming well adapted to these new niches and, hence, from increasing in density and competitive ability. Consequently, they evolve into rare endemics and ultimately become extinct.

FIGURE 14.37 Stages of the taxon cycle as illustrated by birds of the West Indies. Stage I, a widespread, undifferentiated species that presumably has recently colonized from South America, represented here by the gray kingbird *Tyrannus dominicensis*. Stage II, a widespread form with well-differentiated races (letters) on different islands, represented by the finch *Loxigilla noctis*. Stage III, a species with well-differentiated races on only a few islands, represented by the warbler *Dendroica adelaidae*. Stage IV, a narrowly endemic species, represented by the finch *Melanospiza richardsoni*, confined to Saint Lucia. (After Ricklefs and Cox 1972.)

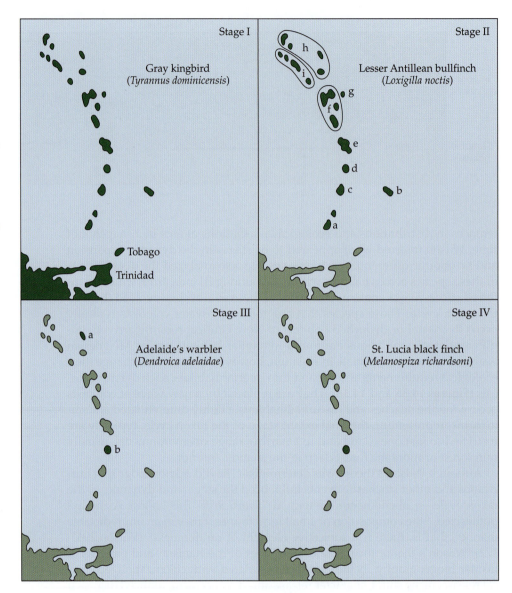

ecological and evolutionary shifts in West Indian birds that showed successive stages of what they interpreted as a taxon cycle (**Figure 14.37**). Similarly, Roughgarden and his colleagues used the theory of taxon cycles to explain the evolution of body size in West Indian anoles (**Figure 14.38**; Roughgarden 1995; Roughgarden et al. 1987; Roughgarden and Pacala 1989; Roughgarden 1992; see also Erwin 1981; Miles and Dunham 1996).

Ricklefs and Bermingham (2002, 2008) have reviewed this subject and provided a much more explicit, causal explanation for taxon cycles. While immigration and the many factors influencing this fundamental process explain the expansion phase, the contraction phase appears to result from subsequent selective pressures of insular environments—including interspecific interactions. For ants of Melansia and birds and *Anolis* lizards of the Lesser Antilles, ecological and evolutionary changes may be driven by coevolution between these species and their predators, competitors, and possibly parasites as well (Ricklefs and Cox 1972; Apanius et al. 2000; Ricklefs and Bermingham 2002; Ricklefs and Fallon 2002; Fallon et al. 2003; Ricklefs 2010). New colonists enter the expansion phase because they have escaped these mainland enemies, but new competitors begin to expand from the beachfront, and insular pred-

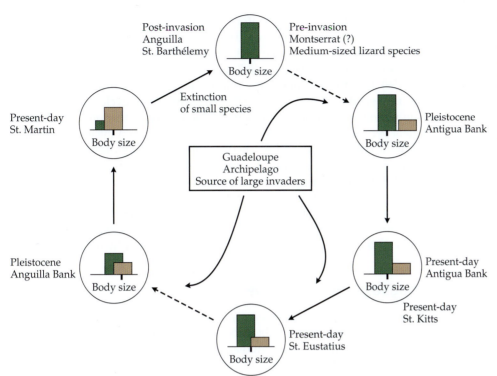

FIGURE 14.38 A proposed taxon cycle for the *Anolis* lizard species in the northern Lesser Antilles of the Caribbean. An island with just one resident species (green bar) evolves a "characteristic" or "optimum" body size (top circle; height of bars indicates relative population density). This island is then invaded by a second and larger species (tan bar), which then evolves toward the size of the original resident, which in turn becomes smaller. The range of the initial resident begins to contract until it eventually becomes extinct. The "invader" now becomes the specialized resident, and its size and range continue to decrease as the island is again colonized by another, larger species, to reinitiate the cycle. (After Roughgarden et al. 1989.)

ators and parasites subsequently learn to exploit these new prey and hosts, eventually triggering a contraction phase. More recently, these and other biogeographers investigating the distributional and evolutionary dynamics of insular biotas have provided evidence that is in direct opposition to one of Wilson's important assertions—that taxon cycles are unidirectional, mainland to islands but never the reverse. That is, colonists and their descendants are derived from mainland populations, but because they eventually become entangled by ecological specialization and evolutionary traps, island endemics should never initiate taxon cycles on the mainland. So it was quite surprising that, in a review of phylogeographic studies published between 1988 and 2008, Bellemain and Ricklefs (2008) found evidence for reverse colonization (island to mainland) in more than a third of these studies. Mainland colonizations were detected across a broad range of archipelagoes and variety of species, including short-faced bats, a variety of birds (a bananaquit, flycatchers, monarch flycatchers, parrots, Darwin's finches, catbirds, and orioles), frogs, lizards, turtles, and drosophilid flies (see also Bellemain and Ricklefs 2007). They have also occurred in higher plants (Carine et al. 2004).

Despite the initial attraction of the concept of the taxon cycle, some researchers questioned both the generality and integrative, ecological explanations for putative taxon cycles, instead offering alternative explanations for observed patterns in distributions, niche shifts, and morphological differences (Losos 1992; Taper and Case 1992; Losos et al. 1993; see also Pielou 1979; Pregill and Olson 1981). Wilson's taxon cycle theory has attracted surprisingly little attention from most biogeographers and ecologists, and many have dismissed it as an interesting, but failed, attempt. However, this lack of progress to some extent reflected the difficulty of amassing the data necessary to test the ideas in a rigorous critical framework. Recent advances in molecular biology and their application to numerous insular radiations now provide the opportunity to revisit and build on the taxon cycle theory. In fact, more recent studies by Ricklefs and his colleagues using phylogenetic analyses coupled with detailed information on the ecology and distributions of Lesser Antillean birds have confirmed most of the tenets of the theory, at least for this insular fauna (see Ricklefs and Bermingham 2002, 2007, 2008; Losos and Ricklefs 2009; Ricklefs 2010). In fact, Ricklefs and Bermingham (2002: 359) assert that "alternating phases of expansion and contraction are nearly universal and that it is possible to study taxon cycles analytically in a wide variety of groups and regions." Much earlier, Diamond (1977) drew parallels between what Wilson and others saw as taxon cycles and similar patterns of colonization and extinctions of insular populations of human populations across the Pacific—"colonization cycles in man and beast."

Yet, despite the conceptually compelling nature of Wilson's early theory and, in particular, its ability to integrate distributional, ecological, and evolutionary phenomena, the theory of taxon cycles was soon to be eclipsed and far overshadowed by a new theory—ironically, one codeveloped by Wilson and intended to build on his central concept of "a biogeography of the species."

A "biogeography of the species" lost

> There is a need for a "biogeography of the species," oriented with respect to the broad background of biogeographic theory but drawn at the species level and correlated with studies on ecology, speciation, and genetics. (Wilson 1959: 122)

This concept of a *biogeography of the species* was integral to Wilson's theory of taxon cycles. The dynamics in distributions, densities, morphology, behavior, niches, endemism, and extinction of insular biotas are strongly influenced by differences among species in their abilities to colonize, maintain populations, evolve, and dominate other species on islands. Students of island biogeography theory may then be surprised to learn that a stated goal of MacArthur and Wilson's equilibrium theory—with its model that views species as being largely equivalent, noninteracting, and not evolving—was to further Wilson's concept of a biogeography of the species (see MacArthur and Wilson 1967: 5–6, 183). A decade earlier, Wilson's collaboration with William L. Brown produced what would become the seminal paper on character displacement (Brown and Wilson 1956), so it is not surprising that ecological character displacement would form an integral part of his future concepts of taxon cycles and a biogeography of the species.

MacArthur and Wilson's 1967 monograph did indeed describe a very general theory of island biogeography, which included extensive discussions of colonization strategies, r/K selection, demography, niche dynamics, dispersal and biotic exchange, and evolution. Yet, even as its influence grew, the conceptual domain of their very general theory contracted to that of the equilibrium model, which could be visualized in a compellingly simple graphical form (see Figure 13.9) and could explain some very common patterns in species richness and turnover. The very integrative theory that came to Wilson while seated on the floor and pouring over distribution maps in his office had been replaced by a theory that was irresistibly simple, not least in its representation of geography and ecological communities.

Despite the tragic loss of Robert MacArthur to cancer in 1972 and Wilson's redirected energies and great success in developing other research programs and entirely new disciplines (including sociobiology, biodiversity research, and conservation biology), their equilibrium theory continued to develop and reign as the paradigm of island biogeography, strongly influencing other disciplines of ecology, evolution and biogeography, and conservation biology as well. In contrast, the taxon cycle theory received much less attention, and

Wilson's concept of a biogeography of the species was seldom if ever mentioned. As we observed at the end of the previous chapter, however, these concepts have quietly but increasingly served to guide an emerging campaign for a reintegration and the development of a more general theory of the ecological and evolutionary assembly of insular biotas (e.g., see Heaney 2000; Lomolino 2000b; Hawkins et al. 2003; Kalmar and Currie 2006; Heaney 2007; Stuessy 2007; Whittaker et al. 2008, 2010; Lomolino and Brown 2009; Lomolino et al. 2010).

15

ECOLOGICAL GEOGRAPHY OF CONTINENTAL AND OCEANIC BIOTAS

Islands, the subject of the previous two chapters, are often inhabited by evolutionary marvels, and they provide fascinating and perhaps unrivaled insights into the forces influencing the ecology, evolution, and biogeography of biotas. On the other hand, islands cover just a small fraction of Earth's surface, and their physical and environmental characteristics are, almost by definition, quite unlike that of most ecosystems (i.e., islands are isolated, relatively small, and characterized by "unusual" environments and biotic communities). It is important, therefore, to ask to what extent the patterns and processes known to occur on islands also hold true in geographically less isolated and biologically more diverse settings—the world's continents and ocean basins (see Crowell 1983).

Biogeography's first and most fundamental pattern, Buffon's law (see Chapter 2), describes differences in species composition among continental biotas—a pattern that Wallace, Darwin, and the following generations of biogeographers ultimately attributed to the equally fundamental processes of evolution, dispersal (immigration), and extinction. Throughout the Age of European Exploration, equally distinguished naturalists and biogeographers such as Forster (1778) and von Humboldt (1808) not only documented the generality of Buffon's law but also discovered many related patterns in the geography of nature. During their historic voyages, they were struck, not just by the complexity of nature, but also by its pervasive geographic signal. The diversity of the biotas they studied, and the size and shape of the individuals in those biotas, varied in a highly regular manner along gradients of latitude, elevation, and depth. While these clines in diversity and composition of ecological communities were traditionally viewed as the realm of "ecological biogeography," modern scientists now realize that they result from the combined effects of processes occurring over ecological to geological time (e.g., predation, competition, and dispersal, along with climate change, tectonic events, adaptation, and speciation).

Thus, the ecological biogeography of continents and ocean basins is indeed a complex and challenging subject, but one which holds the promise of great rewards for anyone interested in the origins, distributions, and diversity of life. In this, the final chapter before we turn to applying the lessons of biogeography for conserving biological diversity, we focus on two of the field's persistent themes (see Table 2.1):

- explaining geographic variation in the characteristics of individuals and populations, and
- explaining differences in numbers (diversity) of species among geographic areas or along geographic gradients.

Because species richness actually represents the overlap of species distributions, we first explore patterns in the sizes and shapes of biogeography's fundamental unit—the geographic range. We then review the earliest accounts of so-called **ecogeographic rules**—very regular geographic gradients in body size, shape, and other characteristics of individuals across the continents and ocean basins. We conclude this chapter with an overview of geographic gradients in diversity and the great wealth of explanations for patterns in the geography of nature.

While the complexity of all of these broadscale patterns—from geographic variation among individuals and populations, to variation in the size of geographic ranges and diversity among regional biotas—may seem overwhelming, each of these patterns should ultimately be explicable within the central framework of modern biogeography. Environmental factors vary in a very regular manner across the geographic template, and species respond to those environmental conditions by dispersing, adapting in situ (and sometimes diversifying), or suffering extinction. Thus, the very regular geographic variation of the natural world is not, in retrospect, that remarkable. These broad scale patterns in biological diversity, however, do constitute some of science's most fascinating and challenging phenomena.

The Geographic Range: Areography and Macroecology

Given that one of the most fundamental units of biogeography is the geographic range, an equally fundamental question is, What determines species distributions? This, of course, is a question biogeographers have been addressing since the origins of the field (see Chapter 4). Drawing on distributional information collected by naturalists over the past three centuries, modern biogeographers have made great strides in describing the patterns and underlying processes influencing the sizes, shapes, and location of geographic ranges. During the 1970s, this research program began to coalesce into what is recognized today as one of biogeography's most active and intriguing subdisciplines—**areography**. This field of research also became recognized as one of the most ambitious and challenging. As we have remarked elsewhere, use of experimental manipulations at spatial scales appropriate for studying geographic ranges are rendered either logistically infeasible or unethical. Areographers are therefore limited to applying the comparative method, but they have shown some great ingenuity and creativity. In fact, their attempts to investigate the influences of processes operating over multiple spatial and temporal scales stimulated the development of a new approach to exploring patterns and processes in ecology and biogeography—**macroecology**.

Macroecology has been described as a multi-scale approach to investigating the assembly and structure of biotas (Brown and Maurer 1987, 1989;

Brown 1995; Gaston and Blackburn 1999, 2000; Blackburn and Gaston 2003). In essence, macroecology attempts to gain "insights that come from applying the questions posed by ecologists to the spatial and temporal scales normally studied by biogeographers and macroevolutionists" (Brown and Maurer 1989: 1149). Because it is a potentially powerful statistical approach to understanding abundance and distribution of species, the macroecological approach has provided many important insights for both ecologists and biogeographers.

The essence of macroecology is that it tries to identify general patterns and to understand the underlying mechanisms by focusing on the statistical distributions of variables across spatial and temporal scales, and among large numbers of equivalent (but not identical) ecological "particles." These particles can be almost anything: individual organisms within a local population or entire species, replicated sample areas or patches of habitat, or species within local communities or larger biotas (for examples, see Brown 1995; Gaston and Blackburn 2000; Blackburn and Gaston 2003). The premise of macroecology is that although these particles may form seemingly random collections at one scale, their interactions and statistical properties (e.g., means, variances, frequency distributions) yield patterns that become emergent at broader spatial or longer temporal scales. Appropriately, Gaston and Blackburn (2000: 16) attribute the early seeds of the macroecological approach to Robert MacArthur (1972), who observed that "we may reveal patterns in the whole that are not evident at all in its separate parts."

Macroecologists continue to discover some fascinating emergent patterns in the statistical distributions of abundances and geographic ranges, and to apply and advance the approach to explore general mechanistic processes involved across different scales of space, time, and biological complexity. As many of these studies focus on questions with only a minor spatial context, we refer our readers to general works by Brown (1995), Gaston and Blackburn (2000), and Blackburn and Gaston (2003) for comprehensive discussions of the subject (see also Kuhn et al. 2008; Smith et al. 2008; Fisher et al. 2009; Terribile et al. 2009; Witman and Roy 2009); we instead limit our discussion here to patterns at geographic scales.

Areography: Sizes, shapes, and overlaps of ranges

Sizes of ranges

Fundamental is, of course, not synonymous with *simple*: And so it is with biogeography's most fundamental unit—the geographic range, which is far from a straightforward and easily estimated characteristic of all but the most geographically restricted yet well-studied species (Chapter 4). One principle reason for the challenges associated with measuring and comparing the sizes of geographic ranges over time or among species and across geographic regions is that studies often confuse or fail to distinguish between two alternative concepts of the range. The **extent of occurrence** is the area that lies within the marginal limits of the species' occurrence, whereas **the area of occupancy** is the area within this extent that is actually occupied by the species during a particular period (Gaston and Fuller 2009; see also Guasp et al. 1997). Thus, the area of occupancy will almost always be less than the extent of occurrence (**Figure 15.1**). Gaston and Fuller (2009) provide a thorough description of these alternative concepts and the various methods for estimating geographic ranges, along with some guidance on choosing the appropriate measure for particular applications, especially those in conservation planning.

Regardless of which of these alternative concepts and measures is used, however, there is enormous variation in the sizes of geographic ranges among

FIGURE 15.1 The area of occupancy is the area within the distributional limits of a species where its populations actually occur and, thus, should seldom exceed the species' extent of occurrence (the area bounded by the most "marginal," i.e., broadly dispersed, occurrences of this species). Exceptional cases for birds reported here (i.e., symbols above the diagonal line of equality) are those for two narrowly distributed, nonmigratory species—the Calayan rail (*Gallirallus calayanensis*) and the Bahia antwren (*Herpsilochmus pileatus*). (After Gaston and Fuller 2009.)

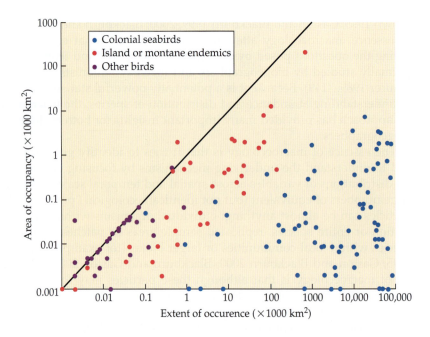

species. Among the smallest ranges are the natural distributions of the Soccoro isopod (*Thermosphaeroma thermophilum*) and the Devil's Hole pupfish (*Cyprinodon diabolis*), each of which occurs in a single freshwater spring with a surface area of less than 100 m^2 (see Figure 4.21). Among the largest ranges are those of several marine organisms, such as the blue whale (*Balaenoptera musculus*), whose ranges include most of the world's unfrozen oceans, an area on the order of 300 million km^2. In addition to whales and a variety of other marine mammals, cosmopolitan species of the marine realm include bluefish and numerous species of planktonic invertebrates and microbes. Among terrestrial organisms, species with very large native ranges include the peregrine falcon, barn owl, osprey, painted lady butterfly (*Vanessa cardui*), and common reed (*Phragmites australis*), all of which are widely distributed over all of the continents except Antarctica. Of course, modern *Homo sapiens* is now one of the most widely distributed species, and humans have carried several species of symbionts and parasites with them as they have spread over the entire Earth (see Chapter 16).

A landmark in the development of modern biogeography was the 1982 publication of the English language edition of a fascinating little book entitled *Areography*, by the Argentine ecologist and biogeographer Eduardo Rapoport (Rapoport 1975). Working largely in isolation in Latin America, Rapoport showed that simple quantitative analyses of spatial distributions of organisms—such as the maps of geographic ranges in Hall's (1981) *Mammals of North America* or Critchfield and Little's (1966) *Geographic Distribution of Pines of the World*—could reveal fascinating patterns and suggest hypotheses about the mechanisms that limit species' distributions and influence community diversity. Rapoport's book inspired a number of ecological biogeographers to undertake conceptually and methodologically similar studies on an intriguing variety of patterns in the size, shape, and overlap among geographic ranges. The result was a renewed emphasis on comparative biogeography that was similar in some ways to the resurgence of research on insular biogeography that was stimulated by MacArthur and Wilson's (1967) seminal monograph about 15 years earlier. Areography remains an active and stimulating research program, exploring all aspects of the structure of geographic ranges (see Gaston 2003; Sexton et al. 2009).

The frequency distribution of range size

One of the most fundamental and interesting areographic patterns is the distinctively shaped frequency distribution of the sizes (areas) of geographic ranges, first documented by Willis (1922) and subsequently confirmed by Rapoport (1982) and others (**Figure 15.2**; see Gaston 1991, 1996; Pagel et al. 1991; Brown 1995; Brown et al. 1996; Gaston and Blackburn 2000; Gaston 2003). Two features of this frequency distribution are of particular interest: (1) it is unimodal, and (2) it is skewed right when plotted on an arithme-

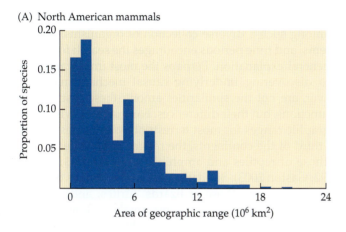

(A) North American mammals

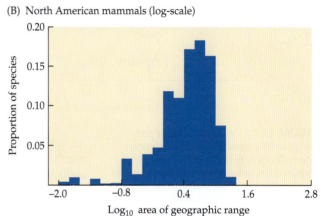

(B) North American mammals (log-scale)

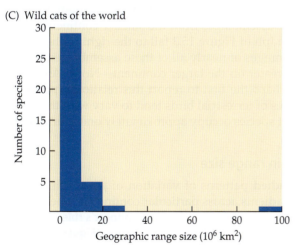

(C) Wild cats of the world

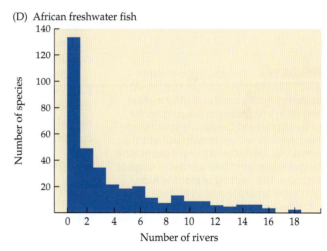

(D) African freshwater fish

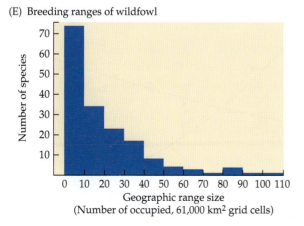

(E) Breeding ranges of wildfowl

FIGURE 15.2 Frequency distribution plots of geographic range sizes reveal that, despite substantial variation, geographic ranges of most species tend to be relatively small. Plotted here are frequency distributions of (A) North American mammals on an arithmetic scale, (B) the same data plotted on a log-transformed scale, and arithmetic-scale plots for (C) wild cats of the world, (D) African freshwater fishes, and (E) breeding ranges of wildfowl. (A,B after Brown 1995; C–E after Gaston 2003.)

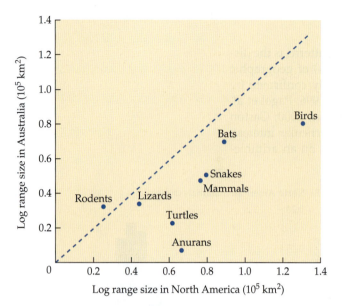

FIGURE 15.3 Geographic range size tends to be a direct function of, among other things, size of the continent that contains the range. For example, the mean geographic range sizes of North American vertebrates (except for rodents) exceed those of vertebrates inhabiting the much smaller continent of Australia (the dashed line is the line of equality). (From Gaston 2003; after Anderson and Marcus 1992.)

FIGURE 15.4 Two examples of Rapoport's rule in marine mollusk species: (A) along the Pacific coast of North America and (B) in tree species in the continental United States and Canada. As species richness declines with increasing latitude, the remaining species tend to have geographic ranges that extend over a broader range of latitudes. This method of plotting Rapoport's rule (after Stevens 1992, 1989) depicts the average breadths of the latitudinal ranges of the species whose ranges are centered on 10° latitudinal bands. (After Brown 1995.)

tic scale (**Figure 15.2A,C–E**). That is, despite substantial variation in range size among species—often spanning three or more orders of magnitude—most species tend to have relatively small geographic ranges. The pattern appears to be a very general one, exhibited by terrestrial, freshwater, and marine assemblages, including modern forms (Diniz-Filho et al. 2005; Hawkins and Diniz-Filho 2006), those from the paleontological record (Jablonski 1987; Roy 1994; Payne and Finnegan 2007), and newly established invasive species as well (Gaston 2003; after Halloy 1999).

The consistent shape of these frequency distributions among such a diverse collection of organisms, ecosystems, and time periods encourages the search for a very general explanation. Perhaps the most important clues to discovering underlying causal mechanisms, however, are not the similarities among these distribution functions, but their differences. For example, while geographic ranges for most terrestrial assemblages tend to occupy less than a third of the continents, the ranges of birds tend to be much larger than those of reptiles and amphibians (**Figure 15.3**; Anderson and Marcus 1992; Gaston 2003). Similarly, geographic ranges of mammals, while typically larger than those of lizards, turtles, and anurans, differ markedly between bats and nonflying mammals such as rodents; see Figure 15.3). Perhaps just as intriguing, geographic range sizes of the same taxa differ considerably and in a predictable manner among the continents, being smaller for assemblages from the smaller continents (note that most points in the graph of Figure 15.3 fall to the right of the line of equality, indicating that ranges of nearly all of these assemblages, except for rodents, tend to be greater on the larger continent—North America). Rapoport (1982) was perhaps the first to report this relationship, noting that the geographic ranges of terrestrial birds tend to vary with the continental size such that most species occupy approximately one-fourth of the continent's land surface.

Geographic gradients in range size

Rapoport (1982) also studied patterns of variation of geographic range size along geographic gradients across particular continents. One pattern seemed so general that it later became known as **Rapoport's rule**—the tendency for geographic range size to increase with latitude (**Figures 15.4** and

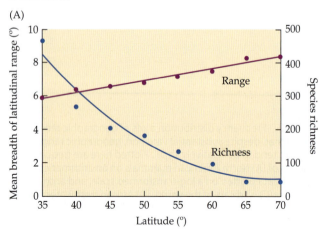

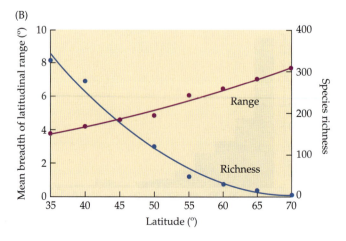

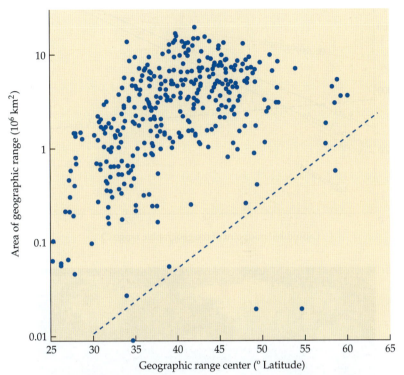

FIGURE 15.5 Variation in size of geographic range with latitude for North American breeding land birds. Although this method of plotting the data appears to show much more variation than that used in Figure 15.4, it also provides an illustration of Rapoport's rule: increasing range size with increasing latitude. The dashed line indicates that this is a constrained, macroecological relationship, suggesting that while geographical ranges of tropical species can vary markedly, those of high-latitude species tend to be much less variable and larger. (After Brown 1995.)

15.5; Stevens 1989; see also earlier report of this pattern by Lutz 1921). In subsequent studies, Stevens (1989, 1992) reported that the pattern is exhibited for many different kinds of organisms, including mammals, birds, mollusks, amphibians, reptiles, and trees, while other biogeographers reported the pattern for such diverse groups as fish, crayfish, amphipods, and beetles (see Gaston 2003: 100–101). Rapoport's rule also has its analogue in elevational gradients in terrestrial environments (**Figure 15.6**) and in depth gradients in the marine realm, but its generality and causality remain uncertain (Stevens 1992, 1996; see Rohde et al. 1993; Rohde 1996; Rohde and Heap 1996; Blackburn and Ruggiero 2001; Smith and Gaines 2003; Fernandez and Vrba 2005a; Hausdorf 2006; Ribas and Schoereder 2006; Beketov 2009).

According to Stevens (1989), the pattern results from the effects of climatic regimes on ecological capacities of species across different regions. To be adapted to their highly variable climatic regions, species occupying

FIGURE 15.6 The elevational equivalent of Rapoport's rule: Species tend to be distributed over a wider range of elevations with ascent up mountains. This pattern is seen in both (A) tree species in Costa Rica and (B) bird species in Venezuela. Compare these patterns with the latitude patterns illustrated in Figure 15.4. (After Stevens 1992.)

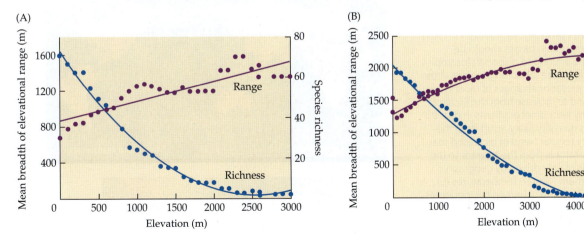

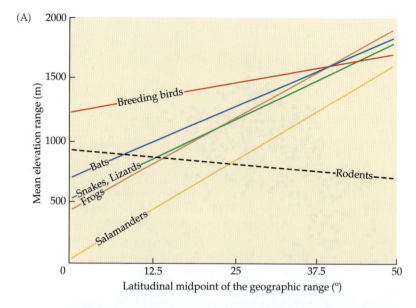

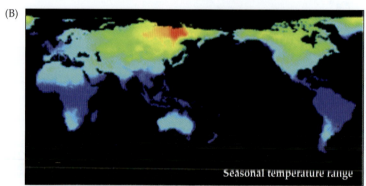

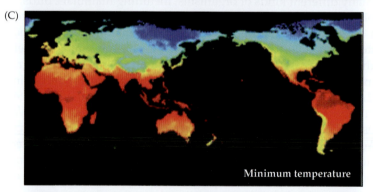

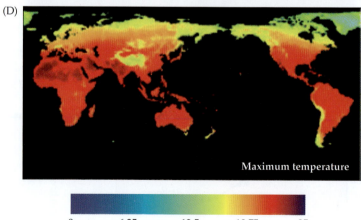

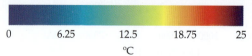

FIGURE 15.7 (A) Consistent with Janzen's (1967) hypothesis that mountain passes should be "higher" (i.e., more limiting to dispersal) for tropical species than for those inhabiting regions in the higher latitudes, elevational ranges of most types of vertebrates tend to increase from the Equator to the poles. Many latitudinal gradients in distributions and species diversity likely result from the effects of the very regular trend of increased seasonality of temperatures from the Equator to the poles (B), which is mainly driven by geographic variation in minimum (winter) temperatures (C). In comparison, maximum (summer) temperatures (D) exhibit much less variation. (A after McCain 2009a; B–D after Ghalambor et al. 2006.)

temperate regions or higher elevations along mountain slopes should have relatively generalized niches and, therefore, greater capacities to disperse across climatic and topographic barriers. Stevens' explanation was based, at least in part, on an intriguing idea published in a paper by Daniel Janzen in 1967—*Why Mountain Passes Are Higher in the Tropics*. In contrast to high-latitude or high-elevation species, those of the tropical lowlands enjoyed less variable environmental conditions. Thus, they could occupy only relatively narrow niches and geographic ranges, with limited pressures to disperse across mountain ranges to persist. The cumulative effect of these differences in selection pressures is that in comparison with species in the tropics, species from temperate regions should be able to cross more barriers and, thus, have larger geographic ranges (**Figure 15.7A**). A review of relevant patterns for ectothermic vertebrates (lizards and amphibians) by Cameron Ghalambor and his colleagues provides strong support for Janzen's hypothesis, verifying that seasonality is indeed much lower in tropical versus temperate regions (**Figure 15.7B**) and that body temperature variation, the ranges of thermal tolerance, and acclimation capacities of species all increase with latitude (Ghalambor et al. 2006).

As with many broad scale patterns, the generality of Rapoport's rule has been challenged on both empirical and conceptual grounds. Many, if not most, species fail to exhibit the predicted pattern, or they exhibit it for some biogeographic regions and not for others (see Gaston 2003). At least one of the alternative explanations for the pattern—that based on a hypothesized link between range size, species diversity, and intense competition of the tropics (Brown 1995)—has been criticized for being circular (see also Smith and Gaines 2003). Are geographic ranges smaller because of competition from a diversity of species, or is it that more species can be packed into tropical regions because their ranges are smaller?

Areographic patterns of primates offer a particularly illustrative set of case studies. As is the very general case for most taxa, diversity of primates declines as we move away from tropical regions (**Figure 15.8A–D**). Geographic ranges of primates (**Figure 15.8E–H**) also appear to exhibit a Rapoport effect—being smallest for tropical species (Harcourt 2000, 2006), which also exhibit more specialized niches (Eely and Foley 1999), thus allowing higher species packing in the lower latitudes (see also Rohde 1992, 1996; Cowlishaw and Hacker 1997; France 1998). But even this group presents some challenging anomalies. Inferences vary as a function of such factors as taxonomic resolution, analytical approach, and region (e.g., primates of the Neotropical region appear to exhibit a pattern, albeit nonsignificant, opposite to that predicted by the Rapoport effect; **Figure 15.8G**). In fact, these mixed results for primates are typical of those for most taxa. As Gaston (2003) concluded in his review of this and related patterns, while occurring in an impressive variety of species, the Rapoport effect is far from universal. On the other hand, the list of species exhibiting this pattern (i.e., increasing range size over either latitudinal or elevational gradients) is quite long and continues to grow (for reports of these patterns in vertebrates, see McCain 2009a; in freshwater invertebrates, see Beketov 2009). Thus, rather than suggesting that we ignore these patterns or become apoplectic over their variation across space, time, and taxa, we concur with Harcourt's (2006) advice to capitalize on this diversity of patterns as raw material for inductive discovery of underlying, causal forces influencing geographic range size, in general.

Geographic range size as a function of body size

In our above discussion of areography, we noted that the sizes of the geographic ranges of species within a continent exhibit a distinctively shaped fre-

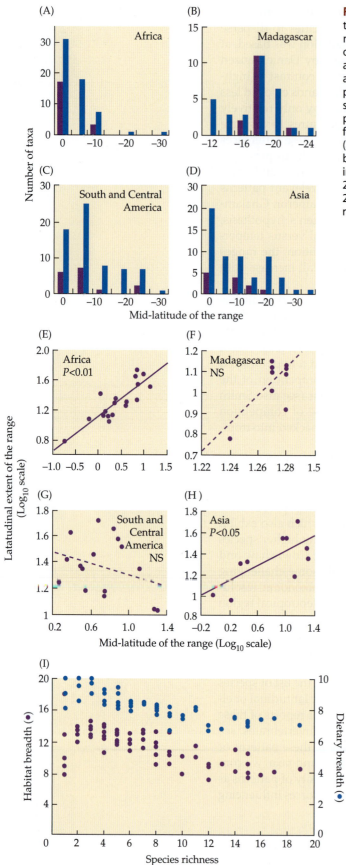

FIGURE 15.8 Although variable among the continents, the tendency for species richness of primates (A–D, expressed as number of taxa whose latitudinal midpoints fall within intervals of latitude) to be highest near the Equator may be explained, at least in part, by the tendency for geographic range size (E–H, as latitudinal extent) to be smallest and, therefore, allow higher packing of species in the tropical regions (dashed lines indicate slopes that were not significantly different from zero after phylogenetic correction). (I) Consistent with this explanation for latitudinal gradients in species richness of primates, niches (as measured by habitat breadths [purple circles] and dietary breadths [blue circles]) of primates tend to be more specialized in high-diversity (tropical) regions. (A–H after Harcourt 2000, 2006; I after Eeley and Foley 1999; see also Bohm and Mayhew 2005 for evidence of a longer history and higher speciation rates of primates in tropical versus extra-tropical regions.)

quency distribution (see Figure 15.2). What are the mechanisms that account for the preponderance of relatively small ranges, and might this be causally related to a similar macroecological pattern in body size? Within nearly all groups of organisms, the frequency distribution of body size takes on a highly skewed, unimodal form very similar to that of geographic range size (see Calder 1974; Peters 1983; Brown and Nicoletto 1991; Brown 1995; Gaston and Blackburn 2000; Blackburn and Gaston 2003; Smith et al. 2004; for a notable exception in African mammals, see Kelt and Meyer 2009). One relatively simple inference is that geographic range size is just a reflection of body size: Species composed of smaller individuals require less energy and less space and, therefore, have smaller geographic ranges. This, however, appears to be too simplistic. Fortunately, there is a more direct and more informative approach to investigating the relationship between body size and size of the geographic range: graphing geographic range size as a function of body size for an assemblage of species (**Figure 15.9**). Given the substantial variation in both of these variables, such plots are typically drawn on logarithmically transformed axes. Along with describing an interesting relationship that links a fundamental property of biogeography (the geographic range) to one equally fundamental to physiology (body size), these graphs also illustrate an innovative contribution of the macroecological approach—constraint lines.

These and many other macroecological plots often fail to suggest a tight, linear relationship between the variables of interest (in this case, range size and body size). The relationship, however, is far from random; it resembles a cloud of points that is constrained to certain regions of the bivariate space. Macroecologists have thus convinced many of us not to focus just on general trends (i.e.,

(A) Land mammals, North America

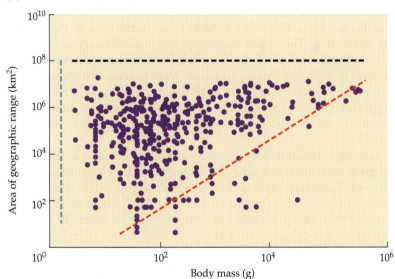

(B) Land birds, North America

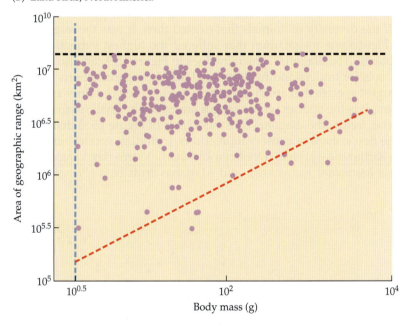

FIGURE 15.9 The relationship between geographic range size and body size on logarithmic axes for North American terrestrial (A) mammals and (B) birds. Note that while there is considerable variation, it is far from random; in particular, there are few large species with small geographic ranges. (After Brown 1995.)

trend lines through a cloud of points), but on the factors and processes constraining the extremes of those observations. Body size and range size plots for North American birds and mammals (see Figure 15.9) are typical of those for most taxa, including native and introduced birds and mammals from other regions (Maurer et al. 1991; Rapoport 1994; Sant'ana and Diniz-Filho 1999; Duncan et al. 2001; Fernandez and Vrba 2005b), fish (Gotelli and Pyron 1991; Pyron 1999; Rosenfield 2002), amphibians (Murray et al. 1998; Murray and Hose 2005), birds (Brown and Maurer 1987), reptiles (Bonfirm et al. 1998), invertebrates (Reaka 1980; Cambefort 1994), and plants (Ruokolainen and Vormisto 2000; see Gaston 2003). The points tend to fall within a roughly triangular space, indicating that range size varies substantially among the smallest species but varies less, and is typically larger, for the larger species. More to the point, the triangular shape of these plots suggests at least three constraint lines.

First, and perhaps easiest to overlook, there seems to be a constraint—perhaps a physiological one—on how small a species can be (the vertical, blue line in Figure 15.9). The other apparent constraint lines in Figure 15.9 are no less interesting. The upper, horizontal constraint line in this figure is most likely a reflection of the effects of the total size of the continent or ocean basin, and so at first glance, this seems trivial. Most species, however, occupy a relatively small proportion of the available space, and the relationship between range size and total area available differs substantially among taxa. Thus, there still remains much to be explained in terms of factors influencing this constraint line.

The remaining constraint line is perhaps the most intriguing one, and it may have great relevance, not only for ecology and biogeography, but for conservation of endangered species as well. Because population density varies inversely with body size (Brown and Maurer 1987; Damuth 1991; Silva and Downing 1995), and probability of extinction varies inversely with absolute population size (MacArthur and Wilson 1967; Leigh 1975; Gilpin and Hanski 1991), species with large body sizes tend to be rare and extinction-prone (**Figure 15.10**). Only if they have large geographic ranges are they likely to have sufficient total numbers of individuals to persist for long periods. This liability of large size was clearly demonstrated by the many waves of anthropogenic extinctions that occurred during the latter part of the Pleistocene. Of the diversity of animals that existed then, it was the largest species (the Pleistocene megafauna) that suffered the highest extinction rates (see Chapter 9). The diagonal constraint lines in Figure 15.9 can also serve as valuable tools for understanding recent and impending extinctions. It is not just the largest species that are in jeopardy—though they certainly are. More to the point, it is the species whose ranges are critically small *relative to their body size* (i.e., below the diagonal constraint lines of Figure 15.9) that should receive the lion's share of conservation efforts, including species ranging in size from Kirtland's warblers (*Dendroica kirtlandii*), Morro Bay kangaroo rats (*Dipodomys heermanni morroensis*), and Marianas fruit bats (*Pteropus* spp.)

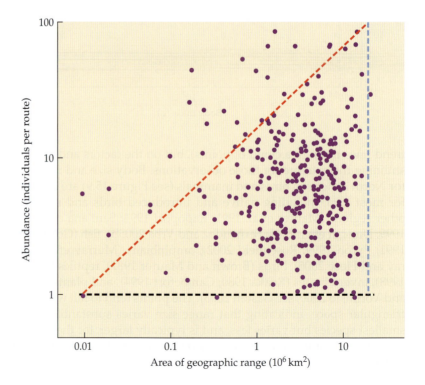

FIGURE 15.10   The relationship between geographic range size and density for local populations of North American land birds. Note that while there is considerable variation, it is far from random. In particular, there is a paucity of species with small geographic ranges and high abundances. (After Brown 1995.)

to lesser prairie chickens (*Tympanuchus pallidicinctus*), black-footed ferrets (*Mustela nigripes*), and giant pandas (*Ailuropoda melanoleuca*). Constraint lines should vary among taxa, *baupläne*, and tropic strategies, and they certainly merit further attention by macroecologists and biogeographers (for an overview of methods that can be used to calculate constraint lines, see Enquist et al. 1995; Garvey et al. 1998; Guo et al. 1998; Cade et al. 1999; Gotelli and Entsminger 2001; Diniz-Filho et al. 2005).

Temporal dynamics of range size

Variability in the size of geographic ranges over time can provide some especially valuable clues to one of the most fundamental questions posed in introductory chapters of this book and, indeed, throughout the history of biogeography: What determines the geographic range? First, it is important to realize that ranges are far from static, but undergo a developmental history analogous to the aging of individuals: They grow, persist for variable lengths of time at or near the maximal level, and then decline, sometimes entering renewed phases of expansion, but ultimately declining to extinction. Thus, alternative conceptual models of range dynamics during the biogeographic and evolutionary development of a species or higher taxon can be distinguished based on the relative length of the three developmental phases— range expansion, stasis, and decline (see Gaston 2008). As Gaston (2003: 81) concluded, "species ranges must ultimately be a product of three processes: speciation, extinction, and the temporal dynamics of the range sizes of species between speciation and extinctions." We shall focus on the latter process (contraction of geographic ranges that may ultimately lead to extinctions) during our discussions of conservation biogeography in the next chapter, but we note here that one of the most consistently significant predictors of taxon persistence in the fossil record, and of viability of contemporary species as well, is geographic range size (Diniz-Filho et al. 2005; Payne and Finnegan 2007; Powell 2007; Foote et al. 2008).

Here, we turn our attention to a complementary phenomenon: the dynamics of geographic ranges during speciation and diversification of a lineage. Webb and Gaston (2000) have addressed this question by investigating the relationship between geographic range size and age of the species. They found that for several monophyletic groups of birds with well-resolved phylogenies, range size tended to expand relatively rapidly at first, but then gradually declined with species age (**Figure 15.11**). Actually, the earlier stages in phylogeographic diversification are entirely analogous to those of range expansion in recently introduced exotic species (see Chapters 9 and 16); that is, once the invasive species become locally established, their populations expand rapidly until they encounter physical or biological barriers, which then slow, redirect, and eventually restrict their range expansion. Gaston's (2008) more recent review of range dynamics of both terrestrial vertebrates and marine invertebrates in the fossil record suggest some emerging patterns in long-term range dynamics: a relatively short period of rapid growth in range size following the appearance of the taxon; a relatively limited period during which it is widespread; and then a prolonged period of decline, often interrupted by responses to climatic events (e.g., those associated with Milankovitch oscillations). Finally, Webb and Gaston (2000) note that for contemporary species, the most threatened species tend to be those whose geographic ranges fall well below that expected for the age of their lineage (i.e., the outliers marked by *x* in Figure 15.11).

In the next chapter, we discuss these and other emerging themes in the relatively newly articulated field of conservation biogeography. It is clear, however, from the foregoing discussion on geographic range size, that spe-

FIGURE 15.11 The geographic range sizes of species within a monophyletic group of birds tend to be correlated with ages of the species (abscissa, in million years), increasing relatively rapidly (in evolutionary time) at first, but then declining over longer time periods. Each circle represents a different species of the monophyletic group that is graphed. The key outliers (marked by x) are those species that have suffered recent collapse of their geographic ranges as a result of anthropogenic factors. (From Gaston 2003; after Webb and Gaston 2000.)

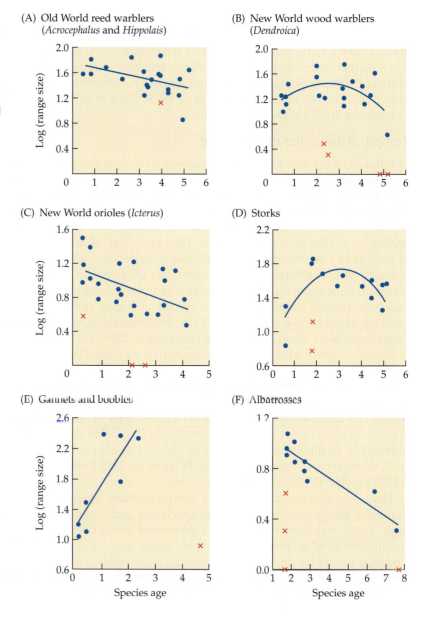

cies distributions result from properties of the species, environmental variation over space and time, and interactions between the two—in particular, the abilities of some species, especially our own, to modify their environments and alter the range dynamics of many others (see Bohning-Gaese et al. 2006; Gaston 2008).

Shapes of ranges

In addition to the impressive and very regular variation in the sizes of geographic ranges, there are also interesting areographic patterns in their shapes. For example, Rapoport (1982) noted that, despite the orders-of-magnitude variation in the areas of the ranges of North American mammals, one measure of shape—their perimeter-to-area ratio—remained relatively constant. That is, when he measured the length of a range boundary and the area encompassed within that boundary, he found that the ratio of the two variables was a nearly constant value of 10. This is surprising because for similarly shaped geometric figures, perimeter-to-area ratios should decrease with range size.

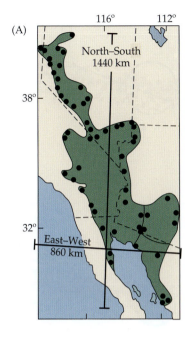

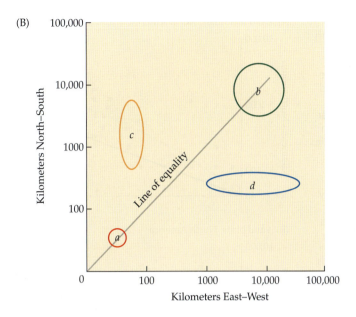

FIGURE 15.12   A simple means of representing both the shapes and sizes of geographic ranges is to measure their maximum extents along the cardinal directions and then plot these measures for various species on one graph to investigate how the shape of a geographic range changes with its size. (A) This map illustrates the geographic range of the desert kangaroo rat (*Dipodomys deserti*). (B) Potential application of this method is shown for four hypothetical species (*a*, *b*, *c*, and *d*); we have added highly simplified representations of range shapes to show whether they were relatively circular (small or large, for species *a* and *b*) or more elliptical ("stretched" north–south or east–west for species *c* and *d*, respectively).

This apparent invariance of perimeter-to-area ratios suggests that large ranges have more convoluted boundaries than do small ranges. While this pattern might suggest something interesting about colonization–extinction dynamics, ecological limiting factors, or some combination of these, it is first important to evaluate an alternative, or null hypothesis; that is, it simply reflects an unintentional bias of mapmakers. We have an inherent tendency to draw maps with a fractal structure, including more detail about boundaries (and other features) for larger areas. In many cases, including Hall's (1981) treatise on North American mammals, small ranges are mapped at greater magnifications than are large ones. On the other hand, if the fractal inclinations of mapmakers could be taken into account, the periphery-to-area ratio could serve as a simple measure of range shape that warrants further study.

Brown and Maurer (1989) used an alternative and relatively simple technique to quantify both the shapes and the orientations of ranges (for other approaches to investigating variation in range shape, see also Maurer 1994; Brown et al. 1996; Ruggiero 2001; Fortin et al. 2005). In order to investigate whether range shape varied in any regular manner with range size, Brown and Mauer (1989) plotted maximum north–south extent across a species' geographic range as a function of maximum east–west extent (**Figure 15.12A**). The result is the kind of graph shown in **Figure 15.12B**. In such a graph, ranges with equal dimensions, such as circles or squares, would fall along the line of equality, with small ranges falling to the lower left and large ranges falling to the upper right of this graph. Ranges longer in a north–south direction would fall above the line of equality, whereas those longer in an east–west direction would fall below the line.

Although there is considerable scatter in such graphs (**Figure 15.13**; Brown and Maurer 1989; Brown 1995), North American mammals, birds, and reptiles all show a consistent trend: Small ranges tend to be oriented north–south, whereas large ranges tend to be oriented east–west. European mammals and birds (see Figure 15.13C) present an interesting contrast, with most ranges being relatively large and oriented east–west. Brown and Maurer suggested that these patterns of range orientation reflect the physical geography of the continents. The east–west orientation of large ranges on both continents likely results from similar orientation of major climatic and vegetation zones

FIGURE 15.13 The shapes of geographic ranges of North American (A) mammals and (B) birds change in a consistent manner—circular for the smallest ranges; elliptical and stretched north–south for species with somewhat larger ranges; and circular and then stretched east–west for species with relatively large ranges (the solid black line is the line of equality and denotes where the ranges are roughly circular, i.e., where a range's north–south extent equals its east–west extent). (C) In contrast to the patterns for North American mammals and birds, geographic ranges of European birds almost all tend to be relatively large and stretched east–west.

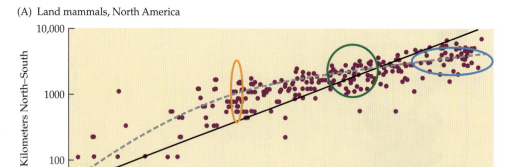

(A) Land mammals, North America

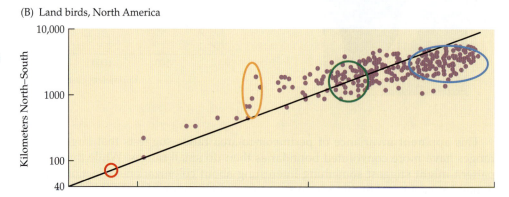

(B) Land birds, North America

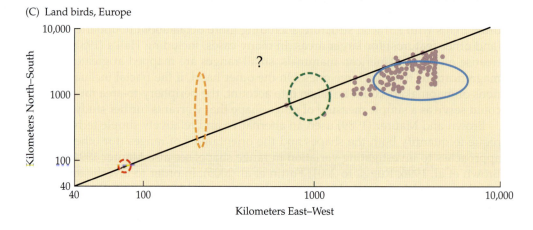

(C) Land birds, Europe

(see Chapters 3 and 5). On the other hand, very small ranges (those of localized endemics) will appear roughly circular or symmetrical because of the cartographic artifact described above. Somewhat larger ranges, however, are likely constrained by broad scale, topographic barriers. As Brown and Maurer (1989) observed, most of the major topographic barriers in North America (e.g., the Atlantic and Pacific coastlines; the Mississippi River; the Rocky Mountains; and the Sierra Nevada, West Coast, and Appalachian Mountains) run roughly north and south. Accordingly, species whose distributions range over a small to intermediate portion of the continent should have ranges with a north–south "stretch." Species with larger ranges are, by definition, not constrained by these barriers. Thus their ranges tend to be more circular or, in those species with the largest ranges, stretched east–west, following the orientation of climatic and vegetation zones or the northern coastlines of the continent.

Why then don't the species of Europe exhibit the same patterns? First, unlike in North America, most of the major geographic barriers of Europe run east–west. Thus, dispersal and range expansion to the south is blocked, first by a broken string of mountain chains (including the Pyrenees, Alps, and Caucasus Mountains) and then by large bodies of water (the Mediterranean, Aegean, and Black Seas) farther south. Northward dispersal is similarly limited by the Norwegian and Baltic seas and by the stresses associated with the subarctic and Arctic climates farther north. This explanation seems plausible, but it does not account for another important difference between European and North American avifaunas (see Figure 15.13, compare B and C). Europe is distinguished by a surprising lack of birds with relatively small geographic ranges. In this case, the explanation is equally as plausible but involves both the climatic dynamics of the Pleistocene and the differences in shapes of the continents. As climates changed and the glaciers expanded across these northern continents, native biotas either adapted by dispersing south of the glaciers, or they suffered extinctions. While the biogeographic dynamics were quite complex (see Chapter 9), it is clear from any topographic map that this was much less challenging for species in North America. In Europe, however, only species capable of dispersing across its formidable geographic barriers (i.e., species with superior dispersal abilities and the largest geographic ranges) survived. On the other hand, geographically restricted species, who by definition had limited dispersal abilities, could not escape the glaciers and severe swings in climatic conditions, and they perished.

Biogeographers and macroecologists have recently begun to investigate the relationship between dispersal abilities and geographic range size (e.g., see Bohning-Gaese et al. 2006; Lester et al. 2007; Beck and Kitching 2009a). The results are, however, inconsistent among studies, which in retrospect may not be surprising given the great challenges associated with studying long-distance dispersal (i.e., extremely rare, albeit important events). As we observed in the closing paragraphs of Chapter 6, all this may be changing with continual advances in our abilities to assess dispersal capacities at the scales most relevant in biogeographic and evolutionary research. Finally, just as comparisons in the dynamics of range shapes for similar taxa among geographic regions provide valuable insights (see Figure 15.13A versus B), comparisons among taxa within the same region are likely to provide additional clues to underlying causal mechanisms. For example, ectotherms differ substantially from birds and mammals in terms of both climatic adaptations and dispersal capacities. Consistent with these differences among taxa, mosquitoes, amphibians and reptiles of North America tend to exhibit dynamics in the sizes and shapes of geographic ranges that differ substantially from those illustrated in Figure 15.13 (Pfrender et al. 1998). Again, it is such differences among focal groups of species and geographic regions that comprise some of the most instructive, scale-appropriate "treatments" for natural experiments in biogeographic research.

The internal structure of geographic ranges

Still further insights into the factors influencing geographic ranges can be obtained by investigating the relationships between characteristics of local populations (e.g., population density or genetic variability) and two biogeographic variables: geographic range size of the species and relative location of each population within the species' geographic range. For example, by plotting average local abundances of species against the sizes of their geographic ranges (see Figure 15.10), macroecologists obtained a triangular-shaped plot, again suggesting a constrained relationship (Brown 1995; Blackburn et al.

1998; Gaston et al. 1998; Gaston and Blackburn 2000; Gaston 2003). The constraint lines include a trivial artifact—the absolute minimum abundance of the species—and a vertical constraint line that again corresponds to the maximum range size for the assemblage (equivalent to the dashed horizontal line in Figure 15.9). The third constraint line (the diagonal one) is certainly the most interesting, suggesting that maximal abundance of local populations increases with the size of the geographic range. This pattern is consistent with the relationship between niche breadth, abundance, and distribution developed in Chapter 4. Species with less restrictive resource requirements and broad tolerances for abiotic and biotic limiting factors are able to be both locally abundant and widely distributed. Furthermore, in part for these reasons, these species are also likely to have a low probability of extinction, even in the face of major environmental change.

In the above examples, the particles of analysis are individual species within continental or oceanic biotas, and the emergent statistical patterns are the constrained plots of Figures 15.9 and 15.10. We can also examine a set of related but more local-scale relationships, in this case with populations serving as the particles and the general patterns emerging at the species level. In fact, we have already described one such pattern in the internal structure of a species' geographic range in Chapter 4 (see Figures 4.2, 4.8, and 4.16). In Rapoport's (1982) classic analysis of aerial photographs, he discovered that the density of a species of a palm (*Copernicia alba*) decreased from sites near the center of its geographic range to those along its periphery. This pattern was subsequently reported for a broad diversity of species, including numerous plants, insects, mollusks, fish, birds, and mammals (Brown 1995; Gaston 2003; McCain 2006). Not only are population densities of many species lower near the edge of a range, but peripheral populations also tend to be more sparsely distributed and more variable over time. The overall result, and one especially relevant to conservation biology as well as biogeography, is that peripheral populations tend to be sinks, experiencing frequent extirpations and, thus, relying heavily on recolonization from more central (source) populations. On the other hand, more recent reviews and analyses suggest that the pattern, although pervasive, is not a universal one: Many species attain their highest population densities near the periphery of their geographic range (e.g., see Sagarin et al. 2006; Fuller et al. 2009; Sexton et al. 2009). It is difficult to assess how much of the discrepancy in conclusions among studies stems from methodological issues (in particular, the difficulty of using a consistent and objective means of locating the range center; Sagarin et al. 2006: 525) versus actual inconsistencies of empirical patterns and underlying forces influencing population densities across the geographic ranges of species. This certainly represents another line of research that is ripe for new discoveries.

Frontiers of areography

We are optimistic that areographers and macroecologists will continue to capitalize on the great legacy of distributional information provided by centuries of biological surveys, and that they will continue to develop more innovative and insightful approaches for investigating the forces that determine one of biogeography's most fundamental units—the geographic range. In most of the studies discussed above, the geographic context of relevant patterns is typically reduced to a single dimension or factor (latitude or elevation). In contrast, modern techniques in GIS, cartography, and spatial data analyses provide more holistic and insightful means of visualizing patterns over multiple dimensions and then assessing underlying causal forces, which may of-

(A) Geographic range size of breeding birds

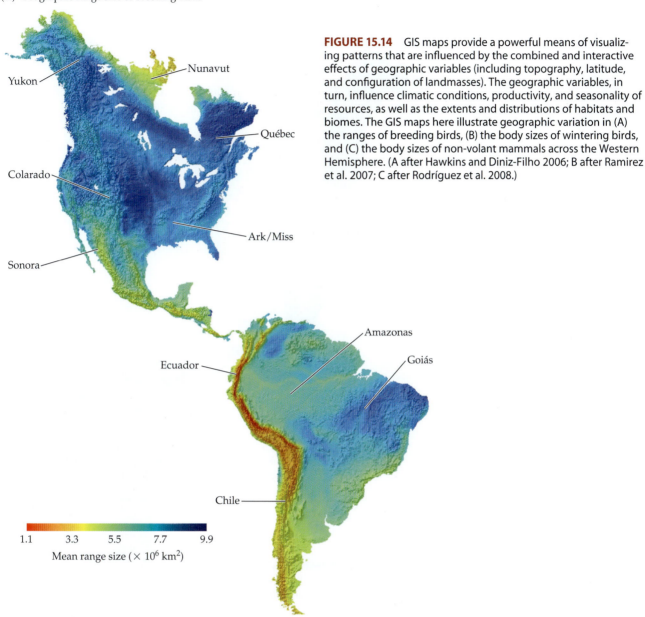

FIGURE 15.14 GIS maps provide a powerful means of visualizing patterns that are influenced by the combined and interactive effects of geographic variables (including topography, latitude, and configuration of landmasses). The geographic variables, in turn, influence climatic conditions, productivity, and seasonality of resources, as well as the extents and distributions of habitats and biomes. The GIS maps here illustrate geographic variation in (A) the ranges of breeding birds, (B) the body sizes of wintering birds, and (C) the body sizes of non-volant mammals across the Western Hemisphere. (A after Hawkins and Diniz-Filho 2006; B after Ramirez et al. 2007; C after Rodríguez et al. 2008.)

Mean range size ($\times 10^6$ km^2)

ten combine to generate complex but now interpretable patterns in variation of geographic ranges over space, time, and taxa.

Hawkins and Diniz-Filho's (2006) research on the ranges of New World birds provides an exemplary case in point. Their approach is based on first principles (essentially Janzen's explanation for differences in elevational ranges of tropical versus temperate species) and utilizes the now readily available techniques for visualizing macroecological patterns with GIS. They complement this with rigorous, statistical methods to assess both simple and interactive effects of driver variables while adjusting for spatial autocorrelation among independent variables (Hawkins and Diniz-Filho 2006; Ruggiero and Hawkins 2006; see also Blackburn and Gaston 2006). The centerpieces of such studies are maps such as that of **Figure 15.14A**, which illustrates the combined and interactive effects of topography and climate (as influenced by elevation and latitude, respectively) on breeding ranges of New World birds. Consistent with Janzen's (1967) hypothesis, the effect of elevation (topography) on geographic range size varies with latitude. That is, geographic ranges of tropical species are much smaller for those inhabiting the slopes

FIGURE 15.14 (*continued*)   (B) Body size of wintering birds

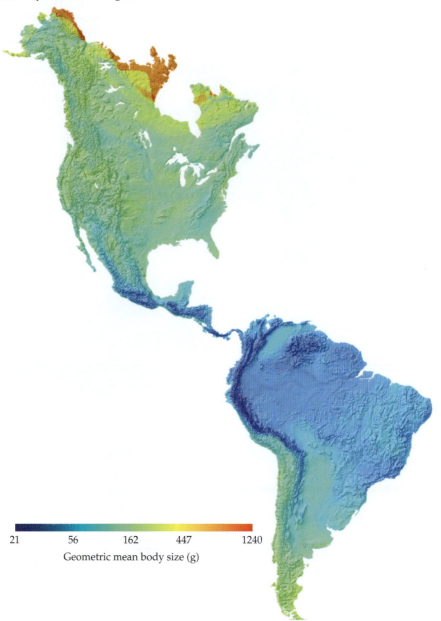

21   56   162   447   1240
Geometric mean body size (g)

of the equatorial Andes than for those of the broad lowlands to the east. In contrast, at higher latitudes to the south and especially across the northern regions of the USA and Canada, geographic range size is little influenced by topography—most species exhibiting consistently broad ranges across the mountains and plains regions of northern North America (note that the ranges of birds in the high-latitude regions south of Chile also exhibit a reduced influence of regional topography, with ranges tending to be intermediate in size and apparently limited by the relatively narrow configuration of this region of South America; see also Ruggiero and Hawkins 2008). In summary, these and similar approaches for rigorously exploring areographic patterns hold great promise, not just for evaluating simple (unidimensional) clines predicted by a given "rule," but for identifying the combination of causal forces that interact to influence the geography of nature, in general.

(C) Body size of non-volant mammals

FIGURE 15.14 (*continued*)

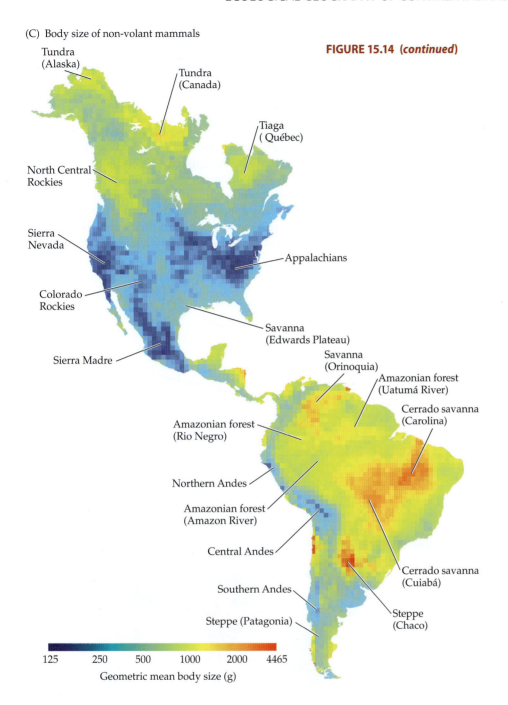

Geometric mean body size (g)

Ecogeographic Rules

Ecogeography of the terrestrial realm

As we noted in the introduction to this chapter, one of biogeography's early and persistent themes is one that explains geographic variation in the characteristics of individuals and populations. While the diversity of such characteristics that could be studied (including genetic, demographic, morphological, physiological, behavioral, and ecological traits) seems overwhelming, some surprisingly regular patterns have emerged. Foremost among these are the so-called ecogeographic rules. As Ernst Mayr (1956) observed, the term *ecogeographic* is less than ideal, because it seems to refer to geographic variation in ecological characteristics or processes (e.g., diversity or predation rates, respectively). The term, however, has been firmly established and is generally understood to refer to very regular geographic gradients in the characteristics of organisms (especially, but not limited to, their morphological traits) across continents and ocean basins.

Body size and Bergmann's rule

Arguably, the most informative of the traits that we could use to characterize an organism is its body size. Almost all other traits, including those influencing its resource requirements, dispersal capacity, reproductive and evolutionary rates, and interactions with other species (e.g., as parasite or host, predator or prey, or dominant competitor) are strongly correlated with body size. Thus, it should not be surprising that the most commonly studied and first articulated ecogeographic rule concerns geographic variation in body size. In 1847, Carl Bergmann noted that among closely related kinds of mam-

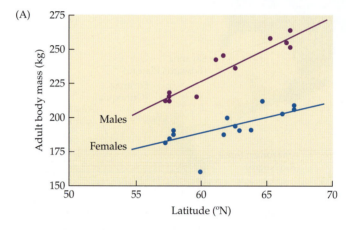

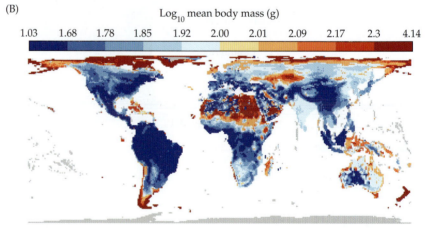

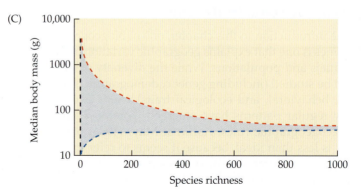

FIGURE 15.15 (A) Body size of moose (*Alces alces*) in Sweden increases with latitude in a manner consistent with Bergmann's rule. (B) This global map of geographic variation in body size for avian species illustrates patterns related to both Bergmann's rule (latitudinal and elevational trends) and the island rule (larger body size of insular birds). (C) Regardless of the particular geographic gradient (i.e., latitudinal, elevational, or otherwise), body size of birds exhibits a consistent relationship with species richness of avian communities. As indicated by the red constraint line, maximum size for species among assemblages of birds decreases with the diversity of its communities. (A after Sand et al. 1995; B and C after Olson et al. 2009.)

mals and birds, the largest species often occurred at higher latitudes (**Figure 15.15A,B**). Bergmann surmised that this was because larger species have lower surface-to-volume ratios, thus giving them an advantage over their smaller relatives in conserving body heat in colder climates.

Since its first articulation, Bergmann's rule has been the subject of frequent study as well as much debate about its actual meaning, generality, and causality (**Table 15.1**). These three properties are, in fact, interrelated. For example, the generality of Bergmann's rule or any other ecogeographic pattern depends on whether it is meant to apply to particular functional groups (e.g., endotherms versus all animals), taxonomic levels (to variation among geographic populations of a species, or to variation among different species and higher taxa), or geographic gradients (e.g., those across latitude, elevation on land, or depth in the oceans). While all sides in such debates have merit, we prefer a more integrative approach that is based on causality and first principles. That is, it may well be that the most insightful approach for understanding Bergmann's rule is one that strives to understand the geography of body size, in general, and is based on a clearly articulated but general explanation. One general hypothesis implicit in many studies referring to Bergmann's rule is that variation in body size results from adaptation to the very regular variation in environmental conditions across the geographic template.

TABLE 15.1 *Selection of Hypotheses Proposed to Explain Bergmann's Rule and Other Patterns of Geographic and Temporal Variation in Body Size of Animals*

Hypothesis	Explanation	Source
Heat conservation (two alternatives)	A. Tolerance to cooler temperatures increases with body size because the ratio of surface area (across which heat is lost) to volume in organisms decreases with their size.	Bergmann 1847; Mayr 1956
	B. Relative thermal conductance (rate of heat lost per mass) and the lower critical temperature for thermoregulation (below which endotherms must increase their metabolism to maintain body temperature) decrease (as relative insulation increases) with body size.	Blackburn et al. 1999b
Heat dissipation; water conservation	The capacity to use evaporative cooling to prevent overheating in hot and moist environments is higher for smaller animals.	Brown and Lee 1969; James 1970
Endurance; thermal independence; starvation resistance; resource (energy or water) availability	The amounts of energy and water stored in the tissues of an organism increase more rapidly with body size than do the rates at which these resources are used; therefore, larger organisms have greater capacities to withstand harsh periods of climatic extremes and shortages of energy and water.	Calder 1974; Lindstedt and Boyce 1985; McNab 2002
Dispersal	The amounts of energy and water required by an organism to travel a given distance (like the energy and water needed to survive, as stated above) increase less rapidly with body size than do the stores of these resources; therefore, larger individuals should have greater capacities for dispersal, which may be adaptive for surviving in low-productivity and highly seasonal environments.	Blackburn and Hawkins 2004
Ecological character release	In species-rich assemblages, small size may allow some species to avoid intense, exploitative competition by specializing on smaller prey, and to avoid interference competition and predation by exploiting smaller refugia. In species-poor (e.g., high-latitude or high-elevation) assemblages, the absence of one or more larger species may result in an increase in size in the otherwise subordinate, smaller species (see Figure 15.16).	McNab 1971; Fuentes and Jaksic 1979; Dayan 1990; Alacantra 1991
Environmental (morphological) plasticity	Observed temporal and geographic trends in body size may simply be nongenetic responses to varying environmental conditions; individuals decline in size under deteriorating conditions and vice versa.	Teplitsky et al. 2008

One advantage of this more inclusive and integrative research program is that it identifies other lines of research for evaluating the general, causal explanation. For example, the above hypothesis would predict that body size should vary with gradients in species richness and over time as well. In fact, there is strong evidence for both of these corollary predictions. Recall from Chapter 14 that one of the fundamental driving forces for body size variation on islands (i.e., the island rule) may be ecological character release. An analogous explanation for latitudinal trends in body size on the continents was proposed by Brian McNab (1971). According to this hypothesis, ecological character release should vary in a regular manner with latitude and, in particular, with the latitudinal gradient in species richness, which we discuss later in this chapter. Low diversity in high latitudes means reduced interspecific interactions, including competition—a condition that often triggers ecological release, especially in the smallest members of an ecological guild (McNab 1971; see also Dayan et al. 1989; Dayan 1990; Iriarte et al. 1990). In species-rich environments, small size may allow some species to avoid intense competition by specializing on smaller prey, and to avoid interference competition and predation by exploiting smaller refugia. When these interspecific pressures are released (e.g., in species-poor communities of the higher latitudes, or on isolated islands), an increase in size in an otherwise small species may be favored because larger individuals tend to dominate intraspecific competitors, and they can exploit a broader range of resources (e.g., larger squirrels can crack large as well as small nuts; larger predators can take large as well as small prey). McNab's (1971) analysis of body size trends in non-volant mammals appears consistent with this ecological explanation for patterns consistent with Bergmann's rule. For a number of guilds of mammals, trends toward increasing body size (whether toward the north or south, or along clines of longitude as well) appear to be associated with prey availability and the absence of a larger member of the guild (**Figure 15.16**; see also Fuentes and Jaksic 1979; King and Moors 1979; Erlinge 1987; Dayan et al. 1989; Dayan 1990; Iriarte et al. 1990; Alacantra 1991). Olson and colleagues' more recent, global analyses of body size variation in birds (see Figure 15.15B; see also Figure 15.14B,C) is consistent with both McNab's hypothesis of interspecific character release and with the more general, integrative explanation of geographic variation in body size, which also includes the effects of climatic conditions (Olson et al. 2009). Larger birds are more likely to be found at higher latitudes and on islands, and the maximum size of species in an avian assemblage over all systems (continental or insular) varies inversely with species richness (**Figure 15.15C**). Olson and colleagues were quick to point out that climatic factors (especially annual means and amplitudes in temperature) and resource productivity were also strongly correlated with avian body size, concluding that global variation in avian body size was driven by interactions among processes operating from the individual (physiological and behavioral) to assemblage and regional (ecological and evolutionary assembly) scales.

This more integrative, process-based explanation for geographic variation in body size thus predicts not just latitudinal, but corollary patterns in body size variation over other geographic gradients associated with underlying factors such as species diversity, climate, and productivity. Indeed, there is much empirical evidence for body size clines across gradients of elevation and depth, in both the terrestrial and marine realms (in mammals: Ashton et al. 2000; Meiri and Dayan 2003; Meiri et al. 2004a,b; in birds: Ashton 2002a; Meiri and Dayan 2003; see also James 1970; Kendeigh 1976; in salamanders: Ashton 2002b; in turtles: Ashton and Feldman 2003; in parasitic flatworms: Poulin 1996; in marine gastropods and ostracods: Olabarria and Thurston

(A) Mustelids

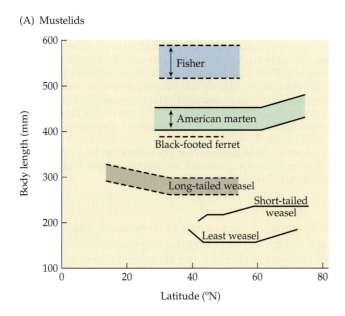

(B) Felids

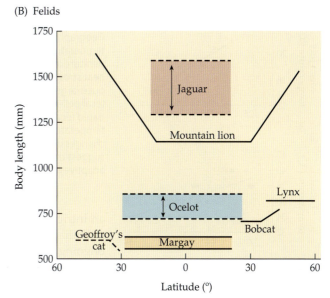

FIGURE 15.16 Geographic trends in body size in ecological guilds (groups of species with similar feeding strategies) may be influenced by interspecific competition. For example, body sizes of smaller species of (A) mustelids and (B) felids sometimes increase along latitudinal gradients (north or south) beyond the range limits for larger members of that guild. (Trend lines drawn here are simplified from actual data presented by McNab 1971.)

2003; Hunt and Roy 2006; in fruit flies: Huey et al. 2000; and in ants: Cushman et al. 1993; Kaspari and Vargo 1995; see also Ray 1960; Van Voorhies 1996).

The accumulating information on geographic variation in body size, however, includes numerous exceptional cases (e.g., in mammals: Fuentes and Jaksic 1979; Erlinge 1987; Alacantra 1991; Thurber and Peterson 1991; Ochocinska and Taylor 2003; in reptiles: Ashton and Feldman 2003; in butterflies: Barlow 1994; Hawkins and Lawton 1995; in land snails: Hausdorf 2003; and in other arthropods: Blanckenhorn and Demont 2004), and others that exhibit significant but conflicting clines along latitudinal, elevational, or bathymetric gradients may ultimately be explicable as the scope of future research expands to simultaneously include these diverse taxa, regional biotas, and geographic clines. Among mammals, the pattern is equivocal for mustelids (weasels and related species) and actually opposite of that expected for heteromyid rodents (Ashton et al. 2000). In addition, some studies indicate that the correlation between body size and latitude differs between large- and small-bodied species (Ashton 2002a; Freckleton et al. 2003; Meiri and Dayan 2003; but see Ashton et al. 2000), between species that are more or less sedentary (e.g., birds: Ashton 2002a, 2004), and between different regions (e.g., tropical vs. temperature latitudes: McNab 1971; but see Ashton et al. 2000). Whether most of the species actually exhibit the pattern also depends on methodological issues. For example, the 2000 review by Ashton et al. found that 78 out of 110 species of mammals (71 percent) exhibited the pattern (i.e., their body sizes increased with latitude), but the intraspecific trends were statistically significant for just 43 (39 percent) of the species—results that are similar to those McNab (1971) reported earlier as evidence against the generality of the rule. Ashton et al. (2000), however, attributed many of these nonsignificant trends to cases that were not sampled extensively. They found that when only the 33 species that were adequately studied were considered, a clear majority (23 of those species, or 70 percent) exhibited significant, positive correlations between body mass and latitude.

Finally, integrative and process-based explanations for the geography of body size also direct us to investigate analogous patterns of variation over time, especially when the time periods are associated with substantial shifts in climate, species diversity, or other putative drivers of body size variation. This line of research, into temporal variation in body size, constitutes

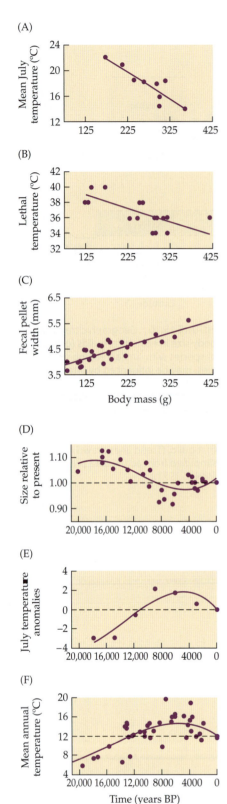

(A)

(B)

(C)

(D)

(E)

(F)

Time (years BP)

FIGURE 15.17 Bergmann's rule and adaptation to temperature in the bushy-tailed wood rat (*Neotoma cinerea*) over both geographic space (A–C) and evolutionary time (D–F). (A) Geographic variation in body size among contemporary populations as a function of July temperature, showing that larger individuals inhabit cooler climates. (B) Upper lethal temperature as a function of body size, showing that larger individuals are more susceptible to heat stress. (C) Fecal pellet size as a function of body size, showing that larger animals produce larger feces, allowing body sizes of past populations to be estimated from fossilized fecal pellets. (D) Relative sizes of past populations, estimated from the sizes of radiocarbon-dated fecal pellets, over the last 20,000 years. (E and F) Two measures of environmental temperature over the last 20,000 years, showing a pattern that almost exactly mirrors the changes in body size. Taken together, these patterns over both space and time strongly suggest that, consistent with some current explanations for Bergmann's rule, smaller size is adaptive in warmer environments. (After Smith et al. 1995.)

an intriguing and rapidly growing field owing to its heuristic value and its potential for providing a means of predicting the response of organisms to ongoing, anthropogenic changes in climates and ecological diversity. Studies retracing the temporal dynamics of a limited but informative sampling of species inhabiting Pleistocene to modern environments have provided some compelling results. As we discussed in Chapter 9, species survived the repeated climatic cycles of the Pleistocene either by dispersing or by adapting to the highly altered local conditions. The fact that those mammals and other species that recolonized exposed regions after the glaciers receded (12,000 to 18,000 years BP) now exhibit the expected trends of body size with latitude indicates relatively rapid adaptation and evolution (Blackburn and Hawkins 2004). For at least some of these species, their body sizes closely tracked the temporal changes in the thermal regime. For example, Felisa Smith and her colleagues (Smith et al. 1995) found that over the past 20,000 years, body size of wood rats (North American rodents of the genus *Neotoma*) varied in a manner entirely consistent with the thermoregulatory explanations for Bergmann's (latitudinal) rule: increasing in size as temperatures cooled, and decreasing in size as temperatures warmed (**Figure 15.17**; see also Smith et al. 2009). Similar patterns of geographic and temporal variation in body size were documented for pocket gophers of North and Central America (Hadley 1997) and for the now extinct moas of New Zealand during climate change of the Late Holocene (Worthy and Holdaway 2002: 140–142; see also Thurber and Peterson 1991). Hunt and Roy (2006) report analogous patterns in the marine realm, with body size of ostracods (class Crustacea) increasing as temperatures cooled during the Cenozoic (0.1 to 40 million years ago) and the magnitude of these changes consistent with the differences among geographically isolated, modern populations.

The 2006 Millien et al. review of ecotypic variation over space and time reveals similar temporal trends, with body size of a diversity of native and introduced species being correlated with changes in local temperatures at particular sites (e.g., Smith et al. 1998; Yom-Tov 2001; Yom-Tov and Yom-Tov 2004, 2005; Yom-Tov et al. 2003, 2006) or with differences in temperature or the number of co-occurring species as species expand their ranges to occupy new sites (see also Schmidt and Jensen 2003, 2005). Studies of introduced species also provide evidence for the surprising rapidity of morphological responses to environmental conditions, whether natural or anthropogenic. For example, Huey et al (2000) found that in just 20 years after being introduced into the New World, the fly *Drosophila subobscura* not only rapidly expanded its exotic range to occupy a broad geographic range, but it also developed a geographic cline in body size similar to that exhibited by populations in its native range—a cline entirely consistent with Bergmann's rule (**Figure 15.18**; see also Thurber and Peterson 1991). In still other studies, Yoram Yom-Tov and his colleagues found that body size of rats (*Rattus exulans* and *R. rattus*) introduced onto islands off the coasts of New Zealand was inversely correlated with the number of co-occurring species of rodents (Yom-Tov et al. 1999). To the degree to which these rapid dynamics in body size represent heritable (i.e., microevolutionary) changes, they have great relevance for predicting responses to ongoing anthropogenic changes in climates and the diversity of native communities (but see Teplitsky et al. 2008). Moreover, because body size influences nearly all traits and capacities of organisms, these rapid anthropogenic changes in life's most fundamental characteristic are likely to cascade through trophic levels and across geographic regions to affect the distributions and diversities of many species.

Allen's rule

Other ecogeographic rules describe additional patterns of geographic variation in morphology. For example, Joel Asaph Allen (1878) noted that among closely related endothermic vertebrates, those forms living in hotter environments (generally, those of the lower latitudes or lower elevations) often tend to have longer appendages. Again, the hypothesized mechanism invoked thermoregulatory advantages: that shorter appendages promote heat conservation in cold environments by reducing the total surface area for heat loss. On the other hand, relatively long appendages increase surface area and thus facilitate heat dissipation where ambient temperatures reach stressfully high levels. Physiological research has shown that the relatively long ears of desert rabbits and elephants, and the long and relatively thin legs of camels, do indeed play important roles in heat dissipation (Schmidt-Nielsen 1963, 1964; see also Brown and Lasiewski 1972). In North America, hares of the genus *Lepus* show a dramatic decline in the length of their ears with decreasing temperature, from the black-tailed and white-sided jackrabbits of the hot, arid Southwest to the varying and Arctic hares of the perpetually cold north (**Figure 15.19**). Studies on a related species, the European rabbit (*Oryctolagus cuniculus*), just over a century after its introduction into Australia also

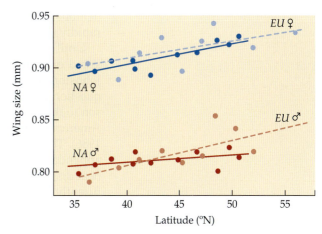

FIGURE 15.18 Within two decades after their introduction into North America (NA), body size for populations of the fruit fly, *Drosophila subobscura*, evolved to the point that the pattern has converged upon that of populations in their native range in Europe (EU), which is consistent with Bergmann's rule. (After Huey et al. 2000.)

FIGURE 15.19 Two classic examples of Allen's rule: variation in the relative size of the ears among hares (A, from left: *Lepus alleni*, *L. californicus*, *L. americanus*, and *L. arcticus*) and foxes (B, from left: *Fennecus zerda*, *Vulpes vulpes*, and *Alopex lagopus*). In each case, the species on the left occurs in a hot desert habitat, the one on the right in a cold tundra habitat, and the one (or two) in the middle in intermediate, temperate environments.

demonstrate that this ecogeographic trend in body shape may develop quite rapidly (Williams and Moore 1989).

Until quite recently, the empirical evidence supporting Allen's rule consisted of a limited number of isolated cases (Hesse et al. 1951; Mayr 1963; Johnston and Selander 1971). The number of exceptions led some authors to question the generality of this ecogeographic rule. For example, in comparison to the cases for North American hares, it is questionable whether North American rabbits (genus *Sylvilagus*) exhibit the same pattern (Stevenson 1986). The presumed mechanistic basis of heat dissipation through enlarged appendages can also be questioned. While long ears can serve as excellent heat dissipators, heat exchange can be regulated much more rapidly and efficiently by other means, including by modifying insulation (e.g., by fluffing up fur or feathers) and by altering blood flow to the extremities. In addition, long ears may be advantageous in warm environments because they aid in sound detection (sound attenuates more rapidly at higher temperatures: Erulkar 1972; Harris 1996).

Some relatively recent studies, however, provide evidence for both the thermoregulatory basis for Allen's rule and ecogeographic variation consistent with the predicted pattern for a variety of species, including shorebirds (Cartar and Morrison 2005; Nudds and Oswald 2007) and kinglets (Chui and Doucet 2009), small mammals (Porter and Kearney 2009), and hominin primates (Weaver and Steudel-Numbers 2005; Tilkens et al. 2005; see also Stegmann 2007; Serrat et al. 2008). We find the latter reports that populations of primates, including our own species, exhibit morpho-geographic trends consistent with both Bergmann's and Allen's rules intriguing, and they certainly merit renewed interests by both biogeographers and anthropologists (see Newman 1953; Roberts 1953, 1978; Houghton 1990; Ruff 1994; Katzmarzyk and Leonard 1995; Bindon and Baker 1997; Lazenby and Smashnuk 1999; see also our discussion on the biogeography of humanity in Chapter 16).

Gloger's rule

A few decades after Bergmann described the pattern of latitudinal variation in body size, the German ornithologist Constantin Wilhelm Lambert Gloger (1883) noted that the coloration of related forms of endotherms was often correlated with the humidity of their environments: that darker colors occurred in more humid, tropical environments. In this case, while some physiological mechanisms have been proposed, they have received little support. Dark colors might be thought to facilitate basking, to absorb solar radiation in cold environments, and light colors might be thought to reflect solar radiation and prevent overheating in hot environments. But much radiant heat exchange occurs in the nonvisible wavelengths, and the thermally effective absorbance and reflectance of fur, feathers, scales, and other body surfaces is often not well correlated with their color. In fact, many animals tend to be white in cooler environments (e.g., polar bears, arctic foxes, arctic hares, and ptarmigans)—just the opposite of what we would expect based on this thermoregulatory argument. For some time, it was claimed that white or hollow fur may be advantageous in cold climates since the hairs might act as fiber-optic conductors of ultraviolet light toward the skin. Detailed analyses of the optical quality of hairs in polar bears, however, clearly demonstrate that this is not the case (Koon 1998).

For many predators such as mustelids, foxes, wolves, and bears exhibiting marked variation in color among their populations, the most likely explanation is selection for crypsis to avoid visual detection by predators or prey (e.g., see Rounds 1987). Humid climates are usually associated with dark soils and dense vegetation that casts deep shadows, whereas arid environments (hot or cold) often have light-colored soils and relatively sparse vegetation. The selective advantages of crypsis also explains variation in pelage color of many prey species, in this case decreasing their likelihood of being detected by visually hunting predators. Kettlewell's (1961) famous study of industrial melanism showed that moths matching the soot-darkened tree trunks and other surfaces in industrial areas were less likely to be discovered and eaten by birds. Less well publicized, but perhaps even more thorough and convincing, are the studies of Sumner (1932), Blair (1943), and Dice (1947) on the adaptive basis of geographic variation in coat color in deer mice (*Peromyscus maniculatus*). They made three key observations:

1. The pattern of geographic variation is such that dorsal pelage of local and regional populations tends to closely match background coloration in the environment.

2. Most of this variation is heritable, controlled by both simple Mendelian genes and more complex polygenic systems.

3. Individuals that do not match their backgrounds are selectively captured by predators (owls) in controlled experiments.

Crypsis, however, is likely not the only explanation for geographic variation in pelage color. A few species of endotherms exhibit geographic clines in pelage color that, although contrary to those predicted by Gloger, may indeed point to at least some thermoregulatory benefits. Color phases in squirrels, especially gray squirrels (*Sciurus carolinensis*), include melanistic individuals whose frequency varies with geographic clines in temperature and forest cover (**Figure 15.20**). At least in historic periods, melanistic forms of this squirrel dominated the more northern regions of its range. With the clearing of native forests of eastern North America, however, selective pres-

FIGURE 15.20 Gray squirrels (*Sciurus carolinensis*) exhibit a strong cline in fur color, which is the opposite of that predicted by Gloger's rule, but apparently is adaptive: Crypsis helps squirrels avoid predator detection, and darker fur assists with thermoregulation (the northern regions of their range being characterized by denser and darker forests, and colder climates). (Top photo © photoGartner/istock; bottom photo © Karla Caspari/istock.)

FIGURE 15.21 Global variation in skin color of indigenous peoples forms a pattern consistent with Gloger's rule—in this case, a strong cline of increasing reflectance (lighter skin) as we move from the Equator toward the poles. Skin color appears to be, in part, an adaptive response to geographic variation in the levels of ultraviolet radiation, which varies with latitude, altitude, precipitation, and humidity. The observed levels of skin pigmentation in local, indigenous populations may reflect a balance between the beneficial effects (mediation of production and storage of vital nutrients) and harmful effects (sunburn and skin cancer) of ultraviolet radiation. (After Chaplin 2004.)

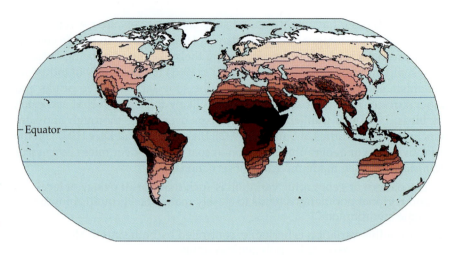

sures changed, and the frequency of melanistic squirrels declined across their range. Although the darker morphs may have been favored for their crypsis in closed-canopy, native forests, those in colder reaches of their range also benefited from enhanced thermoregulatory abilities via basking, increased heat production via **non-shivering thermogenesis**, and overall reductions in heat loss by 18 percent at −10° C (Ducharme et al. 1989; see Rapoport 1969 for a description of an analogous [inverse Gloger's] cline in pigmentation of springtails of South America).

We find the case study of melanistic gray squirrels especially instructive, as it suggests that the ecogeography of color (Gloger's trends or otherwise) remains a largely untapped source of future insights into the geographic variation in selective pressures, both natural and anthropogenic. Future studies that combine advances in spatial data analyses, genetics, thermoregulation, and techniques to objectively describe color variation in relevant visual and thermal spectra are likely to provide some intriguing insights (e.g., see Burt and Ichida 2004; Anderson et al. 2009). Our own species—essentially naked and thus lacking the protective advantages of fur, feathers, or scales—provides a special opportunity to study the geography of coloration. Chaplin's (2004) research is an exemplary case of utilizing the above analytical advances to develop a more comprehensive understanding of geographic variation and adaptive significance of skin coloration in indigenous peoples (**Figure 15.21**).

Ecogeography of the marine realm

As Li (2009) observed, the marine realm encompasses over 70 percent of the Earth's surface and more than 99 percent of its biosphere by volume. Despite the great challenges of studying the planet's last biological frontier, information on geographic variation in the marine realm rapidly accumulated during the latter decades of the nineteenth century to the point where biogeographers began describing some interesting ecogeographic patterns. We review three such patterns for marine biotas here, including Jordan's rule, Thorson's rule, and gradients in a suite of morphological characteristics summarized by Vermeij (1978) in his classic book *Biogeography and Adaptation: Patterns of Marine Life*.

Jordan's rule

In 1892, the American ichthyologist David Starr Jordan observed that the number of vertebrae of marine fish increases along a gradient from the warm waters of the tropics to cooler waters of the high latitudes (Jordan 1891;

Hubbs 1922). Like the other ecogeographic patterns described above, this one seemed so general that it became known as an ecogeographic rule—**Jordan's rule of vertebrae**.

While a comprehensive and systematic review of the generality of this ecogeographic pattern may still be lacking, it is clear that all species do not exhibit the pattern, and the pattern often varies depending on the particular geographic region or the habitats of the populations studied (**Figure 15.22**; McDowall 2003). Although the correlation of vertebrae number with latitude is often highly significant, it is clear that latitude per se cannot be the driving force for this pattern. Instead, ichthyologists and aquatic biogeographers attribute Jordan's rule to natural selection and adaptation of local populations to geographic variation in water temperatures (Lindsey 1975; Lindsey and Arnason 1981; Billerbeck et al. 1997; McDowall 2003). This ecogeographic pattern may also be related to Bergmann's rule. As we have observed for many terrestrial species, body size of many fish also tends to increase with latitude. These and other studies, such as that by Yamahira and Nishida (2009) on the medaka, or Japanese killifish (*Oryzias latipes*), provide additional evidence for the generality of Jordan's rule and its genetic basis, adaptive significance, and relationship to Bergmann's rule. An increased number of vertebrae is directly related to the capacity for rapid growth and large body size in fish inhabiting cold waters with relatively short growing seasons.

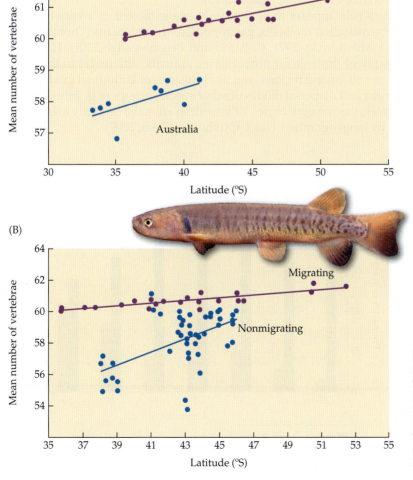

FIGURE 15.22 *Galaxias brevipinnis* is a salmoniform fish that clearly exhibits Jordan's rule: the tendency for number of vertebrae to increase with latitude. Note, however, that the pattern differs not just among species but also among (A) populations of species from different regions and those from different environments or (B) having different migration strategies, as well. (After McDowall 2003; photo courtesy Gunther Schmida.)

McDowall's (2003) study of salmoniform fish provides a tantalizing account of the pattern, demonstrating how it differs across regions and environments, and for species with different migration patterns (see Figure 15.22). This study provides an excellent example of how strategic deconstruction of a broad scale geographic pattern among regions and functional groups can provide especially important insights into the causal mechanisms of that pattern (see the discussion on deconstruction of the latitudinal gradient later in this chapter).

Thorson's rule

Marine invertebrates such as mollusks and gastropods have two principal modes of development: (1) their larvae may become planktonic and disperse with the currents away from their natal range, or (2) they may undergo direct development to adults in their natal home range. It appears that, because of these differences in dispersal abilities and the subsequent effects on isolation and speciation in allopatry, invertebrates with direct development (and limited dispersal abilities) are more likely to become isolated and, thus, exhibit relatively high speciation rates (see Astorga et al. 2003). While the causal link between dispersal mode and speciation rates requires further investigation, it may explain an ecogeographic pattern first reported by Thorson in the 1930s. Along a gradient of increasing latitude and decreasing water temperatures, the relative abundance of r- to K-strategists declines, with high-latitude waters being dominated by invertebrates that produce relatively few, but large, eggs and exhibit higher frequencies of viviparity, ovoviviparity, egg brooding, and direct development (versus pelagic, planktotrophic, widely dispersing forms in tropical waters; **Figure 15.23**).

Thorson's rule appears to be a general one exhibited in a variety of invertebrates, including mollusks, gastropods, and crustaceans (Thorson 1936, 1946, 1950; Mileikovsky 1971; Gallardo and Penchaszadeh 2001). However, the generality of this rule for other marine animals, differences in the trend among taxa and among regions, and its corollaries of latitudinal gradients in development and dispersal modes of other species groups (e.g., other marine invertebrates, marine vertebrates, or terrestrial animals) certainly merit more attention by biogeographers (see Laptikhovsky 2006, 2009).

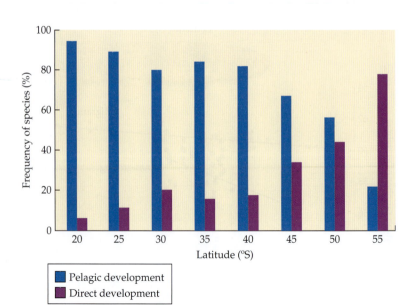

FIGURE 15.23 The principal mode of development in marine mollusks is strongly correlated with latitude, with direct development (those lacking planktonic larvae) being the predominant mode in the higher latitudes. (From Marquet et al. 2004; after Gallardo and Penchaszadeh 2001.)

Geographic gradients in predation, parasitism, and morphological defenses

According to Geerat Vermeij (1978: 108), "antipredatory defenses are developed to a greater degree and are found in an increasing proportion of species along a gradient from high to low latitudes. With qualifications, such gradients have been shown to exist in all the groups for which data are available: gastropods, bivalves, barnacles, sponges, holothurians, and marine sessile photosynthesizers." Vermeij based this on his painstaking examination of the shells and life-forms of many hundreds of specimens of these invertebrates. Shells from lower latitudes tend to be thicker, and more sculpted—often with a higher spire and more narrow aperture, which tends to be equipped with a thicker and more rugged lip and with a sturdier valve (**Figure 15.24**). In addition to comparing intact shells from a wide range of latitudes, Vermeij (1978; see also Vermeij 2004; Vermeij and Williams 2007) inspected shell fragments and damaged shells (i.e., those that were attacked by predators; see also Stachowicz and Hay 2000; Dietl and Kelley 2002; Dietl 2003; Kowalewski et al. 2005; Tomasovych 2008). By doing this, he was able to demonstrate that predation pressures also vary with latitude, being highest in the tropics where the abundance and diversity of predators (and most other types of species) is highest. Thus, Vermeij described what appears to be a general ecogeographic pattern and simultaneously provided a convincing causal explanation for that pattern, in this case linking gradients in morphology with those of interspecific interactions (predation) and species diversity. This hypothesis may well be expanded to include latitudinal gradients in defenses against parasites and infectious disease, which also exhibit strong latitudinal clines in diversity (see Guernier et al. 2004).

Vermeij's **predation escalation hypothesis** has recently been tested in the terrestrial realm. In seed plants, the size of seeds appears to be a measure of predation pressure in perhaps an analogous fashion to shell morphology in the marine realm. Thus, the tendency for seed size to increase from the poles to the Equator (**Figure 15.25**) appears consistent with Vermeij's hypothesis. It is not yet clear, however, whether the actual intensity of seed predation varies with latitude. Therefore, the seed-size–latitudinal gradient may result from some other phenomenon, such as latitudinal variation in climatic conditions, plant growth form, growing period, or seed dispersal. Some of these factors, especially precipitation, temperature, and their effects on soil development, may explain another latitudinal gradient in plant morphology. Although

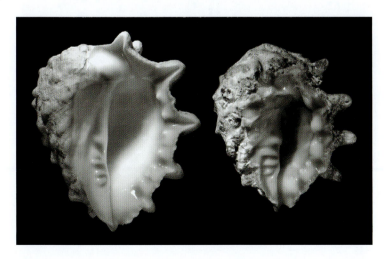

FIGURE 15.24 According to Vermeij's (1978) predation escalation hypothesis, selective pressures from the high diversity of predators in the tropical latitudes (or those from large vs. relatively small and isolated biotas) favor shell morphologies that are thicker and more sculpted, possessing more narrow apertures (often equipped with relatively thick and rugged lips) and sturdy valves; compare the shells for the marine gastropods *Purpura aperta* (left) and *Drupa. clathrata* (right), from low- and high-diversity environments, respectively. (Photo courtesy of G. Vermeij; see Vermeij 2005.)

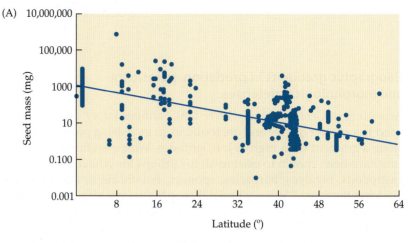

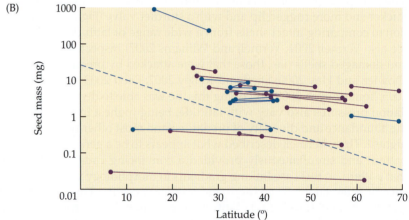

FIGURE 15.25 Seed size tends to increase from the poles to the Equator, both (A) among species and (B) within species. The two points shown for each species are the upper and lower latitudinal limits of the studies; the dashed line represents the slope of the global, multispecies relationship between latitude and seed mass. (After Moles and Westoby 2003.)

highly variable, rooting depth tends to be relatively deep in the tropics, and it decreases along a cline toward the higher latitudes (**Figure 15.26**).

Rensch's rule of geographic variation in life history characteristics

As we saw above, interesting patterns of geographic variation are not confined to morphological traits. There is an extensive literature on geographic variation in life history characteristics such as clutch size and age ratios in birds and other egg-laying animals (e.g., Moreau 1944; Lack 1947; Skutch 1949; Cody 1966; Moll and Legler 1971; MacArthur 1972; Iverson et al. 1993;

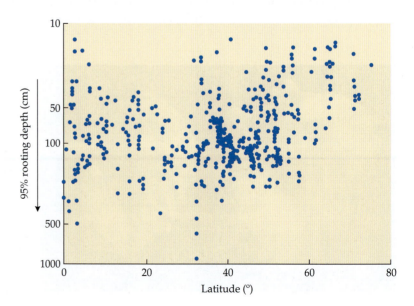

FIGURE 15.26 Rooting depths of plants tend to vary markedly throughout the tropical regions, but then they decrease with increasing latitudes through temperate and subarctic regions (rooting depth calculated as the depth above which 95 percent of the roots are found). (After Schenk and Jackson 2002.)

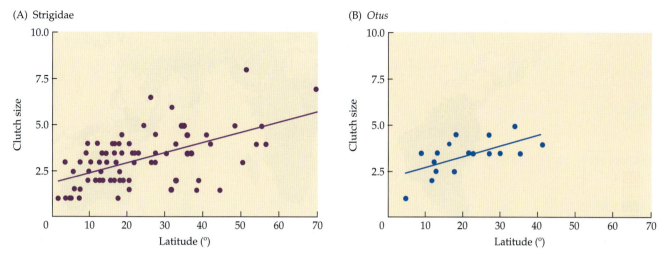

FIGURE 15.27 Clutch size as a function of latitude in owls. Such patterns have been found at three taxonomic scales: (A) among the entire owl family Strigidae; (B) among species of the genus *Otus*, which contains the screech owls and their relatives; and although not shown here, among populations of several widely distributed owl species. The strong positive relationships in these graphs are obvious, although the mechanisms underlying these general patterns are much debated (see also Figure 7.11). (After Cartron et al. 2000.)

Baker 1995; Graves 1997), and there are parallel trends of geographic variation in litter size of mammals (Lord 1960; Cockburn et al. 1983; Glazier 1985; Millar 1989; Cockburn 1991; Reinhardt 1997). Often, the observed patterns are related to latitude, with larger clutches or litters at higher latitudes being the general pattern (e.g., **Figures 15.27** and **15.28A**; see also Figure 7.11). Again, however, there is more consensus about the pattern than about its actual mechanism(s).

Several of the studies listed above suggest that clutch size is influenced by the combined effects of climate (in particular, length of the growing season), food availability, and interspecific interactions. For example, Cody (1966) explained geographic variation in clutch size within the framework of *r*-and *K*-selection (essentially, selection favoring higher population growth rates and productivity vs. selection favoring more efficient utilization of resources, respectively; see also MacArthur and Wilson 1967). Whereas clutch sizes of mainland birds in temperate regions tends to far exceed those of insular populations, mainland and insular clutch sizes are essentially equal in tropical regions (**Figure 15.28B**). Cody reasoned that birds inhabiting seasonally variable (temperate) mainland environments with short growing seasons should be subject to *r*-selection, producing relatively few, but large, clutches. Conversely, on islands of these regions, where local climates tend to be milder and less variable, selective pressures may shift toward more efficient use of available resources, and clutches of fewer—but higher-quality—offspring (i.e., *K*-selection; see Cody 1966; Isenmann 1982). Other studies have reported the same trend for insular salamanders (Anderson 1960) and for microtine rodents (Tamarin 1977; Gliwicz 1980; Ebenhard 1988a,b). These patterns for insular systems may be analogous to those associated with Thorson's rule in marine invertebrates, where life history strategies compensate for the relatively short growing seasons of high-latitude waters by exhibiting more rapid rates of growth and direct development (Lonsdale and Levington 1985; see also Endler 1977; Vermeij 1978). Others have offered alternatives to Cody's explanation for geographic variation in clutch and litter size. For the most part, these hypotheses are based on trade-offs associated with adapting to climatic seasonality, risks of migration, available energy, seasonal pulses in productivity, and risk of predation (e.g., see Lack 1947; Skutch 1949, 1967; Ashmole 1963; Lack and Moreau 1965; Flux 1969; Charnov and Krebs 1974; Dunn et al. 2000; McNamara et al. 2008). As is often the case, these hypotheses are not mutually exclusive, and some combination of them may be necessary to explain the observed geographic clines in clutch and litter size within any particular species, as well as the pattern for terrestrial vertebrates, in general.

(A)

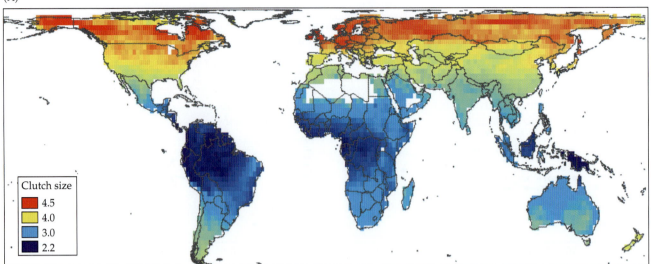

Clutch size
- 4.5
- 4.0
- 3.0
- 2.2

(B)

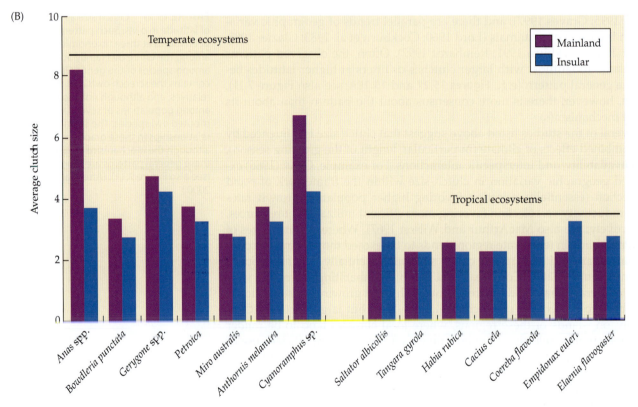

Temperate ecosystems

Tropical ecosystems

Mainland
Insular

Average clutch size

Anas spp.
Bowdleria punctata
Gerygone spp.
Petroica
Miro australis
Anthornis melanura
Cyanoramphus sp.

Saltator albicollis
Tangara gyrola
Habia rubica
Cacius cela
Coereba flaveola
Empidonax euleri
Elaenia flavogaster

FIGURE 15.28 (A) Ecogeographic patterns include, not just those of morphological variation, but also traits such as geographic variation in life history characteristics, including avian clutch size (the number of egg per nest). The empirical patterns illustrated in this map (based on analyses of clutch size across 5290 species, or 56 percent of the world's land birds) confirm observations of early ornithologists such as Rensch (1938), Moreau (1944), and Lack (1947) (see Figures 7.11 and 15.27). (B) Clutch size (number of eggs per nest) in birds tends to vary among geographic regions and, at least in temperate regions, between mainland and insular ecosystems as well. These patterns likely reflect the combined effects of differences in climatic conditions, growing seasons, and diversity of ecological communities on life history traits across geographic regions and types of ecosystems (i.e., mainland versus insular). (A from Jetz et al. 2008; B from MacArthur and Wilson 1967, Cody 1966.)

We find the resurgence of interest and research programs on this and other ecogeographic patterns especially encouraging, and despite the diversity of traits and geographic gradients comprising these patterns, it appears that a more integrative synthesis is now within reach. As Gaston and his colleagues suggest (Gaston et al. 2008), strategies for a more integrative theory of ecogeography should include careful attention to variation in patterns, not just across taxa and geographical gradients, but among levels of biotic organization (intraspecific, interspecific, and assemblage level as well). In addition, the scale dependence, patterns of covariation, and interactions among processes functioning at each of those levels appear to be essential elements to include in this emerging synthesis (see also Lomolino et al. 2006). One final note, or really a question raised by one of our colleagues—Vicki Funk—during a conference on ecogeographic rules held at the 2007 meeting of the International Biogeography Society: Is all this simply a zoological phenomenon? Perhaps more pointedly put, why are nearly all ecogeographic patterns discussed here (save for our mention of geographic variation in seed size and root depth) the domain of zoologists? Here again, we identify a glaring gap in our knowledge, which we offer as a challenge to biogeographers keen on advancing the frontiers of this field.

The Geography of Biotic Diversity

In this section, we expand our focus from patterns of geographic variation in the characteristics of individuals, populations, and species to now consider the geography of entire communities and biotas. As we observed in Chapter 2, comparison of community structures (in particular, the number and types of co-occurring species) among geographic areas has been a persistent theme of biogeography, and one shared with its descendant discipline of community ecology. Indeed, Buffon's law—which proved fundamental to both biogeography and evolution—describes the differences in species composition of regional biotas, including those with similar environmental characteristics but different histories of place and lineages. In Chapter 5 we described the very regular, geographic distributions of types of communities (including biomes and ecoregions) and their dynamic characteristics, including variation in primary productivity among those communities and across geographic gradients of continental and oceanic realms.

Here, we focus on one of the simplest yet most informative indices of community structure—species richness, which has attracted the attention of biogeographers at least since Europeans first developed a global-scale understanding of the natural world. Indeed, when Buffon was describing the discipline's first law, he and other biogeographers of the eighteenth century were discovering an equally compelling signal in the geography of communities—that the number of species found in local and regional biotas varied in a regular manner across the globe. Simultaneously and perhaps irresistibly, those early biogeographers and community ecologists also proposed causal explanations for those patterns in diversity among regional biotas. To their credit, elements of these early, causal explanations can be found in modern theories for the geography of diversity, which we summarize in the final section of this chapter.

Diversity measures and terminology

Species richness and diversity indexes

Despite the variety of measures used to describe the structural characteristics of biological communities and regional biotas (including biomass, trophic web structure, and variation in population levels among species), much if not most research conducted throughout the history of biogeography and community ecology has focused on one index of community structure—**species richness** (the number of species in a sample from a local area or geographic region). In many cases, the rare species may be as interesting and important as the common ones; in fact, many taxa of interest to both historical and ecological biogeographers are both highly restricted in their distributions and uncommon where they do occur (see Chapters 4, 10, and 11). To compare the similarity in biotas among regions, it is often most useful—as well as most convenient—simply to compare lists of species, which can usually be obtained from systematic works, and faunal and floral surveys. Assuming that the samples are relatively complete or that the sampling efforts for each region are at least comparable, species counts (or statistical estimates of the total richness) can provide simple and relatively unambiguous measures of the biological diversity and distinctiveness of different geographic areas.

Although rare species may be important components of ecological systems (see Gaston 1994; Kunin and Gaston 1997), ecologists often deemphasize them and focus on species that dominate the community in terms of abundance, biomass, energy use, cover, or some other estimate of "importance." Consequently, several measures, or indexes, of species diversity have been developed to give greater weight to the dominant species. Most of these measures of community complexity are based on the uncertainty principle, which, as applied by ecologists, asks the following question: Given that an individual is to be drawn randomly from a community or assemblage of species and individuals, what is the likelihood that we can correctly guess the identity of that randomly selective individual? Information theory (developed, in part, to decipher codes and other encrypted information) has proven very useful in this application, with the identity of a particular element or letter in a message being analogous to that of a species in a community (Shannon and Weaver 1949). The toughest codes to crack (i.e., those with the greatest uncertainty) are those that have the greatest number of elements (letters) and those whose elements are used in more equal frequencies. Similarly, most ecologists and biogeographers understand that local diversity increases with both the number of species in the community and the evenness or degree to which population levels of these species are the same.

Many diversity indexes have been proposed, and several are used frequently in the ecological literature. These indexes are defined, and their uses, strengths, and weaknesses discussed, in several reviews (e.g., Pielou 1975; Whittaker 1977; Ludwig and Reynolds 1988; Rosenzweig 1995; Whittaker et al. 2001; Maclaurin and Sterelny 2008; Magurran 2009). Although a thorough understanding of these indexes is important for interpreting many ecological studies, a detailed treatment of them is not necessary here. For our purposes, it is sufficient to know that, in most natural communities, nearly all of the approximately 20 indexes commonly used to quantify species diversity are highly correlated with species richness. For this reason, we will use the term *species diversity* somewhat loosely, and synonymously with *species richness*.

Scales of diversity: Alpha, beta, gamma, and delta

Like most other ecological patterns, species diversity varies with the spatial scale on which it is studied. The continuum of spatial scales of diversity is commonly divided into four convenient categories—two (alpha and beta) at relatively local scales, and two (delta and gamma) at regional to global scales (Cody 1975; Whittaker 1975, 1977; Wilson and Schmida 1984; Magurran 1988; Ricklefs and Schluter 1993). **Alpha diversity** refers to the species richness of a local ecological community—that is, the number of species recorded within some standardized area, such as a hectare, a square kilometer, or some naturally delineated patch of habitat. **Beta diversity** refers to the change (or turnover) in species composition over a relatively small distance, often between recognizably different but adjacent habitats (see Wilson and Schmida 1984; Ricklefs and Schluter 1993; Koleff et al. 2003; Jost 2007). Not surprisingly, beta diversity is often found to be a direct function of topographic (e.g., altitudinal) variation and corresponding dissimilarity in environmental conditions among adjacent sites (**Figure 15.29**). Consistent with Janzen's (1967) hypothesis that "mountain passes are higher" (niches and ranges are more restricted)

(A) Birds (3836 species)

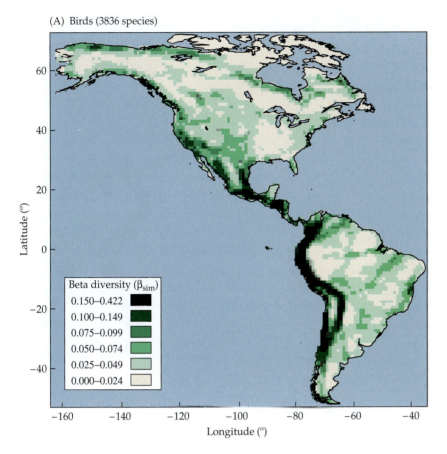

(B) Mammals (1641 species)

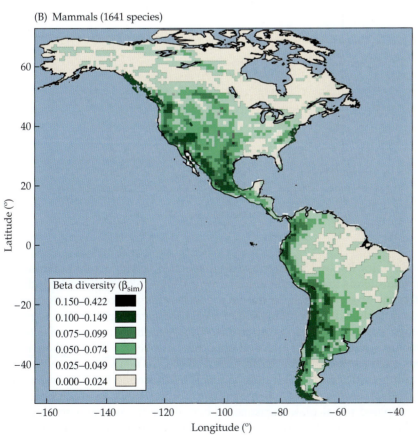

FIGURE 15.29 Beta diversity is a measure of the turnover, or differences in species composition among sites, in this case for Western Hemisphere bird (A) and mammal (B) assemblages (in focal, 1-degree latitude-by-longitude cell in comparison with that of its adjacent cells). Beta diversity appears to be a direct function of elevational (and associated environmental) variation among adjacent cells (C). For North American native and exotic plants (E and F) as well as for these vertebrates (A, B, and D), beta diversity tends to decline with increasing latitude. (A–C after Melo et al. 2009; D after Qian et al. 2009; E after Qian 2009; F after Qian 2008; see also Qian and Ricklefs 2007.)

FIGURE 15.29 (continued)

(C) Topographic variation

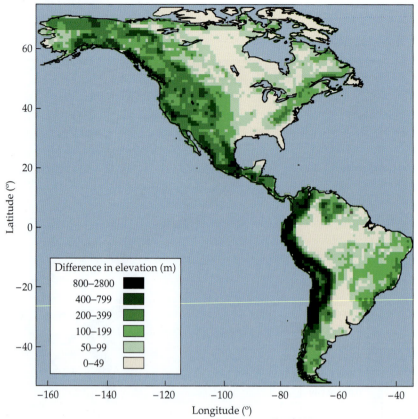

(D) North American mammals

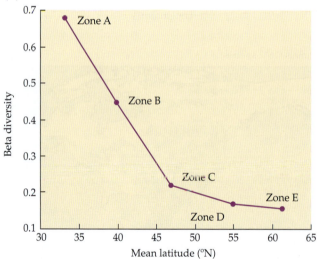

in the tropics, Figure 15.29 also reveals that beta diversity of plants, birds, and mammals tends to be greater among tropical sites than at higher latitudes.

Gamma diversity refers to the total species richness of large geographic areas ranging from a combination of local ecological communities to entire biomes, continents, and ocean basins. Thus, gamma diversity reflects the combined influences of alpha and beta diversity, so it will be highest in regions such as the Amazon basin, where there are many species within local communities and a high turnover of species among habitats and across the landscapes. Finally, just as species composition of local communities can be compared to calculate measures of beta diversity, we can compare species lists for large geographic areas to calculate an analogous, broad scale

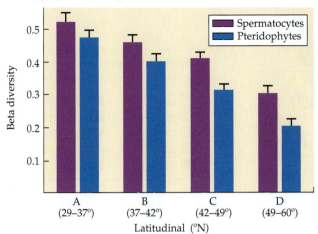

(E) Native flowering plants (Spermatocytes) and cryptogams (Pteridophytes)

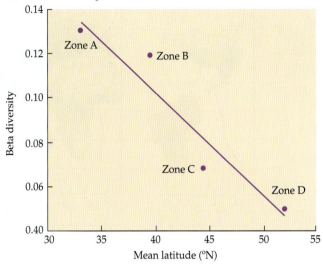

(F) Exotic vascular plants

measure of dissimilarity—**delta diversity**. Among these four measures, delta diversity may be the most fundamental to biogeography and evolution, as it describes the differences in biogeographic regions that were central to Buffon's law (see Chapter 2) and challenged generations of scientists to explain how these broad scale differences in biotas developed. As we have thoroughly discussed diversification among biotas in Chapters 10–12, our focus here will be on geographic variation in alpha and gamma diversity (i.e., species richness at local and regional scales).

The primary data available for numerous taxa and a variety of geographic regions are species counts for small areas. These counts usually are published either by systematists describing the results of biotic surveys or by ecologists summarizing samples of local habitats. An alternative and complementary method is to map the distributions of all species of a particular taxon, and then count the number of geographic ranges that overlap a particular point along some geographic gradient (e.g., latitude, elevation, or depth). Because species lists and samples typically underestimate, to varying degrees, the actual number of species inhabiting particular areas, ecologists have developed statistical methods of standardizing and comparing species richness from samples of different intensity (e.g., see methods available in ecologists' software such as EcoSim©, Acquired Intelligence 1987–2009; see also Gotelli and Colwell 2001). Even when using direct counts to estimate species richness, in order to justify comparisons along geographic gradients or among communities and regions, diversity should be calculated as **species density** (the number of species per standardized sample effort and area).

The latitudinal gradient in species diversity

One of the most striking characteristics of life on Earth is the gradient of increasing species diversity from the poles to the Equator. The earliest explorers/naturalists noticed that the tropics teem with life, the temperate zones have fewer kinds of animals and plants, and the Arctic and Antarctic are stark and barren by comparison (Forster 1778; Humboldt 1808 [see Hawkins 2001]; Wallace 1878; see Chapter 2). This pattern is a very general one in that it holds true, not only for diversity of all organisms combined, but also for most major taxa (classes, orders, and families) of microbes, plants, and animals and for those inhabiting most of the major types of ecosystems in the terrestrial and aquatic realms (**Figure 15.30**; see also Figure 3.25A).

FIGURE 15.29 (*continued*)

(A) All vascular plants

FIGURE 15.30 A–D

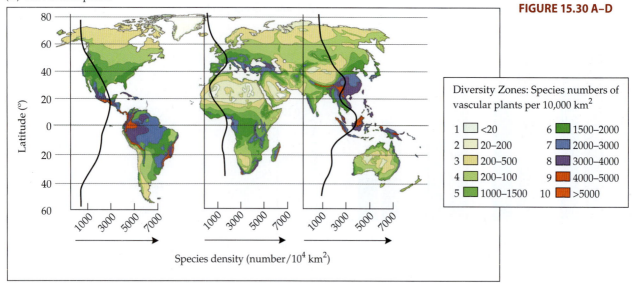

Diversity Zones: Species numbers of vascular plants per 10,000 km^2

1	<20	6	1500–2000
2	20–200	7	2000–3000
3	200–500	8	3000–4000
4	200–100	9	4000–5000
5	1000–1500	10	>5000

Species density (number/10^4 km^2)

(B) Gymnosperms

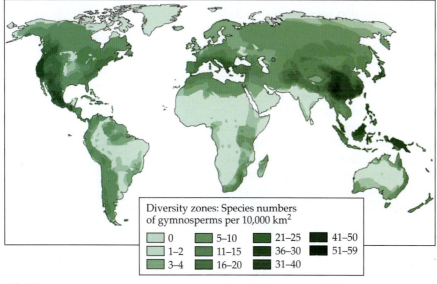

Diversity zones: Species numbers of gymnosperms per 10,000 km^2

0	5–10	21–25	41–50
1–2	11–15	36–30	51–59
3–4	16–20	31–40	

(C) Cacti

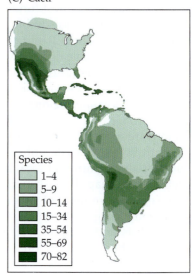

Species
- 1–4
- 5–9
- 10–14
- 15–34
- 35–54
- 55–69
- 70–82

(D) Mosses

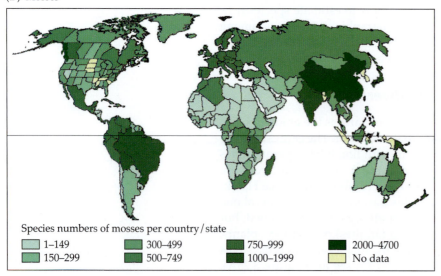

Species numbers of mosses per country/state

1–149	300–499	750–999	2000–4700
150–299	500–749	1000–1999	No data

(E) Terrestrial and marine mammals

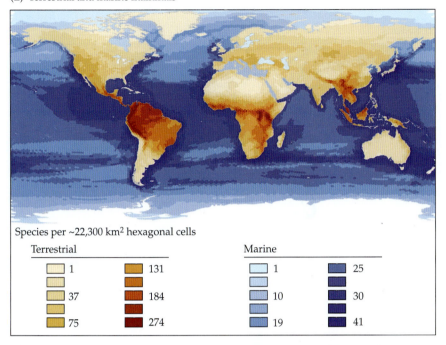

Species per ~22,300 km² hexagonal cells

Terrestrial		Marine	
1	131	1	25
37	184	10	30
75	274	19	41

FIGURE 15.30 The latitudinal gradient in species density (number of species per standardized area of study) is one of biogeography's most general and long-studied patterns. However, these global-scale maps reveal that although the general signal is consistent across a diversity of taxa (with highest diversity in the tropics), other features of the pattern vary substantially among regions and among taxonomic groups, reflecting the influence of other geographic factors, such as topography, bathymetry, and relative position from the coasts or along peninsulas. (A–D after Mutke and Barthlott 2005a,b; E after Schipper et al. 2008; F–I courtesy of John Gittleman and Wes Sechrest; J after Hawkins et al. 2007a; K courtesy of Kurt A. Buhlmann, Savannah River Ecology Laboratory, University of Georgia and the IUCN-SSC and CI/CABS Global Reptile Assessment; L courtesy of Shai Meiri and the Global Assessment of Reptile Distributions workshop, http://www3.imperial.ac.uk/cpb/workshops/globalassessmentofreptiledistributions; M after Buckley and Jetz 2007; N–P from IMAPS World Atlas of Biodiversity, http://bure.unep-wcmc.org/imaps/gb2002/book/viewer.htm, N prepared based on data from Berra 2001, P prepared based on data from Vernon 1995; Q prepared based on data from Spalding 1998; R after Stehli 1968.)

(F) Bats

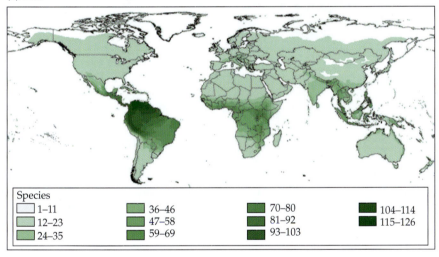

Species			
1–11	36–46	70–80	104–114
12–23	47–58	81–92	115–126
24–35	59–69	93–103	

(G) Primates

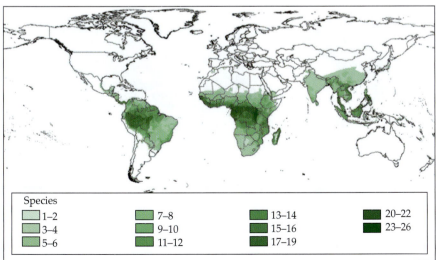

Species			
1–2	7–8	13–14	20–22
3–4	9–10	15–16	23–26
5–6	11–12	17–19	

FIGURE 15.30 (*continued*)   (H) Rodents

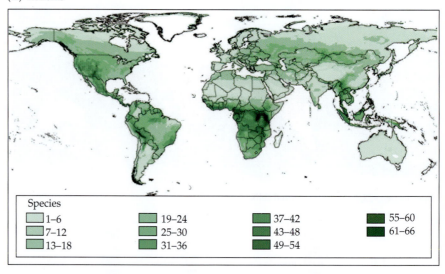

(I) Marsupials

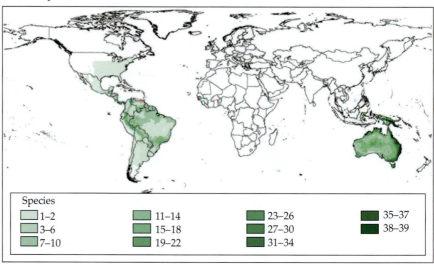

(J) Birds

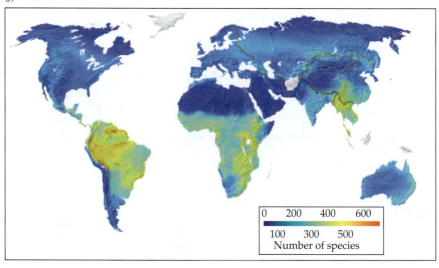

(K) Freshwater turtles

FIGURE 15.30 (*continued*)

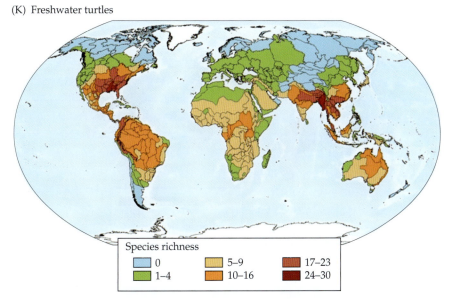

Species richness

0	5–9	17–23
1–4	10–16	24–30

(L) Lizards

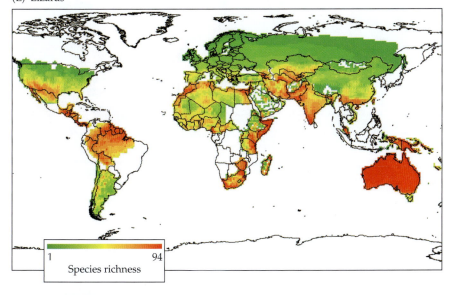

1   94
Species richness

(M) Amphibians

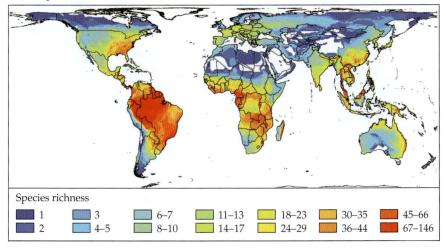

Species richness

1	3	6–7	11–13	18–23	30–35	45–66
2	4–5	8–10	14–17	24–29	36–44	67–146

FIGURE 15.30 (*continued*) (N) Freshwater fish family diversity

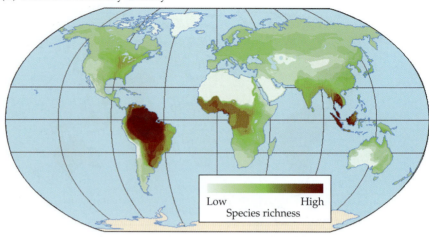

(O) Seagrass

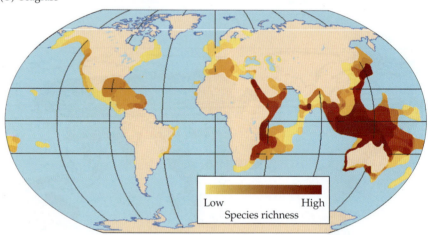

(P) Corals

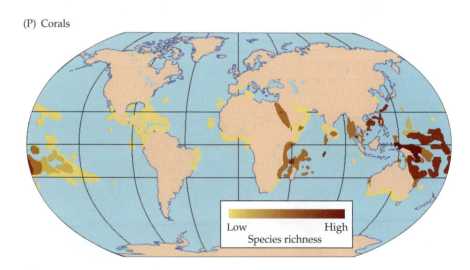

In addition, the latitudinal gradient in diversity appears to be an ancient pattern, although one varying not just among taxa but also within taxa over time. Crame and Lidgard (1989) estimate that there has been a latitudinal gradient of land plant diversity for at least 110 million years. The fossil re-

(Q) Mangroves

FIGURE 15.30 (*continued*)

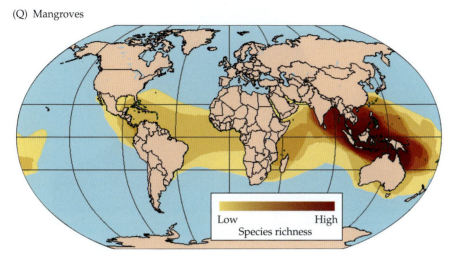

(R) Planktonic foraminifera

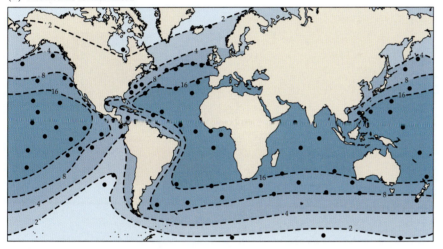

cord for marine animals, such as brachiopods, corals, and single-celled fora-minifera and diatoms, also indicates that the latitudinal gradient in species diversity in the marine realm has been in existence for more than 250 million years in brachiopods and 70 million years in planktonic foraminifera (Stehli et al. 1969). One especially interesting feature of this pattern is that, at least for terrestrial plants and for marine bivalves, the latitudinal gradient has steadily intensified over time (**Figure 15.31**; Crame 2004). According to Cra-me (2001, 2002, 2004), this may have resulted from global cooling following the climatic optimum of the Early Eocene (the so-called Mid Eocene sauna), which marked a period with relatively modest latitudinal variation in ocean temperatures. This cooling appears to have been caused by tectonic events and associated changes in insolation and water currents. Among many other changes, this caused abrupt declines in mean oceanic temperatures by about 2° C at around 45 million years BP, 5° C at around 37 million years BP, and 4° C at around 14 million years BP; all this was followed by more gradual cool-ing thereafter. As a result of simultaneous changes in oceanic currents, latitu-dinal gradients in water temperatures intensified, polar ice caps formed, and the tropical zone contracted as thermally defined climatic zones and marine provinces became more firmly established.

Perhaps just as remarkable as these long-term dynamics in diversity gra-dients is the observation that, at least for some taxa, latitudinal gradients

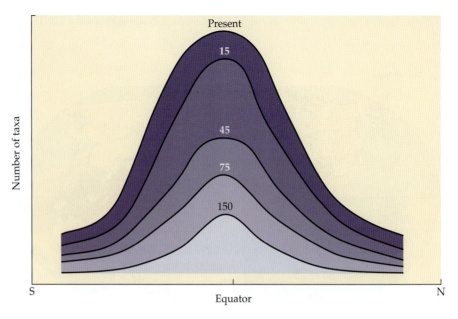

FIGURE 15.31 Latitudinal gradients of species diversity are ancient patterns but, as illustrated here for the marine realm, have intensified over geological time (from 150 million years BP to present), especially during the last 30 million years, when global cooling and altered patterns of ocean circulation intensified latitudinal gradients in temperature. (After Crame 2004.)

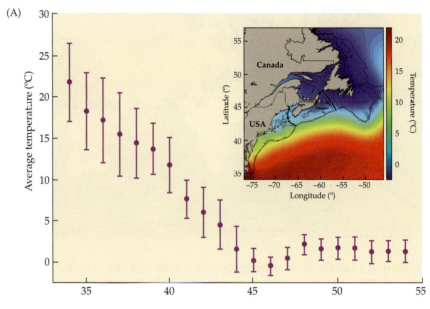

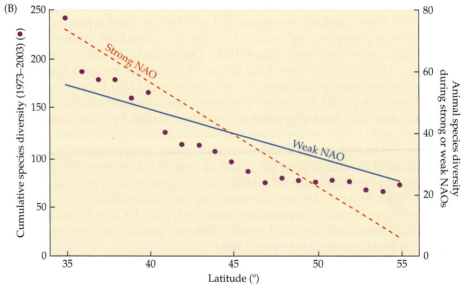

FIGURE 15.32 Latitudinal gradients in species diversity may be surprisingly dynamic over relatively short time periods. For example, the latitudinal gradient in species richness of marine fish off the coasts of northeastern North America is strongly influenced by the intensity of a dominant climatic phenomenon of that region—the North Atlantic Oscillation (NAO; vertical bars in A indicate interannual, standard deviation in temperatures). During years of strong NAO events, latitudinal gradients in water temperatures are intensified and, as a result, so are the gradients of species richness of these marine fish (right-hand axis of graph B). (After Fisher et al. 2008.)

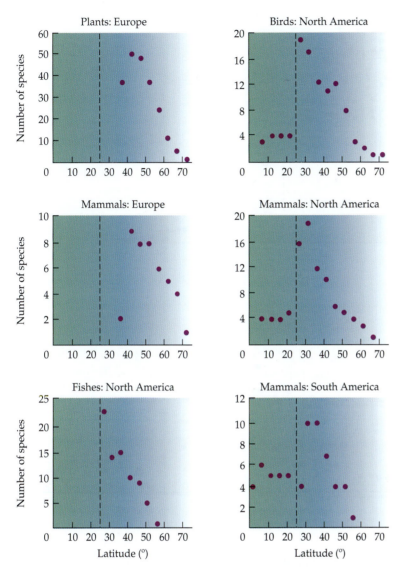

FIGURE 15.33 Species richness gradients for naturalized (established) exotic species of plants and animals in Europe, North America, and South America are similar in many respects to those of native biotas. For some groups of species, however, the gradient appears most prevalent outside of the tropics (dashed line indicates either the Tropic of Cancer or the Tropic of Capricorn; see discussion on the ecology and geography of species invasions in Chapter 16). (After Sax 2001.)

in diversity may vary markedly on an annual basis (**Figure 15.32**) and may become established in an incredibly short amount of time. For example, Sax (2001) reports that within a span of decades to just a few centuries, naturalized (established) exotic species of Europe and North America now exhibit significant latitudinal trends in diversity (**Figure 15.33**; see also Stohlgren et al. 2005; Fridley et al. 2006; Pyšek and Richardson 2006; Fridley 2008).

Nature and complexity of the pattern

Beginning with classic studies by Alfred Fischer (1960) and George Simpson (1964), several authors have quantified latitudinal diversity gradients (for important reviews, see Fischer 1960; Pianka 1966; Stehli 1968; Stehli et al. 1969; Kiester 1971; Stehli and Wells 1971; Arnold 1972; Buzas 1972; Scriber 1973; Wilson 1974; Horn and Allen 1978; Silvertown 1985; Rohde 1992; Stuart and Rex 1994; Rosenzweig 1995; Roy et al. 1998; Willig 2000; Willig et al. 2003; Brown and Sax 2004; Crame 2004; Hillebrand 2004; Turner and Hawkins 2004). For groups of organisms whose distributions are well known, this can be done by counting the species in local areas of approximately equal size, and then graphing these data (species density) as a function of latitude of collection sites.

FIGURE 15.34 When plotted on graphs such as these for (A) South American bats and (B) birds of the world, the latitudinal gradient in species density typically exhibits relatively high, but variable, levels through the tropics, then a rapid decline through subtropical regions, and lowest levels in the higher latitudes. (A after Willig and Selcer 1989; B after Turner and Hawkins 2004.)

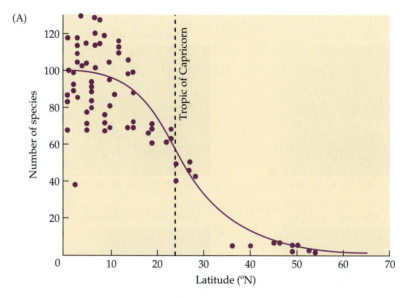

(A)

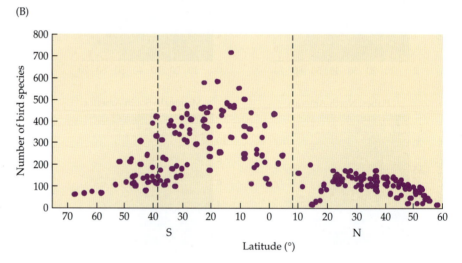

(B)

Graphs of latitudinal gradients in diversity, such as those of **Figure 15.34** and Figure 15.30A have at least two advantages over maps of species density such as those of Figure 15.30. First, although the latter provide a more comprehensive description of the geography of diversity (i.e., variation over a combination of geographic gradients), they of course require much more information. Second, and perhaps most relevant to modern research on, not just patterns, but underlying causes for geographic variation in biological diversity, gradient analyses often provide a much more efficient means of identifying otherwise hidden features of, or features that may be considered secondary to, the general patterns. For example, rather than exhibiting a continuous decline in species density from the Equator to the poles, most taxa exhibit a pattern of relatively high, albeit variable, diversity in the tropics, marked by a rapid decline through the subtropics and much more modest declines through the higher latitudes (see Figure 15.34). Other secondary patterns emerging from these gradient analyses include that the latitudes corresponding to peaks in species diversity vary among regions for the same taxa (e.g., for plants across latitudinal gradients of the Eastern versus Western Hemispheres—see Figure 15.30A) and among taxa as well. In the latter case, while still exhibiting a unimodal latitudinal pattern, diversity of some taxa may actually peak outside the tropics. In addition to those illustrated in **Figure 15.35**, exceptional cases of taxa that exhibit extratropical peaks in diversity include sawflies (Hymenoptera, Symphyta: Kouki et al. 1994), aphids (Dixon et al. 1987), braconid wasps (Janzen 1981; Gauld 1986; Gauld et al. 1992), willows (Myklestad and Birks 1993), helminth parasites of marine mammals (Rohde 1992; but see Rohde 1999), communities of water-filled pitcher plants (Buckley et al. 2003), and birds of peatlands in Finland (Kouki 1999) and eastern deciduous forests of North America (Rabenold 1979). These secondary or exceptional patterns are of special interest to biogeographers, not because they defy the general pattern, but because they may represent responses to long-term and broad scale natural experiments—variation in the characteristics of regions and lineages that influence diversification of biotas.

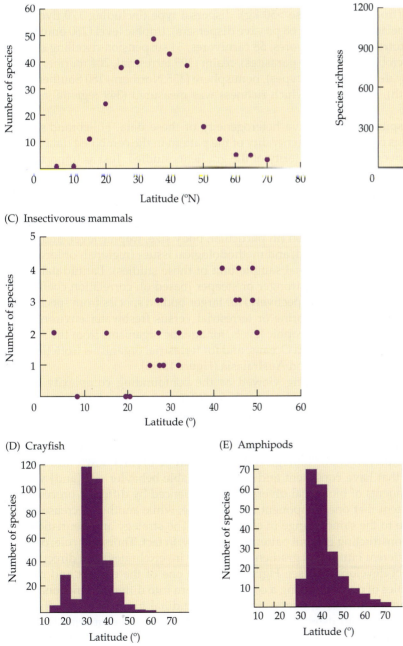

FIGURE 15.35 Despite the impressive generality of the latitudinal gradient in species density, the pattern is not without exceptions. (A) One exceptional group that has been particularly well studied is the plant family Pinaceae—the conifers (pines, spruces, firs, and their relatives). These plants are restricted to the Northern Hemisphere and achieve their highest diversity at midlatitudes in both North America and Eurasia. (B) The Ichneumonidae, a large family of parasitic wasps, are also most diverse in temperate regions, with the greatest species richness in North America occurring at about 40° N latitude. (C) The relationship between species richness of small insectivorous mammals (shrews and moles) and latitude is statistically significant, but anomalous, with diversity for this group peaking in the temperate latitudes. (D and E) Species richness of crayfish and amphipods tends to peak, not at the Equator, but in the subtropical latitudes of the Nearctic Region. (A after Stevens and Enquist 1998; B after Owen and Owen 1974, Janzen 1981; see also Hespenheide 1978; C after Cotgreave and Stockley 1994; D after Stevens 1989; E after France 1992.)

Among the most recent and promising new directions in developing a more thorough understanding of this pattern and its general causes are rigorous statistical analyses across many taxa, and analyses that strategically "deconstruct" complex patterns into those exhibited by functionally different groups of species. Using the former of these approaches, Hillebrand (2004) reviewed nearly 600 studies of latitudinal gradients to statistically assess the generality of this pattern and to explore how the pattern differed among species groups, types of ecosystems, and geographic regions. He applied the methods of meta-analysis (Gurevitch and Hedges 1993; Rosenberg et al. 2000; Rosenberg 2010), which is a rigorous approach for analyzing the results of a collection of different studies and testing for the statistical relationships among variables that influence those results. Hillebrand's data included statistical results from previous analyses of latitudinal gradients for a very diverse assortment of species and environments, including those varying in thermoregulatory strategies (440 ectotherms, 139 endotherms), body mass (from <1 g to those >50 kg), dispersal type (including 170 flyers, 119 with pelagic larvae, and 68 passive dispersers), trophic level (280 omnivores, 87 autotrophs, 65 herbivores, 58 carnivores, 34 suspension-dwelling species, 41 microbivores, and 16 parasites), realm (305 terrestrial, 204 marine, and 69 freshwater), habitat (16 types), hemisphere (335 Northern, 180 Southern, and 64 both), and scale at which richness was measured (349 regional and 222 local).

Despite the impressive heterogeneity of these data, Hillebrand's meta-analysis confirmed the generality of the pattern in the combined data set and in an overwhelming majority of the gradients studied. Just as important, he demonstrated that the strength (correlation coefficient) and the slope of the gradient varied in a systematic manner among the types of species, habitats, and regions studied. That is, *the* pattern is actually many patterns, with the same general signal but varying to different degrees depending on characteristics of the focal taxon (in particular, body size, trophic level, and global richness), region studied, and methodological issues (regional- or local-scale studies; grain, or minimal sample unit of those studies). Latitudinal gradients in diversity were stronger or steeper (based on correlation coefficients or regression slopes, respectively) for larger-bodied species from species-rich taxa, for those from marine or terrestrial versus freshwater environments, for those at the higher trophic levels, for aquatic organisms from the Atlantic versus Pacific and Indian Oceans, and for terrestrial organisms from the New World versus Eurasia and Australasia (**Figure 15.36**).

Rather than just being viewed as the confounding complexity of this large and heterogeneous data set, these differences among functional groups may provide important clues into the causal mechanisms involved. In fact, Hillebrand's analysis is an exemplary case study in *deconstructing* a pattern to explore the underlying causal mechanisms for that pattern. According to Huston (1994: 2), "biological diversity can be broken down into components that have consistent and understandable behavior . . . [and] various components of biological diversity are influenced by different processes, to the extent that one component may increase, while another decreases in response to the same change in conditions." The strategy and heuristic value of deconstructing general patterns is not new. In fact, Thorson's rule, describing different latitudinal gradients in marine invertebrates with direct versus indirect development (see Figure 15.23), is one of the earliest and clearest examples of deconstructing a complex pattern into different functional components (in this case, suggesting that latitudinal gradients are influenced by dispersal and its effects on isolation and speciation of marine lineages). Similarly, gradients in species diversity differ for mammals of different body

(A) Body size, richness scale, habitat

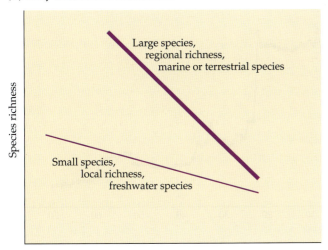

(B) Taxon diversity, grain size

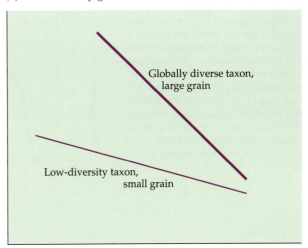

(C) Ocean, New/Old World, trophic level

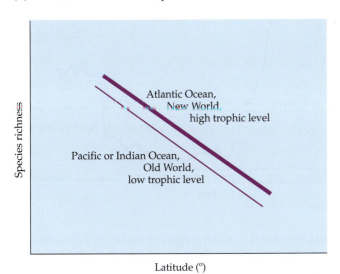

(D) Dispersal mode, thermoregulation,
Northern/Southern Hemisphere

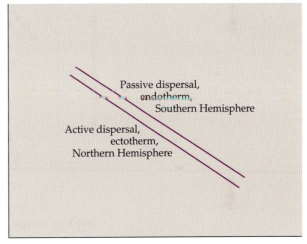

FIGURE 15.36 Hillebrand's (2004) meta-analysis of nearly 600 studies of latitudinal gradients in species diversity not only verified the generality of the pattern but revealed that it varied among functionally different groups of species, geographic regions, and methodological approaches. These graphs summarize the effects of various factors on the steepness and strength of the latitudinal gradient in species richness (as measured by the slopes and correlation coefficients of the relationship; thick lines indicate stronger relationships): The four charts identify factors that have qualitatively different effects on the latitudinal gradient in species diversity, including those that have significant effects (A) on both the slope and strength, (B) on the slope, (C) on the strength, and (D) on neither the slope nor strength. (After Hillebrand 2004.)

size and feeding guilds (Andrews and O'Brien 2000), for carnivorous versus noncarnivorous marine gastropods (**Figure 15.37**), for marine bivalves that burrow into the substratum versus those that live along its surface (infaunal vs. epifaunal species [Roy et al. 2004]), and for groups of plants with different growth forms (Bhattarai and Vetaas 2003).

Marquet et al. (2004: 192) use the term **deconstruction** in its etymological sense, meaning "a 'turning to the roots' of what is being measured, and to 'disaggregate' in order to make apparent what is hidden." They go on to list four reasons for deconstructing patterns in species richness:

1. To understand the underlying, causal mechanism

2. To reconcile seemingly disparate explanations for general patterns in diversity

3. To emphasize the importance of overcoming methodological challenges of analyzing heterogeneous data sets

4. To restate complex questions of biological diversity in a more tractable, comparative framework

FIGURE 15.37 (A) Although the latitudinal gradient of species richness of marine gastropods in the Pacific Ocean follows the general trend, the pattern differs substantially for species with different feeding strategies. (B) In particular, the relative diversity of noncarnivores (measured as the ratio of carnivores to noncarnivores) is highest in the midlatitudes, while that of carnivores is highest in the tropics and, to a lesser degree, in the waters of the higher latitudes. (From Roy et al. 2004; after Valentine et al. 2002.)

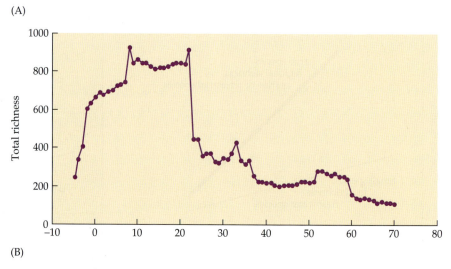

(A)

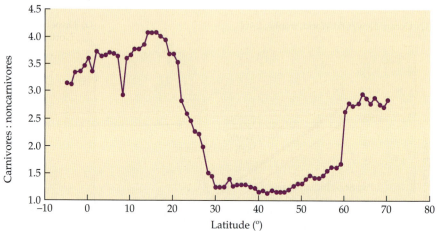

(B)

One key feature of this approach is to strategically separate (deconstruct) heterogeneous data sets into smaller subsets that are internally homogeneous but that differ among each other with respect to putative causal factors and their associated processes. Rather than simply exploring the pattern(s), we would chose subgroups based on the hypothesized underlying causal explanations, selecting groups that differ markedly in their responses to the hypotheses' central factors or processes (see Roy et al. 2000). For example, for evaluation of a hypothesis that is based on energy use, subgroups should be composed of species that are similar in their trophic and energetic characteristics, with different subgroups representing disparate strategies for acquiring or possessing energy (e.g., endotherms vs. ectotherms, carnivores vs. herbivores, or large vs. small species). On the other hand, if we wish to test a hypothesis that latitudinal gradients are ultimately driven by differences in temperature, solar energy, or precipitation, then gradients for terrestrial and shallow water organisms might be compared with those inhabiting the abyssal depths, which are invariably cold, dark, and certainly not water limited. In still other cases, we can compare latitudinal gradients for the same species groups, but across regions with different environmental characteristics (e.g., those with very different climates, currents, and sea surface temperatures; **Figure 15.38**) or different histories (geological age).

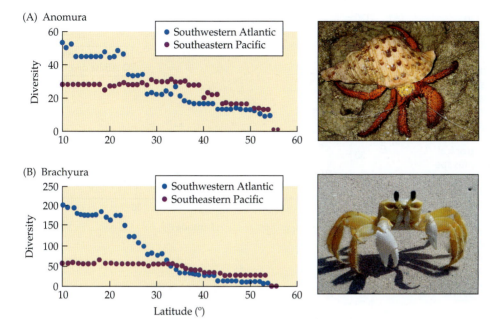

FIGURE 15.38 Latitudinal gradients in diversity of crustaceans differ markedly among species groups—(A) anomurans (false crabs and hermit crabs) and (B) brachyurans (true crabs)—as well as among geographic regions (southwestern Atlantic and southeastern Pacific). (From Marquet et al. 2004; after Astorga et al. 2003; A photo courtesy of Was a bee/Wikimedia; B photo courtesy of Patrick Verdier.)

Latitudinal gradients in other measures of diversity

Up to now, our discussion of latitudinal gradients in species diversity has focused on just one (albeit a very important and intuitively appealing) measure of diversity—species richness (or species density). But, as we indicated earlier, biological diversity is much more than this, as it includes the sum total of variation among life-forms in all of their characteristics (genetic, physiological, morphological, and behavioral) as well as variation in all biological, ecological, and biogeographic processes and across all scales of the biosphere (from organelles and cells to biomes and biogeographic regions). Given that biogeographic studies often analyze combinations of communities, studies of gradients in species diversity may include not just measures of alpha and gamma diversity (i.e., richness of local to regional communities), but also measures of beta and delta (between habitat/region) diversity as well.

Biogeographers of both the terrestrial and marine realms have begun to investigate broad scale patterns in beta diversity. The conceptual basis for such studies was articulated by Robert MacArthur in 1965. He argued that evolution and diversification in the tropics acted across habitats, adding new species to different habitats rather than packing them more tightly into one habitat. Thus, changes in diversity of an area or a latitudinal band composed of a collection of habitats (i.e., gamma diversity) would be due to difference in richness among communities (i.e., beta diversity). Put another way, although localized or within-habitat diversity may vary little across latitudinal gradients, beta diversity should vary and be highest in the tropics. The difficulty is that most studies of latitudinal gradients are conducted at the gamma scale, combining results from a complexity of adjacent habitats, thus rendering MacArthur's hypothesis untestable. Fortunately, sufficient fine-scale data are now available to test this hypothesis, at least for some systems. Clarke and Lidgard's (2000) study of North American bryozoans is an interesting case in point. Their studies indicate that beta diversity is, indeed, correlated with latitude, peaking between 10° and 30° N and then steadily declining in the higher latitudes.

As we reported earlier, bats are one of the many groups of terrestrial organisms that exhibit latitudinal gradients in species richness (see Figure 15.30F

FIGURE 15.39 The latitudinal gradient in diversity is actually many gradients, each differing depending on the particular taxon, region, scale, and measure of diversity employed. (A) Regional species richness (gamma diversity: triangles and dotted line) of New World bats declines much more rapidly than local species richness (alpha diversity: circles and solid line), indicating that (B) beta diversity (turnover in species composition among local communities) is highest in the tropics. These graphs also reveal that, across the tropical latitudes (23.5° S to 23.5° N) beta diversity (as species turnover between communities) tends to be high, but also highly variable and without an obvious latitudinal cline. Thus, the global-scale gradient in diversity of New World bats results primarily from differences between tropical communities and those of subtropical and temperate regions. (C) On the other hand, evenness of populations comprising these communities (Shannon's index of evenness) does not vary in any regular manner with latitude. (After Stevens and Willig 2002.)

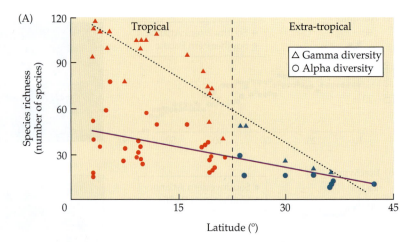

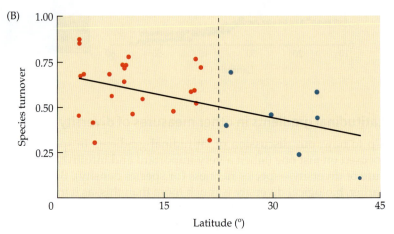

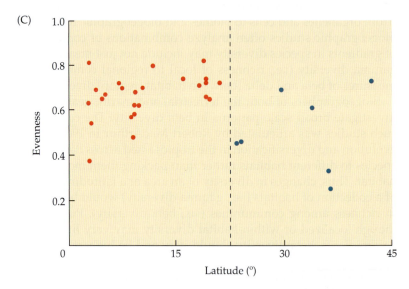

and Figure 15.34A). Stevens and Willig (2002) analyzed latitudinal gradients in 14 measures of community diversity at both local and regional scales. Local species richness (alpha diversity) increased and became more variable along a gradient from the poles to the Equator, whereas evenness of populations did not vary in any systematic manner along the same gradient (**Figure 15.39**). Latitudinal clines in gamma diversity (regional-scale richness) of bats were substantially steeper that those of alpha diversity, indicating that differences in species composition among local communities (i.e., beta diversity) is higher in the tropics (see Figure 15.29). Overall, Stevens and Willig's study indicates that the dramatic increase in gamma diversity of bats toward the tropics is equally influenced by ecological and evolutionary processes that determine differentiation among local communities.

Latitudinal gradients in species richness of marine invertebrates were reported over three decades ago (Stehli 1968; Rex et al. 1993). Only recently, however, have marine biogeographers and ecologists begun to analyze gradients in other measures of diversity. The analysis by Rex et al. (2000) of diversity of epibenthic invertebrates, including bivalves, gastropods, and isopods, from 37° S to 77° N, indicates that species richness of the three groups combined was strongly correlated with latitude. The correlation between latitude and evenness, however, was much steeper in general, but it varied among the three groups (being relatively strong for the isopods, weaker for the gastropods, and not significant at all for the bivalves [see also Rex 2004]).

In a groundbreaking study, Shepherd (1998) examined geographic clines in morphological diversity in 237 species of North American mammals. After accounting for the strong latitudinal gradient in species number, she found that, whereas diversity of body sizes increased with latitude, diversity in body shapes decreased. In their more recent studies of morphological diversity in the marine realm, Roy and his colleagues (Roy et al. 2001; Roy et al. 2004) used principal components analysis to calculate a standardized measure of morphological diversity in strombid gastropods (conchs and related species). As illustrated in the map of **Figure 15.40**, morphological diversity of these species exhibits some interesting geographic clines with both latitude and longitude, trends that are distinctly different from those of species richness.

While our colleagues have only begun to compare geographic gradients in these and other alternative measures of biological diversity, Roy and colleagues' results may be typical: Many, if not most, of the measures (taxonomic, genetic, ecological, functional, or morphological) of diversity will exhibit significant geographic clines, but the trends will differ among measures per-

FIGURE 15.40 Morphological diversity of strombid gastropods (conchs and related species) exhibits a complex but highly nonrandom pattern of geographic variation across gradients of both latitude and longitude (morphological diversity was measured using principal components analyses of physical dimensions of these species). (After Roy et al. 2004; after Roy et al. 2001; photo courtesy of Haplochromis/Wikipedia.)

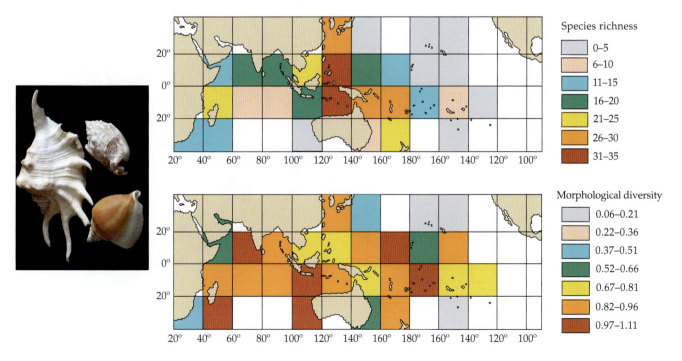

Species richness

- ☐ 0–5
- ☐ 6–10
- ☐ 11–15
- ☐ 16–20
- ☐ 21–25
- ☐ 26–30
- ☐ 31–35

Morphological diversity

- ☐ 0.06–0.21
- ☐ 0.22–0.36
- ☐ 0.37–0.51
- ☐ 0.52–0.66
- ☐ 0.67–0.81
- ☐ 0.82–0.96
- ☐ 0.97–1.11

haps as much as, if not more than, they differ among taxa (see Stevens et al. 2003). Again, comparative methods, and pattern deconstruction in particular, will no doubt be key to deciphering this fascinating variety of diversity gradients and ultimately yielding a much more comprehensive understanding of the processes influencing the geography and complexity of nature.

Other geographic gradients in species richness

While the latitudinal gradient appears to be the primary signal of geographic variation in diversity for most biotas, maps such as those of Figure 15.30, along with the earlier maps of species density developed by G. G. Simpson (1964) and other early biogeographers (**Figure 15.41**), reveal other geographic

(A) Mammals

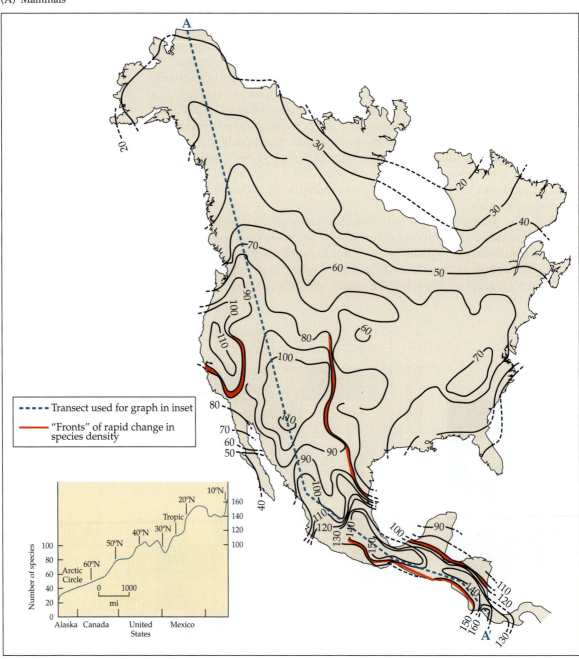

trends that may also inform efforts to understand the key processes influencing biological diversity. Such secondary, or in Simpson's terms "more localized," geographic features include proximity to the coastline, position within peninsulas, and local elevation and depth (topography and bathymetry), each of which may accentuate, obfuscate, or reverse the latitudinal trend in diversity.

Peninsulas

Simpson's (1964) classic map of isograms in species density of mammals led him to hypothesize a **peninsula effect**, which describes the tendency for diversity to decrease from the continental axis toward the terminus of a peninsula

(B) Birds

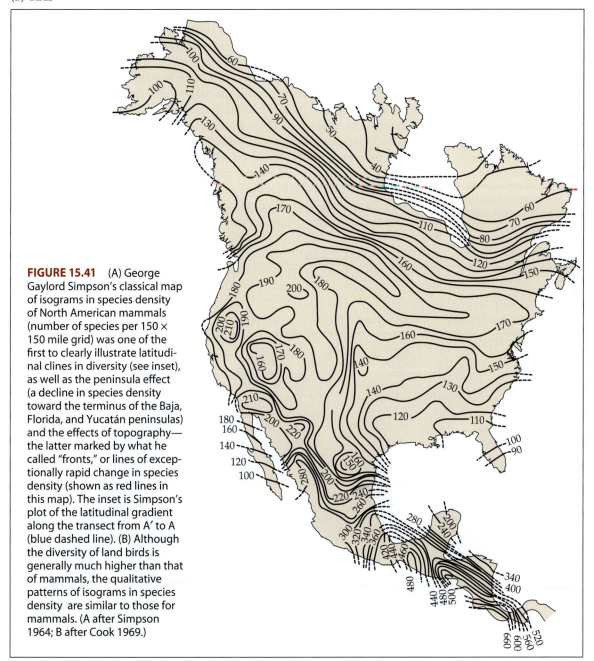

FIGURE 15.41   (A) George Gaylord Simpson's classical map of isograms in species density of North American mammals (number of species per 150 × 150 mile grid) was one of the first to clearly illustrate latitudinal clines in diversity (see inset), as well as the peninsula effect (a decline in species density toward the terminus of the Baja, Florida, and Yucatán peninsulas) and the effects of topography— the latter marked by what he called "fronts," or lines of exceptionally rapid change in species density (shown as red lines in this map). The inset is Simpson's plot of the latitudinal gradient along the transect from A' to A (blue dashed line). (B) Although the diversity of land birds is generally much higher than that of mammals, the qualitative patterns of isograms in species density are similar to those for mammals. (A after Simpson 1964; B after Cook 1969.)

(**Figure 15.42**). This pattern has been reported for a variety of taxa, including plants, invertebrates, and vertebrates, along a variety of peninsulas, including those of Florida, Nova Scotia, Alaska, and Baja California in the Nearctic; the Yucatán in the Neotropics; the Iberian, Balkan, Scandinavian, Italian, and Korean peninsulas in the Palearctic; and the Cape York Peninsula in Australia (Simpson 1964; Cook 1969; Taylor and Regal 1978; Due and Polis 1986; Brown 1987; Means and Simberloff 1987; Brown and Opler 1990; Martin and Gurrea 1990; Barbosa and Benzal 1996; Contoli 2000; Baquero and Telleria 2001; Johnson and Ward 2002; Choi 2004; Peck et al. 2005; Meyer 2008; Choi and Chun 2009; but see Schwartz 1988; Jenkins and Rinne 2008). As Figures 15.41 and 15.42 indicate, peninsula patterns differ among regions and species groups. For the most part, however, richness within the peninsula tends to be substantially less than that of the continental interior and lowest at the terminus of the peninsula. It is also important to note that, for those peninsulas extending toward the lower latitudes (e.g., Cape York, Baja California, Florida, Italy, and Korea), the gradient in species richness actually runs counter to the otherwise very general latitudinal clines in richness (i.e., richness along these peninsulas decreases as we move closer to the Equator).

Explanations for the peninsula effect vary, but nearly all center on the island-like nature of peninsulas—especially their isolation, limited area, maritime climates, and limited habitat diversity. As we discussed in Chapter 13, relatively low species richness is expected for biotas inhabiting isolated and relatively small ecosystems, because immigration rates will be relatively low, while extinction rates should be relatively high (Simpson 1964; MacArthur and Wilson 1967). Area affects not just extinction rates, but also total amount of available resources, diversity of habitats, and the likelihood that the peninsula (or island) will contain topographic barriers that would promote allopatric speciation (e.g., mountain chains and large rivers; Taylor and Regal 1978; Lawlor 1983; Due and Polis 1986; Milne and Forman 1986; Brown 1987; Means and Simberloff 1987). That is, if the peninsula is sufficiently large and not subject to frequent catastrophes, then its isolation may favor

FIGURE 15.42 The peninsula effect describes the tendency for species richness of a taxon to decrease from the axis (mainland connection) toward the most distal point of a peninsula (here illustrated for the peninsula of Baja California, A–C). Along this peninsula, (A) the diversity of heteromyid rodents (kangaroo rats and pocket mice) decreases from its base to tip; (B) that of scorpions does not show a clear pattern but may peak near the center of the peninsula; and (C) that of butterflies is distinctly bimodal, with peaks near the base and tip and with the lowest diversity in the center. (A–C data from Taylor and Regal 1978; Due and Polis 1986; Brown 1987.)

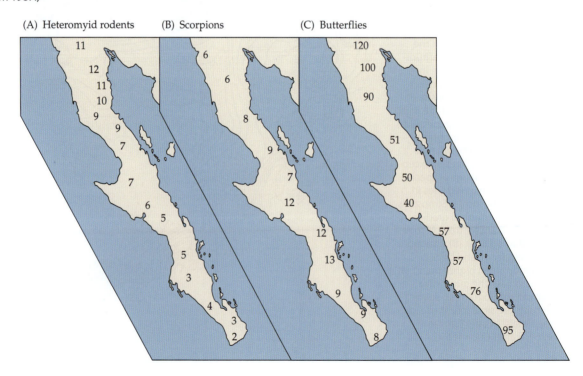

(A) Heteromyid rodents   (B) Scorpions   (C) Butterflies

evolutionary diversification and formation of peninsula endemics. In fact, endemicity is relatively common for biotas of large peninsulas, especially for lineages composed of species with limited dispersal abilities (e.g., scorpions of Baja California: Due and Polis 1986; non-volant mammals of European peninsulas: Baquero and Telleria 2001; ants of Baja California: Johnson and Ward 2002; arachnids of the Balkan Peninsula: Murienne et al. 2010; ground beetles of Florida: Peck et al. 2005, **Figure 15.43**). Cases where biotas exhibit mid-peninsula peaks in richness (e.g., Figures 15.42B and 15.43B) likely result from the combined actions of opposing gradients toward the terminus of the peninsula: (1) declining area, habitat diversity, and immigration rate that, at least in ecological time, confer lower diversity, (2) increasing geographic and genetic isolation, which promotes diversification of lineages with relatively limited dispersal abilities, and (3) for those peninsulas such as Florida, Korea, and Italy that are oriented toward the Equator, the effects of processes contributing to the latitudinal gradient in diversity. We can extend this explanation for mid-peninsula peaks in diversity to a regional scale and apply it to the world's major isthmuses, that is, viewing them as conjoined peninsulas. As Janzen (1967) and others have noted, "the isolation provided by the peninsular situation" of isthmuses may lead "to differentiation and ultimately, speciation," promoting relatively high concentrations of endemics in these regions (e.g., the Isthmus of Panama; see Watson and Peterson 1999) while also serving as biogeographic ecotones of range overlap of species from two distinct continental biotas.

These intriguing complexities once again highlight the heuristic value of strategically exploring patterns, not just in species richness, but those in composition, endemicity, and functional characteristics of component species as well. Each of the alternative hypotheses for the peninsula effect and its variants (reverse gradients and mid-peninsula peaks in diversity) is based on the potential influence of geometry or related geographic factors on immigration, extinction, or speciation. Given that species vary in their abilities to disperse to, survive in, or eventually diversify in such environments,

FIGURE 15.43 Ground beetles (Coleoptera: Carabidae) of Florida exhibit three alternative trends in species richness gradients along peninsulas. (A) Temperate species (i.e., those with affinities to mainland assemblages of southeastern United States) exhibit the predicted pattern of attenuating diversity toward the terminus of the peninsula. On the other hand, species richness of those endemic to Florida (B) peaks at an intermediate point along this gradient, while species with tropical affinities (C) exhibit a reverse peninsula effect (consistent with the latitudinal gradient in diversity). (After Peck et al. 2005.)

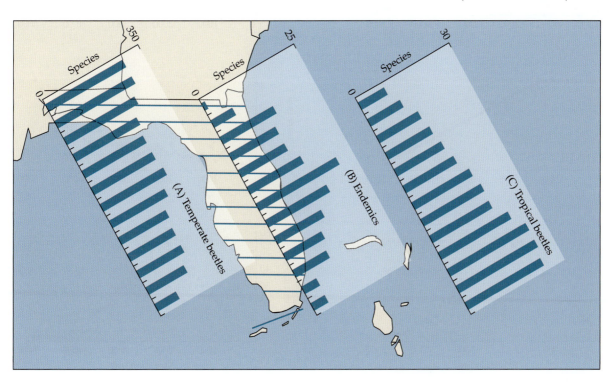

comparisons among peninsulas and among species or functionally different groups of species represent powerful means of discriminating among alternative causal explanations for these patterns. Finally, in the same way that many conservation biologists have applied Simpson's concepts of regional-scale dispersal corridors and filters to more local scales than he intended (i.e., at the scales of habitats and landscapes; see Chapter 16), some are now exploring the relevance of his peninsula effect for understanding the attenuation in densities and diversity of species along linear features of native landscapes (Perault and Lomolino 2000; Tubelis et al. 2006). Thus, even though it

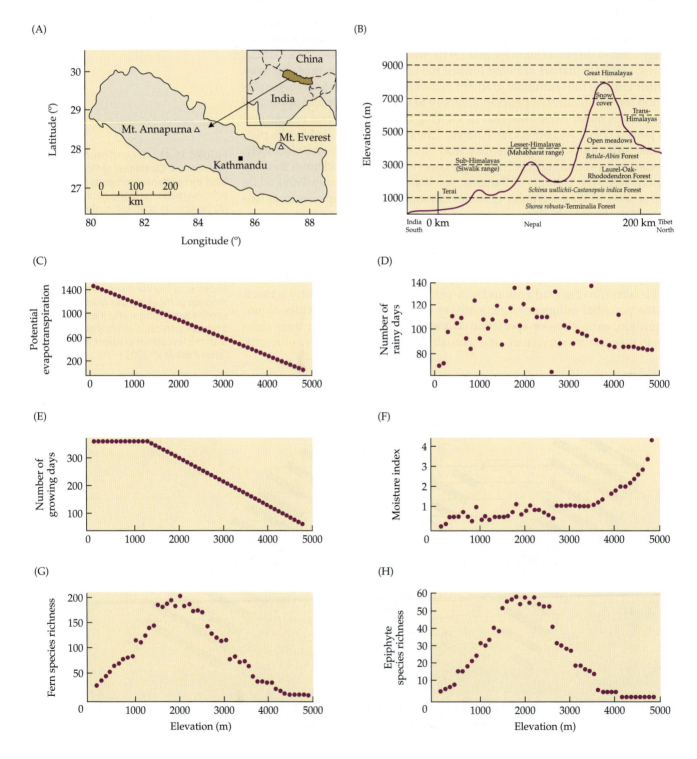

has been over four decades since Simpson (1964) concluded his classic paper on diversity gradients by calling for "a comparison of species ranges, population densities and structures, and related factors in peninsular and non-peninsular species," there remains much work to be done on the generality, causality, and applied relevance of this pattern.

Elevation and topographic relief

Research on montane biotas has played a prominent role in the development of biogeography, from Alexander von Humboldt's (1805) legion of detailed studies along Mount Chimborazo in the Ecuadorian Andes during the 1800s, to those of Charles Darwin (1839, 1859) in the Chilean Andes; from the accounts of Alfred Russel Wallace in Indonesia (1869, 1876), Joseph Dalton Hooker (1877), Asa Gray (1878), and C. H. Merriam (1890) in the North American Rockies during mid to later decades of the nineteenth century, to those of Robert H. Whittaker (1960; Whittaker and Niering 1965) in the Santa Catalina, Great Smoky, and Siskiyou Mountains (southeastern and western North America), and James H. Brown's studies of montane islands of the Great Basin region of North America during this past century (Brown 1971b, 1978). Each of these distinguished scientists capitalized on the great heuristic value of natural experiments offered by elevation gradients along mountain slopes. They detailed the geographic variation in abiotic variables and discovered what appeared to be very general patterns in variation among biological communities along elevational transects. Originally, it appeared that diversity of these communities decreased monotonically from the foothills to the highest elevations (Whittaker 1960, 1977; Yoda 1967; Kikkawa and Williams 1971; Terborgh 1977; Heaney 1991; Heaney et al. 1989; Daniels 1992; Sfenthourakis 1992; Fernandes and Lara 1993; Patterson et al. 1996, 1998). Unfortunately, the apparent patterns may have been influenced to some degree by preferential sampling of some elevations (especially those along the foothills or initial slopes) more than others, and by limitation of surveys to transects along mountain chains, without attention to the coastal plains.

As a result of more rigorous and more extensive surveys during the latter decades of the twentieth century, a somewhat different pattern emerged. Biogeographic surveys of a variety of plants, invertebrates, and vertebrates—especially those conducted along transects extending from sea level to mountain summits—indicate a unimodal, or hump-shaped, pattern with richness peaking at an intermediate elevation (Whittaker 1960, 1977; Whittaker and Niering 1975; Brown 1988; Rosenzweig 1992, 1995; Rahbek 1995, 1997; Fleishman et al. 1998; Hawkins 1999; Heaney 2001; Nor 2001; Rickart 2001; Sanchez-Cordero 2001; Sanders 2002; Sanders et al. 2003; Bhattarai et al. 2004; Lee et al. 2004; McCain 2004; Li et al. 2009; McCain 2009b,c; Rowe and Lidgard 2009). That is, species richness of most groups of terrestrial organisms increases gradually along the coastal plains, rises more rapidly with ascent to the foothills or mid elevations of mountains, and then declines again with approach to the ice-covered summits (**Figures 15.44** and **15.45**).

As you can imagine, over the long history of research on this pattern, there have been many hypothesized causal explanations. The ultimate explanation for this pattern should be based on responses of species to the complex, but predictable, variation in the geographic template along elevational transects.

◄ **FIGURE 15.44** Along a gradient through the central Himalayas from 100 to 4800 m above sea level (A,B), climatic conditions change in a predictable, albeit complex, fashion (C–F). The result is that the most favorable combination of environmental conditions for ferns and epiphytes occurs at intermediate elevations (G,H). (After Bhattarai et al. 2004.)

FIGURE 15.45 Species-elevation patterns vary among species groups and mountain chains, but most elevational gradients tend to exhibit a peak in richness at some intermediate elevation. Note that these surveys were not conducted below 500 m, so it is difficult to determine whether trees of the Siskiyou Mountains also exhibit a unimodal, or hump-shaped, cline in richness (peaking at lower elevations). (Data from Whittaker 1960; Whittaker and Niering 1965.)

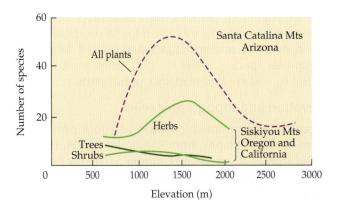

As we described in Chapter 5, species typically respond to geographic clines in environmental conditions in an individualistic fashion, with very few species sharing precise ecological requirements or limits to their geographic ranges. The overlap among those ranges, however, exhibits some very general patterns along elevational gradients. This overlap, of course, is species richness.

Although some earlier explanations for elevational gradients in diversity were based on one putative overriding factor, it is now clear that many factors and processes are involved. That is, species are responding to a combination of factors that vary in a highly regular manner with ascent of elevational gradients: some of these promoting immigration, survival, and speciation, and others having the opposite effect (**Figure 15.46**). For example, as elevation increases, air temperature and air pressure (including partial pressures of oxygen and carbon dioxide) decrease (see McCain and Sanders 2009, and description of temperature-dependent kinetics of organisms and the metabolic theory of ecology; see Table 15.2; see Figure 15.49). Because cooler air cannot hold as much moisture as warmer air, precipitation often increases with ascent up a mountain slope, although at the highest elevations the moisture is frozen and generally unavailable to most organisms (see Fu et al. 2007). Thus, the optimal combination of temperature, gases, and precipitation for most plants and other organisms most likely occurs at intermediate elevations.

It is also likely that the tight juxtaposition and concentration of climatic zones along elevational gradients contributes to the relatively high beta and, ultimately, gamma diversity of these mountainous regions. This, indeed, was Simpson's (1964) explanation for what he termed "fronts," or rapid change in species density of North American mammals (red highlighted lines in Figure 15.41A). Unlike the similar series of biomes that are encountered along extensive latitudinal gradients, geographic ranges of ecologically distinct species are in very close proximity and, therefore, more likely to overlap—or "spill over"—into other communities along the slopes of mountains (see Schmida and Wilson 1985; Pulliam 1988; Grytnes 2002, 2003).

In addition to climatic variables and their influences on montane habitats and landscapes, physiographic features also vary along elevational transects (Korner 2007). Given the roughly conical shape of mountains, the area of any elevational/habitat zone actually decreases as we move from the lowlands to the summit (see Figure 15.46). In addition, communities at the higher elevations are more isolated, both from species assemblages occupying the lowlands and from those occurring in similar alpine communities but in other mountains. Thus, as Robert MacArthur and other biogeographers noted, montane ecosystems can be viewed as archipelagoes composed of ecosys-

(A)

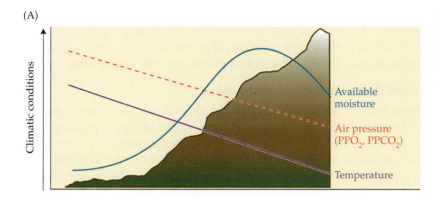

(B)

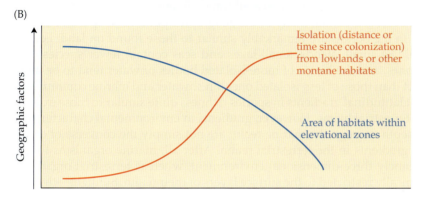

(C)

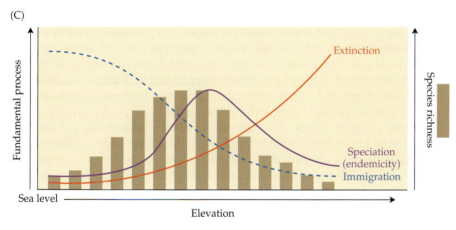

FIGURE 15.46 Transects from sea level to the peaks of mountain ranges are characterized by regular gradients in (A) climatic conditions, (B) geographic variables, and ultimately, (C) the fundamental processes affecting biological diversity. As a result, species richness (bars in part C) of many taxonomic groups often peaks at an intermediate elevation, which corresponds to an optimal combination of conditions for generating and maintaining their species populations.

tems that vary in area and isolation—these factors affecting the rates of extinction and immigration, respectively (see Beck and Kitching 2009b). Based on this montane island model alone, we would predict that species richness should be highest at the lowest elevations because lowland ecosystems occupy the largest area and are the least isolated. However, immigrations and extinctions are also influenced by the climatic factors discussed above, and perhaps just as important, area and isolation also influence the other fundamental process—speciation. Once again, Simpson (1964: 70) anticipated this explanation, observing that "high relief may promote geographic isolation of populations and so increase the likelihood of speciation." As we discussed in Chapters 13 and 14, speciation and endemicity increase with both area and isolation. Because these two physiographic features change in opposite directions along elevational gradients, however, speciation and richness of montane endemics should peak at some intermediate elevation (e.g., see Heaney 2001; Vetaas and Grytnes 2002; Bhattarai et al. 2004; Li et al. 2009).

Again, the overall effect of geographic variation in all of these important environmental factors and their influence on immigration, extinction, and speciation is that species richness peaks at intermediate elevations along an elevational transect between the coastal plains and mountain peaks.

Depth and diversity in the aquatic realm

The marine realm is clearly one of the last frontiers for biogeographers, as well as for most other biologists. Until quite recently, the immense pressures, extreme cold, and complete darkness of deepwater environments limited most of our studies to the top few meters of the oceans. Fortunately, with advances in direct and remote-controlled sampling techniques, we have been able to explore deeper into the oceans and begin to develop a better understanding of the geography of nature across all three dimensions of the marine realm. As we observed earlier, many of the geographic patterns of species richness found in the oceans are remarkably similar to those found on land. For example, like small islands, small lakes and seas contain fewer species than large ones, and the biotas of more-isolated bodies of water tend to be more distinct than those of less-isolated ones. Freshwater and marine communities exhibit latitudinal gradients in species richness quite similar to those exhibited by terrestrial biotas. However, gradients in environmental characteristics with depth through the marine realm are not simply the converse of elevational gradients in the terrestrial realm. After all, marine organisms live in a genuinely three-dimensional environment, whereas terrestrial organisms are typically restricted to the surface layers of their realm. Sunlight—the ultimate source for primary productivity in nearly all communities—attenuates rapidly with depth in the water column, but it is little affected by elevation above the surface and along mountain slopes. Air pressure varies in a regular manner with elevation above sea level, but that gradient seems far from equivalent to the changes in water pressure with depth (increasing at the rate of approximately 1 atmosphere pressure per 32 feet in depth).

Despite these and many more fundamental differences between the aquatic and terrestrial realms, patterns of variation in species richness along gradients of depth appear qualitatively similar to those along gradients of elevation on land with, at least in some cases, the effects of depth dominating those of latitude (see Brandt et al. 2005). For most marine macroinvertebrates and fishes, species richness exhibits a unimodal, or hump-shaped, cline with depth, increasing with descent below sea level and peaking somewhere along the continental shelf, then declining along the continental slope toward the abyssal plain (**Figure 15.47**; Vinogradova 1962; Sanders 1968; Slobodkin and Sanders 1969; Rex 1981; MacPherson and Duarte 1994; Gappa 2000; Smith and Brown 2002; Brandt et al. 2005; Rex et al. 2005; Maciolek and Smith 2009). Freshwater communities of relatively deep lakes exhibit similar clines in species diversity with increasing depth. Some large lakes, including the Great Lakes and the Finger Lakes in northeastern North America, Lake Baikal in Siberia, and several of the large lakes of central Africa, are more than 100 m deep, but few organisms are found in the deepest waters. In Chapter 7, we described the great diversity of fishes, especially in the cichlids, that inhabit the Great Lakes of Africa's Rift Valley. Although Lake Malawi and Lake Tanganyika are 704 and 1470 m deep, respectively, their hundreds of species of fish are most diverse in the surface waters, while none are found below a depth of about 200 m. Their lower depths apparently are inhabited by only a few kinds of anaerobic invertebrates and bacteria.

Thus, the precise pattern and depth of peak diversity varies substantially depending on the particular species group or region studied. The general pattern, along with its explanations, however, is qualitatively similar to those

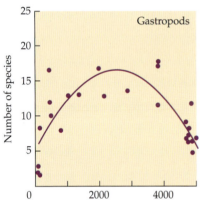

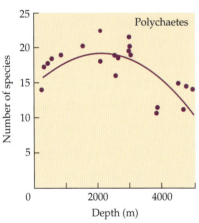

FIGURE 15.47 Similar to elevation gradients in species density (i.e., above sea level), the relationship between diversity of marine species and depth often exhibits a hump-shaped pattern, with highest diversity found at intermediate depths in the heterogeneous environments of the continental slopes. This is probably the typical pattern for most samples of small areas, although at broader spatial scales, diversity also may be relatively high across the abyssal plains. (After Rex 1981.)

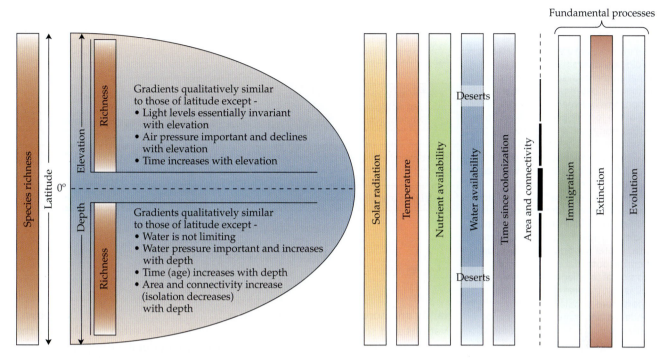

Gradients qualitatively similar
to those of latitude except -
• Light levels essentially invariant
  with elevation
• Air pressure important and declines
  with elevation
• Time increases with elevation

Gradients qualitatively similar
to those of latitude except -
• Water is not limiting
• Water pressure important and increases
  with depth
• Time (age) increases with depth
• Area and connectivity increase
  (isolation decreases)
  with depth

developed to explain diversity gradients on land (**Figure 15.48**). Just as environmental and physiographic factors (e.g., air pressure, temperature, available moisture, area, and isolation) vary in a complex but predictable manner along terrestrial gradients of latitude and elevation, so too do light, nutrients, pressure, temperature, currents, and concentrations of a seemingly endless list of chemicals vary in a regular manner with descent from sea level to the deepest reaches of the oceans. Again, the gradient is bracketed by two harsh boundaries—in this case, the terrestrial environment above and the extreme pressures, darkness, and cold of the abyssal plains below. Surface waters may at first seem to be favorable to many species, given their relatively high temperatures and sunlight, but they also tend to be most subject to frequent, and sometimes intense, disturbance, including erosion and rapid fluxes in temperature and water chemistry. In addition, a quick look at a bathymetric map reveals that deepwater environments cover most of the Earth's surface (i.e., area, which tends to be positively correlated with species diversity, increases with depth in the marine realm). Thus, because of the complex and sometimes opposing clines of factors influencing immigration, speciation, survival, and reproduction of marine organisms, optimal conditions and peak diversity for most species should occur at intermediate depths between surface waters and the abyssal plains.

Toward a general explanation of the geography of diversity

Table 15.2 provides descriptions of a long list of hypotheses that have been proposed to explain one of nature's most general and long-studied patterns—the latitudinal gradient in species richness. We note that this list, while quite long, is not exhaustive. On the other hand, many of these "alternative" hypotheses seem redundant, and most are not mutually exclusive. Null models (of which only two are listed in Table 15.2) may sometimes seem unrealistic, but they are very useful starting points for most investigations in ecology, evolution, and biogeography (see Gotelli and Graves 1996). Explanations based on biological interactions and ecological processes (see Table 15.2) cer-

FIGURE 15.48 Fundamental to biogeography is the tendency for environmental factors to vary in a regular manner across the geographic template. Because the levels of different factors (some promoting, others inhibiting, biodiversity) change in a complex fashion, the result is the creation of particular regions (e.g., the tropics and those at intermediate elevations or intermediate depths) with combinations of conditions that are optimal for generating and maintaining species populations, and for enhancing the number of coexisting species populations in local communities and regional assemblages (through reduced sizes of, or increased overlap among, their geographic ranges). Intensities of the colored bars on the right reflect relative levels of factors influencing biological communities as they vary across the global, latitudinal gradient. As oriented here, these factors also vary in a regular manner along gradients of elevation and depth (charts and lists within the upper and lower portions of the sphere, respectively; see Huston and Wolverton 2009: Figure 3.)

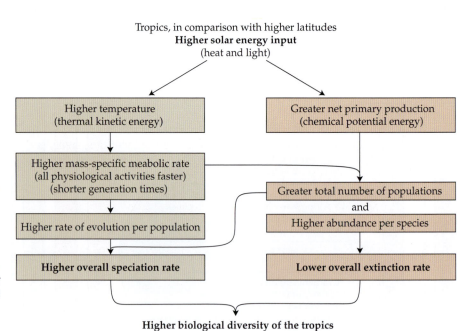

FIGURE 15.49   The metabolic theory of ecology (Allen et al. 2002, 2006; Gillooly and Allen 2007; Stegen et al. 2009) may explain a variety of patterns in biological diversity, including latitudinal gradients in species richness. The theory is based on very general relationships between energy input (here in the form of heat and light from the sun), body size, and metabolic (physiological) processes of all organisms. These processes tend to occur more rapidly for organisms in warmer environments (those with higher "thermal kinetic energy," left-hand side of the diagram) and for those of smaller body size. In addition, the more intense sunlight of tropical environments promotes greater net primary production ("chemical potential energy," right-hand side of the diagram). The overall result is that tropical environments support more populations of higher abundance, and more persistent and rapidly evolving species, than those at higher latitudes. (After Gilloly and Allen 2007.)

tainly are more realistic, for the most part, and are based on empirical observations and sound ecological theory. They, however, often fall short of the broad spatial and temporal contexts requisite for understanding such regional- to global-scale patterns in biological diversity (see Ricklefs 2006a). Even those explanations based on the hypothesized positive relationship between primary productivity and diversity are often challenged, most recently by studies that suggest that annual net primary productivity (NPP) of marine ecosystems, and maximum NPP (during the growing season) of at least some terrestrial ecosystems, peak outside the tropics (Huston and Wolverton 2009). Explanations based on ecological interactions often seem circular, although they do serve to emphasize potentially important feedback mechanisms that may amplify the primary pattern. Many explanations based on the influence of independent abiotic factors also appear to fall short of identifying ultimate

FIGURE 15.50   From ecological and correlative to more integrative explanations for the latitudinal gradient in diversity. (A) Many earlier explanations for the latitudinal gradient in diversity were based on standard ecological approaches (see Table 15.2), on correlations between species richness (represented here by relative lengths of bars along the left-hand surface of this diagram), and on the influence of one or more environmental variables (e.g., sunlight or temperature; see Figure 15.48). In contrast, more integrative explanations for this pattern are based on the combined effects of latitudinal differences in fundamental processes (origination and extinction within, and immigration among, latitudinal zones) and on ecological characteristics of their biotas (e.g., latitudinal differences in niche breadths, size and overlap among their geographic ranges, and propensities and capacities of their species to disperse to other zones). (B) Wiens and Donoghue's (2004) explanation—the niche conservatism model—holds that because tropical climates are relatively benign, aseasonal, and predictable, their species tend to become ecologically specialized and limited in their abilities to disperse to other sites within the tropics and to sites in the higher latitudes. Thus, tropical species are effectively more isolated, and because tropical systems also tend to be larger and older than those in other regions, diversification rates and total number of species accumulated should be higher in the tropics. (After Wiens and Donoghue 2004.)

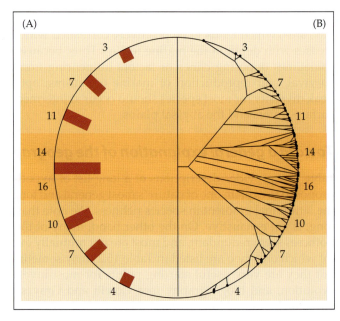

causality and, at least in some cases, focus on isolated factors that apply just to a small subset of all the groups that exhibit this very general pattern.

We believe that the most promising explanations for this pattern are those that are more integrative and take into account the combination of factors that affect the fundamental processes that influence the locations, sizes, and overlap of geographic ranges, that is, evolution, immigration, extinction, and ecological interactions (**Figures 15.49–15.51**; see Figure 15.46). The list of hypotheses in Table 15.2 also demonstrates a succession of approaches toward explaining diversity, in general (see also Scheiner and Willig 2005; Pickett et al. 2007). Whereas many early hypotheses can be characterized as efforts to find the one, overriding factor or driving force responsible for latitudinal clines in diversity, later approaches tended to include a combination of factors (stochastic as well as deterministic) and their interactive as well as independent effects. Ultimately, the list would include genuinely integrative hypotheses and theories of biodiversity and biogeography, including those described at the end of Table 15.2. Three common features of these more integrative explanations for this very general pattern are that they (1) span both historical and ecological biogeography, (2) are more firmly and explicitly based on first principles, and (3) as a result, may be applied to other patterns of species diversity as well. Thus, integrative theories of biogeography and biodiversity, such as the neutral theory of biodiversity (Hubbell 2001), the metabolic theory of energy (Allen et al. 2002, 2006; Gillooly and Allen

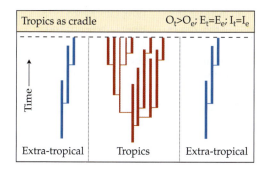

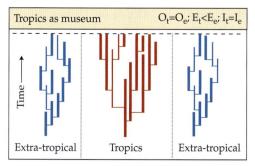

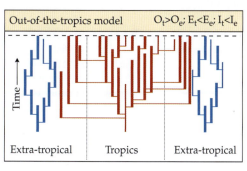

FIGURE 15.51 Jablonski, Roy, and Valentine's out-of-the-tropics model is another integrative explanation for the geography of diversity, based on first principles and the combined effects of differences in fundamental processes among latitudinal zones, that is, on rates of origination (O) and extinction (E) within, and immigration (I) into a region. For species richness to be higher in tropical versus extra-tropical regions, one or more of the following conditions must be satisfied: $O_t > O_e$; $E_t < E_e$; $I_t > I_e$ (where the subscripts t and e refer to tropical and extra-tropical regions, respectively). In addition, the overall sum, which represents net diversification, must be higher for tropical regions. Jablonski et al. (2006; see also Krug et al. 2009) provide evidence that, at least for the marine bivalves they study, the tropics function as "cradles" ($O_t > O_e$), "museums" ($E_t < E_e$), and "immigration pumps" ($I_e > I_t$). This last condition (i.e., that more species move from the tropics into extra-tropical regions than in the reverse direction) is one key distinction between this model and the tropical niche conservatism model (see Figure 15.50); that is, the out-of-the-tropics model emphasizes the *expansion* of niches and geographic ranges over time, both within and among latitudinal and climatic zones. (After Jablonski et al. 2006.)

TABLE 15.2 *Hypotheses for Geography of Diversity[a]*

Hypothesis	Explanation	Source
NULL MODELS		
Mid-domain effect	Random placement of geographic ranges between two presumed hard boundaries (the polar regions) results in the highest overlap of ranges (highest richness) occurring midway between those boundaries (i.e., near the Equator).	Colwell and Hurtt 1994; Colwell et al. 2005, Hawkins et al. 2005; Zapata et al. 2005
Neutral theory of biodiversity and biogeography	Even a set of species that are assumed to be identical with respect to their birth and death rates and dispersal abilities of their individuals, and whose population dynamics is similar to that of random walk, should be comprised of populations that vary over time—changing in population density and some suffering random extinction, while others experience mutations, genetic recombinations, and random speciation. The factors influencing population densities, random extinctions, and random speciation of these equivalent species may favor higher species richness in the tropics.	Hubbell 2001; Hubbell and Lake 2003; Turner and Hawkins 2004
BIOLOGICAL INTERACTIONS AND PROCESSES		
Competition	Competition, especially diffuse competition (that from many species of competitors), tends to hold each population in check, allowing more species to coexist in competitor-rich, tropical communities.	Dobzhansky 1950; Pianka 1966; Huston 1979
Predation	Predators (including herbivores and carnivores) tend to hold prey populations in check by switching and preying most heavily on the most abundant prey, thus preventing prey populations from increasing to levels at which they would exclude each other, especially in the tropics with its high diversity of predators.	Paine 1966; Pianka 1966; Harper 1969; Janzen 1970; Lubchenco 1978, 1980; Lubchenco and Menge 1978
Mutualism	Mutualists, by definition, promote the coexistence of their symbionts and, thus, the empirically observed high frequency of mutualism in the tropics promotes higher diversity of symbionts.	Dobzhansky 1950; Paine 1966; Janzen 1970; Menge and Sutherland 1976
Host diversity— parasitism	High diversity of hosts promotes high diversity of parasites. Therefore, because the tropics are inhabited by more host species, diversity of parasites also is higher in the tropics.	Rohde 1989
Epiphyte load	Diversity of epiphytes is highest where diversity of their host trees are highest (i.e., in the lush forests of the tropics).	Strong 1977
Environmental patchiness (beta diversity)	Tropical habitats tend to be patchily distributed, thus providing more numerous patches to be colonized and occupied by different species.	McCoy and Connor 1980
Habitat (e.g., foliage height) diversity; biotic spatial heterogeneity	Diversity, in general, tends to be higher in more complex and heterogeneous habitats, and the tropics tend to be spatially and vertically more heterogeneous.	MacArthur et al. 1966; Cody 1968, 1974, 1975; MacArthur 1972; Huston 1979; Thiollay 1990
Niche width	Tropical species tend to be more specialized (i.e., have narrower niches) and, therefore, more species can be packed into tropical habitats.	Ben-Eliahu and Safriel 1982; Brown and Gibson 1983
Population growth rate	Tropical species tend to be comprised of populations with higher growth rates, which in turn provides more opportunities for ecological specialization.	Huston 1979
Productivity	Higher levels of primary productivity of tropical plant communities provides more energy, which in turn supports higher populations and more diverse communities of consumers.	Hutchison 1959; Connell and Orians 1964; MacArthur 1965, 1969; Pianka 1966; Wright 1983; Currie and Paquin 1987; Tilman 1988; Currie 1991; Roy et al. 1998

TABLE 15.2 *(continued)*

Hypothesis	Explanation	Source
Canopy/crown morphology (incidence of solar radiation)	Tropical forests receive more direct sunlight, which tends to favor trees with more shallow crowns, also allowing more light to penetrate below the canopy and support multiple layers of trees below and, thus, a greater diversity of trees overall.	Terborgh 1985
Geographic ranges (Rapoport's rule)	Tropical species tend to have smaller geographic ranges, thus allowing more species to coexist in tropical versus temperate regions.	Rapoport 1982; Pianka 1989; Stevens 1989
ABIOTIC/ENVIRONMENTAL FACTORS		
Environmental stability and predictability (over different time periods—see below)	Tropical environments tend to be more stable over both short and long time periods, thus their species are less subject to extinctions and more capable of adapting to—and specializing for—these more predictable environments.	Slobodkin and Sanders 1969; Janzen 1970
Tectonic dynamics	Despite significant changes in the development and drifting of Earth's plates, some areas that are currently in the tropics have been in tropical regions for much of the evolutionary history of their biota.	Scotese 2004
Glacial fluxes	Glacial expansions and climatic fluxes of the Pleistocene (Chapter 9) caused extinctions of high-latitude species, and the subsequent interglacial period has been insufficient to re-establish diversity in these regions.	Fischer 1960; Pianka 1966; Simpson 1974, 1975; Fischer and Arthur 1977; Stanley 1979
Annual stability (seasonality)	Tropical environments tend to be much less variable throughout the year, with little seasonality in temperature or precipitation. Thus, tropical species can become more specialized, allowing more species to coexist in the same amount of space.	Begon et al. 1986
Antiquity of the tropics (evolutionary time)	As indicated above, in comparison to those in higher latitudes, tropical regions tend to have occurred in tropical latitudes for longer periods of time. Thus, tropical regions have accumulated species over longer periods of time than those at higher latitudes.	Pianka 1966, 1988; Whittaker 1969; Mittelbach et al. 2007
Antiquity of the tropics (ecological time)	Tropical species have had more time to disperse to, and colonize, a greater portion of suitable habitats than have their counterparts in temperate regions and higher latitudes (equilibrium levels of species richness between speciation, extinction, and dispersal are higher in the tropics).	Fisher 1960; Pianka 1966
Environmental harshness	Tropical environments tend to be more benign, and thus can be inhabited by more species.	Terborgh 1973b; Brown 1981, 1988; Thiery 1982; Begon et al. 1986
Abiotic rarefaction (intermediate disturbance)	Tropical environments are intermediate with respect to frequency and intensity of storms and other abiotic disturbances that would otherwise reduce population densities. According to this modified form of the intermediate disturbance hypothesis, diversity should thus be higher in the tropics.	Dobzhansky 1950; Connell 1978
Solar energy (species energy)	Tropical environments receive more intense solar energy, supporting higher productivity of plant communities, thus promoting higher diversity of these producers and dependent consumers.	Forster 1778; Willdenow 1805; Wright 1983; Turner 1986; Currie 1991; Roy et al. 1998
Heat and evolutionary rates (see metabolic theory of ecology, below)	Higher environmental temperatures in the tropics promote higher metabolic rates and shorter generation times in tropical species (at least for the ectotherms), allowing more rapid speciation rates and greater overall accumulation of species.	Alekseev 1982; Rohde 1992; Mittelbach et al. 2007; Ricklefs 2006b; Evans and Gaston 2005
Aridity	Because precipitation tends to be higher in the tropics and because water is required by all life forms, species diversity (of terrestrial life forms) should be higher in the tropics.	Begon et al. 1986

continued on next page

TABLE 15.2 *(continued)*

Hypothesis	Explanation	Source
Evapotranspiration	Plants require both water and energy in the form of sunlight. Tropical environments tend to be both wet and warm and, therefore, productivity of plant communities should be higher in the tropics, supporting a greater diversity of producers and consumers.	Wright 1983
Solar radiation and mutation rates (evolutionary speed)	The intensity of solar (ultraviolet) radiation in the tropics causes greater mutation rates, thus providing more raw material for selection and adaptive radiation.	Rensch 1959; Stehli et al. 1969; Rohde 1992; Bromham and Cardillo 2003
Area	Tropical regions tend to occupy a greater portion of Earth's surface. Because diversity tends to increase with area for nearly all types of ecosystems, diversity should be higher in the tropics.	Connor and McCoy 1979; Currie 1991; Chown and Gaston 2000

FUNDAMENTAL PROCESSES AND INTEGRATIVE HYPOTHESES

Equilibrium models of diversity

Hypothesis	Explanation	Source
Speciation, extinction and area (an island/ equilibrium model)	As discussed above, tropical ecosystems tend to occupy a larger portion of the Earth's surface than those in higher latitudes. According to equilibrial models of island biogeography (MacArthur and Wilson 1963, 1967), larger ecosystems should be characterized by relatively high immigration and speciation rates, but low extinction rates. Therefore, at equilibrium, larger ecosystems (i.e., the tropics) should be inhabited by more species.	Rosenzweig 1992, 1995, 1997; Hawkins and Porter 2001; Rohde 1997
Metabolic theory of ecology	Tropic environments receive more solar radiation in the form of heat and light which, in turn support higher rates of metabolism (thermal kinetic energy) and higher rates of primary productivity (chemical potential energy), respectively. The ultimate results include higher evolutionary rates (more populations of rapidly evolving species) and lower extinction rates (resulting from higher abundances per species) and, thus, greater diversity (see Figure 15.49).	Allen et al. 2002, 2006; Algar et al. 2007; Gillooly and Allen 2007; Hawkins et al. 2007b; Stegen et al. 2009

Speciation, extinction, and immigration (non-equilibrial models)

Hypothesis	Explanation	Source
Tropical niche conservatism	In contrast to other regions, the tropics are larger and older and, therefore, should have higher speciation rates and lower extinction rates. In addition, because tropical climates tend to be relatively benign, predictable, and productive, selection in the tropics favors species with relatively narrow niche breadths and limited dispersal abilities.	Wiens and Donoghue 2004; Hawkins et al. 2006; Hawkins 2008
Out-of-the-tropics model	Tropical systems serve as the cradles (high origination rates), museums (high persistence, low extinction rates), and immigration pumps (high rates of immigration to extra-tropical systems) of biological diversity. Niches of tropical species tend to be more dynamic (less conservative), promoting expansion into extra-tropical regions.	Jablonski et al. 2006; Krug et al. 2009

Source: After Rohde 1992.

[a]A diversity of explanations for latitudinal gradients in species richness, many of which may apply to other gradients in the geography of diversity, in general. This list, while long, is not necessarily an exhaustive one. In addition, these hypotheses are not necessarily mutually exclusive.

2007; Stegen et al. 2009), and the theory of niche conservatism (Wiens and Donoghue 2004; Wiens and Graham 2005; Wiens et al. 2006), may eventually be developed to provide an explanation for the geography of diversity, in general (i.e., inclusive of diversity clines along gradients of elevation, depth, area, and isolation, as well as latitude).

While we cannot anticipate the final form of the grand synthesis that may emerge from efforts to advance the frontiers of this field, a general theory of the geography of diversity will likely be based on one of the most fundamental, first principles of biogeography; environmental conditions vary in a regular manner across the geographic template. Because the nature of this variation is complex and inconsistent among environmental factors (some gradients promoting, others inhibiting, diversity), the result will often be the creation of particular regions (the tropics, intermediate elevations, or intermediate depths) with a combination of conditions that are optimal for generating and maintaining populations of most taxa, and for increasing the number of co-existing species populations in local communities and regional assemblages (through reduced sizes of, or increased overlap among, geographic ranges; see Figure 15.48). The latter point is worth additional emphasis: In addition to historical processes (immigration, extinction, and speciation), a general theory of diversity should include ecological interactions that, as Geerat Vermeij (2005) observed, can create positive feedbacks between potential and realized diversity by packing additional, more specialized species into tropical (mid-elevational, etc.) assemblages.

The global analysis by Qian and Ricklefs (2007), utilizing data from nearly 300 terrestrial sites across 157 countries, provides a compelling case for the generality and shared causality of geographic variation in species diversity. The strong concordance they observed in variation of species richness among geographic regions (all continents except Antarctica) and in such disparate taxa (amphibians, reptiles, birds, mammals, and vascular plants) argues strongly for the operation of very general, fundamental processes and basic principles of biology, geography, ecology, and physics. These results, along with those of many others demonstrating some surprisingly common features of diversity patterns over a variety of geographic gradients, taxa, time periods, and ecosystems (as discussed earlier in this chapter; see also Pearson and Carroll 1999; Lamoreux et al. 2006), justifies some optimism that a general theory of the geography of diversity may be within reach.

One final coda to this chapter, and a prelude to the next: As general and ancient as these gradients in species diversity are, they are likely to change—perhaps markedly—during the next few centuries. The geographic ranges of technologically advanced, ecologically significant populations of our own species have expanded to the point that they now threaten regions of peak diversity and endemicity—extending further into the tropics, higher into montane regions, and to much greater depths of the marine realm (see Lomolino 2001; Lee et al. 2004). This is a critical time and one when biogeography, in particular, can provide invaluable insights, not just for understanding patterns in biological diversity, but for conserving them as well.

CONSERVATION AND THE FRONTIERS OF BIOGEOGRAPHY

CONSERVATION BIOGEOGRAPHY AND THE DYNAMIC GEOGRAPHY OF HUMANITY

16

The Biodiversity Crisis and the Geography of Extinctions

Biodiversity is, in the simplest terms, the variety of life. It encompasses the variation among species or other biological elements, including alleles and gene complexes, populations, guilds, communities, ecosystems, landscapes, and biogeographic regions. Biodiversity can be expressed as the variation within a given location, or the variation among elements across geographic units. This variation can include the number of different types of species or elements, their relative frequencies, the degree of variation among these elements, or variation in key processes such as dispersal, gene flow, interspecific interactions, or ecological succession.

Given such a broad and inclusive concept, our hope of conserving biodiversity may at first seem like an impossible dream. Yet, in more pragmatic terms, conserving biodiversity distills down to one simple, yet still challenging, goal: maintaining diversity of species. As Aldo Leopold put it, "The first requisite of intelligent tinkering is to save all the pieces." By *pieces*, of course, he meant native species. With the loss of a species, biodiversity is diminished at all levels, from genetic and local scales to biogeographic and global ones. With each species extinction, associated ecological and evolutionary phenomena such as mass migrations of birds and butterflies, pollination of Hawaii's endemic plants by now-extinct honeycreepers, and a myriad of other interspecific interactions are also lost. Charles Elton's (1958) views were similar to Leopold's, but with a more explicit emphasis on the geographic dimensions of what we now term biodiversity. In his discussion of a "wilderness in retreat," Elton stated that "conservation should mean the keeping or putting in the landscape of the greatest possible ecological variety—in the world, every continent or island, and so far as practicable in every district." We would add to the geographic context of Leopold's initial statement and suggest that the goal of conservation biology is to preserve species distributions—

and, in so doing, preserve the ecological and evolutionary processes required to conserve diversity per se, as well as the natural character of nature.

Our goals in this chapter are fourfold. First, we will summarize the current status of biodiversity and review the geographic signatures of extinctions and endangerment. Second, we will discuss the geographic dynamics of extinction forces, including species invasions, anthropogenic changes in landscapes and seascapes, overharvesting, pollution, and climate change. Because the geography of extinction has much to do with the geography of our own species, we will then present an overview of the biographic dynamics of humanity, before concluding with a description of an emerging synthesis—conservation biogeography—which encapsulates the contribution biogeographic science can make to the conservation, not just of diversity per se, but to the geographic, ecological, and evolutionary context of nature.

Biodiversity and the Linnaean shortfall

The current rate of decline in biodiversity far exceeds that of purely natural conditions. Aldo Leopold, Charles Elton, Edward O. Wilson and other ecologists and evolutionary biologists of this and the past two centuries have repeatedly warned of the ongoing anthropogenic losses of native species and ecosystems. The existence of a biodiversity crisis is documented in the record of historical extinctions and the vulnerable and imperiled status of many extant species (**Figures 16.1** and **16.2**; World Conservation Monitoring Centre 1992; Wildlife Conservation Society 2006; IUCN 2008). Since the early 1600s, biologists have tallied the extinction of 134 species of birds and 78 species of mammals, including such wondrous forms as the great auk (*Pinguinus impennis*), dodo (*Raphus cucullatus*), passenger pigeon (*Ectopistes migratorius*), Tasmanian tiger (*Thylacinus cynocephalus*), Cuban solenodon (*Solenodon cubanis*), and Stellar's seacow (*Hydrodamalis gigas*). Over the same time period, 21 species of reptiles and 38 species of amphibians have vanished, including one amphibian family and 38 genera (most historic extinctions of amphibians have occurred during the last four decades; see Duellman and Trueb 1986; Wake and Vredenburg 2008; see Figure 3.25).

Aquatic species, including freshwater fish, crayfish, and mussels, are also included in this wave of historic extinctions. The rate of extinctions of freshwater fish in North America steadily increased during the past 100 years,

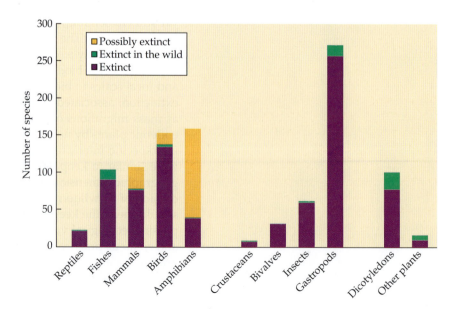

FIGURE 16.1 Extinctions in major groups of animals and plants during the past 500 years. Estimates of the number of species that are possibly extinct, while substantial in some groups (in particular, amphibians), are only available for mammals, birds, and amphibians. (After IUCN 2008.)

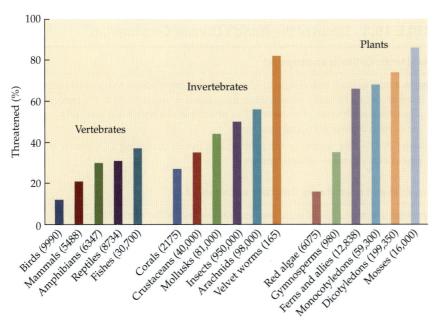

FIGURE 16.2 Status of major groups of animals and plants, as percent of evaluated species that are critically endangered, endangered, or vulnerable to extinction. When expressed as percent of total number of described species (numbers in parentheses in group labels), these values remain essentially unchanged for birds, mammals, amphibians and gymnosperms, but only 11 percent for corals, and less than 5 percent for all other taxa listed here. (After IUCN 2008.)

rising to some 40 species or subspecies lost during the latter decades of the twentieth century (Williams and Miller 1990). In the Great Rift Valley Lakes of East Africa, the diversity of native fish has plummeted just in the past few decades. In Lake Victoria alone, cichlid diversity has declined by as much as 200 species. Well over 10 percent of the freshwater fish of many countries are now threatened, and global extinction rates of these species continue to rise. Worldwide, as much as 20 percent of all freshwater fish species (about 1800 species) are now extinct or in serious jeopardy, with evidence now pointing to equally serious declines in the marine realm, connected with overfishing (especially of large fish species), pollution, and sensitivity to changing global climate (**Table 16.1**).

This wave of extinctions has by no means been restricted to the animal kingdom. In the United States and Canada, 164 native plant species are now recorded as extinct, while another 6217 species are officially listed as imperiled, or at risk (NatureServe 2009). The status of insular plants is especially grave. On the Hawaiian Islands, some 108 endemic plant taxa are now extinct, and another 175 are listed as endangered or vulnerable. On Saint Helena Island in the Atlantic Ocean, 7 endemic plant species are extinct, and all of the remaining 46 endemic species of plants are classified as endangered or threatened.

Even the fungi, an often neglected yet diverse and fascinating kingdom, have experienced a marked pulse of extinctions during recent decades. Today, some 20 percent of the approximate 8000 species of European fungi are threatened with extinction due to the degradation and loss of their habitats (Dahlberg and Croneborg 2003; see also Jaenike 1991; Webster 1997; Koune 2001; Moore et al. 2001; Mueller and Schmidt 2007; Lonsdale et al. 2008). The larger fungi (primarily mushrooms and puffballs) are especially vulnerable to both habitat loss and overharvesting, with 35 percent of their approximate 1000 species being threatened (Cherfas 1991).

As alarming as these statistics may seem, it is likely that they underestimate the actual decline in biodiversity. While scientists have now described over 1.7 million species, this represents only a fraction of the total species thought to occur on Earth. Estimates of total diversity vary but typically range between 5 million and 25 million species (World Conservation Monitoring

TABLE 16.1 *Status of the World's Oceanic Communities*

Coral reefs: Critically endangered

Symptoms: Live coral reduced 50 to 93 percent; fish populations reduced 90 percent; apex predators virtually absent; other megafauna reduced by 90 to 100 percent; population explosions of seaweeds; loss of complex habitat; mass mortality of corals from disease and coral bleaching

Drivers: Overfishing, warming and acidification due to increasing CO_2, runoff of nutrients and toxins, invasive species

Estuaries and coastal seas: Critically endangered

Symptoms: Marshlands, mangroves, sea grasses, and oyster reefs reduced 67 to 91 percent; fish and other shellfish populations reduced 50 to 80 percent; eutrophication and hypoxia, sometimes of entire estuaries, with mass mortality of fish and invertebrates; loss of native species; toxic algal blooms; outbreaks of disease; contamination and infection of fish and shellfish; human disease

Drivers: Overfishing; runoff of nutrients and toxins; warming due to rise of CO_2; invasive species; coastal land use

Continental shelves: Endangered

Symptoms: Loss of complex benthic habitat; fish and sharks reduced 50 to 99 percent; eutrophication and hypoxia in dead zone near river mouths; toxic algal blooms; contamination and infection of fish and shellfish; decreased upwelling of nutrients; changes in plankton communities

Drivers: Overfishing; trophic cascades; trawling; runoff of nutrients and toxins; warming and acidification due to rise of CO_2; introduced species; escape of aquaculture species

Open ocean pelagic: Threatened

Symptoms: Targeted fish reduced 50 to 90 percent; increase in nontargeted fish; increased stratification; changes in plankton communities

Drivers: Overfishing; trophic cascades; warming and acidification due to rise of CO_2

Source: Jackson 2008.

Centre 2000). By almost all estimates, most of what is out there is unknown to us. This deficiency, sometimes called the **Linnaean shortfall**, represents not just a pressing challenge but also a great opportunity for field biologists (see Hopkins and Freckleton 2002; Kozlowski 2008). There are many species waiting to be discovered, some of them so distinct that once described, they become the sole known representatives of new families, orders, or even phyla (**Table 16.2**; see also Staley 1997, and Fenchel and Finlay 2004 for reviews on the global diversity and conservation of microbes).

Given current trends in the specialization and training of taxonomists and systematists, however, the Linnaean shortfall is likely to remain with us for some time. While a great majority of undiscovered animals are thought to be insects, spiders, and other invertebrates, only 30 percent of today's taxonomists specialize in these groups (**Figure 16.3**). The geographic distribution of taxonomists—or, at least, their home institutions—is also biased. Over three-fourths of today's taxonomists are trained in temperate areas of the Nearctic and Palearctic regions—which, as we have seen, are not the world's most species-rich areas. Similarly, a quick review of Table 16.2 reveals that the marine realm is still a great biological frontier. Less than 10 percent of the world's

TABLE 16.2   *The Linnaean Shortfall*

1. Eleven of the 80 extant cetaceans were discovered in the twentieth century, one as recently as 1991.

2. One of the largest shark species, "megamouth," was discovered in 1976.

3. Three new families of flowering plants were discovered in Mexico during the 1990s.

4. Two new *phyla* were discovered during the 1990s, one (Loricefera) in the marine benthos and the other (Cycliophora) clinging to the mouthparts of lobsters.

5. Perhaps 90 percent of tropical forest insects remain unknown.

6. It has been estimated that perhaps 1.5 million fungi remain to be discovered.

7. About 4000 bacterial species have been described; a gram of soil may contain 4000 to 5000 species.

8. Even among mammals, one of the best-studied groups of relatively large organisms, new species continue to be discovered, including some 349 species described between 1992 and 2008.

Source: After Raven and Wilson 1992; Patterson 2000, 2001; IUCN 2008.

oceans have been adequately sampled for biological diversity, and hundreds if not thousands of small and even moderately rare species are easily missed by virtue of their diminutive or otherwise inconspicuous nature (Culotta 1994). For example, the 295 km² reef system off the west coast of New Caledonia (southwestern Pacific) is inhabited by at least 3000 species of mollusks—most of them less than 5 mm long (Vermeij 2004a; after Bouchet et al. 2002).

One of the most serious downsides of the Linnaean shortfall is the likelihood that many of the undiscovered species will go extinct before they are known to science. E. O. Wilson (1992) calls these **Centinelan extinctions**, and they no doubt represent a significant component of the actual loss in biodiversity. As scientists, we are called upon to identify the specific causes of the biodiversity crisis and objectively apply our knowledge to develop effective strategies to minimize the losses. The task is great, and it requires an integrative approach, one involving many disciplines and methods. Therefore, biogeography—being one of science's most holistic disciplines—has much to offer. More than any other scientific discipline, ours focuses on biological

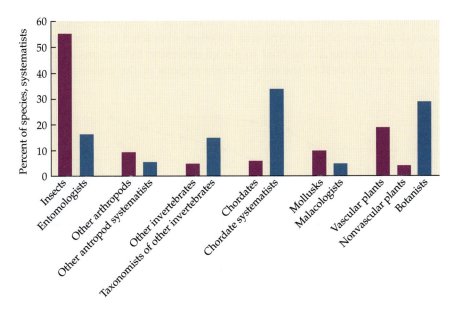

FIGURE 16.3   A comparison between the relative number of species worldwide in each of the principal taxonomic groups and the number of systematists who specialize in them. (After Barrowclough 1992.)

variation from local to global scales, including the geographic dynamics of species and the forces (natural or anthropogenic) that drive those dynamics.

The geography of prehistoric extinctions

Given that diversity varies in a nonrandom manner across geographic gradients and that extinction forces (many of them related to geographic expansions of other species) move across and transform the geographic template in a highly nonrandom manner, we might expect that extinctions will also exhibit a definite geographic signature. Yet, only recently have we had the information and technology to tackle some truly intriguing questions that address this interface between biogeography and conservation biology. Do extinctions vary with latitude, elevation, depth, area, or isolation? Do different taxa exhibit the same geographic trends in susceptibility to anthropogenic extinctions? These are no longer just academic questions. Given the limited time, energy, and financial support available for preserving biodiversity and preventing extinctions, it is imperative that we direct our efforts *where* they will do the most good.

Studies of the fossil record provide some useful clues to some of these questions. Paleontological studies, especially those on marine bivalves, reveal that natural extinctions can be highly nonrandom, but that the selectivity with respect to species traits and variation among geographic regions depends on the intensity of that extinction event. The most severe mass extinctions, such as the one at the end of the Cretaceous Period that saw a 60 to 80 percent drop in global species diversity, tended to be spatially uniform, except of course for the unsurprising concentration of extinctions near the Chicxulub crater where the 10 to 20 km diameter asteroid impacted (see Sharpton et al. 1992). Even these mass extinctions, however, exhibited selectivity among various clades of bivalves—with survivorship highest for the genera with the largest geographic ranges and habitat breadths (**Figure 16.4**; Jablonski 1986, 1989; Raup and Jablonski 1993; Rosenzweig 1995; Jablonski 2008a,b). The selectivity of extinctions was more pronounced, however, for the less severe extinction events and background extinctions throughout the geological record. Extinctions during these "normal periods," in addition to being highest in range-restricted clades, also tended to be positively correlated with body size and to be spatially heterogeneous. In fact, Jablonski's (2008a) research on extinctions and recovery from those events indicates that each of the three fundamental processes—extinction, invasion, and origination—exhibited nonrandom geographic patterns and, together, interacted to determine past and current patterns in regional to global diversity. Extinctions of marine bivalves were greatest in the high-latitude regions of the oceans, which in

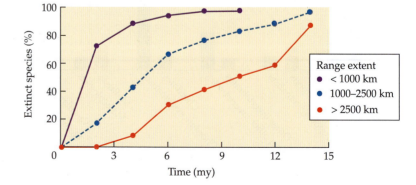

FIGURE 16.4 Extinction rates of marine bivalves and gastropods living along the Atlantic and Gulf coastal plains of North America during the Late Cretaceous were consistently higher for species with relatively small geographic ranges. (After Jablonski 1986.)

turn created opportunities for invasions from regions with relatively high origination rates, that is, the tropics (Jablonski 2008a; Krug et al. 2009; see also Jablonski 2008b). Thus, the spatial dynamics of marine bivalve diversity demonstrates the value of studying both the geography of extinctions and that of the recovery (origination and invasions) from those events.

Plate tectonic processes and, in particular, their influence on geographic isolation and opportunities for dispersal also have played an important role in triggering numerous waves of species invasions and extinctions in the fossil record. As you may recall from Chapter 10, the Great American Interchange is an important case in point. With the formation of the Central American land bridge approximately 3.5 million years ago, the Nearctic and Neotropical regions were connected after a long period of isolation. The ensuing migrations and radiations of immigrants caused a wave of extinctions among native species, especially in South America. As we shall see, this lesson in dynamic paleobiogeography is especially relevant to the current biodiversity crisis. Throughout the fossil record and during recent times as well, whether due to the formation or breakup of Pangaea, the Great American Interchange, or the climate-driven range shifts associated with glacial cycles of the Pleistocene, exchange among biotas that evolved in isolation has typically resulted in waves of extinctions as invaders overwhelmed or outcompeted native species.

The geography of recent extinctions and endangerment

Reviews of the historical record of extinctions have provided us with important clues for understanding the ongoing biodiversity crisis. The thousands of recorded extinctions have a definite geographic signature—one bearing witness to the high endemism and fragility of insular communities. As **Figure 16.5** reveals, a highly disproportionate number of the animal extinctions recorded since 1600 have occurred on islands (see Chapters 13 and 14). This insular bias applies to plants as well (see Figure 16.5A), and although just one in six species of plants are insular, one in three threatened species are endemic to islands (Schemske et al. 1994).

These insular extinctions have resulted from a combination of factors, nearly all related to human activities and the isolation and ecological naiveté of insular biotas. Many insular plants and animals, especially those endemic to more isolated and relatively depauperate and disharmonic islands, have been subjected to the effects of scores of exotic species introduced after humans colonized their islands. On Chiloé Island, located off the coast of Chile, Darwin found the native foxes—later named Darwin's fox (*Pseudalopex fulvipes*)—so naive that he was able to collect the type specimen by hitting it over the head with his geological hammer. Recall from Chapter 14 that, on visiting the Galápagos, he was amazed that the avifauna was so tame that he could prod a hawk off its branch with the butt of his rifle.

> A gun here would be superfluous. What havoc the introduction of any beast of prey must cause in a country before the inhabitants have become adapted to the stranger's craft of power. (Darwin 1860)

Of course, this was one of the many cases in which Darwin's words proved prescient. Two-thirds of the historical extinctions of insular birds have been caused either directly or indirectly by introduced mammals.

Throughout their history, truly remote insular biotas evolved in the absence of large, ground-dwelling mammals. Many species of insular plants, for example, were able to flourish without evolving defenses against mammalian herbi-

FIGURE 16.5 (A) Patterns in recorded extinctions (1600–1990) of plants, terrestrial animals, and mollusks. Extinctions have been far more common among animals inhabiting islands than among their mainland counterparts. Extinctions of two well-documented groups of vertebrates, (B) mammals and (C) birds, illustrate the strong geographic bias in extinctions. While particular distributions of historic extinctions vary among these two faunas, they both reveal the high susceptibility of insular biotas. In addition, the causes for historic extinctions in these two groups differ between them as well as across geographic regions (e.g., extinctions of mammals on the continents, excluding Australia, were primarily due to overexploitation, while those of most islands were caused by a combination of factors, including habitat modification and introduced species). (A after Reid and Miller 1989; World Conservation Monitoring Centre 1992; IUCN 2003; B,C after Heywood, pers. comm.)

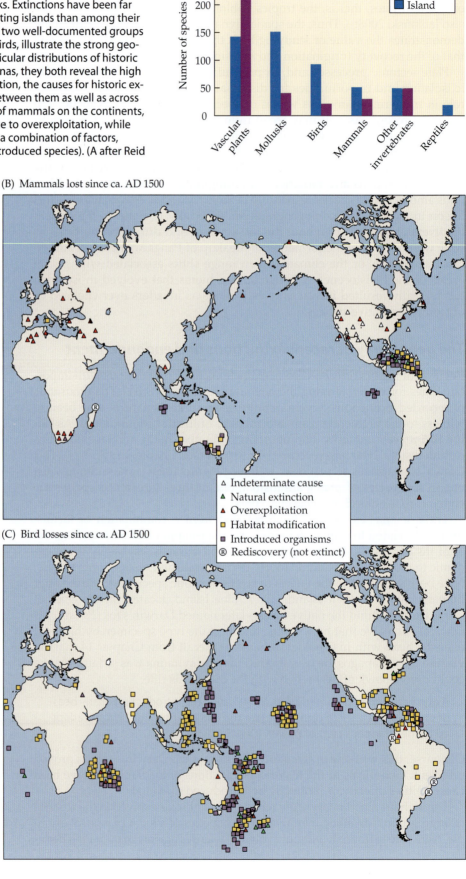

(A)

(B) Mammals lost since ca. AD 1500

△ Indeterminate cause
▲ Natural extinction
▲ Overexploitation
□ Habitat modification
■ Introduced organisms
Ⓡ Rediscovery (not extinct)

(C) Bird losses since ca. AD 1500

vores. In a very real sense, the biological novelty and fragility of insular biotas derives from their isolation. With the introduction of exotic species, whether planned or accidental, this "splendid isolation" was lost. Just as the formation of the Central American land bridge triggered the great faunal interchange and the subsequent extinctions of many species endemic to South America, anthropogenic introductions of exotic species onto islands effectively connected them to the mainland and triggered a wave of extinctions.

Actually, there have been multiple waves of insular extinctions. Europeans were not the only civilization to colonize islands and subsequently devastate the native biota. Subfossil evidence now indicates that insular extinctions typically follow from initial human colonization, and new bursts accompany each new human culture that arrives. Hence, this process has been occurring for millennia, involving people originating in many different regions and cultures across the globe, including the aboriginals of Malaysia, Indonesia, Australia, and Tasmania; the Micronesians, Melanesians, and Polynesians of the isolated archipelagoes across the Pacific (who may have caused extinctions of over 1000 species of birds; Steadman 1995, 2006); and the Amerindians of the West Indies (Morgan and Woods 1986; Woods and Sergile 2001; Steadman and Martin 2003; Steadman 2006). Olson and James (1982a) estimate that the Polynesians may have caused the extinctions of nearly half of the native avifauna of Hawaii. In New Zealand, all of the nine or so species of moas (immense flightless birds that were endemic to these islands) were extirpated by the Maori people and their introduced dogs and rats before European colonization (see Box 14.1; Halliday 1978; Worthy and Holdaway 2002). These extinctions were far from random, with the largest species falling first in a tragic, domino-like wave of extinctions. New Zealand's only surviving ratites are the much smaller kiwis (*Apteryx* spp.).

The historical record of extinctions also reveals that, in addition to the introductions of non-native species (inclusive of diseases), overharvesting and habitat alteration (reduction, fragmentation, and replacement of native habitats) have also taken a heavy toll on native species. For example, hunting, collection, and other forms of overexploitation accounted for nearly a third of the avian extinctions and half of the mammalian extinctions worldwide since 1600 (**Figure 16.6**). Extinction patterns in mollusks, perhaps the hardest-hit group of animals, present a somewhat different picture. Most of their historical extinctions were associated with habitat destruction and pol-

FIGURE 16.6   Causes of historical extinctions (1600–1980) and of current endangerment in selected animal taxa. (After Flather et al. 1994.)

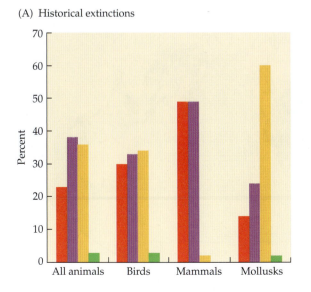

(A) Historical extinctions

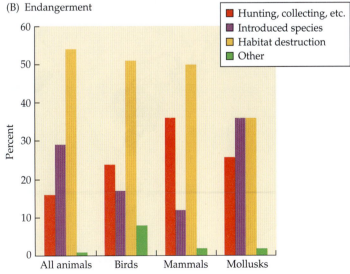

(B) Endangerment

Hunting, collecting, etc.
Introduced species
Habitat destruction
Other

FIGURE 16.7 Historical trends in the relative numbers of insular and mainland extinctions (1600–1990) of animals worldwide. (After World Conservation Monitoring Centre 1992.)

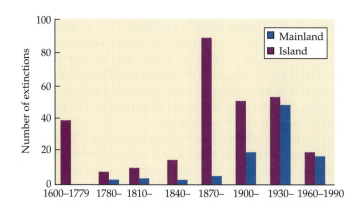

FIGURE 16.8 The clearing of primary forests across the globe has rapidly reduced once-expansive stands of native, forested landscapes to archipelagoes of habitat islands. Insets: (A) Olympic National Forest, Washington, USA; (B) Cadiz Township, Wisconsin, USA; (C) Warwickshire, England; (D) Borneo, Indonesia. (Global map courtesy of Biodiversity and Climate Change Programme, World Conservation Monitoring Centre 2000; inset A after D. R. Perault, unpublished report; inset B after Thomas 1956; inset C after Wilcove et al. 1986; inset D cartography by Hugo Ahlenius, UNEP/GRID-Arendal Maps and Graphics Library 2007.)

lution (60 percent), while introduced species and collecting accounted for a smaller, yet still significant, portion (24 and 14 percent, respectively).

While we have learned much by examining the patterns of past extinctions, it appears that the geographic signature of extinctions and their predominant causal forces are changing (compare the graphs of Figure 16.6A and B). **Figure 16.7** reveals an important temporal–spatial trend in animal extinctions: The extinction front has expanded to include the mainland. Although nearly all animal extinctions recorded between 1600 and 1980 were insular, the number of extinctions of mainland species now rivals that of insular forms. Ironically, the ongoing wave of extinctions may still be an insular phenomenon. Habitat destruction and fragmentation on the continents have transformed the geographic template and reduced once-expansive stands of native habitats to ever-shrinking archipelagoes of habitat islands (**Figure 16.8**). Terrestrial biotas stranded in these remnants of native habitats may undergo a process of ecological relaxation similar to that described in Chapter 13. However, in contrast to the climatic and tectonic changes that have isolated habitats in the past, these anthropogenic habitat changes are occurring much more rapidly and may, therefore, take a heavier toll on the stranded species.

Tropical	Temperate and Boreal
■ Current	■ Current
■ Original	■ Original

Global distribution of original and remaining forests

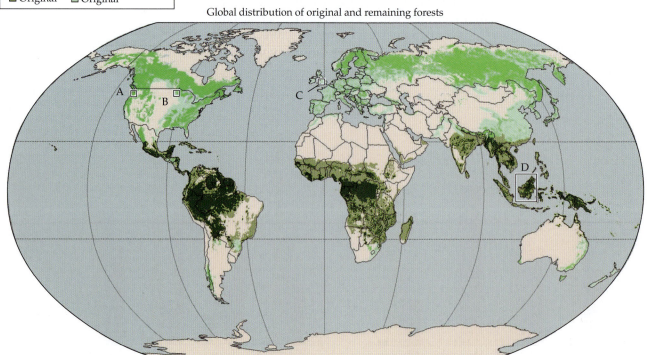

(A) Original forest   1945

1965   1985

10 km

(B) 1831   1882   1902   1950

Forest

Prairie

(C) ca 400   ca 1086   ca 1650   ca 1960

10 km

(D) 1950   1985   2000   2005   2010   2020

FIGURE 16.9 Causes of endangerment for plants of the United States. (After Schemske et al. 1994.)

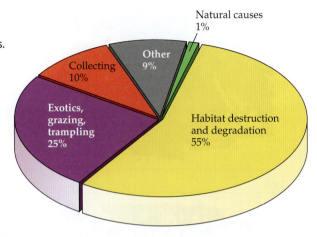

Whereas anthropogenic introductions and hunting were the primary causes of historical extinctions, habitat loss and fragmentation have now become the primary causes of endangerment in terrestrial animals (see Figure 16.6), whilst some scientists now argue that climate change represents an equally great threat for the present century, and one, moreover, that will have synergistic interactions with habitat fragmentation and loss. In mollusks, species introductions and habitat alteration contribute equally to spe-

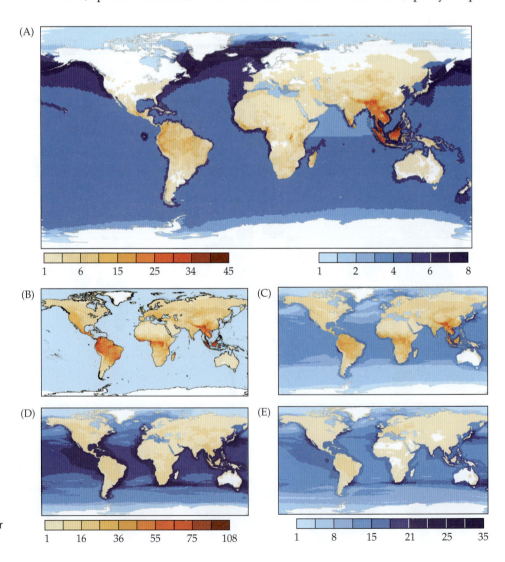

FIGURE 16.10 Geography of endangerment (A) and principal threats to terrestrial and marine mammals from (B) habitat loss, (C) harvesting, (D) accidental mortality, and (E) pollution. (After Schipper et al. 2008.)

cies endangerment. Patterns of endangerment in plants are similar to those reported for terrestrial animals (**Figure 16.9**). Habitat alteration, including development for housing, water control, mining, oil and gas, logging, off-road vehicles, and agriculture, represents the primary cause of endangerment for plant species listed by the U.S. Fish and Wildlife Service under the Endangered Species Act. The effects of introduced species and the effects of grazing and trampling by domestic livestock account for another 25 percent of species endangerment. Perhaps the most sobering finding is that natural causes account for just 1 percent of the endangerment of plants.

Just as we have seen for the spatial patterns of extinctions, both prehistoric and recent, species endangerment also exhibits a strong geographic signature, one ultimately reflecting the nonrandom manner in which we modify native ecosystems. Recent analyses of global patterns of endangerment in vertebrates provides some especially important insights for understanding and developing geographically explicit strategies for conserving biological diversity. Recall from Figure 3.25C that hotspots of threatened amphibians tended to concentrate in tropical and subtropical regions, and especially on relatively large islands or montane regions adjacent to human development. Terrestrial mammals exhibit a similar, albeit more pronounced, latitudinal gradient in endangerment, while the pattern for marine mammals is just the converse of this—that is, with endangerment increasing with latitude, especially in the waters of the Northern Hemisphere (**Figure 16.10**). Levels of endangerment are especially high for insular mammals—in particular, those of the larger islands of Indonesia, and for those of montane regions (e.g., along the Andes, the Mexican Sierras, and the Himalayas). Finally, avian hotspots of endangerment also exhibit some similar geographic trends, but with some unique features reflecting intertaxon differences, including the superior dispersal abilities of birds (**Figure 16.11**). For example, hotspots of threatened birds not shared with those of terrestrial mammals include those of southern Brazil and those extending into more isolated regions of Indonesia and Oceania.

The salient lesson of these exercises in the geography of endangerment is that hotspots of imminent species loss occur where anthropogenic threats

FIGURE 16.11 Geographic variation in species density (species per 1° by 1° grid cell) of threatened birds shares patterns similar to that of mammals (compare to Figure 16.10A), but with some additional, unshared hotspots, including those of southern Brazil and on oceanic islands. (After Grenyer et al. 2006.)

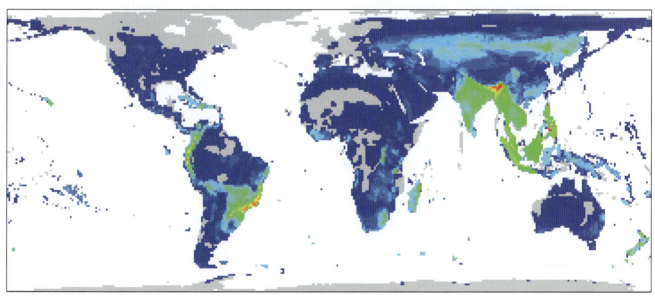

1   31

overlap those of geographically restricted, and often ecologically naive species. Thus, to develop more effective strategies for conserving endangered species, it is paramount that we understand the geographic dynamics of extinction forces and the resultant, geographic dynamics of the rarest species.

Geographic range collapse

> The extinction problem has little to do with the death rattle of its final actor. The curtain in the last act is but a punctuation mark—it is not interesting in itself. What biologists want to know about is the process of decline in range and numbers. (Michael Soulé 1983)

It is axiomatic that all extinctions—historic or prehistoric, natural or anthropogenic—are preceded by declines in geographic ranges. Throughout most of the history of biogeography and conservation biology, however, this process of decline (i.e., geographic range collapse) has largely remained a mystery. When a species' range collapses, does the process exhibit a consistent or predictable pattern? For example, do populations in particular portions of the range (e.g., central vs. peripheral, poleward vs. equatorial, or in high vs. low elevations) tend to fare better than others? These are not just academic questions, but also ones with important implications for guiding conservation strategies. Efforts to reintroduce or translocate populations of endangered species, to locate nature reserves, to establish conservation corridors, or to search for undiscovered remnants of a once-widespread species would be greatly facilitated if we could develop a spatial search image for the locations most likely to yield favorable results.

Early conservation biogeography theory suggested that reintroductions, translocations, and searches for remnant populations should be conducted near the core of a species' historical geographic range. After all, the range boundaries are just that, sites where environmental conditions are so poor that the species cannot maintain its populations without recruitment from other populations. Areas along the periphery of a species' range are often thought to represent sink habitats (sensu Pulliam 1988; see Chapter 4) for individuals emigrating out from more central source populations. In addition, areographic studies indicate that, as we move from the center to the more peripheral areas of a species' geographic range, densities tend to decrease while variability in densities over time tends to increase (Figure 4.16; but see Blackburn et al. 1999b; Sagarin and Gaines 2002). Thus, under the "melting range hypothesis," which assumes an equal and simultaneous decline in all populations across a species' geographic range, dwindling populations should implode, with the final populations persisting near the center of the species' historic range (see also Gaston 2003: 167–177). These predictions of melting and imploding ranges are not new and were, in fact, anticipated by some of the most prominent figures in the history of biogeography, including Alfred Russel Wallace and Philip J. Darlington.

> On a continent, the process of extinction will generally take effect on the circumference of the area of distribution, because it is there that the species comes into contact with such adverse conditions or competing forms as prevent it from advancing further. (A. R. Wallace 1876)

> The actual limits of range will then be determined not by the limits of favorable ground but by a constantly fluctuating equilibrium between tendency to spread at the center of the range and tendency to recede at the margins. (P. J. Darlington Jr. 1957: 548)

Conservation biologists of the latter part of the twentieth century followed the leads of Wallace and Darlington, often viewing the range periphery as the "land of the living dead" and the "domain of zombies"—sites with little value for conserving endangered species (but see Hunter and Hutchinson 1994; Lawton 1995; Lesica and Allendorf 1995; Peterson 2001; Rodriguez 2002). This may, however, be a case in which generic prescriptions can prove misleading. For example, long-term ecological analyses and species distribution modeling indicate that European beech (*Fagus sylvatica*) may in some regions remain in a phase of range expansion from past climate changes. In addition, current and future global climate change is expected to drive a further northward range shift of this tree species, in which case northern range margin populations would be anticipated to continue to do well, while southern range margins may become increasingly marginal for this and other species (compare Lindbladh et al. 2008; Magri 2008). Moreover, setting aside climate change considerations, empirical analyses of geographic range collapse in rare and endangered species of vertebrates, invertebrates, and plants indicate that, contrary to early theory, relictual populations of these now geographically restricted species occur along the periphery, not near the core of the species' historical geographic ranges (Lomolino and Channell 1995, 1998; Channell 1998; see also Safriel et al. 1994; Towns and Daugherty 1994; Laliberte and Ripple 2004). Among the species persisting along the range periphery are at least five "poster species" for the conservation movement: giant panda (*Ailuropoda melanoleuca*), red wolf (*Canis rufus*), black-footed ferret (*Mustela nigripes*), California condor (*Gymnogyps californianus*), and whooping crane (*Grus Americana*; **Figure 16.12** and **16.13**). Analyses of range collapse in other taxa are concordant with these results (**Figure 16.14**). The actual progression of range collapse is typically quite different from that

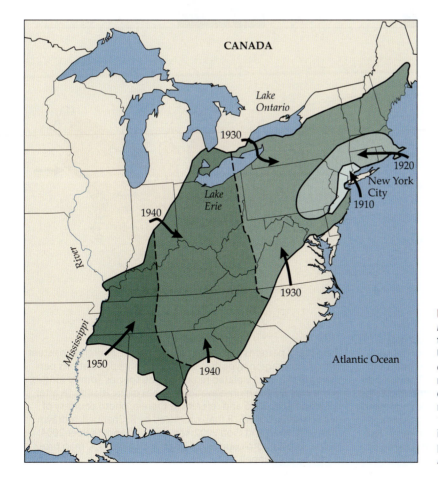

FIGURE 16.12 The American chestnut (*Castanea dentata*) was once one of the most common trees in the eastern deciduous forests of the United States. The chestnut blight, first introduced near New York City around 1910, spread rapidly to cause the decline and range collapse of the chestnut. The solid outline indicates the natural range of the American chestnut prior to 1910; dashed lines and green-shaded regions indicate the sequential collapse of chestnut populations between 1910 and 1950. (After Anderson 1974; Bell and Walker 1992.)

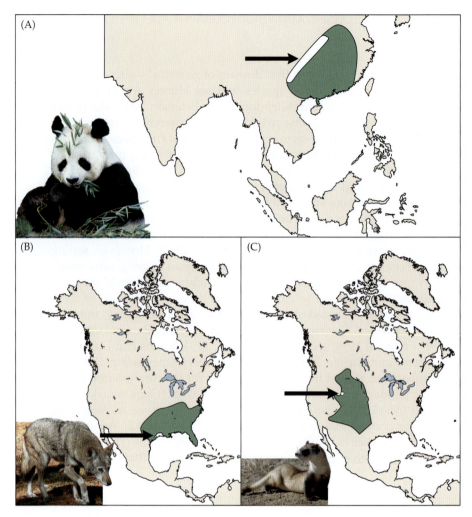

FIGURE 16.13 Patterns of range collapse in three species of terrestrial mammals: the giant panda (*Ailuropoda melanoleuca*), red wolf (*Canis rufus*), and black-footed ferret (*Mustela nigripes*). The locations of remnant populations of these endangered species are indicated in white; the dark green polygons represent the historical ranges of the species. (After Lomolino and Channell 1995.) (Photos: (A) © Gerry Ellis/DigitalVision; (B) courtesy of Tim Ross; (C) courtesy of Ryan Hagerty/USFWS.)

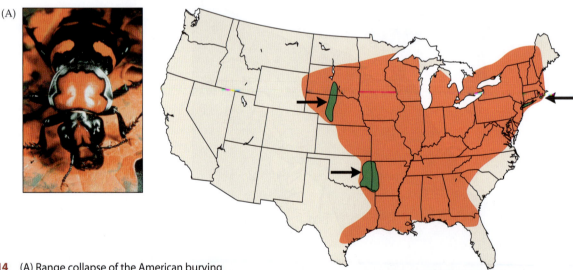

FIGURE 16.14 (A) Range collapse of the American burying beetle (*Nicrophorus americanus*). The historical range of this species (ca 1900–1940) is indicated by red shading, while the extant range is indicated by green polygons and arrows. (B) The Komodo dragon (*Varanus komodoensis*) was once broadly distributed across the island of Flores and nearby islands in Indonesia, but now persists only in peripheral regions of that range, including the island of Komodo and other islands east of Flores (historic ranges in red, extant or final ranges in green). (C) Geographic range collapse of insular animals such as Hawaiian birds may appear anomalous to the general patterns for range collapse, but are entirely consistent with the contagion hypothesis—with the final populations persisting in the most isolated regions of the species' former ranges, in this case on the high-elevation reaches of the islands. (A after Lomolino et al. 1995 and references therein, photo courtesy of Ryan Hagerty/USFWS; B map produced by R. Channell, redrawn from Sastrawan and Ciofi 2002; see also Ciofi and de Boer 2004; Molnar 2004; photo courtesy of Dezidor/Wikipedia; C map produced by R. Channell.)

(B)

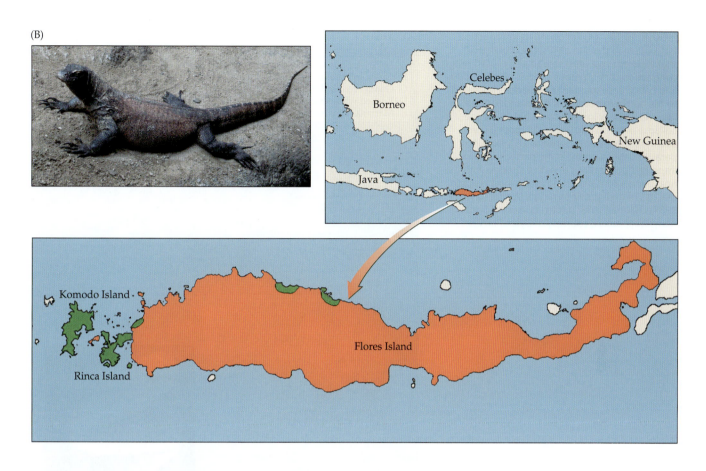

(C)

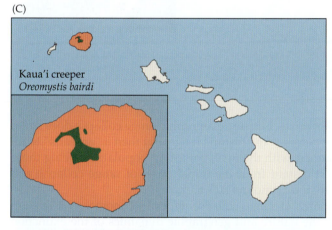

Kaua'i creeper
Oreomystis bairdi

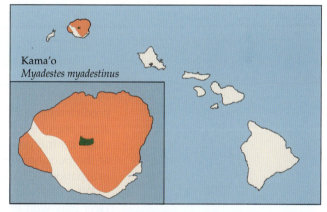

Kama'o
Myadestes myadestinus

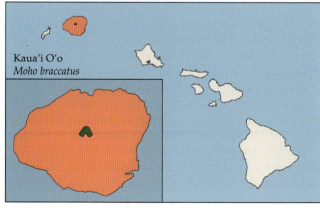

Kaua'i O'o
Moho braccatus

FIGURE 16.15 (A) Following its introduction onto Guam in the late 1940s, the brown tree snake (*Boiga irregularis*) spread and caused range contractions of numerous species of native birds and bats whose remnant populations retreated to the northern portion of the island (i.e., the last area to be colonized by *Boiga*). (B) Range collapse of the exotic browntail moth (*Euproctis chrysorrhoea*) in response to an introduced parasitoid followed the same pattern seen for imperiled populations of native species, with the final populations persisting in the most isolated reaches of their previous range. (A after Savidge 1987; Channell 1998; photo courtesy of Soulgany/Wikipedia; B courtesy of Dylan Parry.)

(A) Guam

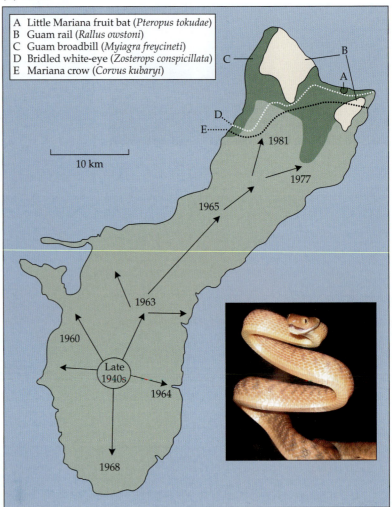

A Little Mariana fruit bat (*Pteropus tokudae*)
B Guam rail (*Rallus owstoni*)
C Guam broadbill (*Myiagra freycineti*)
D Bridled white-eye (*Zosterops conspicillata*)
E Mariana crow (*Corvus kubaryi*)

10 km

predicted by earlier theorists. Initially, geographic range collapse does begin along one of the peripheries of the historic range, but then it "spreads" to include central populations, with the final, persisting populations being those at the periphery most distant from the initial onset of range collapse and extirpations (Channell and Lomolino 2000a,b).

Why do these patterns of range collapse appear so anomalous? The answer may well lie in the nature and geographic dynamics of the extinction forces. Even though the macroecological patterns of population density and variability appear to be very general ones, they may be overwhelmed by anthropogenic disturbance. Indeed, given that we are studying range collapse and extinction, these disturbances must (by definition) be significant enough to render the areographic patterns irrelevant. And here is a second key distinction between the melting range hypothesis and actual processes causing range collapse and extinction: Whereas the melting range hypothesis assumes a ubiquitous extinction force whose effects are nonvarying and simultaneous across space, nearly (if not all) known anthropogenic disturbances are spatially nonrandom and predictably dynamic (**Figures 16.15** and **16.16**). That is, they tend to spread across the landscape like a contagion from the beachheads to the interior, and from the lowlands to mountain peaks. Anthropogenic activities, including agriculture, fragmentation, introduced species, dis-

(B)

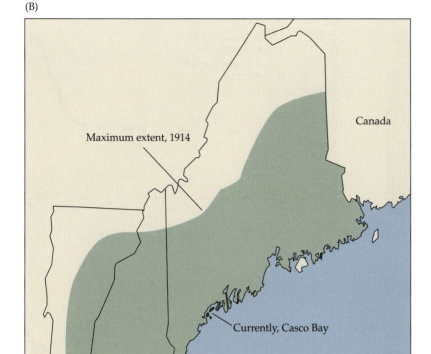

eases, and advanced hunting technologies, tend to move across the landscape like fire across a burning leaf: Regardless of the flash point, the last piece to burn will be along an edge. In these circumstances, isolation may be the key to persistence. Thus, remnant populations subject to contagion-like extinction pressures should occur in isolated regions, on an island, a mountain range, or an otherwise isolated portion of the species' historical range.

An additional, albeit nonexclusive, explanation for persistence of at least some peripheral populations during the latter stages of range collapse derives from the relative number, genetic diversity, and local adaptations of these populations. In contrast to the relatively limited number of central populations, those along the periphery of a species' geographic range are more numerous and may, therefore, exhibit greater variation (physiological, behavioral, ecological, and genetic) among populations. Perhaps just as important, these peripheral populations must adapt to a much greater range of environmental conditions in comparison to those prevailing near the center of the species' range. Therefore, depending of course on the nature of the threatening process, it is more likely that one of the many and diverse peripheral populations will be preadapted to extinction forces that might threaten its populations in the future. Geographic range collapse in the American burying beetle (*Nicrophorus americanus*) illustrates the potential relevance of both the contagion

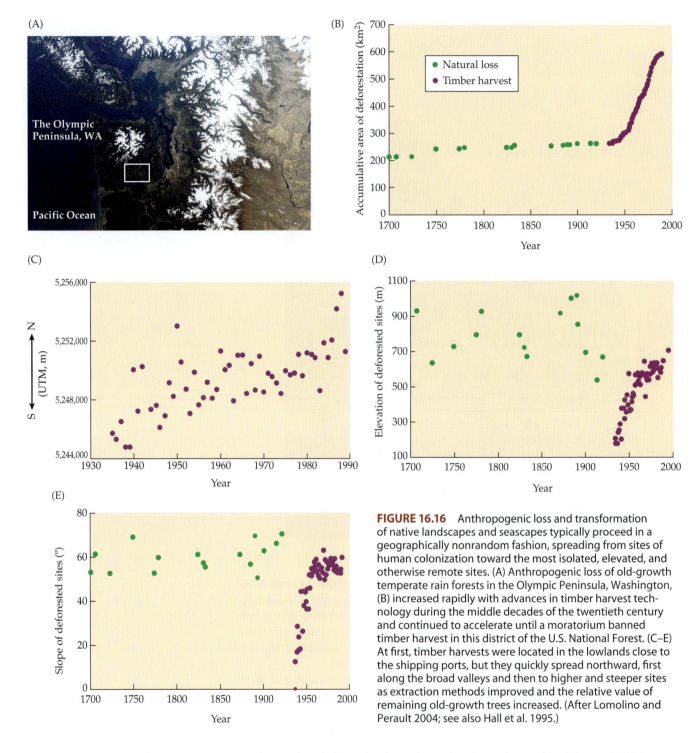

FIGURE 16.16 Anthropogenic loss and transformation of native landscapes and seascapes typically proceed in a geographically nonrandom fashion, spreading from sites of human colonization toward the most isolated, elevated, and otherwise remote sites. (A) Anthropogenic loss of old-growth temperate rain forests in the Olympic Peninsula, Washington, (B) increased rapidly with advances in timber harvest technology during the middle decades of the twentieth century and continued to accelerate until a moratorium banned timber harvest in this district of the U.S. National Forest. (C–E) At first, timber harvests were located in the lowlands close to the shipping ports, but they quickly spread northward, first along the broad valleys and then to higher and steeper sites as extraction methods improved and the relative value of remaining old-growth trees increased. (After Lomolino and Perault 2004; see also Hall et al. 1995.)

and peripheral diversity/preadaptation hypotheses (see Figure 16.14A; see also Thomas et al. 2008). The historic decline of this, the largest member of a guild of carcass-burying beetles, can be attributed to a combination of at least three factors: (1) anthropogenic decline and fragmentation of forested habitats (which had provided optimal substrates for burying carcasses and rearing offspring), (2) range expansion of raccoons, skunks, and other carnivorous mammals that compete for carcasses, and (3) extinction of one of its key sources of carcasses—passenger pigeons (Lomolino et al. 1995; Lomolino and Creighton 1996; Rosenzweig and Lomolino 1997). The locations of the remaining populations of this species, including those along the western margins of its historic range and on an island off the east coast of Rhode Island (USA), are entirely consistent with the contagion hypothesis. Then again, western populations of this species occurred in the transition zone between the deciduous forests of the east and prairies of the Great Plains. Therefore, these populations may also have been preadapted to deforestation that contributed to declines of nearly all of the populations to the east.

Of course, not all cases of range collapse follow these patterns but, as is common in many scientific studies, the anomalies provide invaluable insights into the underlying, causal forces. While over 75 percent of the cases of mammalian range collapse in Australia, North America, Europe, and Asia during historic times were indeed peripheral, the pattern for African species did not differ significantly from random expectations (i.e., a 50-50 ratio of peripheral to central collapse; Channell 1998). The apparently anomalous case for Africa is, however, entirely consistent with the human history of the African continent, which also accounts for the lack of megafaunal extinctions in Africa during historic times (see Chapter 9 and Diamond 1984). Because humans coevolved and were interdispersed with Africa's native plants and animals, extinction fronts, when they occurred, spread from many points across the continent rather than in a single wavelike contagion across the range.

Patterns of range collapse for many insular species also appear anomalous with respect to the general pattern described above; in this case, persistence being highest for the interior ("central") populations (see Figure 16.14C). These patterns are, however, entirely consistent with the contagion hypothesis and patterns of human colonization and expansion across island habitats. Once founding populations of Polynesians established their populations along the beachfronts, subsequent populations spread from these coastal lowlands toward the interior and higher elevations. As a result, the ranges of native plants and animals of Hawaii and many other oceanic islands collapsed away from lower elevations and other areas of anthropogenic disturbance (J. M. Scott, pers. comm.; P. A. V. Borges, pers. comm.), with remnant populations persisting only at the higher elevations (see also Burney et al. 2001). In another illustrative case study, ranges of native birds on the island of Guam collapsed, not along an elevational gradient, but with the well-documented expansion of the range of the introduced brown tree snake (*Boiga irregularis*), which preyed heavily on native birds (see Figure 16.15A). Some bird species now depend for survival on small offshore islands, a pattern in common with many other endangered island vertebrate species. Similarly, range expansion of an exotic parasitoid (the generalist tachinid, *Compsilura concinnata*) introduced into the New England region of North America resulted in range collapse of one of its key target species, the exotic browntail moth (*Euproctis chrysorrhoea*). The pattern of range collapse, this time for an exotic species, followed the pattern observed for most natives, with the final populations of browntail moths persisting in the most isolated reaches of their alien range (see Figure 16.15B). Unfortunately, the parasitoid did not discriminate between exotics and natives, and it simultaneously caused range collapse in giant silk moths (including *Hyalophora cecropia*, *Callosamia promethea*, *Antherea polyphemus*, and *Actias luna*) and numerous other species of native, mega-invertebrates (D. Parry, pers. comm.).

In all these cases, whether populations persist along the edges of once-expansive ranges, on offshore islands, or at high elevations, the common factor is that the final refugia were isolated and, therefore, the last to be affected by the spread of anthropogenic disturbances.

These studies of range collapse have obvious applications for conserving endangered species. In developing effective strategies, conservation biologists must consider the geographic dynamics of extinction forces as well as the ecological and biogeographic characteristics of the species in question, including our own (**Figure 16.17**). Just as important, these studies show that sites along the periphery of a species' historic range should no longer be dismissed as sites with little conservation value. Instead, reintroductions,

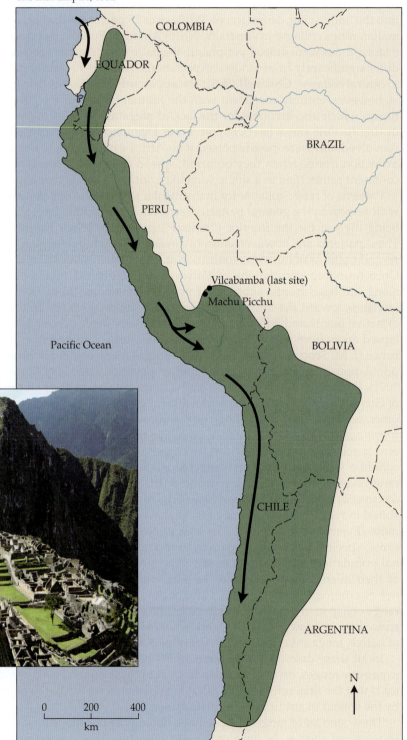

FIGURE 16.17 With the invasions of the Spanish Conquistadores (arrows), the once-vast range of the Incas collapsed to their final stronghold in the isolated and high sites of Machu Picchu and Vilcabamba (elevations 2500 and 3173 m, respectively). The Spaniards passed through the river valleys but failed to discover the last Inca settlements in the mountains. (After Davies 1995; photo © Amy Harris/istock.)

The Inca Empire, 1532

COLOMBIA

EQUADOR

BRAZIL

PERU

Vilcabamba (last site)
Machu Picchu

Pacific Ocean

BOLIVIA

CHILE

ARGENTINA

N

0   200   400
km

Machu Picchu

translocations, locations of nature reserves and corridors, and surveys for relictual populations should be conducted wherever favorable sites can be located throughout the historical range of an endangered species and, perhaps, in the light of climate change projections, outside the known range as well. Of course, conservation biologists must also consider the possibility that a species translocated outside its native range would negatively affect species native to that site. It is also possible that we may apply the lessons of these and future studies of geographic range collapse in reverse—in this case, developing a geographically explicit, range-wide strategy to accelerate and direct range collapse of *exotics* and otherwise problematic species.

The Dynamic Geography of Extinction Forces

The ecology and geography of invasions

Magnitude of the problem

The frequency and diversity of anthropogenic introductions of exotic species are staggering and now far exceed natural invasion rates (Hodkinson and Thompson 1997). Species introductions, whether intentional or accidental, have touched and transformed all types of ecosystems, from oceanic islands and isolated mountaintops to tropical rain forests and the far reaches of the Antarctic (Williamson 1996; Vitousek et al. 1997b; Lonsdale 1999; Mack et al. 2000; Pimentel et al. 2000; **Table 16.3**; **Figure 16.18**). More than any other events, including those associated with plate tectonics or glacial cycles, species introductions have had a demonstrable effect on the diversity and geography of the world's biota, especially those native to islands (McKinney and

TABLE 16.3 *The Number of Introduced, or Nonnative, Species of Birds and Freshwater Fish in Selected Regions*

Region	Breeding birds		Freshwater fish	
	Native	**Nonnative**	**Native**	**Nonnative**
Europe	514	27		74
California			76	42
Alaska			55	1
Canada			177	9
Mexico			275	26
Australia		32	145	22
South Africa	900	14	10,7	20
Peru				12
Brazil	1635	2	517	76
Bermuda		6		
Bahamas	288	4		
Cuba		3		10
Puerto Rico	105	31	3	32
Hawaii	57	38	6	19
New Zealand	155	36	27	30
Japan	248	4		13

Source: Vitousek et al. 1996.

Note: The relative number of exotic species is especially high on islands such as Puerto Rico, Hawaii, and New Zealand, where the number of non-native fish may rival or exceed that of native species.

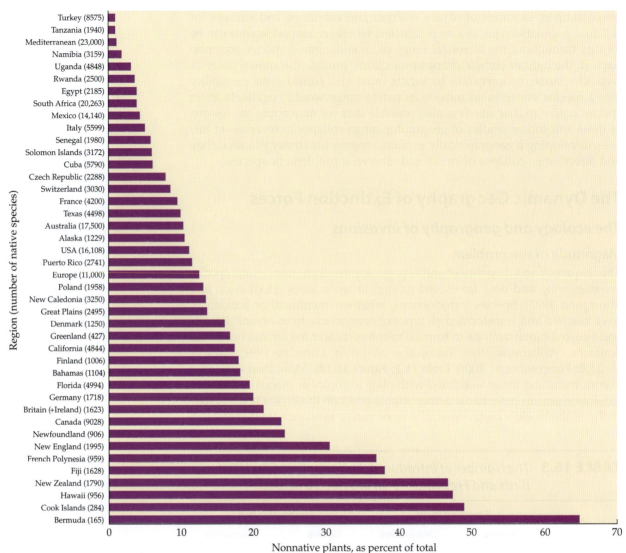

FIGURE 16.18 The percentage of introduced, or nonnative, species of plants varies substantially among geographic regions and tends to be disproportionately high for islands such as Bermuda, Hawaii, the Cook Islands, and New Zealand, where exotics may comprise roughly half of the extant flora. (After Vitousek et al. 1996.)

Lockwood 1999; Lockwood and McKinney 2001; Lockwood 2004; Clavero and Garcia-Berthou 2005; Lockwood et al. 2007; Vellend et al. 2007; Donlan and Wilcox 2008). They have driven many hundreds of geographically restricted, native species to extinction and replaced them with a redundant set (with relatively limited diversity) of species adapted to human-modified habitats.

While it is often overlooked, the list of problematic non-nonnative species threatening the world's biota throughout history, as well as in prehistoric times, includes our own species along with the commensals and pathogens that we spread during our waves of global colonization (Smith et al. 2006; Desprez-Loustau et al. 2007; Snall et al. 2008; Moodley et al. 2009). Thus, the effects of species introductions include two principal extinction forces often considered separately from introduced species—habitat fragmentation and overharvesting by invading populations of humans. As we describe in a later section, because human colonizations and range expansion were highly nonrandom across the geographic template, they resulted in the highly nonrandom geographic signature of range collapse and species extinctions. As with any predator, overharvest by humans was highly selective, resulting in a disproportionate loss of the most vulnerable (e.g., flightless or ecologically naive birds and reptiles, and geographically restricted plants) and the largest and rarest species. The result of these invasions by redundant suites of species, along with extinctions of natives endemic to particular regions has been the ongoing homogenization of nature (sensu McKinney and Lockwood 1999; see also Rahel 2002; Qian and Ricklefs 2006), reducing both diversity of the world's species and the distinctness of regional biotas—the latter, of course, being the basis for biogeography's most fundamental pattern, Buffon's law (see Olden and Poff 2003 for theoretical scenarios of invasion and extinction that may *increase* the distinctiveness among biotas). Even for those species that survive, overharvesting of fish, ungulates, invertebrates, and plants often has long-term, evolutionary consequences that far outpace those of natural selection or other forms of artificial selection (by 300 and 50 percent, respectively; Darimont et al. 2009; Stenseth and Dunlop 2009). Such intense, artificial selection has resulted in downsizing of fish and other wildlife, thus reversing the effects of natural selection by removing the largest, and otherwise fittest, individuals.

Clearly, attempts to understand and ultimately mitigate the effects of invasions by human populations and subsequent species introductions must consider the cascading ecological and evolutionary effects on native biotas and, conversely, how the characteristics of native ecosystems influence the would-be invaders. On the latter point, introduced species may become successfully established in natural (i.e., undisturbed) habitats, especially if they are similar to the invaders' native habitats and if they are inhabited by relatively few species. For example, populations of pigeons (*Columba livia*) are established along many coastal, cliffside habitats of New Zealand (which are similar to their ancestral habitats) and, although restricted to anthropogenic environments on species-rich islands of the Philippines, introduced rodents are often found established in native habitats on species-poor islands (D. F. Sax and L. R. Heaney, pers. comm.). A related observation, from studies of insular birds across a number of well-studied oceanic island archipelagoes, is that the number of species that have become established is generally highly correlated with the number of native birds that were lost to extinctions at some earlier period (Sax et al. 2002).

Taxonomic and geographic biases in introductions and established nonnatives

Although it may be impossible to estimate the total number and diversity of species introductions that have occurred even within historic times, we know that many were conducted as part of "naturalization" programs—deliberate attempts by Europeans to surround themselves with familiar species in exotic lands. Such deliberate programs do at least provide us with useful information on the magnitude, diversity, and geographic patterns of attempted introductions and established nonnative species. It is not just Europeans who do this, of course, and many other human groups have taken useful commensals with them when colonizing new territories, not least the Polynesians.

In 1988, Torbjorn Ebenhard reviewed all available records of historical introductions of birds and mammals (Ebenhard 1988a,b). Again, this has to be viewed as an incomplete record: An inestimable number of introductions went unreported. Yet, Ebenhard's survey is highly instructive. He found published reports documenting 788 mammalian introductions, including 118 species, 30 families, and 8 orders. The most frequently introduced mammals included common rabbits (*Oryctolagus cuniculus*), domestic cats (*Felis catus*), rats (including *Rattus norvegicus*, *R. rattus*, and *R. exulans*), house mice (*Mus musculus*), and domestic pigs, cattle, goats, and dogs. Not surprisingly, the pool of introduced species was not a random sample of the mainland faunas. Carnivores and artiodactyls (deer and related species of ungulates) constituted 19 and 31 percent of the introductions, respectively, whereas they represent less than 7 percent of the mainland species pool. Ebenhard also tallied 771 introductions of birds, including 212 species from 46 families and 16 orders. Again, species introductions were not random: waterfowl, gallinaceous birds, pigeons, and parrots were overrepresented in the sample of introduced species (representing 46 percent of the introduced species versus just 12 percent of all avian species belonging to these orders).

(A)

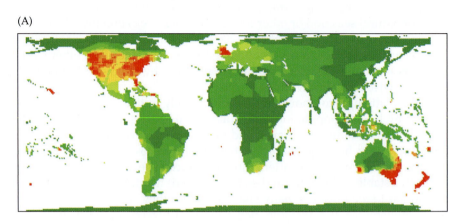

FIGURE 16.19 (A) The richness of established, introduced birds across the globe is highly nonrandom, reflecting patterns of colonization and development by human civilizations of the past two centuries rather than natural patterns of biological diversity. Richness is expressed as number of species per 1° × 1° (approximately 96.5 × 96.5 km) grid cell, with values ranging from 0 (dark green) to 41 (red) species per grid cell. (B–D) The record of historic, human-assisted introductions of birds (including 1378 introduction events of 426 species) reveals strong geographic biases in the native ranges of birds used in introductions and the sites receiving those introductions in comparison with the native distributions of all birds. (E) The diversity of exotic birds established on oceanic islands and the resultant latitudinal gradients in diversity of these species are direct functions of the number of species introduced to those islands (see also Ebenhard 1988b and Blackburn et al. 2004; Cassey et al. 2005). (F) While the diversity of native birds tends to decline with increasing isolation of oceanic archipelagoes, diversity of attempted and established (successful) species introductions exhibits the opposite trend. (After Blackburn et al. 2009, courtesy of Oxford University Press.)

(B) Native ranges of all birds

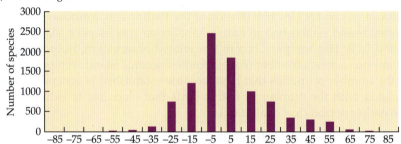

(C) Native ranges of birds used in introductions

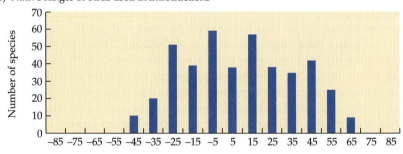

(D) Sites receiving introductions

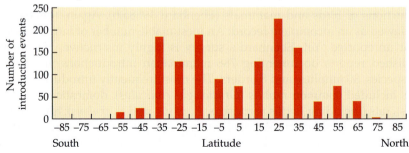

Species introductions of vertebrates such as birds also had a strong geographic bias (**Figure 16.19A**). The most striking feature of the geographic distribution of introductions is that, despite their isolation and relatively small area, oceanic islands have received the bulk of species introductions. Indeed, 60 percent of all documented mammal and bird introductions have occurred on oceanic islands. In an important monograph on species introductions in birds, Long (1981) catalogued 162 species introductions to the Hawaiian Islands, 133 to New Zealand, 56 to Tahiti, and well over 500 species introductions to some 90 other islands or archipelagoes across the globe. As Figure 16.18 illustrates, introductions of exotic plants were similar in their geographic bias, with oceanic islands receiving a highly disproportionate number and diversity of introductions (MacDonald et al. 1986; Drake et al. 1989; Heywood 1989; Atkinson and Cameron 1993).

Blackburn and Duncan's 2001 analyses of 1378 introduction events for 426 bird species provide some interesting and potentially important insights into the geography of human-assisted invasions and, in turn, the geography of extinctions as well. As Ebenhard (1988b) observed earlier, most introductions of birds involve a limited diversity of taxa (mostly from just five families including the pheasants, passerines, parrots, ducks, geese, swans, and pigeons; Blackburn and Duncan 2001). Moreover, avian introductions were highly nonrandom with respect to both the sites of origin and sites of introduction (**Figure 16.19B–D**). As Blackburn and Duncan (2001) observed, although islands cover just 3 percent of Earth's ice-free land surface (i.e., areas where introductions are feasible; Mielke 1989), over half of avian introductions were to islands. Similarly, although the greatest extent of ice-free land and the peak in native bird species diversity occurs in the tropics (Rosenzweig 1992; see Figure 16.19B), there was a strong, temperate-region bias in terms of both the sources and destinations of those introductions (see Figure 16.19C,D; Blackburn et al. 2010). These geographic biases of introduction largely reflect patterns of colonizations by Europeans during the eighteenth to twentieth centuries, with settlement and colonization especially concentrated in particular oceanic archipelagoes (especially Hawaii and New Zealand) and former British colonies (USA and Australia, in particular). Indeed, the record of bird species introductions on temperate zone oceanic islands demonstrates a strong positive relationship between introduction attempts and richness of exotic species (**Figure 16.19E**). Other factors, however, also influence avian introductions and, therefore, the number of exotic birds that become estab-

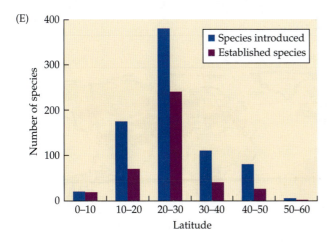

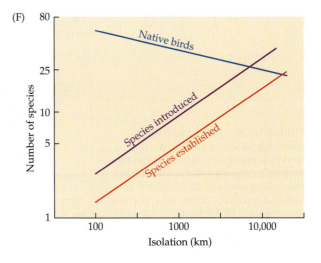

lished on oceanic islands. For example, the diversity of both introduced and naturalized birds varies with island isolation, but in a manner opposite to that predicted by island biogeography theory (**Figure 16.19F**). That is, while natural immigrations and resultant colonizations and native species diversity decline with island isolation, the diversity of human-assisted introductions increases with island isolation. Thus, islands have not only received a highly disproportionate number and diversity of avian introductions, but those introductions have (to a large degree purposefully) targeted the most isolated islands which, as we saw in Chapter 14, are most likely to be inhabited by endemic and ecologically naive species.

Aquatic systems have also been significantly affected by introductions of exotic species (see Lodge 1993; Wonham et al. 2000). In the state of California alone, 48 out of 137 species of fish are exotics. Indeed, 46 of these are either known or suspected to have negatively affected native species. Of the 95 fish species that have bred in Arizona, 67 (71 percent) are not native to the state. Many exotic species that became established in freshwater systems of the United States subsequently invaded Canadian waters (Crossman 1991). In Africa, introduction of the Nile perch (*Lates niloticus*) into Lake Victoria appears to have been one of the primary causes of the extinction of as many as 200 endemic cichlids (see Baskin 1992; Goldschmidt et al. 1993). Although global assessments of the geographic distributions of fish introductions are unavailable, established populations of non-native species of freshwater fish are nonrandomly distributed, reflecting geographic biases in human development and settlement across watersheds (**Figure 16.20**; Leprieur et al. 2008).

Perhaps the most troubling feature of the history of species introductions is that most have occurred in those systems with the greatest endemism, ecological naiveté and vulnerability to invading species: oceanic islands, lakes,

(A) Percentage of fish fauna

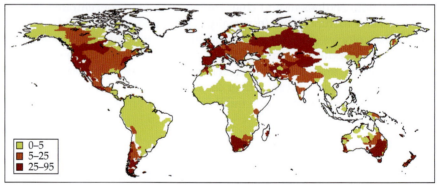

☐	0–5
■	5–25
■	25–95

(B) Number of introduced species

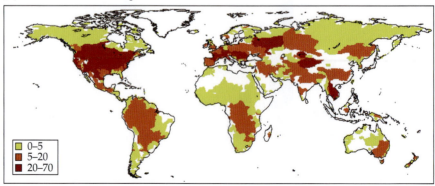

☐	0–5
■	5–20
■	20–70

FIGURE 16.20 The distribution of nonnative freshwater fish across the world's watersheds is highly nonrandom, likely reflecting geographic biases in human settlements and development. (A) Distributions of exotic species expressed as percentage of fish fauna. (B) Species richness of exotics in local watersheds. (Leprieur et al. 2008.)

TABLE 16.4 *Percentage of Threatened Terrestrial Vertebrate Species That Are Affected by Introduced Species*

Taxonomic group	Continental areas within biogeographic realms								All mainland areas	All insular areas
	Eurasia	North America	Africa	Indo-Malaya	Oceania	Antarctica	South America	Australia		
Mammals	16.7 (42)	3.4 (29)	8.0 (100)	12.7 (55)	0.0 (8)	0.0 (0)	10.0 (60)	64.4 (45)	19.4 (283)	11.5 (61)
Birds	4.2 (24)	13.3 (15)	2.5 (118)	0.0 (30)	0.0 (1)	0.0 (0)	4.2 (71)	27.3 (11)	5.2 (250)	38.2 (144)
Reptiles	5.9 (17)	16.7 (24)	25.0 (16)	4.3 (23)	14.3 (7)	0.0 (0)	14.3 (28)	22.2 (9)	15.5 (84)	32.9 (76)
Amphibians	0.0 (8)	6.3 (16)	0.0 (3)	0.0 (0)	0.0 (0)	0.0 (0)	0.0 (1)	0.0 (2)	3.3 (30)	30.8 (13)
Total for all groups considered	9.9 (91)	9.5 (84)	6.3 (237)	7.4 (108)	6.3 (16)	0.0 (0)	8.1 (160)	50.7 (67)	12.7 (647)	31.0 (294)

Source: MacDonald et al. 1989.

Note: The total number of threatened species in each realm is given in parentheses.

and freshwater streams (see Cox and Lima 2006). The prospect for the future is no less troubling. The proportion of threatened terrestrial vertebrates that are affected by introduced species is much higher on oceanic islands (31 percent) and on the "island continent" of Australia (51 percent) than in other continental areas (0 to 10 percent; **Table 16.4**). Biological invasions also threaten the biological diversity of many nature reserves. While we are wise to invest much time and energy in managing and protecting these reserves, most of them have already been significantly impacted by exotic species (MacDonald et al. 1989). Based on a global analysis of the incidence of exotic species in nature reserves in the late 1980s, MacDonald et al. (1989) estimated that introduced species represent from 3 to 28 percent of all vascular plant species found in nature reserves. Again, the situation is much worse on islands, where between 31 percent and 64 percent of all vascular plant species in nature reserves are exotics. Similarly, while introduced species tend to be minor components of avian and mammalian fauna of mainland reserves (averaging 1 to 5 percent for most reserves), they represent significant, if not the dominant, elements in reserves on oceanic islands (20 and 81 percent of the total species for birds and mammals, respectively). In many reserves that include lakes and streams, well over half the species of freshwater fish are exotic. While we cast these statistics in negative terms, it is not the case that all exotic species represent significant threats to the indigenous biota, and some may play important functional roles within ecosystems, including replacing the ecosystem functions (e.g., pollination, dispersal, grazing) of native species that are now extinct.

One final note on the nature of introductions is well worth emphasizing. While historic introductions of birds were indeed biased with respect to characteristics of the species and the locations and types of ecosystems receiving those introductions, these biases may well be changing. The taxonomic bias of avian introductions has shifted from a disproportionate number of purposeful introductions of game birds (especially pheasants and ducks) and those favored by naturalization societies (sparrows and finches) in historic periods, to unplanned releases of caged pets (parrots and parakeets) in recent decades (Blackburn et al. 2010). The source regions of avian introductions also shifted over these periods, from the Nearctic and Palearctic regions to Australian and Neotropical regions. Finally, while birds introduced across both periods tend to be larger in body size and have larger geographic ranges than birds in general, those introduced during recent decades tend to be somewhat smaller and have smaller geographic ranges than those introduced during historic times. Regardless of the reasons for these biases and changes in those biases over time, they are likely to influence both patterns in establishment of introduced populations and the effects of those populations on the diversity of native biotas.

Taxonomic and geographic biases in susceptibility to nonnative species

The insular biases in historic extinctions illustrated in Figure 16.5A are striking, even when we consider the geographic patterns in introduction sites summarized above (i.e., just over half of avian introductions were to islands, but well over 90 percent of historic extinctions of birds were of insular forms). Thus, islands are not just more frequently targeted for human-assisted introductions, but in comparison to mainland biotas, insular species are more susceptible to "the stranger's craft of power." In retrospect, this may not be surprising given the special nature of island environments and their native communities. Isolated oceanic islands tend to be inhabited by relatively depauperate and disharmonic communities that are often composed of novel—yet, ecologically naive—species. Indeed, these defining features of insular biotas—their depauperate and disharmonic nature and, in turn, the naiveté of their inhabitants—are far from random. Again, most of the more rigorous analyses of invasive species focus on the more conspicuous and better documented cases, in particular, avian introductions. Just as with introduction attempts, the ecological effects of avian extinctions also exhibit some highly nonrandom patterns with respect to vulnerability of native species on particular islands. Large-bodied species were especially susceptible to the effects of introduced and invasive species, including humans. For example, analysis of archaeological sites in New Zealand demonstrates that the Maori preyed preferentially on the largest remaining avifauana, which consequently were the first to disappear from the islands (Duncan et al. 2002). A global analysis of biases in avifauna extinctions by Tim Blackburn and his colleagues confirms that the largest birds, along with those that dwell and nest on the ground (in particular rails, ducks, elephant birds, and moas) were most susceptible to the effects of non-native species (Blackburn and Gaston 2005; see also Blackburn et al. 2008a, 2009), although we should also recognize that many island extinctions are attributable to more than one driver, with habitat alteration commonly implicated, alongside the effects of hunting by humans and interactions with non-native species.

Susceptibility to nonnative species also varies with characteristics of the islands, especially those influencing ecological naiveté and the ecological amplitude of insular populations. Accordingly, the proportion of insular avifauna suffering extinctions is highly correlated with island isolation—more isolated islands, presumably those with a greater proportion of ecologically naive species, lost higher proportions of their native species (**Figure 16.21A**). Avifauna extinctions also tended to be more severe on larger and more mountainous islands. Such islands likely included more species that advanced to the latter stages of taxon cycles, occupying smaller ranges and becoming more specialized—factors predisposing these species to the effects of humans and other nonnative species.

The extreme susceptibility of remote islands to invasion is dramatically illustrated by two case studies—the faunas of Hawaii and those of New Zealand. At the time of Charles Elton's classic book on invasion ecology, 95 species of birds inhabited the Hawaiian archipelago. Of these, 57 were native, while 38 (40 percent) were exotics that had been introduced since 1800. Dur-

FIGURE 16.21 (A) The proportion of native avifauna suffering extinctions on oceanic islands exhibits a constrained relationship but generally increases with isolation (distance to nearest mainland region), likely reflecting the naiveté of birds on isolated islands lacking native terrestrial mammals. (B) As is the case for most biotas of isolated archipelagoes, assemblages of the less vagile species (in particular, freshwater fish, land mammals, terrestrial reptiles, and amphibians) in the Hawaiian Islands are dominated by introduced species. (A after Blackburn et al. 2008b; B based on data from Loope and Mueller-Dumbois 1989.)

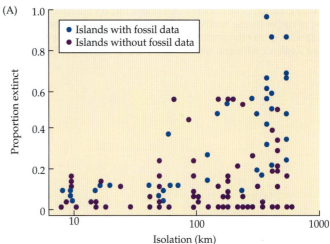

(B) Hawaiian Islands

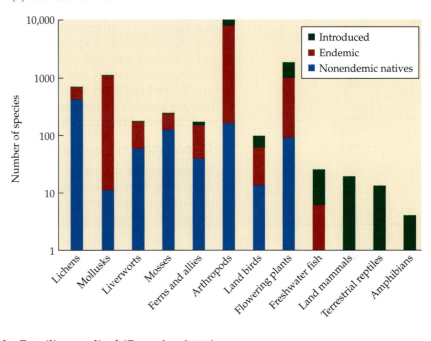

ing the same period, 14 native species became extinct and several others became rare and endangered. Of 94 species known to have been introduced prior to 1940, 53 became established (at least locally and temporarily), and only 41 failed completely. Most of the introduced species were deliberately released to augment the local avifauna with game birds and species of beautiful plumage and song. A strange mixture of species derived from different taxonomic groups and biogeographic provinces now coexists on the archipelago, including the mockingbird (*Mimus polyglottos*), cardinal (*Cardinalis cardinalis*), and California quail (*Callipepla californica*) from North America; the Indian mynah (*Acridotheres tristis*), lace-necked dove (*Streptopelia chinensis*), and Pekin robin (*Leiothrix lutea*—a babbler, not a thrush) from Asia; the house sparrow (*Passer domesticus*), skylark (*Alauda arvensis*), and ring-necked pheasant (*Phasianus colchicus*) from Europe; and the Brazilian cardinal (*Paroaria cristata*) from South America (Elton 1958). A quick review of **Figure 16.21B** reveals that on isolated archipelagoes such as the Hawaiian Islands, assemblages of the less vagile vertebrates, including non-volant mammals, amphibians, reptiles, and freshwater fish, are dominated by exotic species.

The fate of the Hawaiian archipelago is by no means unique. Consider New Zealand, where 50 percent of its extant plants, 43 percent of its extant birds, and some 90 percent of its extant mammals are exotic (Wodzicki 1950; Elton 1958; Falla et al. 1966; Atkinson and Cameron 1993; Sax et al. 2002). The only native terrestrial mammals of New Zealand are three species of bats, which are semifossorial, having converged on the niches of ground-dwelling, forest rodents. It is little wonder that exotic predatory mammals have had a tremendous impact on the native species of lizards and ground-nesting birds, which evolved in the complete absence of mammalian predators.

Given the magnitude and diversity of introductions, it should not be surprising that their effects on native species, while often negative, have involved many different mechanisms. Introduced species can directly compete with or prey on native species. Rats and mongooses are perhaps the most notori-

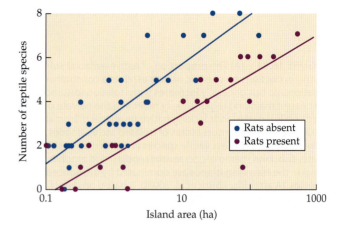

FIGURE 16.22 The effect of introduced rats on the species diversity of native insular reptiles is revealed by consistent differences in species–area relationships for reptiles on islands off the coast of New Zealand, with and without rats. (After Whitaker 1978.)

ous of the predators introduced to oceanic islands (Case and Bolger 1991b; Courchamp et al. 2003; Drake and Hunt 2009). By preying on ground-nesting birds and reptiles, they have decimated many endemic species (**Figures 16.22** and **16.23**). Overgrazing by pigs, goats, sheep, and other mammalian herbivores has affected many insular plants and has indirectly contributed to the demise of insular vertebrates by destroying their food supply or habitat. Other introduced species have served as vectors for pathogens, such as avian malaria and bird pox virus, Dutch elm (*Ophiostoma ulmi*) disease, and American chestnut (*Castanea dentata*) blight (see Jackson 1978; Berger 1981; von Broembsen 1989), which have caused range contractions and extinctions in many animals and plants on continents as well as islands (see also Lafferty et al. 2005; Smith et al. 2006).

Taxon cycles, red queens, and ecogeography of introduced species

The geographic dynamics of introduced species is in many ways difficult to distinguish from those of species that naturally invade biogeographic regions outside their native ranges. Indeed, some of the most valuable insights into the factors influencing range expansion of naturally invasive species come from

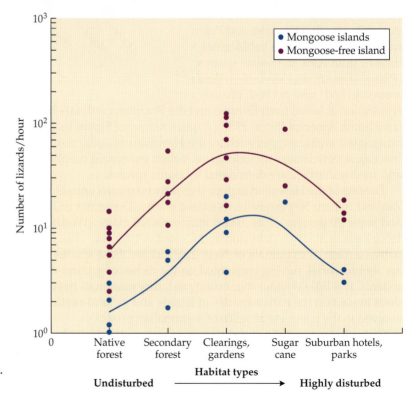

FIGURE 16.23 The effects of introduced mongooses on density and activity (as measured by number of lizards observed per hour) of native lizards inhabiting tropical islands of the Pacific Ocean. (After Case and Bolger 1991b.)

studies of anthropogenic invasions. As we saw in Chapter 6, the first stage of range expansion is frequently marked by a jump dispersal across some existing geographic barrier—in this case aided by human civilizations. Most such introductions, whether naturally or human-assisted, fail, but those that do take hold follow the initial, relatively protracted stage of establishment with one of exponential growth in populations and geographic range size through diffusion (see Kinlan and Hastings 2005; Vermeij 2005). Eventually, however, populations of the focal species encounter physical or biotic barriers, and the process slows or enters a period of relative stasis or, on occasion, one of collapse.

These invasion dynamics, especially when they occur on islands, may be entirely consistent with the regular stages of expansion, specialization and contraction, and eventual extinction that E. O. Wilson described in his classical theory of taxon cycles (see Chapter 14; Ricklefs and Bermingham 2002; Ricklefs 2005). A fundamental assumption of this theory was that species are dynamic and interact with each other to drive the cycle and exclude older, "over-specialized" species as new ones invade. In fact, once the process begins, selective pressures are altered for both invasives and natives, and species not only shift in their ecological associations, but they evolve. The result of this coevolutionary (Red Queen) arms race is that invasive species become integrated into these communities to the point that they may be functionally indistinguishable from native predators, competitors, and parasites. Ecological and evolutionary adaptations of both invasives and natives are especially well documented for pests and biological control agents used in agriculture and animal husbandry (see numerous examples in Cox 2004; Wares et al. 2005). Perhaps most relevant to biogeographers, adaptations of introduced species often take the form of ecogeographic patterns, such as those in body size across climatic, latitudinal, and elevational gradients, or body size evolution on islands consistent with the island rule—these developing with surprising rapidity, often in less than 100 generations. Rapid formation of ecogeographic patterns in introduced species include those of body size in the kiore (*Rattus exulans*) on Pacific islands (Yom-Tov et al. 1999), the small Indian mongoose (*Herpestes javanicus*) in the West Indies (Simberloff et al. 2000), house sparrows (*Passer domesticus*) and tree sparrows (*P. montanus*) introduced into North America (Johnston and Selander 1964; St. Louis and Barlow 1991), brown anoles (*Anolis sagrei*) introduced into central Florida (Campbell and Echternacht 2003), and fruit flies (*Drosophila subobscura*) introduced into North and South America (Huey et al. 2000, 2005; see also Stockwell et al. 2003; Lambrinos 2004; Vellend et al. 2007).

Thus, a comprehensive understanding of the effects of introduced species requires that we continue to study the evolutionary as well as ecological dynamics of both alien and native species (Mooney and Cleland 2001; Strayer et al. 2006). Given that human civilizations are the principal agents of the lion's share of species introductions during historic times, it seems essential that we first acknowledge that we are now unrivaled as ecological and evolutionary engineers of the world's biotas (Palumbi 2001).

Dynamic landscapes and seascapes

Magnitude of the problem

According to many conservation biologists, habitat loss and fragmentation (inclusive of the effects of pollution, eutrophication, and other human activities that degrade natural ecosystems) now represent the most serious threats to biological diversity of many biotas (Wilcox and Murphy 1985; Laurance and Bierregaard 1996; Settele et al. 1996; Birdlife International 2000; Debinski and Holt 2000; Hilton-Taylor 2000; Bissonette and Storch 2002; Haila 2002;

McGarigal and Cushman 2002; Cushman 2005; Lindenmayer and Fischer 2006, 2007; Fischer and Lindenmayer 2007; Prugh et al. 2008). Even on oceanic islands such as Hawaii, where half the plants are already exotics, habitat destruction by humans and their domestic and feral livestock has become the primary threat to ferns and other endangered plants (Wagner 1995). Yet, we need to avoid overstating and simplifying what really is a very complex problem. The relative importance of different extinction forces varies markedly among taxa (see Figures 16.6 and 16.9) and among regions. Species introductions, hunting, pollution, climatic change, and other effects of human activities combine to pose serious threats to most endangered species. That is, almost every endangered species is threatened by a combination of these factors, many of them interacting with the effects of habitat loss and fragmentation. Here, we first summarize the geographic and ecological dynamics of landscapes and seascapes before reviewing their effects on distributions and diversity of native biotas.

The Earth's natural ecosystems are being transformed at an unprecedented rate. Systems affected include some of the world's most diverse assemblages, including those of the tropical rain forest biome, in which just 7 percent of the Earth's surface is home to over 50 percent of its species (Wilson 1988). A single rain forest tree can contain over 40 species of ants from 26 genera, approximately equal to the entire ant fauna of the British Isles. Since 1819, the primary forest of Singapore has been all but completely removed, with less than 1 percent remaining (Corlett 1992). The loss of natural ecosystems has by no means been restricted to tropical rain forests (see Figure 16.8). Prior to the impact of humans, grasslands covered approximately 40 percent of the Earth's ice-free land surface (Clements and Shelford 1939). Now, just half of this amount remains (see World Conservation Monitoring Centre 1992). Pristine prairies once covered approximately 1 million km^2 of North America; now only 10 percent remains. Moreover, less than 1 percent of tallgrass prairie remains, nearly all of it on shallow, rocky soils or in tiny isolated reserves. The loss of wetlands has been no less severe. Estimates are that as much as 50 percent of the world's wetlands have been lost, most of this within the twentieth century, and that 65 percent of biologically significant tropical wetlands are currently threatened (Moyle and Leidy 1992). Freshwater systems are heavily affected by dams and other water control projects, which effectively fragment streams and rivers and their dependent populations.

In the marine realm, coral reefs are the aquatic equivalent of tropical rain forests in both diversity and productivity, yet many are threatened by a variety of human activities, including pollution, sedimentation, overexploitation, and other forms of development. Unfortunately, the threats of human activity on biodiversity of marine systems extend far beyond coral reefs. In fact, recent analyses reveal that over 40 percent of the world's oceans have been strongly affected by human activities (Halpern et al. 2008; see also Jackson 2008; Thrush et al. 2008). The geographic profile of these impacts in the marine realm is, of course, highly nonrandom, with shallow-water ecosystems and those closest to concentrations of human activities (e.g., continental shelf, rocky reef, mangrove, and coral reef ecosystems) being most strongly affected (**Figure 16.24**; Halpern et al. 2008). Some of these effects are so severe that they have created dead zones—oxygen-depleted waters resulting from eutrophication—that now cover over 245,000 km^2 of coastal regions near heavy industrial and agricultural development (Diaz and Rosenberg 2008; **Figure 16.25**).

Given their linear and geographically restricted nature, rivers and other freshwater systems are especially vulnerable to impacts of human societies, including habitat degradation, reduction, and fragmentation. Dams and ar-

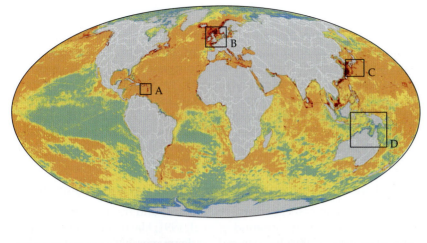

FIGURE 16.24 The geographic profile of environmental impacts on the marine realm is pervasive but highly nonrandom across the globe, consistent with geographic variation in human activities, including pollution, eutrophication, overharvesting, and climate change. (Halpern et al. 2008.)

■ Very low impact (<1.4)
■ Low impact (1.4–4.95)
■ Medium impact (4.95–8.47)
■ Medium high impact (8.47–12)
■ High impact (12–15.52)
■ Very high impact (>15.52)

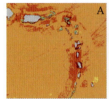

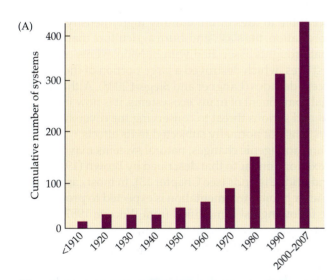

FIGURE 16.25 (A) The so-called dead, or hypoxic, zones of the marine realm (those where dissolved oxygen levels fall below 0.2 ml O_2/liter water) have increased exponentially in number during the past century. (B) Currently, they tend to concentrate in regions of the continental shelves adjacent to areas of high human impact. Here, anthropogenic eutrophication results in increased productivity and biomass of phytoplankton, which then die and fall in the water column, ultimately increasing the oxygen demand of decomposers and creating critical hypoxia of benthic waters. (After Diaz and Rosenberg 2008; see also Jackson 2008.)

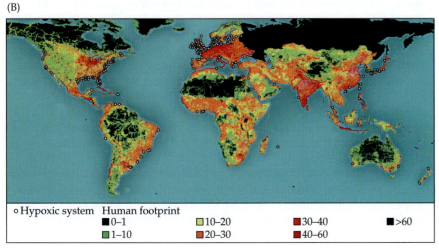

eas receiving input from thermal and other types of pollution can effectively isolate populations of freshwater organisms and destroy spawning migrations of native fish species and other aquatic organisms (e.g., see Gosset et al. 2006; Anderson et al. 2008; Morita et al. 2009). Once again, most endangered species are threatened by a combination of factors, which often confound our attempts to develop effective conservation strategies. An especially challenging case in point for freshwater systems is that of controlling the spread of sea lampreys and other invasive species while maintaining the connectivity of streams and rivers for native species (see Pratt et al. 2009).

Ecological and evolutionary effects of habitat loss and fragmentation

The physical changes associated with habitat loss and fragmentation include the following (see reviews by Saunders et al. 1991; Harrison and Bruna 1999; Ricketts 2001; Villard 2002; Fischer and Lindenmayer 2007):

1. a reduction in the total area, resources, and productivity of native habitats
2. increased isolation of remnant fragments and their local populations
3. significant changes in the environmental characteristics of the fragments, including changes in solar radiation receipt, wind, and water flux

Especially worrisome is the possibility that chronic changes in local climate will trigger feedback effects, which in turn can accelerate the rate of habitat loss or prevent the regeneration of native ecosystems. Deforestation, for example, often leads to drier conditions in remnant patches of forest, which makes them more susceptible to fires of unnaturally high intensity and frequency. In Borneo, deforestation, the slash-and-burn agriculture that replaced the forests, and seasonal droughts associated with El Niño events combined to trigger massive forest fires, which destroyed a large portion of the island's remaining forests in recent decades (Langner and Siegert 2008). Although fire is a natural and essential component of many ecosystems, it is now becoming a much more prevalent and serious threat to conserving the diversity of rain forests and other ecosystems not normally subjected to its effects.

As a result of these anthropogenic changes, natural systems may undergo a type of ecological relaxation similar to that described by Brown (1971b) and Diamond (1972) for land bridge islands (see Chapter 13). In most cases, relaxation or biotic collapse after habitat fragmentation is expected to proceed in a sequence of stages (the following list is modified from Wilcove 1987):

Stage 1   Initial exclusion Some species will be lost from the landscape simply because none of their populations were included in the remnant patches.

Stage 2   Extirpation due to lack of essential resources Species vary tremendously in their resource requirements, many requiring very large areas and/or very rare resources. The likelihood that all of a species' resource requirements can be met decreases as the remaining area decreases.

Stage 3   Perils associated with small populations Small populations are much more susceptible to a host of genetic, demographic, and stochastic problems (see Soulé 1987). As the total area of the remnant patches decreases, these problems become increasingly severe.

Stage 4   Deleterious effects of isolation As discussed in Chapter 13, some populations may be rescued from extinction by migration and recruitment of individuals from other populations. The likelihood of such rescue effects decreases as fragmentation and isolation increase.

Stage 5   Ecological imbalances While communities may not respond as superorganisms or units in the classical Clementsian sense (see Chapter 5), most species are strongly influenced by interactions with other species. Loss of one species during any of the above stages of relaxation may result in the subsequent loss of its predators, parasites, mutualists, or commensals. In addition, habitat disturbance and reductions in community diversity during the earlier stages of relaxation may facilitate the establishment of introduced species, triggering a cascade of subsequent extirpations.

Stage 6   Evolutionary consequences Reduced population levels, increased isolation, and altered environmental conditions and community structure are likely to increase the prevalence of genetic drift, reduce gene flow, and alter selective regimes, thus promoting rapid evolutionary change among populations in fragmented landscapes and in fragmented aquatic ecosystems (see Hoffmeister et al. 2005; Keyghobadi 2007; Aguilar et al. 2008).

Ecological imbalances and the cascading effects of initial losses in biological diversity are fairly well documented within the historical record and appear highly nonrandom with respect to species identities—a phenomenon sometimes termed **community disassembly** (see Chapter 14; Fox 1987; Mikkelson 1993; Lomolino and Perault 2000). For example, as a result of the fragmentation of native forests in North America, populations of brown-headed cowbirds (*Molothrus ater*) underwent dramatic increases in density and expanded their ranges into newly cleared areas and edges of remnant forests. Nest parasitism by cowbirds now takes a heavy toll on many Neotropical songbirds, especially those breeding within 500 m of the forest edge (see Brittingham and Temple 1983; O'Conner and Faaborg 1992). These songbirds suffer from indirect as well as direct effects of deforestation in both their breeding and wintering ranges. In the United States, deforestation has caused significant range contractions among large mammalian predators, including the cougar (*Puma concolor*), gray wolf (*Canis lupus*), and red wolf (*Canis rufus*). The loss of these top carnivores has resulted in compensatory increases, termed **meso-predator release**, in the densities of smaller carnivores (especially coyotes) and omnivores (raccoons, opossums, and skunks), which prey heavily on smaller animals—namely, songbirds and their eggs or nestlings. Fragmentation may also influence community structure by limiting the densities of native decomposers. In the fragmented tropical rain forests north of Manaus, Brazil, the diversity and density of carrion- and dung-feeding beetles was found to be markedly lower in clear-cuts and fragments than in adjacent stands of contiguous forest (Klein 1989; see also Bierregaard et al. 1992; Lomolino et al. 1995; Lomolino and Creighton 1996; Laurance and Gascon 1997). As a result, decomposition and recycling of dead organic material was greatly reduced.

Although ecological imbalances and secondary extinctions represent one of the final stages of relaxation, this stage may be of long duration. Modeling studies by David Tilman and his colleagues (Tilman et al. 1994) suggested that even moderate intensities of habitat destruction may well cause time-delayed, secondary extinctions even among currently dominant species. The cascading effects we have just discussed often take generations to move from one trophic level to another. Because any realistic scenario may involve many steps, today's habitat losses are likely to result in a substantial "extinction debt," a future ecological cost that may take many generations, possibly centuries, before it is fully realized.

Finally, even those species that somehow survive may exhibit evolutionary responses to fragmentation, changes that may be quite similar to those exhibited by populations inhabiting true islands. While research on the evolutionary responses to fragmentation is relatively limited, early indications are that they may include changes in some of the most fundamental traits of organisms, including their body size. Recall from Chapter 14 that the island syndrome includes a suite of ecological and evolutionary changes, among these being evolutionary trends in body size (i.e., the island rule; see Figure 14.26A). Schmidt and Jensen's (2003, 2005) studies of body size trends in mammals and birds inhabiting fragmented forests of northern Europe appear to be the first systematic and broad scale analyses to report morphological trends in response to fragmentation—body size shifts that are entirely consistent with the island rule (i.e., small species tending toward gigantism and large species tending toward dwarfism; see Figure 14.26C; for a report of similar trends in captive and reintroduced Mexican wolves, see Fredrickson and Hedrick 2002; in birds of fragmented African rain forest, see Smith et al. 1997; in skinks of fragmented Australian rain forests, see Sumner et al. 1999; in lynx of now fragmented habitats across the Iberian Peninsula, see Pertoldi et al. 2005; in tropical frogs of fragmented, Amazonian rain forests, see Neckel-Oliveira and Gascon 2006; and in non-volant mammals of fragmented, temperate rain forests of the Pacific Northwest, see Lomolino and Perault 2007).

Biogeography of global climate change

Humanity's influence on the diversity and the geography of nature goes beyond our ability to transform landscapes. It is now clear that we can cause climatic changes at magnitudes perhaps rivaling those associated with the glacial cycles of the Pleistocene. If current trends continue, we will be heating up an interglacial (i.e., an already warm) period by perhaps as much as 6° C over the next century. Most ecologists and biogeographers agree that such a temperature change, if it does occur, will dramatically influence the diversity and distribution of life. Many plants and animals may already be signaling the early effects of global warming. Scientists are currently working to identify which species and which ecosystems will be most affected, and in what ways they may respond. Biogeographers should have some especially relevant insights for predicting geographic dynamics (e.g., shifts, expansions, and contractions of geographic ranges), provided, of course, that reliable information on the magnitude and geographic profile of climate change is forthcoming and that we are able to address some serious shortfalls in our biogeographic knowledge.

Magnitude of the problem

The most comprehensive assessments of climate change are those conducted by the Intergovernmental Panel on Climate Change (IPCC). The IPCC report of 2007 provided the following conclusions regarding observed climatic changes during historical periods (especially the last two centuries), the causes for those changes, and the predicted climate change over the next century. The underlying science is complex, and disentangling different climate change drivers (natural and anthropogenic) is thus not a trivial task, hence the need for a process of sifting and collating evidence from hundreds of studies into the synthetic reviews published by IPCC. The most general conclusion of the 2007 report was that a climate warming trend is underway and is evident in significant increases in global average temperatures over land and sea, in widespread melting of snow and ice, and in rising sea levels (**Figure 16.26**). Although this is a very active area of research, the following summary (extracted from IPCC 2007) provides a useful and very sobering account of the environmental challenges associated with ongoing climate change.

(A) Global average surface temperature

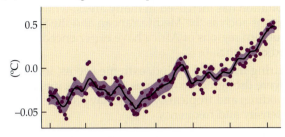

(B) Global average sea level

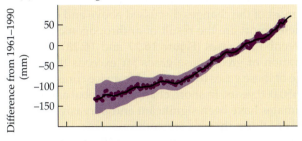

(C) Northern hemisphere snow cover

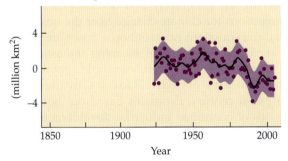

FIGURE 16.26 Recorded changes in (A) global air temperatures, (B) global sea level, and (C) snow cover over the Northern Hemisphere. Circles indicate yearly averages, smooth lines represent decadal averages, and shaded blue areas are 90 percent confidence intervals. (IPCC 2007.)

- Eleven of the 12 years from 1995 to 2006 ranked among the 12 warmest years in the instrumental record, which begins around 1850.

- Average temperatures in the Northern Hemisphere during the latter decades of the twentieth century were warmer than any other period in the last five centuries, and likely the warmest period in the past 1300 years.

- Temperature increases have been widespread, but they have been especially high for the higher northern latitudes, with warming of Arctic regions nearly doubling that of the global average.

- Ocean waters are warming to depths of at least 3000 meters.

- The extent of Arctic sea ice is now decreasing by 2.7 percent per decade, and the total extent of seasonally frozen ground in the Northern Hemisphere has decreased by approximately 7 percent since 1900.

- Thermal expansion of ocean waters, along with addition of waters from melting of glaciers, ice caps, and the polar ice sheets has resulted in increases in global sea levels, accelerating to about 3 cm per decade in recent years. More recent analyses, however, predict that global sea level may rise by over a meter in the next century (see Kopp et al. 2009).

- Increased atmospheric CO_2 has resulted in increased uptake of CO_2 by the oceans, which in turn has increased their acidity. This has led to a 0.1 drop in pH of ocean waters since 1750 and may lead to further declines in pH by 0.14 to 0.35 units over the next century.

• During the latter half of the twentieth century, precipitation has increased in many areas, while other regions such as the Mediterranean and the Sahel and southern regions of Africa have experienced intensified droughts.

• During that same period, extreme events, including severe droughts, heat waves, heavy precipitation and flooding, and tropical cyclones, have increased in intensity and frequency.

The IPCC has also conducted a rigorous assessment of the likely causes of observed climate change, concluding that most of the changes discussed above and elsewhere in their report are the result of anthropogenic increases in greenhouse gases (**Figure 16.27** and **16.28**). The concentration of greenhouse gases far exceeds the natural range of these gases (determined from ice cores) over the last 650,000 years. In short, changes in global and regional climates have been significant and pervasive, and the causes of those changes are anthropogenic.

The most relevant findings of the IPCC's analysis for conserving the geography and diversity of nature are its projections for climate change over the next century. Even if the emissions of greenhouse gases are somehow held at current levels, global warming and associated effects on other climatic conditions would continue, with a predicted increase of from 0.3° to 0.9° C by the end of the twenty-first century. Estimates based on more realistic scenarios of continued industrial development and gas emissions during this period range from an increase of 1.1 to 6.4° C.

The geography of recent and future climate change

One of the primary means of validating climate change models is to evaluate their abilities to predict, or retrodict, both the temporal and geographic dynamics of climate change. The same great engines and gyres in the atmosphere and across the oceans that drive the natural seasonality and geographic variation in climates (see Chapter 3) should also influence the anthropogenic changes in climates. As a result, predicted changes in temperature, precipitation, and major currents across the atmosphere and oceans should have a strong geographic

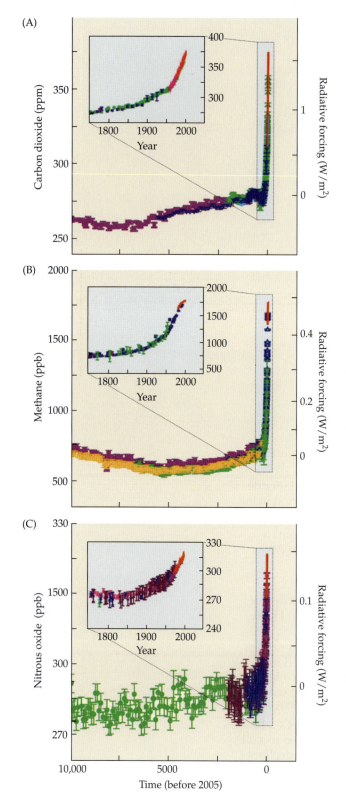

FIGURE 16.27 Atmospheric concentrations of greenhouse gases have increased substantially during the past 10,000 years, with most of this following the advent of the industrial age (i.e., since 1750 in the insets). Red lines in more recent records are from measurement of atmosphere samples, while symbols with error bars are estimates from ice cores (different colors depicting the results of different studies). The right-hand axis, relative forcing, is a standardized measure of the influence of this particular greenhouse gas on global warming. (IPCC 2007.)

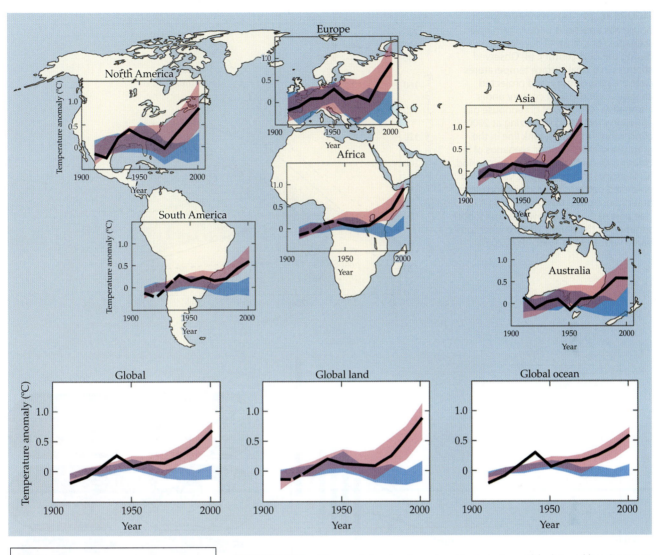

FIGURE 16.28   Observed changes in regional temperatures (black trend lines) are not explained by climate models based solely on naturally generated greenhouse gases (blue-shaded area) but are entirely consistent with the predictions of models that include anthropogenic levels of these gases (pink-shaded area). (IPCC 2007.)

signature (**Figure 16.29**; see also Figure 16.28), and the spatial as well as temporal concordance between those predictions and the observed climatic dynamics provides one way of assessing alternative models. Events such as the infusion of ash and greenhouse gases from major volcanic eruptions have been particularly useful in the assessments and confirm the efficacy of the leading models.

Given successful performance of these models, they can then be used to predict future dynamics of Earth's climate and geographic template—information that is, of course, fundamental to understanding and conserving the distributions and diversity of life over the coming centuries. However, as the climate is a complex and imperfectly understood system, and changes in forcing agents (whether anthropogenic or natural) in the years ahead are very difficult to predict with absolute confidence, all such models have to be viewed as essentially what-if scenarios. Given the importance of developing more accurate predictions, climate scientists are continually at work to gen-

FIGURE 16.29 (A) Changes in greenhouse gas emissions under alternative scenarios for industrial development during the next century. (B) Observed changes in global surface temperatures over the past century, and projected changes over the next century under alternative scenarios of greenhouse gas emissions, including stabilizing at 2000 concentrations (lower, pink line in the graph) and more realistic scenarios of increasing gas emissions during this period. (C) Seasonal and geographic variation in predicted patterns of precipitation under greenhouse gas emission scenario A1B (see part A) during (left) winter and (right) summer for the period 2090 to 2099 compared with 1980 to 1999. (IPCC 2007.)

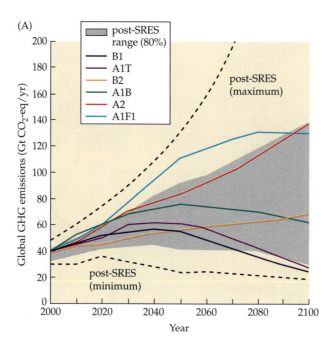

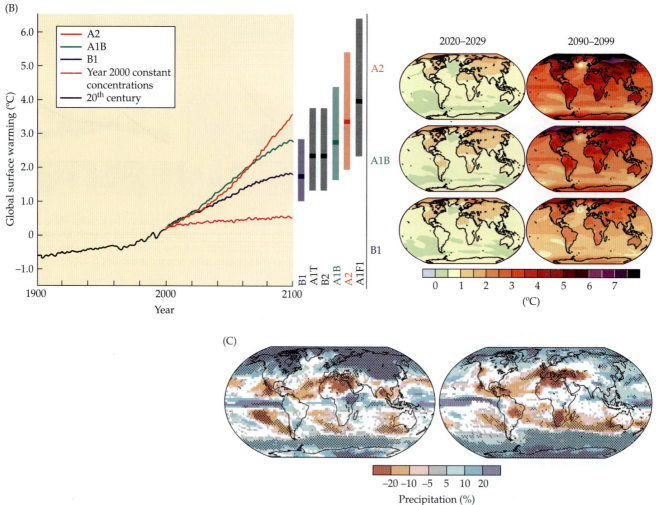

erate and refine alternative models and to develop refined means of choosing between less and more likely outcomes. For discussion of how these approaches can be applied to modeling future distributions and diversity in light of predicted climate change, see Araújo and New (2007).

Impacts of climate change on the geography and diversity of nature

Many studies have already documented ecological, biogeographic, and evolutionary effects of observed climate changes during the latter decades of the twentieth century. Of itself, this is no surprise, as past episodes of climate change have generated similar such responses. Indeed, ecological responses to the end of the last ice age, and to the smaller cooling event of the Little Ice Age, which peaked around 1750, are still ongoing and detectable for some systems and taxa (Lindbladh et al. 2008; Magri 2008). These present-day responses to late twentieth century warming are significant and generally consistent with observed shifts in regional and local climates. They include changing population dynamics of stenothermal species, range contractions, expansions and shifts to higher latitudes and elevations in other species, and a number of cases of phenological, behavioral, physiological, and morphological shifts in characteristics of local populations—at least some of these involving actual evolutionary change (Schneider and Root 2002; Stenseth et al. 2002; Parmesan and Yohe 2003; Root et al. 2003; Thomas et al. 2004; Walther et al. 2005; Araújo et al. 2006; Bradshaw and Holz 2006, 2007; Parmesan 2006; Pounds et al. 2006; Thomas et al. 2006; Musolin 2007; Skelly et al. 2007; Doi and Takahashi 2008; Etges and Levitan 2008; Green et al. 2008; Jackson 2008; NRC 2008; Ling et al. 2009).

A reasonable starting point for predicting climate-driven changes in geographic distributions of a particular species is to define its climate space—the climatic dimensions of its fundamental niche. Given this information, we can then attempt to map out predicted shifts in the species' geographic range under alternative scenarios of future climate change (e.g., see Schaefer et al. 2008; Cheung et al. 2009). There is now a considerable literature that attempts just such forecasting (e.g., Pearson and Dawson 2003; Araújo and Rahbek 2006; Pearson 2006; Araújo and Luoto 2007; Araújo and New 2007). But such models are probably too simplistic and incomplete to allow confident predictions of actual range shifts (e.g., see Pearson and Dawson 2004; Araújo et al. 2005). This is, in part, because we seldom have adequate information for estimating the climate space of most species, especially those of conservation concern, which almost by definition, tend to be the rarest and consequently most difficult to study. Even if we had perfect predictions of future climate and we were also able to accurately delineate the climate space of species, we know that they seldom occupy all regions of that fundamental range, so other factors—in particular, dispersal abilities, physical and biotic barriers to dispersal, interspecific interactions, and possible ecological time lags—must be taken into account when predicting a species' fundamental and realized ranges. Thus, we need at least seven types of information to predict the effects of climatic change on geographic ranges of focal species:

1. The temporal dynamics and geographic profile of future shifts in climates

2. Resulting shifts in biomes, ecoregions, and habitats across terrestrial and aquatic realms (remembering that, as we saw for the major climatic shifts of the Pleistocene [see Chapter 9], novel climatic, edaphic, and water chemistry zones are likely to be created while others may disappear)

3. Estimates of the climate space and habitat affinities and niche breadth of the focal species

4. Predicted distributions and extents of features of future landscapes and seascapes (e.g., expansive tracts of developed lands, rivers, mountains, winds, and oceanic currents) that might serve as immigration barriers, filters, or corridors

5. Estimates of the dispersal capacities of focal species in the context of the above features of projected landscapes and seascapes

6. A much more comprehensive understanding of the abilities of species to influence distributions and dispersal capacities of others through interactions with familiar symbionts as well as with those native to the newly invaded or otherwise novel environments

7. An assessment of ecological and evolutionary capacities of species to shift their fundamental niches and dispersal capacities, and the physiological, behavioral, and morphological traits that underlie these capacities

While we now have at our disposal a powerful arsenal of sophisticated spatial and analytical tools for predicting the biogeographic impacts of global change, whether through the direct actions of human societies (e.g., through habitat loss, fragmentation, overharvesting, or pollution) or through climate-driven changes in terrestrial and aquatic habitats, they each remain beset with uncertainty. To improve the reliability of these estimates and their value for developing effective conservation strategies, we need to redouble our efforts on several fronts, as already commented above. Not the least of these fronts is that we term the **Wallacean shortfall**—the paucity of information on species distributions and on the geographic dynamics of extinction forces, especially the geographic dynamics of human civilizations. In the following section, we present an overview of the current understanding of this latter subject—the biogeography of humanity—and then conclude this chapter with an introduction to an emerging synthesis that addresses the Wallacean shortfall—**conservation biogeography**, which promises to provide insights fundamental to understanding and conserving the diversity and geography of nature

The biogeography of humanity

Perhaps the most fundamental and defining question of sentient beings is, Where did we come from? Typically, the question is assumed to be about evolutionary origins and a succession of species, equal to Who were our ancestors? However, as we have asserted throughout this book, evolution occurs not just over time but across space as well. That is, the full story of our origins as a species and of our evolution to become Earth's most dominant life-form requires an integrative understanding of both evolutionary and geographic dynamics of our ancestors. To paraphrase the distinguished evolutionary biologist Theodosius Dobzhansky (see page 4), the treasure troves of hominid fossils and archaeological artifacts make little sense unless placed within the contexts of place as well as time. It is not a question of establishing just who were our ancestors but also where did they occur, and what were their paths of dispersal and range expansion.

Fortunately, archaeologists, anthropologists, and human biogeographers can now utilize an ever-expanding arsenal of technological advances to reconstruct the historical (evolutionary and biogeographic) development of humanity. In essence, this requires two types of information: (1) the age of the fossils and artifacts of our ancestors, and of significant features of landscapes and seascapes that influenced their subsequent dispersals, and (2) the differentiation among fossils and artifacts across time periods and geographic regions. In fact, we have already discussed many of these technological tools for aging and assessing the similarity among populations and individuals when we discussed the methods employed by historical biogeographers to reconstruct the evolutionary and biogeographic development of other species (see Chapters 10, 11, and 12; see also Chapters 8 and 9). Particularly useful in these applications are recent advances in our abilities to assess genetic similarities and divergence among populations by analyzing the more conservative components of their genomes—mitochondrial DNA and Y-chromosomal DNA, which are not subject to recombination and thus provide a means of reconstructing maternal and paternal lineages, respectively. Genetic analyses can also be used to reconstruct the evolutionary and biogeographic history of humanity by proxy, that is, by genetic analyses of species that our ancestors carried with them as commensals or internal parasites (e.g., the Polynesian rat, *Rattus exulans*, and the human-specific stomach bacterium, *Helicobacter pylori*; see Matisoo-Smith and Robins 2004, 2009; Anderson 2009; Moodley et al. 2009). Analyses of cultural evolution—as evidenced by diversity and divergence of languages and of tools, weapons, pottery, or other artifacts—also provide us with a rich variety of indirect, albeit powerful means of reconstructing the early migrations and geographic and evolutionary history of humanity. In combination, these various lines of evidence are providing an increasingly more detailed and comprehensive, but often surprising, story of the peopling of the world.

Perhaps the most fundamental surprise is that, despite our self-declared uniqueness among the world's biotas, the many waves of human colonization across the globe were relatively unremarkable; at least in a qualitative sense, very similar to the countless episodes of invasions and range expansions of most other species (see Follinsbee and Brooks 2007). As we have observed in this and previous chapters, most invasions fail, and the paths of dispersal and range expansion (whether via jump dispersal, diffusion, or secular migration; see Chapter 6) are strongly influenced by both the capacities of the species and the characteristics of the geographic template. So it was for the repeated waves of geographic expansion of our ancestors, from the earliest primates to the more recent lineages, including known members of the family of man (Hominidae). The fascinating archaeological record of our ancestors reveals that, despite sometimes remarkable bouts of range expansion, the geographic dynamics of these bipedal and largely subtropical species were strongly influenced by geology and geography, which, in turn, shaped the suitability of climates and habitats for their survival and dispersal.

The dynamic biogeography of man

Our focus here is on the genus *Homo* and, in particular, the temporal and spatial dynamics of its best known members (**Figure 16.30**). While we shall never achieve perfect consensus on the details of this captivating story, there is general agreement on some of its salient features, thanks to recent advances in our abilities to reconstruct the past from both archaeological evidence and genetic and anthropological studies of extant *Homo sapiens*—the family's only surviving species.

The following summary of dispersal and range expansions of this lineage was compiled from numerous works on the subject (in particular, those by Aiello and Wells 2002; Detroit et al. 2002; Oppenheimer 2003; Finlayson 2004; Matisoo-Smith and Robins 2004; O'Connell and Allen 2004; Schwartz 2004; Underhill 2004; Finlayson 2005; Trinkaus 2005; Terrell 2006; Follinsbee and Brooks 2007; Palmer 2007; Pope and Terrell 2007; Stoneking 2008; Wilmhurst et al. 2008; Anderson 2009; Haidle et al. 2009; Moodley et al. 2009; and Rongo et al. 2009). Perhaps most fundamental in the emerging consensus on the biogeographic history of hominids is that the homeland of its most successful

FIGURE 16.30 Evolution and geographic expansions of some key members of the genus *Homo*. Numbers in parentheses are millions of years BP for the appearance (numbers near labels) or last record (numbers at ends of arrows) of this hominid. (Modified from Finlayson 2005.)

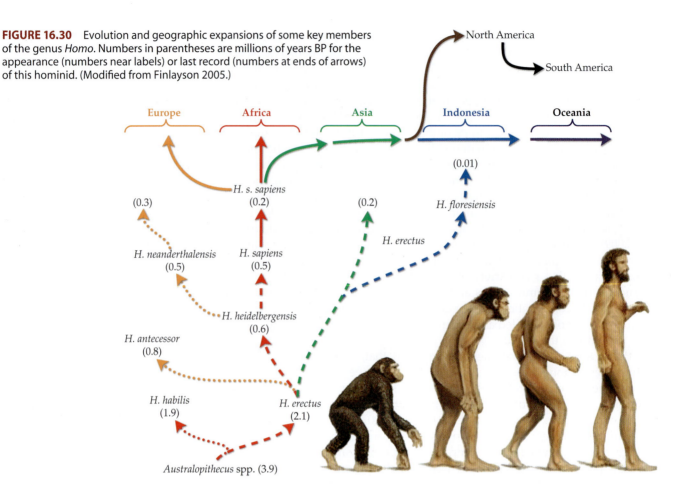

FIGURE 16.31 Maximum geographic range of *Homo erectus* from 1.9 to 0.2 million years BP. Numbers indicate approximate dates of fossil sites. Question marks indicate uncertainty of species identification. (After Finlayson 2005.)

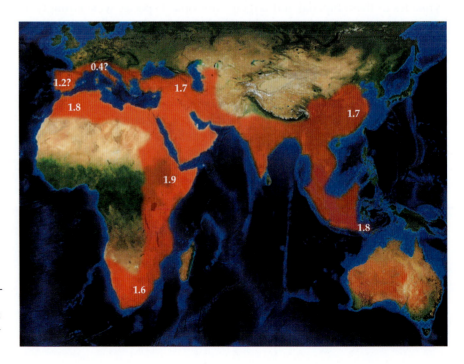

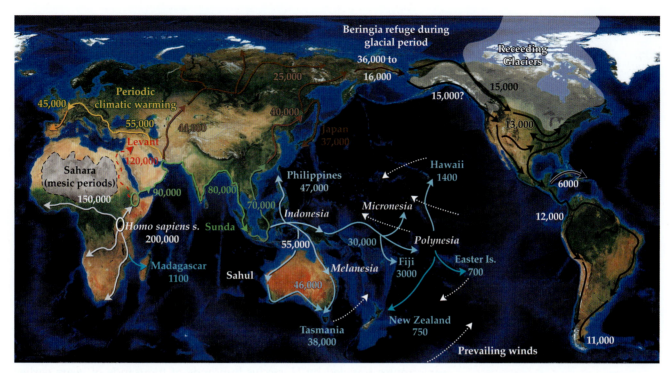

FIGURE 16.32 Geographic range expansion of anatomically modern man (*Homo sapiens sapiens*) from the species' African homeland to its eventual distributions across all continents and most oceanic archipelagoes was, as for most species, strongly influenced by contemporary landscapes, seascapes, and climate (numbers refer to colonization time in years before present). This summary of the biogeographic dynamics of our ancestors is based on a rich diversity of archaeological, anthropological, linguistic, and genetic studies (in particular, see those by Oppenheimer 2003; Finlayson 2004; Matisoo-Smith and Robins 2004; O'Connell and Allen 2004; Underhill 2004; Finlayson 2005; Trinkaus 2005; Follinsbee and Brooks 2007; Palmer 2007; Pope and Terrell 2007; Wilmhurst et al. 2008; Anderson 2009; Moodley et al. 2009).

and broadly distributed species—*Homo erectus* and *H. sapiens*—was Africa (**Figures 16.31** and **16.32**).

Once this is accepted, the compelling question becomes when and, of particular interests to biogeographers, where and how did successive populations of these species disperse into other continental landmasses and across oceanic archipelagoes? Again, geology played a fundamental role, albeit often in an indirect manner. The African and Arabian plates, which shared a long history dating back to the breakup of Pangaea, had collided with the Eurasian plate some 35 million years earlier, well before the first appearance of *H. erectus* in the fossil record, some 2 million years BP. The result was the eventual creation of three potential dispersal routes out of Africa: a northwestern passage across the Gates of Hercules at the Strait of Gibraltar, a northeastern one along the Nile Valley and across the Sinai into the Arabian Peninsula (which by then was welded onto the Eurasian plate), and a southeastern one across the Gates of Grief at the southern terminus of the Red Sea, and then along the southern shores of the Arabian Peninsula and into Asia after crossing the mouth of the Persian Gulf (see Figure 16.32).

Actual dispersal and range expansion across each of these potential routes out of Africa, however, was strongly influenced by the climatic upheavals that marked the 2 million year record of the Pleistocene (e.g., see Scholz et al. 2007; Carto et al. 2009). Although the plates shifted little during this relatively thin slice of the geological record, their positions governed the total amount and regional variation in solar energy absorbed by the planet—factors that, when combined with Milankovitch variation in Earth's orbit, generated the 20 or so glacial-interglacial cycles of the Pleistocene (see Chapter 9). With each climatic reversal came major geographic shifts in habitats across Africa, including the mesic grassland, woodland, and savanna ecosystems of our ancestors, as well as the more xeric habitats that served as formidable barriers to their dispersal. Thus, it appears that throughout this period, the Sahara was an expansive desert that prevented early human populations from reaching the northwestern passage out of Africa. The deserts of the Sahara, however, were much reduced during an earlier interglacial period that brought more mesic conditions to this regions' climate. This created a corridor for dispersal of *H. erectus* and later for *H. sapiens* from their homelands of eastern Africa northward and into the Nile Valley, ultimately expanding their ranges across the Sinai and into the Levant (see Figure 16.32). However, the first excursion of anatomically modern man (*H. sapiens sapiens*) out of Africa—at about 120,000 BP—ended when the next glacial period caused widespread aridification of the Arabian Peninsula and extinction of its earliest humans. Again, as is the very general pattern for most species populations, including those of our ancestors, most episodes of long-distance dispersal fail (see Chapter 6).

This failed, first range expansion out of Africa has been confirmed by recent archaeological and genetic analyses, which trace all current lineages of our species to a dispersal event via a more southerly route out of Africa, that is, across the Gates of Grief when glacial conditions lowered sea levels by approximately 100 meters. The early human colonists likely waded, rafted, and swam for short distances across the shallows and among the islands and reefs across the reduced mouth of the Red Sea, reaching the southern coasts of the Arabian Peninsula by about 90,000 BP. This modest but critical episode of jump dispersal was followed by range expansion through diffusion along the coasts of the Arabian Peninsula until populations of early humans reached and then crossed the shallows and estuarine habitats of the southern shores of the Persian Gulf. From there they may well have walked along the same routes traveled by *H. erectus* hundreds of thousands of years earlier during their own dispersal across this region. In both cases, the rapidity of range expansion from the Arabian Peninsula to Southeast Asia was remarkable. While their path was highly restricted, with little evidence of early expansions into the more arid interiors of the continent, the coastal ecosystems of the Indian Ocean comprised a relatively continuous corridor of equable, mesic climates with ample freshwater and productive terrestrial and aquatic habitats. Thus, by roughly 75,000 years BP, *H. sapiens* had reached Malaysia and Indonesia—spanning the 10,000 km of the great coastline corridor in just 10 to 15 millennia (see Figure 16.32).

Again, glacial cycles played a crucial role in the continuing range expansion of our ancestors once they reached Southeast Asia. Recall that during glacial maxima, sea levels were roughly 100 m lower and, as a result, the mainland stretched out along the continental shelves to include present-day islands of Sumatra, Borneo, and Java (**Figure 16.33**). This allowed expanding populations of early humans to simply walk to these now isolated islands of western Indonesia. The ocean basins east of Java and Borneo, however, are much deeper and thus formed a persistent barrier to range expansion by simple diffusion. Further expansion of *H. sapiens* across Indonesia would require jump dispersals across open waters to reach the more isolated archipelagoes of Wallacea and eastern Indonesia. While occurring at a much slower rate than earlier expansions across the continents and exposed glacial land bridges, this island-hopping would eventually provide access to the Sahul—the glacial-period continent composed of New Guinea, Australia, Tasmania, and their contiguous and then emergent, continental shelves. Perhaps it is not surprising that it was Alfred Russel Wallace who, after comparing anatomical and cultural characteristics of aboriginal tribes across Indonesia, delineated the principal biogeographic division among the peoples of Indonesia along the waters just east of his classic zoogeographic line (i.e., Wallace's line, which roughly corresponded to the boundary between the deep ocean basins that lie east of Java and Borneo; see Figure 16.33; see Vetter 2006).

Archaeological evidence dates invasions of Sahul by *H. sapiens* at roughly 55,000 BP. This was then followed by relatively rapid dispersal and range ex-

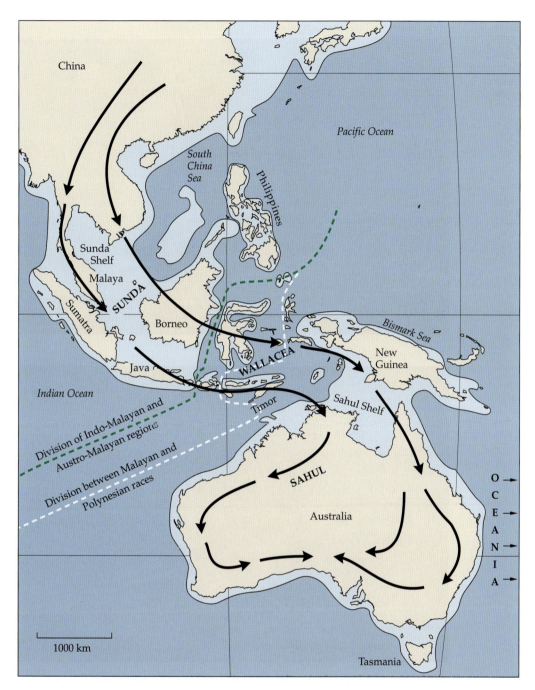

FIGURE 16.33 The lowering of sea levels during glacial maxima of the Pleistocene caused the exposure of continental shelves and the formation of dispersal routes across four regions of the eastern Pacific: Sunda, Wallacea, Sahul, and Oceania. (Light blue areas = land exposed during glacial maxima; dark blue = water > 200 m; possible dispersal routes are indicated by arrows). As Wallace observed, these physiographic divisions of land and sea are reflected in the divisions among regional biotas, including vertebrate species of the Indo-Malayan and Austro-Malayan regions, and geographic races of the peoples of these regions. (After Wallace 1869; Fagan 1990; Guilaine 1991; Vetter 2006.)

pansion along its mesic coastal habitats. As glacial conditions waned, sea levels rose—isolating aboriginal populations of humans on the now separated fragments of Sahul, with aboriginal populations continuing to inhabit their more equable and productive coastal regions. The common association of Australia's aboriginal populations with its xeric interior—the "outback"—is largely a misconception and artifact of European colonization. Having similar requirements for mesic and equable environments, invading Europeans simply displaced aboriginal Australians to the more marginal (interior) regions of the island continent.

Initial colonization of the more isolated archipelagoes of Melanesia, Micronesia, and Polynesia would take many more millennia, re-

FIGURE 16.34 Colonization of the Pacific realm by humans required the development of sophisticated seaworthy sailing vessels many centuries before the Europeans began their "age of exploration." (Artwork © 2010 Herb Kawainui Kane.)

(A) A Pahi of the Tuamotu Archipelago

(B) A Tongiaki of Tonga

quiring development of advanced craft-building, seafaring, and navigational skills, along with a concomitant increased dependence on fish, mollusks, and other marine resources (**Figure 16.34**). These new capabilities to tap the productivity of the marine realm also created significant liabilities, which paradoxically may have served to drive further explorations and range expansions across Oceania. Just as we have observed in recent decades, even minor shifts in climates can cause severe declines in fish populations and other aquatic resources, some of these triggered by outbreaks of disease in native marine organisms (Rongo et al. 2009). Given the variable nature of their aquatic resources, there may have been strong selective pressures for Pacific islanders to continually explore the far reaches of their range—leading to continued range expansion, either purposefully as they discovered new and more productive waters and islands or by accident when their sailing vessels were carried by storms to distant archipelagoes. Thus, range expansions of early humans across Oceania were unpredictable, and strongly dependent on prevailing winds. For example, despite reaching the southeastern coasts of Sahul by about 40,000 BP and western Melanesia by 30,000 BP, the ancestors of Polynesians did not reach Hawaii until about 1400 BP. Curiously, Pacific Islanders did not colonize New Zealand, which lies only 1500 km from the shores of Tasmania and Australia, until 750 BP—just a few decades before they reached Easter Island, which is roughly 2000 km from the Pitcairn Islands, 4500 from French Polynesia, and 9000 km from Australia. Just as we have seen for many other species, geographic distance is often a poor measure of biogeographic isolation. In this case, the prevailing winds would carry any sea canoes leaving Tasmania and southeastern Australia far north of New Zealand (see Figure 16.32, dashed white arrows indicating prevailing winds). Thus, it was not early aboriginal Australians but Pacific Islanders who colonized New Zealand at a much later time, that is, after they had colonized central and eastern Polynesia and were then carried by winds to the shores of North and South Island. As we have discussed earlier in this chapter, this very recent colonization of distant lands devastated many native species of giant birds and other marvels of evolution on these large and long-isolated islands (see Box 14.1).

Much earlier, perhaps roughly around the same time that ancestral aborigines were invading the glacial subcontinent of Sahul, early human populations began to push northward from the coastal regions of the Arabian Peninsula toward the Aral Sea, while others expanded their range toward the northwest along the shores of the Persian Gulf and on to the Tigris and Euphrates valleys, which they reached by roughly 55,000 BP (see Figure 16.32). Subsequent dispersal and range expansion of *H. sapiens* further north into cool temperate regions of Asia and Europe may well have been tied to periods of global warming. As was the case for early colonization of Indonesia and Oceania, continued expansions of our human ancestors into the higher latitudes proceeded at a relatively slow pace, again requiring development of adaptations to new climates and habitats: a process similar in many respects to secular migration, as it required incremental but continual evolution (cultural and biological) en route as populations either adapted to the increasingly harsh conditions or perished (Aiello and Wells 2002; Finlayson 2004; Haidle et al. 2009).

It appears that both *H. erectus* and *H. sapiens* independently developed the abilities not just to capture natural fires, but to create them as well. In comparison to *H. erectus*, however, *H. sapiens* was empowered by a more advanced brain, more sophisticated capacity to develop language, and superior abilities to plan, organize, and coordinate group activities. Early humans manufactured diverse tool kits, each with a variety of implements specialized for different local environments, from forests and open savannas to coastal habitats. Advancing populations of humans continued to develop the ability to manufacture and use fire for protection, warmth, hunting, and preparing food (see Ambrose 2001).

Again, geographic barriers and glacial cycles seem to have had a strong influence on this phase of range expansion by *H. sapiens*. Full glacial conditions persisted throughout Europe for most of the period between 75,000 and 40,000 BP, except for a relatively brief period of warming around 50,000 BP. As Fagan (1990) noted, it may not be just a coincidence that modern humans first appeared in Europe during this brief interstadial. Tools, shelters, and social organization associated with cold adaptation developed rapidly after 40,000 BP, allowing populations of *H. sapiens* to extend their ranges deeper into the higher latitudes of Eurasia and eventually into North America as well. During their migrations, they developed the projectile weapons and group strategies necessary for taking large game. They were then able to use the bones of mammoths and other large mammals to build shelters and fashion tools. Their cultural and ecological adjustments to the cold may have been augmented by morphological and physiological adjustments. Now populations of *H. sapiens* could migrate along with the game and other species that tracked the retreating glaciers and shrub-steppe communities.

By roughly 35,000 BP, populations of *H. sapiens* reached the southern edge of the glaciers in Europe and Siberia. Their colonization of northeastern Siberia by around 36,000 BP set the stage for the great leap into the Western Hemisphere. Again, this major event in human colonization was strongly influenced by climatic cycles and may have required two fundamental steps: colonization and occupation of the glacial subcontinent of Beringia (see Figure 9.9A) from 36,000–16,000 BP, and then their southward dispersal into Alaska and through the remainder of North America and then South America (see Batt and Pollard 1996; Gibbons 1996b; Roosevelt et al. 1996; Oppenheimer 2003; Finlayson 2004, 2005; Kitchen et al. 2007). When glacial conditions prevailed, lowering of sea levels by roughly 100 m exposed the continental shelves of Beringia which, although isolated from North America, was inhabited by a diversity of plants and wildlife derived from episodes of dispersal during earlier periods. The Beringian environments probably varied substantially, ranging from a relatively productive cold steppe that supported a diversity of large mammals (i.e., game species for early humans), to a polar desert (see Chapter 9 and Hoffecker et al. 1993). Subsequent expansion through the remainder of the Americas was probably blocked by the glaciers, which persisted as a continuous ice sheet across northern North America until about 16,000 BP. It is possible that human populations migrated southward along the ice-free coasts or along the chain of nunataks between the Laurentian and Cordilleran glaciers prior to 16,000 BP. Colonization of the New World prior to this is still controversial, however, and a number of paleoecologists maintain that colonization occurred only after the last glacial recession, when a corridor opened between the waning Laurentide and Cordilleran ice sheets (see Batt and Pollard 1996; Josenhans et al. 1997; Marshall 2001a,b; Erlandson 2002; Erlandson et al. 2007; Kitchen et al. 2007; Meltzer 2009).

Following their dispersal south of the ice sheets, human populations expanded through North America and then South America with perhaps unprecedented rapidity (Mosiman and Martin 1975; Kitchen et al. 2007). Barring some taphonomic artifact, it appears that populations of ecologically significant humans expanded across the extent of the Americas (from northern North America to Tierra del Fuego in South America)—some 15,000 km—in just a few thousand years. To understand how this could be so, it may help to review the geography of barriers throughout these regions. In contrast to the continents of the Eastern Hemisphere, where most barriers (e.g., the Sahara, the Mediterranean, the Alps, and the Himalayas) run east to west, dispersal barriers tend to run north to south in the Western Hemisphere, and thus may have actually facilitated human migrations southward from Alaska through the remainder of the Nearctic and Neotropical regions.

Certainly, future paleontological studies will revise many of these dates, but the peopling of the Americas may well continue to represent one of the most rapid and dramatic episodes of range expansion evidenced for any animal taxon. Just as important for understanding the geography of nature, past and present, this remarkably rapid spread of ecologically significant humans across the Western Hemisphere far exceeded the abilities of native wildlife to adapt to what Darwin referred to as "the stranger's craft of power." The result was wholesale extinction of the native megafauna that had developed in the splendid isolation of the last continents to be colonized by humans. Indeed, both Darwin and Wallace had some insightful and sobering observations on the recent, but now lost, species of these new lands.

> It is impossible to reflect on the changed state of the American continent without the deepest astonishment. Formerly it must have swarmed with great monsters: now we find mere pigmies, compared with the antecedent, allied races. (Darwin 1839)

> We live in a zoologically impoverished world, from which all the hugest, and fiercest, and strangest forms have recently disappeared. (Alfred Russel Wallace 1876)

Island biogeography and ecogeography of *H. sapiens sapiens*

As surprising as it may seem, despite our powers to transform landscapes and seascapes along with their native biotas, megafaunal or otherwise, our populations have been and continue to be strongly influenced by geographic variation in environmental conditions and associated selection forces. Granted, we have expanded our niches well beyond the subtropical savannas and other native habitats of our ancestors, but our populations continue to concentrate in lowland regions in close proximity to aquatic resources, and our patterns of dispersal and development follow highly nonrandom paths across regional landscapes (see Figure 16.16; see also Hall et al. 1995). Human populations also exhibit patterns similar to those we have described for other species, including geographic clines in diversity and morphology among populations on islands and across the continents (see Chapters 13–15).

In an insightful account of the colonization of the Pacific Islands, John Terrell (1986) analyzed patterns in the distribution and diversity of humans using many of the principles and models of island biogeographic theory (see also Howells 1973; Terrell 1976; Terrell et al. 1977; Terrell 2006; Boomert and Bright 2007; Fitzpatrick 2007; Pope and Terrell 2007; Anderson 2009). He found, for example, that just as island theory predicts, humans first colonized the less isolated islands and were likely to maintain populations on the more isolated islands only if they also were relatively large. That is, it appears that the minimum island area required to maintain populations of *H. sapiens* increased with isolation (see the discussion of rescue effects and compensatory effects in Chapter 13). In the Atlantic, prehistoric civilizations occupied many small islands relatively close to continents (e.g., the Scilly, Scottish, and Channel islands) by, if not many millennia before, the Bronze Age (5500 to 3000 BP). In contrast, the very isolated islands of the South Atlantic (e.g., Ascension, Saint Helena, Tristan da Cunha, and South Georgia) were uninhabited when Europeans first reached them. Other isolated islands, of Oceania and elsewhere, were colonized by repeated waves of colonists that either supplemented existing insular populations or replaced those that went extinct.

As observed for other insular faunas, persistence of human populations among islands appears to have been positively correlated with island area. For example, during glacial recession and the associated rise in sea levels during the early Holocene, the land bridge connecting Tasmania to the Australian mainland became submerged. As a result, insular populations of humans were extirpated on all but the largest islands of Bass Strait. That is, just as Brown (1971b) and Diamond (1972) reported for the relaxation of other insular biotas, human populations were subject to extinctions following fragmentation of once-expansive landmasses. All of these observations are of course consistent with the fundamental tenets of the equilibrium theory of island biogeography: Immigrations and extinctions tend to be recurrent processes, which are strongly influenced by island characteristics (i.e., isolation and area).

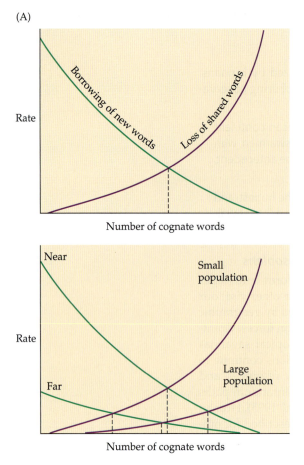

(A)

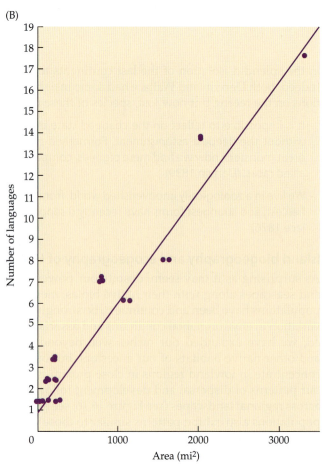

(B)

FIGURE 16.35 (A) Borrowing from island biogeographic theory, John E. Terrell hypothesized that linguistic diversity (number of cognate words) results from a balance between the rate of borrowing of new words between insular societies (which should decrease with increasing distance between islands) and the rate of loss of shared words (which should be higher for smaller islands with smaller populations). Accordingly, linguistic diversity (number of cognate words and number of languages) on islands decreases with isolation and (B) increases with area. The relationship between island area and the number of languages spoken on a given island for the Solomon Islands supports this hypothesis. (After Terrell 1986.)

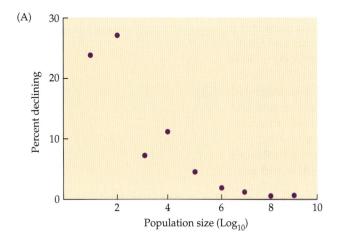

(A)

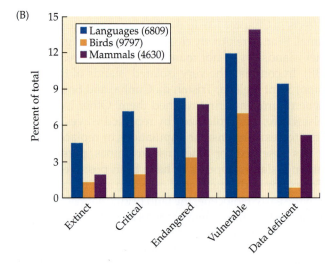

(B)

FIGURE 16.36 Diversity of languages of human populations tends to vary in ways similar to that of species diversity of animals such as birds and mammals. (A) Just as for birds and mammals, endangerment of languages (as percent of languages declining) exhibits an inverse relationship with population size, and (B) languages, bird species and mammal species tend to exhibit qualitatively similar patterns across various categories of threat status, although languages tend to be relatively more threatened than species diversity of these vertebrates. (After Sutherland 2003.)

Terrell and his colleagues took this analogy one step further (see Terrell 1986). Not only does the similarity of languages and other cultural and morphological characteristics among insular populations decrease with isolation, but linguistic diversity (i.e., number of different cognate words) increases with island area (**Figure 16.35B**). Interestingly, and consistent with patterns of species diversity, linguistic diversity on the continents also exhibits a marked latitudinal gradient, increasing from the poles toward the Equator (Mace and Pagel 1995). In addition, the diversity of languages and their levels of endangerment exhibit patterns qualitatively similar to those of species diversity of natural communities (**Figure 16.36**). Sutherland's (2003) global-scale comparison of geographic variation in linguistic diversity and diversity patterns of birds and mammals reveals similar clines—each of these measures increasing with area and maximum elevation of a region, but decreasing with latitude (see Chapter 15). Unlike avian and mammalian diversity, however, linguistic diversity did not decline with isolation. Isolation, as an inverse measure of the rate of addition of new words from surrounding cultures, is only part of the picture, and again it was Terrell who provided a more comprehensive model of linguistic diversity among islands by borrowing directly from island biogeography theory. According to Terrell, the diversity of insular languages can be explained by adapting MacArthur and Wilson's equilibrium model, with linguistic diversity resulting from a balance between rates of borrowing ("immigration") of new words from other insular populations and the loss ("extinction") of shared, extant words (**Figure 16.35A**; see also Wang and Minett 2005; Patriarca and Heinsalu 2009). These rates are functions of both the existing number (pool) of cognate words on an island and its physical characteristics (isolation and area). Similarly, the development and distribution of native linguistic groups across continental regions such as North America appear to have been strongly influenced by the distributions of physical barriers and ecogeographic zones of the Late Pleistocene and Early Holocene (Rogers et al. 1990; see also Fitzhugh 1997; Nettle 1999; Nichols 1999; Mufwene 2004).

Finally, Jared Diamond (1977) suggested that insular human populations may undergo subsequent transformations much like those associated with E. O. Wilson's (1961) taxon cycles (see Chapter 14; see also Flannery 1994; Diamond 1997, 2005). Like classical taxon cycles in other animals, the colonization cycles of humans tend to be unidirectional, with most movements from larger landmasses to smaller ones. After the initial colonists arrive, human populations often expand, not just geographically but ecologically as well, exploiting a broader range of resources and habitats. As population densities increase and resources become saturated, the initial founding population may enter a stage of local adaptation in which selective pressures switch from those favoring colonizing abilities (high dispersal ability, high reproductive rate, and generalized niches) to those favoring efficiency of resource use and competition with conspecifics. Increased competition from newly arriving human populations may then trigger a "retreating stage" in which the niche and distributional range of the initial colonists contract. Finally, just as envisioned by Wilson in his taxon cycle hypothesis, Diamond's colonization cycle of man ends in extinction—that is, with the more ancient and specialized civilizations being replaced by waves of invading populations of generalists (see Figure 16.17).

Collapses of insular populations of humans were often largely self-inflicted—driven by their own destruction of local resources. Again, native biotas of the ecosystems last to be colonized by humans—the Nearctic and Neotropical regions, and oceanic islands—were the most severely affected (**Figure 16.37**; see also Figure 9.30). Studies by David Steadman and his colleagues reveal that the extinctions of as many as two-thirds of the native

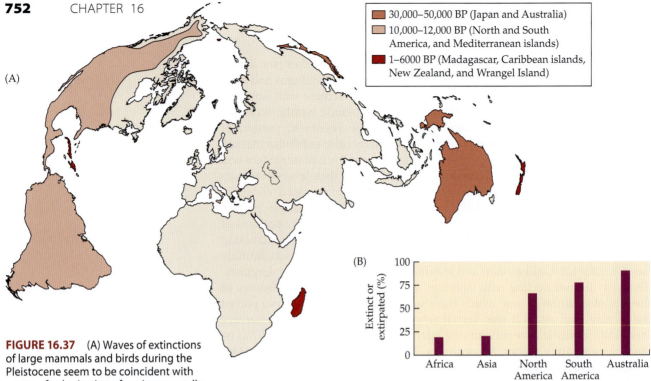

(A)

(B)

FIGURE 16.37 (A) Waves of extinctions of large mammals and birds during the Pleistocene seem to be coincident with waves of colonization of environmentally significant humans (see Figure 9.30). (B) The relative number of Late Pleistocene/Early Holocene (40,000 to 10,000 BP) extinctions or extirpations of the mammalian megafauna (species heavier than 44 kg). Note that the continents long occupied by humans (Africa and Asia) were relatively immune to this wave of extinctions. (A after Martin 1984; B after Martin 1990; see also Martin 2005.)

birds and bats of Oceania's islands were coincident with colonization by humans (Olson and James 1982a, 1984; Koopman and Steadman 1995; Steadman 1995; Burney et al. 2001; Steadman and Martin 2003). In fact, Steadman (1993) suggests that human colonization of oceanic islands influenced insular avifauna more strongly than any tectonic, climatic, or biological event of the past 10,000 years (**Figure 16.38**). Unfortunately, the mishandling of the local resource base of remote islands is far from being a thing of the past, although the processes at work are often now complicated by globalization, that is, by economic, trading, political, and social interconnections with the outside world (e.g., see Whittaker and Fernández-Palacios 2007: their section 12.1).

Provided they persist, human populations across continental and insular ecosystems exhibit some interesting ecogeographic patterns, that is, geographic clines in morphology—in particular, body size. For example, body size variation in preindustrial populations of humans is consistent with Bergmann's rule, increasing with latitude and with decreasing environmental temperatures (Roberts 1953, 1978; Ruff 1991, 1994, 2002) except on islands, where human populations apparently fail to exhibit this pattern (see Bindon and Baker 1997). We have already noted how at least one insular population of early hominids—*Homo floresiensis*—may have undergone a truly remarkable transformation in body size from its ancestors (putatively, the much larger, Indonesian populations of *H. erectus*), but one consistent with the island rule (in this case, dwarfism in an otherwise large mammal). On the other hand, this is just one isolated case, and Pacific Islanders exhibit tremendous geographic variation in body size—from the relatively modest stature among populations in Micronesia and Melanesia to the stature of those in the Hawaiian Islands in Polynesia, who often were larger than the European colonists (Howells 1973). One possible explanation for this apparent cline of increasing body size with increasing isolation across Oceania is immigrant selection (sensu Lomolino 1984, 1985, 1989a), which is equivalent to the anthropologists' hypothesis of selection for "thrifty genotypes" (sensu Neel 1962; Bindon and Baker 1997). Because the abilities of individuals to store energy and water in their tissues increases more rapidly with body size than does their requirement for these resources (per

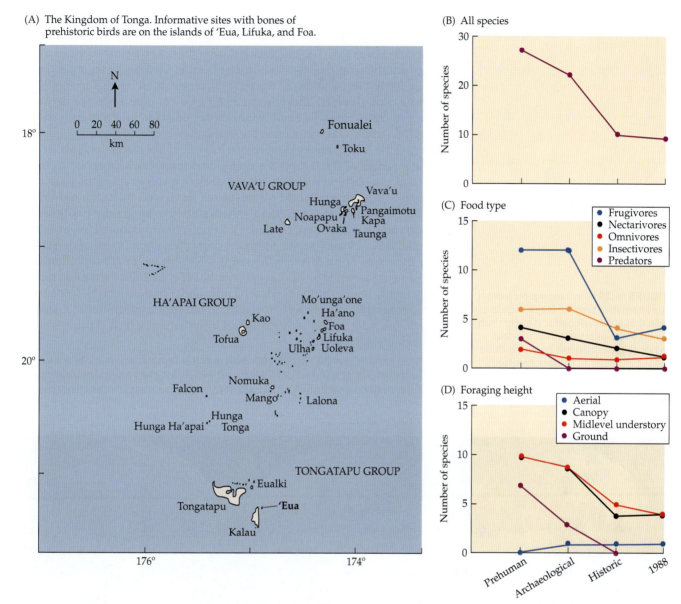

(A) The Kingdom of Tonga. Informative sites with bones of prehistoric birds are on the islands of 'Eua, Lifuka, and Foa.

(B) All species

(C) Food type

(D) Foraging height

FIGURE 16.38 (A,B) The diversity of native forest birds on 'Eua Island in the Kingdom of Tonga has decreased dramatically since the island was colonized by Polynesians. (C,D) Hardest hit were the frugivorous and ground-dwelling species, which were especially susceptible to predation by humans and their ground-dwelling commensals. (After Steadman 1995.)

day or per distance traveled), larger individuals have greater capacities for withstanding periods of shortages in food and water: traits that would seem highly adaptive during long voyages. As a result, the founding populations of the isolated archipelagoes of Polynesia may have been biased in favor of larger individuals, and this selective filter may have repeatedly operated during each period of drought, collapse in aquatic resources, and subsequent colonization.

Although the immigrant selection/thrifty genotype hypothesis remains largely untested, future studies of these and other patterns in ecogeography and biogeography of humanity offer intriguing opportunities for genuinely transformative collaborations and new syntheses among anthropologists and biogeographers.

Conservation Biogeography

In his foreword to a book recounting the voyages of Alfred Russel Wallace, E. O. Wilson (1999) observed that "the vastness of the tropical archipelago [the islands of Malaysia and Indonesia] also provided the knowledge Wallace needed to conceive the biological discipline of biogeography, which has expanded during the late 20th century into a cornerstone of ecology and conservation biology." The linkages between these two fields—conservation biology and biogeography—have in fact been evident from the start (or at least the formal recognition) of the field of conservation biology in the 1970s and 1980s. For example, efforts were being made to derive reserve network design principles from the equilibrium theory of island biogeography in the mid-1970s. Around the same time, the Hungarian biogeographer Miklos Udvardy and the American conservationist Raymond Dasmann developed (under the auspices of the IUCN) the first global framework designed to meet the goal of establishing a worldwide network of reserves encompassing representative areas of the world's ecosystems (reviewed in Ladle and Whittaker 2011). In both cases and many others since, biogeography has been integral to conservation science debates and policy formulation. However, the full potential of biogeographic science to contribute both to an understanding of processes of biodiversity loss and to strategies for conservation and management has yet to be harnessed (Whittaker et al. 2005; for a more recent and thorough review of the field, see Ladle and Whittaker 2011). Conservation biogeography is now recognized as an important programmatic area of research, teaching, and dissemination of information, which is premised on two fundamental assertions:

1. *Success in conserving biological diversity depends heavily on our understanding of the geography of nature.* As Wilson and his colleagues have demonstrated, biogeographers have provided many valuable insights for conserv-

FIGURE 16.39 Anthropogenic biomes are alternative types of major vegetation that are developed and maintained by human civilizations. The principal types of anthropogenic biomes include dense settlements, villages, croplands, rangelands, and managed forests, which together cover roughly 75 percent of the Earth's ice-free land surface. (After Ellis and Ramnakutty 2008.)

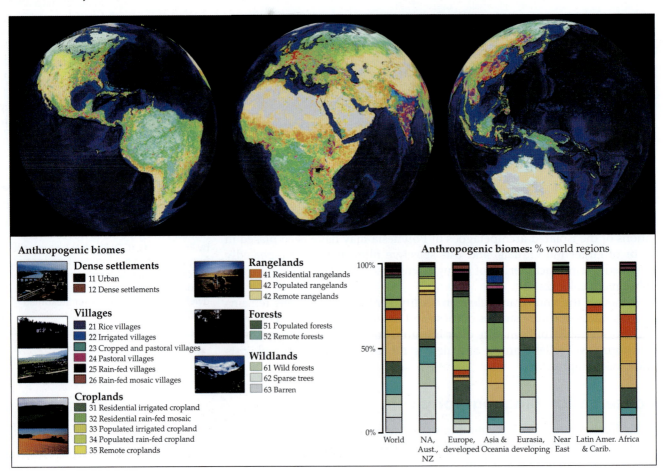

ing biological diversity, tackling the many conservation challenges that require geographic information (e.g., what are the effects of reduced area and increased isolation of habitats; where are the hotspots of diversity and endemism; where will an exotic species spread to in the future; where should we locate nature reserves and conservation corridors; where will native species disperse to under alternative scenarios of climate change; how have human civilizations changed the geography of nature; and where should we be searching for undiscovered species; **Figures 16.39** and **16.40**). The interdependence between biogeographic knowledge and conservation is demonstrated by the case study of the Philippines–one the world's most significant hotspots of diversity and endemicity (**Box 16.1**).

2. *In order to conserve "the hugest, and fiercest, and strangest forms" and the true nature of native species* (including their distinct physiologies, morphologies, behaviors, and ecological interactions), *we need to conserve their distributions.* The lessons from island biogeography and from seemingly unrelated but actually very relevant fields of animal husbandry and horticulture are sobering ones. If we favor and select certain types of individuals (e.g., the tamest, or smallest, or those with the showiest flowers, or

(A)

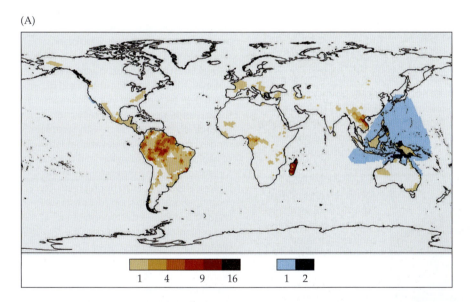

(B)

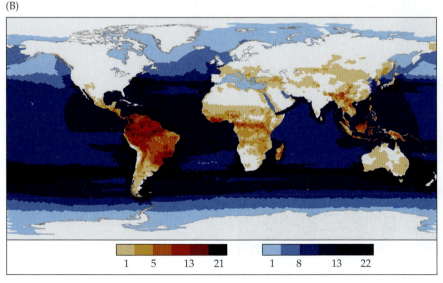

FIGURE 16.40 The locations of (A) newly described species and (B) data-deficient species of terrestrial and marine mammals are highly nonrandom, likely reflecting both the underlying patterns of diversity for these species and the geography of human development across the globe, which by default creates refugia in the less developed areas or those last to be developed. (After Schipper et al. 2008.)

BOX 16.1   *A Case Study in Conservation Biogeography—Biological Diversity of the Philippines*

Science is built up with facts, as a house is with stones. But a collection of facts is no more a science than a heap of stones is a house.

The above quote from the nineteenth century mathematician Jules Henri Poincaré is as applicable to conservation biogeography as it is to any field of science. We are optimistic that the many dedicated scientists and organizations tackling the biodiversity crisis will increase their efforts to conduct biogeographic surveys and lessen the Wallacean shortfall. But even in our optimism, we must admit that it will take many decades to describe the geographic distributions of the majority of today's imperiled species. During that time, and unless we take action to conserve these species now, many of them will have gone extinct. Conservation biology is a science of urgency, but of all the actions we can take, which should receive the highest priority? The solution to this dilemma is that we continue to advance on three important and interdependent fronts:

1. We should increase our efforts to conduct biogeographic surveys to map species distributions and describe the geographic variation in diversity and endemicity.

2. We should apply relevant theory to interpret the results of these surveys.

3. In lieu of more complete atlases on distributions of imperiled species, we should develop predictive, biogeographic models to identify likely hotspots of diversity and endemicity and focus future surveys and conservation initiatives on these sites.

It may appear that we have clouded the distinction between facts and theory, but this is both intentional and instructive. Even maps of geographic distributions—perhaps the most fundamental units of biogeography—are neither pure fact nor pure theory, but a combination of the two (see Chapter 4). As Thomas Kuhn (1970) observed, "Theory and facts are not categorically distinct. Scientific revolutions are driven by the dynamic tension between theory and empiricism."

In biogeography, each distribution map is drawn based on documented locations of occurrence for that species (empirical "facts") and on interpolations and extrapolations based on inferred (i.e., theoretical) associations of the species with

ecological, climatic, and topographic features. These associations and, in turn, predicted distributions must be adjusted when biological surveys provide new distributional data.

The relevance of these lessons for conservation biologists is clearly illustrated in the exemplary case study of conservation in the Philippines, which has been summarized by Heaney (2004). The development of the Philippines Archipelago is quite complex, with some of the oldest geological units developing from tectonic activities that began about 35 million years ago. The three principal units, or plates, that now comprise the Philippines drifted and converged, eventually resulting in the emergence of islands from tectonic uplift and volcanic activities. The Philippines are now composed of over 7000 islands ranging in size from hundreds of tiny islets to the largest island—Luzon, which is just over 100,000 km^2 (**Figures A** and **B**). Given their tectonic and volcanic origins, these islands are topographically and ecologically complex, often with mountain ranges covered with an elevational sere of habitats ranging from lowland tropical forests to cloud and mossy forests at the highest elevations. Even during relatively recent periods—too short for substantial tectonic activities—the topographic and ecological characteristics of these islands, and indeed even their size, isolation, number, and total area, changed dramatically with each of the 20 or so glacial-interglacial episodes and sea level dynamics of the Pleistocene (see Figure B; see also Chapter 9).

This complex history, along with the high insolation, high precipitation, and relatively rich, volcanic soils, created a productive and dynamic evolutionary

FIGURE A   Much of the Philippines was originally covered by a diversity of habitats including lowland forests, montane forests, mossy forests, and others. (Left, top) Mount Lumas, on Leyte. (Right) A tree fern. (Left, below) Mossy forest habitats, which often harbor endemic species of mammals, frogs, orchids, and trees. (Left, top courtesy of Paul D. Heideman; right courtesy of Ben C. Tan; left, below courtesy of Lawrence R. Heaney.)

BOX 16.1   *(continued)*

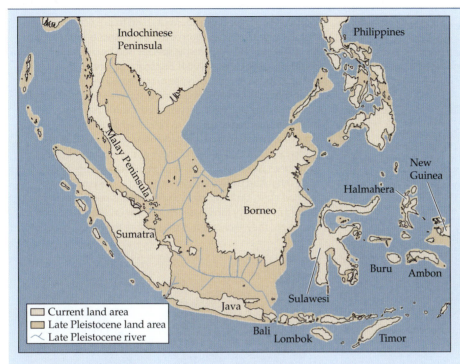

FIGURE D   The Indo-Australian region showing the configuration and locations of its major archipelagoes during recent times and during the previous glacial maximum, when water levels were relatively low and many now-isolated islands and landmasses were connected. (After Heaney 2004.)

arena that allowed the archipelago's biota to diversify to the point that the Philippines rank among the world's richest hotspots of biological diversity. To put this in perspective, we can compare the Philippines to Brazil, which is recognized as one of the world's megadiverse countries. The Philippines has a total of about 952 native land vertebrates, and about 542 of these (57%) are endemic. Brazil has more land vertebrates (3,131 species) and a higher number of endemics (788, 25%), but it is 28 times larger than the Philippines. Thus, on a per unit area basis, the Philippines clearly has the richer fauna, especially in terms of the percentage of endemic species (Heaney 2004; see also Heaney and Regalado 1998). Endemicity is high within each class of terrestrial vertebrates but varies with dispersal abilities, being lowest for the best dispersers (i.e., breeding land birds, where endemicity = 44 percent), intermediate for land mammals and reptiles (i.e., with endemicities of 64 and 65 percent, respectively), and highest for

amphibians (i.e., with endemicity = 73 percent), which are most limited by their inability to osmoregulate while dispersing in saltwater. Both the estimated number of species and percent endemism for all of these groups are increasing rapidly because of very active current research (e.g., Balete et al. 2007a,b, 2008; Brown et al. 2007; Brown and Gonzales 2007; Siler et al. 2009; these and other features of the Philippines hotspot of diversity and endemicity are described in the Vanishing Treasures of the Philippine Rain Forest Web site: http://www.fieldmuseum.org/vanishing_treasures/).

Unfortunately, this is another geographic area where a global hotspot of diversity and endemicity coincides with one of human development. Prior to 1600, mature tropical forests covered about 90 percent of the land surface of the Philippines. By 1898, mature forest cover was reduced to 65 percent, and now just 8 percent of the original forests remain (Kummer 1991; Heaney and Regalado 1998; Environmental Science for Social

Change 1999). Also, just as in other regions where human populations have advanced over hotspots of biological diversity, the geographic progression of deforestation in the Philippines was highly nonrandom. The first forests to be cleared were those that occupied the relatively broad and flat lowlands, but then deforestation advanced to the higher and steeper sites as lowland forests became more scarce, timber technologies advanced, and the human population increased (see also Figure 16.16).

Clearly, the Philippines are in dire need of continued and intensified biogeographic surveys, but they also require immediate conservation action. Given the limited time and financial resources, it is imperative that these efforts be applied where they will do the most good (i.e., on the islands and within the subregions of those islands that harbor the majority of the endemic and imperiled species). Fortunately, the Philippines have been attracting the attention of distinguished naturalists and biogeographers for well over a century (see Heaney and Regalado 1998). Granted, just as for any of the world's hotspots of biological diversity, comprehensive catalogs and atlases of the Philippines' imperiled plants and animals may be many decades away. However, biogeographers have used the available information to identify some very general geographic patterns and, thus, to develop a comprehensive plan for conservation in lieu of complete biogeographic surveys. The most general and relevant patterns are summarized by Heaney (2004):

1. Each deep-water island (or cluster of islands that were united during glacial periods) is a unique center of endemism (**Figure C**).

2. Each isolated mountain range on the larger islands is a subcenter of endemism.

Heaney and his colleagues then applied these patterns to develop quantitative models that could be used to predict the

Continued on next page

BOX 16.1   *(continued)*

FIGURE C   Locations of the regions of endemicity of the Philippines' land mammals, including known centers of endemism along with those predicted based on their isolation (by deep water channels or, in the case of montane assemblages, by lowland habitats). (After Heaney 2004; L. R. Heaney, pers. comm. 2010.)

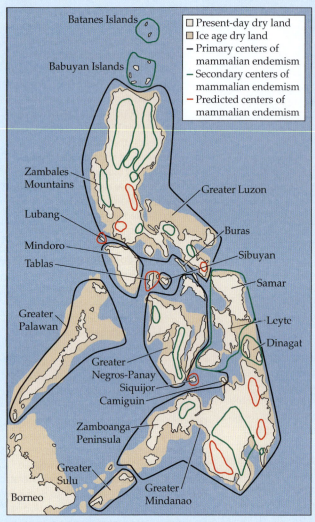

locations of hotspots of endemicity, and to predict (based on area, isolation, and elevation of the island and its mountains) the actual number of endemic species (see Heaney and Regalado 1998; Heaney 2004). Subsequent surveys by these researchers (Rickart et al., in press) confirmed the locations of these hotspots and added several on an even finer geographic scale, but they also typically yielded the same or similar numbers of endemic species as predicted by the model (see Figure C), including some remarkable new species (**Figure D**; Heaney et al. 2009). One of the few and most foreboding exceptions was the Island of Siquijor, which was predicted to be inhabited by two species of endemic land mammals, but had none. Unfortunately, this island has been completely cleared of primary forests, so if it had any endemic land mammals, they likely went extinct before the naturalists could

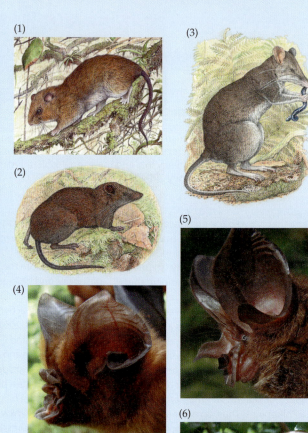

FIGURE D   A sample of some of the diversity of endemic mammalian fauna of the Philippines. *Carpomys phaeurus* (1), the lesser dwarf cloud rat, formerly considered a "possibly extinct species" but rediscovered in 2003. *Archboldomys kalinga* (2), the Cordillera shrew-mouse, described in 2006. *Rhynchomys soricoides* (3), the northern Luzon shrew rat, a member of a previously poorly known genus, with two new species described in 2007. A possibly undescribed member of the *Hipposideros lekaguli* (4) species group, Boonsong's roundleaf bat. One of several apparently undescribed "cryptic species" that are members of the *Rhinolophus philippinensis* (5) species group, the large-eared horseshoe bat. *Musseromys gulantang* (6), the Mount Banahaw tree mouse, a new genus and species discovered in 2004 and formally described in 2009. (1, 2, 3 illustrations by Velizar Simeonovski; 4, 5 photos by Danilo S. Balete; 6 by Lawrence R. Heaney).

BOX 16.1 *(continued)*

survey the island (an apparent case of Centinelan extinctions; see page 701).

Comparisons among patterns for different taxonomic groups revealed that, for the most part, hotspots of endemicity of plants, insects, and land vertebrates occur in the same regions and subregions, with just a few exceptions. In addition to being influenced by island area, isolation, and elevation, endemism of reptiles, amphibians, and plants were also influenced by edaphic characteristics (endemicity of reptiles and amphibians being highest in areas of exposed

limestone, and those of plants being highest in areas of ultrabasic soils; Ong et al. 2002). Thus, the accuracy and generality of the predictive, geographic models can be significantly enhanced with information from geological maps of the islands and knowledge of the ecology of specific groups of organisms.

The success of this mixed strategy, where insights from both biological surveys and biogeographic theory are simultaneously applied to better focus conservation initiatives, is cause for optimism. As Heaney (2004) concluded

in an important paper on conservation biogeography, the destruction of the Philippine forests continues at such a rate that we cannot afford to wait until we inventory the biota of each of its 7000 islands—indeed, there is no longer any need to wait. We can concentrate our limited funds and personnel on the predicted and now verified centers and subcenters of endemicity and protect those sites inhabited by most or, as in the case of such well-studied groups as mammals (see Figure C), virtually all of the Philippines' endemic species. ∎∎∎

the smallest thorns), and if we attempt to "maintain" them only in small, ecologically simplified or otherwise altered environments such as nature reserves, zoos, and botanical gardens, their descendants are not likely to retain the diversity or natural character that we hope to conserve. By the phrase "conserve their distributions" we refer to the importance of retaining core characteristics of the geographic, ecological, and evolutionary context of biogeographic realms, provinces, and assemblages. A core lesson from both historical and ecological biogeography is the essential impermanence of biological systems, species, and assemblages, as species ranges respond to changing environments at coarse scales, and as the organization of local communities ebb and flow in response to fluctuating environments at all scales of analysis. In light of the expected change in global climate in the twenty-first century, it is evident that we need to recognize and accept the dynamic nature of biogeographic distributions, while at the same time recognizing the dangers of biotic homogenization and the "loss of biogeography" that arise from human-assisted long-distance dispersal (e.g., see Trakhtenbrot et al. 2005).

Fittingly, the theme of the second meeting of the International Biogeography Society, held in 2005, was conservation biogeography, and soon thereafter the editorial board of one of the leading journals in the field—*Diversity and Distributions*—adopted conservation biogeography as its unifying theme. By virtue of its genuinely integrative and geographically explicit approach, conservation biogeography continues to rapidly develop as one of the most promising means of addressing some of the greatest challenges of conserving biological diversity. Most fundamental of these challenges is the Wallacean shortfall. As we have noted earlier, for most species, even those that have been formerly described and catalogued, we know precious little about their geographic distributions and even less of the patterns of genetic and phenotypic variation among their populations. Such information seems essential if we are to assess the abilities of endangered species to survive the effects of anthropogenic changes in landscapes, seascapes, and climates, either by local adaptations or by dispersing to track their native habitats or colonize new ones. Because the need for such information will only increase in relevance to both basic and applied scientists, we include a more detailed description of the research agenda for conservation biogeography in our final chapter, The Frontiers of Biogeography (see Lomolino 2004, 2006; Lomolino and Heaney 2004; Whittaker et al. 2005; Ladle and Whittaker, 2011).

THE FRONTIERS OF BIOGEOGRAPHY

From the Foundations to the Frontiers of Biogeography

As we have observed throughout this book, the long and distinguished history of biogeography is tightly interwoven with that of ecology and evolutionary biology. In fact, biogeography draws widely not only from the biological sciences but also from such disparate fields as geology, meteorology, anthropology, and philosophy. The challenge of integrating these fields to develop a comprehensive understanding of the origins and maintenance of global patterns of biodiversity is one of the field's great attractions, but the isolation of people with such diverse interests into disparate disciplines has also represented a major hurdle to progress throughout the historical development of biogeography.

For centuries, the geography of nature provided key insights into the many mysteries of what we now call biological diversity. Indeed, that the natural world varied over space and time was ancient knowledge—information essential to survival of hunting and gathering societies in their struggle for existence. But these ancient observations—that the types and numbers of organisms encountered varied with area searched, distance traveled, elevation of land, and depth of waters—did not coalesce into a true science until the Age of European Exploration. With the procession of broadly trained natural scientists—from Linnaeus, Forster, and Buffon during the eighteenth century to Humboldt, Darwin, Hooker, Sclater, and Wallace in the nineteenth and Wilson, MacArthur, Briggs, Diamond, and their colleagues in the twentieth—our knowledge of the patterns and causal explanations for the geography of nature accumulated and matured into the science of biogeography.

The Age of European Exploration was a period of great discoveries, and it marked the origins of the natural sciences in terrestrial and marine environments, with major contributions from geology, meteorology, systematics, paleontology, evolution, and ecology—all fields practiced by biogeographers (i.e., scientists studying the spatial variation in the Earth and its life-forms, past and present). From these distinguished origins, each of these disciplines evolved, diverged, and eventually flourished (or languished) as increasingly more distinct disciplines. With this growth and diversification of sciences and scientists came a presumed "need" for specialization and splintering, and the grand view—the ultimate synthesis across space and time—became murky and more elusive.

If this abbreviated history of natural science tells us anything, it is that biogeography is not only the foundation but also the frontier of many disciplines. The greatest strides that we can make in unlocking the mysteries and complexities of nature are those from syntheses and collaborations among scientists across the many descendant disciplines, long diverged but now reticulating within a strong spatial context—the *new biogeography*. From the earliest vicariance hypotheses of Joseph Dalton Hooker to the most recent methods of analyzing reticulating phylogenies and phylogeographies, geographic variation over time and space is the key. The unifying question of modern biogeography is How do life-forms from all kingdoms—from the tiniest unicellular organisms to the greatest beasts, and from ancient to current (and into the future)—vary across geographic dimensions? The new emphasis is one of reintegration of divergent conceptual lineages, and new syntheses and bold collaborations among scientists who, in the past, had seldom read each other's work, let alone worked together.

The Frontiers

During the past few decades, biogeography has become an increasingly prominent cornerstone of modern ecology, evolution, and conservation biology. Modern biogeographers explore a great diversity of patterns in the geographic variation of nature, from physiological, morphological, and genetic variation among individuals and populations, to differences in the diversity and composition of biotas along geographic gradients, and to the historical and evolutionary development of those patterns. It is difficult not to be enthusiastic about recent advances in biogeography and its great promise for answering some of science's greatest challenges (i.e., explaining the origins, spread, distribution, and diversification of life across the Earth). A tall order, granted, but our colleagues and our science have made great strides in recent decades, and we are entirely confident that the next generation of biogeographers will not disappoint us.

The frontiers of biogeography are now being defined and driven by the push to develop more general, more complex, and more integrative theories of the distributions and diversity of life—theories based on the dynamics of the Earth, its environments, and its species. We believe that the best means for advancing the frontiers of our science is to foster novel collaborations among complementary research programs. The new syntheses—more complex, scale-variant, and multifactorial views of how the natural world develops and diversifies—may be less appealing to many, but are likely to provide much more realistic and more illuminating accounts of the complexity of nature.

This reintegration of biogeography will continue in earnest largely through the efforts of many broad-thinking scientists who no longer shy away from, but instead embrace, the complexity of nature and who foster collaborations and conceptual syntheses in modern biogeography. New challenges are emerging, and new approaches—and, in some cases, entirely new disciplines—are developing to meet those challenges. We see ten principal strategies and initiatives for advancing the frontiers of this field, but we admit that this list may not be exhaustive. Indeed, we encourage our colleagues and students to redefine the frontiers and surprise us with qualitatively novel contributions to this list.

1. **Scales of space and time**   We should increase attention to scale dependence and to developing process-based linkage across scales of time, space, and biological complexity. As we have noted throughout this book, the influences of many, if not most, processes are emergent at particular scales of space and time, and not at others. Analyses conducted at just the local or global scales are likely to provide only a fragmentary view of the forces influencing geographic variation in biological diversity; developing integrated multi-scale or trans-scalar theories and models remains a key goal and challenge for biogeographers.

2. **The comparative approach**   We should advance and develop more creative applications of the comparative approach, deconstruct complex patterns, and use resultant insights to develop more integrative explanations for the diversity, geographic variation, and distribution of biotas.

3. **Integrations in biogeography**   We should avoid the siren's call to specialize and split into more isolated and independent subdisciplines. Biogeography's greatest distinction and strength is that it is the prototypic interdisciplinary science. As our colleague Stephanie Forrest observed, "The most effective way to do interdisciplinary science is through collaborations among disciplinary scientists."

4. **Biogeography of elusive and novel biotas**   We should address the Wallacean shortfall (i.e., the paucity of information on the geographic variation of nature) and intensify our efforts to understand the biogeography of imperiled species as well as the geographically restricted, less accessible, or otherwise more elusive and novel biotas, including:
   • biotas of remote recesses of the biosphere (e.g., those of the abyssal zones and deep soils);
   • a global-scale, biogeography of freshwater communities;
   • parasites, parasitoids, microbes, fungi, and bryophytes;
   • genetically engineered and other anthropogenically modified species;
   • invasive and now "naturalized" species inhabiting, and in many ways surviving, only in anthropogenic ecosystems (especially zoos, botanical gardens, nature reserves, fragmented landscapes, and the matrix of anthropogenic habitats).

5. **Technological advances**   We should utilize the rapidly expanding databases (e.g., in marine paleontology, palynology, and integrated museum collections data) and increasingly powerful technology (e.g., those in geographic information systems, molecular biology, geology, and geostatistics) in the service of developing a more integrated, conceptually based biogeography.

6. **Evolution**   We should develop an integrated theory of differentiation and diversification of lineages and biotas that includes vicariance, extinction, geodispersal, and reticulating phylogenies (i.e., develop a more comprehensive theory on how the production of living diversity is influenced by geographic circumstances and the dynamics of Earth's geographic template).

7. **Dispersal and immigration**   We should develop a theory of long-distance dispersal, immigration, and gene flow that will provide a more comprehensive understanding of the factors (species and system traits) influencing geographic ranges of species.

8. **Extinction**   We should develop a much more comprehensive understanding of the geography of extinction (i.e., how population persistence varies across the major geographic gradients of area, isolation, latitude, elevation, and depth; across geographic regions in terrestrial and marine realms; and with species traits, including growth form, body size, and the size and relative position of their geographic ranges).

9. **The geography of humanity**   We should advance our understanding of the dynamic geography of our own species: how we, our commensals, diseases, anthropogenic habitat fragmentation, and other anthropogenic environmental disturbances have—and will—expand across and transform native ecosystems.

10. **Conservation biogeography**   Finally, we should apply these approaches and resultant lessons to developing better strategies for conserving, not just isolated samples of imperiled life forms, but their geographic, evolutionary, and ecological context as well (see Richardson and Whittaker 2010; Ladle and Whittaker 2011).

New Dimensions of Biogeography

With technological advances in biogeography such as those in molecular genetics and evolution, we will soon be capable of assaying entire genomes from hundreds of species at a time, elevating the comparative approach to study the evolutionary dynamics of entire biotas. These and other key advances are now complemented by continuing development and availability of geostatistics and geographic information systems (GIS), which provide us with, not just better views and more colorful maps, but new ways of thinking about the patterns and underlying causal processes in the geography of nature. Indeed, biogeography may now be entering a transformative period thanks to our advancing capacities to explore the three-dimensional nature of biodiversity. Despite the myriad of patterns described throughout this book, nearly all of them may be characterized as surface film phenomena, that is, patterns of variation across a very thin slice of the biosphere lying within just a few meters of the surface of the terrestrial and marine realms. Radar, infrared sensitive imagery, ultrasound recordings, and other advances in remote sensing continue to transform our abilities to monitor movements of animals well above Earth's surface. As we observed in Chapter 6, applications of these new tools have spawned a new discipline—aeroecology—and are now providing some fundamental advances in our abilities to directly study long-distance dispersal and immigration. The most transformative and compelling advances in our abilities to expand the dimensions of biogeography, however, are certain to come from the marine realm. Here, organisms don't just pass through, but live and evolve throughout, all layers of this realm. Oceanographers and marine biogeographers continue to make great strides in describing the three-dimensional dynamics of marine environments, how these conditions vary among ocean basins and over time, and ultimately, how they influence the spatial and temporal dynamics of their biotas (see Witman and Roy 2009).

Marine and terrestrial biogeographers have indeed made great strides, even since the last edition of this book, in developing a more integrative science and expanding the physical and conceptual dimensions of biogeography. It has been both a great honor and a wonderful challenge to chronicle these advances, many of which we failed to anticipate. We thus look toward the next edition of this book with great anticipation and with confidence in the present and next generations of biogeographers who will build on existing and rapidly advancing knowledge to explore, advance, and redefine the frontiers of biogeography.

Glossary

abiotic Pertaining to the nonliving components of an ecosystem, such as water, heat, solar radiation, and minerals.

abscissa On a graph, the horizontal (x) axis or the horizontal coordinate of a point.

abyssal plain The relatively flat floor of a deep ocean, mostly between 4 and 6 km beneath the surface.

abyssal zone In deep bodies of water, the zone between 4 and 6 km through which solar radiation does not penetrate (aphotic zone) and in which temperature remains at or slightly below 4° C year-round.

acid rain Precipitation with an extremely low pH. The acid condition is caused by the combination of water vapor in the atmosphere with chemicals such as hydrogen sulfide vapor released from the burning of fossil fuels, producing sulfuric acid.

active dispersal (vagility) The movement of an organism from one point to another by its own motility, such as by active swimming, walking, or flying, rather than by being carried along by some other force; compare with *passive dispersal*.

actualism The philosophical assumption that the physical processes now operating are timeless, and therefore that the fundamental laws of nature have remained unchanged; also called *uniformitarianism*.

adaptation Any feature of an organism that substantially improves its ability to survive and leave more offspring over that of other ancestral forms or coexisting phenotypes.

adaptive radiation The evolutionary divergence of a monophyletic taxon (from a single ancestral condition) into a number of very different forms and lifestyles (adaptive zones).

adaptive zone A way of life, including such properties as ecological preference and mode of feeding, that has been adopted by a group of organisms.

adiabatic cooling The decrease in air temperature as a result of a decrease in air pressure (not a loss of heat to the outside) as warm air rises and expands. The rate of cooling is about 1° C per 100 m for dry air and 0.6° C per 100 m for moist air.

aerial Occurring in the air.

aerial plankton A diverse collection of organisms that are so tiny and light that they are carried by strong winds high above the surface of the Earth and to places far removed from their natal ranges.

aeroecology A relatively new discipline of the natural sciences that utilizes recent advances in remote sensing, including various forms of radar, infrared imagery, and ultrasound detection to visualize the three-dimensional movements of organisms through the aerosphere (the relatively thin portion of the troposphere, closest to the Earth's surface, that supports life).

aestivation A specialized type of animal behavior and physiology in which the organism lives through the summer in a dormant condition.

age and area hypothesis According to Willis, a hypothesis stating that the greater the age of a taxon, the larger its distributional range.

albedo The fraction of solar energy that is reflected from the surface of an object back into the atmosphere, or from the Earth back into space.

allele One of two or more alternative forms of a gene located at a single point (locus) on a chromosome.

allelopathy A type of interspecific interaction in which one species inhibits the growth of another by releasing chemicals into the soil.

Allen's rule Among homeotherms, the ecogeographic (morphogeographic) trend for limbs and extremities to become shorter and more compact in colder climates than in warmer ones.

allochthonous Having originated outside the area in which it now occurs.

allochthonous endemic An endemic taxon that originated in a different location than where it is found today.

allometry The manner in which the relative size of one part of an organism increases in relation to the size increase of the entire organism; also known as *scaling*.

allopatric Occurring in geographically different places; i.e., ranges that are mutually exclusive.

allopatric speciation The formation of new species that occurs when populations are geographically separated.

allopolyploid A hybrid polyploid formed following the union of two gametes, usually from distantly related species, with nonhomologous chromosomes.

alluvial A large, fan-shaped pile of sand, clay and other sediments gradually deposited by moving water along the shores of lakes and estuaries or a river bed that flows onto a flat plain at the foot of a mountain range.

alpha diversity The species richness of a local ecological community, i.e., the number of species recorded within some standardized area, such as a hectare, a square kilometer, or some naturally delineated patch of habitat.

amensalism A pairwise, interspecific interaction in which one species is adversely affected while the other is not directly affected (allelopathy is one example).

amphitropical Occurring in subtropical or temperate areas on opposite sides of the tropics.

anagenesis The process of evolution that produces entirely new levels of structural organization (grades) without branching events. Compare with *cladogenesis*.

ancestor The individual or population that gave rise to some subsequent individuals or populations with different features.

andesite Rocks of volcanic origin, often formed from a blend of sima and sial magma at plate margins such as subduction zones and resulting in, for example, intrusions (dikes) and island arcs.

anemochory Passive dispersal of propagules by winds.

aneuploidy The formation of a new chromosomal arrangement resulting in an increase or decrease of the chromosome number by one pair; often caused by an uneven meiotic division.

anthropochory The intentional transport of disseminules by humans.

anthropogenic biome An alternative type of major vegetation that develops and is maintained by activities of human civilizations. The principal types of anthropogenic biomes include dense settlements, villages, croplands, rangelands, and managed forests, which together cover roughly 75 percent of the Earth's ice-free land surface.

aphotic zone The lower zone in a water column, usually below 50 to 100 m, in which the intensity of solar radiation is too low to permit photosynthesis by plants.

apomixis Reproduction without the union of sexual cells (gametes).

apomorphy In a transformation series, the derived character state.

apterous Wingless; often used to contrast these forms with their primitive ancestors that had wings and the ability to fly.

aquatic Living exclusively or for most of the time in water.

arboreal Living predominantly or entirely in the canopies of trees.

arborescent Treelike.

arctic Pertaining to all nonforested areas north of the coniferous forests of the Northern Hemisphere, especially everything north of the Arctic Circle.

area biogeography The primary goal of an analysis is to reconstruct the biogeographic history of a set of areas of endemism based on the affinities of taxa distributed across those areas.

area cladogram A cladogram of relationships among areas, generated variously from a model of geological history (*geological area cladogram*) or taxon history (see *taxon-area cladogram*).

area of endemism A geographic region containing two or more endemic taxa, defined and diagnosed variously (see Box 12.1).

area of occupancy An alternative conception of the geographic range of a species that attempts to represent or estimate the area within the distributional limits of a species that is actually occupied by its populations. As such, the area of occupancy should almost always be less than the extent of occurrence.

areography The study of the structure of geographic ranges, including variation in their sizes, shapes, and overlap.

arid Exceedingly dry; strictly defined as any region receiving less than 10 cm of annual precipitation.

assembly rules Highly nonrandom patterns in the organization and structure of ecological communities resulting from selective immigrations, extinctions or interactions among species.

asthenosphere A fluid, viscous zone of the upper mantle on which the continental and oceanic plates float (ride) and over which they move.

athenosphere See *asthenosphere*.

austral Pertaining to the temperate and subtemperate zones of the Southern Hemisphere.

Australasia The continental fragments of the original Australian Plate, including Australia, New Zealand, New Guinea, Tasmania, Timor, New Caledonia, and several smaller islands.

Australian Region The biogeographic region including Australia, New Zealand, and New Guinea, other nearby islands, and the Indonesian islands lying east of Wallace's line.

autapomorphy In a transformation series, a derived character state present in a single taxon.

autochthonous Having originated in the area in which it presently occurs.

autochthonous endemic An endemic taxon that differentiated where it is found today.

autocorrelation (spatial and temporal) The tendency for observations that are closer in space or time to be more similar than those that are more removed.

autopolyploid A polyploid possessing more than two sets of homologous chromosomes.

autosome A chromosome that is not a sex chromosome; a somatic chromosome.

autotroph An organism that uses carbon dioxide occurring in the environment as its primary source of cellular carbon.

avifauna All the species of birds inhabiting a specified region.

bajada A broad, sloping depositional surface at the base of a mountain range in deserts, resulting from the coalescing of alluvial fans.

balanced (biota) See *harmonic (biota)*.

barrier Any abiotic or biotic feature that totally or partially restricts the movement (flow) of genes or individuals from one population or locality to another.

basal meristem A plant whose growth (mitotic) tissues are located in or near the soil layer, thus adapting this plant to continual grazing of its apical tissue.

basal metabolic rate A standardized measure of the rate of energy and oxygen requirements of an organism (usually measured at rest and within a range of temperatures at which the organism is not under thermal stress), and indicative of the minimal amount of energy required to maintain vital functions under stress-free conditions.

basin-and-range topography Landscapes characterized by many low, flat areas (basins) that are interrupted by isolated mountain ranges. During pluvial times, many of these basins filled with water.

Batesian mimics Species/individuals that use a form of mimicry in which a harmless and otherwise palatable species avoids predation by resembling a noxious or dangerous species.

bathyal zone The deep sea; in particular, that portion within the aphotic zone but less than 4 km deep.

bathymetry The depth and configuration of the bottom of a body of water.

bauplän (pl *bauplāne*) The body plan or blueprint for how an organism is structured, including such things as its symmetry, number of body segments, and number and relative sizes of limbs, wings, or other appendages.

Beijerinck's Law "Everything is everywhere, but the environment selects." A statement which holds that where the likelihood of dispersal and eventual colonization is relatively high (especially in microscopic organisms with high dispersal capacities), the geographic distributions of these species are largely limited by local environmental conditions and physiological tolerances of the species, not by dispersal.

Benioff zones Zones of high earthquake activity located on the back sides of trenches. The earthquakes, which are caused by the subduction of a plate, are shallow near the trench and progressively deeper at greater distances.

benthic Living at, in, or associated with structures on the bottom of a body of water.

Bergmann's rule The tendency for the average body mass of geographic populations of an animal species to increase with latitude.

Beringia The geographic area of western Alaska, the Aleutians, and eastern Siberia that was connected in the Cenozoic by a landbridge when the Bering Sea and adjacent shallow waters receded.

beta diversity The dissimilarity (turnover) in species composition between local ecological communities.

biogeographic relicts The narrowly endemic descendants of once widespread taxa.

biogeography The science that attempts to document and understand spatial patterns of biological diversity. Traditionally defined as the study of distributions of organisms (both past and present), modern biogeography now includes studies of all patterns of geographic variation in life, from genes to entire communities and ecosystems; elements of biological diversity that vary across geographic gradients including those of area, isolation, latitude, depth and elevation.

biological species A group of potentially interbreeding populations that are reproductively isolated from all other populations.

biomass The total body mass of an organism, population, or community.

biome A major type of natural vegetation that occurs wherever a particular set of climatic and soil conditions prevail, but that may contain different taxa in different regions; e.g., temperate grassland.

biosphere Collectively, all the living things of the Earth and the areas they may inhabit.

biota All species of plants, animals, and microbes inhabiting a specified region.

biotic Pertaining to the components of an ecosystem that are living or came from a once-living form.

biotic components Sets of spatiotemporally integrated taxa that coexist in given areas.

bipedal Using two hindlimbs for locomotion, usually by hopping or jumping, such as a kangaroo or a kangaroo rat.

bipolar Occurring at both poles, in the cold or sub-temperate zones.

bivoltine Breeding twice per year.

bog (mire) A peatland that receives most of its water from precipitation, and is characterized by water that tends to be low in nutrients and acidic, and is often covered with a matt of sphagnum moss.

boreal Occurring in that portion of the temperate and subtemperate zones of the Northern Hemisphere that characteristically contains coniferous (evergreen) forests and some types of deciduous forests.

bottleneck In evolutionary biology, any stressful situation that greatly reduces the size of a population.

brackish Having a salt concentration greater than fresh water (> 0.5%) and less than seawater (35%).

bradytelic Term used by Simpson (1944) to refer to a slow rate of morphological change through time in a lineage with a fossil record.

breeding area In migratory land animals, the area in which populations mate and produce offspring.

Brooks parsimony analysis (BPA) A method in historical biogeography that uses parsimony analyses to construct general area cladograms from a set of taxon-area cladograms.

browser An animal that feeds on plant materials, especially on woody parts of trees and shrubs.

Buffon's law Named in honor of the eighteenth century biogeographer Georges-Louis Leclerc, Comte de Buffon (1707–1788) who observed that different regions, even those with similar environmental characteristics, are inhabited by different assemblages of species.

calcareous In soil biology, pertaining to a soil whose horizons are rich in calcium carbonate and have a basic reaction.

calcification The formation of a soil under continental climatic conditions of relatively low moisture and hot to cool temperatures, resulting in a soil rich in calcium carbonate ($CaCO_3$) because rainfall is not sufficient to leach calcium from the upper soil horizons.

caliche A hard, often rocklike layer of calcium carbonate that forms in soils of arid regions at the level to which the leached calcium salts from the upper soil horizon are precipitated.

canonical distribution A lognormal distribution of the number of individuals or species according to the mathematical formulation of Preston (1962).

carnivore An animal that feeds mostly or entirely on animal prey.

carrying capacity (*K*) An estimate of the population size or biomass of a particular species that can be supported by a given habitat or environment.

cartograms Examples of strategic distortion where mapping units (particular grid cells or polygons such as those representing countries or biogeographic regions) are scaled (and distorted), not according to their surface area but in proportion to another theme such as population density or species diversity.

catastrophic death assemblages Fossil deposits characterized by large numbers of individuals or species preserved at the same time and informative about biological and ecological characterstics of lineages and biotas.

catastrophic extinction See *mass extinction*.

cenocrons Sets of taxa that share the same biogeographic history with a shared distributional and evolutionary history.

census population size The total number of individuals in a population. Compare with *effective population size*.

Centinelan extinctions The loss of species in some unstudied area and time period before they became known to science.

chaparral A type of sclerophyllous scrub vegetation occurring in the southwestern region of North America with a Mediterranean climate.

character A heritable trait with different *character states* that can be recognized by systematists and used to infer evolutionary relationships and classify organisms into taxa.

character displacement The divergence of a feature of two similar species where their ranges overlap so that each uses different resources.

character state One of several alternative forms of a character; e.g., the ancestral form or one of several derived forms.

chorological map A map that attempts to reconstruct and display the paths of range expansion of species or biotas over their evolutionary history.

chorology A term coined by Ernst Haeckel in 1866 to describe the science of the geographic spread of organisms.

chromosome A connected sequence of genetic material on long-stranded deoxyribonucleic acid (DNA) wrapped with proteins.

chronospecies A recognizable stage along a sequence from ancestral to descendant species through non-branching (anagenetic) evolutionary change. See *phyletic speciation*.

circumboreal Occurring in the temperate or subtemperate zones of the New and Old World portions of the Northern Hemisphere.

clade Any monophyletic evolutionary branch in a phylogeny, using derived characters to support genealogical relationships.

cladistic biogeography See *vicariance biogeography*.

cladistics The method of reconstructing the evolutionary history (phylogeny) of a taxon by identifying the branching sequence of differentiation through analysis of shared derived character states. Also called *phylogenetic systematics*. Compare with *evolutionary systematics* and *numerical phenetics*.

cladogenesis The process of evolution that produces a series of branching events.

cladogram A line diagram derived from a cladistic analysis showing the hypothesized branching sequence (genealogy) of a monophyletic taxon and using shared derived character states (synapomorphies) to determine when each branch diverged.

classical species concept See *morphological species concept*.

climax Pertaining to a community that perpetuates itself under the prevailing climatic and soil conditions; therefore, the last stage in secondary succession.

cline A change in one or several heritable characteristics of populations along a geographic transect, attributable to changes in the frequencies of certain alleles and often correlated with a gradual change in the environment.

coalescence In a gene tree, the point in time (absolute time or scaled to generations) at which two allelic lineages diverged from an ancestral lineage.

coenocline A graphical method developed by Robert H. Whittaker to illustrate changes in local abundances or densities of a group of species along an environmental or geographic gradient.

coevolution The simultaneous, interdependent evolution of two unrelated species that have strong ecological interactions, such as a flower and its pollinator or a predator and its prey.

coexistence Living together in the same local community.

cohesion The array of genetic and ecological components that serve to maintain the integrity of a species; particularly relevant to the cohesion species concept.

cohesion species concept A species concept that defines species as "the most inclusive population of individuals having the potential for phenotypic cohesion through intrinsic cohesion mechanisms [genetic and/or demographic exchangeability]" (Templeton 1989, p. 12).

collision zones Regions where two tectonic plates converge, resulting either in the subduction of one of the plates beneath the other or, if both are of equal buoyancy (as in convergence of two continental plates), uplift and formation of mountains.

colonization The immigration of a species into new habitat followed by the successful establishment of a population.

commensalism An interspecific relationship in which one species draws benefits from the association and the other is unaffected.

community An assemblage of organisms that live in a particular habitat and interact with one another.

community assembly hypothesis The assertion that communities develop nonrandom patterns in their structure (primarily in the combination of particular, co-occurring species) as a result of interspecific differences and ecological interactions.

community ecology The study of interactions among co-occurring organisms living in a particular habitat.

comparative phylogeography In phylogeography, the analysis of geographic pattern across multiple co-distributed species or groups of species.

compensatory effects Distributions of populations of a particular, focal species tend to result from the interactive (rather than just the independent) effects of immigration and extinction, such that its populations will occur on isolated islands if their extinction rates on those islands are compensatorily low (i.e., the island is relatively large), and its populations will occur on relatively small islands if their immigration rates are compensatorily high (i.e., the islands are near the mainland or other source of immigrations).

competition Any interaction that is mutually detrimental to both participants. Interspecific competition occurs among species that share requirements for limited resources.

competitive exclusion The principle that when two species with similar resource requirements co-occur, one eventually outcompetes and causes the extinction of the other.

component analysis An approach in historical biogeography that searches for shared cladogenetic events among a set of partially but not completely congruent taxon-area cladograms by using one or more assumptions to develop rules about how to remove the incongruent parts of the cladograms.

congeners Species belonging to the same genus.

consensus area cladogram Summarizes the shared cladogenetic events among a set of taxon-area cladograms.

conservation biogeography A newly articulated discipline that applies lessons from biogeographic theory and patterns to conserve biological diversity, and which emphasizes the need to conserve the geographic, ecological, and evolutionary context of nature. This includes the application of biogeographic principles, theories, and analyses to problems concerning the conservation of biological diversity.

continental drift A model, first proposed by Alfred Wegener, stating that the continents were once united and have since become independent structures that have been displaced over the surface of the globe.

continental island An island that was formed as part of a continent and that has a nucleus of continental (sialic) rocks.

contour maps Maps that use isoclines ("contours" of similar levels of a variable) to illustrate changes in characteristics (e.g., population densities of a species or diversity of ecological communities) across a geographic area.

convergent evolution The development of two or more species with strong superficial resemblances from totally unlike and unrelated ancestors.

Cope's rule A trend in directional evolution (orthogenesis) toward increased body size.

coprolite Fossil excrement.

Coriolis effect A physical consequence of the law of conservation of angular momentum whereby, as a result of the Earth's rotation, a moving object appears to veer to the right in the Northern Hemisphere and to the left in the Southern Hemisphere.

corridor A dispersal route that permits the direct spread of many or most taxa from one region to another.

cosmopolitan Occurring essentially worldwide, as on all habitable landmasses or in all major oceanic regions.

craton The stable crustal nucleus of a continent or continental island, which is older than 600 million years; also called *Precambrian shield*.

crust The outermost rock layer of the Earth, covering the mantle.

cryptic species Species within a genus that are morphologically so similar that they cannot be visually distinguished by superficial features. See *sibling species*.

cryptoturnover Biotic turnover (immigrations and extinctions) of species that may go unrecorded because they occur between survey periods.

Curie point The temperature at which remnant magnetism develops in cooling minerals; e.g., 680° C for hemitite and 580° C for magnetite.

cyclical vicariance model (Also known as the "speciation-pump model.") A model that attributes the high diversity of tropical communities in South America to repeated fragmentation (and allopatric speciation) and reconnection of tropical forests caused by the 20 or so glacial/interglacial cycles of the Pleistocene.

deciduous In plants, having leaves that are shed for at least one season, usually in response to the onset of cold or drought.

declination Dipping of magnetic needles, which orient toward Earth's magnetic poles, which lie deep beneath its crust. Thus, declination can be used to estimate latitude, either under existing conditions or in magnetically active rocks that crystallized in ancient periods, but now drifted to distant positions.

decomposer An organism (usually a bacterium or fungus) capable of metabolically breaking down organic materials into simple organic and inorganic compounds and releasing them into the ecosystem.

deconstruction (pattern deconstruction) An attempt to understand underlying, causal mechanisms for patterns of biodiversity by examining the differences in patterns among functionally different groups of species, or differences among patterns for similar species across different habitats, ecosystems or different regions (see Huston 1994).

deductive reasoning A method of analysis in which one reasons from general constructs to specific cases.

defaunation The elimination of animal life from a particular area.

dehiscence The act of splitting along a natural line to discharge the contents, such as that of an anther to release pollen grains or of a capsule to release seeds.

deletion A form of mutation in which one or more nucleotides are eliminated from a DNA sequence.

delta diversity The dissimilarity (turnover) in species composition between large, geographic areas.

density compensation In island biogeography, an increase in the density of a species inhabiting an island habitat when one or more taxonomically similar competitors are absent.

density enhancement The tendency for densities of populations inhabiting relatively small, species-poor islands to increase with island area.

density inflation The tendency for densities of populations inhabiting islands of intermediate size and species richness to increase beyond the level of this species on the mainland. The accumulated effects of density inflation over many species is referred to as "density overcompensation" (see below).

density overcompensation In island biogeography, a case in which the total densities of a few species inhabiting a small island exceed the combined densities of a much greater number of species of the same taxon occupying similar habitats on a large island or continent.

density stasis On islands of intermediate to relatively large area and high levels of species richness, the combined effects of many species interactions may regulate one another's densities such that few, if any, exhibit any consistent trend of population density with species richness or island area.

desert A general term for an extremely dry habitat, especially one where water is unavailable for plant growth most of the year; in particular, a habitat with long periods of water stress and sparse coverage by plants, often with perennials covering less than 10% of the total area.

desertification The degradation of land in arid, semi arid and dry sub-humid areas into desert.

deterministic Determined or controlled by some regulatory force, such as natural selection. Compare with *stochastic*.

detritivore An organism that feeds solely on detritus.

detritus Freshly dead or partially decomposed organic matter.

dew point The temperature at which air becomes saturated with water vapor and it condenses to form fog or other forms of precipitation.

diadromous Referring to an aquatic animal that must migrate between fresh water and seawater to complete its life cycle, such as certain lampreys and eels.

diapause An arrested state of development in the life cycle, especially of many insects, during which the organism has reduced metabolism and is more resistant to stressful environmental conditions, such as cold, heat, or drought.

diaspore Any part or stage in the life cycle of an organism that is adapted for dispersal.

diffuse competition A type of competition in which one species is negatively affected by numerous other species that collectively cause a significant depletion of shared resources.

diffusion A form of range expansion that is accomplished over generations by individuals spreading out from the margins of the species' range.

dimorphic Having two distinct forms in a population.

dioecious Having individuals with only male or only female reproductive systems.

diploid Having two sets of chromosomes ($2n$).

disassembly (community disassembly) The nonrandom loss of particular species from ecological communities, typically associated with anthropogenic disturbances and resultant, selective extinctions.

discontinuous A common pattern of variation in trait characteristics, where some individuals are very similar to one another but separated by other groups of individuals by distinct gaps.

disharmonic (biota) A biota that is not a random subset of the mainland or source biota, but one biased in favor of species with superior immigration abilities or abilities to survive on islands.

disjunct A taxon whose range is geographically isolated from that of its closest relatives.

disjunction A discontinuous range of a monophyletic taxon in which at least two closely related populations are separated by a wide geographic distance.

dispersal The movement of organisms away from their point of origin.

dispersal biogeography A form of historical biogeography that attempts to account for present-day distributions based on the assumption that they resulted from differences in the dispersal abilities of individual lineages.

dispersal-vicariance analysis (DIVA) An event-based approach in historical biogeography that reconstructs ancestral areas at each node in a taxon-area cladogram and infers events based on an a priori assignment of the "cost" of vicariance, dispersal, and extinction events.

dispersion The spatial distribution of individual organisms within a local population.

disruptive selection Selection that favors extreme and eliminates intermediate phenotypes.

disseminule Any part of a plant that is used for dispersal; occasionally restricted to include only seeds and seed-bearing structures. See also *diaspore*.

distance-decay The tendency for some property (e.g., species richness) to decline with increasing isolation, or for similarity in characteristics of two or more sites to decrease with increasing distance between those sites.

distribution function A method of assessing the relative importance of immigration and extinction as processes influencing community assembly by using a bivariate plot to illustrate the distribution of populations of a focal species on islands of varying isolation and area.

doldrums A narrow equatorial zone characterized by long periods of calm or light shifting winds, caused by the upward movement of air masses from this region of high atmospheric pressure to higher latitudes with relatively lower pressure.

dominant A species having great influence on the composition and structure of a community by virtue of its abundance, size, or aggressive behavior.

dot maps Maps depicting the geographic range of a species by place dots at points of documented occurrence of its individuals and populations.

driftless area A possible glacial refugium located in what is present day southern Wisconsin and adjacent areas of Illinois and Iowa.

eccentricity The degree of ellipticity in Earth's orbit about the sun, which varies in a cyclical manner over a 100,000 year period.

ecogeographic (morphogeographic) rules A regular change in characteristics of organisms (in particular, but not limited to, their morphological characteristics) along geographic gradients (e.g., Bergmann's, Allen's, Jordan's, and Thorson's rules).

ecological biogeography The study of distributions and geographic variation of extant biotas, with special emphasis on the influences of interactions between organisms and their abiotic and biotic environments.

ecological naiveté The tendency for insular biotas to lose the structures and behaviors that enabled their mainland ancestors to avoid predators, parasites, and competitors.

ecological niche See *niche*.

ecological release The tendency for populations inhabiting species-poor environments such as small or isolated islands to occur in a broader range of habitats, feed on a broader variety of prey, or otherwise occupy a broader range of their fundamental niche than they do in species-rich communities.

ecological speciation The development of reproductive isolation between two incipient species as a result of divergent natural selection.

ecological time The period during which a population can interact with its environment and respond to environmental fluctuations without undergoing substantial evolutionary modification; compare with *evolutionary time*.

ecology The study of the abundance and distribution of organisms and of the relationships between organisms and their biotic and abiotic environments.

ecoregions As defined by Robert G. Bailey, a geographic grouping of landscapes, each comprised of a mosaic of ecosystems.

ecosystem The set of biotic and abiotic components in a given environment.

ecosystem engineer A species that significantly alters the functioning and environmental characteristics of an ecosystem.

ecosystem geography The study of the distribution of ecosystems and the processes that have differentiated them in space and time.

ecotone A zone of transition between two habitats or communities.

ectoparasite A parasite that lives on the exterior of its host, such as a louse.

ectotherm An animal whose body temperature is determined largel.

edaphic Pertaining to soil.

edge effects In conservation biology, the potential negative effects of exotic species and disturbances that act along the edges of habitat fragments.

effective population size The number of breeding individuals in a population. Compare with *census population size*.

El Niño Southern Oscillation (ENSO) An approximate 5- to 7-year cycle of regional climatic changes that is caused by variation in sea surface temperatures and oceanic currents in the tropical regions of the Pacific (similar events occur in the tropical Atlantic).

emigration Dispersal of organisms away from a region of interest.

endemic Pertaining to a taxon that is restricted to the geographic area specified, such as a continent, lake, biome, or island.

endemic bird areas Areas containing the ranges of at least two restricted-range species of birds (i.e., those whose breeding ranges are less than 50,000 km^2).

endorheic basin A drainage basin that lacks outflows (either surface or underground), such that all water that it receives is lost only by seepage or evaporation.

endotherm An animal whose body temperature is maintained largely by its own metabolic heat production; compare with *ectotherm*.

endozoochory Dispersal of plant propagules inside the bodies of animals. Some plants adapted for endozoochory require scarification or chemical degradation of their seeds inside the animal's digestive tract for germination.

entropy A measure of the unavailable energy in a closed thermodynamic system.

epeiric sea A large but relatively shallow body of salt water that lies over a part of a continent.

epicenter The point on the Earth's surface directly above the origin (focus) of an earthquake.

epicontinental sea See *epeiric sea*.

epifaunal Living on a substrate, such as a barnacle or coral.

epipelagic Living in open water, mostly within the upper 100 m.

epiphyll Thin layers of mosses, lichens, and algae that grow along the surfaces of trees in rainforests.

epiphyllous Living on a leaf.

epiphyte A plant that usually lives on another plant (is not rooted in soil) and derives its moisture and nutrients from atmospheric precipitation and whatever materials are released by the organisms in the immediate vicinity.

equatorial countercurrent A small current running west to east along the equator (i.e., opposite the major oceanic gyres) in the eastern region of the Pacific Ocean (see Figure 3.5).

equilibrium A condition of balance between opposing forces, such as birth and death rates or immigration and extinction rates.

equilibrium theory of island biogeography The theory proposed by MacArthur and Wilson stating that the number of species on an island results from a dynamic equilibrium between the opposing rates of immigration and extinction.

equilibrium turnover rate The change in species composition per unit of time when immigration equals extinction.

equinox Either of two times (March 21 and September 22) in a year when the sun passes the equator so that day and night are the same length everywhere on Earth.

establishment The successful start or founding of a population.

estuary A body of water where freshwater from rivers mixes with saltwater from the sea.

Ethiopian Region The portion of Africa south of the Sahara Desert plus Madagascar and other nearby islands.

euryhaline Having a tolerance to an extremely wide range of salt concentrations.

eurythermal Having a tolerance to a broad range of temperatures.

eurytopic Having a tolerance to an extremely wide range of habitats and environmental conditions.

eustatic (eustasis) Global fluctuations in sea level resulting from the freezing or melting of great masses of sea ice, thus decreasing or increasing the global volume of liquid water, respectively.

eutherian A placental mammal.

eutrophic lake A lake that is rich in dissolved nutrients and highly productive, but that is usually shallow and seasonally deficient in oxygen.

evapotranspiration The sum total of water lost through evaporation from land and transpiration by plants.

evolution In the strictest sense, any irreversible change in the genetic composition of a population.

evolutionarily significant units Defined in several ways, but one that is popular is as "historically isolated sets of populations for which a stringent and qualitative criterion is reciprocal monophyly for mitochondrial DNA (mtDNA) combined with significant divergence in frequencies of nuclear alleles" (Moritz et al. 1995, p. 249).

evolutionary biogeography An earlier approach in historical biogeography motivated by the search for centers of origin and dispersal over a stable geographic template.

evolutionary species concept A species concept that recognizes a species as "an entity composed of organisms that maintains its identity from other such entities through time and over space and that has its own independent evolutionary fate and historical tendencies" (Wiley and Mayden 2000, p. 73).

evolutionary systematics A method of reconstructing the evolutionary history (phylogeny) of a taxon by analyzing the evolution of major features along with the distribution of both shared primitive and shared derived characteristics. Compare with *cladistics* and *numerical systematics*.

evolutionary time The period during which a population can evolve and become adapted to an environment by means of genetic changes. Compare *ecological time*.

evolvability The capacity of species populations to adapt by mutation and evolution and respond to natural selection.

excess density compensation See *density overcompensation*.

exoskeleton An external skeleton of an animal, such as that of a clam or insect.

exozoochory (epizoochory) Dispersal of plant propagules (e.g., sticky or barbed seeds or fruits) that attach to the skin surfaces of mobile animals.

exploitative competition A negative interaction between species or conspecifics in which individuals use up resources and make them unavailable to others.

exponential growth In ecology, a population that increases in proportion to its population size and, therefore, at a continually accelerating rate.

extant Living at the present time.

extent of occurrence An alternative conception of the geographic range of a species that attempts to represent the distributional limits, measured as the area that lies within the extreme limits of its occurrences (compare with *area of occupancy*).

extinct No longer living.

extratropical Occurring outside the tropics.

facultative relationship An interaction between two organisms that is not essential to the survival of either.

family A taxonomic category above the level of genus and below the level of order.

fault In geology, a weakness in the Earth's crust along which there can be crustal motion and displacement.

fen (fenland) A peatland that receives most of its water from runoff or groundwater. As a result, fens tend to have higher nutrient levels and support more diverse plant and animal communities than bogs.

filter A geographic or ecological barrier that blocks the passage of some forms, but not others.

filter feeder An aquatic animal that feeds on plankton or other minute organic particles by using one of a variety of filtering mechanisms.

first law of thermodynamics Energy is neither created nor destroyed, but can be converted from one form to another.

fitness The ability of a genotype to leave offspring in the next generation or succeeding generations as compared with that of other genotypes.

floristic belts A series of plant communities that are characterized by different growth forms or physiognomies that occur in predictable series (e.g., deserts, savanna, dry woodlands, coniferous forests and tundra) along elevational and latitudinal gradients.

flyway An established air route used by vast numbers of migratory birds.

food chain A diagram or list of species that describes predator and prey relationships and the transfer of energy and nutrients within an ecological community.

forest Any of a variety of vegetation types dominated by trees and usually having a fairly well developed or closed canopy when the trees have leaves.

fossil A remnant, impression, or other trace of a living organism from the past.

fossorial Referring to an animal that lives in and forages on plants from a burrow.

founder effect Genetic drift that occurs when a newly isolated population is founded, such as on an island, by one or a few colonists (founders). The features of the new population may be markedly different from those of the ancestral population because the gene pool of the founders may be a biased and small sample of the source population.

fragment In plate tectonics, a portion of a former landmass.

fresh water In the strictest sense, water that has a salt concentration of less than 0.5%.

frugivore An animal that feeds mainly on juicy fruits.

fundamental geographic range The theoretical distribution that populations of a particular species may achieve based solely on its physiological and abiotic tolerances, assuming that species interactions are unimportant and opportunities of dispersal are unlimited.

fundamental niche The total range of environmental conditions in which a species can survive and reproduce.

fynbos A type of sclerophyllous scrub vegetation occurring in the region of South Africa with a Mediterranean climate.

gamete One of two cells, usually from different parents, that fuse to form a zygote.

gamma diversity The total species richness of large geographic areas ranging from a combination of local ecological communities to those of entire biomes, continents and ocean basins.

gene The small unit of a DNA molecule that codes for a specific protein to produce one of the chemical, physiological, or structural attributes of an organism.

gene conversion A process by which damaged genes can be repaired.

gene flow The movement of alleles within a population or between populations caused by the dispersal of gametes or offspring.

gene frequency (allelic frequency) The proportions of gene forms (alleles) in a population.

gene tree The phylogeny of a particular gene or group of tightly linked genes embedded within the phylogeny of a population or species.

genealogical concordance concept A species concept that proposes that "population subdivisions concordantly identified by multiple independent genetic traits should constitute the population units worthy of recognition as phylogenetic taxa" (Avise and Ball 1990, p. 52).

genealogy The study of the exact sequence of descent from an ancestor to all derived forms.

general area cladogram A cladogram summarizing vicariance (and depending on approach, also dispersal, sympatric speciation, and extinction) events from a set of taxon-area cladograms.

generalized track In panbiogeography, a line drawn on a map representing the coincident distributions of numerous disjunct taxa.

genetic drift Changes in gene frequency within a population caused solely by chance—i.e., which individuals happen to mate and leave offspring in the next generation—without any influence of natural selection.

genome A full set of chromosomes.

genotype The total genetic message found in a cell or an individual.

genus A taxonomic category for classifying species derived from a common ancestor; a level below that of family and tribe.

geo-dispersal The concordant dispersal of several species out of an area of endemism following erosion of a dispersal barrier with subsequent re-emergence of a barrier and a new round of speciation.

geographic coordinate systems A means of using two geographic variables (e.g., latitude and longitude, or easting and northings [UTM system]) to locate points on a map.

geographic information systems (GIS) A system of technologically sophisticated and readily accessible, computer-based tools for visualizing, modifying, and analyzing patterns among spatially referenced observations.

geographic isolation Spatial separation of two potentially interbreeding populations; allopatry.

geographic range The area over which the populations of a species are distributed.

geographic speciation See *allopatric speciation*.

geographic template The highly nonrandom, spatial variation of environmental conditions that forms the foundation for all biogeographic patterns.

glacial-interglacial cycles The 20 or so repeated, global-scale climatic changes that occurred during the Pleistocene Epoch and profoundly affected biogeographic patterns of Earth's biotas.

glacio-pluvial The alternating "ice" to "rain" (glacial to interglacial) stages of the Pleistocene Epoch.

gleization The formation of a soil under moist and cool or cold conditions, resulting in an acidic soil with a large amount of organic matter and iron present in a reduced state (FeO).

Gondwanaland The southern half of the supercontinent Pangaea, consisting of all the southern continental landmasses and India, which were united for at least 1 billion years but broke up during the late Mesozoic.

granivore An animal that feeds mainly on dry seeds and fruits; a subtype of herbivore.

grassland Any of a variety of vegetation types composed mostly of grasses and other herbaceous plants (forbs) but few if any trees and shrubs.

Great American Interchange Term used by the paleontologist George Gaylord Simpson to describe the waves of biotic exchange of terrestrial organisms between Nearctic and Neotropical Regions (North and South America) following the emergence of the Central American landbridge at roughly 3.5 million years BP.

greenhouse effect The retention of heat in the atmosphere when clouds (water vapor) and carbon dioxide absorb the infrared (heat) radiation reradiated from the Earth rather than permitting the heat to escape.

grouping rule Only synapomorphies provide evidence of common ancestry relationships; symplesiomorphies, autapomorphies and homoplasies are uninformative for doing so.

guide fossil A fossil used to date the age of a sedimentary stratum; also called an *index fossil*.

guild Groups of species that exploit the same class of environmental resources in a similar manner (e.g., desert granivores or foliage-gleaning birds).

guyot A flat-topped submarine volcano.

habitat island A geographically isolated patch of habitat, such as a pond, mountaintop, or cave, that can be studied in the same ways as oceanic islands for patterns of colonization and extinction.

habitat selection The preference of an organism for a particular habitat type.

hadal Of or pertaining to the deepest zones of the ocean, below 6 km, which have nearly constant environments with year-round temperatures near 4° C and no light penetration; e.g., in the oceanic trenches.

half-life The amount of time needed for half of the radioactive material in a rock to decay to a stable element.

halomorphic soil Soil that is characterized by very high concentrations of sodium, chlorides, and sulfates and which forms in estuaries and salt marshes, and in arid inland basins where shallow water accumulates and evaporates, leaving behind high concentrations of salts.

halophytic "Salt-loving" plant species that grow in areas of soils with high salt concentrations, including a variety of taxonomic and functional groups, each with special adaptations for dealing with the problem of maintaining osmotic and ionic balance in these environments (e.g., the ability to excrete salts or store them in special cells and tissues).

haploid Having one set of chromosomes (*n*).

haplotype A *haploid* allele for a gene or group of tightly linked genes.

harmonic (biota) A biota that is similar to a random subset of its source pool. See *balanced* (*biota*).

herbivore An animal that feeds mostly or entirely on plants.

hermaphroditic Having both male and female reproductive structures in the same individual.

heterogeneity The state of being mixed in composition, as in genetic or environmental heterogeneity.

heterotroph An organism that uses organic carbon (compounds made by living organisms) as its source of cellular carbon.

historical biogeography The study of the development of lineages and biotas including their origin, dispersal and extinction.

history of lineage The series of changes that have occurred in the intrinsic characteristics of organisms, species, or higher taxa over generations of evolutionary descent.

history of place The past environmental conditions, configurations, and locations of landmasses or ocean basins and other bodies of water.

history of species The environmental and ecological conditions experienced by the ancestors of a particular species or taxon, which have influenced its evolutionary development and geographic distributions, both past and present.

histosols Soils that are characterized by a substantial layer of organic matter (>30%) of more than 40 cm either extending down from the surface or taken cumulatively within the upper 80 cm of the soil. These soils are formed because the production of organic debris exceeds its decay.

Holarctic region The extratropical zone of the Northern Hemisphere, which includes both the Neartic and Palearctic regions.

homeostasis The maintenance of a constant internal state despite fluctuations in the external environment.

homeotherm An animal that maintains a fairly constant body temperature.

homologies Character states shared between two or more taxa because it was inherited from a common ancestor.

homologous chromosomes Two chromosomes (in a diploid organism) that have essentially the same gene sequence and that are similar enough to pair during meiosis and mitosis.

homology A character shared by a group of organisms or taxa due to inheritance from a common ancestor.

homoplasy A character that appears similar among a group of organisms or taxa, but the similarity is due to parallel or convergent evolution rather than inheritance from a common ancestor.

horizon In soils, the major stratifications or zones, each of which has particular structural and chemical characteristics.

horizontal gene transfer The movement of a piece of DNA or a gene from one species, and incorporation into the genome of a new species, often mediated through an intermediate vector (e.g., bacterium or parasite).

horotelic Term used by Simpson (1944) to refer to a moderate rate of morphological change through time in a lineage with a fossil record.

horse latitudes The zones of dry descending air between 30° and 40° N and S latitude, where many deserts of the world are located.

hotspot In plate tectonics, a stationary weak point in the upper mantle that discharges magma (molten rock) as a plate passes over it, producing a narrow chain of islands. In biodiversity studies, an area with a relatively high number of species or high number of endemic species.

humus A brown or black organic substance in soils consisting of partially or wholly decayed vegetable or animal matter that provides nutrients for plants and increases the ability of soil to retain water.

hybridization The production of offspring by parents of two different species or dissimilar populations or genotypes.

hydrochory Passive dispersal of propagules by water.

hyperosmotic Referring to an environment in which water will diffuse from an organism because the external solution has a salt concentration higher than its internal concentration.

hypersaline Having a higher salt concentration than normal seawater.

hypothetico-deductive reasoning A method of analysis in which one starts with a new, tentative hypothesis and then tests the predictions and assumptions following from it one by one in an attempt to falsify it.

ichthyofauna All the species of fishes inhabiting a specified region.

immigrant pattern The tendency for islands along a gradient of increasing isolation to be inhabited by assemblages of species that represent highly nonrandom subsets of less-isolated communities, in this case biased in favor of species with superior immigration abilities.

immigration The arrival of new individuals to an isolated site.

incidence functions A method of exploring community assembly by graphing the proportion of islands inhabited by a given species as a function of the number of other species present.

included niche A niche of a specialized species characterized by a narrow range of conditions and lying entirely within the larger fundamental niche of a more generalized species.

index fossil See *guide fossil*.

inductive reasoning A method of analysis in which one uses specific observations to derive a general principle.

infaunal Living within a substrate, such as clams that bury themselves in sand or mud and that feed by means of a long siphon.

ingroup The focal monophyletic group in a cladistic analysis.

insectivore An animal that feeds mainly on insects.

insertion A form of mutation in which one or more nucleotides are inserted into a DNA sequence.

interference competition A negative interaction between species in which aggressive dominance or active inhibition is used to deny other individuals access to resources.

interglacial A phase during the Quaternary, when glacial ice sheets retreated and the climate became more equable.

interglacials The relatively warm and wet periods that alternated with the glacial periods of the Pleistocene Epoch.

interpolation (spatial and temporal) Procedures that provide estimates of the expected value of a variable at an unmeasured point in space or time, based on statistical models that take into account the values of recorded variables at actual observation sites and times, and their distances from the site (or changes from the time).

intertidal zone The zone above the low tide mark and below the high tide mark of a body of water; the littoral zone.

inversion A form of mutation in which the orientation of a portion of a chromosome becomes reversed with respect to its former orientation.

island rule A graded trend in insular vertebrates from gigantism in the smaller species to dwarfism in the larger species.

isolating mechanism Any structural, physiological, ecological, or behavioral mechanism that blocks or strongly interferes with hybridization or gene exchange between two populations.

isostasy In geology, the term that describes how continental blocks float on a viscous layer of the mantle. While doing this the block may rise or fall to achieve an equilibrium.

isotherm A line on a map connecting all locations with the same mean temperature.

Jordan's rule (law of vertebrae) The tendency for the number of vertebrae in marine fish to increase along a gradient from the warm waters of the tropics to cooler waters of the high latitudes.

jump dispersal Long-distance dispersal that is accomplished by movement of individuals within a relatively short period.

karyotype The morphological appearance and characteristic number of a set of chromosomes in a cell, and generally conserved across a population or species.

kettle lake Lakes that formed during the onset of an interglacial period when large blocks of ice separated from the melting glaciers and formed deep, persistent depressions that later filled with runoff from ice-melt and precipitation.

key innovation Evolution of a new trait that confers a strong adaptive advantage and is often associated with an adaptive radiation.

keystone species A species whose activities have a disproportionate effect on the structure and function of ecological communities.

kriging A statistical means of estimating the value of a particular variable at an unsampled location by calculating a moving-average from data collected at other locations.

K-selection Selection favoring a more efficient utilization of resources, which is more pronounced when the species is at or near carrying capacity (K).

K-strategists Species whose life-history and ecological characteristics are adapted to selective pressures associated with populations near carrying capacity (K) of an environment.

laterite In tropical soils, a hard, rocklike layer composed principally of ferric oxide, produced when this compound accumulates in high concentration in the soil.

laterization The formation of a soil under conditions of abundant moisture, warm temperatures, and high decomposer activity, resulting in a soil from which bases and silica have been removed (leached), leaving behind a clay rich in ferric and aluminum oxides.

Laurasia The northern half of the supercontinent Pangaea, including North America, Europe, and parts of Asia.

leaching In soil science, the removal of soluble substances by water.

lentic Referring to standing freshwater habitats, such as ponds and lakes.

leptokurtic A mathematical distribution characterized by a sharply peaked curve with a long tail.

liana A climbing woody or herbaceous vine that is especially common in wet tropical forests.

Liebig's law of the minimum An early ecological generalization, now discredited or greatly modified, which stated that abundance and distribution are limited by the single factor in the shortest supply. See also *limiting factor*.

life zones The characteristic changes in vegetation composition and form that occur along an elevational or latitudinal gradient.

Lilliputian effect The tendency evidenced in the fossil record for extinctions to be species selective, resulting in differential survival of the smaller life-forms.

limiting factor The resource or environmental parameter that most limits the abundance and distribution of a population.

limnetic zone The offshore waters of a lake that receive light sufficient to support photosynthesis of their plant (macrophyte and phytoplankton) communities.

limnologist A scientist who studies the physical, chemical and biological characteristics of freshwater systems, especially lakes, rivers and ponds.

lineage sorting In a gene tree, the loss of ancestral polymorphism within a population or species as a consequence of stochastic effects influencing the probability of an allele either being retained or lost during genetic transmission across generations.

Linnaean shortfall The disparity between the number of described species and the total number of species in existence.

lithosphere The Earth's crust, exclusive of water (hydrosphere) and living organisms (biosphere).

littoral zone The marginal zone of the sea; i.e., the intertidal zone. In fresh water, the shallow zone that may contain rooted plants.

living fossils An extant taxon that has changed very little morphologically from ancestral taxa in the fossil record.

log-transformed Referring to the change of values in a data set by taking the logarithm of each.

long-distance (jump) dispersal The movement of an organism across inhospitable environments to colonize a favorable distant habitat.

lotic Referring to running freshwater habitats, such as brooks and rapids.

macchia A type of sclerophyllous scrub vegetation occurring in regions of the Old World with a Mediterranean climate.

macroecology A top-down and multi-scale approach to investigating the assembly and structure of biotas, which identifies general patterns and underlying mechanisms by focusing on the statistical distributions of variables across spatial and temporal scales, and among large numbers of equivalent (but not identical) ecological *particles* (e.g., particles can include individual organisms within a local population or entire species, replicated sample areas or patches of habitat, or species within local communities or larger biotas).

macroevolution A general term for evolution above the population level.

magma The molten rock material under the Earth's crust, from which igneous rock is formed by cooling.

magnetic reversals In geology, episodes during which the direction of Earth's magnetic field has been reversed, occurring approximately twice every million years.

magnetic stripes In geology, long alternating stripes of normally and reversely magnetized basaltic rock on the ocean floor.

malacofauna All the species of mollusks inhabiting a specified region.

mangrove swamp A wetland found in tropical climates and along coastal regions that is dominated by mangrove trees and shrubs, particularly red mangroves (*Rhizophora*), black mangroves (*Avicennia*) and white mangroves (*Laguncularia*).

mantle The second and thickest layer of the Earth.

mantle drag One of the forces responsible for continental drift, in this case resulting from the tendency of the crust to ride the gyres of circulating mantle beneath it much like boxes on a conveyor belt.

maquis A type of sclerophyllous scrub vegetation occurring in portions of the Mediterranean region with its characteristic climate.

marginal population A population that has difficulty in surviving as a result of limiting abiotic or biotic factors.

marine Living in salt water.

maritime climate In general, a coastal or island environment with little or no freezing, much cloud cover and fog, and less variance in temperature, and thus a milder year-round climate, than nearby inland or mainland localities.

marsh (moor) A wetland that is dominated by herbaceous vegetation.

mass extinction A major episode of extinction for many taxa, occurring fairly suddenly in the fossil record.

matorral A type of sclerophyllous scrub vegetation occurring in the region of Chile with a Mediterranean climate.

Mediterranean climate A semiarid climate characterized by mild, rainy winters and hot, dry summers.

megafauna A general term for the large terrestrial vertebrates inhabiting a specified region.

mesic Relatively moist and equable.

mesophytes Land plants that grow in environments with an average ("mesic") supply of water.

Messinian crisis The sudden drainage of the Mediterranean Sea in the Cenozoic.

metabolism The sum total of the positive and negative chemical reactions in an organism or a cell that provide the energy and chemical substances necessary for its existence.

metadata Detailed descriptions of methods used to record and store a particular data set, along with the relevant characteristics of those data.

metamorphosis A major change in the form of an individual animal during development; e.g., a caterpillar becomes a moth or a butterfly and a tadpole becomes a frog or a toad.

metapopulation A set or constellation of local populations of a particular species that are linked by dispersal among those populations.

metatherian A marsupial mammal.

microallopatric Refers to spatially small forms of allopatric divergence, such as might occur among different volcanoes on a single oceanic island.

microclimate The fine-scale environmental regime.

microcosm A small community that represents in miniature the components and processes of a larger ecosystem.

microevolution Evolution below the species level, within and among populations (e.g., changes in gene diversity and gene frequencies resulting from mutation, genetic drift, natural selection, and gene flow).

microhabitat The fine-scale environment that often determines the presence or absence of each kind of organism.

microphyllous   Having small, narrow leaves; characteristic of many plants in very dry habitats.

microvicariance   Vicariance at a relatively small spatial scale (e.g., on a single island, or between small peripheral versus large core populations).

midden   A solid mass of collected organic debris left by an animal such as pack rats (Rodentia).

midoceanic ridge   In plate tectonics, a submarine mountain chain, within which seafloor spreading of the oceanic plates occurs.

migration   Dispersal from and return to and from an area, typically a breeding site, by an individual or its immediate descendants.

Milankovitch cycles   Named in honor of their discoverer, Serbian astrophysicist Milutin Milankovitch, the cyclical changes in the characteristics of Earth's orbit about the sun.

mimicry   The marked resemblance of an organism to another organism or a background (e.g., a leaf, tree bark, or sand) to deceive predators or prey by "disappearing" (crypsis) or by causing the predator or prey to confuse the mimic with something it is not.

mitochondrial DNA (mtDNA)   The closed circular DNA of the organelle, transmitted through the cytoplasm from one generation to the next.

mobility   The ability to move or be moved.

molecular phylogenetics (systematics)   A subdiscipline of phylogenetic systematics that employs characters derived from DNA sequences, protein electrophoresis, protein immunology, etc.

monoecious   In botany, having separate male and female flowers on the same plant.

monomorphic   Having only one form in a population.

monophyletic   A group, or clade, of organisms that includes an ancestral taxon and all of its descendant taxa.

monospecific   Having only one species in the genus.

monotypic   Having only one species in the taxon.

monsoon   Predictable and typically intense summer rains that occur in tropical and subtropical regions where heat and rising air masses over the land surface draws cool, moisture laiden air in from the adjacent oceans. As this air rises over land, it cools, reaches its dew point, and results in heavy precipitation.

monsoon forest   Tree-dominated ecosystems, sometimes called the "layperson's jungle," that develop in areas with intense seasonal rainfall (summer monsoons) and are dominated by trees with large leaves and a luxuriant undergrowth of shrubs and small trees.

morphogeographic rule   See *ecogeographic rule*.

morphological species concept   A species concept that proposes that "species are the smallest groups that are consistently and persistently distinct, and distinguishable by ordinary means" (Cronquist 1978, p. 15).

motility   The ability to move under one's own power, as by wings or a flagellum.

mutation   Any change in the genetic information that results in either the alteration in a gene (e.g., point mutation, insertion, or deletion) or a major modification in the karyotype (chromosomal mutation), neither of which is usually reversible in the strictest sense.

mutualism   An interspecific relationship in which both species receive positive benefits from their interaction.

mycorrhiza   A symbiotic relationship between a fungus and a plant root that benefits the plant by providing a source of useful nitrogen, which is manufactured by the fungus.

natural selection   The process of eliminating from a population through differential survival and reproduction those individuals with inferior fitness.

neap tide   A less than average tide occurring at the first and third quarters of the moon.

Nearctic region   The extratropical region of North America.

nectarivore   An animal that feeds on plant nectar.

nekton   Organisms that are free-swimming in the upper zone of open water and strong enough to move against currents.

neoendemic   An endemic that evolved in fairly recent times. Compare with *paleoendemic*.

Neotropical Region   The region from southern Mexico and the West Indies to southern South America.

neritic zone   The shallow water adjoining a seacoast, especially the zone over a continental shelf.

nested clade analysis (NCA)   A method in phylogeography that converts a statistical haplotype network into a hierarchically nested set of clades based on measures of geographical distribution of clades and nested clades relative to one another.

nested clade phylogeographic analysis (NCPA)   See *nested clade analysis*.

nestedness   The tendency for ecological communities from species-poor sites to form proper subsets (include only those species) found in richer sites.

net primary productivity   The rate of formation of plant tissue in an ecological community that represents the rate of energy made available for consumption by herbivores (equal to gross primary productivity minus total respiration of all plants in that community).

niche   The total requirements of a population or species for resources and physical conditions.

niche breadth   The range of resources and physical conditions used by a particular population relative to that of other populations.

niche conservatism   The tendency of species to retain ancestral characteristics.

niche expansion   An increase in the range of habitats or resources used by a population, which may occur when a potential competitor, predator or other interacting species is absent.

nomadic   Having no fixed pattern of migration.

nonadaptive radiation   The process of diversification of a lineage that, although resulting in an increase in the number of descendant species, fails to produce ecological diversification, thus limiting taxa to exclusive (nonoverlapping) distributions.

non-shivering thermogenesis   A process of heat production in certain mammals (especially newborns and small mammals) involving the breakdown of chemicals such as fatty acids found in muscle, liver, and brown adipose tissue (heavily vascularized and rich in mitochondria), which results in the production of large amounts of heat instead of ATP.

nonvolant   An organism that is unable to fly (e.g., rodents versus bats and birds).

nuclear DNA (nucDNA) The vast majority of DNA in a eukaryotic cell, transmitted through the nucleus from one generation to the next.

null hypothesis A statistical hypothesis stating what would be expected by chance alone, which can be tested in order to determine whether an observation could be a result of chance or is instead the result of some directing force.

null phenetics The method of classifying a taxon by using quantitative measures of a large number of characters to assess its overall character similarity to other related taxa. Compare with *evolutionary systematics* and *cladistics (phylogenetic systematics)*.

nunatak An area in a glaciated region that was not covered by an ice sheet; hence, a refugium.

obligate relationship An interaction between species in which at least one of the species cannot survive or reproduce without the other; e.g., many host-parasite relationships, in which a single species of host is required.

obliquity The tilt of the Earth on its axis, which varies between 22.1° to 24.5° over a 41,000-year period.

oceanic In marine ecology, of or pertaining to open ocean with very deep water.

oceanic island An island that was formed de novo from the floor of the ocean through volcanic activity and that has never been attached to a continent.

oceanographer A scientist who studies the physical, chemical and biological characteristics of marine systems.

oligotrophic lake A deep lake with low primary productivity.

omnivore An animal that feeds on both plants and animals.

order A taxonomic category above the level of family and below the level of class.

ordinate On a graph, the vertical (y) axis or the vertical coordinate of a point.

Oriental Region The tropical zone of Southeast Asia eastward to the margin of the continental shelf (Wallace's line).

orogeny The process of mountain building resulting from the upward thrust of Earth's crust due to volcanic or tectonic activity.

orthogenesis The supposed intrinsic tendency of organisms to evolve steadily in a particular direction, e.g., to become larger or smaller.

osmoconformer An organism that does not osmoregulate, but instead has internal salt concentrations in osmotic balance with its environment.

osmoregulation The process of maintaining homeostasis by maintaining a constant internal concentration of body fluids in changing external solutions.

outbreak area In organisms that irrupt, the area into which populations expand during peak population densities.

outcrossing Having gametes that are exchanged between different genotypes.

outgroup A taxon related to the ingroup, used in a cladistic analysis to infer primitive and derived character states in a transformation series.

outline maps A map depicting the geographic limits (range boundary) of a particular population or species.

overkill hypothesis A theory that attributes the mass extinctions of many megafaunal species during the late-Pleistocene and early Holocene to hunting and other forms of over-exploitation by humans.

overturn Vertical mixing of the water column in a lake caused by temperature changes over the seasons.

Pacifica A hypothesized ancient continent that may have existed somewhere in the Pacific basin.

pagility The ability of an organism to be transported by passive dispersal, i.e., where the principal dispersal force is from some force external to the organism (e.g., wind, water currents or other organisms).

Palearctic region The region of extratropical climates in Eurasia and in the coastal area of northernmost Africa.

paleocirculation The ocean currents of the past.

paleoclimatology The study of past climates, as elucidated mainly through the analysis of fossils.

paleoecology The branch of paleontology that attempts to reconstruct the structure of and the processes affecting ancient populations and communities.

paleoendemic An endemic that evolved in the distant past. Compare with *neoendemic*.

paleoflora All the species of plants inhabiting a specified region in the past.

paleomagnetism The magnetism or magnetically induced orientation of microstructures that has existed in a rock since its origin.

paleontology The field of study devoted to describing, analyzing, and explaining the fossil record.

paleotropical Occurring in the Old World tropics and subtropics; i.e., in Africa, Madagascar, India, and Southeast Asia.

panbiogeography An approach in historical biogeography, developed by Croizat, that attempts to reconstruct the events leading to observed distributions of taxa by drawing lines on a map (tracks) connecting known distributions of related taxa. Unlike cladistic vicariance biogeography, panbiogeography does not require phylogenies of the focal taxa.

Pangaea In plate tectonics, the supercontinent of the Permian that was composed of essentially all the present continents and major continental islands.

panmixis The condition whereby interbreeding within a large population is totally random.

Panthalassa The great global ocean that existed during the Permian coincident with the supercontinent Pangaea.

pantropical Occurring in all major tropical areas around the world.

paradigm A unifying principle, pattern or theory, such as MacArthur and Wilson's equilibrium theory of island biogeography, or Darwin's theory of natural selection.

parallel evolution The evolution of species with strong resemblances from fairly closely related ancestors; hence the evolution of two or more taxa in the same direction from related ancestors.

paralogy-free subtrees An approach in historical biogeography that modifies component analysis by removing geographic paralogy (repetition of geographic areas resulting from sympatric or embedded allopatric speciation) prior to analysis of congruence among taxon-area cladograms.

parámo Tropical alpine vegetation, characteristic of high mountains at the equator, that has a low and compact perennial cover as a response to the perpetually wet, cold, and cloudy environment.

parapatric Having contiguous but narrowly overlapping distributions.

parapatric speciation A mode of speciation in which differentiation occurs when two populations have contiguous but narrowly overlapping ranges, often representing two distinct habitat types.

paraphyletic In phylogenetic systematics, referring to taxa that include an ancestral taxon and some, but not all, of its descendant taxa; an artificial taxon.

parasitism An interspecific relationship in which one species (the parasite) draws nutrition from or is somehow dependent for survival on the other species (the host), which is negatively affected by the interaction.

parsimony In phylogenetic analysis, provides a "rule" stating that the tree requiring the fewest number of character state changes is preferred over more complex trees.

parsimony analysis for comparing trees (PACT) A method in historical biogeography conceptually similar to primary and secondary BPA but that employs a different algorithm to incorporate all taxon-area cladogram area relationships into a general area cladogram.

parthenogenesis The development of eggs without fertilization by a male gamete.

passive dispersal (vagility) The movement of an organism from one location to another by means of a stronger force, such as wind or water or via a larger animal.

pattern Nonrandom, repetitive organization.

peatland A freshwater ecosystem that develops in cool temperate regions in sites where drainage is blocked, resulting in saturation of soils and accumulation of partially decomposed plant and other organic material—called "peat."

pedogenic regimes The soil-forming processes; e.g., laterization, podzolization, calcification, and gleization.

pelagic Occurring in open water and away from the bottom.

peninsula (peninsular) effect The hypothesized tendency for species richness to decrease along a gradient from the axis to the most distal point of a peninsula.

peripheral population Any population of a species that occurs along or near the edges of its geographic range, around either the perimeter or the elevational limit. Not synonymous with *marginal population*.

perturbation Any event that greatly upsets the equilibrium of or alters the state or direction of change in a system.

phenetics The study of the overall similarities of organisms. Compare with *phylogeny*.

phenotype The expression of the genetic message of an individual in its morphology, physiology, and behavior.

phoresy The phenomenon where relatively small animals depend on larger animals for dispersal.

photic (euphotic) zone The uppermost zone in a water column where solar radiation is adequate to permit photosynthesis by plants.

photoautotroph An organism that uses light as the energy source and carbon dioxide as the carbon source for its basic metabolism; hence, an organism that is photosynthetic.

photosynthesis The chemical process of using pigments to capture sunlight and then using that energy, along with water and carbon dioxide, to make organic compounds (sugars), releasing oxygen in the process.

phyletic gradualism An evolutionary process whereby a species is gradually transformed over time into a different organism; compare with *punctuated equilibrium*.

phyletic speciation The transformation of one ancestral species into a single descendant species through non-branching (anagenetic) evolutionary change.

PhyloCode A proposed new system of taxonomic classification.

phylogenesis The phylogenetic diversification of a lineage that in some cases, such as on relatively large, isolated, and persistent islands and archipelagoes, can create hotspots of diversity and endemicity, sometimes rivaling or exceeding those found in continental systems. As Heaney (2000) asserts in his review of island biogeography theory, "phylogenesis is not a process that can be treated simply as 'another form of colonization', as it behaves differently than colonization. It interacts in a complex manner with both colonization and extinction, and can generate patterns of species richness almost independently of the other two processes."

phylogenetic biogeography I An earlier approach in historical biogeography that used Hennig's phylogenetic methods, but also incorporated his progression rule under the assumption of a history of sequential dispersal and speciation events as a lineage expanded out of a center of origin.

phylogenetic biogeography II An approach in historical biogeography that searches for shared cladogenetic events among a set of taxon-area cladograms, but also explains incongruent events in terms of post-vicariance dispersal, peripheral isolates speciation, sympatric speciation, and extinction.

phylogenetic systematics See *cladistics*. Compare with *evolutionary systematics* and *numerical phenetics*.

phylogenomics The approach to systematics that samples many genes from across a genome to reconstruct evolutionary histories.

phylogeny The evolutionary relationships between an ancestor and all of its known descendants.

phylogeography An approach in biogeography that studies the geographic distributions of genealogical lineages, within species and among similar species, and attempts to differentiate between historical and ongoing processes leading to the development of observed patterns.

phylogroup A term used in phylogeography to define a population or species whose alleles within a given gene tree form a monophyletic lineage with respect to other populations or species.

physiognomy The external aspect of a landscape; e.g., the topography and other physical characteristics of a land form and its vegetation.

physiological ecology The study of how the physiological characteristics of organisms relate to the abundances and distributions of those organisms in their natural habitats.

phytogeography The study of the distribution of plants.

phytoplankton The collection of small or microscopic plants that float or drift in great numbers in freshwater or marine environments, especially at or near the surface, and serve as food for zooplankton.

phytosociology The quantitative study of the composition of plant communities and how these relate to environmental factors.

placental A mammal that has a placenta connecting the mother to the fetus.

plankton Small organisms (especially tiny plants, small invertebrates, and juvenile stages of larger animals) that inhabit water and are transported mainly by water currents and wave action rather than by individual locomotion.

planktotrophic Referring to aquatic larvae that have no long-term storage of nutrients and so must feed on small organisms in the plankton during their development.

plate In plate tectonics, a portion of the Earth's upper surface, about 100 km thick, that moves over the asthenosphere during seafloor spreading.

plate tectonics The study of the origin, movement, and destruction of plates and how these events have been involved in the evolution of the Earth's crust.

plesiomorphy In a transformation series, the primitive character state.

plunge pool A relatively deep, roughly circular lake which formed as glacial meltwaters flowed over the surface of a glacier and then plunged off its edge to carve a basin in the Earth some 2 to 3 km below.

pluvial Referring to periods of high rainfall and water runoff.

podzolization The formation of a soil under conditions of adequate moisture and low decomposer activity, resulting in a soil in which the bases, humic acids, colloids, and ferric and aluminum oxides have been removed (leached) from the upper horizon.

poikilotherm An animal with a relatively variable body temperature, often ectothermic; i.e., relying on the external environment to control its body temperature.

pollination The transfer of pollen grains to a receptive stigma, usually by wind or flower-visiting animals.

polymorphic Having several distinct forms in a population.

polyphagous Feeding on a variety of different kinds of food.

polyphyletic A group of organisms that does not include their common ancestor; an artificial taxon.

polyploid Any organism or cell that has three or more sets of chromosomes.

population bottleneck Reduction in effective population size, often through the founding of a new population via jump dispersal of one or a few individuals across a barrier; a general result is loss of genetic variation.

Precambrian shield See *craton*.

precession Cyclical changes, or "wandering" of Earth's orientation where the axis of the North Pole shifts from one "north star" (presently Polaris of Ursa Minor) to another (Vega of Lyra) with a periodicity of approximately 22,000 years.

precinctiveness The tendency for organisms of some species to exhibit relatively strong site fidelity and remain in their natal range.

predation The act of feeding on other organisms; an interspecific interaction that has negative effects on the species that is consumed or used.

predation escalation hypothesis (of Vermeij) Morphological structures associated with predatory defense in marine invertebrates tend to increase from polar to tropical waters. Geerat Vermeij hypothesizes that this is a result of the latitudinal cline in predation pressures, which, like the cline in species diversity in general, tend to increase toward the tropics.

predator In a predatory relationship, the species that consumes other organisms.

predator mediated coexistence Coexistence of two intense competitors which results from the actions of a predator that preys most heavily on the more abundant or otherwise dominant competitor, therefore preventing populations of either competitor from increasing to the point where it can exclude the other.

prey In a predatory relationship, the species that is consumed or used by another organism.

prezygotic isolating mechanism An attribute that serves as a form of reproductive isolation between separate *biological species* prior to the formation of a viable zygote.

primary BPA See *Brooks parsimony analysis*.

primary consumer An organism that consumes plants.

primary division freshwater fishes Any fish that is totally intolerant of salt water.

primary production The production of biomass by green plants.

primary succession The gradual transformation of bare rock or another sterile substrate into a soil that supports a living ecological community.

profundal zone The deepwater zone of lakes where light is insufficient to support photosynthesis.

progression rule The idea that a primitive form remains in the center of origin, whereas progressively more derived forms are found at progressively greater distances.

projections (geographic projections) A means of using light projected through plastic models of the Earth (traditionally), or the use of various mathematical formulae to transform spatial data from the curved, 3-dimensional surface of the Earth to a two-dimensional map.

propagule Any part of an organism, or group of organisms, or stage in the life cycle that can reproduce the species and thus establish a new population.

provincialism The coincident occurrence of large numbers of well-differentiated endemic forms in an area; regional or provincial distinctiveness.

pseudo-congruence Superficially congruent taxon-area relationships that in fact have developed at different times in response to different vicariance or dispersal events.

pseudoreplication A source of error in statistical analyses that tends to overestimate the sample size and statistical significance of tests, primarily because of sampling observations that are not independent (e.g., repeated sampling of the same individual or group of related individuals).

pseudoturnover An overestimate of biotic turnover resulting from failure to detect a species that was actually present during a biological survey, and was therefore either recorded as an immigration or an extinction in subsequent surveys.

punctuated equilibrium The hypothesis that evolution occurs during periods of rapid differentiation (often accompanying speciation), which are followed by long periods in which few if any characters evolve. Compare with *phyletic gradualism*.

quadrupedal Having four limbs for locomotion.

radiation In evolutionary biology, a general term for the diversification of a group, implying that many new species have been produced.

radiation zone As MacArthur and Wilson (1967) defined it, a zone near the edges of the distribution of a taxon where immigration from the mainland and other archipelagoes is so rare that speciation and diversification occur easily.

rafting The transport over water of living organisms on large floating mats of debris.

rain-green forest The dominant vegetation of tropical-dry forests characterized by trees that leaf out during the first heavy rains following the dry season.

rain shadow More generally, a "precipitation shadow," which is a relatively dry region typically found along the leeward side of a mountain range where the relative humidity of the descending air decreases as it becomes adiabatically warmed.

range maps Maps depicting the geographic distribution of particular species (includes dot, contour and outline maps).

Rapoport's rule The tendency for geographic range size to increase with latitude.

raster-based GIS An alternative platform for geographic information systems that is composed of a system of cells (typically, rectangular units) that tessellate, each cell having a unique identity so that it can be assigned attributes corresponding to a variety of local characteristics representative of that cell.

realized geographic range The actual distribution of populations of a species, which is restricted to only a subset of the theoretical, *fundamental geographic range* as a result of interactions with other species and limited opportunities for dispersal.

realized niche The actual environmental conditions in which a species survives and reproduces in nature; a subset of the fundamental niche.

reciprocally monophyletic Two populations or species are reciprocally monophyletic if, for a given gene tree, the alleles within each population have coalesced to a monophyletic lineage with respect to the other population.

recombination The exchange of genetic material between chromosomes, resulting in new combinations of genes on the chromosomes of offspring derived from each of the parental chromosomes.

reconciled tree A method in historical biogeography that reconstructs historical events (sympatric speciation, dispersal, vicariance, extinction) in order to "reconcile" phylogenetic and geographic incongruence between co-distributed taxa.

Red Queen hypothesis A hypothesis that states that a species must continually evolve in order to keep pace with an environment that is perpetually changing because all other species are also evolving, altering the availability of resources and the nature of biotic interactions.

Red Queen principle Borrowing from Lewis Carroll's *Through the Looking-Glass* and the Red Queen's warning that "it takes all the running you can do, to keep in the same place," Leigh Van Valen (1973c) developed this principle to emphasize that, rather than evolving toward a static optimal phenotype, species are always changing as they adapt to other species, which themselves must change to keep up.

refugium An area in which climate and vegetation type have remained relatively unchanged while areas surrounding it have changed markedly, and which has thus served as a refuge for species requiring the specific habitat it contains.

reinforcement At a contact zone between differentiated populations, the process of selection for prezygotic mechanisms that completes the reproductive isolation between two new *biological species*.

relative apomorphy rule Homologous characters found within members of a monophyletic group that are also found in the sister group are plesiomorphic, while homologous characters found only in the ingroup are apomorphic.

relative rate test A method used in molecular systematics to determine whether two clades are evolving at a similar rate relative to a more distantly related third clade.

relaxation model An explanation for nonrandom assembly of ecological communities inhabiting ecosystems that have decreased in size over time (due to natural or anthropogenic changes) which holds that as the area of these ecosystems decreased, their communities lost the most resource-intensive species and began to converge on a similar set of species, i.e., those with relatively low resource requirements.

relict A surviving taxon from a group that was once widespread (biogeographic relict) or diverse (taxonomic relict).

relict pattern In island biogeography, Philip Darlington's prediction that ecological communities from relatively small ecosystems would represent a random subset of those found on the mainland or source pool.

remnant magnetism The property of rocks containing iron and titanium oxides to become magnetized as they solidify and cool, thus recording the magnetic fields in their crystalline structure. Such rocks can later serve as a "fossil compasses" indicating both the direction and declination of the magnetic fields at the sites and periods the rocks originally solidified.

remote sensing Any means of collecting data where the recorder or sensing device is not in direct contact with the area or objects of interest. In biogeographic applications, this typically includes the use of technologically sophisticated sensing devices operated from especially remote platforms, including aircrafts, ships, and satellites.

reproductive isolation Inability of individuals from different populations to produce viable offspring.

rescue effect The tendency for extinction rates to be relatively low on less-isolated islands because their relatively high immigration rate tends to supplement declining populations before they suffer extinction.

resident A species that lives year-round in a particular habitat or location.

respiration (community respiration) In ecology, the rate of chemical breakdown of organic material in an ecological community.

reticulate evolution or speciation The formation of new evolutionary lineages of species by the hybridization of dissimilar populations, for example, through interspecific hybridization.

ridge push One of the forces responsible for continental drift, in this case resulting from the rise of hot magma toward and then outward from Earth's surface.

rift zone A region where one or more continental plates are separating, thus forming a low-lying, tectonically active area (spreading zone). Examples include those that in the past created the Red Sea and the great, deep lakes of the Baikal rift zone and the East African Rift Valley.

r-selection Selection that favors high population growth rate ("*r*"), which is more prominent when population size is far below carrying capacity.

r-strategists Species whose life-history and ecological characteristics are adapted to selective pressures associated with relatively low population levels.

saline Having high concentrations of salts and especially ions of chloride and sulfate.

salt marsh A wetland dominated by herbaceous vegetation and characterized by its relatively high salinity and salt adapted inhabitats.

scaling See *allometry*.

scavenger An animal that feeds on carrion.

sclerophyllous Having tough, thick, evergreen leaves.

scrub Any of a wide variety of vegetation types dominated by low shrubs; in exceedingly dry locations, scrub vegetation has few or no trees and widely spaced low shrubs, but in areas of fairly high rainfall, scrub has trees and grades into either woodland or forest.

seafloor spreading In plate tectonics, the process of adding crustal material at a midoceanic ridge and thus displacing older rocks, usually on both sides, away from their point of origin.

seamount A peaked submarine volcano.

second law of thermodynamics As energy is converted into different forms, its capacity to perform useful work diminishes, and the disorder (entropy) of the system increases.

secondary BPA A method in historical biogeography that duplicates areas on a primary BPA tree only as much as required to remove all homoplasy; the resulting tree includes area relationships that are the results of vicariance, post-speciation dispersal, peripheral isolates speciation, and extinction.

secondary consumer An animal that consumes herbivorous animals (i.e., a predator).

secondary division freshwater fishes Any fish that prefers fresh water but can live for short periods in salt water.

secondary succession A series of changes in the vegetational composition of an environment in response to disturbance, involving the gradual and regular replacement of species and ending, at least hypothetically, with the return to a stable state (climax).

secular migration A means of geographic range expansion which is so slow (i.e., over many generations) that it is often accompanied by substantial evolutionary changes in the populations en route, i.e., as they expand to colonize new regions.

selective sweep Rapid differentiation of populations resulting from the fixation of a new mutation with a strong positive fitness effect.

self-incompatible Requiring two individuals to exchange gametes in order to produce offspring.

semiaquatic Living partly in or adjacent to water.

semiarid Having a fairly dry climate with low precipitation, usually 25 to 60 cm per year, and a high evapotranspiration rate, so that potential loss of water to the environment exceeds the input.

semidesert A semiarid habitat characterized by low vegetation; e.g., small, widely spaced shrubs.

seral stage One of the stages in the ecological succession of communities.

sere A series of stages in community transformation during ecological succession.

serpentine A rock or soil type rich in magnesium but deficient in calcium.

sessile Remaining fixed in the same spot throughout adulthood; e.g., most plants and certain benthic aquatic invertebrates.

shrubland Any of a wide variety of vegetation types dominated by shrubs, which may form a fairly solid cover; e.g., sclerophyllous scrub (chaparral), or sparse cover; e.g., desert scrub.

sial Rock rich in silica-aluminum; the principal component of continental rocks.

sibling species Different species that are difficult to delineate with morphological traits (also called *cryptic species*).

siliceous Containing silica.

sima Rock rich in silica-magnesium; the principal component of basalt and of oceanic plates.

similarity index An estimate of the similarity or relatedness of two communities, biota, or taxa.

sink habitats Environments where the birth rate of a particular species is exceeded by its death rate and, therefore, the presence of this species indicates that most individuals are derived from dispersal from other ("source") habitats.

sister group (clade) In a phylogeny, the group or clade most closely related to the focal group, and therefore the most useful outgroup for rooting the phylogeny.

sister taxa In cladistics, the two taxa that are most closely (and therefore most recently) related.

slab pull One of the forces responsible for continental drift, in this case resulting from the tendency for portions of a plate to adhere to the leading, subducting edge of that plate.

small island effect The tendency for species richness to vary independent of island area on relatively small islands.

solstice Either of two times in a year (June 22, December 22) when the sun reaches its highest latitude (23.5° N and S, respectively).

source habitats Environments in which the birth rate of a particular species exceeds its death rate and, therefore, results in a surplus of individuals which contributes to dispersal to other ("sink") habitats.

speciation The process in which two or more contemporaneous species evolve from a single ancestral population.

speciation-pump model See *cyclical vicariance model*.

species The fundamental taxonomic category for organisms; variously defined and diagnosed using different species concepts (see Box 7.1 and Figure 7.3).

species composition The types of species that constitute a given sample or community.

species concept Any one of some 22 criteria and approaches (see Box 7.1 and Figure 7.3) used to delineate separate species.

species density The number of species per standardized sample area.

species pool All the organisms present in neighboring source areas that are theoretically available to colonize a particular habitat or island.

species richness A relatively simple, but important measure of diversity representing the number of species in an ecological community.

species selection An analogue of natural selection at the species level, in which some species with certain characteristics increase while others decrease or become extinct.

species-area relationship A plot of the numbers of species of a particular taxon against the area of islands or other ecosystems. The relationship is often linearized by log-transforming one or both variables.

spreading zone A region where two or more continental plates are separating, often associated with the formation of a new ocean basin (sea-floor spreading) or rift valleys on the continents.

spring tide A greater than average tide occurring during the new and full moons.

stable isotopes Alternative forms of an element that have different masses (i.e., the same number of protons but different numbers of neutrons and are not subject to radioactive decay. These chemicals have played an increasingly important role in biogeography because the combination of stable isotopes varies among regions of the terrestrial and aquatic realms, thus providing a geographic signature when integrated into the tissues of organisms.

stasipatric speciation See *parapatric speciation.*

statistical phylogeography The framing of phylogeographic investigations within a rigorous statistical framework, either through development of alternative hypotheses that are evaluated with coalescence-based statistics; or through a nested clade phylogeographic analysis (see *nested clade analysis*).

stenohaline Having a tolerance for only a narrow range of salt concentrations.

stenothermal Having a tolerance for only a narrow range of temperatures.

stenotypic In ecology, refers to a species with a relatively narrow niche or a highly specialized association with another species.

stochastic Random, expected (statistically) by chance alone; compare with *deterministic.*

stratigraphy The branch of geology dealing with the sequence of deposition of rocks and fossils as well as their composition, origin, and distribution.

subduction In plate tectonics, the movement of one plate beneath another, leading to the heating and subsequent remelting of the lower plate.

subduction zone A region where one relatively dense tectonic plate (oceanic plate, composed sima) slides beneath a less dense (continental, sial) plate.

subfamily A taxonomic category used for grouping genera within a family.

sublittoral zone The coastal marine zone below the intertidal zone and therefore below the point at which the sea bottom is periodically exposed to the atmosphere.

subspecies A taxonomic category used by some systematists to designate a genetically distinct set of populations with a discrete range.

substitution A form of DNA mutation in which one nucleotide base is replaced by one of the three other nucleotide bases.

subterranean Living underground.

supercontinent An ancient landmass formed by the collision and connection of most, if not all, of the present global landmasses (e.g., Pangaea).

superfamily A taxonomic category used for grouping families within a suborder.

super-generalist On remote, species-poor islands, some of the inhabitants may increase their niche breadths and symbiotic capacities to adapt and interact with a relatively high number of species (e.g., as generalist pollinators and seed dispersers) in comparison with their mainland ancestors.

superparámo A vegetation type of high elevations in the equatorial Andes mountains.

superspecies A group of closely related species.

supertramps Species that are relatively common on relatively small and isolated islands (i.e., those with few ecologically similar species), but are absent from islands with more diverse communities.

swamp (marl) A wetland dominated by woody vegetation.

sweepstakes route A severe barrier that results in the partly stochastic dispersal of some elements of a biota, and the establishment of a disharmonic biota.

symbionts Two or more species that occur in the same community and interact as competitors, predators, parasites, mutualists, amensals, or commensals.

symbiosis A long-term interspecific relationship in which two unrelated and unlike organisms live together in close association so that each receives some adaptive benefit.

sympatric In the strictest sense, living in the same local community, close enough to interact; in the more general sense, having broadly overlapping geographic ranges.

sympatric speciation The differentiation of two reproductively isolated species from one initial population within the same local area; hence, speciation that occurs under conditions in which much gene flow potentially could or actually does occur. Compare with *allopatric speciation; parapatric speciation.*

symplesiomorphy In a transformation series, a character state shared by taxa in a focal clade, but also shared by other clades at more basal nodes in a phylogeny and therefore not useful in diagnosing the focal group as being monophyletic. See *grouping rule.*

synapomorphic In a transformation series, a derived character state shared by taxa because of inheritance from a common ancestor and therefore useful in diagnosing a monophyletic group or clade.

synapomorphy In a transformation series, a derived character state shared by taxa due to inheritance from a common ancestor and therefore useful in diagnosing a monophyletic group or clade. See *grouping rule.*

systematic biogeographic maps Maps that attempt to describe the distinctiveness of and/or similarities among biotas from local (or provincial) to global (continental or oceanic) scales.

systematics The study of the evolutionary relationships between organisms. Includes *phylogenetics* and *taxonomy.*

tachytelic Term used by Simpson (1944) to refer to a rapid rate of morphological change through time in a lineage with a fossil record.

taphonomy A subdiscipline of paleontology concerned with the processes by which remains of living things become fossilized, and the ways in which these processes can bias the fossil record or cause problems of interpretation.

taxon (pl. **taxa**) A convenient and general term for any taxonomic category; e.g., a species, genus, family, or order.

taxon-area cladogram An area cladogram generated by replacing the taxa on a taxon cladogram with their areas of occurrence.

taxon biogeography The primary goal of an analysis is to reconstruct the biogeographic history of a single taxon.

taxon cycle A series of ecological and evolutionary changes in insular populations from their colonization of early successional sites along the beachfronts, to their expansion and increased specialization for interior habitats, and their eventual extinction and replacement by subsequent waves of colonists and their descendenats.

taxonomic relicts The sole survivors of once diverse taxonomic groups.

taxonomy In the strictest sense, the study of the names of organisms, but often used for entire process of classification. See *systematics*.

tectonic Referring to any process involved in the production or deformation of the Earth's crust.

terrane An accumulation of fragments of lithosphere that acrete onto other plates.

terrestrial Living on land.

tetrapod Any four-legged vertebrate, including amphibians, reptiles, and mammals.

theory The view, widely held before continental drift was accepted, that the distribution pattern of ocean basins and continents remained relatively constant over the history of the Earth.

thermocline In a water column, the subsurface zone in which the temperature drops sharply.

thorn scrub A relatively xeric ecosystem typically receiving less than 30 cm of annual rainfall, experiencing a 6-month dry season, and dominated by sparse woody vegetation.

three-item statement analysis An approach in historical biogeography that modifies component analysis by converting taxon-area cladograms to subsets of three-area relationships prior to analysis of congruence.

tillites Glacial rock deposits.

timberline The upper elevational limit of trees on mountains.

time dwarfing The tendency for some initially large animals (in particular, large marsupials of Australia; Flannery 1994) to decrease in size over time, especially following significant reduction and isolation of their native habitats.

tokogenetic Reticulate (netlike or interwoven) relationships between individuals within a sexual species.

total evidence An approach in systematics that combines characters drawn from different sources (e.g., morphological and molecular characters) into a single data set for phylogenetic analyses.

trace fossils Taxa known only from their activities (e.g., tracks or burrows).

track In panbiogeography, a line drawn on a map connecting the geographically isolated ranges of the species in a taxon that are closest relatives (vicariants).

trade winds Winds blowing toward the equator between the horse latitudes and the doldrums in the Northern and Southern Hemispheres.

transform fault A fault in an oceanic plate, perpendicular to the midoceanic ridge, that divides the plate into smaller units.

transform zone (Also known as a "strike-slip fault.") A region where two tectonic plates slide against eachother but in opposite directions.

transformation series For a character, two or more *character states*, assumed to be products of evolutionary changes and therefore used by systematists to infer evolutionary relationships and classify organisms into taxa.

translocation A kind of chromosomal mutation in which a segment of one chromosome becomes attached to a different chromosome.

transpiration The loss of water vapor from plants through pores called stomates.

transposable elements Nucleotide sequences that promote their own movement between chromosomal sites.

trench In plate tectonics, an exceedingly deep cut in the ocean floor where the subduction of an oceanic plate is occurring.

triple junction In plate tectonics, a point at which three oceanic plates meet; the position of the junction shifts because each of the plates drifts at a different rate.

trophic level A functional, ecological classification of organism based on their primary source of energy (including primary producers = plants; primary consumers = herbivores; secondary consumers = predators; tertiary and in some cases, higher order consumers; and decomposers).

trophic status The position or role of an organism in the nutritional structure of a community; e.g., primary producer, herbivore, or top carnivore.

tropical alpine scrubland Vegetative communities found above the timberline in equatorial regions and dominated by relatively sparse, low-growing vegetation including tussock grasses and bizarre, erect rosette perennials with thick stems.

Tropical (Inter-Tropical) Convergence Zone The zone along tropical regions of the Earth's surface with most direct sunlight and most intense heating, which is also associated with rising air masses and high precipitation, and shifts with the seasons between the Tropics of Cancer and Capricorn.

turnover The rate of replacement of species in a particular area as some taxa become extinct but others immigrate from outside.

undersaturated Having fewer taxa of a particular kind than expected on the basis of an equilibrium between colonization, speciation, and extinction.

unequal crossover An asymmetric exchange of genetic material between a chromosome pair during recombination.

uniformitarianism See *actualism*.

upwelling The vertical movement of deep water, containing dissolved nutrients from the ocean bottom, to the surface.

vagility The ability to move actively from one place to another.

vector-based GIS An alternative platform for geographic information systems that is composed of a system of points and vectors (lines or curves) and polygons to represent features of interest (e.g., roads, rivers, lakes, and other features to be mapped), each of which has a unique identity so that it can be assigned attributes corresponding to a variety of local characteristics representative of that element (point, vector or polygon).

vicariance biogeography An approach in historical biogeography that attempts to reconstruct the historical events that led to observed distributional patterns based largely on the assumption that these patterns resulted from the splitting (vicariance) of areas and not long-distance dispersal. Compare *dispersal biogeography*.

vicariant events Tectonic, eustatic, climatic, or oceanographic events that result in the geographic isolation of previously connected populations.

vicariants Two disjunct and phylogenetically related species that are assumed to have been created when the initial range of their ancestor was split by some historical event.

Viking funeral ship A term coined by McKenna for a landmass, such as a fragment, containing fossils that were laid down when the land was in one location, but that were transported to a completely different locality via continental drift.

visualizations Illustrations such as maps, graphs, or conceptual diagrams that purposely simplify, exaggerate, or distort relevant features of a subject (landscape, seascape, satellite imagery, ecological or evolutionary model) to convey its salient features in terms of general patterns or underlying causal mechanisms.

waif In dispersal biogeography, a diaspore or any type of individual that is carried passively by waves or air currents to a distant place; e.g., most colonizers of oceanic island beaches.

Wallace's line The most famous biogeographic line, running between Borneo and Celebes and between Bali and Lombok, which marks the boundary between the Oriental and Australian Regions.

Wallacean Shortfall The paucity of information on geographic distributions of species (past and present), and on the geographic dynamics of extinction forces, especially the dynamic geography of humans, their commensals and other anthropogenic threats.

Wegenerism The general idea of continental drift according to Alfred Wegener.

westerlies Prevailing winds in the temperate regions (30° to 60°) of both the Northern and Southern Hemispheres which, as a result of the Coriolis effect, have a strong west-to-east component.

wintering area In migratory land animals, the area where populations spend the cold season and feed, but do not breed.

woodland Any of a variety of vegetation types consisting of small, widely spaced trees with or without substantial undergrowth.

xerophytes Land plants that grow in relatively dry ("xeric") environments.

zonal soils Soils that have distinctive characteristics and are formed by the actions of climate and organisms on the so-called "typical" rocks (sandstone, shale, granite, gneiss and slate).

zoochory Transport of propagules by animals.

zoogeography The study of the distributions of animals.

zooplankton The collection of small or microscopic animals that float or drift in great numbers in fresh or salt water, especially at or near the surface, and serve as food for fish and other larger organisms.

Bibliography

Abatzopoulos, T. J. 2002. *Artemia: Basic and Applied Biology.* Kluwer Academic Publishers.

Abbott, I. 1983. The meaning of z in species/area regressions and the study of species turnover in island biogeography. *Oikos* 41: 385–390.

Abbott, I., L. K. Abbott and P. R. Grant. 1977. Comparative ecology of Galapagos ground finches (*Geospiza* Gould): Evaluation of the importance of floristic diversity and interspecific competition. *Ecological Monographs* 47: 151–184.

Abele, L. G. and W. Kim. 1989. The decapod crustaceans of the Panama canal. *Smithsonian Contributions to Zoology* 482: 1–50. Washington, DC: Smithsonian Institution Press.

Åberg, J., G. Jansson, J. E. Swenson and P. Angelstam. 1995. The effect of matrix on the occurrence of hazel grouse (*Bonasa bonasia*) in isolated habitat fragments. *Oecologia* 103: 265–269.

Adams, C. C. 1902. Southeastern United States as a center of geographical distribution of flora and fauna. *Biological Bulletin* 3: 115–131.

Adams, C. C. 1909. *An Ecological Survey of Isle Royale, Lake Superior.* Lansing, MI: Michigan Biological Survey.

Adams, C. G. and D. V. Ager. 1967. *Aspects of Tethyan Biogeography.* The Systematics Association Publications, no. 7, Wetteren, Universa.

Adams, J., M. Maslin and E. Thomas. 1999. Sudden climate transition during the Quaternary. *Progress in Physical Geography* 23: 1–36.

Adhemar, J. F. 1842. *Les Revolutions de la Mer Deluges Periodiques.* Paris: Carilian et Goeury et V. Dalmont.

Adler, G. H. 1996. The island syndrome in isolated populations of a tropical forest rodent. *Oecologia* 108: 694–700.

Adler, G. H. and R. Levins. 1994. The island syndrome in rodent populations. *The Quarterly Review of Biology* 69: 473–490.

Agapow, P. M., O. R. P. Bininda-Emonds, K. A. Crandall, J. L. Gittleman, G. M. Mace, J. C. Marshall and A. Purvis. 2004. The impact of species concept on biodiversity studies. *Quarterly Review of Biology* 79: 161–179.

Agassiz, L. 1840. Etudes sur les Glaciers/Studies of the glaciers. Neuchatel.

Aguilar, R., M. Quesada, L. Ashworth, Y. Herreria-Diego and J. Lobo. 2008. Genetic consequences of habitat fragmentation in plant populations: Susceptible signals in plant traits and methodological approaches. *Molecular Ecology* 17: 5177–5188.

Agusti, J. and M. Anton. 2002. *Mammoths, Sabertooths and Hominids: 65 Million Years of Mammalian Evolution in Europe.* New York: Columbia University Press.

Ahmed, S., S. G. Compton, R. K. Butlin and P. M. Gilmartin. 2009. Wind-borne insects mediate directional pollen transfer between desert fig trees 160 kilometers apart. *Proceedings of the National Academy of Sciences*, doi: 10.1073.pnas.0902213106.

Aiello, L. C. and J. C. K. Wells. 2002. Energetics and the evolution of the genus *Homo*. *Annual Reviews of Anthropology* 31: 323–338.

Alacantra, M. 1991. Geographical variation in body size of the wood mouse *Apodemus sylvaticus* L. *Mammalian Reviews* 21: 143–150.

Alatalo, R. V. 1982. Bird species distributions in the Galapagos and other archipelagoes: Competition or chance? *Ecology* 63: 881–887.

Albrecht, F. O. 1967. Polymorphisme phasaire et biologie des acridiens migrateurs. Paris: Masson.

Alessa, L. and F. S. Chapin III. 2008. Anthropogenic biomes: A key contribution to Earth-system science. *Trends in Ecology and Evolution* 23: 529–531.

Algar, A. C., J. T. Kerr and D. J. Currie. 2007. A test of metabolic theory as the mechanism underlying broad-scale species-richness gradients. *Global Ecology and Biogeography* 16: 170–178.

Ali, J. R. and J. C. Aitchson. 2008. Gondwana to Asia: Plate tectonics, paleography and the biological connectivity of the Indian sub-continent from the middle Jurassic through latest Eocene (166–35 Ma). *Earth-Science Reviews* 88: 145–166.

Allard, G. O. and V. J. Hurst. 1969. Brazil-Gabon geologic link supports continental drift. *Science* 163: 528–532.

Allen, A. P. and J. F. Gilloly. 2006. Assessing latitudinal gradients in speciation rates and biodiversity at the global scale. *Ecology Letters* 9: 947–954.

Allen, A. P., J. H. Brown and J. F. Gilloly. 2002. Global biodiversity, biochemical kinetics, and the energetic-equivalence rule. *Science* 297: 1545–1548.

Allen, J. A. 1878. The influence of physical conditions in the genesis of species. *Radical Review* 1: 108–140.

Alroy, J. 1998. Cope's rule and the dynamics of body mass evolution in North American fossil mammals. *Science* 280: 731–734.

Alroy, J. 1999. Putting North America's end-Pleistocene extinction in context. In R. MacPhee (ed.), *Extinctions in Near Time*, 105–143. New York: Kluwer Academic/Plenum Publishers.

Alroy, J. 2001. A multispecies overkill simulation of the End-Pleistocene megafaunal mass extinction. *Science* 292: 1893–1896.

Althoff, D. M. and O. Pellmyr. 2002. Examining genetic structure in a bogus yucca moth: A sequential approach to phylogeography. *Evolution* 56: 1632–1643.

Alvarez, L. W., W. Alvarez, F. Asaro and H. V. Michel. 1980. Extraterrestrial cause for the Cretaceous-Tertiary extinction. *Science* 208: 1095–1108.

Alvarez, W., E. G. Kauffman, F. Surlyk, L. W. Alvarez, F. Asaro and H. V. Michel. 1984. Impact theory of mass extinctions and the invertebrate fossil record. *Science* 223: 1135–1141.

Amadon, D. 1950. The Hawaiian honeycreepers (Aves, Drepaniidae). *Bulletin of the American Museum of Natural History* 95: 153–262.

Ambrose, S. H. 2001. Paleolithic technology and human evolution. *Science* 291: 1748–1753.

Amiran, D. H. K. and A. W. Wilson (eds.). 1973. *Coastal Deserts: Their Natural and Human Environments.* Tucson, AZ: University of Arizona Press.

Anderson, A. 2009. The rat and the octopus: Initial human colonization and the prehistoric introduction of domestic animals to Remote Oceania. *Biological Invasions* 11: 1503–1519.

Anderson, D. E., A. S. Goudie and A. G. Parker. 2007. *Global Environments Through the Quaternary: Exploring Environmental Change.* Oxford: Oxford University Press.

Anderson, E. P., C. M. Pringle and M. C. Freeman. 2008. Quantifying the extent of river fragmentation by hydropower dams in the Sarapiqui River Basin, Costa Rica. *Aquatic Conservation* 18: 408–417.

Anderson, I. W. 1974. The chestnut pollen decline as a time horizon in lake sediments in eastern North America. *Canadian Journal of Earth Sciences* 11: 678–685.

Anderson, P. K. 1960. Ecology and evolution in island populations of salamanders in San Francisco Bay region. *Ecological Monographs* 30: 359–385.

Anderson, R. P. and C. O. Handley. 2002. Dwarfism in insular sloths: Biogeography, selection and evolutionary rate. *Evolution* 56: 1045–1058.

Anderson, S. and L. S. Marcus. 1992. Areography of Australian tetrapods. *Australian Journal of Zoology* 40: 627–651.

Anderson, T. M., B. M. vonHoldt, S. I. Candille, M. Musiani, C. Greco, D. R. Stahler, D. W. Smith, B. Padhukasahasram, E. Randi, J. A. Leonard, et al. 2009. Molecular and evolutionary history of melanism in North American gray wolves. *Science* 323: 1339–1343.

Andrewartha, H. G. and L. C. Birch. 1954. *The Distribution and Abundance of Animals.* Chicago: University of Chicago Press.

Andrews, P. and E. M. O'Brien. 2000. Climate, vegetation and predictable gradients in mammal species richness in southern Africa. *Journal of Zoology* (London) 251: 205–231.

Anselin, L., I. Syabri and Y. Kho. 2006. GeoDa: An introduction to spatial data analysis. *Geographical Analysis* 38: 5–22.

Apanius, V., N. Yorinks, E. Bermingham and R. E. Ricklefs. 2000. Island and taxon effects in the prevalence of blood parasites and activity of the immune system in Lesser Antillean birds. *Ecology* 81: 1959–1969.

Aponte, C., G. R. Barreto and J. Terborgh. 2003. Consequences of habitat fragmentation of age structure and life history in a tortoise population. *Biotropica* 35: 550–555.

Araújo, M. B. and C. Rahbek. 2006. How does climate change affect biodiversity? *Science* 313: 1396–1397.

Araújo, M. B. and M. Luoto. 2007. The importance of biotic interactions for modelling species distributions under climate change. *Global Ecology and Biogeography* 16: 743–753.

Araújo, M. B. and M. New. 2007. Ensemble forecasting of species distributions. *Trends in Ecology and Evolution* 22: 42–47.

Araújo, M. B., R. J. Whittaker, R. J. Ladle and M. Erhard. 2005. Reducing uncertainty in projections of extinction risk from climate change. *Global Ecology and Biogeography* 14: 529–538.

Araújo, M. B., W. Thuiller and R. G. Pearson. 2006. Climate warming and the decline of amphibians and reptiles in Europe. *Journal of Biogeography* 33:1712–1728.

Arbogast, B. S. and G. J. Kenagy. 2001. Comparative phylogeography as an integrative approach to historical biogeography. *Journal of Biogeography* 28: 819–825.

Arbogast, B. S., S. V. Edwards, J. Wakeley, P. Beerli and J. B. Slowinski. 2002. Estimating divergence times from molecular data on phylogenetic and population genetic timescales. *Annual Review of Ecology and Systematics* 33: 707–740.

Arim, M., S. R. Abades, P. E. Neill, M. Lima and P. A. Marquet. 2006. Spread dynamics of invasive species. *Proceedings of the National Academy of Sciences, USA* 103: 374–378.

Arnold, E. N. 1979. Indian Ocean giant tortoises: Their systematics and island adaptations. *Philosophical Transactions of the Royal Society of London*, Series B 286: 127–145.

Arnold, M. L. and S. K. Emms. 1998. Paradigm lost: Natural hybridization and evolutionary innovations. In D. J. Howard and S. H. Berlocher (eds.), *Endless Forms: Species and Speciation*, 379–389. New York: Oxford University Press.

Arnold, S. J. 1972. Species densities of predators and their prey. *American Naturalist* 106: 220–236.

Arrhenius, O. 1921. Species and area. *Journal of Ecology* 4: 68–73.

Ashmole, A. P. 1963. The regulation of numbers of tropical ocean birds. *Ibis* 103b: 458–473.

Ashton, K. G. 2002a. Patterns of within-species body size variation in birds: Strong evidence for Bergmann's rule. *Global Ecology and Biogeography* 11: 505–523.

Ashton, K. G. 2002b. Do amphibians follow Bergmann's rule? *Canadian Journal of Zoology* 80: 708–716.

Ashton, K. G. 2004. Sensitivity of intraspecific latitudinal clines of body size for tetrapods to sampling, latitude and body size. *Integrative Comparative Biology* 44: 403–412.

Ashton, K. G. and C. R. Feldman. 2003. Bergmann's rule in non–avian reptiles: Turtles follow it, lizards and snakes reverse it. *Evolution* 57: 1151–1163.

Ashton, K. G., M. C. Tracy and A. de Queiroz. 2000. Is Bergmann's rule valid for mammals? *American Naturalist* 156: 390–415.

Astorga, A., M. Fernandez, E. E. Boschi and N. Lagos. 2003. Two oceans, two taxa and one mode of development: Latitudinal diversity patterns of South American crabs and test for possible causal processes. *Ecology Letters* 6: 420–427.

Athias-Binche, F. 1993. Dispersal in varying environments: The case of phoretic ceropodid mites. *Canadian Journal of Zoology-Revue Canadienne de Zoologie* 71: 1793–1798.

Atkinson, I. A. E. and E. K. Cameron. 1993. Human influence on terrestrial biota and biotic communities of New Zealand. *Trends in Ecology and Evolution* 8: 447–451.

Atmar, W. and B. D. Patterson. 1993. The measure of order and disorder in the distribution of species in fragmented habitats. *Oecologia* 96: 373–382.

Attenborough, D. 1987. *The First Eden: The Mediterranean World and Man.* Boston: Little Brown and Company.

Auffenberg, W. 1981. *The Behavioral Ecology of the Komodo Monitor.* Gainesville: University of Florida Presses.

Austin, M. P. 2007. Species distribution models and ecological theory: A critical assessment and some new approaches. *Ecological Modelling* 200: 1–19.

Austin, M. P., L. Belbin, J. A. Meyers, M. D. Doherty and M. Luoto. 2006. Evaluation of statistical models used for predicting plant species distributions: Role of artificial data and theory. *Ecological Modelling* 199: 197–216.

Avise, J. C. 2000. *Phylogeography: The History and Formation of Species.* Cambridge, MA: Harvard University Press.

Avise, J. C. 2004. *Molecular Markers, Natural History and Evolution.* 2nd edition. Sunderland, MA: Sinauer Associates.

Avise, J. C. and D. E. Walker. 1999. Species realities and numbers in sexual vertebrates: Perspectives from an asexually transmitted genome. *Proceedings of the National Academy of Sciences, USA* 96: 992–995.

Avise, J. C. and R. M. Ball, Jr. 1990. Principles of genealogical concordance in species concepts and biological taxonomy. In D. Futuyma and J. Antonovics (eds.), *Oxford Surveys in Evolutionary Biology*, 45–67. Oxford: Oxford University Press.

Avise, J. C., J. Arnold, R. M. Ball, E. Bermingham, T. Lamb, J. E. Neigel, C. A. Reeb and N. C. Saunders. 1987. Intraspecific phylogeography—The mitochondrial-DNA bridge between population-genetics and systematics. *Annual Review of Ecology and Systematics* 18: 489–522.

Bachraty, C., P. Legendre and D. Desbruyeres. 2009. Biogeographic relationships among deep-sea hydrothermal vent faunas at global scale. *Deep-Sea Research Part I: Oceanographic Research Papers* 56: 1371–1378.

Bahre, C. J. 1995. Human impacts on the grasslands of Southweastern Arizona. In M. P. McClaran and T. R. Van Devender (eds.), *The Desert Grassland*, 230–264. Tucson, AZ: University of Arizona Press.

Bailey, R. G. 2002. *Ecoregion-Based Design for Sustainability.* New York: Springer.

Bailey, R. G. 2009. *Ecosystem Geography: From Ecoregions to Sites*. New York: Springer.

Bak, P. 1996. *How Nature Works: The Science of Self-Organized Criticality*. New York: Copernicus.

Baker, A. J., C. H. Daugherty, R. Colbourne and J. L. McLennan. 1995. Flightless brown kiwis of New Zealand possess extremely subdivided population structure and cryptic species like small mammals. *Proceedings of the National Academy of Sciences, USA* 92: 8254.

Baker, H. G. 1955. Self-compatibility and establishment after 'long-distance' dispersal. *Evolution* 9: 347–349.

Baker, H. G. and G. L. Stebbins (eds.). 1965. *The Genetics of Colonizing Species*. New York: Academic Press.

Baker, M. 1995. Environmental component of latitudinal clutch-size variation in house sparrows (*Passer domesticus*). *The Auk* 112: 249–252.

Baker, R. H. 1968. Habitats and distribution. In J. A. King (ed.), *Biology of Peromyscus*, 98–126. American Society of Mammalogists Special Publication no. 2.

Baker, R. R. 1978. *The Evolutionary Ecology of Animal Migration*. London: Hodder & Stoughton.

Baldwin, B. G. and M. J. Sanderson. 1998. Age and rate of diversification of the Hawaiian silversword alliance (Compositae). *Proceedings of the National Academy of Sciences, USA* 95: 9402–9406.

Baldwin, B. G. and R. H. Robichaux. 1995. Historical biogeography and ecology of the Hawaiian silversword alliance (*Asteraceae*): New molecular phylogenetic perspectives. In W. L. Wagner and V. A. Funk (eds.), *Hawaiian Biogeography: Evolution on a Hot Spot Archipelago*. Washington, DC: Smithsonian Institution Press.

Baldwin, B., P. J. Coney and W. R. Dickinson. 1974. Dilemma of a Cretaceous time scale and rates of sea-floor spreading. *Geology* 2: 267–270.

Balete, D. S., E. A. Rickart and L. R. Heaney. 2007a. A new species of the shrew-mouse, *Archboldomys* (Rodentia: Muridae: Murinae) from the Philippines. *Systematics and Biodiversity* 4: 489–501.

Balete, D. S., E. A. Rickart, R. G. B. Rosell-Ambal, S. Jansa and L. R. Heaney. 2007b. Descriptions of two new species of *Rhynchomys* Thomas (Rodentia: Muridae: Murinae), from Luzon Island, Philippines. *Journal of Mammalogy* 88: 287–301.

Balete, D. S., L. R. Heaney, E. A. Rickart, R. S. Quidlat, and J. C. Ibanez. 2008. A new species of *Batomys* (Mammalia: Muridae) from eastern Mindanao Island, Philippines. *Proceedings of the Biological Society of Washington.* 121: 411–428.

Ballard, H. E., Jr. and K. J. Systsma. 2000. Evolution and biogeography of the woody Hawaiian violets (Viola, Violaceae): Arctic origins, herbaceous ancestry and bird dispersal. *Evolution* 54: 1521–1532.

Banfield, A. W. F. 1954. The role of ice in the distribution of mammals. *Journal of Mammalogy* 35: 104–107.

Banfield, A. W. F. 1961. A revision of the reindeer and caribou, genus *Rangifer*. *Bulletin of the National Museum of Canada* 177: 1–137.

Baquero, R. A. and J. L. Telleria. 2001. Species richness, rarity and endemicity of European mammals: A biogeographical approach. *Biodiversity and Conservation* 10: 29–44.

Barahona, F., S. E. Evans, J. A. Mateo, M. García-Márquez and L. F. López-Jurado. 2000. Endemism, gigantism and extinction in island lizards: The genus *Gallotia* on the Canary Islands. *Journal of Zoology, London* 250: 373–388.

Barber, P. H., S. R. Palumbi, M. V. Erdmann and M. K. Moosa. 2000. Biogeography—A marine Wallace's line? *Nature* 406: 692–693.

Barbosa, A. and J. Benzal. 1996. Diversity and abundance of small mammals in Iberia: Peninsular effect or habitat suitability? *Zeitschrift fur Saugetierkunde* 61: 236–241.

Barbour, C. D. and J. H. Brown. 1974. Fish species diversity in lakes. *American Naturalist* 108: 473–489.

Barker, F. K., A. Cibois, P. A. Schikler, J. Feinstein and J. Cracraft. 2004. Phylogeny and diversification of the largest avian radiation. *Proceedings of the National Academy of Sciences, USA* 101: 11040–11045.

Barker, N. P., P. H. Weston, F. Rutschmann and H. Sauquet. 2007. Molecular dating of the "Gondwanan" plant family Proteaceae is only partially congruent with the timing of the break-up of Gondwana. *Journal of Biogeography* 34: 2012–2027.

Barlow, N. D. 1994. Size distributions of butterfly species and the effect of latitude on species sizes. *Oikos* 71: 326–332.

Barnosky, A. D., P. L. Koch, R. S. Feranec, S. L. Wing and A. B. Shabel. 2004. Assessing the causes of Late Pleistocene extinctions on the continents. *Science* 306: 70–75.

Barraclough, T. G. and A. P. Vogler. 2000. Detecting the geographical pattern of speciation from species-level phylogenies. *American Naturalist* 155: 419–434.

Barrett, K., D. A. Wait and W. B. Anderson. 2003. Small island biogeography in the Gulf of California: Lizards, the subsidized island biogeography hypothesis and the small island effect. *Journal of Biogeography* 30: 1575–1581.

Barrett, S. C. H. 1998. The reproductive biology and genetics of island plants. In P. R. Grant (ed.), *Evolution on Islands*,18–34. New York: Oxford University Press.

Barrowclough, G. F. 1992. Biodiversity and conservation biology. In N. Eldredge (ed.), *Systematics, Ecology and the Biodiversity Crisis*, 121–143. New York: Columbia University Press.

Barton, N. H. and B. Charlesworth. 1984. Genetic revolutions, founder effects and speciation. *Annual Review of Ecology and Systematics* 15: 133–164.

Bascompte, J., P. Jordano, C. J. Melián and J. M. Olesen. 2003. The nested assembly of plant–animal mutualistic networks. *Proceedings of the National Academy of Sciences, USA* 100: 9383–9387.

Baskett, M. L., J. S. Weitz and S. A. Levin. 2007. The evolution of dispersal in reserve networks. *American Naturalist* 170: 59–78.

Baskin, Y. 1992. Africa's troubled waters: Fish introductions and a changing physical profile muddy Lake Victoria's future. *BioScience* 42: 476–481.

Batt, C. M. and A. M. Pollard. 1996. Radiocarbon calibration and the peopling of North America. *American Chemical Society Symposium Series* 625: 415–433.

Baur, B. and J. Bengtsson. 1987. Colonizing ability in land snails on the Baltic uplift archipelagoes. *Journal of Biogeography* 14: 329–341.

Bazzaz, F. A. 1996. Plants in Changing Environments: Linking Physiological, Population and Community Ecology. Cambridge: Cambridge University Press.

Beadle, N. C. W. 1966. Soil phosphate and its role in molding segments of the Australian flora and vegetation, with special reference to xeromorphy and sclerophylly. *Ecology* 47: 991–1007.

Beadle, N. C. W. 1981. *The Vegetation of Australia*. Stuttgart: Gustav Fischer Verlag.

Beard, C. 2002. East of Eden at the Paleocene/Eocene boundary. *Science* 295: 2028–2029.

Beauchamp, G. 2004. Reduced flocking by birds on islands with relaxed predation. *Proceedings of the Royal Society of London, Series B-Biological Sciences* 271: 1039–1042.

Beck, J. and I. J. Kitching. 2009a. Correlates of range size and dispersal ability: A comparative analysis of sphingid moths from the Indo-Australian tropics. *Global Ecology and Biogeography* 16: 341–349.

Beck, J. and I. J. Kitching. 2009b. Drivers of moth species richness on tropical altitudinal gradients: A cross-regional comparison. *Global Ecology and Biogeography* 18: 361–371.

Beckwith, S. L. 1954. Ecological succession on abandoned farm lands and its relationship to wildlife management. *Ecological Monographs* 24: 349–376.

Beerli, P. and J. Felsenstein. 1999. Maximum-likelihood estimation of migration rates and effective population numbers in two populations using a coalescent approach. *Genetics* 152: 763–773.

Beever, E. A., P. F. Brussard and J. Berger. 2003. Patterns of apparent extirpation among isolated populations of pikas (Ochotona princeps) in the Great Basin. *Journal of Mammalogy* 84: 37–54.

Begon, M., J. L. Harper and C. R. Townsend. 1986. *Ecology: Individuals, Populations and Communities.* Oxford: Blackwell Scientific Publications.

Beheregaray, L. B. 2008. Twenty years of phylogeography: The state of the field and the challenges for the Southern Hemisphere. *Molecular Ecology* 17: 3754–3774.

Behle, W. H. 1978. Avian biogeography of the Great Basin and Intermontane Region. *Great Basin Naturalist Memoirs* 2: 55–80.

Behrensmeyer, A. K., J. D. Damuth, W. A. DiMichele, R. Potts, H. Sues and S. L. Wing. 1992. *Terrestrial Ecosystems Through Time: Evolutionary Paleoecology of Terrestrial Plants and Animals.* Chicago: University of Chicago Press.

Beilmann, A. P. and L. G. Brenner. 1951. The recent intrusion of forests in the Ozarks. *Annals of the Missouri Botanical Garden* 38: 261–282.

Beketov, M. A. 2009. The Rapoport effect is detected in a river system and is based on nested organization. *Global Ecology and Biogeography* 18: 498–506.

Belgrano, A., U. M. Scharler, J. Dunne and R. E. Ulanowicz (eds.). 2005. *Aquatic Food Webs: An ecosystem approach.* Oxford: Oxford University Press.

Bell, M. and M. J. C. Walker. 1992. *Late Quaternary Environmental Change: Physical and Human Perspectives.* New York: John Wiley and Sons.

Bellemain, E. and R. E. Ricklefs. 2008. Are islands the end of the colonization road? *Trends in Ecology and Evolution* 23: 461–468.

Bellwood, D. R. and C. P. Meyer. 2009. Searching for heat in a marine biodiversity hotspot. *Journal of Biogeography* 36: 569–576.

Benioff, H. 1954. Orogenesis and deep crustal structure: Additional evidence from seismology. *Bulletin of the Geological Society of America* 65: 385–400.

Benton, M. J. and R. J. Twitchett. 2003. How to kill (almost) all life: The end-Permian extinction event. *Trends in Ecology and Evolution* 18: 358–365.

Bercovici, D. 2003. The generation of plate tectonics from mantle convection. *Earth Planetary Science Letters* 205: 107–121.

Berger, A. and M. F. Loutre. 2002. An exceptionally long interglacial ahead? *Science* 297: 1287–1288.

Berger, A. J. 1981. *Hawaiian Birdlife.* 2nd edition. Honolulu: University of Hawaii Press.

Bergh, N. G. and H. P. Linder. 2009. Cape diversification and repeated out-of-southern-Africa dispersal in paper daisies (Asteraceae-Gnaphalieae). *Molecular Phylogenetics and Evolution* 51: 5–18.

Bergmann, C. 1847. Über die Verhältnisse der Wärmeökonomie der Thiere zu ihren Grösse. *Göttinger Studien* 1: 595–708.

Berlocher, S. H. 1998. Origins: A brief history of research on speciation. In D. J. Howard and S. H. Berlocher (eds.), *Endless Forms: Species and Speciation.* Oxford: Oxford University Press.

Bermingham, E. and A. P. Martin. 1998. Comparative mtDNA phylogeography of neotropical freshwater fishes: Testing shared history to infer the evolutionary landscape of lower Central America. *Molecular Ecology* 7: 499–517.

Bermingham, E. and C. Moritz. 1998. Comparative phylogeography: Concepts and applications. *Molecular Ecology* 7: 367–369.

Bernabo, J. C. and T. Webb. 1977. Changing patterns in the Holocene pollen record from northeastern North America: A mapped summary. *Quaternary Research* 8: 64–96.

Bernatchez, L. and C. C. Wilson. 1998. Comparative phylogeography of nearctic and palearctic fishes. *Molecular Ecology* 7: 431–452.

Bernhardsen, T. 2002. *Geographic Information Systems: An Introduction.* New York: Wiley & Sons.

Berra, T. M. 1981. *An Atlas of Distribution of the Freshwater Fish Families of the World.* Lincoln: University of Nebraska Press.

Berra, T. M. 2001. *Freshwater Fish Distribution.* San Diego and London: Academic Press.

Berry, A. (ed.). 2002. Infinite tropics: An Alfred Russel Wallace anthology. *Reports of the National Center for Science Education* 22: 28–29.

Berry, W. B. N. 1973. Silurian-Early Devonian graptolites. In A. Hallam (ed.), *Atlas of Palaeobiogeography,* 81–87. Amsterdam: Elsevier.

Bertness, M. D. 1989. Competitive and facilitative interactions and acorn barnacle populations in a sheltered habitat. *Ecology* 70: 257–268.

Bertness, M. D. 1991. Interspecific interactions among high marsh perennials in a New England salt marsh. *Ecology* 72: 125–137.

Betancourt, J. L. 2004. Advances in arid lands paleobiogeography: The rodent midden record in the Americas. In M. V. Lomolino and L. R. Heaney (eds.), *Frontiers of Biogeography—New Directions in the Geography of Nature.* London: Cambridge University Press.

Betancourt, J. L., T. R. Van Devender and P. S. Martin. 1990. *Packrat Middens: The Last 40,000 Years of Biotic Exchange.* Tucson, AZ: University of Arizona Press.

Bhagwat, S. A. and K. J. Willis. 2008. Species persistence in northerly glacial refugia of Europe: A matter of chance or biogeographical traits? *Journal of Biogeography* 35: 464–482.

Bhattarai, K. R. and O. R. Vetaas. 2003. Variation in plant species richness of different life forms along a subtropical elevation gradient in the Himalayas, east Nepal. *Global Ecology and Biogeography* 12: 327–340.

Bhattarai, K. R., O. R. Vetaas and J. A. Grytnes. 2004. Fern species richness along a central Himalayan elevational gradient, Nepal. *Journal of Biogeography* 31: 389–400.

Bianchi, C. N. 2007. Biodiversity issues for the forthcoming tropical Mediterranean Sea. *Hydrobiologia* 580: 7–21.

Bierregaard, R. O., Jr., T. E. Lovejoy, V. Kapos, A. A. dos Santos and R. W. Hutchings. 1992. The biological dynamics of tropical rainforest fragments: A prospective comparison of fragments and continuous forest. *Bioscience* 42: 859–866.

Biju-Duval, B. and L. Montadert (eds.). 1977. Structural history of the Mediterranean basins. International Symposium of the 25th Plenary Congress Assembly of the International Commission for the Scientific Exploration of the Mediterranean. Paris: Société des Éditions Technip.

Billerbeck, J. M., G. Ortí and D. O. Conover 1997. Latitudinal variation in vertebrate number has a genetic basis in the Atlantic silverside, *Menidia menidia. Canadian Journal of Fisheries and Aquatic Sciences* 54: 1796–1801.

Bilton, D. T., J. R. Freeland and B. Okamura. 2001. Dispersal in freshwater invertebrates. *Annual Review of Ecology and Systematics* 32: 159–181.

Bindon, J. R. and P. T. Baker. 1997. Bergmann's rule and the thrifty genotype. *American Journal of Physical Anthropology* 104: 201–210.

Bindon, J. R. and P. T. Baker. 1997. Bergmann's rule and the thrifty genotype. *American Journal of Physical Anthropology* 104: 201–210.

BirdLife International. 2000. *Threatened Birds of the World*. Barcelona and Cambridge: Lynx Editions and BirdLife International.

Birge, E. A. and C. Juday. 1911. The inland lakes of Wisconsin. *Bulletin of the Wisconsin Geological Natural History Survey* 22: 1–259.

Birks, H. J. B. 1987. Recent methodological developments in quantitative descriptive biogeography. *Annales Zoologici Fennici* 24: 165–178.

Birks, H. J. B. 1989. Holocene isochrone maps and patterns in tree-spreading in the British Isles. *Journal of Biogeography* 16: 503–540.

Bishop, I. R. and M. J. Delancy. 1963. The ecological distribution of small mammals in the Channel Islands. *Mammalia* 27: 99.

Bissonette, J. A. and I. Storch. 2002. Fragmentation: Is the message clear? *Conservation Ecology* 6(2): 14. Available from: http://www.consecol.org/vol6/iss2/art14.

Bivand, R. S., E. J. Pebesma and V. Gómez-Rubio. 2008. *Applied Spatial Data Analysis with R*. New York: Springer Publications.

Blackburn, T. M. and A. Ruggiero. 2001. Latitude, elevation and body mass variation in Andean passerine birds. *Global Ecology and Biogeography* 10: 245–259.

Blackburn, T. M. and B. Hawkins. 2004. Bergmann's rule and the mammal fauna of northern North America. *Ecography* 27: 715–724.

Blackburn, T. M. and K. J. Gaston (eds.). 2003. *Macroecology: Concepts and Consequences*. Oxford: Blackwell Scientific Publications.

Blackburn, T. M. and K. J. Gaston. 2005. Biological invasions and the loss of birds on islands: Insights into the idiosyncrasies of extinction. In D. F. Sax, J. J. Stachowicz and S. D. Gaines (eds.), *Species Invasions: Insights into Ecology, Evolution, and Biogeography*, 85–134. Sunderland, MA.: Sinauer Associates.

Blackburn, T. M. and K. J. Gaston. 2006. There's more to macroecology than meets the eye. *Global Ecology and Biogeography* 15: 537–540.

Blackburn, T. M. and R. P. Duncan. 2001. Establishment patterns of exotic birds are constrained by nonrandom patterns in introductions. *Journal of Biogeography* 28: 927–939.

Blackburn, T. M., J. L. Lockwood and P. Cassey. 2009. *Avian Invasions: The Ecology and Evolution of Exotic Birds*. Oxford: Oxford University Press.

Blackburn, T. M., K. J. Gaston and M. Parnell. 2010. Changes in non-randomness in the expanding introduced avifauna of the world. *Ecography* 33: 168–174.

Blackburn, T. M., K. J. Gaston and N. Loder. 1999b. Geographic gradients in body size: A clarification of Bergmann's rule. *Diversity and Distributions* 5: 165–174.

Blackburn, T. M., K. J. Gaston, J. J. D. Greenwood and R. D. Gregory. 1998. The anatomy of the interspecific abundance-range size relationship for British avifauna. II. Temporal dynamics. *Ecology Letters* 1: 47–55.

Blackburn, T. M., P. Cassey, K. L. Evans, K. J. Gaston and R. P. Duncan. 2008b. The biogeography of avian extinctions on oceanic islands revisited. *Journal of Biogeography* 36: 1613–1614.

Blackburn, T. M., P. Cassey, R. P. Duncan, K. L. Evans and K. J. Gaston. 2004. Avian extinction and mammalian introductions on oceanic islands. *Science* 305: 1955–1958.

Blackburn, T. M., P. Cassey, R. P. Duncan, K. L. Evans and K. J. Gaston. 2008a. Threats to avifauna on oceanic islands revisited. *Conservation Biology* 22: 492–494.

Blair, W. F. 1943. Activities of the Chihuahuan deer-mouse in relation to light intensity. *Journal of Wildlife Management* 7: 92–97.

Blanckenhorn, W. U. and M. Demont. 2004. Bergmann and converse Bergmann latitudinal clines in arthropods: Two ends of a continuum? *Integrative Comparative Biology* 44: 413–424.

Blondel, J., D. Chessel and B. Frochot. 1988. Bird species impoverishment, niche expansion and density inflation in Mediterranean island habitats. *Ecology* 69: 1899–1917.

Bloom, A. J. 2010. *Global Climate Change: Convergence of Disciplines*. Sunderland, MA: Sinauer Associates.

Boback, S. M. 2003. Body size evolution in snakes: evidence from island populations. *Copeia* 2003: 81–94.

Boback, S. M. and C. Guyer. 2003. Empirical evidence for an optimal body size in snakes. *Evolution* 57: 345–351.

Bocherens, H., J. Michaux, F. G. Talavera and J. Van der Plicht. 2006. Extinction of endemic vertebrates on islands: The case of the giant rat *Canariomys bravoi* (Mammalia, Rodentia) on Tenerife (Canary Islands, Spain). *Comptes Rendus Palevol* 5: 885–891.

Bock, C. E. and L. W. Lepthian. 1976. Synchronous eruptions of boreal seed-eating birds. *American Naturalist* 110: 559–571.

Bohle, U. R., H. H. Hilger and W. F. Martin. 1996. Island colonization and evolution of the insular woody habit in Echium L. (*Boraginaceae*). *Proceedings of the National Academy of Sciences, USA* 93: 11740–11745.

Bohm, M. and P. J. Mayhew. 2005. Historical biogeography and the evolution of the latitudinal gradient of species richness in the Papionini (Primata: Cercopithecidae). *Biological Journal of the Linnean Society* 85: 235–246.

Bohning-Gaese, K., L. I. Gonzalez-Guzman and J. H. Brown. 1998. Constraints on dispersal and the evolution of the avifauna of the Northern Hemisphere. *Evolutionary Ecology* 12: 767–783.

Bohning-Gaese, K., T. Caprano, K. van Ewijk and M. Veith. 2006. Range size: Disentangling current traits and phylogenetic and biogeographic factors. *American Naturalist* 167: 555–567.

Bolnick, D. I. and B. Fitzpatrick. 2007. Sympatric speciation: Theory and empirical data. *Annual Review of Ecology Evolution and Systematics* 38: 459–487.

Bonfirm, F. S., J. A. Diniz-Filho and R. P. Bastos. 1998. Spatial patterns and the macroecology of South American viperid snakes. *Revista Brasileira de Biologia* 58: 97–103.

Bonner, J. T. 2006. *Why Size Matters: From Bacteria to Blue Whales*. Princeton, NJ: Princeton University Press.

Boomert, A. and A. J. Bright. 2007. Island archaeology: In search of a new horizon. *Island Studies Journal* 2: 3–26.

Borodin, A. M., A. G. Bannikov and V. E. Sokolov (eds.). 1984. *Red Data Book of the USSR: Rare and Endangered Species of Animals and Plants*. 2nd edition. Moscow: Forest Industry Publishers.

Bossuyt, F. and M. C. Milinkovitch. 2001. Amphibians as indicators of early Tertiary "out-of-India" dispersal of vertebrates. *Science* 292: 93–95.

Boucher, D. H. 1985. The Biology of Mutualism: Ecology and Evolution. London: Croom Helm.

Boucher, D. H., S. James and K. Kesler. 1984. The ecology of mutualism. *Annual Review of Ecology and Systematics* 13: 315–347.

Bouchet, P., P. Lozouet, P. Maestrati and V. Heros. 2002. Assessing the magnitude of species richness in tropical marine environments: Exceptionally high numbers of mollusks at a New Caledonia site. *Biological Journal of the Linnnean Society* 75: 421–436.

Boucot, A. J. and J. G. Johnson. 1973. Silurian brachiopods. In A. Hallam (ed.), *Atlas of Paleobiogeography*, 59–65. Amsterdam: Elsevier.

Bourliere, F. 1973. The comparative ecology of rain forest mammals in Africa and tropical America: Some introductory remarks. In B. J. Meggers, E. S. Ayensu and W. D. Ducksworth (eds.), *Tropical Forest Ecosystems in Africa and South America: A*

Comparative Review, 279–292. Washington, DC: Smithsonian Institution Press.

Bowen G. J., W. C. Clyde, P. L. Koch, S. Ting, J. Alroy, T. Tsubamoto, Y. Wang and Y. Wang. 2002. Mammalian dispersal at the Paleocene/Eocene boundary. *Science* 295: 2062–2065.

Bowen, L. and D. Van Vuren. 1997. Insular endemic plants lack defenses against herbivores. *Conservation Biology* 11: 1249–1254.

Bowers, M. A. and C. H. Lowe. 1986. Plant-form gradients on Sonoran Desert bajadas. *Oikos* 46: 284–291.

Bowers, M. A. and J. H. Brown. 1982. Body size and coexistence in desert rodents: Chance or community structure. *Ecology* 63: 391–400.

Boyd, E. M. and S. A. Nunneley. 1964. Banding records substantiating the changed status of 10 species of birds since 1900 in the Connecticut Valley. *Bird-Banding* 35: 1–8.

Braba, F., L. Frechilla and G. Orizaola. 1996. Effect of introduced fish on amphibian assemblages in mountain lakes of Northern Spain. *The Herpetological Journal* 6: 145.

Bradley, R. S. 1985. *Quaternary Palaeoclimatology: Methods of Palaeoclimatic Reconstruction*. London: Unwin Hyman.

Bradshaw, W. E. and C. M. Holzapfel. 2006. Evolutionary response to rapid climate change. *Science* 312: 1477.

Bradshaw, W. E. and C. M. Holzapfel. 2007. Genetic response to rapid climate change: It's seasonal timing that matters. *Molecular Ecology* 17: 157–166.

Brandt, S., K. E. Ellingsen, S. Brix and W. Brokeland. 2005. Southern ocean deep-sea isopod species richness (Crustacea, Malacostraca): Influences of depth, latitude and longitude. *Polar Biology* 28: 284–289.

Bremer, K. 1992. Ancestral areas—A cladistic reinterpretation of the center of origin concept. *Systematic Biology* 41: 436–445.

Bremer, K. 1995. Ancestral areas: Optimization and probability. *Systematic Biology* 44: 255–259.

Bremer, K. 2002. Gondwanan evolution of the grass alliance of families (Poales). *Evolution* 56: 1374–1387.

Brenner, W. 1921. Vaxtgeografiska studien: Barosunds skargard. I. Allman del och floran. *Acta Societatis pro Fauna et Flora Fennica* 49: 1–151.

Briggs, D. and S. M. Walters. 1984. *Plant Variation and Evolution*. Cambridge: Cambridge University Press.

Briggs, J. C. 1968. Panama sea-level canal. *Science* 162: 511–513.

Briggs, J. C. 1974. *Marine Zoogeography*. New York: McGraw-Hill.

Briggs, J. C. 1987. *Biogeography and Plate Tectonics*. Amsterdam: Elsevier.

Briggs, J. C. 1994. The genesis of Central America: Biology vs. geophysics. *Global Ecology and Biogeography Letters* 4: 169–172.

Briggs, J. C. 1995. *Global Biogeography*. New York: Elsevier.

Briggs, J. C. 2003a. The biogeographic and tectonic history of India. *Journal of Biogeography* 30: 381–388.

Briggs, J. C. 2003b. Fishes and birds: Gondwana life rafts reconsidered. *Systematic Biology* 52: 548–553.

Briggs, J. C. 2003c. Marine centres of origin as evolutionary engines. *Journal of Biogeography* 30: 1–18.

Briggs, J. C. 2004a. The ultimate expanding earth hypothesis. *Journal of Biogeography* 31: 855–857.

Briggs, J. C. 2004b. Older species: A rejuvenation on coral reefs? *Journal of Biogeography* 31: 525–530.

Briggs, J. C. 2004c. A marine center of origin: Reality and conservation. In M. V. Lomolino and L. R. Heaney (eds.), *Frontiers of Biogeography: New Directions in the Geography of Nature*, Chapter 13. Cambridge University Press.

Briggs, J. C. 2006. Proximate sources of marine biodiversity. *Journal of Biogeography* 33: 1–10.

Briggs, J. C. 2007. Panbiogeography: Its origin, metamorphosis and decline. *Russian Journal of Marine Biology* 33: 273–277.

Brittingham, M. C. and S. A. Temple. 1983. Have cowbirds caused forest songbirds to decline? *BioScience* 33: 31–35.

Bromham, L. and M. Cardillo. 2007. Primates follow the "island rule": Implications for interpreting *Homo floresiensis*. *Biology Letters* 3: 398–400.

Brook, B. W. and D. M. J. S. Bowman. 2002. Explaining the Pleistocene megafaunal extinctions: Models, chronologies and assumptions. *Proceedings of the National Academy of Sciences, USA* 99: 14624–14627.

Brookfield, J. F. Y. 2009. Evolution and evolvability: Celebrating Darwin 200. *Biology Letters* 5: 44–46.

Brooks, D. R. 2004. Reticulations in historical biogeography: The triumph of time over space in evolution. In M. V. Lomolino and L. R. Heaney (eds.), *Frontiers of Biogeography*, 123–144. Sunderland, MA: Sinauer Associates.

Brooks, D. R. and D. A. McLennan. 2002. *The Nature of Diversity: An Evolutionary Voyage of Discovery*. Chicago: University of Chicago Press.

Brooks, D. R., A. P. G. Dowling, M. G. P. van Veller and E. P. Hoberg. 2004. Ending a decade of deception: a valiant failure, a not-so-valiant failure and a success story. *Cladistics-the International Journal of the Willi Hennig Society* 20: 32–46.

Brooks, D. R., M. G. P. Van Veller and D. A. McLennan. 2001. How to do BPA, really. *Journal of Biogeography* 28: 345–358.

Brower, L. P. 1977. Monarch migration. *Natural History* 86(6): 40–53.

Brower, L. P. and S. B. Malcolm. 1991. Animal migrations: Endangered phenomena. *American Zoologist* 31: 265–276.

Brown, D. E. and R. Davis. 1995. One hundred years of vicissitude: Terrestrial bird and mammal distribution changes in the American Southwest. In *Biodiversity and Management of the Madrean Archipelago: The Sky Islands of Southwestern United States and Northwestern Mexico*, 231–244. USDA Forest Service, General Technical Report RM-GTR-264.

Brown, D. E., F. Reichenbacher and S. E. Franson. 1998. *A Classification of North American Biotic Communities*. Salt Lake City: University of Utah Press.

Brown, J. H. 1971b. Mammals on mountaintops: Nonequilibrium insular biogeography. *American Naturalist* 105: 467–478.

Brown, J. H. 1978. The theory of insular biogeography and the distribution of boreal birds and mammals. *Great Basin Naturalist Memoirs* 2: 209–227.

Brown, J. H. 1981. Two decades of homage to Santa Rosalia: Toward a general theory of diversity. *American Zoologist* 21: 877–888.

Brown, J. H. 1988. Species diversity. In A. Myers and R. S. Giller (eds.), *Analytical Biogeography*, 57–89. London: Chapman and Hall.

Brown, J. H. 1995. *Macroecology*. Chicago: University of Chicago Press.

Brown, J. H. 1999. The legacy of Robert MacArthur: From geographical ecology to macroecology. *Journal of Mammalogy* 80: 333–344.

Brown, J. H. and A. K. Lee. 1969. Bergmann's rule and climatic adaptation in woodrats (*Neotoma*). *Evolution* 23: 329–338.

Brown, J. H. and A. Kodric-Brown. 1977. Turnover rates in insular biogeography: Effect of immigration on extinction. *Ecology* 58: 445–449.

Brown, J. H. and A. Kodric-Brown. 1993. Highly structured fish communities in Australian desert springs. *Ecology* 74: 1847–1855.

Brown, J. H. and B. A. Maurer. 1986. Body size, ecological dominance and Cope's rule. *Nature* 324: 248–250.

Brown, J. H. and B. A. Maurer. 1987. Evolution of species assemblages: Effects of energetic constraints and species dynamics on the diversification of the North American avifauna. *American Naturalist* 130: 1–17.

Brown, J. H. and B. A. Maurer. 1989. Macroecology: The division of food and space among species on continents. *Science* 243: 1145–1150.

Brown, J. H. and D. F. Sax. 2004. Gradients in species diversity: Why are there so many species in the tropics? In M. V. Lomolino, D. F. Sax and J. H. Brown (eds.), *Foundations of Biogeography,* 1145–1154. Chicago: University of Chicago Press.

Brown, J. H. and M. A. Bowers. 1984. Patterns and processes in three guilds of terrestrial vertebrates. In D. R. Strong, D. Simberloff, L. G. Abele and A. B. Thistle (eds.), *Ecological Communities: Concepts, Issues and the Evidence,* 282–296. Princeton, NJ: Princeton University Press.

Brown, J. H. and M. A. Bowers. 1985. On the relationship between morphology and ecology: Community organization in hummingbirds. *Auk* 102: 251–269.

Brown, J. H. and M. V. Lomolino. 1989. On the nature of scientific revolutions: Independent discovery of the equilibrium theory of island biogeography. *Ecology* 70: 1954–1957.

Brown, J. H. and P. F. Nicoletto. 1991. Spatial scaling of species assemblages: Body masses of North American land mammals. *American Naturalist* 138: 1478–1512.

Brown, J. H. and R. C. Lasiewski. 1972. Metabolism of weasels: The cost of being long and thin. *Ecology* 53: 939–943.

Brown, J. H., D. W. Mehlman and G. C. Stevens. 1995. Spatial variation in abundance. *Ecology* 76: 2028–2043.

Brown, J. H., G. C. Stevens and D. M. Kaufman. 1996. The geographic range: Size, shape, boundaries and internal structure. *Annual Review of Ecology and Systematics* 27: 597–623.

Brown, J. H., P. A. Marquet and M. L. Taper. 1993. Evolution of body size: Consequences of an energetic definition of fitness. *American Naturalist* 142: 573–584.

Brown, J. M. and D. S. Wilson. 1992. Local specialization of phoretic mites on sympatric carrion beetle hosts. *Ecology* 73: 463–478.

Brown, J. W. 1987. The peninsular effect in Baja California: An entomological assessment. *Journal of Biogeography* 14: 359–365.

Brown, J. W. and P. A. Opler. 1990. Patterns of butterfly species density in peninsular Florida. *Journal of Biogeography* 17: 615–622.

Brown, M. and J. J. Dinsmore. 1988. Habitat islands and the equilibrium theory of island biogeography: Testing some predictions. *Oecologia* 75: 426–429.

Brown, P., T. Sutikna, M. J. Morwood, R. P. Soejono, Jatmiko, E. Wayhu Saptomo and Rokus Awe Due. 2004. A new small-bodied hominin from the Late Pleistocene of Flores, Indonesia. *Nature* 431: 1055.

Brown, R. M, A. C. Diesmos and M. V. Duya. 2007. A new species of Luperosaurus (Squamata: Gekkonidae) from the Sierra Madre of Luzon Island, Philippines. *Raffles Bulletin of Zoology* 55: 167–174.

Brown, R. M. and J. C. Gonzalez. 2007. A new forest frog of the genus *Platymantis* (Amphibia: Anura: Ranidae) from the Bicol Peninsula of Luzon Island, Philippines. *Copeia* 2007: 251–266.

Brown, W. L. and E. O. Wilson. 1956. Character displacement. *Systematic Zoology* 5: 49–64.

Brown, W. M., M. George, Jr. and A. C. Wilson. 1979. Rapid evolution of animal mitochondrial DNA. *Proceedings of the National Academy of Sciences, USA* 76: 1967–1971.

Browne, R. A. 1981. Lakes as islands: Biogeographic distribution, turnover rates and species composition in the lakes of central New York. *Journal of Biogeography* 8: 75–83.

Browne, R. A. and G. H. MacDonald. 1982. Biogeography of the brine shrimp, *Artemia:* Distribution of parthenogenetic and sexual populations. *Journal of Biogeography* 9: 331–338.

Bruhnes, B. 1906. Recherches sur la direction d'aimentation des roches volcaniques (1). *Journal Physique*, 4e Sér., 5: 705–724.

Brundin, L. 1966. Transantarctic relationships and their significance, as evidence by chironomid midges. *Kungliga Svenska Vetenskapsakademiens Handlingar*, 4th ser., II, no. 1: 1–472.

Buckley, H. L., T. E. Miller, A. M. Ellison and N. J. Gotelli. 2003. Reverse latitudinal trends in species richness of pitcher-plant food webs. *Ecology Letters* 6: 825–829.

Buckley, L. B. and W. Jetz. 2007. Environmental and historical constraints on global patterns of amphibian richness. *Proceedings of the Royal Society*, Series B 274: 1167–1173.

Buckley, R. C. and S. B. Knedlhans. 1986. Beachcomber biogeography: Interception of dispersing propagules by islands. *Journal of Biogeography* 13: 69–70.

Buffett, B. A. 2000. Earth's core and the geodynamo. *Science* 288: 2007–2012.

Buffon, G. L. L., Comte de. 1761. *Histoire Naturelle, Generale et Particuliere*, vol. 9. Paris: Imprimerie Royale.

Buffon, G. L. L., Comte de. 1766. *Histoire Naturelle, Generale et Particuliere*. volume 5. Paris: Imprimerie Royale. Trans. into English by W. Smellie (1781, London).

Bullard, E. C., J. E. Everett and A. G. Smith. 1965. Fit of the continents around the Atlantic. In P. M. S. Blackett, E. C. Bullard and S. K. Runcorn (eds.), *A Symposium on Continental Drift,* 41–75. *Philosophical Transactions of the Royal Society of London*, Series A 248.

Bullock, J. M. and R. Kenward. 2001. *Dispersal Ecology.* Malden, MA. Blackwell Publishing.

Bunce, M., T. H. Worthy, M. J. Phillips, R. N. Holdaway, E. Willerslev, J. Haile, B. Shapiro, R. P. Scofield, A. Drummond, P. J. J. Kamp and A. Cooper. 2009. The evolutionary history of the extinct ratite moa and New Zealand Neogene paleography. *Proceedings of the National Academy of Sciences, USA* 106: 20646–20651.

Bureau de Recherches Géologiques et Minières. 1980a. Colloque C5. Géologie des chaines alpines issues de la Téthys. *Mémoirs*, no. 115.

Bureau de Recherches Géologiques et Minières. 1980b. Colloque C6. Géologie de l'Europe du Précambrien aux bassins sedimentaires post-hercyniens. *Mémoirs*, no. 108.

Burke, K., J. F. Dewey and W. S. F. Kidd. 1977. World distribution of sutures: The sites of former oceans. *Tectonophysics* 40: 69–99.

Burney, D. A. and T. F. Flannery. 2005. Fifty millennia of catastrophic extinctions after human conflict. *Trends in Ecology and Evolution* 20: 395–401.

Burney, D. A. and T. F. Flannery. 2006. Response to Wroe et al.: Island extinctions versus continental extinctions. *Trends in Ecology and Evolution* 21: 63–64.

Burney, D. A., G. S. Robinson and L. P. Burney. 2003. *Sporormiella* and the late Holocene extinctions in Madagascar. *Proceedings of the National Academy of Sciences, USA* 100: 10800–10805.

Burney, D. A., H. F. James, L. P. Burney, S. L. Olson, W. Kikuchi, W. L. Wagner, M. Burney, D. McCloskey, D. Kikuchi, F. V. Grady, R. Gage II and R. Nishek. 2001. Fossil evidence for a diverse biota from Kaua'I and its transformation since human arrival. *Ecological Monographs* 71: 615–641.

Burns, K. C., R. P. McHardy and S. Pledger. 2009. The small-island effect: Fact or artefact? *Ecography* 32: 269–276.

Burns, K. J., S. J. Hackett and N. K. Klein. 2002. Phylogenetic relationships and morphological diversity in Darwin's finches and their relatives. *Evolution* 56: 1240–1252.

Burr, B. M. and R. L. Mayden. 1992. Phylogenetics and North American freshwater fishes. In R. L. Mayden (ed.), *Systematics, Historical Ecology and North American Freshwater Fishes.* Stanford, CA: Stanford University Press.

Burrough, P. A. 2001. GIS and geostatistics: Essential partners for spatial analysis. *Environmental and Ecological Statistics* 8: 361–377.

Burt, E. H. Jr. and J. M. Ichida. 2004. Gloger's rule, feather-degrading bacteria, and color variation among song sparrows. *Condor* 106: 681–686.

Bush, A. O., J. C. Fernandez, G. W. Esch and J. R. Seed. 2001. *Parasitism—The diversity and ecology of animal parasites.* London: Cambridge University Press.

Bush, G. L. 1975. Modes of animal speciation. *Annual Review of Ecology and Systematics* 6: 339–364.

Bush, M. B. and R. J. Whittaker. 1991. Krakatau: Colonization patterns and hierarchies. *Journal of Biogeography* 18: 341–356.

Bush, M. B. and R. J. Whittaker. 1993. Non-equilibrium in island theory and Krakatau. *Journal of Biogeography* 20: 453–458.

Bussing, W. A. 1985. Patterns of distribution of the Central American Icthyofauna. In G. G. Stehli and S. D. Webb (eds.), *The Great American Interchange*, 453–473. New York: Plenum Press.

Butler, R. 1995. When did India hit Asia? *Nature* 373: 20–21.

Butlin, R. 1998. What do hybrid zones in general and the *Chorthippus parallelus* zone in particular, tell us about speciation? In D. J. Howard and S. H. Berlocher (eds.), *Endless Forms: Species and Speciation* 367–378. Oxford: Oxford University Press.

Buzas, M. A. 1972. Patterns of species diversity and their explanation. *Taxon* 21: 275–286.

Cabe, P. R. 1993. European starling (*Sturnus vulgaris*). In A. Poole (ed.), *The Birds of North America Online.* Ithaca, NY: Cornell Lab of Ornithology.

Cabrera, A. L. and A. Willink. 1973. *Biogeografía de America Latina.* Washington, DC: Programa Regional de Desarrollo Cientifico y Tecnologico, Departamento Asuntos Cientificos, Secretario General de la Organizacion de los Estados Americanos.

Caccone, A., G. Gentile, J. P. Gibbs, T. H. Fritts, H. L. Snell, J. Betts and J. R. Powell. 2002. Phylogeography and history of giant Galapagos tortoises. *Evolution* 56: 2052–2066.

Caccone, A., J. P. Gibbs, V. Ketmaier, E. Suatoni and J. R. Powell. 1999. Origin and evolutionary relationships of giant Galapagos tortoises. *Proceedings of the National Academy of Sciences, USA* 96: 13223–13228.

Cade, B. S., J. W. Terrell and R. L. Schroeder. 1999. Estimating effects of limiting factors with regression quantiles. *Ecology* 80: 311–323.

Cain, S. A. 1944. *Foundations of Plant Geography.* New York: Harper and Brothers.

Calder, W. A. 1974. Consequences of body size for avian energetics. In R. A. Paynter, Jr. (ed.), *Avian Energetics,* 86–151. Cambridge, MA: Nuttal Orinithology Club, Publ. 15.

Calder, W. A. III. 1984. *Size, Function and Life History.* Cambridge: Harvard University Press.

Calvert, A. J., E. W. Sawyer, W. J. Davis and J. N. Ludden. 1995. Archaean subduction inferred from seismic images of a mantle suture in the Superior Province. *Nature* 375: 670–674.

Cambefort, Y. 1994. Body size, abundance, and geographic distribution of Afrotropical dung beetles (Coleoptera: Scarabaeidae). *Acta Oecologia* 15: 165–179.

Campbell, T. S. and A. C. Echternacht. 2003. Introduced species as moving targets: Changes in body sizes of introduced lizards following experimental introductions and historical invasions. *Biological Invasions* 5: 193–212.

Candolle, A. P. de. 1855. *Géographie Botanique Raisonnée.* 2 vols. Paris: Masson.

Canestrelli, D., R. Cimmaruta, V. Costantini and G. Nascetti. 2006. Genetic diversity and phylogeography of the Apennine yellow-bellied toad *Bombina pachypus,* with implications for conservation. *Molecular Ecology* 15: 3741–3754.

Cannatella, D. C. and L. Trueb. 1988. Evolution of pipoid frogs: Intergeneric relationships of the aquatic frog family Pipidae (Anura). *Zoological Journal of the Linnean Society* 94: 1–38.

Carey, S. W. 1955. The orocline concept of geotectonics, part I. *Proceedings of the Royal Society of Tasmania,* Papers 89: 255–288.

Carey, S. W. 1958. A tectonic approach to continental drift. In *Continental Drift: A Symposium*, 177–355. Hobart: University of Tasmania.

Carine, M. A., S. J. Russell, A. Santos-Guerra and J. Francisco-Ortega. 2004. Relationships of the Macaronesian and Mediterranean floras: Molecular evidence for multiple colonizations into Macaronesia and back-colonization of the continent in *Convolvulus* (Convolvulaceae). *American Journal of Botany* 91: 1070–1085.

Carine, M., A. Guerra, I. R. Guma and J. A. Reyes-Betancort. 2010. Endemism and evolution of the Macaronesian flora. In D. M. Williams and S. K. Knapp (eds.), Beyond Cladistics. Berkeley, CA: University of California Press.

Carlquist, S. 1965. *Island Life.* Garden City, NY: Natural History Press.

Carlquist, S. 1966. The biota of long-distance dispersal. I. Principles of dispersal and evolution. *Quarterly Review of Biology* 41: 247–270.

Carlquist, S. 1974. *Island Biology.* New York: Columbia University Press.

Carlquist, S. 1981. Chance dispersal. *American Scientist* 69: 509–515.

Carnaval, A. C., M. J. Hickerson, C. F. B. Haddad, M. T. Rodrigues and C. Moritz. 2009. Stability predicts genetic diversity in the Brazilian Atlantic Forest Hotspot. *Science* 323: 785–789.

Carpenter, K. E. and V. G. Springer. 2005. The center of the center of marine shore fish biodiversity: The Philippine Islands. *Environmental Biology of Fishes* 72: 467–480.

Carpenter, S. R. 1988. *Complex Interactions in Lake Communities.* New York: Springer-Verlag.

Carpenter, S. R. 2002. Ecological futures: Building an ecology of the long now. *Ecology* 83: 2069–2083.

Carpenter, S. R., J. F. Kitchell and J. R. Hodgson. 1985. Cascading tropic interactions and lake ecosystem productivity. *BioScience* 35: 635–639.

Carpenter, S. R., J. F. Kitchell, J. R. Hodgson, P. A. Cochran, J. J. Elser, M. M. Elser, D. M. Lodge, D. Kretchmer, X. He and C. von Ende. 1987. Regulation of lake primary productivity by food web structure. *Ecology* 68: 1867–1876.

Carr, G. D., B. G. Baldwin and J. L. Strother. 2003. Appendix 1: Accepted names and synonyms for specific taxa in Madiinae. In S. Carlquist, B. G. Baldwin and G. D. Carr (eds.), *Tarweeds and Silverswords: Evolution in the Madiinae (Asteraceae),* 229–243. St. Louis, MO: Missouri Botanical Garden Press.

Carrasco, M. A., A. D. Barnosky and R. W. Graham. 2009. Quantifying the extent of North American mammal extinctions relative to the pre-anthropogenic baseline. *PLoS One* 4: e8331.

Carroll, R. L. 1988. *Vertebrate Paleontology and Evolution.* New York: W.H. Freeman and Company.

Carson, H. L. 1971. Speciation and the founder principle. *University of Missouri, Stadler Symposium* 3: 51–70.

Carson, H. L. and A. R. Templeton. 1984. Genetic revolutions in relation to speciation phenomena: The founding of new populations. *Annual Review of Ecology and Systematics* 15: 97–131.

Carson, H. L. and K. Y. Kaneshiro. 1976. *Drosophila* of Hawaii: Systematics and ecological genetics. *Annual Review of Ecology and Systematics* 7: 311–346.

Carstens, B. C., J. D. Degenhardt, A. L. Stevenson and J. Sullivan. 2005. Accounting for coalescent stochasticity in testing phylogeographical hypotheses: Modeling Pleistocene population structure in the Idaho giant salamander *Dicamptodon aterrimus*. *Molecular Ecology* 14: 255–265.

Carstensen, D. W. and J. M. Olesen. 2009. Wallacea and its nectarivorous birds: Nestedness and modules. *Journal of Biogeography* 36: 1540–1550.

Cartar, R. V. and R. I. G. Morrison. 2005. Metabolic correlates of leg length in breeding arctic shorebirds: The cost of getting high. *Journal of Biogeography* 32: 377–382.

Carto, S. L., A. J. Weaver, R. Hetherington, Y. Lam, E. C. Wiebe. 2009. Out of Africa and into an ice age: On the role of global climate change in the late Pleistocene migration of early modern humans out of Africa. *Journal of Human Evolution* 56: 139–151.

Cartron, J.-L. E., J. F. Kelly and J. H. Brown. 2000. Constraints on patterns of covariation: A case study in strigid owls. *Oikos* 90: 381–389.

Case, T. J. 1975. Species numbers, density compensation and the colonizing ability of lizards on islands in the Gulf of California. *Ecology* 56: 3–18.

Case, T. J. 1978. A general explanation for insular body size trends in terrestrial vertebrates. *Ecology* 59: 1–18.

Case, T. J. 1987. Testing theories of island biogeography. *American Scientist* 75: 402–411.

Case, T. J. and D. T. Bolger. 1991a. The role of interspecific competition in the biogeography of island lizards. *Trends in Ecology and Evolution* 6(4): 135–139.

Case, T. J. and D. T. Bolger. 1991b. The role of introduced species in shaping the distribution and abundance of island reptiles. *Evolutionary Ecology* 5: 272–290.

Case, T. J. and R. Sidell. 1983. Pattern and chance in the structure of model and natural communities. *Evolution* 37: 832–849.

Case, T. J. and T. D. Schwaner. 1993. Island/mainland body size differences in Australian varanid lizards. *Oecologia* 94: 102–109.

Case, T. J., D. T. Bolger and K. Petren. 1994. Invasion and competitive displacement among house geckos in the tropical Pacific. *Ecology* 75: 464–477.

Case, T. J., J. Faaborg and R. Sidell. 1983. The role of body size in the assembly of West Indian bird communities. *Evolution* 37: 1062–1074.

Cassey, P. and T. M. Blackburn. 2004. Body size trends in a Holocene island bird assemblage. *Ecography* 27: 59–67.

Cassey, P., T. M. Blackburn, R. P. Duncan and K. J. Gaston. 2005. Causes of exotic bird establishment across oceanic islands. *Proceedings of the Royal Society of London*, Series B 272: 2059–2063.

Caswell, H. 1978. Predator-mediated coexistence: A nonequilibrium model. *American Naturalist* 112: 127–154.

Censky, E. J., K. Hodge and J. Dudley. 1998. Over-water dispersal of lizards due to hurricanes. *Nature* 395: 556–557.

Cerling, T. E., G. Wittemyer, H. B. Rasmussen, F. Vollrath, C. E. Cerling, T. J. Robinson and I. Douglas-Hamilton. 2006. Stable isotopes in elephant hair document migration patterns and diet changes. *Proceedings of the National Academy of Sciences, USA* 103: 371–373.

Channell, R. 1998. *A Geography of Extinction: Patterns in the Contraction of Geographic Ranges*. Ph.D. dissertation, University of Oklahoma, Norman, OK.

Channell, R. and M. V. Lomolino. 2000a. Dynamic biogeography and conservation of endangered species. *Nature* 403: 84–86.

Channell, R. and M. V. Lomolino. 2000b. Trajectories toward extinction: Dynamics of geographic range collapse. *Journal of Biogeography* 27: 169–179.

Chaplin, G. 2004. Geographic distribution of environmental factors influencing human skin color. *American Journal of Physical Anthropology* 125: 292–302.

Chappell, M. A. 1978. Behavioral factors in the altitudinal zonation of chipmunks (*Eutamias*). *Ecology* 59: 565–579.

Charnov, E. L. and J. R. Krebs. 1974. On clutch size and fitness. *Ibis* 116: 217–219.

Chase, B. M., M. E. Meadows, L. Scott, D. S. G. Thomas, E. Marais, J. Sealy and P. J. Reimer. 2009. A record of rapid Holocene climate change preserved in hyrax middens from southwestern Africa. *Geology* 37: 703–707.

Cheatham, A. H. and J. E. Hazel. 1969. Binary (presence-absence) similarity coefficients. *Journal of Paleontology* 43: 1130–1136.

Cherfas, J. 1991. Disappearing mushrooms: Another mass extinction? *Science* 254: 1458.

Chester, R. H. 1969. Destruction of Pacific corals by the sea star *Acanthaster planci*. *Science* 165: 280–283.

Cheung, W. W. L., V. W. Y. Lam, J. L. Sarmiento, K. Kearney, R. Watson and D. Pauly. 2009. Projecting global marine biodiversity impacts under climate change scenarios. *Fish and Fisheries* 10: 235–251.

Chiba, S. 1998. Synchronized evolution in lineages of land snails in oceanic islands. *Paleobiology* 24: 99–108.

Choi, S. 2004. Trends in butterfly species richness in response to the peninsular effect in South Korea. *Journal of Biogeography* 31: 587–592.

Choi, S. and J. Chun. 2009. Combined effect of environmental factors on distribution of Geometridae (Lepidoptera) in South Korea. *European Journal of Entomology* 106: 69–76.

Chui, C. K. S. and S. M. Doucet. 2009. A test of ecological and sexual selection hypotheses for geographic variation in coloration and morphology of golden-crowned kinglets (*Regulus satrapa*). *Journal of Biogeography* 36: 1945–1957.

Ciofi, C. and M. de Boer. 2004. Distribution and conservation of the Komodo monitor *Varanus komodoensis*. *Herpetological Journal* 14: 99–107.

Clagg, H. B. 1966. Trapping of air-borne insects in the Atlantic-Antarctic area. *Pacific Insects* 8: 455–466.

Clark, J. S. 2008. Beyond neutral science. *Trends in Ecology and Evolution* 24: 8–15.

Clarke, A. and S. Lidgard. 2000. Spatial patterns of diversity in the sea: bryozoan species richness in the North Atlantic. *Journal of Animal Ecology* 69: 799–814.

Clarke, J. A., C. P. Tambussi, J. I. Noriega, G. M. Erickson and R. A. Ketcham. 2005. Definitive fossil evidence for the extant avian radiation in the Cretaceous. *Nature* 433: 305–308.

Clausen, J., D. D. Keck and W. M. Hiesey. 1940. *Experimental Studies on the Nature of Species*. I. *Effect of Varied Environments on Western North American Plants*. Carnegie Institute Publication 520. Washington, DC: Carnegie Institute.

Clausen, J., D. D. Keck and W. M. Hiesey. 1947. Heredity of geographically and ecologically isolated races. *American Naturalist* 81: 114–133.

Clausen, J., D. D. Keck and W. M. Hiesey. 1948. *Experimental Studies on the Nature of Species*. III: *Environmental Responses of Climatic Races of* Achillea. Carnegie Institute Publication 581: 1–129. Washington, DC: Carnegie Institute.

Clavero, M. and E. Garcia-Berthou. 2005. Invasive species are a leading cause of animal extinctions. *Trends in Ecology and Evolution* 20: 110.

Clegg, J. A., M. Almond and P. H. S. Stubbs. 1954. The remnent magnetism of some sedimentary rocks in Britain. *Philosophical Magazine* 45: 583–598.

Clegg, S. 2010. Evolutionary changes following island colonization in birds: Empirical insights into the roles of microevolutionary processes. In J. Losos and R. E. Ricklefs (eds.), *The Theory of Island Biogeography Revisited*, 293–325. Princeton, NJ: Princeton University Press.

Clegg, S. M. and I. P. F. Owens. 2002. The "island rule" in birds: Medium body size and its ecological explanation. *Proceedings of the Royal Society of London,* Series B 269: 1359–1365.

Clegg, S. M., F. D. Frentiu, J. Kikkawa, G. Tavecchia and I. P. Owens. 2008. 4000 years of phenotypic change in an island bird: Heterogeneity of selection over three microevolutionary timescales. *Evolution* 62: 2393–2410.

Clement, M., D. Posada and K. A. Crandall. 2000. TCS: A computer program to estimate gene genealogies. *Molecular Ecology* 9: 1657–1660.

Clements, F. E. 1916. *Plant Succession: An Analysis of the Development of Vegetation.* Carnegie Institute Publication no. 242. Washington, DC: Carnegie Institute.

Clements, F. E. and V. E. Shelford. 1939. *Bio-Ecology.* New York: John Wiley & Sons.

Clobert, J. 2001. *Dispersal.* New York: Oxford University Press.

Cockburn, A. 1991. *An Introduction to Evolutionary Ecology.* London: Blackwell Scientific Publications.

Cockburn, A., A. K. Lee and R. W. Martin. 1983. Macrogeo-graphic variation in litter size in *Antechinus* spp. (Marsupialia: Dasyuridae). *Evolution* 37: 86–95.

Cody, M. L. 1966. The consistency of intra- and inter-continental grassland bird species counts. *American Naturalist* 100: 371–376.

Cody, M. L. 1968. On the methods of resource division in grassland bird communities. *American Naturalist* 102: 107–137.

Cody, M. L. 1973. Parallel evolution and bird niches. In F. di Castri and H. A. Mooney (eds.), *Mediterranean-Type Ecosystems: Origin and Structure,* 307–338. Ecological Studies 7. New York: Springer-Verlag.

Cody, M. L. 1975. Towards a theory of continental species diversity. In M. L. Cody and J. M. Diamond (eds.), *Ecology and Evolution of Communities,* 214–257. Cambridge, MA: Belknap Press.

Cody, M. L. and H. A. Mooney. 1978. Convergence vs. nonconvergence in Mediterranean-climate ecosystems. *Annual Review of Ecology and Systematics* 9: 265–321.

Cody, M. L. and J. M. Diamond (eds.). 1975. *Ecology and Evolution of Communities.* Cambridge, MA: Belknap Press.

Cody, M. L. and J. M. Overton. 1996. Short-term evolution of reduced dispersal ability in island plant populations. *Journal of Ecology* 84: 53–61.

Cody, S., J. E. Richardson, V. Rull, C. Ellis and R. T. Pennington. 2010. The Great American Biotic Interchange revisited. *Ecography,* doi: 10.1111/j.1600-0587.2010.06327.x.

Cogger, H. G. 1975. Sea snakes of Australia and New Guinea. In W. A. Dunson (ed.), *The Biology of Sea Snakes,* 59–139. Baltimore, MD: University Park Press.

Cohan, F. M. 2001. Bacterial species and speciation. *Systematic Biology* 50: 513–524.

Cole, K. L. 1982. Late Quaternary zonation of vegetation in the eastern Grand Canyon. *Science* 217: 1142–1145.

Coleman, B. D., M. A. Mares, M. R. Willig and Y.-H. Hsieh. 1982. Randomness, area and species richness. *Ecology* 63: 1121–1133.

Colinvaux, P. A. 1981. Historical ecology in Beringia: The southland bridge coast at St. Paul Island. *Quaternary Research* 16: 18–36.

Colinvaux, P. A. 1996. Low-down on a landbridge. *Nature* 382: 21–23.

Colinvaux, P. A., P. E. De Oliveira and M. B. Bush. 2000. Amazonian and Neotropical plant communities on glacial time-scales: The failure of the aridity hypothesis. *Quaternary Science Reviews* 19: 141–169.

Collins, N. M. and J. A. Thomas. 1991. *The Conservation of Insects and Their Habitats.* London: Academic Press.

Colwell, R. K. 1973. Competition and coexistence in a simple tropical community. *American Naturalist* 107: 737–760.

Colwell, R. K. 1979. The geographical ecology of hummingbird flower mites in relation to their host plants and carriers. In J. S. Rodrigues (ed.), *Recent Advances in Acarology,* vol. 2, 461–468. New York: Academic Press.

Colwell, R. K., C. R. Rahbek and N. J. Gotelli. 2005. The mid-domain effect: There's a baby in the bathwater. *American Naturalist* 166: E149–E154.

Coney, P. J., D. L. Jones and I. W. H. Monger. 1980. Cordilleran suspect terrains. *Nature* 288: 329–333.

Connell, J. H. 1961. The influence of interspecific competition and other factors on the distribution of the barnacle *Chthamalus stellatus. Ecology* 42: 710–723.

Connell, J. H. 1975. Some mechanisms producing structure in natural communities: A model and evidence from field experiments. In M. L. Cody and J. M. Diamond (eds.), *Ecology and Evolution of Communities,* 460–490. Cambridge, MA: Belknap Press.

Connell, T. H. and R. O. Slatyer. 1977. Mechanisms of succession in natural communities and their role in community stability and organization. *American Naturalist* 111: 1119–1144.

Connor, E. F. and D. Simberloff. 1978. Species number and compositional similarity of the Galápagos flora and avifauna. *Ecological Monographs* 48: 219–248.

Connor, E. F. and E. D. McCoy. 1979. The statistics and biology of the species-area relationship. *American Naturalist* 113: 791–833.

Conrad, C. P. and C. Lithgow-Berteiloni. 2002. How mantle slabs drive plate tectonics. *Science* 298: 207–209.

Contoli, L. 2000. Rodents of Italy species richness maps and forma Italiae. *Hystrix* 11: 39–46.

Cook, J. A., A. L. Bidlack, C. J. Conroy, J. R. Demboski, M. A. Fleming, A. M. Runck, K. D. Stone and S. O. MacDonald. 2001. A phylogeographic perspective on endemism in the Alexander Archipelago of southeast Alaska. *Biological Conservation* 97: 215–227.

Cook, L. G. and M. D. Crisp. 2005a. Directional asymmetry of long-distance dispersal and colonization could mislead reconstructions of biogeography. *Journal of Biogeography* 32: 741–754.

Cook, L. G. and M. D. Crisp. 2005b. Not so ancient: The extant crown group of Nothofagus represents a post-Gondwanan radiation. *Proceedings of the Royal Society, Series B* 272: 2535–2544.

Cook, R. E. 1969. Variation in species density of North American birds. *Systematic Zoology* 18: 63–84.

Cook, R. R. 1995. The relationship between nested subsets, habitat subdivision and species diversity. *Oecologia* 101: 204–210.

Cook, R. R. and J. F. Quinn. 1995. The influence of colonization in nested species subsets. *Oecologia* 102: 413–424.

Cooke, H. B. S. 1972. The fossil mammal fauna of Africa. In A. Keast, F. C. Erk and B. Glass (eds.), *Evolution, Mammals and Southern Continents,* 89–139. Albany: State University of New York Press.

Cooper, A. and D. Penny. 1997. Mass survival of birds across the Cretaceous-Tertiary boundary: Molecular evidence. *Science* 275: 1109–1113.

Cooper, A. and R. A. Cooper. 1995. The Oligocene bottleneck and New Zealand biota: Genetic record of a past environmental crisis. *Proceedings of the Royal Society of London,* Series B 261: 293–302.

Cooper, A., C. Lalueza-Fox, S. Anderson, A. Rambaut, J. Austin and R. Ward. 2001. Complete mitochondrial genome sequences of two extinct moas clarify ratite evolution. *Nature* 409: 704–707.

Cooper, W. S. 1913. The climax forest of Isle Royale, Lake Superior and its development. *Botanical Gazette* 15: 1–44, 115–140, 189–235.

Corlett, R. T. 1992. The ecological transformation of Singapore, 1819–1990. *Journal of Biogeography* 19: 411–420.

Cornelius, J. M. and J. F. Reynolds. 1991. On determining the statistical significance of discontinuities with ordered ecological data. *Ecology* 72: 2057–2070.

Cotgreave, P. and P. Stockley. 1994. Body size, insectivory and abundance in assemblages of small mammals. *Oikos* 71: 89–96.

Courchamp, F., J.-L. Chapuis and M. Pascal. 2003. Mammal invaders on islands: Impact, control and control impact. *Biological Reviews* 78: 347–383.

Cowan, I. McT. and C. J. Guiget. 1956. *The Mammals of British Columbia.* British Columbia Provincial Museum Handbook No. 11.

Cowie, R. H. and B. S. Holland. 2006. Dispersal is fundamental to biogeography and the evolution of biodiversity on oceanic islands. *Journal of Biogeography* 33: 193–198.

Cowie, R. H. and B. S. Holland. 2008. Molecular biogeography and diversification of the endemic terrestrial fauna of the Hawaiian Islands. *Philosophical Transactions of the Royal Society*, Series B 363: 3363–3376.

Cowles, H. C. 1889. The ecological relations of the vegetation of the sand dunes of Lake Michigan. *Botanical Gazette* 27: 95–117, 167–202, 281–308, 361–391.

Cowles, H. C. 1901. The physiographic ecology of Chicago and vicinity: A study of the origin, development and classification of plant societies. *Botanical Gazette* 31: 73–108, 145–182.

Cowlishaw, G. and J. E. Hacker. 1997. Distribution, diversity and latitude in African primates. *American Naturalist* 150: 505–512.

Cox, A. 1973c. *Plate Tectonics and Geomagnetic Reversals.* San Francisco: W. H. Freeman.

Cox, C. B. 1973a. The distribution of Triassic terrestrial tetrapod families. In D. H. Tarling and S. K. Runcorn (eds.), *Implications of Continental Drift to the Earth Sciences,* vol. 1, 369–371. New York: Academic Press.

Cox, C. B. 1973b. Triassic tetrapods. In A. Hallam (ed.), *Atlas of Palaeobiogeography,* 213–223. Amsterdam: Elsevier.

Cox, G. W. 2004. *Alien Species and Evolution: The Evolutionary Ecology of Exotic Plants, Animals, Microbes, and Interacting Native Species.* Washington, DC: Island Press.

Cox, G. W. and R. E. Ricklefs. 1977. Species diversity and ecological release in Caribbean land bird faunas. *Oikos* 28: 113–122.

Cox, G. W., L. C. Contreras and A. V. Milewski. 1994. Role of fossorial animals in community structure and energetics of Mediterranean ecosystems. In M. T. Kilinin de Arroyo, P. H. Zedler and M. D. Fox (eds.), *Ecology of Convergent Ecosystems: Mediterranean-Climate Ecosystems of Chile, California and Australia,* 383–398. New York: Springer-Verlag.

Cox, J. G. and S. L. Lima. 2006. Naiveté and an aquatic-terrestrial dichotomy in the effects of introduced predators. *Trends in Ecology and Evolution* 21: 674–680.

Coyne, J. A. and H. A. Orr. 2004. *Speciation.* Sunderland, MA: Sinauer Associates.

Cracraft, J. 1983. Species concepts and speciation analyses. *Current Ornithology* 1: 159–187.

Cracraft, J. 2001. Avian evolution, Gondwana biogeography and the Cretaceous-Tertiary mass extinction event. *Proceedings of the Royal Society of London,* Series B-Biological Sciences 268: 459–469.

Cracraft, J. and M. J. Donoghue. 2004. *Assembling the Tree of Life.* New York: Oxford University Press.

Craig, M. T., P. A. Hastings and D. J. Pondella. 2004. Speciation in the Central American Seaway: The importance of taxon sampling in the identification of trans-isthmian geminate pairs. *Journal of Biogeography* 31: 1085–1091.

Crame, J. A. 2001. Taxonomic diversity gradients through geologic time. *Diversity and Distributions* 7: 175–189.

Crame, J. A. 2002. Evolution of taxonomic diversity gradients in the marine realm: A comparison of Late Jurassic and Recent bivalve faunas. *Paleobiology* 28: 184–207.

Crame, J. A. 2004. Pattern and process in marine biogeography: A view from the poles. In M. V. Lomolino and L. R. Heaney (eds.), *Frontiers of Biogeography I: New Directions in the Geography of Nature,* 272–292 Sunderland, MA: Sinauer Associates.

Crame, J.A. and B. R. Rosen. 2002. Cenozoic palaeogeography and the rise of modern biodiversity patterns. In J. A. Crame and A. W. Owen (eds.), *Palaeobiogeography and Biodiversity Change: the Ordovician and Mesozoic-Cenozoic Radiations.* Geological Society, London, Special Publications, No. 194, 153–168.

Crandall, K. A. and A. R. Templeton. 1996. Applications of intraspecific phylogenetics. In P. H. Harvey, A. J. Leigh Brown, J. Maynard Smith and S. Nee (eds.), *New Uses for New Phylogenies,* 81–99. New York: Oxford University Press.

Crane, P. R. and S. Lidgard. 1989. Angiosperm diversification and paleolatitudinal gradients in Cretaceous floristic diversity. *Science* 246: 675–678.

Craw, D., C. P. Burridge, P. Upton, D. L. Rowe and J. M. Waters. 2008. Evolution of biological dispersal corridors through a tectonically active mountain range in New Zealand. *Journal of Biogeography* 35: 1790–1802.

Craw, R. C. 1982. Phylogenetics, areas, geology and the biogeography of Croizat: A radical view. *Systematic Zoology* 31: 304–316.

Craw, R. C. 1988. Panbiogeography: Method and synthesis in biogeography. In A. A. Myers and P. S. Giller (eds.), *Analytical Biogeography: An Integrated Approach to the Study of Animal and Plant Distributions,* 405–435. London: Chapman and Hall.

Craw, R. C., J. R. Grehan and M. J. Heads. 1999. *Panbiogeography: Tracking the History of Life.* New York: Oxford University Press.

Creer, K. M., E. Irving and S. K. Runcorn. 1954. The direction of the geomagnetic field in remote epochs in Great Britain. *Journal of Geomagnetism and Geoelectricity* 6: 163–168.

Creer, K. M., E. Irving and S. K. Runcorn. 1957. Geophysical interpretation of palaeomagnetic directions from Great Britain. *Philosophical Transactions of the Royal Society of London,* Series A 250: 144–156.

Cressie, N. 1991. *Statistics for Spatial Data.* New York: John Wiley and Sons.

Crisci, J. V., L. Katinas and P. Posadas. 2003. *Historical Biogeography: An Introduction.* Cambridge, MA: Harvard University Press.

Crisp, M. D., S. Laffan, H. P. Linder and A. Monro. 2001. Endemism in the Australian flora. *Journal of Biogeography* 28: 183–198.

Critchfield, W. B. and E. J. Little. 1966. *Geographic Distributions of Pines of the World.* USDA Forest Service, Miscellaneous Publication 991.

Croizat, L. 1952. *Manual of Phytogeography.* The Hague: Dr. W. Junk.

Croizat, L. 1958. *Panbiogeography.* 2 vol. Caracas: Published by the author.

Croizat, L. 1960. *Principia Botanica.* Caracas: Published by the author.

Croizat, L. 1964. *Space, Time, Form: The Biological Synthesis.* Caracas: Published by the author.

Croizat, L. 1982. Vicariance/vicariism, panbiogeography, "vicariance biogeography," etc: A clarification. *Systematic Zoology* 31: 291–304.

Croizat, L., G. J. Nelson and D. E. Rosen. 1974. Centers of origin and related concepts. *Systematic Zoology* 23: 265–287.

Croll, J. 1875. *Climate and Time in Their Geological Relations.* London: Edward Stanford.

Cronquist, A. 1978. Once again, what is a species? In L. V. Knutson (ed.), *BioSystematics in Agriculture*, 3–20. Allenheld Osmum, Montclair.

Crosby, G. T. 1972. Spread of the cattle egret in the Western Hemisphere. *Bird-Banding* 43: 205–212.

Crossman, E. J. 1991. Introduced freshwater fishes: A review of the North American perspective with emphasis on Canada. *Canadian Journal of Fisheries and Aquatic Sciences* 48 (suppl. 1): 46–57.

Crowell, K. L. 1962. Reduced interspecific competition among the birds of Bermuda. *Ecology* 43: 75–88.

Crowell, K. L. 1983. Islands—insights or artifacts? Population dynamics and habitat utilization in insular rodents. *Oikos* 41: 442–454.

Crowell, K. L. 1986. A comparison of relict vs. equilibrium models for insular mammals of the Gulf of Maine. *Biological Journal of the Linnean Society* 28: 37–64.

Crowell, K. L. and S. L. Pimm. 1976. Competition and niche shifts of mice introduced onto small islands. *Oikos* 27: 251–258.

Cruden, R. W. 1966. Birds as agents of long-distance dispersal for disjunct plant groups of the temperate Western Hemisphere. *Evolution* 20: 517–532.

Cuellar, O. 1977. Animal parthenogenesis. *Science* 197: 837–843.

Culotta, E. 1994. Is marine diversity at risk? *Science* 263: 918–920.

Culver, D. C. 1970. Analysis of simple cave communities. I. Caves as islands. *Evolution* 29: 463–474.

Cumber, R. A. 1953. Some aspects of the biology and ecology of bumblebees bearing upon the yields of red clover seed in New Zealand. *New Zealand Journal of Science and Technology* 34: 227–240.

Cunningham, C. W. and T. M. Collins. 1994. Developing model systems for molecular biogeography: Vicariance and interchange in marine invertebrates. In B. Schierwater, B. Streit, G. P. Wagner and R. DeSalle (eds.), *Molecular Ecology and Evolution: Approaches and Applications*. Basel, Switzerland: Birkhauser.

Cushman, J. H., J. H. Lawton and B. F. J. Manly. 1993. Latitudinal patterns in European ant assemblages: Variation in species richness and body size. *Oecologia* 95: 30–37.

Cushman, S. A. 2005. Effects of habitat loss and fragmentation on amphibians: A review and prospectus. *Biological Conservation* 128: 231–240.

Cutler, A. 1991. Nested faunas and extinction in fragmented habitats. *Conservation Biology* 5: 496–505.

D'Hondt, S. and M. A. Arthur. 1996. Late Cretaceous oceans and the cool tropic paradox. *Science* 271: 1838–1841.

Dahlberg, A. and H. Croneborg. 2003. *Thirty-three Threatened Fungi of Europe*. European Council for Conservation of Fungi, Sweden.

Dalen, L., V. Nystrom, C. Valdiosera, M. Germonpre, M. Sablin, E. Turner, A. Angerbjorn, J. L. Arsuaga and A. Gotherstrom. 2007. Ancient DNA reveals lack of postglacial habitat tracking in the arctic fox. *Proceedings of the National Academy of Sciences, USA* 104: 6726–6729.

Dalrymple, G. B., E. A. Silver and E. D. Jackson. 1973. Origin of the Hawaiian Islands. *American Scientist* 61: 294–308.

Dammermann, K. W. 1948. The fauna of Krakatau, 1883–1933. *Koninklijke Nederlandsche Akademie Wetenschappen Verhandelingen* 44: 1–594.

Damuth, J. 1991. Of size and abundance. *Nature* 351: 268–269.

Damuth, J. D. 1993. Cope's rule, the island rule and the scaling of mammalian population density. *Nature* 365: 748–750.

Dana, J. D. 1853. On an isothermal oceanic chart, illustrating geographic distribution of marine animals. *The American Journal of Science and Arts*, 2nd Series, 66: 153–167, 314–327.

Daniels, R. J. R. 1992. Geographic distribution patterns of amphibians in the western Ghats, India. *Journal of Biogeography* 19: 521–529.

Dansereau, P. M. 1957. *Biogeography: An Ecological Perspective*. New York: Ronald Press.

Darimont, C. T., S. M. Carlson, M. T. Kinnison, P. C. Paquet, T. E. Reimchen and C. C. Wilmers. 2009. Human predators outpace other agents of trait change in the wild. *Proceedings of the National Academy of Sciences, USA* 106: 952–954.

Darlington, P. J. Jr. 1938. The origin of the fauna of the Greater Antilles, with discussion of dispersal of animals over water and through the air. *Quarterly Review of Biology* 13: 274–300.

Darlington, P. J. Jr. 1957. *Zoogeography: The Geographical Distribution of Animals*. New York: John Wiley & Sons.

Darlington, P. J. Jr. 1959a. Darwin and zoogeography. *Proceedings of the American Philosophical Society* 103: 307–319.

Darlington, P. J. Jr. 1959b. Area, climate and evolution. *Evolution* 13: 488–510.

Darlington, P. J. Jr. 1965. *Biogeography of the Southern End of the World: Distribution and History of Far Southern Life and Land, with an Assessment of Continental Drift*. Cambridge, MA: Harvard University Press.

Darwin, C. 1839. *Journal of the Researches into the Geology and Natural History of Various Countries Visited by H.M.S. Beagle, under the Command of Captain Fitzroy, R.N. from 1832 to 1836*. London: Henry Colburn.

Darwin, C. 1859. *On the Origin of Species by Means of Natural Selection or the Preservation of Favored Races in the Struggle for Life*. London: John Murray.

Darwin, C. 1860. *The Voyage of the Beagle*. New Jersey: Doubleday.

Darwin, C. 1862. *On the Various Contrivances by Which British and Foreign Orchids Are Fertilised by Insects, and on the Good Effects of Intercrossing*. London: John Murray.

Daubenmire, R. 1978. *Plant Geography with Special Reference to North America*. New York: Academic Press.

Dávalos, L. M. 2007. Short-faced bats (Phyllostomidae: Stenodermatina): A Caribbean radiation of strict frugivores. *Journal of Biogeography* 34: 364–375.

Davidson, A. D., M. J. Hamilton, A. G. Boyer, J. H. Brown and G. Ceballos. 2009. Multiple ecological pathways to extinction in mammals. *Proceedings of the National Academy of Sciences, USA* 106: 10702–10705.

Davies, N. 1995. *The Incas*. Niwat, CO: University Press of Colorado.

Davies, S. J. J. F. 2003. Moas. In M. Hutchins (ed.), *Grzimek's Animal Life Encyclopedia*, volume 8. (Birds I. Tinamous and Ratites to Hoatzins). 2nd edition. Farmington Hills, MI: Gale Group.

Davis, A. L. V., C. H. Scholtz and S. L. Chown. 1999. Species turnover, community boundaries and biogeographical composition of dung beetle assemblages across an altitudinal gradient in South Africa. *Journal of Biogeography* 26: 1039–1055.

Davis, M. B. 1969. Palynology and environmental history during the Quaternary period. *American Scientist* 57: 317–332.

Davis, M. B. 1976. Pleistocene geography of temperate deciduous forests. *Geoscience and Man* 13: 13–26.

Davis, M. B. 1981. Quaternary history and the stability of forest communities. In D. C. West, H. H. Shugart and D. B. Botkin (eds.), *Forest Succession: Concepts and Application*, 132–153. New York: Springer-Verlag.

Davis, M. B. 1986. Climatic stability, time lags and community disequilibrium. In J. M. Diamond and T. Case (eds.), *Community Ecology*, 269–284. New York: Harper and Row.

Davis, M. B. 1994. Ecology and paleoecology begin to merge. *Trends in Ecology and Evolution* 9: 357–358.

Davis, M. B. and R. G. Shaw. 2001. Range shifts and adaptive responses to Quaternary climate change. *Science* 292: 673–679.

Davis, R. 1996. The Pinalenos as an island in a montane archipelago. In C. A. Istock and R. S. Hoffmann (eds.), *Storm over a Mountain Archipelago: Conservation Biology and the Mt. Graham Affair*, 123–134. Tuscon: University of Arizona Press.

Davis, R. and C. Dunford. 1987. An example of contemporary colonization of montane islands by small, nonflying mammals in the American Southwest. *American Naturalist* 129: 398–406.

Davis, R. and D. E. Brown. 1989. Role of post-Pleistocene dispersal in determining the modern distribution of Albert's squirrel. *Great Basin Naturalist* 49: 425–434.

Davis, R. and J. R. Callahan. 1992. Post-Pleistocene dispersal in the Mexican vole (*Microtus mexicanus*): An example of an apparent trend in the distribution of southwestern mammals. *Great Basin Naturalist* 52: 262–268.

Dawson, W. R. and C. Carey. 1976. Seasonal acclimation to temperature in cardueline finches. *Journal of Comparative Physiology* 112: 317–333.

Dayan, T. 1990. Feline canines: Community-wise character displacement in the small cats of Israel. *American Naturalist* 136: 39–60.

Dayan, T., E. Tchernov, Y. Yom-Tov and D. Simberloff. 1989. Ecological character displacement in Saharo-Arabian *Vulpes*: Outfoxing Bergmann's rule. *Oikos* 55: 263–272.

Dayton, P. K. 1971. Competition, disturbance and community organization: The provision and subsequent utilization of space in a rocky intertidal community. *Ecological Monographs* 41: 351–389.

de Queiroz, K. 2005. The resurrection of oceanic dispersal in historical biogeography. *Trends in Ecology and Evolution* 20: 68–73.

de Queiroz, K. 2007. Species concepts and species delimitation. *Systematic Biology* 56: 879–886.

de Ruiter, P. C., V. Wolters and J. C. Moore (eds.). 2005. *Dynamic Food Webs*, volume 3: *Multispecies Assemblages, Ecosystem Development and Environmental Change*. Burlington, MA: Elsevier Inc.

de Vos, J., L. W. van den Hoek Ostende and G. D. van den Bergh. 2007. Patterns in insular evolution of mammals: A key to island paleogeography. In W. Renema (ed.), *Biogeography, Time and Place: Distributions, Barriers, and Islands*, 315–345. Netherlands: Springer.

de Vries, A. L. 1971. Freezing resistance in fishes. In W. S. Hoar and D. J. Randall (eds.), *Fish Physiology*, vol. 6, 157–190. New York: Academic Press.

de Weerd, D. R. U., D. Schneider and E. Gittenberger. 2005. The provenance of the Greek land snail species *Isabellaria pharsalica*: Molecular evidence of recent passive long-distance dispersal. *Journal of Biogeography* 32: 1571–1581.

de Wet, J. M. J. 1979. Origins of polyploids. In W. H. Lewis (ed.), *Polyploidy: Biological Relevance*, 3–15. New York: Plenum Press.

Debinski, D. M. and R. D. Holt. 2000. A survey and overview of habitat fragmentation experiments. *Conservation Biology* 14: 342–355.

Debruyne, R., G. Chu, C. E. King, K. Bos, M. Kuch, C. Schwarz, P. Szpak, D. R. Grocke, P. Matheus, G. Zazula, et al. 2008. Out of America: Ancient DNA evidence for a new world origin of late quaternary woolly mammoths. *Current Biology* 18: 1320–1326.

DeChaine, E. G. 2008. A bridge or a barrier? Beringia's influence on the distribution and diversity of tundra plants. *Plant Ecology & Diversity* 1: 197–207.

Dehant, V., K. C. Creager, S. Karato and S. Zatman. 2003. *Earth's Core: Dynamics, Structure and Rotation*. Washington, DC: American Geophysical Union.

Delcourt, H. R. and P. A. Delcourt. 1994. Postglacial rise and decline of *Ostrya virginiana* (Mill.) K. Koch and *Carpinus caroliniana* Walt. in eastern North America: Predictable responses of forest species to cyclic changes in seasonality of climate. *Journal of Biogeography* 21: 137–150.

Delcourt, P. A. and H. R. Delcourt. 1977. The Tunica Hills, Louisiana-Mississippi: Late glacial locality for spruce and deciduous forest species. *Quaternary Research* 7: 218–237.

Delcourt, P. A. and H. R. Delcourt. 1996. Quaternary paleoecology of the lower Mississippi Valley. *Engineering Geology* 45: 219–242.

Delsuc, F., H. Brinkmann and H. Philipp. 2005. Phylogenomics and the reconstruction of the tree of life. *National Review of Genetics* 6: 361–375.

Dengler, J. 2009. Which function describes the species–area relationship best? A review and empirical evaluation. *Journal of Biogeography* 36: 728–744.

Deplus, C. 2001. Plate tectonics: Indian Ocean actively deforms. *Science* 292: 1850–1851.

DeSalle, R., T. Freedman, E. M. Prager and A. C. Wilson. 1987. Tempo and mode of sequence evolution in mitochondrial DNA of Hawaiian *Drosophila*. *Journal of Molecular Evolution* 26: 157–164.

Desbruyeres, D., Almeida, A., Biscoito, M., Comtet, T., Khripounoff, A., LeBris, N., Sarradin, P.M., Segonzac, M. 2000. A review of the distribution of hydrothermal vent communities along the northern Mid-Atlantic Ridge: Dispersal vs. environmental controls. *Hydrobiologia* 440(1–3): 201–216.

Desmond, A. 1965. How many people have ever lived on earth? In L. K. Y. Ng and S. Mudd (eds.), *The Population Crisis: Implications and Plans for Action*. Bloomington: Indiana University Press.

Desprez-Loustau, M., C. Robin, M. Buee, R. Courtecuisse, J. Garbaye, F. Suffert, I. Sache and D. M. Rizzo. 2007. The fungal dimension of biological invasions. *Trends in Ecology and Evolution* 22: 472–480.

Detroit, F., E. Dizon, F. Semah, C. Falgueres, S. Hameau, W. Ronquillo and E. Cabanis. 2002. Notes on the morphology and age of the Tabon Cave fossil *Homo sapiens*. *Current Anthropology* 43: 660–666.

Dettmann, M. E., D. T. Pocknall, E. J. Romero and M. C. Zamaloa. 1990. Nothofagidites Erdtman ex Potonié, 1960; a catalogue of species with notes on the paleogeographic distribution of Nothofagus Bl. (Southern Beech). *New Zealand Geological Survey Paleontological Bulletin* 90: 1–79.

Dewey, J. F. 1977. Suture zone complexities: A review. *Tectonophysics* 40: 53–67.

deWit, R. and T. Bouvier. 2006. "Everything is everywhere, but, the environment selects": What did Baas Beking and Beijerinck really say? *Environmental Microbiology* 8: 755–758.

Dexter, R. W. 1978. Some historical notes on Louis Agassiz's lectures on zoogeography. *Journal of Biogeography* 5: 207–209.

Diamond, J. M. 1969. Avifaunal equilibria and species turnover rates on the Channel Islands of California. *Proceedings of the National Academy of Sciences, USA* 64: 57–63.

Diamond, J. M. 1970a. Ecological consequences of island colonization by Southwest Pacific birds. I. Types of niche shifts. *Proceedings of the National Academy of Sciences, USA* 67: 529–536.

Diamond, J. M. 1970b. Ecological consequences of island colonization by Southwest Pacific birds. II. The effect of species diversity on total population density. *Proceedings of the National Academy of Sciences, USA* 67: 1715–1721.

Diamond, J. M. 1972. Biogeographic kinetics: Estimation of relaxation times for avifaunas of Southwest Pacific islands. *Proceedings of the National Academy of Sciences, USA* 69: 3199–3203.

Diamond, J. M. 1973. Distributional ecology of New Guinea birds. *Science* 179: 759–769.

Diamond, J. M. 1974. Colonization of exploded volcanic islands by birds: The supertramp strategy. *Science* 184: 803–806.

Diamond, J. M. 1975a. The island dilemma: Lessons of modern biogeographic studies for the design of natural reserves. *Biological Conservation* 7: 129–146.

Diamond, J. M. 1975b. Assembly of species communities. In M. L. Cody and J. M. Diamond (eds.), *Ecology and Evolution of Communities*, 342–444. Cambridge, MA: Belknap Press.

Diamond, J. M. 1977. Colonization cycles in man and beast. *World Archaeology* 8: 249–261.

Diamond, J. M. 1984. "Normal" extinctions of isolated populations. In M. N. Nitecki (ed.), *Extinctions*, 191–246. Chicago: University of Chicago Press.

Diamond, J. M. 1991a. Did Komodo dragons evolve to eat pygmy elephants? *Nature* 326: 832.

Diamond, J. M. 1991b. A new species of rail from the Solomon Islands and convergent evolution of insular flightlessness. *Auk* 108: 461–470.

Diamond, J. M. 1997. Guns, Germs and Steel: The Fates of Human Societies. New York: W. W. Norton.

Diamond, J. M. 2005. *Collapse: How Societies Chose to Fail or Succeed.* New York: Viking Press.

Diamond, J. M. and R. M. May. 1976. Island biogeography and the design of natural preserves. In R. M. May (ed.), *Theoretical Ecology: Principles and Applications*, 163–186. Philadelphia: W. B. Saunders Co.

Diaz, R. J. and R. Rosenberg. 2008. Spreading dead zones and consequences for marine ecosystems. *Science* 321: 926–929.

Dice, L. R. 1947. Effectiveness of selection by owls of deer mice (*Peromyscus maniculatus*) which contrast in color with their background. *Contributions, Laboratory Vertebrate Biology, University of Michigan* 50: 1–15.

Dice, L. R. and P. M. Blossom. 1937. *Studies of Mammalian Ecology in Southwestern North America with Special Attention to the Colors of Desert Mammals.* Carnegie Institute Publication 485: 1–29. Washington, DC: Carnegie Institute.

Dickison, M. R. 2007. *The allometry of giant flightless birds.* PhD dissertation. Duke University.

Dietl, G. P. 2003. The escalation hypothesis: One long argument. *Palaios* 18: 83–86.

Dietl, G. P. and P. H. Kelley. 2002. The fossil record of predator-prey arms races: Coevolution and escalation hypotheses. In M. Kowalewski and P. H. Kelley (eds.), *The Fossil Record of Predation: The Paleontological Society Papers* 8: 353–374.

Dietz, R. S. 1961. Continental and ocean basin evolution by spreading of the sea floor. *Nature* 190: 854–857.

Diniz-Filho, J. A. F., P. Carvalho, L. M. Bini and N. M. Torres. 2005. Macroecology, geographic range size–body size relationship and minimum viable population analysis for New World carnivore. *Acta Oecologia* 17: 25–30.

Diver, K. C. 2008. Not as the crow flies: Assessing effective isolation for island biogeographical analysis. *Journal of Biogeography* 35: 1040–1048.

Dixon, A. F. G., P. Kindlmann, J. Leps and J. Holman. 1987. Why are there so few species of aphids, especially in the tropics. *American Naturalist* 129: 580–592.

Doak, D. F. and L. S. Mills. 1994. A useful role for theory in conservation. *Ecology* 75: 615–626.

Dobzhansky, T. 1937. *Genetics and the Origin of Species.* New York: Columbia University Press.

Dobzhansky, T. 1950. Evolution in the tropics. *American Scientist* 38: 209–221.

Dobzhansky, T. 1951. *Genetics and the Origin of Species.* 3rd edition. New York: Columbia University Press.

Docters van Leeuwen, W. M. 1936. Krakatau, 1883–1933. *Ann. Jard. Bot. Buitenzorg* 46–47: 1–506.

Dodd, A. P. 1959. The biological control of prickly pear in Australia. In A. Keast (ed.), *Biogeography and Ecology in Australia*, 565–577. Monographiae Biologicae 8. The Hague: Dr. W. Junk.

Doi, H. and M. Takahashi. 2008. Latitudinal patterns in the phonological responses of leaf colouring and leaf fall to climate change in Japan. *Global Ecology and Biogeography* 17: 556–561.

Donlan, C. J. and C. Wilcox. 2008. Diversity, invasive species and extinction in insular ecosystems. *Journal of Applied Ecology* 45: 1114–1123.

Donoghue, M. J. and B. R. Moore. 2003. Toward an integrative historical biogeography. *Integrative and Comparative Biology* 43: 261–270.

Donoghue, M. J. and S. A. Smith. 2004. Patterns in the assembly of temperate forests around the Northern Hemisphere. *Philosophical Transactions of the Royal Society of London*, Series B 359: 1633–1644.

Dorst, J. 1962. *The Migrations of Birds.* Boston: Houghton Mifflin.

Dragoo, J. W., J. A. Lackey, K. E. Moore, E. P. Lessa, J. A. Cook and T. L. Yates. 2006. Phylogeography of the deer mouse (Peromyscus maniculatus) provides a predictive framework for research on hantaviruses. *Journal of General Virology* 87: 1997–2003.

Drake, D. R. and T. L. Hunt. 2009. Invasive rodents on islands: Integrating historical and contemporary ecology. *Biological Invasions* 11: 1483–1487.

Drake, J. A., H. A. Mooney, F. di Castri, R. H. Groves, F. J. Kruger, M. Rejmanek and M. Williamson (eds.). 1989. *Biological Invasions: A Global Perspective.* New York: John Wiley & Sons.

Drezner, T. D. and R. C. Balling Jr. 2008. Regeneration cycles of the keystone species *Carnegiea gigantean* are linked to worldwide volcanism. *Journal of Vegetation Science* 19: 587–596.

Drude, O. 1887. *Atlas der Pflanzenverbreitung.* 5 Abt. Gotha, Berghaus Physikalischer Atlas.

Drummond, A. J. and A. Rambaut. 2007. BEAST: Bayesian evolutionary analysis by sampling trees. *BMC Evolutionary Biology* 7.

Drury, W. H. and I. C. T. Nisbet. 1973. Succession. *Journal of the Arnold Arboretum* 54: 331–368.

du Toit, A. L. 1927. *A Geological Comparison of South America with South Africa.* Publication no. 381, 1–157. Washington, DC: Carnegie Institute.

du Toit, A. L. 1937. *Our Wandering Continents.* Edinburgh: Oliver & Boyd.

Ducharme, M. B., J. Larochelle and D. Richard. 1989. Thermogenic capacity in gray and black morphs of the gray squirrel, *Sciurus carolinensis. Physiological Zoology* 62: 1273.

Due, A. D. and G. A. Polis. 1986. Trends in scorpion diversity along the Baja California Peninsula. *American Naturalist* 128: 460–468.

Duellman, W. E. and L. Trueb. 1986. *Biology of Amphibians.* London, New York: McGraw-Hill.

Duggen, S., K. Hoernle, P. van den Bogaard, L. Rupke and J. P. Morgan. 2003. Deep roots of the Messinian salinity crisis. *Nature* 422: 602–606.

Duncan, R. P., T. M. Blackburn and T. H. Worthy. 2002. Prehistoric bird extinctions and human hunting. *Proceedings of the Royal Society of London*, Series B 269: 517–521.

Duncan, R. P., M. Bomford, D. M. Forsyth and L. Conibear. 2002. High predictability in introduction outcomes and the geographical range size of introduced Australian birds: A role of climate. *Journal of Animal Ecology* 70: 621–632.

Dunn, C. P. and C. Loehle. 1988. Species-area parameter estimation: Testing the null model of lack of relationships. *Journal of Biogeography* 15: 721–728.

Dunn, P. O., K. J. Thusius, K. Kimber and D. W. Winkler. 2000. Geographic and ecological variation in clutch size of tree swallows. *The Auk* 117: 215–221.

Dunne, J. A., R J Williams and N. D.Martinez. 2002. Food-web structure and network theory: The role of connectance and size. *Proceedings of the National Academy of Sciences, USA* 99: 12917–12923.

Dunson, W. A. (ed.) 1975. *The Biology of Sea Snakes*. Baltimore, MD: University Park Press.

Dunson, W. A. and F. J. Mazotti. 1989. Salinity as a limiting factor in the distribution of reptiles in Florida Bay: A theory for the estuarine origin of marine snakes and turtles. *Bulletin of Marine Science* 44: 229–244.

Durham, J. W. and F. S. MacNeil. 1967. Cenozoic migrations of marine invertebrates through the Bering Strait region. In D. M. Hopkins (ed.), *The Bering Land Bridge*, 326–349. Stanford, CA: Stanford University Press.

Duxbury, A. 2000. *An Introduction to the World's Oceans*. 3rd edition. Berkshire: McGraw-Hill.

Dyer, K. R. 1986. *Coastal and Estuarine Sediment Dynamics*. Chichester, UK: John Wiley & Sons.

Dyke, G. J. 2003. The fossil record and molecular clocks: basal radiations within the Neornithes. In P. C. J. Donoghue and M. P. Smith (eds.), *Telling the Evolutionary Time: Molecular Clocks and the Fossil Record*, 263–277. Boca Raton: CRC Press.

Ebach, M. C. and C. J. Humphries. 2002. Cladistic biogeography and the art of discovery. *Journal of Biogeography* 29: 427–444.

Ebach, M. C. and D. Goujet. 2006. The first biogeographical map. *Journal of Biogeography* 33: 761–769.

Ebach, M. C. and J. J. Morrone. 2005. Forum on historical biogeography: What is cladistic biogeography? *Journal of Biogeography* 32: 2179–2183.

Ebenhard, T. 1988a. *Demography and Island Colonization of Experimentally Introduced and Natural Vole Populations*. Ph.D. dissertation, Uppsala University, Sweden.

Ebenhard, T. 1988b. Introduced birds and mammals and their ecological effects. *Swedish Wildlife Research* 13: 1–107.

Echelle, A. A. and T. E. Dowling. 1992. Mitochondrial-DNA variation and evolution of the Death Valley pupfishes (Cyprinodon, Cyprinodontidae). *Evolution* 46: 193–206.

Edgecombe, G. D. 2010. Palaeomorphology: Fossils and the inference of cladistic relationships. *Acta Zoologica* 91: 72–80.

Edwards, G. P., A. R. Pople, K. Saalfeld and P. Caley. 2004. Introduced mammals in Australian rangelands: Future threats and the role of monitoring programmes in management strategies. *Austral Ecology* 29: 40–50.

Edwards, S. V. and P. Beerli. 2000. Perspective: Gene divergence, population divergence and the variance in coalescence time in phylogeography studies. *Evolution* 54: 1839–1854.

Eeley, H. A. C. and Laws, M. J. 1999. Large-scale patterns of species richness and species range size in anthropoid primates. In J. G. Fleagle, C. H. Janson and K. Reed, (eds.), *Primate Communities*, 191–219. Cambridge University Press, Cambridge.

Ehrendorfer, F. 1979. Reproductive biology in island plants. In D. Bramwell (ed.), *Plants and Islands*, 293–306. London: Academic Press.

Ehrlich, P. R. 1961. Intrinsic barriers to dispersal in the checkerspot butterfly. *Science* 134: 108–109.

Ehrlich, P. R. 1964. Some axioms of taxonomy. *Systematic Zoology* 13: 109–123.

Ehrlich, P. R. 1965. The population biology of the butterfly *Euphydryas editha*. II. The structure of the Jaspar Ridge colony. *Evolution* 19: 327–336.

Ehrlich, P. R. and R. W. Holm. 1963. *The Process of Evolution*. New York: McGraw-Hill.

Ehrlich, P. R., R. R. White, M. C. Singer, S. W. McKechnie and L. E. Gilbert. 1975. Checkerspot butterflies: A historical perspective. *Science* 188: 221–228.

Eisenberg, J. F. 1981. *The Mammalian Radiations: An Analysis of Trends in Evolution, Adaptation and Behavior*. Chicago: University of Chicago Press.

Eizirik, E., W. J. Murphy and S. J. O'Brien. 2001. Molecular dating and biogeography of the early placental mammal radiation. *Journal of Heredity* 92: 212–219.

Ekman, S. 1953. *Zoogeography of the Sea*. London: Sidgwick & Jackson.

Eldredge, N. and I. Cracraft. 1980. *Phylogenetic Patterns and the Evolutionary Process*. New York: Columbia University Press.

Eldredge, N. and S. J. Gould. 1972. Punctuated equilibria: An alternative to phyletic gradualism. In T. J. M. Schopf (ed.), *Models in Paleobiology*, 82–115. San Francisco: Freeman, Cooper & Co.

Elena, S. F., V. S. Cooper and R. E. Lenski. 1996. Punctuated evolution caused by selection of rare beneficial mutations. *Science* 272: 1802–1804.

Elewa, A. M. T. (ed.). 2005. *Migration of Organisms: Climate, Geography and Ecology*. New York: Springer.

Elias, S. A. 1997. *The Ice-Age History of Southwestern Parks*. Washington, DC: Smithsonian Institution Press.

Elias, S. A., S. K. Short and C. H. Nelson. 1996. Life and times of the Bering landbridge. *Nature* 382: 60–63.

Elliot, D. H., E. H. Colbert, W. J. Breed, I. A. Jensen and T. S. Powell. 1970. Triassic tetrapods from Antarctica: Evidence for continental drift. *Science* 169: 197–201.

Elliot, G. F. 1951. On the geographical distribution of terebratelloid brachiopods. *Annals and Magazines of Natural History*, Series 12. 4: 305–334.

Ellis, E. C. and N. Ramankutty. 2008. Putting people in the map: Anthropogenic biomes of the world. *Frontiers in Ecology and the Environment* 6: 439–447.

Elmberg, J., P. Nummi, H. Paysa and K. Sjoberg. 1994. Relationship between species number, lake size and resource diversity in assemblages of breeding waterfowl. *Journal of Biogeography* 21: 75–84.

Elton, C. 1927. *Animal Ecology*. New York: Macmillan.

Elton, C. S. 1958. *The Ecology of Invasions by Animals and Plants*. London: Methuen & Co.

Elton, C. S. 1966. *The Patterns of Animal Communities*. London: Methuen & Co.

Embleton, B. J. J. 1973. The palaeolatitude of Australia through Phanerozoic time. *Journal of the Geological Society of Australia* 19: 475–482.

Emerson, B. C. 2002. Evolution on oceanic islands: Molecular phylogenetic approaches to understanding pattern and process. *Molecular Ecology* 11: 951–966.

Emlen, J. T. 1978. Density anomalies and regulation mechanisms in land bird populations on the Florida peninsula. *American Naturalist* 112: 265–286.

Emlen, J. T. 1979. Land bird densities on Baja California islands. *Auk* 96: 152–167.

Emslie, S. D. and G. S. Morgan. 1994. A catastrophic death assemblage and paleoclimatic implications of Pliocene seabirds of Florida. *Science* 264: 684–685.

Emslie, S. D., M. Stiger and E. Wambach. 2005. Packrat middens and late Holocene environmental change in southwestern Colorado. *Southwestern Naturalist* 50: 209–215.

Enckell, P. H., S. A. Bengtson and B. Wiman. 1987. Serf and waif colonization, distribution and dispersal of invertebrate species in Faroe Island settlement areas. *Journal of Biogeography* 14: 89–104.

Endler, J. A. 1977. *Geographic Variation, Speciation and Clines.* Monographs in Population Biology, no. 10. Princeton, NJ: Princeton University Press.

Engel, S. R., K. M. Hogan, J. F. Taylor and S. K. Davis. 1998. Molecular systematics and paleobiogeography of the South American sigmodontine rodents. *Molecular Biology and Evolution* 15: 35–49.

Enghoff, H. 1996. Widespread taxa, sympatry, dispersal and an algorithm for resolved area cladograms. *Cladistics* 12: 349–364.

Enquist, B. J., M. A. Jordan and J. H. Brown. 1995. Connections between ecology, biogeography and paleobiology: Relationship between local abundance and geographic distribution in fossil and recent molluscs. *Evolutionary Ecology* 9: 586–604.

Environmental Science for Social Change. 1999. *Decline of Philippine Forests.* Map. Makati: Bookmark.

Erickson, R. O. 1945. The *Clematis fremontii* var. *riehlii* population in the Ozarks. *Annals of the Missouri Botanical Garden* 32: 413–460.

Erlandson, J. M. 2002. Anatomically modern humans, maritime migrations, and the peopling of the New World. In N. Jablonski (ed.), *The First Americans: The Pleistocene Colonization of the New World,* 59–92. San Francisco, CA: Memoirs of the California Academy of Sciences.

Erlandson, J. M., M. H. Graham, B. J. Bourque, D. Corbett, J. A. Estes and R. S. Steneck. 2007. The kelp highway hypothesis: Marine ecology, the coastal migration theory, and the peopling of the Americas. *Journal of Island and Coastal Archaeology* 2: 161–174.

Erlinge, S. 1987. Why do European stoats *Mustela erminea* not follow Bergmann's rule? *Holarctic Ecology* 10: 33–39.

Erulkar, S. D. 1972. Comparative aspects of spatial localization of sound. *Physiological Reviews* 52: 237–360.

Erwin, D. H. 1993. *The Great Paleozoic Crisis: Life and Death in the Permian.* New York: Columbia University Press.

Erwin, T. C. 1981. Taxon pulses, vicariance and dispersal: An evolutionary synthesis illustrated by carabid beetles. In G. Nelson and D. E. Rosen (eds.), *Vicariance Biogeography: A Critique,* 159–196. New York: Columbia University Press.

Esselstyn, J. A., R. M. Timm and R. M. Brown. 2009. Do geological or climatic processes drive speciation in dynamic archipelagoes? The tempo and mode of diversification in Southeast Asian shrews. *Evolution* 63: 2595–2610.

Estes, J. A. and D. O. Duggins. 1995. Sea otters and kelp forests in Alaska: Generality and variation in a community ecology paradigm. *Ecological Monographs* 65: 75–100.

Estes, R. and A. Baez. 1985. Herptofaunas of North and South America during the Late Cretaceous and Cenozoic: Evidence for interchange? In F. G. Stehli and S. D. Webb (eds.), *The Great American Interchange,* 140–200. New York: Plenum Press.

Etges, W. J. and M. Levitan. 2008. Variable evolutionary response to regional climate change in a polymorphic species. *Biological Journal of the Linnean Society* 95: 702–718.

Evans, K. L. and K. J. Gaston. 2005. Can the evolutionary-rates hypothesis explain species-energy relationships? *Functional Ecology* 19: 899–915.

Facelli, J. and S. T. A. Pickett. 1990. Markovian chains and the role of history in succession. *Trends in Ecology and Evolution* 5: 27–30.

Faeth, S. H. 1984. Density compensation in vertebrates and invertebrates: A review and an experiment. In D. R. Strong, D. Simberloff, L. G. Abele and A. B. Thistle (eds.), *Ecological Communities: Conceptual Issues and the Evidence,* 491–509. Princeton, NJ: Princeton University Press.

Fagan, B. 2009. *The Complete Ice Age: How Climate Change Shaped the World.* London: Thames & Hudson.

Fagan, B. M. 1990. *The Journey from Eden: The Peopling of Our World.* London: Thames and Hudson.

Fairfield, K. N., M. E. Mort and A. Santos. 2004. Phylogenetics and evolution of the Macaronesian members of the genus *Aichryson* (Crassulaceae) inferred from nuclear and chloroplast sequence data. *Plant Systematics and Evolution* 248: 71–83.

Falla, R. A., R. B. Sibson and E. G. Turbott. 1966. *A Field Guide to the Birds of New Zealand and Outlying Islands.* London: William Collins Sons & Co.

Fallaw, W. C. 1979. Trans-North Atlantic similarity among Mesozoic and Cenozoic invertebrates correlated with widening of the ocean basin. *Geology* 7: 398–400.

Fallon, S. M., E. Bermingham and R. E. Ricklefs. 2003. Island and taxon effects in parasitism revisited: Avian malaria in the Lesser Antilles. *Evolution* 57: 606–615.

Fallon, S. M., E. Bermingham and R. E. Ricklefs. 2005. Host specialization and geographic localization of avian malaria parasites: A regional analysis in the Lesser Antilles. *American Naturalist* 165: 466–480.

Farrell, T. M. 1991. Models and mechanisms of succession: An example from a rocky intertidal community. *Ecological Monographs* 61: 95–113.

Felsenstein, J. 1985. Phylogenies and the comparative method. *American Naturalist* 125: 1–15.

Fenchel, T. and B. J. Finlay. 2004. The ubiquity of small species: Patterns of local and global diversity. *BioScience* 54(8): 777–784.

Fernandes, G. W. and A. C. F. Lara. 1993. Diversity of Indonesian gall-forming herbivores along altitudinal gradients. *Biodiversity Letters* 1: 186–192.

Fernandez, M. H. and E. S. Vrba. 2005a. Rapoport effect and biomic specialization in African mammals: Revisiting the climatic variability hypothesis. *Journal of Biogeography* 32: 903–918.

Fernandez, M. H. and E. S. Vrba. 2005b. Body size, biomic specialization and range size of African large mammals. *Journal of Biogeography* 32: 1243–1256.

Fichman, M. 1977. Wallace: Zoogeography and the problem of landbridges. *Journal of the History of Biology* 10: 45–63.

Fiedel, S. 2009. Sudden deaths: The chronology of terminal Pleistocene megafaunal extinction. In G. Haynes (ed.), *American Megafaunal Extinctions at the End of the Pleistocene,* 21–37. Springer.

Fields, P. A., J. B. Graham, R. H. Rosenblatt and G. N. Somero. 1993. Effects of expected global climate change on marine faunas. *Trends in Ecology and Evolution* 8: 361–366.

Figuerola, J. and A. J. Green. 2002. Dispersal of aquatic organisms by waterbirds: A review of past research and priorities for future studies. *Freshwater Biology* 47: 483–494.

Findley, I. S. and C. Jones. 1964. Seasonal distribution of the hoary bat. *Journal of Mammalogy* 45: 461–470.

Findley, J. S. 1969. Biogeography of southwestern boreal and desert mammals. In J. K. Jones Jr. (ed.), *Contributions in Mammalogy* 113–128. Lawrence: University of Kansas, Museum of Natural History Publication no. 51.

Finerty, J. P. 1980. *The Population Ecology of Cycles in Small Mammals: Mathematical Theory and Biological Fact.* New Haven, CT: Yale University Press.

Finlayson, C. 2004. *Neanderthals and Modern Humans: An Ecological and Evolutionary Perspective.* New York: Cambridge University Press.

Finlayson, C. 2005. Biogeography and evolution of the genus *Homo. Trends in Ecology and Evolution* 20: 457–463.

Fiorillo, A. R. 2008. Dinosaurs of Alaska: Implications for the Cretaceous origin of Beringia. In R. B. Blodgett and G. D. Stanley Jr. (eds.), *The Terrane Puzzle: New Perspectives on Paleontology and Stratigraphy from the North American Cordillera,*

313–326. Boulder, CO: Special Paper 442, The Geological Society of America.

Firth, R. and J. W. Davidson. 1945. *Pacific Islands*, vol. 1. General Survey. Naval Intelligence Division, London, 20.3.5.

Fischer, A. G. 1960. Latitudinal variation in organic diversity. *Evolution* 14: 64–81.

Fischer, A. G. 1984. The two Phanerozoic supercycles. In W. A. Berggren and J. A. Van Couvering (eds.), *Catastrophes and Earth History*, 129–150. Princeton, NJ: Princeton University Press.

Fischer, J. and D. B. Lindenmayer. 2002. Treating the nestedness temperature calculator as a "black box" can lead to false conclusions. *Oikos* 99: 193–199.

Fischer, J. and D. B. Lindenmayer. 2007. Landscape modification and habitat fragmentation: A synthesis. *Global Ecology and Biogeography* 16: 265–280.

Fisher, J. A. D., K. T. Frank and W. C. Leggett. 2009. Dynamic macroecology on ecological time-scales. *Global Ecology and Biogeography* 19: 1–15.

Fisher, J. A. D., K. T. Frank, B. Petrie, W. C. Leggett and N. L. Shackell. 2008. Temporal dynamics within a contemporary latitudinal diversity gradient. *Ecology Letters* 11: 883–897.

Fisher, J., K. L. Cole and R. S. Anderson. 2009. Using packrat middens to assess grazing effects on vegetation change. *Journal of Arid Environments* 73: 937–949.

Fitzhugh, W. W. 1997. Biogeographical archaeology in the Eastern North American Arctic. *Human Ecology* 25: 385–418.

Fitzpatrick, S. M. 2007. Archaeology's contribution to island studies. *Island Studies Journal* 2: 77–100.

Flannery, T. 2001. *The Eternal Frontier: An Ecological History of North America and Its Peoples*. New York: Grove Press.

Flannery, T. F. 1994. *The Future Eaters: An Ecological History of the Australasian Lands and People*. Kew, Victoria, N.S.W., Australia: Reed International Books.

Flather, C. H., L. A. Joyce and C. A. Bloomgarden. 1994. *Species Endangerment Patterns in the United States*. USDA Forest Service, General Technical Report RM-241.

Fleischer, R. C. and C. E. McIntosh. 2001. Molecular systematics and biogeography of the Hawaiian avifauna. *Studies of Avian Biology* 22: 51–60.

Fleischer, R. C., C. E. McIntosh and C. L. Tarr. 1998. Evolution on a volcanic conveyor belt: using phylogeographic reconstructions and K-Ar-based ages of the Hawaiian Islands to estimate molecular evolutionary rates. *Molecular Ecology* 7: 533–545.

Fleishman, E., G. T. Austin and A. D. Weiss. 1998. An empirical test of Rapoport's rule: Elevational gradients in montane butterfly communities. *Ecology* 79: 2482–2493.

Fleming, T. H. and A. Estrada (eds.). 1993. *Frugivory and Seed Dispersal: Ecological and Evolutionary Aspects*. Dordrecht, The Netherlands: Kluwer Academic.

Flenley, J. R. 1979a. *The Equatorial Rain Forest: A Geological History*. London: Butterworth.

Flenley, J. R. 1979b. The Late Quaternary vegetational history of the equatorial mountains. *Progress in Physical Geography* 3: 488–509.

Flessa, K. W. 1980. Biological effects of plate tectonics and continental drift. *Bioscience* 30: 518–523.

Flessa, K. W. 1981. The regulation of mammalian faunal similarity among the continents. *Journal of Biogeography* 8: 427–438.

Flessa, K. W., S. G. Barnett, D. B. Cornue, M. A. Lomaga, N. Lombardi, J. M. Miyazaki and A. S. Murer. 1979. Geologic implications of the relationship between mammalian faunal similarity and geographic distance. *Geology* 7: 15–18.

Flint, R. F. 1971. *Glacial and Quaternary Geology*. New York: John Wiley & Sons.

Flohn, H. 1969. *Climate and Weather*. New York: World University Library, McGraw-Hill.

Flux, J. E. C. 1969. Current work on the African hare, *Lepus capensis* L. in Kenya. *Journal of Reproduction and Fertilization* 6 (suppl.): 225–227.

Follinsbee, K. E. and D. R. Brooks. 2007. Miocene hominoid biogeography: Pulses of dispersal and differentiation. *Journal of Biogeography* 34: 383–397.

Foote, M. 2003. Origination and extinction through the Phanerozoic: A new approach. *Journal of Geology* 111: 125–148.

Foote, M., J. S. Crampton, A. G. Beu and R. A. Cooper. 2008. On the bidirectional relationship between geographic range and taxonomic duration. *Paleobiology* 34: 421–433.

Forbes, E. 1844. Report on the Mollusca and Radiata of the Aegean Sea. *Reports of the British Association of Science 1843 (1844)*, 130–193.

Forbes, E. 1856. Map of the distribution of marine life. In A. K. Johnston (ed.), *The Physical Atlas of Natural Phenomena*. Philadelphia: Lea and Blanchard.

Forman, R. T. T. (ed.). 1979. *Pine Barrens: Ecosystem and Landscape*. New York: Academic Press.

Forster, J. R. 1778. *Observations Made during a Voyage Round the World, on Physical Geography, Natural History and Ethic Philosophy*. London: G. Robinson.

Forster, P. 2004. Ice Ages and the mitochondrial DNA chronology of human dispersals: A review. *Philosophical Transactions of the Royal Society of London*, Series B 359: 255–264.

Fortin, M. J., T. H. Keith, B. A. Maurer, M. L. Taper, D. M. Kaufman and T. M. Blackburn. 2005. Species' geographic ranges and distributional limits: Pattern analysis and statistical issues. *Oikos* 108: 7–17.

Fortin, M., P. Drapeau and P. Legendre. 1989. Spatial autocorrelation and sampling design in plant ecology. *Vegetatio* 83: 209–222.

Fortin, M.-J. and M. R. T. Dale. 2005. *Spatial Analysis: A Guide for Ecologists*. Cambridge: Cambridge University Press.

Foster, J. B. 1963. *The Evolution of Native Land Mammals of the Queen Charlotte Islands and the Problem of Insularity*. Ph.D. dissertation, University of British Columbia, Vancouver.

Foster, J. B. 1964. Evolution of mammals on islands. *Nature* 202: 234–235.

Foster, S. A. and J. A. Endler. 1999. *Geographic Variation in Behavior: Perspectives on Evolutionary Mechanisms*. New York: Oxford University Press.

Fox, B. J. 1987. Species assembly and evolution of community structure. *Evolutionary Ecology* 1: 201–213.

France, R. L. 1998. Refining latitudinal gradient analyses: Do we live in a climatically symmetrical world? *Global Ecology and Biogeography Letters* 7: 295–296.

Franken, R. J. and D. S. Hik. 2004. Influence of habitat quality, patch size and connectivity on colonization and extinction dynamics of collared pikas (Ochotona collaris). *Journal of Animal Ecology* 73: 889–896.

Franklin, J. 1995. Predictive vegetation mapping: Geographic modeling of biospatial patterns in relation to environmental gradients. *Progress in Physical Geography* 19: 474–499.

Fratini, S. M. Vannini, S. Cannicci and C. D. Schubart. 2005. Tree-climbing mangrove crabs: A case of convergent evolution. *Evolutionary Ecology Research* 7: 219–233.

Freckleton, R. P., P. H. Harvey and M. Pagel. 2003. Bergmann's rule and body size in mammals. *American Naturalist* 161: 821–825.

Fredrickson, R. and P. Hedrick. 2002. Body size in endangered Mexican wolves: Effects of inbreeding and cross-lineage matings. *Animal Conservation* 5: 39–43.

Free, J. B. 1970. *Insect Pollination of Crops*. New York: Academic Press.

Frey, J. K., M. A. Bogan and T. L. Yates. 2007. Mountaintop island age determines species richness of boreal mammals in the American Southwest. *Ecography* 30: 231–240.

Frick, W. F., J. P. Hayes and P. A. Heady III. 2008. Patterns of island occupancy in bats: Influences of area and isolation on insular incidence of volant mammals. *Global Ecology and Biogeography* 17: 622–632.

Fridley, J. D. 2008. Of Asian forests and European fields: Eastern U.S. plant invasions in a global floristic context. *PLoS ONE* 3: e3630, doi: 10.1371/journal.pone.0003630.

Fridley, J. D., H. Qian, P. S. White and M. W. Palmer. 2006. Plant species invasions along the latitudinal gradient in the United States. *Ecology* 87: 3209–3213.

Fritts, H. C. 1976. *Tree Rings and Climate*. New York: Academic Press.

Fryer, G. and T. D. Iles. 1972. *The Cichlid Fishes of the Great Lakes of Africa: Their Biology and Evolution*. Edinburgh: Oliver & Boyd.

Fu, C., J. Wang, Z. Pu, S. Zhang, H. Chen, B. Zhao, J. Chen and J. Wu. 2007. Elevational gradients of diversity for lizards and snakes in the Hengduan Mountains, China. *Biodiversity and Conservation* 16: 707–726.

Fuentes, E. R. and F. M. Jaksic. 1979. Latitudinal size variation of Chilean foxes: Tests of alternative hypotheses. *Ecology* 60: 43–47.

Fuller, H. L. and A. H. H. Harcourt. 2009. Does the density of primate species decline from centre to edge of their geographic ranges? *Journal of Tropical Ecology* 25: 387–392.

Fuller, H. L., A. H. Harcourt and S. A. Parks. 2009. Does the population density of primate species decline from the centre to edge of their geographic ranges? *Journal of Tropical Ecology* 25: 387–392.

Funk, V. A. 2004. Revolutions in historical biogeography. In M. V. Lomolino, D. F. Sax and J. H. Brown (eds.), *Foundations of Biogeography: Classic Papers with Commentaries* 647–657. Chicago: University of Chicago Press.

Funk, V. A. and W. L. Wagner. 1995. Biogeographic patterns in the Hawaiian Islands. In W. L. Wagner and V. A. Funk (eds.), *Hawaiian Biogeography: Evolution on a Hot Spot Archipelago* 379–419. Washington, DC: Smithsonian Institution Press.

Gadow, H. F. 1913. *The Wanderings of Animals*. Cambridge: Cambridge University Press.

Gaines, S. D. and M. D. Bertness. 1992. Dispersal of juveniles and variable recruitment in sessile marine species. *Nature* 360: 579–580.

Galil, B. S. 2000. A sea under siege—Alien species in the Mediterranean. *Biological Invasions* 2: 177–186.

Gallardo, C. S. and P. E. Penchaszadeh. 2001. Hatching mode and latitude in marine gastropods: Revisiting Thorson's paradigm in the Southern Hemisphere. *Marine Biology* 138: 547–552.

Gallardo, M. H., J. W. Bickham, R. L. Honeycutt, R. A. Ojeda and R. Köhler. 1999. Discovery of tetraploidy in a mammal. *Nature* 401: 341.

Gamble, C., W. Davies, W. P. Pettitt and M. Richards. 2004. Climate change and evolving human diversity in Europe during the last glacial. *Philosophical Transactions of the Royal Society of London*, Series B 359: 243–254.

Gandolfo, M. A., K. C. Nixon and W. L. Crepet. 2008. Selection of fossils for calibration of molecular dating models. *Annals of the Missouri Botanical Garden* 95: 34–42.

Gappa, J. L. 2000. Species richness of marine Bryozoa in the continental shelf and slope of Argentina (south-west Atlantic). *Diversity and Distributions* 6: 15–27.

Garvey, J. E., E. A. Marschall and R. A. Wright. 1998. From star charts to stoneflies: Detecting relationships in continuous bivariate data. *Ecology* 79: 442–447.

Gastner, M. T. and M. E. J. Newman. 2004. Diffusion-based method of producing density-equalizaing maps. *Proceedings of the National Academy of Science, USA* 101: 7499–7504.

Gaston, K. J. 1991. How large is a species' geographic range? *Oikos* 61: 434–438.

Gaston, K. J. 1994. *Rarity*. London: Chapman and Hall.

Gaston, K. J. 1996. Species-range-size distributions: Patterns, mechanisms and implications. *Trends in Ecology and Evolution* 11: 197–201.

Gaston, K. J. 2003. *The Structure and Dynamics of Geographic Ranges*. Oxford: Oxford University Press.

Gaston, K. J. 2008. Biodiversity and extinction: The dynamics of geographic range size. *Progress in Physical Geography* 32: 678–683.

Gaston, K. J. and R. A. Fuller. 2009. The sizes of species' geographic ranges. *Journal of Applied Ecology* 46: 1–9.

Gaston, K. J. and S. F. Matter. 2002. Individuals-area relationships: Comment. *Ecology* 83: 288–293.

Gaston, K. J. and T. M. Blackburn. 1996. The tropics as a museum of biological diversity: Analysis of the New World avifauna. *Proceedings of the Royal Society of London*, Series B-Biological Sciences 263: 63–68.

Gaston, K. J. and T. M. Blackburn. 1999. A critique for macroecology. *Oikos* 84: 353–368.

Gaston, K. J. and T. M. Blackburn. 2000. *Pattern and Process in Macroecology*. Oxford: Blackwell Scientific Publications.

Gaston, K. J., S. L Chown and K. L. Evans. 2008. Ecogeographical rules: Elements of a synthesis. *Journal of Biogeography* 35: 483–500.

Gaston, K. J., T. M. Blackburn and J. I. Spicer. 1998. Rapoport's rule: Time for an epitaph? *Trends in Ecology and Evolution* 13: 70–74.

Gates, D. M. 1993. *Climate Change and Its Biological Consequences*. Sunderland, MA: Sinauer Associates.

Gauld, I. D. 1986. Latitudinal gradients in ichneumonid species-richness in Australia. *Ecological Entomology* 11: 155–161.

Gauld, I. D. K. J. Gaston and D. H. Janzen. 1992. Plant allelochemicals, tritrophic interactions and the anomalous diversity of tropical parasitoids: The "nasty" host hypothesis. *Oikos* 65: 353–357.

Gause, G. F. 1934. *The Struggle for Existence*. Baltimore, MD: Williams & Wilkins.

Gentile, G. and R. Argano. 2005. Island biogeography of the Mediterranean Sea: The species–area relationship for terrestrial isopods. *Journal of Biogeography* 32: 1715–1726.

George, T. L. 1987. Greater land bird densities on island vs. mainland: Relation to nest predation level. *Ecology* 68: 1393–1400.

Ghalambor, C. K., R. B. Huey, P. R. Martin, J. J. Tewksbury and G. Wang. 2006. Are mountain passes higher in the tropics? Janzen's hypothesis revisited. *Integrative and Comparative Biology* 46: 5–17.

Ghebreab, W. 1998. Tectonics of the Red Sea region reassessed. *Earth-Science Reviews* 45: 1–43.

Gibbard, P. 2007. Europe cut adrift. *Nature* 448: 259–260.

Gibbons, A. 1996a. Did Neandertals lose an evolutionary arms race? *Science* 272: 1586–1587.

Gibbons, A. 1996b. The peopling of the Americas. *Science* 274: 31–33.

Gilbert, F. S. 1980. The equilibrium theory of island biogeography: Fact or fiction? *Journal of Biogeography* 7: 209–235.

Gill, A. E. 1976. Genetic divergence of insular populations of deer mice. *Biochemical Genetics* 14: 835–848.

Gillespie, R. 2008. Updating Martin's global extinction model. *Quaternary Science Reviews* 27: 2522–2529.

Gillespie, R. G. 2004. Community assembly through adaptive radiation in Hawaiian spiders. *Science* 303: 356–359.

Gillespie, R. G. and B. G. Baldwin. 2009. Island biogeography of remote archipelagos: Interplay between ecological and evolutionary processes. In J. Losos and R. Ricklefs (eds.) *The Theory of Island Biogeography at 40: Impacts and Prospects*, 358–387. Princeton, NJ: Princeton University Press.

Gillespie, R. G. and G. K. Roderick. 2002. Arthropods on islands: Colonization, speciation and conservation. *Annual Review of Entomology* 47: 595–632.

Gillespie, T. W., G. M. Foody, D. Rocchini, A. P. Giorgi and S. Saatchi. 2008. Measuring and modelling biodiversity from space. *Progress in Physical Geography* 32: 203–221.

Gilloly, J. F. and A. P. Allen. 2007. Linking global patterns in biodiversity to evolutionary dynamics using metabolic theory. *Ecology* 88: 1890–1894.

Gilpin, M. E. and I. Hanski. 1991. *Metapopulation Dynamics: Empirical and Theoretical Investigations*. London: Academic Press.

Gilpin, M. E. and J. M. Diamond. 1976. Calculations of immigration and extinction curves from the species-area distance relation. *Proceedings of the National Academy of Sciences, USA* 73: 4130–4134.

Gil-Romera, G., L. Scott, E. Marais, et al. 2007. Late Holocene environmental change in the northwestern Namib Desert margin: New fossil pollen evidence from hyrax middens. *Palaeogeography, Palaeoclimatology and Palaeoecology* 249: 1–18.

Gittleman, J. L. 1985. Carnivore body size: Ecological and taxonomic correlates. *Oecologia* 67: 540–554.

Givnish, T. J. 1998. Adaptive plant evolution on islands. In P. Grant (ed.), *Evolution on Islands*, 281–304. Oxford: Oxford University Press.

Givnish, T. J., K. J. Sytsma, T. A. Patterson and J. R. Haperman. 1996. Comparison of patterns of geographic speciation and adaptive radiation in *Cyanea* and *Clemontia* (Campanulaceae) based on a cladisitc analysis of DNA sequence and restriction-site data. *American Journal of Botany* 83(Suppl): 159.

Glazier, D. S. 1985. Energetics of litter size in five species of *Peromyscus* with generalizations for other mammals. *Journal of Mammalogy* 66: 629–642.

Gleason, H. A. 1917. The structure and development of the plant association. *Bulletin of the Torrey Botanical Club* 53: 7–26.

Gleason, H. A. 1922. On the relation between species and area. *Ecology* 3: 158–162.

Gleason, H. A. 1926. The individualistic concept of plant associations. *Bulletin of the Torrey Botanical Club* 53: 7–26.

Glick, P. A. 1939. The distribution of insects, spiders and mites in the air. *U.S. Department of Agriculture Technical Bulletin* 673.

Gliwicz, J. 1980. Island populations of rodents: Their organization and functioning. *Biological Reviews* 55: 109–138.

Gloger, C. L. 1883. Das Abandern der Vogel durch Einfluss des Klimas. Breslau: A. Schulz.

Gofas, S. and A. Zenetos. 2003. Exotic molluscs in the Mediterranean basin: Current status and perspectives. *Oceanography and Marine Biology* 41: 237–277.

Goin, F. J. and A. A. Carlini. 1995. An early tertiary microbiotherid marsupial from Antarctica. *Journal of Vertebrate Paleontology* 15: 205–207.

Golani, D. 1993. The sandy shore of the Red Sea—Launching pad for Lessepsian (Suez Canal) migrant fish colonizers of the eastern Mediterranean. *Journal of Biogeography* 20: 579–585.

Goldblatt, P. and J. C. Manning. 2000. Cape Plants. A Conspectus of the Cape Flora of South Africa. *Pretoria: National Botanic Institute.*

Goldman, E. A. 1935. New American mustelids of the genera Martes, Gulo and Lutra. *Proceedings of the Biological Society of Washington* 48: 175–186.

Goldschmidt, T., F. Witte and J. Wanink. 1993. Cascading effects of the introduced Nile perch on the detritivorous/

planktivorous species in the sublittoral areas of Lake Victoria. *Conservation Biology* 7: 686–700.

Gollasch, S., B. S. Galil and A. N. Cohen. 2006. *Bridging Divides: Maritime Canals as Invasion Corridors.* New York: Springer.

Goltsman, M., E. P. Kruchenkova, S. Sergeev, I. Volodin and D. W. Macdonald. 2005. "Island syndrome" in a population of Arctic foxes (*Alopex lagopus*) from Mednyi Island. *Journal of Zoology, London* 267: 405–418.

Good, J. M. and J. Sullivan. 2001. Phylogeography of the red-tailed chipmunk (*Tamias ruficaudus*), a northern Rocky Mountain endemic. *Molecular Ecology* 10: 2683–2695.

Good, R. 1974. *The Geography of the Flowering Plants.* 3rd edition. White Plains, NY: Longman. [First published in 1947.]

Goode, G. B. 1896. *Biobliograph of the Published Writings of Philip Lutley Sclater, FRS Secretary of the Zoological Society of London.* Washington, DC: Government Printing Office.

Goodman, C. S. and B. C. Couglin. 2000. The evolution of evo-devo biology. *Proceedings of the National Academy of Sciences, USA* 97: 4424–4425.

Gordon, K. R. 1986. Insular evolutionary body size trends in *Ursus. Journal of Mammalogy* 67: 395–399.

Gosset, C., J. Rivers and J. Labonne. 2006. Effect of habitat fragmentation on spawning migration of brown trout (*Salmo trutta* L.). *Ecology of Freshwater Fish* 15: 247–254.

Gotelli, N. and R. K. Colwell. 2001. Quantifying biodiversity: Procedures and pitfalls in the measurement and comparison of species richness. *Ecology Letters* 4: 379–391.

Gotelli, N. J. 2000. Null model analysis of species co-occurrence patterns. *Ecology* 81: 2606–2621.

Gotelli, N. J. 2004. Part 7: Assembly rules. In M. V. Lomolino, D. F. Sax and James H. Brown, *Foundations of Biogeography*, 1027–1144. Chicago: Chicago University Press.

Gotelli, N. J. and D. J. McCabe. 2002. Species co-occurrence: A meta–analysis of Diamond's (1975) assembly rules model. *Ecology* 83: 2091–2096.

Gotelli, N. J. and G. L. Entsminger. 2001. *EcoSim: Null models software for ecology.* Version 7.0. Acquired Intelligence Inc. & Kesey-Bear. Available from: http://homepages.together.net/~gentsmin/ecosim.htm.

Gotelli, N. J. and G. R. Graves. 1996. *Null Models in Ecology.* Washington, DC: Smithsonian Institution Press.

Gotelli, N. J. and M. Pyron. 1991. Life history variation in North American freshwater minnows: Effects of latitude and phylogeny. *Oikos* 62: 30–40.

Gould, G. C. and B. J. MacFadden. 2004. Gigantism, dwarfism and Cope's rule: "Nothing in evolution makes sense without phylogeny." *Bulletin of the American Museum of Natural History* 285: 219–237.

Gould, S. J. 1965. Is uniformitarianism necessary? *American Journal of Science* 263: 223–228.

Gould, S. J. 1979. An allometric interpretation of species-area curves: The meaning of the coefficients. *American Naturalist* 114: 335–343.

Gould, S. J. and C. B. Calloway. 1980. Clams and brachiopods: Ships that pass in the night. *Paleobiology* 6: 383–396.

Grace, J. B. and R. G. Wetzel. 1981. Habitat partitioning and competitive displacement in cattails (*Typha*): Experimental field studies. *American Naturalist* 118: 463–474.

Graham, R. W. 1986. Response of mammalian communities to environmental changes during the Late Quaternary. In J. M. Diamond and T. J. Case (eds.), *Community Ecology*, 300–313. New York: Harper and Row.

Graham, R. W. 2001. Late Quaternary biogeography and extinction of Proboscideans in North America. In L. Agenbroad, G. Haynes, E. Johnson and M. R. Palombo (eds,), *The World of Elephants: International Congress*, 707–709. Rome.

Graham, R. W., E. L. Lundelius Jr., M. A. Graham, E. K. Schroeder, R. S. Toomey III, E. Anderson, A. D. Barnosky, J. A. Burns, C. S. Churcher, C. K. Grayson, et al. 1996. Spatial response of mammals to late-Quaternary environmental fluctuations. *Science* 272: 1601–1606.

Grande, L. 1985. The use of paleontology in systematics and biogeography and a time control refinement for historical biogeography. *Paleobiology* 11: 234–243.

Grant, B. R. and P. R. Grant. 1989. *Evolutionary Dynamics of a Natural Population: The Large Cactus Finch of the Galápagos.* Chicago: University of Chicago Press.

Grant, B. S. 2009. Industrial melanism. In M. Ruse and J. Travis (eds.), *Evolution: The First Four Billion Years.* Cambridge, MA: Harvard University Press.

Grant, P. R. 1965. The adaptive significance of some size trends in island birds. *Evolution* 19: 355–367.

Grant, P. R. 1966b. Ecological compatibility of bird species on islands. *American Naturalist* 100: 451–462.

Grant, P. R. 1968. Bill size, body size and the ecological adaptations of bird species to competitive situations on islands. *Systematic Zoology* 17: 319–333.

Grant, P. R. 1971. The habitat preference of *Microtus pennsylvanicus* and its relevance to the distribution of this species on islands. *Journal of Mammalogy* 52: 351–361.

Grant, P. R. 1986. *Ecology and Evolution of Darwin's Finches.* Princeton, NJ: Princeton University Press.

Grant, P. R. 1998. *Evolution on Islands.* New York: Oxford University Press.

Grant, P. R. 2001. Reconstructing the evolution of birds on islands: 100 years of research. *Oikos* 92: 385–403.

Grant, P. R. and I. Abbott. 1980. Interspecific competition, island biogeography and null hypotheses. *Evolution* 34: 332–341.

Grant, P. R. and J. Weiner. 1999. *Ecology and Evolution of Darwin's Finches.* Princeton, NJ: Princeton University Press.

Grant, P. R., B. R. Grant and K. Petren. 2005. Hybridization in the recent past. *American Naturalist* 166: 56–67.

Grant, P. R., B. R. Grant, J. A. Markert, L. F. Keller and K. Petren. 2004. Convergent evolution of Darwin's finches caused by introgressive hybridization and selection. *Evolution* 58: 1588–1599.

Grant, V. 1981. *Plant Speciation.* New York: Columbia University Press.

Graur, D. and W. Martin. 2004. Reading the entrails of chickens: molecular timescales of evolution and the illusion of precision. *Trends in Genetics* 20: 80–86.

Graves, G. R. 1997. Geographic clines of age rations of black-throated blue warblers (*Dendroica caerulescens*). *Ecology* 78: 2524–2531.

Gray, A. 1878. Forest geography and archaeology. *American Journal of Science and Arts* 16, 3rd series: 85–94, 183–196.

Gray, J. and A. J. Boucot (eds.). 1979. *Historical Biogeography, Plate Tectonics and the Changing Environment.* Corvallis: Oregon State University Press.

Gray, J., A. J. Boucot and W. B. N. Berry (eds.). 1981. *Communities of the Past.* Stroudsburg, PA: Hutchinson Ross.

Grayson, D. K. 1981. A mid-Holocene record for the heather vole, *Phenacomys* cf. *intermedius,* in the central Great Basin and its biogeographic significance. *Journal of Mammalogy* 62: 115–121.

Grayson, D. K. 1987a. An analysis of the chorology of late Pleistocene extinctions in North America. *Quaternary Research* 28: 281–289.

Grayson, D. K. 1987b. The biogeographic history of small mammals in the Great Basin: Observations on the last 20,000 years. *Journal of Mammalogy* 68: 359–375.

Grayson, D. K. 1993. *The Desert's Past: A Natural Prehistory of the Great Basin.* Washington, DC: Smithsonian Institution Press.

Grayson, D. K. 2000. Mammalian responses to Middle Holocene climatic change in the Great Basin of the western United States. *Journal of Biogeography* 27: 181–192.

Grayson, D. K. 2001. The archaeological record of human impacts on animal populations. *Journal of World Prehistory* 15: 1–68.

Grayson, D. K. 2005. A brief history of Great Basin pikas. *Journal of Biogeography* 32: 2103–2111.

Grayson, D. K. and D. J. Meltzer. 2002. Clovis hunting and large mammal extinction: A critical review of the evidence. *Journal of World Prehistory* 16: 313–359.

Grayson, D. K. and D. J. Meltzer. 2003. A requiem for North American overkill. *Journal of Archaeological Science* 30: 585–593.

Grayson, D. K. and Madson, D. B. 2000. Biogeographic implications of recent low-elevation recolonization by Neotoma cinerea in the Great Basin. *Journal of Mammalogy* 81: 1100–1105.

Grayson, D. K. and S. D. Livingston. 1993. Missing mammals on Great Basin Mountains: Holocene extinctions and inadequate knowledge. *Conservation Biology* 7: 527–532.

Green, R. E., Y. C. Collingham, S. G. Willis, R. D. Gregory, K. W. Smith and B. Huntley. 2008. Performance of climate envelope models in retrodicting recent changes in bird population size from observed climatic change. *Biology Letters* 4: 599–602.

Greenway, J. C. Jr. 1967. *Extinct and Vanishing Birds of the World.* New York: Dover Publications.

Greenwood, P. H. 1974. The cichlid fishes of Lake Victoria, East Africa: The biology and evolution of a species flock. *Bulletin of the British Museum of Natural History* 6 (suppl.): 1–134.

Greenwood, P. H. 1984. African cichlids and evolutionary theories. In A. A. Echelle and I. Kornfield (eds.), *Evolution of Fish Species Flocks* 141–154. Orono: University of Maine Press.

Grenyer, R., C. D. L. Orme, S. F. Jackson, G. H. Thomas, R. G. Davies, T. J. Davies, K. E. Jones, V. A. Olson, R. S. Ridgely, P. C. Rasmussen, et al. 2006. Global distribution and conservation of rare and threatened vertebrates. *Nature* 444: 93–96.

Gressitt, J. L. (ed.). 1963. *Pacific Basin Biogeography.* Honolulu: Bishop Museum Press.

Gressitt, J. L. 1954. *Insects of Micronesia.* Hawaii: Bishop Museum.

Gressitt, J. L. 1982. Pacific-Asian biogeography with examples from the Coleoptera. *Entomological Genetics* 8: 1–11.

Greve, M., N. J. M. Gremmen, K. J. Gaston and S. L. Chown. 2005. Nestedness of Southern Ocean island biotas: Ecological perspectives on a biogeographical conundrum. *Journal of Biogeography* 32: 155–168.

Griffis, M. R. and R. G. Jaeger. 1998. Competition leads to an extinction-prone species of salamander: Interspecific territoriality in a metapopulation. *Ecology* 79: 2494–2502.

Griffiths, R. C. and S. Tavaré. 1994. Simulating probability distributions in the coalescent. *Theoretical Population Biology* 46: 131–159.

Grinnell, J. 1917. The niche-relationship of the California thrasher. *Auk* 34: 427–433.

Grinnell, J. 1922. The role of the "accidental." *Auk* 39: 373–380.

Grytnes, J. A. 2002. Species richness and altitude, a comparison between simulation models and interpolated plant species richness along the Himalayan altitudinal gradient. *American Naturalist* 159: 294–304.

Grytnes, J. A. 2003. Ecological interpretations of the mid-domain effect. *Ecology Letters* 6: 883–888.

Guasp, A. C., M. I. de Torres and D. E. Gonzalez. 1997. Areography 2.0: A progam to delimit distributional areas of a species. *Environmental Software* 11: 271–275.

Guernier, V., M. E. Hochberg and J.-F. Guegan. 2004. Ecology drives the worldwide distribution of human diseases. *PLoS Biology* 2: 0740–0746.

Guilaine, J. 1991. *Prehistory: The World of Early Man.* New York: Facts on File Publishers.

Guimaraes, P. R. Jr. and P. Guimaraes. 2006. Improving the analyses of nestedness for large sets of matrices. *Environmental Modelling and Software* 21: 1512–1513.

Gupta, S., J. S. Collier, A. Palmer-Felgate and G. Potter. 2007. Catastrophic flooding origin of shelf valley systems in the English Channel. *Nature* 448: 342–345.

Gurevitch, J. and L. V. Hedges. 1993. Meta-analysis: combining the results of independent experiments. In S. M. Scheiner and J. Gurevitch (eds.), *The Design and Analysis of Ecological Experiments*, 378–398. Oxford University Press, New York.

Gurnis, M., R. D. Muller and L. Moresi. 1998. Cretaceous vertical motion of Australia and the Australian–Antarctic discordance. *Science* 279: 1499–1504.

Guthrie, R. D. 1990. *Frozen Fauna of the Mammoth Steppe: The Story of Blue Babe*. Chicago: University of Chicago Press.

Guzman, B. and P. Vargas. 2009. Long-distance colonization of the Western Mediterranean by *Cistus ladanifer* (Cistaceae) despite the absence of special dispersal mechanisms. *Journal of Biogeography* 36: 954–968.

Haddrath, O. and A. J. Baker. 2001. Complete mitochondrial DNA genome sequences of extinct birds: Ratite phylogenetics and the vicariance biogeography hypothesis. *Proceedings of the Royal Society of London, Series B* 268: 939–945.

Hadly, E. A. 1997. Evolutionary and ecological response of pocket gopher (*Thomomys talpoides*) to late-Holocene climatic change. *Biological Journal of the Linnean Society* 60: 277–296.

Hadly, E. A., M. H. Kohn, J. A. Leonard and R. K. Wayne. 1998. A genetic record of population isolation in pocket gophers during Holocene climatic change. *Proceedings of the National Academy of Sciences, USA* 95: 6893–6896.

Hadly, E. A., P. A. Spaeth and C. Li. 2009. Niche conservatism above the species level. *Proceedings of the National Academy of Sciences, USA* 106: 19707–19714.

Haeckel, E. 1876. (later editions in 1907, 1911) The history of creation, or, The development of the Earth and its inhabitants by the action of natural causes: A popular exposition of the doctrine of evolution in general and of that of Darwin, Goethe and Lamarck in particular. New York: D. Appleton.

Haeckel, E. H. P. A. 1866. *Generelle Morphologie der Organismen*. Berlin: Reimer.

Hafner, J. C., N. S. Upham, E. Reddington and C. W. Torres. 2008. Phylogeography of the pallid kangaroo mouse, *Microdipodops pallidus*: A sand-obligate endemic of the Great Basin, western North America. *Journal of Biogeography* 35: 2102–2118.

Haidle, M. N., M. Bolusa, A. A. Brucha, C. Hertlera, A. Kandela, M. Märkera, N. J. Conarda, V. Hochschilda, F. Schrenka and V. Mosbruggera. 2009. The role of culture in early expansions of humans. *Quaternary International*, doi:10.1016/j.quaint.2009.07.011.

Haila, Y. 1986. On the semiotic dimension of ecological theory: The case of island biogeography. *Biology and Philosophy* 1: 377–387.

Haila, Y. 1990. Towards an ecological definition of an island: A northwest European perspective. *Journal of Biogeography* 17: 561–568.

Haila, Y. 2002. A conceptual genealogy of fragmentation research: From island biogeography to landscape ecology. *Ecological Applications* 12: 321–334.

Haila, Y., I. Hanski, O. Jarvinen and E. Ranta. 1982. Insular biogeography: A Northern European perspective. *Acta Oecologica* 3: 303–318.

Haldane, J. B. S. 1926. On being the right size. *Harpers Magazine* March: 424–427.

Halffter, G. 1987. Biogeography of the montane entomofauna of Mexico and Central America. *Annual Review of Entomology* 32: 95–114.

Hall, B. G. 2007. *Phylogenetic Trees Made Easy: A How-to Manual*. 3rd edition. Sunderland, MA: Sinauer Associates.

Hall, C. A. S., T. Hanqin, Y. Qi, G. Pontius and J. Cornell, J. 1995. Modelling spatial and temporal patterns of tropical land use change. *Journal of Biogeography* 22: 753–757.

Hall, E. R. 1981. *The Mammals of North America*. 2nd edition. 2 vols. New York: John Wiley & Sons.

Hall, R. 1998. The plate tectonics of Cenozoic SE Asia and the distribution of land and sea. In R. Hall and J. D. Holloway (eds.), *Biogeography and Geological Evolution of SE Asia*, 99–131. Leiden: Backhuys Publishers.

Hall, R. 2001. Cenozoic reconstructions of SE Asia and the SW Pacific: Changing patterns of land and sea. In I. Metcalfe, J. M. B. Smith, M. Morwood and I. D. Davidson (eds.), *Faunal and Floral Migrations and Evolution in SE Asia–Australasia*, 35–56. Lisse: A. A. Balkema (Swets and Zeitlinger Publishers).

Hall, R. 2002. Cenozoic geological and plate tectonic evolution of SE Asia and the SW Pacific: Computer-based reconstructions, model and animations. *Journal of Asian Earth Sciences* 20: 353–431.

Hallam, A. 1967. The bearing of certain palaeogeographic data on continental drift. *Palaeogeography, Palaeoclimatology and Palaeoecology* 3: 201–224.

Hallam, A. 1973a. Provinciality, diversity and extinction of Mesozoic marine invertebrates in relation to plate movements. In D. H. Tarling and S. K. Runcorn (eds.), *Implications of Continental Drift to the Earth Sciences*, vol. 1, 287–294. New York: Academic Press.

Hallam, A. 1983. Plate tectonics and evolution. In D. S. Bendall (ed.), *Evolution From Molecules to Men*, 367–386. Cambridge: Cambridge University Press.

Hallam, A. 1984. Pre-Quaternary sea-level changes. *Annual Reviews of Earth and Planetary Science* 12: 205–243.

Hallam, A. 1994. *Outline of Phanerozoic Biogeography*. Oxford Biogeography Series, 10. Oxford: Oxford University Press.

Hallam, A. 1997. Estimates of the amount and rate of sea-level change across the Rhaetian–Hettangian and Pliensbachian–Toarcian boundaries (latest Triassic to early Jurassic). *Journal of the Geological Society of London* 154: 773–779.

Hallam, A. 2004. *Catastrophes and Lesser Calamities: The Causes of Mass Extinctions*. Oxford: Oxford University Press.

Hallam, A. and B. W. Sellwood. 1976. Middle Mesozoic sedimentation in relation to tectonics in the British area. *Journal of Geology* 84: 301–321.

Halliday, T. 1978. *Vanishing Birds: Their Natural History and Conservation*. New York: Holt, Rinehart and Winston.

Halloy, S. R. P. 1999. The dynamic contribution of new crops to the agricultural economy: Is it predictable? In J. Janick (ed.), *Perspectives on New Crops and New Uses*, 53–59. Alexandria, VA: ASHS Press.

Hallström, B. M., M. Kullberg, M. A. Nilsson and A. Janke. 2007. Phylogenomic data analyses provide evidence that Xenarthra and Afrotheria are sister groups. *Molecular Biology and Evolution* 24: 2059–2068.

Halpern, B. S., S. Walbridge, K. A. Selkoe, C. V. Kappel, F. Micheli, C. D'Agrosa, J. F. Bruno, K. S. Casey, C. Ebert, H. E. Fox, et al. 2008. A global map of human impact on marine ecosystems. *Science* 319: 948–952.

Hamilton, T. H. and N. E. Armstrong. 1965. Environmental determination of insular variation in bird species abundance in the Gulf of Guinea. *Nature* 207: 148–151.

Hamilton, T. H., R. H. Barth Jr. and I. Rubinoff. 1964. The environmental control of insular variation in bird species abundance. *Proceedings of the National Academy of Sciences, USA* 52: 132–140.

Hansen, D. M. and C. B. Müller. 2009a. The critically endangered enigmatic Mauritian endemic plant *Roussea simplex*

(Rousseaceae): Geckos as pollinators and seed dispersers. *International Journal of Plant Sciences* 170: 42–52.

Hansen, D. M. and C. B. Müller. 2009b. Invasive ants disrupt gecko pollination and seed dispersal of the endangered plant *Roussea simplex* in Mauritius. *Biotropica* 41: 202–208.

Hansen, T. A. 1980. Influence of larval dispersal and geographic distribution on species longevity in neogastropods. *Paleobiology* 6: 193–207.

Hanski, I. 1986. Population dynamics of shrews on small islands accord with the equilibrium model. *Biological Journal of the Linnean Society* 28: 23–36.

Hanski, I. 1992. Inferences from ecological incidence functions. *American Naturalist* 139: 657–662.

Hanski, I. 2004. Island ecology on mainlands: Spatially realistic theory of metapopulation ecology. In J. M. Fernández-Palacios and C. Morici (eds.), *Ecología Insular (Island Ecology)*, 125–146. La Palma, Spain: Cabildo.

Hanski, I. 2005. *Metapopulation Ecology*. Oxford: Oxford University Press.

Hanski, I. 2010. The theories of island biogeography and metapopulation dynamics. In J. Losos and R. E. Ricklefs (eds.), *The Theory of Island Biogeography Revisited*, 186–213. Princeton, NJ: Princeton University Press.

Hanski, I. and A. Peltonen. 1988. Island colonization and peninsulas. *Oikos* 51: 105–106.

Hanski, I. and M. E. Gilpin. 1997. *Metapopulation Biology: Ecology, Genetics and Evolution*. San Diego, CA: Academic Press.

Hanski, I. and O. E. Gaggiotti. 2004. *Ecology, Genetics and Evolution of Metapopulation*. Amsterdam, MA: Elsevier.

Hanski, I. and O. Gaggiotti. 2008. *Ecology, Genetics, and Evolution of Metapopulations*. Amsterdam: Elsevier Academic Press.

Harcourt, A. H. 2000. Latitude and latitudinal extent: A global analysis of the Rapoport effect in tropical mammalian taxon: primates. *Journal of Biogeography* 27: 1169–1182.

Harcourt, A. H. 2006. Rarity in the tropics: Biogeography and macroecology of the primates. *Journal of Biogeography* 33: 2077–2087.

Harcourt, A. H. H. and B. M. Schreier. 2009. Diversity, body mass, and latitudinal gradients in primates. *International Journal of Primatology* 30: 283–300.

Hardin, G. 1960. The competitive exclusion principle. *Science* 131: 1292–1297.

Harold, A. S. and R. D. Mooi. 1994. Areas of endemism— Definition and recognition criteria. *Systematic Biology* 43: 261–266.

Harper, J. L. 1969. The role of predation in vegetational diversity. *Brookhaven Symposia in Biology* 22: 48–62.

Harris, C. M. 1996. Absorption of sound in air vs. humidity and temperature. *Journal of Acoustic Society of America* 40: 148–159.

Harris, P., D. Peschken and J. Milroy. 1969. The status of biological control of the weed *Hypencum perforatum* in British Columbia. *Canadian Entomologist* 101: 1–15.

Harris, S. A. 2002. Global heat budget, plate tectonics and climatic change. *Geografiska Annaler* 84A: 1–9.

Harris, V. T. 1952. An experimental study of habitat selection by prairie and forest races of the deer mouse, *Peromyscus maniculatus*. *Contributions to the Laboratory of Vertebrate Biology, University of Michigan* 56: 1–53.

Harrison, G. L., P. A. McLenachan, M. J. Phillips, K. E. Slack, A. Cooper and D. Penny. 2004. Four new avian mitochondrial genomes help get to basic evolutionary questions in the late Cretaceous. *Molecular Biology and Evolution* 21: 974–983.

Harrison, R. G. 1998. Linking evolutionary pattern and process: The relevance of species concepts for the study of speciation. In D. J. Howard and S. H. Berlocher (eds.), *Endless Forms: Species and Speciation* 19–31. New York: Oxford University Press.

Harrison, S. and E. Bruna. 1999. Habitat fragmentation and large-scale conservation: What do we know for sure? *Ecography* 22: 225–232.

Harshman, J., E. L. Braun, M. J. Braun, C. J. Huddleston, R. C. K. Bowie, J. L. Chojnowski, S. J. Hackett, K.-L. Han, R. T. Kimball, B. D. Marks, et al. 2008. Phylogenomic evidence for multiple losses of flight in ratite birds. *Proceedings of the National Academy of Sciences, USA* 105: 13462–13467.

Hartnett, D. C., K. R. Hickman and L. E. Fischer Walter. 1996. Effects of bison grazing, fire and topography on floristic diversity in tallgrass prairie. *Journal of Range Management* 49: 413–420.

Harvey, F. 2008. *A Primer of GIS: Fundamental Geographic and Cartographic Concepts*. New York: Guilford Publications.

Harvey, P. H. and M. D. Pagel. 1991. *The Comparative Method in Evolutionary Biology*. Oxford: Oxford University Press.

Hastings, J. R. and R. M. Turner. 1965. *The Changing Mile*. Tucson, AZ: University of Arizona Press.

Hastings, J. R., R. M. Turner and D. K. Warren. 1972. *An Atlas of Some Plant Distributions in the Sonoran Desert*. Tucson, AZ: University of Arizona Institute of Atmospheric Physics, Technical Reports of the Meteorology of Arid Lands.

Hatt, R. T. 1928. The relation of the meadow mouse *Microtus pennsylvanicus p.* to the biota of a Lake Champlain island. *Ecology* 9: 88–93.

Hausdorf, B. 1998. Weighted ancestral area analysis and a solution of the redundant distribution problem. *Systematic Biology* 47: 445–456.

Hausdorf, B. 2002. Units in biogeography. *Systematic Biology* 51: 648–652.

Hausdorf, B. 2003. Latitudinal and altitudinal body size variation among north-west European land snail species. *Global Ecology and Biogeography* 12: 389–394.

Hausdorf, B. 2006. Latitudinal and altitudinal diversity patterns and Rapoport effects in north-west European land snails and their causes. *Biological Journal of the Linnean Society* 87: 309–323.

Hausdorf, B. and C. Hennig. 2003. Nestedness of north-west European land snail ranges as a consequence of differential immigration from Pleistocene glacial refuges. *Oecologia* 135: 102–109.

Hawkins, A. F. A. 1999. Altitudinal and latitudinal distribution of east Malagasy forest bird communities. *Journal of Biogeography* 26: 447–458.

Hawkins, B. A. 1994. *Parasitoid Community Ecology*. Oxford: Oxford University Press.

Hawkins, B. A. 2001. Ecology's oldest pattern? *Trends in Ecology and Evolution* 16: 470.

Hawkins, B. A. 2008. Recent progress toward understanding the global diversity gradient. *International Biogeography Society Newsletter* 6: 5–8.

Hawkins, B. A. and J. A. F. Diniz-Filho. 2006. Beyond Rapoport's rule: Evaluating range size patterns of New World birds in a two-dimensional framework. *Global Ecology and Biogeography* 15: 461–469.

Hawkins, B. A. and J. H. Lawton. 1995. Latitudinal gradients in butterfly body sizes: Is there a general pattern? *Oecologia* 102: 31–36.

Hawkins, B. A., F. S. Albuquerque, M. B. Araújo, J. Beck, L. M. Bini, F. J. Cabrero-Sañudo, I. Castro-Parga, J. A. F. Diniz-Filho, D. Ferrer-Castán, R. Field, et al. 2007b. A global evaluation of metabolic theory as an explanation for terrestrial species richness gradients. *Ecology* 88: 1877–1888.

Hawkins, B. A., J. A. F. Diniz-Filho and A. E. Weis. 2005. The mid-domain effect and diversity gradients: Is there anything to learn? *American Naturalist* 166: E140–E143.

Hawkins, B. A., J. A. F. Diniz-Filho, C. A. Jaramillo and S. A. Soeller. 2006. Post-Eocene climate change, niche conservatism, and the latitudinal diversity gradient of New World birds. *Journal of Biogeography* 33: 770–780.

Hawkins, B. A., J. A. F. Diniz-Filho, C. A. Jaramillo and S. A. Soeller. 2007a. Climate, niche conservatism and the global bird diversity gradient. *American Naturalist* 170: S16–S27.

Hawkins, B. A., R. Field, H. V. Cornell, D. J. Currie, J.-F. Guegan, D. M. Kaufman, J. T. Kerr, G. G. Mittelbach, T. Oberdorff, E. M. O'Brien, E. E. Porter and J. R. G. Turner. 2003. Energy, water, and broad-scale geographic patterns of species richness. *Ecology* 84: 3105–3117.

Hay, O. P. 1927. *The Pleistocene of the Western Region of North America and its Vertebrate Animals.* Washington, DC: Carnegie Institute.

Haynes, G. (ed.) 2009b. *American Megafaunal Extinctions at the End of the Pleistocene.* Springer Science.

Haynes, G. 2009a. Introduction to the volume. In G. Haynes (ed.), *American Megafaunal Extinctions at the End of the Pleistocene,* 1–20. Springer Science.

Heads, M. 1984. *Principia Botanica:* Croizat's contribution to botany. *Tuatara* 27(1): 8–13

Heads, M. 2005. Dating nodes on molecular phylogenies: A critique of molecular biogeography. *Cladistics* 21: 62–78.

Heads, M. 2006. Panbiogeography of Nothofagus (Nothofagaceae): Analysis of the main species massings. *Journal of Biogeography* 33: 1066–1075.

Heaney, L. R. 1978. Island area and body size of insular mammals: Evidence from the tri-colored squirrel (*Callosciurus prevosti*) of Southeast Africa. *Evolution* 32: 29–44.

Heaney, L. R. 1984. Mammalian species richness on islands of the Sunda Shelf, Southwest Asia. *Oecologia* 61: 11–17.

Heaney, L. R. 1985. Zoogeographic evidence for middle and late Pleistocene land bridges to the Philippine Islands. *Modern Quaternary Research in Southeast Asia* 9: 127–143.

Heaney, L. R. 1986. Biogeography of mammals in SE Asia: Estimates of rates of colonization, extinction and speciation. In L. R. Heaney and B. D. Patterson (eds.), *Island Biogeography of Mammals,* 127–165. New York: Academic Press.

Heaney, L. R. 1986. Biogeography of the mammals of Southeast Asia: Estimates of colonisation, extinction and speciation. *Biological Journal of the Linnean Society* 28: 127–165.

Heaney, L. R. 1991. An analysis of patterns of distribution and species richness among Philippine fruit bats (Pteropodidae). *Bulletin of the American Museum of Natural History* 206: 145–167.

Heaney, L. R. 2000. Dynamic disequilibrium: A long-term, large-scale perspective on the equilibrium model of island biogeography. *Global Ecology and Biogeography* 9: 59–74.

Heaney, L. R. 2001. Small mammal diversity along elevational gradients in the Philippines: An assessment of patterns and hypotheses. *Global Ecology and Biogeography* 10: 15–40.

Heaney, L. R. 2004. Conservation biogeography in oceanic archipelagoes. In M. V. Lomolino and L. R. Heaney (eds.), *Frontiers of Biogeography: New Directions in the Geography of Nature,* 345–360. Cambridge: Cambridge University Press.

Heaney, L. R. 2007. Is a new paradigm emerging for oceanic island biogeography? *Journal of Biogeography* 34: 753–757.

Heaney, L. R. and G. Vermeij. 2004. Diversification. In M. V. Lomolino, D. F. Sax and J. H. Brown (eds.), *Foundations of Biogeography,* 779–788. Chicago: University of Chicago Press.

Heaney, L. R. and J. C. Regalado, Jr. 1998. *Vanishing Treasures of the Philippine Rain Forest.* Chicago: The Field Museum.

Heaney, L. R., D. S. Balete, E. A. Rickart, M. J. Veluz and S. Jansa. 2009. A new genus and species of small "tree mouse" (Rodentia, Muridae) related to the Philippine giant cloud-rats. *Bulletin of the American Museum of Natural History* 331: 205–229.

Heaney, L. R., E. A. Rickart, R. B. Utzurrum and J. S. F. Klompen. 1989. Elevational zonation of mammals in the central Philippines. *Journal of Tropical Ecology* 5: 259–280.

Heaney, L. R., J. S. Walsh and A. T. Peterson. 2005. The roles of geological history and colonization abilities in genetic differentiation between mammalian populations in the Philippine archipelago. *Journal of Biogeography* 32: 229–247.

Heatwole, H. 1999. *Sea Snakes.* Sydney: University of New South Wales Press.

Hedges, S. B. 1996. Historical biogeography of West Indian vertebrates. *Annual Reviews of Ecology and Systematics* 17: 163–196.

Hedges, S. B. 2001. Biogeography of the West Indies: An overview. In C. A. Woods and F. E. Sergile (eds.), *Biogeography of the West Indies: Patterns and Perspectives* (2nd edition), 15-34. London: CRC Press.

Hedgpeth, J. W. 1957. Classification of marine environments. In J. W. Hedgpeth (ed.), *Treatise on Marine Ecology and Paleoecology,* vol. 1, *Ecology,* 17–18. Memoirs of the Geological Society of America, 67.

Hemmingsen, A. M. 1960. *Energy Metabolism as Related to Body Size and Respiratory Surfaces and Its Evolution.* Reports of the Steno Memorial Hospital and the Nordisk Insulin Laboratorium, Copenhagen, vol. 9(2).

Hendrickson, J. A. Jr. 1981. Community-wide character displacement reexamined. *Evolution* 35: 794–809.

Hengeveld, H. G. 1990. Global climate change: Implications for air-temperature and water supply in Canada. *Transactions of the American Fisheries Society* 119: 176–182.

Hengeveld, R. 1989. *Dynamics of Biological Invasions.* London: Chapman and Hall.

Hennig, W. 1950. Grundzüge einer theorie der phylogenetischen Systematik. Berlin: Deutscher Zentralverlag.

Hennig, W. 1966. *Phylogenetic Systematics.* 3rd edition. Trans. D. D. Davis and R. Zanderl. Urbana: University of Illinois Press.

Henstock, T. J. and A. Levander. 2003. Structure and seismotectonics of the Mendocino Triple Junction. *Journal of Geophysical Research,* 108 (BF-ESE 12) :1–17.

Herbertson, A. J. 1905. The major natural regions: An essay in systematic geography. *Geography Journal* 25: 300–312.

Herczeg, G., A. Gonda and J. Merila. 2009. Evolution of gigantism in nine-spined sticklebacks. *Evolution* 63: 3190–3200.

Hespenheide, H. A. 1978. Are there fewer parasitoids in the tropics? *American Naturalist* 112: 766-769.

Hess, H. H. 1962. History of ocean basins. In A. E. J. Engel, H. L. James and B. F. Leonard (eds.), *Petrological Studies: A Volume in Honor of A. F. Buddington,* 599–620. New York: Geological Society of America.

Hesse, R., W. C. Allee and K. P. Schmidt. 1951. *Ecological Animal Geography.* 2nd edition. New York: John Wiley & Sons.

Hewitt, G. M. 1999. Post-glacial re-colonization of European biota. *Biological Journal of the Linnean Society* 68: 82–108.

Hewitt, G. M. 2001. Speciation, hybrid zones and phylogeography - or seeing genes in space and time. *Molecular Ecology* 10: 537–549.

Hewitt, G. M. 2004. Genetic consequences of climatic oscillations in the Quaternary. *Philosophical Transactions of the Royal Society of London,* Series B-Biological Sciences 359: 183–195.

Hey, J. and R. Nielsen. 2007. Integration within the Felsenstein equation for improved Markov chain Monte Carlo methods in population genetics. *Proceedings of the National Academy of Sciences, USA* 104: 2785–2790.

Heywood, V. H. 1989. Patterns, extents and modes of invasions by terrestrial plants. In J. A. Drake et al. (eds.), *Biological Invasions: A Global Perspective,* 31–60. New York: John Wiley and Sons.

Hickerson, M. J., B. C. Carstens, J. Cavender-Bares, K. A. Crandall, C. H. Graham, J. B. Johnson, L. Rissler, P. F. Victoriano and A. D. Yoder. 2010. Phylogeography's past, present, and future: 10 years after Avise, 2000. *Molecular Phylogenetics and Evolution* 54: 291–301.

Hickerson, M. J., E. A. Stahl and H. A. Lessios. 2006. Test for simultaneous divergence using approximate Bayesian computation. *Evolution* 60: 2435–2453.

Hickerson, M. J., E. Stahl and N. Takebayashi. 2007. msBayes: Pipeline for testing comparative phylogeographic histories using hierarchical approximate Bayesian computation. *BMC Bioinformatics* 8: 268.

Higgins, P. J. 1977. Galápagos iguanas: Models of reptilian differentiation. *BioScience* 28: 512–515.

Hijmans, R. J. and C. H. Graham. 2006. The ability of climate envelope models to predict the effect of climate change on species distributions. *Global Change Biology* 12: 2272–2281.

Hildebrand, S. F. 1939. The Panama Canal as a passageway for fishes, with lists and remarks of the fishes and invertebrates observed. *Zoologica* 24: 15–46.

Hilgard, E. W. 1860. *Report on the Geology and Agriculture of State of Mississippi.* Jackson, Mississippi: Mississippi Printers.

Hillebrand, H. 2004. On the generality of the latitudinal diversity gradient. *American Naturalist* 163: 192–211.

Hilton-Taylor, C. 2000. *IUCN Red List of Threatened Species,* The World Conservation Union. Available from: http://www.redlist.org/ (accessed 2004).

Hnatiuk, S. H. 1979. A survey of germination of seeds of some vascular plants found on Aldabra Atoll. *Journal of Biogeography* 6: 105–114.

Hobbs, R. C., D. J. Rugh, J. M. Waite, J. M. Breiwick and D. P. DeMaster. 2004. Abundance of eastern North Pacific gray whales on the 1995/96 southbound migration. *Journal of Cetacean Research and Management* 6: 115–120.

Hobson, J. A. 2005. Using stable isotopes to trace long-distance dispersal in birds and other taxa. *Diversity and Distribution* 11: 157–164.

Hobson, K. A. 2002. Incredible journeys. *Science* 295: 981–983.

Hobson, K. A. and L. I. Wassenaar (eds.). 2008. *Tracking Animal Migration with Stable Isotopes.* Terrestrial Ecology Series, volume 2. Amsterdam: Elsevier.

Hocker, H. W. Jr. 1956. Certain aspects of climate as related to the distribution of loblolly pine. *Ecology* 37: 824–834.

Hockin, D. C. 1982. Experimental insular zoogeography: Some tests of the equilibrium theory using meiobenthic harpacticoid copepods. *Journal of Biogeography* 9: 487–497.

Hocutt, C. H. and E. O. Wiley (eds.). 1986. *The Zoogeography of North American Freshwater Fishes.* New York: John Wiley & Sons.

Hodell, D. A., M. Brenner and J. H. Curtis. 2005. Terminal classic drought in the northern Maya lowlands inferred from multiple sediment cores in Lake Chichancanab (Mexico). *Quaternary Science Reviews* 24: 1413–1427.

Hodkinson, D. J. and K. Thompson. 1997. Plant dispersal: The role of man. *Journal of Applied Ecology* 34: 1484–1496.

Hoeksema, B. W. 2007. Delineation of the Indo-Malayan centre of maximum marine biodiversity: The coral triangle. In W. Renema (ed.), *Biogeography, Time, and Place: Distributions, Barriers, and Islands,* 117–178. Dordrecht: Springer.

Hoekstra, H. E. and W. F. Fagan. 1998. Body size, dispersal ability and compositional disharmony: The carnivore-dominated fauna of the Kuril Islands. *Diversity and Distributions* 4: 135–149.

Hoekstra, H. E., J. G. Krenz and M. W. Nachman. 2005. Local adaptation in the rock pocket mouse (Chaetodipus intermedius): Natural selection and phylogenetic history of populations. *Heredity* 94: 217–228.

Hoffecker, J. F., W. R. Powers and T. Goebel. 1993. The colonization of Beringia and the peopling of the New World. *Science* 259: 46–52

Hoffmann, R. S. 1971. Relationships of certain Holarctic shrews, genus *Sorex. Zeitschrift fur Saugetierkunde* 36: 193–200.

Hoffmann, R. S. 1981. Different voles for different holes: Environmental restrictions on refugial survival of mammals. In G. E. Scudder and J. L. Reveal (eds.), *Evolution Today: Proceedings of Systematic and Evolutionary Biology,* 24–45. Pittsburgh: Hunt Institute for Botanical Documentation, Carnegie-Mellon University.

Hoffmann, R. S., J. W. Koeppl and C. F. Nadler. 1979. The relationships of the amphiberingian marmots (Mammalia: Sciuridae). *Occasional Papers, Museum of Natural History, University of Kansas* 83: 1–56.

Hoffmeister, T. S., L. E. Vet, A. Biere, K. Holsinger and J. Filser. 2005. Ecological and evolutionary consequences of biological invasion and habitat fragmentation. *Ecosystems* 8: 657–668.

Holdridge, L. R. 1947. Determination of world plant formations from simple climatic data. *Science* 105: 367–368.

Holland, M. M., P. G. Risser and R. J. Naiman (eds.) 1991. *Ecotones: The Role of Landscape Boundaries in the Management and Restoration of Changing Environments.* New York: Chapman and Hall.

Holland, R. F. and S. K. Jain. 1981. Insular biogeography of vernal pools in the central valley of California. *American Naturalist* 117: 24–37.

Holling, C. S. 1965. The functional response of predators to prey density and its role in mimicry and population regulation. *Memoirs of the Entomological Society of Canada* 45: 1–60.

Holloway, T. D. and N. Jardine. 1968. Two approaches to zoogeography: A study based on the distributions of butterflies, birds and bats in the Indo-Australian area. *Proceedings of the Linnean Society of London* 179: 153–188.

Holmgren, C. A., E. Rosello, C. Latorre and J. L. Betancourt. 2008. Late-Holocene fossil rodent middens from the Arica region of northernmost Chile. *Journal of Arid Environments* 72: 677–686.

Holt, R. D. 1977. Predation, apparent competition and the structure of prey communities. *Theoretical Population Biology* 12: 197–229.

Holt, R. D. 2010. Toward a trophic island biogeography: Reflections on the interface of island biogeography and food web ecology. In J. Losos and R. E. Ricklefs (eds.), *The Theory of Island Biogeography Revisited,* 143–185. Princeton, NJ: Princeton University Press.

Hone, D. W. E., G. J. Dyke, M. Haden and M. J. Benton. 2008. Body size evolution in Mesozoic birds. *Journal of Evolutionary Biology* 21: 618–624.

Hone, D. W., T. M. Keesey, D. Pisani and A. Purvis. 2005. Macroevolutionary trends in the Dinosauria: Cope's rule. *Journal of Evolutionary Biology* 18: 587–595.

Hooijer, D. A. 1976. Observations on the pygmy mammoths of the Channel Islands, California. Athlon, Festschrift Loris Russel, Royal Ontario Museum, 220–225.

Hooker, J. D. 1844–1860. The Botany of the Antarctic Voyage of HMS Discovery Ships *Erebus* and *Terror* in the Years 1839–1843, Under the Command of Captain Sir James Clark Ross. London: Reeve Brothers. 3 parts (I. Flora Antarctica II. Flora Novae-Zelandiae III. Flora Tasmaniae) in 6 vols. A.T.: natural history.

Hooker, J. D. 1853. *The Botany of the Antarctic Voyage of H.M.S. Discovery Ships "Erebus" and "Terror" in the Years 1839–1843.* London: Lovell Reeve.

Hooker, J. D. 1866. *Lecture on Insular Floras.* London. Delivered before the British Association for the Advancement of Science at Nottingham, August 27, 1866.

Hooker, J. D. 1877. *Journey to America.* (Unpublished bound volume archived at Royal Botanical Gardens, Kew.)

Hooper, E. T. 1942. An effect on the *Peromyscus maniculatus* Rassenkreis of land utilization in Michigan. *Journal of Mammalogy* 23: 193–196.

Hopkins, D. M. and P. A. Smith. 1981. Dated wood from Alaska and the Yukon: Implications for forest refugia in Beringia. *Quaternary Research* 15: 217–249.

Hopkins, D. M., J. W. Mathews Jr., C. E. Schweger and S. B. Young. 1982. *Paleoecology of Beringia*. New York: Academic Press.

Hopkins, G. W. and R. P. Freckleton. 2002. Declines in the numbers of amateur and professional taxonomists: Implications for conservation. *Animal Conservation* 5: 245–249.

Hopper, S. D., R. J. Smith, M. F. Fay, J. C. Manning and M. W. Chase. 2009. Molecular phylogenetics of Haemodoraceae in the Greater Cape and Southwest Australian Floristic Regions. *Molecular Phylogenetics and Evolution* 51: 19–30.

Horai, S., K. Hayasaka, R. Kondo, K. Tsugane and N. Takahata. 1995. Recent African origin of modern humans revealed by complete sequences of hominoid mitochondrial DNAs. *Proceedings of the National Academy of Sciences, USA* 92: 532–536.

Hormiga, G., M. Arnedo and R. G. Gillespie. 2003. Speciation on a conveyor belt: Sequential colonization of the Hawaiian islands by *Orsonwelles* spiders (Araneae, Linyphiidae). *Systematic Biology* 52: 70–88.

Horn, H. S. 1974. The ecology of secondary succession. *Annual Review of Ecology and Systematics* 5: 25–37.

Horn, H. S. 1975. Markovian processes in forest succession. In M. L. Cody and J. M. Diamond (eds.), *Ecology and Evolution of Communities*, 196–211. Cambridge, MA: Belknap Press.

Horn, H. S. 1981. Succession. In R. M. May (ed.), *Theoretical Ecology*, 253–271. Oxford: Blackwell Scientific Publications.

Horn, J. W. and T. H. Kunz. 2008. Analyzing NEXRAD Doppler radar images to assess nightly dispersal patterns and population trends in Brazilian free-tailed bats (*Tadarida brasiliensis*). *Integrative and Comparative Biology* 48: 24–39.

Horn, J. W., E. B. Arnett and T. H. Kunz. 2008. Behavioral responses of bats to operating wind turbines. *Journal of Wildlife Management* 72: 124–132.

Horn, M. H. and L. G. Allen. 1978. A distributional analysis of California coastal marine fishes. *Journal of Biogeography* 5: 23–42.

Horner, E. 1954. Arboreal adaptations of *Peromyscus* with special reference to use of the tail. *Contributions, Laboratory of Vertebrate Biology, University of Michigan* 61: 1–84.

Horstkotte, J. and U. Strecker. 2005. Trophic differentiation in the phylogenetically young Cyprinodon species flock (Cyprinodontidae, Teleostei) from Laguna Chichancanab (Mexico). *Biological Journal of the Linnean Society* 85: 125–134.

Hostetler, S. W., P. J. Bartlein, P. U. Clarke, E. E. Small and A. M. Solomon. 2000. Simulated influence of Lake Agassiz on the climate of central North America 11,000 years ago. *Nature* 405: 334–337.

Houghton, P. 1990. The adaptive significance of Polynesian body form. *Annals of Human Biology* 17: 19–32.

Howard, D. J. and S. H. Berlocher. 1998. *Endless Forms: Species and Speciation*. Oxford: Oxford University Press.

Howarth, F. G. 1990. Hawaiian terrestrial arthropods: An overview. *Bishop Museum Occasional Papers* 30: 30: 4–26.

Howe, H. F. 1985. Gomphothere fruits: A critique. *The American Naturalist* 125: 853–865.

Howell, A. B. 1917. *Birds of the Islands off the Coast of Southern California*. Pacific Coast Avifauna, no. 12. Hollywood, CA: Cooper Ornithological Club.

Howell, N., I. Kubacka and D. A. Mackey. 1998. How rapidly does the human mitochondrial genome evolve? *American Journal of Human Genetics* 59: 501–509.

Howells, W. 1973. *The Pacific Islanders*. London: Weidenfeld and Nicolson.

Hristov, N. I., M. Betke and T. H. Kunz. 2008. Applications of thermal infrared imaging for research in aeroecology. *Integrative and Comparative Biology* 48: 50–59.

Hsü, K. J. 1972. When the Mediterranean dried up. *Scientific American* 227: 26–36.

Hubbell, S. P. 2001. *The Unified Neutral Theory of Biodiversity and Biogeography*. Princeton, NJ: Princeton University Press.

Hubbs, C. L. 1922. Variations in the number of vertebrae and other meristic characters of fishes correlated with the temperature of water during development. *American Naturalist* 56: 360–372.

Hubbs, C. L. and R. R. Miller. 1948. The zoological evidence: Correlation between fish distribution and hydrographic history in the desert basins of western United States, with emphasis on glacial and postglacial times. In *The Great Basin*, 17–166. *Bulletin of the University of Utah, Biology*: 107.

Huey, R. B., G. W. Gilchrist and A. P. Hendry. 2005. Using invasive species to study evolution: Case studies with *Drosophila* and salmon. In D. F. Sax, J. J. Stachowicz and S. D. Gaines (eds.), *Species Invasions: Insights into Ecology, Evolution, and Biogeography*, 139–164. Sunderland, MA: Sinauer Associates.

Huey, R. B., G. W. Gilchrist, M. L. Carlson, D. Berrigan and L. Serra. 2000. Rapid evolution of a geographic cline in an introduced fly. *Science* 287: 308–310.

Huffaker, C. B. and C. E. Kennett. 1959. A ten-year study of vegetational changes associated with biological control of Klamath weed. *Journal of Range Management* 12: 69–82.

Hugueny, B. 1989. West African rivers as biogeographic islands: Species richness of fish communities. *Oecologia* 79: 236–243.

Hulsey, C. D., F. J. G. de Leon and R. Rodiles-Hernandez. 2006. Micro- and macroevolutionary decoupling of cichlid jaws: A test of Liem's key innovation hypothesis. *Evolution* 60: 2096–2109.

Hulsey, C. D., F. J. García de León, Y. S. Johnson, D. A. Hendrickson and T. J. Near. 2004. Temporal diversification of Mesoamerican cichlid fishes across a major biogeographic boundary. *Molecular Phylogenetics Evolution* 31: 754–764.

Hultén, E. 1937. *Outline of the History of Arctic and Boreal Biota during the Quaternary Period*. Stockholm: Bokforlags Aktiebolaget Thule.

Humboldt, A. von 1808. Ansichten der Natur mit Wissenschaftlichen Erlauterungen. Tubingen, Germany: J. G. Cotta.

Humboldt, A. von and A. Bonpland. 2009. *Essay on the Geography of Plants* (1807). Trans. S. Romanowski; S. T. Jackson ed.; accompanying essays and supplementary material by S. T. Jackson and S. Romanowski. Chicago: University of Chicago Press.

Humboldt, A. von. 1805. Essai sur la geographie des plantes accompagne d'un tableau physique des regions equinoxiales, fonde sur des mesures executees, depuis le dixieme degre de latitude boreale jusqu'au dixieme degre de latitude australe, pendant les annees 1799, 1800, 1801, 1802 et 1803. Paris: Levrault Schoell.

Humboldt, A. von. 1807 (1805–1834). *Le voyage aux régions equinoxiales du Nouveau Continent, fait en 1799–1804, par Alexandre de Humboldt et Aimé Bonpland*. Paris: Libraire grecque-latine-allemande.

Humphries, C. J. 2000. Form, space and time: Which comes first? *Journal of Biogeography* 27: 11–15.

Humphries, C. J. and L. R. Parenti. 1986. *Cladistic Biogeography*. Oxford: Clarendon Press.

Humphries, C. J. and L. R. Parenti. 1999. *Cladistic biogeography: interpreting patterns of plant and animal distributions*, 2nd edition. Oxford University Press.

Hunn, C. A. and P. Upchurch. 2001. The importance of time/space in diagnosing the causality of phylogenetic events: Towards a 'chronobiogeographical' paradigm? *Systematic Biology* 50: 391–407.

Hunt, G. and K. Roy. 2006. Climate change, body size evolution, and Cope's Rule in deep-sea ostracodes. *Proceedings of the National Academy of Sciences, USA* 103: 1347–1352.

Hunter, M. I. and A. Hutchinson. 1994. The virtues and shortcomings of Parochialism: Conserving species that are locally rare, but globally common. *Conservation Biology* 8: 1163–1165.

Huntley, B. and H. J. B. Birks. 1983. *An Atlas of Past and Present Pollen Maps for Europe: 0–12,000 Years Ago*. Cambridge: Cambridge University Press.

Hurley, P. M. 1968. The confirmation of continental drift. *Scientific American* 218(4): 52–62.

Hurley, P. M. and J. R. Rand. 1969. Pre-drift continental nuclei. *Science* 164: 1229–1242.

Huston M. A. 1994. *Biological Diversity: The Coexistence of Species on Changing Landscapes*. Cambridge: Cambridge University Press.

Huston, M. A. and S. Wolverton. 2009. The global distribution of net primary production: Resolving the paradox. *Ecological Monographs* 79: 343–377.

Hutchinson, G. E. 1957. *A Treatise on Limnology*, vol. 1. New York: John Wiley and Sons.

Hutchinson, G. E. 1958. Concluding remarks. *Cold Spring Harbor Symposia on Quantitative Biology* 22: 415–427.

Hutchinson, G. E. 1959. Homage to Santa Rosalia, or why are there so many kinds of animals? *American Naturalist* 93: 145–159.

Hutchinson, G. E. 1967. *A Treatise on Limnology*, vol. 2. New York: John Wiley and Sons.

Hutchinson, G. E. 1975. *A Treatise on Limnology*, vol. 3. New York: John Wiley and Sons.

Hutchinson, G. E. 1978. *An Introduction to Population Ecology*. New Haven, CT: Yale University Press.

Hutchinson, G. E. 1993. *A Treatise on Limnology*, vol. 4. *The Zoobenthos*. New York: John Wiley and Sons.

Hutton, J. 1795. *Theory of the Earth with Proofs and Illustrations*. Edinburgh.

Huybers, P. 2006. Early Pleistocene glacial cycles and the integrated summer insolation forcing. *Science* 313: 508–511.

Hyde, W. T., T. J. Crowley, S. K. Baum and W. R. Peltier. 2000. Neoproterozoic "snowball Earth" simulations with a coupled climate/ice-sheet model. *Nature* 405: 425–429.

Inger, R. F. 1954. Systematics and zoogeography of Philippine Amphibia. *Fieldiana Zoology* 33: 181–531.

IPCC (Intergovernmental Panel on Climate Change). 2007. *Climate Change 2007*. IPCC.

Iriarte, J. A., W. L. Franklin, W. E. Johnson and K. H. Redford. 1990. Biogeographic variation of food habits and body size of the American puma. *Oecologia* 85: 185–190.

Irving, E. 1956. Paleomagnetic and paleoclimatological aspects of polar wandering. *Pure and Applied Geophysics* 33: 23–41.

Irving, E. 1959. Paleomagnetic pole positions. *Journal of the Royal Astronomy Society Geophysics* 2: 51–77.

Irwin, D. E. 2002. Phylogeographic breaks without geographic barriers to gene flow. *Evolution* 56: 2383–2394.

Isenmann, P. 1982. The influence of insularity on fecundity in tits (Aves, Paridae) in Corsica. *Acta Oecologia* 3: 295–301.

Iturralde-Vinent, M. A. 2006. Meso-Cenozoic Caribbean paleogeography: Implications for the historical biogeography of the region. *International Geology Review* 48: 791–827.

IUCN (International Union for the Conservation of Nature and Natural Resources). 2003. *Red List of Threatened Species*. Cambridge: IUCN.

IUCN (International Union for the Conservation of Nature and Natural Resources). 2008. *Wildlife in a Changing World: An Analysis of the 2008 IUCN Red List of Threatened Species*. Gland, Switzerland: IUCN Publication Services.

Ivantsoff, W., P. Unmack, B. Saeed and L. E. L. M. Crowley. 1991. A redfinned blue-eye, a new species and genus of the family Pseudomugilidae from central western Queensland. *Fishes of Sahul* 6: 277–282.

Ivany, L. C., S. Van Simaeys, E. W. Domack and S. D. Samson. 2006. Evidence for an earliest Oligocene ice sheet on the Antarctic Peninsula. *Geology* 34: 377–380.

Iverson, J. B., C. P. Balgooyen, K. K. Byrd and K. K. Lyddan. 1993. Latitudinal variation in egg and clutch size in turtles. *Canadian Journal of Zoology* 71: 2448–2461.

Izawa, T., T. Kawahara and H. Takahashi. 2007. Genetic diversity of an endangered plant, *Cypripedium macranthos* var. *rebunense* (Orchidaceae): Background genetic research for future conservation. *Conservation Genetics* 8: 1369–1376.

Jablonski, D. 1982. Evolutionary rates and modes in Late Cretaceous gastropods: Role of larval ecology. In B. Mamet and M. J. Copeland (eds.), *Proceedings of the Third North American Paleontological Convention, Toronto*.

Jablonski, D. 1986. Background and mass extinctions: The alternation of macroevolutionary regimes. *Science* 231: 129–133.

Jablonski, D. 1987. Heritability at the species level: Analysis of geographic ranges of Cretaceous mollusks. *Science* 238: 360–363.

Jablonski, D. 1989. The biology of mass extinction: A paleontological view. *Philosophical Transactions of the Royal Society of London*, Series B 325: 357–368.

Jablonski, D. 1991. Extinctions: A paleontological perspective. *Science* 253: 754–757.

Jablonski, D. 2000. Micro- and macroevolution: Scale and hierarchy in evolutionary biology and paleobiology. *Paleobiology* 26: 15–52.

Jablonski, D. 2002. Survival without recovery after mass extinctions. *Proceedings of the National Academy of Sciences, USA* 99: 8139–8144.

Jablonski, D. 2008a. Extinction and the spatial dynamics of biodiversity. *Proceedings of the National Academy of Sciences, USA* 105: 11528–11535.

Jablonski, D. 2008b. Species selection: Theory and data. *Annual Reviews of Ecology and Systematics* 39: 501–524.

Jablonski, D. and D. J. Bottjer. 1990. Onshore-offshore trends in marine invertebrate evolution. In R. M. Ross and W. D. Allmon (eds.), *Causes of Evolution: A Paleontological Perspective*, 21–75. Chicago: University of Chicago Press.

Jablonski, D. and D. J. Bottjer. 1991. Environmental patterns in the origins of higher taxa: The post-Paleozoic fossil record. *Science* 252: 1831–1833.

Jablonski, D. and R. A. Lutz. 1980. Molluscan shell morphology: Ecological and paleontological applications. In D. C. Rhoads and R. A. Lutz (eds.), *Skeletal Growth of Aquatic Organisms*, 323–377. New York: Plenum Press.

Jablonski, D., J. J. Sepkoski Jr., D. J. Bottjer and P. M. Sheehan. 1983. Onshore-offshore patterns in the evolution of Phanerozoic shelf communities. *Science* 222: 1123–1125.

Jablonski, D., K. Roy and J. W. Valentine. 2006. Out of the tropics: Evolutionary dynamics of the latitudinal diversity gradient. *Science* 314: 102–106.

Jaccard, P. 1902. Etude comparative de la distribution florale dans une portion des Alpes et du Jura. *Bulletin de la Societe Vaudoise de la Science Naturelle* 37: 547–579.

Jaccard, P. 1908. Nouvelles recherches sur la distribution florale. *Bulletin de la Societe Vaudoise de la Science Naturelle* 44: 223–276.

Jachmann, H., P. S. M. Berry and H. Imae. 1995. Tusklessness in African elephants: A future trend. *African Journal of Ecology* 33: 230.

Jackson, H. H. T. 1919. An apparent effect of winter inactivity upon the distribution of mammals. *Journal of Mammalogy* 1: 58–64.

Jackson, J. A. 1978. Alleviating problems of competition, predation, parasitism and disease in endangered birds: A review. In S. A. Temple (ed.), *Endangered Birds: Management Techniques for Preserving Threatened Species*, 75–112. Madison: University of Wisconsin Press.

Jackson, J. B. C. 1974. Biogeographic consequences of eurytopy and stenotopy among marine bivalves and their evolutionary significance. *American Naturalist* 108: 541–560.

Jackson, J. B. C. 2008. Ecological extinction and evolution in the brave new ocean. *Proceedings of the National Academy of Sciences, USA* 105: 11458–11465.

Jackson, J. B. C. and A. H. Cheetham. 1999. Tempo and mode of speciation in the sea. *Trends in Ecology and Evolution* 14: 72–77.

Jackson, S. T. (ed.). 2009. *Alexander von Humboldt and Aimé Bonpland Essay on the Geography of Plants*. Trans. S. Romanowski. Chicago: University of Chicago Press.

Jackson, S. T. 2004. Quaternary biogeography: Linking biotic responses to environmental variability across timescales. In, M. V. Lomolino and L. R. Heaney (eds.), *Frontiers of Biogeography–New Directions in the Geography of Nature*. London: Cambridge University Press.

Jackson, S. T. and D. R. Whitehead. 1991. Holocene vegetation patterns in the Adirondack Mountains. *Ecology* 72: 641–653.

Jackson, S. T., J. T. Overpeck, T. Webb III, S. E. Keattch and K. H. Anderson. 1997. Mapped plant-macrofossil and pollen records of Late Quaternary vegetation change in eastern North America. *Quaternary Science Reviews* 16: 1–70.

Jaenike, J. 1978. Effect of island area on *Drosophila* population densities. *Oecologia* 36: 327–332.

Jaenike, J. 1991. Mass extinction of European fungi. *Trends in Ecology and Evolution* 6(6): 174–175.

James, F. C. 1970. Geographic size variation in birds and its relationship to climate. *Ecology* 51: 365–390.

James, F. C., R. F. Johnston, N. O. Wamer, G. J. Niemi and W. J. Boecklen. 1984. The Grinellian niche of the wood thrush. *American Naturalist* 124: 17–47.

Jannasch, H. W. and C. O. Wirsen. 1980. Chemosynthetic primary production at East Pacific sea floor spreading center. *Bioscience* 29: 592–598.

Janzen, D. H. 1966. Coevolution of mutualism between ants and acacias in Central America. *Evolution* 20: 249–275.

Janzen, D. H. 1967. Why mountain passes are higher in the tropics. *American Naturalist* 101: 233–249.

Janzen, D. H. 1973. Sweep samples of tropical foliage insects: Effects of seasons, vegetation types, elevation, time of day and insularity. *Ecology* 54: 687–708.

Janzen, D. H. 1981. The peak in North American ichneumonid species richness lies between 38° and 42° N. *Ecology* 62: 532–557.

Janzen, D. H. 1985. The natural history of mutualisms. In D. H. Boucher (ed.), *The Biology of Mutualisms*, 40–99. London: Croom Helm.

Janzen, D. H. and P. S. Martin. 1982. Neotropical anachronisms: The fruits the gomphotheres ate. *Science* 215: 19–27.

Jarrard, R. D. and D. A. Clague. 1977. Implications of Pacific island and seamount ages for the origin of volcanic chains. *Review of Geophysics and Space Physics* 15: 57–76.

Jenkins, D. G. and D. Rinne. 2008. Red herring or low illumination? The peninsula effect revisited. *Journal of Biogeography* 35: 2128–2137.

Jenkins, D. G., C. R. Brescacin, C. V. Duxbury, J. A. Elliott, J. A. Evans, K. R. Grablow, M. Hillegass, B. N. Lyon, G. A. Metzger, M. L. Olandese, et al. 2007. Does size matter for dispersal distance? *Global Ecology and Biogeography* 16: 415–425.

Jetz, W., C. H. Sekercioglu and K. Bohning-Gaese. 2008. The worldwide variation in avian clutch size across species and space. *PLoS Biology* 6: e303, doi: 10.1371/journal.pbio.0060303.

Jiang, L. 2007. Density compensation can cause no effect of biodiversity on ecosystem functioning. *Oikos* 116: 324–334.

Jianu, C.-M. and D. B. Weishampel. 1999. The smallest of the largest: A new look at possible dwarfing in sauropod dinosaurs. *Geologie en Mijnbouw Palaeontology* 36: 361–385.

Johannesson, K. 2003. Evolution in Littorina: Ecology matters. *Journal of Sea Research* 49: 107–117.

Johansson, M. E. and P. A. Keddy. 1991. Intensity and asymmetry of competition between two plant pairs of different degrees of similarity: An experimental study on two guilds of wetland plants. *Oikos* 60: 27–34.

Johnson, D. L. 1978. The origin of island mammoths and the Quaternary land bridge history of the Northern Channel Islands, California. *Quaternary Research* 10: 204–225.

Johnson, D. L. 1980. Problems in the land vertebrate zoogeography of certain islands and the swimming powers of elephants. *Journal of Biogeography* 7: 383–398.

Johnson, D. L. 1981. More comments on the Northern Channel Island mammoths. *Quaternary Research* 15: 105–106.

Johnson, E. A. and K. Miyanishi. 2007. *Plant Disturbance Ecology: The Process and the Response*. Burlington, MA: Elsevier.

Johnson, L. A. S. and B. G. Briggs. 1975. On the Proteaceae: The evolution and classification of a southern family. *Botanical Journal of the Linnean Society* 70(2): 83–182.

Johnson, M. P. and P. H. Raven. 1973. Species number and endemism: The Galápagos Archipelago revisited. *Science* 179: 893–895.

Johnson, M. P., L. G. Mason and P. H. Raven. 1968. Ecological parameters and species diversity. *American Naturalist* 102: 297–306.

Johnson, N. K. 1975. Controls of the number of bird species on montane islands in the Great Basin. *Evolution* 29: 545–567.

Johnson, N. K. and C. Cicero. 2004. New mitochondrial DNA data affirm the importance of Pleistocene speciation in North American birds. *Evolution* 58: 1122–1130.

Johnson, R. A. and P. S. Ward. 2002. Biogeography and endemism of ants (*Hymenoptera: Formicidae*) in Baja California, Mexico: A first overview. *Journal of Biogeography* 29: 1009–1026.

Johnson, T. C., C. A. Scholz, M. R. Talbot, K. Kelts, R. D. Ricketts, G. Ngobi, K. Beuning, I. Ssemmanda and J. W. McGill. 1996. Late Pleistocene desiccation of Lake Victoria and rapid evolution of cichlid fishes. *Science* 273: 1091–1093.

Johnston, M. C. 1963. Past and present grasslands of southern Texas and northeastern Mexico. *Ecology* 44: 456–466.

Johnston, R. F. and R. K. Selander. 1964. House sparrows: Rapid evolution of races in North America. *Science* 144: 548–550.

Johnston, R. F. and R. K. Selander. 1971. Evolution in the house sparrow. II. Adaptive differentiation in North American populations. *Evolution* 25: 1–28.

Johnston, R. F. and W. J. Klitz. 1977. Variation and evolution in a granivorous bird: The house sparrow. In J. Pinowski and S. C. Kendeigh (eds.), *Granivorous Birds in Ecosystems*, 15–51. Cambridge: Cambridge University Press.

Jolivet, L. and C. Faccenna. 2000. Mediterranean extension and the Africa-Eurasia collision. *Tectonics* 19: 1095–1107.

Jones, C. G. and J. H. Lawton. 1994. *Linking Species and Ecosystems*. London: Chapman and Hall.

Jones, H. L. and J. M. Diamond. 1976. Short-time-base studies of turnover in breeding bird populations on the California Channel Islands. *Condor* 78: 526–549.

Jones, J. R. E. 1949. A further study of calcareous streams in the "Black Mountain" district of South Wales. *Journal of Animal Ecology* 18: 142–159.

Jones, M. B. 2003. *Migrations and Dispersal of Marine Organisms*. Dordrecht: Kluwer Academic.

Jonsson, K. A. and J. Fjeldså. 2006. Determining biogeographical patterns of dispersal and diversification in oscine passerine birds in Australia, Southeast Asia and Africa. *Journal of Biogeography* 33: 1155–1165.

Jordan, D. S. 1891. *Temperature and Vertebrae: A Study in Evolution*. New York: Wilder-Quarter Century Books.

Jordan, D. S. 1908. The law of geminate species. *American Naturalist* 42: 73–80.

Jordan, P. 1971. *The Expanding Earth: Some Consequences of Dirac's Gravitational Hypothesis*. New York: Pergamon Press.

Jordan, S., C. Simon, D. Foote and R.A. Englund. 2005. Phylogeographic patterns of Hawaiian Megalagrion damselflies (Odonata: Coenagrionidae) correlate with Pleistocene island boundaries. *Molecular Ecology* 14: 3457–3470.

Josenhans, H. W., D. W. Fedje, R. Pientitz and J. R. Southon. 1997. Early humans and rapidly changing Holocene sea levels in the Queen Charlotte Island – Hecate Strait, British Columbia Canada. *Science* 277: 71–74.

Jost, L. 2007. Partitioning diversity into independent alpha and beta components. *Ecology* 88: 2427–2439.

Kadmon, R. 1995. Nested species subsets and geographic isolation: A case study. *Ecology* 76: 458–465.

Kadmon, R. and O. Allouche. 2007. Integrating the effects of area, isolation, and habitat heterogeneity on species diversity: A unification of island biogeography and niche theory. *American Naturalist* 170: 443–454.

Kalmar, A. and D. J. Currie. 2006. A global model of island biogeography. *Global Ecology and Biogeography* 15: 72–81.

Karanth, K. P., A. Avivi, A. Beharav and E. Nevo. 2004. Microsatellite diversity in populations of blind subterranean mole rats (Spalax ehrenbergi superspecies) in Israel: Speciation and adaptation. *Biological Journal of the Linnean Society* 83: 229–241.

Kark, S., N. Levin and S. Phinn. 2008. Global environmental priorities: Making sense of remote sensing: Reply to TREE Letter: Satellites miss environmental priorities by Loarie et al. (2007). *Trends in Ecology and Evolution* 23: 181–182.

Karl, D. M., C. O. Wirsen and H. W. Jannasch. 1980. Deep-sea primary production at the Galápagos hydrothermal vents. *Science* 207: 1345–1347.

Karr, J. R. 1982 Avian extinction on Barro Colorado Island, Panama: A reassessment. *American Naturalist* 119: 220–239.

Karr, J. R. 1990. Avian survival rates and the extinction process on Barro Colorado Island, Panama. *Conservation Biology* 4: 391–397.

Kaspari, M. and E. Vargo. 1995. Does colony size buffer environmental variation? Bergmann's rule and social insects. *American Naturalist* 145: 610–632

Katsman, C. A., P. C. F. Van der Vaart, H. A. Dijkstra and W. P. M. de Ruijter. 2003. Stability of multilayer ocean vortices: A parameter study including realistic Gulf Stream and Agulhas Rings source. *Journal of Physical Oceanography* 33(6): 1197–1218.

Kattan, G. H., H. Alvarez-Lopez and M. Giraldo. 1994. Forest fragmentation and bird extinctions: San Antonio eighty years later. *Conservation Biology* 8: 138–146.

Kaufman, L. and P. Ochumba. 1993. Evolutionary and conservation biology of cichlid fishes as revealed by faunal remnants in northern Lake Victoria. *Conservation Biology* 7: 719–730.

Kawakami, T., R. K. Butlin, M. Adams, D. J. Paull and S. J. B. Cooper. 2009. Genetic analysis of a chromosomal hybrid zone in the Australian morabine grasshoppers (Vandiemenella, viatica species group). *Evolution* 63: 139–152.

Kearney, M. and W. Porter. 2009. Mechanistic niche modelling: Combining physiological and spatial data to predict species ranges. *Ecology Letters* 12: 334–350.

Keast, A. 1972a. Continental drift and the biota of the mammals on southern continents. In A. Keast, F. C. Erk and B. Glass (eds.), *Evolution, Mammals and Southern Continents*, 23–87. Albany: State University of New York Press.

Keast, A. 1972b. Australian mammals: Zoogeography and evolution. In A. Keast, F. C. Erk and B. Glass (eds.), *Evolution, Mammals and Southern Continents*, 195–246. Albany: State University of New York Press.

Keast, A. 1972c. Comparisons of contemporary mammal faunas of southern continents. In A. Keast, F. C. Erk and B. Glass (eds.), *Evolution, Mammals and Southern Continents*, 433–501. Albany: State University of New York Press.

Keast, A. 1981. Distributional patterns, regional biotas, and adaptations in the Australian biota: A synthesis. In A. Keast (ed.), *Ecological Biogeography of Australia*, 1895–1997. The Hague: Junk.

Keddy, P. A. 1982. Population ecology on an environmental gradient: *Cakile edentula* on a sand dune. *Oecologia* 52: 348–355.

Keddy, P. A. and I. C. Wisheu. 1989. Species richness-standing crop relationships along four lakeshore gradients: Constraints on the general model. *Canadian Journal of Botany* 67: 1609–1617.

Keddy, P. A. and P. MacLelan. 1990. Centrifugal organization in forests. *Oikos* 59: 75–84.

Keeler, M. S. and F. S. Chew. 2008. Escaping an evolutionary trap: Preference and performance of a native insect on an exotic invasive host. *Oecologia* 156: 559–568.

Kellog, C. A. and D. W. Griffin. 2006. Aerobiology and the global transport of desert dust. *Trends in Ecology and Evolution* 21: 638–644.

Kelt, D. A. and D. H. Van Vuren. 2001. The ecology and macroecology of mammalian home range area. *American Naturalist* 157: 637–645.

Kelt, D. A. and D. Van Vuren. 1999. On the relationship between body size and home range area: Consequences of energetic constraints in mammals. *Ecology* 80:337–400.

Kelt, D. A. and J. H. Brown. 1999. Community structure and assembly rules: Confronting conceptual and statistical issues with data on desert rodents. In E. Weiher and P. A. Keddy (eds.), *Ecological Assembly Rules—Perspectives, Advances, Retreats*, 75–107 (Ch. 3). Cambridge: Cambridge University Press.

Kelt, D. A. and M. D. Meyer. 2009. Body size frequency distributions in African mammals are bimodal at all spatial scales. *Global Ecology and Biogeography* 18: 19–29.

Kelt, D. A., J. H. Brown, G. Shenbrot and J. H. Brown. 1999. Patterns in the structure of Asian and North American desert small mammal communities. *Journal of Biogeography* 26: 825–841.

Kendeigh, S. C. 1976. Latitudinal trends in the metabolic adjustments of the House Sparrow. *Ecology* 57: 509–519.

Kent, M., R. A. Moyeed, C. L. Reid, R. Pakeman and R. Weaver. 2006. Geostatistics, spatial rate of change analysis and boundary detection in plant ecology and biogeography. *Progress in Physical Geography* 30: 201–231.

Keogh, J. S., I. A. W. Scott and C. Hayes. 2005. Rapid and repeated origin of insular gigantism and dwarfism in Australian tiger snakes. *Evolution* 59: 226–233.

Kerr, R. A. 1995. Earth's surface may move itself. *Science* 269: 1214–1216.

Kerr, R. A. 2000. An appealing snowball Earth that's still hard to swallow. *Science* 287: 1734–1736.

Kettlewell, H. B. D. 1961. The phenomenon of industrial melanisms in Lepidoptera. *Annual Review of Ecology and Systematics* 6: 245–262.

Keyghobadi, N. 2007. The genetic implications of habitat fragmentation for animals. *Canadian Journal of Zoology* 85: 1049–1064.

Kidd, D. In press. Geophylogenies and the map of life. *Systematic Biology*.

Kidd, D. M. and X. H. Liu. 2008. GEOPHYLOBUILDER 1.0: An ARCGIS extension for creating 'geophylogenies'. *Molecular Ecology Resources* 8: 88–91.

Kidwell, S. M. and S. M. Holland. 2002. The quality of the fossil record: implications for evolutionary analyses. *Annual Review of Ecology and Systematics* 33: 561–588.

Kiester, A. R. 1971. Species density of North American amphibians and reptiles. *Systematic Zoology* 20: 127–137.

Kikkawa, J. and E. E. Williams. 1971. Altitudinal distribution of land birds in New Guinea. *Search* 2: 64–69.

Kikkawa, J. and K. Pearse. 1969. Geographical distribution of land birds in Australia: A numerical analysis. *Australian Journal of Zoology* 17: 821–840.

Kim H.-G., S. C. Keeley, P. S. Vroom and R. K. Jansen. 1998. Molecular evidence for an African origin of the Hawaiian endemic *Hesperomannia* (Asteraceae). *Proceedings of the National Academy of Sciences, USA* 95: 15440–15445.

King, C. M. and P. J. Moors. 1979. On co-existence, foraging strategy and the biogeography of weasels and stoats (*M. nivalis* and *M. erminea*) in Britain. *Oecologia* 39: 129–150.

Kingsolver, J. G. and D. W. Pfennig. 2004. Individual-level selection as a cause of Cope's rule of phyletic size increase. *Evolution* 58: 1608–1612.

Kinlan, B. P. and A. Hastings. 2005. Rates of populations spread and geographic range expansion: What exotic species tell us. In D. F. Sax, S. D. Gaines and J. J. Staichowicz (eds.), *Species Invasions: Insights to Ecology, Evolution and Biogeography*, 381–419. Sunderland, MA: Sinauer Associates.

Kitchen, A., M. M. Miyamoto and C. J. Mulligan. 2007. A three-stage colonization model for the peopling of the Americas. *PLoS ONE* 3: e1596, doi:10:1371.

Kitchener, D. J., A. Chapman, J. Dell, B. G. Muir and M. Palmer. 1980. Lizard assemblage and reserve size and structure in the Western Australian wheatbelt—some implications for conservation. *Biological Conservation* 17: 25–61.

Klein, B. C. 1989. Effects of forest fragmentation on dung and carrion beetle communities in Central Amazon. *Ecology* 70: 1715–1725.

Klicka, J. and R. M. Zink. 1997. The importance of recent ice ages in speciation: A failed paradigm. *Science* 277: 1666–1669.

Klicka, J. and R. M. Zink. 1999. Pleistocene effects on North American songbird evolution. *Proceedings of the Royal Society of London*, Series B-Biological Sciences 266: 695–700.

Knapp, M., K. Stockler, D. Havell, F. Delsuc, F. Sebastiani and P. J. Lockhart. 2005. Relaxed molecular clock provides evidence for long-distance dispersal of Nothofagus (southern beech). *PLoS Biology* 3: 38–43.

Knight, T. M., M. W. McCoy, J. M Chase, K. A. McCoy and R. D. Holt. 2005. Trophic cascades across ecosystems. *Nature* 437: 880–883.

Knoll, A. H. 1986. Patterns of change in plant communities through geological time. In J. Diamond and T. J. Case (eds.), *Community Ecology*, 126–141. New York: Harper and Row.

Knowles, L. L. 2001. Did the Pleistocene glaciations promote divergence? Tests of explicit refugial models in montane grasshoppers. *Molecular Ecology* 10: 691–701.

Knowles, L. L. 2003. The burgeoning field of statistical phylogeography. *Journal of Evolutionary Biology* 17: 1–10.

Knowles, L. L. 2009. Statistical phylogeography. *Annual Review of Ecology Evolution and Systematics* 40: 593–612.

Knowles, L. L. and D. Otte. 2000. Phylogenetic analysis of montane grasshoppers from western North America (Genus *Melanoplus*, Acrididae : Melanoplinae). *Annals of the Entomological Society of America* 93: 421–431.

Knowles, L. L. and W. P. Maddison. 2002. Statistical phylogeography. *Molecular Ecology* 11: 2623–2635.

Knowlton, N. 1993. Sibling species in the sea. *Annual Review of Ecology and Systematics* 24: 189–216.

Knowlton, N. and L. A. Weigt. 1998. New dates and new rates for divergence across the Isthmus of Panama. *Proceedings of the Royal Society of London*, Series B 265: 2257–2263.

Knox, A. K., J. B. Losos and C. Schneider. 2001. Adaptive radiation vs. intraspecific differentiation: Morphological variation in Caribbean Anolis lizards. *Journal of Evolutionary Biology* 14: 904–909.

Knox, E. B., S. R. Downie and J. D. Palmer. 1993. Chloroplast genome rearrangements and the evolution of giant lobelias from herbaceous ancestors. *Molecular Biology and Evolution* 10: 414–430.

Koch, P. L. and A. D. Barnosky. 2006. Late Quaternary extinctions: State of the debate. *Annual Review of Ecology and Systematics* 37: 215–250.

Kocher, T. D. 2004. Adaptive evolution and explosive speciation: The cichlid fish model. *Nature Reviews Genetics* 5: 288–298.

Kodric-Brown, A. and J. H. Brown. 1979. Competition between distantly related taxa and the co-evolution of plants and pollinators. *American Zoologist* 19: 1115–1127.

Köhler, M., S. Moyà-Solà and R. W. Wrangham. 2007. Island rules cannot be broken. *Trends in Ecology and Evolution* 23: 6–7.

Köhler, P., R. Bintanja, H. Fischer, F. Joos, R. Knutti, G. Lohmann and V. Masson-Delmotte. 2010. What caused Earth's temperature variations during the last 800,000 years? Data-based evidence on radiative forcing and constraints on climate sensitivity. *Quaternary Science Reviews* 29: 129–145.

Kohn, A. J. 1978. Ecological shift and release in an isolated population: *Conus miliaris* at Easter Island. *Ecological Monographs* 48: 323–336.

Koleff, P., K. J. Gaston and J. J. Lennon. 2003. Measuring beta diversity for presence absence data. *Journal of Animal Ecology* 72: 367–382.

Koon, D. W. 1998. "Is polar bear hair fiber optic?" *Applied Optics* 37: 3198–3200.

Koopman, K. F. and D. W. Steadman. 1995. Extinction and biogeography of bats on 'Eua, Kingdom of Tonga. *American Museum Novitates* 3125: 1–13.

Koopman, K. F. and J. K. Jones. 1970. Classification of bats. In B. H. Slaughter and D. W. Walton (eds.), *About Bats*, 22–28. Dallas: Southern Methodist University Press.

Kopp, R. E., F. J. Simons, J. X. Mitrovica, A. C. Maloof and M. Oppenheimer. 2009. Probablistic assessment of sea level during the last interglacial stage. *Nature* 462: 863–868.

Koren, I., Y. J. Kaufman, R. Washington, M. C. Todd, Y. Rudich, J. V. Martins and D. Rosenfeld. 2006. The Bodélé depression: A single spot in the Sahara that provides most of the mineral dust to the Amazon forest. *Environmental Research Letters* 1: 014005. doi:10.1088/1748-9326/1/1/014005.

Korner, C. 2007. The use of "altitude" in ecological research. *Trends in Ecology and Evolution* 22: 569–574.

Kornfield, I. and P. F. Smith. 2000. African cichlid fishes: Model systems for evolutionary biology. *Annual Review of Ecology and Systematics* 31: 163–196.

Korobytsina, K. V., D. F. Nadler, N. N. Vorontsov and R. S. Hoffmann. 1974. Chromosomes of the Siberian snow sheep, *Ovis nivicola* and implications concerning the origin of amphiberingian wild sheep. *Quaternary Research* 4: 235–245.

Kotze, D. J., J. Niemela and M. Nieminen. 2000. Colonization success of carabid beetles on Baltic Islands. *Journal of Biogeography* 27: 807–819.

Kouki, J. 1999. Latitudinal gradients in species richness in northern areas: Some exceptional patters. *Ecological Bulletins* 47: 30–37.

Kouki, J., P. Niemelä and M. Viitasaari. 1994. Reversed latitudinal gradient in species richness of sawflies (Hymenoptera, Symphyta). *Annales Zoologica Fennici* 31: 83–88.

Koune, J.-P. 2001. *Threatened Mushrooms in Europe.* Strasbourg: Council of Europe Publishers.

Kowalewski, M., A. P. Hoffmeister, T. K. Baumiller and R. K. Bambach. 2005. Secondary evolutionary escalation between brachiopods and enemies of other prey. *Science* 308: 1774–1777.

Kozak, K. H. and J. J. Wiens. 2006. Does niche conservatism promote speciation? A case study in North American salamanders. *Evolution* 60: 2604–2621.

Kozak, K. H., C. H. Graham and J. J. Wiens. 2008. Integrating GIS-based environmental data into evolutionary biology. *Trends in Ecology and Evolution* 23: 141–148.

Kozlowski, G. 2008. Is the global conservation status assessment of a threatened taxon a utopia? *Biodiversity Conservation* 17: 445–448.

Krasnov, B. R., G. I. Shenbrot, D. Mouillot, I. S. Khokhlova and R. Poulin. 2005. Spatial variation in species diversity and composition of flea assemblages in small mammalian hosts: Geographical distance or faunal similarity? *Journal of Biogeography* 32: 633–644.

Kratter, A. W. 1992. Montane avian biogeography in southern California and Baja California. *Journal of Biogeography* 19: 269–283.

Krebs, C. J., B. L. Keller and R. H. Tamarin. 1969. *Microtus* population biology: Demographic changes in fluctuating populations of *M. ochrogaster* and *M. pennsylvanicus* in southern Indiana. *Ecology* 50: 587–607.

Kreft, H., W. Jetz, J. Mutke, G. Kier and W. Barthlott. 2008. Global diversity of island floras from a macroecological perspective. *Ecology Letters* 11: 116–127.

Kreuzer, M. P. and N. J. Huntly. 2003. Habitat-specific demography: Evidence for source-sink population structure in a mammal, the pika. *Oecologia* 134: 343–350.

Krijgsman, W., F. J. Hilgen, I. Raffi, F. J. Sierro and D. S. Wilson. 1999. Chronology, causes and progression of the Messinian salinity crisis. *Nature* 400: 652–655.

Krug, A. Z., D. Jablonski and J. W. Valentine. 2009. Signature of the End-Cretaceous mass extinction in modern biota. *Science* 323: 767–771.

Krug, A. Z., D. Jablonski, J. W. Valentine and K. Roy. 2009. Generation of Earth's first-order biodiversity pattern. *Astrobiology* 9: 113–124.

Krzanowski, A. 1967. The magnitude of islands and the size of bats (Chiroptera). *Acta Zoologica Cracoviensia* 15, XI: 281–348.

Kuch, M., N. Rohland, J. L. Betancourt, C. Latorre, S. Steppan and H. N. Poinar. 2002. Molecular analysis of an 11,700-year-old rodent midden from the Atacama Desert, Chile. *Molecular Ecology* 11: 913–924.

Kuhn, I., K. Bohning-Gaese, W. Cramer and S. Klotz. 2008. Macroecology meets global change research. *Global Ecology and Biogeography* 17: 3–4.

Kuhn, T. 1970. *The Structure of Scientific Revolutions.* Chicago: University of Chicago Press.

Kuhn, T. S. 1996. *The Structure of Scientific Revolutions.* 3rd edition. Chicago: University of Chicago Press.

Kuhner, M. K. 2006. LAMARC 2.0: maximum likelihood and Bayesian estimation of population parameters. *Bioinformatics* 22: 768–770.

Kuhner, M. K., J. Yamato and J. Felsenstein. 1998. Maximum likelihood estimation of population growth rates based on the coalescent. *Genetics* 149: 429–434.

Kumar, S. 2005. Molecular clocks: Four decades of evolution. *Nature Reviews Genetics* 6: 654–662.

Kumar, S. and S. B. Hedges. 1998. A molecular timescale for vertebrate evolution. *Nature* 392: 917–920.

Kunin, W. E. and K. J. Gaston. 1997. *The Biology of Rarity: Causes and Consequences of Rare-Common Differences.* New York: Chapman and Hall.

Kunz, T. H., S. A. Gauthreaux Jr., N. I. Hristov, J. W. Horn, G. Jones, E. K. V. Kalko, R. P. Larkin, G. F. McCracken, S. M. Swartz, R. B. Srygley, R. Dudley, J. K. Westbrook and M. Wikelski. 2008. Aeroecology: Probing and modeling the aerosphere. *Integrative and Comparative Biology* 48: 1–11.

La Marche, V. C. 1973. Holocene climatic variations inferred from treeline fluctuations in the White Mountains, California. *Quaternary Research* 3: 632–660.

Lack, D. 1947. *Darwin's Finches.* Cambridge: Cambridge University Press.

Lack, D. 1970. Island birds. *Biotropica* 2: 29–31.

Lack, D. 1973. The numbers of species of hummingbirds in the West Indies. *Evolution* 27: 326–337.

Lack, D. 1974. *Evolution Illustrated by Waterfowl.* Oxford: Blackwell Scientific Publications.

Lack, D. 1976. *Island Biology Illustrated by the Land Birds of Jamaica.* Studies in Ecology, vol. 3. Berkeley: University of California Press.

Lack, D. and R. E. Moreau. 1965. Clutch size in tropical passerine birds of forest and savanna. *L'Oiseau* 35: 76–89.

Ladle, R. J. and R. J. Whittaker. 2011. *Conservation Biogeography.* Oxford: Oxford University Press.

Lafferty, K. D., K. F. Smith, M. E. Torchin, A. P. Dobson and A. M. Kuris. 2005. The role of infectious diseases in natural communities: What introduced species tell us. In D. F. Sax, J. J. Stachowicz and S. D. Gaines (eds.), *Species Invasions: Insights into Ecology, Evolution, and Biogeography,* 111–134. Sunderland, MA: Sinauer Associates.

Lai, D. Y. and P. L. Richardson. 1977. Distribution and movement of Gulf Stream rings. *Journal of Physical Oceanography* 7: 670–683.

Laliberte, A. S. and W. J. Ripple. 2004. Range contractions of North American carnivores and ungulates. *BioScience* 54: 123–138.

Lamarck, J. B. P. A. de M. de, and A. P. de Candolle. 1805. *Flore francaise, ou descriptions succinctes de toutes les plantes qui croissent naturellement en France, disposées selon une nouvelle méthode d'analyse, et précédées par un exposé des principes élémentaires de la botanique, 3rd edn.* Paris: Desray Publishers.

Lambeck, K. and J. Chappell. 2001. Sea level change through the last glacial cycle. *Science* 292: 679–686.

Lambert, T. D., G. H. Adler, C. M. Riveros, L. Lopez, R. Ascanio and J. Terborgh. 2003. Rodents on tropical land-bridge islands. *Journal of Zoology* (London) 260: 179–187.

Lambrinos, J. G. 2004. How interactions between ecology and evolution influence contemporary invasion dynamics. *Ecology* 85: 2061–2070.

Lamoreux, J. F., J. C. Morrison, T. H. Ricketts, D. M. Olson, E. Dinerstein, M. W. McNight and H. H. Shugart. 2006. Global tests of biodiversity concordance and the importance of endemism. *Nature* 440: 212–214.

Langner, A. and F. Siegert. 2008. Spatiotemporal fire occurrence in Borneo over a period of 10 years. *Global Change Biology* 14: 48–62.

Lanner, R. M. and T. R. Van Devender. 1998. The recent history of pinyon pines in the American Southwest. In D. M. Richardson (ed.), *Ecology and Biogeography of Pinus*, 171–182. Cambridge: Cambridge University Press.

Laptikhovsky, V. 2006. The rule of Thorson-Rass: One or two independent phenomena? *Russian Journal of Marine Biology* 32: 201–204.

Laptikhovsky, V. 2009. Latitudinal and bathymetric trends in egg size variation: A new look at Thorson's and Rass's rules. *Marine Ecology* 27: 7–14.

Larson, H. K. 1995. A review of the Australian endemic gobiid fish genus Chlamydogobius, with description of five new species. *The Beagle, Records of the Museums and Art Galleries of the Northern Territory* 12: 19–51.

Laskar, J., P. Robutel, F. Joutel, M. Gastineau, A. C. M. Correia and B. Levrard. 2004. A long-term numerical solution for the insolation quantities of the Earth. *Astronomy and Astrophysics* 428: 261–285.

Latorre, C., J. L. Betancourt, K. A. Rylander and J. A. Quade. 2002. Vegetation invasions into Absolute Desert: A 45,000-year rodent midden record from the Calama-Salar de Atacama Basins, Chile. *Geological Society of America Bulletin* 114: 349–366.

Laurance, W. F. 1990. Comparative responses of five arboreal marsupials to tropical forest fragmentation. *Journal of Mammalogy* 71: 641–653.

Laurance, W. F. and C. Gascon. 1997. How to creatively fragment a landscape. *Conservation Biology* 11: 577–580.

Laurance, W. F. and R. O. Bierregaard. 1996. Fragmented tropical forests. *Bulletin of the Ecological Society of America* 77: 34–36.

Lawlor, T. E. 1983. The mammals. In T. J. Case and M. L. Cody (eds.), *Island Biogeography of the Sea of Cortez*, 265–289, 482–500. Berkeley: University of California Press.

Lawlor, T. E. 1986. Comparative biogeography of mammals on islands. *Biological Journal of the Linnean Society* 28: 99–125.

Lawlor, T. E. 1998. Biogeography of great mammals: Paradigm lost? *Journal of Mammalogy* 79: 1111–1130.

Lawton, J. H. 1995. Population dynamic principles. In J. H. Lawton and R. M. May (eds.), *Extinction Rates*, 147–163. Oxford: Oxford University Press.

Lawton, J. H., S. Nee, A. J. Letcher and P. H. Harvey. 1994. Animal distributions: Patterns and processes. In P. J. Edwards, R. M. May and N. R. Webb (eds.), *Large-Scale Ecology and Conservation Biology*, 41–58. London: Blackwell Scientific Publications.

Lawver, L. A. and L. M. Gahagan. 2003. Evolution of Cenozoic seaways in the circum-Antarctic region. *Palaeogeography Palaeoclimatology Palaeoecology* 198: 11–37.

Lazell, J. D. Jr. 1983. Biogeography of the herptofauna of the British Virgin Islands, with description of a new anole (Sauria: Iguanidae). In A. G. J. Rhodin and K. Miyata (eds.), *Advances in Herpetology and Evolutionary Biology*, 99–117. Cambridge, MA: Museum of Comparative Zoology.

Lazenby, R. and A. Smashnuk. 1999. Osteometric variation in Inuit second metacarpal: A test of Allen's rule. *International Journal of Osteoarchaeology* 9: 182–188.

Leathwick, J. R., J. Elith, W. L. Chadderton, D. Rowe and T. Hastie. 2008. Dispersal, disturbance and the contrasting biogeographies of New Zealand's diadromous and non-diadromous fish species. *Journal of Biogeography* 35: 1481–1497.

Lee, P. C. and C. J. Moss. 1995. Statural growth in known-age African elephants (*Loxodonta Africana*). *Journal of Zoology, London* 236: 29–41.

Lee, P.-F., T.-S. Ding, Fu-H. Hsu, G. Shu. 2004. Breeding bird species richness in Taiwan: Distribution on gradients in elevation, primary productivity and urbanization. *Journal of Biogeography* 31: 307–314.

Legendre, P. 1993. Spatial autocorrelation: Problem or new paradigm. *Ecology* 74: 1659–1673.

Leibold, M. A. 1996. A graphical model of keystone predators in food webs: Trophic regulation of abundance, incidence and diversity patterns in communities. *American Naturalist* 147: 784–812.

Leigh, E. 1975. Population fluctuations and community structure. In W. H. Van Dobben and R. H. Lowe-McConnel (eds.), *Unifying Concepts in Ecology*, 67–88. The Hague: Dr. W. Junk.

Leigh, E. G. 1981. The average lifetime of a population in a varying environment. *Journal of Theoretical Biology* 90: 213–239.

Leith, H. 1956. Ein Beitrag zur Frage der korrelation zwischen mittleren klimawerten und vegetationsformationen. *Berichte Deutsche Botanische Gesellschaft* 69: 169–176.

Leprieur, F., O. Beauchard, S. Blanchet, T. Oberdorff and S. Brosse. 2008. Fish invasions in the world's river systems: When natural processes are blurred by human activities. *PLoS* 6: 404–410.

Lesica, P. and F. W. Allendorf. 1995. When are peripheral populations valuable for conservation? *Conservation Biology* 9: 753–760.

Lessios, H. A. 1998. The first stage of speciation as seen in organisms separated by the Isthmus of Panama. In D. J. Howard and S. H. Berlocher (eds.), *Endless Forms: Species and Speciation*, 186–201. New York: Oxford University Press.

Lester, S. E., B. I. Ruttenberg, S. D. Gaines and B. P. Kinlan. 2007. The relationship between dispersal ability and geographic range size. *Ecology Letters* 10: 745–758.

Leston, D. 1957. Spread potential and colonization of the islands. *Systematic Zoology* 6: 41–46.

Leverington, D. and J. Teller. 2003. Paleotopographic reconstructions of the eastern outlets of glacial Lake Agassiz. *Canadian Journal of Earth Sciences* 40: 1259–1278.

Leverington, D., J. Mann and J. Teller. 2002. Changes in the bathymetry and volume of Glacial Lake Agassiz between 9200 and 7700 ^{14}C yr B.P. *Quaternary Research* 244–252.

Levin, D. A. (ed.). 1979. *Hybridization: An Evolutionary Perspective*. Stroudsburg, PA: Dowden, Hutchinson, & Ross.

Levins, R. 1970. Extinction. In M. Gesternhaber (ed.), *Some Mathematical Problems in Biology*, 77–107. Providence, RI: American Mathematical Society.

Lewis, W. H. (ed.). 1979. *Polyploidy: Biological Relevance*. New York: Plenum Press.

Lewontin, R. C. and L. C. Birch. 1966. Hybridization as a source of variation for adaptation to new environments. *Evolution* 20: 315–336.

Li, J., W. He, X. Hua, J. Zhou, H. Xu, J. Chen and C. Fu. 2009. Climate and history explain species richness peak at mid-elevation for *Schizothorax* fishes (Cypriniformes: Cyprinidae) distributed in the Tibetan Plateau and its adjacent regions. *Global Ecology and Biogeography* 28: 264–272.

Li, W. and D. Graur. 1991. *Fundamentals of molecular evolution*. Sunderland, MA: Sinauer Associates.

Li, W. H. 1999. *Molecular Evolution*. Sunderland, MA: Sinauer Associates.

Li, W. K. W. 2009. Plankton populations and communities. In J. D. Witman and K. Roy (eds.), *Marine Macroecology*. Chicago: University of Chicago Press.

Lidicker, W. Z. Jr. 1988. Solving the enigma of microtine "cycles." *Journal of Mammalogy* 69: 225–235.

Lieberman, B. S. 2000. *Paleobiogeography: Using Fossils to Study Global Change, Plate Tectonics and Evolution.* New York: Kluwer Academic/Plenum Publishers.

Lieberman, B. S. 2002. Phylogenetic biogeography with and without the fossil record: Gauging the effects of extinction and paleontological incompleteness. *Palaeogeography Palaeoclimatology Palaeoecology* 178: 39–52.

Lieberman, B. S. 2003. Paleobiogeography: The relevance of fossils to biogeography. *Annual Review of Ecology Evolution and Systematics* 34: 51–69.

Lieberman, B. S. 2004. Range expansion, extinction and biogeographic congruence: A deep time perspective. In M. V. Lomolino and L. R. Heaney (eds.), *Frontiers of Biogeography,* 111–124. Sunderland, MA: Sinauer Associates.

Lieth, H. 1973. Primary production: Terrestrial ecosystems. *Human Ecology* 1: 303–332.

Lillegraven, J. A. 1972. Ordinal and familial diversity of Cenozoic mammals. *Taxon* 21: 261–274.

Lindbladh, M., M. Niklasson, M. Karlsson, L. Björkman and M. Churski. 2008. Close anthropogenic control of *Fagus sylvatica* establishment and expansion in a Swedish protected landscape: Implications for forest history and conservation. *Journal of Biogeography* 35: 682–697.

Lindeman, R. 1942. The trophic-dynamic aspect of ecology. *Ecology* 23: 399–418.

Lindenmayer, D. B. and J. Fischer. 2006. *Habitat Fragmentation and Landscape Change: An Ecological and Conservation Synthesis.* Washington, DC: Island Press.

Lindenmayer, D. B. and J. Fischer. 2007. Tackling the habitat fragmentation panchreston. *Trends in Ecology and Evolution* 22: 127–132.

Linder, H. P. 2001. On areas of endemism, with an example from the African Restionaceae. *Systematic Biology* 50: 892–912.

Linder, H. P. 2003. The radiation of the Cape flora, southern Africa. *Biological Reviews* 78: 597–638.

Lindsey, C. C. 1975. Peomerism, the widespread tendency among related fish species for vertebral number to be correlated with maximum body length. *Journal of the Fisheries Research Board of Canada* 28: 2453–2469.

Lindsey, C. C. and A. N. Arnason. 1981. A model for responses for vertebral number in fish to environmental influences during development. *Canadian Journal of Fisheries and Aquatic Sciences* 38: 334–347.

Lindstedt, S. L. and M. S. Boyce. 1985. Seasonality, body size and survival time in mammals. *American Naturalist* 125: 873–878.

Ling, S. D., C. R. Johnson, K. Ridgway, A. J. Hobday and M. Haddon. 2009. Climate-driven range extension of a sea urchin: Inferring future trends by analysis of recent population dynamics. *Global Change Biology* 15: 719–731.

Linnaeus, C. 1781. On the increase of the habitable earth. *Amonitates Academicae* 2: 17–27.

Lisiecki, L. E. and M. E. Raymo. 2005. A Pliocene-Pleistocene stack of 57 globally distributed benthic d18O records. *Paleoceanography* 20: 1–17.

Lister, A. and P. Bahn. 1994. *Mammoths.* New York: MacMillan Press.

Lister, A. M. 1989. Rapid dwarfing of red deer on Jersey in the last interglacial. *Nature* 342: 539–542.

Lister, A. M. 1993. Mammoths in miniature. *Nature* 362: 188–189.

Lister, B. C. 1976a. The nature of niche expansion in West Indian *Anolis* lizards. I. Ecological consequences of reduced competition. *Evolution* 30: 659–676.

Lister, B. C. 1976b. The nature of niche expansion in West Indian *Anolis* lizards. II. Evolutionary consequences. *Evolution* 30: 677–692.

Lithgow-Bertelloni, C. and M. A. Richards. 1995. Cenozoic plate driving forces. *Geophysical Research Letters* 22: 1317–1320.

Liu, H. P., R. Hershler and K. Clift. 2003. Mitochondrial DNA sequences reveal extensive cryptic diversity within a western American springsnail. *Molecular Ecology* 12: 2771–2782.

Livezey, B. C. 1993. An ecomorphological review of the dodo (*Raphus cucullatus*) and solitaire (*Pezophaps solitaria*), flightless Columbiformes of the Macarene Islands. *Journal of Zoology* (London) 230: 247.

Loarie, S. R., L. N. Joppa and S. L. Pimm. 2007. Satellites miss environmental priorities. *Trends in Ecology and Evolution* 22: 630–632.

Lockwood, J. L. 2004. How do biological invasions alter diversity patterns? A biogeographic perspective. In M. V. Lomolino and L. R. Heaney (eds.), *Frontiers of Biogeography,* Chapter 15. Sunderland, MA: Sinauer Associates.

Lockwood, J. L. and M. L. McKinney. 2001. *Biotic Homogenization.* New York: Kluwer Academic/Plenum Publishers.

Lockwood, J. L. and M. P. Moulton. 1994. Ecomorphological pattern in Bermuda birds: The influence of competition and implications for nature preserves. *Evolutionary Ecology* 8: 53–60.

Lockwood, J. L., M. Hoopes and M. Marchetti. 2007. *Invasion Ecology.* Hoboken, NJ: Wiley-Blackwell.

Lodge, D. M. 1993. Biological invasions: Lessons for ecology. *Trends in Ecology and Evolution* 8: 133–137.

Lomolino, M. V. 1982. Species-area and species-distance relationships of terrestrial mammals in the Thousand Island Region. *Oecologia* 54: 72–75.

Lomolino, M. V. 1983. *Island Biogeography, Immigrant Selection and Mammalian Body Size on Islands.* Ph.D. dissertation, Department of Biology, State University of New York at Binghamton.

Lomolino, M. V. 1984. Immigrant selection, predatory exclusion and the distributions of *Microtus pennsylvanicus* and *Blarina brevicauda* on islands. *American Naturalist* 123: 468–483.

Lomolino, M. V. 1985. Body size of mammals on islands: The island rule re-examined. *American Naturalist* 125: 310–316.

Lomolino, M. V. 1986. Mammalian community structure on islands: Immigration, extinction and interactive effects. *Biological Journal of the Linnean Society* 28: 1–21.

Lomolino, M. V. 1988. Winter immigration abilities and insular community structure of mammals in temperate archipelagoes. In J. F. Downhower (ed.), *Biogeography of the Island Region of Western Lake Erie,* 185–196. Columbus: Ohio State University Press.

Lomolino, M. V. 1989a. Bioenergetics of cross-ice movements of *Microtus pennsylvanicus, Peromyscus leucopus* and *Blarina brevicauda. Holarctic Ecology* 12: 213–218.

Lomolino, M. V. 1989b. Interpretation and comparisons of constants in the species-area relationship: An additional caution. *American Naturalist* 133: 71–75.

Lomolino, M. V. 1990. The target area hypothesis: The influence of island area on immigration rates of non-volant mammals. *Oikos* 57: 297–300.

Lomolino, M. V. 1993a. Matching of granivorous mammals of the Great Basin and Sonoran deserts on a species-for-species basis. *Journal of Mammalogy* 74: 863–867.

Lomolino, M. V. 1993b. Winter filtering, immigrant selection and species composition of insular mammals of Lake Huron. *Ecography* 16: 24–30.

Lomolino, M. V. 1994b. Species richness patterns of mammals inhabiting nearshore archipelagoes: Area, isolation and immigration filters. *Journal of Mammalogy* 75: 39–49.

Lomolino, M. V. 1996. Investigating causality of nestedness of insular communities: Selective immigrations or extinctions? *Journal of Biogeography* 23: 699–703.

Lomolino, M. V. 1999. A species-based, hierarchical model of island biogeography. In E. A. Weiher and P. A. Keddy (eds.), *The Search for Assembly Rules in Ecological Communities*. New York: Cambridge University Press.

Lomolino, M. V. 2000a. A call for a new paradigm of island biogeography. *Global Ecology and Biogeography* 9: 1–6.

Lomolino, M. V. 2000b. A species-based theory of insular zoogeography. *Global Ecology and Biogeography* 9: 39–58.

Lomolino, M. V. 2000c. Ecology's most general, yet protean pattern: The species–area relationship. *Journal of Biogeography* 27: 555–557.

Lomolino, M. V. 2001. Elevation gradients of species-density: Historical and prospective views. *Global Ecology and Biogeography* 10: 3–12.

Lomolino, M. V. 2002. "There are areas too small and areas too large to show clear diversity patterns …" R. H. MacArthur (1972:191). *Journal of Biogeography* 29: 555–557.

Lomolino, M. V. 2004. Introduction to conservation biogeography. In M. V. Lomolino and L. R. Heaney (eds.), *Frontiers of Biogeography*, 294–296. Sunderland, MA: Sinauer Associates.

Lomolino, M. V. 2005. Body size evolution in insular vertebrates: Generality of the island rule. *Journal of Biogeography* 32: 1683–1699.

Lomolino, M. V. 2006. Space, time and conservation biogeography. In J. M. Scott, D. D. Goble and F. W. Davis (eds.), *The Endangered Species Act at 30: Conserving Biodiversity in Human-Dominated Landscapes*, 61–69. Washington, DC: Island Press.

Lomolino, M. V. and D. R. Perault. 2000. Assembly and disassembly of mammal communities in a fragmented temperate rainforest. *Ecology* 81: 1517–1532.

Lomolino, M. V. and D. R. Perault. 2004. Geographic gradients of deforestation and mammalian communities in a fragmented, temperate rainforest landscape. *Global Ecology and Biogeography* 13: 55–64.

Lomolino, M. V. and D. R. Perault. 2007. Body size variation of mammals in a fragmented, temperate rainforest. *Conservation Biology* 21: 1059–1069.

Lomolino, M. V. and G. A. Smith. 2004. Prairie dog towns as islands: Applications of island biogeography and landscape ecology for conserving non-volant terrestrial vertebrates. *Global Ecology and Biogeography* 12: 275–286.

Lomolino, M. V. and J. C. Creighton. 1996. Habitat selection and breeding success of the endangered American burying beetle (*Nicrophorus americanus*). *Biological Conservation* 77: 235–241.

Lomolino, M. V. and J. H. Brown. 2009. The reticulating phylogeny of island biogeography theory. *Quarterly Review of Biology* 84: 357–390.

Lomolino, M. V. and L. R. Heaney. 2004. *Frontiers of Biogeography*. Sunderland, MA: Sinauer Associates.

Lomolino, M. V. and M. D. Weiser. 2001. Towards a more general species-area relationship: Diversity on all islands, great and small. *Journal of Biogeography* 28: 431–445.

Lomolino, M. V. and R. Channell. 1995. Splendid isolation: Patterns of range collapse in endangered mammals. *Journal of Mammalogy* 76: 335–347.

Lomolino, M. V. and R. Channell. 1998. Range collapse, reintroductions and biogeographic guidelines for conservation. *Conservation Biology* 12: 481–484.

Lomolino, M. V. and R. Davis. 1997. Biogeographic scale and biodiversity of mountain forest mammals of western North America. *Global Ecology and Biogeography Letters* 6: 57–76.

Lomolino, M. V., D. F. Sax, B. R. Riddle and J. H. Brown. 2006. The island rule and a research agenda for studying ecogeographic patterns. *Journal of Biogeography* 33: 1503–1512.

Lomolino, M. V., J. C. Creighton, G. D. Schnell and D. L. Certain. 1995. Ecology and conservation of the endangered American burying beetle (*Nicrophorus americanus*). *Conservation Biology* 9: 605–614.

Lomolino, M. V., J. H. Brown and D. F. Sax. 2009. Island biogeography theory: Reticulations and re-integration of "a biogeography of the species." In J. Losos and R. E. Ricklefs (eds.), *The Theory of Island Biogeography Revisited*, 13–51. Princeton, NJ: Princeton University Press.

Lomolino, M. V., J. H. Brown and R. Davis. 1989. Island biogeography of montane forest mammals in the American Southwest. *Ecology* 70: 180–194.

Lomolino, M. V., R. Channell, D. R. Perault and G. A. Smith. 2001. Downsizing nature: Anthropogenic dwarfing of species and ecosystems. In M. McKinney and J. Lockwood (eds.), *Biotic Homogenization: The Loss of Diversity Through Invasion and Extinction*. 33–56. New York: Kluwer Academic/Plenum Publishers.

Lomolino, M. V., R. Channell, D. R. Perault and G. A. Smith. 2001. Downsizing nature: Anthropogenic dwarfing of species and ecosystems. In M. McKinney and J. Lockwood (eds.), *Biotic Homogenization: The loss of diversity through invasion and extinction*, 223–243. New York: Kluwer Academic/Plenum Publishers.

Long, J. 1981. *Introduced Birds of the World*. London: David and Charles.

Long, J. D., G. C. Trussell and T. Elliman. 2009. Linking invasions and biogeography: Isolation differentially affects exotic and native plant diversity. *Ecology* 90: 863–868.

Lonsdale, D. J. and J. S. Levington. 1985. Latitudinal differentiation in copepod growth: An adaptation to temperature. *Ecology* 66: 1397–1407.

Lonsdale, D., M. Pautasso and O. Holdenrieder. 2008. Wood-decaying fungi in the forest: Conservation needs and management options. *European Journal of Forest Research* 127: 1–22.

Lonsdale, W. M. 1999. Global patterns of plant invasions and the concept of invisibility. *Ecology* 80: 1522–1536.

Loo, S. E., R. MacNally and P. Quinn. 2002. An experimental examination of colonization as a generator of biotic nestedness. *Oecologia* 132: 118–124.

Loope, L. L. and D. Mueller-Dombois. 1989. Characteristics of invaded islands, with special reference to Hawaii. In J. A. Drake et al. (eds.), *Biological Invasions: A Global Perspective*, 257–280. New York: John Wiley and Sons.

Lord, R. D. Jr. 1960. Litter size and latitude in North American mammals. *American Midland Naturalist* 64: 488–499.

Losos, J. 2009. *Lizards in an Evolutionary Tree: Ecology and Adaptive Radiation of Anoles*. Berkeley, CA: California University Press.

Losos, J. and R. E. Ricklefs. 2009. Adaptation and diversification on islands. *Nature* 457: 830–836.

Losos, J. and R. Ricklefs (eds.). 2009. *The Theory of Island Biogeography at 40: Impacts and Prospects*. Princeton: Princeton University Press.

Losos, J. B. 1992. A critical comparison of the taxon cycle and character displacement models of size evolution in *Anolis* lizards in the lesser Antilles. *Copeia* 1991: 279–288.

Losos, J. B. and C. J. Schneider. 2009. *Anolis* lizards. *Current Biology* 19: R316–R318.

Losos, J. B. and D. Schluter. 2000. Analysis of an evolutionary species–area relationship. *Nature* 408: 847–850.

Losos, J. B. and K. de Queiroz. 1997. Evolutionary consequences of ecological release in Caribbean Anolis lizards. *Biological Journal of the Linnean Society* 61: 459–483.

Losos, J. B. and R. E. Ricklefs (eds.). 2010. *The Theory of Island Biogeography Revisited*. Princeton, NJ: Princeton University Press.

Losos, J. B., J. C. Marks and T. W. Schoener. 1993. Habitat use and ecological interactions of an introduced and a native species of Anolis lizard on Grand Cayman, with a review of the outcomes of anole introductions. *Oecologia* 95: 525–532.

Lourens, L. J., J. Becker, R. Bintanja, F. J. Hilgen, E. Tuenter, R. S. W. van de Wal and M. Ziegler. 2010. Linear and non-linear response for the Milankovitch theory. *Quaternary Science Reviews* 29: 352–365.

Lourie, S. A. and A. C. J. Vincent. 2004. A marine fish follows Wallace's Line: The phylogeography of the three-spot seahorse (*Hippocampus trimaculatus*, Syngnathidae, Teleostei) in Southeast Asia. *Journal of Biogeography* 31: 1975–1985.

Lourie, S. A., D. M. Green and A. C. J. Vincent. 2005. Dispersal, habitat preferences and comparative phylogeography of Southeast Asian seahorses (*Syngnathidae: Hippocampus*). *Molecular Ecology* 14: 1073–1094.

Lovejoy, T. E., R. O. Birregaard, Jr., A. B. Rylands, J. R. Malcolm, C. E. Quintela, L. H. Harper, K. S. Brown, Jr., A. H. Powell, G. V. N. Powell, H. O. R. Schubart and M. B. Hays. 1986. Edge and other effects of isolation on Amazon forest fragments. In M. E. Soulé (ed.), *Conservation Biology: The Science of Scarcity and Diversity*, 257–285. Sunderland, MA: Sinauer Associates.

Lovette, I. J. 2005. Glacial cycles and the tempo of avian speciation. *Trends in Ecology and Evolution* 20: 57–59.

Lowrey, T. K. 1995. Phylogeny, adaptive radiation, and biogeography of Hawaiian Tetramolopium (Asteraceae, Astereae). In W. L. Wagner and V. A. Funk (eds.), *Hawaiian biogeography: Evolution on a hot spot archipelago*, 195–220. Washington D.C.: Smithsonian Institution Press.

Lowrie, W. 1997. *Fundamentals of Geophysics*. Cambridge: Cambridge University Press.

Lubchenco, J. 1980. Algal zonation in the New England rocky intertidal community: An experimental analysis. *Ecology* 61: 333–344.

Lubchenco, J. and B. A. Menge. 1978. Community development and persistence in a low rocky intertidal zone. *Ecological Monographs* 48: 67–94.

Lubick, N. 2002. Snowball fights. *Nature* 417: 12–13.

Ludwig, J. A. and J. F. Reynolds. 1988. *Statistical Ecology: A Primer on Methods and Computing*. New York: John Wiley and Sons.

Lukoschek, V. and J. S. Keogh. 2006. Molecular phylogeny of sea snakes reveals a rapidly diverged adaptive radiation. *Biological Journal of the Linnean Society* 89: 523–539.

Lundberg, J. G. and B. Chernoff. 1992. A Miocene fossil of the Amazonian fish *Arapaima* (Teleostei, Arapaimidae) from the Magdalena river region of Colombia: Biogeographic and evolutionary implications. *Biotropica* 24: 2–14.

Lundberg, J. G., A. Machado-Allison and R. F. Kay. 1986. Miocene characid fishes from Columbia: Evolutionary stasis and extirpation. *Science* 234: 208–209.

Luo, Z.-X. 2007. Transformation and diversification in early mammal evolution. *Nature* 450: 1011–1019.

Lutz, F. E. 1921. Geographic average, a suggested method for the study of distribution. *American Museum Novitates* 5: 1–7.

Lyell, C. 1834 (3rd of 6 editions). *Principles of Geology, Being an Attempt to Explain the Former Changes of the Earth's Surface, by Reference to Causes Now in Operation*. London: John Murray.

Lynch, J. D. and N. V. Johnson. 1974. Turnover and equilibria in insular avifaunas, with special reference to the California Channel Islands. *Condor* 76: 370–384.

Lyons, S. K. 2003. A quantitative assessment of the range shifts of Pleistocene mammals. *Journal of Mammalogy* 84: 385–402.

Lyons, S. K. 2005. A quantitative model for assessing community dynamics of Pleistocene mammals. *American Naturalist* 165: E168–185.

Lyons, S. K., F. A. Smith, P. J. Wagner, E. P. White and J. H. Brown. 2004. Was a "hyperdisease" responsible for the late Pleistocene megafaunal extinction. *Ecology Letters* 7: 859–868.

Lyras, G. A., M. D. Dermitzakis, A. A. E. Van der Geer, S. B. Van der Geer and J. De Vos. 2009. The origin of *Homo floresiensis* and its relation to evolutionary processes under isolation. *Anthropological Science* 117: 33–43.

MacArthur, R. H. 1958. Population ecology of some warblers of northeastern coniferous forests. *Ecology* 39: 599–619.

MacArthur, R. H. 1965. Patterns of species diversity. *Biological Review* 40: 510–533.

MacArthur, R. H. 1972. *Geographical Ecology: Patterns in the Distributions of Species*. New York: Harper & Row.

MacArthur, R. H. and E. O. Wilson. 1963. An equilibrium theory of insular zoogeography. *Evolution* 17: 373–387.

MacArthur, R. H. and E. O. Wilson. 1967. *The Theory of Island Biogeography*. Monographs in Population Biology, no. 1. Princeton, NJ: Princeton University Press.

MacArthur, R. H. and T. H. Connell. 1966. *The Biology of Populations*. New York: John Wiley and Sons.

MacArthur, R. H., J. M. Diamond and J. Karr. 1972. Density compensation in island faunas. *Ecology* 53: 330–342.

MacArthur, R. H., J. MacArthur, D. MacArthur and A. MacArthur. 1973. The effect of island area on population densities. *Ecology* 54: 657–658.

MacDonald, I. A. W., F. J. Kruger and A. A. Ferrar. 1986. *The Ecology and Management of Invasions in Southern Africa*. Cape Town: Oxford University Press.

MacDonald, I. A. W., L. L. Loope, M. B. Usher and O. Hamann. 1989. Wildlife conservation and the invasion of nature reserves by introduced species: A global perspective. In J. A. Drake et al. (eds.), *Biological Invasions: A Global Perspective*, 215–256. New York: John Wiley and Sons.

Mace, R. and M. Pagel. 1995. A latitudinal gradient in the density of human languages in North America. *Proceedings of the Royal Society of London*, Series B 261: 117–121.

MacFadden, B. J. 2005. Fossil horses: Evidence for evolution. *Science* 307: 1728–1730.

MacFadden, B. J. 2006. Extinct mammalian biodiversity of the ancient New World tropics. *Trends in Ecology and Evolution* 21: 157–165.

Maciolek, N. J. and W. K. Smith. 2009. Benthic species diversity along a depth gradient: Boston Harbor to Lydonia Canyon. *Deep Sea Research II* 56: 1763–1774.

Mack, R. N., D. Simberloff, W. M. Lonsdale, H Evans, M. Clout and F. A. Bazzaz. 2000. Biotic invasions: Causes, epidemiology, global consequences and control. *Ecological Applications* 10: 689–710.

Mackey, B. G., S. L. Berry and T. Brown. 2008. Reconciling approaches to biogeographical regionalization: A systematic and generic framework examined with a case study of the Australian continent. *Journal of Biogeography* 35: 213–229.

Maclaurin, J. and K. Sterelny. 2008. *What Is Biodiversity?* Chicago: University of Chicago Press.

MacPhee, R. D. E. 2009. *Insulae infortunatae*: Establishing a Chronology for Late Quaternary Mammal Extinctions in the West Indies. In G. Haynes (ed.), *American Megafaunal Extinctions at the End of the Pleistocene*, 169–193. Berlin: Springer.

MacPhee, R. D. E. and P. A. Marx. 1997. The 40,000-year plague: Humans, hyperdisease and first-contact extinctions. In S. Goodman and B. D. Patterson (eds.), *Human Impact and Natural Change in Madagascar*, 169–217. Washington, DC: Smithsonian Institution Press.

MacPherson, E. and C. M. Duarte. 1994. Patterns in species richness, size and latitudinal range of East Atlantic fishes. *Ecography* 17: 242–248.

Maddison, W. P. and D. R. Maddison. 2000. *Mesquite: A modular programming system for evolutionary analysis.* Available from: http://mesquiteproject.org. (Accessed 8 June 2005.)

Magnanou, E., R. Fons, J. Blondel and S. Morand. 2005. Energy expenditure in Crocidurinae shrews (Insectivora): Is metabolism a key component of the insular syndrome? *Comparative Biochemistry and Physiology*, Part A 142: 276–285.

Magri, D. 2008. Patterns of post-glacial spread and the extent of glacial refugia of European beech (*Fagus sylvatica*). *Journal of Biogeography* 35: 450–463.

Magurran, A. E. 1988. *Ecological Diversity and Its Measurement.* Princeton, NJ: Princeton University Press.

Magurran, A. E. 2009. *Measuring Biological Diversity.* Somerset, NJ: John Wiley and Sons.

Maloney, B. K. 1980. Pollen analytical evidence for early forest clearance in North Sumatra. *Nature* 287: 324–326.

Malte, E. C. and D. F. Goujet. 2006. The first biogeographical map. *Journal of Biogeography* 33: 761–769.

Manly, B. F. J. 1991. *Randomization and Monte Carlo Methods in Biology.* London: Chapman and Hall.

Mares, M. A. 1976. Convergent evolution in desert rodents: Multivariate analysis and zoogeographic implications. *Paleobiology* 2: 39–63.

Mares, M. A. 1993a. Desert rodents, seed consumption and convergence. *BioScience* 43: 373–379.

Mares, M. A. 1993b. Heteromyids and their ecological counterparts: A pandesertiv view of rodent ecology and evolution. In H. H. Genoways and J. H. Brown (eds.), *Biology of the Heteromyidae*, 652–719. American Society of Mammalogists, Special Publication no. 10.

Mares, M. A. and T. E. Lacher Jr. 1987. Ecological, morphological and behavioral convergence in rock-dwelling mammals. In H. H. Genoways (ed.), *Current Mammalogy*, vol. 1, 307–347. New York: Plenum Press.

Markgraf, V., M. McGlone and G. Hope. 1995. Neogene paleoenvironmental and paleoclimatic change in southern temperate ecosystems: A southern perspective. *Trends in Ecology and Evolution* 10: 143–147.

Marko, P. B. 2002. Fossil calibration of molecular clocks and the divergence times of geminate species pairs separated by the Isthmus of Panama. *Molecular Biology and Evolution* 19: 2005–2021.

Marko, P. B. and A. L. Moran. 2009. Out of sight, out of mind: High cryptic diversity obscures the identities and histories of geminate species in the marine bivalve subgenus Acar. *Journal of Biogeography* 36: 1861–1880.

Marks, G. and W. K. Beatty. 1976. *Epidemics.* New York: Charles Scribner's Sons.

Marquet, P. A. and M. L. Taper. 1998. On size and area: Patterns of mammalian body size extremes across landmasses. *Evolutionary Ecology* 12: 127–139.

Marquet, P. A., M. Fernández, S. A. Navarrete and C. Valdovinos. 2004. Diversity emerging: Towards a deconstruction of biodiversity patterns. In, M. V. Lomolino and L. R. Heaney, *Frontiers of Biogeography: New Directions in the Geography of Nature.* Sunderland, MA: Sinauer Associates.

Marshall, E. 2001a. Clovis first. *Science* 291: 1732.

Marshall, E. 2001b. Pre-clovis sites fight for acceptance. *Science* 291: 1730.

Marshall, L. G. 1979. Evolution of metatherian and eutherian (mammalian) characters: A review based on cladistic methodology. *Zoological Journal of the Linnean Society* 66: 369–410.

Martens, K., B. Godderis and G. Coulter. 1994. *Speciation in Ancient Lakes.* Stuttgart: E. Schweizerbart'sche Verlagsbuchhandlund.

Martin, B. A. and M. K. Saiki. 2005. Relation of desert pupfish abundance to selected environmental variables in natural and manmade habitats in the Salton Sea Basin. *Environmental Biology of Fish* 73: 97–107.

Martin, J. and P. Gurrea. 1990. The peninsular effect in Iberian butterflies (Lepidoptera: Papilionoidea and Hesperioidea). *Journal of Biogeography* 17: 85–96.

Martin, J. L. 1992. Niche expansion in an insular bird community: An autecological perspective. *Journal of Biogeography* 19: 375–381.

Martin, P. S. 1967. Prehistoric overkill. In P. S. Martin and H. E. Wright Jr. (eds.), *Pleistocene Extinctions: The Search for a Cause*, 75–120. New Haven, CT: Yale University Press.

Martin, P. S. 1984. Prehistoric overkill. In P. S. Martin and R. G. Klein (eds.), *Quaternary Extinctions: A Prehistoric Revolution*, 354–403. Tucson, AZ: University of Arizona Press.

Martin, P. S. 1990. 40,000 years of extinction on the "planet of doom." *Palaeogeography, Palaeoclimatology, Palaeoecology* 82: 187–201.

Martin, P. S. 1995. Mammoth extinction: Two continents and Wrangel Island. *Radiocarbon* 37: 1–6.

Martin, P. S. 2005. *Twilight of the Mammoths: Ice-Age Extinctions and the Rewilding of America.* Berkeley, CA: University of California Press.

Martin, P. S. and R. G. Klein (eds.). 1984. *Quaternary Exctinctions: A Prehistoric Revolution.* Tucson, AZ: University of Arizona Press.

Martin, T. E. 1981. Species-area slopes and coefficients: A caution on their interpretation. *American Naturalist* 118: 823–837.

Mast, A. R. and R. Nyffeler. 2003. Using a null model to recognize significant co-occurrence prior to identifying candidate areas of endemism. *Systematic Biology* 52: 271–280.

Mathys, B. A. and J. L. Lockwood. 2009. Rapid evolution of great kiskadees on Bermuda: An assessment of the ability of the island rule to predict the direction of contemporary evolution in exotic vertebrates. *Journal of Biogeography* 36: 2204–2211.

Matisoo-Smith, E. and J. H. Robins. 2004. Origins and dispersals of Pacific peoples: evidence from mtDNA phylogenies of the Pacific rat. *Proceedings of the National Academy of Sciences, USA* 101: 9167–9172.

Matisoo-Smith, E. and J. H. Robins. 2009. Mitochondrial DNA evidence for the spread of Pacific rats through Oceania. *Biological Invasions* 11: 1521–1527.

Matthew, W. D. 1915. Climate and evolution. *Annals of the New York Academy of Sciences* 24: 171–318.

Maurer, B. A. 1994. *Geographic Population Analysis: Tools for Analysis of Biodiversity.* London: Blackwell Scientific Publications.

Maurer, B. A., H. A. Ford and E. H. Rapoport. 1991. Extinction rate, body size, and avifaunal diversity. *Acta XX Congressus Internationalis Ornithologici* 826–834.

Maurer, B. A., J. H. Brown and R. D. Rusler. 1992. The micro and macro of body size evolution. *Evolution* 46: 939–953.

Maxson, L. R. and J. D. Roberts. 1984. Albumin and Australian frogs—Molecular-data a challenge to speciation model. *Science*, 225: 957–958.

Mayden, R. L. (ed.) 1992b. *Systematics, Historical Ecology and North American Freshwater Fishes.* Stanford, CA: Stanford University Press.

Mayden, R. L. 1992a. Explorations into the past and the dawn of systematics and historical ecology. In R. L. Mayden (ed.), *Systematics, Historical Ecology and North American Freshwater Fishes.* 3–17. Stanford, CA: Stanford University Press.

Mayden, R. L. 1997. A hierarchy of species concepts: The denouement in the saga of the species problem. In M. F. Clardige, H. A. Dawah and M. R. Wilson (eds.), *Species: The Units of Biodiversity*, 439. London: Chapman and Hall.

Mayden, R. L. and R. M. Wood. 1995. Systematics, species concepts and the evolutionarily significant unit in biodiversity and conservation biology. *American Fisheries Society Symposium* 17: 58–113.

Mayr, E. 1942. *Systematics and the Origin of Species*. New York: Columbia University Press.

Mayr, E. 1944a. Wallace's Line in the light of recent zoogeographic studies. *Quarterly Review of Biology* 19: 1–14.

Mayr, E. 1944b. The birds of Timor and Sunda. *Bulletin of the American Museum of Natural History* 83: 127–194.

Mayr, E. 1954. Changes in genetic environment and evolution. In J. Huxley, A. C. Hardy and E. B. Ford (eds.), *Evolution as a Process*, 157–180. London: Allen and Unwin.

Mayr, E. 1956. Geographical character gradients and climatic adaptation. *Evolution* 10: 105–108.

Mayr, E. 1963. *Animal Species and Evolution*. Cambridge, MA: Harvard University Press.

Mayr, E. 1965a. The nature of colonization in birds. In H. G. Baker and G. L. Stebbins (eds.), *The Genetics of Colonizing Species*, 3047. New York: Academic Press.

Mayr, E. 1965b. Avifauna: Turnover on islands. *Science* 150: 1587–1588.

Mayr, E. 1969. *Principles of Systematic Zoology*. New York: McGraw-Hill.

Mayr, E. 1970. *Populations, Species and Evolution*. Cambridge, MA: Harvard University Press.

Mayr, E. 1974. Cladistic analysis or cladistic classification? *Z. Zool. Syst. Evolut.-forsch.* 12: 94–128.

Mayr, E. 1982. *The Growth of Biological Thought*. Cambridge, MA: Harvard University Press.

Mayr, E. 1988. The why and how of species. *Biology and Philosophy* 3: 431–441.

Mayr, E. and J. M. Diamond. 2001. *The Birds of Northern Melanesia: Speciation, Ecology and Biogeography*. New York: Oxford University Press.

Mayr, E. and P. D. Ashlock. 1991. *Principles of Systematic Zoology*. New York: McGraw-Hill.

McAtee, W. L. 1947. Distribution of seeds by birds. *American Midland Naturalist* 38: 214–223.

McAuliffe, J. R. 1994. Landscape evolution, soil formation and ecological processes in Sonoran Desert bajadas. *Ecological Monographs* 64: 111–148.

McCain, C. M. 2004. The mid-domain effect applied to elevational gradients: Species richness of small mammals in Costa Rica. *Journal of Biogeography* 31: 19–31.

McCain, C. M. 2006. Do elevational range size, abundance, and body size patterns mirror those documented for geographic ranges? A case study using Costa Rican rodents. *Evolutionary Ecology Research* 8: 435–454.

McCain, C. M. 2009a. Vertebrate range sizes indicate that mountains may be "higher" in the tropics. *Ecology Letters* 12: 550–560.

McCain, C. M. 2009b. Global analysis of bird elevational diversity. *Global Ecology and Biogeography* 18: 346–360.

McCain, C. M. 2009c. Global analysis of reptile elevational diversity. *Global Ecology and Biogeography* 19: 541–553.

McCain, C. M. and N. J. Sanders. 2009. Metabolic theory of elevational diversity of vertebrate ectotherms. Ecology in press.

McCall, R. A., S. Nee and P. H. Harvey. 1998. The role of wing length in the evolution of avian flightlessness. *Evolutionary Ecology* 12: 569–580.

McClain, C. R., A. G. Boyer and G. Rosenberg. 2006. The island rule and the evolution of body size in the deep sea. *Journal of Biogeography* 33: 1578–1584.

McClanahan, T. and G. Branch (eds.). 2008. *Food Webs and the Dynamics of Marine Reefs*. Oxford: Oxford University Press.

McClure, H. E. 1974. *Migration and Survival of the Birds of Asia*. Bangkok: Applied Scientific Research Corporation of Thailand.

McCook, L. J. 1994. Understanding ecological succession: Causal models and theories, a review. *Vegetatio* 110: 115–147.

McCook, L. J. and A. R. O. Chapman. 1997. Patterns and variations in natural succession following massive ice-scour of a rocky intertidal seashore. *Journal of Experimental Marine Biology and Ecology* 214: 121–147.

McCoy, E. D., S. S. Bell and K. Walters. 1986. Identifying biotic boundaries along environmental gradients. *Ecology* 67: 749–759.

McDowall, R. M. 2003. Variation in vertebral number in galaxiid fishes (Teleostei: Galaxiidae): A legacy of life history, latitude and length. *Environmental Biology of Fishes* 66: 361–381.

McDowell, S. B. 1969. Notes on the Australian sea snake *Ephalophis greyi* M. Smith (Serpentes: Elapidae, Hydrophiinae) and the origin and classification of sea snakes. *Zoological Journal of the Linnean Society* 48: 333–349.

McDowell, S. B. 1972. The genera of sea snakes of the *Hydrophis* group (Serpentes: Elapidae). *Transactions of the Zoological Society of London* 32: 189–247.

McDowell, S. B. 1974. Additional notes on the rare and primitive sea snake, *Ephalophis greyi*. *Journal of Herpetology* 8: 123–128.

McElhinny, M. W. 1973a. *Paleomagnetism and Plate Tectonics*. Cambridge: Cambridge University Press.

McFarlane, D. A. 1989. Patterns of species co-occurrence in the Antillean bat fauna. *Mammalia* 53: 59–66.

McGarigal, K. and S. A. Cushman. 2002. Comparative evaluation of experimental approaches to the study of habitat fragmentation studies. *Ecological Applications* 12(2): 335–345.

McGlone, M. S. 2005. Goodby Gondwana. *Journal of Biogeography.* 32: 739–740.

McIntosh, R. P. 1967. The continuum concept of vegetation. *Botanical Review* 33: 130–187.

McIntosh, R. P. 1981. Succession in ecological theory. In D. C. West, H. H. Shugart and D. B. Botkin (eds.), *Forest Succession: Concepts and Applications*, 10–23. New York: Springer-Verlag.

McIntosh, R. P. 1999. The succession of succession: A lexical chronology. *Bulletin of the Ecological Society of America* 80: 256–264.

McKenna, M. C. 1972a. Eocene final separation of the Eurasian and Greenland-North American landmasses. *24th International Geological Congress, Section 7*: 275–281.

McKenna, M. C. 1972b. Was Europe connected directly to North America prior to the middle Eocene? In Th. Dobzkansky, M. K. Hecht and W. C. Steere (eds.), *Evolutionary Biology 6*: 179–188. New York: Appleton-Century-Crofts.

McKenna, M. C. 1973. Sweepstakes, filters, corridors, Noah's Arks and beached Viking funeral ships in paleogeography. In D. H. Tarling and S. K. Runcorn (eds.), *Implications of Continental Drift to the Earth Sciences*, vol. 1, 295–308. New York: Academic Press.

McKenna, M. C. and S. K. Bell. 1997. *Classification of Mammals above the Species Level*. New York: Columbia University Press.

McKenzie, D. P. and W. J. Morgan. 1969. The evolution of triple junctions. *Nature* 226: 239–243.

McKinney, M. L. and J. L. Lockwood. 1999. Biotic homogenization: A few winners replacing many losers in the next mass extinction. *Trends in Ecology and Evolution* 14: 450–453.

McKitrick, M. C. and R. M. Zink. 1988. Species Concepts in Ornithology. *Condor* 90: 1–14.

McLaughlin, J. F. and J. Roughgarden. 1989. Avian predation on *Anolis* lizards in the northeastern Caribbean: An inter-island contrast. *Ecology* 70: 617–628.

McLaughlin, S. P. 1989. Natural floristic area of the western United States. *Journal of Biogeography* 16: 239–248.

McLaughlin, S. P. 1992. Are floristic areas hierarchically arranged? *Journal of Biogeography* 19: 21–32.

McMaster, R. B. and E. L. Usery. 2005. *A Research Agenda for Geographic Information Science.* Boca Raton, FL: CRC Press.

McMillan, W. O. and S. R. Palumbi. 1995. Concordant evolutionary patterns among Indo-West pacific butterflyfishes. *Proceedings of the Royal Society of London,* Series B-Biological Sciences 260: 229–236.

McNab, B. K. 1963. Bioenergetics and the determination of home range size. *American Naturalist* 97: 113–140.

McNab, B. K. 1971. On the ecological significance of Bergmann's rule. *Ecology* 52: 845–854.

McNab, B. K. 1994a. Energy conservation and the evolution of flightlessness in birds. *American Naturalist* 144: 628–642.

McNab, B. K. 1994b. Resource use and the survival of land and freshwater vertebrates on oceanic islands. *American Naturalist* 144: 643–660.

McNab, B. K. 2002. Minimizing energy expenditure facilitates vertebrate persistence on oceanic islands. *Ecology Letters* 5: 693–704.

McNamara, J. M., Z. Barta, M. Wikelski and A. I. Houston. 2008. A theoretical investigation of the effect of latitude on avian life histories. *American Naturalist* 172: 331–345.

McNeill, W. H. 1976. *Plagues and Peoples.* Garden City, NY: Anchor Press.

McPeek, M. A. and J. M. Brown. 2007. Clade age and not diversification rate explains species richness among animal taxa. *American Naturalist* 169: E97–E106.

McPhail, J. D. 1994. Speciation and the evolution of reproductive isolation in the sticklebacks (*Gasterosteus*) of southwestern British Columbia. In M. A. Bell and S. A. Foster (eds.), *Evolutionary Biology of the Threespine Stickleback,* 399–437. Oxford: Oxford University Press.

McPhail, J. D. and C. C. Lindsey. 1986. Zoogeography of the freshwater fishes of Cascadia (the Columbia River north of the Stikine). In C. H. Hocutt and E. O. Wiley (eds.), *Zoogeography of North American Freshwater Fishes,* 615–637. New York: John Wiley and Sons, Inc.

Meadows, M. E. 2001. Biogeography: Does theory meet practice? *Progress in Physical Geography* 25: 134–142.

Means, D. B. and D. Simberloff. 1987. The peninsula effect: Habitat-correlated species decline in Florida's herptofauna. *Journal of Biogeography* 14: 551–568.

Meiri, S. 2007. Size evolution in island lizards. *Global Ecology and Biogeography* 16: 702–708.

Meiri, S. 2008. Evolution and ecology of lizard body sizes. *Global Ecology and Biogeography* 17: 724–734.

Meiri, S. and T. Dayan. 2003. On the validity of Bergmann's rule. *Journal of Biogeography* 30: 331–351.

Meiri, S., E. Meijaard, S. A. Wich, C. P. Groves and K. M. Helgen. 2008. Mammals of Borneo: Small size on a large island. *Journal of Biogeography* 35: 1087–1094.

Meiri, S., T. Dayan and D. Simberloff. 2004a. Body size of insular carnivores: Little support for the island rule. *American Naturalist* 163: 469–479.

Meiri, S., T. Dayan and D. Simberloff. 2004b. Carnivores, biases and Bergmann's rule. *Biological Journal of the Linnean Society* 81: 579–598.

Meiri, S., T. Dayan, D. Simberloff and R. Grenyer. 2009. Life on the edge: Carnivore body size variation is all over the place. *Proceedings of the Royal Society of London,* Series B 276: 1469–1476.

Melo, A. S., T. F. L. V. B. Rangel and J. A. F. Diniz-Filho. 2009. Environmental drivers of beta-diversity patterns in New-World birds and mammals. *Ecography* 32: 226–236.

Meltzer, D. J. 2009. *First Peoples in a New World: Colonizing Ice Age America.* Berkeley, CA: University of California Press.

Menge, B. A. and J. Lubchenco. 2001. Essays on ecological classics—On the genesis of "community development and persistence in a low rocky intertidal zone." *Bulletin of the Ecological Society of America* 82: 124–125.

Menge, B. A. and J. P. Sutherland. 1976. Species diversity gradients: Synthesis of the roles of predation, competition and temporal heterogeneity. *American Naturalist* 110: 351–369.

Menge, B. A., E. L. Berlow, C. A. Blanchette, S. A. Navarrete and S. B. Yamada. 1994. The keystone species concept: Variation in interaction strength in a rocky intertidal habitat. *Ecological Monographs* 64: 249–286.

Mercer, J. M. and V. L. Roth. 2003. The effects of Cenozoic global change on squirrel phylogeny. *Science* 299: 1568–1572.

Merriam, C. H. 1890. Results of a biological survey of the San Francisco Mountain region and the desert of the Little Colorado, Arizona. *North American Fauna* 3: 1–136.

Merriam, C. H. 1892. The geographical distribution of life in North America with special reference to the Mammalia. *Proceedings of the Biological Society of Washington* 7: 1–64.

Merriam, C. H. 1894. Laws of temperature control of the geographic distribution of terrestrial animals and plants. *National Geographic* 6: 229–238.

Mertens, R. 1934. Die Inseleidenchsen des Golfes von Salerno. *Senckenbergiana Biologica* 42: 31–40.

Meschede, M. and W. Frisch. 1998. A plate-tectonic model for the Mesozoic and Early Cenozoic history of the Caribbean plate. *Tectonophysics* 296: 269–291

Metcalf, H. and J. F. Collins. 1911. The control of the chestnut bark disease. *Farmer's Bulletin of the U. S. Deptartment of Agriculture* 467: 1–24.

Metcalfe, I. 1999. *Gondwana Dispersion and Asian Accretion.* Rotterdam, Netherlands: A. A. Balkema.

Meyer, A. 1993. Phylogenetic relationships and evolutionary processes in East African cichlid fishes. *Trends in Ecology and Evolution* 8: 279–284.

Meyer, C. P., J. B. Geller and G. Paulay. 2005. Fine scale endemism on coral reefs: archipelagic differentiation in turbinid gastropods. *Evolution* 59: 113–125.

Meyer, H. A. 2008. Distribution of tardigrades in Florida. *Southeastern Naturalist* 7: 91–100.

Meyers, G. S. 1953. Ability of amphibians to cross sea barriers, with especial reference to Pacific zoogeography. *Proceedings of the Seventh Pacific Science Congress* 4: 19–17.

Mielke, H. W. 1989. *Patterns of Life: Biogeography of a Changing World.* Boston: Unwin Hyman.

Mikkelson, G. M. 1993. How do food webs fall apart? A study of changes in tropic structure during relaxation on habitat fragments. *Okios* 67: 539–547.

Milankovitch, M. 1920. *Théorie Mathématique des Phénomènes Thermiques produits par la Radiation Solaire.* Paris: Gauthier-Villars.

Milankovitch, M. 1930. *Mathematische Klimalehre und Astronomische Theorie der Klimaschwankungen, Handbuch der Klimalogie Band 1.* Berlin: Teil A Borntrager.

Mileikovsky, S. A. 1971. Types of larval development in marine bottom invertebrates, their distribution and ecological significance: A re-evaluation. *Marine Biology* 10: 193–213.

Miles, D. B. and A. E. Dunham. 1996. The paradox of the phylogeny: Character displacement of analyses of body size in island *Anolis. Evolution* 50: 594–603.

Miles, J. 1987. Vegetation and succession: Past and present perceptions. In A. J. Gray, M. J. Crawley and P. J. Edwards (eds.), *Colonisation, Succession and Stability,* 1–30. Oxford: Blackwell Scientific Publications.

Millar, J. S. 1989. Reproduction and development. In G. L. Kirkland Jr. and J. N. Layne (eds.), *Advances in the Study of Peromyscus (Rodentia)*, 169–232. Lubbock, TX: Texas Tech University Press.

Miller, A. I. 1989. Spatio-temporal transitions in Paleozoic Bivalvia: An analysis of North America fossil assemblages. *Historical Biology* 1: 251–273.

Miller, B., G. Ceballos and R. Reading. 1994. The prairie dog and biotic diversity. *Conservation Biology* 8: 677–681.

Miller, R. R. 1948. The cyprinodont fishes of the Death Valley system of eastern California and southwestern Nevada. *University of Michigan Museum of Zoology Miscellaneous Publications* 42: 1–80.

Miller, R. R. 1961a. Man and the changing fish fauna of the American Southwest. *Papers of the Michigan Academy of Science, Arts and Letters* 46: 365–404.

Miller, R. R. 1961b. Speciation rates in some freshwater fishes of western North America. In W. F. Blair (ed.), *Vertebrate Speciation*, 537–560. Austin, TX: University of Texas Press.

Miller, R. R. 1966. Geographical distribution of Central American freshwater fishes. *Copeia* (4): 773–802.

Miller, R. S. 1967. Pattern and process in competition. *Advances in Ecological Research* 4: 1–74.

Millien, V. 2004. Relative effects of climate change, isolation and competition on body-size evolution in the Japanese field mouse, *Apodemus argenteus*. *Journal of Biogeography* 31: 1267–1276.

Millien, V. 2006. Morphological evolution is accelerated among island mammals. *PLoS Biology* 4(10): e321.

Millien, V. and J. Damuth. 2004. Climate change and size evolution in an island rodent species: New perspectives on the island rule. *Evolution* 58: 1353–1360.

Millien, V., S. K. Lyons, L. Olson, F. A. Smith, A. B. Wilson and Y. Yom-Tov. 2006. Ecotypic variation in the context of global climate change: Revisiting the rules. *Ecology Letters* 9: 853–869.

Millington, A., S. Walsh and P. E. Osborne (eds.). 2002. *GIS and Remote Sensing Applications in Biogeography and Ecology*. New York: Springer/Kluwer.

Milne, B. T. and R. T. Forman. 1986. Peninsulas in Maine: Woody plant diversity, distance and enviromental patterns. *Ecology* 67: 967–974.

Mittelbach, G. G., D. W. Schemske, H. V. Cornell, A. P. Allen, J. M. Brown, M. B. Bush, S. P. Harrison, A. H. Hurlbert, N. Knowlton, H. A. Lessios, et al. 2007. Evolution and the latitudinal diversity gradient: Speciation, extinction and biogeography. *Ecology Letters* 10: 315–331.

Moen, D. S., S. A. Smith and J. J. Wiens. 2009. Community assembly through evolutionary diversification and dispersal in Middle American treefrogs. *Evolution* 63: 3228–3247.

Moilanen, A., I. Hanski and A. T. Smith. 1998. Long-term dynamics in a metapopulation of the American pika. *American Naturalist* 152: 530–542.

Moles, A. T. and W. Westoby. 2003. Latitude, seed predation and seed mass. *Journal of Biogeography* 30: 105–128.

Moll, E. O. and J. M. Legler. 1971. The life history of a Neotropical slider turtle, *Pseudomys scripta* (Schoepff), in Panama. *Bulletin of the Los Angeles County Museum of Natural History of Science*, no. 11.

Molles, M. C., Jr. 1978. Fish species diversity on model and natural reef patches: Experimental insular biogeography. *Ecological Monographs* 48: 289–305.

Molnar, R. E. 2004. *Dragons in the Dust: The Paleobiology of the Giant Monitor Lizard Megalania*. Bloomington, IN: University of Indiana Press.

Monk, C. D. 1968. Successional and environmental relationships of the forest vegetation of north central Florida. *American Midland Naturalist* 79: 441–457.

Moodley, Y., B. Linz, Y. Yamaoka, H. M. Windsor, S. Breurec, J. Wu, A. Maady, S. Bernhoft, J. Thiberge, S. Phuanukoonnon, et al. 2009. The peopling of the Pacific from a bacterial perspective. *Science* 323: 527–530.

Mooney, H. A. (ed.). 1977. *Convergent Evolution in Chile and California: Mediterranean Climate Ecosystems*. US/IBP Synthesis Series 5. Stroudsburg, PA: Dowden, Hutchinson, & Ross.

Mooney, H. A. and E. E. Cleland. 2001. The evolutionary impact of invasive species. *Proceedings of the National Academy of Sciences, USA* 98: 5446–5451.

Mooney, H. A., J. Kummerow, A. W. Johnson, D. J. Parsons, D. Kelley, A. Hoffman, R. I. Hays, J. Giliberto and C. Chu. 1977. The producers: Their resources and adaptive responses. In H. A. Mooney (ed.), *Convergent Evolution in Chile and California*, 85–143. Stroudsburg, PA: Dowden, Hutchinson and Ross.

Moore, P. D. 2004. Isotopic biogeography. *Progress in Physical Geography* 28: 145–151.

Moore, W. S. 1995. Inferring phylogenies from mtdna variation—Mitochondrial-gene trees vs. nuclear-gene trees. *Evolution* 49: 718–726.

Mora, C., P. M. Chittaro, P. F. Sale, J. P. Kritzer and S. A. Ludsin. 2003. Patterns and processes in reef fish diversity. *Nature* 421: 933–936.

Moreau, R. E. 1944. Clutch size: A comparative study with special reference to African birds. *Ibis* 86: 286–347.

Morgan, G. S. and C. A. Woods. 1986. Extinction and zoogeography of West Indian land mammals. *Biological Journal of the Linnean Society* 28: 167–203.

Morgan, W. J. 1972a. Deep mantle convection plumes and plate motion. *Bulletin of the American Association of Petroleum Geologists* 56: 203–213.

Morgan, W. J. 1972b. Plate motions and deep mantle convection. *Memoirs of the Geological Society of America* 132: 7–22.

Morin, P. J and S. P. Lawler. 1995. Food web architecture and population dydnamics: Theory and empirical evidence. *Annual Review of Ecology and Systematics* 26: 505–530.

Morita, S. H., K. Morita and S. Yamamoto. 2009. Effects of habitat fragmentation by damming on salmonid fish: Lessons from white-spotted charr in Japan. *Ecological Research* 24: 711–722.

Moritz, C., J. L. Patton, C. J. Schneider and T. B. Smith. 2000. Diversification of rainforest faunas: An integrated molecular approach. *Annual Review of Ecology and Systematics* 31: 533–563.

Moritz, C., S. Lavery and R. Slade. 1995. Using allele frequency and phylogeny to define units for conservation and management. *American Fisheries Society Symposium* 17: 249–262.

Morrone, J. J. 1994. On the identification of areas of endemism. *Systematic Biology* 43: 438–441.

Morrone, J. J. 2002. Biogeographical regions under track and cladistic scrutiny. *Journal of Biogeography* 29: 149–152.

Morrone, J. J. 2006. Biogeographic areas and transition zones of Latin America and the Caribbean islands based on panbiogeographic and cladistic analyses of the entomofauna. *Annual Review of Entomology* 51: 467–494.

Morrone, J. J. 2009. *Evolutionary Biogeography: An Integrative Approach with Case Studies*. New York: Columbia University Press.

Morrone, J. J. and J. V. Crisci. 1995. Historical biogeography: Introduction to methods. *Annual Review of Ecology and Systematics* 26: 373–401.

Morton, E. S. 1978. Avian arboreal folivores: Why not? In G. G. Montgomery (ed.), *The Ecology of Arboreal Folivores*, 123–130. Washington, DC: Smithsonian Institution Press.

Morwood, M. J., R. P. Soejono, R. G. Roberts, T. Sutikna, C. S. M. Turney, K. E. Westaway, W. J. Rink, J.-X. Zhao, G. D. van den Bergh, R. A. Due, D. R. Hobbs, M. W. Moore, M. I. Bird and L.

K. Fifield. 2004. Archaeology and age of a new hominin from Flores in eastern Indonesia. *Nature* 431: 1087–1091.

Mosiman, J. E. and P. S. Martin. 1975. Simulating overkill by paleoindians. *American Scientist* 63: 304–313.

Moulton, M. P. 1993. The all-or-none pattern in introduced Hawaiian passeriforms: The role of competition sustained. *American Naturalist* 141: 105–119.

Moulton, M. P. and S. L. Pimm. 1983. The introduced Hawaiian avifauna: Biogeographical evidence for competition. *American Naturalist* 121: 669–690.

Moulton, M. P. and S. L. Pimm. 1986. The extent of competition in shaping an introduced avifauna. In J. M. Diamond and T. Case (eds.), *Community Ecology*, 80–97. New York: Harper and Row.

Moulton, M. P. and S. L. Pimm. 1987. Morphological assortment and introduced Hawaiian passerines. *Evolutionary Ecology* 1: 113–124.

Moyal, A. 2001. *Platypus: The Extraordinary Story of How a Curious Creature Baffled the World*. Washington, DC: Smithsonian Institution Press.

Moyle, P. B. and R. A. Leidy. 1992. Loss of biodiversity in aquatic ecosystems: Evidence from fish faunas. In P. L. Fiedler and S. K. Jain (eds.), *Conservation Biology: The Theory and Practice of Nature Conservation, Preservation and Management*, 127–170. New York: Chapman and Hall.

Mueller, G. M. and J. P. Schmit. 2007. Fungal biodiversity: What do we know? What can we predict? *Biodiversity and Conservation* 16: 1–5.

Mufwene, S. S. 2004. Language birth and death. *Annual Reviews of Anthropology* 33: 201–222.

Mullen, L. M. and H. E. Hoekstra. 2008. Natural selection along an environmental gradient: A classic cline in mouse pigmentation. *Evolution* 62: 1555–1570.

Muller, P. 1974. *Aspects of Zoogeography*. The Hague: Dr. W. Junk.

Muller, R. D., M. Sdrolias, C. Gaina and W. R. Roest. 2008. Age, spreading rates, and spreading asymmetry of the world's ocean crust. *Geochemistry Geophysics Geosystems* 9 (Q04006).

Munroe, E. 1996. Distributional patterns of Lepidoptera in the Pacific Islands. In A. Keast and S. E. Miller (eds.), *The Origin and Evolution of Pacific Island Biotas. New Guinea to Eastern Polynesia: Patterns and Processes*, 275–295. Amsterdam: SPB Academic.

Munroe, E. G. 1948. *The Geographical Distribution of Butterflies in the West Indies*. Ph.D. dissertation, Cornell University, Ithaca, NY.

Munroe, E. G. 1953. The size of island faunas. In *Proceedings of the Seventh Pacific Science Congress of the Pacific Science Association*, vol. IV, Zoology, 52–53. Auckland, New Zealand: Whitcome and Tombs.

Murienne, J., I. Karaman and G. Giribet. 2010. Explosive evolution of an ancient group of Cyphophthalmi (Arachnida: Opiliones) in the Balkan Peninsula. *Journal of Biogeography* 37: 90–102.

Murray, B. A. and G. C. Hose. 2005. The interspecific range size–body size relationship in Australian frogs. *Global Ecology and Biogeography* 14: 339–345.

Murray, B. R., C. R. Fonesca and M. Westoby. 1998. The macroecology of Australian frogs. *Journal of Animal Ecology* 67: 567–579.

Murray, J. 1895. A summary of scientific results. In *Challenger Report Summary*. 2 volumes. London: Nell and Company.

Musolin, D. L. 2007. Insects in a warmer world: Ecological, physiological and life-history responses of true bugs (Heteroptera) to climate change. *Climate Change Biology* 13: 1565–1585.

Mutke, J. and W. Barthlott. 2005a. Patterns of vascular plant diversity at continental to global scales. In I. Friis and H. Balsey (eds.), *Plant Diversity and Complexity patterns: Local,* *Regional, and Global Dimensions*. 521–537. The Royal Danish Academy of Sciences and Letters, Copenhagen.

Mutke, J. and W. Barthlott. 2005b. Patterns of vascular plant diversity at continental to global scales. *Biologiske Skrifter* 55: 521–531.

Myers, E. M. and D. C. Adams. 2008. Morphology is decoupled from interspecific competition in plethodon salamanders in the Shenandoah Mountains, USA. *Herpetologica* 64: 281–289.

Myklestad, A. and H. J. B. Birks. 1993. A numerical analysis of the distribution patterns of Salix L. species in Europe. *Journal of Biogeography* 20: 1–32.

Nadler, C. F., M. Zhurkevich, R. S. Hoffmann, A. I. Kozlovskii, L. Deutsch and D. F. Nadler Jr. 1978. Biochemical relatedness of some Holarctic voles of the genera *Microtus*, *Arvicola* and *Clethrionomys* (Rodentia: Arvicolae). *Canadian Journal of Zoology* 56: 1564–1575.

Nadler, C. F., N. N. Vorontsov, R. S. Hoffmann, I. I. Formichova and C. F. Nadler Jr. 1973. Zoogeography of transferins in arctic and long-tailed ground squirrel populations. *Comparative Biochemistry and Physiology* 44B: 33–40.

Nason, J. D., J. L. Hamrick and T. H. Fleming. 2002. Historical vicariance and postglacial colonization effects on the evolution of genetic structure in Lophocereus, a Sonoran Desert columnar cactus. *Evolution* 56: 2214–2226.

Nathan, R. 2005. Long-distance dispersal research: Building a network of yellow brick roads. *Diversity and Distribution* 11: 125–130.

Nathan, R., F. M. Schurr, O. Spiegel, O. Steinitz, A. Trakhtenbrot and A. Tsoar. 2008. Mechanisms of long-distance seed dispersal. *Trends in Ecology and Evolution* 23: 638–647.

NatureServe. 2009. NatureServe Explorer, version 7.1. http://www.natureserve.org/explorer/sumplan.htm.

Navarro, A. and N. H. Barton. 2003. Accumulating postzygotic isolation genes in parapatry: A new twist on chromosomal speciation. *Evolution* 57: 447–459.

Neall, V. E. and S. A. Trewick. 2008. The age and origin of the Pacific islands: A geological overview. *Philosophical Transactions of the Royal Society*, Series B 363: 3293–3308.

Neckel-Oliveira, S. and C. Gascon. 2006. Abundance, body size and movement patterns of a tropical treefrog in continuous and fragmented forests in the Brazilian Amazon. *Biological Conservation* 128: 308–315.

Neckel-Oliveira, S. and C. Gascon. 2006. Abundance, body size and movement patterns of a tropical tree frog in continuous and fragmented forests in the Brazilian Amazon. *Biological Conservation* 128: 308–316.

Neel, J. V. 1962. Diabetes mellitus: A "thrifty" genotype rendered detrimental by "progress"? *American Journal of Human Genetics* 14: 353–362.

Nei, M. 1987. *Molecular Evolutionary Genetics*. New York: Columbia University Press.

Neill, W. T. 1958. The occurrence of amphibians and reptiles in saltwater areas and the bibliography. *Bulletin of Marine Science of the Gulf and Caribbean* 8: 1–97.

Nelson, G. and P. Y. Ladiges. 1992. *TAS and TAX: MSDos computer programs for cladistics*. New York and Melbourne: published by the authors.

Nelson, G. and P. Y. Ladiges. 1995. *TASS (Three Area Subtrees)*. New York and Melbourne: published by the authors.

Nelson, G. and P. Y. Ladiges. 1996. Paralogy in cladistic biogeography and analysis of paralogy-free subtrees. *American Museum Novitates* 3167: 1–58.

Nelson, G. J. 1974. Historical biogeography: An alternative formalization. *Systematic Zoology* 23: 555–558.

Nelson, G. J. and D. E. Rosen (eds.). 1981. *Vicariance Biogeography: A Critique*. New York: Columbia University Press.

Nelson, G. J. and N. Platnick. 1981. *Systematics and Biogeography: Cladistics and Vicariance.* New York: Columbia University Press.

Nettle, D. 1999. *Linguistic Diversity.* Oxford: Oxford University Press.

Nevo, E. and H. Bar-El. 1976. Hybridization and speciation in fossorial mole rats. *Evolution* 30: 831–840.

Newman, M. T. 1953. The application of ecological rules to the racial anthropology of the aboriginal New World. *American Anthropology* 55: 311–327.

Nichols, J. 1999. *Linguistic Diversity in Space and Time.* Chicago: University of Chicago Press.

Nielson, M., K. Lohman and J. Sullivan. 2001. Phylogeography of the tailed frog (*Ascaphus truei*): Implications for the biogeography of the Pacific Northwest. *Evolution* 55: 147–160.

Niering, W. A. 1963. Terrestrial ecology of Kapingamarangi Atoll, Caroline Islands. *Ecological Monographs* 33: 131–160.

Niethammer, G. 1958. Tiergeographie (Bericht uber die Jahren 1950–56). *Fortschr. Zool.* 11: 35–141.

Niklas, K. J., B. H. Tiffney and A. H. Knoll. 1983. Apparent changes in the diversity of fossil plants. In M. C. Hecht, W. C. Steere and B. Wallace (eds.), *Evolutionary Biology,* 12: 1–89. New York: Plenum Press.

Nilsson, I. N. and S. G. Nilsson. 1982. Turnover of vascular plant species on small islands in Lake Möckeln, South Sweeden 1976–1980. *Oecologia* 53: 128–133.

Nilsson, I. N. and S. G. Nilsson. 1995. Experimental estimates of census efficiency and pseudo-turnover on islands: Error trend and between-observer variation when recording vascular plants. *Journal of Ecology* 73: 65–70.

Nilsson, S. G. 1977. Density compensation and competition among birds on small islands in a south Swedish lake. *Oikos* 28: 170–176.

Niven, J. 2007. Brains, islands and evolution: Breaking all the rules. *Trends in Ecology and Evolution* 22: 57–59.

Nixon, K. and Q. Wheeler, Q. 1990. An Amplification of the Phylogenetic Species Concept. *Cladistics* 6: 211–213.

Nobel, P. S. 1978. Surface temperarures of cacti: Influences of environmental and morphological factors. *Ecology* 59: 986–996.

Nobel, P. S. 1980a. Morphology, nurse plants and minimum apical temperatures for young *Carnegiea gigantea. Botanical Gazette* 141: 188–191.

Nobel, P. S. 1980b. Morphology, surface temperatures and northern limits of columnar cacti in the Sonoran Desert. *Ecology* 61: 1–7.

Noonan, B. P. and P. T. Chippindale. 2006. Vicariant origin of Malagasy reptiles supports Late Cretaceous Antarctic land bridge. *American Naturalist* 168: 730–741.

Norell, M. A. and X. Xu. 2005. Feathered dinosaurs. *Annual Review of Earth and Planetary Sciences* 33: 277–299.

Nores, M. 1995. Insular biogeography of birds on mountain-tops in north Argentina. *Journal of Biogeography* 22: 61–70.

NRC (National Research Council). 2008. *Ecological impacts of climate change.* NRC Committee on Ecological Impacts of Climate Change, National Academy Press, http://www.nap.edu/catalog/12491.html.

Nudds, R. L. and S. A. Oswald. 2007. An interspecific test of Allen's rule: Evolutionary implications for endothermic species. *Evolution* 61: 2839–2848.

Nur, A. and Z. Ben-Avraham. 1977. Lost Pacifica continent. *Nature* 270: 41–43.

Nylander, J. A. A., U. Olsson, P. Alstrom and I. Sanmartín. 2008. Accounting for phylogenetic uncertainty in biogeography: A Bayesian approach to dispersal-vicariance analysis of the thrushes (Aves: Turdus). *Systematic Biology* 57: 257–268.

O'Connell, J. F. and J. Allen. 2004. Dating the colonization of Sahul (Pleistocene Australia–New Guinea): A review of recent research. *Journal of Archaeological Science* 31: 835–853.

O'Conner, R. J. and J. Faaborg. 1992. The relative abundance of the brown-headed cowbird (*Molothrus ater*) in relation to exterior and interior edges in forests of Missouri. *Transactions of the Missouri Academy of Science* 26: 1–9.

O'Grady, P. and R. DeSalle. 2008. Out of Hawaii: The origin and biogeography of the genus *Scaptomyza* (Diptera: Drosophilidae). *Biology Letters* 4: 195–199.

O'Malley, M. A. 2008. "Everything is everywhere: But the environment selects": Ubiquitous distribution and ecological determinism in microbial biogeography. *Studies in History and Philosophy of Biological and Biomedical Sciences* 39: 314–325.

Ochocinska, D. and J. R. E. Taylor. 2003. Bergmann's rule in shrews: Geographical variation of body size in Palearctic *Sorex* species. *Biological Journal of the Linnean Society* 78: 365–381.

Odening, W. R., B. R. Strain and W. C. Oechel. 1974. The effects of decreasing water potential on net CO_2 exchange of intact desert shrubs. *Ecology* 55: 1086–1095.

Odum, E. P. 1969. The strategy of ecosystem development. *Science* 164: 262–270.

Odum, E. P. 1971. *Fundamentals of Ecology.* 3rd edition. Philadelphia: W. B. Saunders.

Odum, H. T. 1957. Trophic structure and productivity of Silver Springs, Florida. *Ecological Monographs* 27: 55–112.

Olabarria, C. and M. H. Thurston. 2003. Latitudinal and bathymetric trends in body size of the deep-sea gastropod *Troschelia berniciensis* (King). *Marine Biology* 143: 723–730.

Olden, J. D. and N. L. Poff. 2003. Toward a mechanistic understanding and prediction of biotic homogenization. *American Naturalist* 162: 442–460.

Olden, J. D., J. J. Lawler and N. L. Poff. 2008. Machine learning methods without tears: A primer for ecologists. *Quarterly Review of Biology* 83: 171–193.

Olesen, J. M. and A. Valido. 2003. Lizards as pollinators and seed dispersers: An island phenomenon. *Trends in Ecology and Evolution* 18: 177–181.

Olesen, J. M. and P. Jordano. 2002. Geographical patterns in plant-pollinator mutualistic networks. *Ecology* 83: 2416–2424.

Olesen, J. M., L. I. Eskildsen and S. Venkatasamy. 2002. Invasion of pollination networks on oceanic islands: Importance of invader complexes and endemic super-generalists. *Diversity and Distributions* 8: 181–192.

Oliva, M. E. and M. T. Gonzalez. 2005. The decay of similarity over geographical distance in parasite communities of marine fishes. *Journal of Biogeography* 32: 1327–1332.

Olson, D. M., E. Dinerstein, E. D. Wikramanayake, N. D. Burgess, G. V. N. Powell, E. C. Underwood, J. A. D'Amico, I. Itoua, H. E. Strand, J. C. Morrison, et al. 2001. Terrestrial ecoregions of the world: A new map of life on Earth. *BioScience* 51: 933–938.

Olson, S. L. 1973. Evolution of the rails of the South Atlantic Islands. *Smithsonian Contributions to Zoology* 152: 1–43.

Olson, S. L. 1976. Oligocene fossils bearing on the origins of the Totidae and Momotidae (Aves: Coraciiformes). *Smithsonian Contributions to Paleobiology* 27: 111–119.

Olson, S. L. and H. F. James. 1982a. Fossil birds from the Hawaiian Islands: Evidence for wholesale extinction by man before Western contact. *Science* 217: 633–635.

Olson, S. L. and H. F. James. 1984. The role of Polynesians in the extinction of the avifauna of the Hawaiian Islands. In P. S. Martin and R. G. Klein (eds.), *Quarterly Extinctions,* 768–780. Tucson, AZ: University of Arizona Press.

Olson, V. A., R. G. Davies, D. L. Orme, G. H. Thomas, S. Meiri, T. M. Blackburn, K. J. Gaston, I. P. F. Owens and P. M. Bennett. 2009. Global biogeography and ecology of body size in birds. *Ecology Letters* 12: 249–259.

Omi, P. N., L. C. Wensel and J. L. Murphy. 1979. An application of multivariate statistics to land-use planning: Classifying land units into homogeneous zones. *Forest Science* 25: 399–414.

Ong, P., L. E. Afuang and R. G. Rosell-Ambal (eds.). 2002. *Philippine Biodiversity Conservation Priorities: A Second Iteration of the National Biodiversity Strategy and Action Plan.* Philippine Department of the Environment and Natural Resources, Quezon City. xviii, 113.

Oppenheimer, C. 2003. *Out of Eden: The Peopling of the World.* London: Constable.

Oreskes, N. (ed.). 2001. *Plate Tectonics: An Insider's History of the Modern Theory of the Earth.* Cambridge, MA: Westview Press.

Orti, G. and A. Meyer. 1997. The radiation of characiform fishes and the limits of resolution of mitochondrial ribosomal DNA sequences. *Systematic Biology* 46: 75–100.

Ortmann, A. E. 1896. *Grundzuge der marinen tiergeographie.* Jena: G. Fisher.

Osborn, S. 2007. Distribution of the cattle egret (*Bubulcus ibis*) in North America. In A. Poole (ed.), *The Birds of North America Online.* Ithaca, NY: Cornell Lab of Ornithology.

Ospovat, D. 1977. Lyell's theory of climate. *Journal of the History of Biology* 10: 317–339.

Ostfeld, R. S. 1994. The fence effect reconsidered. *Oikos* 70: 340–348.

Ovaskainen, O. and I. Hanski. 2003. The species-area relationship derived from species-specific incidence functions. *Ecology Letters* 6: 903–909.

Overpeck, J. T., R. S. Webb and T. Webb III. 1992. Mapping eastern North American vegetation changes of the past 18 ka: No-analogs and the future. *Geology* 20: 1071–1074.

Owen, D. F. and J. Owen. 1974. Species diversity in temperate and tropical Ichneumonidae. *Nature* 249: 583–584.

Owen-Smith, N. 1987. Pleistocene extinctions: The pivotal role of megaherbivores. *Paleobiology* 13: 351–362.

Owen-Smith, N. 1989. Megafaunal extinctions: The conservation message from 11,000 years B.P. *Conservation Biology* 3: 405–412.

Pacala, S. and J. Roughgarden. 1985. Population experiments with the *Anolis* lizards of St. Maarten and St. Eustatius. *Ecology* 66: 128–141.

Paetkau, D., G. F. Shields and C. Strobeck. 1998. Gene flow between insular, coastal and interior populations of brown bears in Alaska. *Molecular Ecology* 7: 1283–1292.

Pafilis, P., S. Meiri, J. Foufopoulos and E. Valakos. 2009. Intraspecific competition and high food availability are associated with insular gigantism in a lizard. *Naturwissenschaften* 96: 1432–1904.

Page, R. D. M. 1987. Graphs and generalized tracks: Quantifying Croizat panbiogeography. *Systematic Zoology* 36: 1–17.

Page, R. D. M. 1989. *COMPONENT Users Manual*, Release 1.5. Auckland: Author.

Page, R. D. M. 1990. Tracks and trees in the antipodes: A reply. *Systematic Zoology* 39: 288–299.

Page, R. D. M. 1994. Parallel phylogenies: Reconstructing the history of host-parasite assemblages. *Cladistics* 10: 155–173.

Page, R. D. M. 2003. *Treemap 2.02. Program and User's Manual.* Division of Environmental and Evolutionary Biology, Institute of Biomedical and Life Sciences, University of Glasgow, Glasgow, U.K.

Pagel, M. D., R. M. May and A. R. Collie. 1991. Ecological aspects of the geographical distribution and diversity of mammalian species. *American Naturalist* 137: 791–815.

Paine, R. T. 1966. Food web complexity and species diversity. *American Naturalist* 100: 65–76.

Paine, R. T. 1974. Intertidal community structure: Experimental studies on the relationship between a dominant competitor and its principal predator. *Oecologia* 15: 93–120.

Paine, R. T. 1995. A conversation on refining the concept of keystone species. *Conservation Biology* 9: 962–965.

Palkovacs, E. P. 2003. Explaining adaptive shifts in body size on islands: A life history approach. *Oikos* 103: 37–44.

Palmer, D. 2007. *How Did We Get Here? The Complex Story of Human Evolution.* London: New Holland Publishers.

Palmer, M. A., R. F. Ambrose and N. L. Poff. 1997. Ecological theory and community restoration ecology. *Restoration Ecology* 5: 291–300.

Palombo M. R. 2009. Body size structure of the Pleistocene Mammalian communities from Mediterranean islands. *Integrative Zoology* 4: 341–356.

Palombo, M. R. 2007. How can endemic proboscideans help us understand the "island rule"? A case study of Mediterranean islands. *Quaternary International* 169–170: 105–124.

Palombo, M. R. 2008. Insularity and its effects. *Quaternary International* 182: 1–5.

Palumbi, S. R. 1997. Molecular biogeography of the Pacific. *Coral Reefs* 16: S47–S52.

Palumbi, S. R. 2001. Humans as the world's greatest evolutionary force. *Science* 293: 1786–1790.

Panero, J. L., J. Francisco-Ortega, R. K. Jansen and A. Santos-Guerra. 1999. Molecular evidence for multiple origins of woodiness and a New World biogeographic connection of the Macaronesian Islands endemic *Pericalis* (Asteracea: Senecioneae). *Proceedings of the National Academy of Sciences, USA* 96: 13886–13891.

Panitsa, M., D. Tzanoudakis, K. A. Triantis and S. Sfenthourakis. 2006. Patterns of species richness on very small islands: The plants of the Aegean archipelago. *Journal of Biogeography* 33: 1223–1234.

Parent, C. E., A. Caccone and K. Petren. 2008. Colonization and diversification of Galápagos terrestrial fauna: A phylogenetic and biogeographical synthesis. *Philosophical Transactions of the Royal Society of London B*: Biological Sciences 363: 3347–3361.

Parenti, L. R. 1981. Discussion [of C. Patterson, *Methods of Paleo-biogeography*]. In G. Nelson and D. E. Rosen (eds.), *Vicariance Biogeography: A Critique*, 490–497. New York: Columbia University Press.

Parker, V. T. 2002. Conceptual problems and scale limitations of defining ecological communities: A critique of the CI concept (community of individuals). *Perspectives in Plant Ecology, Evolution and Systematics* 4: 80–96.

Parmesan, C. 1996. Climate and species range. *Nature* 382: 765–766.

Parmesan, C. 2006 Ecological and evolutionary responses to recent climate change. *Annual Reviews of Ecology and Systematics* 37: 637–669.

Parmesan, C. and G. Yohe. 2003. A globally coherent fingerprint of climate change impacts across natural systems. *Nature* 421: 37–42.

Parrish, T. 2002. *Krakatau: Genetic Consequences of Island Colonization.* Herteren, Netherlands: University of Utrecht and Netherlands Institute of Ecology, Heteren.

Parsons, T. J. and M. M. Holland. 1998. Mitochondrial mutation rate revisited: Hot spots and polymorphism—Response, *Nature Genetics* 18: 110.

Pascual, R., M. Archer, E. O. Jaureguizar, J. L. Prado, H. Godthelp and S. J. Hand. 1992. First discovery of monotremes in South America. *Nature* 356: 704–706.

Paterson, H. E. H. 1993. *Evolution and the Recognition Concept of Species.* Baltimore: Johns Hopkins University Press.

Patriarca, M. and E. Heinsalu. 2009. Influence of geography on language competition. *Physica* 388: 174–186.

Patterson, B. D. 1980. Montane mammalian biogeography in New Mexico. *Southwestern Naturalist* 25: 33–40.

Patterson, B. D. 1981a. Morphological shifts of some isolated populations of *Eutamias* (Rodentia: Sciuridae) in different congeneric assemblages. *Evolution* 35: 53–66.

Patterson, B. D. 1984. Mammalian extinction and biogeography in the southern Rocky Mountains. In M. H. Nitecki (ed.), *Extinctions*, 247–294. Chicago: University of Chicago Press.

Patterson, B. D. 1987. The principle of nested subsets and its implications for biological conservation. *Conservation Biology* 1: 323–334.

Patterson, B. D. 1990. On the temporal development of nested subset patterns of species composition. *Oikos* 59: 330–342.

Patterson, B. D. 1995. Local extinctions and the biogeographic dynamics of boreal mammals in the southwest. In C. A. Istock and R. S. Hoffman (eds.), *Storm over a Mountain Island: Conservation Biology and the Mount Graham Affair*. Tucson, AZ: University of Arizona Press.

Patterson, B. D. 2000. Patterns and trends in the discovery of new Neotropical mammals. *Diversity and Distributions* 6: 145–151.

Patterson, B. D. 2001. Fathoming tropical biodiversity: The continuing discovery of Neotropical mammals. *Diversity and Distributions* 7: 191–196.

Patterson, B. D. and J. H. Brown. 1991. Regionally nested patterns of species composition in granivorous rodent assemblages. *Journal of Biogeography* 18: 395–402.

Patterson, B. D. and R. Pascual. 1972. The fossil mammal fauna of South America. In A. Keast, F. C. Erk and B. Glass (eds.), *Evolution, Mammals and Southern Continents*, 247–309. Albany: State University of New York Press.

Patterson, B. D. and W. Atmar. 1986. Nested subsets and the structure of insular mammalian faunas and archipelagoes. *Biological Journal of the Linnean Society* 28: 65–82.

Patterson, B. D., D. F. Stotz, S. Solari, J. W. Fitzpatrick and V. Pacheco. 1998. Contrasting patterns of elevational zonation for birds and mammals in the Andes of southeastern Peru. *Journal of Biogeography* 25: 583–607.

Patterson, B. D., V. Pacheco and S. Solari. 1996. Distribution of bats along an elevational gradient in the Andes of south-east Peru. *Journal of the Zoological Society of London* 240: 637–658.

Patterson, C. 1981. Methods of paleobiogeography. In G. Nelson and D. E. Rosen (eds.), *Vicariance Biogeography: A Critique*, 446–489. New York: Columbia University Press.

Patton, J. L. 1969. Chromosomal evolution in the pocket mouse, *Perognathus goldmani* Osgood. *Evolution* 23: 645–662.

Patton, J. L. 1972. Patterns of geographic variation in karyotype in the pocket gopher, *Thomomys bottae* (Eydoux and Gervais). *Evolution* 25: 574–586.

Patton, J. L. 1985. Population structure and the genetics of speciation in pocket gophers, genus *Thomomys*. *Acta Zoologica Fennica* 170: 109–114.

Patton, J. L., B. Berlin and E. A. Berlin. 1982. Aboriginal perspectives of a mammal community in amazonian Perú: Knowledge and utilization patterns among the Aguaruna Jivaro. In M. A. Mares and H. H. Genoways (eds.), *Mammalian Biology in South America*, vol. 6, 111–128. Linesville, PA: University of Pittsburgh.

Pauly, G. B., O. Piskurek and H. B. Shaffer. 2007. Phylogeographic concordance in the southeastern United States: The flatwoods salamander, *Ambystoma cingulatum*, as a test case. *Molecular Ecology* 16: 415–429.

Payne, J. L. and S. Finnegan. 2007. The effect of geographic range on extinction risk during background and mass extinction. *Proceedings of the National Academy of Sciences, USA* 104: 10506–10511.

Pearson, D. L. and S. S. Carroll. 1999. The influence of spatial scale on cross-taxon congruence patterns and prediction accuracy of species richness. *Journal of Biogeography* 26: 1079–1090.

Pearson, R. G. 2006. Climate change and the migration capacity of species. *Trends in Ecology and Evolution* 21: 111–113.

Pearson, R. G. and T. P. Dawson. 2003. Predicting the impacts of climate change on the distribution of species: Are bioclimate envelope models useful? *Global Ecology and Biogeography* 12: 361–371.

Pearson, S. and J. L. Betancourt.2002. Understanding arid environments using fossil rodent middens. *Journal of Arid Environments* 50: 499–511.

Pearson, T. G. (ed.). 1936. *Birds of America*. Garden City, NY: Garden City Publishing Co.

Peck, S. B. 2006. *The Beetles of the Galápagos Islands, Ecuador: Evolution, Ecology, and Diversity (Insecta: Coleoptera)*. Ottawa: NRC Research Press.

Peck, S. B., M. Larivee and J. Browne. 2005. Biogeography of ground beetles of Florida (Coleoptera: Carabidae): The peninsula effect and beyond. *Annals of the Entomological Society of America* 98: 951–959.

Pellmyr, O. and H. W. Krenn. 2002. Origin of a complex key innovation in an obligate insect–plant mutualism. *Proceedings of the National Academy of Sciences, USA* 99: 5498–5502.

Peltonen, A. and I. Hanski. 1991. Patterns of island occupancy explained by colonization and extinction rates in shrews. *Ecology* 72: 1698–1708.

Peltonen, A., S. Peltonen, P. Vilpas and A. Beloff. 1989. Distributional ecology of shrews in three archipelagoes in Finland. *Annales Zoologici Fennici* 26: 381–387.

Peng, C. H., J. Guiot, E. VanCamp and R. Cheddar. 1995. Temporal and spatial variations of terrestrial biomes and carbon storage since 13,000 yr B.P. in Europe: Reconstruction from pollen data and statistical models. *Water, Air and Soil Pollution* 82: 375–390.

Pennington, R. T., M. Lavin and A. Oliveira. 2009. Woody plant diversity, evolution, and ecology in the tropics: Perspectives from seasonally dry tropical forests. *Annual Review of Ecology, Evolution, and Systematics* 40: 437–457.

Perault, D. R. and M. V. Lomolino. 2000. Corridors and mammal community structure across a fragmented, old-growth forest landscape. *Ecological Monographs* 70: 401–422.

Perez-del-Olmo, A., M. Fernandez, J. A. Raga, A. Kostadinova and S. Morand. 2009. Not everything is everywhere: The distance decay of similarity in a marine host–parasite system. *Journal of Biogeography* 36: 200–209.

Perrie, L. and P. Brownsey. 2007. Molecular evidence of long-distance dispersal in the New Zealand pteridophyte flora. *Journal of Biogeography* 34: 2028–2038.

Pertoldi, C., R. Garcia-Perea, J. A. Godoy, M. Delibes and V. Loeschcke. 2005. Morphological consequences of range fragmentation and population decline on the endangered Iberian lynx (*Lynx pardinus*). *Journal of Zoology* 268: 73–86.

Pesole, G., C. Gissi, A. De Chirico and C. Saccone. 1999. Nucleotide substitution rate of mammalian mitochondrial genomes. *Journal of Molecular Evolution* 48: 427–434.

Peters, R. H. 1983. *The Ecological Implications of Body Size*. Cambridge: Cambridge University Press.

Peterson, A. T. 2001. Endangered species and peripheral populations: Cause for reflection. *Endangered Species Update* 18: 30–31.

Peterson, C. H. 1991. Intertidal zonation of marine invertebrates in sand and mud. *American Scientist* 79: 236–249.

Peterson, R. L. 1955. *North American Moose*. Toronto: University of Toronto Press.

Petren, K. and T. J. Case. 1997. A phylogenetic analysis of body size evolution and biogeography in chuckwallas (*Sauromalus*) and other iguanines. *Evolution* 51: 206–219.

Petren, K., D. T. Bolger and T. J. Case. 1993. Mechanisms in the competitive success of an invading sexual gecko over an asexual native. *Science* 259: 354–358.

Petren, K., P. R. Grant, B. R. Grant and L. F. Keller. 2005. Comparative landscape genetics and the adaptive radiation of Darwin's finches: The role of peripheral isolation. *Molecular Ecology* 14: 2943–2957.

Petuch, E. J. 1995. Molluscan diversity in the late Neogene of Florida: Evidence for a two-staged mass extinction. *Science* 270: 275–277.

Pfrender, M. E., W. E. Bradshaw and C. A. Kleckner. 1998. Patterns in the geographical range sizes of ectotherms in North America. *Oecologia* 115: 439–444.

Phillips, B. L., G. P. Brown, J. K. Webb and R. Shine. 2006. Invasion and the evolution of speed in toads. *Nature* 439: 803.

Phillips, B. L., G. P. Brown, J. M. J. Travis and R. Shine. 2008. Reid's paradox revisited: The evolution of dispersal kernels during range expansion. *American Naturalist* 172: S34–S48.

Pianka, E. R. 1966. Latitudinal gradients in species diversity: A review of concepts. *American Naturalist* 100: 33–46.

Pianka, E. R. 1986. *Ecology and Natural History of Desert Lizards*. Princeton, NJ: Princeton University Press.

Pianka, E. R. 1995. Evolution of body size: Varanid lizards as a model system. *American Naturalist* 146: 398–414.

Pickett, S. T. A. and P. S. White. 1985. *The Ecology of Natural Disturbance and Patchiness*. New York: Academic Press.

Pickett, T. A., J. Kolasa and C. G. Jones. 2007. *Ecological Understanding: The Nature of Theory and the Theory of Nature*. 2nd edition. Burlington, MA: Academic Press.

Pielou, E. C. 1975. *Ecological Diversity*. New York: John Wiley and Sons.

Pielou, E. C. 1977a. *Mathematical Ecology*. New York: John Wiley and Sons.

Pielou, E. C. 1977b. The latitudinal spans of seaweed species and their patterns of overlap. *Journal of Biogeography* 4: 299–311.

Pielou, E. C. 1979. *Biogeography*. New York: John Wiley and Sons.

Pielou, E. C. 1991. *After the Ice Age*. Chicago: University of Chicago Press.

Pijl, L. van den. 1972. *Principles of Dispersal in Higher Plants*. 2nd edition. New York: Springer-Verlag.

Pimentel, D., L. Lach, R. Zuniga and D. Morrison. 2000. Environmental and economic costs of nonindigenous species in the United States. *BioScience* 50: 53–65.

Pimm, S. L. 1991. *The Balance of Nature? Ecological Issues in the Conservation of Species and Communities*. Chicago: University of Chicago Press.

Pimm, S. L., H. L. Jones and J. M. Diamond. 1988. On the risk of extinction. *American Naturalist* 132: 757–785.

Plath, M. and U. Strecker. 2008. Behavioral diversification in a young species flock of pupfish (Cyprionodon spp.): Shoaling and aggressive behavior. *Behavioral Ecology and Sociobiology* 62: 1727–1737.

Platnick, N. I. 1991. On areas of endemism. *Australian Systematic Botany* 4(unnumbered).

Platnick, N. I. and G. Nelson. 1978. A method of analysis for historical biogeography. *Systematic Zoology* 27: 1–16.

Platt, W. J. 1975. The colonization and formation of equilibrium plant species associations on badger disturbances in a tall-grass prairie. *Ecological Monographs* 45: 285–305.

Poinar, G. O., G. M. Thomas and B. Lighthart. 1990. Bioassay to determine the effect of commercial preparations of bacillus-thuringiensis on entomogeneous rhabditoid nematodes. *Agriculture Ecosystems and Environment* 30: 195–202.

Poinar, H. N., M. Hofreiter, W. G. Spaulding, P. S. Martin, B. A. Stankiewicz, H. Bland, R. P. Evershed, G. Possnert and S. Paabo. 1998. Molecular coproscopy: Dung and diet of the extinct ground sloth *Nothrotheriops shastensis*. *Science* 281: 402–406.

Pole, M. 1994. The New Zealand flora—Entirely long-distance dispersal? *Journal of Biogeography* 21: 625–635.

Polis, G. A. and D. S. Hurd, 1995. Extraordinarily high spider densities on islands: Flow of energy from the marine to terrestrial food webs and the absence of predation. *Proceedings of the National Academy of Sciences, USA* 92: 4382–4386.

Pope, K. O. and J. E. Terrell. 2007. Environmental setting of human migrations in the circum-Pacific region. *Journal of Biogeography* 35: 1–21.

Popper, K. R. 1968a. *The Logic of Scientific Discovery*. 2nd edition. New York: Harper & Row.

Popper, K. R. 1968b. *Conjectures and Refutations*. New York: Harper & Row.

Por, F. D. 1971. One hundred years of Suez Canal: A century of Lessepsian migration: Retrospect and viewpoints. *Systematic Zoology* 20: 138–159.

Por, F. D. 1975. Pleistocene pulsation and preadaptation of biotas in Mediterranean seas: Consequences for Lessepsian migration. *Systematic Zoology* 24: 72–78.

Porter, J. W. 1972. Ecology and species diversity of coral reefs on opposite sides of the Isthmus of Panama. *Bulletin of the Biological Society of Washington* 2: 89–116.

Porter, J. W. 1974. Community structure and coral reefs on opposite sides of the Isthmus of Panama. *Science* 186: 543–545.

Porter, W. P. and M. Kearney. 2009. Size, shape, and the thermal niche of endotherms. *Proceedings of the National Academy of Sciences, USA* 106: 19666–19672.

Posada, D. and K. A. Crandall. 1998. MODELTEST: Testing the model of DNA substitution. *Bioinformatics* 14: 817–818.

Posada, D., K. A. Crandall and A. R. Templeton. 2000. GeoDis: A program for the cladistic nested analysis of the geographical distribution of genetic haplotypes. *Molecular Ecology* 9: 487–488.

Poulin, R. 1996. How many parasite species are there: Are we closest to answers? *International Journal for Parasitology* 26: 1127–1129.

Poulson, T. L. and W. B. White. 1969. The cave environment. *Science* 165: 971–981.

Pounds, J. A., M. R. Bustamante, L. A. Coloma, J. A. Consuegra, M. P. L. Fogden, P. N. Foster, E. La Marca, K. L. Masters, A. Merino-Viteri, R. Puschendorf, et al. 2006. Widespread amphibian extinctions from epidemic disease driven by global warming. *Nature* 439: 161–167.

Powell, M. G. 2007. Geographic range and genus longevity of late Paleozoic brachiopods. *Paleobiology* 33: 530–546.

Power, D. M. 1972. Numbers of bird species on the California Islands. *Evolution* 26: 451–463.

Power, M. E., D. Tilman, J. A. Estes, B. A. Menge, W. J. Bond, L. S. Mills, G. Daily, J. C. Castilla, J. Lubchenco and R. T. Paine. 1996. Challenges in the quest for keystones. *Biosience* 46: 609–620.

Powledge, F. 2003. Island biogeography's lasting impact. *BioScience* 53: 1032–1038.

Pratt, H. D. 2003. *The Hawaiian Honeycreepers: Drepanidinae*. Oxford: Oxford University Press.

Pratt, T. C., L. M. O'Connor, A. G. Hallett, R. L. McLaughlin, C. Katopodis, D. B. Hayes and R. A. Bergstedt. 2009. Balancing aquatic habitat fragmentation and control of invasive species: Enhancing selective fish passage at sea lamprey control barriers. *Transactions of the American Fisheries Society* 138: 652–666.

Pregill, G. K. and S. L. Olson. 1981. Zoogeography of West Indian vertebrates in relation to Pleistocene climatic cycles. *Annual Review of Ecology and Systematics* 12: 75–98.

Presley, S. J. and M. R. Willig. 2008. Composition and structure of Caribbean bat (Chiroptera) assemblages: Effects of inter-island distance, area, elevation and hurricane-induced disturbance. *Global Ecology and Biogeography* 17: 747–757.

Preston, F. W. 1962a. The canonical distribution of commonness and rarity: part I. *Ecology* 43: 185–215.

Preston, F. W. 1962b. The canonical distribution of commonness and rarity: part II. *Ecology* 43: 410–432.

Preston, F. W. 1957. Analysis of Maryland statewide bird counts. *Maryland Birdlife [Bulletin of the Maryland Ornithological Society]* 13: 63–65.

Price, G. J. 2008. Is the modern koala (*Phascolarctos cinereus*) a derived dwarf of a Pleistocene giant? Implications for testing megafauna extinction hypotheses. *Quaternary Science Reviews* 27: 2516–2521.

Price, J. P. and D. A. Clague. 2002. How old is the Hawaiian biota? Geology and phylogeny suggest recent divergence. *Proceedings of the Royal Society of London,* Series B 269: 2429–2435.

Price, J. P. and D. Elliott-Fisk. 2004. Topographic history of the Maui Nui complex, Hawai'i, and its implications for biogeography. *Pacific Science* 58: 27–45.

Price, N. J. 2001. *Major Impacts and Plate Tectonics.* New York: Routledge.

Price, P. W. 1980. *Evolutionary Biology of Parasites.* Monographs in Population Biology, no. 15. Princeton, NJ: Princeton University Press.

Price, T. D. and A. B. Phillimore. 2007. Reduced major axis regression and the island rule. *Journal of Biogeography* 34: 1998–1999.

Primack, R. 1998. *Essentials of Conservation Biology.* 2nd edition. Sunderland, MA: Sinauer Associates.

Pritchard, J. K., M. Stephens and P. Donnelly. 2000. Inferences of population structure using multilocus genotype data. *Genetics* 155: 945–959.

Prothero, D. R., L. C. Ivany and E. A. Nesbitt (eds.). 2002. *From Greenhouse to Icehouse: The Marine Eocene-Oligocene Transition.* New York: Columbia University Press.

Provan, J. and K. D. Bennett. 2008. Phylogeographic insights into cryptic glacial refugia. *Trends in Ecology and Evolution* 23: 564–571.

Prugh, L. R., K. E. Hodges, A. R. E. Sinclair and J. S. Brashares. 2008. Effects of habitat area and isolation on fragmented animal populations. *Proceedings of the National Academy of Sciences, USA* 105: 20770–20775.

Pruvot, G. 1896. *Essai sur les fonds et la faune de la Manche occidentale (cotes de Bretagne) compares a ceux du golfe du Lion.* Paris: Schleicher Frères.

Pulliam, H. R. 1988. Sources, sinks and population regulation. *American Naturalist* 132: 652–661.

Pyron, M. 1999. Relationships between geographical range size, body size, local abundance and habitat breadth in North American suckers and sunfishes. *Journal of Biogeography* 26: 549–558.

Pyšek, P. and D. M. Richardson. 2006. The biogeography of naturalization in alien plants. *Journal of Biogeography* 33: 2040–2050.

Qian, H. 2008. A latitudinal gradient of beta diversity for exotic vascular plant species richness in North America. *Diversity and Distributions* 14: 556–560.

Qian, H. 2009. Beta diversity in relation to dispersal ability for vascular plants in North America. *Global Ecology and Biogeography* 18: 327–332.

Qian, H. and R. E. Ricklefs. 2006. The role of exotic species in homogenizing the North American flora. *Ecology Letters* 9: 1293–1298.

Qian, H. and R. E. Ricklefs. 2007. A latitudinal gradient in large-scale beta diversity for vascular plants in North America. *Ecology Letters* 10: 737–744.

Qian, H., C. Badgley and D. L. Fox. 2009. The latitudinal gradient of beta diversity in relation to climate and topography for mammals of North America. *Global Ecology and Biogeography* 18: 111–133.

Rabenold, K. N. 1979. A reversed latitudinal diversity gradient in avian communities of eastern deciduous forests. *American Naturalist* 114: 275–286.

Rabinowitz, D., S. Cairns and T. Dillon. 1986. Seven forms of rarity and their frequency in the flora of the British Isles. In M. E. Soulé (ed.), *Conservation Biology: The Science of Scarcity and Diversity,* 182–204. Sunderland, MA: Sinauer Associates.

Rahbek, C. 1995. The elevational gradient of species richness: A uniform pattern? *Ecography* 18: 200–205.

Rahbek, C. 1997. The relationship among area, elevation and regional species richness in Neotropical birds. *American Naturalist* 149: 875–902.

Rahel, F. J. 2002. Homogenization of freshwater faunas. *Annual Review of Ecology and Systematics* 33: 291–315.

Rai, S. N., D. V. Ramana and A. Manglik. 2002. *Dynamics of Earth's Fluid System.* Rotterdam, Netherlands: A. A. Balkema., Netherlands.

Raikow, R. J. 1976. The origin and evolution of the Hawaiian honeycreepers (Drepaniidae). *Living Bird* 15: 95–117.

Ramakrishnan, U. and E. A. Hadly, 2009. Using phylochronology to reveal cryptic population histories. *Molecular Ecology* 18: 1310–1330.

Ramirez, L., J. A. F. Diniz-Filho and B. A. Hawkins. 2007. Partitioning phylogenetic and adaptive components of the geographic body size pattern of New World birds. *Global Ecology and Biogeography* 17: 100–110.

Rangel, T. F. L. V. B., J. A. F. Diniz-Filho and L. M. Bini. 2006. Towards an integrated computational tool for spatial analysis in macroecology and biogeography. *Global Ecology and Biogeography* 15: 321–327.

Rapoport, E. H. 1969. Gloger's rule and pigmentation of Collembola. *Evolution* 23: 622–626.

Rapoport, E. H. 1975. *Areografia: Estrategias Geograficas de las Especies.* Fondo de Cultura Economica.

Rapoport, E. H. 1982. *Areography: Geographical Strategies of Species.* New York: Pergamon Press.

Rapoport, E. H. 1994. Remarks on marine and continental biogeography: An aerographical viewpoint. *Philosophical Transactions of the Royal Society,* Series B 343: 71–78.

Rass, T. S. 1986. Vicariance icthyogeography of the Atlantic ocean pelagial. In *Pelagic Biogeography,* 237–241. UNESCO Technical Papers in Marine Science, 49.

Raunkiaer, C. 1934. *The Life Forms of Plants and Statistical Plant Geography.* Oxford: Clarendon Press.

Raup, D. M. 1976. Species diversity in the Phanerozoic: An interpretation. *Paleobiology* 2: 289–297.

Raup, D. M. 1979. Size of the Permo-Triassic bottleneck and its evolutionary implications. *Science* 206: 217–218.

Raup, D. M. 1994. The role of extinction in evolution. *Proceedings of the National Academy of Sciences, USA* 91: 6758–6763.

Raup, D. M. and D. Jablonski. 1993. Geography of end-Cretaceous marine bivalve extinctions. *Science* 260: 971–973.

Raup, D. M. and J. J. Sepkoski Jr. 1982. Mass extinctions in the marine fossil record. *Science* 215: 1501–1503.

Raup, D. M., S. J. Gould, T. J. M. Schopf and D. Simberloff. 1973. Stochastic models of phylogeny and the evolution of diversity. *Journal of Geology* 81: 525–542.

Raven, P. H. and E. O. Wilson. 1992. A fifty-year plan for biodiversity surveys. *Science* 258: 1099–1100.

Rawlinson, P. A., R. A. Zann, S. van Balen and I. W. B. Thornton. 1992. Colonization of the Krakatau islands by vertebrates. *Geojournal* 28: 225–31.

Raxworthy, C. J., M. R. J. Forstner and R. A. Nussbaum. 2002. Chameleon radiation by oceanic dispersal. *Nature* 415: 784–786.

Ray, C. 1960. The application of Bergmann's and Allen's rule to the poikilotherms. *Journal of Morphology* 106: 85–109.

Ray, N. and J. M. Adams. 2001. A GIS-based vegetation map of the world at the last glacial maximum (25,000–15,000 BP). *Internet Archaeology* 11: 1–44.

Reaka, M. L. 1980. Geographic range, life history patterns, and body size in a guild of coral-dwelling mantis shrimps. *Evolution* 34: 1019–1030.

Ree, R. H. and I. Sanmartín. 2009. Prospects and challenges for parametric models in historical biogeographical inference. *Journal of Biogeography* 36: 1211–1220.

Ree, R. H. and S. A. Smith. 2008. Maximum likelihood inference of geographic range evolution by dispersal, local extinction, and cladogenesis. *Systematic Biology* 57: 4–14.

Ree, R. H., B. R. Moore, C. O. Webb and M. J. Donoghue. 2005. A likelihood framework for inferring the evolution of geographic range on phylogenetic trees. *Evolution* 59: 2299–2311.

Reichman, O. J. and S. C. Smith. 1985. Impact of pocket gopher burrows on overlying vegetation. *Journal of Mammalogy* 66: 720–725.

Reid, W. V. and R. K. Miller. 1989. *Keeping Options Alive—The Scientific Basis for Conserving Biodiversity.* Washington, DC: World Resources Institute.

Reig, O. A. 1989. Karyotypic repatterning as one triggering factor in cases of explosive speciation. In A. Fondevila (ed.), *Evolutionary Biology of Transient Unstable Populations*, 246–289. Berlin: Springer-Verlag.

Reinhardt, K. 1997. Breeding success of southern hemisphere skuas Catharacta spp.: The influence of latitude. *Ardea* 85: 73–82.

Renner, S. S. 2004. Multiple Miocene Melastomataceae dispersal between Madagascar, Africa and India. *Philosophical Transactions of the Royal Society of London*, Series B 359: 1485–1494.

Rensch, B. 1938. Some problems of geographical variation and species-formation. *Journal of the Proceedings of the Linnean Society* 150: 275–285.

Rensch, B. 1960. *Evolution above the Species Level.* New York: Columbia University Press.

Rex, M. A. 1981. Community structure in the deep-sea benthos. *Annual Review of Ecology and Systematics* 12: 331–354.

Rex, M. A., C. R. McClain, N. A. Johnson, R. J. Etter, J. A. Allen, P. Bouchet and A. Waren. 2005. A source-sink hypothesis of abyssal biodiversity. *American Naturalist* 165: 163–178.

Rex, M. A., C. T. Stuart, R. R. Hessler, J. A. Allen, H. L. Sanders and G. D. F. Wilson. 1993. Global-scale patterns of species diversity in the deep sea benthos. *Nature* 365: 636–639.

Rey, J. R. 1981. Ecological biogeography of arthropods on *Spartina* islands in northwest Florida. *Ecological Monographs* 51: 237–265.

Rhode, D. 2001. Packrat middens as a tool for reconstructing historic ecosystems. In D. Egan and E. Howell (eds.), *Historical Ecology Handbook: A Restorationist's Guide to Reference Ecosystems*, 257–293. Covelo, CA: Island Press.

Ribas, C. R. and J. H. Schoereder. 2006. Is the Rapoport effect widespread? Null models revisited. *Global Ecology and Biogeography* 15: 614–624.

Richards, C. L., B. C. Carstens and L. L. Knowles. 2007. Distribution modelling and statistical phylogeography: An integrative framework for generating and testing alternative biogeographical hypotheses. *Journal of Biogeography* 34: 1833–1845.

Richards, M. A., R. G. Gordon and R. D. Van der Hilst. 2000. Plate tectonics and mantle convection: thirty years later. In M. A. Richards, R. G. Gordon and R. D. Van der Hilst (eds.), *History and Dynamics of Global Plate Motions, Geophysical Monograph 121.* 1–4. Washington, DC: American Geophysical Union.

Richards, P. W. 1996 *The Tropical Rainforest: An Ecological Study.* 2nd edition. Cambridge: Cambridge University Press.

Richardson, D. M., and R. J. Whittaker. 2010. Conservation biogeography–foundations, concepts and challenges. *Diversity and Distributions* 16: 313–320.

Rickart, E. A. 2001. Elevational diversity gradients, biogeography and the structure of montane mammal communities in the intermountain region of North America. *Global Ecology and Biogeography* 10: 77–100.

Rickart, E. A., L. R. Heaney, D. S. Balete and B. R. Tabaranza Jr. In press. Small mammal diversity along an elevational gradient in northern Luzon, Philippines. *Mammalian Biology*, doi:10.1016/j.mambio.2010.01.006.

Ricketts, J. H. 2001. The matrix matters: Effective isolation in fragmented landscapes. *American Naturalist* 158: 87–99.

Ricklefs, R. E. 1980. Geographical variation in clutch size among passerine birds: Ashmole's hypothesis. *Auk* 97: 38–49.

Ricklefs, R. E. 2005. Taxon cycles: Insights from invasive species. In D. F. Sax, J. J. Stachowicz and S. D. Gaines (eds.), *Species Invasions: Insights into Ecology, Evolution and Biogeography*, 165–200. Sunderland, MA: Sinauer Associates.

Ricklefs, R. E. 2006a. Evolutionary diversification and the origin of the diversity-environment relationship. *Ecology* 87: S3–S13.

Ricklefs, R. E. 2006b. Global variation in the diversification rate of passerine birds. *Ecology* 87: 2468–2478.

Ricklefs, R. E. 2008. Disintegration of the ecological community. *The American Naturalist* 172: 741–750.

Ricklefs, R. E. 2010. Dynamics of colonization and extinction on islands: Insights from Lesser Antillean birds. In J. Losos and R. E. Ricklefs (eds.), *The Theory of Island Biogeography Revisited*, 388–414. Princeton, NJ: Princeton University Press.

Ricklefs, R. E. and D. Schluter (eds.). 1993. *Species Diversity in Ecological Communities: Historical and Geographical Perspectives.* Chicago: University of Chicago Press.

Ricklefs, R. E. and E. Bermingham. 2002. The concept of the taxon cycle in biogeography. *Global Ecology and Biogeography* 11: 353–361.

Ricklefs, R. E. and E. Bermingham. 2004. History and the species–area relationship in Lesser Antillean birds. *American Naturalist* 163: 227–239.

Ricklefs, R. E. and E. Bermingham. 2007. The causes of evolutionary radiations in archipelagoes: Passerine birds in the Lesser Antilles. *American Naturalist* 169: 285–297.

Ricklefs, R. E. and E. Bermingham. 2008. The West Indies as a laboratory of biogeography and evolution. *Philosophical Transactions of the Royal Society of London*, Series B 363: 2393–2413.

Ricklefs, R. E. and G. W. Cox. 1972. Taxon cycles of the West Indian avifauna. *American Naturalist* 106: 295–219.

Ricklefs, R. E. and G. W. Cox. 1978. Stage of taxon cycle, habitat distribution and population density in the avifauna of the West Indies. *American Naturalist* 112: 875–895.

Ricklefs, R. E. and S. M. Fallon. 2002. Diversification and host switching in avian malaria parasites. *Proceedings of the Royal Society of London*, Series B 269: 885–892.

Ricklefs, R. E., S. M. Fallon and E. Bermingham. 2004. Evolutionary relationships, cospeciation, and host switching in avian malaria parasites. *Systematic Biology* 53: 111–119.

Rico-Gray, V. and P. S. Oliveira. 2007. *The Ecology and Evolution of Ant–Plant Interactions.* Chicago: University of Chicago Press.

Riddle, B. R. 1996. The molecular phylogenetic bridge between deep and shallow history in continental biotas. *Trends in Ecology and Evolution* 11: 207–211.

Riddle, B. R. 2005. Is biogeography emerging from its identity crisis? *Journal of Biogeography* 32: 185–186.

Riddle, B. R. and D. J. Hafner. 2004. The past and future roles of phylogeography in historical biogeography. In M. V. Lomolino and L. R. Heaney (eds.), *Frontiers of Biogeography*, 93–110. Sunderland, MA: Sinauer Associates.

Riddle, B. R. and D. J. Hafner. 2006a. A step-wise approach to integrating phylogeographic and phylogenetic biogeographic perspectives on the history of a core North American warm deserts biota. *Journal of Arid Environments* 65: 435–461.

Riddle, B. R. and D. J. Hafner. 2006b. Phylogeography in historical biogeography: Investigating the biogeographic histories of populations, species, and young biotas. In M. C. Ebach and R. S. Tangney (eds.), *Biogeography in a Changing World*, 161–176. Boca Raton, FL: CRC Press.

Riddle, B. R., D. J. Hafner, L. F. Alexander and J. R. Jaeger. 2000. Cryptic vicariance in the historical assembly of a Baja California Peninsular Desert biota. *Proceedings of the National Academy of Sciences, USA* 97: 14438–14443.

Riddle, B. R., M. N. Dawson, E. A. Hadly, D. J. Hafner, M. J. Hickerson, S. J. Mantooth and A. D. Yoder. 2008. The role of molecular genetics in sculpting the future of integrative biogeography. *Progress in Physical Geography* 32: 173–202.

Ridley, H. N. 1930. *The Dispersal of Plants throughout the World*. Ashford, England: L. Reeve & Co.

Rieppel, O. 2002. A case of dispersing chameleons. *Nature* 415: 744–745.

Rissler, L. J. and J. J. Apodaca. 2007. Adding more ecology into species delimitation: Ecological niche models and phylogeography help define cryptic species in the Black Salamander (*Aneides flavipunctatus*). *Systematic Biology* 56: 924–942.

Roberts, D. F. 1953. Body weight, race and climate. *American Journal of Physical Anthropology* 11: 533–558.

Roberts, D. F. 1978. *Climate and Human Variability*. 2nd edition. Menlo Park, CA: C. A Cummings.

Roberts, R. G. and B. W. Brook. 2010. Turning back the clock on extinction of megafauna of Australia. *Quaternary Science Reviews* 29: 593–595.

Robinove, C. J. 1979. *Integrated Terrain Mapping with Digital Landsat Images in Queensland, Australia*. Professional Paper 1102. Washington, DC: U.S. Geological Survey. 39 pp.

Robinson, G. R., R. D. Holt, M. S. Gaines, S. P. Hamburg, M. L. Johnson, H. S. Fitch, E. A. Martinko. 1992. Diverse and contrasting effects of habitat fragmentation. *Science* 257: 524–526.

Robinson, W. D. 1999. Long-term changes in the avifauna of Barro Colorado Island, Panama, a tropical forest isolate. *Conservation Biology* 13: 85–97.

Roca, A. L., N. Georgiadis and S. J. O'Brien. 2005. Cytonuclear genomic dissociation in African elephant species. *Nature Genetics* 37: 96–100.

Rocha, S., M. A. Carretero, M. Vences, F. Glaw and D. J. Harris. 2005. Deciphering patterns of transoceanic dispersal: The evolutionary origin and biogeography of coastal lizards (*Cryptoblepharus*) in the western Indian Ocean region. *Journal of Biogeography* 33: 13–22.

Rodda, G. H. and K. Dean-Bradley. 2002. Excess density compensation of island herpetofaunal assemblages. *Journal of Biogeography* 29: 623–632.

Roderick, G. K. 1997. Herbivorous insects and the Hawaiian silversword alliance: Coevolution or cospeciation? *Pacific Science* 51: 440–449.

Rodriguez, J. P. 2002. Range contraction in declining North American bird populations. *Ecological Applications* 12: 238–248.

Rodriguez, J., M. T. Alberdi, B. Azanza and J. L. Prado. 2004. Body size structure in north-western Mediterranean Plio-Pleistocene mammalian faunas. *Global Ecology and Biogeography* 13: 163–176.

Rodríguez, M. Á., M. Á. Olalla-Tárraga and B. A. Hawkins. 2008. Bergmann's rule and the geography of mammal body size in the Western Hemisphere. *Global Ecology and Biogeography* 17: 274–283.

Roff, D. A. 1986. The evolution of wing dimorphism in insects. *Evolution* 40: 1009–1020.

Roff, D. A. 1990. The evolution of flightlessness in insects. *Ecological Monographs* 60: 389–421.

Rogers, R. A., L. A. Rogers, R. S. Hoffmann and L. D. Martin. 1991. Native American biological diversity and the biogeographic influence of Ice Age refugia. *Journal of Biogeography* 18: 623–630.

Rogers, R. A., L. D. Martin and T. D. Nicklas. 1990. Ice-age geography and the distribution of native North American languages. *Journal of Biogeography* 17: 131–143.

Rohde, K. 1992. Latitudinal gradients in species diversity: The search for the primary cause. *Oikos* 65: 514–527.

Rohde, K. 1996. Rapoport's rule is a local phenomenon and cannot explain latitudinal gradients in species diversity. *Biodiversity Letters* 3: 10–13.

Rohde, K. 1999. Latitudinal gradients in species diversity and Rapoport's rule revisited: A review of recent work and what can parasites teach us about the causes of the gradients? *Ecography* 22: 593–613.

Rohde, K. and M. Heap. 1996. Latitudinal ranges of teleost fish in the Atlantic and Indo-Pacific oceans. *American Naturalist* 147: 659–665.

Rohde, K., M. Heap and D. Heap. 1993. Rapoport's rule does not apply to marine teleosts and cannot explain latitudinal gradients in species richness. *American Naturalist* 142: 1–16.

Romer, A. S. 1966. *Vertebrate Paleontology*. Chicago: Chicago University Press.

Romm, J. 1992. *The Edges of the Earth in Ancient Thought: Geography, Exploration, and Fiction*. Princeton, NJ: Princeton University Press.

Ronce, O. 2007. How does it feel to be like a rolling stone? Ten questions about dispersal evolution. *Annual Reviews of Ecology, Evolution and Systematics* 38: 231–253.

Rongo, T., M. Bush and R. van Woesik. 2009. Did ciguatera prompt the late Holocene Polynesian voyages of discovery? *Journal of Biogeography* 36: 1423–1432.

Ronquist, F. 1997. Dispersal-vicariance analysis: A new approach to the quantification of historical biogeography. *Systematic Biology* 46: 195–203.

Ronquist, F. 2002. Parsimony analysis of coevolving species associations. In R. D. M. Page (ed.), *Cospeciation*, 22–64. Chicago: University of Chicago Press.

Ronquist, F. and S. Nylin. 1990. Process and pattern in the evolution of species association. *Systematic Zoology* 39: 323–344.

Roosevelt, A. C., M. Lima deCosta, C. Lopes Machado, M. Michab, N. Mercier, H. Vallada, J. Feathers, W. Barnett, M. Imazio da Silveira and K. Schick. 1996. Paleoindian cave dwellers in the Amazon: The peopling of the Americas. *Science* 272: 373–384.

Root, T. 1988a. *Atlas of Wintering North American Birds: An Analysis of Christmas Bird Count Data*. Chicago: University of Chicago Press.

Root, T. 1988b. Energy constraints on avian distributions and abundances. *Ecology* 69: 330–339.

Root, T. 1988c. Environmental factors associated with avian distributional boundaries. *Journal of Biogeography* 15: 489–505.

Root, T. L., J. T. Price, K. R. Hall, S. H. Schnieder, C. Rosenzweig and J. A. Pounds. 2003. Fingerprints of global warming on wild animals and plants. *Nature* 421: 57–60.

Rosen, B. R. 1984. Reef coral biogeography and climate through the Late Cainozoic: Just islands in the sun or a critical pattern of islands. In P. J. Brenchley (ed.), *Fossils and Climate*. New York: Wiley.

Rosen, B. R. 1988. Progress, problems and patterns in the biogeography of reef corals and other tropical marine organisms. *Helgolander Meeresuntersuchungen* 42: 269–301.

Rosen, D. E. 1975. A vicariance model of Caribbean biogeography. *Systematic Zoology* 24: 431–464.

Rosen, D. E. 1978. Vicariant patterns and historical explanations in biogeography. *Systematic Zoology* 27: 159–188.

Rosen, D. E. 1979. Fishes from the uplands and intermontane basins of Guatemala: Revisionary studies and comparative geography. *Bulletin of the American Museum of Natural History* 162: 267–376.

Rosenberg, M. S. 2010. A generalized formula for converting chi-square tests to effect sizes for meta-analysis. *PLoS ONE* 5: e10059.

Rosenberg, M. S., D. C. Adams and J. Gurevitch. 2000. *MetaWin. Statistical software for meta-analysis. Version 2.0.* Sinauer Associates, Sunderland, Massachusetts.

Rosenfield, J. A. 2002. Pattern and process in the geographical ranges of freshwater fishes. *Global Ecology and Biogeography* 11: 323–332

Rosenzweig, M. L. 1966. Community structure in sympatric Carnivora. *Journal of Mammalogy* 47: 602–612.

Rosenzweig, M. L. 1968. Net primary productivity of terrestrial communities: Predictions from climatological data. *American Naturalist* 102: 67–74.

Rosenzweig, M. L. 1978. Competitive speciation. *Biological Journal of the Linnaean Society* 10: 275–289.

Rosenzweig, M. L. 1992. Species diversity gradients: We know more and less than we thought. *Journal of Mammalogy* 73: 715–730.

Rosenzweig, M. L. 1995. *Species Diversity in Space and Time.* New York: Cambridge University Press.

Rosenzweig, M. L. and M. V. Lomolino. 1997. Rarity and community ecology: Who gets the tiny bits of the broken stick. In W. E. Kunin and K. J. Gaston (eds.), *The Biology of Rarity: Causes and Consequences of Rare-Common Differences,* 63–90. London: Chapman and Hall.

Roth, V. L. 1990. Insular dwarf elephants: A case study in body mass estimation and ecological inference. In J. Damuth and B. J. MacFadden (eds.), *Body Size in Mammalian Paleobiology: Estimation and Biological Implications,* 151–179. New York: Cambridge University Press.

Roughgarden, J. 1972. Evolution of niche width. *American Naturalist* 106: 683–718.

Roughgarden, J. 1974. Niche width: Biogeographic patterns among *Anolis* lizard populations. *American Naturalist* 108: 429–442.

Roughgarden, J. 1992. Comments on the paper by Losos: Character displacement vs. taxon loop. *Copeia* 1992: 288–295.

Roughgarden, J. 1995. Anolis *Lizards of the Caribbean: Ecology, Evolution and Plate Tectonics.* Oxford: Oxford University Press.

Roughgarden, J. and E. R. Fuentes. 1977. The environmental determinants of size in solitary populations of West Indian *Anolis* lizards. *Oikos* 29: 44–51.

Roughgarden, J. and M. Feldman. 1975. Species packing and predation pressure. *Ecology* 56: 489–492.

Roughgarden, J., D. Heckel and E. R. Fuentes. 1983. Coevolutionary theory and the biogeography and community

structure of *Anolis*. In R. B. Huey, E. R. Pianka and T. W. Schoener (eds.), *Lizard Ecology: Studies of a Model Organism,* 371–410. Cambridge, MA: Harvard University Press.

Roughgarden, J., R. M. May and S. A. Levin. 1989. *Perspectives in Ecological Theory.* Princeton, NJ: Princeton University Press.

Roughgarden, J., S. D. Gaines and S. Pacala. 1987. Supply side ecology: The role of physical transport processes. In P. Giller and J. Gee (eds.), *Organization of Communities: Past and Present,* 491–518. Oxford: Blackwell Scientific Publications.

Rounds, R. C. 1987. Distribution and analysis of colourmorphs of the black bear (*Ursus americanus*). *Journal of Biogeography* 14: 521–538.

Rowe, J. S. 1980. The common denominator in land classification in Canada: An ecological approach to mapping. *Forestry Chronicle* 56: 19–20.

Rowe, R. J. and S. Lidgard. 2009. Elevational gradients and species richness: Do methods change pattern perception? *Global Ecology and Biogeography* 18: 163–177.

Roy, K. 1994. Effects of the Mesozoic Marine Revolution on the taxonomic, morphologic and biogeographic evolution of a group: Aporrhaid gastropods during the Mesozoic. *Paleobiology* 20: 274–296.

Roy, K., D. Jablonski and J. W. Valentine. 2001b. Climate change, species range limits and body size in marine bivalves. *Ecology Letters* 4: 366–370.

Roy, K., D. Jablonski and J. W. Valentine. 2004. Beyond species richness: Biogeographic patterns and biodiversity dynamics using other metrics of diversity. In M. V. Lomolino and L. R. Heaney, *Frontiers of Biogeography: New Directions in the Geography of Nature,* 151–170. Sunderland, MA: Sinauer Associates.

Roy, K., D. Jablonski and K. K. Martien. 2000. Invariant size-frequency distributions along a latitudinal gradient in marine bivalves. *Proceedings of the National Academy of Sciences, USA* 97: 13150–12155.

Roy, K., D. Jablonski, J. W. Valentine and G. Rosenberg. 1998. Marine latitudinal gradients: Tests of causal hypotheses. *Proceedings of the National Academy of Sciences, USA* 95: 3699–3702.

Roy, K., D. P. Balch and M. E. Hellberg. 2001. Spatial patterns of morphological diversity across the Indo-Pacific: Analysis using strombid gastropods. *Proceedings of the Royal Society of London,* Series B 268: 1–6.

Rubinoff, I. 1968. Central American sea-level canal: Possible biological effects. *Science* 161: 857–861.

Rubinoff, R. W. and I. Rubinoff. 1971. Geographic and reproductive isolation in Atlantic and Pacific populations of Panamanian *Bathygobius*. *Evolution* 25: 88–97.

Ruff, C. B. 1991. Climate, body size and body shape in hominid evolution. *Journal of Human Evolution* 21: 81–105.

Ruff, C. B. 1994. Morphological adaptation to climate in modern and fossil hominids. *Yearbook of Physical Anthropology* 37: 65–107.

Ruff, C. B. 2002. Variation in human body size and shape. *Annual Review of Anthropology* 31: 211–232.

Ruffieux L., J. Elouard, and M. Sartori. 1998. Flightlessness in mayflies and its relevance to hypotheses on the origin of insect flight. *Proceedings of the Royal Society B,* Biological Sciences 265: 2135–2140.

Ruggiero, A. 2001. Size and shape of the geographical ranges of Andean passerine birds: Spatial patterns in environmental resistance and anisotropy. *Journal of Biogeography* 28: 1281–1294.

Ruggiero, A. and B. A. Hawkins. 2006. Mapping macroecology. *Global Ecology and Biogeography* 15: 433–437.

Ruggiero, A. and B. A. Hawkins. 2008. Why do mountains support so many species of birds? *Ecography* 31: 306–315.

Runcorn, S. K. 1956. Paleomagnetic comparisons between Europe and North America. *Proceedings of the Geological Association of Canada* 8: 77–85.

Runcorn, S. K. 1962. Paleomagnetic evidence for continental drift and its geophysical cause. In S. K. Runcorn (ed.), *Continental Drift*, 1–40. International Geophysics Series 3. New York: Academic Press.

Ruokolainen, K. and J. Vormisto. 2000. The most widespread Amazonian palms tend to be tall and habitat generalists. *Basic and Applied Ecology* 1: 97–108.

Rushton, S. P., S. J. Ormerod and G. Kerby. 2004. New paradigms for modelling species distributions? *Journal of Applied Ecology* 41: 193–200.

Russello, M. A., S. Glaberman, J. P. Gibbs, C. Marquez, J. R. Powell and A. Caccone. 2005. A cryptic taxon of Galapagós tortoise in conservation peril. *Biology Letters* 1: 287–290.

Russo, C. A. M., N. Takezaki and M. Nei. 1995. Molecular phylogeny and divergence times of drosophilid species. *Molecular Biology and Evolution* 12: 391–404.

Rutschmann, F. 2006. Molecular dating of phylogenetic trees: A brief review of current methods that estimate divergence times. *Diversity and Distributions* 12: 35–48.

Ryder, O. A. 1986. Species conservation and systematics: The dilemma of subspecies. *Trends in Ecology & Evolution* 1: 9–10.

Saarma, U., S. Y. W. Ho, O. G. Pybus, M. Kaljuste, I. L. Tumanov, I. Kojola, A. A. Vorobiev, N. I. Markov, A. P. Saveljev, H. Valdmann, et al. 2007. Mitogenetic structure of brown bears (*Ursus arctos* L.) in northeastern Europe and a new time frame for the formation of European brown bear lineages. *Molecular Ecology* 16: 401–413.

Safriel, U. N., S. Volis and S. Kark. 1994. Core and peripheral populations and global climate change. *Israel Journal of Plants Sciences* 42: 331–345.

Sagarin, R. D. and S. D. Gaines. 2002. The 'abundant centre' distribution: To what extent is it a biogeographical rule? *Ecology Letters* 5: 137–147.

Sagarin, R. D., S. D. Gaines and B. Gaylord. 2006. Moving beyond assumptions to understand abundance distributions across the ranges of species. *Trends in Ecology and Systematics* 21: 524–530.

Sakai, A., W. L. Wagner, D. M. Ferguson and D. R. Herbst. 1995. Origins of dioecy in the Hawaiian flora. *Ecology* 76: 2517–2529.

Saleeby, J. B. 1983. Accretionary tectonics of the North American cordillera. *Annual Review of Earth and Planetary Sciences* 11: 45–73.

Salzburger, W., T. Mack, E. Verheyen and A. Meyer. 2005. Out of Tanganyika: Genesis, explosive speciation, key-innovations and phylogeography of the haplochromine cichlid fishes. *BMC Evolutionary Biology* 1: 17.

Sanchez-Cordero, V. 2001. Elevation gradients of diversity for rodents and bats on Oaxaca, Mexico. *Global Ecology and Biogeography* 10: 63–76.

Sand, H., G. Cederlund and Kjell Danell. 1995. Geographical and latitudinal variation in growth patterns and adult body size of Swedish moose (*Alces alces*). *Oecologia* 102: 433–442.

Sander, P. M., O. Mateus, T. Laven and N. Knotschke. 2006. Bone histology indicates insular dwarfism in a new Late Jurassic sauropod dinosaur. *Nature* 441: 739–741.

Sanders, H. L. 1968. Marine benthic diversity: A comparative study. *American Naturalist* 102: 243–282.

Sanders, J. J. 2002. Elevational gradients in ant species richness: Area, geometry and Rapoport's rule. *Ecography* 25: 25–32.

Sanders, K. L., M. S. Y. Lee, R. Leys, R. Foster and J. S. Keogh. 2008. Molecular phylogeny and divergence dates for Australasian elapids and sea snakes (Hydrophiinae):

Evidence from seven genes for rapid evolutionary radiations. *Journal of Evolutionary Biology* 21: 682–695.

Sanders, N. J., J. Moss and D. Wagner. 2003. Patterns of ant species richness along elevational gradients in an arid ecosystem. *Global Ecology and Biogeography* 12: 93–102.

Sanderson, M. J. 2002. Estimating absolute rates of molecular evolution and divergence times: A penalized likelihood approach. *Molecular Biology and Evolution* 19: 101–109.

Sanmartín, I. and F. Ronquist. 2004. Southern Hemisphere biogeography inferred by event-based models: Plant vs. animal patterns. *Systematic Biology* 53: 216–243.

Sanmartin, I., H. Enghoff and F. Ronquist. 2001. Patterns of animal dispersal, vicariance and diversification in the Holarctic. *Biological Journal of the Linnean Society* 73: 345–390.

Sanmartín, I., L. Wanntorp and R. C. Winkworth. 2007. West Wind Drift revisited: Testing for directional dispersal in the Southern Hemisphere using event-based tree fitting. *Journal of Biogeography* 34: 398–416.

Sanmartín, I., P. Van der Mark and F. Ronquist. 2008. Inferring dispersal: A Bayesian approach to phylogeny-based island biogeography, with special reference to the Canary Islands. *Journal of Biogeography* 35: 428–449.

Sant'ana, C. E. R. and J. A. Diniz-Filho. 1999. Macroecologigia de corujas (Aves, Strigiformes) de America do Sul. *Ararajuba* 7: 3–11.

Santini, F. and R. Winterbottom. 2002. Historical biogeography of Indo-western Pacific coral reef biota: Is the Indonesian region a centre of origin? *Journal of Biogeography* 29: 189–205.

Sarich, V. M. and A. C. Wilson. 1967. Immunological time scale for hominid evolution. *Science* 158: 1200–1203.

Sarich, V. M. and A. C. Wilson. 1973. Generation time and genomic evolution in primates. *Science* 179: 1144–1147.

Sastrawan, P. and C. Ciofi. 2002. Distribution and home range: In J. Murphy, C. Ciofi, C. De La Panouse and T. Walsh (eds.), *Komodo dragons: Biology and conservation*, 42–77. Washington, DC: Smithsonian Institution Press.

Sastre, P., P. Roca, M. Lobo and EDIT co-workers. 2009. A geoplatform for improving accessibility to environmental cartography. *Journal of Biogeography* 36: 568.

Sauer, J. 1969. Oceanic islands and biogeographic theory: A review. *Geographical Review* 59: 582–593.

Sauer, J. D. 1988. *Plant Migration: The Dynamics of Geographic Patterning in Seed Plants*. Berkeley: University of California Press.

Saunders, D. A., R. J. Hobbs and C. R. Margules. 1991. Biological consequences of ecosystem fragmentation: A review. *Conservation Biology* 5: 18–32.

Sauquet, H., P. H. Weston, N. P. Barker, C. L. Anderson, D. J. Cantrill and V. Savolainen. 2009. Using fossils and molecular data to reveal the origins of the Cape proteas (subfamily Proteoideae). *Molecular Phylogenetics and Evolution* 51: 31–43.

Savage, J. M. 1973. The geographic distribution of frogs: Patterns and predictions. In J. L. Vial (ed.), *Evolutionary Biology of the Anurans*, 351–445. Columbia: University of Missouri Press.

Savidge, J. A. 1987. Extinction of an island forest avifauna by an introduced snake. *Ecology* 68: 660–668.

Savilov, A. I. 1961. The distribution of the ecological forms of the by-the-wind sailor, *Velella lata*, Ch. and Eys. and the Portugese man-of-war *Physalia utriculus* (La Martiniere) Esch., in the North Pacific. *Transactions Institute Okenologyi, Akademis Nauk. SSSR* 45: 223–239.

Savolainen, V., M. C. Anstett, C. Lexer, I. Hutton, J. J. Clarkson, M. V. Norup, M. P. Powell, D. Springate, N. Salamin and W. J. Baker. 2006. Sympatric speciation in palms on an oceanic island. *Nature* 441: 210–213.

Sax, D. F., J. J. Stachowicz and S. D. Gaines. 2005. *Species Invasions: Insights into Ecology, Evolution and Biogeography.* Sunderland, MA: Sinauer Associates.

Sax, D. F., J. J. Stachowicz, J. H. Brown, J. F. Bruno, M. N. Dawson, S. D. Gaines, R. K. Grosberg, A. Hastings, R. D. Holt, M. M. Mayfield, M. I. O'Connor and W. R. Rice. 2007. Ecological and evolutionary insights from species invasions. *Trends in Ecology and Evolution* 22: 465–471.

Sax, D. F., S. D. Gaines and J. H. Brown. 2002. Species invasions exceed extinctions on islands worldwide: A comparative study of plants and birds. *American Naturalist* 160: 766–783.

Sax, D.F. 2001. Latitudinal gradients and geographic ranges of exotic species: implications for biogeography. *Journal of Biogeography* 28: 139–150.

Scanlon, J. D. and M. S. Y. Lee. 2004. Phylogeny of Australasian venomous snakes (*Colubroidea, Elapidae, Hydrophiinae*) based on phenotypic and molecular evidence. *Zoologica Scripta* 33: 335–366.

Schaefer, H.-C., W. Jetz and K. Böhning-Gaese. 2008. Impact of climate change on migratory birds: Community reassembly versus adaptation. *Global Ecology and Biogeography* 17: 38–49.

Scheiner, S. M. 2003. Six types of species–area curves. *Global Ecology and Biogeography* 12: 441–447.

Scheiner, S. M. and M. R. Willig. 2005. Developing unified theories in ecology as exemplified with diversity gradients. *American Naturalist* 166: 458–469.

Scheiner, S. M. and M. R. Willig. 2008. A general theory of ecology. *Theoretical Ecology* 1: 21–28.

Schemske, D. W., B. C. Husband, M. H. Ruckelshaus, C. Goodwillie, I. M. Parker and J. G. Bishop. 1994. Evaluating approaches to the conservation of rare and endangered plants. *Ecology* 75: 584–606.

Schenk, H. J. and R. B. Jackson. 2002. The global biogeography of roots. *Ecological Monographs* 72: 311–328.

Schimper, A. F. W. 1898. *Pflanzengeographie auf physiologischer Grundlage.* 1st edition. Jena.

Schimper, A. F. W. 1903. *Plant-Geography upon a Physiological Basis.* Trans. W. R. Fisher. Oxford: Clarendon Press.

Schipper, J. et al. 2008. The status of the world's land and marine mammals: Diversity, threat and knowledge. *Science* 322(5899): 225–230, doi: 10.1126/science.1165115.

Schlaepfer, M. A., M. C. Runge and P. W. Sherman. 2002. Ecological and evolutionary traps. *Trends in Ecology and Evolution* 17: 478–480.

Schliewen, U. K. and B. Klee. 2004. Reticulate speciation in Cameroonian crater lake cichlids. *Frontiers in Zoology* 1: 1–12.

Schlinger, E. I. 1974. Continental drift, *Nothofagus* and some ecologically associated insects. *Annual Review of Entomology* 19: 323–343.

Schlotfeldt, B. E. and S. Kleindorfer. 2006. Adaptive divergence in the superb fairy-wren (*Malurus cyaneus*): A mainland versus island comparison of morphology and foraging behaviour. *Emu* 106: 309–319.

Schluter D. 1996a. Ecological causes of adaptive radiation. *American Naturalist* 148: S40–S64.

Schluter, D. 2000. *The Ecology of Adaptive Radiation.* Oxford: Oxford University Press.

Schluter, D. and G. L. Conte. 2009. Genetics and ecological speciation. *Proceedings of the National Academy of Sciences USA* 106: 9955–9962.

Schluter, D. and P. R. Grant. 1984. The distribution of *Geospiza difficilis* in relation to *G. fuliginosa* in the Galápagos Islands: Tests of three hypotheses. *Evolution* 36: 1213–1226.

Schluter, D., T. D. Price and P. R. Grant. 1985. Ecological character displacement in Darwin's finches. *Science* 227: 1056–1059.

Schmida, A. and M. V. Wilson. 1985. Biological determinants of species diversity. *Journal of Biogeography* 12: 1–20.

Schmidt, N. M. and P. M. Jensen. 2003. Changes in mammalian body length over 175 years—Adaptations to a fragmented landscape? *Conservation Ecology* 7: 6.

Schmidt, N. M. and P. M. Jensen. 2005. Concomitant patterns in avian and mammalian body length changes in Denmark. *Ecology and Society* 10(2): 5. URL: http://www.ecologyandsociety.org/vol10/iss2/art5/

Schmidt-Nielsen, K. 1963. Osmotic regulation in higher vertebrates. *Harvey Lectures*, ser. 58: 53–93.

Schmidt-Nielsen, K. 1964. *Desert Animals: Physiological Problems of Heat and Water.* New York: Oxford University Press.

Schmitt, R. J. 1987. Indirect interactions between prey: Apparent competition, predator aggregation and habitat segregation. *Ecology* 68: 1887–1897.

Schneider, S. H. and T. L. Root. 2002. *Wildlife Responses to Climate Change: North American Case Studies.* Washington, DC: Island Press.

Schoener, T. W. 1968a. The *Anolis* lizards of Bimini: Resource partitioning in a complex fauna. *Ecology* 49: 704–726.

Schoener, T. W. 1968b. Sizes of feeding territories among birds. *Ecology* 49: 123–141.

Schoener, T. W. 1970. Size patterns in West Indian *Anolis* lizards. II. Correlations with the sizes of particular sympatric species: Displacement and divergence. *American Naturalist* 104: 155–174.

Schoener, T. W. 1975. Presence and absence of habitat shift in some widespread lizard species. *Ecological Monographs* 45: 233–258.

Schoener, T. W. 1976. The species-area relationship within archipelagoes: Models and evidence from island birds. *Proceedings of the XVI International Ornithological Congress* 6: 629–642.

Schoener, T. W. 1983. Rate of species turnover decreases from lower to higher organisms: A review of the data. *Oikos* 41: 372–377.

Schoener, T. W. 1986. Patterns in terrestrial vertebrate vs. arthropod communities: Do systematic differences in regulation exist? In J. Diamond and T. J. Case (eds.), 556–586. *Community Ecology.* New York: Harper and Row.

Schoener, T. W. 1988. The ecological niche. In J. M. Cherrett (ed.), *Ecological Concepts*, 79–113. Oxford: Blackwell Scientific Publications.

Schoener, T. W. 1989. Food webs from the small to the large. *Ecology* 70: 1559–1589.

Schoener, T. W. 2010. The MacArthur–Wilson equilibrium model: A chronicle of what it said and how it was tested. In J. B. Losos and R. E. Ricklefs (eds.), *The Theory of Island Biogeography Revisited*, 52–87. Princeton, NJ: Princeton University Press.

Schoener, T. W. and A. Schoener. 1983. Distribution of vertebrates on some very small islands. I. Occurrence sequences of individual species. *Journal of Animal Ecology* 52: 209–235.

Schoener, T. W. and D. A. Spiller. 1987. High population persistence in a system with high turnover. *Nature* 330: 474–477.

Schoener, T. W. and G. H. Adler. 1991. Greater resolution of distributional complementarities by controlling for habitat affinities: A study of Bahamian lizards. *American Naturalist* 137: 669–692.

Scholz, C. A., T. C. Johnson, A. S. Cohen, J. W. King, J. A. Peck, J. T. Overpeck, M. R. Talbot, E. T. Brown, L. Kalindekafe, P. Y. O. Amoako, et al. 2007. East African megadroughts between 135 and 75 thousand years ago and bearing on early-modern human origins. *Proceedings of the National Academy of Sciences, USA* 104: 16416–16421.

Schopf, J. M. 1970a. Gondwana paleobotany. *Antarctic Journal of the U.S.* 5: 62–66.

Schopf, J. M. 1970b. Relation of floras of the Southern Hemisphere to continental drift. *Taxon* 19: 657–674.

Schopf, J. M. 1974. Permo-Triassic extinctions: Relation to sea-floor spreading. *Journal of Geology* 82: 129–144.

Schopf, J. M. 1976. Morphologic interpretation of fertile structures in glossopterid gymnosperms. *Review of Palaeobotany and Palynology* 21: 25–64.

Schopf, J. W., A. B. Kudryavtsev, A. D. Czaja and A. B. Tripathi. 2007. Evidence of archean life: Stromatolites and microfossils. *Precambrian Research* 158: 141–155.

Schouw, F. 1823. Grunzüge einer algemeinen Pflanzengeographie. Berlin.

Schreiber, A. 2009. Comparative allozyme genetics and range history of the European river barbel (Teleostei, Cyprinidae: *Barbus barbus*) in the Rhine/upper Danube contact area. *Journal of Zoological Systematics and Evolutionary Research* 47: 149–159.

Schuster, M. F. and S. McDaniel. 1973. A vegetative analysis of a Black Belt prairie relict site near Aliceville, Alabama. *Journal of the Mississippi Academy of Science* 19: 153–159.

Schwartz, J. H. 2004. Getting to know *Homo erectus*. *Science* 305: 53–55.

Schwartz, M. W. 1988. Species diversity patterns in woody flora of three North American peninsulas. *Journal of Biogeography* 15: 759–774.

Schwarzbach, M. 1980. *Alfred Wegener: The Father of the Continental Drift*. Madison, WI: Science Technology.

Sclater, J. G. and C. Tapscott. 1979. The history of the Atlantic. *Scientific American* 240: 156–174.

Sclater, P. L. 1858. On the general geographical distribution of the members of the class Aves. *Journal of the Linnean Society, Zoology* 2: 130–145.

Sclater, P. L. 1897. On the distribution of marine mammals. *Proceedings of the Zoological Society of London* 41: 347–359.

Scotese, C. R. 2004. Cenozoic and Mesozoic paleogeography: Changing terrestrial biogeographic pathways. In M. V. Lomolino and L. R. Heaney (eds.), *Frontiers of Biogeography: New Directions in the Geography of Nature,* Chapter 1. Cambridge: Cambridge University Press.

Scott, J. M., P. J. Heglund, M. L. Morrison, J. B. Haufler, M. G. Raphael, W. A. Wall and F. B. Samson. (eds). 2002. *Predicting Species Occurrences: Issues of Accuracy and Scale*. Washington, DC: Island Press.

Scott, S. N., S. M. Clegg, S. P. Blomberg, J. Kikkawa and I. P. F. Owens. 2003. Morphological shifts in island-dwelling birds: The roles of generalist foraging and niche expansion. *Evolution* 57: 2147–2156.

Scriber, J. M. 1973. Latitudinal gradients in larval feeding specialization of the world Papilionidae (Lepidoptera). *Psyche* 80: 355–373.

Seehausen, O. 2002. Patterns in fish radiation are compatible with Pleistocene desiccation of Lake Victoria and 14,600 year history for its cichlid species flock. *Proceedings of the Royal Society of London: Biological Sciences* 269: 491–497.

Seelanan, T., A. Schnabel and J. F. Wendel. 1997. Congruence and consensus in the cotton tribe (Malvaceae). *Systematic Botany* 22: 259–290.

Segurado, P. and M. B. Araujo. 2004. An evaluation of methods for modeling species distributions. *Journal of Biogeography* 31: 1555–1568.

Selander, R. K., M. H. Smith, S. Y. Yang, W. E. Johnson and J. B. Gentry. 1971. Biochemical polymorphism and systematics in the genus *Peromyscus*. II. Genic heterozygosity and genetic similarity among populations of the old-field mouse (*Peromyscus polionotus*). *Studies in Genetics VI, University of Texas Publication,* 7103: 49–90.

Sequeira, F., J. Alexandrino, S. Weiss and N. Ferrand. 2008. Documenting the advantages and limitations of different classes of molecular markers in a well-established phylogeographic context: Lessons from the Iberian endemic Golden-striped salamander, *Chioglossa lusitanica* (Caudata: Salamandridae). *Biological Journal of the Linnean Society* 95: 371–387.

Serrat, M. A., D. King and C. O. Lovejoy. 2008. Temperature regulates limb length in homeotherms by directly modulating cartilage growth. *Proceedings of the National Academy of Sciences, USA* 105: 19348–19353.

Settele, J., C. R. Margules, P. Poschlod and K. Henle, editors. 1996. *Species Survival in Fragmented Landscapes*. Dordrecht, The Netherlands: Kluwer Academic.

Sexton, J. P., P. J. McIntyre, A. L. Angert and K. J. Rice. 2009. Evolution and ecology of species range limits. *Annual Reviews of Ecology and Systematics* 40: 415–436.

Sexton, P. F. and R. D. Norris. 2008. Dispersal and biogeography of marine plankton: Long-distance dispersal of the foraminifer *Truncorotalia truncatulinoides*. *Geology* 36: 899–902.

Sfenthourakis, S. and K. A. Triantis. 2009. Habitat diversity, ecological requirements of species and the small island effect. *Diversity and Distributions* 15: 131–140.

Shane, C. K. and E. J. Cushing. 1991. *Quaternary Landscapes*. Minneapolis: University of Minnesota Press.

Shannon, C. E. and W. Weaver. 1949 (1964). *The Mathematical Theory of Communication*. Urbana, IL: University of Illinois Press.

Shapiro, B., A. J. Drummond, A. Rambaut, M. C. Wilson, P. E. Matheus, A. V. Sher, O. G. Pybus, M. T. P. Gilbert, I. Barnes, J. Binladen, et al. 2004. Rise and fall of the Beringian steppe bison. *Science* 306: 1561–1565.

Sharpton, V. L., G. B. Dalrymple, L. E. Marin, G. Ryder and B. C. Schuraytz. 1992. New links between the Chicxulub impact structure and the Cretaceous/Tertiary boundary. *Nature* 359: 819.

Shefer, S. A. Abelson, O. Mokady and E. Geffen. 2004. Red to Mediterranean Sea bioinvasion: Natural drift through the Suez Canal, or anthropogenic transport? *Molecular Ecology* 13: 2333–2343.

Shepherd, U. L. 1998. A comparison of species diversity and morphological diversity across the North American latitudinal gradient. *Journal of Biogeography* 25: 19–29.

Shilton, L. A. and R. J. Whittaker. 2010. The role of pteropodid bats (Megachiroptera) in re-establishing tropical forests on Krakatau. In T. H. Fleming and P. A. Racey (eds.), *Island Bats: Ecology, Evolution, and Conservation*, 176–215. Chicago: University of Chicago Press.

Shilton, L. A., J. D. Altringham, S. G. Compton and R. J. Whittaker. 1999. Old World fruit bats can be long-distance seed dispersers through extended retention of viable seed in the gut. *Proceedings of the Royal Society of London*, Series B 266: 219–223.

Shipley, B. and P. A. Keddy. 1987. The individualistic and community-unit concepts as falsifiable hypothesis. *Vegetatio* 69: 47–55.

Shreve, F. 1942. The desert vegetation of North America. *The Botanical Review* 8: 195–246.

Shrode, J. B. 1975. Developmental temperature tolerance of a Death Valley pupfish (*Cyprirodon nevadensis*). *Physiological Zoology* 48: 378–389.

Sibley, C. G. and J. E. Ahlquist. 1985. The Phylogeny and Classification of the Australo-Papuan Passerine Birds. *Emu* 85: 1–14.

Sibley, C. G. and J. E. Ahlquist. 1990. *Phylogeny and Classification of Birds*. New Haven, CT: Yale University Press.

Siler, C. D., J. D. McVay, A. C. Diesmos and R. M. Brown. 2009. A new species of fanged frog, genus *Limnonectes* (Amphibia:

Anura: Dicroglossidae) from Southeast Mindanao Island, Philippines. *Herpetologica* 65: 105–114.

Silkas, B., S. L. Olson and R. C. Fleischer. 2001. Rapid, independent evaluation of flightlessness in four species of Pacific Island rails (Rallidae): An analysis based on mitochondrial sequence data. *Journal of Avian Biology* 32: 5–14.

Silva, M. and J. A. Downing. 1995. The allometric scaling of density and body size: A non-linear relationship for terrestrial mammals. *American Naturalist* 141: 704–727.

Silvertown, J. 2004. The ghost of competition past in the phylogeny of island endemic plants. *Journal of Ecology* 92: 168–173.

Silvertown, J., J. Francisco-Ortega and M. Carine. 2005. The monophyly of island radiations: An evaluation of niche pre-emption and some alternative explanations. *Journal of Ecology* 93: 653–657.

Simard, M. A., S. D. Côté, R. B. Weladji and L. J. Huot. 2008. Feedback effects of chronic browsing on life-history traits of a large herbivore. *Journal of Animal Ecology* 77: 678–686.

Simberloff, D. S. 1974a. Equilibrium theory of island biogeography and ecology. *Annual Review of Ecology and Systematics* 5: 161–182.

Simberloff, D. S. 1974b. Permo-Triassic extinctions: Effects of area on biotic equilibrium. *Journal of Geology* 82: 267–274.

Simberloff, D. S. 1978. Using island biogeographic distributions to determine if colonization is stochastic. *American Naturalist* 112: 713–726.

Simberloff, D. S. and E. O. Wilson. 1969. Experimental zoogeography of islands: The colonization of empty islands. *Ecology* 50: 278–296.

Simberloff, D. S. and E. O. Wilson. 1970. Experimental zoogeography of islands: A two-year record of colonization. *Ecology* 51: 934–937.

Simberloff, D. S. and J. Martin. 1991. Nestedness of insular avifaunas: Simple summary statistics masking complex species patterns. *Ornis Fennica* 68: 178–192.

Simberloff, D. S. and M. D. Collins. 2010. Birds of the Solomon Islands: The domain of the dynamic equilibrium theory and assembly rules, with comments on the taxon cycle. In J. Losos and R. E. Ricklefs (eds.), *The Theory of Island Biogeography Revisited*, 237–263. Princeton, NJ: Princeton University Press.

Simberloff, D. S. and W. Boecklen. 1991. Patterns of extinction in the introduced Hawaiian avifauna: A re-examination of the role of competition. *American Naturalist* 138: 300–327.

Simberloff, D. S., K. L. Heck, E. D. McCoy and E. F. Conner. 1981a. There have been no statistical tests of cladistic biogeographical hypothesis. In G. Nelson and D. E. Rosen (eds.), *Vicariance Biogeography: A Critique*, 40–64. New York: Columbia University Press.

Simberloff, D. S., K. L. Heck, E. D. McCoy and E. F. Conner. 1981b. Response. In G. Nelson and D. E. Rosen (eds.), *Vicariance Biogeography: A Critique*, 85–93. New York: Columbia University Press.

Simberloff, D., T. Dayan, C. Jones and G. Ogura. 2000. Character displacement and release in the small Indian mongoose, *Herpestes javanicus*. *Ecology* 81: 2086–2099.

Simmons, N. B. and J. H. Geisler. 1998. Phylogenetic relationships of *Icaronycteris*, *Archaeonycteris*, *Hassianycteris* and *Palaeochiropteryx* to extant bat lineages, with comments on the evolution of echolocation and foraging strategies in Microchiroptera. *Bulletin of the American Musuem of Natural History* 4: 1–182.

Simmons, N. B., K. L. Seymour, J. Habersetzer and G. F. Gunnell. 2008. Primitive Early Eocene bat from Wyoming and the evolution of flight and echolocation. *Nature* 451: 818–822.

Simpson, G. G. 1936. Data on the relationships of local and continental mammal faunas. *Journal of Paleontology* 10: 410–414.

Simpson, G. G. 1940. Mammals and land bridges. *Journal of the Washington Academy of Science* 30: 137–163.

Simpson, G. G. 1944. *Tempo and Mode in Evolution*. New York: Columbia University Press.

Simpson, G. G. 1945. The principles of classification and a classification of mammals. *Bulletin of the American Museum of Natural History* 85: i–xvi, 1–350.

Simpson, G. G. 1950. History of the fauna of Latin America. *American Scientist* 38: 361–389.

Simpson, G. G. 1952a. Periodicity in vertebrate evolution. *Journal of Paleontology* 26: 359–370.

Simpson, G. G. 1953. *The Major Features of Evolution*. New York: Columbia University Press.

Simpson, G. G. 1961a. *Principles of Animal Taxonomy*. New York: Columbia University Press.

Simpson, G. G. 1961b. Historical zoogeography of Australian mammals. *Evolution* 15: 413–446.

Simpson, G. G. 1964. Species density of North American Recent mammals. *Systematic Zoology* 13: 57–73.

Simpson, G. G. 1969. South American mammals. In E. J. Fittkau, J. Illies, H. Klinge, G. H. Schwabe and H. Sioli (eds.), *Biogeography and Ecology in South America*, vol. 2, Monographiae Biologicae 20: 879–909. The Hague: Dr. W. Junk.

Simpson, G. G. 1970. Uniformitarianism: An inquiry into principle, theory and method in geohistory and biohistory. In M. K. Hecht and W. C. Steere (eds.), *Essays in Evolution and Genetics in Honor of Theodosius Dobzhansky*, 43–96. New York: Appleton-Century-Crofts.

Simpson, G. G. 1978. Early mammals in South America: Fact, controversy and mystery. *Proceedings of the American Philosophical Society* 122: 318–328.

Sinclair, W. A. 1964. Comparisons of recent declines of white ash, oaks and sugar maple in northeastern woodlands. *Cornell Plantations* 20: 62–67.

Sismondo, S. 2000. Island biogeography and the multiple domains of models. *Biology and Philosophy* 15: 239–258.

Sites, J. W. and J. C. Marshall. 2003. Delimiting species: A Renaissance issue in systematic biology. *Trends in Ecology and Evolution* 18: 462–470.

Skarpaas, O. and K. Shea. 2007. Dispersal patterns, dispersal mechanisms, and invasion wave speeds for invasive thistles. *American Naturalist* 170: 421–430.

Skelly, D. K, J. L. Joseph, H. P. Possingham, L. K. Freidenburg, T. J. Farrugia, M. T. Kinnison and A. P. Hendry. 2007. Evolutionary responses to climate change. *Conservation Biology* 21: 1353–1355.

Skelton, P. and A. Smith. 2002. *Cladistics: A Practical Primer on CD-ROM*. Cambridge: Cambridge University Press.

Skutch, A. F. 1949. Do tropical birds rear as many young as they can nourish? *Ibis* 91: 430–455.

Skutch, A. F. 1967. Adaptive limitation of the reproductive rate of birds. *Ibis* 109: 579–599.

Slatkin, M. 1973. Gene flow and selection in a cline. *Genetics* 75: 733–756.

Slatyer, C., D. Rosauer and F. Lemckert. 2007. An assessment of endemism and species richness patterns in the Australian Anura. *Journal of Biogeography* 34: 583–596.

Slobodkin, L. B. and H. L. Sanders. 1969. On the contribution of environmental predictability to species diversity. In G. M. Woodwell and H. H. Smith (eds.), *Diversity and Stability in Ecological Systems*, 82–93. Brookhaven Symp. Biol. 22.

Smith, A. G. and A. Hallam. 1970. The fit of the southern continents. *Nature* 225: 139–144.

Smith, A. T. 1974. The distribution and dispersal of pikas: Consequences of insular population structure. *Ecology* 55: 1112–1119.

Smith, A. T. 1980. Temporal changes in insular populations of the pika (*Ochotona princeps*). *Ecology* 61: 8–13.

Smith, B. T. and J. Klicka 2010. The profound influence of the Late Pliocene Panamanian uplift on the exchange, diversification, and distribution of New World birds. *Ecography*, doi: 10.1111/j.1600-0587.2009.06335.x.

Smith, C. H. 1983. A system of world mammal faunal regions I: Logical and statistical derivation of regions. *Journal of Biogeography* 10: 455–466.

Smith, F. A., D. L. Crawford, L. E. Harding, H. M. Lease, I. W. Murray, A. Raniszweski and K. M. Youberg. 2009. A tale of two species: Extirpation and range expansion during the late Quaternary in an extreme environment. *Global Planetary Change* 65: 122–133.

Smith, F. A., H. Browning and U. L. Shepherd. 1998. The influence of climate change on the body mass of woodrats *Neotoma* in an arid region of New Mexico, USA. *Ecography* 21: 104–148.

Smith, F. A., J. H. Brown, J. P. Haskell, S. K. Lyons, J. Alroy, E. L. Charnov, T. Dayan, B. J. Enquist, S. K. M. Ernest, E. A. Hadly, et al. 2004. Similarity of mammalian body size across the taxonomic hierarchy and across space and time. *American Naturalist* 163: 672–691.

Smith, F. A., J. L. Betancourt and J. H. Brown. 1995. Evolution of body size in the woodrat over the past 25,000 years of climate change. *Science* 270: 2012–2014.

Smith, F. A., S. K. Lyons, S. K. M. Ernest and J. H. Brown. 2008. Macroecology: More than the division of food and space among species on continents. *Progress in Physical Geography* 21: 115–138.

Smith, F.A., H. Browning and U. L. Shepherd. 1998. The influence of climatic change on the body mass of woodrats Neotoma in an arid region of New Mexico, USA. *Ecography* 21: 140–148.

Smith, G. A. and M. V. Lomolino. 2004. Black-tailed prairie dogs and the structure of avian communities on the shortgrass plains. *Oecologia*. 138 [4]:592–602.

Smith, G. B. 1979. Relationships of eastern Gulf of Mexico reef-fish communities to the species equilibrium theory of insular biogeography. *Journal of Biogeography* 6: 49–61.

Smith, G. R. 1978. Biogeography of intermountain fishes. *Great Basin Naturalist Memoirs* 2: 17–42.

Smith, G. R. 1981. Late Cenozoic freshwater fishes of North America. *Annual Review of Ecology and Systematics* 12: 163–193.

Smith, G. R., T. E. Dowling, K. W. Gobalet, T. Lugaski, D. K. Shiozawa and R. P. Evans. 2002. Biogeography and timing of evolutionary events among Great Basin fishes. In R. Hershler, D. B. Madsen and D. R. Currey (eds.), *Great Basin Aquatic Systems History*, 175–234. Washington, DC: Smithsonian Institution Press.

Smith, H. G. and D. M. Wilkinson. 2007. Not all free-living microorganisms have cosmopolitan distributions: The case of *Nebela* (*Apodera*) *vas Certes* (Protozoa: Amoebozoa: Arcellinida). *Journal of Biogeography* 34: 1822–1831.

Smith, J. M. B. 1994. Patterns of disseminule dispersal by drift in the northwest Coral Sea. *New Zealand Journal of Botany* 30: 57–67.

Smith, J. M. B., H. Heatwole, M. Jones and B. M. Waterhouse. 1990. Drift disseminules on cays of the Swain Reefs, Great Barrier Reef, Australia. *Journal of Biogeography* 17: 5–17.

Smith, K. F. and J. H. Brown. 2002. Patterns of diversity, depth range and body size among pelagic fishes along a gradient in depth. *Global Ecology and Biogeography* 11: 313–322.

Smith, K. F. and S. D. Gaines. 2003. Rapoport's bathymetric rule and the latitudinal species diversity gradient for Northeast Pacific fishes and Northwest Atlantic gastropods: Evidence against a causal link. *Journal of Biogeography* 30: 1153–1159.

Smith, K. F., D. F. Sax and K. D. Lafferty. 2006. Evidence for the role of infectious disease in species extinction and endangerment. *Conservation Biology* 20: 1349–1357.

Smith, T. B., R. K. Wayne, D. J. Birman and M. W. Bruford. 1997. A role for ecotones in generating rainforest biodiversity. *Science* 276: 1855–1857.

Smithson, P., K. Addison and K. Atkinson. 2002. *Fundamentals of the Physical Environment*. 3rd edition. London: Taylor and Francis.

Snall, T., R. B. O'Hara, C. Ray and S. K. Collinge. 2008. Climate-driven spatial dynamics of plague among prairie dog colonies. *American Naturalist* 171: 238–248.

Snead, R. E. 1980. *World Atlas of Geomorphic Features*. Huntington, NY: Krieger.

Sneath, P. H. A. and R. R. Sokal. 1973. *Numerical Taxonomy*. San Francisco: W. H. Freeman.

Snider-Pellegrini, A. 1858. *La Creation et ses mysteres*. Paris: Frank and Dentu.

Solbrig, O. T. 1972. The floristic disjunctions between the "Monte" in Argentina and the "Sonoran Desert" in Mexico and the United States. *Annals of the Missouri Botanical Garden* 59: 218–223.

Solem, A. 1959. Zoogeography of the land and freshwater mollusca of the New Herbrides. *Fieldiana Zoology* 43: 239–343.

Solem, A. 1981. Land-snail biogeography: A true snail's pace of change. In G. Nelson and D. E. Rosen (eds.), *Vicariance Biogeography: A Critique*, 197–221. New York: Columbia University Press.

Soltis, D. E., A. B. Morris, J. S. McLachlan, P. S. Manos and P. S. Soltis. 2006. Comparative phylogeography of unglaciated eastern North America. *Molecular Ecology* 15: 4261–4293.

Soltis, D. E., M. A. Gitzendanner, D. D. Strenge and P. S. Soltis. 1997. Chloroplast DNA intraspecific phylogeography of plants from the Pacific Northwest of North America. *Plant Systematics and Evolution*, 206: 353–373.

Soltz, D. L. and R. J. Naiman. 1978. *The Natural History of Native Fishes in the Death Valley System*. Los Angeles, CA: Natural History Museum of Los Angeles County.

Sondaar, P. Y. 1977. Insularity and its effect on mammal evolution. In M. K. Hecht, P. C. Goody and B. M. Hecht (eds.), *Major Patterns in Vertebrate Evolution*, 671–707. New York: Plenum Press.

Sorenson et al. 1999. Relationships of the extinct moa-nalos, flightless Hawaiian waterfowl, based on ancient DNA. *Proceedings of the Royal Society*, Series B 266: 2187–2194.

Soulé, M. E. 1966. Trends in insular radiation of a lizard. *American Midland Naturalist* 100: 47–64.

Soulé, M. E. 1983. What do we really know about extinction? In C. M. Schonewald-Cox, S. M. Chambers, B. MacBryde and W. L. Thomas (eds.), *Genetics and Conservation: A Reference for Managing Wild Animal and Plant Populations*, 111–124. London: Benjamin/Cummings Publishing Company.

Soulé, M. E. 1987. *Viable Populations for Conservation*. New York: Cambridge University Press.

Souza, W. P., S. C. Schroeter and S. D. Gaines. 1981. Latitudinal variation in intertidal algal community structure. *Oecologia* 48(3): 297–303.

Spalding, M. D. 1998. *Biodiversity Patterns in Coral Reefs and Mangrove Forests: Global and Local Scales*. Unpublished Ph.D. dissertation. University of Cambridge.

Spalding, M. D., H. E. Fox, G. R. Allen, N. Davidson, Z. A. Ferdana, M. Finlayson, B. S. Halpern, M. A. Jorge, A. Lombana, S. A. Lourie, et al. 2007. Marine ecoregions of the world: A bioregionalization of coastal and shelf areas. *BioScience* 57: 573–583.

Sparks, J. S. and W. L. Smith. 2005. Freshwater fishes, dispersal ability and nonevidence: "Gondwana Life Rafts" to the rescue. *Systematic Biology* 54: 158–165.

Spaulding, W. G. and L. J. Graumlich. 1986. The plast pluvial climatic episodes in the deserts of Southwestern North America. *Nature* 320: 441–444.

Spencer, W. B. 1896. On the faunal subregions of Australia. *Proceedings of the Royal Society of Victoria* 28: 139–148.

Spiller, D. A., J. B. Losos and T. W. Schoener. 1998. Impact of a catastrophic hurricane on island populations. *Science* 281: 695–696.

Spolsky, C. and T. Uzzell. 1986. Evolutionary history of the hybridogenetic hybrid frog *Rana esculenta* as deduced from mtDNA analyses. *Molecular Biology and Evolution* 3: 44–56.

St. Louis, V. L. and J. C Barlow. 1991. Morphometric analyses of introduced and ancestral populations of the Eurasian tree sparrow. *Wilson Bulletin* 103: 1–12.

Stachowicz, J. J. and M. E. Hay. 2000. Geographic variation in camouflage specialization by a decorator crab. *American Naturalist* 156: 59–71.

Stadler, B. and A. F. G. Dixon. 2008. *Mutualism: Ants and Their Insect Partners*. Cambridge: Cambridge University Press.

Staley, J. T. 1997. Biodiversity: Are microbial species threatened? *Current Opinion in Biotechnology* 8: 340–345.

Stanley, S. M. 1979. *Macroevolution: Pattern and Process*. San Francisco: W. H. Freeman.

Stanley, S. M. 1984. Marine mass extinctions: A dominant role for temperature. In R. H. Nitecki (ed.), *Extinctions*, 69–117. Chicago: University of Chicago Press.

Stanley, S. M. 1987. *Extinction*. New York: Scientific American Books, Inc.

Stanley, S. M. 2009. *Earth System History*. 3rd edition. New York: W. H. Freeman.

Steadman, D. W. 1986. Two new species of rails (Aves: Rallidae) from Mangaia, southern Cook Islands. *Pacific Science* 40: 27–43.

Steadman, D. W. 1989. Extinction of birds in eastern Polynesia: A review of the record and comparisons with other Pacific islands groups. *Journal of Archaeological Science* 16: 177–205.

Steadman, D. W. 1993. Biogeography of Tongan birds before and after human contact. *Proceedings of the National Academy of Sciences* 90: 818–822.

Steadman, D. W. 1995. Prehistoric extinctions of Pacific island birds: Biodiversity meets zooarchaeology. *Science* 267: 1123–1131.

Steadman, D. W. 2006. *Extinction and Biogeography of Tropical Pacific Birds*. Chicago: University of Chicago Press.

Steadman, D. W. and P. S. Martin. 2003. The late Quaternary extinction and future resurrection of birds on Pacific islands. *Earth-Science Reviews* 61: 133–147.

Steadman, D. W. and S. L. Olson. 1985. Bird remains from an archaeological site on Henderson Island, South Pacific. *Acta XX Congresus Internationalis Ornithologici* 2: 361–382.

Stebbins, G. L. 1971a. Adaptive radiation of reproductive characteristics in angiosperms. 2. Seeds and seedlings. *Annual Review of Ecology and Systematics* 2: 237–260.

Stebbins, G. L. 1971b. *Chromosomal Evolution in Higher Plants*. London: Edward Arnold.

Steegman, A. T. Jr. 2007. Human cold adaptation: An unfinished agenda. *American Journal of Human Biology* 19: 218–227.

Steenbergh, W. F. and C. H. Lowe. 1976. Ecology of the saguaro. I. The role of freezing weather in a warm desert plant population. In *Research in the Parks: National Park Service Symposium*, ser. no. 1: 49–92. Washington, DC: U.S. Government Printing Office.

Steenbergh, W. F. and C. H. Lowe. 1977. Ecology of the saguaro. II. Reproduction, germination, establishment, growth and survival of the young plant. *National Park Service Scientific Monographs*, ser. no. 8. Washington, DC: U.S. Government Printing Office.

Stegen, J. C., B. C. Enquist and R. Ferriere. 2009. Advancing the metabolic theory of biodiversity. *Ecology Letters* 12: 1001–1015.

Stehli, F. G. 1968. Taxonomic diversity gradients in pole locations: The recent model. In E. T. Drake (ed.), *Evolution and Environment*, 163–227. Peabody Museum Centennial Symposium. New Haven, CT: Yale Unversity Press.

Stehli, F. G. and J. W. Wells. 1971. Diversity and age patterns in hermatypic corals. *Systematic Zoology* 20: 115–126.

Stehli, F. G. and S. D. Webb (eds.). 1985. *The Great American Biotic Interchange*. New York: Plenum Press.

Stehli, F. G., R. G. Douglas and N. D. Newell. 1969. Generation and maintenance of gradients in taxonomic diversity. *Science* 164: 947–949.

Steinauer, E. M. and S. L. Collins. 1996. Prairie ecology: The tallgrass prairie. In F. B. Samson and F. L. Knopf (eds.), *Prairie Conservation*, 39–52. Washington, DC: Island Press.

Steinitz, O., J. Heller, A. Tsoar, D. Rotem and R. Kadmon. 2006. Environment, dispersal and patterns of species similarity. *Journal of Biogeography* 33: 1044–1054.

Stenseth, N. C. and E. S. Dunlop. 2009. Unnatural selection. *Nature* 457: 803–804.

Stenseth, N. C. and R. A. Ims. 1993. *Biology of Lemmings*. London: Academic Press.

Stenseth, N. C. and W. Z. Lidicker, Jr. 1992. *Animal Dispersal: Small Mammals as a Model*. New York: Chapman and Hall.

Stenseth, N. C., A. Mysterud, G. Ottersen, J. W. Hurrell, K. Chan and M. Lima. 2002. Ecological effects of climate fluctuations. *Science* 297: 1292–1295.

Stenthouarskis, S. 1992. Altitudinal effect on species richness of Oniscidea (Crustacea; Isopoda) on three mountains in Greece. *Global Ecology and Biogeography Letters* 2: 157–164.

Stephens, D. W. and J. Gardner. 1999. Brine shrimp in Great Salt Lake, Utah. *U.S. Geological Survey*, Utah District.

Sterba, G. 1966. *Freshwater Fishes of the World*. London: Studio Vista.

Stevens, G. C. 1989. The latitudinal gradients in geographic range: How so many species coexist in the tropics. *American Naturalist* 132: 240–256.

Stevens, G. C. 1992. The elevational gradient in altitudinal range: An extension of Rapoport's latitudinal rule to altitude. *American Naturalist* 140: 893–911.

Stevens, G. C. 1996. Extending Rapoport's rule to Pacific marine fishes. *Journal of Biogeography* 23: 149–154.

Stevens, G. C. and B. J. Enquist. 1998. Macroecological limits to the abundance and distribution of *Pinus*. In D. M. Richardson (ed.), *Ecology and Biogeography of the Genus* Pinus, 183–190. Cambridge: Cambridge University Press.

Stevens, G. C. and J. F. Fox. 1991. The causes of treeline. *Annual Review of Ecology and Systematics* 22: 177–191.

Stevens, R. D. and M. R. Willig. 2002. Geographical ecology at the community level: Perspectives on the diversity of new world bats. *Ecology* 83: 545–560.

Stevens, R. D., S. B. Cox, R. E. Strauss and M. R. Willig. 2003. Patterns of functional diversity across an extensive environmental gradient: Vertebrate consumers, hidden treatments and latitudinal trends. *Ecology Letters* 6: 1099–1108.

Stevenson, R. D. 1986. Allen's rule in North American rabbits (*Sylvilagus*) and hares (*Lepus*) is an exception, not a rule. *Journal of Mammalogy* 67: 312–316.

Stewart, J. R. and A. M. Lister. 2001. Cryptic northern refugia and the origins of the modern biota. *Trends in Ecology and Evolution* 16: 608–613.

Stigall, A. L. and B. S. Lieberman. 2006. Quantitative palaeobiogeography: GIS, phylogenetic biogeographical analysis, and conservation insights. *Journal of Biogeography* 33: 2052–2060.

Stockwell, C. A., A. P. Hendry and M. T. Kinnison. 2003. Contemporary evolution meets conservation biology. *Trends in Ecology and Evolution* 18: 94–101.

Stoddart, D. R. 1973. Coral reefs: The last two million years. *Geographical Association Journal* 58: 313–323.

Stohlgren, T. J., D. Barnett, C. Flather, J. Kartesz and B. Peterjohn. 2005. Plant species invasions along the latitudinal gradient in the United States. *Ecology* 86: 2298–2309.

Stoneking, M. 2008. Human origins: The molecular perspective. *EMBO Reports* 9: 546–550.

Stoneking, M., S. T. Sherry, A. J. Redd and L. Vigilant. 1992. New approaches to dating suggest a recent age for the human mtDNA ancestor. *Philisophical Transactions of the Royal Society of London,* Series B 337: 167–175.

Storey, B. C. 1995. The role of mantle plumes in continental break-up: Case histories from Gondwanaland. *Nature* 377: 301–308.

Strahler, A. N. 1975. *Physical Geography*, 4th edition. New York: John Wiley and Sons.

Strahler, A. N. 1998. *Plate Tectonics*. Cambridge, MA: Geo-Books.

Strahler, A. N. and A. H. Strahler. 1973. *Environmental Geoscience.* Santa Barbara: Hamilton.

Strauss, S. Y. 1991. Indirect effects in community ecology: Their definition, study and importance. *Trends in Ecology and Evolution* 6: 206–210.

Strayer, D. L., V. T. Eviner, J. M. Jaschke and M. L. Pace. 2006. Understanding the long-term effects of species invasions. *Trends in Ecology and Evolution* 21: 645–651.

Strecker, U. 2006. Genetic differentiation and reproductive isolation in a Cyprinodon fish species flock from Laguna Chichancanab, Mexico. *Molecular Phylogenetics and Evolution* 39: 865–872.

Streelman, J. T. and P. D. Danley. 2003. The stages of vertebrate evolutionary radiation. *Trends in Ecology & Evolution* 18: 126–131.

Strong, D. R. Jr. and J. R. Rey. 1982. Testing for MacArthur-Wilson equilibrium with the arthropods of the miniature *Spartina* archipelago at Oyster Bay, Florida. *American Zoologist* 22: 355–360.

Stuart, C. T. and M. A. Rex. 1994. The relationship between developmental pattern and species diversity in deep-sea prosobranch snails. In C. M. Young and K. J. Eckelbarger (eds.), *Reproduction, Larval Biology and Recruitment of the Deep-Sea Benthos*, 119–136. New York: Columbia University Press.

Stuessy, T. F. 2007. Evolution of specific and genetic diversity during ontogeny of island floras: The importance of understanding process for interpreting island biogeographic patterns. In M. C. Ebach and R. S. Tangney (eds.), *Biogeography in a Changing World*, 117–134. New York: CRC Press.

Stuessy, T. F., G. Jakubowsky, R. Salguero Gómez, M. Pfosser, P. M. Schlüter, T. Fer, B.-Y. Sun and H. Kato. 2006. Anagenetic evolution in island plants. *Journal of Biogeography* 33: 1259–1265.

Stute, M., M. Forster, H. Frischkorn, A. Serejo, J. F. Clark, P. Schlosser, W. S. Broeker and G. Bonani. 1995. Cooling of tropical Brazil: 5°C during the last glacial maximum. *Science* 269: 379–380.

Sugihara, G. 1981. $S = CAz, z = 1/4$: A reply to Connor and McCoy. *American Naturalist* 117: 790–793.

Sukumar, R. 2003. *The Living Elephants: Evolutionary Ecology, Behavior and Conservation*. New York: Oxford University Press.

Sullivan, J. and P. Joyce. 2005. Model selection in phylogenetics. *Annual Review of Ecology, Evolution, and Systematics* 36: 445–466.

Sumner, F. B. 1932. Genetic, distributional and evolutionary studies of the subspecies of deer mice (*Permoscus*). *Bibliographia Genetica* 9: 1–106.

Sumner, J., C. Moritz and R. Shine. 1999. Shrinking forest shrinks skink: Morphological change in response to rainforest fragmentation in the prickly forest skink (*Gnypetoscincus queenslandiae*)—A review. *Biological Conservation* 91: 159–167.

Surovell, T., N. Waguespack and P. J. Brantingham. 2005. Global archaeological evidence for proboscidean overkill. *Proceedings of the National Academy of Sciences, USA* 102: 6231–6236.

Sutherland, W. J. 2003. Parallel extinction risk and global distribution of languages and species. *Nature* 423: 276–279.

Swetnam, T. W., C. D. Allen and J. Betancourt. 1999. Applied historical ecology: Using the past to manage for the future. *Ecological Applications* 9:1189-1206.

Swift, J. 1726 (amended 1735). *(Gulliver's Travels) Travels into Several Remote Nations of the World, in Four Parts. By Lemuel Gulliver, First a Surgeon, and Then a Captain of Several Ships.* Worcester, MA : Printed by I. Thomas, June 1802.

Swisher, C. C. III, J. M. Grajales-Nishimura, A. Montanari, S. V. Margolis, P. Claeys, W. Alvarez, P. Renne, E. Cedillo-Pardo, F. J.-M. R. Maurrasse, G. H. Curtis, J. Smit and M. O. McWilliams. 1992. Coeval 40AR/39AR ages of 65.0 million years ago from Chicxulub crater melt rock and Cretaceous-Tertiary boundary tektites. *Science* 257: 954–958.

Szumik, C. A. and Goloboff. P. A. 2004. Areas of endemism: An improved optimality criterion. *Systematic Biology* 53: 968–977.

Taberlet, P. and J. Bouvet. 1994. Mitochondrial DNA polymorphism, phylogeography, and conservation genetics of the brown bear *Ursus arctos* in Europe. *Proceedings of the Royal Society of London,* Series B 255: 195–200.

Takhtajan, A. 1986. *Floristic Regions of the World*. Berkeley: University of California Press.

Tamarin, R. H. 1977. Dispersal in island and mainland voles. *Ecology* 58: 1044–1054.

Tamura, K. and M. Nei. 1993. Estimation of the number of nucleotide substitutions in the control region of mitochondrial DNA in humans and chimpanzees. *Molecular Biology and Evolution* 10: 512–526.

Tansley, A. G. 1935. The use and abuse of vegetational concepts and terms. *Ecology* 16: 284–307.

Taper, M. L. and T. J. Case. 1992. Models of character displacement and the theoretical robustness of taxon cycles. *Evolution* 46: 317–333.

Tarling, D. H. 1962. Tentative correlation of Samoan and Hawaiian Islands using "reversals" of magnetism. *Nature* 196: 882–883.

Tarr, C. L. and R. C. Fleischer. 1995. Evolutionary relationships of the Hawaiian honeycreepers (Aves, Drepanidinae). In W. L. Wagner and V. A. Funk (eds.), *Hawaiian Biogeography: Evolution on a Hot Spot Archipelago*. Washington, DC: Smithsonian Institution Press.

Taylor, B. 1991. Investigating species incidence over habitat fragments of different areas: A look at error estimation. *Biological Journal of the Linnean Society* 42: 477–491.

Taylor, C. M. 1996. Abundance and distribution within a guild of benthic stream fishes: Local processes and regional patterns. *Freshwater Biology* 36: 385–396.

Taylor, F. B. 1910. Bearing of the Tertiary mountain belt on the origin of the earth's plan. *Geological Society of America Bulletin* 21: 179–226.

Taylor, F. B. 1928. Sliding continents and tidal and rotational forces. In W. A. Van Der Gracht and J. M. Waterschoot (eds.), *Theory of Continental Drift*, 158-177. Tulsa, OK: American Association of Petroleum Geologists.

Taylor, R. J. 1987. The geometry of colonization: 1. Islands. *Oikos* 48: 225–231.

Taylor, R. J. and P. J. Regal. 1978. The peninsular effect on species diversity and the biogeography of Baja California. *American Naturalist* 112: 583–593.

Teal, J. M. 1957. Community metabolism in a temperate cold spring. *Ecological Monographs* 27: 283–302.

Teal, J. M. 1962. Energy flow in the salt marsh ecosystem of Georgia. *Ecology* 43: 614–624.

Tegelstrom, H. and L. Hansson. 1987. Evidence for long distance dispersal in the common shrew (*Sorex araneus*). *Sonderdunk aus Zeitschrift fur Saugetierkunde* 52: 52–54.

Teller, J. T., D. W. Leverington and J. D. Mann. 2002. Freshwater outbursts to the oceans from glacial Lake Agassiz and their role in climate change during the last deglaciation. *Quaternary Science Reviews* 21: 879–887.

Templeton, A. R. 1980a. The theory of speciation via the founder principle. *Genetics* 94: 1011–1038.

Templeton, A. R. 1981. Mechanisms of speciation: A population genetic approach. *Annual Review of Ecology and Systematics* 12: 23–48.

Templeton, A. R. 2002. Out of Africa again and again. *Nature* 416: 45–51.

Templeton, A. R. 2004. Statistical phylogeography: Methods of evaluating and minimizing inference errors. *Molecular Ecology* 13: 789–809.

Templeton, A. R., E. Routman and C. A. Phillips. 1995. Separating population-structure from population history—A cladistic-analysis of the geographical-distribution of mitochondrial-DNA haplotypes in the tiger salamander, *Ambystoma-Tigrinum*. *Genetics* 140: 767–782.

Teplitsky, C., J. A. Mills, J. S. Alho, J. W. Yarrall and J. Merila. 2008. Bergmann's rule and climate change revisited: Disentangling environmental and genetic responses in a wild bird population. *Proceedings of the National Academy of Sciences, USA* 105: 13492–13496.

Terborgh, J. 1974. Preservation of natural diversity: The problem of extinction prone species. *Bioscience* 24: 715–722.

Terborgh, J. 1977. Bird species diversity on an Andean elevational gradient. *Ecology* 58: 1007–1019.

Terborgh, J. 1986. Keystone plant resources in tropical forest. In M. E. Soulé (ed.), *Conservation Biology: The Science of Scarcity and Diversity*, 330–344. Sunderland, MA: Sinauer Associates.

Terborgh, J. 2010. Trophic cascades on islands. In J. Losos and R. E. Ricklefs (eds.), *The Theory of Island Biogeography Revisited*, 116–142. Princeton, NJ: Princeton University Press.

Terborgh, J. and B. Winter. 1980. Some causes of extinction. In M. E. Soulé and B. A. Wilcox, (eds.), *Conservation Biology: An Ecological-Evolutionary Perspective*, 119–134. Sunderland, MA: Sinauer Associates.

Terborgh, J. and J. A. Estes (eds.) 2010. *Trophic Cascades: Predators, Prey, and the Changing Dynamics of Nature*. Island Press.

Terborgh, J. and J. Faaborg. 1973. Turnover and ecological release in the avifauna of Mona Island, Puerto Rico. *Auk* 90: 759–779.

Terborgh, J., L. Lopez and J. Tellos. 1997a. Bird communities in transition: The Lago Guri Islands. *Ecology* 78: 1494–1501.

Terborgh, J., L. Lopez, J. Tello, D. Yu and A. R. Bruni. 1997b. Transitory states in relaxing bridge islands. In W. F. Laurance and R. O. Bierregaard Jr. (eds.), *Tropical Forest Remnants: Ecology, Management, and Conservation of Fragmented Communities*, 256–274. Chicago: University of Chicago Press.

Terborgh, J., L. Lopez, V. P. Nunez, M. Rao, G. Shahababuddin, G. Orihuela, M. Riveros, R. Ascanio, G. H. Adler, T. D. Lambert and L. Balbas. 2001. Ecological meltdown in predator-free forest fragments. *Science* 294: 1923–1926.

Terrell, J. 1976. Island biogeography and man in Melanesia. *Archaeology and Physical Anthropology in Oceania* 11: 1–17.

Terrell, J. 1986. *Prehistory in the Pacific Islands: A Study of Variation in Language, Customs and Human Biology*. Cambridge: Cambridge University Press.

Terrell, J. E. 2006. Human biogeography: Evidence of our place in nature. *Journal of Biogeography* 33: 2088–2098.

Terrell, J., M. Miller and D. Roe. 1977. *Human Biogeography*. World Archaeology, Vol. 8. Henley-on-Thames: Routledge and Kegan Paul.

Terribile, L. C., J. A. F. Diniz-Filho, M. A. Rodriguez and T. F. L. V. B. Rangel. 2009. Richness patterns, species distributions and the principle of extreme deconstruction *Global Ecology and Biogeography* 18: 123–136.

Thiollay, J. 1998. Distribution patterns and insular biogeography of South Asian raptor communities. *Journal of Biogeography* 25: 57–72.

Thomas, C. D., A. Cameron, R. E. Green, M. Bakkenes, L. J. Beaumont, Y. C. Collingham, B. F. N. Erasmus, M. F. De Siqueira, A. Grainger, L. Hannah, et al. 2004. Extinction risk from climate change. *Nature* 427: 145–148.

Thomas, C. D., A. M. A. Franco and J. K. Hill. 2006. Range retractions and extinction in the face of climate warming. *Trends in Ecology and Evolution* 21: 415–416.

Thomas, C. D., C. R. Bulman and R. J. Wilson. 2008. Where within a geographical range do species survive best? A matter of scale. *Insect Conservation and Diversity* 1: 2–8.

Thomas, F., J. Guegan and F. Renaud. 2009. *Ecology and Evolution of Parasitism: Hosts to Ecosystems*. Oxford: Oxford University Press.

Thomas, G. H., S. Meiri and A. B. Phillimore. 2009. Body size diversification in *Anolis*: Novel environment and island effects. *Evolution* 63: 2017–2030.

Thomas, W. L. 1956. *Man's Role in Changing the Face of the Earth*. Chicago: Univeristy of Chicago Press.

Thompson, D. B. 1990. Different spatial scales of adaptation in the climbing behavior of *Permyscus maniculatus*: Geographic variation, natural selection and gene flow. *Evolution* 44: 952–965.

Thompson, D. Q., R. L. Stuckey and E. B. Thompson. 1987. *Spread, Impact and Control of Purple Loosestrife* (Lythrum saclicaria) *in North American Wetlands*. Washington, DC: U.S. Fish and Wildlife Service, Research 2.

Thompson, R. S. and J. I. Mead. 1982. Late Quaternary environments and biogeography in the Great Basin. *Quaternary Research* 17: 39–55.

Thornton, I. 1996. *Krakatau: The Destruction and Reassembly of an Island Ecosystem*. Cambridge, MA: Harvard University Press.

Thornton, I. W. B., R. A. Zann and S. van Balen. 1993. Colonization of Rakata (Krakatau Is.) by non-migrant land birds from 1883 to 1992 and implications for the value of island biogeography theory. *Journal of Biogeography* 20: 441–452.

Thornton, I. W. B., T. R. New, R. A. Zann and P. A. Rawlinson. 1990. Colonization of the Krakatau Islands by animals: Perspective from the 1980s. *Philosophical Transactions of the Royal Society of London*, Series B 328: 131–165.

Thorson, G. 1936. The larval development, growth and metabolism of Arctic marine bottom invertebrates compared with those of other seas. *Meddelelser om Gronland* 100: 1–155.

Thorson, G. 1946. Reproduction and larval development of Danish marine bottom invertebrates, with special reference to the planktonic larvae in the Sound (Oresund). *Meddelelser fra Kommissionen for Danmarks Fiskeri– og Havundersogelser, Serie Plankton* 4: 1–523.

Thorson, G. 1950. Reproductive and larval ecology of marine bottom invertebrates. *Biological Reviews* 25: 1–45.

Thrush, S. F., J. Halliday, J. E. Hewitt and A. M. Lohrer. 2008. The effects of habitat loss, fragmentation, and community homogenization on resilience in estuaries. *Ecological Applications* 18: 12–22.

Thurber, J. M. and R. O. Peterson. 1991. Changes in body size associated with range expansion in the coyote (*Canis latrans*). *Journal of Mammalogy* 72: 750–755.

Tiffney, B. H. 1985. Perspectives on the origin of the floristic similarity between eastern Asia and eastern North America. *Journal of the Arnold Arboretum* 66: 73–94.

Tiffney, B. H. and S. R. Manchester. 2001. The use of geological and paleontological evidence in evaluating plant phylogeographic hypotheses in the Northern Hemisphere Tertiary. *International Journal of Plant Science* 162: S41–S52.

Tilkens, M. J., C. Wall-Scheffler, T. D. Weaver, K. Steudel-Numbers. 2005. The effects of body proportions on thermoregulation: An experimental assessment of Allen's rule. *Journal of Human Evolution* 53: 286–291.

Tilman, D., R. M. May, C. L. Lehman and M. A. Nowak. 1994. Habitat destruction and the extinction debt. *Nature* 371: 65–66.

Tjørve, E. 2009. Shapes and functions of species–area curves (II): A review of new models and parameterizations. *Journal of Biogeography* 36: 1435–1445.

Tjørve, E. and W. R. Turner. 2009. The importance of samples and isolates for species–area relationships. *Ecography* 32: 391–400.

Tokita, M., T. Okamoto and T. Hikida. 2005. Evolutionary history of African lungfish: A hypothesis from molecular phylogeny. *Molecular Phylogenetics and Evolution* 35: 281–286.

Tomasovych, A. 2008. Evaluating neutrality and the escalation hypothesis in brachiopod communities from shallow, high-productivity habitats. *Evolutionary Ecology Research* 10: 667–698.

Tonn, W. M. and J. J. Magnunson. 1982. Patterns in the species composition and richness of fish assemblages in northern Wisconsin lakes. *Ecology* 63: 1149–1166.

Towns, D. R. and C. H. Daugherty 1994. Patterns of range contractions and extinctions in the New Zealand herpetofauna following human colonization. *New Zealand Journal of Zoology* 21: 325–339.

Trakhtenbrot, A., R. Nathan, G. Perry and D. M. Richardson. 2005. The importance of long-distance dispersal in biodiversity conservation. *Diversity and Distributions* 11: 173–181.

Trewick, S. A. 1996. Morphology and evolution of two takahe: Flightless rails of New Zealand. *Journal of Zoology* (London) 238: 221.

Triantis, K. A., K. Vardinoyannis, E. Tsolaki, I. Botsaris, K. Lika and M. Mylonas. 2006. Re-approaching the small island effect. *Journal of Biogeography* 33: 914–923.

Triantis, K. A., M. Mylonas and R. J. Whittaker. 2008. Evolutionary species–area curves as revealed by single-island endemics: Insights for the inter-provincial species–area relationship. *Ecography* 31: 401–407.

Triantis, K. A., M. Mylonas, K. Lika and K. Vardinoyannis. 2003. A model for the species–area–habitat relationship. *Journal of Biogeography* 30: 19–27.

Trinkaus, E. 2005. Early modern humans. *Annual Reviews of Anthropology* 34: 207–230.

Trusty, J. L., R. G. Olmstead, A. Santos-Guerra, S. Sá-Fontinha and J. Francisco-Ortega. 2005. Molecular phylogenetics of the Macaronesian-endemic genus *Bystropogon* (Lamiaceae): Palaeo-islands, ecological shifts and interisland colonizations. *Molecular Ecology* 14: 1177–1189.

Tubelis, D. P., D. B. Lindenmayer and A. Cowling. 2006. The peninsula effect on bird species in native eucalypt forests in a wood production landscape in Australia. *Journal of Zoology* 27: 11–18.

Tull, D. S. and K. Bohning-Gaese. 1993. Patterns of drilling predation on gastropods of the family Turritellidae in the Gulf of California. *Paleobiology* 19: 476–486.

Turner, G. F. 2007. Adaptive radiation of cichlid fish. *Current Biology* 17: R827–R831.

Turner, J. R. G. and B. A. Hawkins. 2004. The global diversity gradient. In M. V. Lomolino and L. R. Heaney (eds.), *Frontiers of Biogeography: New Directions in the Geography of Nature*, 171–190. Sunderland, MA: Sinauer Associates.

Turner, J. T. 1981. Latitudinal patterns of calanoid and cyclopoid copepod diversity in estuarine waters of eastern North America. *Journal of Biogeography* 8: 369–382.

Turner, R. M., J. E. Bowers and T. L. Burgess. 1995. *Sonoran Desert Plants: An Ecological Atlas*. Tucson, AZ: University of Arizona Press.

Turner, W., S. Spector, N. Gardiner, M. Fladeland, E. Sterling and M. Steininger. 2003. Remote sensing for biodiversity science and conservation. *Trends in Ecology and Evolution* 18: 306–314.

Turrill, W. B. 1953. *Pioneer Plant Geography: The Phytogeography Researches of Sir Joseph Dalton Hooker*. The Hague: Martinus Nijhoff.

Twitchett, R. J. 2006. The palaeoclimatology, palaeoecology and palaeoenvironmental analysis of mass extinction events. *Palaeogeography, Palaeoclimatology, Palaeoecology* 232: 190–213.

Twitchett, R. J. 2007. The Lilliput effect in the aftermath of the end-Permian extinction event. *Palaeogeography, Palaeoclimatology, Palaeoecology* 252: 132–144.

Udvardy, M. D. F. 1969. *Dynamic Zoogeography*. New York: Van Nostrand Reinhold.

Ulrich, W., M. Almeida-Neto and N. J. Gotelli. 2009. A consumer's guide to nestedness analysis. *Oikos* 118: 3–17.

Underhill, P. A. 2004. A synopsis of the extant Y chromosome diversity in East Asia and Oceania. In L. Sagart, R. Blench and A. Sanchez-Mazas (eds.), *The Peopling of East Asia: Putting Together Archaeology, Linguistics and Genetics*, 301–319. London: Routledge Curzon.

Upchurch, P., C. A. Hunn and D. B. Norman. 2002. An analysis of dinosaurian biogeography: Evidence for the existence of vicariance and dispersal patterns caused by geological events. *Proceedings of the Royal Society of London*, Series B 269: 613–621.

Upton, G. and B. Fingleton. 1990. *Spatial Data Analysis by Example*. New York: John Wiley and Sons.

Urban, M. C., B. L. Phillips, D. K. Skelly and R. Shine. 2008. A toad more traveled: The heterogeneous invasion dynamics of cane toads in Australia. *American Naturalist* 171: E134–E148.

Urquhart, F. A. 1960. *The Monarch Butterfly*. Toronto: University of Toronto Press.

Usher, M. B. 1979. Markovian approaches to ecological succession. *Journal of Animal Ecology* 48: 413–426.

Uy, J. A. C., R. G. Moyle and C. E. Filardi. 2009. Plumage and song differences mediate species recognition between incipient flycatcher species of the Solomon Islands. *Evolution* 63: 153–164.

Vagvolgyi, J. 1975. Body size, aerial dispersal and origin of the Pacific land snail fauna. *Systematic Zoology* 24: 465–488.

Valdiosera, C. E., N. Garcia, C. Anderung, L. Dalen, E. Cregut-Bonnoure, R. D. Kahlke, M. Stiller, M. Brandstrom, M. G. Thomas, J. L. Arsuaga, A. Gotherstrom and I. Barnes. 2007. Staying out in the cold: Glacial refugia and mitochondrial DNA phylogeography in ancient European brown bears. *Molecular Ecology* 16: 5140–5148.

Valentine, J. W. 1961. Paleoecologic molluscan geography of the Californian Pleistocene. *University of California Publications in Geological Science* 34: 309–442.

Valentine, J. W. 1966. Numerical analysis of marine molluscan ranges on the extratropical northeastern Pacific shelf. *Limnology and Oceanography* 11: 198–211.

Valentine, J. W. 1989. Phanerozoic marine faunas and the stability of the Earth system. *Global and Planetary Change* 1: 137–155.

Valentine, J. W. and D. Jablonski. 1982. Larval strategies and patterns of brachiopod diversity in space and time. *Geological Society of America Abstracts* 14: 241.

Valentine, J. W., K. Roy and D. Jablonski. 2002. Carnivore-noncarnivore ratios in northeastern Pacific marine gastropods. *Marine Ecology-Progress Series* 228: 153–163.

Van Den Bosch, F., R. Hengeveld and J. A. J. Metz. 1992. Analyzing the velocity of animal range expansion. *Journal of Biogeography* 19: 135–150.

Van der Hammen, T., T. A. Wijmstra and W. H. Zagwijn. 1971. The floral record of the late Cenozoic of Europe. In K. K. Turekian (ed.), *The Late Cenozoic Glacial Ages*, 391–424. New Haven, CT: Yale University Press.

Van Devender, T. R. 1977. Holocene woodlands in the southwestern deserts. *Science* 198: 189–192.

Van Devender, T. R. and W. G. Spaulding. 1979. Development of vegetation and climate in the southwestern United States. *Science* 204: 701–710.

Van Oosterzee, P. and M. Morwood. 2007. *A New Human: The Startling Discovery and Strange Story of the "Hobbits" of Flores, Indonesia*. London: Collins.

Van Riper, C., S. G. Van Riper, M. L. Goff and M. Laird. 1986. The epizootiology and ecological significance of malaria in Hawaiian islands. *Ecological Monographs* 56: 327–344.

van Tuinen, M. and S. B. Hedges. 2001. Calibration of avian molecular clocks. *Molecular Biology and Evolution* 18: 206–213.

Van Valen, L. 1973a. Body size and the number of plants and animals. *Evolution* 27: 27–35.

Van Valen, L. 1973b. A new evolutionary law. *Evolutionary Theory* 1: 1–33.

van Veller, M. G. P. and D. R. Brooks. 2001. When simplicity is not parsimonious: A priori and a posteriori approaches in historical biogeography. *Journal of Biogeography* 28: 1–11.

van Veller, M. G. P., D. R. Brooks and M. Zandee. 2003. Cladistic and phylogenetic biogeography: The art and science of discovery. *Journal of Biogeography* 30: 319–329.

van Veller, M. G. P., M. Zandee and D. J. Kornet. 1999. Two requirements for obtaining valid common patterns under different assumptions in vicariance biogeography. *Cladistics* 15: 393–406.

Van Voorhies, W. A. 1996. Bergmann size clines: A simple explanation for their occurrence in ectotherms. *Evolution* 50: 1259–1264.

van Vuren, D. and L. Bowen. 1999. Reduced defenses in insular endemic plants: An evolutionary framework. *Conservation Biology* 13: 211–212.

Vanni, M. J. and D. L. Findlay. 1990. Trophic cascades and phytoplankton community structure. *Ecology* 71: 921–937.

Vanzolini, P. E. and W. R. Heyer. 1985. The American herptofauna and the interchange. In F. G. Stehli and S. D. Webb (eds.), *The Great American Interchange*, 475–483. New York: Plenum Press.

Varela-Romero, A., G. Ruiz-Campos, L. M. Yépiz-Velázquez and J. Alaníz-García. 2002. Distribution, habitat and conservation status of desert pupfish (*Cyprinodon macularius*) in the Lower Colorado River Basin, Mexico. *Reviews in Fish Biology and Fisheries* 12, no. 2 (2002): 157–165.

Varley, G. C. 1970. The concept of energy flow applied to a woodland community. In A. Watson (ed.), *Animal Populations in Relation to Their Food Resources*, 389–405. London: Blackwell Scientific Publications.

Vartanyan, S. L., K. A. Arslanov, T. V. Tertychnaya and S. B. Chernov. 1995. Radiocarbon dating evidence for mammoths on Wrangel Island, Arctic Ocean, until 2000 B.C. *Radiocarbon* 37: 1–6.

Veizer, J., D. Ala, K. Azmy, P. Bruckschen, D. Buhl, F. Bruhn, G. A. F. Carden, A. Diener, S. Ebneth, Y. Godderis, T. Jasper, C. Korte, F. Pawellek, O. Podlaha and H. Strauss. 1999. 87Sr / 86Sr, δ13C and δ18O evolution of Phanerozoic seawater. *Chemical Geology* 161: 59–88.

Vellend, M. and J. L. Orrock. 2010. Ecological and genetic models of diversity: Lessons across disciplines. In J. Losos and R. E. Ricklefs (eds.), *The Theory of Island Biogeography Revisited*, 439–462. Princeton, NJ: Princeton University Press.

Vellend, M., L. J. Harmon, J. L. Lockwood, M. M. Mayfield, A. R. Hughes, J. P. Wares and D. F. Sax. 2007. Effects of exotic species on evolutionary diversification. *Trends in Ecology and Evolution* 22: 481–488.

Vences, M., D. R. Vieites, F.Glaw, H. Brinkmann, J. Kosuch, M. Veith and A. Meyer. 2003. Multiple overseas dispersal in amphibians. *Proceedings of the Royal Society of London*, Series B 270: 2435–2442.

Verboom, G. A., J. K. Archibald, F. T. Bakker, D. U. Bellstedt, F. Conrad, L. L. Dreyer, F. Forest, C. Galley, P. Goldblatt, J. F. Henning, et al. 2009. Origin and diversification of the Greater Cape flora: Ancient species repository, hot-bed of recent radiation, or both? *Molecular Phylogenetics and Evolution* 51: 44–53.

Verheyen, E., W. Salzburger, J. Snoeks and A. Meyer. 2003. Origin of the superflock of cichlid fishes from Lake Victoria, East Africa. *Science* 300: 325–329.

Vermeij, G. 2005. From phenomenology to first principles: Toward a theory of diversity. *Proceedings of the California Academy of Sciences* 56(Suppl. I): 12–23.

Vermeij, G. and S. T. Williams. 2007. Predation and the geography of opercular thickness in turbinid gastropods. *Journal of Molluscan Studies* 73: 67–73.

Vermeij, G. J. 1974. Marine faunal dominance and molluscan shell form. *Evolution* 28: 656–664.

Vermeij, G. J. 1978. *Biogeography and Adaptation: Patterns of Marine Life*. Cambridge, MA: Harvard University Press.

Vermeij, G. J. 1991a. Anatomy of an invasion: The trans-Arctic interchange. *Paleobiology* 17: 281–307.

Vermeij, G. J. 1991b. When biotas meet: Understanding biotic interchange. *Science* 253: 1099–1103.

Vermeij, G. J. 2004a. Island life: A view from the sea. In M. V. Lomolino and L. R. Heaney (eds.), *Frontiers of Biogeography: New Directions in the Geography of Nature*, 239–254. Sunderland, MA: Sinauer Associates.

Vermeij, G. J. 2005. Invasions as expectation: A historical fact of life. In D. F. Sax, J. J. Stachowicz and S. D. Gaines (eds.), *Species Invasions: Insights into Ecology, Evolution and Biogeography*, 315–340. Sunderland, MA: Sinauer Associates.

Vernon, J. E. N. 1995. *Corals in Space and Time: The Biogeography of Scleractinia*. Sydney: University of New South Wales Press.

Vetaas, O. R., J.-A. Grytnes. 2002. Distribution of vascular plant species richness and endemic richness along the Himalayan elevational gradient in Nepal. *Global Ecology and Biogeography* 11: 291–301.

Vetter, J. 2006. Wallace's other line: Human biogeography and field practice in the eastern colonial tropics. *Journal of the History of Biology* 39: 89–123.

Villard, M.-A. 2002. Habitat fragmentation: Major conservation issue or intellectual attractor? *Ecological Applications* 12(2): 319–320.

Vine, F. J. and D. H. Matthews. 1963. Magnetic anomalies over a young oceanic ridge. *Nature* 199: 947–949.

Vinogradova, N. G. 1962. Vertical zonation in the distribution of deep-sea benthic fauna in the ocean. *Deep-Sea Research* 8: 245–250.

Vitousek, P. M., C. M. D'Antonio, L. L. Loope and R. West-brooks. 1996. Biological invasions as a global environmental challenge. *American Scientist* 84: 468–478.

Vitousek, P. M., C. M. D'Antonio, L. L. Loope, M. Rejmanek and R. Westbrooks. 1997b. Introduced species: A significant component of human-caused global change. *New Zealand Journal of Ecology* 21: 1–16.

Vogiatzakis, I. N. 2003. *GIS-Based Modelling and Ecology: A Review of Tools and Methods*. Reading, England: University of Reading Press.

Voigt, F. A., R. Arafeh, N. Farwig, E. M. Griebeler and K. Bohning-Gaese. 2009. Linking seed dispersal and genetic structure of trees: A biogeographic approach. *Journal of Biogeography* 36: 242–254.

von Broembsen, S. L. 1989. Invasions of natural ecosystems by plant pathogens. In J. A. Drake et al. (eds.), *Biological Invasions: A Global Perspective*, 77–84. New York: John Wiley and Sons.

Von Holle, B., H.R. Delcourt and D. Simberloff. 2003. The importance of biological inertia in plant community resistance to invasion. *Journal of Vegetation Science* 14: 425–432.

Voris, H. K. 1977. A phylogeny of the sea snakes (Hydrophiidae). *Fieldiana Zoology* 70(4): 79–166.

Vrba, E. S. 1999. Habitat theory in relation to the evolution of African Neogene biota and hominids. In T. G. Bromage and F. Schrenk (eds.), *African Biogeography, Climate Change, and Early Hominid Evolution*, 19–34. Oxford University Press, Oxford.

Vuilleumier, F. 1970. Insular biogeography in continental species. I. The Northern Andes of South America. *American Naturalist* 104: 373–388.

Vuilleumier, F. 1973. Insular biogeography in continental regions. II. Cave faunas from Tesin, southern Switzerland. *Systematic Zoology* 22: 64–76.

Vuilleumier, F. 1984. Faunal turnover and development of fossil avifaunas in South America. *Evolution* 38: 1384–1396.

Vuilleumier, F. 1985. Fossil and recent avifaunas and the Interamerican interchange. In F. G. Stehli and S. D. Webb (eds.), *The Great American Biotic Interchange*, 387–424. New York: Plenum Press.

Wagner, D. L. and J. K. Liebherr. 1992. Flightlessness in insects. *Trends in Ecology and Evolution* 7: 216–220.

Wagner, W. H. Jr. 1995. Evolution of Hawaiian ferns and fern allies in relation to their conservation status. *Pacific Science* 49: 31–41.

Wagner, W. L. and V. A. Funk (eds.). 1995. *Hawaiian Biogeography: Evolution on a Hot Spot Archipelago*. Washington, DC: Smithsonian Institution Press.

Wainright, P. C., D. R. Bellwood and M. W. Westneat. 2002. Ecomorphology of locomotion in labrid fishes. *Environmental Biology of Fishes* 65: 47–62.

Wake, D. B. 1966. Comparative osteology and evolution of the lungless salamanders, family Plethodontidae. *Memoirs of the Southern California Academy of Sciences* 4: 1–111.

Wake, D. B. and V. T. Vredenburg. 2008. Are we in the midst of the sixth mass extinction? A view from the world of amphibians. *Proceedings of the National Academy of Science, USA* 105: 11466–11473.

Walker, L. R. and R. Moral. 2003. *Primary succession and ecosystem rehabilitation*. London: Cambridge University Press.

Walker, L. R., J. Walker and R. J. Hobbs (eds.). 2007. *Linking Restoration and Ecological Succession*. New York: Springer Science.

Wallace, A. R. 1857. On the natural history of the Aru Islands. *Annals and Magazine of Natural History*, Supplement to Volume 20, December.

Wallace, A. R. 1860. On the zoological geography of the Malay Archipelago. *Journal of the Linnaean Society of London* 4: 172–184.

Wallace, A. R. 1869. *The Malay Archipelago: The Land of the Orangutan and the Bird of Paradise*. New York: Harper.

Wallace, A. R. 1876. *The Geographical Distribution of Animals*. 2 vols. London: Macmillan.

Wallace, A. R. 1878. *Tropical Nature and Other Essays*. New York: Macmillan.

Wallace, A. R. 1880. *Island Life, or the Phenomena and Causes of Insular Faunas and Floras*. London: Macmillan.

Wallace, A. R. 1893. Mr. H. O. Forbes's discoveries in the Chatam Islands. *Nature* (May 11, 1893): 27.

Waloff, Z. 1966. The upsurges and recessions of the desert locust plague: An historical survey. *Anti-Locust Memoirs* 8: 1–111.

Waltari, E. and R. P. Guralnick. 2009. Ecological niche modelling of montane mammals in the Great Basin, North America: Examining past and present connectivity of species across basins and ranges. *Journal of Biogeography* 36: 148–161.

Waltari, E., E. P. Hoberg, E. P. Lessa and J. A. Cook. 2007. Eastward ho: Phylogeographical perspectives on colonization of hosts and parasites across the Beringian nexus. *Journal of Biogeography* 34: 561–574.

Waltari, E., R. J. Hijmans, A. T. Peterson, A. S. Nyari, S. L. Perkins and R. P. Guralnick. 2007. Locating Pleistocene refugia: Comparing phylogeographic and ecological niche model predictions. *PLoS One* 2: e563.

Walter, H. S. 1998. Driving forces of island biodiversity: An appraisal of two theories. *Physical Geography* 19: 351–377.

Walter, H. S. 2000. Stability and change in the bird communities of the Channel Islands. In D. R. Browne, K. L. Mithcell and H. W. Chaney (eds.), *Proceedings of the 5th California Islands Symposium*, 307–314. U. S. Department of the Interior, Mineral Management Service (OCS Study MMS 99–0038).

Walter, H. S. 2004. The mismeasure of islands: Biogeographic theory and the conservation of nature. *Journal of Biogeography* 31: 177–197.

Walther, G.-R., L. Hughes, P. Vitouset and M. Stenseth. 2005. Consensus on climate change. *Trends in Ecology and Evolution* 20: 645–646.

Wang, W. S.-Y. and J. W. Minett. 2005. The invasion of language: Emergence, change and death. *Trends in Ecology and Evolution* 20: 263–269.

Waples, R. S. 1991. Pacific salmon, Oncorhynchus spp., and the definition of a "species" under the Endangered Species Act. *Marine Fisheries Review* 53: 11–22.

Ward, 2001. Sudden productivity collapse associated with the Triassic-Jurassic boundary mass extinction. *Science* 292: 1148–1151.

Wares, J. P. 2002. Community genetics in the Northwestern Atlantic intertidal. *Molecular Ecology* 11: 1131–1144.

Wares, J. P., A. R. Hughes and R. K. Grosberg. 2005. Mechanisms that drive evolutionary change: Insights from species introductions and invasions. In D. F. Sax, J. J. Stachowicz and S. D. Gaines (eds.), *Species Invasions: Insights into Ecology, Evolution and Biogeography*, 229–257. Sunderland, MA: Sinauer Associates.

Waser, N. M., L. Chittka, M. V. Price, N. M. Williams and J. Ollerton. 1996. Generalization in pollination systems and why it matters. *Ecology* 77: 1043–1060.

Waters, J. M. 2008. Driven by the West Wind Drift? A synthesis of southern temperate marine biogeography, with new directions for dispersalism. *Journal of Biogeography* 35: 417–427.

Waters, J. M. and D. Craw. 2006. Goodbye Gondwana? New Zealand biogeography, geology, and the problem of circularity. *Systematic Zoology* 55: 351–356.

Waters, J. M., A. Lopez and G. P. Wallis. 2000. Molecular phylogenetics and biogeography of galaxiid fishes (Osteichthys: Galaxiida): Dispersal, vicariance and position of Lexidogalaxias salamandroides. *Systematic Biology* 49: 777–795.

Watson, D. M. and A. T. Peterson. 1999. Determinants of diversity in a naturally fragmented landscape: Humid montane forest avifaunas of Mesoamerica. *Ecography* 22: 582–589.

Watson, H. C. 1859. *Cybele Britannica, or British Plants and their Geographical Relations.* London: Longman and Company.

Watson, J. E. M., R. J. Whittaker and D. Freudenberger. 2005. Bird community responses to habitat fragmentation: How consistent are they across landscapes? *Journal of Biogeography* 32: 1353–1370.

Watters, G. T. 1992. Unionids, fishes and the species-area curve. *Journal of Biogeography* 19: 481–490.

Weaver, T. and K. Steudel-Numbers. 2005. Does climate or mobility explain differences in body proportions between Neanderthals and their Upper Paleolithic successors? *Evolutionary Anthropology* 14: 218–223.

Webb, C. O., D. D. Ackerly, M. A. McPeek and M. J. Donoghue. 2002. Phylogenies and community ecology. *Annual Review of Ecology and Systematics* 33: 475–505.

Webb, S. D. 1991. Ecogeography and the great American interchange. *Paleobiology* 17: 266–280.

Webb, S. D. and L. G. Marshall. 1982. Historical biogeography of recent South American land mammals. In M. A. Mares and H. H. Genoways (eds.), *Mammalian Biology in South America,* 39–52. Special Publication Series 6, Pymatuning Laboratory of Ecology, University of Pittsburgh.

Webb, T. III. 1987. The appearance and disappearance of major vegetational assemblages: Long-term vegetational dynamics in eastern North America. *Vegetatio* 69: 177–187.

Webb, T. J. and K. J. Gaston. 2000. Geographic range size and evolutionary age in birds. *Proceedings of the Royal Society of London,* Series B-Biological Sciences 267: 1843–1850.

Webeck, K. and S. Pearson. 2005. Stick-nest rat middens and a late-Holocene record of White Range, central Australia. *The Holocene* 15: 466–471.

Weber, F. R., T. D. Hamilton, D. M. Hopkins, C. A. Repenning and H. Haas. 1981. Canyon Creek: A Late Pleistocene vertebrate locality in interior Alaska. *Quaternary Research* 16: 167–180.

Webster, J. 1997. Fungal biodiversity. *Biodiversity and Conservation* 6: 657.

Webster, R. and M. A. Oliver. 2007. *Geostatistics for Environmental Scientists (Statistics in Practice).* 2nd edition. Sussex, England: Wiley and Sons.

Wecker, S. C. 1963. The role of early experience in habitat selection by the prairie deer-mouse, *Peromyscus maniculatus bairdi. Ecological Monographs* 33: 307–325.

Wecker, S. C. 1964. Habitat selection. *Scientific American* 211: 109–116.

Wegener, A. 1912a. Die Entstehung der Kontinente. *Petermanns Geogr. Mitt.* 58: 185–195, 253–256, 305–308.

Wegener, A. 1912b. Die Entstehung der Kontinente. *Geol. Rundsch.* 3: 276–292.

Wegener, A. 1915. *Die Entstehung der Kontinente und Ozeane.* Braunschweig: Vieweg. [Other editions 1920, 1922, 1924, 1929, 1936.]

Wegener, A. 1966. *The Origin of Continents and Oceans.* New York: Dover Publications. [Translation of 1929 edition by J. Biram.]

Weigert, R. G. and D. F. Owen. 1971. Trophic structure, available resources and populations densities in terrestrial ecosystems versus aquatic ecosystems. *Journal of Theoretical Biology* 30: 69–81.

Weiher, E. and P. Keddy (eds.). 1999. *Ecological Assembly Rules: Perspectives, Advances, Retreats.* Cambridge: Cambridge University Press.

Weir, J. T. and D. Schluter. 2004. Ice sheets promote speciation in boreal birds. *Proceedings of the Royal Society of London* Series B-Biological Sciences 271: 1881–1887.

Weksler, M., H. C. Lanier and L. E. Olson. 2010. Eastern Beringian biogeography: historical and spatial genetic structure of singing voles in Alaska. *Journal of Biogeography,* doi: 10.1111/j.1365-2699.2010.02310.x.

Weller, M. W. 1980. *The Island Waterfowl.* Ames: Iowa State University Press.

Wells, P. V. 1976. Macrofossil analysis of wood rat (*Neotoma*) middens as a key to the Quaternary vegetational history of arid America. *Quaternary Research* 6: 223–248.

Wells, P. V. 1979. An equable glaciopluvial in the West: Pleni-glacial evidence of increased precipitation on a gradient from the Great Basin to the Sonoran and Chihuahuan Deserts. *Quaternary Research* 12: 311–325.

Wells, P. V. and C. D. Jorgensen. 1964. Pleistocene wood rat middens and climatic change in Mojave Desert: A record of juniper woodlands. *Science* 143: 1171–1174.

Wells, P. V. and R. Berger. 1967. Late Pleistocene history of coniferous woodland in the Mohave Desert. *Science* 155: 1640–1647.

Wenner, A. M. and D. L. Johnson. 1980. Land vertebrates on the California Channel Islands: Sweepstakes or bridges? In D. M. Power (ed.), *The California Islands: Proceedings of a Multi-Disciplinary Symposium,* 497–530. Santa Barbara, CA: Museum of Natural History.

Werger, M. J. A. and A. C. van Bruggen (eds.). 1978. *Biogeography and Ecology of Southern Africa.* 2 vols. Monographiae Biologicae 31. The Hague: Dr. W. Junk.

Westing, A. H. 1966. Sugar maple decline: An evaluation. *Economic Botany* 20: 196–212.

Wetzel, R. G. 1975. *Limnology.* Philadelphia: W. B. Saunders.

Wheeler, Q. D. and R. Meier. 2000. *Species Concepts and Phylogenetic Theory: A Debate.* New York: Columbia University Press.

Whitaker, R. J., D. W. Grogan and J. W. Taylor. 2003. Geographic barriers isolate endemic populations of hyperthermophilic Archaea. *Science* 301: 976–978.

White, J. L. and B. C. Harvey. 2001. Effects of an introduced piscivorous fish on native benthic fishes in a coastal river. *Freshwater Biology* 46, no. 7 (2001): 987–995.

White, J. L. and R. D. E. MacPhee. 2001. The sloths of the West Indies: A systematic and phylogenetic review. In C. A. Woods and F. E. Sergile (eds.), *Biogeography of the West Indies: Patterns and Perspectives,* 2nd Edition, 201–236. London: CRC Press.

White, M. J. D. 1973. *Animal Cytology and Evolution.* 3rd edition. Cambridge: Cambridge University Press.

White, M. J. D. 1978. *Modes of Speciation.* San Francisco: W. H. Freeman.

White, T. A. and J. B. Searle. 2006. Factors explaining increased body size in common shrews (*Sorex araneus*) on Scottish islands. *Journal of Biogeography* 34: 356–363.

White, T. C. R. 1976. Weather, food and plagues of locusts. *Oecologia* 22: 119–134.

Whitehead, D. R. and C. E. Jones. 1969. Small islands and the equilibrium theory of insular biogeography. *Evolution* 23: 171–179.

Whiteside, D. I. and J. E. A. Marshall. 2008. The age, fauna and paleoenvironment of the Late Triassic fissure deposits of Tytherington, South Gloucestershire, UK. *Geological Magazine* 145: 105–147.

Whittaker, A. H. 1978. The effects of rodents on reptiles and amphibians. In P. R. Dingwall and I. A. Atkinson (eds.), *The*

Ecology and Control of Rodents in New Zealand's Nature Reserves, 75–88. Department of Lands and Survey Information Series no. 4.

Whittaker, R. H. 1956. Vegetation of the Great Smoky Mountains. *Ecological Monographs* 22: 1–44.

Whittaker, R. H. 1960. Vegetation of the Siskiyou Mountains, Oregon and California. *Ecological Monographs* 30: 279–338.

Whittaker, R. H. 1967. Gradient analysis of vegetation. *Biological Review* 42: 207–264.

Whittaker, R. H. 1975. *Communities and Ecosystems*. 2nd edition. New York: Macmillan.

Whittaker, R. H. 1977. Evolution of species diversity in land communities. In M. K. Hecht, W. C. Steere and B. Wallace (eds.), *Evolutionary Biology* 10, 1–67. New York: Plenum Press.

Whittaker, R. H. and G. E. Likens. 1973. Carbon in the biota. In G. M. Woodwell and E. V. Pecan (eds.), *Carbon and the Biosphere*, 281–300. Conf. 72501. Springfield, VA: National Technical Information Service.

Whittaker, R. H. and W. A. Niering. 1965. Vegetation of the Santa Catalina Mountains, Arizona: A gradient analysis of the south slope. *Ecology* 46: 429–452.

Whittaker, R. H. and W. A. Niering. 1968. Vegetation of the Santa Catalina Mountains, Arizona. IV. Limestone and acid soils. *Journal of Ecology* 56: 523–544.

Whittaker, R. H. and W. A. Niering. 1975. Vegetation of the Santa Catalina Mountains, Arizona. V. Biomass, production and diversity along the elevation gradient. *Ecology* 56: 771–790.

Whittaker, R. J. 1995. Disturbed island ecology. *Trends in Ecology and Evolution* 10: 421–425.

Whittaker, R. J. 1998. *Island Biogeography: Ecology, Evolution and Conservation*. New York: Oxford University Press.

Whittaker, R. J. 2000. Scale, succession and complexity in island biogeography: Are we asking the right questions? *Global Ecology and Biogeography* 9: 75–86.

Whittaker, R. J. 2004. Dynamic hypotheses of richness on islands and continents. In M. V. Lomolino and L. R. Heaney (eds.), *Frontiers of Biogeography: New Directions in the Geography of Nature*, 211–232. Sunderland, MA: Sinauer Associates.

Whittaker, R. J. and J. M. Fernández-Palacios. 2007. *Island Biogeography: Ecology, Evolution, and Conservation*, 2nd edition. Oxford: Oxford University Press.

Whittaker, R. J. and S. H. Jones. 1994. The role of frugivorous bats and birds in the rebuilding of a tropical forest ecosystem, Krakatau, Indonesia. *Journal of Biogeography* 21: 245–258.

Whittaker, R. J., K. A. Triantis and R. J. Ladle. 2008. A general dynamic theory of oceanic island biogeography. *Journal of Biogeography* 35: 977–994.

Whittaker, R. J., K. A. Triantis and R. J. Ladle. 2010. A general dynamic theory of oceanic island biogeography: Extending the MacArthur–Wilson theory to accommodate the rise and fall of volcanic islands. In J. B. Losos and R. E. Ricklefs (eds.), *The Theory of Island Biogeography Revisited*, 88–115. Princeton, NJ: Princeton University Press.

Whittaker, R. J., K. J. Willis and R. Field. 2001. Scale and species richness: Towards a general, hierarchical theory of species diversity. *Journal of Biogeography* 28: 453–470.

Whittaker, R. J., M. B. Araujo, P. Jepson, R. J. Ladle, J. E. M. Watson and K. J. Willis. 2005. Conservation biogeography: Assessment and prospect. *Diversity and Distributions* 11: 3–23.

Whittaker, R. J., M. B. Bush and K. Richards. 1989. Plant recolonization and vegetation succession on the Krakatau Islands, Indonesia. *Ecological Monographs* 59: 59–123.

Whittaker, R. J., R. Field and T. Partomihardjo. 2000. How to go extinct: Lessons from the lost plants of Krakatau. *Journal of Biogeography* 27: 1049–1064.

Whittaker, R. J., S. H. Jones and T. Partomihardjo. 1997. The rebuilding of an isolated rain forest assemblage: How disharmonic is the flora of Krakatau? *Biodiversity and Conservation* 6: 1671–1696.

Whittier, T. R. and T. M. Kincaid. 1999. Introduced fish in northeastern USA lakes: Regional extent, dominance and effect on native species richness. *Transactions of the American Fisheries Society* 128: 769–784.

Wiens, J. J. 2009. Paleontology, genomics and combined-data phylogenetics: Can molecular data improve phylogeny estimation for fossil taxa? *Systematic Biology* 58: 87–99.

Wiens, J. J. and C. H. Graham. 2005. Niche conservatism: Integrating evolution, ecology, and conservation biology. *Annual Review of Ecology, Evolution, and Systematics* 36: 519–539.

Wiens, J. J. and M. J. Donoghue. 2004. Historical biogeography, ecology and species richness. *Trends in Ecology and Evolution* 19: 639–644.

Wiens, J. J., C. H. Graham, D. S. Moen, S. A. Smith and T. W. Reeder. 2006. Evolutionary and ecological causes of the latitudinal diversity gradient in hylid frogs: Treefrog trees unearth the roots of high tropical diversity. *American Naturalist* 168: 579–596.

Wikelski, M. 2005. Evolution of body size in Galápagos marine iguanas. *Proceedings of the Royal Society of London*, Series B 272: 1985–1993.

Wilcove, D. S. 1987. From fragmentation to extinction. *Natural Areas Journal* 7: 23–29.

Wilcove, D. S., C. H. McLellan and A. P. Dobson. 1986. Habitat fragmentation in the temperate zone. In M. E. Soulé (ed.), *Conservation Biology: The Science of Scarcity and Diversity*, 237–256. Sunderland, MA: Sinauer Associates.

Wilcox, B. A. 1978. Supersaturated island faunas: A species-age relationship for lizards on post-Pleistocene land-bridge islands. *Science* 199: 996–998.

Wilcox, B. A. 1980. Insular ecology and conservation. In M. E. Soulé and B. A. Wilcox, (eds.), *Conservation Biology: An Ecological-Evolutionary Perspective*, 95–117. Sunderland, MA: Sinauer Associates.

Wilcox, B. A. and D. D. Murphy. 1985. Conservation strategy: The effects of fragmentation on extinction. *American Naturalist* 125: 879–887.

Wildlife Conservation Society. 2006. *State of the Wild: A Global Portrait of Wildlife, Wildlands, and Oceans*. Washington, DC: Island Press.

Wildman, D. E., M. Uddin, J. C. Opazo, G. Liu, V. Lefort, S. Guindon, O. Gascuel, L. I. Grossman, R. Romero and M. Goodman. 2007. Genomics, biogeography, and the diversification of placental mammals. *Proceedings of the National Academy of Sciences, USA* 104: 14395–14400.

Wiley, E. O. 1981. *Phylogenetics: The Theory and Practice of Phylogenetic Systematics*. New York: Wiley-Interscience.

Wiley, E. O. 1988. Vicariance biogeography. *Annual Review of Ecology and Systematics* 19: 513–542.

Wiley, E. O. and R. L. Mayden. 2000. The evolutionary species concept. In Q. D. Wheeler and R. Meier (eds.), *Species Concepts and Phylogenetic Theory: A Debate*, 70–89. New York: Columbia University Press.

Wiley, E. O., D. J. Siegel-Causey, D. R. Brooks and V. A. Funk. 1991. *The Compleat Cladist: A Primer of Phylogenetic Procedures*. Lawrence, KS: Museum of Natural History, University of Kansas.

Wilkinson, D. M. 1998. Mycorrhizal fungi and Quaternary plant migrations. *Global Ecology and Biogeography* 7: 137–140.

Williams, C. B. 1953. The relative abundance of different species in a wild animal population. *Journal of Animal Ecology* 22: 14–31.

Williams, C. B. 1964. *Patterns in the Balance of Nature and Related Problems in Quantitative Ecology*. New York: Academic Press.

Williams, C. K. and Moore, R. J. 1989. Genetic divergence in fecundity of Australian wild rabbits *Oryctolagus cuniculus*. *Journal of Animal Ecology* 58: 495–507.

Williams, E. E. 1976. West Indian anoles: A taxonomic and evolutionary summary. I. Introduction and a species list. *Breviora* no. 440.

Williams, E. E. 1983. Ecomorphs, faunas, island size and diverse end points in island radiations of *Anolis*. In R. B. Huey, E. R. Pianka and T. W. Schoener (eds.), *Lizard Ecology: Studies on a Model Organism*, 326–370. Cambridge, MA: Belknap Press.

Williams, J. E. and R. R. Miller. 1990. Conservation status of North American fish fauna in fresh water. *Journal of Fish Biology* 37: 79–85.

Williams, J. W., B. N. Shuman, T. Webb III, P. J. Bartlein and P. Leduc. 2003. Late Quaternary vegetation dynamics in North America: Scaling from taxa to biomes. *Ecological Monographs* 74: 309–334.

Williamson, M. 1981. *Island Populations*. Oxford: Oxford University Press.

Williamson, M. 1989. The equilibrium theory today: True but trivial. *Journal of Biogeography* 16: 3–4.

Williamson, M. 1996. *Biological Invasions*. London: Chapman and Hall.

Williamson, M., K. J. Gaston and W. M. Lonsdale. 2001. The species–area relationship does not have an asymptote! *Journal of Biogeography* 28: 827–830.

Williamson, M., K. J. Gaston and W. M. Lonsdale. 2002. An asymptote is an asymptote and not found in species–area relationships. *Journal of Biogeography* 29: 1713.

Williamson, P. G. 1981b. Paleontological documentation of speciation in Cenozoic molluscs from Turkana basin. *Nature* 293: 437–443.

Willig, M. R. 2000. Latitude, Trends with. In S. Levin (ed.), *Encyclopedia of Biodiversity*, 701–714. San Diego, CA: Academic Press.

Willig, M. R. and K. W. Selcer. 1989. Bat species density gradient in the New World: A statistical assessment. *Journal of Biogeography* 16: 189–195.

Willig, M. R., D. M. Kaufman and R.D. Stevens. 2003. Latitudinal gradients of biodiversity: Pattern, process, scale and synthesis. *Annual Review of Ecology, Evolution and Systematics* 34: 272–309.

Willis, E. O. 1974. Populations and local extinctions of birds on Barro Colorado Island, Panama. *Ecological Monographs* 44: 153–169.

Willis, J. C. 1922. *Age and Area*. Cambridge: Cambridge University Press.

Willis, K. J. and R. J. Whittaker. 2000. The refugial debate. *Science* 287: 1406–1407.

Willis, K. J., K. D. Bennett and D. Walker. 2004. Introduction. *Philosophical Transactions of the Royal Society of London*, Series B 359: 157–158.

Wilmhurst, J., A. J. Anderson, T. F. G. Highman and T. H. Worthy. 2008. Dating the late prehistoric dispersal of Polynesians to New Zealand using the commensal Pacific rat. *Proceedings of the National Academy of Sciences, USA* 105: 7676–7680.

Wilson, D. S. 1975. The adequacy of body size as a niche difference. *American Naturalist* 109: 769–784.

Wilson, E. O. (ed.) 1988. *Biodiversity*. Washington, DC: National Academy Press.

Wilson, E. O. 1959. Adaptive shift and dispersal in a tropical ant fauna. *Evolution* 13: 122–144.

Wilson, E. O. 1961. The nature of the taxon cycle in the Melanesian ant fauna. *American Naturalist* 95: 169–193.

Wilson, E. O. 1969. The species equilibrium. *Brookhaven Symposia in Biology* 22: 38–47.

Wilson, E. O. 1992. *The Diversity of Life*. Cambridge, MA: Belknap Press.

Wilson, E. O. 1994. *Naturalist*. Washington, DC: Island Press.

Wilson, E. O. 1999. Prologue. In G. Daws and M. Fujita (eds.), *Archipelago: The Islands of Indonesia*. Berkeley: University of California Press.

Wilson, E. O. and D. S. Simberloff. 1969. Experimental zoogeography of islands: Defaunation and monitoring techniques. *Ecology* 50: 267–278.

Wilson, I., M. Weale and D. Balding. 2003. Inferences from DNA data: Population histories, evolutionary processes and forensic math probabilities. *Journal of the Royal Statistical Society* Series A 166: 155–188.

Wilson, J. T. 1963a. A possible origin of the Hawaiian Islands. *Canadian Journal of Physics* 41: 863–870.

Wilson, J. T. 1963b. Evidence from islands on the spreading of the ocean floors. *Nature* 197: 536–538.

Wilson, J. W. III. 1974. Analytical zoogeography of North American mammals. *Evolution* 28: 124–140.

Wilson, M. V. and A. Schmida. 1984. Measuring beta diversity with presence-absence data. *Journal of Ecology* 72: 1055–1064.

Winchester, S. 2001. *The Map That Changed the World: William Smith and the Birth of Modern Geology*. New York: Harper Collins.

Windley, B. F. 1977. *The Evolving Continents*. London: John Wiley and Sons.

Winemiller, K. O. 1990. Spatial and temporal variation in tropical fish trophic networks. *Ecological Monographs* 60: 331–367.

Winemiller, K. O., E. R. Pianka, L. J. Vitt. and A. Joern. 2001. Food web laws or niche theory? Six independent empirical tests. *American Naturalist* 158: 193–200.

Winfield, I. J. and C. Hollingworth. 2001. Nonindigenous fishes introduced into inland waters of the United States (American Fisheries Society, Special Publication 27) *Fish and Fisheries* 2: 172–173.

Witman, J. D. and K. Roy. 2009. *Marine Macroecology*. Chicago: University of Chicago Press.

Wodzicki, K. A. 1950. Introduced mammals of New Zealand: An ecological and economic survey. *Bulletin Department of Scientific and Industrial Research* 98: 1–255.

Wojcicki, M. and D. R. Brooks. 2004. Escaping the matrix: A new algorithm for phylogenetic comparative studies of co-evolution. *Cladistics* 20: 341–361.

Wojcicki, M. and D. R. Brooks. 2005. PACT: An efficient and powerful algorithm for generating area cladograms. *Journal of Biogeography* 32: 755–774.

Wolfe, J. A. 1975. Some aspects of plant geography in the Northern Hemisphere during the late Cretaceous and Tertiary. *Annals of the Missouri Botanical Garden* 62: 264–279.

Wolfe, J. A. 1979. *Temperature Parameters of Humid to Mesic Forests of Eastern Asia and Relation to Forests of Other Regions of the Northern Hemisphere and Australasia*. U.S. Geological Survey Professional Paper no. 1106. Washington, DC: U.S. Government Printing Office.

Wollaston, T. V. 1877. *Coleoptera Sanctae-Helenae*. London: John Van Voorst.

Wonham, M. J., J. T. Carlton, G. M. Ruiz and L. D. Smith. 2000. Fish and ships: Relating dispersal frequency to success in biological invasions. *Marine Biology* 136: 1111–1121.

Woodburne, M. O. and J. A. Case. 1996. Dispersal, vicariance and the late Cretaceous to early Tertiary land mammal biogeography from South America to Australia. *Journal of Mammalian Evolution* 3: 121–161.

Woodburne, M. O. and W. J. Zinmeister. 1982. Fossil land mammals from Antarctica. *Science* 218: 284–286.

Woodring, W. P. 1966. The Panama canal landbridge as a sea barrier. *Proceedings of the American Philosophical Society* 110: 425–433.

Woods, C. A. and F. E. Sergile. 2001. Biogeography of the West Indies: Patterns and Perspectives, 2nd edition. London: CRC Press.

Woods, K. D. and M. D. Davis. 1989. Paleoecology of range limits: Beech in the upper peninsula of Michigan. *Ecology* 70: 681–696.

Wootton, J. T. 1992. Indirect effects, prey susceptibility and habitat selection: Impacts of birds on limpets and algae. *Ecology* 73: 981–991.

World Conservation Monitoring Centre. 1992. *Global Biodiversity: Status of the Earth's Living Resources*. New York: Chapman and Hall.

World Conservation Monitoring Centre. 2000. Freshwater biodiversity. *World Conservation Monitoring Centre Biodiversity Series No. 8*: PP-PP.

World Conservation Monitoring Centre. 2000. *Global Diversity: Earth's Living Resources in the 21st Century*. B.Groombridge and M. D. Jenkins (eds.) World Conservation Press, Cambridge, UK.

Worthy, T. H. and R. N. Holdaway. 2002. *The Lost World of the Moa: Prehistoric Life of New Zealand*. Bloomington, IN: Indiana University Press.

Wright, D. H. and J. H. Reeves. 1992. On the meaning and measurement of nestedness of species assemblages. *Oecologia* 92: 416–428.

Wright, D. H., B. D. Patterson, G. M. Mikkelson, A. Cutler and W. Atmar. 1996. A comparative analysis of nested subset patterns of species composition. *Oecologia* 113: 1–20.

Wright, H. E. 1976. Ice retreat and revegetation of the western Great Lakes area. In W. C. Mahaney (ed.), *Quaternary Stratigraphy of North America*, 119–132. Stroudsburg, PA: Dowden, Hutchison and Ross.

Wright, S. 1978. *Variability Within and Among Natural Populations*. Chicago: University of Chicago Press.

Wright, S. D., C. G. Yong, J. W. Dawson, D. J. Whittkacr and R. C. Gardner. 2001. Stepping stones to Hawaii: A trans-equatorial dispersal pathway for Metrosideros (Myrtaceae) inferred from nrDNA (ITS plus ETS). *Journal of Biogeography* 28: 769–774.

Wright, S. J. 1980. Density compensation in island avifaunas. *Oecologia* 45: 385–389.

Wright, S. J. 1981. Inter-archipelago vertebrate distributions: The slope of the species-area relation. *American Naturalist* 118: 726–748.

Wright, S. J. 1985. How isolation affects rates of turnover of species on islands. *Oikos* 44: 331–340.

Wroe, S., J. Field and D. K. Grayson. 2006. Megafaunal extinction: Climate, humans and assumptions. *Trends in Ecology and Evolution* 21: 61–62.

Wulff, E. V. 1943. *An Introduction to Historical Plant Geography*. Trans. E. Brissenden. Waltham, MA: Chronica Botanica.

Wyatt, R. E. 1992. *Ecology and Evolution of Plant Reproduction*. New York: Chapman and Hall.

Xu, X., Z. H. Zhou, X. L. Wang, X. W. Kuang, F. C. Zhang and X. K. Du. 2003. Four-winged dinosaurs from China. *Nature* 421: 335–340.

Yamahira, K. and T. Nishida. 2009. Latitudinal variation in axial patterning of the medaka (Actinopterygii: Adrianichthyidae): Jordan's rule is substantiated by genetic variation in abdominal vertebral number. *Biological Journal of the Linnean Society* 96: 856–866.

Yeaton, R. I. 1974. An ecological analysis of chaparral and pine forest bird communities on Santa Cruz Island and mainland California. *Ecology* 55: 959–973.

Yeaton, R. I. 1981. Seedling morphology and the altitudinal distribution of pines in the Sierra Nevada of central California: A hypothesis. *Madroño* 28: 67–77.

Yoda, K. 1967. A preliminary survey of the forest vegetation of eastern Nepal. II. General description, structure and floristic composition of sample plots chosen from different vegetation zones. *Journal of the College of Art and Science. Chi'ba University National Science* 5: 99–140.

Yoder, A. D. and M. D. Nowak. 2006. Has vicariance or dispersal been the predominant biogeographic force in Madagascar? Only time will tell. *Annual Review of Ecology, Evolution, and Systematics* 37: 405–431.

Yoder, A. D., M. M. Burns, S. Zehr, T. Delefosse, G. Veron, S. M. Goodman and J. J. Flynn. 2003. Single origin of Malagasy Carnivora from an African ancestor. *Nature* 421: 734–737.

Yokoyama, Y., K. Lambeck, P. de Dekker, P. Johnston and L. K. Fifield. 2000. Timing of the last glacial maximum from observed sea level minima. *Nature* 406: 713–716.

Yom-Tov, Y. 2001. Global warming and body mass decline in Israeli passerine birds. *Proceedings of the Royal Society of London*, Series B 268: 947–952.

Yom-Tov, Y. and S. Yom-Tov. 2004. Climatic change and body size in two species of Japanese rodents. *Biological Journal of the Linnean Society* 82: 263–267.

Yom-Tov, Y. and S. Yom-Tov. 2005. Global warming, Bergmann's rule and body size in the masked shrew *Sorex cinereus* in Alaska. *Journal of Animal Ecology* 74: 803–808.

Yom-Tov, Y., S. Yom-Tov and H. Baagoe. 2003. Increase of skull size in the red fox (*Vulpes vulpes*) and Eurasian badger (*Meles meles*) in Denmark during the twentieth century: An effect of improved diet. *Evolutionary Ecology Research* 5: 1037–1048.

Yom-Tov, Y., S. Yom-Tov and H. Moller. 1999. Competition, coexistence, and adaptation amongst rodent invaders to Pacific and New Zealand islands. *Journal of Biogeography* 26: 947–958.

Yom-Tov, Y., S. Yom-Tov, J. Wrigth, C. J. R. Thorne and R DuFeu. 2006. Recent changes in body weight and wing length among some British passerine birds. *Oikos* 112: 91–101.

Zachos, J., M. Pagani, L. Sloan, E. Thomas and K. Billups. 2001. Trends, rhythms, and aberrations in global climate 65 Ma to present. *Science* 292: 686–693.

Zakharov, E. V., C. R. Smith, D. C. Lees, A. Cameron, R. I. Vane-Wright and F. A. H. Sperling. 2004. Independent gene phylogenies and morphology demonstrate a Malagasy origin for a wide-ranging group of swallowtail butterflies. *Evolution* 58: 2763–2782.

Zapata, F. A., K. J. Gaston and S. L. Chown. 2005. The mid-domain effect revisited. *American Naturalist* 166: E144–E148.

Zaret, T. M. 1980. *Predation and Freshwater Communities*. New Haven, CT: Yale University Press.

Zaret, T. M. and R. T. Paine. 1973. Species introduction in a tropical lake. *Science* 182: 449–455.

Zavodna, M., P. Arens, P. J. van Dijk, T. Partomihardjo, B. Vosman and J. M. M. van Damme. 2005. Pollinating fig wasps: Genetic consequences of island recolonization. *Journal of Evolutionary Biology* 18: 1234–1243.

Zazula, G. D., D. G. Froeseb, S. A. Eliasc, S. Kuzminac and R. W. Mathewes. 2007. Arctic ground squirrels of the mammoth-steppe: paleoecology of Late Pleistocene middens (24000–29450 14C yr BP), Yukon Territory, Canada. *Quaternary Science Reviews* 26: 979–1003.

Zeh, D. W. and J. A. Zeh. 1992. Failed predation or transportation? Causes and consequences of phoretic behavior in the pseudoscorpion *Dinocheirus arizonensis* (Pseudoscorpionida: Chernetidae). *Journal of Insect Behavior* 5: 37–50.

Zink, R. M. 1996. Comparative phylogeography in North American birds. *Evolution* 50: 308–317.

Zink, R. M., A. E. Kessen, T. V. Line and R. C. Blackwell-Rago. 2001. Comparative phylogeography of some aridland bird species. *Condor* 103: 1–10.

Zink, R. M., J. Klicka and B. R. Barber. 2004. The tempo of avian diversification during the Quaternary. *Philosophical Transactions of the Royal Society of London,* Series B-Biological Sciences 359: 215–219.

Zink, R. M., R. C. Blackwell-Rago and F. Ronquist. 2000. The shifting roles of dispersal and vicariance in biogeography. *Proceedings of the Royal Society of London,* Series B 267: 497–503.

Zuckerkandl, E. and L. Pauling. 1965. Evolutionary divergence and convergence in proteins. In V. Bryson and H. J. Vogel (eds.), *Evolving Genes and Proteins,* 97–166. New York: Academic Press.

Index

Ecological character displacement
body size and, *599*
Red Queens and, 592
See also Character displacement
Ecological differentiation, 236–238
Ecological diversification, body sizes of insular populations and, 594–595
Ecological geography
geographic range, 622–641 (*see also* Geographic range)
geography of biodiversity (*see* Biodiversity)
of the marine realm, 650–657
of the terrestrial realm, 641–650
See also Ecogeography
Ecological guilds, geographic trends in body size, *645*
Ecological naiveté, 592, *593*
Ecological niche
multidimensional, 89–91
See also Niches
Ecological niche modeling (ENM), 118, 441, *442*, 491, 550
Ecological "particles," 623
Ecological pyramids, 128–129
Ecological release
body size and, *599, 643,* 644
characteristics in insular populations, 574
isolated biotas and, 569
Red Queens and, 592
Ecological selection, insular communities, 574–585
Ecological speciation, 224
Ecological succession, 133–134
See also Succession
"Ecological time," 164
Ecology
Ernst von Haeckel and, 31
metabolic theory of, *688, 692*
origin of, 15–16
Ecoregions, 122, 159, 161–163, 164
*Eco*RI enzyme, 432
Ecosystem engineers, 49
Ecosystem geography, 159, 161–164
Ecosystems, 123–124
Ecotones, 131

Ecotypes
cohesion species concept and, 213
defined, 215
Ectopistes migratorius, 246, 334, 698
Edentates, 399, 408
Edgbaston goby, 362, *363*
Effective population size, 218
Egretta intermedia, 576, *577*
Eichhornia crassipes, 178
"Eigenplace," 527
Einstein, Albert, 10
Ekman, Sven, 40
El Niño, 57
El Niño Southern Oscillation (ENSO), 57, 65
Elapidae, 461, *568*
Eldredge, Niles, 250
Electrophoresis, 432
Elephant birds, 391, 396, 587, 588
Elephant seals, 246
Elephants
haplotype networks, *439*
swimming, *183, 184,* 570
tusklessness in modern populations, 612
Elephas, 439
E. falconeri, 601
Elevation
as a barrier to dispersal, 190, *191*
cooling effect of, 51
Elevational gradients, *682, 683*–686
Elms, *180, 336*
Elrathii kingii, 262
Elton, Charles, 697, 726
Elusive biotas, future biogeographic studies and, 763
Emberizinae, 464
Emberizine buntings, 403
Emeus, 603
Emperor seamount, *303, 304*
Emus, 364, 396, 401, 601–602
Endangered Species Act of 1973 (U.S.), 215
Endangerment
causes of, *705*
overview, 708–710
Endeavor (HMS), 20, *21*
Endemicity analysis, *475*
Endemics
distributions, 362–364
fossil record of past distributions, 447
of New Zealand and Madagascar, 366–367
origins of, 368–370
provincialism (*See* Provincialism)

of South America and Australia, 376
Endemism
defining and delineating areas of, 472–473
elevational gradients in, 685
hierarchical patterns, 365–366
in Holarctic temperate deciduous forests, 502–503
in insular biotas, 559–562
peninsula effect and, 681
in the Philippines, 757–759
reasons for, 365
Endler, John, 116
Endoheric basins, 301
Endotherms, 128
Endothia parasitica, 246
Endozoochory, 187–188, 591
Endurance, body size and, *643*
English Channel, 343, *344*
Enhydra lutris, 246
Enlightenment, 22
Entosphenus, 109
Environment, impact on the geography of diversity, 691–692
Environmental patchiness, impact on the geography of diversity, *690*
Environmental plasticity, body size and, *643*
Envisat, 79
Eocene Optimum, *314, 317*
Eoraptor, 445
Epeiric seas, 301, 309
Ephalophis, 461, 462
Ephemeroptera, 588
Ephydra cinerea, 101
Epicontinental seas, 301
Epiphylls, 140
Epiphyte load, *690*
Epiphytes, *139, 140, 682*
Equator, wind patterns at, 52
Equatorial countercurrent, 57
Equidae, *595*
Equilibrium theory of island biogeography, 42
"biogeography of the species" concept, 618
described, 520, 522–524
estimates of turnover on land bridge islands, 530, 532
experimental defaunation, 539–541
founders of, 510–511, 512
graphical model of, 75–76
historical background to, 511–512

"independent discovery" of, 519
Krakatau Islands and, 524–525, 532–538
nonequilibrium biotas, 546–553
premises, 524
rescue effect, 541–542, *543*
small island effect, 544–546
species equivalence assumption, 562
statement of the equilibrium condition in, 525
strength and weaknesses, 525–529
target area effect, 542–544
terms, *520*
tests of, 529–541
turnover on recently created anthropogenic islands, 538–539
Equinoxes, 50–51, *319*
Erebus (HMS), 30
Erica, 460
E. arborea, 416
Erinaceus europaeus, 504
Eschrichtius robustus, 246
Essay on the Geography of Plants (Humboldt), 22–23, 75
Estuaries, 156, *700*
Etheostoma, 242, 379
Ethiopian Region, 371
'Eua Island, *753*
Eubalaena glacialis, 387
Eucalyptus, 147
Euphorbia, 141, 144
E. polygonifolia, 327, 329
Euphorbiaceae, 141, 416, 609
Euphrates valley, 746
Euphydryas editha, 108, 398
Euproctis chrysorrhoea, 714, 715, 717
Eurasian Plate, 282, 284, 299
Eurasian tree sparrow, 400
Europasaurus, 607
Europe, human dispersal to, 748
European beech, 711
European rabbit, 173, *176,* 647–648
Euryhaline species, 67, 190
Eurypateryx, 603
Eurythermal species, 190
Eustatic changes, 326–327
Eutamias. See Tamias
Eutrophic lakes, 155
Evapotranspiration, *692*
Evergreen forests
coniferous, 122–123
subtropical, 145–146
Evergreens, acidic soils and, 62

ARCTIC OCEAN

Queen
Elisabeth
Islands

Ellesmere Island

Melville I.

Devon I.

GREENLAND

BEAUFORT SEA

Banks
Island

Baffin Bay

N O

Victoria
Island

Prince of
Wales I.

ARCTIC CIRCLE

St. Lawrence
Island

ALASKA

Great Bear
Lakes

Baffin Island

Iceland

Faroe Isla

BERING
SEA

Mt. McKinley

Mackenzie
Mts.

Great Slave
Lakes

Hudson
Bay

LABRADOR SEA

Shetland Islan
Orkney Islands

Nunivak I.

Gulf of Alaska

Canadian Shield

Labrador

Hebrides

Brita

Pribilof Is.

Kodiak I.
Alexander Archipelago

Lake
Winnipeg

James
Bay

Ireland

Aleutian Islands

Queen Charlotte Is.

R
O
C
K
Y
M
O
U
N
T
A
I
N
S

G
r
e
a
t

P
l
a
i
n
s

Superior

L. Michigan

L. Huron

St. Lawrence

Island of
Newfoundland

English Cha

Vancouver
Island

Cascade Range

NORTH
AMERICA

Ontario

L. Erie

Gulf of
St. Lawrence

Nova Scotia

Olympic Peninsula

Great Salt
Lake

Mississippi

Appalachian Mountains

Chesapeake Bay

Azores

Strait of Gibraltar

Iberia
Penins

Sierra Nevada

Great
Basin

Death
Valley

Grand
Canyon

Bermuda
Islands

Madeira
Islands

Atlas

Channel Is.

Rio Grande

Florida

ATLANTIC OCEAN

Canary
Islands

Guadalupe I.

Sierra Madre Occidental

Baja California

Gulf of California

Sierra Madre
Oriental

GULF OF
MEXICO

Bahama
Islands

Cape Verde
Islands

Senegal

Midway Is.

Hawaiian Islands

TROPIC OF CANCER

Yucatan
Peninsula

Greater Antilles

Cuba

Jamaica

Hispaniola

Puerto Rico

Hawaii

Revillagigedo
Islands

CENTRAL
AMERICA

CARIBBEAN SEA

Lesser Antilles

Trinidad

P
O
L
Y
N
E
S
I
A

PACIFIC OCEAN

Isthmus of
Panama

Orinoco

Guiana Highlands

Cocos Island

Negro

Amazon

Ascension

Kiritimati
(Christmas I.)

Galápagos
Islands

Amazon
Basin

EQUATOR

Marquesas
Islands

Caroline I.

Samoa
Is.

Scilly

Society Is.

Tahiti

Tuamotu Archipelago

A
N
D
E
S

SOUTH
AMERICA

Lake Titicaca

M
I
D

A
T
L
A
N
T
I
C

R
I
D
G
E

Fiji Is.

Tonga Is.

Cook Islands

Trindade

Si

Rarotonga

A
N
D
E
S

Gran Chaco

TROPIC OF CAPRICORN

Pitcairn I.

Austral Islands

Easter Island

P
a
t
a
g
o
n
i
a

P
a
m
p
a
s

Tristan
Group

Chatham Islands

Juan Fernándes
Islands

Isla Grande
de Chiloé

Falkland Islands

South Georgia

ANTARCTIC CIRCLE

Tierra del Fuego

Cape Horn

South Orkney
Islands

Antarctic
Peninsula

WEDDELL
SEA

ROSS SEA

Mt. Sidley

Ellsworth Mts.

Ross Ice Shelf

A N T A R C T I C A